Dictionary & Thesaurus
of
Environment,
Health & Safety

U.S. Department of Energy
Office of Environment, Safety and
Health

CRC Press
Taylor & Francis Group
Boca Raton London New York

CRC Press is an imprint of the
Taylor & Francis Group, an **informa** business

CRC Press
Taylor & Francis Group
6000 Broken Sound Parkway NW, Suite 300
Boca Raton, FL 33487-2742

© 1992 by Taylor & Francis Group, LLC
CRC Press is an imprint of Taylor & Francis Group, an Informa business

First issued in paperback 2019

No claim to original U.S. Government works

ISBN 13: 978-0-367-45025-0 (pbk)
ISBN 13: 978-0-87371-876-9 (hbk)

Visit the Taylor & Francis Web site at
http://www.taylorandfrancis.com

and the CRC Press Web site at
http://www.crcpress.com

CONTENTS

PREFACE

The *Dictionary & Thesaurus of Environment, Health & Safety,* in addition to including standard thesaurus-type information, has several unique features. This preface, along with the scope note and field explanation pages, provides basic information to assist the user in effectively using the document.

In addition to special data fields found within the individual word blocks, the most noteworthy features of the document are the three appendices following the main body of the thesaurus. These appendices include: (1) a listing of all thesaurus acronyms and their reciprocal phrases; (2) a listing of all thesaurus terms under broader subject categories; and (3) a separate mini-thesaurus for the DOE FRASE (Factor Relationship and Sequence of Events) vocabulary used on the Safety Performance Measurement System (SPMS).

In developing the *Dictionary & Thesaurus of Environment, Health & Safety* we followed the ANSI (American National Standards Institute) Standard for Thesaurus Construction, but with several notable exceptions to meet unique project requirements (e.g., inclusion of phrases as main terms, etc.). When available, definitions were also included for each term, with acronyms indicating the source(s) of the definition(s).

This thesaurus attempts to provide a semantic structure for environment, health, and safety terminology drawn from a variety of sources, used for a variety of different purposes: a difficult but critical task in the management of increasingly complex ES&H programs and information systems. Terminology usage and structure are dynamic by their nature. Given this fact, thesauri are "living" documents that attempt to reflect *consensus* agreement among subject professionals concerning their vocabulary.

The *Dictionary & Thesaurus of Environment, Health & Safety* is a first attempt at capturing the usage and structure of selected ES&H terminology. It is by no means a comprehensive or final document. Continual editing and updating will be necessary to ensure that the dictionary/thesaurus correctly reflects vocabulary usage by professionals in the field.

A final note of appreciation to Robert W. Eicher (who initiated this project), the staff of the Department's System Safety Development Center, Laura Cunningham, and the many reviewers who shared their expertise through comments and suggestions.

D. Charlynn Clayton
Technical Editor

SCOPE NOTE ACRONYMS

An explanation of acronyms found in the scope notes and definitions for each Main Term.

This list also indicates books and other sources from which terms were taken for the *Dictionary & Thesaurus of Environment, Health & Safety*.

ANL   Argonne National Laboratory. *Glossary of Hazardous, Solid and Radioactive Waste Terms*. September 1990.

ARPA   Archaeological Resources Protection Act of 1979.

BNA   Bureau of National Affairs. *Environment Reporter*.

CAA   Clean Air Act.

CWA   Clean Water Act.

CERCLA   Comprehensive Environmental Response, Compensation, and Liability Act.

CFR   *Code of Federal Regulations*.

DOE Order   Definition sections from pertinent DOE Orders.

DSTT   Lapedes, Daniel N., Editor-in-Chief. *McGraw-Hill Dictionary of Scientific and Technical Terms*, 2nd Edition. McGraw-Hill Book Company: New York, 1978.

EM   U.S. Department of Energy. Office of Restoration and Waste Management. *Applied Research, Development, Demonstration, Testing and Evaluation Plan*. November 1989.

      Environmental Restoration and Waste Management: Five-Year Plan. August 1989.

EMER   Training Resources and Data Exchange (TRADE), Emergency Preparedness Special Interest Group, Glossary Task Force. *Glossary and Acronyms of Emergency Preparedness Terms*. ORAU 89/B-80, Oak Ridge Associated Universities: Oak Ridge, TN, 1988.

EPA   U.S. Environmental Protection Agency. Office of Information Resources Management (OIRM). *Information Resources Directory (IRD):* Glossary. 1989.

ESA   Endangered Species Act.

ESH   U.S. Department of Energy. Office of the Assistant Secretary for Environment, Safety and Health. *Environmental Audit Manual*, Volumes I-III.

FIFRA   Federal Insecticide, Fungicide, and Rodenticide Act.

FRASE   EG&G Idaho, Inc. System Safety Development Center. Safety Performance Measurement System (SPMS). Selection of terms from the DOE FRASE (Factor Relationship and Sequence of Events) vocabulary which is currently used as the searching vocabulary on SPMS. FRASE terms include the numeric key and FRASE key in the Scope Note.

FWCA	Fish and Wildlife Coordination Act.
HMTA	Hazardous Materials Transportation Act.
IAEA	International Atomic Energy Agency. *Radiation Protection Glossary.* Safety Series No. 76. Vienna, 1986.
LLRWPA	Low Level Radioactive Waste Policy Act.
MORT	Management Oversight and Risk Tree.
NCRP	National Council on Radiation Protection and Measurement. Report of NCRP SC-46-2. Draft of July 1990. Appendix A: Glossary. NSR/SC-46-2/ljm.
NEPA	National Environmental Protection Act.
NFI	U.S. Department of Energy. Office of the Assistant Secretary for Environment, Safety and Health. *Nuclear Facilities Incidents Database Reference Manual:* Appendix E: Synonyms Used in Text. April 1990.
NIH	U.S. Department of Health and Human Services. Public Health. National Institutes of Health. *National Institutes of Health Radiation Safety Guide:* Glossary. 1979.
NPDWA	National Primary Drinking Water Act.
NWPA	Nuclear Waste Policy Act.
OSHA	U.S. Department of Labor. Occupational Safety and Health Administration. *OSHA Thesaurus.*
RCRA	Resource Conservation and Recovery Act.
RHA	Rivers and Harbors Act.
SARA	Superfund Amendments and Reauthorization Act.
SDWA	Safe Drinking Water Act.
SEA	Structural Engineers Association of California. Seismology Committee. *Recommended Lateral Force Requirements and Tentative Commentary.* Section 1B. Definitions. San Francisco. 1976.
SMRP	T. Theodore Fujita. *Workbook of Tornadoes and High Winds for Engineering Applications.* SMRP Research Paper 165. University of Chicago. Department of Geophysical Sciences. Satellite and Mesometeorological Research Project. September 1978.
SSDC	EG&G Idaho, Inc., System Safety Development Center. *Glossary of SSDC Terms and Acronyms.* October 1984.
SWDA	Solid Waste Disposal Act.
TSA	U.S. Department of Energy. Office of the Assistant Secretary for Environment, Safety and Health. Glossaries and acronyms lists from various *Technical Safety Appraisals.*
TSCA	Toxic Substances Control Act.
TTGM	U.S. Department of Energy. Office of the Assistant Secretary for Environment, Safety and Health. Office of Special Projects. *Tiger Team Guidance Manual. Environment, Safety and Health, and Management and Organization Assessment.* February 1990.

USGS U.S. Geological Survey. *National Earthquake Hazards Reduction Program: Overview.* Report of the United States Congress Geological Survey. Circular 918.

WEBSTER Woolf, Henry B., Editor-in-Chief. *Webster's New Collegiate Dictionary.* G&C Merriam Company: Springfield, MA, 1977.

WTID Gove, Philip B., Editor-in-Chief. *Webster's Third New International Dictionary of the English Language.* G&C Merriam Company: Springfield, MA, 1976.

NOTE: For DOE FRASE vocabulary terms included in the Dictionary/Thesaurus, the Scope Note Field includes the numeric key and FRASE key for each term.

FIELD EXPLANATION

An explanation of the "typical" fields found in the ES&H *Dictionary & Thesaurus of Environment, Health & Safety.*

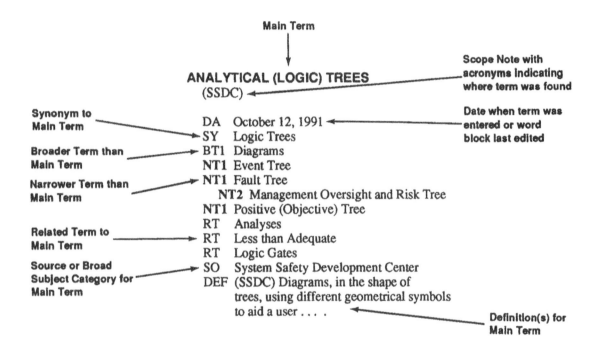

NOTE: A listing of all the Source terms (SO) or broad subject categories is included in the appendices. The SOs are shown as bolded Main Terms with all the referenced terms shown as **Source To.**

Dictionary & Thesaurus
of
Environment,
Health & Safety

Environment, Safety and Health Thesaurus/Dictionary

Safety Performance Measurement System

A-SCALE SOUND LEVELS
(EPA)
DA October 12, 1990
RT Decibel
RT Noise Pollution Abatement
SO Environmental Protection Agency
    Glossary
DEF (EPA)Measurements of sound
    approximating the sensitivity of the
    human ear, used to note the
    intensity or annoyance of sounds.

A/E
DA October 12, 1990
SEE Architect Engineer
SO Acronyms

AACS
DA October 12, 1990
SEE Airborne Activity Confinement
    Systems
SO Acronyms

ABANDONED AREAS
(CFR)
DA October 12, 1990
BT1 Sites/Areas
NT1 Temporarily Abandoned Areas
RT Inactive Mines
RT Land Reclamation
SO Air Pollution
DEF (CFR)Deserted mine areas in which
    work has ceased and in which
    further work is not intended. Areas
    which function as escapeways and
    areas formerly used as
    lunchrooms, shops, and
    transformer or pumping stations
    are not considered abandoned
    areas. Except for designated
    ventilation passageways designed
    to minimize the distance to vents,
    worked-out mine areas are
    considered abandoned areas for
    the purpose of this subpart (40
    CFR 61.21).

ABANDONED WELLS
(SDWA; CFR)
DA October 12, 1990
BT1 Wells

SO Environmental Protection Agency
    Glossary
SO Water Pollution
DEF (EPA)Wells whose use has been
    permanently discontinued or which
    are in a state of disrepair such
    that they cannot be used for their
    intended purpose.

ABATEMENT
(EPA)
DA October 12, 1990
BT1 Pollution Recovery Processes
 BT2 Processes
NT1 Noise Pollution Abatement
RT Pollutant Discharge Elimination
    System
RT Reclamation
SO Environmental Protection Agency
    Glossary
DEF (EPA)Reducing the degree or
    intensity of, or eliminating,
    pollution.

ABDOMEN
(1115 ABDOMEN)
DA November 28, 1990
BT1 Trunk
 BT2 Human Body Parts
RT Groin
SO DOE FRASE VOCABULARY

ABEL
(EPA)
DA October 12, 1990
BT1 Computer Codes
RT Modeling
SO Environmental Protection Agency
    Glossary
DEF (EPA)EPA's computer model for
    analyzing a violator's ability to pay
    a civil penalty.

ABLE
DA October 12, 1990
SEE Scram
SO Acronyms

ABNORMAL CONDITION CONTROL
DA January 8, 1991
SF ACC

BT1 Controls
SO Nuclear Facilities Incident Database

ABNORMAL EXPOSURE CONDITIONS
(IAEA)
DA October 12, 1990
BT1 Conditions
BT1 Emergencies
 BT2 Reportable Occurrences
  BT3 Occurrences
RT Emergency Exposure
SO Radiation
DEF (IAEA)Conditions in which a source
    or the radiation from it is not under
    control.

ABOSFn
DA October 12, 1990
SEE Nominal Automatic Burnout Safety
    Factor
SO Acronyms

ABOVEGROUND RELEASES
(SWDA; RCRA)
DA October 19, 1990
BT1 Releases
RT Airborne Releases
SO Wastes
DEF Releases to the surface of the land
    or to surface water. This includes,
    but is not limited to, releases from
    the aboveground portion of an
    underground storage tank (UST)
    system and aboveground releases
    associated with overfills and
    transfer operations as the
    regulated substance moves to or
    from a UST system.

ABOVEGROUND TANKS
(SWDA; RCRA; CFR)
DA October 12, 1990
BT1 Tanks
 BT2 Facility Components
SO Environmental Management
SO Wastes
DEF Devices situated in such a way that
    the entire surface area of the tank
    is completely above the plane of
    the adjacent surrounding surface
    and the entire surface area of the

SY-Synonymous Terms   SO-Source/Subject Category   SF-See From

1

tank (including the tank bottom) is
able to be visually inspected.

ABRASION
(1300 AB)
DA November 28, 1990
BT1 Injuries
RT Razor(s)
SO DOE FRASE VOCABULARY

ABS-SC
(NFI)
DA October 12, 1990
SEE Automatic Backup Shutdown of the
    Safety Computer
SO Acronyms
DEF (NFI) Automatic Backup Shutdown
    of the Safety Computer.

ABSORBED DOSE
(DOE Orders 5480.11, 5400.5; ESH;
  NIH; NCRP; IAEA; EMER)
DA October 12, 1990
SF D
BT1 Doses
BT1 Radiation Units
 BT2 Units of Measure
RT Dose Equivalents
RT Gray
RT Ionizing Radiation
RT Irradiation
RT Relative Biological Effectiveness
SO Emergency Preparedness
SO Industrial Hygiene
SO Radiation
DEF (ESH) The energy imparted to
    matter by ionizing radiation per
    unit mass of irradiated material at
    the place of interest in that
    material. The absorbed dose is
    expressed in units of rad (or gray).
    (1 rad = 0.01 gray.) (NIH) The
    amount of energy imparted to
    matter by ionizing radiation per
    unit mass of irradiated material.

ABSORBED DOSE RATE
(IAEA)
DA October 12, 1990
BT1 Rates
RT Dose Rate Meters
RT Ionizing Radiation
SO Radiation
DEF (IAEA)The energy from ionizing
    radiation absorbed per unit mass
    is called the absorbed dose. The
    unit of absorbed dose is the gray
    (1 Joule/kg) or, historically , the
    rad which is equal numerically to
    10^{-2} Joule/kg (100 erg/g).

ABSORPTION (CHEMICAL)
(EPA; NIH)
DA October 12, 1990
BT1 Chemical Processes
 BT2 Processes
RT Sorption

SO Environmental Protection Agency
    Glossary
DEF (EPA) (1) Adhesion of molecules of
    gas, liquid or dissolved solids to a
    surface. (2) The passage of one
    substance into or through another;
    e.g., an operation in which one or
    more soluble components of a gas
    mixture are dissolved in a liquid.

ABSORPTION (RADIATION)
DA February 28, 1991
RT Radiation Protection
RT Radiation Shields
RT Radiation Exposure
SO Environmental Protection Agency
    Glossary
SO Radiation
DEF (EPA) The phenomenon by which
    radiation imparts some or all of its
    energy to any material through
    which it passes.

ABSORPTION (WASTE)
(EPA)
DA February 28, 1991
BT1 Wastewater Treatment Processes
 BT2 Treatment
  BT3 Waste Management Processes
 BT4 Processes
RT Sorption
SO Environmental Protection Agency
    Glossary
SO Wastes
DEF (EPA) An advanced method of
    treating wastes in which activated
    carbon removes organic matter
    from wastewater.

AC
DA October 12, 1990
SEE Alternating Current
SO Acronyms

3 AC FLOW
(NFI)
DA October 12, 1990
RT Coolants
SO Nuclear Facilities Incident Database
DEF Refers to the flow conditions when
    3 AC motors of the primary
    coolant circuit are working.

AC Motors
DA October 12, 1990
SEE Allis-Chalmers Motors
SO Acronyms

ACC
DA October 12, 1990
SEE Abnormal Condition Control
SO Acronyms

ACCELERATOR PRODUCED
RADIOACTIVE MATERIALS
(DOE Order 5820.2A; ANL)
DA May 24, 1991

RT Naturally Occurring Radioactive
    Material
SO Radiation

ACCELERATORS
(EPA)
DA October 12, 1990
BT1 Devices
RT Radiation
SO Environmental Protection Agency
    Glossary
DEF (EPA) In radiation science, devices
    that speed up charged particles
    such as electrons or protons.

ACCEPTABLE DAILY INTAKE
(EPA)
DA October 12, 1990
SF ADI
BT1 Intake
 BT2 Measurements
BT1 Standards
 BT2 Codes, Standards, and
    Regulations
RT Hazardous Chemicals
SO Environmental Protection Agency
    Glossary
DEF (EPA) An estimate similar in
    concept to the Reference Dose
    (RfD), however, derived using a
    less rigorously defined
    methodology. RfDs have replaced
    the ADI as the Agency's preferred
    value for use in evaluating
    potential noncarcinogenic health
    effects resulting from exposure to
    a chemical.

ACCEPTABLE INTAKE FOR CHRONIC
EXPOSURE
(EPA)
DA October 12, 1990
SF AIC
BT1 Standards
 BT2 Codes, Standards, and
    Regulations
RT Hazardous Chemicals
RT Intake
SO Environmental Protection Agency
    Glossary
DEF (EPA) An estimate similar in
    concept to the Reference Dose
    (RfD), however, derived using a
    less rigorously defined
    methodology. RfDs have replaced
    AICs as the Agency's preferred
    value for use in evaluating
    potential noncarcinogenic health
    effects resulting from exposure to
    a chemical.

ACCEPTABLE INTAKE FOR
SUBCHRONIC EXPOSURE
(EPA)
DA October 12, 1990
SF AIS
BT1 Standards
 BT2 Codes, Standards, and
    Regulations

BT-Broader Term   NT-Narrower Term   RT Related Term

RT Hazardous Chemicals
RT Intake
SO Environmental Protection Agency Glossary
DEF (EPA) An estimate similar in concept to a subchronic reference dose (RfD), however, derived using a less rigorously defined methodology. Subchronic RfDs have replaced AICs as the Agency's preferred value for use in evaluating potential noncarcinogenic health effects resulting from exposure to a chemical.

ACCEPTABLE RISKS
(SSDC)
DA October 12, 1990
SY Residual Risks
BT1 Risks
RT Assumed Risks
SO System Safety Development Center Glossary
DEF (SSDC) Residual risks remaining after controls have been applied to associated hazards that have been identified, quantified to the maximum extent practicable, analyzed, communicated to the proper level of management and accepted after proper evaluation.

ACCESSIBLE ENVIRONMENT
(ANL; CFR)
DA May 22, 1991
BT1 Environment
RT Atmosphere
RT Controlled Areas
RT Surface Waters
SO Wastes
DEF (CFR) The atmosphere, land surface, surface water, oceans, and the portion of the lithosphere that is outside the controlled area.

ACCIDENT PRONE SITUATION
(SSDC; IAEA)
DA October 12, 1990
BT1 Conditions
RT Accidents
SO System Safety Development Center Glossary
DEF (SSDC) A condition in which accidents are predisposed due to the existence of unwanted energy flows around potential targets in the absence of adequate barriers.

ACCIDENT RESPONSE CAPABILITIES COORDINATING COMMITTEE
(DOE Order 5500.1A Attach. II, 2)
DA January 24, 1991
BT1 Committees
BT2 Administrative Organizations
BT3 Organizations
SO Environmental Management
DEF (DOE Order 5500.1A) An advisory body that assists and advises the manager of the Albuquerque Operations Office in matters relating to the overall management and coordination of DOE's nuclear weapons accident response.

ACCIDENT RESPONSE GROUPS
(DOE Order 5500.1A Attach. II, 3)
DA January 24, 1991
BT1 Groups
BT2 Administrative Organizations
BT3 Organizations
SO Environmental Management
DEF (DOE Order 5500.1A) Groups of technical and scientific experts composed of DOE and DOE-contractor personnel assigned responsibility for providing DOE assistance to peacetime accidents and significant incidents involving nuclear weapons anywhere in the world.

ACCIDENT SITES
(EPA)
DA October 12, 1990
BT1 Sites/Areas
RT Accidents
RT Facilities
RT Incidents
RT Occurrences
SO Environmental Protection Agency Glossary
DEF (EPA) Locations of unexpected occurrences, failures, or losses, either at a plant or along a transportation route, resulting in a release of hazardous materials.

ACCIDENT TYPES
(SSDC)
DA October 12, 1990
SY Accidental Occurrences
SY Accidents
RT Accidents
SO System Safety Development Center Glossary
DEF (SSDC) Classification of accidents according to cause, such as the energy or environmental conditions involved, e.g., electrical accidents, radiation accidents.

ACCIDENTAL OCCURRENCES
(SWDA; RCRA; CFR)
DA October 12, 1990
SY Accident Types
BT1 Occurrences
NT1 Non-Sudden Accidental Occurrences
NT1 Sudden Accidental Occurrences
NT2 Pressure Releases
NT3 Sudden Releases of Pressure
RT Events and Causal Factors
SO Wastes
DEF Accidents, including continuous or repeated exposure to conditions, which result in bodily injury or property damage neither expected nor intended from the standpoint of the insured.

ACCIDENTS
(DOE Order 5500.1A Attach. II, 1; FIFRA; CFR; SSDC; IAEA; EMER; MORT)
DA October 12, 1990
SY Accident Types
SY Events
SY Incidents
SY Mishaps
SY Occurrences
NT1 Accidents (explosive)
NT1 Aircraft Accidents
NT1 Core Melt Accidents
NT1 Credible Accidents
NT2 Maximum Credible Accident
NT1 Criticality Accidents
NT1 Design Basis Accidents
NT2 Design Basis Earthquakes
NT3 Safe Shutdown Earthquake 135
NT2 Design Basis Fires
NT2 Design Basis Floods
NT2 Design Basis Tornadoes
NT2 Operational DBA
NT1 Failures
NT2 Catastrophic Collapses
NT2 Common Cause Failure
NT2 Containment Failure
NT2 Failed Element Monitors
NT2 Failed Fuel Element
NT2 Flow Failures
NT2 Ground Failures
NT3 Differential Settlement
NT3 Flow Failures
NT3 Landslides
NT3 Lateral Spreads
NT2 Malfunctions
NT2 No Containment Failure
NT2 PCB Transformer Rupture
NT2 Single Failure
NT2 Steam Generator Tube Rupture
NT2 Structural Collapse
NT1 Loss of Coolant Accident
NT2 Small Break LOCA
NT1 Loss of Target Accident
NT1 Minor Accidents
NT1 Nuclear Weapons Accidents
NT1 Operating Basis Accident
NT1 Operational Accidents
NT1 Radioactive Material Transportation Accidents
NT1 Radiological Accidents
RT Accident Prone Situation
RT Accident Types
RT Accident Sites
RT Accidents (explosive)
RT Amelioration
RT Barriers
RT Computerized Accident/Incident Reporting System
RT Contingency Plans
RT Controls
RT Corrective Actions
RT Emergency Response
RT Events
RT Fires
RT Hazards

RT Human Factors Engineering
RT Incidents
RT Occurrences
RT Radiation Protection
RT Safety
RT Specific Control Factors
RT Targets
RT Witnesses
SO Emergency Preparedness
SO Environmental Management
SO Radiation
SO System Safety Development Center Glossary
DEF (MORT) In MORT analysis an accident occurs when an unwanted energy flow or environmental condition that results in adverse consequences reaches persons and/or objects. MORT combines this concept and others into a functional accident definition as follows...(SSDC) Unwanted transfers of energy or environmental conditions which, due to the absence or failure of barriers and/or controls, produces injury to persons, property, or process. (IAEA) In the context of nuclear safety or radiation protection, events which lead or could lead to abnormal exposure conditions.

ACCIDENTS (EXPLOSIVE)
(DOE Order 6430.1A)
DA October 12, 1990
BT1 Accidents
RT Accidents
SO Construction
DEF (DOE Order 6430.1A) Incidents or occurrences that result in an uncontrolled chemical reaction involving explosives.

ACCLIMATIZATION
(EPA)
DA October 12, 1990
SY Adaptations
BT1 Processes
RT Organisms
SO Environmental Protection Agency Glossary
DEF (EPA) The physiological and behavioral adjustments of an organism to changes in its environment.

ACCREDITATION
(DOE Orders 5480.15, 5480.18)
DA October 16, 1990
BT1 Administrative Processes
BT2 Processes
RT Accrediting Boards
RT Accreditation Maintenance Reports
RT Accreditation Review Teams
RT Performance-Based Training
RT Training Programs
RT Training Program Accreditation Plan
DEF (DOE Order 5480.15) The process

of evaluating a program which, through the use of radiation dosimeters, measures and records dose equivalents received by radiation workers. The accreditation process consists of performance tests of personnel dosimeters and site visits by assessors. (DOE Order 5480.18) A process to formally recognize reactor and non-reactor facility training programs as meeting established accreditation objectives and criteria.

ACCREDITATION COORDINATORS
(DOE Order 5480.18)
DA October 16, 1990
BT1 Personnel
RT Training Accreditation Program Staff
DEF (DOE Order 5480.18) Individuals appointed within the program office and the field organization who are responsible for reviewing accreditation documents and maintaining the communication between DOE and the contractor concerning all accreditation activities associated with the office.

ACCREDITATION MAINTENANCE REPORTS
(DOE Order 5480.18)
DA October 12, 1990
BT1 Reports
RT Accreditation
DEF (DOE Order 5480.18) Reports written 2 years after accreditation or renewal of accreditation which describe changes in the accredited training programs since the last accreditation review.

ACCREDITATION REVIEW TEAMS
(DOE Order 5480.18; EMER)
DA October 12, 1990
BT1 Management Teams
BT2 Teams
BT3 Administrative Organizations
BT4 Organizations
RT Accreditation
RT Training Accreditation Program Staff
SO Standards
DEF (DOE Order 5480.18) Groups of individuals representing the Training Accreditation Program with collective expertise in nuclear facility or reactor operations, nuclear facility training, instructional processes, and training program evaluation. These teams review the facility's Contractor Self-Evaluation Report, visit the facility, evaluate training, and prepare a report of conclusions and recommendations.

ACCREDITED
(DOE Order 5480.15)
DA October 12, 1990
RT Personnel Dosimetry Programs
SO Standards
DEF (DOE Order 5480.15) The status conferred upon DOE and DOE contractor dosimetry programs that have undergone the accreditation process and met or exceeded the applicable criteria of DOE/EH-0026 and DOE/EH-0027. Programs are accredited for a period of 2 years from the date of notification.

ACCREDITING BOARDS
(DOE Order 5480.18; EPA)
DA October 12, 1990
BT1 Boards
BT2 Administrative Organizations
BT3 Organizations
RT Accreditation
SO Standards
DEF (DOE Order 5480.18) Independent groups of individuals responsible for making the decision to award or defer accreditation. The Accrediting Board consists of five members with collective expertise in nuclear facility and reactor operations, nuclear and non-nuclear industrial training, instructional processes, and educational accreditation.

ACCURACY
(ESH)
DA October 12, 1990
BT1 Measurements
RT Calibration
RT Inspections
SO Quality Assurance
DEF (ESH) The nearness of a result or the mean of a set of results to the true value. Accuracy is assessed by means of reference sample and percent recoveries.

ACETYLCHOLINE
(EPA)
DA October 12, 1990
BT1 Organic Chemicals
BT2 Chemical Substances
SO Environmental Protection Agency Glossary
DEF (EPA) A drug used in the human body having important neurotransmitter effects on various internal systems; often used as a bronchoconstrictor.

ACH
DA October 12, 1990
SEE Air Changes Per Hour
SO Acronyms

ACID DEPOSITION
(EPA)

DA   October 12, 1990
NT1 Acid Rain
RT   Air Pollution
RT   Combustion
RT   Flue Gases
RT   Fossil Fuels
RT   Manufacturing Processes
RT   Nitric Acid Plants
RT   Nitrogen Oxides (NO$_x$)
RT   Sulfuric Acid Plants
RT   Sulfur Dioxide
SO   Environmental Protection Agency
     Glossary
DEF (EPA) A complex chemical and
     atmospheric phenomenon that
     occurs when emissions of sulfur
     and nitrogen compounds, and
     other substances are transformed
     by chemical processes in the
     atmosphere, often far from the
     original sources, and then
     deposited on earth in either a wet
     or dry form. The wet forms,
     popularly called "acid rain," can fall
     as rain, snow, or fog. The dry
     forms are acidic gases or
     particulates.

ACID RAIN
(EPA)
DA   October 12, 1990
BT1 Acid Deposition
RT   Manufacturing Processes
SO   Environmental Protection Agency
     Glossary
DEF (EPA) A complex chemical and
     atmospheric phenomenon that
     occurs when emissions of sulfur
     and nitrogen compounds and
     other substances are transformed
     by chemical processes in the
     atmosphere, often far from the
     original sources, and then
     deposited on earth in the wet form
     of acid rain.

ACID SUIT
(2650 ACID SUIT)
DA   January 3, 1991
BT1 Anticontamination Clothing
 BT2. Clothing
 BT2 Personal Protective Equipment
  BT3 Equipment/Parts - Personal
       Protective (DOE FRASE
       Vocabulary)
   BT4 Equipment
SO   DOE FRASE VOCABULARY

ACLs
DA   October 12, 1990
SEE Alternate Concentration Limits
SO   Acronyms

ACO
DA   June 18, 1991
SEE Administrative Consent Order
SO   Acronyms

ACQUISITION OF THE EQUIVALENT
(CFR)
DA   November 15, 1990
SY   Replacement
RT   Functionally Equivalent
     Components
DEF (CFR) The substitution for an
     injured resource with a resource
     that provides the same or
     substantially similar services,
     when such substitutions are in
     addition to any substitutions made
     or anticipated as part of response
     actions and when such
     substitutions exceed the level of
     response actions determined
     appropriate to the site pursuant to
     the NCP.

ACR
DA   October 12, 1990
SEE Area Control Room
SO   Acronyms

ACRONYMS
(EPA; ESH; TSA; EM)
DA   October 12, 1990
DEF An alphanumeric abbreviation for a
     complex term, name, or concept.
     NOTE: There are over 900
     acronyms in this thesaurus.
     Acronyms, and their corresponding
     phrases, are found in their correct
     alphabetical position in the main
     body of the thesaurus. A concise
     listing of all acronyms and their
     reciprocal phrases is included in
     the appendices.

ACRR
DA   October 12, 1990
SEE Annular Core Research Reactor
SO   Acronyms

ACRS
DA   October 12, 1990
SEE Advisory Committee on Reactor
     Safety
SO   Acronyms

ACT OF GOD
(ANL; CFR)
DA   May 22, 1991
SY   Act of Nature
SO   Natural Phenomenon
DEF (CFR) An unanticipated grave
     natural disaster or other natural
     phenomenon of an exceptional,
     inevitable, and irresistible
     character, the effects of which
     could not have been prevented or
     avoided by the exercise of due
     care or foresight.

ACT OF NATURE
(CECLA; USC; SSDC)
DA   October 12, 1990
SY   Act of God
SY   Natural Phenomenon

NT1 Divergent Windstorms
 NT2 Downbursts
  NT3 Microbursts
 NT2 Downslope Wind
 NT2 Gust Fronts
NT1 Rotational Windstorms
 NT2 Dust Devils
 NT2 Tornadoes
 NT2 Waterspouts
NT1 Tectonic Deformations
SO   Natural Phenomenon
SO   System Safety Development Center
     Glossary
DEF (SSDC) An act exclusively by
     violence of nature and without the
     interference of any human agency.
     A natural occurrence that cannot
     be foreseen or prevented. Also
     called an act of God.

ACTION AREAS
(ESA; CFR)
DA   October 12, 1990
BT1 Sites/Areas
RT   Actions
SO   Endangered Species
DEF (CFR) All areas to be affected
     directly or indirectly by the Federal
     action and not merely the
     immediate area involved in the
     action.

ACTION CHARLIE
(NFI)
DA   October 12, 1990
SY   Manual Reactor Scram
SO   Nuclear Facilities Incident Database
SO   Radiation
DEF (NFI) Procedural shutdown, manual
     scram.

ACTION DESCRIPTION MEMORANDA
(DOE Order 5440.1D)
DA   May 14, 1991
SF   ADM
SO   Management
DEF (DOE Order 5440.1D) An internal
     DOE document (normally, not
     more than 5 pages) containing a
     concise description of a proposed
     action and discussion of relevant
     potential environmental issues, to
     assist DOE in determining the
     appropriate level of NEPA
     document for a proposed action.

ACTION LEVELS
(TSCA; CFR; EPA)
DA   October 12, 1990
RT   Hazardous Substances
RT   Pesticides
RT   Response Actions
RT   Superfund
RT   Tolerances
SO   Environmental Management
SO   Environmental Protection Agency
     Glossary
DEF (EPA) Regulatory levels
     recommended by EPA for

enforcement by FDA and USDA when pesticide residues occur in food or feed commodities for reasons other than the direct application of the pesticide. As opposed to "tolerances" which are established for residues occurring as a direct result of proper usage, action levels are set for inadvertent residues resulting from previous legal use or accidental contamination. (2) In the Superfund program, the existence of a contaminant concentration in the environment high enough to warrant action or trigger a response under Superfund Amendments and Reauthorization Act of 1986 (SARA) and the National Oil and Hazardous Substances Contingency Plan. The term can be used similarly in other regulatory programs.

ACTION PLANS
(TTGM)
DA May 6, 1991
BT1 Plans
RT Tiger Team Assessments
RT Tiger Team Assessment Reports
SO Management
DEF (TTGM) The Secretary of DOE requests the Program Senior Official to prepare an action plan which addresses each of the assessment findings and root causes identified in a Tiger Team Assessment Report. The action plan is intended to set forth specific actions which the site will undertake to remedy deficiencies noted in the Tiger Team Assessment Report. The plan will include a timetable and funding requirements for the implementation of the planned actions.

ACTION PROPENSITY
(SSDC)
DA October 12, 1990
SO System Safety Development Center Glossary
DEF (SSDC) The ability to make things happen. In an organization, the general manager has a much higher action propensity than those in staff positions, such as safety.

ACTIONS
(ESA; CFR)
DA October 12, 1990
BT1 Responses
NT1 Agency Actions
NT1 Alternative Courses of Action
NT1 Assessment Actions
NT1 Automatic Incident Actions
NT1 Categorical Exclusions
NT1 Corrective Actions

NT1 Emergency Actions
NT1 Incident Actions
NT1 Interim Actions
NT1 Proposed Actions
NT1 Protective Actions
 NT2 Denials
NT1 Reasonable and Prudent Alternatives
NT1 Reasonable and Prudent Measures
NT1 Recovery Actions
NT1 Remedial Actions
 NT2 Remedial Design
 NT2 Source Control Remedial Actions
 NT2 Uranium Mill Tailings Remedial Action
NT1 Remedial Measures
NT1 Remedies
NT1 Removal Actions
NT1 Replacement
NT1 Response Actions
 NT2 Operable Units
NT1 Restoration
RT Action Areas
RT Effects of the Action
SO Endangered Species
DEF (CFR) All activities or programs of any kind authorized, funded, or carried out, in whole or in part, by Federal agencies in the United States or upon the high seas. Examples include, but are not limited to:(a) actions intended to conserve listed species or their habitat; (b) the promulgation of regulations; (c) the granting of licenses, contracts, leases, easements, rights-of-way, permits, or grants-in-aid; or (d) actions directly or indirectly causing modifications to the land, water, or air.

ACTIVATED CARBON
(EPA)
DA October 12, 1990
RT Carbon Adsorbers
RT Granular Activated Carbon Treatment
RT Sorption
RT Wastewater Treatment Processes
SO Environmental Protection Agency Glossary
DEF (EPA) A highly adsorbent form of carbon used to remove odors and toxic substances from liquid or gaseous emissions. In waste treatment it is used to remove dissolved organic matter from waste water. It is also used in motor vehicle evaporative control systems.

ACTIVATED SLUDGES
(EPA)
DA October 12, 1990
BT1 Sludge
 BT2 Wastes
RT Mixed Liquor
RT Wastewater Treatment Processes

SO Environmental Protection Agency Glossary
DEF (EPA) Sludges that result when primary effluent is mixed with bacteria-laden sludge and then agitated and aerated to promote biological treatment. This speeds breakdown of organic matter in raw sewage undergoing secondary waste treatment.

ACTIVATION (EMERGENCY)
(CERCLA; CFR; IAEA; ANL)
DA October 12, 1990
BT1 Administrative Processes
 BT2 Processes
RT Emergency Situations
RT National Response Teams
RT Regional Response Teams
SO Compensation and Liability
SO Environmental Management
SO Radiation
DEF (CFR) Notification by telephone or other expeditious manner or, when required, the assembly of some or all appropriate members of the National Response Team or the Regional Response Team.

ACTIVATION (NUCLEAR)
(IAEA)
DA November 20, 1990
RT Induced Radioactivity
SO Radiation
DEF (IAEA) The production of induced activity by nuclear reactions.

ACTIVE INGREDIENTS
(FIFRA; CFR; USC; EPA)
DA October 12, 1990
RT Pesticides
SO Environmental Management
SO Environmental Protection Agency Glossary
DEF (EPA) In any product, components which kill, or otherwise control, target pests. Pesticides are regulated primarily on the basis of active ingredients. (USC) (1) in the case of a pesticide other than a plant regulator, defoliant, or desiccant, ingredients which will prevent, destroy, repel, or mitigate any pest;(2)in the case of a plant regulator, ingredients which, through physiological action, will accelerate or retard the rate of growth or rate of maturation or otherwise alter the behavior of ornamental or crop plants or product thereof; (3) in the case of a defoliant, ingredients which will cause the leaves or foliage to drop from a plant; and (4)in the case of desiccant, ingredients which will artificially accelerate the drying of plant tissue.

ACTIVE LIFE
(SWDA; RCRA)
DA October 12, 1990
BT1 Time Designations
RT Hazardous Waste Facilities
SO Wastes
DEF (SWDA) For a facility, means the
     period from the initial receipt of
     hazardous waste at the facility until
     the regional administrator receives
     certification of final closure.

ACTIVE MAINTENANCE
(ANL; CFR)
DA May 22, 1991
BT1 Maintenance
 BT2 Activities
SO Wastes
DEF (CFR) Any significant remedial
     activity needed during the period
     of institutional control to maintain a
     reasonable assurance that the
     performance objectives in 10 CFR
     61.41 and 61.42 are met. Such
     active maintenance includes
     ongoing activities such as the
     pumping and treatment of water
     from a disposal unit or one-time
     measures such as replacement of
     a disposal unit cover. Active
     maintenance does not include
     custodial activities such as repair
     of fencing, repair or replacement
     of monitoring equipment,
     revegetation, minor additions to
     soil cover, minor repair of disposal
     unit covers, and general disposal
     site upkeep such as mowing
     grass.

ACTIVE MINES
(CFR)
DA October 12, 1990
BT1 Conventional Mines
 BT2 Sites/Areas
SO Air Pollution
DEF (CFR) Underground uranium mines
     from which ore or waste material
     is currently removed by
     conventional methods.

ACTIVE PORTIONS
(SWDA; RCRA; TSCA; CFR)
DA October 12, 1990
BT1 Sites/Areas
RT Closed Portions
RT Inactive Portions
RT Treatment, Storage, and Disposal
     Facilities
SO Environmental Management
SO Wastes
DEF (SWDA) Portions of a facility where
     treatment, storage, or disposal
     operation are being or have been
     conducted and which are not
     closed portions.

ACTIVITIES
DA February 11, 1991

NT1 Activity Types (DOE FRASE
      Vocabulary)
 NT2 Building/Equip Maint/Repair
       Activity
 NT2 Classified Activity
 NT2 Construction Activity
 NT2 Decommissioning Activity
 NT2 Decontamination Activity
 NT2 Emergency Response Activity
 NT2 Equipment Installation Activity
 NT2 Food Service Activity
 NT2 Fuel Handling Activity
 NT2 Grounds Maintenance Activity
 NT2 Inspection/Monitoring Activity
 NT2 Janitorial/Housekeeping Activity
 NT2 Material Handling Activity
 NT2 Mining/Drilling Activity
  NT3 Strip Mining
 NT2 No Activity
 NT2 Office Activity
 NT2 Other Non-Task Activity
 NT2 Physical Fitness Training Activity
 NT2 Pre Start-Up/ Calibration Activity
 NT2 Production/Operation Activity
 NT2 Reactor Refueling Activity
 NT2 Recreation/Break Activity
 NT2 Research/Testing Activity
 NT2 Security Activity
 NT2 Training/Education Activity
 NT2 Transportation Activity
 NT2 Travel Activity
 NT2 Unknown/Undetermined Activity
 NT2 Vehicle Maint/Repair Activity
NT1 Atomic Energy Defense Activities
NT1 Civilian Nuclear Activities
NT1 Commercial Activities
NT1 Explosives Activities
NT1 Maintenance
 NT2 Active Maintenance
 NT2 Building/Equip Maint/Repair
       Activity
 NT2 Corrective Repair Maintenance
 NT2 Grounds Maintenance Activity
 NT2 Predictive Maintenance
 NT2 Preventive Maintenance
 NT2 Reliability-Centered Maintenance
 NT2 Source Control Maintenance
       Measures
 NT2 Surveillance and Maintenance
 NT2 Vehicle Maint/Repair Activity
 NT2 Wastewater Operations and
       Maintenance
NT1 Major Construction Activities
NT1 Monitoring
 NT2 Air Monitoring
 NT2 Biomonitoring
 NT2 Environmental Monitoring
  NT3 Effluent Monitoring
  NT3 Environmental Surveillance
 NT2 Field Monitoring
 NT2 Health Monitoring
 NT2 Interstitial Monitoring
 NT2 Radiological Monitoring
  NT3 Smear
 NT2 Vibration and Acoustic Monitoring
 NT2 Well Monitoring
NT1 Nuclear Regulatory Commission
      Licensed Activities
NT1 Operations
 NT2 Aviation Operations
  NT3 Charter Operations

 NT2 Burial Operations
  NT3 Shallow Land Burial
 NT2 Cleanup Operations
 NT2 Department of Energy Operations
 NT2 DOE Operations
 NT2 Emergency Renovation
       Operations
 NT2 Headquarters Operations
 NT2 Planned Renovation Operations
 NT2 Reactor Operations
 NT2 Wastewater Operations and
       Maintenance
 NT2 Waste Management Operations
NT1 Operations and Maintenance
NT1 Post Removal Site Control
NT1 Post-Incident Activities
NT1 Protection of the Public Health and
      Welfare
NT1 Regulated Activities
NT1 Research, Development,
      Demonstration, Testing and
      Evaluation
NT1 Research and Development
 NT2 Applied Research
 NT2 Basic Research
NT1 Vital Activities
NT1 Waterborne Activities
RT Graded Approach
RT Processes
SO Management
DEF (WTID) Natural or normal functions
     or operations; an occupation,
     pursuit or recreation in which an
     individual participates.

ACTIVITY (NUCLEAR)
(NIH; IAEA)
DA November 20, 1990
SY Radioactivity
BT1 Measurements
NT1 Gross Alpha Particle Activity
NT1 Gross Beta Particle Activity
NT1 Specific Activity
 NT2 Low Specific Activity
RT Hazards
SO Radiation
DEF (NIH) The number of nuclear
     disintegrations occurring in a given
     quantity of material per unit time.
     Sometimes the former special unit
     Curie (Ci) is still used. (IAEA) A
     special name for the unit of activity
     is becquerel (Bq).

ACTIVITY DATA SHEETS
DA January 8, 1991
SF ADS (Activity Data Sheets)
BT1 Reports
RT NEPA Status Reports

**ACTIVITY-MEDIAN AERODYNAMIC
DIAMETER**
(CFR; IAEA)
DA October 12, 1990
SF AMAD
BT1 Measurements
RT Aerosols
SO Radiation
DEF (IAEA) The diameter of a unit
     density sphere with the same

terminal settling velocity in air as that of the aerosol particle whose activity is the median for the entire aerosol considered.

ACTIVITY PER TON
DA January 8, 1991
SF *G/T*
BT1 Units of Measure

ACTIVITY TYPES (DOE FRASE VOCABULARY)
(DOE FRASE Vocabulary Numeric Keys 1225-1299)
DA November 28, 1990
BT1 Activities
NT1 Building/Equip Maint/Repair Activity
NT1 Classified Activity
NT1 Construction Activity
NT1 Decommissioning Activity
NT1 Decontamination Activity
NT1 Emergency Response Activity
NT1 Equipment Installation Activity
NT1 Food Service Activity
NT1 Fuel Handling Activity
NT1 Grounds Maintenance Activity
NT1 Inspection/Monitoring Activity
NT1 Janitorial/Housekeeping Activity
NT1 Material Handling Activity
NT1 Mining/Drilling Activity
 NT2 Strip Mining
NT1 No Activity
NT1 Office Activity
NT1 Other Non-Task Activity
NT1 Physical Fitness Training Activity
NT1 Pre Start-Up/ Calibration Activity
NT1 Production/Operation Activity
NT1 Reactor Refueling Activity
NT1 Recreation/Break Activity
NT1 Research/Testing Activity
NT1 Security Activity
NT1 Training/Education Activity
NT1 Transportation Activity
NT1 Travel Activity
NT1 Unknown/Undetermined Activity
NT1 Vehicle Maint/Repair Activity
RT DOE FRASE Categories
SO DOE FRASE VOCABULARY
DEF A subject category used with the DOE FRASE Vocabulary.

ACTS
(SWDA; RCRA; ESA; CFR; ESH)
DA October 12, 1990
BT1 Statutes and Regulations
NT1 Administrative Procedures Act
 NT2 Individual Plant Examination
NT1 Archaeological Resources Protection Act of 1979
NT1 Atomic Energy Act
NT1 Clean Air Act
NT1 Clean Water Act
NT1 Comprehensive Environmental Response, Compensation, etc.
NT1 Endangered Species Act
NT1 National Environmental Policy Act
NT1 Nuclear Waste Policy Act
NT1 Resource Conservation and Recovery Act

NT1 Safe Drinking Water Act
NT1 Solid Waste Disposal Act
NT1 Superfund Amendments and Reauthorization Act of 1986
NT1 Toxic Substances Control Act
SO Air Pollution
SO Endangered Species
SO Wastes
SO Water Pollution
DEF The formal products of a legislative body. Statutes: Clean Air Act; Solid Waste Disposal Act; Resource Conservation and Recovery Act; Endangered Species Act; Public Health Service Act; Clean Water Act; Safe Drinking Water Act; Superfund Amendments and Reauthorization Act.

ACTUAL CASH VALUE
(SSDC)
DA October 12, 1990
BT1 Costs
RT Property Loss
RT Replacement Values
SO System Safety Development Center Glossary
DEF (SSDC) Replacement cost of property at the time of loss, less depreciation based on age, condition, time in use, and obsolescence.

ACTUARIAL
(SSDC)
DA October 12, 1990
RT Known Precedents
SO System Safety Development Center Glossary
DEF (SSDC) Using data from actual past experience to project future performance.

ACUTE DERMAL LD 50
(FIFRA; CFR)
DA January 24, 1991
BT1 Doses
SO Environmental Management
DEF (CFR) A single dermal dose of a substance, expressed as milligrams per kilogram of body weight, that is lethal to 50% of the test population of animals under test conditions as specified in the Registration Guidelines.

ACUTE EFFECTS
(EMER)
DA February 1, 1991
BT1 Effects
RT Risk Assessment
SO Emergency Preparedness
DEF (EMER) Symptoms of exposure to a hazardous material; normally the result of a short-term exposure which comes quickly to a crisis.

ACUTE EXPOSURE
(EPA; EMER)
DA October 12, 1990
BT1 Exposure
RT Acute Oral LD 50
RT Acute Toxicity
SO Emergency Preparedness
SO Environmental Protection Agency Glossary
DEF (EPA) A single exposure to a toxic substance which results in severe biological harm or death. Acute exposures are usually characterized as lasting no longer than a day.

ACUTE LD 50
(FIFRA; CFR)
DA January 24, 1991
BT1 LD 50
 BT2 Doses
SO Environmental Management
DEF (CFR) A concentration of a substance, expressed as parts per million parts of medium, that is lethal to 50% of the test population of animals under test conditions as specified in the Registration Guidelines.

ACUTE ORAL LD 50
(FIFRA; CFR)
DA January 24, 1991
BT1 LD 50
 BT2 Doses
BT1 Lethal Dose
RT Acute Exposure
SO Environmental Management
DEF (CFR) A single orally administered dose of a substance, expressed as milligrams per kilogram of body weight, that is lethal to 50 percent of the test population of animals under test conditions as specified in the Registration Guidelines.

ACUTE TOXICITY
(FIFRA; CFR; EPA; EMER)
DA October 12, 1990
BT1 Toxicity
RT Acute Exposure
SO Emergency Preparedness
SO Environmental Management
SO Environmental Protection Agency Glossary
DEF (EPA) The ability of a substance to cause poisonous effects resulting in severe biological harm or death soon after a single exposure or dose. Also, any severe poisonous effect resulting from a single short-term exposure to a toxic substance.

ADAPTATIONS
(EPA)
DA October 12, 1990
SY Acclimatization
BT1 Biological Processes

BT-Broader Term   NT-Narrower Term   RT Related Term

BT2 Processes
RT Ecology
RT Natural Selection
RT Organisms
SO Environmental Protection Agency
   Glossary
DEF (EPA) Changes in an organism's
   structure that help it adjust to its
   surroundings.

ADAPTER(S)
(2375 ADAPTER)
DA January 3, 1991
BT1 Equipment/Parts - Electrical (DOE
   FRASE Vocabulary)
BT2 Equipment
SO DOE FRASE VOCABULARY

ADC
DA October 12, 1990
SEE Analog to Digital Converter
SO Acronyms

ADD ON CONTROL DEVICES
(EPA)
DA October 12, 1990
BT1 Control Devices
BT2 Devices
SO Environmental Protection Agency
   Glossary
DEF (EPA) An air pollution control
   device such as carbon absorber or
   incinerator which reduces the
   pollution in an exhaust gas. The
   control device usually does not
   affect the process being controlled
   and thus is "add on " technology,
   as opposed to a scheme to control
   pollution through making some
   alteration to the basic process.

ADDITIVES
(TSCA; CFR)
DA January 24, 1991
BT1 Chemical Substances
NT1 Biological Additives
NT1 Burning Agents
NT1 Sinking Agents
SO Environmental Management
DEF (CFR) A chemical substance that is
   intentionally added to another
   chemical substance to improve its
   stability or impart some other
   desirable quality.

ADEQUATELY WETTED
(CFR)
DA October 12, 1990
RT Air Pollutants
RT Dust
SO Air Pollution
DEF (CFR) Sufficiently mixed or coated
   with water or an aqueous solution
   to prevent dust emissions.

ADHESION
(EPA)
DA October 12, 1990
BT1 Processes

RT Agglutination
SO Environmental Protection Agency
   Glossary
DEF (EPA) Molecular attraction which
   holds the surfaces of two
   substances in contact.

ADI
DA October 12, 1990
SEE Acceptable Daily Intake
SO Acronyms

ADM
DA May 14, 1991
SEE Action Description Memoranda
SO Acronyms

**ADMIN. SUPPORT/CLERICAL
EMPLOYEE**
(0450 OFFICE;CLERK)
DA November 28, 1990
BT1 Occupations
BT1 Personnel
NT1 Firefighter
NT1 Food Service Employee
NT1 Janitor
NT1 Misc Service Employee
NT1 Security Guard
RT Office Activity
SO DOE FRASE VOCABULARY

ADMINISTRATIVE CONSENT ORDER
DA June 18, 1991
SF ACO
SO Management

ADMINISTRATIVE ORDERS
(EPA)
DA October 12, 1990
RT Enforcement Decision Documents
SO Environmental Protection Agency
   Glossary
DEF (EPA) Legal documents signed by
   EPA directing an individual,
   business, or other entity to take
   corrective action or refrain from an
   activity. They describe the
   violation and actions to be taken,
   and can be enforced in court.
   Such orders may be issued, for
   example, as a result of an
   administrative complaint whereby
   the respondent is ordered to pay a
   penalty for violations of a statute.

ADMINISTRATIVE ORGANIZATIONS
DA February 5, 1991
BT1 Organizations
NT1 Agencies
 NT2 Federal Agencies
  NT3 Atomic Energy Commission
  NT3 Cognizant Federal Agencies
  NT3 DOE Design Agency
  NT3 DOE Production Agency
  NT3 Federal Aviation Administration
  NT3 Lead Agencies
  NT3 National Aeronautic and Space
   Administration
  NT3 Nuclear Regulatory Commission

   NT4 NRC Office of Nuclear Reactor
    Regulation
   NT4 NRC Office of Nuclear
    Regulatory Research
  NT3 Occupational Safety and Health
   Administration
  NT3 Office of Management and
   Budget
  NT3 Procuring Agencies
  NT3 U.S. Department of Energy
   NT4 Headquarters Operations
   NT4 Naval Reactors
   NT4 Operations Offices
    NT5 Albuquerque Operations Office
    NT5 Chicago Operations Office
    NT5 Idaho Operations Office
    NT5 Nevada Operations Office
    NT5 Oak Ridge Operations Office
    NT5 Richland Operations Office
    NT5 San Francisco Operations
     Office
    NT5 Savannah River Operations
     Office
   NT4 Program Offices
    NT5 Office of the Assistant
     Secretary for Nuclear Energy
    NT5 Office of the Assistant
     Secretary for Fossil Energy
    NT5 Office of the Assistant
     Secretary for Defense
     Programs
    NT5 Office of the Assistant
     Secretary for Conservation
     et.al.
    NT5 Office of the Assistant
     Secretary for Environment, et.
     al.
     NT6 Office of Special Projects
    NT5 Office of Environmental
     Restoration and Waste
     Management
    NT5 Office of New Production
     Reactors
    NT5 Office of Civilian Radioactive
     Waste Management
    NT5 Office of Energy Research
     NT6 Office of Basic Energy
      Sciences
     NT6 Office of Health and
      Environmental Research
  NT3 U.S. Department of Justice
  NT3 U.S. Department of Labor
  NT3 U.S. Department of
   Transportation
   NT4 Research and Special Programs
    Administration
  NT3 U.S. Environmental Protection
   Agency
   NT4 Regional Office
  NT3 User Agencies
 NT2 Local Agencies
  NT3 Implementing Agencies
  NT3 Intermunicipal Agencies
  NT3 Local Governments
  NT3 Local Educational Agencies
 NT2 State Agencies
  NT3 Air Pollution Control Agencies
   NT4 Interstate Air Pollution Control
    Agencies
  NT3 Colorado Department of Health
  NT3 Division of the State Fire Marshall

NT3 Implementing Agencies
NT3 Interstate Agencies
NT4 Interstate Air Pollution Control
Agencies
NT3 Missouri Department of Natural
Resources
NT3 New Mexico Environmental
Improvement Division
NT3 Ohio Bureau of Underground
Storage Tank Regulation
NT3 Ohio Department of Commerce
NT3 Ohio Department of Health
NT3 Ohio Environmental Protection
Agency
NT3 Procuring Agencies
NT3 State Routing Agencies
NT3 State Emergency Response
Commission
NT3 State Authority
NT3 Tennessee Department of
Commerce
NT4 Bureau of Environment
NT3 Utah Department of Health
NT2 Support Agencies
NT1 Boards
NT2 Accrediting Boards
NT2 Energy Research Advisory Board
NT2 GDC Planning Board
NT2 Regional Water Quality Control
Board
NT1 Committees
NT2 Accident Response Capabilities
Coordinating Committee
NT2 Advisory Committee on Reactor
Safety
NT2 Emergency Management
Coordination Committee
NT2 Interagency Committee on
Standards Policy
NT2 Local Emergency Planning
Committees
NT2 Metrication Operating Committee
NT2 Metric Transition Committee
NT2 Plant Oversight Review Committee
NT2 Regional Assistance Committees
NT2 Safety Review Committee
NT2 Subcommittee on Federal
Response
NT2 Waste Acceptance Criteria
Committee
NT1 Divisions
NT2 Computer Operations Division
NT2 Health, Safety, and Environment
Division
NT2 New Mexico Environmental
Improvement Division
NT2 Occupational Health Division
NT1 Groups
NT2 Accident Response Groups
NT2 Carcinogen Risk Assessment
Verification Endeavor Workgroup
NT2 Emergency Management
Coordination Committee
Secretariat
NT2 Interagency Group on Energy
Vulnerability
NT2 Purchasing Group
NT2 Reactor Materials Control Group
NT2 Risk Retention Group
NT2 Standards-developing Groups

NT2 State and Tribal Government
Working Group
NT1 Indian Governing Bodies
NT1 Offices
NT2 Grand Junction Projects Office
NT2 NRS Office for Analysis and
Evaluation of Operational Data
NT2 NRS Office of Inspection and
Enforcement
NT2 Office of Defense Waste and
Transportation Management
NT2 Office of Management and Budget
NT2 Office of Technology Development
NT2 Office of Weapons Production
NT2 Operations Offices
NT3 Albuquerque Operations Office
NT3 Chicago Operations Office
NT3 Idaho Operations Office
NT3 Nevada Operations Office
NT3 Oak Ridge Operations Office
NT3 Richland Operations Office
NT3 San Francisco Operations Office
NT3 Savannah River Operations
Office
NT2 Program Offices
NT3 Office of the Assistant Secretary
for Nuclear Energy
NT3 Office of the Assistant Secretary
for Fossil Energy
NT3 Office of the Assistant Secretary
for Defense Programs
NT3 Office of the Assistant Secretary
for Conservation et.al.
NT3 Office of the Assistant Secretary
for Environment, et. al.
NT4 Office of Special Projects
NT3 Office of Environmental
Restoration and Waste
Management
NT3 Office of New Production
Reactors
NT3 Office of Civilian Radioactive
Waste Management
NT3 Office of Energy Research
NT4 Office of Basic Energy Sciences
NT4 Office of Health and
Environmental Research
NT2 Program Enrichment Office
NT2 Regional Coordinating Offices
NT2 Sample Management Offices
NT1 Operating Organizations
NT1 Program Organizations
NT1 Teams
NT2 Management Teams
NT3 Accreditation Review Teams
NT3 Continuity of Government
Emergency Management Teams
NT3 Crisis Management Teams
NT3 Emergency Management Teams
NT4 Energy Emergency
Management Teams
NT4 Operational Emergency
Management Teams
NT3 Headquarters Coordinating
Teams
NT2 Response Teams
NT3 Airborne Response Teams
NT3 Emergency Support Teams
NT3 Emergency Response Teams
NT3 Energy Emergency Teams
NT3 Environmental Response Teams

NT3 Explosive Ordnance Disposal
Teams
NT3 Hazardous Materials Response
Teams
NT3 Hostage Negotiation Teams
NT3 National Response Teams
NT3 Nuclear Emergency Search
Teams
NT3 Radiological Assistance Teams
NT3 Regional Response Teams
NT3 Special Response Teams
NT3 Tactical Response Teams
NT2 Tiger Teams
RT Commercial Organizations
RT Educational Organizations
RT Research and Development
Organizations

ADMINISTRATIVE PROCEDURES ACT
(EPA)
DA October 12, 1990
BT1 Acts
BT2 Statutes and Regulations
NT1 Individual Plant Examination
RT Procedures
SO Environmental Protection Agency
Glossary
DEF (EPA) Laws that spell out
procedures and requirements
related to the promulgation of
regulations.

ADMINISTRATIVE PROCESSES
DA February 26, 1991
BT1 Processes
NT1 Accreditation
NT1 Activation (Emergency)
NT1 Arbitration
NT1 Assessments
NT2 Biological Assessments
NT2 Consequence Assessments
NT2 Credibility Assessments
NT2 Damage Assessments
NT3 Natural Resource Damage
Assessment
NT3 Natural Resource Damage
Preassessment Screens
NT2 Data Quality Assessments
NT2 Endangered Assessment
NT2 Environmental Assessments
NT2 Exposure Assessment
NT2 Operation Assessment and
Readiness
NT2 Performance Assessments
NT2 Preliminary Assessment
NT2 Process Waste Assessment
NT2 RCRA Facility Assessment
NT2 Risk Assessment
NT3 Probabilistic Risk Assessment
NT2 Seabrook Station Probabilistic
Safety Assessment
NT2 Self-Assessment
NT2 Systematic Assessment of
Licensee Performance
NT2 Tiger Team Assessments
NT3 Environment Assessments
NT3 Management and Organization
Assessment
NT3 Safety and Health Assessments
NT2 Type A Assessments

NT2 Type B Assessments
NT2 Unified Dose Assessments
NT1 Audits (SSDC)
NT2 Environmental Audits
NT2 Field Audits
NT2 Internal Audits
NT2 System Audits
NT1 Classification
NT1 Closure
NT2 Current Closure Cost Estimate
NT2 Current Post-Closure Cost Estimate
NT2 Final Closure
NT2 Partial Closure
NT2 Site Closure and Stabilization
NT1 Comprehensive Planning
NT1 Construction Project Planning
NT1 Coordination Processes
NT1 Cost Recovery
NT1 Critical Incident Technique
NT1 Decommissioning
NT1 Final Authorizations
NT1 Formal Consultation
NT1 General Design Process
NT1 Hazards Identification
NT1 Innovation Diffusion
NT1 Inspections
NT2 Compliance Inspections
NT2 Condition Assessment Surveys
NT2 Failed Instrument Component Inspection
NT2 Maintenance Team Inspection
NT2 Safety System Functional Inspection
NT2 Safety System Outage Modification Inspection
NT2 Site Inspections
NT1 Inspections (Nuclear)
NT1 Inspection and Evaluation
NT1 Investigations
NT2 Remedial Investigations
NT3 RCRA Remedial Investigation
NT2 Special Health Supervision
NT1 Issue Identification
NT1 Modification
NT2 Major Modification
NT2 Major PSD Modification
NT1 Program Evaluation and Review Technique
NT1 Project Planning and Control
NT1 Reconstruction
NT1 Release of Property
NT1 Renovation
NT2 Emergency Renovation Operations
NT1 Reviews
NT2 Environmental Reviews
NT2 Operational Readiness Reviews
NT2 Safety Reviews
NT3 Backfit Reviews
NT3 Independent (Safety) Reviews
NT2 Safety Program Reviews
NT2 Special Reviews
NT1 Safety Analysis Process
NT1 Siting
NT1 State Coordination
NT1 State Notification
NT1 Surveillance
NT2 Environmental Surveillance
NT1 Test Authorizations
NT1 Tiering

NT1 Upstream Processes
NT1 Validation
NT1 Verification
NT2 Calibration Checks
NT2 Verification of Training and Retraining
NT1 Work Processes

ADMINISTRATORS
(SWDA; RCRA; CERCLA; TSCA; SDWA; CFR; EPA; ESH)
DA   October 12, 1990
BT1 Personnel
NT1 Assistant Administrator for Fisheries
NT1 Regional Administrator
RT   Authorized Representatives
SO   Air Pollution
SO   Compensation and Liability
SO   Hazardous Materials
SO   Wastes
SO   Water Pollution
DEF (CFR) The Administrator of the U.S. Environmental Protection Agency, or an authorized representative. (ESH) The Administrator of the Environmental Protection Agency, or any office or employee of the Agency to whom authority has heretofore been delegated, or to whom authority may hereafter be delegated, to act in his stead.

ADS (Activity Data Sheets)
DA   October 12, 1990
SEE Activity Data Sheets
SO   Acronyms

ADS (automatic depressurization system)
DA   October 12, 1990
SEE Automatic Depressurization System
SO   Acronyms

ADULTERANTS
(EPA)
DA   October 12, 1990
BT1 Chemical Substances
RT   Impurities
SO   Environmental Protection Agency Glossary
DEF (EPA) Chemical impurities or substances that by law do not belong in a food or in a pesticide.

ADVANCED LIGHT SOURCE
DA   January 8, 1991
SF   ALS

ADVANCED TEST REACTOR
(SSDC)
DA   May 24, 1991
SY   Test Reactor Area
BT1 Reactors
SO   Radiation
SO   System Safety Development Center Glossary

ADVANCED WASTE WATER TREATMENT
(EPA)
DA   October 12, 1990
BT1 Wastewater Treatment Processes
BT2 Treatment
BT3 Waste Management Processes
BT4 Processes
SO   Environmental Protection Agency Glossary
DEF (EPA) Any treatment of sewage that goes beyond the secondary or biological water treatment stage and includes the removal of nutrients such as phosphorus and nitrogen and a high percentage of suspended solids.

ADVERSE IMPACT ON VISIBILITY
(CAA; CFR; ESH)
DA   October 12, 1990
RT   Visibility Impairments
SO   Air Pollution
DEF (CFR) Denotes visibility impairment which interferes with the management, protection, preservation, or enjoyment of the visitor's visual experience of the Federal Class I area. This determination must be made on a case-by-case basis taking into account the geographic extent, intensity, duration, frequency and time of visibility impairments, and how these factors correlate with (1) times of visitor use of the Federal Class I area, and (2) the frequency and timing of natural conditions that reduce visibility. This term does not include effects on integral vistas.

ADVERSE MODIFICATIONS
(ESA; CFR)
DA   October 19, 1990
SY   Destruction
RT   Critical Habitats
SO   Endangered Species
DEF (CFR) Direct or indirect alterations that appreciably diminish the value of critical habitat for both the survival and recovery of a listed species. Such alterations include, but are not limited to, alterations adversely modifying any of those physical or biological features that were the basis for determining the habitat to be critical.

ADVISORY
(EPA)
DA   October 12, 1990
BT1 Reports
RT   Risk Management
SO   Environmental Protection Agency Glossary
DEF (EPA) A non-regulatory document that communicates risk information to persons who may have to make risk management decisions.

SY-Synonymous Terms   SO-Source/Subject Category   SF-See From

ADVISORY COMMITTEE ON REACTOR SAFETY
DA   January 8, 1991
SF   *ACRS*
BT1   Committees
  BT2   Administrative Organizations
  BT3   Organizations
RT   Reactors

AEA
DA   October 12, 1990
SEE Atomic Energy Act
SO   Acronyms

AEC
DA   October 12, 1990
SEE Atomic Energy Commission
SO   Acronyms

AEOD
DA   October 12, 1990
SEE NRS Office for Analysis and
      Evaluation of Operational Data
SO   Acronyms

AERATION
(EPA)
DA   October 12, 1990
BT1   Biological Processes
  BT2   Processes
NT1   Diffused Air Process
NT1   Mechanical Aeration
RT   Aeration Tanks
RT   Aerobic Treatment
RT   Air
RT   Biological Treatment
RT   Composting
SO   Environmental Protection Agency
     Glossary
DEF (EPA) A process which promotes
     biological degradation of organic
     water. The process may be
     passive (as when waste is
     exposed to air) or active (as when
     a mixing or bubbling device
     introduces the air).

AERATION TANKS
(EPA)
DA   October 12, 1990
BT1   Tanks
  BT2   Facility Components
RT   Aeration
SO   Environmental Protection Agency
     Glossary
DEF (EPA) Chambers used to inject air
     into water.

AERATION ZONES
(SWDA; RCRA)
DA   October 19, 1990
SY   Unsaturated Zones
BT1   Zones
  BT2   Sites/Areas
RT   Saturated Zones
SO   Wastes
DEF (CFR) The zones between the land
     surface and the water table.

AERIAL MEASURING SYSTEMS
(DOE Order 5500.1A Attach. II, 4)
DA   January 24, 1991
BT1   Systems
RT   Radiological Monitoring
SO   Environmental Management
DEF (DOE 5500.1A) Aerial detection
     systems with the capability of
     measuring extremely low levels of
     gamma radiation and locating and
     tracking multi-spectral sensing
     capabilities.

AEROBIC
(EPA)
DA   October 12, 1990
RT   Anaerobic
RT   Organisms
SO   Environmental Protection Agency
     Glossary
DEF (EPA) Life or processes that
     require, or are not destroyed by,
     the presence of oxygen.

AEROBIC TREATMENT
(EPA)
DA   October 12, 1990
BT1   Biological Treatment
  BT2   Treatment
  BT3   Waste Management Processes
  BT4   Processes
RT   Aeration
RT   Microorganisms
SO   Environmental Protection Agency
     Glossary
DEF (EPA) Process by which microbes
     decompose complex organic
     compounds in the presence of
     oxygen and use the liberated
     energy for reproduction and
     growth. Types of aerobic
     processes include extended
     aeration, trickling filtration, and
     rotating biological contactors.

AEROSOLS
(EPA)
DA   October 12, 1990
RT   Activity-Median Aerodynamic
     Diameter
RT   Atomize
RT   Dust
RT   Inhalation
RT   Particulates
SO   Environmental Protection Agency
     Glossary
DEF (EPA) A suspension of liquid or
     solid particles in a gas.

AFB
DA   October 12, 1990
SEE Air Force Base
SO   Acronyms

AFESC
DA   October 12, 1990
SEE Air Force Engineering and Services
      Center
SO   Acronyms

AFFECTED FACILITIES
(CAA; ESH)
DA   October 12, 1990
BT1   Facilities
RT   Stationary Sources
SO   Air Pollution
DEF (ESH) Affected facilities means,
     with reference to a stationary
     source, apparatuses to which a
     New Source Performance
     Standard is applicable.

AFFECTED PERSONS
(EMER)
DA   February 1, 1991
SO   Emergency Preparedness
DEF (EMER) Individuals who have been
     exposed and/or injured as a result
     of an accident involving any type
     of hazardous material, to a degree
     requiring special attention (i.e.,
     decontamination, first aid, or
     medical service).

AFTERBURNERS
(EPA)
DA   October 12, 1990
BT1   Control Devices
  BT2   Devices
RT   Combustion
RT   Incinerators
SO   Environmental Protection Agency
     Glossary
DEF (EPA) In incinerator technology,
     burners located so that the
     combustion gases are made to
     pass through their flame in order
     to remove smoke and odors. They
     may be attached to or be
     separated from the incinerator
     proper.

AFW
DA   October 12, 1990
SEE Auxiliary Feedwater
SO   Acronyms

AGENCIES
(SDWA; CFR; USC; ESH; EMER)
DA   October 12, 1990
BT1   Administrative Organizations
  BT2   Organizations
NT1   Federal Agencies
  NT2   Atomic Energy Commission
  NT2   Cognizant Federal Agencies
  NT2   DOE Design Agency
  NT2   DOE Production Agency
  NT2   Federal Aviation Administration
  NT2   Lead Agencies
  NT2   National Aeronautic and Space
        Administration
  NT2   Nuclear Regulatory Commission
    NT3   NRC Office of Nuclear Reactor
          Regulation
    NT3   NRC Office of Nuclear Regulatory
          Research
  NT2   Occupational Safety and Health
        Administration
  NT2   Office of Management and Budget

BT-Broader Term   NT-Narrower Term   RT Related Term

NT2 Procuring Agencies
NT2 U.S. Department of Energy
NT3 Headquarters Operations
NT3 Naval Reactors
NT3 Operations Offices
NT4 Albuquerque Operations Office
NT4 Chicago Operations Office
NT4 Idaho Operations Office
NT4 Nevada Operations Office
NT4 Oak Ridge Operations Office
NT4 Richland Operations Office
NT4 San Francisco Operations Office
NT4 Savannah River Operations
Office
NT3 Program Offices
NT4 Office of the Assistant Secretary
for Nuclear Energy
NT4 Office of the Assistant Secretary
for Fossil Energy
NT4 Office of the Assistant Secretary
for Defense Programs
NT4 Office of the Assistant Secretary
for Conservation et.al.
NT4 Office of the Assistant Secretary
for Environment, et. al.
NT5 Office of Special Projects
NT4 Office of Environmental
Restoration and Waste
Management
NT4 Office of New Production
Reactors
NT4 Office of Civilian Radioactive
Waste Management
NT4 Office of Energy Research
NT5 Office of Basic Energy
Sciences
NT5 Office of Health and
Environmental Research
NT2 U.S. Department of Justice
NT2 U.S. Department of Labor
NT2 U.S. Department of Transportation
NT3 Research and Special Programs
Administration
NT2 U.S. Environmental Protection
Agency
NT3 Regional Office
NT2 User Agencies
NT1 Local Agencies
NT2 Implementing Agencies
NT2 Intermunicipal Agencies
NT2 Local Governments
NT2 Local Educational Agencies
NT1 State Agencies
NT2 Air Pollution Control Agencies
NT3 Interstate Air Pollution Control
Agencies
NT2 Colorado Department of Health
NT2 Division of the State Fire Marshall
NT2 Implementing Agencies
NT2 Interstate Agencies
NT3 Interstate Air Pollution Control
Agencies
NT2 Missouri Department of Natural
Resources
NT2 New Mexico Environmental
Improvement Division
NT2 Ohio Bureau of Underground
Storage Tank Regulation
NT2 Ohio Department of Commerce
NT2 Ohio Department of Health

NT2 Ohio Environmental Protection
Agency
NT2 Procuring Agencies
NT2 State Routing Agencies
NT2 State Emergency Response
Commission
NT2 State Authority
NT2 Tennessee Department of
Commerce
NT3 Bureau of Environment
NT2 Utah Department of Health
NT1 Support Agencies
SO Air Pollution
SO Emergency Preparedness
SO Hazardous Materials
SO Water Pollution
DEF (WEBSTER) Official bodies through
which power is exerted or an end
is achieved; an administrative
division of the government.

AGENCY ACTIONS
(ESA; CFR)
DA October 12, 1990
BT1 Actions
BT2 Responses
RT Federal Agencies
SO Endangered Species
DEF (CFR) All actions of any kind
authorized, funded or carried out,
in whole or in part by Federal
agencies, including, in the
instance of an application for a
permit or license, the underlying
activity for which the permit or
license is sought.

AGENCY LEAD OFFICIALS
(EMER)
DA February 1, 1991
BT1 Personnel
SO Emergency Preparedness
DEF (EMER) The designated officials in
each participating agency
authorized to direct that agency's
response to a radiological
emergency.

AGENT ORANGE
(EPA)
DA October 12, 1990
BT1 CERCLA Hazardous Substances
BT2 Hazardous Substances
BT1 Defoliants
BT2 Herbicides
BT3 Pesticides
BT4 Hazardous Substances
RT Halogenated Organic Compounds
SO Environmental Protection Agency
Glossary
DEF (EPA) A toxic herbicide and
defoliant which was used in the
Vietnam conflict. It contains
2,4,5-trichlorophenoxyactic acid
(2,4,5-T) and 2-4
dichlorophenoxyacetic acid (2,4-D)
with trace amounts of dioxin.

AGGLOMERATION
(EPA)
DA October 12, 1990
BT1 Processes
RT Air Pollution
RT Flocculation
RT Sedimentation
SO Environmental Protection Agency
Glossary
DEF (EPA) The process by which
precipitation particles grow larger
by collision or contact with cloud
particles or other precipitation
particles.

AGGLUTINATION
(EPA)
DA October 12, 1990
BT1 Processes
RT Adhesion
SO Environmental Protection Agency
Glossary
DEF (EPA) The process of uniting solid
particles coated with a thin layer of
adhesive material or of arresting
solid particles by impact on a
surface coated with an adhesive.

AGITATOR
(2100 AGITATOR)
DA December 10, 1990
BT1 Machines (DOE FRASE
Vocabulary)
BT2 Equipment
SO DOE FRASE VOCABULARY

AGREEMENT STATES
(CFR; EMER)
DA October 12, 1990
BT1 States
RT Agreements
RT Nuclear Regulatory Commission
SO Air Pollution
SO Emergency Preparedness
DEF (CFR) States with which the Atomic
Energy Commission or the
Nuclear Regulatory Commission
has entered into an effective
agreement under subsection
274(b) of the Atomic Energy Act of
1954, as amended.

AGREEMENTS
(ESA; CFR)
DA October 12, 1990
NT1 Consent Order and Compliance
Agreement
NT1 Cooperative Agreements
NT1 Federal Facilities Agreement
NT2 Federal Facilities Compliance
Agreement
NT1 Interagency Agreement
NT2 Federal Facilities Compliance
Agreement
NT1 Mutual Assistance Agreements
NT1 State/EPA Agreements
NT2 Superfund Memorandum of
Agreement
RT Agreement States

SY-Synonymous Terms   SO-Source/Subject Category   SF-See From

SO Endangered Species
DEF (CFR) Signed documented
   statements of the actions to be
   taken by the State(s) and the
   Director in furthering certain
   purposes of the Act. They include:
   (1) A Cooperative Agreement
   entered into pursuant to section
   6(c) of the Act and, where
   appropriate, containing provisions
   found in section 6(d)(2) of the act.
   (2) A Grant-In-Aid Award which
   includes a statement of the actions
   to be taken in connection with the
   conservation of endangered or
   threatened species receiving
   Federal financial assistance,
   objectives and costs of such
   action and costs to be borne by
   the Federal Government and by
   the State(s).

AGRICULTURAL MACHINE
(2101 AGRICULTURAL)
DA   December 10, 1990
BT1 Machines (DOE FRASE
       Vocabulary)
 BT2 Equipment
SO DOE FRASE VOCABULARY

AGRICULTURAL POLLUTION
(EPA)
DA   October 12, 1990
BT1 Pollution
BT1 Wastes
RT Agricultural Solid Wastes
RT Air Pollution
RT Methane
RT Pesticides
RT Water Pollution
SO Environmental Protection Agency
       Glossary
DEF (EPA) The liquid and solid wastes
       from farming, including runoff and
       leaching of pesticides and
       fertilizers, erosion and dust from
       plowing, animal manure and
       carcasses, crop residues, and
       debris.

AGRICULTURAL SOLID WASTES
(RCRA; CFR)
DA   January 24, 1991
BT1 Solid Wastes
 BT2 Wastes
RT Agricultural Pollution
SO Environmental Management
DEF (CFR) The solid waste that is
       generated by the rearing of
       animals and the producing and
       harvesting of crops or trees.

AGRICULTURE PERSONNEL
DA   November 29, 1990
BT1 Occupations
BT1 Personnel
NT1 Forest Worker
NT1 Groundskeeper

NT1 Misc Agriculture Employee
SO DOE FRASE VOCABULARY

AIA
DA   October 12, 1990
SEE Automatic Incident Actions
SO Acronyms

AIC
DA   October 12, 1990
SEE Acceptable Intake for Chronic
       Exposure
SO Acronyms

AIF
DA   October 12, 1990
SEE Atomic Industrial Forum
SO Acronyms

AIR
(CFR)
DA   November 15, 1990
SY Air Resources
NT1 Ambient Air
NT1 Indoor Air
NT1 Outside Air
NT1 Overfire Air
NT1 Stable Air
NT1 Underfire Air
RT Aeration
RT Atmosphere
RT Inhalation
RT Ventilation
DEF (CFR) Those naturally occurring
       constituents of the atmosphere,
       including those gases essential for
       human, plant, and animal life.

AIR CHANGES PER HOUR
DA   January 14, 1991
SF ACH
BT1 Measurements
DEF (EPA) The movement of a volume
       of air in a given time; if a house
       has one air change per hour, it
       means all of the air in the house
       will be replaced in a one-hour time
       period.

AIR COMPRESSOR
(2102 AIR COMPRESS)
DA   December 10, 1990
BT1 Compressor
 BT2 Machines (DOE FRASE
       Vocabulary)
 BT3 Equipment
SO DOE FRASE VOCABULARY

AIR CONDITIONER
(2475 AIR CONDITIO)
DA   January 3, 1991
SO DOE FRASE VOCABULARY

AIR CONTAMINANTS
(EPA)
DA   October 12, 1990
SY Air Pollutants
BT1 Contaminants

RT Air Pollution
RT Particulates
SO Environmental Protection Agency
       Glossary
DEF (EPA) Particulate matter, gases, or
       combination thereof, other than
       water vapor or natural air.

AIR CURTAINS
(EPA)
DA   October 12, 1990
RT Oil Spills
SO Environmental Protection Agency
       Glossary
DEF (EPA) A method of containing oil
       spills. Air bubbling through a
       perforated pipe causes an upward
       water flow that slows the spread of
       oil. It can also be used to stop fish
       from entering polluted water.

AIR DRYER
(2103 AIR DRYER)
DA   December 10, 1990
BT1 Machines (DOE FRASE
       Vocabulary)
 BT2 Equipment
SO DOE FRASE VOCABULARY

AIR FORCE BASE
DA   January 8, 1991
SF AFB
BT1 Facilities

AIR FORCE ENGINEERING AND
SERVICES CENTER
DA   January 8, 1991
SF AFESC
BT1 Centers
BT1 Facilities

AIR HOIST
(2300 AIR HOIST)
DA   December 10, 1990
BT1 Hoist(s)
 BT2 Hoisting Apparatus
  BT3 Material Handling Device
   BT4 Devices
   BT4 Equipment/Parts - Material
         Handling (DOE FRASE
         Vocabulary)
    BT5 Equipment
SO DOE FRASE VOCABULARY

AIR LOCK ROOM
(1675 AIR LOCK ROO)
DA   December 10, 1990
BT1 Room
 BT2 Sites/Areas
SO DOE FRASE VOCABULARY

AIR MASK
(2651 AIR MASK)
DA   January 3, 1991
BT1 Personal Protective Equipment
 BT2 Equipment/Parts - Personal
       Protective (DOE FRASE
       Vocabulary)

BT-Broader Term   NT-Narrower Term   RT Related Term

BT3 Equipment
RT Dust Mask
RT Nose
SO DOE FRASE VOCABULARY

AIR MASS
(EPA)
DA October 12, 1990
RT Inversion
RT Smog
SO Environmental Protection Agency
  Glossary
DEF (EPA) A widespread body of air that
  gains certain meteorological or
  polluted characteristics, e.g., a
  heat inversion or smogginess
  while set in one location. The
  characteristics can change as it
  moves away.

AIR MONITORING
(EPA)
DA October 12, 1990
BT1 Monitoring
 BT2 Activities
RT Air Pollution
RT Alternate Methods
RT Monitoring Systems
RT Reference Methods
RT Run
SO Environmental Protection Agency
  Glossary
DEF (EPA) Periodic or continuous
  surveillance or testing to
  determine the level of compliance
  with statutory requirements and/or
  pollutant levels in the atmosphere.

AIR POLLUTANTS
(EPA; USC)
DA October 12, 1990
SY Air Contaminants
NT1 Designated Pollutants
NT1 Dust
 NT2 Soot
NT1 Emissions
 NT2 Effluents
  NT3 Radioactive Effluents
 NT2 Exhaust Gases
 NT2 Fugitive Emissions
  NT3 Uncontrolled Total Arsenic
    Emissions
 NT2 Fumes
 NT2 Particulate Matter Emissions
  NT3 Allowable Emissions
  NT3 Excess Emissions
  NT3 PM_{10} Emissions
  NT3 Visible Emissions
 NT2 Process Emissions
 NT2 Reactor Opening Loss
 NT2 Secondary Emissions
NT1 Flue Gases
 NT2 Condenser Stack Gases
NT1 Fly Ash
NT1 Hazardous Air Pollutants
 NT2 Volatile Hazardous Air Pollutant
    (VHAP)
  NT3 Vinyl Chloride
NT1 Particulates
 NT2 Airborne Particulates

 NT3 Particulate Matter
  NT4 Particulate Asbestos Materials
  NT4 PM_{10}
NT1 Photochemical Smog
 NT2 Oxidants
NT1 Plumes
 NT2 Vapor Plumes
NT1 Smog
NT1 Smoke
NT1 Toxic Clouds
NT1 Toxic Air Pollutants
RT Adequately Wetted
RT Air Pollution
RT Alternate Methods
RT Area Sources
RT Emission Factor
RT Emission Inventory
RT Emission Limitation
SO Environmental Protection Agency
  Glossary
DEF (EPA)Any substance in air which
  could, if in high enough
  concentration, harm man, other
  animals, vegetation, or material.
  Pollutants may include almost any
  natural or artificial composition of
  matter capable of being airborne.
  They may be in the form of solid
  particles, liquid droplets, gases, or
  in combinations of these forms.
  Generally, they fall into two main
  groups: (1) those emitted directly
  from identifiable sources; and (2)
  those produced in the air by
  interaction between two or more
  primary pollutants, or by reaction
  with normal atmospheric
  constituents, with or without
  photoactivation. Exclusive of
  pollen, fog, and dust, which are of
  natural origin, about 100
  contaminants have been identified
  and fall into these categories:
  solids, sulfur compounds, volatile
  organic chemicals, nitrogen
  compounds, oxygen compounds,
  halogen compounds, radioactive
  compounds, and odors.

AIR POLLUTION
(EPA)
DA October 12, 1990
BT1 Pollution
NT1 Hazardous Air Pollution
NT1 Indoor Air Pollution
NT1 Man-Made Air Pollution
NT1 Photochemical Smog
 NT2 Oxidants
NT1 Smog
NT1 Wood-Burning Stove Pollution
RT Acid Deposition
RT Agglomeration
RT Agricultural Pollution
RT Air Contaminants
RT Air Monitoring
RT Air Pollutants
RT Air Pollution Episodes
RT Air Pollution Control Agencies
RT Air Quality Standards
RT Background Levels

RT Banking (Air Pollution)
RT Emissions
RT Emission Standards
RT Episodes (Pollution)
RT Equivalent Methods
RT Excess Emissions
RT Exhaust Gases
RT Particulates
RT Plumes
RT Pollutant Standard Index
RT Reclamation
RT Soot
SO Environmental Protection Agency
  Glossary
DEF (EPA) The presence of contaminant
  or pollutant substances in the air
  that do not disperse properly and
  interfere with human health or
  welfare, or produce other harmful
  environmental effects.

AIR POLLUTION CONTROL AGENCIES
(USC)
DA November 15, 1990
BT1 State Agencies
 BT2 Agencies
  BT3 Administrative Organizations
   BT4 Organizations
NT1 Interstate Air Pollution Control
    Agencies
RT Air Pollution
RT Municipalities
RT State Agencies
DEF (USC) Any of the following: (1) A
  single State agency designated by
  the Governor of that State as the
  official State air pollution control
  agency for purposes of this Act;
  (2) An agency established by two
  or more States and having
  substantial powers or duties
  pertaining to the prevention and
  control of air pollution; (3) A city,
  county, or other local government
  health authority, or, in the case of
  any city, county, or other local
  government in which there is an
  agency other than the health
  authority charged with
  responsibility for enforcing
  ordinances or laws relating to the
  prevention and control of air
  pollution, such other agency; or (4)
  An agency of two or more
  municipalities located in the same
  State or in different States and
  having substantial powers or
  duties pertaining to the prevention
  and control of air pollution.

AIR POLLUTION EPISODES
(EPA)
DA October 12, 1990
BT1 Incidents
RT Air Pollution
SO Environmental Protection Agency
  Glossary
DEF (EPA) Periods of abnormally high
  concentration of air pollutants,
  often due to low winds and

SY-Synonymous Terms   SO-Source/Subject Category   SF-See From

temperature inversion, that can cause illness and death.

AIR QUALITY CONTROL REGIONS
(EPA)
DA   October 12, 1990
BT1 Sites/Areas
NT1 Federal Class I areas
 NT2 Mandatory Class I Federal Areas
NT1 Regions
RT   Air Quality Standards
RT   Bay Area Air Quality Management District
SO   Environmental Protection Agency Glossary
DEF (EPA) Areas, designated by the federal government, in which communities share a common air pollution problem. Sometimes several states are involved.

AIR QUALITY CRITERIA
(EPA)
DA   October 12, 1990
RT   Air Quality Standards
RT   Anti-Degradation Clause
SO   Environmental Protection Agency Glossary
DEF (EPA) The levels of pollution and lengths of exposure above which adverse health and welfare effects may occur.

AIR QUALITY STANDARDS
(EPA)
DA   October 12, 1990
BT1 Standards
 BT2 Codes, Standards, and Regulations
NT1 Ambient Air Quality Standards
 NT2 National Ambient Air Quality Standards
  NT3 Primary Standards
  NT3 Secondary Standards
RT   Air Pollution
RT   Air Quality Control Regions
RT   Air Quality Criteria
RT   Control Strategies
RT   Emission Limitation
RT   Emission Standards
RT   National Standards (Air Pollution)
SO   Environmental Protection Agency Glossary
DEF (EPA) The level of pollutants prescribed by regulations that may not be exceeded during a specified time in a defined area.

AIR RESOURCES
(CFR)
DA   November 15, 1990
SY   Air
BT1 Natural Resources
DEF (CFR) Those naturally occurring constituents of the atmosphere, including those gases essential for human, plant, and animal life.

AIR TAMPER
(2950 AIR TAMPER)
DA   January 3, 1991
BT1 Tools - Powered
 BT2 Tools (DOE FRASE Vocabulary)
  BT3 Equipment
SO   DOE FRASE VOCABULARY

AIRBORNE ACTIVITY CONFINEMENT SYSTEMS
DA   January 8, 1991
SF   AACS
BT1 Confinement Systems
 BT2 Systems

AIRBORNE PARTICULATES
(EPA)
DA   October 12, 1990
BT1 Particulates
 BT2 Air Pollutants
NT1 Particulate Matter
 NT2 Particulate Asbestos Materials
 NT2 PM_{10}
RT   Airborne Radioactive Materials
SO   Environmental Protection Agency Glossary
DEF (EPA) Total suspended particulate matter found in the atmosphere as solid particles or liquid droplets. The chemical composition of particulates varies widely, depending on location and time of year. Airborne particulates include windblown dust, emissions from industrial processes, smoke from the burning of wood and coal, and the exhaust of motor vehicles.

AIRBORNE RADIOACTIVE MATERIALS
(AEA; CFR)
DA   January 24, 1991
BT1 Radioactive Materials
 BT2 Materials
RT   Airborne Particulates
SO   Environmental Management
DEF (CFR) Radioactive materials dispersed in the air in the form of dust, fume, mist, vapor, or gas.

AIRBORNE RELEASES
(EPA; EMER)
DA   October 12, 1990
BT1 Releases
RT   Aboveground Releases
SO   Emergency Preparedness
SO   Environmental Protection Agency Glossary
DEF (EPA) Releases of any chemical into the air.

AIRBORNE RESPONSE TEAMS
(EMER)
DA   February 1, 1991
BT1 Response Teams
 BT2 Teams
  BT3 Administrative Organizations
   BT4 Organizations
SO   Emergency Preparedness
DEF (EMER) Groups of one or more

armed security inspectors who are specially trained in airborne tactics and equipment to respond to security incidents.

AIRCRAFT
(DOE Order 5480.13)
DA   October 16, 1990
NT1 Cargo Aircraft Only
NT1 Civil Aircraft
NT1 Modern Aircraft
NT1 Passenger-carrying Aircraft
NT1 Public Aircraft
RT   Aircraft Pilot
RT   Aircraft Accidents
RT   Aircraft Incidents
RT   Airports
RT   Aviation Operations
RT   Jet Thrust Units
RT   Passengers (Aircraft)
RT   Starter Cartridges
RT   Targets
SO   Aviation Safety
DEF (DOE Order 5480.13) A device that is used or intended to be used for flight in the air, including heavier-than-air and lighter-than-air aircraft, airplanes, gliders, helicopters, rigid and nonrigid airships, and balloons.

AIRCRAFT ACCIDENTS
(DOE Order 5480.13 5B)
DA   October 12, 1990
BT1 Accidents
RT   Aircraft
SO   Aviation Safety
SO   Environmental Management
DEF (DOE Order 5480.13) An occurrence associated with the operation of an aircraft which takes place between the time a person boards the aircraft with the intention of flight until such time as all persons have disembarked and in which any person suffers death or serious injury as a result of being in or on the aircraft or anything attached thereto, or in which the aircraft receives substantial damage.

AIRCRAFT INCIDENTS
(DOE Order 5480.13 5C)
DA   October 12, 1990
BT1 Incidents
RT   Aircraft
SO   Aviation Safety
SO   Environmental Management
DEF (DOE Order 5480.13) Deviations from the normal, planned, or expected aviation operation, if deviations have adverse safety, health, or environmental effects or potential effects and are not classified as accidents.

AIRCRAFT PILOT
(0825 AIR PILOT)

DA   November 28, 1990
BT1 Transport Personnel
 BT2 Occupations
 BT2 Personnel
RT   Aircraft
SO   DOE PHASE VOCABULARY

AIRPORTS
(DOE Order 5480.13)
DA   October 12, 1990
NT1 Heliports
RT   Aircraft
SO   Aviation Safety
DEF (DOE Order 5480.13) An area of
     land or water that is used or
     intended to be used for the landing
     and takeoff of aircraft, including its
     buildings and facilities, if any.

AIS
DA   October 12, 1990
SEE Acceptable Intake for Subchronic
     Exposure
SO   Acronyms

ALACHLOR
(EPA)
DA   October 12, 1990
BT1 Herbicides
 BT2 Pesticides
  BT3 Hazardous Substances
SO   Environmental Protection Agency
     Glossary
DEF (EPA) An herbicide, marketed
     under the trade name Lasso, used
     mainly to control weeds in corn
     and soybean fields.

ALAP
DA   October 12, 1990
SEE As Low As Practicable
SO   Acronyms

ALAR
(EPA)
DA   October 12, 1990
BT1 Organic Chemicals
 BT2 Chemical Substances
SO   Environmental Protection Agency
     Glossary
DEF (EPA) Trade name for daminozide,
     a plant growth regulator,that
     makes apples redder, firmer, and
     less likely to drop off trees before
     growers are ready to pick them. It
     is also used to a lesser extent on
     peanuts, tart cherries, concord
     grapes, and other fruits.

ALARA
DA   October 12, 1990
SEE As Low As Reasonably Achievable
SO   Acronyms

ALARMS
(EMER)
DA   February 1, 1991
BT1 Devices

NT1 Attack Warning Signals
NT1 Attention Signals
NT1 Criticality Alarm System
NT1 Criticality Alarms
NT1 Fire Alarm System
NT1 High Temperature Alarm
NT1 Intrusion Alarm System (Perimeter
     or Interior)
NT1 Kanne Alarm
NT1 Level Alarm
NT1 Radiation Alarm System
RT   Alert Signals
RT   Central Alarm Stations
RT   Controls
RT   Defense Readiness Conditions
SO   Emergency Preparedness
DEF (EMER) Series of alerting devices
     which may include effluent or
     stack monitor alarms, duress
     alarms, vault alarms, gas alarms,
     entry alarms, radiation monitor
     alarms, or other alarms.

ALASKA POWER ADMINISTRATION
DA   May 15, 1991
BT1 Power Marketing Administrations
SO   Management
DEF (U.S. Government Manual) The
     Administration is responsible for
     operating and marketing power for
     two Federal hydroelectric projects
     in Alaska. Power operations and
     marketing functions involving the
     Eklutna and Snettisham
     Hydroelectric Projects include the
     projects' transmission systems
     serving the Anchorage and
     Juneau areas.

ALASKAN NATIVES
(ESA; CFR)
DA   October 12, 1990
RT   Indian Tribes
SO   Endangered Species
DEF (CFR) A person defined in the
     Alaska Native Claims Settlement
     Act (43 USC 1603(b); 85 Stat.
     588) as a citizen of the United
     States who is of one-fourth degree
     or more Alaska Indian (including
     Tsimshian Indians enrolled or not
     enrolled in the Metlaktla Indian
     Community), Eskimo, or Aleut
     blood, or combination thereof. The
     term includes any Native, as so
     defined, either or both of whose
     adoptive parents are not Natives.
     It also includes, in the absence of
     proof of a minimum blood
     quantum, any citizen of the United
     States who is regarded as an
     Alaska Native by the Native village
     or town of which he claims to be a
     member and whose father or
     mother is (or, if deceased, was)
     regarded as Native by any Native
     village or Native town. Any citizen
     enrolled by the Secretary pursuant
     to section 5 of the Alaska Native
     Claims Settlement Act shall be

conclusively presumed to be an
Alaskan Native for purposes of 50
CFR 17.3.

ALBUQUERQUE OPERATIONS OFFICE
DA   January 11, 1991
BT1 Operations Offices
 BT2 Offices
  BT3 Administrative Organizations
   BT4 Organizations
 BT2 U.S. Department of Energy
  BT3 Federal Agencies
   BT4 Agencies
    BT5 Administrative Organizations
     BT6 Organizations
RT   Allied-Signal Inc.
RT   EG&G Mound Applied
     Technologies, Inc.
RT   General Electric
RT   Lovelace Medical Foundation
RT   Mason and Hanger-Silas Mason
     Co.
RT   Ross Aviation, Inc.
RT   Sandia Corporation AT&T
     Technologies, Inc.
RT   University of California
RT   Westinghouse Electric Corp.
DEF (Capsule Review of DOE Research
     and Development and Field
     Facilities, 1986) The principal
     mission of the Albuquerque
     Operations Office is national
     security. Established in 1946, AL
     operates an extensive weapons
     laboratory and production complex
     from Florida to California. AL
     directs weapons research,
     development and production and
     provides program execution and
     project management functions for
     assigned energy research and
     development activities. AL
     operates a system for safe and
     secure transport of all
     Government-owned Special
     Nuclear Material of strategic
     quantities, coordinates and
     administers nuclear test detection
     activities and maintains
     capabilities for radiological
     assistance and nuclear weapons
     accident response.

ALDICARB
(EPA)
DA   October 12, 1990
BT1 CERCLA Hazardous Substances
 BT2 Hazardous Substances
BT1 Hazardous Constituents
BT1 Insecticides
 BT2 Pesticides
  BT3 Hazardous Substances
SO   Environmental Protection Agency
     Glossary
DEF (EPA) An insecticide sold under the
     trade name Temik. It is made from
     ethyl isocyanate.

ALERT
(EMER)

DA   February 1, 1991
BT1 Operational Emergency Response
     Levels
 BT2 Emergency Response Levels
RT   Defense Readiness Conditions
SO   Emergency Preparedness
DEF (EMER) An emergency response
     level which represents an event in
     progress or having occurred which
     involves an actual or potential
     substantial reduction of the level of
     safety of the facility.

ALERT SIGNALS
(EMER)
DA   February 1, 1991
RT   Alarms
RT   Emergencies
SO   Emergency Preparedness
DEF (EMER) Three- to five-minute
     steady tones, sounded strictly at
     the option of and on the authority
     of local government officials. The
     signals may be activated for
     natural or man-made disasters as
     local authorities may determine
     and may also be used to call
     attention to essential emergency
     information. Use of the attention or
     alert signals should always be
     accompanied by a public
     explanation and instructions to the
     public over local broadcast
     stations or by other means.

ALGAE
(EPA)
DA   October 12, 1990
BT1 Phytoplankton
 BT2 Plankton
BT1 Plants
RT   Eutrophication
SO   Environmental Protection Agency
     Glossary
DEF (EPA) Simple rootless plants that
     grow in sunlit waters in relative
     proportion to the amounts of
     nutrients available. They can affect
     water quality adversely by
     lowering the dissolved oxygen in
     the water. They are food for fish
     and small aquatic animals.

ALGAE BLOOM
(SSDC)
DA   May 24, 1991
SY   Red Tide
SY   Water Bloom
BT1 Natural Phenomenon
SO   Water Pollution

ALGAL BLOOMS
(EPA)
DA   October 12, 1990
RT   Water Pollution
SO   Environmental Protection Agency
     Glossary
DEF (EPA) Sudden spurts of algal
     growth, which can affect water

quality adversely and indicate
potentially hazardous changes in
local water chemistry.

ALI
DA   October 12, 1990
SEE Annual Limit on Intake
SO   Acronyms
SO   Industrial Hygiene
SO   Radiation

ALIQUOTS
(ESH)
DA   October 12, 1990
BT1 Samples
RT   Replicate Samples
RT   Standard Sample
SO   Management
DEF (ESH) Fractions of a field sample
     taken for complete processing
     through an analytical procedure (a
     "laboratory sample" of a field
     sample).

ALLEGATIONS
(TSCA; CFR)
DA   October 12, 1990
RT   Complaints
SO   Hazardous Materials
DEF (CFR) Statements, made without
     formal proof or regard for
     evidence, that a chemical
     substance or mixture has caused
     a significant adverse reaction to
     health or to the environment.

ALLEN WRENCH
(3025 ALLEN WRENCH)
DA   January 3, 1991
BT1 Wrench(s)
 BT2 Tools (DOE FRASE Vocabulary)
 BT3 Equipment
SO   DOE FRASE VOCABULARY

ALLEY
(1600 ALLEY)
DA   December 10, 1990
BT1 Site (DOE FRASE Vocabulary)
 BT2 Sites/Areas
SO   DOE FRASE VOCABULARY

ALLIED-SIGNAL INC.
DA   January 11, 1991
BT1 Companies
 BT2 Commercial Organizations
 BT3 Organizations
BT1 DOE Contractors
 BT2 Potentially Responsible Parties
RT   Albuquerque Operations Office
RT   Kansas City Plant

ALLIS-CHALMERS MOTORS
DA   January 8, 1991
SF   *AC Motors*
BT1 Equipment

ALLOWABLE EMISSIONS
(CAA; ESH)

DA   October 12, 1990
BT1 Particulate Matter Emissions
 BT2 Emissions
 BT3 Air Pollutants
RT   State Implementation Plans
RT   Stationary Sources
SO   Air Pollution
DEF (ESH) The emissions rate of a
     stationary source calculated using
     the maximum rated capacity of the
     source (unless the source is
     subject to federally enforceable
     limits which restrict the operating
     rate, or hours of operation, or
     both) and the most stringent of the
     following: (1) The applicable
     standards set forth in 40 CFR 60
     or 61; (2) any applicable State
     Implementation Plan emissions
     limitation including those with a
     future compliance date; or (3) the
     emissions rate specified as a
     federally enforceable permit
     condition, including those with a
     future compliance date.

ALLOWABLE OUTAGE TIME
DA   January 8, 1991
SF   *AOT*
BT1 Time Designations
RT   Time Delay

**ALLOWABLE SOIL BEARING
CAPACITY**
(DOE Order 6430.1A)
DA   October 12, 1990
SO   Construction
DEF (DOE Order 6430.1A) The
     maximum permissible pressure on
     foundation soils under which the
     settlements of various footings will
     not exceed a reasonable value.

ALPHA
DA   February 26, 1991
SY   Safeguards and Security Alert I
     (Code Designator: Alpha)
BT1 Safeguards and Security Alerts
 BT2 Safeguards and Security
      Emergencies
 BT3 Emergencies
 BT4 Reportable Occurrences
 BT5 Occurrences
SO   Emergency Preparedness
DEF (EMER) A maximum alert action
     which shall be effected when the
     head of a field office determines
     that conditions warrant maximum
     security measures at U.S.
     Department of Energy (DOE) or
     DOE contractor facilities.

ALPHA DECAY RADIOISOTOPES
(EDB)
DA   March 29, 1991
BT1 Radionuclides
 BT2 CERCLA Hazardous Substances
 BT3 Hazardous Substances
 BT2 Nuclides

NT1 Radium 226
NT1 Radon 222
NT1 Thorium 230
NT1 Uranium 238
SO Radiation

ALPHA PARTICLES
(EPA; NIH; NCRP; EMER)
DA October 12, 1990
BT1 Charged Particles
RT Alpha Rays
RT Charged Particle Equilibrium
RT Directly Ionizing Particles
RT Gross Alpha Particle Activity
RT Ionizing Radiation
RT Radioactive Decay
SO Emergency Preparedness
SO Environmental Protection Agency
   Glossary
SO Radiation
DEF (EPA) Positively charged particles
   composed of 2 neutrons and 2
   protons released by some atoms
   undergoing radioactive decay. The
   particles are identical to the
   nucleus of a helium atom. (NIH) A
   strongly ionizing particle emitted
   from the nucleus during
   radioactive decay having a mass
   and charge equal in magnitude to
   a helium nucleus, consisting of 2
   protons and 2 neutrons with a
   double positive charge. The
   nucleus of a helium atom which is
   ejected from some radionuclides
   during radioactive decay.

ALPHA RAYS
(NIH)
DA October 12, 1990
BT1 Non-Penetrating Radiation
 BT2 Radiation
RT Alpha Particles
RT Ionizing Radiation
SO Radiation
DEF (NIH) Streams of fast-moving
   helium nuclei (alpha particles), a
   strongly ionizing and weakly
   penetrating radiation.

ALS
DA October 12, 1990
SEE Advanced Light Source
SO Acronyms

ALTERNATE CONCENTRATION LIMITS
DA January 8, 1991
SF ACLs
BT1 Limits

ALTERNATE EMERGENCY
OPERATIONS CENTER
(DOE Order 5500.1A Attach. II, 5)
DA January 24, 1991
BT1 Centers
RT Emergency Operations Centers
SO Environmental Management
DEF (DOE 5599.1A) An alternate facility
   to the designated Emergency

Operations Center from which the
emergency management team can
carry out emergency response
activities in the event the
designated primary Emergency
Operations Center cannot be
used.

ALTERNATE METHODS
(EPA)
DA October 12, 1990
RT Air Monitoring
RT Air Pollutants
RT Equivalent Methods
RT Reference Methods
SO Environmental Protection Agency
   Glossary
DEF (EPA) Methods of sampling and
   analyzing for an air pollutant which
   are not reference or equivalent
   methods but which have been
   demonstrated in specific cases to
   EPA's satisfaction to produce
   results adequate for compliance.

ALTERNATE REMOVAL SYSTEMS
(NFI)
DA October 12, 1990
SF ARS
BT1 Systems
RT D Machines
RT Reactor Operations
SO Radiation
DEF (NFI) Systems for discharging fuel
   assemblies from D machines.

ALTERNATE ROD INSERTION
DA January 8, 1991
SF ARI

ALTERNATING CURRENT
DA January 8, 1991
SF AC
NT1 Volts Alternating Current
RT Electrical Equipment

ALTERNATIVE COURSES OF ACTION
(ESA; CFR; USC)
DA October 12, 1990
SY Reasonable and Prudent
   Alternatives
BT1 Actions
 BT2 Responses
RT DOE Alternative
RT Record of Decision (NEPA)
SO Endangered Species
DEF All reasonable and prudent
   alternatives, including both no
   action and alternatives extending
   beyond original project objectives
   and acting agency jurisdiction.

ALTERNATIVE WATER SUPPLIES
(CERCLA; USC)
DA October 12, 1990
RT Drinking Water Supplies
SO Compensation and Liability
DEF (CERCLA) Includes, but is not

limited to, drinking water and
household water supplies.

AMAD
DA October 12, 1990
SEE Activity-Median Aerodynamic
   Diameter
SO Acronyms

AMBIENT
(DOE Order 6430.1A)
DA October 12, 1990
RT Environment
SO Construction
DEF (DOE Order 6430.1A) Surrounding
   environmental conditions.

AMBIENT AIR
(CAA; CFR; EPA)
DA October 12, 1990
SY Outside Air
BT1 Air
SO Environmental Management
SO Environmental Protection Agency
   Glossary
DEF (EPA) Any unconfined portion of
   the atmosphere: open air,
   surrounding air. (CFR) That
   portion of the atmosphere,
   external to buildings, to which the
   general public has access.

AMBIENT AIR QUALITY STANDARDS
(SWDA; CFR)
DA November 5, 1990
BT1 Air Quality Standards
 BT2 Standards
  BT3 Codes, Standards, and
      Regulations
NT1 National Ambient Air Quality
   Standards
 NT2 Primary Standards
 NT2 Secondary Standards
DEF (EPA) Outside air quality standards
   established by the EPA.

AMELIORATION
(SSDC)
DA October 12, 1990
SY Corrective Actions
NT1 Mitigation
NT1 Off-Site Federal Support
NT1 On-Site Federal Support
NT1 Protective Actions
 NT2 Denials
NT1 Protective Measures
NT1 Recovery Plans
 NT2 Recovery Actions
RT Accidents
RT Incidents
RT Occurrences
RT Protection of the Public Health and
   Welfare
SO System Safety Development Center
   Glossary
DEF (SSDC) Those things that are done
   immediately following an accident
   to limit its consequences and to

reduce the sensitivity of those consequences.

AMERICAN NATIONAL METRIC COUNCIL
(DOE Order 5900.2)
DA October 12, 1990
BT1 Research and Development
   Organizations
 BT2 Organizations
RT Metric System
RT Metrication
SO Quality Assurance
DEF (DOE Order 5900.2) A private nonprofit organization that serves as a planning, coordinating, and information center for metric activities in the United States.

AMERICAN NATIONAL STANDARDS INSTITUTE
DA January 8, 1991
SF ANSI
BT1 Institutes
 BT2 Research and Development
   Organizations
  BT3 Organizations
RT Standards

AMERICAN NUCLEAR SOCIETY
DA January 8, 1991
SF ANS
BT1 Research and Development
   Organizations
 BT2 Organizations

AMERICAN PETROLEUM INSTITUTE
DA January 8, 1991
SF API (American Petroleum Institute)
BT1 Institutes
 BT2 Research and Development
     Organizations
  BT3 Organizations

AMERICAN SOCIETY OF MECHANICAL ENGINEERS
DA January 8, 1991
SF ASME
BT1 Professional Bodies
 BT2 Organizations

AMMETER(S)
(2750 AMMETER)
DA January 3, 1991
BT1 Equipment/Parts -
    Instrumentation/Measuring (DOE
    FRASE Voc.)
 BT2 Equipment
BT1 Indicator(s)
 BT2 Instrument(s)
  BT3 Equipment/Parts -
     Instrumentation/Measuring
     (DOE FRASE Voc.)
   BT4 Equipment
SO DOE FRASE VOCABULARY

AMMUNITION
(DOE Order 5480.16)

DA October 16, 1990
BT1 Munitions
NT1 Chemical Ammunition
 NT2 Smoke Grenades
NT1 Rocket Ammunition
NT1 Small Arms Ammunition
RT Small Arms
SO Firearms
DEF (DOE Order 5480.16) As associated with small arms, is the assembled cartridge, including primer, powder, case, and projectile.

AMPLIFIER(S)
(2376 AMPLIFIER)
DA January 3, 1991
BT1 Equipment/Parts - Electrical (DOE
    FRASE Vocabulary)
 BT2 Equipment
SO DOE FRASE VOCABULARY

AMPUTATION
(1301 AMPUTATION)
DA November 28, 1990
BT1 Injuries
RT Avulsion
SO DOE FRASE VOCABULARY

ANADROMOUS FISH
(EPA)
DA October 12, 1990
BT1 Fish
 BT2 Animals
SO Environmental Protection Agency
    Glossary
DEF (EPA) Fish that spend their adult life in the sea but swim upriver to fresh water spawning grounds to reproduce.

ANAEROBIC
(EPA)
DA October 12, 1990
RT Aerobic
RT Anaerobic Digestion
SO Environmental Protection Agency
    Glossary
DEF (EPA) A life or process that occurs in, or is not destroyed by, the absence of oxygen.

ANAEROBIC DIGESTION
(DOE Order 6430.1A)
DA October 12, 1990
BT1 Digestion
 BT2 Processes
RT Anaerobic
RT Biological Oxidation
RT Denitrification
RT Microorganisms
SO Construction
DEF (DOE Order 6430.1A) Biological stabilization of domestic wastewater sludge by microorganisms that function in the absence of oxygen.

ANALOG TO DIGITAL CONVERTER
DA January 8, 1991
SF ADC
BT1 Equipment

ANALYSES
(SSDC; MORT)
DA October 12, 1990
NT1 Change Analyses
NT1 Cost-Effective Analysis
NT1 Differential Cost Benefit Analysis
NT1 Failure Mode and Effect Analysis
NT1 Fault Tree Analysis
NT1 Hazard Analysis
NT1 Matrix/Spike-Duplicate Analysis
NT1 Proper Job Analysis
NT1 Regional Frequency Analysis
NT1 Risk Analysis
NT1 Root Cause Analysis
NT1 Safety Analysis
 NT2 Job Safety Analysis
 NT2 Preliminary Safety Analysis
 NT2 System (Safety) Analyses
NT1 Safety (Hazard) Analysis
NT1 Single Failure Analysis
NT1 Special Analytical Services
NT1 Supplement Analyses
NT1 Task Analysis
NT1 Time Loss (T/I) Analysis
NT1 Vulnerability Analyses
RT Analytical Batches
RT Analytical (Logic) Trees
RT Assessments
RT Communication
RT Corrective Action Triggers
RT Failure Reporting Analysis and
    Corrective Action System
RT Fire Hazard Analysis Report
RT Knowledge
RT Monitoring
RT NRS Office for Analysis and
    Evaluation of Operational Data
RT Nuclear Safety Analysis Center
RT Predictive Maintenance
RT Routine Analytical Services
RT Safety Analysis Reports
RT Safety Analysis Process
RT Sampling and Analysis Plans
RT Sampling and Analysis
RT Technical Information
RT Trending
SO System Safety Development Center
    Glossary
DEF (SSDC) Examinations of a complex, its elements, and their relationships. The use of methods and techniques of arranging data to (a) assist in deciding what additional facts are needed; (b) establish consistency, validity, and logic; (c) establish necessary and sufficient events for causes; and (d) guide and support inferences and judgments. (MORT) MORT analysis asks: were the data collection and analysis procedures adequate? were there analyses made of the data? did the analyses provide the proper risk assessment information to the

decision maker responsible for the risk assumption? Related issues include: risk projection analysis, diagnostic statistics analysis, status and predictive statistics, etc.

ANALYTES
(EPA)
DA   October 12, 1990
RT   Chemical Substances
RT   Sampling
SO   Environmental Protection Agency Glossary
DEF (EPA) The chemicals for which a sample is analyzed.

ANALYTICAL (LOGIC) TREES
(SSDC)
DA   October 12, 1990
SY   Logic Trees
BT1 Diagrams
NT1 Event Tree
NT1 Fault Tree
  NT2 Management Oversight and Risk Tree
NT1 Positive (objective) Trees
RT   Analyses
RT   Less than Adequate
RT   Logic Gates
SO   System Safety Development Center Glossary
DEF (SSDC) Diagrams, in the shapes of trees, using different geometrical symbols to aid a user in systematically portraying information in a logical sequence and showing relationships between elements of the tree. Trees may be positive or negative (fault tree).

ANALYTICAL BATCHES
(ESH)
DA   October 12, 1990
BT1 Batches
 BT2 Samples
RT   Analyses
SO   Quality Assurance
DEF (ESH) Samples which are analyzed together with the same method sequence and the same lots of reagents and with the manipulations common to each sample within the same time period or in continuous sequential time periods. Samples in each batch should be of similar composition.

ANCILLARY EQUIPMENT
(SWDA; RCRA; CFR)
DA   October 12, 1990
BT1 Equipment
RT   Underground Storage Tanks
SO   Environmental Management
SO   Wastes
DEF Any devices including, but not limited to, such devices as piping, fittings, flanges, valves, and

pumps used to distribute, meter, or control the flow of regulated substances to and from an underground storage tank (UST).

AND GATE
(SSDC)
DA   October 12, 1990
BT1 Logic Gates (Boolean)
SO   System Safety Development Center Glossary
DEF (SSDC) A logic gate that produces an output only when all input events occur. May contain the identifying word "AND".

ANIMAL BITE
(1302 ANIMAL BITE)
DA   November 28, 1990
BT1 Injuries
NT1 Snake Bite
SO   DOE FRASE VOCABULARY

ANIMALS
(USC)
DA   November 15, 1990
NT1 Fish
  NT2 Anadromous Fish
  NT2 Catanadramous Fish
  NT2 Game Fish
  NT2 Rough Fish
NT1 Herbivores
NT1 Insects
NT1 Nematodes
NT1 Sea Turtles
RT   Captivity-bred
RT   Captivity
RT   Ecology
RT   Targets
RT   Threatened Species
RT   Wildlife
DEF (USC) All vertebrate and invertebrate species, including but not limited to man and other mammals, birds, fish, and shellfish.

ANKLE(S)
(1126 ANKLE)
DA   November 28, 1990
BT1 Joint(s)
 BT2 Human Body Parts
RT   Foot/Feet
RT   Leg(s)
SO   DOE FRASE VOCABULARY

ANKLE PROTECTION
(2652 ANKLE PROTEC)
DA   January 3, 1991
BT1 Foot Protection
 BT2 Personal Protective Equipment
  BT3 Equipment/Parts - Personal Protective (DOE FRASE Vocabulary)
   BT4 Equipment
SO   DOE FRASE VOCABULARY

ANL-E
DA   October 12, 1990

SEE Argonne National Laboratory-East (Chicago)
SO   Acronyms

ANL-W
DA   October 12, 1990
SEE Argonne National Laboratory-West (At INEL)
SO   Acronyms

ANNIHILATION (ELECTRON)
(NIH)
DA   October 12, 1990
RT   Electrons
SO   Radiation
DEF (NIH) An interaction between a positive and negative electron; their energy, including rest energy, being converted into electromagnetic radiation (annihilation radiation).

ANNUAL DOCUMENT LOGS
(TSCA)
DA   October 19, 1990
BT1 Reports
RT   Annual Reports
RT   PCB Wastes
RT   Reports
SO   Hazardous Materials
DEF (TSCA) Detailed information maintained by a facility on the PCB waste handling at that facility.

ANNUAL DOSE EQUIVALENT
(DOE Order 5480.11)
DA   October 16, 1990
BT1 Dose Equivalents
 BT2 Radiation Units
  BT3 Units of Measure
NT1 Annual Effective Dose Equivalent
  NT2 Cumulative Annual Effective Dose Equivalent
RT   Annual Dose Equivalent Limit
DEF (DOE Order 5480.11) The dose equivalent received in a year. Annual dose equivalent is expressed in units of rem (or sievert).

ANNUAL DOSE EQUIVALENT LIMIT
(IAEA)
DA   October 12, 1990
BT1 Dose Limit
 BT2 Limits
RT   Annual Dose Equivalent
RT   Radiation Safety
SO   Radiation
DEF (IAEA) The value of the annual dose equivalent that must not be exceeded, according to the ICRP system of dose limitation. It is regarded as the lower boundary of an unacceptable dose region.

ANNUAL EFFECTIVE DOSE EQUIVALENT
(DOE Order 5480.11)
DA   October 12, 1990

BT1 Annual Dose Equivalent
 BT2 Dose Equivalents
  BT3 Radiation Units
   BT4 Units of Measure
BT1 Effective Dose Equivalent
 BT2 Dose Equivalents
  BT3 Radiation Units
   BT4 Units of Measure
NT1 Cumulative Annual Effective Dose
   Equivalent
SO Industrial Hygiene
DEF (DOE Order 5480.11) The effective
   dose equivalent received in a
   year. The annual effective dose
   equivalent is expressed in units of
   rem (or sievert).

ANNUAL LIMIT ON INTAKE
(DOE Order 5480.11; NCRP; IAEA;
 EMER)
DA October 16, 1990
SF *ALI*
BT1 Intake
 BT2 Measurements
BT1 Radiation Standards
 BT2 Standards
  BT3 Codes, Standards, and
     Regulations
RT Derived Air Concentration
RT Dose Commitments
RT Occupational Exposure
RT Radiation Safety
RT Reference Man
SO Emergency Preparedness
SO Environmental Management
SO Industrial Hygiene
SO Radiation
DEF (ICRP) The quantity of a single
   radionuclide which, if inhaled or
   ingested in 1 year, would irradiate
   a person, represented by
   Reference Man (ICRP Publication
   23) to the limiting value for control
   of the workplace (paragraph 9j(2)).

ANNUAL REPORTS
(TSCA; CFR)
DA October 19, 1990
BT1 Reports
RT Annual Document Logs
SO Hazardous Materials
DEF (CFR) Written documents submitted
   each year by each disposer and
   commercial storer of PCB waste to
   the appropriate EPA Regional
   Administrator. The annual report is
   a brief summary of the information
   included in the annual document
   log.

ANNUAL SITE ENVIRONMENTAL
REPORTS
(DOE Order 5400.1)
DA May 24, 1991
BT1 Reports
RT Environmental Monitoring
DEF (DOE Order 5400.1) The purpose
   of these reports is to present
   summary environmental data so
   as to characterize site

environmental management
performance, confirm compliance
with environmental standards and
requirements, and highlight
significant programs and efforts.
Reports shall be prepared for all
sites that conduct significant
environmental protection
programs. The breadth and detail
should reflect the size and extent
of any program at a particular site.

ANNULAR CORE RESEARCH
REACTOR
DA January 8, 1991
SF *ACRR*
BT1 Reactors

ANO-1
 DA October 12, 1990
 SEE Arkansas Nuclear One-1
 SO Acronyms

ANS
 DA October 12, 1990
 SEE American Nuclear Society
 SO Acronyms

ANSI
 DA October 12, 1990
 SEE American National Standards
    Institute
 SO Acronyms

ANTAGONISM
(EPA)
DA October 12, 1990
RT Chemical Substances
RT Neutralization
SO Environmental Protection Agency
   Glossary
DEF (EPA) The interaction of two
   chemicals having an opposing, or
   neutralizing, effect on each other,
   or given some specific biological
   effect a chemical interaction that
   appears to have an opposing or
   neutralizing effect over what might
   otherwise be expected.

ANTARCTIC "OZONE HOLE"
(EPA)
DA October 12, 1990
BT1 Ozone Depletion
RT Ozone Depletion
SO Environmental Protection Agency
   Glossary
DEF (EPA) Refers to the seasonal
   depletion of ozone in a large area
   over Antarctica.

ANTENNA
(2377 ANTENNA)
DA January 3, 1991
BT1 Equipment/Parts - Electrical (DOE
   FRASE Vocabulary)
 BT2 Equipment
SO DOE FRASE VOCABULARY

ANTI-DEGRADATION CLAUSE
(EPA)
DA October 12, 1990
BT1 Requirements
RT Air Quality Criteria
RT Pollution
RT Water Quality Criteria
SO Environmental Protection Agency
   Glossary
DEF (EPA) Part of federal air quality and
   water quality requirements
   prohibiting deterioration where
   pollution levels are above the legal
   limit.

ANTI-MICROBIAL AGENTS
(FIFRA; CFR)
DA January 24, 1991
SO Environmental Management
DEF (CFR) Includes all substances or
   mixtures of substances intended
   for inhibiting the growth of or
   destroying any bacteria, fungi
   pathogenic to man and other
   animals, or viruses declared to be
   pests. These include but are not
   limited to: disinfectants, sanitizers,
   bacteriostats, sterilizers, fungicides
   and fungistats, and commodity
   preservatives and protectants.
   These do not include (a)
   substances or mixtures of
   substances intended to inhibit the
   growth of, inactivate or destroy
   fungi, bacteria, or viruses in or on
   living man or other animals; and
   (b) substances or mixtures of
   substances intended to inhibit the
   growth of, inactivate or destroy
   fungi, bacteria, or viruses in or on
   processed food, beverages, or
   pharmaceuticals including
   cosmetics.

ANTIBODIES
(EPA)
DA October 12, 1990
BT1 Proteins
 BT2 Organic Chemicals
  BT3 Chemical Substances
NT1 Monoclonal Antibodies
RT Antigens
SO Environmental Protection Agency
   Glossary
DEF (EPA) Proteins produced in the
   body by immune system cells in
   response to antigens, and capable
   of combining with antigens.

ANTICIPATED OPERATIONAL
OCCURRENCES
(DOE Order 6430.1A)
DA October 12, 1990
BT1 Occurrences
RT Design Capacity
RT Fires
RT Spills
SO Construction
DEF (DOE Order 6430.1A) Abnormal
   events that are expected to occur

once or more during the lifetime of the facility (e.g. small radioactive materials spills, small fires, etc.).

ANTICIPATED PROCESSES AND EVENTS
(ANL; CFR)
DA May 22, 1991
BT1 Natural Phenomenon
BT1 Processes
SO Wastes
DEF (CFR) Those natural processes and events that are reasonably likely to occur during the period the intended performance objective must be achieved. To the extent reasonable in the light of the geologic record, it shall be assumed that those processes operating in the geologic setting during the Quaternary Period continue to operate but with the perturbations caused by the presence of emplaced radioactive waste superimposed thereon.

ANTICIPATED TRANSIENT WITHOUT SCRAM
DA January 8, 1991
SF ATWS
RT Reactor Shutdown

ANTICONTAMINATION CLOTHING
(2695 ANTI C CLOTH)
DA January 3, 1991
BT1 Clothing
BT1 Personal Protective Equipment
BT2 Equipment/Parts - Personal Protective (DOE FRASE Vocabulary)
BT3 Equipment
NT1 Acid Suit
NT1 Lab Coat
NT1 Radiation Suit
SO DOE FRASE VOCABULARY

ANTIGENS
(EPA)
DA October 12, 1990
RT Antibodies
RT Vaccines
SO Environmental Protection Agency Glossary
DEF (EPA) Substances that cause production of antibodies when introduced into animal or human tissue.

ANTIMONY 125
(EDB)
DA March 29, 1991
BT1 Beta-Minus Decay Radioisotopes
BT2 Beta Decay Radioisotopes
BT3 Radionuclides
BT4 CERCLA Hazardous Substances
BT5 Hazardous Substances
BT4 Nuclides
BT1 Years Living Radioisotopes

BT2 Radionuclides
BT3 CERCLA Hazardous Substances
BT4 Hazardous Substances
BT3 Nuclides
SO Radiation

AO
DA October 12, 1990
SEE Auxiliary Orifices
SO Acronyms

AOT
DA October 12, 1990
SEE Allowable Outage Time
SO Acronyms

API
DA October 12, 1990
SEE Axial Power Indicator
SO Acronyms

API (American Petroleum Institute)
DA October 12, 1990
SEE American Petroleum Institute
SO Acronyms

APM
DA October 12, 1990
SEE Axial Power Monitors
SO Acronyms

APPLIANCE(S)
(2378 APPLIANCE)
DA January 3, 1991
BT1 Equipment/Parts - Electrical (DOE FRASE Vocabulary)
BT2 Equipment
NT1 Heating Appliance(s)
SO DOE FRASE VOCABULARY

APPLICABLE OR RELEVANT AND APPROPRIATE REQUIREMENTS
(EM)
DA November 5, 1990
SF ARARs
BT1 Standards
BT2 Codes, Standards, and Regulations
RT Superfund
DEF (EM) Requirements, including cleanup standards, standards of control, and other substantive environmental protection requirements and criteria for hazardous substances as specified under Federal and State law and regulations, that must be met when complying with the Comprehensive Environmental Response, Compensation, and Liability Act (From the Superfund Amendments and Reauthorization Act).

APPLICABLE STANDARDS AND LIMITATIONS
(SWDA; RCRA; CFR)
DA October 19, 1990

BT1 Standards
BT2 Codes, Standards, and Regulations
SO Wastes
DEF (CFR) All State, interstate, and federal standards and limitations to which a "discharge," a "sludge use or disposal practice" or a related activity is subject under the CWA (Clean Water Act), including "standards for sewage sludge use or disposal," "effluent limitations," water quality standards, standards of performance, toxic effluent standards or prohibitions, "best management practices," and pretreatment standards under sections 301, 302, 303, 304, 306, 307, 308, 403, and 405 of the Clean Water Act.

APPLICATION (PESTICIDE)
DA February 25, 1991
BT1 Processes
NT1 Band Application
NT1 Basal Application
NT1 Broadcast Application
RT Drift

APPLICATION FOR FEDERAL ASSISTANCE
(ESA; CFR)
DA October 12, 1990
RT Federal Agencies
SO Endangered Species
DEF (CFR) A description of work to be accomplished, including objectives and needs, expected results and benefits, approach, cost, location and time required for completion.

APPLIED RESEARCH
(DOE Order 4700.1)
DA January 24, 1991
BT1 Research and Development
BT2 Activities
RT Basic Research
SO Environmental Management
DEF (DOE 4700.1) Systematic study directed towards fuller scientific knowledge or understanding for direct use in fulfilling specific energy requirements.

APPRAISALS
(ESH)
DA October 12, 1990
NT1 Management Appraisals
NT2 Functional Appraisals
NT2 Internal Appraisals
NT1 Multidiscipline Technical Safety Assurance Appraisal
NT1 Safety Appraisals
NT2 Technical Safety Appraisals
SO Quality Assurance
DEF (DOE Order 5700.6B; ESH) A planned and documented activity performed in accordance with procedures to determine, by

SY-Synonymous Terms SO-Source/Subject Category SF-See From

examination and evaluation of objective evidence, the adequacy of and extent to which applicable elements of the quality assurance program have been developed, documented, and effectively implemented in accordance with specified requirements. Audits can be internal examinations of programs or activities under an organization's control and within its organizational structure or external examinations of programs or activities of another organization.

APPROPRIATE ACT AND REGULATIONS
(SWDA; RCRA; CWA; SDWA)
DA   October 12, 1990
NT1 Clean Water Act
NT1 Resource Conservation and Recovery Act
NT1 Safe Drinking Water Act
NT1 Solid Waste Disposal Act
RT   Codes, Standards, and Regulations
RT   Statutes and Regulations
SO   Wastes
SO   Water Pollution
DEF The Clean Water Act (CWA); the Solid Waste Disposal Act, as amended by the Resource Conservation Recovery Act (RCRA); or Safe Drinking Water Act (SDWA), whichever is applicable; and applicable regulations promulgated under those statutes. In the case of an "approved State program," appropriate Act and regulations includes program requirements.

APPROVALS NECESSARY TO BEGIN PHYSICAL CONSTRUCTION
(CFR)
DA   November 9, 1990
BT1 Licenses
BT1 Permits
RT   Construction
RT   Requirements
DEF (CFR) Permits and approvals required under Federal, State or local hazardous waste control statutes, regulations or ordinances.

APPROVED MEDICAL PRACTITIONERS
(IAEA)
DA   October 12, 1990
BT1 Technically Qualified Individuals
BT2 Personnel
RT   Medical Treatment
RT   Occupational Medicine
SO   Radiation
DEF (IAEA) Medical practitioners responsible for the medical surveillance of occupationally exposed workers whose capacity to act in this respect is recognized by the competent authority.

APPROVED PROGRAMS
(SWDA; RCRA; CFR)
DA   October 12, 1990
BT1 Programs
RT   PSD Permits
RT   State Agencies
SO   Environmental Management
SO   Wastes
DEF (SWDA) State implementation plans providing for issuance of PSD permits which have been approved by EPA under the Clean Air Act and 40 CFR 51. An "approved State" is one administering an approved program. State Director as used in 40 CFR 124.4 means the person(s) responsible for issuing PSD permits under an approved program, or that person's delegated representative. Approved program or approved State means a State which has been approved or authorized by EPA under 40 CFR 271.

APPROVED STATE PRIMACY PROGRAM
(CFR)
DA   October 19, 1990
BT1 Programs
NT1 State Program Revisions
RT   Primary Drinking Water Regulations
RT   State Agencies
RT   State Primary Drinking Water Regulation
SO   Water Pollution
DEF (CFR) Those program elements listed in 40 CFR 142.11(a) that were submitted with the initial State application for primary enforcement authority and approved by the EPA Administrator and all State program revisions thereafter that were approved by the EPA Administrator.

APPROVED STATES
(SWDA; RCRA)
DA   October 19, 1990
BT1 States
SO   Wastes
DEF (SWDA; RCRA) States which have been approved or authorized by EPA under 40 CFR 271.

APPROVED STORAGE CONTAINERS
(DOE Order 6430.1A)
DA   October 12, 1990
BT1 Containers
SO   Construction
DEF (DOE Order 6430.1A) Containers that are fabricated from noncombustible material(s) that satisfy container integrity criteria developed from the safety analysis for the particular form(s) of stored material under normal storage conditions, design basis fire and

other design basis accident conditions, and that are approved for their intended use by the responsible DOE operating contractor and the responsible DOE field organization.

AQUATIC ECOSYSTEM
(CWA; RHA)
DA   January 11, 1991
SY   Aquatic Environment
BT1 Environment

AQUATIC ENVIRONMENT
(CWA; RHA)
DA   October 12, 1990
SY   Aquatic Ecosystem
BT1 Environment
NT1 Benthic Region
RT   Eutrophication
RT   Filling
RT   Limnology
RT   Special Aquatic Sites
RT   Surface Waters
RT   Unacceptable Adverse Effects
SO   Water Pollution
DEF (CFR) Waters of the United States, including wetlands, that serve as habitat for interrelated and interacting communities and populations of plants and animals.

AQUIFERS
(DOE Order 6430.1A; SWDA; RCRA; SDWA; CFR; ESH)
DA   October 12, 1990
BT1 Formations
NT1 Confined Aquifers
NT1 Exempted Aquifers
NT1 Principal Source Aquifers
NT1 Semi-Confined Aquifers
NT1 Significant Source of Groundwater
NT1 Sole Source Aquifer
NT1 Underground Sources of Drinking Water
NT1 Uppermost Aquifers
RT   Confining Beds
RT   Confining Zones
RT   Groundwater
SO   Construction
SO   Environmental Management
SO   Environmental Protection Agency Glossary
SO   Wastes
SO   Water Pollution
DEF (CFR) An underground geological formation, or group of formations, containing usable amounts of groundwater that can supply wells and springs.

ARARs
DA   October 12, 1990
SEE Applicable or Relevant and Appropriate Requirements
SO   Acronyms

ARBITRATION
(EPA)

DA   October 12, 1990
BT1 Administrative Processes
 BT2 Processes
RT   Work Processes
RT   Work Environment
SO   Environmental Protection Agency
       Glossary
DEF (EPA) A process for the resolution
       of disputes. Decisions are made
       by an impartial arbitrator selected
       by the parties. These decisions
       are usually legally binding.

ARCHAEOLOGICAL RESOURCES
(ARPA)
DA   April 24, 1991
RT   Archaeological Resources
       Protection Act of 1979
RT   Federal Land Managers
RT   Indian Tribes
RT   Indian Lands
RT   Natural Resources
RT   Public Lands
RT   Targets
SO   Environmental Management
DEF (ARPA) Any material remains of
       past human life or activities which
       are of archaeological interest, as
       determined under uniform
       regulations promulgated pursuant
       to the Archaeological Resources
       Protection Act of 1979. Such
       regulations containing such
       determination shall include, but not
       be limited to: pottery, basketry,
       bottles, weapons, weapon
       projectiles, tools, structures, or
       portions of structures, pit houses,
       rock paintings, rock carvings,
       intaglios, graves, human skeletal
       materials, or any portion or piece
       of any of the foregoing items.
       Nonfossilized and fossilized
       paleontological specimens, or any
       portion or piece thereof, shall not
       be considered archaeological
       resources, under the regulations
       under ARPA, unless found in an
       archaeological context. No item
       shall be treated as an
       archaeological resource under
       regulations of this act unless such
       item is at least 100 years of age.

ARCHAEOLOGICAL RESOURCES
PROTECTION ACT OF 1979
DA   April 24, 1991
SF   ARPA
BT1 Acts
 BT2 Statutes and Regulations
RT   Archaeological Resources
RT   Federal Land Managers
RT   Indian Lands
RT   Natural Resources
RT   Public Lands
SO   Environmental Management
DEF (ARPA) The purpose of this Act is
       to secure, for the present and
       future benefit of the American
       people, the protection of

archaeological resources and sites
which are on public lands and
Indian lands, and to foster
increased cooperation and
exchange of information between
governmental authorities, the
professional archaeological
community, and private individuals
having collections of
archaeological resources and data
which were obtained before the
date of the enactment of this act.

ARCHITECT ENGINEER
DA   January 8, 1991
SF   A/E
BT1 Personnel

AREA
(1601 AREA)
DA   December 10, 1990
BT1 Sites/Areas
NT1 Canal Area
NT1 Construction Area
NT1 Customer Service Area
NT1 Data Processing Area
NT1 Decontamination Areas
NT1 Eating Area
 NT2 Cafeteria
NT1 Field Area
NT1 Maintenance Area
NT1 Office Area
NT1 Pit Area
NT1 Production/Operations Area
NT1 Pump Area
NT1 Recreation Area
NT1 Sand Blasting Area
NT1 Storage Area
 NT2 Stock Room
 NT2 Warehouse
NT1 Sump Area
NT1 Test Area
 NT2 Fast Flux Test Facility
 NT2 Nevada Test Site
 NT2 Rocket Motor Test Sites
 NT2 Test Reactor Facility
 NT2 Test Stations
 NT2 Test Reactor Area
 NT2 Tonopah Test Range
NT1 Vehicle Service Area
NT1 Yard Area
SO   DOE FRASE VOCABULARY

AREA AROUND EYE
(1083 AREA EYE)
DA   November 28, 1990
BT1 Face
 BT2 Head
  BT3 Human Body Parts
NT1 Eyelid(s)
RT   Eye Protection
RT   Eye(s)
SO   DOE FRASE VOCABULARY

AREA CONTROL ROOM
DA   January 8, 1991
SF   ACR

AREA OF REVIEW
(SDWA; CFR; EPA)
DA   October 12, 1990
BT1 Sites/Areas
RT   Injection Wells
SO   Environmental Protection Agency
       Glossary
DEF (EPA) In the UIC (Underground
       Injection Control) program, the
       area surrounding an injection well
       that is reviewed during the
       permitting process to determine
       whether the injection operation will
       induce flow between aquifers.
       (CFR) The area surrounding an
       injection well described according
       to the criteria set forth in 40 CFR
       146.06 or in the case of an area
       permit, the project area plus a
       circumscribing area the width of
       which is either 1/4 of a mile or a
       number calculated according to
       the criteria set forth in section
       146.06.

AREA SOURCES
(CAA; CFR; EPA)
DA   October 12, 1990
RT   Air Pollutants
RT   Mobile Sources
RT   Point Sources
SO   Air Pollution
SO   Environmental Management
SO   Environmental Protection Agency
       Glossary
DEF (EPA) Small sources of nonnatural
       air pollution that are released over
       a relatively small area but which
       cannot be classified as a point
       source. Such sources may include
       vehicles and other small fuel
       combustion engines. (CFR) Any
       small residential, governmental,
       institutional, commercial, or
       industrial fuel combustion
       operations; onsite solid waste
       disposal facility; motor vehicles,
       aircraft vessels, or other
       transportation facilities or other
       miscellaneous sources identified
       through inventory techniques
       similar to those described in the
       AEROS Manual series, Vol. II
       AEROS User's Manual,
       EPA-450/2-76- 029, December
       1976.

ARGONNE NATIONAL LABORATORY
DA   June 13, 1991
BT1 Government-Owned
       Contractor-Operated Facilities
 BT2 Federal Facilities
  BT3 Facilities
BT1 Laboratories
 BT2 Research and Development
       Organizations
  BT3 Organizations
NT1 Argonne National Laboratory-East
       (Chicago)

SY-Synonymous Terms   SO-Source/Subject Category   SF-See From

NT1 Argonne National Laboratory-West
   (At INEL)
RT Chicago Operations Office
SO Management
DEF (Capsule Review of DOE Research
   and Development and Field
   Facilities, 1986) ANL, established
   by the Atomic Energy Act of 1946,
   conducts applied research and
   engineering development in
   nuclear fission and other energy
   technologies and scientific
   research in basic physical and life
   sciences. R&D at ANL links
   technology base research with
   engineering development, from
   concept stages to application.
   ANL's role is to develop and
   operate research facilities for
   members of the scientific
   community, maintain close
   interaction with personnel in
   universities and industry and aid in
   the education of scientists and
   engineers.

**ARGONNE NATIONAL
LABORATORY-EAST (CHICAGO)**
DA   January 8, 1991
SF   *ANL-E*
BT1 Argonne National Laboratory
 BT2 Government-Owned
    Contractor-Operated Facilities
  BT3 Federal Facilities
  BT4 Facilities
 BT2 Laboratories
  BT3 Research and Development
    Organizations
  BT4 Organizations
BT1 Laboratories
 BT2 Research and Development
    Organizations
 BT3 Organizations
RT University of Chicago

**ARGONNE NATIONAL
LABORATORY-WEST (AT INEL)**
DA   January 8, 1991
SF   *ANL-W*
BT1 Argonne National Laboratory
 BT2 Government-Owned
    Contractor-Operated Facilities
  BT3 Federal Facilities
  BT4 Facilities
 BT2 Laboratories
  BT3 Research and Development
    Organizations
  BT4 Organizations
BT1 Laboratories
 BT2 Research and Development
    Organizations
 BT3 Organizations
RT University of Chicago

ARI
DA   October 12, 1990
SEE Alternate Rod Insertion
SO Acronyms

ARKANSAS NUCLEAR ONE-1
DA   January 8, 1991
SF   *ANO-1*
BT1 Reactors

ARM(S)
(1095 ARM)
DA   November 28, 1990
BT1 Human Body Parts
NT1 Forearm(s)
NT1 Upper Arm
RT Elbow(s)
RT Forearm Protection
RT Hand(s)
SO DOE FRASE VOCABULARY

ARMORERS
(DOE Order 5480.16)
DA   October 12, 1990
BT1 Personnel
RT Defective Firearm
RT Firearms
RT Misfires
RT Munitions
RT Protective Force Personnel
RT Tagging
SO Firearms
DEF (DOE Order 5480.16) Individuals
   who by schooling, experience, and
   assignment are trained to operate,
   maintain, and repair weapons
   used by protective force
   personnel.

ARPA
DA   April 24, 1991
SEE Archaeological Resources
   Protection Act of 1979
SO Acronyms

ARS
DA   October 12, 1990
SEE Alternate Removal Systems
SO Acronyms

ARSENIC CONTAINING GLASS TYPES
(CFR)
DA   November 15, 1990
RT Theoretical Arsenic Emissions
   Factor
SO Air Pollution
DEF (CFR) Any glass that is
   distinguished from other glass
   solely by the weight percent of
   arsenic added as a raw material
   and by the weight percent of
   arsenic in the glass produced. Any
   two or more glasses that have the
   same weight percent of arsenic in
   the raw materials as well as in the
   glass produced shall be
   considered to belong to one
   arsenic-containing glass type,
   without regard to the recipe used
   or any other characteristics of the
   glass or the method of production.

ARSENIC KITCHENS
(CFR)

DA   October 12, 1990
BT1 Control Devices
 BT2 Devices
RT Inorganic Arsenic
RT Theoretical Arsenic Emissions
   Factor
SO Air Pollution
DEF (CFR) Baffled brick chambers
   where inorganic arsenic vapors
   are cooled, condensed, and
   removed in a solid form.

ARTICLES
(TSCA; CFR)
DA   October 12, 1990
NT1 Consumer Products
NT1 PCB Articles
NT1 Radioactive Articles
SO Compensation and Liability
SO Environmental Management
DEF (CFR) Manufactured items (1)
   which are formed to a specific
   shape or design during
   manufacture; (2) which have end
   use functions dependent in whole
   or in part upon their shape or
   design during end use; and (3)
   which do not release a toxic
   chemical under normal conditions
   of processing or use of those
   items at the facility or
   establishments.

ARTIFICIAL REEFS
(CWA; RHA; CFR)
DA   October 12, 1990
BT1 Structures (DOE FRASE
   Vocabulary)
RT Fish
SO Water Pollution
DEF (CFR) Structures that are
   constructed or placed in the
   navigable waters of the United
   States or in the waters overlying
   the outer continental shelf for the
   purpose of enhancing fishery
   resources and commercial and
   recreational fishing opportunities.
   The term does not include
   activities or structures such as
   wing deflectors, bank stabilization,
   grade stabilization structures, or
   low flow key ways, all of which
   may be useful to enhance
   fisheries resources.

AS EXPEDITIOUSLY AS PRACTICABLE
(CAA)
DA   October 12, 1990
SO Air Pollution
DEF (CAA) Means as expeditiously as
   practicable but in no event later
   than five years after the date of
   approval of a plan revision.

AS LOW AS PRACTICABLE
(SSDC)
DA   October 12, 1990
SF   *ALAP*

BT1 Radiation Standards
 BT2 Standards
  BT3 Codes, Standards, and
      Regulations
 RT As Low As Reasonably Achievable
 SO System Safety Development Center
    Glossary
 DEF (SSDC) This has been replaced by
     "As Low as Reasonably
     Achievable (ALARA)." This was a
     DOE guideline on limiting
     personnel exposures to ionizing
     radiation ALAP in terms of
     economic and technical feasibility.

AS LOW AS REASONABLY
ACHIEVABLE
(DOE Orders 5400.5, 5480.11; SSDC)
 DA October 12, 1990
 SF ALARA
 BT1 Radiation Standards
  BT2 Standards
   BT3 Codes, Standards, and
       Regulations
 RT As Low As Practicable
 RT Dose Equivalents
 RT Radiation Protection
 RT Radiation Safety
 SO Construction
 SO Environmental Management
 SO Industrial Hygiene
 SO System Safety Development Center
    Glossary
 DEF (DOE Order 5400.5) A phrase
     (acronym) used to describe an
     approach to radiation protection to
     control or manage exposures
     (both individual and collective to
     the work force and the general
     public) and releases of radioactive
     material to the environment as low
     as social, technical, economic,
     practical, and public policy
     considerations permit. As used in
     this Order, ALARA is not a dose
     limit, but rather it is a process that
     has as its objective the attainment
     of dose levels as far below the
     applicable limits of the Order as
     practicable. (SSDC) Order DOE
     5480.1, Chapter 1, outlines
     requirements for radiation
     protection for DOE and DOE
     contractor operations. This
     includes keeping worker and
     public exposures to ionizing
     radiation as low as reasonably
     achievable by techniques outlined
     in the chapter.

ASBESTOS
(TSCA; CFR; EPA)
 DA October 12, 1990
 BT1 Carcinogens
  BT2 Hazardous Substances
 BT1 CERCLA Hazardous Substances
  BT2 Hazardous Substances
 NT1 Commercial Asbestos
 NT1 Friable Asbestos
 RT Asbestos Materials

 SO Air Pollution
 SO Environmental Management
 SO Environmental Protection Agency
    Glossary
 SO Hazardous Materials
 DEF (EPA) A mineral fiber that can
     pollute air or water and cause
     cancer or asbestosis when
     inhaled. EPA has banned or
     severely restricted its use in
     manufacturing and construction.
     (CFR) Asbestiform varieties of
     chrysotile (serpentine), crocidolite
     (riebeckite), amosite
     (cummingtonite-grunerite),
     anthophyllite, tremolite, or
     actinolite.

ASBESTOS-CONTAINING MATERIALS
(CFR)
 DA October 12, 1990
 BT1 Asbestos Materials
  BT2 Materials
 NT1 Friable Asbestos-Containing
     Materials
 RT EPA Guidance Document
 SO Hazardous Materials
 DEF (CFR) Materials which contain more
     than 1 percent asbestos by weight.

ASBESTOS-CONTAINING WASTE
MATERIALS
(TSCA; CAA; CFR; EPA)
 DA November 15, 1990
 BT1 Asbestos Materials
  BT2 Materials
 BT1 Wastes
 NT1 Asbestos Tailings
 NT1 Asbestos Waste from Control
     Devices
 RT Inactive Waste Disposal Sites
 RT Renovation
 SO Air Pollution
 SO Environmental Management
 DEF (CFR) Any waste that contains
     commercial asbestos and is
     generated by a source subject to
     the provisions of this subpart. This
     term includes asbestos mill
     tailings, asbestos waste from
     control devices, friable asbestos
     waste material, and bags or
     containers that previously
     contained commercial asbestos.
     However, as applied to demolition
     and renovation operations, this
     term includes only friable asbestos
     waste and asbestos waste from
     control devices.

ASBESTOS MATERIALS
(CFR)
 DA October 12, 1990
 BT1 Materials
 NT1 Asbestos-containing Materials
  NT2 Friable Asbestos-Containing
      Materials
 NT1 Asbestos-Containing Waste
     Materials
  NT2 Asbestos Tailings

 NT2 Asbestos Waste from Control
     Devices
 NT1 Friable Asbestos Materials
 NT1 Particulate Asbestos Materials
 RT Asbestos
 SO Air Pollution
 SO Environmental Management
 DEF (CFR) Any materials containing
     asbestos.

ASBESTOS MILLS
(CAA; CFR)
 DA October 12, 1990
 BT1 Facilities
 RT Commercial Asbestos
 SO Air Pollution
 SO Environmental Management
 DEF (CFR) Facilities engaged in
     converting, or in any intermediate
     step in converting, asbestos ore
     into commercial asbestos. Outside
     storage of asbestos material is not
     considered a part of the asbestos
     mill.

ASBESTOS TAILINGS
(CFR)
 DA October 12, 1990
 BT1 Asbestos-Containing Waste
     Materials
  BT2 Asbestos Materials
   BT3 Materials
  BT2 Wastes
 BT1 Tailings
  BT2 Solid Wastes
   BT3 Wastes
 SO Air Pollution
 DEF (CFR) Solid wastes that contain
     asbestos and are products of
     asbestos mining or milling
     operations.

ASBESTOS WASTE FROM CONTROL
DEVICES
(CAA; CFR)
 DA October 12, 1990
 BT1 Asbestos-Containing Waste
     Materials
  BT2 Asbestos Materials
   BT3 Materials
  BT2 Wastes
 RT Control Devices
 RT Friable Asbestos
 SO Air Pollution
 SO Environmental Management
 DEF (CFR) Any waste material that
     contains asbestos and is collected
     by a pollution control device.

ASBESTOSIS
(EPA)
 DA October 12, 1990
 BT1 Diseases
 BT1 Illnesses
 RT Occupational Illnesses
 RT Occupational Exposure
 SO Environmental Protection Agency
    Glossary
 DEF (EPA) A disease associated with

SY-Synonymous Terms   SO-Source/Subject Category   SF-See From

chronic exposure to and inhalation
of asbestos fibers. The disease
makes breathing progressively
more difficult and can lead to
death.

ASHES
(EPA)
DA   October 12, 1990
BT1 Wastes
NT1 Bottom Ash
NT1 Fly Ash
RT   Combustion
RT   Smoke
SO   Environmental Protection Agency
     Glossary
DEF (EPA) The mineral content of a
     product remaining after complete
     combustion.

ASME
DA   October 12, 1990
SEE American Society of Mechanical
     Engineers
SO   Acronyms

ASPHYXIA
(1303 ASPHYXIA)
DA   November 28, 1990
BT1 Injuries
SO   DOE FRASE VOCABULARY

ASPHYXIANTS
(EMER)
DA   February 1, 1991
BT1 Hazardous Chemicals
 BT2 Hazardous Substances
RT   Ventilation
SO   Emergency Preparedness
DEF (EMER) Chemical vapors or gases
     which replace the oxygen in air
     and can cause death by
     suffocation. Asphyxiants are
     especially hazardous when
     present in confined spaces.

ASPHYXIATION
(SSDC)
DA   October 12, 1990
SO   System Safety Development Center
     Glossary
DEF (SSDC) Human deprivation of
     oxygen by chemical or physical
     means. Chemical asphyxiants
     prevent oxygen transfer from the
     blood to body cells. Physical
     asphyxiants prevent oxygen from
     reaching the blood.

ASSESSMENT ACTIONS
(EMER)
DA   February 1, 1991
BT1 Actions
 BT2 Responses
RT   Assessments
SO   Emergency Preparedness
DEF (EMER) Those actions taken during
     or immediately after an incident or
     emergency to gather and process

the information necessary to make
decisions and to implement
specific emergency measures.

ASSESSMENT AREA
(CFR)
DA   November 15, 1990
BT1 Sites/Areas
RT   Assessments
RT   Hazardous Substances
RT   Natural Resources
RT   Oil Spills
DEF (CFR) The area or areas within
     which natural resources have
     been affected directly or indirectly
     by the discharge of oil or release
     of a hazardous substance and that
     serves as the geographic basis for
     the injury assessment.

ASSESSMENT PLANS
(TTGM)
DA   May 7, 1991
BT1 Plans
RT   Tiger Team Assessments
RT   Tiger Team Leaders
SO   Management
DEF (TTGM) Following the
     pre-assessment site visit and
     review of information, the Tiger
     Team Leader, Assessment
     component team leaders, and
     team members will prepare an
     Assessment Plan which will outline
     the key issues, general approach,
     and specific on- site activities. The
     plan should include: 1) issue
     identification; 2) an agenda; 3) a
     list of the records, files, and other
     documents that team members
     intend to review on site; and 4) a
     determination of a) the presence
     of classified projects/wastes and
     b) the adequacy of oversight.

ASSESSMENT TEAM LEADERS
(TTGM)
DA   May 7, 1991
BT1 Personnel
RT   Tiger Team Assessments
RT   Tiger Teams
RT   Tiger Team Leaders
RT   Tiger Team Administrators
SO   Management
DEF (TTGM) Assessment Team Leaders
     are primarily responsible for the
     detailed technical conduct and
     results of their respective
     assessment teams, and provide
     direct supervision of the
     day-to-day activities of their
     individual team members.

ASSESSMENTS
(CFR; EMER)
DA   November 15, 1990
BT1 Administrative Processes
 BT2 Processes
NT1 Biological Assessments

NT1 Consequence Assessments
NT1 Credibility Assessments
NT1 Damage Assessments
 NT2 Natural Resource Damage
      Assessment
 NT2 Natural Resource Damage
      Preassessment Screens
NT1 Data Quality Assessments
NT1 Endangered Assessment
NT1 Environmental Assessments
NT1 Exposure Assessment
NT1 Operation Assessment and
     Readiness
NT1 Performance Assessments
NT1 Preliminary Assessment
NT1 Process Waste Assessment
NT1 RCRA Facility Assessment
NT1 Risk Assessment
 NT2 Probabilistic Risk Assessment
NT1 Seabrook Station Probabilistic
     Safety Assessment
NT1 Self-Assessment
NT1 Systematic Assessment of
     Licensee Performance
NT1 Tiger Team Assessments
 NT2 Environment Assessments
 NT2 Management and Organization
      Assessment
 NT2 Safety and Health Assessments
NT1 Type A Assessments
NT1 Type B Assessments
NT1 Unified Dose Assessments
RT   Analyses
RT   Assessment Actions
RT   Assessment Area
RT   Carcinogen Risk Assessment
     Verification Endeavor Workgroup
RT   Comprehensive Environmental
     Assessment and Response
     Program
RT   Federal Radiological Monitoring
     and Assessment Centers
RT   Federal Radiological Monitoring
     and Assessment Plans
RT   Inspection and Evaluation
RT   Multimedia Environmental Pollutant
     Assessment System
RT   Natural Resources
RT   Remedial Action Assessment
     System
RT   Solid Waste Assessment Test
SO   Emergency Preparedness
SO   Management
DEF (CFR) The processes of collecting,
     compiling, and analyzing
     information, statistics, or data
     through prescribed methodologies.

ASSETS
(SWDA; RCRA; CFR)
DA   October 12, 1990
NT1 Current Assets
RT   Capital Expenditure
RT   Liabilities
SO   Wastes
SO   Water Pollution
DEF All existing and all probable future
     economic benefits obtained or
     controlled by a particular entity.

BT-Broader Term   NT-Narrower Term   RT Related Term

ASSIGNED RISKS
(SSDC)
DA   October 12, 1990
BT1 Risks
RT   States
SO   System Safety Development Center
     Glossary
DEF (SSDC) Risks which are assigned,
     by State law, to an insurer from a
     pool of insurers (usually all those
     licensed in the State) who would
     not otherwise accept it.

ASSIMILATION
(EPA)
DA   October 12, 1990
BT1 Pollution Recovery Processes
 BT2 Processes
RT   Pollutants
RT   Surface Waters
SO   Environmental Protection Agency
     Glossary
DEF (EPA) The ability of a body of water
     to purify itself of pollutants.

**ASSISTANT ADMINISTRATOR FOR
FISHERIES**
(ESA; CFR)
DA   October 12, 1990
BT1 Administrators
 BT2 Personnel
SO   Endangered Species
DEF (CFR) The Assistant Administrator
     for Fisheries, National Oceanic
     and Atmospheric Administration,
     Department of Commerce, or his
     authorized delegate. The Assistant
     Administrator for Fisheries is in
     charge of the National Marine
     Fisheries Service.

ASSOCIATED UNIVERSITIES, INC.
DA   January 11, 1991
BT1 Companies
 BT2 Commercial Organizations
  BT3 Organizations
RT   Brookhaven National Laboratory
RT   Chicago Operations Office

ASSUMED RISKS
(SSDC; MORT)
DA   October 12, 1990
BT1 Risks
RT   Acceptable Risks
RT   Events
RT   Oversights and Omissions
RT   Specific Control Factors
SO   System Safety Development Center
     Glossary
DEF (SSDC) Specific, analyzed residual
     risks accepted at an appropriate
     level of management. Ideally, the
     risk has had analysis of
     alternatives for increasing control
     and evaluation of significance of
     consequences. (MORT) MORT
     analysis asks the following
     questions: What are the assumed
     risks? Are they specific, named

events? Are they analyzed, and
where possible, calculated
(quantified)? Was there a specific
decision to assume each risk?
Was it made by a person who had
management delegated authority
to assume the risk.

ATMOSPHERE
(EPA)
DA   October 12, 1990
NT1 Stratosphere
NT1 Troposphere
RT   Accessible Environment
RT   Air
RT   Atmosphere Gases
RT   Biosphere
RT   Contrails
RT   Lithosphere
RT   Pressure (Surface)
SO   Environmental Protection Agency
     Glossary
DEF (EPA) (1) A standard unit of
     pressure representing the
     pressure exerted by a 29.92-inch
     column of mercury at sea level at
     45' latitude and equal to 1000
     grams per square centimeter. (2)
     The whole mass of air surrounding
     the earth, composed largely of
     oxygen and nitrogen.

ATMOSPHERE GASES
(SWDA; RCRA; ESH)
DA   October 12, 1990
BT1 Gases
RT   Atmosphere
SO   Hazardous Materials
DEF Gases that are commercially
     derived through an air separation
     process.

**ATMOSPHERIC RELEASE ADVISORY
CAPABILITY**
(DOE Order 5500.1A, Attach. II, 6)
DA   January 24, 1991
RT   Emissions
RT   Information Systems
SO   Environmental Management
DEF (DOE 5500.1A) A computer based
     system that provides rapid
     predictions of the transport,
     diffusion, and deposition of
     radioactive nuclides or other toxic
     materials released to the
     atmosphere and dose projections
     to people and the environment.

ATOMIC ENERGY ACT
(Doe Order Definitions)
DA   October 12, 1990
SF   AEA
BT1 Acts
 BT2 Statutes and Regulations
RT   Extraordinary Nuclear Occurrences
RT   Inconsistencies
SO   Industrial Hygiene
DEF The Act (1954) which placed
     production and control of nuclear

materials within a civilian agency,
originally the Atomic Energy
Commission.

ATOMIC ENERGY ACT FACILITIES
(DOE Order 5480.2)
DA   October 16, 1990
SY   Government-Owned
     Contractor-Operated Facilities
BT1 Facilities
RT   Atomic Energy Defense Activities
SO   Environmental Management
DEF (DOE Order 5480.2) Those DOE
     facilities operated under authority
     of the Atomic Energy Act of 1954
     as amended.

ATOMIC ENERGY COMMISSION
DA   January 8, 1991
SF   AEC
BT1 Federal Agencies
 BT2 Agencies
  BT3 Administrative Organizations
   BT4 Organizations
RT   Energy Research and Development
     Administration
RT   Extraordinary Nuclear Occurrences
RT   U.S. Department of Energy
DEF Federal agency created in 1946 to
     manage the development, use and
     control of nuclear energy for
     military and civilian application.
     Abolished by the Energy
     Reorganization Act of 1974 and
     succeeded by the Energy
     Research and Development
     Administration (now part of the
     U.S. Department of Energy) and
     the U.S. Nuclear Regulatory
     Commission.

**ATOMIC ENERGY DEFENSE
ACTIVITIES**
(ANL; NWPA)
DA   May 22, 1991
SY   Atomic Energy Defense Activities of
     the Secretary
BT1 Activities
RT   Atomic Energy Act Facilities
RT   Civilian Nuclear Activities
SO   Management
DEF (NWPA) Activities of the Secretary
     of Energy performed in whole or in
     part in carrying out any of the
     following functions: naval reactors
     development; weapons activities
     including defense inertial
     confinement fusion; verification
     and control technology; defense
     nuclear material production;
     defense nuclear waste and
     materials by-product management;
     defense nuclear materials security
     and safeguards and security
     investigations; and defense
     research and development.

ATOMIC ENERGY DEFENSE ACTIVITIES OF THE SECRETARY
(ANL; LLRWPA)
DA   May 24, 1991
SY   Atomic Energy Defense Activities
DEF (LLRWPA) Those activities and facilities of the Department of Energy carrying out the function of 1) naval reactors development and propulsion; 2) weapons activities, verification and control technology; 3) defense materials production; 4) defense waste management; and 5) defense nuclear materials security and safeguards (all as included in the Department of Energy appropriations account in any fiscal year for atomic energy defense activities).

ATOMIC INDUSTRIAL FORUM
DA   January 8, 1991
SF   AIF
BT1 Professional Bodies
 BT2 Organizations

ATOMIC WEAPONS
(AEA; CFR)
DA   January 24, 1991
RT   Nevada Test Site
RT   Nuclear Winter
SO   Environmental Management
DEF (CFR) Any device utilizing atomic energy, exclusive of the means for transporting or propelling the device (where such means are a separable and divisible part of the device), the principal purpose of which is for use as, or for development of, a weapon prototype, or a weapon test device.

ATOMIZE
(EPA)
DA   October 12, 1990
BT1 Processes
RT   Aerosols
RT   Particulates
SO   Environmental Protection Agency Glossary
DEF (EPA) To divide a liquid into extremely minute particles, either by impact with a jet of steam or compressed air, or by passage through some mechanical device.

ATOMS
(NIH)
DA   October 12, 1990
SO   Radiation
DEF (NIH) Smallest particles of an element which are capable of entering into a chemical reaction.

ATS
DA   October 12, 1990
SEE Automatic Transfer Switch
SO   Acronyms

ATTACK WARNING SIGNALS
(EMER)
DA   February 1, 1991
BT1 Alarms
 BT2 Devices
SO   Emergency Preparedness
DEF (EMER) Three- to five-minute wavering tones on sirens or short blasts on horns or other devices, repeated as deemed necessary. It means that an actual attack against this country has been detected and that protective action should be taken immediately. As a matter of national civil defense policy, attack warning signals shall be used for no other purposes and have no other meanings.

ATTAINMENT AREAS
(EPA)
DA   October 12, 1990
BT1 Sites/Areas
RT   National Ambient Air Quality Standards
SO   Environmental Protection Agency Glossary
DEF (EPA) Areas considered to have air quality as good as or better than the national ambient air quality standards as defined in the Clean Air Act. An area may be an attainment area for one pollutant and a nonattainment area for others.

ATTENTION SIGNALS
(EMER)
DA   February 1, 1991
BT1 Alarms
 BT2 Devices
SO   Emergency Preparedness
DEF (EMER) Three- to five-minute steady tones, sounded strictly at the option of and on the authority of local government officials. The signals may be activated for natural or man-made disasters as local authorities may determine and may also be used to call attention to essential emergency information. Use of the attention or alert signals should always be accompanied by a public explanation and instructions to the public over local broadcast stations or by other means.

ATTENUATION
(EPA)
DA   October 12, 1990
BT1 Processes
RT   Degradation
SO   Environmental Protection Agency Glossary
DEF (EPA) The process by which a compound is reduced in concentration over time, through adsorption, degradation, dilution, and/or transformation.

ATTRACTANTS
(EPA)
DA   November 5, 1990
BT1 Chemical Agents
RT   Insects
RT   Pests
SO   Environmental Management
SO   Environmental Protection Agency Glossary
DEF (EPA) Chemicals or agents that lure insects or other pests by stimulating their sense of smell.

ATTRACTIVE NUISANCE
(SSDC)
DA   October 12, 1990
SO   System Safety Development Center Glossary
DEF (SSDC) Conditions of a property that tend to attract children; for example, a sand pile or structures that can be climbed. Owners must take reasonable precautions to protect children, even though they illegally trespass on the property.

ATTRITION
(EPA)
DA   October 12, 1990
BT1 Processes
RT   Dust
SO   Environmental Protection Agency Glossary
DEF (EPA) Wearing or grinding down of a substance by friction. A contributing factor in air pollution.

ATWS
DA   October 12, 1990
SEE Anticipated Transient Without Scram
SO   Acronyms

AUDITS (SSDC)
(SSDC)
DA   October 19, 1990
SY   Internal Audits
BT1 Administrative Processes
 BT2 Processes
NT1 Environmental Audits
NT1 Field Audits
NT1 Internal Audits
NT1 System Audits
RT   Functional Units
RT   Laboratory Audits
SO   System Safety Development Center Glossary
DEF (SSDC) Reviews and evaluations by an organization of its own internal safety/loss control program, program plan implementation, and operations under its direct control. They may consist of management audits, functional audits or comprehensive audits, which are internal counterparts of the appraisals by the same names. They also may be monitored self audits or

independent audits conducted by
safety disciplines or other
specialists, or by technical or
managerial peers.

AUTHORIZED INSPECTORS
DA   June 3, 1991
BT1 Personnel
RT   Inspections
SO   Hazardous Materials
DEF Inspectors who are currently
     commissioned by the National
     Board of Boiler and Pressure
     Vessel Inspectors and employed
     as an Inspector by an Authorized
     Inspection Agency.

AUTHORIZED LIMITS
(IAEA)
DA   November 20, 1990
SY   Prescribed Limits
BT1 Limits
RT   Competent Authorities
SO   Radiation
DEF (IAEA) Limits of any quantity
     specified by the competent
     authority for a given radiation
     practice or source. These are
     generally lower than the primary,
     secondary or derived limits.

AUTHORIZED OFFICERS
(ESA; CFR)
DA   October 12, 1990
BT1 Personnel
RT   Authorized Officials
RT   Federal Agencies
RT   State Agencies
SO   Endangered Species
DEF (CFR) Any commissioned, warrant,
     or petty officer of the United
     States Coast Guard, or any officer
     or agent designated by the
     Director of the U.S. Fish and
     Wildlife Service, the Secretary of
     the Interior, the Secretary of
     Commerce, or the Secretary of the
     Treasury, or any officer designated
     by the head of a Federal or State
     agency which has entered into an
     agreement with the Secretary of
     the Interior, Secretary of
     Commerce, Secretary of the
     Treasury, or Secretary of
     Transportation to enforce the Acts,
     or any Coast Guard personnel
     accompanying and acting under
     the direction of a person included
     above in this definition.

AUTHORIZED OFFICIALS
DA   November 15, 1990
BT1 Personnel
NT1 Lead Authorized Officials
RT   Authorized Officers
RT   Indian Tribes
DEF (CFR) Federal or State officials to
     whom is delegated the authority to
     act on behalf of the Federal or

State agency designated as
trustee, or officials designated by
Indian tribes, pursuant to section
126(d) of CERCLA, to perform a
natural resource damage
assessment. As used in this part,
authorized official is equivalent to
the phrase "authorized official or
lead authorized official," as
appropriate.

AUTHORIZED REPRESENTATIVES
(SWDA; RCRA)
DA   October 12, 1990
BT1 Personnel
RT   Administrators
RT   Facilities
RT   Operators
RT   Project Managers
SO   Environmental Management
SO   Wastes
DEF (SWDA; RCRA) People responsible
     for the overall operation of a
     facility or an operational unit (i.e.,
     part of a facility); for example,
     plant managers, superintendents
     or people of equivalent
     responsibility.

AUTHORIZED WEAPONS
(DOE Order 5480.16)
DA   October 12, 1990
BT1 Firearms
RT   Protective Force Personnel
SO   Firearms
DEF (DOE Order 5480.16) Weapons
     authorized by the Department and
     issued by the responsible
     contractor or Departmental
     element to be used by protective
     force personnel in the
     performance of their duties.

AUTOCLAVE
(2476 AUTOCLAVE)
DA   January 3, 1991
BT1 Heater(s)
BT2 Equipment/Parts - Heating (DOE
     FRASE Vocabulary)
BT3 Equipment
SO   DOE FRASE VOCABULARY

**AUTOMATIC BACKUP SHUTDOWN OF
THE SAFETY COMPUTER**
DA   January 28, 1991
SF   ABS-SC
RT   Reactor Shutdown
SO   Nuclear Facilities Incident Database

**AUTOMATIC DEPRESSURIZATION
SYSTEM**
DA   January 8, 1991
SF   ADS (automatic depressurization
     system)
BT1 Systems
RT   Depressurization

AUTOMATIC INCIDENT ACTIONS
(NFI)

DA   October 12, 1990
SF   AIA
BT1 Actions
BT2 Responses
RT   Process Water
SO   Radiation
DEF (NFI) Mitigates large process water
     leaks automatically.

AUTOMATIC REACTOR SCRAM
(1543 ARS)
DA   November 29, 1990
BT1 Nature of Property Damage
BT1 Scram
BT2 Reactor Shutdown
RT   Manual Reactor Scram
RT   Nuclear Facility
RT   Unscheduled Shutdown
RT   Violations
SO   DOE FRASE VOCABULARY

AUTOMATIC RIFLES
(DOE Order 5480.16)
DA   October 12, 1990
BT1 Rifles
BT2 Small Arms
BT3 Firearms
RT   Magazines
SO   Firearms
DEF (DOE Order 5480.16) Firearms that
     employ either gas pressure or
     recoil force and mechanical spring
     action in ejecting the empty
     cartridge case after the first shot,
     loading the next cartridge from the
     magazine, firing and ejecting that
     cartridge, and repeating the above
     cycle as long as the pressure on
     the trigger is maintained or until
     the ammunition is exhausted.

AUTOMATIC TRANSFER SWITCH
DA   January 8, 1991
SF   ATS

AUTORADIOGRAPHS
(NIH)
DA   October 12, 1990
RT   Radiological Monitoring
SO   Radiation
DEF (NIH) Records of radiation from
     radioactive material in an object,
     made by placing the object in
     close proximity to a photographic
     emulsion.

AUTOTROPHIC
(EPA)
DA   October 12, 1990
RT   Organisms
SO   Environmental Protection Agency
     Glossary
DEF (EPA) Producing food from
     inorganic substances.

AUXILIARY AIR UNITS
(DOE Order 6430.1A)
DA   October 12, 1990
BT1 Equipment

RT Hood Capture Efficiency
SO Construction
DEF (DOE Order 6430.1A)
Factory-fabricated options or
additions to a fume hood that
introduce some portion of the
make-up air directly at the hood
with features that do not minimize
the performance of the hood nor
create operator discomfort.

AUXILIARY FEEDWATER
DA January 8. 1991
SF *AFW*
BT1 Feedwater
DEF Backup feedwater supply used
during nuclear plant startup and
shutdown; also know as
emergency feedwater.

AUXILIARY ORIFICES
DA January 8, 1991
SF *AO*

AVIATION OPERATIONS
(DOE Order 5480.13)
DA October 12, 1990
BT1 Operations
BT2 Activities
NT1 Charter Operations
RT Aircraft
RT Aviation Safety
RT Bird Strike
RT Flight Crewmembers
RT Ground Crews
RT Heliports
SO Aviation Safety
DEF (DOE Order 5480.13) Any
operations of aircraft or airports, or
the provision of any aviation
support services thereto.

AVIATION SAFETY
(DOE Order 5480.13)
DA October 19, 1990
BT1 Safety
RT Aviation Operations
RT Bird Strike
DEF (DOE order 5480.13) Establishes
procedures and provides guidance
to ensure that DOE and DOE
contractor aviation operations are
conducted in the safest manner
possible, and that, to the extent
possible, passenger and
hazardous cargo air carrying
operations maintain a level of
safety equivalent to that attained
by United States air carriers
operating under 14 CFR 121.

AVULSION
(1304 AV)
DA November 28, 1990
BT1 Injuries
RT Amputation
SO DOE FRASE VOCABULARY

AX(S)
(3026 AX)
DA January 3, 1991
BT1 Tools - Manual
BT2 Tools (DOE FRASE Vocabulary)
BT3 Equipment
RT Hatchet(s)
SO DOE FRASE VOCABULARY

AXIAL POWER INDICATOR
DA January 8, 1991
SF *API*
BT1 Equipment

AXIAL POWER MONITORS
(NFI)
DA October 12, 1990
SF *APM*
BT1 Monitors
BT2 Equipment
SO Radiation
DEF (NFI) Provide roof-top-ratio and
axial flux information.

B&RC
DA June 18, 1991
SEE Budget and Reporting Code
SO Acronyms

B/O
DA October 12, 1990
SEE Blackout
SO Acronyms

BAAQMD
DA October 12, 1990
SEE Bay Area Air Quality Management
District
SO Acronyms

BACK
(1107 BACKP)
DA November 28, 1990
BT1 Trunk
BT2 Human Body Parts
NT1 Lower Back
NT1 Upper Back
RT Ruptured Disk
RT Spinal Cord
SO DOE FRASE VOCABULARY

BACK PRESSURE VALVE
DA January 8, 1991
SF *BPV*
BT1 Valves
BT2 Devices

BACKFIT REVIEWS
(SSDC)
DA October 12, 1990
BT1 Safety Reviews
BT2 Reviews
BT3 Administrative Processes
BT4 Processes
SO System Safety Development Center
Glossary
DEF (SSDC) Safety reviews of a
system, process, procedure or

plant already in existence or in
process. Such reviews may be
conducted due to changing
conditions or requirements or, due
to the fact that an adequate
system safety review had not been
conducted previously.

BACKGROUND LEVELS
(EPA)
DA October 12, 1990
NT1 Ubiquitous Background Levels
RT Air Pollution
RT Toxic Substances
SO Environmental Protection Agency
Glossary
DEF (EPA) In air pollution control,
concentrations of air pollutants in
a definite area during a fixed time
prior to the starting up or on the
stoppage of a source of emission
under control. In toxic substances
monitoring, the average presence
in the environment, originally
referring to naturally occurring
phenomena.

BACKGROUND RADIATION
(NIH)
DA October 12, 1990
SY Natural Background Radiation
BT1 Ionizing Radiation
BT2 Radiation
SO Radiation
DEF (NIH) Ionizing radiation arising from
radioactive material other than the
one directly under consideration.
Background radiation due to
cosmic rays and natural
radioactivity is always present.
There may also be background
radiation due to the presence of
radioactive substances in other
parts of the building, in the
building material itself, etc.

BACKGROUND SOIL PH
(RCRA; CFR)
DA January 24, 1991
BT1 pH
BT2 Measurements
RT Soils
SO Environmental Management
DEF (CFR) "Background soil pH" means
the pH of the soil prior to the
addition of substances that alter
the hydrogen ion concentration.

BACKUP SYSTEMS
(NFI)
DA October 12, 1990
BT1 Emergency Systems
BT2 Systems
RT Reactor Operations
SO Nuclear Facilities Incident Database
SO Radiation
DEF (NFI) Systems that are in backup
status when a malfunction is
detected.

BT-Broader Term NT-Narrower Term RT Related Term

BACT
(CAA; EPA; ESH)
DA   November 5, 1990
SEE Best Available Control Technology
SO   Acronyms
DEF An emission limitation based on the maximum degree of emission reduction which (considering energy, environmental, and economic impacts and other costs) is achievable through application of production processes and available methods, systems, and techniques. In no event does BACT permit emissions in excess of those allowed under any applicable Clean Air Act provisions. Use of the BACT concept is allowable on a case by case basis for major new or modified emissions sources in attainment areas and applies to each regulated pollutant.

BACTERIA
(EPA)
DA   October 12, 1990
BT1 Microorganisms
 BT2 Organisms
NT1 Coliform Organisms
 NT2 Fecal Coliform Bacteria
NT1 Legionella
NT1 Recombinant Bacteria
RT   Confluent Growth
RT   Inocula
RT   Vectors
SO   Environmental Protection Agency Glossary
DEF (EPA) Microscopic living organisms that can aid in pollution control by consuming or breaking down organic matter in sewage, or by similarly acting on oil spills or other water pollutants. Bacteria in soil, water or air can also cause human, animal and plant health problems. The singular form of bacteria is bacterium.

BAFFLE CHAMBERS
(EPA)
DA   October 12, 1990
RT   Combustion Products
RT   Fly Ash
RT   Incinerators
SO   Environmental Protection Agency Glossary
DEF (EPA) In incinerator design, chambers designed to promote the settling of fly ash and coarse particulate matter by changing the direction and/or reducing the velocity of the gases produced by the combustion of the refuse or sludge.

BAGHOUSE FILTERS
(EPA)
DA   November 5, 1990
BT1 Filters

BT2 Devices
RT   Particulate Matter
SO   Environmental Protection Agency Glossary
DEF (EPA) Large fabric bags, usually made of glass fibers, used to eliminate intermediate and large (greater than 20 microns in diameter) particles. These devices operate in a way similar to the bag of an electric vacuum cleaner, passing the air and smaller particulate matter, while entrapping the larger particulates.

BALANCE OF PLANT
DA   January 8, 1991
SF   *BOP*
RT   Facilities

BALANCED BIOLOGICAL COMMUNITIES
DA   January 8, 1991
SF   *BBC*
RT   Biological Resources
RT   Ecology

BALCONY
(1677 BALCONY)
DA   December 10, 1990
BT1 Site (DOE FRASE Vocabulary)
 BT2 Sites/Areas
SO   DOE FRASE VOCABULARY

BALERS
(RCRA; CFR)
DA   January 24, 1991
BT1 Equipment
RT   Baling
SO   Environmental Management
DEF (CFR) Machines used to compress solid waste, primary material or recoverable material, with or without binding, to a density or form which will support handling and transportation as a material unit rather than requiring a disposable or reusable container. This specifically excludes briquetters and stationary compaction equipment which is used to compact materials into disposable or reusable containers.

BALING
(EPA)
DA   October 12, 1990
RT   Balers
RT   Compaction
SO   Environmental Protection Agency Glossary
DEF (EPA) Compacting solid waste into blocks to reduce volume and simplify handling.

BALL MILL
(2104 BALL MILL)
DA   December 10, 1990
BT1 Milling Machine

BT2 Machines (DOE FRASE Vocabulary)
BT3 Equipment
SO   DOE FRASE VOCABULARY

BALLAST (RAILROADS)
(DOE Order 6430.1A)
DA   October 12, 1990
SO   Construction
DEF (DOE Order 6430.1A) Crushed stone used in a railroad bed to support the ties, hold the track in line, and help drainage.

BALLISTIC SEPARATORS
(EPA)
DA   October 12, 1990
BT1 Equipment
RT   Composting
RT   Waste Treatment Plants
SO   Environmental Protection Agency Glossary
DEF (EPA) A machine that sorts organic from inorganic matter for composting.

BAND APPLICATION
(EPA)
DA   October 12, 1990
BT1 Application (Pesticide)
 BT2 Processes
RT   Basal Application
RT   Broadcast Application
RT   Pesticides
SO   Environmental Protection Agency Glossary
DEF (EPA) In pesticides, the spreading of chemicals over, or next to, each row of plants in a field.

BAND SAW
(2105 BAND SAW)
DA   December 10, 1990
BT1 Machines (DOE FRASE Vocabulary)
 BT2 Equipment
BT1 Saws
SO   DOE FRASE VOCABULARY

BANKING (AIR POLLUTION)
(EPA)
DA   October 12, 1990
BT1 Pollution Recovery Processes
 BT2 Processes
RT   Air Pollution
RT   Emissions Trading
SO   Air Pollution
SO   Environmental Protection Agency Glossary
DEF (EPA) A system for recording qualified air emission reductions for later use in bubble, offset, or netting transactions.

BAR SCREENS
(EPA)
DA   October 12, 1990
BT1 Devices
RT   Solids

SY-Synonymous Terms   SO-Source/Subject Category   SF-See From

RT Wastewater Treatment Processes
SO Environmental Protection Agency
   Glossary
DEF (EPA) In wastewater treatment,
   devices used to remove large
   solids.

BARGES
(SWDA; RCRA; ESH)
DA October 12, 1990
BT1 Vessels
SO Hazardous Materials
DEF Vessels that are not self-propelled.

BARRELS
(CERCLA; USC)
DA October 12, 1990
BT1 Containers
SO Compensation and Liability
SO Environmental Management
DEF (CERCLA) Forty-two United States
   gallons at sixty degrees
   Fahrenheit.

BARRIER COATING(S)
(EPA)
DA October 12, 1990
BT1 Barriers
SO Environmental Protection Agency
   Glossary
DEF (EPA) Layers of a material that act
   to obstruct or prevent passage of
   something through a surface that
   is to be protected, e.g., grout,
   caulk, or various sealing
   compounds. Sometimes used with
   polyurethane membranes to
   prevent corrosion or oxidation of
   metal surfaces, chemical impacts
   on various materials, or, for
   example, to prevent soil-gas-borne
   radon from moving through walls,
   cracks, or joints in a house.

BARRIERS
(AEA; CFR; SSDC; IAEA; MORT)
DA October 12, 1990
NT1 Barrier Coating(s)
NT1 Control Barriers
NT1 Engineered Barriers
NT1 Intruder Barriers
NT1 Liners
NT1 Natural Barriers
NT1 Protective Barriers
NT1 Radionuclide Barriers (Natural or
   Engineered)
 NT2 Bulkheads
NT1 Safety Barriers
RT Accidents
RT Class A Equipment
RT Containers
RT Controls
RT Equipment/Parts - Personal
   Protective (DOE FRASE
   Vocabulary)
RT Hazards
RT Load Securing Device
RT Plutonium Process Hood

RT Protection of the Public Health and
   Welfare
RT Safety Limits
RT Tanks
RT Targets
SO Environmental Management
SO System Safety Development Center
   Glossary
DEF (SSDC) Anything used to control,
   prevent, or impede energy flows.
   Types of barriers include physical,
   equipment design, warning
   devices, procedures and work
   processes, knowledge and skills,
   and supervision. Barriers may be
   control or safety barriers or act as
   both. (MORT) MORT analysis
   asks: were adequate barriers in
   place to prevent vulnerable
   persons and objects from being
   exposed to harmful energy flows
   and/or environmental conditions?
   were barriers designed to prevent
   harmful energy flows or
   environmental conditions from
   reaching vulnerable people and
   objects? were barriers designed to
   prevent vulnerable people and
   objects from encountering harmful
   energy flows and environmental
   conditions?

BART
DA November 15, 1990
SEE Best Available Retrofit Technology
SO Acronyms
DEF An emission limitation based on the
   degree of reduction achievable
   through the application of the best
   system of continuous emission
   reduction for each pollutant which
   is emitted by an existing stationary
   facility. The emission limitation
   must be established, on a
   case-by-case basis, taking into
   consideration the technology
   available, the costs of compliance,
   the energy and nonair quality
   environmental impacts of
   compliance, any pollution control
   equipment in use or in existence
   at the source, the remaining useful
   life of the source, and the degree
   of improvement in visibility which
   may reasonably be anticipated to
   result from the use of such
   technology.

BASAL APPLICATION
(EPA)
DA October 12, 1990
BT1 Application (Pesticide)
 BT2 Processes
RT Band Application
RT Broadcast Application
RT Pesticides
SO Environmental Protection Agency
   Glossary
DEF (EPA) In pesticides, the application

of a chemical on plant stems or
tree trunks just above the soil line.

BASE COURSE
(DOE Order 6430.1A)
DA October 12, 1990
RT Roadways
SO Construction
DEF (DOE Order 6430.1A) The first
   layer of underlying material
   installed prior to the placement of
   a roadway pavement wearing
   surface.

BASE SHEAR
(SEA)
DA October 12, 1990
RT Bases
RT Facilities
SO Construction
DEF (SEA) The total design lateral force
   or shear at the base of a structure.

BASELINE
(CFR)
DA November 15, 1990
RT Control Resources
DEF (CFR) The condition or conditions
   that would have existed at the
   assessment area had the
   discharge of oil or release of the
   hazardous substance under
   investigation not occurred.

BASELINE CONCENTRATION
(CAA; USC)
DA October 12, 1990
RT Emission Limitation
RT Pollutants
RT Potential to Emit
SO Air Pollution
SO Environmental Management
DEF (CAA) With respect to a pollutant,
   the ambient concentration levels
   which exist at the time of the first
   application for a permit in an area
   subject to 42 USCS 7470 et seq.,
   based on air quality data available
   in the Environmental Protection
   Agency or a State air pollution
   control agency and on such
   monitoring data as the permit
   applicant is required to submit.
   Such ambient concentration levels
   shall take into account all
   projected emissions in, or which
   may affect, such area from any
   major emitting facility on which
   construction commenced prior to
   January 6, 1975, but which has
   not begun operation by the date of
   the baseline air quality
   concentration determination.
   Emissions of sulfur oxides and
   particulate matter from any major
   emitting facility on which
   construction commenced after
   January 6, 1975, shall not be
   included in the baseline and shall

be counted against the maximum allowable increases in pollutant concentrations established under 42 USCS 7470 et seq.

BASEMENT
(1678 BASEMENT)
DA   December 10, 1990
BT1 Site (DOE FRASE Vocabulary)
 BT2 Sites/Areas
SO DOE FRASE VOCABULARY

BASES
(SEA)
DA   October 12, 1990
RT   Base Shear
RT   Earthquake Magnitude
RT   Facilities
SO   Construction
DEF (SEA) The level at which the
    earthquake motions are
    considered to be imparted to the
    structure or the level at which the
    structure as a dynamic vibrator is
    supported.

BASIC CAUSES
(SSDC)
DA   October 12, 1990
SY   Root Causes
BT1 Causes
SO   System Safety Development Center
    Glossary
DEF (SSDC) Root causes.

BASIC RESEARCH
(DOE Order 4700.1)
DA   January 24, 1991
BT1 Research and Development
 BT2 Activities
RT   Applied Research
DEF (DOE 4700.1) Systematic,
    fundamental study directed
    towards fuller scientific knowledge
    or understanding of subjects
    bearing on national energy needs.

BAT
DA   October 19, 1990
SEE Best Available Technology
SO   Acronyms

BATCHES
(TSCA; CFR)
DA   October 19, 1990
BT1 Samples
NT1 Analytical Batches
RT   Control Substances
RT   Test Substances
SO   Environmental Management
SO   Hazardous Materials
DEF Specific quantities or lot of test or
    control substances that have been
    characterized according to 40
    CFR 792.105(a).

**BATTELLE COLUMBUS
LABORATORIES**
DA   January 8, 1991
SF   *BCL*
BT1 DOE Contractors
 BT2 Potentially Responsible Parties
BT1 Laboratories
 BT2 Research and Development
    Organizations
  BT3 Organizations

BATTELLE MEMORIAL INSTITUTE
DA   January 11, 1991
BT1 DOE Contractors
 BT2 Potentially Responsible Parties
BT1 Institutes
 BT2 Research and Development
    Organizations
  BT3 Organizations
RT   Pacific Northwest Laboratory
RT   Richland Operations Office

BATTERY(S)
(2379 BATTERY)
DA   January 3, 1991
BT1 Equipment/Parts - Electrical (DOE
    FRASE Vocabulary)
 BT2 Equipment
SO   DOE FRASE VOCABULARY

**BAY AREA AIR QUALITY
MANAGEMENT DISTRICT**
DA   January 8, 1991
SF   *BAAQMD*
RT   Air Quality Control Regions
RT   Municipalities

BBC
DA   October 12, 1990
SEE Balanced Biological Communities
SO   Acronyms

BCL
DA   October 12, 1990
SEE Battelle Columbus Laboratories
SO   Acronyms

BDAT
DA   October 12, 1990
SEE Best Demonstrated Available
    Technology
SO   Acronyms

BEARING CAPACITY
(DOE Order 6430.1A)
DA   October 12, 1990
RT   Structural Members
SO   Construction
DEF (DOE Order 6430.1A) A loading
    intensity that the bearing materials
    can sustain without such
    deformation as would result in
    settlement damaging to the
    structure.

BEARING WALL SYSTEMS
(SEA)
DA   October 12, 1990

BT1 Building Frame Systems
 BT2 Space Frames
  BT3 Structures (DOE FRASE
    Vocabulary)
 BT2 Systems
BT1 Structural Members
SO   Construction
DEF (SEA) Structural systems without
    complete vertical load carrying
    space frames.

**BECHTEL PETROLEUM OPERATION,
INC.**
DA   January 11, 1991
BT1 Companies
 BT2 Commercial Organizations
  BT3 Organizations
BT1 DOE Contractors
 BT2 Potentially Responsible Parties
RT   Headquarters Operations

BECQUEREL
(NCRP; IAEA)
DA   October 12, 1990
SF   *Bq*
BT1 Radiation Units
 BT2 Units of Measure
RT   Radioactivity
SO   Radiation
DEF (IAEA; NCRP) A unit of
    radioactivity. One becquerel is one
    nuclear transformation per second.

BEETLES
(NFI)
DA   October 12, 1990
BT1 Leak Detection Systems
 BT2 Systems
SO   Nuclear Facilities Incident Database
SO   Radiation
DEF (NFI) Tank top leak detectors.

BEF
DA   October 12, 1990
SEE Bottom End Fitting
SO   Acronyms

BEHAVIORAL STEREOTYPES
(SSDC)
DA   October 12, 1990
RT   Human Performance
RT   Stimulus-Mediation-Response
SO   System Safety Development Center
    Glossary
DEF (SSDC) The "normal" or fixed
    patterns in which a person thinks
    and acts or reacts in a given
    situation. These patterns may be
    altered through practice, but under
    stress a person will likely return to
    the original pattern of behavior. An
    example of a behavioral
    stereotype is the way in which a
    person crosses his arms. We
    consistently place the same arm
    on top of the other and it takes a
    conscious effort to cross the arms
    with the other one on top.

SY-Synonymous Terms   SO-Source/Subject Category   SF-See From

BELOW REGULATORY CONCERN
(DOE Order 5820.2A)
DA January 8, 1991
SF *BRC*
RT Low Level Wastes
SO Environmental Management
SO Wastes
DEF (DOE Order 5820.2A) A definable
    amount of low level waste that can
    be deregulated with minimal risk to
    the public.

BELOWGROUND RELEASES
(SWDA; RCRA; CFR)
DA October 12, 1990
SY Underground Releases
BT1 Releases
SO Wastes
DEF Releases to the subsurface of the
    land and to groundwater. This
    includes, but is not limited to,
    releases from the belowground
    portions of an underground portion
    of an underground storage tank
    system and belowground releases
    associated with overfills and
    transfer operations as the
    regulated substance moves to or
    from an underground storage tank.

BEN
(EPA)
DA October 12, 1990
BT1 Computer Codes
SO Environmental Protection Agency
    Glossary
DEF (EPA) EPA's computer model for
    analyzing a violator's economic
    gain from not complying with the
    law.

BENCH MARKS
(DOE Order 6430.1A)
DA October 12, 1990
RT Construction
RT Maps
RT Monumentation
SO Construction
DEF (DOE Order 6430.1A) Survey
    control monuments installed to
    provide vertical control for
    construction purposes. Official
    elevation marker used in
    topographic surveys by the
    U.S.G.S. (United States
    Geological Survey).

BENEFACTION
(NCRP)
DA October 12, 1990
BT1 Processes
RT Extraction Plants
SO Radiation
DEF (NCRP) Preliminary conditioning of
    an ore for refinement.

BENEFICIAL ORGANISMS
(FIFRA; CFR)
DA January 24, 1991

BT1 Organisms
RT Biological Controls
DEF (CFR) The term "beneficial
    organism" means any pollinating
    insect or any pest predator,
    parasite, pathogen or other
    biological control agent which
    functions naturally or as part of an
    integrated pest management
    program to control another pest.

BENIGN AND UNSPECIFIED NEOPLASM
(1376 BENIGN AND U)
DA November 28, 1990
BT1 Illnesses
SO DOE FRASE VOCABULARY

BENTHIC ORGANISMS (BENTHOS)
(EPA)
DA October 12, 1990
BT1 Organisms
RT Benthic Region
SO Environmental Protection Agency
    Glossary
DEF (EPA) Forms of aquatic plant or
    animal life that is found on or near
    the bottom of a stream, lake, or
    ocean.

BENTHIC REGION
(EPA)
DA October 12, 1990
BT1 Aquatic Environment
 BT2 Environment
RT Benthic Organisms (Benthos)
SO Environmental Protection Agency
    Glossary
DEF (EPA) The bottom layer of a body
    of water.

BENTONITE CLAYS
(DOE Order 6430.1A)
DA October 12, 1990
BT1 Soils
SO Construction
DEF (DOE Order 6430.1A) Particular
    types of colloidal clay that swell
    when wet and form a gel
    membrane.

BERYLLIUM
(CAA; CFR; EPA)
DA October 12, 1990
BT1 CERCLA Hazardous Substances
 BT2 Hazardous Substances
NT1 Beryllium 7
RT Beryllium Ores
RT Beryllium-Containing Wastes
SO Air Pollution
SO Environmental Management
SO Environmental Protection Agency
    Glossary
DEF (EPA) An airborne metal that can
    be hazardous to human health
    when inhaled. It is discharged by
    machine shops, ceramic and
    propellant plants, and foundries.
    (CFR) The element beryllium.

Where weights or concentrations
are specified, such weights or
concentrations apply to beryllium
only, excluding the weight or
concentration of any associated
elements.

BERYLLIUM 7
(EDB)
DA March 29, 1991
BT1 Beryllium
 BT2 CERCLA Hazardous Substances
  BT3 Hazardous Substances
BT1 Days Living Radioisotopes
 BT2 Radionuclides
  BT3 CERCLA Hazardous Substances
   BT4 Hazardous Substances
  BT3 Nuclides
SO Radiation

BERYLLIUM ALLOYS
(CAA; CFR)
DA October 12, 1990
RT Foundries
SO Air Pollution
SO Environmental Management
DEF (CFR) Metals to which beryllium
    has been added in order to
    increase its beryllium content and
    which contain more than 0.1
    percent beryllium by weight.

BERYLLIUM-CONTAINING WASTES
(CAA; CFR)
DA October 12, 1990
BT1 Wastes
RT Beryllium
SO Air Pollution
SO Environmental Management
DEF (CFR) Materials contaminated with
    beryllium and/or beryllium
    compounds used or generated
    during any process or operation
    performed by a source subject to
    this subpart (40 CFR 61.31).

BERYLLIUM ORES
(CAA; CFR)
DA October 12, 1990
RT Beryllium
RT Extraction Plants
SO Air Pollution
SO Environmental Management
DEF (CFR) Any naturally occurring
    materials mined or gathered for
    their beryllium content.

BERYLLIUM PROPELLANTS
(CFR)
DA October 12, 1990
BT1 Propellants
 BT2 Fuels
SO Air Pollution
DEF (CFR) Propellants incorporating
    beryllium.

BEST AVAILABLE CONTROL TECHNOLOGY
(CAA; USC; EPA; ESH)

BT-Broader Term

NT-Narrower Term

RT Related Term

DA   October 12, 1990
SF   *BACT*
BT1 Best Available Technology
BT1 Control Systems
 BT2 Controls
 BT2 Systems
RT Emissions
RT Reasonably Available Control
      Technology
SO Air Pollution
SO Environmental Management
DEF Treatment technologies that have
     been shown through actual use to
     yield the greatest environmental
     benefit among competing
     technologies that are practically
     available. (See 42 USC 7491
     section 169 for additional
     definition.)

BEST AVAILABLE RETROFIT TECHNOLOGY
(CAA; CFR)
DA   October 12, 1990
SF   *BART*
BT1 Best Available Technology
BT1 Control Systems
 BT2 Controls
 BT2 Systems
RT Emissions
RT Major Modification
SO Air Pollution
DEF An emission limitation based on the
     degree of reduction achievable
     through the application of the best
     system of continuous emission
     reduction for each pollutant which
     is emitted by an existing stationary
     facility. The emission limitation
     must be established, on a
     case-by-case basis, taking into
     consideration the technology
     available, the costs of compliance,
     the energy and nonair quality
     environmental impacts of
     compliance, any pollution control
     equipment in use or in existence
     at the source, the remaining useful
     life of the source, and the degree
     of improvement in visibility which
     may reasonably be anticipated to
     result from the use of such
     technology.

BEST AVAILABLE TECHNOLOGY
(DOE Orders 6430.1A, 5400.5; CFR;
ESH)
DA   October 12, 1990
SF   *BAT*
NT1 Best Available Control Technology
NT1 Best Available Retrofit Technology
NT1 Best Demonstrated Available
     Technology
RT Categorical Pretreatment Standards
RT Equipment
SO Construction
SO Industrial Hygiene
SO Water Pollution
DEF (DOE Order 6430.1A) The
     preferred technology for treating a

particular process liquid waste,
selected from among others after
taking into account factors related
to technology, economics, public
policy, and other parameters. As
used in this Order, BAT is not a
specific level of treatment, but the
conclusion of a selection process
that includes several treatment
alternatives.

BEST DEMONSTRATED AVAILABLE TECHNOLOGY
DA   January 8, 1991
SF   *BDAT*
BT1 Best Available Technology

BEST MANAGEMENT PRACTICE FINDINGS
(TTGM)
DA   April 26, 1991
BT1 Findings
RT Best Management Practices
RT Compliance Findings
RT Noteworthy Practices
RT Tiger Team Assessments
RT Tiger Teams
SO Management
DEF (TTGM) Within a Tiger Team
     Assessment, these findings are
     derived from regulatory agency
     guidance, DOE draft Orders,
     accepted industry practices, and
     professional judgement.

BEST MANAGEMENT PRACTICES
(ESH)
DA   October 12, 1990
SF   *BMP*
BT1 Management
RT Best Management Practice
      Findings
SO Hazardous Materials
DEF (ESH) Methods, measures, or
     practices selected by an agency to
     meet its nonpoint source control
     needs. BMPs include but are not
     limited to structural and
     nonstructural controls and
     operation and maintenance
     procedures applied before, during,
     or after pollution-producing
     activities to reduce or eliminate the
     introduction of pollutants into
     receiving waters.

BEST PRACTICAL TECHNOLOGY
DA   June 18, 1991
SF   *BPT*
SO Management

BETA DECAY
(NCRP)
DA   October 12, 1990
BT1 Radioactive Decay
NT1 Electron Capture
RT Beta Particles
SO Radiation
DEF (NCRP) Radioactive decay in which

a beta particle is emitted or in
which orbital electron capture
occurs.

BETA DECAY RADIOISOTOPES
(EDB)
DA   March 29, 1991
BT1 Radionuclides
 BT2 CERCLA Hazardous Substances
  BT3 Hazardous Substances
 BT2 Nuclides
NT1 Beta-Minus Decay Radioisotopes
NT2 Antimony 125
NT2 Cerium 144
NT2 Cerium 141
NT2 Cesium 137
NT2 Cesium 134
NT2 Cobalt 60
NT2 Cobalt 58
NT2 Iodine 131
NT2 Radium 228
NT2 Ruthenium 103
NT2 Strontium 90
NT2 Tellurium 132
NT2 Thorium 232
NT2 Tritium
NT2 Zinc 65
NT2 Zirconium 95
SO Radiation

BETA-MINUS DECAY RADIOISOTOPES
(EDB)
DA   March 29, 1991
BT1 Beta Decay Radioisotopes
 BT2 Radionuclides
  BT3 CERCLA Hazardous Substances
   BT4 Hazardous Substances
  BT3 Nuclides
NT1 Antimony 125
NT1 Cerium 144
NT1 Cerium 141
NT1 Cesium 137
NT1 Cesium 134
NT1 Cobalt 60
NT1 Cobalt 58
NT1 Iodine 131
NT1 Radium 228
NT1 Ruthenium 103
NT1 Strontium 90
NT1 Tellurium 132
NT1 Thorium 232
NT1 Tritium
NT1 Zinc 65
NT1 Zirconium 95
SO Radiation

BETA PARTICLES
(EPA; NIH; NCRP; EMER)
DA   October 12, 1990
BT1 Charged Particles
RT Beta Decay
RT Beta Rays
RT Charged Particle Equilibrium
RT Electrons
RT Gross Beta Particle Activity
RT Ionizing Radiation
RT Man-made Beta Particle and
      Photon Emitters
SO Emergency Preparedness

SO Environmental Protection Agency
    Glossary
SO Radiation
DEF (EPA) Elementary particles emitted
    by radioactive decay that may
    cause skin burns. They are halted
    by a thin sheet of paper. (NIH)
    Charged particles emitted from the
    nucleus of an atom, having mass
    and charge equal in magnitude to
    electrons. An electron, of either
    positive or negative charge, which
    has been emitted by an atomic
    nucleus or neutron in a nuclear
    transformation.

BETA RAYS
(NIH)
DA October 12, 1990
BT1 Non-Penetrating Radiation
 BT2 Radiation
RT Beta Particles
RT Ionizing Radiation
SO Radiation
DEF (NIH) Streams of high speed
    electrons or positrons of nuclear
    origin more penetrating but less
    ionizing than alpha rays.

BFA
DA October 12, 1990
SEE Blank Fire Adapter (BFA)
SO Acronyms

BFI
DA October 12, 1990
SEE Bottom Fitting Insert
SO Acronyms

BG (Blanket Gas)
DA October 12, 1990
SEE Blanket Gas
SO Acronyms

BG (Burial Ground)
DA October 12, 1990
SEE Burial Grounds
SO Acronyms

BIA (BUREAU OF INDIAN AFFAIRS)
(CFR)
DA October 19, 1990
SO Water Pollution
DEF (CFR) The Bureau of Indian Affairs,
    United States Department of
    Interior.

BIF
DA June 4, 1991
SEE Boiler/Industrial Furnaces
SO Acronyms

BIOACCUMULATIVE
(EPA)
DA October 12, 1990
RT Biological Magnification

SO Environmental Protection Agency
    Glossary
DEF (EPA) Substances that increase in
    concentration in living organisms
    (that are very slowly metabolized
    or excreted) as they breathe
    contaminated air, drink
    contaminated water, or eat
    contaminated food.

BIOASSAY PROCEDURES
(NCRP)
DA October 12, 1990
BT1 Procedures
RT Bioassays
RT Body Content
SO Radiation
DEF (NCRP) Procedures used to
    determine the kind, quality,
    location and/or retention or
    radionuclides in the body by direct
    (in vivo) measurements or by in
    vitro analysis of material excreted
    or removed from the body.

BIOASSAYS
(EPA)
DA October 12, 1990
BT1 Biological Processes
 BT2 Processes
NT1 Direct Bioassays
RT Bioassay Procedures
SO Environmental Protection Agency
    Glossary
DEF (EPA) Using living organisms to
    measure the effect of a substance,
    factor, or condition by comparing
    before and after data. Term is
    often used to mean cancer
    bioassays.

BIOAVAILABILITY
(TSCA; CFR)
DA January 24, 1991
RT Exposure Pathways
DEF (CFR) Refers to the rate and extent
    to which the administered
    compound is absorbed, i.e.,
    reaches the systemic circulation.

BIOCHEMICAL OXYGEN DEMAND
(EPA)
DA October 12, 1990
SY Biological Oxygen Demand
BT1 Measurements
RT BOD5
RT CBOD5
RT Chemical Oxygen Demand
RT Kinetic Rate Coefficient
SO Environmental Protection Agency
    Glossary
DEF (EPA) A measure of the amount of
    oxygen consumed in the biological
    processes that break down
    organic matter in water. The
    greater the BOD, the greater the
    degree of pollution.

BIODEGRADABLE
(EPA)
DA October 12, 1990
RT Biological Oxidation
RT Decomposition
SO Environmental Protection Agency
    Glossary
DEF (EPA) The ability to break down or
    decompose rapidly under natural
    conditions and processes.

BIOLOGICAL ADDITIVES
(CFR)
DA October 12, 1990
BT1 Additives
 BT2 Chemical Substances
RT Microorganisms
RT Oil Spills
SO Compensation and Liability
DEF (CFR) Microbiological cultures,
    enzymes, or nutrient additives that
    are deliberately introduced into an
    oil discharge for the specific
    purpose of encouraging
    biodegradation to mitigate the
    effects of the discharge.

BIOLOGICAL ASSESSMENTS
(ESA; CFR)
DA October 12, 1990
BT1 Assessments
 BT2 Administrative Processes
  BT3 Processes
RT Effects of the Action
RT Listed Species
RT Proposed Species
SO Endangered Species
DEF (CFR) Refers to the information
    prepared by or under the direction
    of the Federal agency concerning
    listed and proposed species and
    designated and proposed critical
    habitat that may be present in the
    action area and the evaluation
    potential effects of the action on
    such species and habitat.

BIOLOGICAL CLEARANCE RATE
(IAEA)
DA October 12, 1990
BT1 Rates
RT Lung Classes
SO Radiation
DEF (IAEA) The fractional change per
    unit time in the number of atoms
    of a stable chemical element in a
    tissue, an organ, or the whole
    body occurring when the removal
    of that element follows an
    approximately exponential
    function.

BIOLOGICAL CONTROLS
(EPA)
DA October 12, 1990
BT1 Controls
RT Beneficial Organisms
RT Pests

SO Environmental Protection Agency
   Glossary
DEF (EPA) In pest control, the use of
   animals and organisms that eat or
   otherwise kill or out-compete
   pests.

BIOLOGICAL HALF-LIFE
(IAEA; NIH)
DA October 12, 1990
SY Biological Half-Time
BT1 Half-Life
 BT2 Time Designations
RT Body Content
SO Radiation
DEF (IAEA; NIH) The time required for
   the body to eliminate one-half of
   an administered dose of any
   substance by the regular
   processes of elimination. This time
   is approximately the same for both
   stable and radionuclides of a
   particular element.

BIOLOGICAL HALF-TIME
(EMER)
DA February 1, 1991
SY Biological Half-Life
BT1 Time Designations
RT Effective Half-Life
SO Emergency Preparedness
DEF The time required for a biological
   system to eliminate, by natural
   processes, half the amount of a
   substance (e.g., radioactive
   material) that has entered it. Also
   called biological half-life.

BIOLOGICAL MAGNIFICATION
(EPA)
DA October 12, 1990
BT1 Biological Processes
 BT2 Processes
RT Bioaccumulative
SO Environmental Protection Agency
   Glossary
DEF (EPA) Refers to the process
   whereby certain substances such
   as pesticides or heavy metals
   move up the food chain, work their
   way into a river or lake and are
   eaten by aquatic organisms such
   as fish, which in turn are eaten by
   large birds, animals, or humans.
   The substances become
   concentrated in tissues or internal
   organs as they move up the chain.

BIOLOGICAL OPINION
(ESA; CFR)
DA October 12, 1990
BT1 Reports
NT1 Preliminary Biological Opinion
RT Critical Habitats
RT Formal Consultation
RT Listed Species
SO Endangered Species
DEF (CFR) A document that states an
   opinion as to whether or not a

Federal action is likely to
jeopardize the continued existence
of listed species or result in the
destruction or adverse
modification of critical habitat.

BIOLOGICAL OXIDATION
(EPA)
DA October 12, 1990
BT1 Biological Processes
 BT2 Processes
RT Anaerobic Digestion
RT Biodegradable
RT Biological Treatment
RT Decomposition
SO Environmental Protection Agency
   Glossary
DEF (EPA) The way bacteria and
   microorganisms feed on and
   decompose complex organic
   materials. Used in self-purification
   of water bodies and in activated
   sludge wastewater treatment.

BIOLOGICAL OXYGEN DEMAND
DA January 8, 1991
SY Biochemical Oxygen Demand
SF *BOD (Biological Oxygen Demand)*
BT1 Measurements

BIOLOGICAL PROCESSES
DA February 26, 1991
BT1 Processes
NT1 Adaptations
NT1 Aeration
 NT2 Diffused Air Process
 NT2 Mechanical Aeration
NT1 Bioassays
 NT2 Direct Bioassays
NT1 Biological Magnification
NT1 Biological Oxidation
NT1 Cloning
NT1 Composting
NT1 Decomposition
NT1 Denitrification
NT1 Destruction
NT1 Differentiation
NT1 Evapotranspiration
NT1 Genetic Engineering
NT1 Infiltration
NT1 Natural Selection
NT1 Osmosis
 NT2 Reverse Osmosis
NT1 Photosynthesis
NT1 Regeneration
NT1 Sterilization
NT1 Transpiration

BIOLOGICAL RESOURCES
(CFR)
DA November 15, 1990
BT1 Natural Resources
RT Balanced Biological Communities
RT Biomass
RT Biota
DEF (CFR) Those natural resources
   referred to in section 101(16) of
   CERCLA as fish and wildlife and
   other biota. Fish and wildlife

include marine and freshwater
aquatic and terrestrial species;
game, nongame, and commercial
species; and threatened,
endangered, and State sensitive
species. Other biota encompass
shellfish, terrestrial and aquatic
plants, and other living organisms
not otherwise listed in this
definition.

BIOLOGICAL TREATMENT
(EPA)
DA October 12, 1990
BT1 Treatment
 BT2 Waste Management Processes
  BT3 Processes
NT1 Aerobic Treatment
RT Aeration
RT Biological Oxidation
RT Composting
SO Environmental Protection Agency
   Glossary
DEF (EPA) A treatment technology that
   uses bacteria to consume waste.
   This treatment breaks down
   organic materials.

BIOMASS
(EPA)
DA October 12, 1990
SY Biota
NT1 Buffer Strips
RT Biological Resources
RT Loading
RT Plants
SO Environmental Protection Agency
   Glossary
DEF (EPA) All of the living material in a
   given area; often refers to
   vegetation. Also called "biota."

BIOMONITORING
(EPA)
DA October 12, 1990
BT1 Monitoring
 BT2 Activities
RT Effluent Monitoring
RT Test Systems
SO Environmental Protection Agency
   Glossary
DEF (EPA) (1) The use of living
   organisms to test the suitability of
   effluents for discharge into
   receiving waters and to test the
   quality of such waters downstream
   from the discharge. (2) Analysis of
   blood, urine, tissues, etc., to
   measure chemical exposure in
   humans.

BIOSPHERE
(EPA)
DA October 12, 1990
SY Ecosphere
RT Atmosphere
RT Ecosystems
RT National Environmental Policy Act

SY-Synonymous Terms   SO-Source/Subject Category   SF-See From

SO Environmental Protection Agency
Glossary
DEF (EPA) The portion of Earth and its
atmosphere that can support life.

BIOSTABILIZERS
(EPA)
DA October 12, 1990
BT1 Machines (DOE FRASE
Vocabulary)
BT2 Equipment
RT Composting
SO Environmental Protection Agency
Glossary
DEF (EPA) Machines that convert solid
waste into compost by grinding
and aeration.

BIOTA
(EPA)
DA November 5, 1990
SY Biomass
RT Biological Resources
RT Plants
DEF (EPA) All of the living material in a
given area; often refers to
vegetation.

BIOTECHNOLOGY
(EPA)
DA October 12, 1990
RT Cloning
RT Designer Bugs
SO Environmental Protection Agency
Glossary
DEF (EPA) Techniques that use living
organisms or parts of organisms to
produce a variety of products
(from medicines to industrial
enzymes) to improve plants or
animals or to develop
microorganisms for specific uses
such as removing toxics from
bodies of water, or as pesticides.

BIOTIC COMMUNITIES
(EPA)
DA October 12, 1990
RT Ecosystems
RT Populations
SO Environmental Protection Agency
Glossary
DEF (EPA) Naturally occurring
assemblages of plants and
animals that live in the same
environment and are mutually
sustaining and interdependent.

BIRD STRIKE
(DOE Order 6430.1A)
DA October 12, 1990
RT Aviation Operations
RT Aviation Safety
SO Construction
DEF (DOE Order 6430.1A) Airspace
conflict between aircraft flight
patterns and birds or waterfowl.

BIRDCAGE
(2550 BIRDCAGE)
DA January 3, 1991
BT1 Equipment/Parts - Nuclear (DOE
FRASE Vocabulary)
BT2 Equipment
BT2 Reactor Components
SO DOE FRASE VOCABULARY

BIT
DA October 12, 1990
SEE Boron Injection Tank
SO Acronyms

BLACK LUNG
(EPA)
DA October 12, 1990
BT1 Diseases
RT Coal
RT Inhalation
SO Environmental Protection Agency
Glossary
DEF (EPA) A disease of the lungs
caused by habitual inhalation of
coal dust.

BLACKOUT
DA January 8, 1991
SF *B/O*

BLACKWATER
(EPA)
DA October 12, 1990
BT1 Wastewater
BT2 Wastes
BT2 Water
SO Environmental Protection Agency
Glossary
DEF (EPA) Water that contains animal,
human, or food wastes.

BLANK AMMUNITION
(DOE Order 5480.16)
DA October 12, 1990
NT1 ESS Blanks
RT Blank Fire Adapter (BFA)
RT Firearms
RT Live Round Excluders
RT Munitions
RT Weapon Simulators
SO Firearms
DEF (DOE Order 5480.16) A cartridge
loaded with powder but containing
no projectile or ammunition that is
deemed by the manufacturer to be
incapable of firing a projectile that
will kill, wound, or otherwise harm
any individual at a distance
greater than 10 feet (3 meters).

BLANK FIRE ADAPTER (BFA)
(DOE Order 5480.16)
DA October 12, 1990
SF *BFA*
BT1 Devices
RT Blank Ammunition
RT Firearms
SO Firearms
DEF (DOE Order 5480.16) A mechanical

device attached to a firearm for
the purpose of adapting it for use
with blank ammunition.

BLANKET ASSEMBLY
(NFI)
DA October 12, 1990
SO Nuclear Facilities Incident Database
SO Radiation
DEF (NFI) Natural convection cooled;
reduces neutron irradiation to the
reactor walls; lithium assemblies.

BLANKET GAS
DA January 8, 1991
SY Inert Atmosphere
SF *BG (Blanket Gas)*

BLANKS
(ESH)
DA October 12, 1990
BT1 Samples
RT Calibration Checks
SO Quality Assurance
DEF (ESH) Artificial samples designed
to monitor the introduction of
artifacts into the process. For
aqueous samples, reagent water
is used as a blank matrix;
however, a universal blank matrix
does not exist for solid samples,
and therefore, no matrix is used.
The blank is taken through the
appropriate steps of the process.

BLASTING AGENTS
(CFR)
DA October 12, 1990
BT1 Explosives
SO Hazardous Materials
DEF (CFR) Blasting agents are materials
designed for blasting and found to
be so insensitive that there is very
little probability of accidental
initiation to explosion. Ammonium
nitrate fuel oil mixtures, containing
only prilled ammonium nitrate and
fuel oil, are examples of blasting
agents.

BLISTER(S)
(1305 BLISTER)
DA November 28, 1990
BT1 Injuries
SO DOE FRASE VOCABULARY

BLOOD
(1140 BLOOD)
DA November 28, 1990
BT1 Circulatory System
BT2 Body System(s)
BT3 Human Body Parts
RT Blood Clot
SO DOE FRASE VOCABULARY

BLOOD CLOT
(1306 BLOOD CLOT)
DA November 28, 1990

BT-Broader Term

NT-Narrower Term

RT Related Term

BT1 Injuries
RT Blood
SO DOE FRASE VOCABULARY

BLOWING
(CFR)
DA October 12, 1990
BT1 Processes
RT Copper Converters
SO Air Pollution
DEF (CFR) The injection of air or oxygen-enriched air into a molten converter bath.

BLVR
DA October 12, 1990
SEE Building Power Low Voltage Relay
SO Acronyms

BMP
DA October 12, 1990
SEE Best Management Practices
SO Acronyms

BNL
DA October 12, 1990
SEE Brookhaven National Laboratory
SO Acronyms

BOARDS
DA February 1, 1991
BT1 Administrative Organizations
 BT2 Organizations
NT1 Accrediting Boards
NT1 Energy Research Advisory Board
NT1 GDC Planning Board
NT1 Regional Water Quality Control Board
RT Centers
RT Committees
RT Groups
RT Offices
SO Management
DEF (WEBSTER) A group of persons having managerial, supervisory, or investigatory powers.

BOD (Biological Oxygen Demand)
DA October 12, 1990
SEE Biological Oxygen Demand
SO Acronyms

BOD5
(EPA)
DA October 12, 1990
RT Biochemical Oxygen Demand
SO Environmental Protection Agency Glossary
DEF (EPA) The amount of dissolved oxygen consumed in five days by biological processes breaking down organic matter.

BODILY INJURIES
(CFR)
DA November 9, 1990
BT1 Injuries
DEF (CFR) Has the meaning given the

term by applicable State law. However, this term does not include those liabilities which, consistent with standard industry practices, are excluded from coverage in liability policies for bodily injury and property damage. The Agency intends the meanings of other terms used in the liability insurance requirements to be consistent with their common meanings within the insurance industry.

BODY CONTENT
(IAEA)
DA October 12, 1990
BT1 Measurements
RT Bioassay Procedures
RT Biological Half-Life
RT Direct Bioassays
RT Radionuclides
RT Radioactive Contamination
RT Retention
SO Radiation
DEF (IAEA) The total amount (which may be expressed as activity) of a specified radionuclide in a human or animal body (formerly called body burden).

BODY OF WATER
(1603 BODY OF WATE)
DA December 10, 1990
BT1 Site (DOE FRASE Vocabulary)
 BT2 Sites/Areas
RT Dam
RT Drowning
RT Pond
RT River/Creek
SO DOE FRASE VOCABULARY

BODY SYSTEM(S)
(1135 BODY SYSTEM)
DA November 28, 1990
BT1 Human Body Parts
NT1 Circulatory System
 NT2 Blood
NT1 Digestive System
 NT2 Mouth
  NT3 Tooth/Teeth
 NT2 Rectum
NT1 Excretory System
NT1 Nervous System
 NT2 Brain
 NT2 Spinal Cord
NT1 Respiratory System
 NT2 Bronchial Epithelium
 NT2 Nose
 NT2 Throat
SO DOE FRASE VOCABULARY

BODY WAVE MAGNITUDE
(USGS)
DA October 12, 1990
SY Local Magnitude (M_L)
SF M_b
BT1 Units of Measure
RT Earthquake Magnitude

RT Secondary Body Waves
RT Surface Wave Magnitude
SO Natural Phenomenon
DEF (USGS) Measure of seismic body waves, primary (P) and Secondary (S), which have periods usually from 1 to 10 seconds.

BOEING PETROLEUM SERVICES
DA January 8, 1991
SF *BPS*
BT1 Companies
 BT2 Commercial Organizations
  BT3 Organizations
RT Oak Ridge Operations Office
RT Strategic Petroleum Reserve

BOGS
(EPA)
DA October 12, 1990
BT1 Wetlands
 BT2 Sites/Areas
 BT2 Surface Water Resources
  BT3 Natural Resources
RT Dystrophic Lakes
RT Humus
RT Swamps
SO Environmental Protection Agency Glossary
DEF (EPA) Wetlands that accumulate appreciable peat deposits. Bogs depend primarily on precipitation for their water source and are usually acidic and rich in plant residue with a conspicuous mat of living green moss.

BOILER(S)
(2479 BOILER)
DA January 3, 1991
BT1 Heater(s)
 BT2 Equipment/Parts - Heating (DOE FRASE Vocabulary)
  BT3 Equipment
NT1 Boiler Part(s)
RT Boiler/Industrial Furnaces
SO DOE FRASE VOCABULARY

BOILER PART(S)
(2478 BOILER PART)
DA January 3, 1991
BT1 Boiler(s)
 BT2 Heater(s)
  BT3 Equipment/Parts - Heating (DOE FRASE Vocabulary)
   BT4 Equipment
RT Boiler/Industrial Furnaces
SO DOE FRASE VOCABULARY

BOILER/INDUSTRIAL FURNACES
DA June 4, 1991
SF *BIF*
BT1 Furnace(s)
 BT2 Heater(s)
  BT3 Equipment/Parts - Heating (DOE FRASE Vocabulary)
   BT4 Equipment
RT Boiler Part(s)
RT Boiler(s)

SY-Synonymous Terms   SO-Source/Subject Category   SF-See From

RT Heater(s)
SO Construction

BOILING WATER REACTOR
DA January 8, 1991
SF *BWR*
BT1 Reactors
DEF (NRC Glossary of Terms: Nuclear Power and Radiation) A reactor in which water, used as both coolant and moderator, is allowed to boil in the core. The resulting steam can be used directly to drive a turbine and electrical generator.

BOMB INCIDENTS
(EMER)
DA February 1, 1991
BT1 Emergencies
 BT2 Reportable Occurrences
  BT3 Occurrences
BT1 Incidents
SO Emergency Preparedness
DEF (EMER) Incidents involving the threatened, attempted, or actual use of conventional explosives in a malevolent manner (including threatened, attempted, or actual use of flammable, corrosive, or hazardous substances).

BOMB THREATS
(EMER)
DA February 1, 1991
BT1 Emergencies
 BT2 Reportable Occurrences
  BT3 Occurrences
BT1 Incidents
BT1 Threats
 BT2 Emergencies
  BT3 Reportable Occurrences
   BT4 Occurrences
RT Explosives
SO Emergency Preparedness
DEF (EMER) Expressions of intention to destroy or damage facilities by use of explosives.

BONE(S)
(1136 BONE)
DA November 28, 1990
BT1 Human Body Parts
NT1 Rib(s)
NT1 Spine
 NT2 Coccyx
RT Fracture
SO DOE FRASE VOCABULARY

BONE SEEKERS
(IAEA)
DA October 12, 1990
BT1 Radionuclides
 BT2 CERCLA Hazardous Substances
  BT3 Hazardous Substances
 BT2 Nuclides
SO Radiation
DEF (IAEA) Any radionuclides that are incorporated more readily into bone than into other living tissue.

BONNEVILLE POWER ADMINISTRATION
DA May 15, 1991
BT1 Power Marketing Administrations
SO Management
DEF (U.S. Government Manual) The Administration markets electric power and energy from Federal hydroelectric projects in the Pacific Northwest. Through interregional connections, it sells surplus power to areas outside the Pacific Northwest region and participates in exchanges of power. The Administration is responsible for energy conservation, renewable resource development, and fish and wildlife enhancement under the provisions of the Pacific Northwest Electric Power Planning and Conservation Act of 1980. In cooperation with the Corps of Engineers, it represents the United States in implementing the provisions of the Columbia River Treaty with Canada. States served include Washington, Oregon, Nevada, Idaho, Montana, and Wyoming.

BOOM
(2301 BOOM)
DA December 10, 1990
BT1 Equipment/Parts - Material Handling (DOE FRASE Vocabulary)
 BT2 Equipment
RT Derrick
SO DOE FRASE VOCABULARY

BOOM CRANE
(2302 BOOM CRANE)
DA December 10, 1990
BT1 Crane(s)
 BT2 Material Handling Device
  BT3 Devices
  BT3 Equipment/Parts - Material Handling (DOE FRASE Vocabulary)
   BT4 Equipment
SO DOE FRASE VOCABULARY

BOOMS
(EPA)
DA October 12, 1990
RT Equipment
RT Oil Spills
SO Environmental Protection Agency Glossary
DEF (EPA) (1) Floating devices used to contain oil on a body of water. (2) Pieces of equipment used to apply pesticides from ground equipment such as a tractor or truck.

BOP
DA October 12, 1990
SEE Balance of Plant
SO Acronyms

BOR
DA October 12, 1990
SEE Burnout Risk
SO Acronyms

BORATED WATER STORAGE TANK
DA January 8, 1991
SF *BWST*

BORING MACHINE
(2106 BORING MACHI)
DA December 10, 1990
BT1 Machines (DOE FRASE Vocabulary)
 BT2 Equipment
SO DOE FRASE VOCABULARY

BORINGS
(DOE Order 6430.1A)
DA October 12, 1990
RT Sampling
SO Construction
DEF (DOE Order 6430.1A) Boreholes drilled to collect soil samples as part of subsurface investigations conducted for the purpose of structural foundation design.

BORON INJECTION TANK
DA January 8, 1991
SF *BIT*
BT1 Tanks
 BT2 Facility Components

BOSF
DA October 12, 1990
SEE Burnout Safety Factor
SO Acronyms

BOSFn
DA October 12, 1990
SEE Nominal BOSF
SO Acronyms

BOTANICAL PESTICIDES
(EPA)
DA October 12, 1990
BT1 Pesticides
 BT2 Hazardous Substances
SO Environmental Protection Agency Glossary
DEF (EPA) Pesticides whose active ingredient is a plant-produced chemical such as nicotine or strychnine.

BOTTLE BILL
(EPA)
DA October 12, 1990
BT1 Statutes and Regulations
RT Pollution
SO Environmental Protection Agency Glossary
DEF (EPA) Proposed or enacted legislation that requires a returnable deposit on beer or soda containers and provides for retail store or other redemption centers.

BT-Broader Term

NT-Narrower Term

RT Related Term

Such legislation is designed to discourage use of throwaway containers.

BOTTLES
DA   June 3, 1991
BT1 Containers
SO   Hazardous Materials
DEF Containers having a neck of relatively smaller cross section than the body and an opening capable of holding a closure for retention of the contents.

BOTTOM ASH
(SWDA; RCRA; CFR; ESH)
DA   October 12, 1990
BT1 Ashes
 BT2 Wastes
RT   Solid Wastes
SO   Environmental Management
SO   Wastes
DEF (CFR) The solid material that remain on a hearth or fall off the grate after thermal processing is complete.

BOTTOM END FITTING
(NFI)
DA   February 12, 1991
SF   *BEF*
SO   Nuclear Facilities Incident Database

BOTTOM FITTING INSERT
(NFI)
DA   January 8, 1991
SF   *BFI*
SO   Nuclear Facilities Incident Database

BOTTOMLAND HARDWOODS
(EPA)
DA   October 12, 1990
BT1 Wetlands
 BT2 Sites/Areas
 BT2 Surface Water Resources
 BT3 Natural Resources
SO   Environmental Protection Agency Glossary
DEF (EPA) Forested freshwater wetlands adjacent to rivers in the southeastern United States. They are especially valuable for wildlife breeding and nesting and habitat areas.

BOUNDARY ELEMENTS
(SEA)
DA   October 12, 1990
NT1 Diaphragm Chord
RT   Diaphragms
RT   Shear Walls
SO   Construction
DEF (SEA) Elements at edges of openings or at perimeters of shear walls or diaphragms.

BPS
DA   October 12, 1990

SEE Boeing Petroleum Services
SO   Acronyms

BPS TECHNICAL REPRESENTATIVE
DA   January 8, 1991
SF   *BTR*
BT1 Personnel

BPT
DA   June 18, 1991
SEE Best Practical Technology
SO   Acronyms

BPV
DA   October 12, 1990
SEE Back Pressure Valve
SO   Acronyms

Bq
DA   October 12, 1990
SEE Becquerel
SO   Acronyms

BRACED FRAME
(SEA)
DA   October 12, 1990
BT1 Building Frame Systems
 BT2 Space Frames
  BT3 Structures (DOE FRASE Vocabulary)
 BT2 Systems
NT1 Concentric Braced Frames
NT1 Eccentric Braced Frames
RT   Dual Systems
SO   Construction
DEF (SEA) An essentially vertical truss system of the concentric or eccentric type which is provided to resist lateral forces.

BRACKISH WATERS
(EPA)
DA   October 12, 1990
BT1 Surface Waters
 BT2 Water
BT1 Surface Water Resources
 BT2 Natural Resources
SO   Environmental Protection Agency Glossary
DEF (EPA) A mixture of fresh and salt waters.

BRAIN
(1079 BRAIN)
DA   November 28, 1990
BT1 Nervous System
 BT2 Body System(s)
  BT3 Human Body Parts
RT   Head
RT   Skull (Frase)
SO   DOE FRASE VOCABULARY

BRAIN DAMAGE
(1307 BRAIN DAMA)
DA   November 28, 1990
BT1 Injuries
NT1 Concussion
NT1 Contusion(S)

RT   Skull (Frase)
SO   DOE FRASE VOCABULARY

BRAVO
DA   February 26, 1991
SY   Safeguards and Security Alert II (Code Designator: Bravo)
BT1 Safeguards and Security Alerts
 BT2 Safeguards and Security Emergencies
  BT3 Emergencies
   BT4 Reportable Occurrences
    BT5 Occurrences
SO   Emergency Preparedness
DEF (EMER) A substantial alert action which shall be put into effect when the head of a field office determines that conditions or information received warrants more that the preparatory safeguards actions under Safeguards and Security Alert III.

BRC
DA   October 12, 1990
SEE Below Regulatory Concern
SO   Acronyms
SO   Environmental Management
SO   Wastes

BREAK-BULK
(SWDA; RCRA; ESH)
DA   October 12, 1990
BT1 Packaging
 BT2 Packages
RT   Transportation
SO   Hazardous Materials
DEF Packages of hazardous materials that are handled individually, palletized, or unitized for purposes of transportation as opposed to bulk and containerized freight.

BREAKTHROUGH
(SSDC)
DA   October 12, 1990
BT1 Controls (Reactor)
 BT2 Controls
SO   System Safety Development Center Glossary
DEF (SSDC) The process of developing and attaining a new, high standard for control of anything.

BREAST(S)
(1110 BREAST)
DA   November 28, 1990
BT1 Human Body Parts
RT   Chest
SO   DOE FRASE VOCABULARY

BREMSSTRAHLUNG
(NIH)
DA   October 12, 1990
BT1 Radiation
RT   Mass Energy Absorption Coefficient
SO   Radiation
DEF (NIH) Electromagnetic (x-ray) radiation associated with the

deceleration of charged particles passing through matter. Usually associated with energetic beta emitters, e.g., phosphorus-32.

BRIDGE
(1850 BRIDGE)
DA   December 10, 1990
BT1 Structures (DOE FRASE Vocabulary)
SO   DOE FRASE VOCABULARY

BRIDGE CRANE
(2303 BRIDGE CRANE)
DA   December 10, 1990
BT1 Crane(s)
 BT2 Material Handling Device
  BT3 Devices
  BT3 Equipment/Parts - Material Handling (DOE FRASE Vocabulary)
  BT4 Equipment
BT1 Equipment/Parts - Instrumentation/Measuring (DOE FRASE Voc.)
 BT2 Equipment
SO   DOE FRASE VOCABULARY

BRITISH THERMAL UNIT
DA   January 8, 1991
SF   BTU
BT1 Units of Measure

BROADCAST APPLICATION
(EPA)
DA   October 12, 1990
BT1 Application (Pesticide)
 BT2 Processes
RT   Band Application
RT   Basal Application
SO   Environmental Protection Agency Glossary
DEF (EPA) In pesticides, the spreading of chemicals over an entire area.

BRONCHIAL EPITHELIUM
(NCRP)
DA   October 12, 1990
BT1 Respiratory System
 BT2 Body System(s)
  BT3 Human Body Parts
RT   Inhalation
SO   Radiation
DEF (NCRP) The surface layer of cells lining the conducting airways.

BROOKHAVEN NATIONAL LABORATORY
DA   January 8, 1991
SF   BNL
BT1 Laboratories
 BT2 Research and Development Organizations
  BT3 Organizations
RT   Associated Universities, Inc.
DEF (Capsule Review of DOE Research and Development and Field Facilities, 1986) BNL, established in 1947, conceives, develops,

constructs and operates complex research facilities for the study of fundamental properties of matter. BNL conducts basic and applied research in technology base areas, supports research facilities and establishes important new directions for research. Major disciplinary strengths at BNL are high energy, nuclear and solid state physics, chemistry and biology. Research programs include the exploration of the fundamental constituents of matter, properties and interactions.

BROOM(S)
(3027 BROOM)
DA   January 3, 1991
BT1 Tools - Manual
 BT2 Tools (DOE FRASE Vocabulary)
  BT3 Equipment
SO   DOE FRASE VOCABULARY

BRUSH(S)
(3028 BRUSH)
DA   January 3, 1991
BT1 Tools - Manual
 BT2 Tools (DOE FRASE Vocabulary)
  BT3 Equipment
SO   DOE FRASE VOCABULARY

BTDR
DA   October 12, 1990
SEE Building Time Delay Relay
SO   Acronyms

BTR
DA   October 12, 1990
SEE BPS Technical Representative
SO   Acronyms

BTU
DA   October 12, 1990
SEE British Thermal Unit
SO   Acronyms

BUBBLE
(EPA)
DA   October 12, 1990
RT   Bubble Policy
RT   Emissions
SO   Environmental Protection Agency Glossary
DEF (EPA) A system under which existing emissions sources can propose alternate means to comply with a set of emissions limitations. Under the bubble concept, sources can control more than required at one emission point where control costs are relatively low in return for a comparable relaxation of controls at a second emission point where costs are higher.

BUBBLE POLICY
(EPA)
DA   October 12, 1990
BT1 Policies
RT   Bubble
RT   Emissions
SO   Environmental Protection Agency Glossary
DEF EPA policy that allows a plant complex with several facilities to decrease pollution from some facilities while increasing it form others, so long as total results are equal to or better than previous limits. Facilities where this is done are treated as if they exist in a bubble in which total emissions are averaged out. Complexes that reduce emissions substantially may "bank" their "credits" or sell them to other industries.

BUCKLED ZONES
(NFI)
DA   October 12, 1990
SF   BZ
BT1 Zones
 BT2 Sites/Areas
RT   Reactor Fuel
SO   Radiation
DEF (NFI) Regions of highest neutron flux.

BUDDY SYSTEM
(OSHA; CFR)
DA   January 24, 1991
BT1 Systems
RT   Industrial Safety
DEF (CFR) A system of organizing employees into work groups in such a manner that each employee of the work group is designated to be observed by at least one other employee in the work group. The purpose of the buddy system is to provide rapid assistance to employees in the event of an emergency.

BUDGET AND REPORTING CODE
DA   June 18, 1991
SF   B&RC
SO   Management

BUFFER STRIPS
(EPA)
DA   October 12, 1990
BT1 Biomass
RT   Plants
SO   Environmental Protection Agency Glossary
DEF (EPA) Strips of grass or other erosion-resisting vegetation between or below cultivated strips or fields.

BUFFER ZONE
(DOE Order 5820.2A)
DA   January 24, 1991

BT-Broader Term   NT-Narrower Term   RT Related Term

BT1 Zones
BT2 Sites/Areas
DEF (DOE 5820.2A) The smallest region
beyond the disposal unit that is
required as controlled space for
monitoring and for taking mitigative
measures, as may be required.

BUILDING (DOE FRASE VOCABULARY)
(1775 BUILDING; DOE Order 4330.4A)
DA   December 10, 1990
BT1 Facilities and Buildings (DOE
FRASE Vocabulary)
 BT2 Facilities
NT1 Containment Building
NT1 Demineralizer Plant
NT1 Fire Station
NT1 Garage
NT1 Guard Station
NT1 Hospital
NT1 Hot Shop
NT1 Machine Shop
NT1 Maintenance Shop
NT1 Motel/Hotel
NT1 Research & Development
Laboratory
NT1 Restaurant
NT1 Sewage Plant
NT1 Steam Plant
SO   DOE FRASE VOCABULARY
DEF (DOE Order 4330.4A) A roofed
structure that is suitable for
housing people, material, or
equipment. Also included are
sheds and other roofed structures
that provide partial protection from
the weather.

BUILDING ACQUISITIONS
(DOE Order 6430.1A)
DA   October 12, 1990
SO   Construction
DEF (DOE Order 6430.1A) New
pre-engineered metal buildings,
other semipermanent or temporary
facilities such as in-plant-
fabricated modular-relocatable
buildings and trailer units, and
other buildings to be acquired.

BUILDING FRAME SYSTEMS
(SEA)
DA   October 12, 1990
BT1 Space Frames
 BT2 Structures (DOE FRASE
Vocabulary)
BT1 Systems
NT1 Bearing Wall Systems
NT1 Braced Frame
 NT2 Concentric Braced Frames
 NT2 Eccentric Braced Frames
NT1 Horizontal Bracing Systems
NT1 Lateral Force Resisting System
NT1 Shoring
RT   P-Delta Effect
SO   Construction
DEF (SEA) Essentially complete space
frames that provide support for
gravity loads.

BUILDING POWER LOW VOLTAGE RELAY
DA   January 8, 1991
SF   BLVR
BT1 Relay(s)
 BT2 Equipment/Parts - Electrical (DOE
FRASE Vocabulary)
 BT3 Equipment

BUILDING TIME DELAY RELAY
DA   January 8, 1991
SF   BTDR
BT1 Relay(s)
 BT2 Equipment/Parts - Electrical (DOE
FRASE Vocabulary)
 BT3 Equipment

BUILDING/EQUIP MAINT/REPAIR ACTIVITY
(1228 BEM ACTIVITY)
DA   November 28, 1990
BT1 Activity Types (DOE FRASE
Vocabulary)
 BT2 Activities
BT1 Maintenance
 BT2 Activities
SO   DOE FRASE VOCABULARY

BUILDUP FACTOR
(NFI; IAEA)
DA   October 12, 1990
BT1 Ratios
RT   Radiation
SO   Radiation
DEF (IAEA) A dimensionless coefficient
equal to the ratio of a given
radiation quantity characterizing
the total scattered and unscattered
radiation field at some point in a
medium through which the
radiation is passing to the quantity
characterizing the unscattered
field along at that point.

BULK CONTAINERS
(RCRA; CFR)
DA   January 24, 1991
BT1 Containers
DEF (CFR) "Bulk container" means a
large container that can either be
pulled or lifted mechanically onto a
service vehicle or emptied
mechanically into a service
vehicle.

BULK PACKAGING
(SWDA; RCRA; CFR; ESH)
DA   October 12, 1990
BT1 Packaging
 BT2 Packages
RT   Freight Containers
RT   Non-bulk Packaging
SO   Hazardous Materials
DEF A packaging, other than a vessel or
a barge, including a transport
vehicle or freight container, in
which hazardous materials are
loaded with no intermediate form
of containment and which has (1)

An internal volume greater than
450 liters (118.9 gallons) as a
receptacle for a liquid; (2) a
capacity greater than 400
kilograms (881.8 pounds) as a
receptacle for a solid; or (3) a
water capacity greater than 1000
pounds (453.6 kilograms) as a
receptacle for a gas.

BULK RESINS
(CFR)
DA   October 12, 1990
BT1 Types of Resin
 BT2 Resin Grade
SO   Air Pollution
DEF (CFR) Resins that are produced by
a polymerization process in which
no water is used.

BULKHEADS
(CFR)
DA   October 12, 1990
BT1 Radionuclide Barriers (Natural or
Engineered)
 BT2 Barriers
RT   Containment
SO   Air Pollution
DEF (CFR) Air-restraining barriers
constructed for long-term control
of radon-222 and radon-222 decay
product levels in mine air.

BULKY WASTES
(RCRA; CFR)
DA   January 24, 1991
BT1 Solid Wastes
 BT2 Wastes
DEF (CFR) "Bulky waste" means large
items of solid waste such as
household appliances, furniture,
large auto parts, trees, branches,
stumps, and other oversize wastes
whose large size precludes or
complicates their handling by
normal solid wastes collection,
processing, or disposal methods.

BULLET CONTAINMENT DEVICES
(DOE Order 5480.16)
DA   October 12, 1990
BT1 Devices
NT1 Clearing Barrels
SO   Firearms
DEF (DOE Order 5480.16) A device
used to point a weapon at or into
during the loading or unloading
process that will contain any
inadvertently discharged round.

BUMP CAP
(2653 BUMP CAP)
DA   January 3, 1991
BT1 Head Protection
 BT2 Personal Protective Equipment
 BT3 Equipment/Parts - Personal
Protective (DOE FRASE
Vocabulary)

BT4 Equipment
SO DOE FRASE VOCABULARY

BUREAU OF ENVIRONMENT
DA   June 4, 1991
BT1 Tennessee Department of
     Commerce
 BT2 State Agencies
  BT3 Agencies
   BT4 Administrative Organizations
    BT5 Organizations

BUREAU OF EXPLOSIVES
(SWDA: RCRA; CFR; ESH)
DA   October 12, 1990
RT   Explosives
SO   Hazardous Materials
DEF The Bureau of Explosives (B of E)
     of the Association of American
     Railroads.

BURIAL BOX(S)
(2551 BURIAL BOX)
DA   January 3, 1991
BT1 Equipment/Parts - Nuclear (DOE
     FRASE Vocabulary)
 BT2 Equipment
 BT2 Reactor Components
SO DOE FRASE VOCABULARY

BURIAL GROUNDS
(EPA)
DA   October 12, 1990
SF   BG (Burial Ground)
BT1 Sites/Areas
RT   Radioactive Waste Management
SO   Environmental Protection Agency
     Glossary
DEF (EPA) Disposal sites for radioactive
     waste materials that use earth or
     water as a shield.

BURIAL OPERATIONS
DA   January 24, 1991
SY   Trenching
BT1 Operations
 BT2 Activities
BT1 Storage
 BT2 Waste Management Processes
  BT3 Processes
NT1 Shallow Land Burial
RT   Waste Management
SO   Environmental Management
DEF (CFR) Any method, technique or
     process, including storage for
     radioactive decay, designed to
     change the physical, chemical or
     biological characteristics or
     composition of any waste in order
     to render the waste for transport,
     storage or disposal, amendable to
     recovery, convertible to another
     usable material or reduced in
     volume.

BURN(S)
(1308 BURN;N)
DA   November 28, 1990
BT1 Injuries

NT1 Chemical Burn(s)
NT1 Electrical Burn(s)
NT1 Flash Burn(s)
SO DOE FRASE VOCABULARY

BURNER(S)
(2480 BURNER)
DA   January 3, 1991
BT1 Equipment/Parts - Heating (DOE
     FRASE Vocabulary)
 BT2 Equipment
BT1 Heating Equipment
SO DOE FRASE VOCABULARY

BURNING AGENTS
(CFR)
DA   October 12, 1990
BT1 Additives
 BT2 Chemical Substances
RT   Combustion
SO   Compensation and Liability
DEF (CFR) Those additives that,
     through physical or chemical
     means, improve the combustibility
     of the materials to which they are
     applied.

BURNOUT RISK
DA   January 8, 1991
SF   BOR

BURNOUT SAFETY FACTOR
DA   January 8, 1991
SF   BOSF
NT1 Nominal Automatic Burnout Safety
     Factor
NT1 Nominal BOSF

BURSITIS
DA   November 28, 1990
BT1 Injuries
SO DOE FRASE VOCABULARY

BUS DRIVER
(0821 BUS DRIVER)
DA   November 28, 1990
BT1 Transport Personnel
 BT2 Occupations
 BT2 Personnel
SO DOE FRASE VOCABULARY

BUSINESS CONFIDENTIALITY
(CFR)
DA   November 15, 1990
SY   Confidential Business Information
DEF (CFR) Includes the concept of trade
     secrecy and other related legal
     concepts which give (or may give)
     a business the right to preserve
     the confidentiality of business
     information and to limit its use or
     disclosure by others in order that
     the business may obtain or retain
     business advantages it derives
     from its right in the information.
     The definition is meant to
     encompass any concept which
     authorizes a Federal agency to

withhold business information
under 5 U.S.C. 552(b)(4), as well
as any concept which requires
EPA to withhold information from
the public for the benefit of a
business under 18 U.S.C. 1905.

BUSTR
DA   October 12, 1990
SEE Ohio Bureau of Underground
     Storage Tank Regulation
SO   Acronyms

BUTTOCK(S)
(1119 BUTTOCK)
DA   November 28, 1990
BT1 Trunk
 BT2 Human Body Parts
RT   Hip(s)
SO DOE FRASE VOCABULARY

BWR
DA   October 12, 1990
SEE Boiling Water Reactor
SO   Acronyms

BWST
DA   October 12, 1990
SEE Borated Water Storage Tank
SO   Acronyms

BY-PRODUCT MATERIALS
(EMER)
DA   February 1, 1991
BT1 By-products
 BT2 Materials
NT1 Uranium By-Product Materials
SO   Emergency Preparedness
DEF (EMER) 1. Radioactive materials
     (except special nuclear material)
     yielded in or made radioactive by
     exposure to the radiation incident
     to the process of producing or
     utilizing special nuclear material.
     2. The tailings or wastes produced
     by the extraction or concentration
     of uranium or thorium from any
     ore processed primarily for its
     source material content.

BY-PRODUCTS
(TSCA; RCRA; CFR)
DA   October 12, 1990
BT1 Materials
NT1 By-Product Materials
 NT2 Uranium By-Product Materials
RT   Coproducts
RT   Process (Toxic Chemical)
SO   Environmental Management
SO   Environmental Protection Agency
     Glossary
SO   Hazardous Materials
DEF (EPA) Materials, other than the
     principal products, that are
     generated as a consequence of an
     industrial process. (CFR)
     Chemical substances produced
     without separate commercial intent
     during the manufacturing or

processing of another chemical substance(s) or mixture(s).

BYPASSES
(NFI)
DA   October 12, 1990
RT   Electrical Equipment
SO   Nuclear Facilities Incident Database
SO   Radiation
DEF (NFI) Applies to electric circuits as well as cooling circuits.

BZ
DA   October 12, 1990
SEE Buckled Zones
SO   Acronyms

C&D
DA   October 12, 1990
SEE Charge and Discharge
SO   Acronyms

Ca
DA   October 12, 1990
SEE Calcium
SO   Acronyms

CAA
DA   October 12, 1990
SEE Clean Air Act
SO   Acronyms

CAD
DA   October 12, 1990
SEE Computer-Assisted Design
SO   Acronyms

CADMIUM
(EPA)
DA   October 12, 1990
SF   Cd
BT1 CERCLA Hazardous Substances
 BT2 Hazardous Substances
BT1 Heavy Metals
SO   Environmental Protection Agency Glossary
DEF (EPA) A heavy metal element that accumulates in the environment.

CAFETERIA
(1680 CAFETERIA)
DA   December 10, 1990
BT1 Eating Area
 BT2 Area
  BT3 Sites/Areas
BT1 Site (DOE FRASE Vocabulary)
 BT2 Sites/Areas
RT   Kitchen
RT   Restaurant
SO   DOE FRASE VOCABULARY

CAIRS
DA   October 12, 1990
SEE Computerized Accident/Incident Reporting System
SO   Acronyms
SO   Environmental Management

CAISSON FOUNDATIONS
(DOE Order 6430.1A)
DA   October 12, 1990
RT   Construction
SO   Construction
DEF (DOE Order 6430.1A) Shafts of concrete placed under a building column or wall that extend down to rock or solid substratum (also known as pier foundations).

CALCINERS
(CFR)
DA   October 12, 1990
RT   Nodulizing Kilns
RT   Pyrolysis
SO   Air Pollution
DEF (CFR) Units in which phosphate rock is heated to high temperatures to remove organic material and/or to convert it to nodular form. Calciners and nodulizing kilns are considered to be similar units.

CALCIUM
DA   January 8, 1991
SF   Ca

CALIBRATION
(NIH)
DA   October 12, 1990
RT   Accuracy
RT   Inspections
RT   Measurements
SO   Radiation
DEF (NIH) Determination of variation from standard, or accuracy, of a measuring instrument to ascertain necessary correction factors.

CALIBRATION CHECKS
(ESH)
DA   October 12, 1990
BT1 Verification
 BT2 Administrative Processes
  BT3 Processes
RT   Blanks
SO   Quality Assurance
DEF (ESH) Verification of the ratio of instrument response to analyte amount, a calibration check, is done by analyzing for analyte standards in an appropriate solvent. Calibration check solutions are made from a stock solution which is different from the stock used to prepare standards.

CALIFORNIUM TARGETS
DA   January 8, 1991
SF   CFTs

CAMS
DA   October 12, 1990
SEE Continuous Air Monitors
SO   Acronyms
SO   Emergency Preparedness

CANAL AREA
(1604 CANAL AREA)
DA   December 10, 1990
BT1 Area
 BT2 Sites/Areas
SO   DOE FRASE VOCABULARY

CANCELLATIONS
(EPA)
DA   October 12, 1990
RT   Registration
SO   Environmental Protection Agency Glossary
DEF (EPA) Refers to Section 6(b) of the Federal Insecticide, Fungicide and Rodenticide Act (FIFRA), which authorizes cancellation of a pesticide registration if unreasonable adverse effects to the environment and public health develop when a product is used according to widespread and commonly recognized practice, or if its labeling or other material required to be submitted does not comply with FIFRA provisions.

CANDIDATES
(ESA)
DA   October 12, 1990
BT1 Species
RT   Endangered Species
RT   Threatened Species
SO   Endangered Species
DEF (CFR) Species being considered for listing as an endangered or a threatened species, but not yet the subject of a proposed rule.

CANDIS COMPUTER TAPES
(NFI)
DA   October 12, 1990
SO   Nuclear Facilities Incident Database
SO   Radiation
DEF (NFI) Computer tapes that operate C&D machines automatically; contain listing of x,y coordinates.

CANTILEVER FOOTINGS
(DOE Order 6430.1A)
DA   October 12, 1990
SO   Construction
DEF (DOE Order 6430.1A) Footings used to support a wall column near its edge without causing nonuniform soil pressure.

CAPABLE FAULT
(AEA; CFR)
DA   January 24, 1991
BT1 Faults
DEF (CFR) A "capable fault" is a fault which has exhibited one or more of the following characteristics: (1) Movement at or near the ground surface at least once within the past 35,000 years or movement of a recurring nature within the past 5000,000 years.

SY-Synonymous Terms   SO-Source/Subject Category   SF-See From

CAPACITOR(S)
(2380 CAPACITOR)
DA January 3, 1991
BT1 Equipment/Parts - Electrical (DOE
    FRASE Vocabulary)
 BT2 Equipment
SO DOE FRASE VOCABULARY

CAPACITORS
(CFR; ESH)
DA October 12, 1990
BT1 Electrical Equipment
 BT2 Equipment
NT1 Large High Voltage Capacitors
NT1 Large Low Voltage Capacitors
NT1 Small Capacitors
RT Dielectric Materials
SO Hazardous Materials
DEF (ESH) Devices for accumulating
    and holding a charge of electricity
    and consisting of conducting
    surfaces separated by a dielectric.

CAPACITY FACTOR
(CFR)
DA October 12, 1990
BT1 Load Factor
BT1 Ratios
RT Equipment
SO Air Pollution
DEF (CFR) The ratio of the average load
    on a machine or equipment for the
    period of time considered to the
    capacity rating of the machine or
    equipment.

CAPILLARY WATER
(DOE Order 6430.1A)
DA October 12, 1990
BT1 Groundwater
 BT2 Water
SO Construction
DEF (DOE Order 6430.1A) Soil moisture
    held as a continuous adsorbed
    film around soil particles and in
    interstices between the soil
    particles due to surface attraction.

CAPITAL EXPENDITURE
(CFR)
DA October 12, 1990
BT1 Costs
RT Assets
RT Fixed Capital Costs
SO Air Pollution
DEF (CFR) An expenditure for a
    physical or operational change to
    a stationary source which exceeds
    the product of the applicable
    "annual asset guideline repair
    allowance percentage" specified in
    the latest edition of Internal
    Revenue Service (IRS) Publication
    534 and the stationary source's
    basis, as defined by section 1012
    of the Internal Revenue Code.
    However, the total expenditure for
    a physical or operational change
    to a stationary source must not be

reduced by any "excluded
additions" as defined for stationary
sources constructed after
December 31, 1981, in IRS
Publication 534, as would be done
for tax purposes. In addition,
"annual asset guideline repair
allowance" may be used even
though it is excluded for tax
purposes in IRS Publication 534.

CAPITAL IMPROVEMENT PROJECT
DA January 8, 1991
SF CIP
BT1 Projects

CAPS
(EPA)
DA October 12, 1990
RT Landfills
SO Environmental Protection Agency
    Glossary
DEF (EPA) Layers of clay, or other highly
    impermeable materials, installed
    over the top of a closed landfill to
    prevent entry of rainwater and
    minimize production of leachate.

CAPTAIN OF THE PORT
(SWDA; RCRA; CFR; ESH)
DA October 12, 1990
BT1 Personnel
SO Hazardous Materials
DEF The officer of the Coast Guard,
    under the command of a District
    Commander, so designated by the
    Commandant for the purpose of
    giving immediate direction to
    Coast Guard law enforcement
    activities within his assigned area
    or, with respect to remaining areas
    in his District not assigned to
    officers designated by the
    Commandant, or the District
    Commander.

CAPTIVITY
(SWDA; ESA; CFR)
DA October 12, 1990
RT Animals
RT Captivity-bred
SO Endangered Species
DEF (CFR) That living wildlife is held in
    a controlled environment that is
    intensively manipulated by man for
    the purpose of producing wildlife of
    the selected species, and that has
    boundaries designed to prevent
    animal, eggs or gametes of the
    selected species from entering or
    leaving the controlled environment.
    General characteristics of captivity
    may include but are not limited to
    artificial housing, waste removal,
    health care, protection from
    predators, and artificially supplied
    food.

CAPTIVITY-BRED
(ESA; CFR)
DA October 12, 1990
RT Animals
RT Captivity
SO Endangered Species
DEF (CFR) Refers to wildlife, including
    eggs, born or otherwise produced
    in captivity from parents that
    mated or otherwise transferred
    gametes in captivity, if
    reproduction is sexual, or from
    parents that were in captivity when
    development of the progeny
    began, if development is asexual.

CAPTURE EFFICIENCY
(EPA)
DA October 12, 1990
BT1 Ratios
RT Control Devices
SO Environmental Protection Agency
    Glossary
DEF (EPA) The fraction of all organic
    vapors generated by a process
    that are directed to an abatement
    or recovery device.

CAR
DA June 18, 1991
SEE Corrective Action Reporting
SO Acronyms

CARBON ADSORBERS
(EPA)
DA October 12, 1990
BT1 Control Devices
 BT2 Devices
RT Activated Carbon
RT Sorption
SO Environmental Protection Agency
    Glossary
DEF (EPA) Added-on control devices
    that use activated carbon to
    absorb volatile organic compounds
    (VOCs) from a gas stream. The
    VOCs are later recovered from the
    carbon.

CARBON DIOXIDE
(EPA)
DA October 12, 1990
SF CO_2
BT1 Gases
BT1 Inorganic Chemicals
 BT2 Chemical Substances
SO Environmental Protection Agency
    Glossary
DEF (EPA) A colorless, odorless,
    nonpoisonous gas that results
    from fossil fuel combustion and is
    normally a part of the ambient air.

CARBON DIOXIDE SPACES
(NFI)
DA October 12, 1990
SO Nuclear Facilities Incident Database
SO Radiation
DEF (NFI) Annular spaces between

BT-Broader Term   NT-Narrower Term   RT Related Term

main tanks and carbon steel liners
filled with carbon dioxide.

CARBON MONOXIDE
(EPA)
DA October 12, 1990
SF *CO (Carbon Monoxide)*
BT1 Criteria Pollutants
 BT2 Pollutants
BT1 Gases
BT1 Inorganic Chemicals
 BT2 Chemical Substances
SO Environmental Protection Agency
 Glossary
DEF (EPA) A colorless, odorless,
 poisonous gas produced by
 incomplete fossil fuel combustion.

CARBON TETRAFLUORIDE
DA January 8, 1991
SF *CF₄*

CARBOXYHEMOGLOBIN
(EPA)
DA October 12, 1990
BT1 Proteins
 BT2 Organic Chemicals
 BT3 Chemical Substances
SO Environmental Protection Agency
 Glossary
DEF (EPA) Hemoglobin in which the iron
 is associated with carbon
 monoxide (CO). The affinity of
 hemoglobin for CO is about 300
 times greater than for oxygen.

CARC
DA October 12, 1990
SEE Containment Air Recirculation and
 Cooling
SO Acronyms

**CARCINOGEN RISK ASSESSMENT
VERIFICATION ENDEAVOR
WORKGROUP**
(EPA)
DA October 12, 1990
BT1 Groups
 BT2 Administrative Organizations
 BT3 Organizations
RT Assessments
RT Risk Assessment
SO Environmental Protection Agency
 Glossary
DEF (EPA) An EPA workgroup formed to
 validate Agency carcinogen risk
 assessments and resolve
 conflicting potency values among
 various program offices.

CARCINOGENS
(EPA)
DA October 12, 1990
BT1 Hazardous Substances
NT1 Asbestos
 NT2 Commercial Asbestos
 NT2 Friable Asbestos
NT1 Polychlorinated Biphenyls
 NT2 High Concentration PCBs

NT2 Low Concentration PCBs
NT2 Recycled PCBs
SO Environmental Protection Agency
 Glossary
DEF (EPA) Substances that can cause
 or contribute to the production of
 cancer.

CARFLOATS
(SWDA; RCRA; CFR; ESH)
DA October 12, 1990
BT1 Vessels
RT Transport (Hazardous Substances)
SO Hazardous Materials
DEF (ESH) Vessels that operate on a
 short run on an irregular basis and
 serve one or more points in a port
 area as an extension of a rail line
 or highway over water, and do not
 operate in ocean, coastwise, or
 ferry service.

CARGO AIRCRAFT ONLY
(SWDA; RCRA; CFR; ESH)
DA October 12, 1990
BT1 Aircraft
RT Transport (Hazardous Substances)
SO Hazardous Materials
DEF (ESH) An aircraft that is used to
 transport cargo and is not
 engaged in carrying passengers.

CARGO TANKS
(SWDA; RCRA; ESH)
DA October 12, 1990
BT1 Tanks
 BT2 Facility Components
RT Motor Vehicles
SO Hazardous Materials
DEF (ESH) Any tank permanently
 attached to or forming a part of
 any motor vehicle or any bulk
 liquid or compressed gas
 packaging not permanently
 attached to any motor vehicle
 which by reason of its size,
 construction, or attachment to a
 motor vehicle. Any packaging
 fabricated under specifications for
 cylinders is not a cargo tank.

CARGO VESSELS
(SWDA; RCRA; CFR; ESH)
DA October 12, 1990
BT1 Vessels
NT1 Containerships
SO Hazardous Materials
DEF (ESH) (1) Vessels other than
 passenger vessels; and, (2)ferries
 being operated under authority of
 a change of character certificate
 issued by a Coast Guard
 Officer-in-Charge, Marine
 Inspection.

CARPENTER
(0642 CARPENTER)
DA November 28, 1990
BT1 Repair/Construction Personnel

BT2 Occupations
BT2 Personnel
SO DOE FRASE VOCABULARY

CARRIER FREE
(NIH)
DA October 12, 1990
RT Radionuclides
SO Radiation
DEF (NIH) An adjective applied to one
 or more radionuclides of an
 element in minute quantity,
 essentially undiluted with stable
 isotope carrier.

CARRIERS
(DOE Orders 5480.3, 1540.1 5.B;
 SWDA; RCRA; CFR; ESH)
DA October 12, 1990
RT Transportation
SO Environmental Management
SO Hazardous Materials
DEF Any person engaged in the
 transportation of passengers or
 property as common, contract, or
 private charter, or freight
 forwarder, as those terms are
 used in the Interstate Commerce
 Act, as amended, or by the U.S.
 Postal Service. (ESH) Any person
 engaged in the transportation of
 passengers or property by (1) land
 or water, as a common, contract,
 or private carrier, or (2) civil
 aircraft.

CARRYING CAPACITY
(EPA)
DA October 12, 1990
BT1 Measurements
SO Environmental Protection Agency
 Glossary
DEF (EPA) (1) In recreation
 management, the amount of use a
 recreation area can sustain
 without deterioration of its quality.
 (2) In wildlife management, the
 maximum number of animals an
 area can support during a given
 period of the year.

CART(S)
(2304 CART)
DA December 10, 1990
BT1 Hand Truck(s)
 BT2 Material Handling Device
 BT3 Devices
 BT3 Equipment/Parts - Material
 Handling (DOE FRASE
 Vocabulary)
 BT4 Equipment
SO DOE FRASE VOCABULARY

CAS
DA June 4, 1991
SEE Condition Assessment Surveys
SO Acronyms

CASED EXPLOSIVES
(DOE Order 6430.1A)
DA   October 12, 1990
BT1 Explosives
SO   Construction
DEF (DOE Order 6430.1A) Explosives that are enclosed in a physical protective covering that will retain the explosives securely and will offer significant protection against accidental detonation during approved handling and intraplant transportation operations.

CASINGS
(SDWA; CFR)
DA   October 12, 1990
NT1 Surface Casings
RT   Wells
SO   Water Pollution
DEF (CFR) Pipes or tubings of appropriate material, of varying diameter and weight, lowered into a borehole during or after drilling in order to support the sides of the hole and thus prevent the walls from caving; to prevent loss of drilling mud into porous ground; or to prevent water, gas, or other fluid from entering or leaving the hole.

CASKS
(EPA)
DA   October 12, 1990
SY   Coffins
SY   Pigs
BT1 Containers
SO   Environmental Protection Agency Glossary
DEF (EPA) Thick-walled containers (usually lead) used to transport radioactive material. Also called coffins.

CATALOGS
(SSDC)
DA   October 12, 1990
RT   Vocabulary
SO   System Safety Development Center Glossary
DEF (SSDC) In computer usage, lists of vocabulary words which are systematically grouped.

CATALYTIC CONVERTERS
(EPA)
DA   October 12, 1990
BT1 Devices
RT   Motor Vehicles
RT   Transportation Control Measures
SO   Environmental Protection Agency Glossary
DEF (EPA) Air pollution abatement devices that remove pollutants from motor vehicle exhaust, either by oxidizing them into carbon dioxide and water or reducing them to nitrogen and oxygen.

CATALYTIC INCINERATORS
(EPA)
DA   October 12, 1990
BT1 Control Devices
 BT2 Devices
BT1 Incinerators
 BT2 Equipment/Parts - Heating (DOE FRASE Vocabulary)
 BT3 Equipment
 BT2 Heating Equipment
RT   Volatile Organic Compounds
SO   Environmental Protection Agency Glossary
DEF (EPA) Control devices which oxidize volatile organic compounds (VOCs) by using a catalyst to promote the combustion process. Catalytic incinerators require lower temperatures than conventional thermal incinerators, with resultant fuel and cost savings.

CATANADRAMOUS FISH
(EPA)
DA   October 12, 1990
BT1 Fish
 BT2 Animals
SO   Environmental Protection Agency Glossary
DEF (EPA) Fish that swim downstream to spawn.

CATASTROPHIC COLLAPSES
(SDWA; CFR)
DA   October 12, 1990
BT1 Failures
 BT2 Accidents
RT   Liquefaction
RT   Subsidence
RT   Surface Faulting
SO   Water Pollution
DEF (CFR) Sudden and utter failures of overlying strata caused by removal of underlying materials.

CATEGORICAL EXCLUSIONS
(DOE Order 5440.1D; NEPA; CFR; EPA)
DA   October 12, 1990
SF   CX
BT1 Actions
 BT2 Responses
RT   Environmental Impact Statements
RT   National Environmental Policy Act Documents
SO   Environmental Management
SO   Environmental Protection Agency Glossary
DEF (EPA) Classes of actions which either individually or cumulatively would not have a significant effect on the human environment and therefore would not require preparation of an environmental assessment or environmental impact statement under the National Environmental Policy Act (NEPA). (DOE Order) A category of actions, as defined at 40 CFR 1508.4 and listed in Section D of the DOE NEPA Guidelines, that do not individually or cumulatively have a significant effect on the human environment and for which neither an environmental assessment (EA) nor an environmental impact statement (EIS) is normally required.

CATEGORICAL PRETREATMENT STANDARDS
(EPA)
DA   October 12, 1990
BT1 Limits
BT1 Standards
 BT2 Codes, Standards, and Regulations
RT   Best Available Technology
SO   Environmental Protection Agency Glossary
DEF (EPA) Technology-based effluent limitations for an industrial facility that discharges into a municipal sewer system. Analogous in stringency to Best Availability Technology (BAT) for direct dischargers.

CATEGORY A REACTORS
(DOE Order 5480.6)
DA   October 12, 1990
BT1 Reactors
SO   Environmental Management
SO   Industrial Safety
DEF (DOE Order 5480.6) A DOE designation based on power level (e.g., 20 MW steady state), potential fission product inventory, and experimental capability. Category A reactors are listed in Attachment 1, paragraph 3. All other DOE-owned reactors (excluding reactors assigned to the Deputy Assistant Secretary for Naval Reactors, NE-60) are designated Category B.

CATEGORY I SERIOUSNESS
(ESH; TSA)
DA   October 12, 1990
BT1 Seriousness
 BT2 Conditions
DEF (ESH) Addresses a situation for which a clear and present danger exists to workers or member of the public. A concern in this category is to be immediately conveyed to the managers of the facility for action. At this point, consideration shall be given to whether a "clear and present danger" exists such that the facility shutdown authority of the Assistant Secretary (EH- 1) should be exercised. If so, the Assistant Secretary or his designee is informed immediately.

CATEGORY II SERIOUSNESS
(ESH; TSA)
DA   October 12, 1990

BT-Broader Term   NT-Narrower Term   RT Related Term

BT1 Seriousness
 BT2 Conditions
DEF (ESH) Addresses a significant risk (but does not involve a situation for which a clear and present danger exists to workers or members of the public) or substantial noncompliance with DOE Orders. A concern in this category is to be conveyed to the manager of the facility no later than the appraisal closeout meeting for immediate attention. Category II concerns have a significance and urgency such that the necessary field response should not be delayed until the preparation of a final report and the routine development of an action plan. Any issues surrounding the concern or the suggested response should be addressed during the appraisal or immediately thereafter. Again, consideration should be given to whether facility shutdown is warranted under the circumstances.

CATEGORY III SERIOUSNESS
(ESH; TSA)
DA October 12, 1990
BT1 Seriousness
 BT2 Conditions
DEF (ESH) Addresses significant noncompliance with DOE Orders or suggests significant improvements in the margin of safety, but is not of sufficient urgency to require immediate attention.

CATHODIC PROTECTION
(SWDA; RCRA)
DA October 12, 1990
BT1 Processes
RT Cathodic Protection Testers
RT Corrosion
SO Environmental Protection Agency Glossary
SO Wastes
DEF A technique to prevent corrosion of a metal surface by making that surface the cathode of an electrochemical cell. For example, a tank system can be cathodically protected through the application of either galvanic anodes or impressed current.

CATHODIC PROTECTION TESTERS
(SWDA; RCRA; CFR)
DA October 12, 1990
BT1 Personnel
RT Cathodic Protection
SO Wastes
DEF People who demonstrate an understanding of the principles and measurements of all common types of cathodic protection

systems as applied to buried or submerged metal piping and tank systems. At a minimum, such persons must have education and experience in soil resistivity, stray current, structure-to-soil potential, and component electrical isolation measurements of buried metal piping and tank systems.

CATION EXCHANGE CAPACITY
(RCRA; CFR)
DA January 24, 1991
BT1 Measurements
RT Soils
DEF (CFR) "Cation exchange capacity" means the sum of exchangeable cations a soil can absorb expressed in milli-equivalents per 100 grams of soil as determined by sampling the soil to the depth of cultivation or solid waste placement, whichever is greater, and analyzing by the summation method for distinctly acid soils or the sodium acetate method for neutral, calcareous or saline soils.

CATS
DA October 12, 1990
SEE Computer Assisted Tracking System
SO Acronyms
SO Environmental Management

CATWALK
(1681 CATWALK)
DA December 10, 1990
BT1 Site (DOE FRASE Vocabulary)
 BT2 Sites/Areas
SO DOE FRASE VOCABULARY

CAUSES
(SSDC)
DA October 12, 1990
NT1 Basic Causes
NT1 Contributing Causes
NT1 Immediate Causes
NT1 Probable Causes
NT1 Proximate Causes
NT1 Root Causes
RT Compensation and Liability
RT Reasonably Attributable
SO System Safety Development Center Glossary
DEF (SSDC) In system safety, causes are anything that contributes to an accident or incident. In analysis or investigation, avoid the use of cause as a singular term; prefer, causal factors. Any "probable cause" statement required in an investigation report should be the immediate, proximate cause of the primary, major damage and should be accompanied by succeeding statements of contributing causes of energy buildup and release and

of inadequacies of plans, operations, detection, and control.

CAUSTIC SODA
(EPA)
DA October 12, 1990
RT Detergents
SO Environmental Protection Agency Glossary
DEF (EPA) Sodium hydroxide, a strong alkaline substance used as the cleaning agent in some detergents.

CBOD5
(EPA)
DA October 12, 1990
BT1 Measurements
RT Biochemical Oxygen Demand
SO Environmental Protection Agency Glossary
DEF (EPA) The amount of dissolved oxygen consumed in 5 days from the carbonaceous portion of biological processes breaking down in an effluent. The test methodology is the same as for BOD5, except that nitrogen demand is suppressed.

CCDF
DA October 12, 1990
SEE Complementary Cumulative Distribution Function
SO Acronyms

CCF
DA October 12, 1990
SEE Common Cause Failure
SO Acronyms

CCR
DA October 12, 1990
SEE Central Control Room
SO Acronyms

CCTV
DA October 12, 1990
SEE Crane Control, Close Circuit TV
SO Acronyms

CCW
DA October 12, 1990
SEE Component Cooling Water
SO Acronyms

Cd
DA October 12, 1990
SEE Cadmium
SO Acronyms

CDC
DA October 12, 1990
SEE Center for Disease Control
SO Acronyms

CDF
DA   October 12, 1990
SEE Core Damage Frequency
SO   Acronyms

CDH
DA   October 12, 1990
SEE Colorado Department of Health
SO   Acronyms

CDPM
DA   October 12, 1990
SEE Crane Drip Pan Monitor
SO   Acronyms

CDR
DA   October 12, 1990
SEE Conceptual Design Report
SO   Acronyms

CE
DA   October 12, 1990
SEE Combustion Engineering
SO   Acronyms

CEARP
DA   October 12, 1990
SEE Comprehensive Environmental
       Assessment and Response
       Program
SO   Acronyms

CEL
DA   October 12, 1990
SEE Channel Effluent Limit
SO   Acronyms

CELL
(1682 CELL)
DA   December 10, 1990
BT1 Site (DOE FRASE Vocabulary)
 BT2 Sites/Areas
SO DOE FRASE VOCABULARY
SO Environmental Management

CELL ROOMS
(CFR)
DA   October 12, 1990
RT Mercury Chlor-alkali Cells
SO Air Pollution
DEF (CFR) Structure(s) housing one or
       more mercury electrolytic
       chlor-alkali cells.

CELLS
(RCRA; CFR; EPA)
DA   October 12, 1990
NT1 Landfill Cells
RT Daily Cover
RT Solid Waste Disposal
SO Environmental Protection Agency
       Glossary
DEF (EPA) (1) In solid waste disposal,
       holes where waste is dumped,
       compacted, and covered with
       layers of dirt on a daily basis.

CEMENTING
(SDWA; CFR)
DA   October 12, 1990
BT1 Processes
RT Plugging
RT Portland Cement
SO Water Pollution
DEF (CFR) The operation whereby a
       cement slurry is pumped into a
       drilled hole and/or forced behind
       the casing.

CENTER FOR DISEASE CONTROL
DA   January 8, 1991
SF   *CDC*
BT1 Centers

CENTERS
DA   February 4, 1991
NT1 Air Force Engineering and Services
       Center
NT1 Alternate Emergency Operations
       Center
NT1 Center for Disease Control
NT1 DOE Emergency Operations Center
NT1 Emergency Briefing Centers
NT1 Emergency Control Centers
NT1 Emergency Operations Centers
NT1 Energy Technology Engineering
       Center (Canoga Park)
NT1 Federal Radiological Monitoring
       and Assessment Centers
NT1 Federal Response Centers
NT1 Feed Materials Production Center
NT1 Joint Information Center
NT1 Joint Nuclear Accident Coordination
       Center
NT1 Manned Control Centers
NT1 Motor Control Center
NT1 National Response Centers
NT1 Nuclear Safety Analysis Center
NT1 Secure Communications Centers
NT1 Security Communications Control
       Center
NT1 Stanford Linear Accelerator Center
NT1 U.S. Army Toxic and Hazardous
       Materials Center
NT1 University of Tennessee Center for
       Biotechnology
RT Boards
RT Committees
RT Divisions
RT Groups
RT Offices
DEF (WTID) Places, areas, people,
       groups, or concentrations marked
       significantly or dominantly by an
       indicated activity, pursuit, interest,
       or appeal.

CENTIMETERS PER SECOND
DA   January 8, 1991
SF   *CM/sec*
BT1 Units of Measure

CENTRAL ALARM STATIONS
(EMER)
DA   February 1, 1991
BT1 Sites/Areas

RT Alarms
RT Emergencies
SO Emergency Preparedness
DEF (EMER) Physical locations to which
       physical security system alarms,
       radiation alarms, fire alarms, and
       hazardous substance alarms are
       electronically communicated.

CENTRAL CONTROL ROOM
DA   January 8, 1991
SF   *CCR*
BT1 Room
 BT2 Sites/Areas

**CENTRAL SERVICE WORKS
ENGINEERING**
DA   January 8, 1991
SF   *CSWE*

CENTRAL TRAINING ACADEMY
(DOE Order 5480.16)
DA   October 12, 1990
SF   *CTA*
BT1 Educational Organizations
 BT2 Organizations
RT Engagement Simulation Systems
RT Firearms
RT Firearms Ranges
RT Protective Force Personnel
SO Firearms
DEF (DOE Order 5480.16) Located at
       Kirtland Air Force Base East,
       Albuquerque, New Mexico.
       Program responsibility is to the
       Director of Safeguards and
       Security (DP-34).

CENTRIFUGAL COLLECTORS
(EPA)
DA   October 12, 1990
BT1 Systems
SO Environmental Protection Agency
       Glossary
DEF (EPA) Mechanical systems using
       centrifugal force to remove
       aerosols from a gas stream or to
       de-water sludge.

CENTRIFUGE
(2110 CENTRIFUGE)
DA   December 10, 1990
BT1 Machines (DOE FRASE
       Vocabulary)
 BT2 Equipment
SO DOE FRASE VOCABULARY

CEQ
DA   October 12, 1990
SEE Council on Environmental Quality
SO Acronyms

CERAMIC PLANTS
(CFR)
DA   October 12, 1990
BT1 Facilities

SO Air Pollution
DEF (CFR) Manufacturing plants
    producing ceramic items.

CERCLA
(Acronyms and Abbreviations)
DA October 12, 1990
SEE Comprehensive Environmental
    Response, Compensation, etc.
SO Acronyms
DEF (CERCLA) To provide for liability,
    compensation, cleanup, and
    emergency response for
    hazardous substances released
    into the environment and the
    cleanup of inactive hazardous
    waste disposal sites.

CERCLA HAZARDOUS SUBSTANCES
(CFR)
DA November 15, 1990
SY Listed Hazardous Substances
BT1 Hazardous Substances
NT1 Agent Orange
NT1 Aldicarb
NT1 Asbestos
 NT2 Commercial Asbestos
 NT2 Friable Asbestos
NT1 Beryllium
 NT2 Beryllium 7
NT1 Cadmium
NT1 Chlorinated Hydrocarbons
 NT2 Chlorofluorocarbons (CFCs)
 NT2 DDT
 NT2 Heptachlor
 NT2 Polychlorinated Biphenyls
  NT3 High Concentration PCBs
  NT3 Low Concentration PCBs
  NT3 Recycled PCBs
 NT2 Polyvinyl Chloride
 NT2 Trichloroethylene (TCE)
 NT2 Vinyl Chloride
NT1 Chromium
NT1 DES
NT1 Diazinon
NT1 Dinoseb
NT1 Ethylene Dibromide
NT1 Formaldehyde
NT1 Lead
NT1 Nitrogen Oxides (NO$_x$)
 NT2 Nitric Oxides
 NT2 Nitrogen Dioxide (NO$_2$)
NT1 Phenols
NT1 Phosphorus
NT1 Radionuclides
 NT2 Alpha Decay Radioisotopes
  NT3 Radium 226
  NT3 Radon 222
  NT3 Thorium 230
  NT3 Uranium 238
 NT2 Beta Decay Radioisotopes
  NT3 Beta-Minus Decay Radioisotopes
  NT4 Antimony 125
  NT4 Cerium 144
  NT4 Cerium 141
  NT4 Cesium 137
  NT4 Cesium 134
  NT4 Cobalt 60
  NT4 Cobalt 58
  NT4 Iodine 131

NT4 Radium 228
NT4 Ruthenium 103
NT4 Strontium 90
NT4 Tellurium 132
NT4 Thorium 232
NT4 Tritium
NT4 Zinc 65
NT4 Zirconium 95
NT2 Bone Seekers
NT2 Days Living Radioisotopes
NT3 Beryllium 7
NT3 Cerium 144
NT3 Cerium 141
NT3 Cesium 137
NT3 Cobalt 58
NT3 Iodine 131
NT3 Manganese 54
NT3 Radon 222
NT3 Selenium 75
NT3 Zinc 65
NT3 Zirconium 95
NT2 Delayed Neutron Precursors
NT2 Delayed Proton Precursors
NT2 Element 104 Isotopes
NT2 Element 105 Isotopes
NT2 Element 106 Isotopes
NT2 Element 107 Isotopes
NT2 Element 108 Isotopes
NT2 Element 109 Isotopes
NT2 Heavy Ion Decay Radioisotopes
NT2 Hours Living Radioisotopes
 NT3 Cesium 134
 NT3 Cobalt 58
NT2 Internal Conversion Radioisotopes
 NT3 Cesium 134
 NT3 Cobalt 60
 NT3 Cobalt 58
NT2 Isomeric Transition Isotopes
 NT3 Cesium 134
 NT3 Cobalt 60
 NT3 Cobalt 58
NT2 Microsec Living Radioisotopes
NT2 Millisec Living Radioisotopes
NT2 Minutes Living Radioisotopes
 NT3 Cobalt 60
 NT3 Thorium 232
NT2 Nanosec Living Radioisotopes
NT2 Neutron-Deficient Isotopes
NT2 Neutron-Rich Isotopes
NT2 Proton Decay Radioisotopes
NT2 Radon Progeny
 NT3 Radon 222
NT2 Seconds Living Radioisotopes
NT2 Transuranic Radionuclides
NT2 Years Living Radioisotopes
 NT3 Antimony 125
 NT3 Cesium 134
 NT3 Cobalt 60
 NT3 Radium 226
 NT3 Radium 228
 NT3 Ruthenium 103
 NT3 Strontium 90
 NT3 Thorium 230
 NT3 Tritium
 NT3 Uranium 238
RT Chemical Substances
RT Extremely Hazardous Substances
RT Regulated Substances
RT Reportable Quantities
RT Toxic Substances
DEF (See hazardous substances)

CERIUM 141
(EDB)
DA March 29, 1991
BT1 Beta-Minus Decay Radioisotopes
 BT2 Beta Decay Radioisotopes
  BT3 Radionuclides
   BT4 CERCLA Hazardous
    Substances
    BT5 Hazardous Substances
   BT4 Nuclides
BT1 Days Living Radioisotopes
 BT2 Radionuclides
  BT3 CERCLA Hazardous Substances
   BT4 Hazardous Substances
  BT3 Nuclides
SO Radiation

CERIUM 144
(EDB)
DA March 29, 1991
BT1 Beta-Minus Decay Radioisotopes
 BT2 Beta Decay Radioisotopes
  BT3 Radionuclides
   BT4 CERCLA Hazardous
    Substances
    BT5 Hazardous Substances
   BT4 Nuclides
BT1 Days Living Radioisotopes
 BT2 Radionuclides
  BT3 CERCLA Hazardous Substances
   BT4 Hazardous Substances
  BT3 Nuclides
SO Radiation

CERTIFICATION
(FIFRA; SWDA; RCRA; CFR; ESH)
DA October 12, 1990
RT Certifying Officials
RT Pesticides
SO Environmental Management
SO Hazardous Materials
SO Wastes
DEF (ESH) The recognition by a
    certifying agency that a person is
    competent and thus authorized to
    use or supervise the use of
    restricted use pesticides.

CERTIFIED APPLICATORS
(FIFRA; CFR; USC)
DA November 15, 1990
BT1 Technically Qualified Individuals
 BT2 Personnel
RT Commercial Applicators
RT Private Applicators
SO Environmental Management
DEF (USC) Individuals who are certified
    under section 4 [7 USCS 136b] as
    authorized to use or supervise the
    use of any pesticide which is
    classified for restricted use. Any
    applicator who holds or applies
    registered pesticides, or uses
    dilutions of registered pesticides
    consistent with section 2(ee) of
    this Act, only to provide a service
    of controlling pests without
    delivering any unapplied pesticide
    to any person so served is not
    deemed to be a seller or

SY-Synonymous Terms   SO-Source/Subject Category   SF-See From

distributor of pesticides under this
Act [7 USCS 136 et seq.].

CERTIFIED WASTES
(DOE Order 5820.2A)
DA January 24, 1991
BT1 Wastes
DEF (DOE 5828.2A) Waste that has
    been confirmed to comply with
    disposal site waste acceptance
    criteria (e.g., the Waste Isolation
    Pilot Plant-Waste Acceptance
    Criteria for transuranic waste)
    under an approved certification
    program.

CERTIFYING OFFICIALS
(DOE Order 1540.2 6.J)
DA January 24, 1991
BT1 Personnel
RT Certification
DEF (DOE 1540.2) Certifying official is
    the designated Headquarters
    official responsible for
    administering the DOE program
    for the design review of DOE
    packagings and issuance of a
    certificate of compliance upon
    approval.

CESIUM (CS)
(EPA)
DA October 12, 1990
BT1 Heavy Metals
SO Environmental Protection Agency
    Glossary
DEF (EPA) A silver-white, soft, ductile
    element of the alkali metal group
    that is the most electropositive
    element known. Used especially in
    photoelectric cells.

CESIUM 134
(EDB)
DA March 29, 1991
BT1 Beta-Minus Decay Radioisotopes
 BT2 Beta Decay Radioisotopes
  BT3 Radionuclides
   BT4 CERCLA Hazardous
       Substances
    BT5 Hazardous Substances
   BT4 Nuclides
BT1 Hours Living Radioisotopes
 BT2 Radionuclides
  BT3 CERCLA Hazardous Substances
   BT4 Hazardous Substances
  BT3 Nuclides
BT1 Internal Conversion Radioisotopes
 BT2 Radionuclides
  BT3 CERCLA Hazardous Substances
   BT4 Hazardous Substances
  BT3 Nuclides
BT1 Isomeric Transition Isotopes
 BT2 Radionuclides
  BT3 CERCLA Hazardous Substances
   BT4 Hazardous Substances
  BT3 Nuclides
BT1 Years Living Radioisotopes
 BT2 Radionuclides

  BT3 CERCLA Hazardous Substances
   BT4 Hazardous Substances
  BT3 Nuclides
SO Radiation

CESIUM 137
(EDB)
DA March 29, 1991
BT1 Beta-Minus Decay Radioisotopes
 BT2 Beta Decay Radioisotopes
  BT3 Radionuclides
   BT4 CERCLA Hazardous
       Substances
    BT5 Hazardous Substances
   BT4 Nuclides
BT1 Days Living Radioisotopes
 BT2 Radionuclides
  BT3 CERCLA Hazardous Substances
   BT4 Hazardous Substances
  BT3 Nuclides
SO Radiation

CF
DA October 12, 1990
SEE Containment Failure
SO Acronyms

CFEE
DA May 20, 1991
SEE Conference of Federal
      Environmental Engineers
SO Acronyms

CFM
DA October 12, 1990
SEE Cubic Feet Per Minute
SO Acronyms

CFR
DA October 12, 1990
SEE Code of Federal Regulations
SO Acronyms

CFS
DA October 12, 1990
SEE Core Flood System
SO Acronyms

CFT
DA October 12, 1990
SEE Core Flood Tanks
SO Acronyms

CFTs
DA October 12, 1990
SEE Californium Targets
SO Acronyms

CF$_4$
DA October 12, 1990
SEE Carbon Tetrafluoride
SO Acronyms

CH (Chicago Operations Office)
DA October 12, 1990
SEE Chicago Operations Office
SO Acronyms

CH (Component Handling)
DA October 12, 1990
SEE Component Handling
SO Acronyms

CH (contact handled)
DA October 12, 1990
SEE Contact Handled
SO Acronyms

CHAIN OF COMMAND
(EMER)
DA February 1, 1991
RT Management
SO Emergency Preparedness
DEF (EMER) A hierarchy of command
    structures utilized in emergency
    operations centers for the purpose
    of defining a structure for the
    resolution of emergencies.

CHAIN SAW
(2951 CHAIN SAW)
DA January 3, 1991
BT1 Saws
BT1 Tools - Powered
 BT2 Tools (DOE FRASE Vocabulary)
  BT3 Equipment
SO DOE FRASE VOCABULARY

CHAMBER
(1683 CHAMBER)
DA December 10, 1990
BT1 Site (DOE FRASE Vocabulary)
 BT2 Sites/Areas
RT Room
SO DOE FRASE VOCABULARY

CHANGE
(SSDC)
DA October 12, 1990
NT1 Temporary Procedure Change
RT Change Analyses
SO System Safety Development Center
    Glossary
DEF (SSDC) In MORT, change can be
    thought of as stress on a system
    which was previously in a state of
    dynamic equilibrium, or anything
    which disturbs the planned or
    normal functioning of a system.

CHANGE ANALYSES
(SSDC)
DA October 12, 1990
BT1 Analyses
RT Change
SO System Safety Development Center
    Glossary
DEF (SSDC) Methods of accident
    investigation wherein accident-free
    reference bases are established,
    and then changes and differences
    relative to accident cases and
    situations are systematically
    searched out. In change analysis,
    all changes are considered
    including those considered to be
    trivial or obscure.

BT-Broader Term   NT-Narrower Term   RT Related Term

CHANGED USE PATTERN
(FIFRA; CFR)
DA   January 24, 1991
RT   Pesticides
RT   Polychlorinated Biphenyls
RT   Registration
SO   Environmental Management
DEF (CFR) The term "changed use
pattern" means a significant
change from a use pattern
approved in connection with the
registration of a pesticide product.
Examples of significant changes
include, but are not limited to,
changes from nonfood to food
use, outdoor to indoor use, ground
to aerial application, terrestrial to
aquatic use, and nondomestic to
domestic use.

CHANNEL EFFLUENT LIMIT
DA   January 8, 1991
SF   CEL
BT1 Limits

CHANNELIZATION
(EPA)
DA   October 12, 1990
BT1 Processes
RT   Dredging
RT   Marshes
SO   Environmental Protection Agency
      Glossary
DEF (EPA) Straightening and deepening
streams so water will move faster.
A flood-reduction or
marsh-drainage tactic that can
interfere with waste assimilation
capacity and disturb fish and
wildlife habitats.

CHARGE AND DISCHARGE
DA   January 8, 1991
SF   C&D

CHARGED PARTICLE EQUILIBRIUM
(IAEA)
DA   October 12, 1990
RT   Alpha Particles
RT   Beta Particles
SO   Radiation
DEF (IAEA) The condition existing at a
point within a medium under
irradiation, when, for every
charged particle leaving a volume
element surrounding the point,
another particle of the same kind
and energy enters.

CHARGED PARTICLES
(EDB)
DA   January 29, 1991
NT1 Alpha Particles
NT1 Beta Particles
NT1 Directly Ionizing Particles
NT1 Electrons
NT1 Ions
NT1 Protons

RT   Hazards
DEF (DSTT) A particle whose charge is
not zero; the charge of a particle
is added to its designation as a
superscript, with particles of
charge +1 and -1 (in terms of the
charge of the proton) denoted by
+ and - respectively.

CHARGING
(CFR)
DA   October 12, 1990
BT1 Processes
RT   Converter Arsenic Charging Rate
RT   Copper Converters
SO   Air Pollution
DEF (CFR) The addition of a molten or
solid material to a copper
converter.

CHARLIE
DA   February 26, 1991
SY   Safeguards and Security Alert III
      (Code Designator: Charlie)
BT1 Safeguards and Security Alerts
 BT2 Safeguards and Security
      Emergencies
  BT3 Emergencies
   BT4 Reportable Occurrences
    BT5 Occurrences
SO   Emergency Preparedness
DEF (EMER) A preparatory alert action
which shall be effected when the
head of a field office determines
that existing preemergency
conditions warrant increased
safeguards and security measures
at facilities under his/her
jurisdiction; however, the Office of
the Assistant Secretary for
Defense Programs may establish
a general U.S. Department of
Energy-wide or a locally confined
Safeguards and Security Alert III
without prior consultation with the
head of field offices.

CHARTER OPERATIONS
(DOE Order 5480.13)
DA   October 12, 1990
BT1 Aviation Operations
 BT2 Operations
  BT3 Activities
SO   Aviation Safety
DEF (DOE Order 5480.13) The carrying
in air commerce of any persons or
property for compensation or hire,
and the use of special mission
aircraft.

CHECK SAMPLES
(ESH)
DA   October 12, 1990
BT1 Samples
RT   Sampling
SO   Quality Assurance
DEF (ESH) Artificial samples that have
been spiked with the analyte(s)
from an independent source in

order to monitor the execution of
the analytical method are called
check samples. The level of the
spike shall be at the regulatory
action level when applicable.
Otherwise, the spike shall be at 5
times the estimate of the
quantification limit. The matrix
used shall be phase matched with
the samples and well
characterized: for an example,
reagent grade water is appropriate
for an aqueous sample.

CHECK VALVES
(NFI)
DA   October 12, 1990
BT1 Valves
 BT2 Devices
SO   Nuclear Facilities Incident Database
SO   Radiation
DEF (NFI) These prevent moderator
backflow from the reactor, e.g., to
SSS ink supply.

CHEMICAL AGENTS
(CFR)
DA   October 12, 1990
NT1 Attractants
NT1 Desiccants
NT1 Dispersants
NT1 Soft Detergents
NT1 Surface Collecting Agents
RT   Surfactants
SO   Compensation and Liability
DEF (CFR) In general, are those
elements, compounds, or mixtures
that coagulate, disperse, dissolve,
emulsify, foam, neutralize,
precipitate, reduce, solubilize,
oxidize, concentrate, congeal,
entrap, fix, make the pollutant
mass more rigid or viscous, or
otherwise facilitate the mitigation
of deleterious effects or removal of
the pollutant from the water.

CHEMICAL AMMUNITION
(HMTA; CFR)
DA   May 20, 1991
BT1 Ammunition
 BT2 Munitions
NT1 Smoke Grenades
SO   Hazardous Materials
DEF (CFR) Chemical ammunition used
in warfare is all shells, bombs,
grenades, etc., loaded with toxic,
tear, or other gas, smoke or
incendiary agent, also such
miscellaneous apparatus as
cloud-gas cylinders, smoke
generators, etc., that may be
utilized to project chemicals.

CHEMICAL BURN(S)
(1310 CHEMICAL BUR)
DA   November 28, 1990
BT1 Burn(s)

BT2 Injuries
SO DOE FRASE VOCABULARY

CHEMICAL CONTAMINATION
(1525 CHEMICAL CON)
DA November 28, 1990
BT1 Contamination
RT Environmental Release
RT Hazardous Spill
SO DOE FRASE VOCABULARY

CHEMICAL HAZARDS EMERGENCY MANAGEMENT SYSTEM
(SSDC)
DA January 8, 1991
SF *CHEMS*
BT1 Safety Performance Measurement System
 BT2 Information Systems
  BT3 Security Interests
  BT3 Systems
RT Hazardous Chemicals
SO Environmental Management
SO Management
DEF (SSDC) This system was developed for DOE by the Center for Assessment of Chemical and Physical Hazards at Brookhaven National Laboratory. The purpose of the database is to provide health, safety and environmental information to DOE and its contractors for chemicals and materials of interest to them.

CHEMICAL OXYGEN DEMAND
(EPA)
DA October 12, 1990
SF *COD (Chemical Oxygen Demand)*
BT1 Measurements
RT Biochemical Oxygen Demand
SO Environmental Protection Agency Glossary
DEF (EPA) A measure of the oxygen required to oxidize all compounds in water, both organic and inorganic.

CHEMICAL PROCESSES
DA February 26, 1991
BT1 Processes
NT1 Absorption (Chemical)
NT1 Corrosion
 NT2 External Corrosion
NT1 Dechlorination
NT1 Degradation
NT1 Desalinization
NT1 Fermentation
NT1 Hydrolysis
NT1 In Situ Volatization
NT1 In-Situ Extraction
NT1 Neutralization
NT1 Nitrification
NT1 Oil Fingerprinting
NT1 Oxidation
 NT2 Combustion
NT1 Precipitation
NT1 Process (Toxic Chemical)
 NT2 Process for Commercial Purposes

NT1 Pyrolysis
NT1 Reduction
NT1 Sorption

CHEMICAL REACTION
(1377 CHEMICAL REA)
DA November 28, 1990
BT1 Illnesses
RT Insulin Reaction
SO DOE FRASE VOCABULARY

CHEMICAL SUBSTANCES
(TSCA; CFR; USC; ESH)
DA October 12, 1990
NT1 Additives
 NT2 Biological Additives
 NT2 Burning Agents
 NT2 Sinking Agents
NT1 Adulterants
NT1 Chemicals of Potential Concern
NT1 Coproducts
NT1 Impurities
NT1 Inorganic Chemicals
 NT2 Carbon Dioxide
 NT2 Carbon Monoxide
 NT2 Hydrogen Sulfide
 NT2 Nitrogen Oxides (NO_x)
  NT3 Nitric Oxides
  NT3 Nitrogen Dioxide (NO_2)
NT1 Intermediates
 NT2 Nitrites
 NT2 Non-Isolated Intermediates
 NT2 Site-Limited Intermediates
NT1 New Chemical Substances
NT1 Non-detects
NT1 Organic Chemicals
 NT2 Acetylcholine
 NT2 Alar
 NT2 Dioxins
 NT2 DNA
  NT3 Plasmids
  NT3 Recombinant DNA
 NT2 Formaldehyde
 NT2 Halogenated Organic Compounds
  NT3 Chlorinated Hydrocarbons
   NT4 Chlorofluorocarbons (CFCs)
   NT4 DDT
   NT4 Heptachlor
   NT4 Polychlorinated Biphenyls
    NT5 High Concentration PCBs
    NT5 Low Concentration PCBs
    NT5 Recycled PCBs
   NT4 Polyvinyl Chloride
   NT4 Trichloroethylene (TCE)
   NT4 Vinyl Chloride
  NT3 Fluorocarbons
   NT4 Chlorofluorocarbons (CFCs)
  NT3 Trihalomethane
 NT2 High-Density Polyethylene
 NT2 Hydrocarbons
  NT3 Methane
 NT2 Nitrilotriacetic Acid
 NT2 Organic Peroxides
 NT2 Phenols
 NT2 Proteins
  NT3 Antibodies
   NT4 Monoclonal Antibodies
  NT3 Carboxyhemoglobin
  NT3 Restriction Enzymes
 NT2 Ribonucleic Acid

 NT2 Synthetic Organic Chemicals
  NT3 Polyelectrolytes
  NT3 Volatile Synthetic Organic Chemicals
 NT2 Volatile Organic Compounds
  NT3 Volatile Synthetic Organic Chemicals
RT Analytes
RT Antagonism
RT CERCLA Hazardous Substances
RT Listed Hazardous Substances
RT Retailers
RT Small Manufacturers
RT Testing Requirements Rule
RT Testing Requirements
SO Environmental Management
SO Hazardous Materials
DEF (CFR) Organic or inorganic substances of a particular molecular identity, including any combination of such substances occurring in whole or part as a result of a chemical reaction or occurring in nature, and any element or uncombined radical. (2) This term does not include any mixture; pesticide (as defined in FIFRA) when manufactured, processed, or distributed in commerce for use as a pesticide; tobacco or any tobacco product; source material, special nuclear material, or byproduct material; article the sale of which is subject to the tax imposed by §4181 of the Internal Revenue Code of 1954; and any food, food additive, drug, cosmetic, or device.

CHEMICAL TREATMENTS
(EPA)
DA October 12, 1990
BT1 Treatment
 BT2 Waste Management Processes
  BT3 Processes
RT Neutralization
SO Environmental Protection Agency Glossary
DEF (EPA) Any one of a variety of technologies that use chemicals or a variety of chemical processes to treat waste.

CHEMICAL VOLUME AND CONTROL SYSTEM
DA January 8, 1991
SF *CVCS*
BT1 Control Systems
 BT2 Controls
 BT2 Systems

CHEMICAL WASTE LANDFILLS
(CFR; ESH)
DA October 12, 1990
BT1 Landfills
 BT2 Land Disposal Units
  BT3 Disposal Units
   BT4 Corrective Action Management Units
    BT5 Sites/Areas

SO Hazardous Materials
DEF (ESH) Landfills at which protection against risk of injury to health or the environment from migration of PCBs to land, water, or the atmosphere is provided from PCBs and PCB items deposited therein by locating, engineering, and operating the landfill as specified in 40 CFR 761.75.

CHEMICALS OF POTENTIAL CONCERN
(EPA)
DA October 12, 1990
BT1 Chemical Substances
RT Hazardous Chemicals
SO Environmental Protection Agency Glossary
DEF (EPA) Chemicals that are potentially site-related and whose data are of sufficient quality for use in the quantitative risk assessment.

CHEMOSTERILANTS
(EPA)
DA October 12, 1990
RT Sterilization
SO Environmental Protection Agency Glossary
DEF (EPA) Chemicals that control pests by preventing reproduction.

CHEMS
DA October 12, 1990
SEE Chemical Hazards Emergency Management System
SO Acronyms
SO Environmental Management

CHEST
(1109 CHEST)
DA November 28, 1990
BT1 Trunk
BT2 Human Body Parts
RT Breast(s)
RT Respiratory System
RT Rib(s)
SO DOE FRASE VOCABULARY

CHICAGO OPERATIONS OFFICE
DA January 8, 1991
SF CH (Chicago Operations Office)
BT1 Operations Offices
BT2 Offices
BT3 Administrative Organizations
BT4 Organizations
BT2 U.S. Department of Energy
BT3 Federal Agencies
BT4 Agencies
BT5 Administrative Organizations
BT6 Organizations
RT Argonne National Laboratory
RT Associated Universities, Inc.
RT Iowa State University
RT Midwest Research Institute
RT Trustees of Princeton University
RT University of Chicago

RT Universities Research Association, Inc.
DEF (Capsule Review of DOE Research and Development and Field Facilities, 1986) The Chicago Operations Office, an offspring of the Manhattan Engineer District, was initially established in 1946 as one of the Atomic Energy Commission's first field offices. In 1977, as part of the newly established DOE, the Office focused its efforts on long-term, high-risk research and development in conjunction with business and industry as well as the academic community. Major missions include: institutional management of major government-owned, contractor-operated laboratories and facilities; scientific and technical management of programs and projects including nuclear waste management, magnetic fusion and coal gasification; and business and technical management of contracts and grants, including fiscal and construction management as well responsibilities for safety, environmental protection and quality assurance.

CHILLING EFFECT
(EPA)
DA October 12, 1990
BT1 Effects
RT Nuclear Winter
SO Environmental Protection Agency Glossary
DEF (EPA) The lowering of the Earth's temperature because of increased particles in the air blocking the sun's rays.

CHIN
(1091 CHIN)
DA November 28, 1990
BT1 Head
BT2 Human Body Parts
RT Jaw
SO DOE FRASE VOCABULARY

CHIN STRAP
(2654 CHIN STRAP)
DA January 3, 1991
BT1 Personal Protective Equipment
BT2 Equipment/Parts - Personal Protective (DOE FRASE Vocabulary)
BT3 Equipment
SO DOE FRASE VOCABULARY

CHISEL(S)
(3029 CHISEL)
DA January 3, 1991
BT1 Tools - Manual
BT2 Tools (DOE FRASE Vocabulary)

BT3 Equipment
SO DOE FRASE VOCABULARY

CHLORINATED HYDROCARBONS
(EPA)
DA October 12, 1990
BT1 CERCLA Hazardous Substances
BT2 Hazardous Substances
BT1 Halogenated Organic Compounds
BT2 Halogenated
BT2 Organic Chemicals
BT3 Chemical Substances
NT1 Chlorofluorocarbons (CFCs)
NT1 DDT
NT1 Heptachlor
NT1 Polychlorinated Biphenyls
NT2 High Concentration PCBs
NT2 Low Concentration PCBs
NT2 Recycled PCBs
NT1 Polyvinyl Chloride
NT1 Trichloroethylene (TCE)
NT1 Vinyl Chloride
SO Environmental Protection Agency Glossary
DEF (EPA) These include a class of persistent, broad-spectrum insecticides that linger in the environment and accumulate in the food chain. Among them are DDT, aldrin, dieldrin, heptachlor, chlordane, lindane, endrin, mirex, hexachloride, and toxaphene. Another example is TCE, used as an industrial solvent.

CHLORINATED SOLVENTS
(EPA)
DA October 12, 1990
BT1 Solvents
SO Environmental Protection Agency Glossary
DEF (EPA) Organic solvents containing chlorine atoms, e.g., methylene chloride and 1,1,1-trichloromethane, which are used in aerosol spray containers and in traffic paint.

CHLORINATION
(EPA)
DA October 12, 1990
BT1 Processes
RT Chlorinators
RT Chlorine-Contact Chambers
RT Dechlorination
RT Disinfection
SO Environmental Protection Agency Glossary
DEF (EPA) The application of chlorine to drinking water, sewage, or industrial waste to disinfect or to oxidize undesirable compounds.

CHLORINATORS
(EPA)
DA October 12, 1990
RT Chlorination

SO Environmental Protection Agency
Glossary
DEF (EPA) Devices that add chlorine, in
gas or liquid form, to water or
sewage to kill infectious bacteria.

CHLORINE
DA January 8, 1991
SF Cl

CHLORINE-CONTACT CHAMBERS
(EPA)
DA October 12, 1990
RT Chlorination
RT Water Supply System
SO Environmental Protection Agency
Glossary
DEF (EPA) Those parts of a water
treatment plant where effluent is
disinfected by chlorine.

CHLORINE TRIFLUORIDE
DA January 8, 1991
SF ClF_3

CHLOROFLUOROCARBONS (CFCS)
(EPA)
DA October 12, 1990
BT1 Chlorinated Hydrocarbons
BT2 CERCLA Hazardous Substances
BT3 Hazardous Substances
BT2 Halogenated Organic Compounds
BT3 Halogenated
BT3 Organic Chemicals
BT4 Chemical Substances
BT1 Fluorocarbons
BT2 Halogenated Organic Compounds
BT3 Halogenated
BT3 Organic Chemicals
BT4 Chemical Substances
RT Hazardous Constituents
SO Environmental Protection Agency
Glossary
DEF (EPA) A family of inert, nontoxic,
and easily liquefied chemicals
used in refrigeration, air
conditioning, packaging, insulation,
or as solvents and aerosol
propellants. Because CFCs are
not destroyed in the lower
atmosphere, they drift into the
upper atmosphere where their
chlorine components destroy
ozone.

CHLOROSIS
(EPA)
DA October 12, 1990
RT Plants
SO Environmental Protection Agency
Glossary
DEF (EPA) Discoloration of normally
green plant parts that can be
caused by disease, lack of
nutrients, or various air pollutants.

CHR
DA October 12, 1990

SEE Confinement Heat Removal
SO Acronyms

CHROMIUM
(EPA)
DA October 12, 1990
SF Cr
BT1 CERCLA Hazardous Substances
BT2 Hazardous Substances
BT1 Heavy Metals
SO Environmental Protection Agency
Glossary
DEF (EPA) A heavy metal.

CHROMIUM-HEXAVALENT
DA January 8, 1991
SF Cr^{+6}

CHRONIC EFFECTS
(EMER)
DA February 1, 1991
BT1 Effects
RT Exposure
RT Hazardous Materials
SO Emergency Preparedness
DEF (EMER) Effects of exposure to a
hazardous material which develop
slowly after many exposures or
which recur often.

CHRONIC EXPOSURE
(EMER)
DA February 1, 1991
BT1 Exposure
SO Emergency Preparedness
DEF (EMER) Repeated exposure or
contact with a toxic substance
over a long period of time.

CHRONIC RFD
(EPA)
DA October 12, 1990
BT1 Reference Dose (RfD)
BT2 Doses
RT Subchronic RfD (RfD)
SO Environmental Protection Agency
Glossary
DEF (EPA) An estimate (with uncertainty
spanning perhaps an order of
magnitude or greater) of a lifetime
daily exposure level for the human
population, including sensitive
subpopulations, that is likely to be
without an appreciable risk of
deleterious effects. A chronic RfD
is specifically developed to be
protective for long-term exposure
to a compound (7 years to
lifetime).

CHRONIC TOXICITY
(FIFRA; CFR; EPA)
DA October 12, 1990
BT1 Toxicity
SO Emergency Preparedness
SO Environmental Management
SO Environmental Protection Agency
Glossary
DEF (EPA) The capacity of a substance

to cause long-term poisonous
human health effects.

CIC
DA October 12, 1990
SEE Compensated Ion Chamber
SO Acronyms

CIF
DA October 12, 1990
SEE Consolidated Incinerator Facility
SO Acronyms

CIL
DA October 12, 1990
SEE Critical Items List
SO Acronyms

CIL (Computer)
DA October 12, 1990
SEE Computer Inoperative Limits
SO Acronyms

CILRT
DA October 12, 1990
SEE Containment Integrated Leak Rate
Test
SO Acronyms

CIP
DA October 12, 1990
SEE Capital Improvement Project
SO Acronyms

CIRCUIT(S)
(2382 CIRCUIT)
DA January 3, 1991
BT1 Equipment/Parts - Electrical (DOE
FRASE Vocabulary)
BT2 Equipment
SO DOE FRASE VOCABULARY

CIRCUIT BREAKER(S)
(2381 CIRCIUT BREA)
DA January 3, 1991
BT1 Equipment/Parts - Electrical (DOE
FRASE Vocabulary)
BT2 Equipment
SO DOE FRASE VOCABULARY

CIRCUIT TO PRESSURE CONVERTER
DA January 8, 1991
SF I/P

CIRCULATORY SYSTEM
(1141 CIRCULATORY)
DA November 28, 1990
BT1 Body System(s)
BT2 Human Body Parts
NT1 Blood
SO DOE FRASE VOCABULARY

CIT
DA October 12, 1990
SEE Critical Incident Technique
SO Acronyms

CIVIL AIRCRAFT
(DOE Order 5480.13)
DA   October 12, 1990
BT1 Aircraft
SO   Aviation Safety
DEF (DOE Order 5480.13) All aircraft
     other than public aircraft.

CIVILIAN NUCLEAR ACTIVITIES
(ANL; NWPA)
DA   May 22, 1991
BT1 Activities
RT   Atomic Energy Defense Activities
SO   Management
DEF (NWPA) Any atomic energy
     activities other than atomic energy
     defense activities.

CI
DA   October 12, 1990
SEE Chlorine
SO   Acronyms

CLADDING
(2552 CLADDING)
DA   January 3, 1991
BT1 Fuel Plate
 BT2 Fuel Element(s)
  BT3 Equipment/Parts - Nuclear (DOE
      FRASE Vocabulary)
  BT4 Equipment
  BT4 Reactor Components
SO   DOE FRASE VOCABULARY
DEF (NRC Glossary of Terms: Nuclear
     Power and Radiation) The
     thin-walled metal tube that forms
     the outer jacket of a nuclear fuel
     rod. It prevents corrosion of the
     fuel by the coolant and the release
     of fission products into the coolant.
     Aluminum, stainless steel and
     zirconium alloys are common
     cladding materials.

CLAIMANT
(CERCLA; CFR; USC)
DA   October 12, 1990
RT   Claims
SO   Compensation and Liability
DEF A person submitting a claim of
     trade secrecy to EPA in connection
     with a chemical otherwise required
     to be disclosed in a report or other
     filing made under Title III of the
     Superfund Amendments and
     Reauthorization Act of 1986.

CLAIMS
(CERCLA; CFR; USC)
DA   October 12, 1990
RT   Claimant
SO   Compensation and Liability
SO   Environmental Management
DEF As defined by section 101(4) of
     CERCLA, demands in writing for a
     sum certain.

CLARIFICATION
(EPA)

DA   October 12, 1990
BT1 Wastewater Treatment Processes
 BT2 Treatment
  BT3 Waste Management Processes
  BT4 Processes
RT   Clarifiers
RT   Sedimentation
SO   Environmental Protection Agency
     Glossary
DEF (EPA) Clearing action that occurs
     during wastewater treatment when
     solids settle out. This is often
     aided by centrifugal action and
     chemically induced coagulation in
     wastewater.

CLARIFIERS
(EPA)
DA   October 12, 1990
BT1 Tanks
 BT2 Facility Components
RT   Clarification
SO   Environmental Protection Agency
     Glossary
DEF (EPA) A tank in which solids are
     settled to the bottom and are
     subsequently removed as sludge.

CLASS 1E ELECTRICAL EQUIPMENT
DA   January 8, 1991
SF   1E (Electrical Equipment)
BT1 Equipment

CLASS A EQUIPMENT
(DOE Order 5000.3A)
DA   December 14, 1990
BT1 Equipment
RT   Barriers
RT   Primary Environmental Monitors
RT   Safety Device/System
DEF (DOE Order 5000.3A) Any active or
     passive safety device/system or
     any primary environmental
     monitors.

CLASS A EXPLOSIVES
(CFR)
DA   October 12, 1990
BT1 Explosives
SO   Hazardous Materials
DEF (CFR) Class A Explosives are
     detonating or otherwise a
     maximum hazard. Black powder,
     explosive boosters, blasting caps,
     and explosive bombs are
     examples of Explosives Class A.

CLASS B EQUIPMENT
(DOE Order 5000.3A)
DA   December 14, 1990
BT1 Equipment
RT   Safety Device/System
RT   Secondary Environmental Monitors
DEF (DOE Order 5400.3A) Any
     device/system which, although not
     primarily safety related, will result
     in facility shutdown or degradation
     of operating parameters (i.e. less
     throughput, longer processing

times, shorter worker endurance
periods, etc.), or any secondary
environmental monitors.

CLASS B EXPLOSIVES
(CFR)
DA   October 12, 1990
BT1 Explosives
NT1 Starter Cartridges
SO   Hazardous Materials
DEF (CFR) Class B Explosives are
     those explosives which function by
     rapid combustion, rather than
     detonation. Special fireworks, jet
     thrust units, rocket ammunition
     with solid projectile are examples
     of Class B Explosives.

CLASS C EXPLOSIVES
(CFR)
DA   October 12, 1990
BT1 Explosives
RT   Small Arms Ammunition
SO   Hazardous Materials
DEF (CFR) Class C Explosives include
     certain types of manufactured
     articles which contain Class A or
     Class B explosives, or both, as
     components, but in restricted
     quantities. This Class of explosive
     also includes certain types of
     fireworks specifically identified as
     Class C Explosives. Explosive
     release devices, fuse igniters, time
     fuses, and igniter cords are
     examples of Class C Explosives.

CLASS C WASTES
(AEA; CFR)
DA   January 24, 1991
BT1 Wastes
DEF (CFR) Waste that will not decay to
     levels which present an
     unacceptable hazard to an
     intruder within 100 years.

CLASS II WELLS
(SDWA; CFR)
DA   October 12, 1990
BT1 Injection Wells
 BT2 Wells
NT1 Existing Class II Wells
NT1 New Class II Wells
SO   Water Pollution
DEF Wells that inject fluids (a) that are
     brought to the surface in
     connection with conventional oil or
     natural gas production and may be
     commingled with waste waters
     from gas plants that are an
     integral part of production
     operations, unless those waters
     would be classified as a
     hazardous waste at the time of
     injection; (b) for enhanced
     recovery of oil or natural gas; and
     (c) for storage of hydrocarbons
     that are liquid at standard
     temperature and pressure.

CLASSIFICATION
(EMER)
DA February 1, 1991
BT1 Administrative Processes
 BT2 Processes
RT Classified Activity
RT Classified Information
RT Classified Interests
RT Classified Matter
RT Confidential
SO Emergency Preparedness
DEF (EMER) A means of identifying
 information concerning the
 national defense and foreign
 relations of the United States that
 requires protection against
 unauthorized disclosure.

CLASSIFIED ACTIVITY
(1229 CLASSIFIED A)
DA November 28, 1990
BT1 Activity Types (DOE FRASE
 Vocabulary)
 BT2 Activities
RT Classification
SO DOE FRASE VOCABULARY

CLASSIFIED INFORMATION
(DOE Orders 4330.4A and 6430.1A)
DA October 12, 1990
BT1 Classified Matter
 BT2 Security Interests
NT1 National Security Information
NT1 Sensitive Compartmented
 Information
NT1 Sensitive Nuclear Material
 Production Information
RT Classification
RT Classified Telecommunications
 Facilities
RT Confidential
RT Couriers
RT Crypto
RT National Security
RT Protect as Restricted Data
SO Construction
DEF (DOE Order 6430.1A) Top Secret,
 Secret, and Confidential Restricted
 Data, Formerly Restricted Data,
 and National Security Information,
 for which the Department of
 Energy is responsible and that
 requires safeguarding in the
 interest of national security and
 defense.

CLASSIFIED INTERESTS
(DOE Order 6430.1A)
DA October 12, 1990
BT1 Security Interests
RT Classification
RT National Security
SO Construction
DEF (DOE Order 6430.1A) Classified
 documents, information, or
 material including classified
 special nuclear material
 possessed by the Department, a
 contractor of the Department, a
 Departmental facility, or any other

facility under the Department's
jurisdiction.

CLASSIFIED MATTER
(DOE Order 6430.1A; EMER)
DA October 12, 1990
BT1 Security Interests
NT1 Classified Information
 NT2 National Security Information
 NT2 Sensitive Compartmented
 Information
 NT2 Sensitive Nuclear Material
 Production Information
RT Classification
RT Couriers
RT Exclusion Areas
SO Construction
SO Emergency Preparedness
SO Environmental Management
DEF (DOE Order 6430.1A) Classified
 information, documents, parts
 components, or other material.

CLASSIFIED TELECOMMUNICATIONS FACILITIES
(DOE Order 6430.1A)
DA October 12, 1990
BT1 Facilities
RT Classified Information
SO Construction
DEF (DOE Order 6430.1A) Facilities that
 contains both crypto equipment
 and input/output equipment for
 the electronic transmission,
 receipt, or processing of classified
 information. The crypto equipment
 and input/output equipment may
 either be installed in the same
 area and share common security
 measures or be installed in
 different parts of the same security
 area connected by a protected
 distribution system, with each area
 having its own security measures.

CLASSIFIED WASTES
(RCRA; CFR)
DA January 24, 1991
BT1 Wastes
DEF (CFR) "Classified waste" means
 waste material that has been
 given security classification in
 accordance with 50 U.S.C. 401
 and Executive Order 11652.

CLDTMP
(NFI)
DA October 12, 1990
BT1 Computer Codes
SO Nuclear Facilities Incident Database
DEF (NFI) A computer code to calculate
 cladding temperature limit.

CLEAN AIR ACT
DA January 8, 1991
SF CAA
BT1 Acts
 BT2 Statutes and Regulations
DEF (CAA) To protect and enhance the

quality of the Nation's air
resources so as to promote the
public health and welfare and the
productive capacity of its
population; to initiate and
accelerate a national research and
development program to achieve
the prevention and control of air
pollution; to provide technical and
financial assistance to state and
local governments in connection
with the development and
execution of their air pollution
prevention and control programs;
and to encourage and assist the
development and operation of
regional air pollution control
programs.

CLEAN ROOMS
(TSCA; CFR)
DA January 24, 1991
BT1 Room
 BT2 Sites/Areas
DEF (CFR) "Clean room" means an
 uncontaminated room having
 facilities for the storage of
 employees; street clothing and
 uncontaminated materials and
 equipment.

CLEAN WATER ACT
(CWA)
DA October 19, 1990
SF CWA
BT1 Acts
 BT2 Statutes and Regulations
BT1 Appropriate Act and Regulations
SO Wastes
SO Water Pollution
DEF Pub. L. 95-217 and as amended;
 33 USC 1251 et seq.

CLEANUP
(EPA)
DA October 12, 1990
SY Remedial Actions
SY Removal Actions
SY Response Actions
RT Decontamination
RT Double Wash/Rinse
RT Removal
RT Sinking Agents
RT Standard Wipe Test
RT Surface Collecting Agents
SO Environmental Protection Agency
 Glossary
DEF (EPA) Actions taken to deal with a
 release or threat of release of a
 hazardous substance that could
 affect humans and/or the
 environment. The term "cleanup" is
 sometimes used interchangeably
 with the terms remedial action,
 removal action, response action,
 or corrective action.

CLEANUP OPERATIONS
(OSHA; CFR)

DA January 24, 1991
BT1 Operations
 BT2 Activities
BT1 Waste Management
 BT2 Processes
RT Decontamination
RT Remedial Actions
DEF (CFR) An operation where
    hazardous substances are
    removed, contained, incinerated,
    neutralized, stabilized, cleared up,
    or in any other manner processed
    or handled with the ultimate goal
    of making the site safer for people
    or the environment.

CLEAR CUT
(EPA)
DA October 12, 1990
RT Silviculture
SO Environmental Protection Agency
    Glossary
DEF (EPA) A forest management
    technique that involves harvesting
    all the trees in one area at one
    time. Under certain soil and slope
    conditions it can contribute
    sediment to water pollution.

CLEARING BARRELS
(DOE Order 5480.16)
DA October 12, 1990
BT1 Bullet Containment Devices
 BT2 Devices
SO Firearms
DEF (DOE Order 5480.16) Devices used
    to point a weapon at or into during
    the loading and unloading process
    that will contain any inadvertently
    discharged round.

ClF_3
DA October 12, 1990
SEE Chlorine Trifluoride
SO Acronyms

CLONING
(EPA)
DA October 12, 1990
BT1 Biological Processes
 BT2 Processes
RT Biotechnology
SO Environmental Protection Agency
    Glossary
DEF (EPA) In biotechnology, obtaining a
    group of genetically identical cells
    from a single cell. This term has
    assumed a more general meaning
    that includes making copies of a
    gene.

CLOSE REFLECTION BY WATER
(DOE Order 5480.3)
DA October 12, 1990
RT Criticality
SO Hazardous Materials
DEF (DOE Order 5480.3) Immediate
    contact by water of sufficient

thickness to reflect a maximum
number of neutrons.

CLOSED-LOOP RECYCLING
(EPA)
DA October 12, 1990
BT1 Recycling
 BT2 Waste Management Processes
  BT3 Processes
BT1 Reuse
 BT2 Uses
SO Environmental Protection Agency
    Glossary
DEF (EPA) Reclaiming or reusing
    wastewater for nonpotable
    purposes in an enclosed process.

CLOSED PORTIONS
(SWDA; RCRA; CFR)
DA October 12, 1990
RT Active Portions
RT Inactive Portions
RT Treatment, Storage, and Disposal
    Facilities
SO Environmental Management
SO Wastes
DEF (SWDA; RCRA) Portions of a
    facility which an owner or operator
    has closed in accordance with the
    approved facility closure plan and
    all applicable closure
    requirements.

CLOSED VENT SYSTEMS
(CFR)
DA November 15, 1990
BT1 Systems
NT1 Double Block and Bleed System
RT Control Devices
SO Air Pollution
DEF (CFR) Systems that are not open to
    atmosphere and that are
    composed of piping, connections,
    and, if necessary, flow- inducing
    devices that transport gas or vapor
    from a piece or pieces of
    equipment to a control device.

CLOSURE
(DOE Order 5480.2A; SWDA; RCRA;
CFR)
DA October 19, 1990
BT1 Administrative Processes
 BT2 Processes
NT1 Current Closure Cost Estimate
NT1 Current Post-Closure Cost Estimate
NT1 Final Closure
NT1 Partial Closure
NT1 Site Closure and Stabilization
RT Closure Plans
RT Hazardous Waste Management
    Facilities
RT Shutdown (Emissions)
SO Environmental Management
SO Wastes
DEF (SWDA; RCRA) The act of securing
    a Hazardous Waste Management
    facility.

CLOSURE PERIOD
(ANL; CFR)
DA May 22, 1991
BT1 Time Designations
RT Closure Plans
RT Waste Management
SO Wastes
DEF (CFR) The period of time beginning
    with the cessation, with respect to
    waste impoundment of uranium
    ore processing operations and
    ending with completion of
    requirements specified under a
    closure plan.

CLOSURE PLANS
(SWDA; RCRA)
DA October 12, 1990
BT1 Plans
RT Closure
RT Closure Period
RT Current Closure Cost Estimate
RT Current Post-Closure Cost Estimate
RT Site Closure and Stabilization
SO Wastes
DEF (SWDA; RCRA) Documentation
    prepared to guide the deactivation,
    stabilization, and surveillance of a
    waste management unit or facility
    under the Resource Conservation
    and Recovery Act. (ESH) DOE
    uses this term to mean the closure
    of an operating facility. The plans
    contain a complete survey of the
    environmental problems at a DOE
    facility, budget data which
    indicates the cost of environmental
    restoration and remediation at the
    facility, and a discussion of the
    proposed cleanup schedule.

CLOTHING
DA January 30, 1991
NT1 Anticontamination Clothing
 NT2 Acid Suit
 NT2 Lab Coat
 NT2 Radiation Suit
NT1 Flame Retardant Clothing
RT Equipment/Parts - Personal
    Protective (DOE FRASE
    Vocabulary)
SO DOE FRASE VOCABULARY

CLP
DA October 12, 1990
SEE Contract Laboratory Program
SO Acronyms

CM
DA October 12, 1990
SEE Core Melt
SO Acronyms

CM/sec
DA October 12, 1990
SEE Centimeters per second
SO Acronyms

CMA
DA   October 12, 1990
SEE Crane Maintenance Area
SO   Acronyms

CMS
DA   October 12, 1990
SEE RCRA Corrective Measures Study
SO   Acronyms

CMT
DA   October 12, 1990
SEE Crisis Management Teams
SO   Acronyms

CNHR
DA   October 12, 1990
SEE Containment Atmosphere Heat
      Removal
SO   Acronyms

CNRR
DA   October 12, 1990
SEE Containment Radioactivity Removal
SO   Acronyms

CO (Carbon Monoxide)
DA   October 12, 1990
SEE Carbon Monoxide
SO   Acronyms

CO (Contracting Officer)
DA   October 12, 1990
SEE Contracting Officers
SO   Acronyms

COAGULATION
(SDWA; CFR)
DA   October 12, 1990
BT1 Waste Management Processes
 BT2 Processes
RT   Flocculation
RT   Precipitates
RT   Sedimentation
SO   Environmental Protection Agency
       Glossary
SO   Water Pollution
DEF (EPA) A clumping of particles in
      wastewater to settle out impurities.
      It is often induced by chemicals
      such as lime, alum, and iron salts.
      (CFR) A process using coagulant
      chemicals and mixing by which
      colloidal and suspended materials
      are destabilized and agglomerated
      into flocs.

COAL
(CFR)
DA   October 12, 1990
BT1 Fossil Fuels
 BT2 Fuels
 BT2 Geologic Resources
  BT3 Natural Resources
NT1 Ultra Clean Coal
RT   Black Lung
RT   Gasification
RT   Soot

SO   Air Pollution
DEF (CFR) All solid fuels classified as
      anthracite, bituminous,
      subituminous, or lignite by the
      American Society and Testing and
      Materials, Designation D388-77
      (incorporated by reference -- see
      40 CFR 60.17).

COAL REFUSE
(CFR)
DA   October 12, 1990
BT1 Refuse
RT   Tailings
SO   Air Pollution
DEF (CFR) Waste-products of coal
      mining, cleaning, and coal
      preparation operations (e.g., culm,
      gob, etc.) containing coal, matrix
      material, clay, and other organic
      and inorganic material.

COASTAL WATERS
(CWA; CFR)
DA   October 12, 1990
BT1 Water
RT   Coastal Zones
RT   Tidal Waters
SO   Compensation and Liability
SO   Environmental Management
DEF (CFR) For the purposes of
      classifying the size of discharges,
      means the waters of the coastal
      zone except for the Great Lakes
      and specified ports and harbors on
      inland rivers.

COASTAL ZONES
(CERCLA; CFR; EPA)
DA   October 12, 1990
BT1 Zones
 BT2 Sites/Areas
RT   Coastal Waters
RT   Territorial Seas
RT   Tidal Marshes
SO   Compensation and Liability
SO   Environmental Management
SO   Environmental Protection Agency
       Glossary
DEF (EPA) Lands and waters adjacent
      to the coast that exert an influence
      on the uses of the sea and its
      ecology, or, inversely, whose uses
      and ecology are affected by the
      sea. (CFR) All U.S. waters subject
      to the tide, U.S. waters of the
      Great Lakes, specified ports and
      harbors on the inland rivers,
      waters of the contiguous zone,
      other waters of the high seas
      subject to this Plan, and the land
      surface and land substrata,
      ground waters, and ambient air
      proximal to those waters. The term
      coastal zone delineates an area of
      Federal responsibility for response
      action. Precise boundaries are
      determined by EPA/USCG
      agreements and identified in

Federal regional contingency
plans.

COBALT 58
(EDB)
DA   March 29, 1991
BT1 Beta-Minus Decay Radioisotopes
 BT2 Beta Decay Radioisotopes
  BT3 Radionuclides
   BT4 CERCLA Hazardous
        Substances
    BT5 Hazardous Substances
   BT4 Nuclides
BT1 Days Living Radioisotopes
 BT2 Radionuclides
  BT3 CERCLA Hazardous Substances
   BT4 Hazardous Substances
  BT3 Nuclides
BT1 Hours Living Radioisotopes
 BT2 Radionuclides
  BT3 CERCLA Hazardous Substances
   BT4 Hazardous Substances
  BT3 Nuclides
BT1 Internal Conversion Radioisotopes
 BT2 Radionuclides
  BT3 CERCLA Hazardous Substances
   BT4 Hazardous Substances
  BT3 Nuclides
BT1 Isomeric Transition Isotopes
 BT2 Radionuclides
  BT3 CERCLA Hazardous Substances
   BT4 Hazardous Substances
  BT3 Nuclides
SO   Radiation

COBALT 60
(EDB)
DA   March 29, 1991
BT1 Beta-Minus Decay Radioisotopes
 BT2 Beta Decay Radioisotopes
  BT3 Radionuclides
   BT4 CERCLA Hazardous
        Substances
    BT5 Hazardous Substances
   BT4 Nuclides
BT1 Internal Conversion Radioisotopes
 BT2 Radionuclides
  BT3 CERCLA Hazardous Substances
   BT4 Hazardous Substances
  BT3 Nuclides
BT1 Isomeric Transition Isotopes
 BT2 Radionuclides
  BT3 CERCLA Hazardous Substances
   BT4 Hazardous Substances
  BT3 Nuclides
BT1 Minutes Living Radioisotopes
 BT2 Radionuclides
  BT3 CERCLA Hazardous Substances
   BT4 Hazardous Substances
  BT3 Nuclides
BT1 Years Living Radioisotopes
 BT2 Radionuclides
  BT3 CERCLA Hazardous Substances
   BT4 Hazardous Substances
  BT3 Nuclides
SO   Radiation

COCA
DA   October 12, 1990

SEE Consent Order and Compliance
Agreement
SO Acronyms

COCCYX
(1116 COCCYX)
DA   November 28, 1990
BT1 Spine
BT2 Bone(s)
BT3 Human Body Parts
SO   DOE FRASE VOCABULARY

COD (Chemical Oxygen Demand)
DA   October 12, 1990
SEE Chemical Oxygen Demand
SO Acronyms

COD (Computer Operations Division)
DA   October 12, 1990
SEE Computer Operations Division
SO Acronyms

CODE OF FEDERAL REGULATIONS
DA   January 8, 1991
SF   *CFR*
BT1 Codes, Standards, and Regulations
BT1 Statutes and Regulations
NT1 Federal Aviation Regulation
NT1 Prevention of Significant
Deterioration Regulations
NT1 Primary Drinking Water Regulations
NT1 Secondary Drinking Water
Regulations
NT1 State Primary Drinking Water
Regulation
SO Management

**CODES, STANDARDS, AND
REGULATIONS**
(SSDC)
DA   October 12, 1990
SF   *CS&R*
NT1 Code of Federal Regulations
NT2 Federal Aviation Regulation
NT2 Prevention of Significant
Deterioration Regulations
NT2 Primary Drinking Water
Regulations
NT2 Secondary Drinking Water
Regulations
NT2 State Primary Drinking Water
Regulation
NT1 Ohio Administrative Code
NT1 Ohio Revised Code
NT1 Standards
NT2 Acceptable Daily Intake
NT2 Acceptable Intake for Chronic
Exposure
NT2 Acceptable Intake for Subchronic
Exposure
NT2 Air Quality Standards
NT3 Ambient Air Quality Standards
NT4 National Ambient Air Quality
Standards
NT5 Primary Standards
NT5 Secondary Standards
NT2 Applicable Standards and
Limitations

NT2 Applicable or Relevant and
Appropriate Requirements
NT2 Categorical Pretreatment
Standards
NT2 Commercial Standards
NT2 Emission Standards
NT3 National Emissions Standards for
Hazardous Air Pollutants
NT3 New Source Performance
Standards
NT2 Environmental Protection
Standards
NT2 Fuel Economy Standard
NT2 Government Standards
NT2 Mandatory Standards
NT2 National Standards (Air Pollution)
NT2 Radiation Standards
NT3 Annual Limit on Intake
NT3 As Low As Reasonably
Achievable
NT3 As Low As Practicable
NT2 Reference Standards
NT2 Technology-Based Standards
NT2 Voluntary Standards
NT2 Water Quality Standards
RT   Appropriate Act and Regulations
RT   Guidelines
RT   Statutes and Regulations
SO   System Safety Development Center
Glossary
DEF (SSDC) Nonspecific terminology
meaning all of the guidelines,
requirements and laws of federal,
state or local origin by which a
company or organization is
controlled in their operation. These
could include all aspects, such as
specifications of materials,
performance, design, or
operations; measurements of
quality control; required
documentation; and qualification of
personnel.

COEFFICIENT OF HAZE
(EPA)
DA   October 12, 1990
SF   *COH*
BT1 Measurements
RT   Turbidity
RT   Visibility Impairments
SO   Environmental Protection Agency
Glossary
DEF (EPA) A measurement of visibility
interference in the atmosphere.

COFFINS
(EPA)
DA   January 11, 1991
SY   Casks
SY   Pigs
SO   Environmental Protection Agency
Glossary

COGEMT
DA   February 13, 1991
SEE Continuity of Government
Emergency Management Teams
SO Acronyms
SO Emergency Preparedness

COGNIZANT DOE AUTHORITY
(DOE Order 6430.1A)
DA   October 12, 1990
RT   Cognizant Federal Agency Officials
RT   Department of Energy Sites
SO   Construction
DEF (DOE Order 6430.1A) An entity in
the DOE field organization unless
otherwise stated.

COGNIZANT FEDERAL AGENCIES
(DOE Order 5500.1 Attach. II, 7; EMER)
DA   January 24, 1991
BT1 Federal Agencies
BT2 Agencies
BT3 Administrative Organizations
BT4 Organizations
RT   Cognizant Federal Agency Officials
RT   Radiation Safety
SO   Emergency Preparedness
SO   Environmental Management
DEF (DOE 5500.1) The federal agency
that owns, authorizes, regulates,
or is otherwise deemed
responsible for the radiological
activity causing the emergency
and that has the authority to take
whatever action is necessary to
stabilize the accident.

**COGNIZANT FEDERAL AGENCY
OFFICIALS**
(DOE Order 5500.1A Attach. II, 8;
EMER)
DA   January 24, 1991
BT1 Personnel
RT   Cognizant Federal Agencies
RT   Cognizant DOE Authority
SO   Emergency Preparedness
SO   Environmental Management
DEF (DOE Order 5500.1A) For DOE, the
senior DOE official with jurisdiction
over the facility and activities
involved in the accident for or with
DOE is the Cognizant Federal
Agency Official.

COH
DA   October 12, 1990
SEE Coefficient of Haze
SO Acronyms

COLD TRAP
(2553 COLD TRAP)
DA   January 3, 1991
BT1 Equipment/Parts - Nuclear (DOE
FRASE Vocabulary)
BT2 Equipment
BT2 Reactor Components
SO   DOE FRASE VOCABULARY

COLIFORM INDEX
(EPA)
DA   October 12, 1990
RT   Fecal Coliform Bacteria
SO   Environmental Protection Agency
Glossary
DEF (EPA) A rating of the purity of water
based on a count of fecal bacteria.

SY-Synonymous Terms   SO-Source/Subject Category   SF-See From

COLIFORM ORGANISMS
(ESH)
DA   November 5, 1990
BT1 Bacteria
 BT2 Microorganisms
  BT3 Organisms
NT1 Fecal Coliform Bacteria
RT   Standard Sample
DEF (EPA) Microorganisms found in the
    intestinal tract of humans and
    animals. Their presence in water
    indicates fecal pollution and
    potentially dangerous bacterial
    contamination by disease causing
    microorganisms.

COLLECTIVE DOSE EQUIVALENT
(DOE Orders 5480.11, 5400.5; ESH;
EMER)
DA   October 12, 1990
BT1 Dose Equivalents
 BT2 Radiation Units
  BT3 Units of Measure
NT1 Collective Effective Dose Equivalent
SO   Emergency Preparedness
SO   Industrial Hygiene
DEF (EMER) The sum of the dose
    equivalents or effective dose
    equivalents of all individuals in an
    exposed population within an
    80-km radius, and expressed in
    units of person-rem (or
    person-sievert). When the
    collective dose equivalent of
    interest is for a specific organ, the
    units would be organ-rem (or
    organ-sievert). The 80-km
    distance shall be measured from a
    point located centrally with respect
    to major facilities or DOE program
    activities. (ES&H) The sums of the
    dose equivalents of all individuals
    in an exposed population within 50
    mi. It is expressed in units of
    person-rem or person-sieverts (1
    person-Sv = 100 person-rem).

**COLLECTIVE EFFECTIVE DOSE
EQUIVALENT**
(DOE Orders 5480.11, 5400.5; ESH;
 IAEA)
DA   October 12, 1990
BT1 Collective Dose Equivalent
 BT2 Dose Equivalents
  BT3 Radiation Units
   BT4 Units of Measure
BT1 Effective Dose Equivalent
 BT2 Dose Equivalents
  BT3 Radiation Units
   BT4 Units of Measure
RT   Collective Effective Dose
     Equivalent Commitment
RT   Collective Effective Dose
     Equivalent Rate
SO   Industrial Hygiene
SO   Radiation
DEF The sum of the dose equivalents or
    effective dose equivalents of all
    individuals in an exposed
    population within an 80-km radius,

expressed in units of person-rem
(or person-sievert). When the
collective dose equivalent of
interest is for a specific organ, the
units would be organ-rem (or
organ-sievert). The 80-km
distance shall be measured from a
point located centrally with respect
to major facilities or DOE program
activities.(ES&H) The sum of the
effective dose equivalents of all
individuals in an exposed
population within 50 mi. It is
expressed in units of person-rem
or person-sieverts (1 person = 100
person- rem).

**COLLECTIVE EFFECTIVE DOSE
EQUIVALENT COMMITMENT**
(IAEA)
DA   October 12, 1990
RT   Collective Effective Dose Equivalent
SO   Radiation
DEF (IAEA) For any specified event,
    decision or defined finite portion of
    a practice, the infinite time integral
    of the collective effective dose
    equivalent rate as a function of
    time, caused by that event
    decision or defined finite practice.

**COLLECTIVE EFFECTIVE DOSE
EQUIVALENT RATE**
(IAEA)
DA   October 12, 1990
BT1 Rates
RT   Collective Effective Dose Equivalent
SO   Radiation
DEF (IAEA) The integrated product of
    the effective dose equivalent rate
    and the number of individuals in
    the population. The total collective
    effective dose equivalent rate from
    a given source is obtained by
    including all individuals irradiated
    by the source, and is a function of
    time.

COLLECTORS
(SEA)
DA   October 12, 1990
RT   Diaphragm Strut
RT   Lateral Force Resisting System
SO   Construction
DEF (SEA) Members or elements
    provided to transfer lateral forces
    from a portion of a structure to
    vertical elements of the lateral
    force resisting system.

**COLORADO DEPARTMENT OF
HEALTH**
DA   January 8, 1991
SF   CDH
BT1 State Agencies
 BT2 Agencies
  BT3 Administrative Organizations
   BT4 Organizations

COMBINED SEWERS
(EPA)
DA   October 12, 1990
BT1 Sewers
SO   Environmental Protection Agency
     Glossary
DEF (EPA) Sewer systems that carry
    both sewage and stormwater
    runoff. Normally, its entire flow
    goes to a waste treatment plant,
    but during a heavy storm, the
    stormwater volume may be so
    great as to cause overflows. When
    this happens, untreated mixtures
    of stormwater and sewage may
    flow into receiving waters.
    Stormwater runoff may also carry
    toxic chemicals from industrial
    areas or streets into the sewer
    system.

COMBUSTIBLE LIQUIDS
(CFR)
DA   October 12, 1990
BT1 Liquids
 BT2 Fluids
RT   Flammable Liquids
RT   Flash Point
SO   Hazardous Materials
DEF (CFR) Combustible liquids are
    liquids that do not meet any other
    classification of hazardous
    materials and that have a flash
    point at or above 100°F and
    below 200°F, except any mixture
    having one component or more
    with a flash point at 200°F or
    higher that makes up at least 99%
    of the volume of the mixture.

COMBUSTION
(EPA)
DA   October 12, 1990
BT1 Oxidation
 BT2 Chemical Processes
  BT3 Processes
RT   Acid Deposition
RT   Afterburners
RT   Ashes
RT   Burning Agents
RT   Combustion Products
RT   Fires
RT   Incinerators
SO   Environmental Protection Agency
     Glossary
DEF (EPA) Burning, or rapid oxidation,
    accompanied by release of energy
    in the form of heat and light. A
    basic cause of air pollution.

COMBUSTION ENGINEERING
DA   January 8, 1991
SF   CE

COMBUSTION PRODUCTS
(EPA)
DA   October 12, 1990
RT   Baffle Chambers
RT   Combustion

RT Flue Gases
SO Environmental Protection Agency
Glossary
DEF (EPA) Substances produced during
the burning or oxidation of a
material.

COMMAND POSTS
(EPA)
DA October 12, 1990
BT1 Facilities
SO Environmental Protection Agency
Glossary
DEF (EPA) Facilities located at a safe
distance upwind from an accident
site, where the on-scene
coordinator, responders, and
technical representatives can
make response decisions, deploy
manpower and equipment,
maintain liaison with news media,
and handle communications.

COMMENT PERIODS
(EPA)
DA October 12, 1990
BT1 Time Designations
RT Members of the Public
SO Environmental Protection Agency
Glossary
DEF (EPA) Time periods provided for the
public to review and comment on
a proposed EPA action or
rulemaking after it is published in
the Federal Register.

COMMERCIAL ACTIVITIES
(ESA; USC)
DA October 12, 1990
BT1 Activities
RT Consumer Commodities
RT Foreign Commerce
SO Endangered Species
DEF (ESA) All activities of industry and
trade, including, but not limited to,
the buying or selling of
commodities and activities
conducted for the purpose of
facilitating such buying and selling,
provided, however, that it does not
include exhibition of commodities
by museums or similar cultural or
historical organizations.

COMMERCIAL APPLICATORS
(FIFRA; CFR; USC)
DA November 15, 1990
RT Certified Applicators
RT Private Applicators
SO Environmental Management
DEF (USC) An applicator (whether or
not he is a private applicator with
respect to some uses) who uses
or supervises the use of any
pesticide which is classified for
restricted use for any purpose or
on any property other than as
provided by paragraph (2). (4)
Under the direct supervision of a

certified applicator. Unless
otherwise prescribed by its
labeling, a pesticide shall be
considered to be applied under the
direct supervision of a certified
applicator if it is applied by a
competent person acting under
the instructions and control of a
certified applicator who is available
if and when needed, even though
such certified applicator is not
physically present at the time and
place the pesticide is applied.

COMMERCIAL AREAS
(CFR)
DA November 9, 1990
BT1 Sites/Areas
RT Residential Areas
DEF (CFR) Those areas where people
work in other than manufacturing
or farming industries. Commercial
areas are typically accessible to
both members of the general
public and employees and include
public assembly properties,
institutional properties, stores,
office buildings, and transportation
centers.

COMMERCIAL ARSENIC
(CFR)
DA October 12, 1990
BT1 Poisons
RT Inorganic Arsenic
SO Air Pollution
DEF (CFR) Any form of arsenic that is
produced by extraction from any
arsenic-containing substance and
is intended for sale or for
intentional use in a manufacturing
process. Arsenic that is a naturally
occurring trace constituent of
another substance is not
considered "commercial arsenic."

COMMERCIAL ASBESTOS
(CAA; CFR)
DA October 12, 1990
BT1 Asbestos
BT2 Carcinogens
BT3 Hazardous Substances
BT2 CERCLA Hazardous Substances
BT3 Hazardous Substances
RT Asbestos Mills
RT Fabricating
SO Air Pollution
SO Environmental Management
DEF (CFR) Any asbestos that is
extracted from asbestos ore.

COMMERCIAL ORGANIZATIONS
DA February 5, 1991
BT1 Organizations
NT1 Companies
NT2 Allied-Signal Inc.
NT2 Associated Universities, Inc.
NT2 Bechtel Petroleum Operation, Inc.
NT2 Boeing Petroleum Services

NT2 EG&G Energy Measurements, Inc.
NT2 EG&G Mound Applied
Technologies, Inc.
NT2 EG&G Idaho, Inc.
NT3 System Safety Development
Center
NT2 General Electric
NT2 Goodyear Atomic Corporation
NT2 International Business Machines
NT2 John Brown, E&C, Inc.
NT2 Kaiser Engineers Hanford, Co.
NT2 Lawrence-Allison & Associates
NT2 Martin Marietta Energy Systems
NT2 Mason and Hanger-Silas Mason
Co.
NT2 Mississippi Power and Light
NT2 MK-Ferguson
NT2 MK-Ferguson of Idaho Co.
NT2 Ohio Valley Electric Company
NT2 Pan American World Services,
INC.
NT2 Raytheon Services Nevada
NT2 Reactive Metals, Inc.
NT2 Reynolds Electrical and
Engineering Company
NT2 Rockwell International Corp.
NT2 Ross Aviation, Inc.
NT2 Sandia Corporation AT&T
Technologies, Inc.
NT2 Science Applications International
Corporation
NT2 United Electric
NT2 Universities Research Association,
Inc.
NT2 Wackenhut Services Inc.
NT2 West Valley Nuclear Services Co.,
Inc.
NT2 Westinghouse Electric Corp.
NT2 Westinghouse Idaho Nuclear Co.,
Inc.
NT2 Westinghouse Savannah River
Company
NT2 Westinghouse Hanford Company
NT2 Westinghouse Materials Company
of Ohio
NT1 Retailers
RT Administrative Organizations
RT Educational Organizations
RT Research and Development
Organizations

COMMERCIAL SOLID WASTES
(RCRA; CFR)
DA January 24, 1991
BT1 Solid Wastes
BT2 Wastes
DEF (CFR) "Commercial solid waste"
means all types of solid wastes
generated by stores, offices,
restaurants, warehouses, and
other non-manufacturing activities,
excluding residential and industrial
wastes.

COMMERCIAL STANDARDS
(DOE Order 1300.2)
DA October 16, 1990
SY Voluntary Standards
BT1 Standards

BT2 Codes, Standards, and
Regulations
RT Standards-developing Groups
RT Voluntary Standards Bodies
DEF (DOE Order 1300.2) Those
standards that are established
generally by private sector bodies
and are available for use by any
person or organization, private or
governmental. Voluntary standards
are also referred to as "industry
standards" as well as "consensus
standards" (standards developed
under due process procedures)
but do not include professional
standards of personal conduct,
private standards of individual
firms, standards mandated by law,
or standards of individual
organizations for their internal use.

COMMERCIAL STORERS OF PCB WASTE
(TSCA)
DA October 19, 1990
RT PCB Wastes
SO Hazardous Materials
DEF (TSCA) The owner or operator of
each facility which is subject to the
PCB storage facility standards of
40 CFR 761.65, and who engages
in storage activities involving PCB
waste generated by others, or
PCB waste that was removed
while servicing the equipment
owned by others and brokered for
disposal.The receipt of a fee or
any other form of compensation
for storage services is not
necessary to qualify as a
commercial storer of PCB waste.It
is sufficient under this definition
that the facility stores PCB waste
generated by others or the facility
removed the PCB waste while
servicing equipment owned by
others.A generator who stores
only the generator's own waste is
subject to the storage
requirements of 40 CFR 761.65,
but is not required to seek
approval as a commercial storer.If
a facility's storage of PCB waste
at no time exceeds 500 gallons of
PCBs, the owner or operator is not
required to seek approval as a
commercial storer of PCB waste.

COMMINUTERS
(EPA)
DA October 12, 1990
BT1 Machines (DOE FRASE
Vocabulary)
BT2 Equipment
RT Comminution
SO Environmental Protection Agency
Glossary
DEF (EPA) Machines that shred or
pulverize solids to make waste
treatment easier.

COMMINUTION
(EPA)
DA October 12, 1990
BT1 Waste Management Processes
BT2 Processes
RT Comminuters
RT Solid Waste Management
RT Wastewater Treatment Processes
SO Environmental Protection Agency
Glossary
DEF (EPA) Mechanical shredding or
pulverizing of waste. Used in both
solid waste management and
wastewater treatment.

COMMITTED DOSE EQUIVALENT
(DOE Orders 5480.11, 5400.5; IAEA;
EMER)
DA October 12, 1990
BT1 Dose Equivalents
BT2 Radiation Units
BT3 Units of Measure
NT1 Committed Effective Dose
Equivalent
SO Emergency Preparedness
SO Industrial Hygiene
SO Radiation
DEF (DOE Order 5400.5) The predicted
total dose equivalent to a tissue or
organ over a 50-year period after
a known intake of a radionuclide
into the body. It does not include
contributions from external dose.
Committed dose equivalent is
expressed in units of rem (or
sievert).

COMMITTED EFFECTIVE DOSE EQUIVALENT
(DOE Order 5400.5; IAEA; EMER)
DA October 12, 1990
BT1 Committed Dose Equivalent
BT2 Dose Equivalents
BT3 Radiation Units
BT4 Units of Measure
BT1 Effective Dose Equivalent
BT2 Dose Equivalents
BT3 Radiation Units
BT4 Units of Measure
SO Emergency Preparedness
SO Industrial Hygiene
DEF (DOE Order 5400.5) The sum of
the committed dose equivalents to
various tissues in the body, each
multiplied by the appropriate
weighting factor. Committed
effective dose equivalent is
expressed in units of rem (or
sievert). (IAEA) The effective dose
equivalent in an individual that will
be accumulated during the fifty
years following an intake of
radioactive material into the body.

COMMITTED USES
(CFR)
DA November 15, 1990
BT1 Uses

RT Irreversible Commitment of
Resources
DEF (CFR) Either current public uses or
planned public uses of a natural
resource for which there is a
documented legal, administrative,
budgetary, or financial
commitment established before
the discharge of oil or release of a
hazardous substance is detected.

COMMITTEES
DA February 1, 1991
BT1 Administrative Organizations
BT2 Organizations
NT1 Accident Response Capabilities
Coordinating Committee
NT1 Advisory Committee on Reactor
Safety
NT1 Emergency Management
Coordination Committee
NT1 Interagency Committee on
Standards Policy
NT1 Local Emergency Planning
Committees
NT1 Metrication Operating Committee
NT1 Metric Transition Committee
NT1 Plant Oversight Review Committee
NT1 Regional Assistance Committees
NT1 Safety Review Committee
NT1 Subcommittee on Federal
Response
NT1 Waste Acceptance Criteria
Committee
RT Boards
RT Centers
RT Groups
RT Offices
SO Management
DEF (WEBSTER) A body of persons
delegated to consider, investigate,
take action on or report on some
specific matter.

COMMON CAUSE FAILURE
DA January 8, 1991
SF CCF
BT1 Failures
BT2 Accidents

COMMON EXPOSURE ROUTES
(FIFRA; CFR)
DA January 24, 1991
BT1 Routes (Exposure)
RT Exposure Pathways
RT Food Chains
DEF (CFR) The term "common exposure
route" means a likely way (oral)
contaminants may reach and/or
enter an organism.

COMMON LABORATORY CONTAMINANTS
(EPA)
DA October 12, 1990
BT1 Contaminants
RT Organic Chemicals

BT-Broader Term

NT-Narrower Term

RT Related Term

SO Environmental Protection Agency
    Glossary
DEF (EPA) Certain organic chemicals
    (considered by EPA to be acetone,
    2-butanone, methylene chloride,
    toluene, and the phthalate esters)
    that are commonly used in the
    laboratory and thus may be
    introduced into a sample from
    laboratory cross- contamination,
    not from the site.

COMMUNICATION
(MORT)
DA   April 3, 1991
RT Analyses
RT Corrective Action Triggers
RT Knowledge
RT Monitoring
RT Technical Information
RT Trending
DEF (MORT) MORT analysis asks: was
    the exchange or transmittal of
    knowledge adequate (relative to a
    potential unwanted energy
    transfer)? was internal
    communication adequate? was
    external communication
    adequate? External
    communication relates to the
    interface between the in-house
    (internal) information system and
    national information systems, such
    as the National Safety Council,
    NASA, etc. Was the method of
    searching, retrieving, and
    processing relevant information
    adequate?

COMMUNICATORS
(EMER)
DA   February 1, 1991
BT1 Personnel
SO Emergency Preparedness
DEF (EMER) Individuals who provide for
    the timely and accurate flow of
    information among emergency
    response staffs. These individuals
    are normally found at emergency
    control centers but may also be
    located in control rooms or field
    command posts, as needed.
    Communicators shall maintain
    staff journals which contain a
    chronology of events involving
    their areas or facilities.

COMMUNITY AWARENESS AND EMERGENCY RESPONSE PROGRAMS
(EMER)
DA   February 1, 1991
BT1 Programs
RT Hazardous Materials
RT Releases
SO Emergency Preparedness
DEF (EMER) Programs developed by
    the Chemical Manufacturers
    Association (CMA) to assist
    chemical plant managers in taking
    the initiative in cooperating with
    local communities to develop
    integrated (community/industry)
    plans for responding to releases of
    hazardous materials.

COMMUNITY RELATIONS
(EPA)
DA   October 12, 1990
SO Environmental Protection Agency
    Glossary
DEF (EPA) The EPA effort to establish
    two-way communication with the
    public to create understanding of
    EPA programs and related actions,
    to assure public input into
    decision-making processes related
    to affected communities, and to
    make certain that the Agency is
    aware of and responsive to public
    concerns. Specific community
    relations activities are required in
    relation to Superfund remedial
    actions.

COMMUNITY WATER SYSTEMS
(SDWA; CFR; EPA; ESH)
DA   October 12, 1990
BT1 Public Water Systems
 BT2 Water Systems
  BT3 Publicly Owned Treatment Works
   BT4 Treatment Facilities
    BT5 Facilities
  BT3 Systems
SO Environmental Protection Agency
    Glossary
SO Water Pollution
DEF Public water systems which serve
    at least 15 service connections
    used year-round by residents or
    regularly serve at least 25
    year-round residents.

COMPACTION
(EPA)
DA   October 12, 1990
BT1 Solid Waste Management
 BT2 Waste Management
  BT3 Processes
RT Baling
SO Environmental Protection Agency
    Glossary
DEF (EPA) Reduction of the bulk of solid
    waste by rolling and tamping.

COMPACTOR
(2112 COMPACTOR)
DA   December 10, 1990
BT1 Machines (DOE FRASE
    Vocabulary)
 BT2 Equipment
SO DOE FRASE VOCABULARY

COMPANIES
DA   February 1, 1991
BT1 Commercial Organizations
 BT2 Organizations
NT1 Allied-Signal Inc.
NT1 Associated Universities, Inc.
NT1 Bechtel Petroleum Operation, Inc.
NT1 Boeing Petroleum Services
NT1 EG&G Energy Measurements, Inc.
NT1 EG&G Mound Applied
    Technologies, Inc.
NT1 EG&G Idaho, Inc.
 NT2 System Safety Development
    Center
NT1 General Electric
NT1 Goodyear Atomic Corporation
NT1 International Business Machines
NT1 John Brown, E&C, Inc.
NT1 Kaiser Engineers Hanford, Co.
NT1 Lawrence-Allison & Associates
NT1 Martin Marietta Energy Systems
NT1 Mason and Hanger-Silas Mason
    Co.
NT1 Mississippi Power and Light
NT1 MK-Ferguson
NT1 MK-Ferguson of Idaho Co.
NT1 Ohio Valley Electric Company
NT1 Pan American World Services, INC.
NT1 Raytheon Services Nevada
NT1 Reactive Metals, Inc.
NT1 Reynolds Electrical and
    Engineering Company
NT1 Rockwell International Corp.
NT1 Ross Aviation, Inc.
NT1 Sandia Corporation AT&T
    Technologies, Inc.
NT1 Science Applications International
    Corporation
NT1 United Electric
NT1 Universities Research Association,
    Inc.
NT1 Wackenhut Services Inc.
NT1 West Valley Nuclear Services Co.,
    Inc.
NT1 Westinghouse Electric Corp.
NT1 Westinghouse Idaho Nuclear Co.,
    Inc.
NT1 Westinghouse Savannah River
    Company
NT1 Westinghouse Hanford Company
NT1 Westinghouse Materials Company
    of Ohio
RT Contractors
RT Parent Corporations
RT Small Manufacturers
SO Management
DEF (WEBSTER) An association of
    persons for carrying on a
    commercial or industrial
    enterprise.

COMPARTMENTALIZED VEHICLES
(RCRA; CFR)
DA   January 24, 1991
BT1 Transport Vehicles
DEF (CFR) "Compartmentalized vehicle"
    means a collection vehicle which
    has two or more compartments for
    placement of solid wastes or
    recyclable materials. The
    compartments may be within the
    main truck body or on the outside
    of that body or in the form of
    metal racks.

COMPATIBLE
(SWDA; RCRA; ESA; CFR)
DA   October 12, 1990
SO   Wastes
DEF The ability of two or more
     substances to maintain their
     respective physical and chemical
     properties upon contact with one
     another for the design life of the
     tank system under conditions
     likely to be encountered in the
     underground storage tank (UST).

COMPENSATED ION CHAMBER
DA   January 8, 1991
SF   *CIC*

COMPENSATION AND LIABILITY
(CERCLA: USC)
DA   October 19, 1990
RT   Causes
RT   Contract Specifications
RT   Contractual Liability
RT   Indemnify
RT   Liabilities
RT   Natural Resource Damage
     Assessment
RT   Product Liability
RT   Responses
DEF A legal phrase which embraces the
     idea of placement of fault for an
     unwanted occurrence, including
     appropriate restitution by the party
     or parties responsible for the
     unwanted occurrence.

COMPETENT AUTHORITIES
(SWDA; RCRA; IAEA)
DA   October 12, 1990
BT1 Technically Qualified Individuals
 BT2 Personnel
RT   Authorized Limits
RT   Radiation Safety
RT   Transportation
SO   Hazardous Materials
SO   Radiation
DEF (IAEA) Authorities designated or
     otherwise recognized by a
     government for specific purposes
     in connection with radiation
     protection and/or nuclear safety.
     (CFR) A national agency
     responsible under its national law
     for the control or regulation of a
     particular aspect of the
     transportation of hazardous
     materials (dangerous goods).

COMPLAINTS
(DOE Order 5483.1A)
DA   October 12, 1990
RT   Allegations
SO   Industrial Safety
DEF (DOE Order 5483.1A) Oral or
     written communications by an
     employee or representative
     thereof, alleging that there are
     conditions in the work environment
     which are in violation of the

DOE-prescribed OSHA standards
or which pose safety or health
hazards to employees.

**COMPLEMENTARY CUMULATIVE
DISTRIBUTION FUNCTION**
DA   January 8, 1991
SF   *CCDF*

COMPLIANCE COATINGS
(EPA)
DA   November 5, 1990
SO   Environmental Protection Agency
     Glossary
DEF (EPA) Coatings whose volatile
     organic compound content do not
     exceed that allowed by regulation.

COMPLIANCE CONSIDERATIONS
(CAA; CWA; CFR; ESH; TSA)
DA   October 12, 1990
NT1 Level 1 Compliance
NT1 Level 2 Compliance
NT1 Level 3 Compliance
RT   Concerns
RT   Environment Assessments
RT   Maintenance Management
RT   Technical Safety Appraisals
SO   Environmental Management
SO   Standards
DEF (ESH) Three separate levels of
     compliance (Level 1, Level 2, and
     Level 3) are determined for rating
     concerns.

COMPLIANCE FINDINGS
(TTA)
DA   April 26, 1991
BT1 Findings
RT   Best Management Practice
     Findings
RT   Noteworthy Practices
RT   Tiger Team Assessments
RT   Tiger Teams
SO   Management
DEF (TTGM) Within a Tiger Team
     Assessment, conditions that, in
     the judgement of the assessment
     team, may not satisfy applicable
     environmental or safety and health
     regulations, DOE Orders (including
     internal DOE memoranda, where
     referenced), enforcement actions,
     agreements with regulatory
     agencies, or permit conditions.

COMPLIANCE INSPECTIONS
(DOE Order 5483.1A)
DA   October 16, 1990
BT1 Inspections
 BT2 Administrative Processes
  BT3 Processes
SO   Industrial Safety
DEF (DOE Order 5483.1A) Documented
     visits to and evaluations of a
     GOCO facility, to include an
     examination of the equipment,
     physical plant, methods,
     operations, procedures, and

processes, and to assess and
assure the contractor's
conformance with the
DOE-prescribed OSHA standards.

COMPLIANCE SCHEDULES
(CFR: EPA)
DA   October 12, 1990
SY   Schedule of Compliance
BT1 Time Designations
RT   Control Strategies
RT   Delayed Compliance Orders
RT   Increments of Progress
SO   Air Pollution
SO   Environmental Protection Agency
     Glossary
DEF (EPA) Negotiated agreements
     between a pollution source and a
     government agency that specify
     dates and procedures by which a
     source will reduce emissions and,
     thereby, comply with a regulation.
     (CFR) The date or dates by which
     a source or category of sources is
     required to comply with specific
     emission limitations contained in
     an implementation plan and with
     any increments of progress toward
     such compliance.

**COMPLICATIONS PECULIAR TO MED.
CARE**
(1378 COMPLICATION)
DA   November 28, 1990
BT1 Illnesses
SO   DOE FRASE VOCABULARY

COMPONENT COOLING WATER
DA   January 8, 1991
SF   *CCW*

COMPONENT HANDLING
DA   January 8, 1991
SF   *CH (Component Handling)*

COMPONENTS
(SWDA; RCRA; CFR)
DA   October 12, 1990
NT1 Functionally Equivalent
     Components
NT1 One-Inch Components
RT   Equipment
SO   Environmental Management
SO   Wastes
DEF (SWDA; RCRA) Any constituent
     parts of a unit or any group of
     constituent parts of a unit which
     are assembled to perform a
     specific function (e.g., a pump
     seal, pump, kiln liner, kiln
     thermocouple). Also, either the
     tank or ancillary equipment of a
     tank system.

COMPOST
(EPA)
DA   October 12, 1990
BT1 Soil Conditioner
 BT2 Organic Matter

RT Composting
RT Humus
SO Environmental Protection Agency
Glossary
DEF (EPA) A mixture of garbage and
degradable trash with soil in which
certain bacteria in the soil break
down the garbage and trash into
organic fertilizer.

COMPOSTING
(EPA)
DA October 12, 1990
BT1 Biological Processes
BT2 Processes
RT Aeration
RT Ballistic Separators
RT Biological Treatment
RT Biostabilizers
RT Compost
RT Decomposition
SO Environmental Protection Agency
Glossary
DEF (EPA) The natural biological
decomposition of organic material
in the presence of air to form a
humus-like material. Controlled
methods of composting include
mechanical mixing and aerating,
ventilating the materials by
dropping them through a vertical
series of aerated chambers, or
placing the compost in piles out in
the open air and mixing or turning
it periodically.

COMPREHENSIVE ENVIRONMENTAL ASSESSMENT AND RESPONSE PROGRAM
DA January 8, 1991
SF CEARP
BT1 Programs
RT Assessments

COMPREHENSIVE ENVIRONMENTAL RESPONSE, COMPENSATION, ETC.
(CERCLA)
DA January 10, 1991
SF CERCLA
BT1 Acts
BT2 Statutes and Regulations
RT In Situ Volatization
RT Post Removal Site Control
RT Support Agencies
RT Trustees
SO Compensation and Liability
SO Emergency Preparedness
SO Environmental Management
DEF (CERCLA) To provide for liability,
compensation, cleanup, and
emergency response for
hazardous substances released
into the environment and the
cleanup of inactive hazardous
waste disposal sites.

COMPREHENSIVE PLANNING
(USC)
DA November 15, 1990

BT1 Administrative Processes
BT2 Processes
RT Resource Conservation
RT Resource Recovery
DEF (USC) Includes planning or
management respecting resource
recovery and resource
conservation.

COMPRESSED GASES
(CFR)
DA October 12, 1990
BT1 Gases
NT1 Compressed Gases in Solution
NT1 Flammable Compressed Gases
NT1 Liquefied Compressed Gases
NT1 Non-Flammable Compressed
Gases
NT1 Non-Liquefied Compressed Gases
RT Cryogenic Liquids
RT Cylinders
SO Hazardous Materials
DEF (CFR) Includes any materials or
mixtures having a container
pressure exceeding 40 psi(pounds
per square inch absolute) at 70
degrees F, or 104 psi at 130
degrees F. Compressed gases are
further defined as flammable or
nonflammable.

COMPRESSED GASES IN SOLUTION
DA June 3, 1991
BT1 Compressed Gases
BT2 Gases
RT Non-Liquefied Compressed Gases
SO Hazardous Materials
DEF Non-liquefied compressed gases
which are dissolved in a solvent.

COMPRESSIONAL WAVES
(USGS)
DA October 12, 1990
SY Primary Body Waves
SO Natural Phenomenon
DEF (USGS) Body waves, which along
with shear waves, mainly cause
high-frequency (greater than 1
Hertz) vibrations, which are more
efficient than low-frequency waves
in causing low buildings to vibrate.

COMPRESSOR
(2113 COMPRESSOR)
DA December 10, 1990
BT1 Machines (DOE FRASE
Vocabulary)
BT2 Equipment
NT1 Air Compressor
SO DOE FRASE VOCABULARY

COMPUTER(S)
(2384 COMPUTER)
DA January 3, 1991
BT1 Equipment/Parts - Electrical (DOE
FRASE Vocabulary)
BT2 Equipment
RT Data Processing Area
SO DOE FRASE VOCABULARY

COMPUTER-ASSISTED DESIGN
DA January 8, 1991
SF CAD

COMPUTER ASSISTED TRACKING SYSTEM
(SSDC)
DA January 8, 1991
SF CATS
BT1 Safety Performance Measurement
System
BT2 Information Systems
BT3 Security Interests
BT3 Systems
RT Findings
SO Environmental Management
DEF (SSDC) A system developed by
DOE to track and analyze
individually identifiable concerns,
findings, actions, verifications, and
noteworthy practices for
appraisals, audits, and surveys
conducted by the Department's
Office of Environment, Safety and
Health (EH).

COMPUTER CODES
(Prentice-Hall Standard Glossary of
Computer Terminology)
DA January 8, 1991
NT1 ABEL
NT1 BEN
NT1 CLDTMP
NT1 FLOCHK
NT1 FRANTIC
NT1 IRRAS
NT1 MACCS
NT1 MSCANS
NT1 NSPKTR
NT1 PILOT
NT1 SETS
NT1 SQUIMP
NT1 TSCANS
RT Software
DEF Systems of symbols and rules for
representing data to a computer.
For example, the most common
codes used for the handling of
data within a computer and for
transmitting data are ASCII, BCD,
and EBCDIC.

COMPUTER INOPERATIVE LIMITS
(NFI)
DA October 12, 1990
SF CIL (Computer)
BT1 Limits
RT Reactor Operations
SO Radiation
DEF (NFI) Reactor power levels.

COMPUTER OPERATIONS DIVISION
DA January 8, 1991
SF COD (Computer Operations
Division)
BT1 Divisions
BT2 Administrative Organizations
BT3 Organizations

COMPUTER STORES
(SSDC)
DA   October 12, 1990
SO   System Safety Development Center
      Glossary
DEF (SSDC) Information recorded in the
      computer which can be drawn
      from as needed.

COMPUTER TERMINAL(S)
(2383 COMPUTER TER)
DA   January 3, 1991
BT1 Equipment/Parts - Electrical (DOE
      FRASE Vocabulary)
 BT2 Equipment
RT   Data Processing Area
SO   DOE FRASE VOCABULARY

COMPUTERIZED ACCIDENT/INCIDENT REPORTING SYSTEM
(SSDC)
DA   January 8, 1991
SF   CAIRS
BT1 Safety Performance Measurement
      System
 BT2 Information Systems
  BT3 Security Interests
  BT3 Systems
RT   Accidents
RT   Incidents
RT   Occurrence Reporting and
      Processing System
RT   Reports
SO   Environmental Management
DEF (SSDC) The CAIRS database
      contains all of the type A, B, and
      C accident investigations
      submitted on DOE for 5484x.

COMVAN
(EMER)
DA   February 1, 1991
BT1 Motor Vehicles
 BT2 Transport Vehicles
SO   Emergency Preparedness
DEF (EMER) An emergency
      communications van.

CONCENTRIC BRACED FRAMES
(SEA)
DA   October 12, 1990
BT1 Braced Frame
 BT2 Building Frame Systems
  BT3 Space Frames
   BT4 Structures (DOE FRASE
      Vocabulary)
  BT3 Systems
SO   Construction
DEF (SEA) Braced frames in which the
      members are subjected primarily
      to axial forces.

CONCEPT OF OPERATIONS
(EMER)
DA   February 1, 1991
RT   Emergencies
RT   Management
SO   Emergency Preparedness
DEF (EMER) A term describing the
organizational structure and
general requirements for a
coordinated federal response to
an emergency.

CONCEPTS AND REQUIREMENTS
(MORT)
DA   April 3, 1991
RT   Design and Development Plans
RT   Hazard Analysis
RT   Requirements
DEF (MORT) MORT analysis asks: are
      the concepts and requirements of
      hazard analysis adequately
      defined? Related issues/topics
      include: safety goals/risks;
      performance goals/risks; safety
      analysis criteria including change
      analysis, analytical methods,
      scaling mechanisms, safety
      precedence sequence; procedures
      criteria; specification of safety
      requirements; information search;
      and life cycle analysis.

CONCEPTUAL DESIGN REPORT
DA   January 8, 1991
SF   CDR
BT1 Reports
RT   Conceptual Designs

CONCEPTUAL DESIGNS
(DOE Order 4700.1)
DA   January 24, 1991
RT   Conceptual Design Report
RT   Preliminary Designs
RT   Project Design Criteria
RT   Technical Support
DEF (DOE Order 4700.1) Conceptual
      design encompasses those efforts
      to: (A) Develop a project scope
      that will satisfy program needs; (B)
      Assure project feasibility and
      attainable performance levels; (C)
      Develop reliable cost estimates
      and realistic schedules in order to
      provide a complete description of
      the project for congressional
      consideration; and (D) Develop
      project criteria and design
      parameters for all engineering
      disciplines, identification of
      applicable codes and standards;
      quality assurance requirements,
      environmental studies, materials of
      construction, space allowances,
      energy conservation features,
      health safety, safeguards, and
      security requirements and any
      other features or requirements
      necessary to describe the project.

CONCERNS
(TTA)
DA   April 26, 1991
RT   Compliance Considerations
RT   Findings
RT   Issue Identification
RT   Performance Objectives
RT   Potential Hazards
RT   Root Causes
RT   Safety and Health Assessments
RT   Seriousness
RT   Tiger Team Assessments
RT   Tiger Teams
RT   Tiger Team Assessment Reports
SO   Management
DEF (TTGM) Within the scope of a Tiger
      Team Assessment, concerns focus
      primarily on safety and health
      issues. A concern addresses a
      situation that, in the judgement of
      the Tiger Team, meets one or
      more of the following three criteria:
      1) does not comply with a DOE
      safety and health requirement or
      mandatory safety standard; 2)
      threatens to compromise the safe
      operation of the facility; or 3) if
      properly addressed, would
      substantially improve that
      particular situation, even though
      that part of the operation was
      judged to have a currently
      acceptable margin of safety. Each
      concern is supported by several
      findings and has the
      characteristics of being explicit,
      identifying the problem, being
      measurable (auditable), and being
      justifiable. Each concern is
      categorized by its seriousness,
      potential hazard consideration,
      and compliance consideration.

CONCRETE ENCASEMENT
(DOE Order 6430.1A)
DA   October 12, 1990
RT   Leaks
RT   Sewers
SO   Construction
DEF (DOE Order 6430.1A) Placement of
      concrete around a sewer at its
      point of intersection with a potable
      waterline to provide a leakage
      barrier.

CONCRETE SAW
(2114 CONCRETE SAW)
DA   December 10, 1990
BT1 Machines (DOE FRASE
      Vocabulary)
 BT2 Equipment
BT1 Saws
SO   DOE FRASE VOCABULARY

CONCUSSION
(1311 CONCUSSION)
DA   November 28, 1990
BT1 Brain Damage
 BT2 Injuries
RT   Contusion(S)
RT   Head
RT   Head Protection
RT   Loss of Consciousness
RT   Skull (Frase)
SO   DOE FRASE VOCABULARY

BT-Broader Term   NT-Narrower Term   RT Related Term

**COND OF RESPIRATORY SYS.
NON-TOXIC**
(1379 COND OF RESP)
DA   November 28, 1990
BT1 Diseases
BT1 Illnesses
NT1 Upper Respiratory Condition
NT1 Upper Respiratory Disease
NT2 Tuberculosis
RT   Respiratory System
SO   DOE FRASE VOCABULARY

CONDENSATE STORAGE TANK
DA   January 8, 1991
SF   *CST*
BT1 Tanks
BT2 Facility Components

CONDENSER(S)
(2385 CONDENSER)
DA   January 3, 1991
BT1 Equipment/Parts - Electrical (DOE
     FRASE Vocabulary)
BT2 Equipment
SO   DOE FRASE VOCABULARY

CONDENSER STACK GASES
(CFR)
DA   October 12, 1990
BT1 Flue Gases
BT2 Air Pollutants
BT2 Gases
RT   Mercury Ore Processing Facilities
SO   Air Pollution
DEF (CFR) The gaseous effluents
     evolved from the stack of
     processes utilizing heat to extract
     mercury metal from mercury ore.

CONDITION ASSESSMENT SURVEYS
(DOE Order 4330.4A)
DA   June 4, 1991
SF   *CAS*
BT1 Inspections
BT2 Administrative Processes
BT3 Processes
BT1 Surveys
RT   Maintenance
SO   Management
DEF (DOE Order 4330.4A) Periodic
     inspections of property using
     universally accepted methods and
     standards.

CONDITIONAL AND GATE
(SSDC)
DA   October 12, 1990
BT1 Logic Gates (Boolean)
BT1 Logic Gates
SO   System Safety Development Center
     Glossary
DEF (SSDC) Input produces the output
     provided the conditions written in
     the ELLIPSE are satisfied.
     Example, PRIORITY AND gate
     specifying order of input event
     occurrence.

CONDITIONAL OR GATE
(SSDC)
DA   October 12, 1990
BT1 Logic Gates (Boolean)
BT1 Logic Gates
SO   System Safety Development Center
     Glossary
DEF (SSDC) Input produces output
     provided the constraint conditions
     are met. Example, EXCLUSIVE
     OR gate enabling an output to
     occur only if a single input is
     present.

CONDITIONAL PROBABILITY
(SSDC)
DA   October 12, 1990
BT1 Probability
RT   Coupling
SO   System Safety Development Center
     Glossary
DEF (SSDC) The probability of an event
     (which requires the satisfying of
     two conditions) after one condition
     has occurred.

CONDITIONAL REGISTRATION
(EPA)
DA   October 12, 1990
BT1 Registration
SO   Environmental Protection Agency
     Glossary
DEF (EPA) Under special circumstances,
     the Federal Insecticide, Fungicide,
     and Rodenticide Act (FIFRA)
     permits registration of pesticide
     products that is "conditional" upon
     the submission of additional data.
     These special circumstances
     include a finding by the EPA
     Administrator that a new product
     or use of existing pesticide will not
     significantly increase the risk of
     unreasonable adverse effects. A
     product containing a new
     (previously unregistered) active
     ingredient may be conditionally
     registered only if the Administrator
     finds that such conditional
     registration is in the public interest,
     that a reasonable time for
     conducting the additional studies
     has not elapsed, and the use of
     the pesticide for the period of
     conditional registration will not
     present an unreasonable risk.

CONDITIONS
(DOE Order 5000.3A)
DA   December 14, 1990
NT1 Abnormal Exposure Conditions
NT1 Accident Prone Situation
NT1 Danger
NT2 Imminent Danger
NT1 Defense Readiness Conditions
NT1 Emergency Conditions
NT1 Hazards
NT2 Fire Hazards
NT2 Health Hazards

NT3 Delayed (Chronic) Health
     Hazards
NT3 Immediate (Acute) Health
     Hazards
NT2 Imminent Hazard
NT2 Potential Hazards
NT3 Level 1 Potential Hazards
NT3 Level 2 Potential Hazards
NT3 Level 3 Potential Hazards
NT1 Immediately Dangerous to Life or
     Health Values
NT1 Imminent and Substantial
     Endangerment
NT1 Limiting Factors
NT1 Limiting Condition for Operation
NT1 Natural Conditions
NT1 Operable
NT1 Operational Readiness
NT1 Potential to Emit
NT1 Safe
NT1 Seriousness
NT2 Category I Seriousness
NT2 Category II Seriousness
NT2 Category III Seriousness
NT1 Standby
NT1 Station Blackout
NT1 SWP Conditions
NT1 Uncertainty
NT1 Working Conditions
NT2 Working Condition A
NT2 Working Condition B
DEF (DOE Order 5000.3A) Any as-found
     state(s), whether or not resulting
     from an event, which may have
     adverse safety, health, quality
     assurance, security, operational,
     or environmental implications. A
     condition is more programmatic in
     nature; for example, an error in
     analysis or calculation, an
     anomaly associated with design or
     performance, or an item indicating
     a weakness in the management
     process are all conditions.

CONDUCTOR(S)
(2386 CONDUCTOR)
DA   January 3, 1991
BT1 Equipment/Parts - Electrical (DOE
     FRASE Vocabulary)
BT2 Equipment
SO   DOE FRASE VOCABULARY

CONDUIT(S)
(2387 CONDUIT)
DA   January 3, 1991
BT1 Equipment/Parts - Electrical (DOE
     FRASE Vocabulary)
BT2 Equipment
SO   DOE FRASE VOCABULARY

**CONFERENCE OF FEDERAL
ENVIRONMENTAL ENGINEERS**
DA   May 20, 1991
SF   *CFEE*
BT1 Professional Bodies
BT2 Organizations
SO   Environmental Management

CONFERENCES
(ESA; CFR)
DA   October 12, 1990
RT   Informal Consultation
SO   Endangered Species
DEF (CFR) A process which involves informal discussions between a Federal agency and the Service under section 7(a)(4) of the Endangered Species Act regarding the impact of an action on proposed species or proposed critical habitat and recommendations to minimize or avoid the adverse effects.

CONFIDENCE
(SSDC)
DA   October 12, 1990
RT   Probability
SO   System Safety Development Center Glossary
DEF (SSDC) The chance of likelihood that a specified value is part of the population (if a specific value lies outside ± three standard deviations from the mean, then we have 99.7% confidence that this observation is different from the population since 99.7% of the population lies in that range).

CONFIDENTIAL
(EMER)
DA   February 1, 1991
RT   Classification
RT   Classified Information
RT   Security Interests
SO   Emergency Preparedness
DEF (EMER) A classification level which is applied to classified information, the unauthorized disclosure of which reasonably could be expected to cause damage to the national security.

CONFIDENTIAL BUSINESS INFORMATION
(CFR)
DA   November 15, 1990
SY   Business Confidentiality
DEF (CFR) Includes the concept of trade secrecy and other related legal concepts which give (or may give) a business the right to preserve the confidentiality of business information and to limit its use or disclosure by others in order that the business may obtain or retain business advantages it derives from its right in the information. The definition is meant to encompass any concept which authorizes a Federal agency to withhold business information under 5 U.S.C. 552(b)(4), as well as any concept which requires EPA to withhold information from the public for the benefit of a business under 18 U.S.C. 1905.

CONFIGURATION CONTROL
(SSDC)
DA   October 12, 1990
BT1 Controls
SO   System Safety Development Center Glossary
DEF (SSDC) As defined in MORT, having the right people in the right place at the right time, working with the proper hardware, and in accordance with the proper procedures and management controls.

CONFINED AQUIFERS
(SWDA; RCRA; EPA; CFR)
DA   October 12, 1990
BT1 Aquifers
 BT2 Formations
SO   Environmental Management
SO   Environmental Protection Agency Glossary
SO   Wastes
DEF (EPA) Aquifers in which groundwater is confined under pressure that is significantly greater than atmospheric pressure.

CONFINEMENT AREAS
(DOE Order 6430.1A)
DA   October 12, 1990
BT1 Sites/Areas
RT   Confinement Systems
RT   Operating Area Compartment
SO   Construction
DEF (DOE Order 6430.1A) Areas having structures or systems from which releases of hazardous materials are controlled. The primary confinement systems are the process enclosures (glove boxes, conveyors, transfer boxes, other spaces normally containing hazardous materials), which are surrounded by one or more secondary confinement areas (operating area compartments).

CONFINEMENT HEAT REMOVAL
DA   January 8, 1991
SF   CHR

CONFINEMENT PROTECTION LIMITS
DA   January 8, 1991
SF   CPL
BT1 Limits

CONFINEMENT SYSTEMS
(DOE Order 6430.1A; AEA; CFR)
DA   October 12, 1990
BT1 Systems
NT1 Airborne Activity Confinement Systems
NT1 Primary Confinement Systems
 NT2 Enclosures
RT   Confinement Areas
RT   Containment Systems
RT   Containment

RT   Design Basis Accidents
RT   Radiation Protection
SO   Construction
SO   Environmental Management
DEF (DOE Order 6430.1A) The barrier and its associated systems (including ventilation) between areas containing hazardous materials and the environment or other areas in the facility that are normally expected to have levels of hazardous materials lower than allowable concentration limits.

CONFINING BEDS
(SDWA; CFR)
DA   October 12, 1990
RT   Aquifers
RT   Permeability
RT   Strata
SO   Water Pollution
DEF Bodies of impermeable or distinctly less permeable material stratigraphically adjacent to one or more aquifers.

CONFINING ZONES
(SDWA; CFR)
DA   October 12, 1990
BT1 Formations
BT1 Zones
 BT2 Sites/Areas
RT   Aquifers
RT   Strata
SO   Water Pollution
DEF A geological formation, group of formations, or part of a formation that is capable of limiting fluid movement above an injection zone.

CONFLUENT GROWTH
(SDWA; CFR)
DA   October 12, 1990
RT   Bacteria
RT   Sanitary Surveys
SO   Water Pollution
DEF (CFR) A continuous bacterial growth covering the entire filtration area of a membrane filter, or a portion thereof, in which bacterial colonies are not discrete.

CONJUNCTIVITIS
(1380 CONJUNCTIVIT)
DA   November 28, 1990
BT1 Contagious or Infectious Disease
 BT2 Diseases
 BT2 Illnesses
BT1 Illnesses
RT   Eyelid(s)
SO   DOE FRASE VOCABULARY

CONNECTED PIPING
(SWDA; RCRA; CFR)
DA   October 12, 1990
BT1 Piping
 BT2 Facility Components

BT-Broader Term   NT-Narrower Term   RT Related Term

SO Wastes
DEF All underground piping including valves, elbows, joints, flanges, and flexible connectors attached to a tank system through which regulated substances flow. For the purpose of determining how much piping is connected to any individual underground storage tank (UST) system, the piping that joins two UST systems should be allocated equally between them.

CONNECTORS
(CFR)
DA   October 12, 1990
BT1 Electrical Equipment
 BT2 Equipment
RT   Piping
SO   Air Pollution
DEF (CFR) Flanged, screwed, welded, or other joined fittings used to connect two pipe lines or a pipe line and a piece of equipment. For the purpose of reporting and recordkeeping, connector means flanged fittings that are not covered by insulation or other materials that prevent location of the fittings.

CONSENT DECREES
(EPA)
DA   October 12, 1990
RT   Remedial Actions
SO   Environmental Protection Agency Glossary
DEF (EPA) Legal documents, approved by a judge, that formalize an agreement reached between EPA and potentially responsible parties (PRPs) through which PRPs will conduct all of part of a cleanup action at a Superfund site, cease or correct actions or processes that are polluting the environment, or otherwise comply with regulations where the PRP's failure to comply caused EPA to initiate regulatory enforcement actions. The consent decree describes the actions PRPs will take and may be subject to a public comment period.

CONSENT ORDER AND COMPLIANCE AGREEMENT
DA   January 8, 1991
SF   *COCA*
BT1 Agreements

CONSEQUENCE ASSESSMENTS
(EMER)
DA   February 1, 1991
BT1 Assessments
 BT2 Administrative Processes
  BT3 Processes
RT   Hazard Analysis

SO   Emergency Preparedness
DEF (EMER) The evaluation and interpretation of radiological or other hazardous substance measurements and other information to provide a basis for decision making. Consequence assessment can include projections of off-site impact.

CONSEQUENCES
(EMER)
DA   February 1, 1991
SY   Effects
SO   Emergency Preparedness
DEF (EMER) The results or effects (especially projected doses or dose rates) of a release of radioactive material to the environment.

CONSEQUENTIAL LOSSES
(SSDC)
DA   October 12, 1990
BT1 Losses
RT   Loss Ratio
SO   System Safety Development Center Glossary
DEF (SSDC) Losses not directly due to a peril, but caused indirectly as a consequence of that peril; for example, spoilage of frozen foods is a loss consequent upon power failure.

CONSERVATION
(ESA; CFR; EPA)
DA   October 12, 1990
NT1 Resource Conservation
RT   Environmental Management
RT   Land Reclamation
RT   Resource Conservation and Recovery Act
SO   Environmental Protection Agency Glossary
DEF (EPA)Avoiding waste of, and renewing when possible, human and natural resources. The protection, improvement, and use of natural resources according to principles that will assure their highest economic or social benefits.

CONSERVATION RECOMMENDATIONS
(ESA; CFR)
DA   October 12, 1990
BT1 Recommendations
RT   Critical Habitats
SO   Endangered Species
DEF (CFR) Suggestions regarding discretionary measures to minimize or avoid adverse effects of a proposed action on listed species or critical habitat or regarding the development of information.

CONSOLIDATED INCINERATOR FACILITY
DA   January 8, 1991
SF   *CIF*
BT1 Facilities

CONSOLIDATED PRE-MANUFACTURE NOTICES
(TSCA; CFR)
DA   January 24, 1991
BT1 Pre-Manufacture Notices
 BT2 Section 5 Notices
SO   Environmental Management
DEF (CFR) Any Pre-Manufacture Notices submitted to EPA which cover more than one chemical substance (each being assigned a separate PMN number by EPA) as a result of a pre-notice agreement with EPA.

CONSTANT AIR MONITOR
(2751 CONSTANT AIR)
DA   January 3, 1991
BT1 Equipment/Parts - Instrumentation/Measuring (DOE FRASE Voc.)
 BT2 Equipment
BT1 Monitors
 BT2 Equipment
SO   DOE FRASE VOCABULARY

CONSTITUENTS
(SWDA; RCRA)
DA   October 19, 1990
BT1 Wastes
DEF (SWDA; RCRA) Those constituents that are listed in Appendix VIII to 40 CFR 261.

CONSTRAINTS
(SSDC)
DA   October 12, 1990
BT1 Requirements
RT   Logic Gates (Boolean)
RT   Logic Gates
SO   System Safety Development Center Glossary
DEF (SSDC) Restrictions affecting freedom to act; boundaries or conditions which may dictate performance in other than the desired or ideal manner. (SSDC) Conditional events that apply conditions or constraints to a basic logic gate or output event. Imposed condition is written in an ELLIPSE.

CONSTRUCTION
(SWDA; RCRA; CFR; USC)
DA   October 12, 1990
BT1 Processes
NT1 Physical Construction
NT1 Substantial Construction
RT   Approvals Necessary to Begin Physical Construction
RT   Bench Marks
RT   Caisson Foundations

SY-Synonymous Terms   SO-Source/Subject Category   SF-See From

RT Construction and Demolition
    Wastes
RT Construction Joints
RT Construction Projects
RT Construction Project Planning
RT General Purpose Facilities Projects
RT Monumentation
RT Significant Modification
RT Stationary Sources
RT Structural Members
RT Subslabs
RT Water Tight
SO Air Pollution
SO Environmental Management
SO Wastes
DEF With respect to any project of
    construction under this chapter,
    means (A) the erection or building
    of new structures and acquisition
    of lands or interests therein, or the
    acquisition, replacement,
    expansion, remodeling, alteration,
    modernization, or extension of
    existing structures, and (B) the
    acquisition and installation of initial
    equipment of, or required in
    connection with, new or newly
    acquired structures or the
    expanded, remodeled, altered,
    modernized or extended part of
    existing structures (including
    trucks and other motor vehicles,
    and tractors, cranes, and other
    machinery) necessary for the
    proper utilization and operation of
    the facility after completion of the
    project; and includes...(See 41
    USCS 6903 (2) for continuation of
    definition).

CONSTRUCTION ACTIVITY
(1227 CON ACTIVITY)
DA November 28, 1990
BT1 Activity Types (DOE FRASE
    Vocabulary)
 BT2 Activities
RT Jackhammer
RT Misc Repair/Construction
    Employee
SO DOE FRASE VOCABULARY

**CONSTRUCTION AND DEMOLITION
WASTES**
(RCRA; CFR)
DA January 24, 1991
BT1 Wastes
RT Construction
DEF (CFR) "Construction and demolition
    waste" means the waste building
    materials, packaging, and rubble
    resulting from construction,
    remodeling, repair, and demolition
    operations on pavements, houses,
    commercial buildings and other
    structures.

CONSTRUCTION AREA
(1605 CONST AREA)
DA December 10, 1990
BT1 Area

 BT2 Sites/Areas
SO DOE FRASE VOCABULARY

CONSTRUCTION JOINTS
(DOE Order 6430.1A)
DA October 12, 1990
RT Construction
SO Construction
DEF (DOE Order 6430.1A) Vertical or
    horizontal concrete surfaces where
    construction can be temporarily
    interrupted and continued later.

CONSTRUCTION PROJECT PLANNING
(DOE Order 6430.1A)
DA October 12, 1990
BT1 Administrative Processes
 BT2 Processes
RT Construction
RT Construction Projects
RT Contract Specifications
RT Projects
SO Construction
DEF (DOE Order 6430.1A) All activities
    that are performed, after the initial
    identification of a project, for the
    purposes of developing the project
    concept, reliable cost estimates,
    realistic performance schedules,
    and methods of performance.

CONSTRUCTION PROJECTS
(DOE Order 6430.1A)
DA October 12, 1990
BT1 Projects
NT1 Major Construction Activities
RT Construction
RT Construction Project Planning
RT Department of Energy Resident
    Construction Contractors
RT GDC Planning Board
SO Construction
DEF (DOE Order 6430.1A) New
    facilities, facility additions, and
    facility alteration projects where
    engineering and design are
    required in their performance.

**CONSULTANT FIRE PROTECTION
SURVEY PROGRAM**
(DOE Order 5480.7)
DA October 12, 1990
BT1 Programs
RT Environment, Safety, and Health
    (ES&H) Program
SO Fires
DEF (DOE Order 5480.7) The program
    under which fire protection
    surveys of principal DOE facilities
    are conducted for the Office of
    Operational Safety by fire
    protection engineers of selected
    contractors administered by this
    organization.

CONSUMER COMMODITIES
DA June 3, 1991
SY Consumer Products
RT Commercial Activities

SO Hazardous Materials
DEF Materials that are packaged and
    distributed in a form intended or
    suitable for sale through a retail
    sales agencies or instrumentalities
    for consumption by individuals for
    the purposes of personal care or
    household use. This term also
    includes drugs and medicines.

CONSUMER PRODUCTS
(CFR)
DA October 12, 1990
SY Consumer Commodities
BT1 Articles
SO Compensation and Liability
DEF (CERCLA) Articles, or component
    parts thereof, produced or
    distributed (i) for sale to a
    consumer for use in or around a
    permanent or temporary
    household or residence, a school,
    in recreation, or otherwise, or (ii)
    for the personal use, consumption
    or enjoyment of a consumer in or
    around a permanent or temporary
    household or residence, a school,
    in recreation, or otherwise; see 15
    USC 2052 for exceptions.

CONSUMPTIVE USE
(SWDA; RCRA; CFR)
DA October 12, 1990
BT1 Uses
RT Heating Oils
SO Wastes
DEF With respect to heating oil means
    consumed on the premises.

CONTACT HANDLED
DA January 8, 1991
SF CH (contact handled)

**CONTACT-HANDLED TRANSURANIC
WASTES**
(DOE Order 5820.2A)
DA January 24, 1991
BT1 Transuranic Wastes (TRU Waste)
 BT2 Radioactive Wastes
  BT3 Wastes
DEF (DOE 5820.2A) Packaged
    transuranic waste whose external
    surface dose rate does not exceed
    200 mrem per hour.

CONTACT PESTICIDES
(EPA)
DA October 12, 1990
BT1 Pesticides
 BT2 Hazardous Substances
RT Dermal Toxicity
SO Environmental Protection Agency
    Glossary
DEF (EPA) Chemicals that kill pests
    when they touch them, rather than
    by being eaten (stomach poison).
    Also, soil that contains the minute
    skeletons of certain algae that
    scratches and dehydrates

BT-Broader Term   NT-Narrower Term   RT Related Term

waxy-coated insects
(diatomaceous earth).

CONTACT RATE
(EPA)
DA   October 12, 1990
BT1 Rates
RT   Exposure Rate
SO   Environmental Protection Agency
      Glossary
DEF (EPA) Amount of medium (e.g.,
      ground water, soil) contacted per
      unit time or event (e.g. liters of
      water ingested per day).

CONTAGIOUS OR INFECTIOUS
DISEASE
(1382 CONTAGIOUS O)
DA   November 28, 1990
BT1 Diseases
BT1 Illnesses
NT1 Conjunctivitis
SO   DOE FRASE VOCABULARY

CONTAINERS
(FIFRA; SWDA; RCRA; CFR; ESH)
DA   October 12, 1990
NT1 Approved Storage Containers
NT1 Barrels
NT1 Bottles
NT1 Bulk Containers
NT1 Casks
NT1 Copper Converters
NT1 Cylinders
NT1 Denuders
NT1 Drums
NT1 Dustfall Jars
NT1 End Boxes
NT1 Freight Containers
 NT2 Unit Load Devices
NT1 Non-Reuseable Containers
NT1 Outside Container
 NT2 Strong Outside Containers
NT1 PCB Containers
NT1 Pigs
NT1 Single Strip Containers
NT1 Waste Containers
RT   Barriers
RT   Containment Vessels
RT   Containment
RT   Disposal Packages
RT   Inner Liners
RT   Packaging
RT   Storage
RT   Tanks
RT   Transport (Hazardous Substances)
SO   Environmental Management
SO   Wastes
DEF Portable devices in which a
      material is stored, transported,
      treated, disposed of, or otherwise
      handled. (ESH) Any package, can,
      bottle, bag, barrel, drum, tank, or
      other containing device (excluding
      spray applicator tanks) used to
      enclose a pesticide or
      pesticide-related waste.

CONTAINERSHIPS
(SWDA; RCRA)
DA   October 12, 1990
BT1 Cargo Vessels
 BT2 Vessels
RT   Freight Containers
RT   Portable Tanks
RT   Transport (Hazardous Substances)
SO   Hazardous Materials
DEF (CFR) Cargo vessels designed and
      constructed to transport, within
      specifically designed cells,
      portable tanks and freight
      containers that are lifted on and
      off with their contents intact.

CONTAINMENT
(SSDC; IAEA; EMER)
DA   October 12, 1990
BT1 Processes
NT1 Filtered Vented Containment
RT   Bulkheads
RT   Confinement Systems
RT   Containers
RT   Containment Systems
RT   Containment Storage Tanks
RT   Leak Detection Systems
RT   Liners
RT   Radiation Protection
RT   Radioactive Materials
SO   Emergency Preparedness
SO   Environmental Management
SO   Radiation
SO   System Safety Development Center
      Glossary
DEF (SSDC) Control of the expansion or
      propagation of accidental loss;
      commonly used in fire control.
      (IAEA) A term signifying either the
      confinement of radioactive material
      in such a way that it is prevented
      from being dispersed into the
      environment or is only released at
      a specified rate, or the device
      used to effect such confinement.

CONTAINMENT AIR RECIRCULATION
AND COOLING
DA   January 8, 1991
SF   CARC

CONTAINMENT ATMOSPHERE HEAT
REMOVAL
DA   January 8, 1991
SF   CNHR

CONTAINMENT BUILDING
(1776 CONTAIN BLDG)
DA   December 10, 1990
BT1 Building (DOE FRASE Vocabulary)
 BT2 Facilities and Buildings (DOE
      FRASE Vocabulary)
 BT3 Facilities
SO   DOE FRASE VOCABULARY

CONTAINMENT COMMANDERS
(EMER)
DA   February 1, 1991
BT1 On-Scene Commanders

BT2 Personnel
SO   Emergency Preparedness
DEF (EMER) Individuals working under
      the direction of the on-scene
      commander who is responsible for
      immediate containment of the
      emergency area.

CONTAINMENT FAILURE
DA   January 8, 1991
SF   CF
BT1 Failures
 BT2 Accidents

CONTAINMENT INTEGRATED LEAK
RATE TEST
DA   January 8, 1991
SF   CILRT

CONTAINMENT RADIOACTIVITY
REMOVAL
DA   January 8, 1991
SF   CNRR

CONTAINMENT SPRAY
DA   January 8, 1991
SF   CS

CONTAINMENT SPRAY SYSTEM
DA   January 8, 1991
SF   CSS
BT1 Containment Systems
 BT2 Emergency Systems
 BT3 Systems
NT1 Containment Spray System
      (Post-accident Injection Phase)
NT1 Containment Spray System
      (Post-accident Recirculation
      Phase)

CONTAINMENT SPRAY SYSTEM
(POST-ACCIDENT INJECTION PHASE)
DA   January 8, 1991
SF   CSSI
BT1 Containment Spray System
 BT2 Containment Systems
 BT3 Emergency Systems
 BT4 Systems

CONTAINMENT SPRAY SYSTEM
(POST-ACCIDENT RECIRCULATION
PHASE)
DA   January 8, 1991
SF   CSSR
BT1 Containment Spray System
 BT2 Containment Systems
 BT3 Emergency Systems
 BT4 Systems

CONTAINMENT STORAGE TANKS
(NFI)
DA   October 12, 1990
BT1 Tanks
 BT2 Facility Components
RT   Containment
RT   Contaminated Water Storage
SO   Nuclear Facilities Incident Database

SY-Synonymous Terms   SO-Source/Subject Category   SF-See From

SO Radiation
DEF (NFI) Contaminated water storage
tanks.

CONTAINMENT SYSTEMS
(AEA; CFR)
DA January 24, 1991
BT1 Emergency Systems
BT2 Systems
NT1 Containment Spray System
NT2 Containment Spray System
(Post-accident Injection Phase)
NT2 Containment Spray System
(Post-accident Recirculation
Phase)
RT Confinement Systems
RT Containment
DEF (CFR) The components of the
packaging intended to retain the
radioactive material during
transport.

CONTAINMENT VESSELS
(DOE Order 5480.3)
DA October 12, 1990
RT Containers
RT Maximum Normal Operating
Pressure
RT Optimum Interspersed
Hydrogenous Moderation
SO Hazardous Materials
DEF (DOE Order 5480.3) Receptacles in
which principal reliance is placed
to retain the radioactive material
during transport.

CONTAMINANT CARRIERS
(CWA; RHA)
DA October 12, 1990
RT Dredged Materials
RT Fill Materials
SO Water Pollution
DEF (CFR) Dredged or fill materials that
contain contaminants.

CONTAMINANTS
(SDWA; CFR; USC; EPA; ESH)
DA October 12, 1990
NT1 Air Contaminants
NT1 Common Laboratory Contaminants
RT Hazards
RT Maximum Contaminant Levels
RT Maximum Contaminant Level Goal
RT Pollutants
RT Pollutant or Contaminant
RT Secondary Maximum Contaminant
Levels
SO Compensation and Liability
SO Environmental Management
SO Environmental Protection Agency
Glossary
SO Hazardous Materials
SO Water Pollution
DEF Physical, chemical, biological, or
radiological substances or matter
that have an adverse affect on air,
water, or soil. (ESH) A chemical or
biological substance in a form that
can be incorporated into, onto or

be injested by and that harms
aquatic organisms, or users of the
aquatic environment, and includes
but is not limited to the substances
on the 40 CFR 307(a)(1) list of
toxic pollutants.

CONTAMINATED WATER STORAGE
(NFI)
DA October 12, 1990
SF CWS
BT1 Storage
BT2 Waste Management Processes
BT3 Processes
RT Containment Storage Tanks
SO Radiation
DEF (NFI) Part of containment systems,
500,000 gallon tank located in 50
million gallon basin.

CONTAMINATION
(DOE Order 5480.14; NCRP; EMER)
DA October 12, 1990
SY Radioactive Contamination
NT1 Chemical Contamination
NT1 Radioactive Contamination
NT2 Non-fixed Radioactive
Contamination
RT Impurities
RT Normal Form
RT Pollutants
SO Emergency Preparedness
SO Environmental Management
SO Radiation
DEF (NCRP) Deposition of radioactive
material in any place where it may
make products or equipment
unsuitable for some specific use.
The presence of any unwanted,
biological, physical, chemical or
radiological substance or matter
that has an adverse effect on air,
water, or soil.

CONTIGUOUS ZONES
(CERCLA; CFR)
DA October 12, 1990
BT1 Zones
BT2 Sites/Areas
RT Territorial Seas
SO Compensation and Liability
SO Environmental Management
DEF (CFR) The zone of the high seas,
established by the United States
under Article 24 of the Convention
on the Territorial Sea and
Contiguous Zone, which is
contiguous to the territorial sea
and which extends nine miles
seaward from the outer limit of the
territorial sea.

CONTINGENCY PLANNING ZONES
(EMER)
DA February 1, 1991
BT1 Zones
BT2 Sites/Areas
SO Emergency Preparedness
DEF (EMER) Provide precautionary

emergency planning for prompt
and effective actions beyond the
emergency planning zones.

CONTINGENCY PLANS
(SWDA; RCRA; CFR; EPA)
DA October 12, 1990
BT1 Emergency Plans
BT2 Plans
RT Accidents
RT Procedures
SO Environmental Management
SO Environmental Protection Agency
Glossary
SO Wastes
DEF (EPA) Documents setting out an
organized, planned, and
coordinated courses of action to
be followed in case of fires,
explosions, or other accidents that
release toxic chemicals,
hazardous wastes, or radioactive
materials which threaten human
health or the environment.

CONTINGENT LIABILITY
(SSDC)
DA October 12, 1990
BT1 Liabilities
SO System Safety Development Center
Glossary
DEF (SSDC) Liability incurred because
of negligence of a person engaged
by the insured to perform work; for
example, a contractor's
responsibility for work of a
subcontractor.

CONTINUITY OF GOVERNMENT
(DOE Order 5500.1A Attach. A, 15 (b))
DA January 24, 1991
BT1 Emergency Response Levels
RT Continuity of Government
Emergency Management Teams
RT Vital Records
SO Environmental Management
DEF (DOE 5500.1A) A condition caused
by the domestic or enemy attack
involving a national security threat
to the continuity of the federal
government.

CONTINUITY OF GOVERNMENT
EMERGENCY MANAGEMENT TEAMS
(DOE Order 5500.1A Attach. II, 9;
EMER)
DA January 24, 1991
SF COGEMT
BT1 Management Teams
BT2 Teams
BT3 Administrative Organizations
BT4 Organizations
RT Continuity of Government
RT Emergency Response Teams
SO Emergency Preparedness
SO Environmental Management
DEF (DOE 5500.1A) The DOE team
predesignated to ensure the
performance of the DOE essential

functions at designated locations and coordinate response to a condition caused by domestic or enemy attack involving a national security threat to the continuity of the federal government.

CONTINUOUS AIR MONITORS
(EMER)
DA   January 8, 1991
SF   *CAMS*
BT1 Monitors
 BT2 Equipment
SO   *Emergency Preparedness*
DEF (EMER) Instruments which collect airborne contamination and continuously count the activity collected on a filter medium. As concentrations rise above a set point, an audible and visual alarm is activated which continues until reset.

CONTINUOUS DISPOSAL
(CFR)
DA   October 12, 1990
BT1 Disposal
 BT2 Waste Management Processes
 BT3 Processes
RT Tailings
SO Air Pollution
DEF (CFR) A method of tailings management and disposal in which tailings are dewatered by mechanical methods immediately after generation. The dried tailings are then placed in trenches or other disposal areas and immediately covered to Federal standards.

CONTOUR PLOWING
(EPA)
DA   October 12, 1990
BT1 Cultivating
RT Erosion
RT Run-off
RT Strip Cropping
SO Environmental Protection Agency Glossary
DEF (EPA) Farming methods that break ground following the shape of the land in a way that discourages erosion.

CONTRACT LABORATORY PROGRAM
(EPA)
DA   October 12, 1990
SF   *CLP*
BT1 Programs
RT Contract Labs
RT Sample Management Offices
SO Environmental Protection Agency Glossary
DEF (EPA) Analytical program developed for Superfund waste site samples to fill the need for legally defensible analytical results

supported by a high level of quality assurance and documentation.

CONTRACT LABS
(EPA)
DA   October 12, 1990
BT1 Facilities
RT Contract Laboratory Program
SO Environmental Protection Agency Glossary
DEF (EPA) Laboratories under contract to EPA, which analyze samples taken from wastes, soil, air, and water or carry out research projects.

CONTRACT-REQUIRED QUANTITATION LIMITS
(EPA)
DA   October 12, 1990
BT1 Limits
RT Sampling
SO Environmental Protection Agency Glossary
DEF (EPA) Chemical-specific levels that a CLP (Contract Laboratory Program) laboratory must be able to routinely and reliably detect and quantitate in specified sample matrices. May or may not equal to the reported quantitation limit of a given chemical in a given sample.

CONTRACT SPECIFICATIONS
(RCRA; CFR)
DA   January 24, 1991
BT1 Specifications
 BT2 Requirements
RT Compensation and Liability
RT Construction Project Planning
RT Guide Specifications
DEF (CFR) "Contract specifications" means the set of specifications prepared for an individual construction project, which contains design, performance, and material requirements for that project.

CONTRACTING OFFICER'S REPRESENTATIVE
(DOE Order 5483.1A)
DA   October 12, 1990
BT1 Personnel
RT Contracting Officers
SO Industrial Safety
DEF (DOE Order 5483.1A) A DOE employee designated in writing by the contracting officer to represent the contracting officer for administrative and technical functions regarding the contract between DOE and the contractor.

CONTRACTING OFFICERS
(DOE Orders 5483.1A, 5484.1)
DA   October 12, 1990
SF   *CO (Contracting Officer)*
BT1 Personnel

RT Contracting Officer's Representative
SO Industrial Safety
SO Management
DEF DOE officials designated by Headquarters to enter into or administer contracts between DOE and contractors, and make contract-related determinations and findings.

CONTRACTOR EMPLOYEES
(DOE Order 5483.1A)
DA   October 12, 1990
BT1 Personnel
RT Contractors
RT Employees' Representative
RT Full-time Employees
SO Industrial Safety
DEF (DOE Order 5483.1A) Persons who is employed by a contractor.

CONTRACTOR SELF-EVALUATION REPORTS
(DOE Orders 5480.18, 5480.18 - Accreditation)
DA   October 12, 1990
BT1 Reports
BT1 Self-Evaluation
RT Training Programs
RT Training Program Accreditation Plan
SO Standards
DEF (DOE Order 5480.18) Formal reports prepared by the contractor summarizing the comparison of a training program to each accreditation objective and its supporting criteria.

CONTRACTORS
(DOE Orders 5480.4, 5483.1A; ESH)
DA   October 12, 1990
SY   DOE Contractors
RT Companies
RT Contractor Employees
RT Operating Level
SO Firearms
SO Industrial Safety
SO Management
SO Standards
DEF For purposes of DOE Order 5480.4, includes DOE prime contractors or subcontractors subject to the contractual provisions of DOE PR-50.704.2(a) or (b), or specific negotiated contract provisions indicating DOE's decision to enforce environmental protection, safety, and health protection requirements.

CONTRACTUAL LIABILITY
(SSDC)
DA   October 12, 1990
BT1 Liabilities
RT Compensation and Liability

SO System Safety Development Center
Glossary
DEF (SSDC) Liability assumed by
contract or agreement, and which
would not otherwise exist.

CONTRACTUAL RELATIONSHIPS
(CERCLA; USC)
DA October 12. 1990
RT Substantial Business Relationships
SO Compensation and Liability
DEF (USC) For the purpose of 42 USCS
9607(b)(3), includes, but is not
limited to, land contracts, deeds or
other instruments transferring title
or possession, unless the real
property on which the facility
concerned is located was acquired
by the defendant after the disposal
or placement of the hazardous
substance on, in, or at the facility,
and one or more of the
circumstances described in clause
(i), (ii), or (iii) is also established
by the defendant by a
preponderance of the evidence...
(See 42 USCS 9601 (35)(A) for
continuation of definition).

CONTRAILS
(EPA)
DA October 12, 1990
RT Atmosphere
SO Environmental Protection Agency
Glossary
DEF (EPA) Long, narrow clouds caused
when high-flying jet aircraft disturb
the atmosphere.

CONTRIBUTING CAUSES
(SSDC)
DA October 12, 1990
BT1 Causes
RT Root Causes
SO System Safety Development Center
Glossary
DEF (SSDC) Root causes or other
factors contributing to the
immediate cause of an accident or
incident.

CONTROL BARRIERS
(SSDC)
DA October 12, 1990
BT1 Barriers
SO System Safety Development Center
Glossary
DEF (SSDC) Those barriers used to
control wanted energy flows, such
as the insulation on an electrical
cord.

CONTROL DEVICES
(CFR)
DA October 12, 1990
BT1 Devices
NT1 Add On Control Devices
NT1 Afterburners
NT1 Arsenic Kitchens

NT1 Carbon Adsorbers
NT1 Catalytic Incinerators
NT1 Cyclone Collectors
NT1 Packed Towers
NT1 Precipitators
 NT2 Electrostatic Precipitators
NT1 Primary Emission Control Systems
NT1 Secondary Hood Systems
NT1 Vapor Capture Systems
 NT2 Floor Sweeps
RT Asbestos Waste from Control
Devices
RT Capture Efficiency
RT Closed Vent Systems
RT Curtail
RT Dispersion Techniques
SO Air Pollution
DEF (CFR) The air pollution control
equipment used to collect
particulate matter emissions. An
enclosed combustion device,
vapor recovery system, or flare.

CONTROL PANEL(S)
(2554 CONTROL PANE)
DA January 3, 1991
BT1 Equipment/Parts - Nuclear (DOE
FRASE Vocabulary)
 BT2 Equipment
 BT2 Reactor Components
SO DOE FRASE VOCABULARY

CONTROL RESOURCES
(CFR)
DA November 15, 1990
RT Baseline
DEF (CFR) Resources unaffected by the
discharge of oil or release of the
hazardous substance under
investigation. A control area or
resource is selected for its
comparability to the assessment
area or resource and may be used
for establishing the baseline
condition and for comparison to
injured resources.

CONTROL ROD(S)
(2556 CONTROL ROD)
DA January 3, 1991
BT1 Equipment/Parts - Nuclear (DOE
FRASE Vocabulary)
 BT2 Equipment
 BT2 Reactor Components
NT1 Gangs
NT1 Raincoats
RT Control Rod Drive Assembly
RT Core(s)
SO DOE FRASE VOCABULARY

CONTROL ROD DRIVE
DA January 8, 1991
SF CRD
BT1 Control Rod Drive Control System
 BT2 Control Systems
  BT3 Controls
  BT3 Systems
 BT2 Reactor Components
NT1 Pistol Grip Switches

CONTROL ROD DRIVE ASSEMBLY
(2555 CRDA)
DA January 3, 1991
BT1 Control Rod Drive Control System
 BT2 Control Systems
  BT3 Controls
  BT3 Systems
 BT2 Reactor Components
BT1 Equipment/Parts - Nuclear (DOE
FRASE Vocabulary)
 BT2 Equipment
 BT2 Reactor Components
RT Control Rod(s)
RT Rod Drive
SO DOE FRASE VOCABULARY

CONTROL ROD DRIVE CONTROL
SYSTEM
DA January 8, 1991
SF CRDCS
BT1 Control Systems
 BT2 Controls
 BT2 Systems
BT1 Reactor Components
NT1 Control Rod Drive Assembly
NT1 Control Rod Drive
 NT2 Pistol Grip Switches
NT1 Control Rod Drive Mechanism

CONTROL ROD DRIVE MECHANISM
DA January 8, 1991
SF CRDM
BT1 Control Rod Drive Control System
 BT2 Control Systems
  BT3 Controls
  BT3 Systems
 BT2 Reactor Components

CONTROL ROD FAULT
(NFI)
DA October 12, 1990
SO Nuclear Facilities Incident Database
SO Radiation
DEF (NFI) Control rod position differs
from its demand position by 4
veeder units.

CONTROL ROOM
(1684 CONTROL ROOM)
DA December 10, 1990
BT1 Room
 BT2 Sites/Areas
SO DOE FRASE VOCABULARY
DEF The area in a nuclear power plant
from which most of the plant
power production and emergency
safety equipment can be operated
by remote control.

CONTROL STRATEGIES
(CAA; CFR; ESH)
DA October 12, 1990
NT1 Transportation Control Measures
RT Air Quality Standards
RT Compliance Schedules
RT Controls (Reactor)
RT Schedule of Compliance
SO Air Pollution
DEF (CFR) A combination of measures

BT-Broader Term   NT-Narrower Term   RT Related Term

designated to achieve the aggregate reduction of emissions necessary for attainment and maintenance of national standards. See 40 CFR 51.100 for types of measures included.

CONTROL SUBSTANCES
(TSCA; CFR)
DA   October 19, 1990
RT   Batches
RT   Test Substances
SO   Environmental Management
SO   Hazardous Materials
DEF (CFR) Chemical substances or mixtures or any other material other than a test substance that are administered to the test system in the course of a study for the purpose of establishing a basis for comparison with the test substance.

CONTROL SYSTEMS
(SSDC)
DA   October 12, 1990
BT1 Controls
BT1 Systems
NT1 Best Available Control Technology
NT1 Best Available Retrofit Technology
NT1 Chemical Volume and Control System
NT1 Control Rod Drive Control System
NT2 Control Rod Drive Assembly
NT2 Control Rod Drive
NT3 Pistol Grip Switches
NT2 Control Rod Drive Mechanism
NT1 Energy Monitoring and Control Systems
NT1 Intermittent Control System
NT1 Primary Emission Control Systems
NT1 Remote Detection and Control Systems
NT1 Safety Injection Control System
NT1 Standby Liquid Control System
RT   Emergency Systems
RT   Energy Management Systems
RT   Information Systems
RT   Water Systems
SO   Environmental Management
SO   Radiation
SO   System Safety Development Center Glossary
DEF (SSDC) All elements comprising a system, including management's plans and policy, procedures, personnel, hardware, and facilities, as they relate to the control of safety within the system.

CONTROL TECHNIQUE GUIDELINES
(EPA)
DA   October 12, 1990
SF   CTG
BT1 Guidelines
SO   Environmental Protection Agency Glossary
DEF (EPA) A series of EPA documents designed to assist states in defining reasonable available

control technology (RACT) for major sources of volatile organic compounds (VOCs).

CONTROL WIDTH
(AEA; CFR)
DA   January 24, 1991
BT1 Measurements
RT   Faults
DEF (CFR) The "control width" of a fault is the maximum width of the zone containing mapped fault traces, including all faults which can be reasonably inferred to have experienced differential movement during Quaternary times and which join or can reasonably be inferred to join the main fault trace, measured within 10 miles along the fault's trend in both directions from the point of nearest approach to the site.

CONTROLLED AREAS
(DOE Order 5480.11; AEA; CFR; IAEA; EMER)
DA   October 12, 1990
BT1 Sites/Areas
NT1 Disturbed Zones
RT   Accessible Environment
RT   Non-Restricted Access Areas
RT   Radiation Safety
RT   Special Source of Ground Water
SO   Emergency Preparedness
SO   Environmental Management
SO   Industrial Hygiene
SO   Radiation
DEF (DOE Order 5480.11) Areas to which access is controlled in order to protect individuals from exposure to radiation and radioactive materials. (IAEA) Areas where workers might receive doses in excess of three-tenths of the occupational dose equivalent limits during the anticipated working period and where appropriate controls (such as restricted access, individual assessment of dose and special health supervision) are accordingly applied.

CONTROLLED COPIES
(DOE Order 5500.1A Attach. II, 10; EMER)
DA   January 24, 1991
SO   Emergency Preparedness
SO   Environmental Management
DEF (EMER) Documents that are maintained on current basis by means of a formal transmittal and filing system.

CONTROLLERS
(DOE Order 5480.16; EMER)
DA   October 12, 1990
BT1 Personnel
NT1 Senior Controllers

SO   Emergency Preparedness
SO   Firearms
DEF (DOE Order 5480.16) Individuals trained in firearms activities who help to ensure that training exercises are conducted safely and that all participants are following the rules.

CONTROLS
(MORT)
DA   February 25, 1991
NT1 Abnormal Condition Control
NT1 Biological Controls
NT1 Configuration Control
NT1 Controls (Reactor)
NT2 Breakthrough
NT1 Control Systems
NT2 Best Available Control Technology
NT2 Best Available Retrofit Technology
NT2 Chemical Volume and Control System
NT2 Control Rod Drive Control System
NT3 Control Rod Drive Assembly
NT3 Control Rod Drive
NT4 Pistol Grip Switches
NT3 Control Rod Drive Mechanism
NT2 Energy Monitoring and Control Systems
NT2 Intermittent Control System
NT2 Primary Emission Control Systems
NT2 Remote Detection and Control Systems
NT2 Safety Injection Control System
NT2 Standby Liquid Control System
NT1 Incident Control
NT1 Institution Control
NT1 Instrumentation and Control
NT1 Interface Control Unit
NT1 Maintenance Information and Control
NT1 Post Removal Site Control
NT1 Quality Control
NT2 Quality Assurance Overviews
NT3 Quality Assurance Plans
NT4 Site Quality Assurance Plan
NT2 Quality Inspection Control
NT2 Quality Verification
NT1 Reactor Volume Control
NT1 Remote Detection and Control
NT1 Underground Injection Control
RT   Accidents
RT   Alarms
RT   Barriers
RT   First Line Supervision
RT   Hazards
RT   Higher Supervision
RT   Inspections
RT   Load Securing Device
RT   Maintenance
RT   Monitors
RT   Operability
RT   Personnel Monitoring Equipment
RT   Protection of the Public Health and Welfare
RT   Radiation Detectors
RT   Redundance
RT   Systems
RT   Systems (DOE FRASE Vocabulary)
RT   Targets

SY-Synonymous Terms   SO-Source/Subject Category   SF-See From

RT Technical Information
RT Valves
DEF (MORT) MORT analysis asks: were adequate controls in place to prevent vulnerable persons and objects from being exposed to harmful energy flows and/or environmental conditions? were controls designed to prevent harmful energy flows or environmental conditions from reaching vulnerable people and objects? were controls designed to prevent vulnerable people and objects from encountering harmful energy flows and environmental conditions?

CONTROLS (REACTOR)
(DOE Order 5480.6; SSDC)
DA   October 12, 1990
BT1 Controls
NT1 Breakthrough
RT Control Strategies
SO Industrial Safety
SO System Safety Development Center Glossary
DEF (DOE Order 5480.6) When used with respect to nuclear reactors, means apparatus and mechanisms that, when manipulated, directly or indirectly affect the reactivity or power level of a reactor or status of an engineered safety feature. (SSDC) The implementation of management policies, standards, and procedures in achieving desired system performance and efficient work processes. Controls can be thought of as administrative barriers in both the prevention of accidents and incidents, and in their analysis and investigation.

CONTUSION(S)
(1312 CO)
DA   November 28, 1990
BT1 Brain Damage
 BT2 Injuries
RT Concussion
RT Head
RT Head Protection
RT Loss of Consciousness
RT Skull (Frase)
SO DOE FRASE VOCABULARY

CONVENTIONAL FILTRATION TREATMENT
(SDWA; CFR)
DA   October 12, 1990
BT1 Treatment
 BT2 Waste Management Processes
  BT3 Processes
RT Filtration
RT Flocculation
RT Particulates
RT Sedimentation
RT Wastewater Treatment Processes

SO Water Pollution
DEF (CFR) A series of processes including coagulation, flocculation, sedimentation, and filtration resulting in substantial particulate removal.

CONVENTIONAL MINES
(SDWA; CFR)
DA   October 12, 1990
BT1 Sites/Areas
NT1 Active Mines
NT1 Inactive Mines
NT1 Underground Uranium Mines
RT Strip Mining
SO Water Pollution
DEF (CFR) An open pit or underground excavation for the production of minerals.

CONVENTIONAL POLLUTANTS
(EPA; ESH)
DA   October 12, 1990
BT1 Pollutants
SO Environmental Protection Agency Glossary
SO Hazardous Materials
DEF (ESH) Statutorily listed pollutants which are understood well by scientists. These may be in the form of organic waste, sediment, acid, bacteria and viruses, nutrients, oil and grease, or heat.

CONVENTIONAL SYSTEMS
(EPA)
DA   October 12, 1990
BT1 Systems
RT Waste Water Collection Systems
SO Environmental Protection Agency Glossary
DEF (EPA) Systems that have been traditionally used to collect municipal wastewater in gravity sewers and convey it to a central primary or secondary treatment plant prior to discharge to surface waters.

CONVERTER ARSENIC CHARGING RATE
(CFR)
DA   October 12, 1990
BT1 Rates
RT Charging
RT Copper Converters
RT Copper Mattes
SO Air Pollution
DEF (CFR) The hourly rate at which arsenic is charged to the copper converters in the copper converter department based on the arsenic content of the copper matte and of any lead matte that is charged to the copper converters.

CONVEYOR(S)
(2310 CONVEYOR)
DA   December 10, 1990

BT1 Material Handling Device
 BT2 Devices
 BT2 Equipment/Parts - Material Handling (DOE FRASE Vocabulary)
 BT3 Equipment
SO DOE FRASE VOCABULARY

COOLANT RETURN PUMP
(2557 CR PUMP)
DA   January 3, 1991
BT1 Equipment/Parts - Nuclear (DOE FRASE Vocabulary)
 BT2 Equipment
 BT2 Reactor Components
BT1 Pump(s)
 BT2 Machines (DOE FRASE Vocabulary)
 BT3 Equipment
SO DOE FRASE VOCABULARY

COOLANT RETURN TANK
DA   January 8, 1991
SF   *CRT*
BT1 Tanks
 BT2 Facility Components

COOLANTS
(EPA)
DA   October 12, 1990
NT1 Primary Coolants
RT 3 AC Flow
RT Cooling Towers
SO Environmental Protection Agency Glossary
DEF (EPA) Liquids or gases used to reduce the heat generated by power production in nuclear reactors, electric generators, various industrial and mechanical processes, and automobile engines.

COOLING SYSTEM
(2026 COOLING SYST)
DA   December 10, 1990
BT1 Systems (DOE FRASE Vocabulary)
 BT2 Systems
SO DOE FRASE VOCABULARY

COOLING TOWERS
(EPA)
DA   October 12, 1990
BT1 Structures (DOE FRASE Vocabulary)
RT Coolants
SO Environmental Protection Agency Glossary
DEF (EPA) Structures that help remove heat from water used as a coolant, e.g., in electric power generating plants. (NRC Glossary of Terms: Nuclear Power and Radiation) A heat exchanger designed to aid in the cooling of water that was used to cool exhaust steam exiting the turbines of a power plant. Cooling towers transfer exhaust heat into

the air instead of into a body of water.

COOLING WATER GAMMA MONITOR
DA January 8, 1991
SF *CWGM*
BT1 Monitors
 BT2 Equipment
BT1 Radiation Detectors
 BT2 Equipment
RT Non-Contact Cooling Water

COOPER E
(DOE Order 6430.1A)
DA October 12, 1990
SO Construction
DEF (DOE Order 6430.1A) The recommended live load in pounds per axle and the uniform trailing load for each track.

COOPERATIVE AGREEMENTS
(ANL; CFR)
DA May 24, 1991
BT1 Agreements
RT U.S. Environmental Protection Agency
SO Environmental Protection Agency Glossary
SO Wastes
DEF (CFR) A legal instrument EPA uses to transfer money, property, services, or anything of value to a recipient to accomplish a public purpose in which substantial EPA involvement is anticipated during the performance of the project.

COORDINATION PROCESSES
(ESH)
DA October 12, 1990
BT1 Administrative Processes
 BT2 Processes
RT Significant Environmental Compliance Issues
SO Management
DEF (ESH) The coordination processes are the means by which significant environmental compliance issues will be resolved or disseminated to ensure timely development and consistent application of Departmental environmental policy and guidance.

COPPER
DA January 8, 1991
SF *Cu*

COPPER CONVERTER DEPARTMENT
(CFR)
DA October 12, 1990
SO Air Pollution
DEF (CFR) All copper converters at a primary copper smelter.

COPPER CONVERTERS
(CFR)

DA October 12, 1990
BT1 Containers
RT Blowing
RT Charging
RT Converter Arsenic Charging Rate
RT Copper Mattes
RT Lead Mattes
RT Pouring
RT Primary Copper Smelter
SO Air Pollution
DEF (CFR) Vessels in which copper matte is charged and is oxidized to copper.

COPPER MATTES
(CFR)
DA October 12, 1990
BT1 Solutions
 BT2 Mixtures
RT Converter Arsenic Charging Rate
RT Copper Converters
RT Lead Mattes
SO Air Pollution
DEF (CFR) Molten solutions of copper and iron sulfides produced by smelting copper sulfide ore concentrates or calcines.

COPRODUCTS
(TSCA; CFR)
DA January 24, 1991
BT1 Chemical Substances
RT By-products
DEF (CFR) A chemical substance produced for a commercial purpose during the manufacture, processing, use, or disposal of another chemical substance or mixture.

CORE(S)
(2558 CORE)
DA January 3, 1991
BT1 Equipment/Parts - Nuclear (DOE FRASE Vocabulary)
 BT2 Equipment
 BT2 Reactor Components
RT Control Rod(s)
RT Fuel Element(s)
RT Reactor(s)
SO DOE FRASE VOCABULARY

CORE DAMAGE FREQUENCY
DA January 8, 1991
SF *CDF*

CORE FLOOD SYSTEM
DA January 8, 1991
SF *CFS*
BT1 Systems

CORE FLOOD TANKS
DA January 8, 1991
SF *CFT*
BT1 Tanks
 BT2 Facility Components

CORE MELT
DA January 8, 1991
SF *CM*

CORE MELT ACCIDENTS
(EMER)
DA February 1, 1991
BT1 Accidents
RT Reactor Shutdown
RT Reactor Fuel
SO Emergency Preparedness
DEF (EMER) Reactor accidents in which fuel melts because of significant overheating.

CORRECTIVE ACTION MANAGEMENT UNITS
(RCRA)
DA January 29, 1991
BT1 Sites/Areas
NT1 Disposal Units
 NT2 Land Disposal Units
  NT3 Landfill Cells
  NT3 Landfills
   NT4 Chemical Waste Landfills
   NT4 Sanitary Landfills
   NT4 Specially Designated Landfills
NT1 Hazardous Waste Management Units
 NT2 Miscellaneous Units
NT1 Solid Waste Management Units
RT Corrective Actions
RT Waste Management
SO Environmental Management
DEF (CFR) Large, land-based units which allow for the removal of waste from a smaller unit to another larger unit (if they are next to each other and there is no uncontaminated soil between them) without first undergoing treatment.

CORRECTIVE ACTION REPORTING
DA June 18, 1991
SF *CAR*
SO Management

CORRECTIVE ACTION TRIGGERS
(MORT)
DA April 3, 1991
RT Analyses
RT Communication
RT Hazard Analysis
RT Knowledge
RT Monitoring
RT Risk Assessment
RT Technical Information
RT Trending
DEF (MORT) MORT analysis asks: were triggers (stimuli) for the initiation of hazard analysis adequate? were they utilized to obtain early safety participation and review in planned or unplanned changes? Related issues/topics include: one-on-one fixes; priority problem fixes; planned change controls;

SY-Synonymous Terms   SO-Source/Subject Category   SF-See From

unplanned change controls; and
new information uses.

CORRECTIVE ACTIONS
(RCRA; USC; EMER; MORT)
DA   January 24, 1991
SY   Amelioration
BT1 Actions
 BT2 Responses
BT1 Recommendations
RT   Accidents
RT   Corrective Action Management
       Units
RT   Recovery Actions
RT   Relations
RT   Remedial Actions
RT   Specific Control Factors
SO   Emergency Preparedness
SO   Environmental Management
DEF (USC) EPA is required to
       promulgate regulations that require
       evidence of financial responsibility
       for corrective action and corrective
       action must encompass affected
       areas beyond the boundary of the
       facility. All permits must address
       releases of hazardous waste or
       constituents regardless of when
       the source materials were
       emplaced. Corrective actions can
       be required of interim status
       facilities. (MORT) In MORT
       analysis, once an accident has
       occurred was there adequate
       amelioration on the part of all
       concerned parties? The intent is
       to limit the consequences of what
       has immediately occurred and to
       reduce the sensitivity of those
       consequences whenever possible.

CORRECTIVE REPAIR MAINTENANCE
(DOE Order 4330.4A)
DA   June 4, 1991
SF   CRM (Corrective Repair
       Maintenance)
BT1 Maintenance
 BT2 Activities
SO   Management
DEF (DOE Order 4330.4A) The repair of
       failed or malfunctioning equipment,
       systems, or facilities to restore the
       intended function or design
       condition. This maintenance does
       not result in a significant extension
       of the expected useful life.

CORRIDOR/HALL
(1685 CORRIDOR)
DA   December 10, 1990
BT1 Site (DOE FRASE Vocabulary)
 BT2 Sites/Areas
SO   DOE FRASE VOCABULARY

CORROSION
(EPA)
DA   October 12, 1990
BT1 Chemical Processes
 BT2 Processes

NT1 External Corrosion
RT   Cathodic Protection
RT   Corrosion Experts
RT   Corrosive Materials
RT   Oxidation
SO   Environmental Protection Agency
       Glossary
DEF (EPA) The dissolving and wearing
       away of metal caused by a
       chemical reaction such as
       between water and the pipes that
       the water contacts, chemicals
       touching a metal surface, or
       contact between two metals.

CORROSION EXPERTS
(SWDA; RCRA; CFR)
DA   October 12, 1990
BT1 Technically Qualified Individuals
 BT2 Personnel
RT   Corrosion
SO   Environmental Management
SO   Wastes
DEF (CFR) People who, by reason of
       thorough knowledge of the
       physical sciences and the
       principles of engineering and
       mathematics acquired by a
       professional education and related
       practical experience, are qualified
       to engage in the practice of
       corrosion control on buried or
       submerged metal piping systems
       and metal tanks. Such people
       must be accredited or certified as
       being qualified by the National
       Association of Corrosion
       Engineers or be registered
       professional engineers who have
       certification or licensing that
       includes education and experience
       in corrosion control of buried or
       submerged metal piping systems
       and metal tanks.

CORROSIVE MATERIALS
(HMTA; CFR)
DA   October 12, 1990
BT1 Hazardous Materials
 BT2 Materials
RT   Corrosion
SO   Emergency Preparedness
SO   Hazardous Materials
DEF (CFR) Corrosive materials are
       liquids or solids that cause visible
       destruction or irreversible
       alterations in human skin tissue at
       the site of contact, or in the case
       of leakage from its packaging, a
       liquid that has a severe corrosion
       rate on steel. Bromine, chemical
       kits, and solid chloroacetic acid
       are examples of a corrosive
       material.

COST-EFFECTIVE ANALYSIS
(IAEA)
DA   October 12, 1990
BT1 Analyses

SO   Radiation
DEF (IAEA) A procedure that is used to
       determine the most effective
       protection obtainable from fixed
       resources or, alternatively, to
       determine the least expensive
       protection for a given level of
       exposure.

COST PLUS AWARDS FEE
DA   January 8, 1991
SF   CPAF

COST RECOVERY
(EPA)
DA   October 12, 1990
BT1 Administrative Processes
 BT2 Processes
SO   Environmental Protection Agency
       Glossary
DEF (EPA) A legal process by which
       potentially responsible parties who
       contributed to contamination at a
       Superfund site can be required to
       reimburse the Trust Fund for
       money spent during any cleanup
       actions by the federal government.

COSTS
DA   February 25, 1991
NT1 Actual Cash Value
NT1 Capital Expenditure
NT1 Current Closure Cost Estimate
NT1 Current Post-Closure Cost Estimate
NT1 Current Plugging and
       Abandonment Cost Estimates
NT1 Eligible Costs
NT1 Fixed Capital Costs
NT1 Legal Defense Costs
NT1 Life Cycle Costs
NT1 Program Significant Cost
NT1 Reasonable Costs
NT1 Replacement Values

**COUNCIL ON ENVIRONMENTAL
QUALITY**
(NEPA)
DA   January 8, 1991
SF   CEQ
BT1 Professional Bodies
 BT2 Organizations
RT   National Environmental Policy Act
DEF (NEPA) Title II of the National
       Environmental Policy Act create
       the Council on Environmental
       Quality and defines the scope as:
       1) preparing an annual report on
       environmental quality and on
       existing and proposed federal
       efforts to improve environmental
       quality; 2) appraising programs
       and activities of the Federal
       Government in the light of the
       national environmental policy; 3)
       formulating and recommending
       national policies to promote
       environmental improvement; 4)
       gathering and analyzing
       environmental information, making

relevant studies, including monitoring trends in environmental quality; and 5) advising the president.

COUNT
(2780 COUNT)
DA   January 3, 1991
BT1 Equipment/Parts -
     Instrumentation/Measuring (DOE
     FRASE Voc.)
 BT2 Equipment
SO   DOE FRASE VOCABULARY

COUNT (RADIATION MEASUREMENT)
(NIH)
DA   October 12, 1990
BT1 Measurements
RT   External Fission Counter
RT   Scintillation Counters
SO   Radiation
DEF (NIH) The external indication of a
     device designed to enumerate
     ionizing events. It may refer to a
     single detected event or to the
     total registered in a given period of
     time. The term is often erroneously
     used to designate a disintegration,
     ionizing event, or voltage pulse.

COUNT RATE METER
DA   January 8, 1991
SF   *CRM (Count Rate Meter)*
RT   Radiation Detectors

COUPLING
(SSDC)
DA   October 12, 1990
RT   Conditional Probability
SO   System Safety Development Center
     Glossary
DEF (SSDC) The situation wherein if a
     person makes an error, there is an
     increased probability of an error
     immediately following. This can
     lead to chains of errors occurring,
     called conditional probability.

COURIERS
(EMER)
DA   February 1, 1991
BT1 Personnel
RT   Classified Information
RT   Classified Matter
SO   Emergency Preparedness
DEF (EMER) Department of Energy
     (DOE) employees or members of
     the armed forces assigned to and
     performing duties under the
     direction and control of the DOE,
     who are specifically designated for
     armed protection in transit of top
     secret or other matter which, in
     the opinion of the responsible
     head of a department element,
     requires such protection. Couriers
     are required to carry credential
     identification.

COVER MATERIALS
(RCRA; CFR; EPA)
DA   October 12, 1990
BT1 Materials
NT1 Daily Cover
NT1 Final Cover
NT1 Intermediate Cover
RT   Sanitary Landfills
SO   Environmental Management
SO   Environmental Protection Agency
     Glossary
DEF (EPA) Soil used to cover
     compacted solid waste in sanitary
     landfills.

COVERS
(EPA)
DA   October 12, 1990
SO   Construction
SO   Environmental Protection Agency
     Glossary
DEF (EPA) Vegetation or other material
     providing protection as ground
     covers.

COVERT THREATS
(EMER)
DA   February 1, 1991
BT1 Threats
 BT2 Emergencies
  BT3 Reportable Occurrences
   BT4 Occurrences
SO   Emergency Preparedness
DEF (EMER) Threats to a U.S.
     Department of Energy security
     interest caused by one or more
     individuals wishing to keep
     unfriendly actions and their
     identities from detection through
     the use of such tactics as stealth,
     guile, or deceit.

CO₂
DA   October 12, 1990
SEE Carbon Dioxide
SO   Acronyms

CPAF
DA   October 12, 1990
SEE Cost Plus Awards Fee
SO   Acronyms

CPL
DA   October 12, 1990
SEE Confinement Protection Limits
SO   Acronyms

Cr
DA   October 12, 1990
SEE Chromium
SO   Acronyms

CRANE(S)
(2313 CRANE)
DA   December 10, 1990
BT1 Material Handling Device
 BT2 Devices
 BT2 Equipment/Parts - Material

Handling (DOE FRASE
Vocabulary)
 BT3 Equipment
NT1 Boom Crane
NT1 Bridge Crane
NT1 Mobile Crane
NT1 Overhead Crane
RT   Derrick
SO   DOE FRASE VOCABULARY

CRANE BRIDGE
(2312 CRANE BRIDGE)
DA   December 10, 1990
SO   DOE FRASE VOCABULARY

CRANE CONTROL, CLOSE CIRCUIT TV
DA   January 8, 1991
SF   *CCTV*

CRANE CONTROL ROOM
DA   January 8, 1991
SF   *KCR*

CRANE DRIP PAN MONITOR
DA   January 8, 1991
SF   *CDPM*
BT1 Monitors
 BT2 Equipment

CRANE MAINTENANCE AREA
DA   January 8, 1991
SF   *CMA*
BT1 Sites/Areas

CRAWL SPACES
(EPA)
DA   October 12, 1990
BT1 Sites/Areas
SO   Environmental Protection Agency
     Glossary
DEF (EPA) In some types of houses,
     which are constructed so that the
     floor is raised slightly above the
     ground, areas beneath the floor
     which allow access to utilities and
     other services. This is in contrast
     to slab-on-grade or basement
     construction houses.

CRD
DA   October 12, 1990
SEE Control Rod Drive
SO   Acronyms

CRDCS
DA   October 12, 1990
SEE Control Rod Drive Control System
SO   Acronyms

CRDM
DA   October 12, 1990
SEE Control Rod Drive Mechanism
SO   Acronyms

CREDIBILITY ASSESSMENTS
(EMER)
DA   February 1, 1991

SY-Synonymous Terms   SO-Source/Subject Category   SF-See From

BT1 Assessments
 BT2 Administrative Processes
  BT3 Processes
RT Threats
SO Emergency Preparedness
DEF (EMER) Actions undertaken to determine the reliability or believability of expressed threats.

CREDIBLE ACCIDENTS
(DOE Order 6430.1A)
DA October 12, 1990
BT1 Accidents
NT1 Maximum Credible Accident
SO Construction
DEF (DOE Order 6430.1A) Those accidents with an estimated probability of occurrence greater than ten raised to the negative sixth power per year. Natural phenomena use separate probability criteria as stated in "Design and Evaluation Guidelines for Department of Energy Facilities Subjected to Natural Phenomena Hazards" (UCRL-15910).

CRISIS MANAGEMENT TEAMS
DA January 8, 1991
SF *CMT*
BT1 Management Teams
 BT2 Teams
  BT3 Administrative Organizations
   BT4 Organizations

CRISIS MANAGERS
(EMER)
DA February 1, 1991
BT1 Personnel
RT Duty Officers
RT Emergency Management Teams
SO Emergency Preparedness
DEF (EMER) The senior persons designated to assume command of operations in a particular emergency operations center (EOC) and authorized to direct the emergency management team assigned to that EOC. May also be called by other names such as team leader.

CRITERIA
(DOE Order 5482.1B; RCRA; CFR; EPA)
DA October 12, 1990
RT Management Appraisals
SO Construction
SO Environmental Management
SO Environmental Protection Agency Glossary
SO Management
DEF (DOE Order 5482.1B) Rules or tests against which the quality of performance can be measured. They are most effective when expressed quantitatively. Fundamental criteria are contained in policies and objectives, as well as codes, standards, regulations,

and recognized professional practices that DOE and DOE contractors are required to observe. (EPA) Descriptive factors taken into account by EPA in setting standards for various pollutants. These factors are used to determine limits on allowable concentration levels and to limit the number of violations per year. When issued by EPA, the criteria provide guidance to the states on how to establish their standards.

CRITERIA POLLUTANTS
(EPA)
DA October 12, 1990
BT1 Pollutants
NT1 Carbon Monoxide
NT1 Lead
NT1 Nitrogen Oxides (NO_x)
 NT2 Nitric Oxides
 NT2 Nitrogen Dioxide (NO_2)
NT1 Ozone
NT1 Sulfur Dioxide
NT1 Total Suspended Particulates
SO Environmental Protection Agency Glossary
DEF (EPA) The 1970 amendments to the Clean Air Act required EPA to set National Ambient Air Quality Standards for certain pollutants known to be hazardous to human health. EPA has identified and set standards to protect human health and welfare for six pollutants: ozone, carbon monoxide, total suspended particulates, sulfur dioxide, lead, and nitrogen oxide. The term, "criteria pollutants" derives from the requirement that EPA must describe the characteristics and potential health and welfare effects of these pollutants. It is on the basis of these criteria that standards are set or revised.

CRITICAL AREAS
(DOE Order 6430.1A)
DA October 12, 1990
BT1 Sites/Areas
RT Critical Facilities
SO Construction
DEF (DOE Order 6430.1A) Those structures and enclosures containing safety class items whose continued integrity is essential to ensure the operability of those safety class items in the event of a DBA (Design Basis Accident).

CRITICAL FACILITIES
(DOE Order 6430.1A)
DA October 12, 1990
BT1 Facilities
RT Critical Areas
RT Emergency Control Centers
RT Safety Class Items

SO Construction
DEF (DOE Order 6430.1A) Facilities such as those for radioactive material handling, processing, or storage and those facilities having high replacement value or vital importance to DOE programs.

CRITICAL GROUP
(IAEA)
DA October 12, 1990
SY Critical Population Group
RT Critical Pathways
RT Members of the Public
RT Public Doses
SO Radiation
DEF (IAEA) For a given radiation source, the members of the public whose exposure is reasonably homogeneous and is typical of individuals receiving the highest effective dose equivalent or dose equivalent (whichever is relevant) from the source.

CRITICAL HABITATS
(ESA; CFR; USC)
DA October 12, 1990
BT1 Habitats
NT1 Proposed Critical Habitats
RT Adverse Modifications
RT Biological Opinion
RT Conservation Recommendations
RT Destruction
RT Wildlife Refuges
SO Endangered Species
DEF (CFR) (1) The specific areas within the geographical area currently occupied by a species, at the time it is listed in accordance with the Act, on which are found those physical or biological features (i) essential to the conservation of the species and (ii) that may require special management considerations or protection, and (2) specific areas outside the geographical area occupied by a species at the time it is listed upon a determination by the secretary that such areas are essential for the conservation of the species.

CRITICAL INCIDENT TECHNIQUE
(SSDC)
DA October 12, 1990
SF *CIT*
BT1 Administrative Processes
 BT2 Processes
RT Hazard Analysis
RT Industrial Safety
RT Reported Significant Observations
SO System Safety Development Center Glossary
DEF (SSDC) A method of identifying human errors and unsafe conditions which contribute to incidents and accidents within a given plant or system group. It involves interviewing a

representative sample of employees exposed to various hazards and having them describe unsafe behavior or conditions which they have observed in the past. The data is then tabulated into hazard categories to identify accident problem areas.

CRITICAL ITEMS LIST
DA January 8, 1991
SF *CIL*

CRITICAL JOBS (TASK)
(SSDC)
DA October 12, 1990
RT Industrial Safety
RT Standard Job Procedures
SO System Safety Development Center Glossary
DEF (SSDC) Jobs or tasks within an occupation that have been associated with major loss more frequently than others. They also could be jobs with the potential for a major loss.

CRITICAL MASS
(DOE Order 5480.5; EMER)
DA October 12, 1990
BT1 Measurements
RT Criticality
SO Emergency Preparedness
SO Environmental Management
SO Standards
DEF (DOE Order 5480.5) The smallest mass of fissionable material that will support a self-sustaining chain reaction under specified conditions.

CRITICAL ORGANS
(SDWA; CFR; NIH)
DA October 12, 1990
BT1 Whole Body
RT Irradiation
RT Known Human Effects
RT Maximum Permissible Concentration
RT Weighting Factors
SO Air Pollution
SO Management
SO Radiation
DEF Those organs or tissues, the irradiation of which will result in the greatest hazard to the health of the individual or his descendants. (40 CFR 61.91) The most exposed human organ or tissue exclusive of the integumentary system (skin) and the cornea.

CRITICAL PATHWAYS
(ESH; IAEA)
DA October 12, 1990
BT1 Routes (Exposure)
RT Critical Group
RT Exposure Pathways

SO Radiation
DEF (ESH) The specific routes of transfer of radionuclides form one environmental component to another (e.g., from one trophic level to another) that results in the greatest fraction of an applicable dose limit to a population group or an individual's whole body, organ or tissue. (IAEA) The dominant environmental pathways through which given radionuclides reach critical groups.

CRITICAL POPULATION GROUP
(Key Definitions under Management Controls)
DA October 12, 1990
SY Critical Group
SO Management
DEF (ESH) Population group showing the greatest fraction of an applicable radiation dose limit as a result of site releases.

CRITICAL TEMPERATURE RATIO
DA January 8, 1991
SF *CTR*
BT1 Ratios

CRITICALITY
(IAEA; EMER)
DA October 12, 1990
SY Nuclear Criticality
NT1 Reactor Subcriticality
RT Close Reflection By Water
RT Critical Mass
RT Criticality Excursions
RT Delta K Rods
SO Emergency Preparedness
SO Radiation
DEF (IAEA) The conditions in which a system is capable of sustaining a nuclear chain reaction.

CRITICALITY ACCIDENTS
(IAEA; EMER)
DA October 12, 1990
BT1 Accidents
RT Criticality Incidents
RT Nuclear Criticality Safety
SO Emergency Preparedness
SO Radiation
DEF (IAEA) Accidents resulting from criticality excursions. (EMER) Unplanned incidents where a series of faults and/or errors cause a very short-lived, uncontrolled nuclear chain reaction.

CRITICALITY ALARM SYSTEM
(2027 CAS)
DA December 10, 1990
BT1 Alarms
BT2 Devices
BT1 Emergency Systems
BT2 Systems
BT1 Systems (DOE FRASE Vocabulary)

BT2 Systems
RT Violation of Criticality Specifictn
SO DOE FRASE VOCABULARY

CRITICALITY ALARMS
(EMER)
DA February 1, 1991
BT1 Alarms
BT2 Devices
SO Emergency Preparedness
DEF (EMER) Devices incorporating a radiation detector and alarm circuitry placed at locations where significant quantities of fissionable material are handled or stored. At a preset value, the devices trigger alarms indicating high radiation is present at the detector probe and that a criticality accident may have occurred.

CRITICALITY EXCURSIONS
(IAEA)
DA October 12, 1990
BT1 Processes
RT Criticality
SO Radiation
DEF (IAEA) Processes characterized by short releases of energy produced by uncontrolled nuclear chain reactions.

CRITICALITY INCIDENTS
(DOE Order 6430.1A)
DA October 12, 1990
BT1 Incidents
RT Criticality Accidents
SO Construction
DEF (DOE Order 6430.1A) Accidental, self-sustained atomic chain reactions.

CRM (Corrective Repair Maintenance)
DA June 4, 1991
SEE Corrective Repair Maintenance
SO Acronyms

CRM (Count Rate Meter)
DA October 12, 1990
SEE Count Rate Meter
SO Acronyms

CROSSING FROGS
(DOE Order 6430.1A)
DA October 12, 1990
BT1 Devices
SO Construction
DEF (DOE Order 6430.1A) Devices that enable the wheels of a train to cross the rail of an intersecting track.

CROWBAR(S)
(3030 CROWBAR)
DA January 3, 1991
BT1 Tools - Manual
BT2 Tools (DOE FRASE Vocabulary)

SY-Synonymous Terms   SO-Source/Subject Category   SF-See From

BT3 Equipment
SO DOE FRASE VOCABULARY

CRT
DA October 12, 1990
SEE Coolant Return Tank
SO Acronyms

CRUSHING MACHINE
(2115 CRUSHING MAC)
DA December 10, 1990
BT1 Machines (DOE FRASE
    Vocabulary)
 BT2 Equipment
SO DOE FRASE VOCABULARY

CRYOGENIC LIQUIDS
DA June 3, 1991
BT1 Liquids
 BT2 Fluids
RT Compressed Gases
SO Hazardous Materials
DEF Refrigerated liquefied gases having
    boiling points colder than 130
    degrees F. (-90 degrees C.) at
    one atmosphere, absolute.
    Materials meeting this definition
    are subject to requirements of
    whether they meet the definition of
    a compressed gas in paragraph
    (a) of this section. The materials
    are partially described as "...,
    refrigerated liquid (cryogenic
    liquid)" in @172.101 of this
    subchapter.

CRYPTO
(DOE Order 6430.1A)
DA October 12, 1990
RT Classified Information
SO Construction
DEF (DOE Order 6430.1A) A
    designation or marking applied to
    classified and unclassified
    telecommunications keying
    material indicating that it requires
    special accounting and
    safeguarding.

Cr⁺⁶
DA October 12, 1990
SEE Chromium-hexavalent
SO Acronyms

CS
DA October 12, 1990
SEE Containment Spray
SO Acronyms

CS&R
DA October 12, 1990
SEE Codes, Standards, and Regulations
SO Acronyms

CSR
DA October 12, 1990
SEE Scram Relay
SO Acronyms

CSS
DA October 12, 1990
SEE Containment Spray System
SO Acronyms

CSSI
DA October 12, 1990
SEE Containment Spray System
    (Post-accident Injection Phase)
SO Acronyms

CSSR
DA October 12, 1990
SEE Containment Spray System
    (Post-accident Recirculation
    Phase)
SO Acronyms

CST
DA October 12, 1990
SEE Condensate Storage Tank
SO Acronyms

CSWE
DA October 12, 1990
SEE Central Service Works Engineering
SO Acronyms

CT
DA October 12, 1990
SEE Current Transformer
SO Acronyms

**CT (PUBLIC WATER DISINFECTION
FORMULA)**
(CFR)
DA November 9, 1990
SY CTcalc
RT Disinfectant Contact Time
RT Residual Disinfectant Concentration
DEF (CFR) The product of "residual
    disinfectant concentration" (C) in
    mg/l determined before or at the
    first customer, and the
    corresponding "disinfectant contact
    time" (T) in minutes, i.e., "C" x "T".
    See 40 CFR 141.2 for continued
    definition.

CTA
DA October 12, 1990
SEE Central Training Academy
SO Acronyms

CTCALC
(CFR)
DA October 19, 1990
SY CT (Public Water Disinfection
    Formula)
RT Disinfectant Contact Time
RT Public Water Systems
RT Residual Disinfectant Concentration
SO Water Pollution
DEF (CFR) The product of "residual
    disinfectant concentration" (C) in
    mg/1 determined before or at the
    first customer, and the
    corresponding "disinfectant contact

time" (T) in minutes, i.e., "C" x "T".
See 40 CFR 141.2 for additional
definition.

CTG
DA October 12, 1990
SEE Control Technique Guidelines
SO Acronyms

CTR
DA October 12, 1990
SEE Critical Temperature Ratio
SO Acronyms

Cu
DA October 12, 1990
SEE Copper
SO Acronyms

CUBIC FEET PER MINUTE
(EPA)
DA October 12, 1990
SF *CFM*
BT1 Units of Measure
SO Environmental Protection Agency
    Glossary
DEF (EPA) A measure of the volume of
    a substance flowing through air
    within a fixed period of time. With
    regard to indoor air, refers to the
    amount of air, in cubic feet, that is
    exchanged with indoor air in a
    minute's time, or an air exchange
    rate.

CUBIC YARDS
DA January 8, 1991
SF *Yards³*
BT1 Units of Measure

CULLET
(CFR)
DA October 12, 1990
RT Glass Melting Furnaces
RT Recycling
SO Air Pollution
DEF (CFR) Waste glass recycled to a
    glass melting furnace.

CULTIVATING
(CWA; RHA; CFR)
DA October 12, 1990
NT1 Contour Plowing
NT1 Strip Cropping
RT Irrigation
SO Water Pollution
DEF (CFR) Physical methods of soil
    treatment employed within
    established farming, ranching, and
    silviculture lands on farm, ranch,
    or forest crops to aid and improve
    their growth, quality or yield.

CULTURAL EUTROPHICATION
(EPA)
DA October 12, 1990
BT1 Eutrophication
 BT2 Processes

BT-Broader Term   NT-Narrower Term   RT Related Term

SO Environmental Protection Agency
    Glossary
DEF (EPA) Increasing rate at which
    water bodies "die" by pollution
    from human activities.

CULTURAL RESOURCE SITES
(DOE Order 6430.1A)
DA October 12, 1990
BT1 Sites/Areas
SO Construction
DEF (DOE Order 6430.1A)
    Human-associated ruins of
    archaeologic significance.

**CUMULATIVE ANNUAL EFFECTIVE
DOSE EQUIVALENT**
(DOE Order 5480.11)
DA October 12, 1990
BT1 Annual Effective Dose Equivalent
 BT2 Annual Dose Equivalent
  BT3 Dose Equivalents
   BT4 Radiation Units
    BT5 Units of Measure
 BT2 Effective Dose Equivalent
  BT3 Dose Equivalents
   BT4 Radiation Units
    BT5 Units of Measure
SO Industrial Hygiene
DEF (DOE Order 5480.11) The sum of
    the annual effective dose
    equivalents recorded for an
    individual for each year of
    employment at DOE or DOE
    contractor facility since the
    effective date of this Order.

**CUMULATIVE FREQUENCY
(PROBABILITY)**
(SSDC)
DA October 12, 1990
RT Probability
SO System Safety Development Center
    Glossary
DEF (SSDC) The frequency or
    probability which includes (or
    accumulates) all observations
    above or below a specified value.

CUMULATIVE IMPACT
(NEPA; CFR)
DA January 24, 1991
RT Ecological Impacts
DEF (CFR) The impact on the
    environment which results from
    the incremental impact of the
    action when added to other past,
    present, and reasonably
    foreseeable future actions
    regardless of what agency or
    person undertakes such other
    actions. Cumulative impacts can
    result from individually minor but
    collectively significant actions
    taking place over a period of time.

**CUMULATIVE WORKING LEVEL
MONTHS**
(EPA)

DA October 12, 1990
SF CWLM
BT1 Measurements
BT1 Working Level Month
 BT2 Radiation Units
  BT3 Units of Measure
SO Environmental Protection Agency
    Glossary
DEF (EPA) The sum of lifetime exposure
    to radon working levels expressed
    in total working level months.

CURB INLETS
(DOE Order 6430.1A)
DA October 12, 1990
RT Sewers
SO Construction
DEF (DOE Order 6430.1A) Inlets to a
    subsurface stormwater
    conveyance system.

CURB RETURN
(DOE Order 6430.1A)
DA October 12, 1990
SO Construction
DEF (DOE Order 6430.1A) The end
    point of a curb radius.

CURIE
(CFR; NIH; NCRP; EMER)
DA October 12, 1990
BT1 Radiation Units
 BT2 Units of Measure
RT Radioactivity
SO Air Pollution
SO Emergency Preparedness
SO Environmental Protection Agency
    Glossary
SO Radiation
DEF A quantitative measure of
    radioactivity equal to 3.7×10^2
    disintegrations per second. (NIH)
    The quantity of any radioactive
    material in which the number of
    disintegrations is 3.700×10^{10} per
    second. Abbreviated Ci. Millicurie:
    One-thousandth of a curie ($3.7 \times$
    10^7 disintegrations per second).
    Abbreviated mCi. Microcurie: One
    millionth of a curie (3.7×10^4
    disintegrations per second).
    Abbreviated μCi. Picocurie: One
    millionth of a microcurie ($3.7 \times$
    10^2 disintegrations per second or
    2.22 disintegrations per minute).
    Abbreviated pCi.

CURRENT ASSETS
(SWDA; RCRA; CFR)
DA October 12, 1990
BT1 Assets
RT Current Liabilities
RT Net Working Capital
SO Wastes
SO Water Pollution
DEF (CFR) Cash or other assets or
    resources commonly identified as
    those which are reasonably
    expected to be realized in cash or

sold or consumed during the
    normal operating cycle of the
    business.

CURRENT CLOSURE COST ESTIMATE
(SWDA; RCRA; CFR)
DA October 12, 1990
BT1 Closure
 BT2 Administrative Processes
  BT3 Processes
BT1 Costs
RT Closure Plans
SO Wastes
DEF (CFR) The most recent of the
    estimates prepared in accordance
    with 40 CFR Part 264.142 (a), (b),
    and (c).

CURRENT LIABILITIES
(SWDA; RCRA; CFR)
DA October 12, 1990
BT1 Liabilities
RT Current Assets
RT Net Working Capital
SO Wastes
SO Water Pollution
DEF (CFR) Obligations whose
    liquidation is reasonably expected
    to require the use of existing
    resources properly classifiable as
    current assets or the creation of
    other current liabilities.

**CURRENT PLUGGING AND
ABANDONMENT COST ESTIMATES**
(SWDA; RCRA; CFR)
DA October 19, 1990
BT1 Costs
RT Net Working Capital
RT Plugging
RT Wells
SO Wastes
SO Water Pollution
DEF (CFR) The most recent of the
    estimates prepared in accordance
    with 40 CFR 144.62 (a), (b) and
    (c).

**CURRENT POST-CLOSURE COST
ESTIMATE**
(SWDA; RCRA; CFR)
DA October 19, 1990
BT1 Closure
 BT2 Administrative Processes
  BT3 Processes
BT1 Costs
RT Closure Plans
SO Wastes
DEF (CFR) The most recent of the
    estimates prepared in accordance
    with 40 CFR Part 264.144(a), (b),
    and (c).

CURRENT TRANSFORMER
DA January 8, 1991
SF CT

CURTAIL
(CFR)

DA   October 12, 1990
RT   Control Devices
RT   Shutdown (Emissions)
SO   Air Pollution
DEF (CFR) To cease operations to the
         extent technically feasible to
         reduce emissions.

CUSTOMER SERVICE AREA
(1687 CS AREA)
DA   December 10, 1990
BT1 Area
  BT2 Sites/Areas
SO DOE FRASE VOCABULARY

CUTIE PIES
(EPA; NFI)
DA   October 12, 1990
BT1 Radiation Detectors
  BT2 Equipment
RT Scintillation Counters
SO Environmental Protection Agency
        Glossary
SO Nuclear Facilities Incident Database
SO Radiation
DEF (NFI) Hand-held radiation activity
         monitors.

CUTTER(S)
(3031 CUTTER)
DA   January 3, 1991
BT1 Tools - Manual
  BT2 Tools (DOE FRASE Vocabulary)
    BT3 Equipment
RT Scissors
RT Snips
SO DOE FRASE VOCABULARY

CVCS
DA   October 12, 1990
SEE Chemical Volume and Control
        System
SO Acronyms

CWA
DA   October 12, 1990
SEE Clean Water Act
SO Acronyms

CWGM
DA   October 12, 1990
SEE Cooling Water Gamma Monitor
SO Acronyms

CWLM
DA   October 12, 1990
SEE Cumulative Working Level Months
SO Acronyms

CWS
DA   October 12, 1990
SEE Contaminated Water Storage
SO Acronyms

CX
DA   May 14, 1991
SEE Categorical Exclusions
SO Acronyms

CYCLONE COLLECTORS
(EPA)
DA   October 12, 1990
BT1 Control Devices
  BT2 Devices
BT1 Inertial Separators
  BT2 Devices
RT Scrubbers
SO Environmental Protection Agency
        Glossary
DEF (EPA) Devices that use centrifugal
         force to pull large particles from
         polluted air.

CYLINDERS
(SWDA; RCRA; CFR; ESH)
DA   October 12, 1990
BT1 Containers
RT Compressed Gases
SO Hazardous Materials
DEF (CFR) Pressure vessels designed
         for pressures higher than 40 psia
         and having a circular cross
         section. They do not include
         portable tanks, multi-unit tank car
         tanks, cargo tanks, or tank cars.

$C_2C_{12}F_4$
DA   October 12, 1990
SEE Freon-114
SO Acronyms

D
DA   October 12, 1990
SEE Absorbed Dose
SO Acronyms

D&D
DA   October 12, 1990
SEE Decontamination and
        Decommissioning
SO Acronyms

D&E
DA   October 12, 1990
SEE Discharge and Exit
SO Acronyms

D&E CANAL
(NFI)
DA   October 12, 1990
SO Nuclear Facilities Incident Database
SO Radiation
DEF (NFI) Discharge or Deposit and Exit
         canal, connects process room and
         disassembly area.

D MACHINES
(NFI)
DA   October 12, 1990
NT1 Outer Chuck
RT Alternate Removal Systems
RT Dog Houses
SO Nuclear Facilities Incident Database
SO Radiation
DEF (NFI) Reactor Discharging
         Machines

d/m
DA   October 12, 1990
SEE Drips per Minute
SO Acronyms

D/M/G
DA   October 12, 1990
SEE Disintegrations per minute per gram
SO Acronyms

DA
DA   October 12, 1990
SEE DOE Design Agency
SO Acronyms

DAC
DA   October 12, 1990
SEE Derived Air Concentration
SO Acronyms

DAILY COVER
(SWDA; RCRA; CFR; ESH)
DA   October 12, 1990
BT1 Cover Materials
  BT2 Materials
RT Cells
RT Final Cover
RT Intermediate Cover
RT Sanitary Landfills
SO Environmental Management
SO Wastes
DEF (ESH) Cover material that is spread
         and compacted on the top and
         side slopes of compacted solid
         waste at least at the end of each
         operating day to control vectors,
         fire, moisture, and erosion and to
         ensure an aesthetic appearance.

DAILY INSTRUCTION LOGS
(NFI)
DA   October 12, 1990
SF *DIL*
RT Procedures
SO Radiation
DEF (NFI) Contain temporary
         procedures before they become
         official.

DAM
(1852 DAM)
DA   December 10, 1990
BT1 Structures (DOE FRASE
        Vocabulary)
RT Body of Water
RT Pond
SO DOE FRASE VOCABULARY

DAMAGE
(1526 DAMAGE;N)
DA   November 28, 1990
BT1 Nature of Property Damage
NT1 Electronic Damage
NT1 Fire Damage
NT1 Heat Damage
NT1 Mechanical Damage
NT1 Smoke Damage
NT1 Structural Damage

NT1 Water Damage
NT1 Wind Damage
SO DOE FRASE VOCABULARY

DAMAGE ASSESSMENTS
(DOE Order 5500.1A Attach. II, 11;
EMER)
DA January 24, 1991
BT1 Assessments
 BT2 Administrative Processes
 BT3 Processes
NT1 Natural Resource Damage
      Assessment
NT1 Natural Resource Damage
      Preassessment Screens
RT Emergencies
RT Nature of Programmatic Impact
RT Nature of Property Damage
SO Emergency Preparedness
SO Environmental Management
DEF (DOE 5500.1A) The capability to
      estimate the damage to DOE
      operations, facilities, and
      equipment as the result of an
      emergency.

DAMAGE TO PROSTHETIC DEVICE
(1313 DAMAGE TO PR)
DA November 28, 1990
BT1 Injuries
SO DOE FRASE VOCABULARY

DAMS
(CFR)
DA October 19, 1990
BT1 Hydraulic Structures
 BT2 Structures (DOE FRASE
      Vocabulary)
SO Water Pollution
DEF (CFR) Impoundment structures that
      completely span a navigable water
      of the United States and that may
      obstruct interstate waterborne
      commerce. The term does not
      include a weir. Weirs are regulated
      pursuant to section 10 of the
      Rivers and Harbors Act of 1899.
      (See 33 CFR Part 322.)

DANGER
(SSDC)
DA October 12, 1990
BT1 Conditions
NT1 Imminent Danger
SO System Safety Development Center
      Glossary
DEF (SSDC) Potential for physical harm
      to people or damage to property.

DATA CALL-IN
(EPA)
DA October 12, 1990
RT Registration Standards
SO Environmental Protection Agency
      Glossary
DEF (EPA) A part of the Office of
      Pesticide Programs (OPP)
      process of developing key
      required test data, especially on

the long- term, chronic effects of
existing pesticides, in advance of
scheduled Registration Standard
reviews. Data Call-In is an adjunct
of the Registration Standards
program intended to expedite
reregistration and involves the
"calling in" of data from
manufacturers.

DATA PROCESSING AREA
(1688 DP AREA)
DA December 10, 1990
BT1 Area
 BT2 Sites/Areas
RT Computer Terminal(s)
RT Computer(s)
SO DOE FRASE VOCABULARY

DATA QUALITY ASSESSMENTS
(ESH)
DA October 12, 1990
BT1 Assessments
 BT2 Administrative Processes
 BT3 Processes
RT Quality Assurance
SO Quality Assurance
DEF (ESH) The requirements for data
      precision, accuracy,
      representativeness, completeness
      and comparability in Superfund
      are very strict. All monitoring
      entities (Federal Agencies, PRPs,
      states, etc.) should perform data
      reduction and validation in
      accordance with accepted
      REM/FIT procedures or as
      specified in EPA's CLP (Contract
      Laboratory Program) where
      applicable. Aspects of data quality
      which will be addressed are
      precision, accuracy, traceability of
      standards, traceability of data,
      methodology, reference or spiked
      samples, performance audits, and
      representativeness, comparability,
      and completeness.

DATUM
(DOE Order 6430.1A)
DA October 12, 1990
RT Monumentation
SO Construction
DEF (DOE Order 6430.1A) A direction,
      level or position from which
      angles, heights or distances are
      conveniently measured.

DAYS LIVING RADIOISOTOPES
(EDB)
DA March 29, 1991
BT1 Radionuclides
 BT2 CERCLA Hazardous Substances
 BT3 Hazardous Substances
 BT2 Nuclides
NT1 Beryllium 7
NT1 Cerium 144
NT1 Cerium 141
NT1 Cesium 137

NT1 Cobalt 58
NT1 Iodine 131
NT1 Manganese 54
NT1 Radon 222
NT1 Selenium 75
NT1 Zinc 65
NT1 Zirconium 95
SO Radiation

db (Decibel)
DA October 12, 1990
SEE Decibel
SO Acronyms

DB (Dumbbell)
DA October 12, 1990
SEE Dumbbell
SO Acronyms

DBA
DA October 12, 1990
SEE Design Basis Accidents
SO Acronyms

DBE
DA October 12, 1990
SEE Design Basis Earthquakes
SO Acronyms

DBF
DA October 12, 1990
SEE Design Basis Fires
SO Acronyms

DBFL
DA October 12, 1990
SEE Design Basis Floods
SO Acronyms

DBT
DA October 12, 1990
SEE Design Basis Tornadoes
SO Acronyms

DC
DA October 12, 1990
SEE Direct Current
SO Acronyms

DCG
DA October 12, 1990
SEE Derived Concentration Guide
SO Acronyms

DCH
DA October 12, 1990
SEE Direct Containment Heating
SO Acronyms

DCR
DA October 12, 1990
SEE Design Change Request
SO Acronyms

DDT
(EPA)
DA October 12, 1990

BT1 Chlorinated Hydrocarbons
 BT2 CERCLA Hazardous Substances
  BT3 Hazardous Substances
 BT2 Halogenated Organic Compounds
  BT3 Halogenated
  BT3 Organic Chemicals
   BT4 Chemical Substances
BT1 Hazardous Constituents
BT1 Insecticides
 BT2 Pesticides
  BT3 Hazardous Substances
SO Environmental Protection Agency
    Glossary
DEF (EPA) The first chlorinated
    hydrocarbon insecticide (chemical
    name: Dichloro-Diphenyl-
    Trichloroethane). It has a half-life
    of 15 years and can collect in fatty
    tissues of certain animals. EPA
    banned registration and interstate
    sale of DDT for virtually all but
    emergency uses in the United
    States in 1972 because of its
    persistence in the environment
    and accumulation in the food
    chain.

DE MINIMIS
(IAEA)
DA October 12, 1990
RT De Minimus Violations
SO Radiation
DEF (IAEA) Part of the maxim "de
    minimis non curat lex" (the law
    does not concern itself with trifles),
    sometimes used with reference to
    sources of radiation which a
    competent authority may decide to
    exempt from defined regulatory
    requirements because individual
    and collective effective dose
    equivalents received from them
    are both so low that they may be
    ignored.

DE MINIMUS VIOLATIONS
(SSDC)
DA October 12, 1990
BT1 Violations
RT de minimis
SO System Safety Development Center
    Glossary
DEF (SSDC) "Those that have no
    immediate or direct relationship to
    safety or health" (OSHA).

DEAD LOADS
(DOE Order 6430.1A)
DA October 12, 1990
RT Live Loads
SO Construction
DEF (DOE Order 6430.1A) Non-varying
    loads exerted by the weight of a
    mass at rest.

DEATH
(1314 DEATH; TSCA; CFR)
DA November 28, 1990

BT1 Injuries
SO DOE FRASE VOCABULARY

DECAY HEAT REMOVAL
DA January 8, 1991
SF *DHR*

DECHLORINATION
(EPA)
DA October 12, 1990
BT1 Chemical Processes
 BT2 Processes
RT Chlorination
SO Environmental Protection Agency
    Glossary
DEF (EPA) Removal of chlorine from a
    substance by chemically replacing
    it with hydrogen or hydroxide ions
    in order to detoxify the substances
    involved.

DECIBEL
(EPA)
DA October 12, 1990
SF *db (Decibel)*
RT A-Scale Sound Levels
SO Environmental Protection Agency
    Glossary
DEF (EPA) A unit of sound
    measurement. In general, a sound
    doubles in loudness for every
    increase of ten decibels.

DECOMMISSIONING
(DOE Orders 6430.1A, 5820.2A)
DA October 12, 1990
BT1 Administrative Processes
 BT2 Processes
RT Nuclear Facilities
RT Remedial Actions
SO Construction
SO Environmental Management
DEF (DOE Order 6430.1A) The process
    of closing and securing a nuclear
    facility, or nuclear materials
    storage facility so as to provide
    adequate protection from radiation
    exposure and to isolate
    radioactive contamination from the
    human environment.

DECOMMISSIONING ACTIVITY
(1250 DEC ACTIVITY)
DA November 28, 1990
BT1 Activity Types (DOE FRASE
    Vocabulary)
 BT2 Activities
RT Nuclear Facility
SO DOE FRASE VOCABULARY

DECOMPOSITION
(EPA)
DA October 12, 1990
BT1 Biological Processes
 BT2 Processes
RT Biodegradable
RT Biological Oxidation
RT Composting
RT Degradation

RT Putrescible
SO Environmental Protection Agency
    Glossary
DEF (EPA) The breakdown of matter by
    bacteria and fungi. It changes the
    chemical makeup and physical
    appearance of materials.

DECONTAMINATION
(DOE Orders 6430.1A, 5820.2A; SARA;
    OSHA; CFR; IAEA; EMER)
DA October 12, 1990
BT1 Processes
RT Cleanup Operations
RT Cleanup
RT Decontamination Factor
RT Detergents
RT Double Wash/Rinse
RT Radiation Protection
RT Radiation Safety
RT Remedial Actions
SO Construction
SO Emergency Preparedness
SO Environmental Management
SO Radiation
DEF (EMER) The removal of unwanted
    material (typically radioactive
    material) from facilities, soils, or
    equipment by washing, chemical
    action, mechanical cleaning, or
    other techniques.

DECONTAMINATION ACTIVITY
(1255)
DA November 28, 1990
BT1 Activity Types (DOE FRASE
    Vocabulary)
 BT2 Activities
RT Nuclear Facility
SO DOE FRASE VOCABULARY

**DECONTAMINATION AND
DECOMMISSIONING**
DA January 8, 1991
SF *D&D*

DECONTAMINATION AREAS
(1689 DECON AREA; TSCA; CFR)
DA December 10, 1990
BT1 Area
 BT2 Sites/Areas
SO DOE FRASE VOCABULARY
SO Environmental Management

DECONTAMINATION FACTOR
(IAEA)
DA October 12, 1990
BT1 Ratios
RT Decontamination
SO Radiation
DEF (IAEA) The ratio of the initial level
    of contaminating radioactive
    material to the residual level
    achieved through a
    decontamination process.

DEDICATED FIRE WATER SYSTEMS
(DOE Order 6430.1A)
DA October 12, 1990

BT1 Fire Protection Systems
BT2 Emergency Systems
BT3 Systems
NT1 Sprinkler System
SO Construction
DEF (DOE Order 6430.1A) Water
    storage and distribution systems
    that are available for and used
    solely for fire protection purposes,
    as opposed to a combined system
    that may be used for potable and
    process water supply in addition to
    fire protection.

DEEP DOSE EQUIVALENT
(DOE Order 5400.5)
DA October 12, 1990
BT1 Dose Equivalents
BT2 Radiation Units
BT3 Units of Measure
NT1 Deep Eye Dose Equivalent
SO Industrial Hygiene
DEF (DOE Order 5400.5) As used in this
    Order, means the dose equivalent
    in tissue at a depth of 1 cm
    deriving from external
    (penetrating) radiation.

DEEP EYE DOSE EQUIVALENT
(DOE Order 5480.11)
DA October 16, 1990
BT1 Deep Dose Equivalent
BT2 Dose Equivalents
BT3 Radiation Units
BT4 Units of Measure
DEF (DOE Order 5480.11) The dose
    equivalent at the respective
    depths of 0.007 cm, 1.0 cm, and
    0.3 cm in tissue.

DEFECTIVE FIREARM
(DOE Order 5480.16)
DA October 12, 1990
BT1 Firearms
RT Armorers
RT Misfires
SO Firearms
DEF (DOE Order 5480.16) A firearm
    that, because of improper
    assembly, excessive wear, or
    broken or missing parts, does not
    function according to design
    specifications.

DEFECTS
(SSDC)
DA October 12, 1990
SO System Safety Development Center
    Glossary
DEF (SSDC) Substandard physical
    conditions, either inherent in the
    material or created through
    another action or event.

**DEFENSE FACILITY
DECOMMISSIONING PROGRAM**
DA January 8, 1991
SF DFDP
BT1 Programs

DEFENSE PROGRAMS
DA January 8, 1991
SY Office of the Assistant Secretary for
    Defense Programs
SF DP (Defense Programs)
BT1 DOE Programs
BT2 Programs

DEFENSE READINESS CONDITIONS
(EMER)
DA February 1, 1991
BT1 Conditions
RT Alarms
RT Alert
RT Emergencies
SO Emergency Preparedness
DEF (EMER) Graded warning codes,
    which when accompanied by a
    number, signify a different degree
    of defense readiness from normal
    operating status to maximum
    readiness.

**DEFENSE WASTE MANAGEMENT
PLAN**
DA January 8, 1991
SF DWMP
BT1 Plans

**DEFENSE WASTE PROCESSING
FACILITY**
DA January 8, 1991
SF DWPF
BT1 Facilities

DEFLAGRATION
(DOE Order 6430.1A)
DA October 12, 1990
RT Explosives
RT Fires
SO Construction
DEF (DOE Order 6430.1A) A rapid
    chemical reaction in which the
    output of heat is sufficient to
    enable the reaction to proceed
    and be accelerated without input
    of heat from another source.
    Deflagration is a surface
    phenomenon, with the reaction
    products flowing away from the
    unreacted material along the
    surface at subsonic velocity. The
    effect of a true deflagration under
    confinement is an explosion.
    Confinement of the reaction
    increases pressure, rate of
    reaction and temperature, and
    may cause transition into a
    detonation.

DEFLECTION ANGLES
(DOE Order 6430.1A)
DA October 12, 1990
SO Construction
DEF (DOE Order 6430.1A) The angles
    measured between a foresight and
    a prolongation of the backsight.

DEFOLIANTS
(FIFRA; CFR; USC; EPA)
DA October 12, 1990
BT1 Herbicides
BT2 Pesticides
BT3 Hazardous Substances
NT1 Agent Orange
SO Environmental Management
SO Environmental Protection Agency
    Glossary
DEF (EPA) Herbicides that remove
    leaves from trees and growing
    plants. (USC) Substances or
    mixtures of substances intended
    for causing the leaves or foliage to
    drop from a plant, with or without
    causing abscission.

DEGRADATION
(EPA)
DA October 12, 1990
BT1 Chemical Processes
BT2 Processes
RT Attenuation
RT Decomposition
SO Environmental Protection Agency
    Glossary
DEF (EPA) The process by which a
    chemical is reduced to a less
    complex form.

DEGRADATION PRODUCTS
(FIFRA; CFR)
DA January 24, 1991
RT Pesticides
DEF (CFR) The term "degradation
    product" means a substance
    resulting from the transformation
    of a pesticide by physicochemical
    or biochemical means.

DEIONIZED
DA January 8, 1991
SF DI

DEIONIZED WATER
(NFI)
DA October 12, 1990
BT1 Water
SO Radiation
DEF (NFI) Water for thermal shields.

**10-MINUTE DELAY LINE GAMMA
MONITOR**
DA January 8, 1991
SF 10 MDLM
BT1 Monitors
BT2 Equipment

DELAY OF EXPERIMENT
(1575 DELAY OF EXP)
DA November 28, 1990
BT1 Delays
BT1 Nature of Programmatic Impact
SO DOE FRASE VOCABULARY

DELAY OF RESPONSE
(1576 DELAY OF RES)

DA November 28, 1990
BT1 Delays
BT1 Nature of Programmatic Impact
SO DOE FRASE VOCABULARY

DELAY OF SAMPLING
(1577 DELAY OF SAM)
DA November 28, 1990
BT1 Delays
BT1 Nature of Programmatic Impact
SO DOE FRASE VOCABULARY

DELAYED (CHRONIC) HEALTH HAZARDS
(CFR)
DA October 12, 1990
BT1 Hazard Categories
BT1 Health Hazards
 BT2 Hazards
  BT3 Conditions
DEF (CFR) Includes "carcinogens" (as defined by 29 CFR 1910.1200) and other hazardous chemicals that cause an adverse effect to a target organ and which effect generally occurs as a result of long term exposure and is of long duration.

DELAYED COMPLIANCE ORDERS
(CAA; USC)
DA October 12, 1990
RT Compliance Schedules
RT Implementation Plans
SO Air Pollution
DEF (USC) Orders issued by the State or by the Administrator to an existing stationary source, postponing the date required under an applicable implementation plan for compliance by such source with any requirement of such plan.

DELAYED NEUTRON PRECURSORS
(EDB)
DA March 29, 1991
BT1 Radionuclides
 BT2 CERCLA Hazardous Substances
  BT3 Hazardous Substances
 BT2 Nuclides
SO Radiation

DELAYED PROTON PRECURSORS
(EDB)
DA March 29, 1991
BT1 Radionuclides
 BT2 CERCLA Hazardous Substances
  BT3 Hazardous Substances
 BT2 Nuclides
SO Radiation

DELAYS
DA January 30, 1991
NT1 Delay of Experiment
NT1 Delay of Response
NT1 Delay of Sampling
NT1 Program Significant Delay

RT Nature of Programmatic Impact
SO DOE FRASE VOCABULARY

DELEGATED STATES
(EPA)
DA October 12, 1990
BT1 States
SO Environmental Protection Agency Glossary
DEF (EPA) States (or other governmental entities) which have applied for, and received authority to administer, within its territory, its state regulatory program as the federal program required under a particular federal statute. As used in connection with NPDES (National Pollutant Discharge Elimination System), UIC (Underground Injection Control), and PWS programs, the term does not connote any transfer of federal authority to a state.

DELIST
(EPA)
DA October 12, 1990
SO Environmental Protection Agency Glossary
DEF (EPA) Use of the petition process to have a facility's toxic designation rescinded.

DELISTING PETITIONS
(RCRA; CFR)
DA January 24, 1991
RT Waste Management
DEF (CFR) A petition to exclude a waste produced at a particular generating facility. The waste must not meet any of the criteria under which the waste was listed as hazardous waste or an acutely hazardous waste. Even after application of the petition to the Administrator and the waste is excluded, it still may be a hazardous waste by operation of Part 261 Subpart C. Procedures for application of a delisting petition are explained in 40 CFR 260.22. The regulatory intent of a delisting petition is to enable EPA to consider all factors in delisting wastes, not just those for which the waste was originally listed.

DELPHI PROCESS
(SSDC)
DA October 12, 1990
SO System Safety Development Center Glossary
DEF (SSDC) Means for providing a (hazard) consensus from a group of people who are familiar with the (hazards) subject.

DELTA K RODS
(NFI)

DA October 12, 1990
RT Criticality
SO Nuclear Facilities Incident Database
SO Radiation
DEF (NFI) K=neutron multiplication factor; criticality when K=1.

DELTA P
(NFI)
DA October 12, 1990
SO Nuclear Facilities Incident Database
SO Radiation
DEF (NFI) Pressure difference.

DELTA T
(NFI)
DA October 12, 1990
SO Nuclear Facilities Incident Database
SO Radiation
DEF (NFI) Temperature difference.

DELUGE SYSTEM
(2028 DELUGE SYSTE)
DA December 10, 1990
BT1 Systems (DOE FRASE Vocabulary)
 BT2 Systems
SO DOE FRASE VOCABULARY

DEMINERALIZER PLANT
(1796 DEMINERALIZE)
DA December 10, 1990
BT1 Building (DOE FRASE Vocabulary)
 BT2 Facilities and Buildings (DOE FRASE Vocabulary)
  BT3 Facilities
SO DOE FRASE VOCABULARY

DEMOLITION
(TSCA; CFR)
DA October 12, 1990
RT Renovation
RT Structural Members
SO Air Pollution
SO Environmental Management
DEF (CFR) The wrecking or taking out of any load-supporting structural member of a facility together with any related handling operations.

DEMONSTRATION
(SWDA; RCRA; USC)
DA October 12, 1990
SO Wastes
DEF (USC) The initial exhibition of a new technology process or practice or a significantly new combination or use of technologies, processes or practices, subsequent to the development stage, for the purpose of proving technological feasibility and cost effectiveness.

DENIALS
(EMER)
DA February 1, 1991
BT1 Protective Actions
 BT2 Actions
  BT3 Responses

BT2 Amelioration
SO Emergency Preparedness
DEF (EMER) 1. Protective actions that deny an adversary access to the intended target. 2. Protection strategies of the same name.

DENITRIFICATION
(EPA)
DA October 12, 1990
BT1 Biological Processes
 BT2 Processes
RT Anaerobic Digestion
RT Nitrates
RT Nitrification
SO Environmental Protection Agency Glossary
DEF (EPA) The anaerobic biological reduction of nitrate nitrogen to nitrogen gas.

DENSITY FUNCTION
(SSDC)
DA October 12, 1990
RT Distribution
SO System Safety Development Center Glossary
DEF (SSDC) The value of the y-axis on a probability distribution curve. It is a measure of frequency of stated values on the x- axis.

DENT
(1527 DENT)
DA November 28, 1990
BT1 Nature of Property Damage
SO DOE FRASE VOCABULARY

DENTAL INJURY
(1315 DENTAL INJUR)
DA November 28, 1990
BT1 Injuries
RT Mouth
RT Tooth/Teeth
SO DOE FRASE VOCABULARY

DENUDERS
(CFR)
DA October 12, 1990
BT1 Containers
RT Hydrogen Gas Streams
RT Mercury Chlor-alkali Cells
SO Air Pollution
DEF (CFR) Horizontal or vertical containers which are part of a mercury chlor-alkali cell and in which water and alkali metal amalgams are converted to alkali metal hydroxide, mercury, and hydrogen gas in a short-circuited, electrolytic reaction.

DENY
(CFR)
DA October 19, 1990
RT Prohibit Specification
SO Water Pollution
DEF (CFR) To deny or restrict the use of any area for the present or future

discharge of any dredged or fill material.

DEPARTMENT'S METRIC COORDINATOR
(DOE Order 5700.6B)
DA October 12, 1990
BT1 Personnel
SO Quality Assurance
DEF (ESH) A person designated by the Assistant Secretary for Environment to act as the Department's central point of contact for metrication matters.

DEPARTMENT'S PROCUREMENTS
(DOE Order 5900.2)
DA October 12, 1990
SO Quality Assurance
DEF (DOE Order 5900.2) Contracts, grants, or cooperative agreements awarded by or for the Department of Energy.

DEPARTMENT OF ENERGY CONTRACTORS
(DOE Order 5484.1)
DA October 12, 1990
SY DOE Contractors
SO Management
DEF (DOE Order 5484.1) Includes any prime contractors or subcontractors subject to the contractual provisions of 48 CFR 923.70, 48 CFR 970.23, or other contractual provisions where DOE has elected to enforce environment, safety, and health requirements by specific negotiated contract provisions.

DEPARTMENT OF ENERGY OPERATIONS
(DOE Order 5484.1)
DA October 16, 1990
BT1 Operations
 BT2 Activities
SO Industrial Safety
DEF (DOE Order 5484.1) Those activities funded by DOE for which DOE has authority to enforce for environmental protection, safety, and health protection requirements.

DEPARTMENT OF ENERGY PROJECT CONSTRUCTION CONTRACTORS
(DOE Order 5480.9; TSA)
DA October 16, 1990
BT1 DOE Contractors
 BT2 Potentially Responsible Parties
RT Projects
SO Construction
DEF (DOE Order 5480.9) Any DOE prime contractors or subcontractors engaged in construction activities exempt from, or not subject to, Nuclear Regulatory Commission licensing,

but subject to the contractual provisions of DEAR or modifications thereof. These contractors may make modifications to existing facilities or construct new facilities for the Department but they are not considered a permanent construction force. Their site tenure may be for short or long periods depending on the nature of the project.

DEPARTMENT OF ENERGY RESIDENT CONSTRUCTION CONTRACTORS
(DOE Order 5480.9)
DA October 12, 1990
BT1 DOE Contractors
 BT2 Potentially Responsible Parties
RT Construction Projects
SO Construction
DEF (DOE Order 5480.9) Any DOE prime contractor or subcontractor exempt from, or not subject to, Nuclear Regulatory Commission licensing but subject to the contractual provisions of DEAR or modifications thereof, who is in residence and considered to be permanent. Field organizations that have such contractors include Nevada, Richland, Oak Ridge, Savannah River, and Albuquerque. The tenures of resident construction contractors on site are usually from 3 to 5 years and may be extended.

DEPARTMENT OF ENERGY SITES
(DOE Order 5484.1)
DA October 12, 1990
BT1 Sites/Areas
RT Cognizant DOE Authority
RT DOE Operations
SO Environmental Management
SO Management
DEF (DOE Order 5484.1) Either tracts owned by DOE or tracts leased or otherwise made available to the Federal Government under terms that afford to the Department of Energy rights of access and control substantially equal to those that the Department of Energy would possess if it were the holder of the fee (or pertinent interest therein) as agent of and on behalf of the Government. One or more DOE operations program activities are carried out within the boundaries of the described tract.

DEPARTMENT OF ENERGY WASTES
(DOE Order 5820.2A)
DA January 24, 1991
BT1 Wastes
DEF (DOE 5820.2A) Radioactive waste generated by activities of the Department (or its predecessors), waste for which the Department is

responsible under law or contract, or other waste for which the Department is responsible. Such waste may be referred to as DOE waste.

DEPARTMENTAL-APPROVED EQUIPMENT
(DOE Order 6430.1A)
DA   October 12, 1990
BT1 Equipment
RT   Safeguards
SO   Construction
DEF (DOE Order 6430.1A) Equipment (e.g. alarm, assessment, monitoring, detection) used in conjunction with all or other elements of a site-specific safeguards and security system as described in the site-specific safeguards and security plan (after such plan is approved by the Departmental element).

DEPARTMENTAL ELEMENTS
(DOE Order 6430.1A)
DA   October 12, 1990
NT1 Field Organizations
SO   Construction
DEF (DOE Order 6430.1A) DOE Headquarters and field organizations.

DEPLETED URANIUM
(HMTA; CFR)
DA   May 20, 1991
BT1 Uranium
 BT2 Geologic Resources
  BT3 Natural Resources
 BT2 Heavy Metals
SO   Hazardous Materials
DEF (CFR) Depleted uranium means uranium containing less uranium-235 than the naturally occurring distribution of uranium isotopes.

DEPLETION CURVES
(EPA)
DA   October 12, 1990
SO   Environmental Protection Agency Glossary
DEF (EPA) In hydraulics, graphical representations of water depletion from storage-stream channels, surface soil, and ground water. Depletion curves can be drawn for base flow, direct runoff, or total flow.

DEPOSITION
(NFI; IAEA)
DA   October 12, 1990
BT1 Measurements
RT   Retention
RT   Uptake
SO   Radiation
DEF (IAEA) The amount of radioactive

material incorporated into tissues and organs.

DEPOSITION (PROCESS)
DA   February 26, 1991
BT1 Waste Management Processes
 BT2 Processes
NT1 Sedimentation

DEPOSITORIES
(ANL; CFR)
DA   May 24, 1991
BT1 Sites/Areas
RT   Waste Disposal Site
SO   Environmental Management
SO   Wastes
DEF (CFR) A disposal site (other than a processing site). See 40 CFR 192.01(e).

DEPRESSURIZATION
(EPA)
DA   October 12, 1990
RT   Automatic Depressurization System
SO   Environmental Protection Agency Glossary
DEF (EPA) A condition that occurs when the air pressure inside a structure is lower than the air pressure outside. Depressurization can occur when household appliances that consume or exhaust house air, such as fireplaces or furnaces, are not supplied with enough makeup air. Radon-containing soil gas may be drawn into a house more rapidly under depressurized conditions.

DERIVATIVE ULTRAVIOLET ABSORBTION SPECTROSCOPY
DA   January 8, 1991
SF   DUVAS

DERIVED AIR CONCENTRATION
(DOE Order 5480.11; NCRP; IAEA; EMER)
DA   October 12, 1990
SF   DAC
RT   Annual Limit on Intake
RT   Reference Man
SO   Emergency Preparedness
SO   Environmental Management
SO   Industrial Hygiene
SO   Radiation
DEF (DOE Order 5480.11) Quantity obtained by dividing the ALI (Annual Limit on Intake) for any given radionuclide by the volume of air breathed by an average worker (Reference Man) during a working year (2.4 x 10sup 3msup 3).

DERIVED CONCENTRATION GUIDE
(ESH; EMER)
DA   October 12, 1990
SF   DCG
BT1 Measurements

RT   Effective Dose Equivalent
RT   Reference Man
SO   Emergency Preparedness
SO   Industrial Hygiene
DEF (ESH) The lesser of the concentrations of a radionuclide in air or water that, under conditions of continuous exposure by one exposure mode (i.e., ingestion of water, submersion in air, or inhalation) for one year for a "reference man", would result in either an effective dose equivalent of 100 mrem (1 mSv) or a dose equivalent of 5 rem (50 mSv) to any tissue, including the skin and lens of the eye.

DERIVED LIMITS
(IAEA)
DA   November 20, 1990
BT1 Limits
SO   Radiation
DEF (IAEA) Values of quantities related to the primary or secondary limits by a defined model such that if the derived limits are not exceeded, it is most unlikely that the primary limits will be exceeded.

DERIVED PROJECTION
(SSDC)
DA   October 12, 1990
RT   Extreme Value Projection
RT   Risk Assessment
SO   System Safety Development Center Glossary
DEF (SSDC) Projecting risk from sources or calculations other than actual past experience.

DERMAL TOXICITY
(EPA)
DA   October 12, 1990
BT1 Toxicity
RT   Contact Pesticides
SO   Environmental Protection Agency Glossary
DEF (EPA) The ability of a pesticide or toxic chemical to poison people or animals by contact with the skin.

DERMATITIS
(1382 DERMATITIS)
DA   November 28, 1990
BT1 Illnesses
RT   Other Skin Conditions
RT   Skin
SO   DOE FRASE VOCABULARY

DERRICK
(2314 DERRICK)
DA   December 10, 1990
BT1 Hoisting Apparatus
 BT2 Material Handling Device
  BT3 Devices
  BT3 Equipment/Parts - Material Handling (DOE FRASE Vocabulary)

BT-Broader Term   NT-Narrower Term   RT Related Term

BT4 Equipment
RT Boom
RT Crane(s)
SO DOE FRASE VOCABULARY

DES
(EPA)
DA October 12, 1990
BT1 CERCLA Hazardous Substances
 BT2 Hazardous Substances
SO Environmental Protection Agency
    Glossary
DEF (EPA) A synthetic estrogen,
    diethylstilbestrol is used as a
    growth stimulant in food animals.
    Residues in meat are thought to
    be carcinogenic.

DESALINIZATION
(EPA)
DA October 12, 1990
BT1 Chemical Processes
 BT2 Processes
RT Distillation
RT Electrodialysis
RT Ion Exchange
SO Environmental Protection Agency
    Glossary
DEF (EPA) Removing salt from ocean or
    brackish water.

DESICCANTS
(USC; EPA)
DA October 12, 1990
BT1 Chemical Agents
SO Environmental Protection Agency
    Glossary
DEF (EPA) Chemical agents that absorb
    moisture; some desiccants are
    capable of drying out plants or
    insects, causing death. (USC)
    Substances or mixtures of
    substances intended for artificially
    accelerating the drying of plant
    tissue.

DESIGN AND DEVELOPMENT PLANS
(MORT)
DA April 3, 1991
BT1 Plans
RT Concepts and Requirements
RT General Design Process
RT Hazard Analysis
RT Human Factors
RT Inspection Plans
RT Maintenance Plans
RT Operational Specifications
RT Project Design Criteria
DEF (MORT) MORT analysis asks: does
    the development phase provide for
    the use of the major safety results
    of the Concepts and
    Requirements Phase? is the
    design a true representation of the
    developed criteria, definitions,
    specifications, and requirements?
    Related issues/topics include:
    energy control procedures, human

factors review, maintenance and
inspection plans, etc.

DESIGN BASIS ACCIDENTS
(DOE Order 6430.1A)
DA October 12, 1990
SF *DBA*
BT1 Accidents
NT1 Design Basis Earthquakes
 NT2 Safe Shutdown Earthquake 135
NT1 Design Basis Fires
NT1 Design Basis Floods
NT1 Design Basis Tornadoes
NT1 Operational DBA
RT Confinement Systems
RT Safety Class Items
SO Construction
DEF (DOE Order 6430.1A) Postulated
    accidents, or natural forces, and
    resulting conditions for which the
    confinement structure, systems,
    components and equipment must
    meet their functional goals. These
    safety class items are those
    necessary to assure the capability:
    to safely shut down operations,
    maintain the plant in a safe
    shutdown condition, and maintain
    integrity of the final confinement
    barrier of radioactive or other
    hazardous materials; to prevent or
    mitigate the consequences of
    accidents; or to monitor releases
    that could result in potential offsite
    exposures.

DESIGN BASIS EARTHQUAKES
(DOE Order 6430.1A)
DA October 12, 1990
SF *DBE*
BT1 Design Basis Accidents
 BT2 Accidents
NT1 Safe Shutdown Earthquake 135
SO Construction
DEF (DOE Order 6430.1A) Earthquakes
    that are the most severe design
    basis accident of this type and that
    produce the vibratory ground
    motion for which safety class items
    are designed to remain functional.

DESIGN BASIS FIRES
(DOE Order 6430.1A)
DA October 12, 1990
SF *DBF*
BT1 Design Basis Accidents
 BT2 Accidents
RT Fire Protection Systems
RT Fires
SO Construction
DEF (DOE Order 6430.1A) Fires that are
    the most severe design basis
    accidents of this type. In
    postulating such fires, failure of
    automatic and manual fire
    suppression provisions shall be
    assumed except for those safety
    class items/systems that are
    specifically designed to remain

available (structurally or
functionally) through the event.

DESIGN BASIS FLOODS
(DOE Order 6430.1A)
DA October 12, 1990
SF *DBFL*
BT1 Design Basis Accidents
 BT2 Accidents
RT Maximum Probable Flood
SO Construction
DEF (DOE Order 6430.1A) Floods that
    are the most severe design basis
    accidents of that type applicable to
    the area under consideration.

DESIGN BASIS TORNADOES
(DOE Order 6430.1A)
DA October 12, 1990
SF *DBT*
BT1 Design Basis Accidents
 BT2 Accidents
RT Tornadoes
SO Construction
DEF (DOE Order 6430.1A) Tornadoes
    that are the most severe design
    basis accidents of that type
    applicable to the area under
    consideration.

DESIGN CAPACITY
(RCRA; CFR)
DA January 24, 1991
RT Anticipated Operational
    Occurrences
RT Engineered Safety Features
DEF (CFR) Anticipated operational
    occurrences (such as the loss of
    coolant flow or a reactivity
    excursion) which are used to
    determine the specific design
    requirements for the reactor safety
    system.

DESIGN CHANGE REQUEST
DA January 8, 1991
SF *DCR*

DESIGN FLOODS
(DOE Order 6430.1A)
DA October 12, 1990
SO Construction
DEF (DOE Order 6430.1A) The floods,
    (either observed or synthetic)
    chosen as the basis for the design
    of a hydraulic structure.

DESIGNATED FACILITIES
(SWDA; RCRA; CFR; ESH)
DA October 12, 1990
BT1 Treatment, Storage, and Disposal
    Facilities
 BT2 Facilities
RT Hazardous Waste Facilities
SO Hazardous Materials
SO Wastes
DEF Hazardous waste treatment,
    storage, or disposal facilities that

have been designated on the manifest by the generator.

DESIGNATED NON-FEDERAL REPRESENTATIVES
(ESA; CFR)
DA   October 12, 1990
BT1 Personnel
RT   Informal Consultation
SO   Endangered Species
DEF (CFR) Persons designated by the Federal agency as its representatives to conduct informal consultation and/or to prepare any biological assessment.

DESIGNATED POLLUTANTS
(EPA)
DA   October 12, 1990
BT1 Air Pollutants
SO   Environmental Protection Agency Glossary
DEF (EPA) Air pollutants which are neither criteria nor hazardous pollutants, as described in the Clean Air Act, but for which new sources performance standards exist. The Clean Air Act does require states to control these pollutants, which include acid mist, total reduced sulfur (TRS), and fluorides.

DESIGNATED USES
(EPA)
DA   October 12, 1990
BT1 Uses
SO   Environmental Protection Agency Glossary
DEF (EPA) Those water uses identified in state water quality standards which must be achieved and maintained as required under the Clean Water Act. Uses can include cold water fisheries, public water supply, agriculture, etc.

DESIGNER BUGS
(EPA)
DA   October 12, 1990
RT   Biotechnology
SO   Environmental Protection Agency Glossary
DEF (EPA) Popular term for microbes developed through biotechnology that can degrade specific toxic chemicals at their source in toxic waste dumps or in ground water.

DESTRUCTION
(ESA; CFR)
DA   October 19, 1990
SY   Adverse Modifications
BT1 Biological Processes
 BT2 Processes
RT   Critical Habitats
SO   Endangered Species
DEF (CFR) The total and irreversible

loss of a natural resource. A direct or indirect alteration that appreciably diminishes the value of critical habitat for both the survival and recovery of a listed species. Such alterations include, but are not limited to, alterations adversely modifying any of those physical or biological features that were the basis for determining the habitat to be critical.

DESTRUCTION REMOVAL EFFICIENCY
DA   January 8, 1991
SF   DRE

DESULFURIZATION
(EPA)
DA   October 12, 1990
BT1 Pollution Recovery Processes
 BT2 Processes
RT   Flue Gas Desulfurization
SO   Environmental Protection Agency Glossary
DEF (EPA) Removal of sulfur from fossil fuels to reduce pollution.

DETAILED OPERATING PROCEDURES
(NFI; SSDC)
DA   October 12, 1990
SF   DOP
BT1 Operating Procedures
 BT2 Procedures
RT   Working Level
SO   Management
SO   System Safety Development Center Glossary
DEF (SSDC) Working level documents to perform a task or job. Used to help avoid oversights and omissions and to assure a proper sequence of work steps.

DETECTION
(DOE Order 6430.1A; EMER)
DA   October 12, 1990
NT1 Release Detection
RT   Detection Equipment
RT   Leak Detection Systems
RT   Physical Protection (physical security)
SO   Construction
SO   Emergency Preparedness
DEF (DOE Order 6430.1A) The positive assessment that a specific object is the cause of an alarm.

DETECTION EQUIPMENT
(DOE Order 6430.1A)
DA   October 12, 1990
BT1 Equipment
NT1 Intrusion Alarm System (Perimeter or Interior)
RT   Detection
SO   Construction
DEF (DOE Order 6430.1A) Any equipment or system that is designed to provide a high

probability of positive assessment of intrusion.

DETECTION LIMITS
(EPA)
DA   October 12, 1990
BT1 Limits
NT1 Environmental Detection Limits
NT1 Lower Limit of Detection
SO   Environmental Protection Agency Glossary
DEF (EPA) The lowest amount that can be distinguished from the normal electronic noise of an analytical instrument.

DETERGENTS
(EPA)
DA   October 12, 1990
NT1 Soft Detergents
RT   Caustic Soda
RT   Decontamination
RT   Surfactants
RT   Xenobiotic
SO   Environmental Protection Agency Glossary
DEF (EPA) Synthetic washing agents that help to remove dirt and oil. Some contain compounds which kill useful bacteria and encourage algae growth when they are in wastewater that reaches receiving waters.

DETONATIONS
(DOE Order 6430.1A)
DA   October 19, 1990
RT   Explosives
RT   Hazards
RT   Initiation Stimuli
SO   Construction
DEF (DOE Order 6430.1A) Violent chemical reactions within a chemical compound or mechanical mixture evolving heat and pressure. They are reactions that proceed through the reacted material toward the unreacted material at supersonic velocity. The result of the chemical reactions is the exertion of extremely high pressure on the surrounding medium, forming a propagating shock wave of supersonic velocity. For the purposes of these criteria the terms detonation and explosion will be used interchangeably regardless of the velocity of the reaction or propagating shock wave.

DETONATORS
(DOE Order 6430.1A)
DA   October 12, 1990
BT1 Electroexplosive Devices
 BT2 Devices
RT   Explosives

SO Construction
DEF (DOE Order 6430.1A) The
    explosive devices that are used to
    initiate the detonation of other
    explosives.

DETRIMENT
(IAEA)
DA October 12, 1990
BT1 Risks
SO Radiation
DEF (IAEA) The mathematical
    expectation of the harm (damage
    to health and other effects)
    incurred from the exposure of
    individuals or groups of persons in
    a human population to a radiation
    source, taking into account not
    only the probabilities but also the
    severity of each type of
    deleterious effect.

DEVELOPERS
(EPA)
DA October 12, 1990
SO Environmental Protection Agency
    Glossary
DEF (EPA) Persons, government units,
    or companies that propose to build
    a hazardous waste treatment,
    storage, or disposal facility.

DEVELOPMENTAL RFD
(EPA)
DA October 12, 1990
BT1 Reference Dose (RfD)
 BT2 Doses
SO Environmental Protection Agency
    Glossary
DEF (EPA) An estimate (with uncertainty
    spanning perhaps an order of
    magnitude or greater) of an
    exposure level for the human
    population, including sensitive
    subpopulations, that is likely to be
    without an appreciable risk of
    developmental effects.
    Developmental RfDs are used to
    evaluate the effects of a single
    event (generally 1 day) exposure.

DEVICES
DA January 30, 1991
NT1 Accelerators
NT1 Alarms
 NT2 Attack Warning Signals
 NT2 Attention Signals
 NT2 Criticality Alarm System
 NT2 Criticality Alarms
 NT2 Fire Alarm System
 NT2 High Temperature Alarm
 NT2 Intrusion Alarm System (Perimeter
     or Interior)
 NT2 Kanne Alarm
 NT2 Level Alarm
 NT2 Radiation Alarm System
NT1 Bar Screens
NT1 Blank Fire Adapter (BFA)
NT1 Bullet Containment Devices

 NT2 Clearing Barrels
NT1 Catalytic Converters
NT1 Control Devices
 NT2 Add On Control Devices
 NT2 Afterburners
 NT2 Arsenic Kitchens
 NT2 Carbon Adsorbers
 NT2 Catalytic Incinerators
 NT2 Cyclone Collectors
 NT2 Packed Towers
 NT2 Precipitators
  NT3 Electrostatic Precipitators
 NT2 Primary Emission Control Systems
 NT2 Secondary Hood Systems
 NT2 Vapor Capture Systems
  NT3 Floor Sweeps
NT1 Crossing Frogs
NT1 Drinking Water Coolers
NT1 Electroexplosive Devices
 NT2 Detonators
NT1 Elementary Neutralization Units
NT1 Emergency Shutdown Device
NT1 Fall Protection Device
 NT2 Safety Belt
 NT2 Safety Line
 NT2 Safety Net
NT1 Filters
 NT2 Baghouse Filters
 NT2 Fabric Filters
 NT2 Filters (Radiology)
 NT2 Hepa Filter
 NT2 High Efficiency Particulate Air
     Filters
 NT2 Sand Filters
 NT2 Trickling Filters
NT1 Flares
NT1 Golf Bags
NT1 Graphic Panel
NT1 Grapper Pick Up
NT1 Grinder Pumps
NT1 Inertial Separators
 NT2 Cyclone Collectors
NT1 Lifting Station
NT1 Live Round Excluders
NT1 Load Securing Device
NT1 Marine Sanitation Devices
NT1 Material Handling Device
 NT2 Conveyor(s)
 NT2 Crane(s)
  NT3 Boom Crane
  NT3 Bridge Crane
  NT3 Mobile Crane
  NT3 Overhead Crane
 NT2 Earth Moving Equipment
  NT3 Dredge(s)
 NT2 Hand Truck(s)
  NT3 Cart(s)
  NT3 Dolly
 NT2 Hoisting Apparatus
  NT3 Derrick
  NT3 Elevator
  NT3 Forklift(s)
  NT3 Hoist(s)
   NT4 Air Hoist
  NT3 Lift Bucket
  NT3 Manlift(s)
  NT3 Scissor Lift
NT1 Ozonators
NT1 Packers
NT1 Point-of-entry Treatment Devices
NT1 Point-of-use Treatment Devices

NT1 Pumping Stations
NT1 Reset Device
NT1 Safety Device/System
NT1 Scrubbers
NT1 Sensors
NT1 Sludge Dryer
NT1 Suspicious Devices
NT1 Torpedoes
NT1 Turbidimeters
NT1 Valves
 NT2 Back Pressure Valve
 NT2 Check Valves
 NT2 Main Steam Isolation Valve
 NT2 Motor Operated Valves
 NT2 Open Ended Valve
 NT2 Pressure Regulator Valve
 NT2 Relief Valves
  NT3 Emergency Relief Valve
  NT3 Inadvertently Opened Relief
      Valve
  NT3 Power Operated Relief Valve
  NT3 Safety/Relief Valve
  NT3 Stuck-Open Relief Valve
 NT2 Turbine Block Valve
NT1 Wastewater Treatment Units
 NT2 Digesters
 NT2 Wastewater Treatment Tanks
RT Equipment
RT Production Facilities
SO DOE FRASE VOCABULARY

DEWATERED
(CFR)
DA October 12, 1990
RT Tailings
SO Air Pollution
DEF (CFR) To remove the water from
    recently produced tailings by
    mechanical or evaporative
    methods such that the water
    content of the tailings does not
    exceed 30 percent by weight.

DFDP
DA October 12, 1990
SEE Defense Facility Decommissioning
    Program
SO Acronyms

DFM
DA October 12, 1990
SEE Flow Monitor
SO Acronyms

DG
DA October 12, 1990
SEE Diesel Generator
SO Acronyms

DHR
DA October 12, 1990
SEE Decay Heat Removal
SO Acronyms

DI
DA October 12, 1990
SEE Deionized
SO Acronyms

SY-Synonymous Terms   SO-Source/Subject Category   SF-See From

DIAGNOSIS OF MULTIPLE ALARMS SYSTEM
DA   January 8, 1991
SF   *DMA*
BT1 Emergency Systems
 BT2 Systems

DIAGRAMS
(SSDC)
DA   October 12, 1990
NT1 Analytical (Logic) Trees
 NT2 Event Tree
 NT2 Fault Tree
  NT3 Management Oversight and Risk Tree
 NT2 Positive (objective) Trees
NT1 Flow Charts
NT1 Histograms
NT1 Logic Trees
NT1 Piping and Instrumentation Drawing
NT1 Ringlemann Chart
NT1 Safety Assurance System Summary
NT1 Schematics
RT   Drafting Break
RT   Maps
SO   System Safety Development Center Glossary
DEF (SSDC) Geometric drawings used to explain a fact, a process, the sequence of an activity, or the composition of an element as associated with an accident or incident.

DIAPHRAGM CHORD
(SEA)
DA   October 12, 1990
BT1 Boundary Elements
RT   Diaphragms
SO   Construction
DEF (SEA) The boundary element of a diaphragm or shear wall which is assumed to take axial stresses analogous to the flanges of a beam.

DIAPHRAGM STRUT
(SEA)
DA   October 12, 1990
RT   Collectors
RT   Diaphragms
SO   Construction
DEF (SEA) The element of a diaphragm parallel to the applied load which collects and transfers diaphragm shear to vertical resisting elements or distributes loads within the diaphragm. Such members may take axial tension or compression.

DIAPHRAGMS
(SEA)
DA   October 12, 1990
NT1 Horizontal Bracing Systems
RT   Boundary Elements
RT   Diaphragm Chord
RT   Diaphragm Strut

SO   Construction
DEF (SEA) Horizontal or nearly horizontal systems acting to transmit lateral forces to the vertical resisting elements. The term "diaphragm" includes horizontal bracing systems.

DIATOMACEOUS EARTH (DIATOMITE)
(EPA)
DA   October 12, 1990
RT   Diatomaceous Earth Filtration
SO   Environmental Protection Agency Glossary
DEF (EPA) A chalk-like material (fossilized diatoms) used to filter out solid waste in waste-water treatment plants, also used as an active ingredient in some powdered pesticides.

DIATOMACEOUS EARTH FILTRATION
(SDWA; CFR)
DA   October 12, 1990
BT1 Filtration
 BT2 Pollution Recovery Processes
  BT3 Processes
BT1 Wastewater Treatment Processes
 BT2 Treatment
  BT3 Waste Management Processes
   BT4 Processes
RT   Diatomaceous Earth (Diatomite)
SO   Water Pollution
DEF (CFR) A process resulting in substantial particulate removal in which (1) a precoat cake of diatomaceous earth filter media is deposited on a support membrane (septum), and (2) while the water is filtered by passing through the cake on the septum, additional filter media known as body feed is continuously added to the feed water to maintain the permeability of the filter cake.

DIAZINON
(EPA)
DA   October 12, 1990
BT1 CERCLA Hazardous Substances
 BT2 Hazardous Substances
BT1 Insecticides
 BT2 Pesticides
  BT3 Hazardous Substances
SO   Environmental Protection Agency Glossary
DEF (EPA) An insecticide. In 1986, EPA banned its use on open areas such as sod farms and golf courses because it posed a danger to migratory birds who gathered on them in large numbers. The ban did not apply to its use in agriculture, or on lawns of homes and commercial establishments.

DICOFOL
(EPA)

DA   October 12, 1990
BT1 Insecticides
 BT2 Pesticides
  BT3 Hazardous Substances
SO   Environmental Protection Agency Glossary
DEF (EPA) A pesticide used on citrus fruits.

DIELECTRIC MATERIALS
(SWDA; RCRA; CFR)
DA   October 12, 1990
BT1 Materials
RT   Capacitors
SO   Wastes
DEF (CFR) Materials that do not conduct direct electrical current. Dielectric coatings are used to electrically isolate UST (Underground Storage Tank) systems from the surrounding soils. Dielectric bushings are used to electrically isolate portions of the UST system (e.g., tank from piping).

DIESEL GENERATOR
DA   January 8, 1991
SF   *DG*
BT1 Equipment
NT1 Standby Diesel Generator

DIFFERENCE DEVIATION
(SSDC)
DA   October 12, 1990
BT1 Measurements
SO   System Safety Development Center Glossary
DEF (SSDC) Departure from a norm.

DIFFERENTIAL COST BENEFIT ANALYSIS
(IAEA)
DA   October 12, 1990
BT1 Analyses
RT   Radiation Protection
SO   Radiation
DEF (IAEA) A procedure for optimization of radiation protection used to determine the point at which exposures have been decreased so far that any further decrease is considered less important than the additional necessary effort required to achieve it.

DIFFERENTIAL IN HOURS
(NFI)
DA   October 12, 1990
SF   *DIH*
RT   Ink
SO   Radiation
DEF (NIH) Effective neutron absorption of ink.

DIFFERENTIAL PRESSURE
DA   January 10, 1991
SF   *dP (Differential Pressure)*
BT1 Pressure
DEF (NRC Glossary of Terms: Nuclear

BT-Broader Term   NT-Narrower Term   RT Related Term

Power and Radiation) The difference in pressure between two points of a system, such as between the inlet and outlet of a pump.

DIFFERENTIAL SETTLEMENT
(USGS)
DA   October 12, 1990
BT1 Ground Failures
 BT2 Failures
  BT3 Accidents
 BT2 Liquefaction
  BT3 Processes
RT   Subsidence
SO   Natural Phenomenon
DEF (USGS) Nonuniform settlement of land or the uneven lowering of it. A principal cause of damage and casualties resulting from earthquakes. (See "Glossary of Geology".)

DIFFERENTIATION
(EPA)
DA   October 12, 1990
BT1 Biological Processes
 BT2 Processes
SO   Environmental Protection Agency Glossary
DEF (EPA) The process by which single cells grow into particular forms of specialized tissue, e.g., root, stem, leaf.

DIFFUSED AIR PROCESS
(EPA)
DA   October 12, 1990
BT1 Aeration
 BT2 Biological Processes
  BT3 Processes
SO   Environmental Protection Agency Glossary
DEF (EPA) A type of aeration that forces oxygen into sewage by pumping air through perforated pipes inside a holding tank and bubbling it through the sewage.

DIGESTERS
(ESA; EPA)
DA   November 6, 1990
BT1 Wastewater Treatment Units
 BT2 Devices
SO   Environmental Protection Agency Glossary
DEF (EPA) In wastewater treatment, closed tanks; in solid waste conversion, units in which bacterial action is induced and accelerated in order to break down organic matter and establish the proper carbon to nitrogen ratio.

DIGESTION
(EPA)
DA   October 12, 1990
BT1 Processes
NT1 Anaerobic Digestion

RT   Ingestion
SO   Environmental Protection Agency Glossary
DEF (EPA) The biochemical decomposition of organic matter, resulting in partial gasification, liquefaction, and mineralization of pollutants.

DIGESTIVE SYSTEM
(1142 DIGESTIVE SY)
DA   November 28, 1990
BT1 Body System(s)
 BT2 Human Body Parts
NT1 Mouth
 NT2 Tooth/Teeth
NT1 Rectum
SO   DOE FRASE VOCABULARY

DIH
DA   October 12, 1990
SEE Differential in Hours
SO   Acronyms

DIKES
(SWDA; RCRA; CWA; RHA; CFR; EPA)
DA   October 12, 1990
BT1 Hydraulic Structures
 BT2 Structures (DOE FRASE Vocabulary)
NT1 Terracing
SO   Environmental Management
SO   Environmental Protection Agency Glossary
SO   Wastes
SO   Water Pollution
DEF (CFR) Embankments or ridges of either natural or man-made materials used to prevent the movement of liquids, sludges, solids, or other materials. Low walls that can act as barriers to prevent a spill from spreading.

DIL
DA   October 12, 1990
SEE Daily Instruction Logs
SO   Acronyms
SO   Nuclear Facilities Incident Database

DILUTION RATIO
(EPA)
DA   October 12, 1990
BT1 Ratios
SO   Environmental Protection Agency Glossary
DEF (EPA) The relationship between the volume of water in a stream and the volume of incoming water. It affects the ability of the stream to assimilate waste.

DINOCAP
(EPA)
DA   October 12, 1990
BT1 Fungicides
 BT2 Pesticides
  BT3 Hazardous Substances

SO   Environmental Protection Agency Glossary
DEF (EPA) A fungicide used primarily by apple growers to control summer diseases. EPA, in 1986, proposed restrictions on its use when laboratory tests found it caused birth defects in rabbits.

DINOSEB
(EPA)
DA   October 12, 1990
BT1 CERCLA Hazardous Substances
 BT2 Hazardous Substances
BT1 Herbicides
 BT2 Pesticides
  BT3 Hazardous Substances
SO   Environmental Protection Agency Glossary
DEF (EPA) A herbicide that is also used as a fungicide and insecticide. It was banned by EPA in 1986 because it posed the risk of birth defects and sterility.

DIODE(S)
(2388 DIODE)
DA   January 3, 1991
BT1 Equipment/Parts - Electrical (DOE FRASE Vocabulary)
 BT2 Equipment
SO   DOE FRASE VOCABULARY

DIOXINS
(EPA)
DA   October 12, 1990
BT1 Organic Chemicals
 BT2 Chemical Substances
SO   Environmental Protection Agency Glossary
DEF (EPA) Any of a family of compounds known chemically as dibenzo-p-dioxins. Concern about them arises from their potential toxicity and contaminants in commercial products. Tests on laboratory animals indicate that it is one of the more toxic man-made chemicals known.

DIP-SLIP
(USGS)
DA   October 12, 1990
BT1 Surface Faulting
 BT2 Faults
RT   Subsidence
SO   Natural Phenomenon
DEF (USGS) A type of surface faulting; in a fault, the component of the movement or slip that is parallel to the dip of the fault. The faulting may be normal or reverse displacement.

DIRECT BIOASSAYS
(NCRP)
DA   October 12, 1990
BT1 Bioassays
 BT2 Biological Processes

SY-Synonymous Terms   SO-Source/Subject Category   SF-See From

BT3 Processes
RT Body Content
SO Radiation
DEF (NCRP) Assessments of
    radioactive material deposited in
    the body by detection of radiation
    emitted by the material in the body
    (in vivo measurement).

DIRECT CONTAINMENT HEATING
DA January 8, 1991
SF *DCH*

DIRECT CURRENT
DA January 8, 1991
SF *DC*
NT1 Volts Direct Current

DIRECT DISCHARGERS
(EPA)
DA October 12, 1990
BT1 Facilities
RT Point Sources
SO Environmental Protection Agency
    Glossary
DEF (EPA) Municipal or industrial
    facilities which introduce pollution
    through a defined conveyance or
    system; a point source.

DIRECT FILTRATION
(SDWA; CFR)
DA October 12, 1990
BT1 Filtration
 BT2 Pollution Recovery Processes
  BT3 Processes
SO Water Pollution
DEF (CFR) A series of processes
    including coagulation and filtration
    but excluding sedimentation
    resulting in substantial particulate
    removal.

DIRECTLY IONIZING PARTICLES
(IAEA)
DA October 12, 1990
BT1 Charged Particles
RT Alpha Particles
RT Electrons
RT Indirectly Ionizing Particles
RT Ionization
SO Radiation
DEF (IAEA) Charged particles
    (electrons, protons, alpha
    particles, etc) having sufficient
    kinetic energy to produce
    ionization by collision.

DIRECTOR OF EMERGENCY OPERATIONS
(DOE Order 5500.1A Attach. II, 12;
 EMER)
DA January 24, 1991
BT1 Personnel
RT Emergency Management Teams
SO Emergency Preparedness
SO Environmental Management
DEF (DOE 5500.1A) The DOE official
    with responsibility to coordinate

the planning, development, and
implementation of the overall DOE
Emergency Management System.

DIRECTORS
(SWDA; RCRA; ESA; CFR)
DA October 12, 1990
BT1 Personnel
RT Regional Administrator
RT State Director
SO Endangered Species
SO Wastes
SO Water Pollution
DEF The Regional Administrator or the
    State Director, the context
    requires, or an authorized
    representative. When there is no
    approved State program, and
    there is an EPA administered
    program, Director means the
    Regional Administrator. When
    there is an approved State
    program, Director normally means
    the State Director. In some
    circumstances, however, EPA
    retains the authority to take certain
    actions even when there is an
    approved State program. In such
    cases, the term Director means
    the Regional Administrator and not
    the State Director. The Director of
    the National Marine Fisheries
    Service, National Oceanic and
    Atmospheric Administration,
    Department of Commerce, or his
    authorized designee.

DISABILITIES
(SSDC)
DA October 12, 1990
NT1 Permanent Partial Disabilities
NT1 Permanent Total Disabilities
RT Disabling Injuries
SO System Safety Development Center
    Glossary
DEF (SSDC) Any illnesses or injuries
    which prevent a person from
    normal human activity, either
    temporarily or permanently.

DISABLING INJURIES
(SSDC)
DA October 12, 1990
BT1 Injuries
RT Disabilities
RT Occupational Injuries
SO System Safety Development Center
    Glossary
DEF (SSDC) Injuries which prevent a
    person from performing his/her
    regularly established job for one
    full day beyond the day of the
    accident (ANSI Z-16.1); also
    called a lost-time injury.

DISCARDED MATERIALS
(SWDA; RCRA; CFR)
DA October 12, 1990
BT1 Materials

RT Recycling
RT Solid Wastes
SO Wastes
DEF (CFR) Solid wastes are discarded
    materials that are not excluded by
    40 CFR Part 261.4(a) of that is not
    excluded by variance granted
    under Parts 260.30 and 260.31. A
    discarded material is any material
    which is (i) abandoned, as
    explained in Part 261.2(b); or (ii)
    recycled, as explained in Part
    261.2(c); or (iii) considered
    inherently wastelike, as explained
    in Part 261.2(d). 40 CFR Parts
    261.2(b) through (f) define
    abandoned, recycled, and
    inherently abandoned, and
    materials that are and are not
    solid wastes in the recycling of
    discarded materials.

DISCHARGE AND EXIT
DA January 8, 1991
SF *D&E*

DISCHARGE MACHINE
(2116 DISCHARGE MA)
DA December 10, 1990
BT1 Machines (DOE FRASE
    Vocabulary)
 BT2 Equipment
SO DOE FRASE VOCABULARY

DISCHARGE OF DREDGED MATERIAL
(CWA; RHA; CFR)
DA October 12, 1990
BT1 Discharges
RT Discharge Point
RT Dredged Materials
SO Water Pollution
DEF (CFR) Any addition of dredged
    material into the waters of the
    United States. The term includes,
    without limitation, the addition of
    dredged material to a specified
    discharge site located in waters of
    the United States and the runoff or
    overflow from a contained land or
    water disposal area. Discharges of
    pollutants into waters of the United
    States resulting from the onshore
    subsequent processing of dredged
    material that is extracted for any
    commercial use (other than fill) are
    not included within this term and
    are subject to section 402 of the
    Clean Water Act even though the
    extraction and deposit of such
    material may require a permit from
    the Corps of Engineers. The term
    does not include plowing,
    cultivating, seeding and harvesting
    for the production of food, fiber,
    and forest products (See 40 CFR
    323.4 for the definition of these
    terms). The term does not include
    de minimis, incidental soil
    movement occurring during normal
    dredging operations.

BT-Broader Term

NT-Narrower Term

RT Related Term

DISCHARGE OF FILL MATERIAL
(CWA; RHA; CFR)
DA October 12, 1990
BT1 Discharges
RT Discharge Point
RT Fill Materials
SO Water Pollution
DEF (CFR) The addition of fill material into waters of the United States. The term generally includes, without limitation, the following activities: Placement of fill that is necessary for the construction of any structure in a water of the United States; the building of any structure or impoundment requiring rock, sand, dirt, or other material for its construction; site-development fills for recreational, industrial, commercial, residential, and other uses; causeways or road fills; dams and dikes; artificial islands; property protection and/or reclamation devices such as riprap, groins, seawalls, breakwaters, and revetments; beach nourishment; levees; fill for structures such as sewage treatment facilities, intake and outfall pipes associated with power plants and subaqueous utility lines; and artificial reefs. The term does not include plowing, cultivating, seeding and harvesting for the production of food, fiber, and forest products (See 40 CFR 323.4 for the definition of these terms)

DISCHARGE OF POLLUTANTS
(CWA; CFR; ESH)
DA October 12, 1990
BT1 Discharges
RT Pollutants
RT Pollutant Discharge Elimination System
RT Waste Load Allocations
RT Water Pollution
SO Environmental Management
SO Hazardous Materials
DEF (ESH) (1)The addition of pollutants to navigable waters from any point source and (2)any addition of pollutants to the waters of the contiguous zone or the ocean from any point source, other than from a vessel or other floating craft being used as a means of transportation. The term discharge includes either the discharge of a single pollutant or the discharge of multiple pollutants.

DISCHARGE POINT
(CWA; RHA; CFR)
DA October 12, 1990
BT1 Sites/Areas
RT Discharge of Dredged Material
RT Discharge of Fill Material
SO Water Pollution
DEF (CFR) The point within the disposal site at which the dredged or fill material is released.

DISCHARGES
(SWDA; RCRA; CFR)
DA October 12, 1990
NT1 Discharge of Pollutants
NT1 Discharge of Dredged Material
NT1 Discharge of Fill Material
NT1 Effluents
NT2 Radioactive Effluents
NT1 Hazardous Waste Discharge
NT1 Indirect Discharges
NT1 Interferences
NT1 Major Discharges
NT1 Medium Discharges
NT1 Minor Discharges
NT1 On-Site Discharges
NT1 Plumes
NT2 Vapor Plumes
NT1 Relief Valve Discharge
NT1 Unauthorized Discharges
RT Effluent Limitations
RT Emissions
RT Management of Migrations
RT Mixing Zones
RT Releases
RT Spills
SO Compensation and Liability
SO Environmental Management
SO Wastes
DEF (CFR) As defined by section 311(a)(2) the of Clean Water Act, includes, but is not limited to, any spilling, leaking, pumping, pouring, emitting, emptying, or dumping of oil. For purposes of this Plan, discharge shall also mean substantial threat of discharge.

DISCRETIONARY AUTHORITY
(CWA; RHA; CFR)
DA October 12, 1990
SO Water Pollution
DEF The authority delegated to division engineers to override provisions of nationwide permits, to add regional conditions, or to require individual permit application.

DISCRIMINATION
(DOE Order 5483.1A)
DA October 12, 1990
SO Industrial Safety
DEF (DOE Order 5483.1A) Discharge, demotion, reduction in pay, coercion, restraint, threats, or other negative actions taken against a contractor employee by a contractor, as a result of the employee's exercise of occupational safety and health rights set forth in this Order.

DISEASE OF CENTRAL NERVOUS SYSTEM
(1384 DISEASE OF C)

DA November 28, 1990
BT1 Diseases
BT1 Illnesses
SO DOE FRASE VOCABULARY

DISEASES
DA February 14, 1991
NT1 Asbestosis
NT1 Black Lung
NT1 Cond of Respiratory Sys. Non-Toxic
NT2 Upper Respiratory Condition
NT2 Upper Respiratory Disease
NT3 Tuberculosis
NT1 Contagious or Infectious Disease
NT2 Conjunctivitis
NT1 Disease of Central Nervous System
NT1 Diseases of Blood
NT2 Hepatitis
RT Hazards
RT Illnesses
RT Injuries

DISEASES OF BLOOD
(1385 DISEASES OF)
DA November 28, 1990
BT1 Diseases
BT1 Illnesses
NT1 Hepatitis
SO DOE FRASE VOCABULARY

DISINFECTANT CONTACT TIME
(SDWA; CFR)
DA October 12, 1990
BT1 Time Designations
RT CT (Public Water Disinfection Formula)
RT CTcalc
RT Point of Disinfectant Application
RT Residual Disinfectant Concentration
SO Water Pollution
DEF (CFR) "T" in CT calculations; means the time in minutes that it takes for water to move from the point of disinfectant applications or the previous point of disinfectant residual measurement to a point before or at the point where residual disinfectant concentration ("C") is measured. See 40 CFR 141.2 for continued explanation.

DISINFECTANTS
(SDWA; CFR; EPA; ESH)
DA October 12, 1990
BT1 Germicides
RT Disinfection
RT Pesticides
RT Trihalomethane
SO Environmental Protection Agency Glossary
SO Water Pollution
DEF (CFR) Chemical or physical agents that kill pathogenic organisms in water. Chlorine is often used to disinfect sewage treatment effluent, water supplies, wells, and swimming pools.

SY-Synonymous Terms   SO-Source/Subject Category   SF-See From

DISINFECTION
(SDWA; CFR)
DA   October 12, 1990
BT1 Purification Processes
 BT2 Processes
RT   Chlorination
RT   Disinfectants
RT   Oxidation
RT   Pathogens
RT   Point of Disinfectant Application
RT   Residual Disinfectant Concentration
SO   Water Pollution
DEF (CFR) A process which inactivates
     pathogenic organisms in water by
     chemical oxidants or equivalent
     agents.

DISINTEGRATIONS PER MINUTE PER GRAM
DA   January 8, 1991
SF   *D/M/G*
BT1 Units of Measure

DISK SANDER(S)
(2952 DISK SANDER)
DA   January 3, 1991
BT1 Tools - Powered
 BT2 Tools (DOE FRASE Vocabulary)
 BT3 Equipment
SO   DOE FRASE VOCABULARY

DISLOCATION
(1316 DISLOCATION)
DA   November 28, 1990
BT1 Injuries
SO   DOE FRASE VOCABULARY

DISPERSANTS
(CFR; EPA)
DA   October 12, 1990
BT1 Chemical Agents
RT   Surfactants
SO   Compensation and Liability
SO   Environmental Protection Agency
     Glossary
DEF (CFR) Chemical agents that
     emulsify, disperse, or solubilize oil
     into the water column or promote
     the surface spreading of oil slicks
     to facilitate dispersal of the oil into
     the water column.

DISPERSION RESINS
(CFR)
DA   October 12, 1990
BT1 Types of Resin
 BT2 Resin Grade
SO   Air Pollution
DEF (CFR) Resins manufactured in such
     a way as to form fluid dispersions
     when dispersed in a plasticizer or
     plasticizer/diluent mixtures.

DISPERSION TECHNIQUES
(CFR)
DA   October 12, 1990
NT1 Intermittent Control System
RT   Control Devices
RT   Plumes

RT   Smoke
RT   Stacks
SO   Air Pollution
DEF (CFR) Techniques which attempt to
     affect the concentration of a
     pollutant in the ambient air by (i)
     using that portion of a stack which
     exceeds good engineering
     practice stack height (ii) varying
     the rate of emission of a pollutant
     according to atmospheric
     conditions or ambient
     concentrations of that pollutant; or
     (iii) increasing final exhaust gas
     plume rise by manipulating source
     process parameters, exhaust gas
     parameters, stack parameters, or
     combining exhaust gases from
     several existing stacks into one
     stack, or other selective handling
     of exhaust gas streams so as to
     increase the exhaust gas plume
     rise. See 40 CFR 51.100 (hh)(2)
     for exclusions to this definition.

DISPOSAL
(SWDA; RCRA; TSCA; AEA; CFR; USC;
 EPA; ESH)
DA   October 12, 1990
BT1 Waste Management Processes
 BT2 Processes
NT1 Continuous Disposal
NT1 Land Disposal
 NT2 Near Surface Disposal
 NT3 Shallow Land Burial
NT1 Land Farming (of waste)
NT1 Solid Waste Disposal
 NT2 Phased Disposal
NT1 Water Dumping
RT   Disposal Facilities
RT   Disposal Sites
RT   Disposal Wells
RT   Disposal Units
RT   Treatment, Storage, and Disposal
RT   Waste Containers
RT   Waste Packages
SO   Environmental Management
SO   Environmental Protection Agency
     Glossary
SO   Hazardous Materials
SO   Wastes
DEF Waste emplacement designed to
     ensure isolation of waste from the
     biosphere, with no intention of
     retrieval for the foreseeable future,
     and that requires deliberate action
     to regain access to the waste.
     (EPA) Final placement or
     destruction of toxic, radioactive, or
     other wastes; surplus or banned
     pesticides or other chemicals;
     polluted soils; and drums
     containing hazardous materials
     from removal actions or accidental
     releases. Disposal may be
     accomplished through use of
     approved secure landfills, surface
     impoundments, land farming, deep
     well injection, ocean dumping, or
     incineration. (ESH) Disposal

means intentionally or accidentally
to discard, throw away, or
otherwise complete or terminate
the useful life of PCBs and PCB
items. Disposal includes spills,
leaks, and other uncontrolled
discharges of PCBs as well as
actions related to containing,
transporting, destroying,
degrading, decontaminating, or
confining PCBs and PCB items.

DISPOSAL AREAS
DA   May 24, 1991
BT1 Piles (Wastes)
 BT2 Hazardous Wastes
 BT3 Hazardous Materials
 BT4 Materials
 BT3 Wastes
BT1 Sites/Areas
SO   Wastes
DEF (CFR) The regions within the
     perimeter of impoundments or
     piles containing uranium
     by-product materials to which the
     post- closure requirements apply.
     See 40 CFR 192.31(f).

DISPOSAL FACILITIES
(DOE Order 5820.2A; SWDA; RCRA;
 CFR; ESH)
DA   October 12, 1990
BT1 Hazardous Waste Management
     Facilities
 BT2 Hazardous Waste Facilities
 BT3 Facilities
NT1 Land Disposal Facilities
 NT2 Near Surface Disposal Facilities
NT1 Repositories
NT1 Test and Evelation Facilities
RT   Disposal
RT   Disposal Units
RT   Geologic Repository Operations
     Areas
RT   Geologic Repositories
RT   Sanitary Landfills
SO   Wastes
DEF (CFR) Facilities or parts of a facility
     at which hazardous waste is
     intentionally placed into or on the
     land or water, and at which
     hazardous waste will remain after
     closure.

DISPOSAL PACKAGES
(ANL; NWPA)
DA   May 24, 1991
BT1 Packages
RT   Containers
RT   Radioactive Materials
RT   Radioactive Contents
SO   Hazardous Materials
SO   Wastes
DEF (NWPA) The primary containers
     that hold, and are in contact with,
     solidified high-level radioactive
     waste, spent nuclear fuel, or other
     radioactive materials, and any
     other overpacks that are emplaced
     at a repository.

DISPOSAL SITES
(DOE Order 5820.2A; AEA; CEA; RHA;
CFR)
DA   October 12, 1990
BT1 Sites/Areas
NT1 Inactive Waste Disposal Sites
 NT2 Inactive Hazardous Waste
      Disposal Sites
RT   Disposal
RT   Migration
RT   Prohibit Specification
SO   Environmental Management
SO   Water Pollution
DEF Those portions of the "waters of the
     United States" where specific
     disposal activities are permitted
     and consist of a bottom surface
     area and any overlying volume of
     water. In the case of wetlands on
     which surface water is not present,
     the disposal site consists of the
     wetland surface area.

DISPOSAL SYSTEMS
(ANL; CFR)
DA   May 24, 1991
BT1 Systems
RT   Engineered Barriers
RT   Natural Barriers
RT   Waste Management
SO   Industrial Safety
SO   Wastes
DEF (CFR) Consists of any combination
     of engineered and natural barriers
     that isolate spent nuclear fuel or
     radioactive wastes after disposal.

DISPOSAL UNITS
(DOE Order 5820.2A; AEA; CFR)
DA   January 24, 1991
BT1 Corrective Action Management
     Units
 BT2 Sites/Areas
NT1 Land Disposal Units
 NT2 Landfill Cells
 NT2 Landfills
  NT3 Chemical Waste Landfills
  NT3 Sanitary Landfills
  NT3 Specially Designated Landfills
RT   Disposal Facilities
RT   Disposal
DEF (DOE 5820.2A) A discrete portion
     of the disposal site into which
     waste is placed for disposal. For
     near-surface disposal the unit is
     usually a trench.

DISPOSAL WELLS
(SDWA; CFR; ESH)
DA   October 12, 1990
BT1 Wells
RT   Disposal
SO   Water Pollution
DEF (CFR) Wells used for the disposal
     of waste into a subsurface stratum.

DISPOSER OF PCB WASTE
(TSCA)
DA   October 19, 1990

RT   PCB Wastes
SO   Hazardous Materials
DEF (TSCA) Any person who owns or
     operates a facility approved by the
     EPA for the disposal of PCB waste
     which is regulated for disposal
     under the requirements of 40 CFR
     761 Subpart D.

DISSOLVED OXYGEN
(EPA)
DA   October 12, 1990
RT   Kinetic Rate Coefficient
SO   Environmental Protection Agency
     Glossary
DEF (EPA) The oxygen freely available
     in water. Dissolved oxygen is vital
     to fish and other aquatic life and
     for the prevention of odors.
     Traditionally, the level of dissolved
     oxygen has been accepted as the
     single most important indicator of
     a water body's ability to support
     desirable aquatic life. Secondary
     and advanced waste treatment are
     generally designed to protect DO
     in waste-receiving waters.

DISSOLVED SOLIDS
(EPA)
DA   October 12, 1990
BT1 Leachates
BT1 Solids
 BT2 Materials
SO   Environmental Protection Agency
     Glossary
DEF (EPA) Disintegrated organic and
     inorganic materials contained in
     water. Excessive amounts make
     water unfit to drink or use in
     industrial processes.

DISTILLATION
(EPA)
DA   October 12, 1990
BT1 Purification Processes
 BT2 Processes
RT   Desalinization
RT   Product Accumulator Vessels
RT   Vaporization
RT   Volatility
SO   Environmental Protection Agency
     Glossary
DEF (EPA) The act of purifying liquids
     through boiling, so that the steam
     condenses to a pure liquid and the
     pollutants remain in a
     concentrated residue.

DISTRIBUTION
(SSDC)
DA   October 12, 1990
NT1 Frequency Distribution
 NT2 Joint Frequency Distribution
RT   Density Function
RT   Histograms
RT   Outlier
RT   Range (Statistical)

SO   System Safety Development Center
     Glossary
DEF (SSDC) In statistics, the frequency
     or manner in which observations
     of different value are distributed
     over the range of values. These
     values can be numbers,
     frequency, size, cost, etc.

DISTRICT COMMANDER
(SWDA; RCRA; CFR; ESH)
DA   October 12, 1990
BT1 Personnel
SO   Hazardous Materials
DEF The District Commander of the
     Coast Guard, or his authorized
     representative, who has
     jurisdiction in the particular
     geographical area.

DISTURBED ZONES
(ANL; CFR)
DA   May 24, 1991
BT1 Controlled Areas
 BT2 Sites/Areas
BT1 Zones
 BT2 Sites/Areas
SO   Construction
SO   Wastes
DEF (CFR) Those portions of controlled
     areas, the physical or chemical
     properties of which have changed
     as a result of underground facility
     construction or as a result of heat
     generated by the emplaced
     radioactive wastes such that the
     resultant change of properties may
     have a significant effect on the
     performance of the geologic
     repository.

DIV
DA   October 12, 1990
SEE Divisions
SO   Acronyms

DIVERGENT WINDSTORMS
(SMRP)
DA   October 12, 1990
BT1 Act of Nature
NT1 Downbursts
 NT2 Microbursts
NT1 Downslope Wind
NT1 Gust Fronts
RT   Rotational Windstorms
SO   Natural Phenomenon
DEF (SMRP) Includes downbursts, gust
     fronts, and downslope winds, and
     are characterized predominantly
     by a divergent flow field.

DIVERSIONARY DEVICES
(DOE Order 5480.16)
DA   October 12, 1990
NT1 Flash Grenades
NT1 Smoke Grenades
SO   Firearms
DEF (DOE Order 5480.16) Special
     purpose pyrotechnic munitions

SY-Synonymous Terms   SO-Source/Subject Category   SF-See From

(known as flash bangs, stun grenades, and flash grenades) intended to give protective forces nonlethal force options whenever possible.

DIVISION OF THE STATE FIRE MARSHALL
DA   January 8, 1991
SF   *DSFM*
BT1 State Agencies
 BT2 Agencies
  BT3 Administrative Organizations
   BT4 Organizations

DIVISIONS
DA   January 8, 1991
SF   *DIV*
BT1 Administrative Organizations
 BT2 Organizations
NT1 Computer Operations Division
NT1 Health, Safety, and Environment Division
NT1 New Mexico Environmental Improvement Division
NT1 Occupational Health Division
RT   Centers
SO   Management
DEF (WTID) Subordinate units of a political or commercial body; typically embracing other subordinate units (e.g. departments, branches, etc.).

DMA
DA   October 12, 1990
SEE Diagnosis of Multiple Alarms System
SO   Acronyms

DNA
(EPA)
DA   October 12, 1990
BT1 Organic Chemicals
 BT2 Chemical Substances
NT1 Plasmids
NT1 Recombinant DNA
RT   DNA Hybridization
RT   Ribonucleic Acid
RT   Shotgun
SO   Environmental Protection Agency Glossary
DEF (EPA) Deoxyribonucleic acid, the molecule in which the genetic information for most living cells is encoded. Viruses, too, can contain RNA.

DNA HYBRIDIZATION
(EPA)
DA   October 12, 1990
BT1 Processes
RT   DNA
SO   Environmental Protection Agency Glossary
DEF (EPA) Use of a segment of DNA, called a DNA probe, to identify its complementary DNA; used to detect specific genes. This

process takes advantage of the ability of a single strand of DNA to combine with a complimentary strand.

DNO
DA   October 12, 1990
SEE Do Not Operate
SO   Acronyms
SO   Nuclear Facilities Incident Database

DO NOT OPERATE
DA   January 8, 1991
SF   *DNO*

DO NOT OPERATE TAGS
(NFI)
DA   October 12, 1990
SO   Radiation
DEF (NFI) Identification tags for maintenance job plans.

DOCTOR/NURSE
(0260 DOCTOR;NURSE)
DA   November 28, 1990
BT1 Professional Personnel
 BT2 Occupations
 BT2 Personnel
RT   Hospital
SO   DOE FRASE VOCABULARY

DOE
DA   October 12, 1990
SEE U.S. Department of Energy
SO   Acronyms

DOE ACCIDENT RESPONSE GROUP TEAM LEADERS
(EMER)
DA   February 1, 1991
BT1 Personnel
SO   Emergency Preparedness
DEF (EMER) U.S. Department of Energy (DOE) senior officials who are responsible for all DOE field operations involved in responding to a nuclear weapon accident or significant incident in which the Department of Defense is the cognizant federal agency and a military person is the on-scene commander.

DOE ALTERNATIVE
(DOE Order 1540.2)
DA   January 24, 1991
BT1 Variances
RT   Alternative Courses of Action
DEF (DOE 1540.2) DOE alternative is an administrative relief from DOE regulations that meets and provides equivalent health and safety protection.

DOE CONTRACTORS
(DOE Orders 5480.1B, 5482.1B, 1540.2; 5400.1; ESH; TSA)
DA   October 12, 1990

SY   Contractors
SY   Department of Energy Contractors
BT1 Potentially Responsible Parties
NT1 Allied-Signal Inc.
NT1 Battelle Memorial Institute
NT1 Battelle Columbus Laboratories
NT1 Bechtel Petroleum Operation, Inc.
NT1 Department of Energy Resident Construction Contractors
NT1 Department of Energy Project Construction Contractors
NT1 EG&G Energy Measurements, Inc.
NT1 EG&G Mound Applied Technologies, Inc.
NT1 EG&G Idaho, Inc.
 NT2 System Safety Development Center
NT1 General Electric
NT1 Hanford Environmental Health Foundation
NT1 John Brown, E&C, Inc.
NT1 Kaiser Engineers Hanford, Co.
NT1 Lawrence-Allison & Associates
NT1 Lovelace Medical Foundation
NT1 Martin Marietta Energy Systems
NT1 Mason and Hanger-Silas Mason Co.
NT1 MK-Ferguson
NT1 MK-Ferguson of Idaho Co.
NT1 Oak Ridge Associated Universities
NT1 Pan American World Services, INC.
NT1 Raytheon Services Nevada
NT1 Reynolds Electrical and Engineering Company
NT1 Rockwell International Corp.
NT1 Ross Aviation, Inc.
NT1 Sandia Corporation AT&T Technologies, Inc.
NT1 Science Applications International Corporation
NT1 Southeastern Universities Research Association, Inc.
NT1 Stanford University
NT1 Trustees of Princeton University
NT1 TRW
NT1 Universities Research Association, Inc.
NT1 University of Georgia Research Foundation, Inc.
NT1 University of California
NT1 Wackenhut Services Inc.
NT1 West Valley Nuclear Services Co., Inc.
NT1 Westinghouse Electric Corp.
NT1 Westinghouse Idaho Nuclear Co., Inc.
NT1 Westinghouse Savannah River Company
NT1 Westinghouse Hanford Company
NT1 Westinghouse Materials Company of Ohio
RT   DOE Operations
RT   Field Organizations
SO   Construction
SO   Environmental Management
SO   Industrial Hygiene
SO   Management
DEF Includes prime contractors or subcontractors subject to the contractual provisions of 48 CFR 923.70, 48 CFR 970.23, or other

BT-Broader Term
NT-Narrower Term
RT Related Term

contractual provisions where DOE has elected to enforce ES&H requirements by specific negotiated contract provisions.

DOE DESIGN AGENCY
DA January 8, 1991
SF *DA*
BT1 Federal Agencies
 BT2 Agencies
  BT3 Administrative Organizations
   BT4 Organizations

DOE EMERGENCY OPERATIONS CENTER
(EMER)
DA February 1, 1991
BT1 Centers
SO Emergency Preparedness
DEF (EMER) The center located at DOE HQ through which DOE's emergency management team coordinates the departmental response to an emergency.

DOE ENERGY MANAGEMENT COORDINATORS
(DOE Order 6430.1A)
DA October 12, 1990
BT1 Personnel
SO Construction
DEF (DOE Order 6430.1A) The DOE site representatives designated responsible for energy management.

DOE FACILITY REPRESENTATIVES
(DOE Order 5000.3A)
DA December 14, 1990
BT1 Personnel
SO Management
DEF (DOE Order 5000.3A) For each major facility or group of lesser facilities, individuals or his or her designees assigned responsibility by the Head of the Field Organization for monitoring the performance of the facility and its operations. These individuals shall be the primary point of contact with the contractor and will be responsible to the appropriate Program Senior Official and Head of Field Organization for implementing the Requirements of DOE Order 5000.3A.

DOE FIRE PROTECTION AUTHORITIES
(DOE Order 6430.1A)
DA October 12, 1990
BT1 Personnel
RT Fire Protection
SO Construction
DEF (DOE Order 6430.1A) The DOE site representatives responsible for fire protection.

DOE FRASE CATEGORIES
DA January 31, 1991

RT Activity Types (DOE FRASE Vocabulary)
RT Equipment/Parts - Electrical (DOE FRASE Vocabulary)
RT Equipment/Parts - Heating (DOE FRASE Vocabulary)
RT Equipment/Parts - Instrumentation/Measuring (DOE FRASE Voc.)
RT Equipment/Parts - Material Handling (DOE FRASE Vocabulary)
RT Equipment/Parts - Nuclear (DOE FRASE Vocabulary)
RT Equipment/Parts - Personal Protective (DOE FRASE Vocabulary)
RT Facilities and Buildings (DOE FRASE Vocabulary)
RT FRASE Vocabulary
RT Human Body Parts
RT Illnesses
RT Injuries
RT Machines (DOE FRASE Vocabulary)
RT Nature of Programmatic Impact
RT Nature of Property Damage
RT Occupations
RT Sites/Areas
RT Structures (DOE FRASE Vocabulary)
RT Systems (DOE FRASE Vocabulary)
RT Tools - Manual
RT Tools - Powered
SO DOE FRASE VOCABULARY
DEF Categories under which FRASE (Factor Relationship And Sequence of Events) terms and phrases are grouped according to broader concepts or things. See individual category entries for a list (and structure) of terms and phrases under that heading.

DOE FRASE VOCABULARY
DA November 28, 1990
SY Factor Relationship and Sequence of Events Vocabulary
SY FRASE Vocabulary
BT1 Vocabulary
DEF A controlled vocabulary designed for searching the narrative description fields of records on the Safety Performance Measurement System (SPMS).

DOE HEADQUARTERS
DA January 8, 1991
SF *HQ*

DOE NUCLEAR WEAPONS COMPLEX
DA January 8, 1991
SF *NWC*
BT1 Facilities

DOE OPERATIONS
(DOE Orders 5400.1, 5480.1B, 5480.2, 5481.1B)
DA October 12, 1990

BT1 Operations
 BT2 Activities
RT Department of Energy Sites
RT DOE Contractors
RT Operational Emergencies
SO Environmental Management
SO Industrial Hygiene
SO Management
DEF Those DOE-funded activities for which DOE has assumed responsibility for the environment, safety, and health programs.

DOE PRODUCTION AGENCY
DA January 8, 1991
SF *PA*
BT1 Federal Agencies
 BT2 Agencies
  BT3 Administrative Organizations
   BT4 Organizations

DOE PROGRAMS
(DOE Order 5700.6B)
DA October 12, 1990
BT1 Programs
NT1 Defense Programs
NT1 Environment, Safety, and Health (ES&H) Program
SO Quality Assurance
DEF (DOE Order 5700.6B) Organized sets of activities within a resource area having common objectives based on strategy set forth to meet assigned goals. It may include one or more projects and research and development activities in support of new, improved, or more efficient supply, or conservation systems or procedures.

DOE PROPERTY
(EMER)
DA February 1, 1991
BT1 Facilities
BT1 Security Interests
BT1 Sites/Areas
RT Property
SO Emergency Preparedness
DEF (EMER) All land, buildings, and structures (real property) and equipment, records, and supplies (personnel property) owned, or rented and leased from commercial sources by the U.S. Government and subject to the administrative custody or jurisdiction of DOE.

DOE REPRESENTATIVES
(ESH)
DA October 12, 1990
BT1 Personnel
SO Management
SO Standards
DEF (ESH) Employees approved by the DOE Designating Official (1) to work on standards committee assignments by reason of individual professional or technical

expertise to further technical programmatic objectives of the Department or (2) to serve as an official spokesperson for the Department on boards of directors governing as policy-developing bodies, including, for example, management boards of standards developing organizations.

DOE RESERVATION
(DOE Order 5820.2A)
DA   January 24, 1991
BT1 Sites/Areas
DEF (DOE 5820.2A) A location consisting of a DOE-controlled land area including DOE-owned facilities (e.g., the Oak Ridge Reservation) in some cases referred to as a Site, such as the Nevada Test Site or the Hanford Site; or as a Laboratory, such as the Idaho National Engineering Laboratory; or as a Plant, such as Rocky Flats Plant; or as a Center, such as the Feed Materials Production Center.

DOE SAFEGUARDS AND SECURITY COORDINATORS
(DOE Order 6430.1A)
DA   October 12, 1990
BT1 Personnel
RT   Safeguards and Security Emergencies
RT   Safeguards
RT   Security
SO   Construction
DEF (DOE Order 6430.1A) The DOE site representatives designated responsible for safeguards and security.

DOE SPECIFICATION PACKAGING
(DOE Order 1540.2)
DA   January 24, 1991
BT1 Packaging
 BT2 Packages
DEF (DOE 1540.2) DOE specification packaging is a general packaging designed to meet requirements established by DOT for hazardous materials.

DOG HOUSES
(NFI)
DA   October 12, 1990
BT1 Sites/Areas
RT   D Machines
SO   Nuclear Facilities Incident Database
SO   Radiation
DEF (NFI) Alcoves in the Process Area for temporary parking of D machine.

DOJ
DA   October 12, 1990
SEE U.S. Department of Justice
SO   Acronyms

DOL
DA   October 12, 1990
SEE U.S. Department of Labor
SO   Acronyms

DOLLY
(2315 DOLLY)
DA   December 10, 1990
BT1 Hand Truck(s)
 BT2 Material Handling Device
  BT3 Devices
  BT3 Equipment/Parts - Material Handling (DOE FRASE Vocabulary)
   BT4 Equipment
SO   DOE FRASE VOCABULARY

DOP
DA   October 12, 1990
SEE Detailed Operating Procedures
SO   Acronyms

DOP TESTED
(NFI)
DA   October 12, 1990
SO   Radiation
DEF (NFI) Di Octyl Phythillate tested.

DOSE COMMITMENTS
(DOE Order 5484.1)
DA   October 12, 1990
BT1 Measurements
RT   Annual Limit on Intake
RT   Dose Equivalents
SO   Environmental Management
SO   Management
DEF (DOE Order 5484.1) Dose equivalent (rems) received by specified organs during a period of 1 calendar year that was the result of an uptake of a radionuclide by a person occupationally exposed.

DOSE EQUIVALENT INDEX
(IAEA)
DA   October 12, 1990
RT   Dose Equivalents
SO   Radiation
DEF (IAEA) For the purposes of radiation protection either the deep dose equivalent index, shallow dose equivalent index, or unrestricted dose equivalent index.

DOSE EQUIVALENTS
(DOE Orders 5480.11, 5400.5; CAA; CFR; ESH; NIH; NCRP; IAEA; EMER)
DA   October 12, 1990
SF   H
BT1 Radiation Units
 BT2 Units of Measure
NT1 Annual Dose Equivalent
 NT2 Annual Effective Dose Equivalent
  NT3 Cumulative Annual Effective Dose Equivalent
NT1 Collective Dose Equivalent
 NT2 Collective Effective Dose Equivalent
NT1 Committed Dose Equivalent

 NT2 Committed Effective Dose Equivalent
NT1 Deep Dose Equivalent
 NT2 Deep Eye Dose Equivalent
NT1 Effective Dose Equivalent
 NT2 Annual Effective Dose Equivalent
  NT3 Cumulative Annual Effective Dose Equivalent
 NT2 Collective Effective Dose Equivalent
 NT2 Committed Effective Dose Equivalent
NT1 Lens of Eye Dose Equivalents
NT1 Shallow Eye Dose Equivalents
NT1 Shallow Dose Equivalents
RT   Absorbed Dose
RT   As Low As Reasonably Achievable
RT   Dose Commitments
RT   Dose Equivalent Index
RT   Dose Limit
RT   Dose Rate Meters
RT   Doses
RT   Dosimeters
RT   Quality Factors
RT   Sievert
RT   Tissue Equivalents
SO   Air Pollution
SO   Emergency Preparedness
SO   Environmental Management
SO   Industrial Hygiene
SO   Management
SO   Radiation
SO   Water Pollution
DEF (EMER) The product of absorbed dose (D) in rad (or gray) in tissue, a quality factor (Q), and other modifying factors (N) to account for differences in biological effectiveness due to the quality of radiation and its distribution in the body. Dose equivalent is expressed in units of rem (or sievert).

DOSE LIMIT
(IAEA)
DA   October 12, 1990
BT1 Limits
NT1 Annual Dose Equivalent Limit
RT   Dose Equivalents
RT   Dose Upper Bounds
RT   Maximum Permissible Dose
RT   Planned Special Exposure
RT   Radiation Units
RT   Radiation Safety
RT   Working Condition A
RT   Working Condition B
SO   Radiation
DEF The value of a quantity of radiation which must not be exceeded.

DOSE RATE METERS
(IAEA)
DA   October 12, 1990
BT1 Radiation Detectors
 BT2 Equipment
RT   Absorbed Dose Rate
RT   Dose Equivalents
SO   Radiation
DEF (IAEA) Devices, instruments, or

systems which can be used to measure or evaluate any quantity that can be related to the determination of either absorbed dose rate or dose equivalent rate.

DOSE-RESPONSE EVALUATION
(EPA)
DA   October 12, 1990
BT1 Processes
RT   Lowest-Observed-Adverse-Effect Level
RT   No-Observed-Effect-Level
RT   Non-Observed Adverse Effect Level
SO   Environmental Protection Agency Glossary
DEF (EPA) The process of quantitatively evaluating the toxicity information and characterizing the relationship between the dose of the contaminant administered or received and the incidence of adverse health effects in the exposed population. From the quantitative dose-response relationship, toxicity values are derived that are used in the risk characterization step to estimate the likelihood of adverse effects occurring in humans at different exposure levels.

DOSE UPPER BOUNDS
(IAEA)
DA   October 12, 1990
RT   Dose Limit
RT   Doses
RT   Radiation Safety
SO   Radiation
DEF (IAEA) Dose levels established by a competent authority to constrain the optimization of protection for a given source or source type.

DOSES
(EPA; NIH; IAEA; EMER)
DA   October 12, 1990
NT1 Absorbed Dose
NT1 Acute Dermal LD 50
NT1 Effective Dose
NT1 LD 0
NT1 LD 50
 NT2 Acute LD 50
 NT2 Acute Oral LD 50
NT1 LD L0
NT1 Lethal Dose Low
NT1 Maximum Permissible Dose
NT1 Median Lethal Dose
NT1 Occupational Doses
NT1 Projected Dose
NT1 Public Doses
NT1 Recorded Doses
NT1 Reference Dose (RfD)
 NT2 Chronic RfD
 NT2 Developmental RfD
 NT2 Subchronic RfD (RfD)
NT1 RFD
NT1 Whole Body Dose
RT   Dose Equivalents
RT   Dose Upper Bounds

RT   Dosimeters
RT   Exposure
RT   Measurements
RT   Protective Action Guides
RT   Radiation Safety
RT   Units of Measure
SO   Emergency Preparedness
SO   Environmental Protection Agency Glossary
SO   Radiation
DEF (EPA) The amounts of substances penetrating the exchange boundaries of organisms after contact. Doses are calculated from the intake and the absorption efficiency, and usually are expressed as mass of a substance absorbed into the body per unit body weight per unit time (e.g., mg/kg-day). Also, n radiology, the quantity of energy or radiation absorbed. (NIH) A general term denoting the quantity of radiation or energy absorbed in a specified mass. For special purposes it must be appropriately qualified, e.g., absorbed dose.

DOSIMETERS
(EPA; IAEA)
DA   October 12, 1990
BT1 Radiation Detectors
 BT2 Equipment
NT1 Nuclear Accident Dosimeter
NT1 Personnel Dosimeters
NT1 Thermoluminescent Dosimeters
RT   Dose Equivalents
RT   Doses
RT   Efficiency (Counters)
RT   Radiation Detectors
RT   Scintillation Counters
SO   Emergency Preparedness
SO   Environmental Protection Agency Glossary
SO   Radiation
DEF (EPA) Instruments that measure exposure to radiation.

DOSIMETRY
(EMER)
DA   February 1, 1991
SO   Emergency Preparedness
DEF (EMER) The theory and application of the principles and techniques involved in measuring and recording radiation doses.

DOSIMETRY PROCESSORS
(AEA; CFR)
DA   January 24, 1991
RT   Personnel Dosimeters
DEF (CFR) "Dosimetry processor" means an individual or an organization that processes and evaluates personnel monitoring equipment in order to determine the radiation dose delivered to the equipment.

DOT
DA   October 12, 1990
SEE U.S. Department of Transportation
SO   Acronyms

DOUBLE ACTION SEMIAUTOMATIC PISTOLS
(DOE Order 5480.16)
DA   October 12, 1990
BT1 Pistols
 BT2 Handguns
  BT3 Small Arms
   BT4 Firearms
BT1 Semiautomatic Firearms
 BT2 Small Arms
  BT3 Firearms
RT   Magazines
SO   Firearms
DEF (DOE Order 5480.16) Magazine-fed and can be selectively fired with a single pull of the trigger with the hammer in either the cocked or uncocked positions.

DOUBLE BLOCK AND BLEED SYSTEM
(CFR)
DA   October 12, 1990
BT1 Closed Vent Systems
 BT2 Systems
BT1 Systems
SO   Air Pollution
DEF (CFR) Two block valves connected in series with a bleed valve or line that can vent the line between the two block valves.

DOUBLE-SHELL TANK
DA   January 8, 1991
SF   DST
BT1 Tanks
 BT2 Facility Components

DOUBLE WASH/RINSE
(TSCA; CFR)
DA   October 12, 1990
RT   Cleanup
RT   Decontamination
RT   Solvents
RT   Triple rinse
SO   Hazardous Materials
DEF (EPA) A minimum requirement to cleanse solid surfaces (both impervious and nonimpervious) two times with an appropriate solvent or other material in which PCBs are at least 5 percent soluble (by weight). A volume of PCB-free fluid sufficient to cover the contaminated surface completely must be used in each wash/rinse. The wash/rinse requirement does not mean the mere spreading of solvent or other fluid over the surface, nor does the requirement mean a once-over wipe with a soaked cloth. Precautions must be taken to contain any runoff resulting from the cleansing and to dispose

SY-Synonymous Terms   SO-Source/Subject Category   SF-See From

properly of wastes generated
during the cleansing.

DOWNBURSTS
(SMRP)
DA   October 12, 1990
BT1 Divergent Windstorms
 BT2 Act of Nature
NT1 Microbursts
RT   Tornadoes
SO   Natural Phenomenon
DEF (SMRP) A strong downward current
     of air which induces an outward
     burst of damaging wind on or near
     the ground. Practically all
     downbursts occur beneath
     cumulonimbus clouds during their
     precipitation stages. Downburst
     winds are highly divergent,
     covering an area up to 30 miles
     long and 10 miles wide. The peak
     windspeed of downbursts is less
     than mid-F3 (180 mph). Due to a
     rapidly sinking motion, the
     descending air successively hits
     the surface to burst out. As a
     result, the gustiness factor of
     downbursting winds is only 0.1 to
     0.3. Most downbursts are highly
     divergent but irrotational. However,
     some downbursts are rotational
     with a curved airflow identified as
     the "twisting downburst." Small
     tornadoes frequently form on the
     edge of a twisting downburst.

DOWNSLOPE WIND
(SMRP)
DA   October 12, 1990
BT1 Divergent Windstorms
 BT2 Act of Nature
RT   Dust Devils
SO   Natural Phenomenon
DEF (SMRP) A hot (warm), dry
     downslope wind descending from
     its source region at high
     elevations. Usually downslope
     winds are non-divergent and
     irrotational. Can form over ice
     and/or snow.

DP (Defense Programs)
DA   October 12, 1990
SEE Defense Programs
SO   Acronyms

dP (Differential Pressure)
DA   October 12, 1990
SEE Differential Pressure
SO   Acronyms

DRAFT PERMITS
(SWDA; RCRA; CFR)
DA   October 12, 1990
BT1 Permits
SO   Environmental Management
SO   Wastes
DEF Documents prepared under 40 CFR
     124.6 indicating the Director's

tentative decision to issue or deny,
modify, revoke and reissue,
terminate, or reissue a permit. A
notice of intent to terminate a
permit and a notice of intent to
deny a permit as discussed in 40
CFR 124.5, are types of draft
permits. A denial of a request for
modification, revocation and
reissuance or termination, as
discussed in Part 124.5, is not a
draft permit. A proposal permit is
not a draft permit.

DRAFTING BREAK
(SSDC)
DA   October 12, 1990
RT   Diagrams
SO   System Safety Development Center
     Glossary
DEF (SSDC) The transferral of a whole
     section of an analytical diagram or
     tree to another location, simply as
     a space convenience.

DRE
DA   October 12, 1990
SEE Destruction Removal Efficiency
SO   Acronyms

DREDGE(S)
(2316 DREDGE)
DA   December 10, 1990
BT1 Earth Moving Equipment
 BT2 Material Handling Device
  BT3 Devices
  BT3 Equipment/Parts - Material
       Handling (DOE FRASE
       Vocabulary)
   BT4 Equipment
SO   DOE FRASE VOCABULARY

DREDGED MATERIALS
(CWA; RHA; CFR)
DA   October 12, 1990
BT1 Materials
RT   Contaminant Carriers
RT   Discharge of Dredged Material
RT   Dredging
SO   Water Pollution
DEF (CFR) Materials that are excavated
     or dredged from waters of the
     United States.

DREDGING
(EPA)
DA   November 6, 1990
BT1 Removal
 BT2 Strip
  BT3 Processes
 BT2 Waste Management Processes
  BT3 Processes
RT   Channelization
RT   Dredged Materials
SO   Environmental Protection Agency
     Glossary
DEF (EPA) Removal of mud from the
     bottom of water bodies using a
     scooping machine. This disturbs

the ecosystem and causes silting
that can kill aquatic life. Dredging
of contaminated muds can expose
aquatic life to heavy metals and
other toxics. Dredging activities
may be subject to regulation under
Section 404 of the Clean Water
Act.

DRIFT
(FIFRA; CFR)
DA   January 24, 1991
RT   Application (Pesticide)
RT   Pesticides
SO   Air Pollution
DEF (CFR) "Drift" means movement of a
     pesticide during or immediately
     after application or use through air
     to a site other than the intended
     site of application or use.

DRILL RIG(S)
(2317 DRILL RIG)
DA   December 10, 1990
BT1 Equipment/Parts - Material
     Handling (DOE FRASE
     Vocabulary)
 BT2 Equipment
SO   DOE FRASE VOCABULARY

DRILLING MACHINE
(2117 DRILLING MAC)
DA   December 10, 1990
BT1 Machines (DOE FRASE
     Vocabulary)
 BT2 Equipment
SO   DOE FRASE VOCABULARY

DRILLS
(EMER)
DA   February 1, 1991
BT1 Exercises
SO   Emergency Preparedness
DEF (EMER) Activities, either
     announced or unannounced,
     which test limited portions of an
     emergency plan. A drill limits play
     to specific components of the
     emergency response team.

DRINKING WATER COOLERS
(SDWA; USC)
DA   October 12, 1990
BT1 Devices
RT   Lead Free
RT   Potable Water
SO   Water Pollution
DEF (USC) Mechanical devices affixed
     to drinking water supply plumbing
     which actively cools water for
     human consumption.

DRINKING WATER SUPPLIES
(CERCLA; CFR; USC)
DA   October 12, 1990
RT   Alternative Water Supplies
RT   Public Water Systems
RT   Safe Drinking Water Act
RT   Special Source of Ground Water

BT-Broader Term   NT-Narrower Term   RT Related Term

SO Compensation and Liability
DEF As defined by section 101(7) of
    CERCLA, any raw or finished
    water source that is or may be
    used by a public water system (as
    defined in the Safe Drinking Water
    Act) or as drinking water by one or
    more individuals.

DRINKING WATER SYSTEM
(2029 DRINKING WAT)
DA December 10, 1990
BT1 Public Water Systems
 BT2 Water Systems
  BT3 Publicly Owned Treatment Works
   BT4 Treatment Facilities
    BT5 Facilities
  BT3 Systems
BT1 Systems (DOE FRASE Vocabulary)
 BT2 Systems
RT Raw Water System
RT Waste Water System
SO DOE FRASE VOCABULARY

DRIPS PER MINUTE
DA January 8, 1991
SF d/m
BT1 Units of Measure

DROWNING
(1317 DROWNING)
DA November 28, 1990
BT1 Injuries
RT Body of Water
RT Pool
SO DOE FRASE VOCABULARY

DRUMS
DA May 20, 1991
BT1 Containers
SO Environmental Management
DEF (DSTT) Hollow, cylindrical
    containers. A typically metal
    cylindrical shipping container
    having a capacity of 12-110
    gallons (45-416 liters) of liquid.
    Within the context of chemical
    engineering drums are tower or
    vessels in a refinery into which
    heated products are conducted so
    that volatile portions can separate.

DRY CAVES
(NFI)
DA October 12, 1990
BT1 Sites/Areas
RT Storage
SO Nuclear Facilities Incident Database
SO Radiation
DEF (NFI) Long term storage areas for
    irradiated elements that do not
    require cooling.

DRY FIRING
(DOE Order 5480.16)
DA October 12, 1990
RT Firearms
RT Firearms Ranges

SO Firearms
DEF (DOE Order 5480.16) A training
    procedure to improve proficiency
    that uses a weapon without blank
    or live ammunition.

DRYING MACHINE
(2119 DRYING MACHI)
DA December 10, 1990
BT1 Machines (DOE FRASE
    Vocabulary)
 BT2 Equipment
SO DOE FRASE VOCABULARY

DRYWELL
DA January 8, 1991
SF DW

DSFM
DA October 12, 1990
SEE Division of the State Fire Marshall
SO Acronyms

DST
DA October 12, 1990
SEE Double-shell Tank
SO Acronyms

DUAL SYSTEMS
(SEA)
DA October 12, 1990
BT1 Systems
RT Braced Frame
RT Intermediate Moment Resisting
    Space Frame
RT Shear Walls
RT Special Moment Resisting Space
    Frame
SO Construction
DEF (SEA) Combinations of Special or
    Intermediate Moment Resisting
    Space Frames and Shear Walls or
    Braced Frames.

DUDS
(DOE Order 5480.16)
DA October 12, 1990
BT1 Munitions
SO Firearms
DEF (DOE Order 5480.16) Bombs,
    grenades, or shells that fail to
    explode.

DUMBBELL
DA January 8, 1991
SF DB (Dumbbell)

DUMPS
(EPA)
DA October 12, 1990
BT1 Sites/Areas
NT1 Open Dumps
SO Environmental Protection Agency
    Glossary
DEF (EPA) Sites used to dispose of solid
    wastes without environmental
    controls.

DURESS SYSTEMS
(DOE Order 6430.1A; EMER)
DA October 12, 1990
BT1 Emergency Systems
 BT2 Systems
SO Construction
SO Emergency Preparedness
DEF (DOE Order 6430.1A) Systems that
    can covertly communicate a
    situation of duress to a security
    control center or other personnel
    who can notify a security control
    center.

DUST
(EPA)
DA October 12, 1990
BT1 Air Pollutants
NT1 Soot
RT Adequately Wetted
RT Aerosols
RT Attrition
RT Dust Devils
RT Fly Ash
RT Particulates
SO Environmental Protection Agency
    Glossary
DEF (EPA) Particles light enough to be
    suspended in air.

DUST DEVILS
(SMRP)
DA October 12, 1990
BT1 Rotational Windstorms
 BT2 Act of Nature
RT Downslope Wind
RT Dust
RT Waterspouts
SO Natural Phenomenon
DEF (SMRP) Rotating columns of air
    which form over dry ground heated
    by strong solar radiation. The
    direction of rotation is not unique,
    affected by the environmental
    flow field during the formation stage.
    Most dust devils are F0 (72 mph)
    or weaker. Occasionally F1
    (73-112 mph) dust devils damage
    outbuildings or garages. The
    central region of a dust devil is
    characterized by a descending
    motion and relatively clear air. Can
    form over ice and/or snow with
    snow rotating (snow devils).

DUST MASK
(2655 DUST MASK)
DA January 3, 1991
BT1 Personal Protective Equipment
 BT2 Equipment/Parts - Personal
    Protective (DOE FRASE
    Vocabulary)
  BT3 Equipment
RT Air Mask
RT Nose
SO DOE FRASE VOCABULARY

DUSTFALL JARS
(EPA)

DA   October 12, 1990
BT1 Containers
SO   Environmental Protection Agency
     Glossary
DEF (EPA) Open containers used to
     collect large particles from the air
     for measurement and analysis.

DUTY OFFICERS
(EMER)
DA   February 1, 1991
BT1 Personnel
RT   Crisis Managers
SO   Emergency Preparedness
DEF (EMER) Personnel with appropriate
     knowledge of emergency
     procedures to act as an
     intermediate crisis manager. Duty
     officers generally have a
     prescribed period of assigned
     duties.

DUVAS
DA   October 12, 1990
SEE Derivative Ultraviolet Absorbtion
     Spectroscopy
SO   Acronyms

DW
DA   October 12, 1990
SEE Drywell
SO   Acronyms

DWELLING
(1729 DWELLING)
DA   December 10, 1990
BT1 Site (DOE FRASE Vocabulary)
 BT2 Sites/Areas
SO   DOE FRASE VOCABULARY

DWMP
DA   October 12, 1990
SEE Defense Waste Management Plan
SO   Acronyms

DWPF
DA   October 12, 1990
SEE Defense Waste Processing Facility
SO   Acronyms

DWTM
DA   October 12, 1990
SEE Office of Defense Waste and
     Transportation Management
SO   Acronyms

DYSTROPHIC LAKES
(EPA)
DA   October 12, 1990
BT1 Lakes
 BT2 Inland Waters
 BT2 Surface Waters
  BT3 Water
 BT2 Surface Water Resources
  BT3 Natural Resources
RT   Bogs
RT   Eutrophic Lakes
RT   Muck Soils

SO   Environmental Protection Agency
     Glossary
DEF (EPA) Shallow bodies of water that
     contain much humus and/or
     organic matter, that contain many
     plants but few fish and are highly
     acidic.

E&CF
DA   October 12, 1990
SEE Events and Causal Factors
SO   Acronyms

E&S
DA   October 12, 1990
SEE Error and Sensitivity
SO   Acronyms

E-MAD
DA   October 12, 1990
SEE Engine Maintenance Assembly and
     Disassembly
SO   Acronyms

EA (Emergency Actions)
DA   October 12, 1990
SEE Emergency Actions
SO   Acronyms

EA (Environmental Assessment)
DA   October 12, 1990
SEE Environmental Assessments
SO   Acronyms

EAR(S)
(1085 EAR)
DA   November 28, 1990
BT1 Head
 BT2 Human Body Parts
RT   Ear Muffs
RT   Ear Plug(s)
RT   Hearing Impairment
RT   Hearing Protection
SO   DOE FRASE VOCABULARY

EAR MUFFS
(2656 EAR MUFFS)
DA   January 3, 1991
RT   Ear Plug(s)
RT   Ear(s)
RT   Hearing Impairment
RT   Hearing Protection
SO   DOE FRASE VOCABULARY

EAR PLUG(S)
(2657 EAR PLUG)
DA   January 3, 1991
BT1 Hearing Protection
 BT2 Personal Protective Equipment
  BT3 Equipment/Parts - Personal
       Protective (DOE FRASE
       Vocabulary)
   BT4 Equipment
RT   Ear Muffs
RT   Ear(s)
RT   Hearing Impairment
SO   DOE FRASE VOCABULARY

EARTH-LINED CHANNELS
(DOE Order 6430.1A)
DA   October 12, 1990
BT1 Structures (DOE FRASE
     Vocabulary)
RT   Flumes
SO   Construction
DEF (DOE Order 6430.1A) Open
     channel conveyance structures
     with sides and bottom constructed
     of naturally occurring earth
     materials.

EARTH MOVING EQUIPMENT
(2318 EARTH MOVING)
DA   December 10, 1990
BT1 Material Handling Device
 BT2 Devices
 BT2 Equipment/Parts - Material
      Handling (DOE FRASE
      Vocabulary)
  BT3 Equipment
NT1 Dredge(s)
SO   DOE FRASE VOCABULARY

EARTHQUAKE MAGNITUDE
(USGS)
DA   October 12, 1990
RT   Bases
RT   Body Wave Magnitude
RT   Local Magnitude (M_L)
RT   Love Seismic Waves
RT   Moment Magnitude
RT   Primary Body Waves
RT   Rayleigh Seismic Waves
RT   Richter Magnitude (M_L)
RT   Surface Wave Magnitude
RT   Tectonic Deformations
RT   Tsunamis
SO   Natural Phenomenon
DEF (USGS) A measure of the strength
     of an earthquake, or the strain
     energy released by it as calculated
     from the instrumental record made
     by the event on a calibrated
     seismograph.

EAS
DA   October 12, 1990
SEE Engineering Assistance Section
SO   Acronyms

EATING AREA
(1691 EATING AREA)
DA   December 10, 1990
BT1 Area
 BT2 Sites/Areas
NT1 Cafeteria
RT   Food Service Activity
RT   Kitchen
RT   Restaurant
SO   DOE FRASE VOCABULARY

EBF
DA   October 12, 1990
SEE Eccentric Braced Frames
SO   Acronyms

EBWR
DA October 12, 1990
SEE Experimental Boiling Water Reactor
SO Acronyms

ECC
DA October 12, 1990
SEE Emergency Control Centers
SO Acronyms

ECCENTRIC BRACED FRAMES
(SEA)
DA October 12, 1990
SF EBF
BT1 Braced Frame
 BT2 Building Frame Systems
  BT3 Space Frames
   BT4 Structures (DOE FRASE
       Vocabulary)
  BT3 Systems
SO Construction
DEF (SEA) Steel braced frame designs.

ECCI
DA October 12, 1990
SEE Emergency Core Cooling Injection
SO Acronyms

ECCS
DA October 12, 1990
SEE Emergency Core Cooling System
SO Acronyms

ECOLOGICAL IMPACTS
(EPA)
DA October 12, 1990
BT1 Effects
RT Cumulative Impact
RT Ecology
RT Environment
RT Environmental Impact Statements
SO Environmental Protection Agency
 Glossary
DEF (EPA) Effects that a man-made or
 natural activity has on living
 organisms and their non-living
 (abiotic) environment.

ECOLOGY
(EPA)
DA October 12, 1990
RT Adaptations
RT Animals
RT Balanced Biological Communities
RT Ecological Impacts
RT Ecosystems
SO Environmental Protection Agency
 Glossary
DEF (EPA) The relationship of living
 things to one another and their
 environment, or the study of such
 relationships.

ECONOMIC POISONS
(FIFRA; CFR; EPA)
DA October 12, 1990
BT1 Poisons
RT Pesticides

SO Environmental Management
SO Environmental Protection Agency
 Glossary
DEF (EPA) Chemicals used to control
 pests and to defoliate cash crops
 such as cotton.

ECOSPHERE
(EPA)
DA October 12, 1990
SY Biosphere
BT1 Environment
SO Environmental Protection Agency
 Glossary
DEF (EPA) The "bio-bubble" that
 contains life on earth, in surface
 waters, and in the air.

ECOSYSTEMS
(EPA)
DA October 12, 1990
NT1 Tundra
RT Biosphere
RT Biotic Communities
RT Ecology
RT Environment
RT Pesticides
RT Populations
RT Soils
RT Targets
SO Environmental Protection Agency
 Glossary
DEF (EPA) Interacting systems of
 biological communities and their
 non-living environmental
 surroundings.

ECP
DA October 12, 1990
SEE Engineering Change Proposal
SO Acronyms

ECS (Emergency Control Station)
DA October 12, 1990
SEE Emergency Control Stations
SO Acronyms

ECS (Emergency Cooling System)
DA October 12, 1990
SEE Emergency Cooling System
SO Acronyms

ECT
DA October 12, 1990
SEE Evaporator Condensate Tank
SO Acronyms

ECW
DA October 12, 1990
SEE Effluent Cooling Water
SO Acronyms

ECWD
DA October 12, 1990
SEE Effluent Cooling Water Drainage
SO Acronyms

ED CHARGE
(NFI)
DA October 12, 1990
SO Nuclear Facilities Incident Database
SO Radiation
DEF (NFI) Enriched/depleted charge.

EDB
DA October 12, 1990
SEE Ethylene Dibromide
SO Acronyms

EDG
DA October 12, 1990
SEE Emergency Diesel Generator
SO Acronyms

EDUCATIONAL ORGANIZATIONS
DA February 5, 1991
BT1 Organizations
NT1 Central Training Academy
NT1 Universities
 NT2 Iowa State University
 NT2 Massachusetts Institute of
     Technology
 NT2 North Carolina State University
 NT2 Stanford University
 NT2 University of Chicago
 NT2 University of California
 NT2 University of California-Irvine
RT Administrative Organizations
RT Commercial Organizations
RT Research and Development
 Organizations

EED (construction)
DA October 12, 1990
SEE Electroexplosive Devices
SO Acronyms

**EED (Equipment Engineering
Department)**
DA October 12, 1990
SEE Equipment Engineering Department
SO Acronyms

EEM
DA October 12, 1990
SEE Essential Equipment Monitor
SO Acronyms

EEMT MEMBER
(DOE Order 5500.5)
DA January 24, 1991
BT1 Personnel
RT Energy Emergency Teams
DEF (DOE 5500.5.h) The EEMT is
 composed of individuals appointed
 by their respective DOE offices
 who are knowledgeable in their
 policy areas and who are
 empowered to speak for and
 commit resources of their
 respective organizations.

EF
DA October 12, 1990

SEE End Fitting
SO Acronyms

EFC
DA   October 12, 1990
SEE External Fission Counter
SO Acronyms

EFFECTIVE DATE
(CAA; CFR)
DA   October 12, 1990
BT1 Time Designations
RT Primary Standard Attainment Date
SO Air Pollution
DEF (CFR) The date of promulgation in
     the Federal Register of an
     applicable standard or other
     regulation.

EFFECTIVE DOSE
DA   January 8, 1991
SF   *HE (Effective Dose)*
BT1 Doses

EFFECTIVE DOSE EQUIVALENT
(DOE Orders 5400.5, 5480.5, and
   6430.1A; CAA; CFR; IAEA)
DA   October 12, 1990
BT1 Dose Equivalents
 BT2 Radiation Units
  BT3 Units of Measure
NT1 Annual Effective Dose Equivalent
 NT2 Cumulative Annual Effective Dose
      Equivalent
NT1 Collective Effective Dose Equivalent
NT1 Committed Effective Dose
      Equivalent
RT Derived Concentration Guide
RT Effective Dose Equivalent
      Commitment
RT Weighting Factors
RT Whole Body
SO Air Pollution
SO Construction
SO Emergency Preparedness
SO Environmental Management
SO Industrial Hygiene
SO Radiation
DEF The summation of the products of
     the dose equivalent received by
     specified tissues of the body and a
     tissue-specific weighting factor.
     This sum is a risk-equivalent value
     and can be used to estimate the
     health-effects risk of the exposed
     individual. The tissue-specific
     weighting factor represents the
     fraction of the total health risk
     resulting from uniform whole-body
     irradiation that would be
     contributed by that particular
     tissue. The effective dose
     equivalent includes the committed
     effective dose equivalent from
     internal deposition of radionuclides
     and the effective dose equivalent
     due to penetrating radiation from
     sources external to the body.
     Effective dose equivalent is

expressed in units of rem (or
sievert).

EFFECTIVE DOSE EQUIVALENT COMMITMENT
(IAEA)
DA   October 12, 1990
RT Effective Dose Equivalent
RT Sievert
SO Radiation
DEF (IAEA) The infinite time integral of
     the per capita effective dose
     equivalent rate resulting from a
     given event, decision or defined
     finite portion of a practice for a
     specified population.

EFFECTIVE HALF-LIFE
(EMER)
DA   October 12, 1990
BT1 Half-Life
 BT2 Time Designations
RT Biological Half-Time
SO Emergency Preparedness
SO Radiation
DEF (EMER) Time required for a
     radioactive nuclide in a system to
     be diminished 50 percent as a
     result of the combined action of
     radioactive decay and biological
     elimination. Effective half-life =
     Biological half-life x Radioactive
     half-life/Biological half-life +
     Radioactive half-life.

EFFECTS
DA   January 30, 1991
SY   Consequences
NT1 Acute Effects
NT1 Chilling Effect
NT1 Chronic Effects
NT1 Ecological Impacts
NT1 Effects on Welfare
NT1 Effects of the Action
NT1 Frostbite/Other Low Temp Effects
NT1 Genetic Radiation Effects
NT1 Greenhouse Effect
NT1 Heat Stroke/Other High Temp
     Effect
NT1 Heat Island Effects
NT1 Ionizing Radiation Effects
NT1 Known Human Effects
NT1 Non-Stochastic Effects
 NT2 Non-Stochastic Radiation Effects
NT1 Orthogonal Effects
NT1 P-Delta Effect
NT1 Somatic Radiation Effects
NT1 Stack Effect
NT1 Stochastic Effects
 NT2 Radiation Induced Hereditary
      Effects
 NT2 Radiation Induced Genetic Effects
 NT2 Stochastic Radiation Effects
NT1 Systemic Effects
NT1 Toxic Effects to Single System
NT1 Unacceptable Adverse Effects
RT Illnesses
RT Injuries
SO DOE FRASE VOCABULARY

EFFECTS OF THE ACTION
(ESA; CFR)
DA   October 12, 1990
BT1 Effects
RT Actions
RT Biological Assessments
SO Endangered Species
DEF (CFR) The direct and indirect
     effects of an action on the species
     or critical habitat, together with the
     effects of other activities that are
     interrelated or interdependent with
     that action, that will be added to
     the environmental baseline. The
     environmental baseline includes
     the past and present impacts of all
     Federal, State, or private actions
     and other human activities in the
     action area, the anticipated
     impacts of all proposed Federal
     projects in the action area that
     have already undergone formal or
     early section 7 consultation, and
     the impact of State or private
     actions that are contemporaneous
     with the consultation in process.
     Indirect effects are those that are
     caused by the proposed action
     and are later in time, but still are
     reasonably certain to occur.
     Interrelated actions are those that
     are part of larger action and
     depend on larger action for their
     justification. Interdependent
     actions are those that have no
     independent utility apart from
     action under consideration.

EFFECTS ON WELFARE
(CAA)
DA   October 12, 1990
BT1 Effects
RT Known Human Effects
SO Air Pollution

EFFICIENCY (COUNTERS)
(IAEA; NIH)
DA   October 12, 1990
RT Dosimeters
RT Scintillation Counters
SO Radiation
DEF A measure of the probability that a
     count will be recorded when
     radiation is incident on a detector.
     Usage varies considerably so it is
     well to make sure which factors
     (window, transmission, sensitive
     volume, energy dependence, etc.)
     are included in a given case.

EFFLUENT COOLING WATER
DA   January 8, 1991
SF   *ECW*

EFFLUENT COOLING WATER DRAINAGE
DA   January 8, 1991
SF   *ECWD*

EFFLUENT LIMITATION GUIDELINES
(CWA; CFR; ESH)
DA   October 12, 1990
BT1 Guidelines
RT   Effluent Limitations
SO   Environmental Management
SO   Hazardous Materials
DEF (ESH) Any effluent issued by the
     Administrator pursuant to §304(b)
     of the Act.

EFFLUENT LIMITATIONS
(CWA; CFR; EPA; ESH)
DA   October 12, 1990
BT1 Limits
RT   Discharges
RT   Effluent Limitation Guidelines
SO   Environmental Protection Agency
     Glossary
SO   Hazardous Materials
DEF (EPA) Restrictions established by a
     State or EPA on quantities, rates,
     and concentrations in wastewater
     discharges. (ESH) Any restriction
     established by the Administrator of
     EPA on quantities, discharge
     rates, and concentrations of
     chemical, physical, biological, and
     and other constituents which are
     discharged from point sources,
     other than new sources, into
     navigable waters, the waters of
     the contiguous zone or the ocean.

EFFLUENT MONITORING
(DOE Orders 5400.1, 5484.1, and
 5400.5)
DA   October 12, 1990
BT1 Environmental Monitoring
 BT2 Monitoring
  BT3 Activities
RT   Biomonitoring
RT   Effluents
SO   Environmental Management
SO   Management
SO   Wastes
DEF The collection and analysis of
     samples, or measurements of
     liquid and gaseous effluents for
     the purpose of characterizing and
     quantifying contaminants,
     assessing radiation exposures of
     members of the public, providing a
     means to control effluents at or
     near the point of discharge, and
     demonstrating compliance with
     applicable standards and permit
     requirements.

EFFLUENTS
(DOE Orders 5400.1, 5484.1)
DA   October 12, 1990
BT1 Discharges
BT1 Emissions
 BT2 Air Pollutants
NT1 Radioactive Effluents
RT   Effluent Monitoring
RT   Pollution
RT   Pollutants
RT   Solid Wastes

SO   Construction
SO   Environmental Management
SO   Environmental Protection Agency
     Glossary
SO   Management
DEF (DOE) Treated or untreated air
     emissions or liquid discharges at a
     DOE site or from a DOE facility.
     Airborne and liquid wastes
     deliberately discharged from a
     DOE site or facility following such
     engineered waste treatment and
     all effluent controls, including
     onsite retention and decay, as
     may be provided. This term does
     not include solid wastes, wastes
     for shipment offsite, wastes that
     are contained (e.g., underground
     nuclear test debris) or stored (e.g.,
     in tanks) or wastes that are to
     remain onsite through treatment or
     disposal. (EPA) Wastewater,
     treated or untreated, that flows out
     of a treatment plant, sewer, or
     industrial outfall. Generally refers
     to wastes discharged into surface
     waters.

EFS
DA   October 12, 1990
SEE Emergency Feedwater System
SO   Acronyms

EFΔP
DA   October 12, 1990
SEE End Fitting Delta Pressure
SO   Acronyms

EG&G
DA   October 12, 1990
SO   Acronyms

EG&G ENERGY MEASUREMENTS, INC.
DA   January 11, 1991
BT1 Companies
 BT2 Commercial Organizations
  BT3 Organizations
BT1 DOE Contractors
 BT2 Potentially Responsible Parties
RT   Nevada Test Site
RT   Nevada Operations Office

EG&G IDAHO, INC.
DA   January 11, 1991
BT1 Companies
 BT2 Commercial Organizations
  BT3 Organizations
BT1 DOE Contractors
 BT2 Potentially Responsible Parties
NT1 System Safety Development Center
RT   Idaho Operations Office
RT   Idaho National Engineering
     Laboratory

EG&G MOUND APPLIED TECHNOLOGIES, INC.
DA   January 11, 1991
BT1 Companies

 BT2 Commercial Organizations
  BT3 Organizations
BT1 DOE Contractors
 BT2 Potentially Responsible Parties
RT   Albuquerque Operations Office

EG (Emergency generators)
DA   October 12, 1990
SEE Emergency (Diesel) Generators
SO   Acronyms

EG (Engine-Generators)
DA   October 12, 1990
SEE Engine-Generators
SO   Acronyms

EGRESS
(DOE Order 6430.1A)
DA   October 12, 1990
RT   Entry Control Points
SO   Construction
DEF (DOE Order 6430.1A) The act of
     departing from a point of access.

EH
DA   October 12, 1990
SEE Office of the Assistant Secretary for
     Environment, et. al.
SO   Acronyms

EIS
DA   October 12, 1990
SEE Environmental Impact Statements
SO   Acronyms

EIS IMPLEMENTATION PLANS
(DOE Order 5440.1D)
DA   May 14, 1991
SF   IP
BT1 Implementation Plans
 BT2 Plans
RT   Environmental Impact Statements
SO   Environmental Management
DEF (DOE Order) Brief written plans that
     provide guidance for the
     preparation of a DOE
     Environmental Impact Statement
     (EIS) (including a Supplemental
     EIS). The plans record the results
     of the scoping process and outline
     the procedures by which an EIS is
     to be prepared.

ELBOW(S)
(1097 ELBOW)
DA   November 28, 1990
BT1 Joint(s)
 BT2 Human Body Parts
RT   Arm(s)
RT   Forearm Protection
SO   DOE FRASE VOCABULARY

ELECTRIC POWER RESEARCH INSTITUTE
DA   January 8, 1991
SF   EPRI
BT1 Institutes

BT2 Research and Development
   Organizations
BT3 Organizations
NT1 Nuclear Safety Analysis Center

ELECTRIC SHOCK
(1318 ELECTRIC SHO)
DA November 28, 1990
BT1 Injuries
RT Electrical Burn(s)
RT Electrocution
RT Electrical Short
SO DOE FRASE VOCABULARY

ELECTRICAL APPARATUS
(2389 ELEC APPARAT)
DA January 3, 1991
BT1 Equipment/Parts - Electrical (DOE
   FRASE Vocabulary)
BT2 Equipment
SO DOE FRASE VOCABULARY

ELECTRICAL BURN(S)
(1319 ELEC BURN)
DA November 28, 1990
BT1 Burn(s)
BT2 Injuries
RT Electric Shock
RT Electrocution
SO DOE FRASE VOCABULARY

ELECTRICAL BUS
(2390 ELEC BUS)
DA January 3, 1991
BT1 Equipment/Parts - Electrical (DOE
   FRASE Vocabulary)
BT2 Equipment
SO DOE FRASE VOCABULARY

ELECTRICAL EQUIPMENT
(SWDA; RCRA; ESA; CFR)
DA October 12, 1990
BT1 Equipment
NT1 Capacitors
NT2 Large High Voltage Capacitors
NT2 Large Low Voltage Capacitors
NT2 Small Capacitors
NT1 Connectors
NT1 PCB Transformers
NT2 Mineral Oil PCB Transformers
RT Alternating Current
RT Bypasses
RT Megger Testing
RT Power Losses
SO Wastes
DEF Underground equipment that
   contains dielectric fluid that is
   necessary for the operation of
   equipment such as transformers
   and buried electrical cable.

ELECTRICAL INSULATOR(S)
(2391 ELEC INSULAT)
DA January 3, 1991
BT1 Equipment/Parts - Electrical (DOE
   FRASE Vocabulary)
BT2 Equipment
SO DOE FRASE VOCABULARY

ELECTRICAL OFFICE MACHINE
(2120 ELEC OFFICE)
DA December 10, 1990
BT1 Machines (DOE FRASE
   Vocabulary)
BT2 Equipment
SO DOE FRASE VOCABULARY

ELECTRICAL SHORT
(1528 ELEC SHORT)
DA November 28, 1990
BT1 Nature of Property Damage
RT Electric Shock
RT Electronic Damage
SO DOE FRASE VOCABULARY

ELECTRICIAN
(0643 ELECTRICIAN)
DA November 28, 1990
BT1 Repair/Construction Personnel
BT2 Occupations
BT2 Personnel
SO DOE FRASE VOCABULARY

ELECTROCUTION
(1320 ELECTROCUTIO)
DA November 28, 1990
BT1 Injuries
RT Electric Shock
RT Electrical Burn(s)
SO DOE FRASE VOCABULARY

ELECTRODE(S)
(2392 ELECTRODE)
DA January 3, 1991
BT1 Equipment/Parts - Electrical (DOE
   FRASE Vocabulary)
BT2 Equipment
SO DOE FRASE VOCABULARY

ELECTRODIALYSIS
(EPA)
DA October 12, 1990
BT1 Processes
RT Desalinization
SO Environmental Protection Agency
   Glossary
DEF (EPA) A process that uses
   electrical current applied to
   permeable membranes to remove
   minerals from water. Often used to
   desalinize salty or brackish water.

ELECTROEXPLOSIVE DEVICES
(DOE Order 6430.1A)
DA October 12, 1990
SF EED (construction)
BT1 Devices
NT1 Detonators
SO Construction
DEF (DOE Order 6430.1A) Devices
   containing some reaction mixture
   (explosive or pyrotechnic) that is
   electrically initiated. The output of
   the initiation is heat, shock, or
   mechanical action.

ELECTROMECHANICAL
MANIPULATOR(S)
(2559 EM)
DA January 3, 1991
BT1 Manipulator
BT2 Equipment/Parts - Nuclear (DOE
   FRASE Vocabulary)
BT3 Equipment
BT3 Reactor Components
SO DOE FRASE VOCABULARY

ELECTRON CAPTURE
(NIH)
DA October 12, 1990
BT1 Beta Decay
BT2 Radioactive Decay
RT Electrons
SO Radiation
DEF (NIH) A mode of radioactive decay
   involving the capture of an orbital
   electron by its nucleus. Capture
   from the particular electron shell is
   designated as "K-electron
   capture," "L-electron capture," etc.

ELECTRON VOLT
(IAEA; NIH)
DA October 12, 1990
RT Volts Alternating Current
RT Volts Direct Current
SO Radiation
DEF A unit of energy equivalent to the
   amount of energy gained by an
   electron in passing through a
   potential difference of 1 volt.
   Abbreviated eV. Larger multiple
   units of the electron volt frequently
   used are: keV for thousand or
   kiloelectron volts, MeV for million
   electron volts and BeV for billion
   electron volts.

ELECTRONIC DAMAGE
(1529 ELECTRONIC D)
DA November 28, 1990
BT1 Damage
BT2 Nature of Property Damage
RT Electrical Short
SO DOE FRASE VOCABULARY

ELECTRONS
(NIH)
DA October 12, 1990
BT1 Charged Particles
RT Annihilation (Electron)
RT Beta Particles
RT Directly Ionizing Particles
RT Electron Capture
SO Radiation
DEF (NIH) Negatively charged
   elementary particles which are a
   constituent of every neutral atom.
   Its unit of negative electricity
   equals 4.8×10^{-19} coulombs. Its
   mass is 0.00549 atomic mass
   units.

ELECTROSTATIC PRECIPITATORS
(EPA)

DA   October 12, 1990
SF   *ESP*
BT1 Precipitators
 BT2 Control Devices
  BT3 Devices
SO   Environmental Protection Agency
     Glossary
DEF (EPA) Air pollution control devices
     that remove particles from a gas
     stream (smoke) after combustion
     occurs. The ESP imparts an
     electrical charge to the particles,
     causing them to adhere to metal
     plates inside the precipitator.
     Rapping on the plates causes the
     particles to fall into a hopper for
     disposal.

ELEMENT 104 ISOTOPES
(EDB)
DA   March 29, 1991
BT1 Radionuclides
 BT2 CERCLA Hazardous Substances
  BT3 Hazardous Substances
 BT2 Nuclides
SO   Radiation

ELEMENT 105 ISOTOPES
(EDB)
DA   March 29, 1991
BT1 Radionuclides
 BT2 CERCLA Hazardous Substances
  BT3 Hazardous Substances
 BT2 Nuclides
SO   Radiation

ELEMENT 106 ISOTOPES
(EDB)
DA   March 29, 1991
BT1 Radionuclides
 BT2 CERCLA Hazardous Substances
  BT3 Hazardous Substances
 BT2 Nuclides
SO   Radiation

ELEMENT 107 ISOTOPES
(EDB)
DA   March 29, 1991
BT1 Radionuclides
 BT2 CERCLA Hazardous Substances
  BT3 Hazardous Substances
 BT2 Nuclides
SO   Radiation

ELEMENT 108 ISOTOPES
(EDB)
DA   March 29, 1991
BT1 Radionuclides
 BT2 CERCLA Hazardous Substances
  BT3 Hazardous Substances
 BT2 Nuclides
SO   Radiation

ELEMENT 109 ISOTOPES
DA   March 29, 1991
BT1 Radionuclides
 BT2 CERCLA Hazardous Substances
  BT3 Hazardous Substances

BT2 Nuclides
SO   Radiation

ELEMENTAL PHOSPHORUS PLANTS
(CFR)
DA   October 12, 1990
BT1 Facilities
RT   Phosphate Rocks
RT   Phosphorus
SO   Air Pollution
DEF (CFR) Facilities that process
     phosphate rock to produce
     elemental phosphorus using
     pyrometallurgical techniques.

**ELEMENTARY NEUTRALIZATION
UNITS**
(SWDA; RCRA; CFR)
DA   October 12, 1990
BT1 Devices
RT   Neutralization
SO   Environmental Management
SO   Wastes
DEF (CFR) Devices which (a) are used
     for neutralizing wastes only
     because they exhibit the
     corrosivity characteristic defined in
     40 CFR 261.22, or are listed in
     Subpart D of Part 261 of this
     chapter only for this reason; and
     (b) meets the definition of tank,
     tank system, container, transport
     vehicle, or vessel in 40 CFR
     260.10.

ELEVATOR
(2320 ELEVATOR)
DA   December 10, 1990
BT1 Hoisting Apparatus
 BT2 Material Handling Device
  BT3 Devices
  BT3 Equipment/Parts - Material
      Handling (DOE FRASE
      Vocabulary)
   BT4 Equipment
RT   Escalator
SO   DOE FRASE VOCABULARY

ELIGIBLE COSTS
(EPA)
DA   October 12, 1990
BT1 Costs
RT   Wastewater Treatment Units
SO   Environmental Protection Agency
     Glossary
DEF (EPA) The construction costs for
     waste-water treatment works upon
     which EPA grants are based.

EMERGENCIES
(DOE Orders 5000.3A, 5500.1A Attach.
II, 15; EMER)
DA   December 14, 1990
BT1 Reportable Occurrences
 BT2 Occurrences
NT1 Abnormal Exposure Conditions
NT1 Bomb Incidents
NT1 Bomb Threats
NT1 Emergency Situations

NT1 Energy Emergencies
NT1 General Emergencies
NT1 Hazardous Materials Emergencies
NT1 National Emergencies
NT1 Natural Phenomena Emergencies
NT1 Operational Emergencies
NT1 Safeguards and Security
     Emergencies
 NT2 Safeguards and Security Alerts
  NT3 Alpha
  NT3 Bravo
  NT3 Charlie
  NT3 Safeguards and Security Alert I
      (Code Designator: Alpha)
  NT3 Safeguards and Security Alert II
      (Code Designator: Bravo)
  NT3 Safeguards and Security Alert III
      (Code Designator: Charlie)
 NT2 Special Nuclear Materials
     Emergencies
NT1 Site Emergencies
NT1 Threats
 NT2 Bomb Threats
 NT2 Covert Threats
 NT2 Nuclear Threat Incidents
 NT2 Terrorist Threats
RT   Alert Signals
RT   Central Alarm Stations
RT   Concept of Operations
RT   Damage Assessments
RT   Defense Readiness Conditions
RT   Emergency Situations
SO   Emergency Preparedness
SO   Environmental Management
DEF (DOE Order 5000.3A) Emergencies
     are the most serious occurrences
     and require an increased alert
     status for onsite personnel and, in
     specified cases, for offsite
     authorities. The detailed definitions
     and classifications of emergencies
     and appropriate emergency
     responses to be taken are
     provided in DOE Order 5500.2A
     The types of occurrences that are
     to be categorized as emergencies
     are specified in DOE Order
     5000.3A.

EMERGENCY (DIESEL) GENERATORS
DA   January 8, 1991
SF   *EG (Emergency generators)*
BT1 Equipment

EMERGENCY ACTIONS
(MORT)
DA   January 8, 1991
SF   *EA (Emergency Actions)*
BT1 Actions
 BT2 Responses
RT   Recovery Actions
RT   Relations

**EMERGENCY AND HAZARDOUS
CHEMICAL INVENTORY FORMS**
(EMER)
DA   February 1, 1991
BT1 Reports
SO   Emergency Preparedness
DEF (EMER) Information on the

inventory of hazardous chemicals required to be submitted by facility owners and operators pursuant to Section 312 of the Emergency Planning and Community Right-to-Know Act of 1986.

EMERGENCY BRIEFING CENTERS
(EMER)
DA   February 1, 1991
BT1 Centers
SO   Emergency Preparedness
DEF (EMER) Facilities located within DOE HQ for the purpose of briefing of management officials on emergencies.

EMERGENCY BROADCAST SYSTEM
(EMER)
DA   February 1, 1991
BT1 Emergency Systems
 BT2 Systems
SO   Emergency Preparedness
DEF (EMER) Broadcast stations and interconnecting facilities which have been authorized by the Federal Communications Commission to operate in a controlled manner during a war, state of public peril or disaster, or other national emergency as provided by the EBS plan.

EMERGENCY CONDITIONS
(FIFRA; CFR)
DA   January 24, 1991
BT1 Conditions
RT   Imminent Hazard
RT   Occurrences
RT   Pesticides
DEF (CFR) The term "emergency condition'" means an urgent, nonroutine situation that requires the use of a pesticide(s) and shall be deemed to exist when: (1) No effective pesticides are available under the Act that have labeled uses registered for control of the pest under the conditions of the emergency; and (2) No economically or environmentally feasible alternative practices which provide adequate control are available.

EMERGENCY CONTROL CENTERS
(DOE Order 6430.1A; EMER)
DA   October 12, 1990
SF   ECC
BT1 Centers
RT   Critical Facilities
RT   Emergency Control Stations
RT   Emergency Operations Centers
RT   Essential Facilities
RT   Field Command Posts
SO   Construction
SO   Emergency Preparedness
DEF (DOE Order 6430.1A) A facility from which designated management

can immediately direct the response to an emergency. The ECC may be an office, conference room, or other predesignated location having communication and informational materials appropriate to carry on the necessary supportive functions of directing an emergency response.

EMERGENCY CONTROL STATIONS
(DOE Orders 6430.1A, 5500.1A Attach. A, 17; EMER)
DA   October 12, 1990
SF   ECS (Emergency Control Station)
BT1 Sites/Areas
RT   Emergency Control Centers
RT   Emergency Operations Centers
SO   Construction
SO   Emergency Preparedness
SO   Environmental Management
DEF (DOE Order 6430.1A) Locations within or near a designated critical facility or plant area for the purpose of maintaining control, orderly shutdown, and/or surveillance of operations and equipment during an emergency.

EMERGENCY COOLING SYSTEM
DA   January 8, 1991
SF   ECS (Emergency Cooling System)
BT1 Emergency Systems
 BT2 Systems

EMERGENCY CORE COOLING INJECTION
DA   January 8, 1991
SF   ECCI

EMERGENCY CORE COOLING SYSTEM
(2030 ECCS)
DA   December 10, 1990
SF   ECCS
BT1 Emergency Systems
 BT2 Systems
BT1 Reactor Protection System
 BT2 Emergency Systems
  BT3 Systems
BT1 Systems (DOE FRASE Vocabulary)
 BT2 Systems
SO   DOE FRASE VOCABULARY
DEF Reactor system components (pumps, valves, heat exchangers, tanks and piping) that are specifically designed to remove residual heat from the reactor fuel rods should the normal core cooling system fail.

EMERGENCY DIESEL GENERATOR
(2121 EDG)
DA   December 10, 1990
SF   EDG
BT1 Machines (DOE FRASE Vocabulary)
 BT2 Equipment

RT   Emergency Power System
SO   DOE FRASE VOCABULARY

EMERGENCY EQUIPMENT
(DOE Order 5500.1A Attach. A, 19; EMER)
DA   January 24, 1991
BT1 Equipment
SO   Emergency Preparedness
SO   Environmental Management
DEF (DOE 5500.1A) Any equipment that may be required to measure, control, or mitigate the consequences of, or in any way be involved in, an emergency.

EMERGENCY EXPOSURE
(IAEA)
DA   October 12, 1990
BT1 Exposure
RT   Abnormal Exposure Conditions
SO   Radiation
DEF (IAEA) An incurred exposure received during abnormal exposure conditions in the interests of preventing serious injury or saving life or valuable property.

EMERGENCY FEEDWATER SYSTEM
DA   January 8, 1991
SF   EFS
BT1 Emergency Systems
 BT2 Systems
RT   Feedwater

EMERGENCY MANAGEMENT COORDINATION COMMITTEE
(DOE Order 5500.1A Attach. A, 18; EMER)
DA   January 24, 1991
BT1 Committees
 BT2 Administrative Organizations
  BT3 Organizations
RT   Management
SO   Emergency Preparedness
SO   Environmental Management
DEF (DOE 5500.1A) A group of senior-level representatives from appropriate organizations who collectively provide executive oversight and coordination of the Emergency Management System. The Emergency Management Coordination Committee is chaired by the Under Secretary.

EMERGENCY MANAGEMENT COORDINATION COMMITTEE SECRETARIAT
(EMER)
DA   February 1, 1991
BT1 Groups
 BT2 Administrative Organizations
  BT3 Organizations
SO   Emergency Preparedness
DEF (EMER) A group of Department of Energy staff members designated by each principal of the EMCC

and operations office emergency
management
directors/coordinators who assist
in the development of policy and
technical decisions and provide
necessary technical and
administrative support to the
EMCC and director of operations.

**EMERGENCY MANAGEMENT
INFORMATION SYSTEM**
(EMER)
DA   February 1, 1991
BT1 Emergency Systems
 BT2 Systems
BT1 Information Systems
 BT2 Security Interests
 BT2 Systems
SO   Emergency Preparedness
DEF (EMER) A system designed under
     U.S. Department of Energy
     contract by Sandia National
     Laboratories for the processing
     and display of emergency
     information.

**EMERGENCY MANAGEMENT
SYSTEMS**
(EMER)
DA   February 1, 1991
BT1 Emergency Systems
 BT2 Systems
RT   Protection of the Public Health and
     Welfare
SO   Emergency Preparedness
DEF (EMER) Comprising a Department
     of Energy (DOE) program for the
     development, coordination, and
     direction of emergency planning,
     preparedness, response, and
     readiness assurance. 1) Planning.
     The development and preparation
     of emergency plans and
     procedures and the determination
     of availability of resources to
     provide an effective response. 2)
     Preparedness. The training of
     personnel, acquisition of resources
     and facilities, and testing of
     emergency plans and procedures
     to ensure an effective response. 3)
     Response. The action(s) taken to
     cope with and minimize the effects
     of any emergency. 4) Readiness
     Assurance. The actions taken to
     provide assurance that
     headquarters, field elements, and
     facility contractors implement
     appropriate aspects of DOE
     emergency management program
     policies and requirements as
     established by DOE orders.

**EMERGENCY MANAGEMENT TEAM
DIRECTORS**
(DOE Order 4400.1A Attach. II, 13;
 EMER)
DA   January 24, 1991
BT1 Personnel
RT   Emergency Management Teams

SO   Emergency Preparedness
SO   Environmental Management
DEF (DOE 550.1A) A predesignated
     senior official identified to direct an
     emergency management team.

EMERGENCY MANAGEMENT TEAMS
(DOE Order 5500.1A Attach. A, 16;
 EMER)
DA   January 24, 1991
BT1 Management Teams
 BT2 Teams
  BT3 Administrative Organizations
   BT4 Organizations
NT1 Energy Emergency Management
     Teams
NT1 Operational Emergency
     Management Teams
RT   Crisis Managers
RT   Director of Emergency Operations
RT   Emergency Management Team
     Directors
RT   Emergency Response
SO   Emergency Preparedness
SO   Environmental Management
DEF (EMER) U.S. Department of Energy
     (DOE) teams designated to
     manage activities during
     emergencies involving DOE or
     requiring DOE assistance.

EMERGENCY MEDICAL TECHNICIAN
DA   January 8, 1991
SF   *EMT*
BT1 Personnel

**EMERGENCY OPERATING
PROCEDURES**
DA   January 8, 1991
SF   *EOP*
BT1 Operating Procedures
 BT2 Procedures

EMERGENCY OPERATIONS CENTERS
(DOE Order 6430.1A; EMER)
DA   October 12, 1990
SF   *EOC*
BT1 Centers
RT   Alternate Emergency Operations
     Center
RT   Emergency Control Centers
RT   Emergency Control Stations
RT   Essential Facilities
SO   Construction
SO   Emergency Preparedness
DEF (EMER) Facilities from which
     management and support
     personnel carry out emergency
     response activities. The EOC may
     be a dedicated facility or office,
     conference room, or other
     predesignated location having
     appropriate communications and
     informational materials to carry out
     the assigned emergency response
     mission and located where
     possible in a secure and protected
     location.

EMERGENCY PERMITS
(RCRA; CFR)
DA   January 24, 1991
BT1 Permits
RT   Permit (RCRA)
DEF (CFR) Emergency permit means a
     RCRA permit issued in
     accordance with 270.61.

**EMERGENCY PLANNING,
PREPAREDNESS, AND RESPONSE
PROGRAM**
(EMER)
DA   February 1, 1991
BT1 Programs
SO   Emergency Preparedness
DEF (EMER) An emergency program
     consisting of the following parts:
     1) planning. The development and
     preparation of emergency plans
     and procedures and the
     determination of availability of
     resources to provide an effective
     response. 2) preparedness. The
     training of personnel, acquisition
     of resources and facilities, and
     testing of emergency plans and
     procedures to ensure an effective
     response. 3) response. The
     action(s) taken to cope with and
     minimize the effects of an
     emergency.

EMERGENCY PLANNING ZONES
(DOE Order 6430.1A; EMER)
DA   October 12, 1990
BT1 Zones
 BT2 Sites/Areas
SO   Construction
SO   Emergency Preparedness
DEF (DOE Order 6430.1A) Areas for
     which planning is done to ensure
     that prompt and effective actions
     can be taken to protect the
     environment and the health and
     safety of on-site personnel and the
     public in the event of a major
     emergency.

EMERGENCY PLANS
(EMER)
DA   February 1, 1991
BT1 Plans
NT1 Contingency Plans
NT1 Field Facility/Building Emergency
     Plans
NT1 Field Site Emergency Plans
NT1 Operational Emergency
     Preparedness Management Plans
NT1 Primary Emergency Plans
NT1 Spill Prevention Control and
     Countermeasures Plan
NT1 Terrorist Response Plans
RT   Procedures
SO   Emergency Preparedness
DEF (EMER) Consist of a brief, clear,
     and concise description of the
     overall emergency organization,
     designation of responsibilities, and
     descriptions of the procedures,

including notifications, involved in coping with any or all aspects of a potential credible emergency.

EMERGENCY POWER
(DOE Order 6430.1A)
DA   October 12, 1990
RT   Standby Power
RT   Uninterruptible Power Supplies
SO   Construction
DEF (DOE Order 6430.1A) Auxiliary power systems that provide power to safety and security related equipment during periods of partial or total power failure of associated primary power system.

EMERGENCY POWER SYSTEM
(2031 EPS)
DA   December 10, 1990
BT1 Emergency Systems
 BT2 Systems
BT1 Systems (DOE FRASE Vocabulary)
 BT2 Systems
RT   Emergency Diesel Generator
SO   DOE FRASE VOCABULARY

EMERGENCY PREPAREDNESS
(EMER)
DA   January 23, 1991
RT   Emergency Response
RT   Interagency Group on Energy Vulnerability
DEF The training of personnel, acquisition of resources and facilities, and testing of emergency plans and procedures to ensure effective response to emergency situations.

EMERGENCY PROCEDURES
(EMER)
DA   February 1, 1991
BT1 Procedures
SO   Emergency Preparedness
DEF (EMER) Detailed instructions and guidance for carrying out emergency response actions.

EMERGENCY PROJECTS
(TSCA; CFR)
DA   January 24, 1991
BT1 Projects
RT   Friable Asbestos
RT   Sudden Accidental Occurrences
DEF (CFR) "Emergency project" means a project involving the removal, enclosure, or encapsulation off friable asbestos- containing material that was not planned but results from a sudden unexpected event.

EMERGENCY RAW COOLING WATER SYSTEM
DA   January 8, 1991
SF   ERCWS
BT1 Emergency Systems
 BT2 Systems

EMERGENCY REFERENCE LEVELS
(IAEA)
DA   November 20, 1990
SY   Intervention Levels
SY   Protective Action Guides
SF   ERL
BT1 Reference Levels
SO   Radiation
DEF (IAEA) Intervention levels: Levels usually specified in advance by the competent authority or management for use in abnormal situations: if the value of the quantity of interest exceeds or is predicted to exceed a particular level, the appropriate remedial action may have to be taken.

EMERGENCY RELIEF VALVE
DA   January 8, 1991
SF   ERV
BT1 Relief Valves
 BT2 Valves
 BT3 Devices

EMERGENCY RENOVATION OPERATIONS
(CAA; CFR)
DA   October 12, 1990
BT1 Operations
 BT2 Activities
BT1 Renovation
 BT2 Administrative Processes
 BT3 Processes
SO   Air Pollution
DEF (CFR) Renovation operations that were not planned but result from a sudden, unexpected event. This term includes operations necessitated by nonroutine failures of equipment.

EMERGENCY RESOURCES
(EMER)
DA   February 1, 1991
SO   Emergency Preparedness
DEF (EMER) Individuals, items of equipment or instrumentation, and specialized services that have been assembled, organized, or developed for the purpose of assisting in alleviating the consequences of an emergency.

EMERGENCY RESPONSE
(EMER; SARA; OSHA)
DA   January 24, 1991
BT1 Responses
RT   Accidents
RT   Emergency Preparedness
RT   Emergency Management Teams
RT   Post Emergency Responses
SO   Emergency Preparedness
SO   Environmental Management
DEF (CFR) A response effort by employees from outside the immediate release area or by other designated responders (i.e., mutual-aid groups, local fire

departments, etc.) to an occurrence which results, or is likely to result, in an uncontrolled release of a hazardous substance. Responses to incidental releases of hazardous substances where the substance can be absorbed, neutralized, or otherwise controlled at the time of release by employees in the immediate release area or by maintenance personnel are not considered to be emergency responses within the scope of this standard. Responses to releases of hazardous substances where there is a potential safety or health hazard (i.e., fire, explosion, or chemical exposure) are not considered to be emergency responses.

EMERGENCY RESPONSE ACTIVITY
(1230 ER ACTIVITY)
DA   November 28, 1990
BT1 Activity Types (DOE FRASE Vocabulary)
 BT2 Activities
SO   DOE FRASE VOCABULARY

EMERGENCY RESPONSE LEVELS
(EMER)
DA   February 1, 1991
NT1 Continuity of Government
NT1 Energy Emergencies
NT1 Operational Emergencies
NT1 Operational Emergency Response Levels
 NT2 Alert
NT1 Unusual Events
SO   Emergency Preparedness
DEF (EMER) Response levels associated with the following DOE-categorized emergencies: operational, energy, and continuity of government. Each of the levels reflects the severity of the accident, based on the consequences or potential consequences of the accident. These levels are to be used by field elements in emergency planning and for reporting emergencies to DOE HQ. Sometimes referred to as emergency action levels.

EMERGENCY RESPONSE TEAMS
(EMER)
DA   January 8, 1991
SF   ERT
BT1 Response Teams
 BT2 Teams
 BT3 Administrative Organizations
 BT4 Organizations
RT   Continuity of Government Emergency Management Teams
RT   Hazardous Materials Response Teams
SO   Emergency Preparedness
DEF (EMER) Federal Emergency

Management Agency (FEMA) teams deployed to a radiological emergency scene by the FEMA director to make an initial assessment of the situation and then provide FEMA's primary response capability.

EMERGENCY SERVICE WATER
DA   January 8, 1991
SF   *ESW*

EMERGENCY SHUTDOWN DEVICE
DA   January 8, 1991
SF   *ESD*
BT1 Devices

EMERGENCY SITUATIONS
(TSCA)
DA   October 19, 1990
BT1 Emergencies
 BT2 Reportable Occurrences
  BT3 Occurrences
RT   Activation (Emergency)
RT   Emergencies
RT   Power Outage
SO   Hazardous Materials
DEF (TSCA) Exist when (1) neither a non-PCB transformer nor a PCB-contaminated transformer is currently in storage for reuse or readily available (i.e., available within 24 hours) for installation. (2) Immediate replacement is necessary to continue service to power users.

EMERGENCY SPRAY WATER SYSTEM
(NFI)
DA   October 12, 1990
BT1 Emergency Systems
 BT2 Systems
BT1 Reactor Coolant System
 BT2 Reactor Components
 BT2 Systems
SO   Nuclear Facilities Incident Database
SO   Radiation
DEF (NFI) Sprays into Process Area to cool dropped fuel assemblies.

EMERGENCY SUPPORT TEAMS
(EMER)
DA   February 1, 1991
BT1 Response Teams
 BT2 Teams
  BT3 Administrative Organizations
   BT4 Organizations
SO   Emergency Preparedness
DEF (EMER) Federal Emergency Management Agency (FEMA) headquarters teams that carry out notification, activation, and coordination procedures from the FEMA Emergency Information and Coordination Center (EICC). The emergency support teams are responsible for federal agency headquarters coordination, staff support of the FEMA director, and

support of the senior FEMA official.

EMERGENCY SYSTEMS
DA   February 12, 1991
BT1 Systems
NT1 Backup Systems
NT1 Containment Systems
 NT2 Containment Spray System
  NT3 Containment Spray System (Post-accident Injection Phase)
  NT3 Containment Spray System (Post-accident Recirculation Phase)
NT1 Criticality Alarm System
NT1 Diagnosis of Multiple Alarms System
NT1 Duress Systems
NT1 Emergency Core Cooling System
NT1 Emergency Power System
NT1 Emergency Spray Water System
NT1 Emergency Broadcast System
NT1 Emergency Management Information System
NT1 Emergency Management Systems
NT1 Emergency Cooling System
NT1 Emergency Feedwater System
NT1 Emergency Raw Cooling Water System
NT1 Energy Emergency System
NT1 Explosion Suppression System
NT1 Fire Protection Systems
 NT2 Dedicated Fire Water Systems
  NT3 Sprinkler System
 NT2 Fire Alarm System
 NT2 Fire Suppression System
NT1 Intrusion Alarm System (Perimeter or Interior)
NT1 National Warning System
NT1 Plant Protection System
NT1 Radiation Alarm System
NT1 Reactor Protection System
 NT2 Emergency Core Cooling System
NT1 Safe Shutdown System
 NT2 Sparger Assemblies
NT1 Supplementary Safety System
NT1 Surveillance Systems
NT1 Surveillance and Nuclear Detection Systems
RT   Control Systems
RT   Information Systems
RT   Water Systems
SO   Emergency Preparedness
SO   Industrial Safety
SO   Management
SO   Radiation

EMERGENCY TELECOMMUNICATIONS SERVICES
(EMER)
DA   February 1, 1991
SO   Emergency Preparedness
DEF (EMER) Telecommunications services directly supporting federal government activity responding to a presidentially declared disaster or emergency as defined in the Disaster Relief Act (42 U.S. Code of Federal Regulations Sec. 5122) and directly resulting from any of

the following circumstances: 1) state of crisis declared by the national command authorities 2) efforts to protect endangered U.S. personnel or property 3) enemy action, civil disturbance, natural disaster, or any other unpredictable occurrence that has damaged facilities whose uninterrupted operation is essential to national security emergency preparedness or the management of other ongoing crises 4) certification by the head or director of a federal agency, commander of a unified/specified command, chief of a military service, or commander or major chief of a military command that a telecommunications service is so critical to protection of life and property or to the national security that it must be processed immediately

EMINENT DOMAIN
(EPA)
DA   October 12, 1990
SO   Environmental Protection Agency Glossary
DEF (EPA) Government taking – or forced acquisition – of private land for public use, with compensation paid to the landowner.

EMISSION FACTOR
(EPA)
DA   October 12, 1990
BT1 Ratios
NT1 Theoretical Arsenic Emissions Factor
RT   Air Pollutants
SO   Environmental Protection Agency Glossary
DEF (EPA) The relationship between the amount of pollution produced and the amount of raw material processed. For example, an emission factor for a blast furnace making iron would be the number of pounds of particulates per ton of raw materials.

EMISSION INVENTORY
(EPA)
DA   October 12, 1990
RT   Air Pollutants
RT   Emission Standards
SO   Environmental Protection Agency Glossary
DEF (EPA) A listing, by source, of the amount of air pollutants discharged into the atmosphere of a community. It is used to establish emission standards.

EMISSION LIMITATION
(CAA; CFR)
DA   November 15, 1990

SY Emission Standards
BT1 Limits
BT1 Requirements
RT Air Pollutants
RT Air Quality Standards
RT Baseline Concentration
RT National Emissions Standards for
      Hazardous Air Pollutants
SO Air Pollution
DEF (CFR) A requirement established
      by the State or the Administrator
      which limits the quantity, rate, or
      concentration of emissions of air
      pollutants on a continuous basis,
      including any requirement relating
      to the operation or maintenance of
      a source to assure continuous
      emission reduction.

EMISSION STANDARDS
(CAA; CFR; ESH)
DA November 15, 1990
SY Emission Limitation
BT1 Requirements
BT1 Standards
  BT2 Codes, Standards, and
        Regulations
NT1 National Emissions Standards for
      Hazardous Air Pollutants
NT1 New Source Performance
      Standards
RT Air Pollution
RT Air Quality Standards
RT Emission Inventory
DEF (USC) Requirements established
      by a State, local government, or
      the Administrator which limit the
      quantity, rate, or concentration of
      emissions of air pollutants on a
      continuous basis, including any
      requirements which limit the level
      of opacity, prescribe equipment,
      set fuel specifications, or prescribe
      operation or maintenance
      procedures for a source to assure
      continuous emission reduction.
      The maximum amount of air
      polluting discharge legally allowed
      from a single source, mobile or
      stationary. (ESH) Legally
      enforceable regulation setting forth
      an allowable rate of emissions into
      the atmosphere, or prescribing
      equipment specifications for
      control of air pollution emissions.

EMISSIONS
(RCRA; CFR; EPA)
DA October 12, 1990
BT1 Air Pollutants
NT1 Effluents
  NT2 Radioactive Effluents
NT1 Exhaust Gases
NT1 Fugitive Emissions
  NT2 Uncontrolled Total Arsenic
        Emissions
NT1 Fumes
NT1 Particulate Matter Emissions
  NT2 Allowable Emissions
  NT2 Excess Emissions

NT2 PM$_{10}$ Emissions
NT2 Visible Emissions
NT1 Process Emissions
NT1 Reactor Opening Loss
NT1 Secondary Emissions
RT Air Pollution
RT Atmospheric Release Advisory
      Capability
RT Best Available Control Technology
RT Best Available Retrofit Technology
RT Bubble
RT Bubble Policy
RT Discharges
RT Releases
SO Environmental Management
SO Environmental Protection Agency
      Glossary
DEF (EPA) Pollution discharged into the
      atmosphere from smokestacks,
      other vents, and surface areas of
      commercial or industrial facilities;
      from residential chimneys; and
      from motor vehicle, locomotive, or
      aircraft exhausts.

EMISSIONS TRADING
(EPA)
DA October 12, 1990
BT1 Policies
RT Banking (Air Pollution)
SO Environmental Protection Agency
      Glossary
DEF (EPA) EPA policy that allows a plant
      complex with several facilities to
      decrease pollution from some
      facilities while increasing it from
      others, so long as total results are
      equal to or better than previous
      limits. Facilities where this is done
      are treated as if they exist in a
      bubble in which total emissions
      are averaged out. Complexes that
      reduce emissions substantially
      may "bank" their "credits" or sell
      them to other industries.

EMPIRICAL CORRELATES
(SSDC)
DA October 12, 1990
RT Human Factors
RT Logical Correlates
SO System Safety Development Center
      Glossary
DEF (SSDC) Information which may
      appear to correlate with human
      performance, but without logical
      reason. An outstanding current
      example is the postulated
      association of biorhythms with
      human performance. Some
      correlates may become "logical"
      with further study.

EMPLOYEES' REPRESENTATIVE
(Doe Order 5483.1A)
DA October 12, 1990
BT1 Personnel
RT Contractor Employees
SO Industrial Safety
DEF (DOE Order 5483.1A) A person

chosen by contractor employees to
represent their occupational safety
and health related views, interests,
and concerns. For purposes of
access to an employee's bioassay,
monitoring, or radiation exposure
records, if the representative is not
the recognized/certified collective
bargaining agent, then he or she
must have the employee's written
authorization for such access.

EMPLOYERS
(TSCA; CFR)
DA January 24, 1991
RT Personnel
DEF (CFR)"Employer" means the public
      department, agency, or entity
      which hires an employee. The
      term includes, but is not limited to,
      any state, country, city, or other
      local governmental entity which
      operates or administers schools, a
      department of health or human
      services, a library, a police
      department, a fire department, or
      similar public service agencies or
      offices.

EMT
DA October 12, 1990
SEE Emergency Medical Technician
SO Acronyms

EN ROUTE
(1606 EN ROUTE)
DA December 10, 1990
BT1 Site (DOE FRASE Vocabulary)
  BT2 Sites/Areas
SO DOE FRASE VOCABULARY

ENCLOSED PROCESSES
(TSCA; CFR)
DA January 24, 1991
BT1 Processes
RT Releases
DEF (CFR) "Enclosed process" means a
      manufacturing or processing
      operation that is designed and
      operated so that there is no
      intentional release into the
      environment of any substance
      present during the operation. An
      operation with fugitive, inadvertent,
      or emergency pressure relief
      releases remains an enclosed
      process so long as measures are
      taken to prevent worker exposure
      to and environmental
      contamination from the releases.

ENCLOSURES
(DOE Order 6430.1A)
DA October 12, 1990
BT1 Primary Confinement Systems
  BT2 Confinement Systems
    BT3 Systems
RT Radiation Protection

SO Construction
DEF (DOE Order 6430.1A) Primary
confinement systems such as
process systems, glove boxes,
conveyors, hotcells, and canyons.

END BOX VENTILATION SYSTEM
(CAA; CFR)
DA November 15, 1990
BT1 Systems
BT1 Ventilation System
 BT2 Systems (DOE FRASE
 Vocabulary)
 BT3 Systems
RT End Boxes
SO Air Pollution
DEF (CFR) A ventilation system which
collects mercury emissions from
the end boxes, the mercury sump
pumps, and their water collection
systems.

END BOXES
(CAA; CFR)
DA October 12, 1990
BT1 Containers
RT End Box Ventilation System
RT Mercury Chlor-alkali Electrolyzers
SO Air Pollution
DEF (CFR) A container(s) located on
one or both ends of a mercury
chlor-alkali electrolyzer which
serves as a connection between
the electrolyzer and denuder for
rich and stripped amalgam.

END FITTING
DA January 8, 1991
SF *EF*

END FITTING DELTA PRESSURE
DA January 8, 1991
SF *EFΔP*
BT1 Pressure

ENDANGERED ASSESSMENT
(EPA)
DA October 12, 1990
BT1 Assessments
 BT2 Administrative Processes
 BT3 Processes
RT Remedial Investigations
SO Environmental Protection Agency
 Glossary
DEF (EPA) A study conducted to
determine the nature and extent of
contamination at a site on the
National Priorities List and the risk
posed to public health or the
environment. EPA or the state
conduct the study when a legal
action is to be taken to direct
potentially responsible parties to
clean up a site or pay for the
cleanup. An endangered
assessment supplements a
remedial investigation.

ENDANGERED SPECIES
(ESA; CFR)
DA October 12, 1990
BT1 Species
NT1 Listed Species
RT Candidates
RT Proposed Species
RT Threatened Species
SO Environmental Protection Agency
 Glossary
DEF (CFR) Animals, birds, fish, plants,
or other living organisms
threatened with extinction by
man-made or natural changes in
their environment. Requirements
for declaring a species
endangered are contained in the
Endangered Species Act.

ENDANGERED SPECIES ACT
DA February 12, 1991
SF *ESA*
BT1 Acts
 BT2 Statutes and Regulations
SO Environmental Management
DEF (ESA) The Endangered Species
Conservation Act of 1973 provides
for an evaluation of declining
species and the institution of a
recovery plan. A species can be
listed as "endangered" (threatened
with extinction) throughout all or a
significant part of its range, or
"threatened" (likely to become
endangered in the foreseeable
future).

ENERGY
(SSDC)
DA October 12, 1990
NT1 Nuclear Energy
NT1 Radiant Energy
RT Energy Flow
SO System Safety Development Center
 Glossary
DEF (SSDC) The capacity to do work
and overcome resistance. Energy
exists in many forms, including
acoustic, electrical, kinetic,
thermal, biological, chemical and
radiation (both ionizing and
monitoring).

ENERGY EMERGENCIES
(DOE Orders 5500.1A Attach. A, 15(c),
5500 5.e; EMER)
DA January 24, 1991
BT1 Emergencies
 BT2 Reportable Occurrences
 BT3 Occurrences
BT1 Emergency Response Levels
RT Energy Emergency System
RT Energy Emergency Teams
SO Emergency Preparedness
DEF (DOE 5500.1A) A condition or
potential affecting the supply of
energy or the energy infrastructure
with significant potential impact on
the national economy, national

security, defense preparedness, or
health and safety.

ENERGY EMERGENCY MANAGEMENT
TEAMS
(EMER)
DA February 1, 1991
BT1 Emergency Management Teams
 BT2 Management Teams
 BT3 Teams
 BT4 Administrative Organizations
 BT5 Organizations
SO Emergency Preparedness
DEF (EMER) Department of Energy
(DOE) teams predesignated to
coordinate response to energy
emergencies involving DOE or
requiring DOE assistance.

ENERGY EMERGENCY SYSTEM
(DOE Order 5500 5.f)
DA January 24, 1991
BT1 Emergency Systems
 BT2 Systems
RT Energy Emergencies
DEF (DOE 5500 5.f) The Energy
Emergency System (EEMS) is the
component of DOE's Emergency
Management System (EMS) which
defines the functional
requirements of energy
emergency management
responsibilities assigned to the
Office of International Affairs and
Energy Emergencies (IE).

ENERGY EMERGENCY TEAMS
(DOE Order 5500 5.g; EMER)
DA January 24, 1991
BT1 Response Teams
 BT2 Teams
 BT3 Administrative Organizations
 BT4 Organizations
RT EEMT Member
RT Energy Emergencies
DEF (DOE 5500.5g) The Energy
Emergency Teams (EEMT) are
established in accordance with
DOE Order 5500.1A to manage
and coordinate energy emergency
responses. The EEMTs are
composed of predesigned DOE
individuals and chaired by the
Deputy Assistant Secretary for
Energy Emergencies (DAS/EE).
EEMTs are activated at the outset
of an energy emergency and
serve as the focal point for the
development and coordination of
energy emergency activities.

ENERGY FLOW
(SSDC)
DA October 12, 1990
RT Energy
RT Energy Trace
RT Linear Energy Transfer

SO System Safety Development Center
    Glossary
DEF (SSDC) The transfer of energy from
    its source to some other point. In
    safety we are concerned with two
    types of energy flows: wanted
    energy flows-controlled (to do
    work) and unwanted energy
    flows-uncontrolled.

ENERGY FLUENCE
(IAEA)
DA October 12, 1990
BT1 Measurements
RT Energy Fluence Rate
RT Radiant Energy
SO Radiation
DEF (IAEA) Radiant energy incident
    divided by the cross-sectional area
    of an elementary sphere.

ENERGY FLUENCE RATE
(IAEA)
DA October 12, 1990
BT1 Rates
RT Energy Fluence
SO Radiation
DEF (IAEA) The increment of energy
    fluence in a specified time interval.

ENERGY FLUX
(IAEA)
DA October 12, 1990
BT1 Measurements
RT Radiant Energy
SO Radiation
DEF (IAEA) The quotient of dR by dt,
    where dR is the increment of
    radiant energy in the time interval
    dt.

ENERGY MANAGEMENT SYSTEMS
(DOE Order 6430.1A)
DA October 12, 1990
SY Energy Monitoring and Control
    Systems
BT1 Systems
RT Control Systems
SO Construction
DEF (DOE Order 6430.1A) An
    automated system for monitoring
    and controlling energy-related
    systems and devices.

ENERGY MONITORING AND
CONTROL SYSTEMS
(DOE Order 6430.1A)
DA October 12, 1990
SY Energy Management Systems
BT1 Control Systems
 BT2 Controls
 BT2 Systems
BT1 Monitoring Systems
 BT2 Systems
SO Construction
DEF (DOE Order 6430.1A) An
    automated system for monitoring
    and controlling energy-related
    systems and devices.

ENERGY RESEARCH ADVISORY
BOARD
DA January 8, 1991
SF *ERAB*
BT1 Boards
 BT2 Administrative Organizations
 BT3 Organizations

ENERGY RESEARCH AND
DEVELOPMENT ADMINISTRATION
DA January 8, 1991
SF *ERDA*
RT Atomic Energy Commission
RT U.S. Department of Energy

ENERGY TECHNOLOGY
ENGINEERING CENTER (CANOGA
PARK)
DA January 8, 1991
SF *ETEC*
BT1 Centers
BT1 Government-Owned
    Contractor-Operated Facilities
 BT2 Federal Facilities
 BT3 Facilities
RT Rockwell International Corp.
DEF (Capsule Review of DOE Research
    and Development and Field
    Facilities, 1986) Founded in 1966,
    the ETEC provides management,
    engineering, testing, consultation
    and project monitoring services for
    a wide range of DOE programs.
    The ETEC manages a reactor
    component testing program, in
    addition to providing engineering
    support, technical management
    and monitoring for a number of
    DOE solar, conservation,
    geothermal and fusion energy
    programs.

ENERGY TRACE
(SSDC)
DA October 12, 1990
RT Energy Flow
SO System Safety Development Center
    Glossary
DEF (SSDC) Tracking an energy flow
    from its origin to any other place
    or object.

ENFORCEMENT DECISION
DOCUMENTS
(EPA)
DA October 12, 1990
SY Record of Decision (NEPA)
BT1 Reports
RT Administrative Orders
SO Environmental Protection Agency
    Glossary
DEF (EPA) A document that provides an
    explanation to the public of EPA's
    selection of the cleanup alternative
    at enforcement sites on the
    National Priorities List. Similar to a
    Record of Decision.

ENGAGEMENT SIMULATION SYSTEMS
(Doe Order 5480.16)
DA October 12, 1990
SF *ESS*
BT1 Simulators
NT1 Multiple Integrated Laser
    Engagement System
RT Central Training Academy
RT ESS Blanks
RT Weapon Simulators
SO Firearms
DEF (DOE Order 5480.16) a combat
    simulation system that has three
    elements: (1) devices that provide
    weapon effects; (2) a control
    system; and (3) a training support
    package.

ENGINE-GENERATORS
DA January 8, 1991
SF *EG (Engine-Generators)*

ENGINE MAINTENANCE ASSEMBLY
AND DISASSEMBLY
DA January 8, 1991
SF *E-MAD*

ENGINEER
(0160 ENGINEER)
DA November 28, 1990
BT1 Professional Personnel
 BT2 Occupations
 BT2 Personnel
SO DOE FRASE VOCABULARY

ENGINEERED BARRIER SYSTEMS
(ANL; CFR)
DA May 24, 1991
BT1 Systems
NT1 Engineered Barriers
SO Wastes
DEF (CFR) The waste packages and the
    underground facility.

ENGINEERED BARRIERS
(AEA; CFR)
DA January 24, 1991
BT1 Barriers
BT1 Engineered Barrier Systems
 BT2 Systems
RT Disposal Systems
DEF (CFR) A man-made structure or
    device that is intended to improve
    the land disposal facility's ability to
    meet the performance objectives
    in Subpart C.

ENGINEERED SAFETY FEATURES
(DOE Order 6430.1A)
DA October 12, 1990
SF *ESF*
RT Design Capacity
RT General Design Process
RT Safety Class Items
SO Construction
SO Industrial Safety
DEF (DOE Order 6430.1A) Systems or
    design characteristics that are
    provided to prevent or mitigate the

potential consequences of postulated design basis accidents. An engineered-safety-feature system is a safety class system.

ENGINEERED SAFETY FEATURES (SHUTDOWN)
(DOE Order 5484.1)
DA October 16, 1990
SF *ESF*
RT Reactor Shutdown
DEF (DOE Order 5484.1) Components or equipment designed to: (a) provide the capability to shutdown the facility and maintain it in a safe shutdown condition; (b) ensure the integrity of the process system which provides a boundary against release of radioactive material (e.g., the integrity of the coolant pressure boundary in a reactor or a glove box in a process facility); (c) prevent or mitigate the consequences of events or accidents that could result in potentially measurable offsite exposures.

ENGINEERED SAFETY FEATURES ACTUATION SYSTEM
DA January 8, 1991
SF *ESFAS*
BT1 Systems

ENGINEERING ASSISTANCE SECTION
DA January 8, 1991
SF *EAS*

ENGINEERING CHANGE PROPOSAL
DA January 8, 1991
SF *ECP*
SO Management

ENGINEERING CONTROL ZONES
(SWDA; RCRA)
DA October 12, 1990
BT1 Zones
 BT2 Sites/Areas
RT Hazardous Wastes
RT Source Control Remedial Actions
SO Environmental Management
SO Wastes
DEF (CFR) Areas under the control of the owner/operator that, upon detection of a hazardous waste release, can be readily cleaned up prior to the release of hazardous waste or hazardous constituents to ground water or surface water.

ENGINEERING TECHNICIAN
(0370 ENGR TECH)
DA November 28, 1990
BT1 Technicians
 BT2 Professional Personnel
  BT3 Occupations
  BT3 Personnel
 SO DOE FRASE VOCABULARY

ENGINES
(SWDA; RCRA; ESH)
DA October 12, 1990
SO Hazardous Materials
DEF (CFR) Locomotives propelled by any form of energy and used by a railroad.

ENRICHED URANIUM
(AEA; CFR)
DA January 24, 1991
BT1 Uranium
 BT2 Geologic Resources
  BT3 Natural Resources
 BT2 Heavy Metals
NT1 Unirradiated Enriched Uranium
DEF (CFR) Uranium containing more uranium-235 than the naturally occurring distribution of uranium isotopes.

ENRICHMENT
(EPA)
DA October 12, 1990
BT1 Processes
RT Eutrophic Lakes
RT Eutrophication
SO Environmental Protection Agency Glossary
DEF (EPA) The addition of nutrients (e.g., nitrogen, phosphorus, carbon compounds) from sewage effluent or agricultural runoff to surface water. This process greatly increases the growth potential for algae and aquatic plants.

ENTRY CONTROL POINTS
(DOE Order 6430.1A)
DA October 12, 1990
BT1 Sites/Areas
RT Egress
RT Ingress
RT Security Areas
SO Construction
DEF (DOE Order 6430.1A) Controlled access entry points to a site or a secured area.

ENVIRONMENT
(CAA; RCRA; CWA; ESA; FWCA; NPDWA; CERCLA; TSCA; FIFRA; CFR; EPA; RHA; SARA; SWDA)
DA October 12, 1990
NT1 Accessible Environment
NT1 Aquatic Ecosystem
NT1 Aquatic Environment
 NT2 Benthic Region
NT1 Ecosphere
NT1 General Environment
NT1 Human Environment
NT1 Indoor Climate
NT1 Media
NT1 Work Environment
RT Ambient
RT Ecological Impacts
RT Ecosystems
RT Environmental Impact Statements
RT Habitats

RT Lithosphere
RT Pollution
RT Significant Adverse Reactions
RT Targets
SO Compensation and Liability
SO Environmental Management
SO Environmental Protection Agency Glossary
SO Hazardous Materials
DEF The sum of all external conditions affecting the life, development and survival of an organism. (ESH) Refers to the navigable waters, the waters of the contiguous zone, and the ocean waters of which the natural resources are under the exclusive management and authority of the United States under the Fishery Conservation and Management Act of 1976; and any other surface water, ground water, drinking water supply, land surface or subsurface strata, or ambient air within the United States or under the jurisdiction of the United States. Includes water, air, and land and the interrelationship which exists among and between water, air, and land and all living things.

ENVIRONMENT, SAFETY, AND HEALTH
DA January 8, 1991
SY Office of the Assistant Secretary for Environment, et. al.
SF *ES&H*

ENVIRONMENT, SAFETY, AND HEALTH (ES&H) PROGRAM
(DOE Orders 5480.1B and 5482.1B; ESH)
DA October 12, 1990
SF *ES&H*
BT1 DOE Programs
 BT2 Programs
RT Consultant Fire Protection Survey Program
RT Environment, Safety, and Health Overview
RT Health Physics
RT Industrial Hygiene
RT Occupational Medicine
SO Emergency Preparedness
SO Industrial Hygiene
SO Management
DEF Encompasses those DOE requirements, activities, and functions in the conduct of all DOE and DOE-controlled operations that are concerned with: controlling air, water, and soil pollution; limiting the risks to the well being of both operating personnel and the general public to acceptably low levels; and protecting property adequately against accidental loss and damage. Typical activities and functions related to this program

include, but are not limited to, the following: environmental protection, occupational safety, fire protection, industrial hygiene, health physics, occupational medicine, process and facilities safety, nuclear safety, emergency preparedness, quality assurance, and radioactive and hazardous waste management.

ENVIRONMENT, SAFETY, AND HEALTH OVERVIEW
(Doe Order 5480.1B)
DA   October 12, 1990
RT   Environment, Safety, and Health (ES&H) Program
SO   Industrial Hygiene
DEF (DOE Order 5480.1B) An organized set of activities performed as independent functions. Its purpose is to assure that all aspects of environment, safety and health-related activities at the program, project, and contractor level are adequately addressed. Such activities include: (1) establishing Department-wide environment, safety, and health policies, requirements and standards; (2) periodic and timely reviews of program and project documents, activities, actions, and plans; (3) appraising the implementation of environment, safety and health programs at the Headquarters, field, and contractor level as appropriate; and (4) providing support, assistance, and guidance to Headquarters program offices and field organizations.

ENVIRONMENT ASSESSMENTS
(TTGM)
DA   May 6, 1991
BT1 Tiger Team Assessments
  BT2 Assessments
   BT3 Administrative Processes
    BT4 Processes
RT   Compliance Considerations
RT   Environmental Assessments
RT   Findings
RT   Management and Organization Assessment
RT   Potential Hazards
RT   Safety and Health Assessments
RT   Seriousness
SO   Management
DEF (TTGM) One of the three major assessments conducted as part of a Tiger Team Assessment. It is concerned with both regulatory (i.e., externally enforced) requirements and DOE Orders (i.e., internally enforced requirements). Environment Assessments specifically examine the following topic areas: air; soils/sediments/biota; surface water; groundwater; waste management; Toxic and Chemical Materials; quality assurance environmental monitoring; radiation/radiologic materials; inactive waste sites; NEPA; and special issues (e.g. natural/cultural resources acts and executive orders.

ENVIRONMENTAL ASSESSMENTS
(DOE Order 5440.1D; NEPA; CFR; EPA)
DA   October 12, 1990
SY   Environmental Reviews
SF   *EA (Environmental Assessment)*
BT1 Assessments
  BT2 Administrative Processes
   BT3 Processes
RT   Environmental Surveys
RT   Environmental Audits
RT   Environmental Impact Statements
RT   Environment Assessments
RT   Mitigation Action Plans
RT   National Environmental Policy Act Documents
RT   Programmatic NEPA Documents
RT   Site-Wide NEPA Documents
RT   State Coordination
SO   Environmental Management
SO   Environmental Protection Agency Glossary
SO   Management
DEF (DOE Order) Documents defined at 40 CFR 1508.9, that assess whether a proposed action is a "major Federal action significantly affecting the quality of the human environment," and which serve as the basis for determining whether to prepare an environmental impact statement or a Finding of No Significant Impact. (EPA) Written environmental analyses which are prepared pursuant to the National Environmental Policy Act to determine whether a federal action would significantly affect the environment and thus require preparation of a more detailed environmental impact statement. (NOTE: see Environment Assessment if term usage is within the area of Tiger Team Assessments)

ENVIRONMENTAL AUDITS
(DOE Orders 5480.1B, 5482.1B; EPA; ESH)
DA   October 12, 1990
BT1 Audits (SSDC)
  BT2 Administrative Processes
   BT3 Processes
RT   Environmental Reviews
RT   Environmental Surveys
RT   Environmental Assessments
SO   Construction
SO   Environmental Management
SO   Environmental Protection Agency Glossary
SO   Industrial Hygiene

SO   Management
DEF (DOE Orders) Documented assessments of a facility to monitor the progress of necessary corrective actions, to assure compliance with environmental laws and regulations, and to evaluate field organization practices and procedures. (EPA) 1) An independent assessment of the current status of a party's compliance with applicable environmental requirements. 2) An independent evaluation of a party's environmental compliance policies, practices, and controls.

ENVIRONMENTAL DETECTION LIMITS
(ESH)
DA   October 12, 1990
BT1 Detection Limits
  BT2 Limits
RT   Radionuclides
SO   Management
DEF (ESH) The smallest levels at which a radionuclide in an environmental medium can be unambiguously distinguished for a given confidence level using a particular combination of sampling and measurement procedures, sample volume, analytical detection limit, and processing procedure.

ENVIRONMENTAL EXPOSURE
(NCRP)
DA   October 12, 1990
BT1 Exposure
RT   Ionizing Radiation
SO   Radiation
DEF (NCRP) Exposure to radiation in nonoccupational situations.

ENVIRONMENTAL IMPACT STATEMENTS
(DOE Order 5440.1D; NEPA; CFR; EPA)
DA   October 12, 1990
SF   *EIS*
NT1 Supplemental EIS
RT   Categorical Exclusions
RT   Ecological Impacts
RT   EIS Implementation Plans
RT   Environment
RT   Environmental Information Documents
RT   Environmental Assessments
RT   Finding of No Significant Impact
RT   Implementation Plans
RT   Interim Actions
RT   Mitigation Action Plans
RT   National Environmental Policy Act Documents
RT   Notice of Availability
RT   Notice of Intent
RT   Programmatic NEPA Documents
RT   Site-Wide NEPA Documents
RT   Supplement Analyses
RT   Tiering
SO   Environmental Management

BT-Broader Term   NT-Narrower Term   RT Related Term

SO Environmental Protection Agency
    Glossary
SO Management
DEF (DOE Order) Documents defined at
    40 CFR 1508.11 or a
    Supplemental EIS, and prepared
    in accordance with the
    requirements of section 102(2)(C)
    or NEPA, the CEQ Regulations,
    and the DOE NEPA Guidelines.

**ENVIRONMENTAL INFORMATION
DOCUMENTS**
(NEPA; CFR)
DA January 24, 1991
BT1 Reports
RT Environmental Impact Statements
DEF (CFR) Any written analysis
    prepared by an applicant, grantee
    or contractor describing the
    environmental impacts of a
    proposed action.

ENVIRONMENTAL MANAGEMENT
DA January 23, 1991
BT1 Management
RT Conservation
RT Waste Management
DEF The management of resources,
    property, and all associated
    programs with an emphasis on
    environmental conservation and
    enhancement. Also, may be used
    to include those actions necessary
    to remediate situations where
    environmental conditions have
    already been negatively affected.

ENVIRONMENTAL MONITORING
(DOE Orders 5400.1, 5484.1)
DA October 12, 1990
BT1 Monitoring
 BT2 Activities
NT1 Effluent Monitoring
NT1 Environmental Surveillance
RT Annual Site Environmental Reports
SO Environmental Management
SO Management
DEF The collection and analysis of
    samples or direct measurements
    of environmental media.
    Environmental monitoring consists
    of two major activities: effluent
    monitoring and environmental
    surveillance.

ENVIRONMENTAL OCCURRENCES
(DOE Order 5400.1; ESH)
DA October 12, 1990
BT1 Occurrences
RT Unusual Occurrences
SO Management
DEF (DOE Order 5400.1) Any sudden or
    sustained deviations from a
    regulated or planned performance
    at a DOE operation that has
    environmental protection and
    compliance significance.

**ENVIRONMENTAL PROTECTION
AGENCY GLOSSARY**
(EPA)
DA October 19, 1990
DEF The Environmental Protection
    Agency Glossary as included in
    the Agency's Information
    Resources Directory, an annual
    EPA publication which gives
    general reference and referral
    information of relevance to EPA's
    programs.

**ENVIRONMENTAL PROTECTION
STANDARDS**
(DOE Order 5400.1; ESH)
DA October 12, 1990
BT1 Standards
 BT2 Codes, Standards, and
     Regulations
SO Environmental Management
SO Management
DEF Specified sets of rules or conditions
    concerned with delineation of
    procedures, definition of terms,
    specification of performance,
    design, or operations, or
    measurements that define the
    quantity of emissions, discharges,
    or releases to the environment
    and the quality of the environment.

ENVIRONMENTAL QUALIFICATION
DA January 8, 1991
SF EQ

ENVIRONMENTAL RELEASE
(1563 ENV RELEASE)
DA November 29, 1990
BT1 Nature of Property Damage
BT1 Releases
RT Chemical Contamination
RT Hazardous Spill
RT Leak Detector
SO DOE FRASE VOCABULARY

ENVIRONMENTAL RESPONSE TEAMS
(EPA; EMER)
DA October 12, 1990
BT1 Response Teams
 BT2 Teams
  BT3 Administrative Organizations
  BT4 Organizations
SO Environmental Protection Agency
    Glossary
DEF (EPA) EPA experts located in
    Edison, NJ, and Cincinnati, OH,
    who can provide around-the-clock
    technical assistance to EPA
    regional offices and states during
    all types of emergencies involving
    hazardous waste sites and spills
    of hazardous substances.

ENVIRONMENTAL REVIEWS
(NEPA; CFR)
DA January 24, 1991
SY Environmental Assessments
BT1 Reviews

BT2 Administrative Processes
BT3 Processes
RT Environmental Surveys
RT Environmental Audits
DEF (CFR) The process whereby an
    evaluation is undertaken by EPA
    to determine whether a proposed
    Agency action may have a
    significant impact on the
    environment and therefore require
    the preparation of the EIS.

ENVIRONMENTAL SURVEILLANCE
(DOE Orders 5400.1, 5400.5)
DA October 12, 1990
BT1 Environmental Monitoring
 BT2 Monitoring
  BT3 Activities
BT1 Surveillance
 BT2 Administrative Processes
  BT3 Processes
SO Environmental Management
SO Management
SO Wastes
DEF The collection and analysis of
    samples, or direct measurements,
    of air, water, soil, foodstuff, biota,
    and other media from DOE sites
    and their environs for the purpose
    of determining compliance with
    applicable standards and permit
    requirements, assessing radiation
    exposures of members of the
    public and assessing the effects, if
    any, on the local environment.

ENVIRONMENTAL SURVEYS
(DOE Orders 5480.1B, 5482.1B; ESH)
DA October 12, 1990
BT1 Surveys
RT Environmental Reviews
RT Environmental Audits
RT Environmental Assessments
RT Technical Safety Appraisals
SO Construction
SO Industrial Hygiene
SO Management
DEF Documented, multidiscipline
    assessments (with sampling and
    analysis) of a facility to determine
    environmental conditions and to
    identify environmental problem
    areas of environmental risk
    requiring corrective action.

EOC
DA October 12, 1990
SEE Emergency Operations Centers
SO Acronyms
SO Environmental Management

EOD
DA October 12, 1990
SEE Explosives Ordnance Disposal
SO Acronyms

EOP
DA October 12, 1990
SEE Emergency Operating Procedures

SY-Synonymous Terms   SO-Source/Subject Category   SF-See From

SO Acronyms
SO Management

EP
DA   October 12, 1990
SEE Essential Power
SO Acronyms

EP Toxic
DA   October 12, 1990
SEE Extraction Procedure Toxic
SO Acronyms

EPA
DA   October 12, 1990
SEE U.S. Environmental Protection
   Agency
SO Acronyms
SO Air Pollution
SO Environmental Protection Agency
   Glossary
SO Hazardous Materials
SO Wastes
SO Water Pollution

EPA GUIDANCE DOCUMENT
(TSCA; CFR)
DA   October 12, 1990
BT1 Reports
RT Asbestos-containing Materials
RT Procedures
SO Hazardous Materials
DEF (CFR) EPA Guidance Document.
   The term "Guidance for Controlling
   Asbestos-Containing Material in
   Buildings" means the
   Environmental Protection Agency
   document with such title as in
   effect on March 31, 1985.

EPA HAZARDOUS WASTE NUMBERS
(SWDA; RCRA; CFR)
DA   October 12, 1990
RT Hazardous Wastes
SO Environmental Management
SO Wastes
DEF (SWDA; RCRA) Numbers assigned
   by the Environmental Protection
   Agency and used by industry;
   solid wastes are listed hazardous
   wastes from non-specific sources
   in 40 CFR 261.31; solid wastes
   are listed hazardous wastes from
   specific sources in 40 CFR 261.32
   unless they are excluded under 40
   CFR 260.20 and 260.22 and are
   listed in Part 261 - Appendix IX.
   (Exclusions are for hazardous
   wastes from non-specific and
   specific sources.)

EPA IDENTIFICATION NUMBER
(SWDA; RCRA; TSCA; CFR)
DA   October 12, 1990
RT Treatment, Storage, and Disposal
   Facilities
SO Hazardous Materials
SO Wastes
DEF (CFR) The number assigned by the

Environmental Protection Agency
   to each generator, transporter,
   and treatment, storage, or
   disposal facility.

EPA REGION
(SWDA; RCRA)
DA   October 12, 1990
BT1 Sites/Areas
RT States
RT United States
SO Wastes
DEF (SWDA; RCRA) The states and
   territories found in one of the ten
   regions of the United States.

EPIDEMICS
(EPA)
DA   October 12, 1990
BT1 Occurrences
NT1 Waterborne Disease Outbreaks
RT Epidemiology
SO Environmental Protection Agency
   Glossary
DEF (EPA) Widespread outbreaks of
   disease, or a large number of
   cases of a disease in a single
   community or relatively small area.

EPIDEMIOLOGY
(EPA)
DA   October 12, 1990
RT Epidemics
RT Populations
SO Environmental Protection Agency
   Glossary
DEF (EPA) The study of diseases as
   they affect population, including
   the distribution of disease, or other
   health-related states and events in
   human populations, the factors
   (e.g., age, sex, occupation,
   economic status) that influence
   this distribution, and the
   application of this study to control
   health problems.

EPISODES (POLLUTION)
(EPA)
DA   October 12, 1990
BT1 Incidents
RT Air Pollution
SO Environmental Protection Agency
   Glossary
DEF (EPA) Air pollution incidents in a
   given area caused by a
   concentration of atmospheric
   pollution reacting with
   meteorological conditions that may
   result in a significant increase in
   illnesses or deaths. Although most
   commonly used in relation to air
   pollution, the term may also be
   used in connection with other
   kinds of environmental events
   such as a massive water pollution
   situation.

EPRI
DA   October 12, 1990
SEE Electric Power Research Institute
SO Acronyms

EQ
DA   October 12, 1990
SEE Environmental Qualification
SO Acronyms

EQUILIBRIUM
(EPA; NCRP)
DA   October 12, 1990
RT Radioactive Series
SO Environmental Protection Agency
   Glossary
SO Radiation
DEF (EPA)In relation to radiation, the
   state at which the radioactivity of
   consecutive elements within a
   radioactive series is neither
   increasing nor decreasing. (NCRP)
   Equilibrium exists when the
   activity of all the progeny within a
   decay series is equal to the parent
   activity. For radon progeny,
   equilibrium is rarely achieved and
   the progeny activities are usually
   less than the radon activity.

EQUIPMENT
(CAA; CFR; DSTT)
DA   October 12, 1990
NT1 Allis-Chalmers Motors
NT1 Analog to Digital Converter
NT1 Ancillary Equipment
NT1 Auxiliary Air Units
NT1 Axial Power Indicator
NT1 Balers
NT1 Ballistic Separators
NT1 Class A Equipment
NT1 Class B Equipment
NT1 Class 1E Electrical Equipment
NT1 Departmental-Approved Equipment
NT1 Detection Equipment
   NT2 Intrusion Alarm System (Perimeter
     or Interior)
NT1 Diesel Generator
   NT2 Standby Diesel Generator
NT1 Electrical Equipment
   NT2 Capacitors
     NT3 Large High Voltage Capacitors
     NT3 Large Low Voltage Capacitors
     NT3 Small Capacitors
   NT2 Connectors
   NT2 PCB Transformers
     NT3 Mineral Oil PCB Transformers
NT1 Emergency Equipment
NT1 Emergency (Diesel) Generators
NT1 Equipment/Parts - Electrical (DOE
     FRASE Vocabulary)
   NT2 Adapter(s)
   NT2 Amplifier(s)
   NT2 Antenna
   NT2 Appliance(s)
     NT3 Heating Appliance(s)
   NT2 Battery(s)
   NT2 Capacitor(s)
   NT2 Circuit Breaker(s)
   NT2 Circuit(s)

NT2 Computer Terminal(s)
NT2 Computer(s)
NT2 Condenser(s)
NT2 Conductor(s)
NT2 Conduit(s)
NT2 Diode(s)
NT2 Electrical Apparatus
NT2 Electrical Bus
NT2 Electrical Insulator(s)
NT2 Electrode(s)
NT2 Fuse(s)
NT2 Ground Fault Interrupter(s)
NT2 Heat Coil
NT2 Insulator
NT2 Jumper
NT2 Laser
NT2 Light(s)
NT2 Magnetic/Electrolytic Apparatus
NT2 Motor(s)
NT2 Outlet/Receptacle(s)
NT2 Power Cord(s)
NT2 Power Key(s)
NT2 Power Lead
NT2 Power Line(s)
NT2 Power Pole(s)
NT2 Power Supply
 NT3 Uninterruptible Power Supplies
NT2 Preamplifier(s)
NT2 Pull Box
NT2 Radio
NT2 Rectifier(s)
NT2 Regulator
NT2 Relay(s)
 NT3 Building Power Low Voltage
   Relay
 NT3 Building Time Delay Relay
 NT3 Phase Failure Relays
 NT3 Scram Relay
NT2 Reset Device
NT2 Resistor(s)
NT2 Rheostat(s)
NT2 Shut-Down Circuit(s)
NT2 Solar Collector(s)
NT2 Solenoid(s)
NT2 Starter
NT2 Switch Box
NT2 Switchboard
NT2 Switchgear
NT2 Thermocouple(s)
NT2 Transducer(s)
NT2 Transformer(s)
NT2 Transmitter(s)
NT2 Trouble Light
NT2 Ultraviolet Equipment
NT2 Voltage
NT2 Voltage Regulator
NT2 Wiring
NT2 X-Ray Equipment
NT1 Equipment/Parts - Heating (DOE
   FRASE Vocabulary)
NT2 Burner(s)
NT2 Evaporator(s)
NT2 Heater(s)
 NT3 Autoclave
 NT3 Boiler(s)
  NT4 Boiler Part(s)
 NT3 Feed Water Heater(s)
 NT3 Furnace(s)
  NT4 Boiler/Industrial Furnaces
 NT3 Oven(s)
 NT3 Stove(s)

NT2 Incinerators
 NT3 Catalytic Incinerators
 NT3 Qualified Incinerators
NT2 Pilot Light(s)
NT2 Retort
NT1 Equipment/Parts -
   Instrumentation/Measuring (DOE
   FRASE Voc.)
NT2 Ammeter(s)
NT2 Bridge Crane
NT2 Constant Air Monitor
NT2 Count
NT2 Fast Response Gamma
   Thermometer
NT2 Instrument(s)
 NT3 Gauge(s)
  NT4 Scale(s)
  NT4 Straight Edge Ruler
  NT4 Thermostat(s)
  NT4 Time Clock
 NT3 Graphitar(s)
 NT3 Indicator(s)
  NT4 Ammeter(s)
  NT4 Flow Meter(s)
  NT4 Ohmmeter(s)
  NT4 Potentiometer
  NT4 Spectrometer(s)
  NT4 Tachometer(s)
  NT4 Thermoluminescent Dosimeters
  NT4 Thermometer(s)
  NT4 Volt Meter
 NT3 Level Alarm
 NT3 Log N Recorder
 NT3 Ram(s)
 NT3 Recorder
 NT3 Testing Equipment
  NT4 Heat Detector
  NT4 Leak Detector
  NT4 Oscilloscope(s)
  NT4 Scintillation Probe(s)
  NT4 Sensor(s)
NT1 Equipment/Parts - Material
   Handling (DOE FRASE
   Vocabulary)
NT2 Boom
NT2 Drill Rig(s)
NT2 Heavy Mobile Equipment
NT2 Highway Construction Equipment
NT2 Hook(s)
NT2 Load Securing Device
NT2 Manipulator Tape
NT2 Material Handling Device
 NT3 Conveyor(s)
 NT3 Crane(s)
  NT4 Boom Crane
  NT4 Bridge Crane
  NT4 Mobile Crane
  NT4 Overhead Crane
 NT3 Earth Moving Equipment
  NT4 Dredge(s)
 NT3 Hand Truck(s)
  NT4 Cart(s)
  NT4 Dolly
 NT3 Hoisting Apparatus
  NT4 Derrick
  NT4 Elevator
  NT4 Forklift(s)
  NT4 Hoist(s)
   NT5 Air Hoist
  NT4 Lift Bucket
  NT4 Manlift(s)

  NT4 Scissor Lift
NT2 Reel(s)
NT2 Scraper
NT2 Sling(s)
NT2 Tine(s)
NT1 Equipment/Parts - Nuclear (DOE
   FRASE Vocabulary)
NT2 Birdcage
NT2 Burial Box(s)
NT2 Cold Trap
NT2 Control Panel(s)
NT2 Control Rod Drive Assembly
NT2 Control Rod(s)
 NT3 Gangs
 NT3 Raincoats
NT2 Coolant Return Pump
NT2 Core(s)
NT2 Exhaust Stack
NT2 Flux Channel(s)
NT2 Fuel Element(s)
 NT3 Fuel Plate
  NT4 Cladding
  NT4 Plutonium Fuel Plate(s)
 NT3 Fuel Rod(s)
 NT3 Piles (Nuclear)
NT2 Fuel Handling Equipment
 NT3 Fuel Canister(s)
 NT3 Fuel Drawer(s)
NT2 Fuel Tool(s)
NT2 Glove Box
NT2 Heat Exchanger
 NT3 Low Pressure Recirculation
   System Heat Exchanger
 NT3 Shutdown Heat Exchanger
NT2 Hepa Filter
NT2 Ion Exchange Column
NT2 Ion Pump
NT2 Log N Channel
NT2 Manipulator
 NT3 Electromechanical Manipulator(s)
NT2 Neutron Source
NT2 Neutron(s)
NT2 Plutonium Process Hood
NT2 Plutonium Recovery Process
NT2 Pressure Vessel Part
NT2 Pressure Vessel(s)
NT2 Reactor Fuel
NT2 Reactor Waste
NT2 Reactor(s)
NT2 Rod Drive
NT2 Shield Plug(s)
NT2 Sodium Scrubber(s)
NT2 Test Train
NT2 Transient Operation
NT2 Transient Test
NT1 Equipment/Parts - Personal
   Protective (DOE FRASE
   Vocabulary)
NT2 Personal Protective Equipment
 NT3 Air Mask
 NT3 Anticontamination Clothing
  NT4 Acid Suit
  NT4 Lab Coat
  NT4 Radiation Suit
 NT3 Chin Strap
 NT3 Dust Mask
 NT3 Eye Protection
  NT4 Goggles
  NT4 Safety Glasses
   NT5 Safety Glasses W Side Shields
   NT5 Tinted Safety Glasses

SY-Synonymous Terms   SO-Source/Subject Category   SF-See From

NT4 Side Shields
NT3 Fall Protection Device
NT4 Safety Belt
NT4 Safety Line
NT4 Safety Net
NT3 Flame Retardant Clothing
NT3 Foot Protection
NT4 Ankle Protection
NT4 Metatarsal Protection
NT4 Safety Boots
NT4 Safety Shoe(s)
NT4 Shoe Cover(s)
NT5 Metal Shoe Cover
NT3 Hand Protection
NT4 Glove(s)
NT4 Wrist Band
NT3 Head Protection
NT4 Bump Cap
NT4 Faceshield
NT4 Hard Hat
NT4 Helmet
NT4 Sand Blaster's Hood
NT4 Welder's Hood
NT3 Hearing Protection
NT4 Ear Plug(s)
NT3 Leggings
NT3 Padding
NT3 Respirator(s)
NT3 Seat Belt(s)
NT1 Fuel Distribution Analyzer
NT1 Installed Equipment
NT1 Isolation Condenser
NT1 Machines (DOE FRASE
   Vocabulary)
NT2 Agitator
NT2 Agricultural Machine
NT2 Air Dryer
NT2 Band Saw
NT2 Biostabilizers
NT2 Boring Machine
NT2 Centrifuge
NT2 Comminuters
NT2 Compactor
NT2 Compressor
NT3 Air Compressor
NT2 Concrete Saw
NT2 Crushing Machine
NT2 Discharge Machine
NT2 Drilling Machine
NT2 Drying Machine
NT2 Electrical Office Machine
NT2 Emergency Diesel Generator
NT2 Grinding Machine
NT2 Hammermills
NT2 Lathe
NT2 Milling Machine
NT3 Ball Mill
NT3 Roll Mill
NT2 Mixing Machine
NT2 Polishing Machine
NT2 Press
NT3 Isostatic Press
NT2 Printing Machine
NT2 Pump(s)
NT3 Coolant Return Pump
NT3 Ion Pump
NT3 Reactor Coolant Pump
NT3 Sump Pump
NT2 Recirculator
NT2 Sanding Machine
NT2 Slicing Machine

NT2 Stitching/Sewing Machine
NT2 Table Saw
NT2 Turbine
NT2 Vacuum Cleaner
NT1 Measurement and Test Equipment
NT1 Monitors
NT2 10-minute Delay Line Gamma
   Monitor
NT2 Axial Power Monitors
NT2 Constant Air Monitor
NT2 Continuous Air Monitors
NT2 Cooling Water Gamma Monitor
NT2 Crane Drip Pan Monitor
NT2 Essential Equipment Monitor
NT2 Failed Element Monitors
NT2 Flow Monitor
NT2 Gang Temperature Monitor
NT2 High Level Flux Monitor
NT2 Internal Gamma Flux Monitor
NT2 Low Energy Gamma Monitor
NT2 Nuclear Incident Monitor
NT2 Power Density Monitor
NT2 Primary Environmental Monitors
NT2 Process Water Gamma Monitor
NT2 Radial Power Monitor
NT2 Secondary Environmental Monitor
NT2 Secondary Environmental
   Monitors
NT2 Selection and Monitoring Chassis
NT2 Stack Tritium Monitor
NT2 Temperature Alarm Monitor
NT2 Temperature Scram Circuit Monitor
NT2 Traveling Wire Flux Monitor
NT1 Motor Generator
NT1 Motor Generator-Motor Alternator
NT1 Multiplexor
NT2 TRAnsient Multiplexor
NT1 Other Equipment
NT1 PCB Equipment
NT1 Personnel Monitoring Equipment
NT2 Personnel Dosimeters
NT1 Process Units
NT1 Programmatic Equipment
NT1 Radiation Detectors
NT2 Cooling Water Gamma Monitor
NT2 Cutie Pies
NT2 Dose Rate Meters
NT2 Dosimeters
NT3 Nuclear Accident Dosimeter
NT3 Personnel Dosimeters
NT3 Thermoluminescent Dosimeters
NT2 External Fission Counter
NT2 Geiger-Mueller Counters
NT2 Geiger Counters
NT2 Internal Fission Counter
NT2 Internal Gamma Flux Monitor
NT2 Ionization Chambers
NT2 Kanne Alarm
NT2 Low Energy Gamma Monitor
NT2 Process Water Gamma Monitor
NT2 Scintillation Counters
NT2 Vibrating Reed Electrometers
NT2 Whole Body Counter
NT1 Rasps
NT1 Slip Gauges
NT1 Steam Generator
NT2 Fossil Fuel and Wood
   Residue-Fired Steam Generating
   Unit
NT1 Strippers
NT1 Tools (DOE FRASE Vocabulary)

NT2 Leader Seater
NT2 Tools - Manual
NT3 Ax(s)
NT3 Broom(s)
NT3 Brush(s)
NT3 Chisel(s)
NT3 Crowbar(s)
NT3 Cutter(s)
NT3 File(s)
NT3 Hack Saw
NT3 Hammer
NT4 Sledgehammer(s)
NT3 Hand Iron(s)
NT3 Hand Saw
NT3 Hand Stapler
NT3 Hatchet(s)
NT3 Hoe(s)
NT3 Knife(s)
NT4 Pocket Knife(s)
NT3 Mop(s)
NT3 Non-Powered Handtool(s)
NT3 Pick(s)
NT3 Plane(s)
NT3 Pliers
NT3 Punch
NT3 Rake(s)
NT3 Razor(s)
NT3 Scalpel(s)
NT3 Scissors
NT3 Screwdriver(s)
NT3 Shear(s)
NT3 Shovel(s)
NT3 Snips
NT3 Snubber(s)
NT3 Spatula(s)
NT3 Threader(s)
NT3 Tongs
NT3 Torch(s)
NT3 Tweezers
NT3 Vise
NT3 Visegrips
NT2 Tools - Powered
NT3 Air Tamper
NT3 Chain Saw
NT3 Disk Sander(s)
NT3 Jackhammer
NT3 Mixer(s)
NT3 Nail Gun
NT3 Post Driver(s)
NT3 Power Actuated Tool(s)
NT3 Power Buffer(s)
NT3 Power Chisel(s)
NT3 Power Cutter(s)
NT3 Power Drill(s)
NT3 Power File(s)
NT3 Power Grinder(s)
NT3 Power Hammer(s)
NT3 Power Impact Wrench(s)
NT3 Power Polisher(s)
NT3 Power Riveter(s)
NT3 Power Sandblaster(s)
NT3 Power Screwdriver(s)
NT3 Power Spray Gun(s)
NT3 Power Stapler(s)
NT3 Power Waxer(s)
NT3 Powered Handtool(s)
NT3 Router(s)
NT3 Soldering Iron(s)
NT3 Spray Gun(s)
NT3 Weed Eater
NT2 Wrench(s)

BT-Broader Term   NT-Narrower Term   RT Related Term

NT3 Allen Wrench
NT3 Impact Wrench
NT3 Power Impact Wrench(s)
NT3 Ratchet Wrench(s)
NT3 Torque Wrench(s)
NT1 Turbidity Removal Evaporator
NT1 Vital Equipment
NT1 ZENER Diodes
RT Best Available Technology
RT Booms
RT Capacity Factor
RT Components
RT Devices
RT Installations
RT Personal Property
RT Production Facilities
RT Real Property
RT Related Personal Property
RT Targets
SO Air Pollution
DEF (DOE Order 4330.4A) The systems
and devices used throughout DOE
and commonly referred to as
equipment are divided into three
categories for the purpose of DOE
Order 4330.4A. It is the intent of
this definition to separately identify
the installed equipment that can
logically be considered as an
integral part of a real property
improvement from other types of
equipment. The purpose of such a
determination is to provide a
uniform basis for analysis of
various maintenance and repair
costs.

1E (Electrical Equipment)
DA October 12, 1990
SEE Class 1E Electrical Equipment
SO Acronyms

EQUIPMENT ENGINEERING DEPARTMENT
DA January 8, 1991
SF EED (Equipment Engineering
Department)

EQUIPMENT INSTALLATION ACTIVITY
(1231 EI ACTIVITY)
DA November 28, 1990
BT1 Activity Types (DOE FRASE
Vocabulary)
BT2 Activities
SO DOE FRASE VOCABULARY

EQUIPMENT OPERATOR
(0830 EQUIPMENT OP)
DA November 28, 1990
BT1 Transport Personnel
BT2 Occupations
BT2 Personnel
SO DOE FRASE VOCABULARY

EQUIPMENT ROOM
(1693 EQUIPMENT RO)
DA December 10, 1990
BT1 Room
BT2 Sites/Areas

SO DOE FRASE VOCABULARY
SO Environmental Management

EQUIPMENT/PARTS - ELECTRICAL (DOE FRASE VOCABULARY)
(DOE FRASE Vocabulary Numeric Keys
2375-2474)
DA January 17, 1991
BT1 Equipment
NT1 Adapter(s)
NT1 Amplifier(s)
NT1 Antenna
NT1 Appliance(s)
 NT2 Heating Appliance(s)
NT1 Battery(s)
NT1 Capacitor(s)
NT1 Circuit Breaker(s)
NT1 Circuit(s)
NT1 Computer Terminal(s)
NT1 Computer(s)
NT1 Condenser(s)
NT1 Conductor(s)
NT1 Conduit(s)
NT1 Diode(s)
NT1 Electrical Apparatus
NT1 Electrical Bus
NT1 Electrical Insulator(s)
NT1 Electrode(s)
NT1 Fuse(s)
NT1 Ground Fault Interrupter(s)
NT1 Heat Coil
NT1 Insulator
NT1 Jumper
NT1 Laser
NT1 Light(s)
NT1 Magnetic/Electrolytic Apparatus
NT1 Motor(s)
NT1 Outlet/Receptacle(s)
NT1 Power Cord(s)
NT1 Power Key(s)
NT1 Power Lead
NT1 Power Line(s)
NT1 Power Pole(s)
NT1 Power Supply
 NT2 Uninterruptible Power Supplies
NT1 Preamplifier(s)
NT1 Pull Box
NT1 Radio
NT1 Rectifier(s)
NT1 Regulator
NT1 Relay(s)
 NT2 Building Power Low Voltage Relay
 NT2 Building Time Delay Relay
 NT2 Phase Failure Relays
 NT2 Scram Relay
NT1 Reset Device
NT1 Resistor(s)
NT1 Rheostat(s)
NT1 Shut-Down Circuit(s)
NT1 Solar Collector(s)
NT1 Solenoid(s)
NT1 Starter
NT1 Switch Box
NT1 Switchboard
NT1 Switchgear
NT1 Thermocouple(s)
NT1 Transducer(s)
NT1 Transformer(s)
NT1 Transmitter(s)
NT1 Trouble Light

NT1 Ultraviolet Equipment
NT1 Voltage
NT1 Voltage Regulator
NT1 Wiring
NT1 X-Ray Equipment
RT DOE FRASE Categories
SO DOE FRASE VOCABULARY
DEF A subject category used with the
DOE FRASE Vocabulary.

EQUIPMENT/PARTS - HEATING (DOE FRASE VOCABULARY)
(DOE FRASE Vocabulary Numeric Keys
2475-2549)
DA January 17, 1991
SY Heating Equipment
BT1 Equipment
NT1 Burner(s)
NT1 Evaporator(s)
NT1 Heater(s)
 NT2 Autoclave
 NT2 Boiler(s)
 NT3 Boiler Part(s)
 NT2 Feed Water Heater(s)
 NT2 Furnace(s)
 NT3 Boiler/Industrial Furnaces
 NT2 Oven(s)
 NT2 Stove(s)
NT1 Incinerators
 NT2 Catalytic Incinerators
 NT2 Qualified Incinerators
NT1 Pilot Light(s)
NT1 Retort
RT DOE FRASE Categories
RT Heating Appliance(s)
SO DOE FRASE VOCABULARY
DEF A subject category used with the
DOE FRASE Vocabulary.

EQUIPMENT/PARTS - INSTRUMENTATION/MEASURING (DOE FRASE VOC.)
(DOE FRASE Vocabulary Numeric Keys
2750-2824)
DA January 17, 1991
BT1 Equipment
NT1 Ammeter(s)
NT1 Bridge Crane
NT1 Constant Air Monitor
NT1 Count
NT1 Fast Response Gamma
Thermometer
NT1 Instrument(s)
 NT2 Gauge(s)
 NT3 Scale(s)
 NT3 Straight Edge Ruler
 NT3 Thermostat(s)
 NT3 Time Clock
 NT2 Graphitar(s)
 NT2 Indicator(s)
 NT3 Ammeter(s)
 NT3 Flow Meter(s)
 NT3 Ohmmeter(s)
 NT3 Potentiometer
 NT3 Spectrometer(s)
 NT3 Tachometer(s)
 NT3 Thermoluminescent Dosimeters
 NT3 Thermometer(s)
 NT3 Volt Meter
 NT2 Level Alarm

SY-Synonymous Terms   SO-Source/Subject Category   SF-See From

NT2 Log N Recorder
NT2 Ram(s)
NT2 Recorder
NT2 Testing Equipment
NT3 Heat Detector
NT3 Leak Detector
NT3 Oscilloscope(s)
NT3 Scintillation Probe(s)
NT3 Sensor(s)
RT   DOE FRASE Categories
SO   DOE FRASE VOCABULARY
DEF A subject category used with the
    DOE FRASE Category.

**EQUIPMENT/PARTS - MATERIAL
HANDLING (DOE FRASE
VOCABULARY)**
(DOE FRASE Vocabulary Numeric Keys
2300-2374)
DA   January 17, 1991
BT1 Equipment
NT1 Boom
NT1 Drill Rig(s)
NT1 Heavy Mobile Equipment
NT1 Highway Construction Equipment
NT1 Hook(s)
NT1 Load Securing Device
NT1 Manipulator Tape
NT1 Material Handling Device
 NT2 Conveyor(s)
 NT2 Crane(s)
  NT3 Boom Crane
  NT3 Bridge Crane
  NT3 Mobile Crane
  NT3 Overhead Crane
 NT2 Earth Moving Equipment
  NT3 Dredge(s)
 NT2 Hand Truck(s)
  NT3 Cart(s)
  NT3 Dolly
 NT2 Hoisting Apparatus
  NT3 Derrick
  NT3 Elevator
  NT3 Forklift(s)
  NT3 Hoist(s)
   NT4 Air Hoist
  NT3 Lift Bucket
  NT3 Manlift(s)
  NT3 Scissor Lift
NT1 Reel(s)
NT1 Scraper
NT1 Sling(s)
NT1 Tine(s)
RT   DOE FRASE Categories
SO   DOE FRASE VOCABULARY
DEF A subject category used with the
    DOE FRASE Vocabulary.

**EQUIPMENT/PARTS - NUCLEAR (DOE
FRASE VOCABULARY)**
(DOE FRASE Vocabulary Numeric Keys
2550-2649)
DA   January 17, 1991
BT1 Equipment
BT1 Reactor Components
NT1 Birdcage
NT1 Burial Box(s)
NT1 Cold Trap
NT1 Control Panel(s)
NT1 Control Rod Drive Assembly

NT1 Control Rod(s)
NT2 Gangs
NT2 Raincoats
NT1 Coolant Return Pump
NT1 Core(s)
NT1 Exhaust Stack
NT1 Flux Channel(s)
NT1 Fuel Element(s)
 NT2 Fuel Plate
  NT3 Cladding
  NT3 Plutonium Fuel Plate(s)
 NT2 Fuel Rod(s)
 NT2 Piles (Nuclear)
NT1 Fuel Handling Equipment
 NT2 Fuel Canister(s)
 NT2 Fuel Drawer(s)
NT1 Fuel Tool(s)
NT1 Glove Box
NT1 Heat Exchanger
 NT2 Low Pressure Recirculation
      System Heat Exchanger
 NT2 Shutdown Heat Exchanger
NT1 Hepa Filter
NT1 Ion Exchange Column
NT1 Ion Pump
NT1 Log N Channel
NT1 Manipulator
 NT2 Electromechanical Manipulator(s)
NT1 Neutron Source
NT1 Neutron(s)
NT1 Plutonium Process Hood
NT1 Plutonium Recovery Process
NT1 Pressure Vessel Part
NT1 Pressure Vessel(s)
NT1 Reactor Fuel
NT1 Reactor Waste
NT1 Reactor(s)
NT1 Rod Drive
NT1 Shield Plug(s)
NT1 Sodium Scrubber(s)
NT1 Test Train
NT1 Transient Operation
NT1 Transient Test
RT   DOE FRASE Categories
SO   DOE FRASE VOCABULARY
DEF A subject category used with the
    DOE FRASE Vocabulary.

**EQUIPMENT/PARTS - PERSONAL
PROTECTIVE (DOE FRASE
VOCABULARY)**
(DOE FRASE Vocabulary Numeric Keys
2650-2749)
DA   January 17, 1991
BT1 Equipment
NT1 Personal Protective Equipment
 NT2 Air Mask
 NT2 Anticontamination Clothing
  NT3 Acid Suit
  NT3 Lab Coat
  NT3 Radiation Suit
 NT2 Chin Strap
 NT2 Dust Mask
 NT2 Eye Protection
  NT3 Goggles
  NT3 Safety Glasses
   NT4 Safety Glasses W Side Shields
   NT4 Tinted Safety Glasses
  NT3 Side Shields
 NT2 Fall Protection Device

  NT3 Safety Belt
  NT3 Safety Line
  NT3 Safety Net
 NT2 Flame Retardant Clothing
 NT2 Foot Protection
  NT3 Ankle Protection
  NT3 Metatarsal Protection
  NT3 Safety Boots
  NT3 Safety Shoe(s)
  NT3 Shoe Cover(s)
   NT4 Metal Shoe Cover
 NT2 Hand Protection
  NT3 Glove(s)
  NT3 Wrist Band
 NT2 Head Protection
  NT3 Bump Cap
  NT3 Faceshield
  NT3 Hard Hat
  NT3 Helmet
  NT3 Sand Blaster's Hood
  NT3 Welder's Hood
 NT2 Hearing Protection
  NT3 Ear Plug(s)
 NT2 Leggings
 NT2 Padding
 NT2 Respirator(s)
 NT2 Seat Belt(s)
RT   Barriers
RT   Clothing
RT   DOE FRASE Categories
SO   DOE FRASE VOCABULARY
DEF A subject category used with the
    DOE FRASE Vocabulary.

EQUIVALENCE DATA
DA   January 24, 1991
DEF (CFR) Chemical data or biological
    test data intended to show that
    two substances or mixtures are
    equivalent.

EQUIVALENT METHODS
(SWDA; CAA; RCRA; CFR; EPA)
DA   October 12, 1990
RT   Air Pollution
RT   Alternate Methods
RT   Functional Equivalent
RT   Reference Methods
RT   Sampling
SO   Air Pollution
SO   Environmental Management
SO   Environmental Protection Agency
     Glossary
SO   Wastes
DEF (CFR) Methods of sampling and
    analyzing for air pollution which
    has been demonstrated to the
    EPA Administrator's satisfaction to
    be, under specific conditions, an
    acceptable alternative to the
    normally used reference methods.

ER (Environmental Restoration)
DA   October 12, 1990
SEE Office of Environmental Restoration
     and Waste Management
SO   Acronyms

BT-Broader Term   NT-Narrower Term   RT Related Term

ER (Office of Energy Research)
DA   October 12, 1990
SEE Office of Energy Research
SO   Acronyms

ERAB
DA   October 12, 1990
SEE Energy Research Advisory Board
SO   Acronyms

ERCWS
DA   October 12, 1990
SEE Emergency Raw Cooling Water
      System
SO   Acronyms

ERDA
DA   October 12, 1990
SEE Energy Research and Development
      Administration
SO   Acronyms

ERGONOMICS
(SSDC)
DA   October 12, 1990
SY   Human Factors
SY   Human Factors Engineering
SO   System Safety Development Center
      Glossary
DEF (SSDC) English term for human
      factors or human factors
      engineering.

ERL
DA   November 20, 1990
SEE Emergency Reference Levels
SO   Acronyms

EROSION
(EPA)
DA   October 12, 1990
BT1 Processes
RT   Contour Plowing
RT   Groundcovers
RT   Run-off
SO   Environmental Protection Agency
      Glossary
DEF (EPA) The wearing away of land
      surface by wind or water. Erosion
      occurs naturally from weather or
      runoff but can be intensified by
      land-clearing practices related to
      farming, residential or industrial
      development, road building, or
      timber- cutting.

ERROR AND SENSITIVITY
DA   January 8, 1991
SF   *E&S*

ERROR SAMPLING
(SSDC)
DA   October 12, 1990
BT1 Sampling
 BT2 Sampling and Analysis
RT   Human Error

SO   System Safety Development Center
      Glossary
DEF (SSDC) A defined, structured
      process employing experienced
      observers, defined deviations and
      reproducible patterns of
      observation.

ERT
DA   October 12, 1990
SEE Emergency Response Teams
SO   Acronyms

ERV
DA   October 12, 1990
SEE Emergency Relief Valve
SO   Acronyms

ES&H
DA   October 12, 1990
SEE Environment, Safety, and Health
      (ES&H) Program
OR   Environment, Safety, and Health
SO   Acronyms
SO   Industrial Hygiene
SO   Management

ESA
DA   February 12, 1991
SEE Endangered Species Act
SO   Acronyms

ESCALATOR
(1694 ESCALATOR)
DA   December 10, 1990
BT1 Site (DOE FRASE Vocabulary)
 BT2 Sites/Areas
RT   Elevator
SO   DOE FRASE VOCABULARY

ESCAPE HATCH
(1695 ESCAPE HATCH)
DA   December 10, 1990
BT1 Site (DOE FRASE Vocabulary)
 BT2 Sites/Areas
SO   DOE FRASE VOCABULARY

ESD
DA   October 12, 1990
SEE Emergency Shutdown Device
SO   Acronyms

ESF
DA   October 12, 1990
SEE Engineered Safety Features
OR   Engineered Safety Features
      (Shutdown)
SO   Acronyms

ESFAS
DA   October 12, 1990
SEE Engineered Safety Features
      Actuation System
SO   Acronyms

ESP
DA   October 12, 1990

SEE Electrostatic Precipitators
SO   Acronyms

ESS
DA   October 12, 1990
SEE Engagement Simulation Systems
SO   Acronyms

ESS BLANKS
(Doe Order 5480.16)
DA   October 12, 1990
BT1 Blank Ammunition
RT   Engagement Simulation Systems
SO   Firearms
DEF (DOE Order 5480.16) Blank
      cartridges that are used in ESS
      equipment; including ordinary
      blank cartridges and special
      charges (e.g., LAW simulator
      charges).

ESSENTIAL EQUIPMENT MONITOR
DA   January 8, 1991
SF   *EEM*
BT1 Monitors
 BT2 Equipment

**ESSENTIAL EXPERIMENTAL
POPULATIONS**
(ESA; CFR)
DA   October 12, 1990
BT1 Experimental Populations
 BT2 Populations
SO   Endangered Species
DEF (CFR) Experimental populations
      whose loss would be likely to
      appreciably reduce the likelihood
      of the survival of the species in
      the wild. All other experimental
      populations are to be classified as
      "nonessential."

ESSENTIAL FACILITIES
(SEA)
DA   October 12, 1990
BT1 Facilities
RT   Emergency Control Centers
RT   Emergency Operations Centers
SO   Construction
DEF (SEA) Structures that are
      necessary for emergency post-
      earthquake operations.

ESSENTIAL POWER
DA   January 8, 1991
SF   *EP*

ESTUARIES
(EPA)
DA   October 12, 1990
BT1 Surface Waters
 BT2 Water
BT1 Surface Water Resources
 BT2 Natural Resources
SO   Environmental Protection Agency
      Glossary
DEF (EPA) Regions of interaction
      between rivers and nearshore

ocean waters, where tidal action and river flow create a mixing of fresh and salt water. These areas may include bays, mouths of rivers, salt marshes, and lagoons. These brackish water ecosystems shelter and feed marine life, birds, and wildlife.

ESW
DA October 12, 1990
SEE Emergency Service Water
SO Acronyms

ET
DA October 12, 1990
SEE Event Tree
SO Acronyms

ETEC
DA October 12, 1990
SEE Energy Technology Engineering Center (Canoga Park)
SO Acronyms

ETHYLENE DIBROMIDE
(EPA)
DA October 12, 1990
SF *EDB*
BT1 CERCLA Hazardous Substances
 BT2 Hazardous Substances
BT1 Hazardous Constituents
RT Fumigants
RT Halogenated Organic Compounds
SO Environmental Protection Agency Glossary
DEF (EPA) A chemical used as an agricultural fumigant and in certain industrial processes. Extremely toxic and found to be a carcinogen in laboratory animals, EDB has been banned for most agricultural uses in the United States.

ETHYLENE DICHLORIDE PLANTS
(CAA; CFR)
DA October 12, 1990
BT1 Facilities
RT Ethylene Dichloride Purification
SO Air Pollution
DEF (CFR) Includes any plant which produces ethylene dichloride by reaction of oxygen and hydrogen chloride with ethylene.

ETHYLENE DICHLORIDE PURIFICATION
(CAA; CFR)
DA October 12, 1990
BT1 Purification Processes
 BT2 Processes
RT Ethylene Dichloride Plants
RT Impurities
SO Air Pollution
DEF (CFR) Includes any part of the process of ethylene dichloride production which follows ethylene dichloride formation, excluding

product storage following the final finishing column.

ETIOLOGIC AGENTS
(CFR)
DA October 12, 1990
RT Hazards
RT Microorganisms
SO Hazardous Materials
DEF (CFR) The Hazardous Materials Regulations of the Department of Transportation concerning etiologic agents are supplemental to the regulations of the Department of Health, Education, and Welfare. For the purposes of controlling the transport of such agents in commerce, the DOT (Department of Transportation) defines etiologic agent as a viable microorganism, or its toxin, which causes or may cause human disease.

EUTROPHIC LAKES
(EPA)
DA October 12, 1990
BT1 Lakes
 BT2 Inland Waters
 BT2 Surface Waters
  BT3 Water
 BT2 Surface Water Resources
  BT3 Natural Resources
RT Dystrophic Lakes
RT Enrichment
RT Oligotrophic Lakes
RT Senescence
SO Environmental Protection Agency Glossary
DEF (EPA) Shallow, murky bodies of water that have excessive concentrations of plant nutrients causing excessive algal production.

EUTROPHICATION
(EPA)
DA October 12, 1990
BT1 Processes
NT1 Cultural Eutrophication
NT1 Senescence
RT Algae
RT Aquatic Environment
RT Enrichment
RT Fertilizers
RT Lakes
RT Nutrients
RT Water Pollution
SO Environmental Protection Agency Glossary
DEF (EPA) The slow aging process during which a lake, estuary, or bay evolves into a bog or marsh and eventually disappears. During the later stages of eutrophication the water body is choked by abundant plant life as the result of increased amounts of nutritive compounds such as nitrogen and phosphorus. Human activities can accelerate the process.

EVAPORATION PONDS
(EPA)
DA October 12, 1990
BT1 Sites/Areas
RT Holding Ponds
RT Impoundment
RT Sewage Sludge
SO Environmental Protection Agency Glossary
DEF (EPA) Areas where sewage sludge is dumped and allowed to dry out.

EVAPORATOR(S)
(2481 EVAPORATOR)
DA January 3, 1991
BT1 Equipment/Parts - Heating (DOE FRASE Vocabulary)
 BT2 Equipment
BT1 Heating Equipment
SO DOE FRASE VOCABULARY

EVAPORATOR CONDENSATE TANK
DA January 8, 1991
SF ECT
BT1 Tanks
 BT2 Facility Components

EVAPOTRANSPIRATION
(EPA)
DA October 12, 1990
BT1 Biological Processes
 BT2 Processes
RT Plants
RT Transpiration
SO Environmental Protection Agency Glossary
DEF (EPA) The loss of water from the soil both by evaporation and by transpiration from the plants growing in the soil.

EVENT TREE
DA January 8, 1991
SF ET
BT1 Analytical (Logic) Trees
 BT2 Diagrams
RT Events

EVENTS
(DOE Order 5000.3A; SSDC, MORT)
DA October 12, 1990
SY Accidents
NT1 Exposure Event
NT1 Officially Reportable Events
NT1 Security Events
NT1 Trigger Event
NT1 Unusual Events
RT Accidents
RT Assumed Risks
RT Event Tree
RT Incidents
RT Licensee Event Report
RT Occurrences
RT Oversights and Omissions
SO System Safety Development Center Glossary
DEF (DOE Order 5000.3A) Real-time occurrences (e.g. pipe break, valve failure, loss of power, etc.).

BT-Broader Term   NT-Narrower Term   RT Related Term

(SSDC) Occurrences, happenings. Accidents involve a sequence of events that occur in the course of good-intentioned work activity, but that culminate in unintentional injury or damage. (MORT) MORT analysis of an event asks: What happened? Why? and What were the losses e.g. number and type of injuries, amount of property damage, production downtime, product degradation, reduction in employee morale, program impact, negative publicity, etc.

EVENTS AND CAUSAL FACTORS
(SSDC)
DA   October 12, 1990
SF   *E&CF*
RT   Accidental Occurrences
RT   Root Causes
SO   System Safety Development Center Glossary
DEF (SSDC) In MORT, these terms are used together and an E&CF chart (diagram) is used to depict in logical sequence the necessary and sufficient events and causal factors for accident occurrence.

EVIDENCE
(SSDC)
DA   October 12, 1990
SO   System Safety Development Center Glossary
DEF (SSDC) Something that tends to prove; grounds for belief; in law, something legally presented before a court which bears on or establishes the point in question. The kinds of evidence are direct, circumstantial, and real.

EXCAVATION ZONE
(SWDA; RCRA; CFR)
DA   October 12, 1990
BT1 Zones
 BT2 Sites/Areas
RT   Tank Systems
SO   Wastes
DEF (CFR) The volume containing the tank system and backfill material bounded by the ground surface, walls, and floor of the pit and trenches into which the UST (Underground Storage Tank) system is placed at the time of installation.

EXCEPTIONS
(DOE Orders 5480.1B, 5480.18, and 5483.1A; ESH)
DA   October 12, 1990
SY   Variances
RT   Exemptions
RT   Limited Quantity
SO   Industrial Hygiene
SO   Industrial Safety
SO   Management

SO   Standards
DEF Interim releases from a standard of the type specified under the Occupational Safety and Health Act. Exceptions is processed in accordance with DOE 5483.1A. Exceptions shall not exceed 180 days and are not renewable.

EXCESS EMISSIONS
(CAA; CFR)
DA   October 12, 1990
BT1 Particulate Matter Emissions
 BT2 Emissions
  BT3 Air Pollutants
RT   Air Pollution
RT   Standard of Performance
SO   Air Pollution
DEF (CFR) Emissions of an air pollutant in excess of an emission standard.

EXCLUSION AREAS
(DOE Order 6430.1A; AEA; CFR)
DA   October 12, 1990
BT1 Security Areas
 BT2 Sites/Areas
RT   Classified Matter
SO   Construction
SO   Environmental Management
DEF (DOE Order 6430.1A) Security areas for the protection of classified matter where mere access to the area would result in access to classified matter.

EXCLUSIONARY
(EPA)
DA   October 12, 1990
SO   Environmental Protection Agency Glossary
DEF (EPA) Any form of zoning ordinance that tends to exclude specific classes of persons or businesses from a particular district or area.

EXCRETORY SYSTEM
(1143 EXCRETORY SY)
DA   November 28, 1990
BT1 Body System(s)
 BT2 Human Body Parts
SO   DOE FRASE VOCABULARY

EXEMPT OR LIMITED QUANTITIES
(EMER)
DA   February 1, 1991
BT1 Quantities
RT   Hazardous Materials
SO   Emergency Preparedness
DEF (EMER) Quantities of hazardous material which are so limited in magnitude that they are exempt from most requirements specified in the federal regulations.

EXEMPT SOLVENTS
(EPA)
DA   October 12, 1990
BT1 Solvents

SO   Environmental Protection Agency Glossary
DEF (EPA) Specific organic compounds that are not subject to requirements of regulation because they have been deemed by EPA to be of negligible photochemical reactivity.

EXEMPTED AQUIFERS
(SDWA; CFR)
DA   October 12, 1990
BT1 Aquifers
 BT2 Formations
RT   Underground Sources of Drinking Water
SO   Water Pollution
DEF Underground bodies of water defined in the Underground Injection Control program as aquifers that are sources of drinking water (although they are not being used as such) and that are exempted from regulations barring underground injection activities. (CFR) An aquifer or its portion that meets the criteria in the definition of "underground source of drinking water" but which has been exempted according to the procedures of 40 CFR 144.3.

EXEMPTIONS
(DOE Orders 5480.2, 5480.18; TSCA; CFR)
DA   October 12, 1990
NT1 Field Level Exemptions
NT1 Generic Exemptions
NT1 Permanent Exemptions
NT1 Temporary Exemptions
RT   Exceptions
SO   Environmental Management
SO   Standards
DEF (DOE Order 5480.2) Releases from the requirements of this Order or any selected requirements of this Order.

EXERCISES
DA   February 13, 1991
NT1 Drills
NT1 Federal Field Exercises
NT1 Force-on-Force Exercises
NT1 Full-Scale Exercises
NT1 Tabletop
SO   Emergency Preparedness
DEF (WTID) Actions characterized by maneuvers, drills, and other repetitive operations; and carried out for testing, training, or discipline.

EXHAUST GASES
(CFR)
DA   October 12, 1990
BT1 Emissions
 BT2 Air Pollutants
BT1 Gases
RT   Air Pollution

RT Stacks
SO Air Pollution
DEF (CFR) Any offgases (the
    constituents of which may consist
    of any fluids, either as a liquid
    and/or gas) discharged directly or
    ultimately to the atmosphere that
    were initially contained in or were
    in direct contact with the
    equipment for which exhaust gas
    limits are prescribed in 40 CFR
    61.62 (a) and (b); 61.63(a); 61.64
    (a)(1), (a)(2), (b), (c), and (d);
    61.65(b) (1)(ii), (b)(2), (b)(5),
    (b)(6)(ii) and (b)(9)(ii).

EXHAUST STACK
(2560 EXHAUST STAC)
DA   January 3, 1991
SY   Stack Effect
BT1 Equipment/Parts - Nuclear (DOE
     FRASE Vocabulary)
 BT2 Equipment
 BT2 Reactor Components
SO   DOE FRASE VOCABULARY

EXISTING CLASS II WELLS
(SDWA; CFR)
DA   October 12, 1990
BT1 Class II Wells
 BT2 Injection Wells
  BT3 Wells
SO   Water Pollution
DEF Wells that were authorized by BIA
     (Bureau of Indian Affairs) and
     constructed and completed before
     the effective date of this program.

EXISTING FACILITIES
(SWDA; ESH)
DA   November 9, 1990
SY   Existing Hazardous Waste
     Management (HWM) Facilities
BT1 Facilities
NT1 Reconstructed Sources
SO   Wastes
DEF Facilities which were in operation or
     for which construction commenced
     on or before November 19, 1980.
     A facility has commenced
     construction if (a) the owner or
     operator has obtained the Federal,
     State and local approvals or
     permits necessary to begin
     physical construction; and either
     (b)(1) a continuous on-site,
     physical construction program has
     begun; or (2) the owner or
     operator has entered into
     contractual obligations which
     cannot be cancelled or modified
     without substantial loss – for
     physical construction of the facility
     to be completed within a
     reasonable time.

EXISTING HAZARDOUS WASTE
MANAGEMENT (HWM) FACILITIES
(SWDA; RCRA; CFR; ESH)

DA   October 12, 1990
SY   Existing Facilities
BT1 Hazardous Waste Management
     Facilities
 BT2 Hazardous Waste Facilities
  BT3 Facilities
SO   Environmental Management
SO   Wastes
DEF (CFR) Facilities which were in
     operation or for which construction
     commenced on or before
     November 19, 1980. A facility has
     commenced construction if: (a) the
     owner or operator has obtained
     the Federal, State and local
     approvals or permits necessary to
     begin physical construction; and
     either (b)(1) a continuous on-site,
     physical construction program has
     begun; or (2) the owner or
     operator has entered into
     contractual obligations which
     cannot be cancelled or modified
     without substantial loss – for
     physical construction of the facility
     to be completed within a
     reasonable time.

EXISTING INJECTION WELLS
(SDWA; CFR)
DA   October 12, 1990
BT1 Injection Wells
 BT2 Wells
SO   Water Pollution
DEF An injection well other than a new
     injection well.

EXISTING SOURCES
(CAA; CFR; USC)
DA   October 12, 1990
SY   Existing Stationary Facilities
BT1 Stationary Sources
 BT2 Facilities
RT   Reasonably Available Control
     Technology
SO   Air Pollution
SO   Environmental Management
DEF (CFR) Stationary sources that are
     not new sources.

EXISTING STATIONARY FACILITIES
(CAA; CFR)
DA   October 12, 1990
SY   Existing Sources
BT1 Stationary Sources
 BT2 Facilities
SO   Air Pollution
DEF (CFR) Any of the following
     stationary sources of air pollutants,
     including any reconstructed
     source, which was not in operation
     prior to August 7, 1962, and was
     in existence on August 7, 1977,
     and has the potential to emit 250
     tons per year or more of any air
     pollutant. In determining potential
     to emit, fugitive emissions, to the
     extent quantifiable, must be
     counted. See listing at 40 CFR
     51.301 (e) et seq.

EXISTING TAILINGS PILES
(CAA; CFR)
DA   October 12, 1990
RT   New Tailings
RT   Piles (Wastes)
RT   Tailings
SO   Air Pollution
DEF (CFR) Tailings piles that are in
     operation on the effective date of
     this rule (40 CFR 61).

EXISTING TANK SYSTEMS
(SWDA; RCRA; ESA; CFR)
DA   October 12, 1990
BT1 Tank Systems
SO   Environmental Management
SO   Wastes
DEF (CFR) Tank systems used to
     contain an accumulation of
     regulated substances or for which
     installation has commenced on or
     before December 22, 1988.
     Installation is considered to have
     commenced if (a) the owner or
     operator has obtained all federal,
     state, and local approvals or
     permits necessary to begin
     physical construction of the site or
     installation of the tank system; and
     if, (b)(1) either a continuous
     on-site physical construction or
     installation program has begun; or,
     (2) the owner or operator has
     entered into contractual obligations
     – which cannot be cancelled or
     modified without substantial loss –
     for physical construction at the site
     or installation of the tank system
     to be completed within a
     reasonable time.

EXIT
(1696 EXIT;N)
DA   December 10, 1990
BT1 Site (DOE FRASE Vocabulary)
 BT2 Sites/Areas
SO   DOE FRASE VOCABULARY

EXPANSION JOINTS
(DOE Order 6430.1A)
DA   October 12, 1990
SO   Construction
DEF (DOE Order 6430.1A) Joints
     between parts of a structure to
     avoid distortion when subjected to
     temperature change.

EXPERIMENTAL BOILING WATER
REACTOR
DA   January 8, 1991
SF   *EBWR*
BT1 Reactors

EXPERIMENTAL POPULATIONS
(ESA; CFR)
DA   October 12, 1990
BT1 Populations
NT1 Essential Experimental Populations

SO Endangered Species
DEF (CFR) Introduced and/or designated populations (including any offspring arising solely therefrom) that have been so designated in accordance with the procedures of this subpart but only when, and at such times as the populations are wholly separate geographically from nonexperimental populations of the same species. Where part of an experimental population overlaps with natural populations of the same species on a particular occasion, but is wholly separate at other times, specimens of the experimental population will not be recognized as such while in the area of overlap. That is, experimental status will only be recognized outside the areas of overlap. Thus, such a population shall be treated as experimental only when the times of geographic separation are reasonably predictable; e.g., fixed migration patterns, natural or man-made barriers. A population is not treated as experimental if total separation will occur solely as a result of random and unpredictable events.

EXPERIMENTAL START DATE
(TSCA)
DA October 19, 1990
BT1 Time Designations
SO Hazardous Materials
DEF (TSCA) The first date the test substance is applied to the test system.

EXPERIMENTAL TERMINATION DATE
(TSCA)
DA October 19, 1990
BT1 Time Designations
SO Hazardous Materials
DEF (TSCA) The last date on which data are collected directly from the study.

EXPLOSION SUPPRESSION SYSTEM
(2032 ESS)
DA December 10, 1990
BT1 Emergency Systems
 BT2 Systems
BT1 Systems (DOE FRASE Vocabulary)
 BT2 Systems
SO DOE FRASE VOCABULARY

EXPLOSIVE BOMBS
(HMTA; CFR)
DA May 20, 1991
BT1 Munitions
SO Hazardous Materials
DEF (CFR) Explosive bombs are metal or other containers filled with explosives. They are used in warfare and include aeroplane bombs and depth bombs

EXPLOSIVE MINES
(HMTA; CFR)
DA May 20, 1991
BT1 Munitions
SO Hazardous Materials
DEF (CFR) Explosive mines are metal or composition containers filled with a high explosive.

EXPLOSIVE ORDNANCE DISPOSAL TEAMS
(EMER)
DA February 1, 1991
BT1 Response Teams
 BT2 Teams
 BT3 Administrative Organizations
 BT4 Organizations
RT Explosives
RT Explosives Ordnance Disposal
SO Emergency Preparedness
DEF (EMER) Teams of trained explosive ordnance personnel. Responsibilities include identification, examination, and disposal of explosive ordnance.

EXPLOSIVE PROJECTILES
(HMTA; CFR)
DA May 20, 1991
BT1 Munitions
RT Explosives
SO Hazardous Materials
DEF (CFR) Explosive projectiles are shells, projectiles, warheads, or rocket heads, loaded with explosives or bursting charges, with or without other materials, for use in cannons, guns, tubes, mortars or other firing or launching devices.

EXPLOSIVE TORPEDOES
(HMTA; CFR)
DA May 20, 1991
BT1 Munitions
SO Hazardous Materials
DEF (CFR) Explosive torpedoes, such as are used in warfare, are metal devices containing a means of propulsion and a quantity of high explosives.

EXPLOSIVES
(DOE Order 6430.1A)
DA October 12, 1990
NT1 Blasting Agents
NT1 Cased Explosives
NT1 Class A Explosives
NT1 Class B Explosives
 NT2 Starter Cartridges
NT1 Class C Explosives
NT1 High Explosives
 NT2 Insensitive High Explosives
RT Bomb Threats
RT Bureau of Explosives
RT Deflagration

RT Detonators
RT Detonations
RT Explosive Ordnance Disposal Teams
RT Explosives Buildings
RT Explosives Hazard Classes
RT Explosive Projectiles
RT Hazards
RT Ignitability
RT Initiation Stimuli
RT Munitions
SO Construction
DEF (DOE Order 6430.1A) Chemical compounds or mechanical mixtures that, when subjected to heat, impact, friction, shock, or other suitable initiation stimulus, undergo a very rapid chemical change with the evolution of large volumes of highly heated gases that exert pressures in the surrounding medium. The term applies to materials that either detonate or deflagrate.

EXPLOSIVES ACTIVITIES
(DOE Order 6430.1A)
DA October 12, 1990
BT1 Activities
RT Explosives Hazard Classes
RT Magazine Vessels
RT Support Buildings
SO Construction
DEF (DOE Order 6430.1A) Functions (storage, handling, and processing) involving explosives from the manufacture or receipt of the explosives through the final shipping, configuration, including final storage but excluding the movement of explosives between explosives areas.

EXPLOSIVES BAYS
(DOE Order 6430.1A)
DA October 12, 1990
BT1 Explosives Buildings
 BT2 Structures (DOE FRASE Vocabulary)
NT1 Staging Bays (in-process)
SO Construction
DEF (DOE Order 6430.1A) Locations (room, cubicle, cell, work area) containing a single type of explosives activity that affords the requirement protection for the appropriate hazard classification (Class I, II, II, IV) of the explosives activity involved. Examples of such explosives activities are machining, pressing, meltcasting, nondestructive testing, and assembly operations.

EXPLOSIVES BUILDINGS
(DOE Order 6430.1A)
DA October 12, 1990
BT1 Structures (DOE FRASE Vocabulary)
NT1 Explosives Bays

NT2 Staging Bays (in-process)
RT Explosives
RT Explosives Hazard Classes
RT Inhabited Building Distance
RT Occupied Area (Explosives)
SO Construction
DEF (DOE Order 6430.1A) Structures containing one or more explosives bays.

EXPLOSIVES HAZARD, CLASS I

(DOE Order 6430.1A)
DA October 12, 1990
BT1 Explosives Hazard Classes
SO Construction
DEF (DOE Order 6430.1A) Class I consists of those explosives activities involving a high potential for an accident that is unacceptable for the exposure of any personnel, thus requiring remote operations. In general, this would include activities where the energies that may interface with the explosives are approaching the upper limits of safety, and/or loss of control of the energy is likely to exceed the safety limits for the explosives involved. This category includes those research and development activities where the safety implications have not been fully characterized. Examples of Class I activities are screening, blending, pressing, extrusion, drilling of holes, dry machining, some wet machining, machining explosives and metal in combination, development of some new explosives or explosives processing methods, and explosives disposal.

EXPLOSIVES HAZARD, CLASS II

(DOE Order 6430.1A)
DA October 12, 1990
BT1 Explosives Hazard Classes
SO Construction
DEF (DOE Order 6430.1A) Class II consists of those explosives activities that involve a moderate potential for an accident because of type of explosives, condition of the explosives and/or nature of the operations involved. This category consists of activities where the accident potential is greater than Class III, but exposure of personnel performing contact operations is acceptable. Included are activities where energies that do or may interface with the explosives are normally well within the safety boundaries for the explosives involved but where loss of control of these energies might approach the safety limits of the explosives. Examples of Class II activities involving HE are weighing, some

wet machining, assembly and disassembly, and environmental testing (exposure of explosives samples to variations in temperature, humidity, etc.).

EXPLOSIVES HAZARD, CLASS III

(DOE Order 6430.1A)
DA October 12, 1990
BT1 Explosives Hazard Classes
SO Construction
DEF (DOE Order 6430.1A) Class III consists of those explosives activities that represent a low potential for an accident because of the type of explosives, the conditions of the explosives and/or the nature of the activity involved. Class III includes explosives activities where the accident potential of the operation being performed is not significantly different from explosives storage. Examples are normal handling, storage, packaging, unpackaging, and some inspection and nondestructive testing.

EXPLOSIVES HAZARD, CLASS IV

(DOE Order 6430.1A)
DA October 12, 1990
BT1 Explosives Hazard Classes
SO Construction
DEF (DOE Order 6430.1A) Class IV consists of those explosives activities with insensitive high explosives (IHE) or IHE subassemblies that, although mass detonating, are so insensitive that there is negligible probability for accidental initiation or transition from burning to detonation. Explosions will be limited to pressure ruptures of containers heated in a fire. Although the fire hazards of IHE and IHE subassemblies are not as great as those of other explosives, it is classified as hazard class/division 1.3 (mass fire) to be consistent with DOD 6055.9. Most processing and storage activities with IHE and IHE subassemblies are class IV. However, the following are examples of explosive activities with IHE or IHE subassemblies that remain class I: pressing, some machining (see DOE/EV-06194), dry blending, dry milling, and dry screening.

EXPLOSIVES HAZARD CLASSES

(DOE Order 6430.1A)
DA October 12, 1990
NT1 Explosives Hazard, Class I
NT1 Explosives Hazard, Class II
NT1 Explosives Hazard, Class III
NT1 Explosives Hazard, Class IV
RT Explosives
RT Explosives Activities

RT Explosives Buildings
SO Construction
DEF (DOE Order 6430.1A) The level of protection required for any specific explosives activity, based on the hazard class (accident potential) for the explosives activity involved.

EXPLOSIVES ORDNANCE DISPOSAL

DA January 8, 1991
SF *EOD*
RT Explosive Ordnance Disposal Teams

EXPOSURE

(EPA; NIH; IAEA)
DA October 12, 1990
NT1 Acute Exposure
NT1 Chronic Exposure
NT1 Emergency Exposure
NT1 Environmental Exposure
NT1 Irradiation
NT1 Medical Exposure
NT1 Occupational Exposure
NT1 Planned Special Exposure
NT1 Positive Exposure
RT Chronic Effects
RT Doses
RT Exposure Rate
RT Exposure Assessment
RT Exposure Event
RT Exposure Pathways
RT Exposure Point
RT Exposure Route
RT Film Badges
RT Ingestion Exposure Pathways
RT Ionizing Radiation
RT Lethal Concentration Low
RT Lethal Dose Low
RT Lethal Dose of Radiation
RT Level of Concern
RT Median Lethal Concentration
RT Radiation Areas
RT Risks
RT Routes (Exposure)
RT Slope Factor
RT Working Level Month
SO Environmental Protection Agency Glossary
SO Radiation
DEF (1) The amount of radiation or pollutant present in an environment which represents a potential health threat to the living organisms in that environment. (2) Contact of an organism with a chemical or physical agent. Exposure is quantified as the amount of the agent available at the exchange boundaries of the organism (e.g., skin, lungs, gut) and available for absorption. (NIH) A measure of the ionization produced in air by x or gamma radiation. It is the sum of the electrical charges on all ions of one sign produced in air when all electrons liberated by photons in a volume element of air are completely stopped in air, divided

by the mass of air in the volume element. The special unit of exposure is the roentgen.

EXPOSURE ASSESSMENT
(EPA)
DA   October 12, 1990
BT1 Assessments
 BT2 Administrative Processes
 BT3 Processes
RT   Exposure
SO   Environmental Protection Agency
      Glossary
DEF (EPA) The determination or
      estimation (qualitative or
      quantitative) of the magnitude,
      frequency, duration, and route of
      exposure.

EXPOSURE EVENT
(EPA)
DA   October 12, 1990
BT1 Events
RT   Exposure
RT   Exposure Route
SO   Environmental Protection Agency
      Glossary
DEF (EPA) An incident of contact with a
      chemical or physical agent. An
      exposure event can be defined by
      time (e.g., day, hour) or by the
      incident (e.g., eating a single meal
      of contaminated fish).

EXPOSURE PATHWAYS
(EPA; IAEA)
DA   October 12, 1990
BT1 Routes (Exposure)
NT1 Ingestion Exposure Pathways
NT1 Plume Exposure Pathways
RT   Bioavailability
RT   Common Exposure Routes
RT   Critical Pathways
RT   Exposure
RT   Exposure Route
RT   Food Chains
SO   Environmental Protection Agency
      Glossary
SO   Radiation
DEF (EPA) The course a chemical or
      physical agent takes from the
      source to the exposed organism.
      An exposure pathway describes a
      unique mechanism by which an
      individual or population is exposed
      to chemicals or physical agents at
      or originating from the site. Each
      exposure pathway includes a
      source or release from a source,
      an exposure point, and an
      exposure route. If the exposure
      point differs from the source, a
      transport/exposure medium (e.g.,
      air) or media (in cases of
      intermedia transfer) also is
      included.

EXPOSURE POINT
(EPA)

DA   October 12, 1990
NT1 Total Exposure Points
RT   Exposure
RT   Exposure Point Concentration
SO   Environmental Protection Agency
      Glossary
DEF (EPA) A point of potential contact
      between an organism and a
      chemical or physical agent.

EXPOSURE POINT CONCENTRATION
(EPA)
DA   October 12, 1990
BT1 Measurements
RT   Exposure Point
SO   Environmental Protection Agency
      Glossary
DEF (EPA) The concentration of a
      chemical at the exposure point.

EXPOSURE RATE
(IAEA)
DA   October 12, 1990
BT1 Rates
RT   Contact Rate
RT   Exposure
SO   Radiation
DEF (IAEA) The increment of exposure
      in a specified time interval.

EXPOSURE ROUTE
(EPA)
DA   October 12, 1990
BT1 Routes (Exposure)
RT   Exposure
RT   Exposure Event
RT   Exposure Pathways
SO   Environmental Protection Agency
      Glossary
DEF (EPA) The way a chemical or
      physical agent comes in contact
      with an organism (i.e., by
      ingestion, inhalation, or dermal
      contact).

EXTERNAL CORROSION
(DOE Order 6430.1A)
DA   October 12, 1990
BT1 Corrosion
 BT2 Chemical Processes
 BT3 Processes
SO   Construction
DEF (DOE Order 6430.1A) Corrosion of
      that portion of a metal structure
      (i.e., pipe) that is exposed to
      external elements such as air,
      water, or soil.

EXTERNAL FISSION COUNTER
DA   January 8, 1991
SF   EFC
BT1 Radiation Detectors
 BT2 Equipment
RT   Count (Radiation Measurement)

EXTRACTION PLANTS
(CAA; CFR)
DA   October 12, 1990
BT1 Facilities

RT   Benefaction
RT   Beryllium Ores
SO   Air Pollution
DEF (CFR) Facilities chemically
      processing beryllium ore to
      beryllium metal, alloy, or oxide, or
      performing any of the intermediate
      steps in these processes.

EXTRACTION PROCEDURE TOXIC
DA   January 8, 1991
SF   EP Toxic
BT1 Procedures

EXTRACTION SITES
(CWA; CFR)
DA   October 12, 1990
BT1 Sites/Areas
SO   Water Pollution
DEF (CFR) The places from which the
      dredged or fill material proposed
      for discharge is to be removed.

EXTRAORDINARY NUCLEAR OCCURRENCES
(AEA; ANL)
DA   May 24, 1991
BT1 Nuclear Incidents
 BT2 Occurrences
RT   Atomic Energy Act
RT   Atomic Energy Commission
SO   Radiation
SO   Wastes
DEF (AEA) Any event causing a
      discharge or dispersal of source,
      special nuclear, or byproduct
      material from its intended place of
      confinement in amounts offsite, or
      causing radiation levels offsite,
      which the Commission determines
      to be substantial and which the
      Commission determines has
      resulted or will probably result in
      substantial damages to persons
      offsite or property offsite. Any
      determination by the Commission
      that such an event has or has not,
      occurred shall be final and
      conclusive, and no other official or
      any court shall have power or
      jurisdiction to review any such
      determination. The Commission
      shall establish criteria in writing
      setting forth the basis upon which
      such determination shall be made.
      As used in this "subsection,"
      "offsite" means away from the "the
      location" or the contract location
      as defined in the applicable
      Commission indemnity agreement,
      entered into pursuant to section
      170.

EXTREME VALUE
(SSDC)
DA   October 12, 1990
BT1 Measurements
RT   Extreme Value Projection

SO System Safety Development Center
       Glossary
DEF (SSDC) The largest observation
       during a given period of
       observation.

EXTREME VALUE PROJECTION
(SSDC)
DA   October 12, 1990
BT1 Measurements
RT   Derived Projection
RT   Extreme Value
RT   Risk Assessment
SO   System Safety Development Center
       Glossary
DEF (SSDC) Risk projection based upon
       the worst event each period over a
       number of periods, such as the
       worst property damage case each
       6 months over a period of 5 years.

EXTREMELY HAZARDOUS
SUBSTANCES
(CERCLA; CFR; EMER)
DA   October 12, 1990
BT1 Hazardous Substances
BT1 Hazardous Chemicals
 BT2 Hazardous Substances
RT   CERCLA Hazardous Substances
RT   Listed Hazardous Substances
RT   Threshold Planning Quantities
SO   Compensation and Liability
SO   Emergency Preparedness
DEF (EMER) Any of 406 chemicals
       identified by EPA on the basis of
       toxicity, and listed under SARA
       (Superfund Amendments and
       Reauthorization Act of 1986) Title
       III. The list is subject to revision.

EXTREMELY HAZARDOUS WASTES
(EMER)
DA   February 1, 1991
BT1 Hazardous Wastes
 BT2 Hazardous Materials
  BT3 Materials
 BT2 Wastes
SO   Emergency Preparedness
DEF (EMER) Any dangerous wastes
       which will persist in a hazardous
       form for several years or more at
       a disposal site. It presents a
       significant environmental hazard,
       may be concentrated by living
       organisms through a food chain,
       or may affect the genetic makeup
       of humans or wildlife. It is highly
       toxic to humans or wildlife if
       disposed of in such quantities as
       would present an extreme hazard
       to humans or the environment.

EXTREMITIES
(Doe Order 5480.11)
DA   October 12, 1990
BT1 Human Body Parts
RT   Whole Body
SO   Industrial Hygiene
DEF (DOE Order 5480.11) Extremities

include hands and arms below the
elbow or feet and legs below the
knee.

EYE(S)
(1084 EYE)
DA   November 28, 1990
BT1 Face
 BT2 Head
  BT3 Human Body Parts
RT   Area Around Eye
RT   Eye Protection
RT   Eyelid(s)
RT   Goggles
RT   Safety Glasses
RT   Safety Glasses W Side Shields
RT   Side Shields
SO   DOE FRASE VOCABULARY

EYE PROTECTION
(2658 EYE PROTECTI)
DA   January 3, 1991
BT1 Personal Protective Equipment
 BT2 Equipment/Parts - Personal
       Protective (DOE FRASE
       Vocabulary)
  BT3 Equipment
NT1 Goggles
NT1 Safety Glasses
 NT2 Safety Glasses W Side Shields
 NT2 Tinted Safety Glasses
NT1 Side Shields
RT   Area Around Eye
RT   Eye(s)
RT   Eyelid(s)
SO   DOE FRASE VOCABULARY

EYELID(S)
(1082 EYELID)
DA   November 28, 1990
BT1 Area Around Eye
 BT2 Face
  BT3 Head
   BT4 Human Body Parts
RT   Conjunctivitis
RT   Eye Protection
RT   Eye(s)
SO   DOE FRASE VOCABULARY

F-V
DA   October 12, 1990
SEE Fussell-Vesely
SO   Acronyms

F/S
DA   October 12, 1990
SEE Frequency/Severity
SO   Acronyms

FAA
DA   October 12, 1990
SEE Federal Aviation Administration
SO   Acronyms

FABRIC FILTERS
(EPA)
DA   October 12, 1990
BT1 Filters

 BT2 Devices
SO   Environmental Protection Agency
       Glossary
DEF (EPA) Cloth devices that catch dust
       particles from industrial emissions.

FABRICATING
(CAA; CFR)
DA   October 12, 1990
BT1 Manufacturing Processes
 BT2 Processes
RT   Commercial Asbestos
SO   Air Pollution
DEF (CFR) Any processing of a
       manufactured product that
       contains commercial asbestos,
       with the exception of processing at
       temporary sites for the
       construction or restoration of
       facilities.

FACE
(1081 FACE)
DA   November 28, 1990
BT1 Head
 BT2 Human Body Parts
NT1 Area Around Eye
 NT2 Eyelid(s)
NT1 Eye(s)
NT1 Lip(s)
NT1 Nose
RT   Faceshield
RT   Mouth
SO   DOE FRASE VOCABULARY

FACESHIELD
(2659 FACESHIELD)
DA   January 3, 1991
BT1 Head Protection
 BT2 Personal Protective Equipment
  BT3 Equipment/Parts - Personal
       Protective (DOE FRASE
       Vocabulary)
   BT4 Equipment
RT   Face
RT   Forehead
SO   DOE FRASE VOCABULARY

FACILITIES
(DOE Order 5000.3; SWDA; CERCLA;
 CAA; CFR; ESH; EMER)
DA   October 12, 1990
NT1 Affected Facilities
NT1 Air Force Base
NT1 Air Force Engineering and Services
       Center
NT1 Asbestos Mills
NT1 Atomic Energy Act Facilities
NT1 Ceramic Plants
NT1 Classified Telecommunications
       Facilities
NT1 Command Posts
NT1 Consolidated Incinerator Facility
NT1 Contract Labs
NT1 Critical Facilities
NT1 Defense Waste Processing Facility
NT1 Direct Dischargers
NT1 DOE Nuclear Weapons Complex
NT1 DOE Property

NT1 Elemental Phosphorus Plants
NT1 Essential Facilities
NT1 Ethylene Dichloride Plants
NT1 Existing Facilities
 NT2 Reconstructed Sources
NT1 Extraction Plants
NT1 Facilities and Buildings (DOE
     FRASE Vocabulary)
 NT2 Building (DOE FRASE Vocabulary)
  NT3 Containment Building
  NT3 Demineralizer Plant
  NT3 Fire Station
  NT3 Garage
  NT3 Guard Station
  NT3 Hospital
  NT3 Hot Shop
  NT3 Machine Shop
  NT3 Maintenance Shop
  NT3 Motel/Hotel
  NT3 Research & Development
       Laboratory
  NT3 Restaurant
  NT3 Sewage Plant
  NT3 Steam Plant
 NT2 Facility (DOE FRASE Vocabulary)
  NT3 Fuel Fabrication Facility
  NT3 Fusion Facility
  NT3 Hot Cell Facility
  NT3 Irradiated Fuel Process Facility
  NT3 Irradiated Fuel Storage Facility
  NT3 Low Power Reactor Facility
  NT3 Nuclear Facility
  NT3 Particle Accelerator Facility
  NT3 Production Reactor Facility
  NT3 Rad Waste Treatment/Storage
       Facility
  NT3 Test Reactor Facility
  NT3 Waste Monitoring Facility
NT1 Fast Flux Test Facility
NT1 Federal Facilities
 NT2 Federal Class I areas
  NT3 Mandatory Class I Federal Areas
 NT2 Government-Owned
     Contractor-Operated Facilities
  NT3 Argonne National Laboratory
   NT4 Argonne National
        Laboratory-East (Chicago)
   NT4 Argonne National
        Laboratory-West (At INEL)
  NT3 Energy Technology Engineering
       Center (Canoga Park)
  NT3 Feed Materials Production Center
  NT3 Idaho Chemical Processing Plant
  NT3 Idaho National Engineering
       Laboratory
   NT4 Power Burst Facility
  NT3 Inhalation Toxicology Research
       Institute
  NT3 Kansas City Plant
  NT3 Lawrence Livermore National
       Laboratory
  NT3 Lawrence Berkeley Laboratory
  NT3 Los Alamos National Laboratory
  NT3 Nevada Test Site
  NT3 Oak Ridge Gaseous Diffusion
       Plant
  NT3 Oak Ridge National Laboratory
  NT3 Pacific Northwest Laboratory
  NT3 Paducah Gaseous Diffusion Plant
       (Paducah)

  NT3 Portsmouth Gaseous Diffusion
       Plant
  NT3 Rocky Flats Plant
  NT3 Sandia National Laboratories
   NT4 Sandia National
        Laboratories-Alburquerque
   NT4 Sandia National
        Laboratories-Livermore
  NT3 Savannah River Site
  NT3 Stanford Linear Accelerator
       Center
  NT3 Strategic Petroleum Reserve
  NT3 Waste Isolation Pilot Plant
  NT3 Y-12 Plant
NT1 Fixed Sources
NT1 Foundries
NT1 Gas Centrifuge Enrichment Plant
NT1 Gaseous Diffusion Plant
NT1 Grand Gulf Nuclear Station
NT1 Hazardous Waste Facilities
 NT2 Hazardous Waste Management
     Facilities
  NT3 Disposal Facilities
   NT4 Land Disposal Facilities
    NT5 Near Surface Disposal Facilities
   NT4 Repositories
   NT4 Test and Evalution Facilities
  NT3 Existing Hazardous Waste
       Management (HWM) Facilities
  NT3 New Hazardous Waste
       Management (HWM) Facilities
  NT3 Off-Site Facilities
  NT3 On-Site Facilities
 NT2 Totally Enclosed Treatment
     Facilities
NT1 Inactive Facilities
NT1 Industrial Buildings
NT1 Land Treatment Facilities
NT1 Major Facilities
NT1 Mercury Ore Processing Facilities
NT1 Model Plants
NT1 Natural Gas Processing Plants
NT1 New Facilities
NT1 Nitric Acid Plants
NT1 Non-Fractionating Plants
NT1 Non-Nuclear Facilities
NT1 NRC-Licensed Facilities
NT1 Nuclear Facilities
 NT2 Fixed Nuclear Facilities
 NT2 New Storage Facilities
 NT2 Nuclear Power Plants
 NT2 Plutonium Processing and
     Handling Facilities
 NT2 Plutonium Storage Facilities
 NT2 Reactor Facilities
  NT3 Low Power Reactor Facility
  NT3 Production Reactor Facility
  NT3 Test Reactor Facility
 NT2 Waste Isolation Pilot Plant
NT1 Offshore Facilities
NT1 On-Shore Facilities
NT1 Pipeline Facilities
NT1 Polyvinyl Chloride Plants
 NT2 Vinyl Chloride Plants
NT1 Portsmouth Uranium Enrichment
     Complex
NT1 Pretreatment Facilities
NT1 Process Experimental Pilot Plant
NT1 Production Facilities
NT1 Propellant Plants
NT1 Radio Repeater Stations

NT1 Resource Recovery Facilities
NT1 Retrieval Containment Building
NT1 Security Facilities
NT1 Shelters
NT1 Significant Municipal Facilities
NT1 Smelters
 NT2 Primary Copper Smelter
NT1 Sodium Components Test
     Installation
NT1 Solid Waste Management Facilities
NT1 Stationary Sources
 NT2 Existing Stationary Facilities
 NT2 Existing Sources
 NT2 Major Stationary Source
  NT3 Major Emitting Facilities
  NT3 Major PSD Stationary Source
 NT2 Modified Sources
 NT2 New Sources
 NT2 Point Sources
NT1 Sulfuric Acid Plants
NT1 Support Buildings
NT1 Surplus Facilities
NT1 Transfer Facilities
NT1 Transuranic Waste Facility
NT1 Transuranic Waste Treatment and
     Storage Facility
NT1 Treatment Facilities
 NT2 Grout Treatment Facility
 NT2 Land Treatment Facilities
 NT2 Publicly Owned Treatment Works
  NT3 Water Systems
   NT4 Public Water Systems
    NT5 Community Water Systems
    NT5 Drinking Water System
    NT5 Non-Community Water
         Systems
    NT5 Non-Transient Non-Community
         Water System
    NT5 Water Supply System
   NT4 Storm Water Collection Systems
   NT4 Waste Water System
    NT5 Waste Water Collection
         Systems
 NT2 Totally Enclosed Treatment
     Facilities
NT1 Treatment, Storage, and Disposal
     Facilities
 NT2 Designated Facilities
 NT2 Totally Enclosed Treatment
     Facilities
NT1 Turbine Building
NT1 Utilization Facilities
NT1 Vital Facilities
NT1 Waste Treatment Plants
 NT2 Waste Experimental Reduction
     Facility
 NT2 Waste Handling and Packaging
     Plant
 NT2 Waste Receiving and Processing
     Plant
  NT3 Mixed Waste Management
       Facility
 RT Accident Sites
 RT Authorized Representatives
 RT Balance of Plant
 RT Base Shear
 RT Bases
 RT Facility Components
 RT Facility Authority
 RT Facility Boundaries
 RT Facility Mailing Lists

RT   Functional Units
RT   Real Property
RT   Self-Assessment
RT   Sites/Areas
RT   Siting
RT   Structures (DOE FRASE
     Vocabulary)
RT   Targets
SO   Air Pollution
SO   Compensation and Liability
SO   Construction
SO   Emergency Preparedness
SO   Environmental Management
SO   Hazardous Materials
SO   Management
SO   Wastes
SO   Water Pollution
DEF (ESH) Systems, buildings, utilities,
     services, and related activities
     whose use is directed to a
     common purpose at a single
     location. Examples include
     accelerators, storage areas, test
     loops, nuclear reactors, coal
     conversion plants,
     magnetohydrodynamics (MHD)
     experiments, windmills, radioactive
     waste disposal systems and burial
     grounds, testing laboratories,
     research laboratories, and
     accommodations for analytical
     examinations of irradiated and
     unirradiated components. Also
     includes pipelines, ponds,
     impoundments, landfills and the
     like, and motor vehicles, rolling
     stock, and aircraft./ Also see any
     appropriate statute or regulation
     for specific definition.

**FACILITIES AND BUILDINGS (DOE
FRASE VOCABULARY)**
(DOE FRASE Vocabulary Numeric Keys
1775-1849)
DA   January 17, 1991
BT1 Facilities
NT1 Building (DOE FRASE Vocabulary)
NT2 Containment Building
NT2 Demineralizer Plant
NT2 Fire Station
NT2 Garage
NT2 Guard Station
NT2 Hospital
NT2 Hot Shop
NT2 Machine Shop
NT2 Maintenance Shop
NT2 Motel/Hotel
NT2 Research & Development
     Laboratory
NT2 Restaurant
NT2 Sewage Plant
NT2 Steam Plant
NT1 Facility (DOE FRASE Vocabulary)
NT2 Fuel Fabrication Facility
NT2 Fusion Facility
NT2 Hot Cell Facility
NT2 Irradiated Fuel Process Facility
NT2 Irradiated Fuel Storage Facility
NT2 Low Power Reactor Facility
NT2 Nuclear Facility

NT2 Particle Accelerator Facility
NT2 Production Reactor Facility
NT2 Rad Waste Treatment/Storage
     Facility
NT2 Test Reactor Facility
NT2 Waste Monitoring Facility
RT   DOE FRASE Categories
RT   Targets
SO   DOE FRASE VOCABULARY
DEF A subject category used with the
     DOE FRASE Vocabulary.

FACILITY (DOE FRASE VOCABULARY)
(1778 Facility)
DA   January 31, 1991
BT1 Facilities and Buildings (DOE
     FRASE Vocabulary)
BT2 Facilities
NT1 Fuel Fabrication Facility
NT1 Fusion Facility
NT1 Hot Cell Facility
NT1 Irradiated Fuel Process Facility
NT1 Irradiated Fuel Storage Facility
NT1 Low Power Reactor Facility
NT1 Nuclear Facility
NT1 Particle Accelerator Facility
NT1 Production Reactor Facility
NT1 Rad Waste Treatment/Storage
     Facility
NT1 Test Reactor Facility
NT1 Waste Monitoring Facility
SO   DOE FRASE VOCABULARY

FACILITY AUTHORITY
(DOE Order 6430.1A)
DA   October 12, 1990
BT1 Personnel
RT   Facilities
SO   Construction
DEF (DOE Order 6430.1A) The
     individual, designated by the DOE
     project manager, developing
     specific project criteria not
     contained in the DOE 6430.1A.

FACILITY BOUNDARIES
(DOE Order 6430.1A)
DA   October 12, 1990
BT1 Structures (DOE FRASE
     Vocabulary)
RT   Facilities
RT   Tension Wires
SO   Construction
DEF (DOE Order 6430.1A) Fences or
     other barriers that surround and
     prevent uncontrolled access to the
     facility or facilities.

FACILITY COMPONENTS
(CAA; CFR)
DA   October 12, 1990
NT1 Industrial Furnaces
NT2 Glass Melting Furnaces
NT3 Pot Furnaces
NT1 New Tank Components
NT1 Piping
NT2 Connected Piping
NT2 Vitrified Clay Pipe
NT1 Reactors (Chemical)

NT1 Tanks
NT2 Aboveground Tanks
NT2 Aeration Tanks
NT2 Boron Injection Tank
NT2 Cargo Tanks
NT2 Clarifiers
NT2 Condensate Storage Tank
NT2 Containment Storage Tanks
NT2 Coolant Return Tank
NT2 Core Flood Tanks
NT2 Double-shell Tank
NT2 Evaporator Condensate Tank
NT2 Farm Tanks
NT2 Flow-Through Process Tanks
NT2 Fuel Tanks
NT2 Hydraulic Lift Tanks
NT2 Inground Tanks
NT2 On Ground Tanks
NT2 Portable Tanks
NT3 Intermodal Portable Tanks
NT2 Quench Tanks
NT2 Refueling Water Storage Tank
NT2 Residential Tanks
NT2 Safety Injection Tank
NT2 Sedimentation Tanks
NT2 Settling Tanks
NT2 Single Shell Tanks
NT2 Underground Tanks
NT3 Underground Storage Tanks
NT4 Non-Operational Storage Tanks
NT4 Septic Tanks
NT2 Wastewater Treatment Tanks
NT1 Turbine
RT   Facilities
SO   Air Pollution
DEF (CFR) Any pipe, duct, boiler, tank,
     reactor, turbine, or furnace at or in
     a facility; or any structural member
     of a facility.

FACILITY MAILING LISTS
(SWDA; RCRA)
DA   October 19, 1990
RT   Facilities
SO   Wastes
DEF (SWDA; RCRA) The mailing lists
     for facilities maintained by EPA.

FACILITY MANAGERS
(DOE Order 5000.3A)
DA   December 14, 1990
BT1 Personnel
SO   Management
DEF (DOE Order 5000.3A) Those
     individuals, or their designees
     usually but not always contractors,
     who have direct line responsibility
     for operation of a facility or group
     of related facilities, including
     authority to direct physical
     changes to the facility.

FACT
(SSDC)
DA   October 12, 1990
SO   System Safety Development Center
     Glossary
DEF (SSDC) (A) Actuality, actual
     existence, event, objective reality.

BT-Broader Term   NT-Narrower Term   RT Related Term

(b) That which is perceived with the senses.

FACTOR RELATIONSHIP AND SEQUENCE OF EVENTS VOCABULARY
DA February 28, 1991
SY DOE FRASE VOCABULARY
SY FRASE Vocabulary
BT1 Vocabulary
RT Safety Performance Measurement System
SO Environmental Management
DEF A controlled vocabulary designed for searching the narrative description fields of records on the Safety Performance Measurement System (SPMS).

FACTORY MUTUAL
(SSDC)
DA October 12, 1990
SF FM
RT Insurance
SO System Safety Development Center Glossary
DEF (SSDC) A group of mutual insurers, underwriting large highly protected properties. They place strong emphasis in and are experts on loss prevention engineering.

FAIL-SAFE
(DOE Order 6430.1A)
DA October 12, 1990
RT Vital Equipment
SO Construction
DEF (DOE Order 6430.1A) A design characteristic by which a unit or system will become safe and remain safe if a system or component fails or loses its activation energy.

FAILED ELEMENT MONITORS
(NFI)
DA October 12, 1990
BT1 Failures
 BT2 Accidents
BT1 Monitors
 BT2 Equipment
RT HARP
SO Nuclear Facilities Incident Database
SO Radiation
DEF (NFI) Low Energy Gamma Monitors; Ten-Minute Delay Monitors; Blanket Gas Activity Monitors; Gas Chromatographs.

FAILED FUEL ELEMENT
DA January 8, 1991
SF FFE
BT1 Failures
 BT2 Accidents

FAILED INSTRUMENT COMPONENT INSPECTION
DA January 8, 1991
SF FICI

BT1 Inspections
 BT2 Administrative Processes
 BT3 Processes

FAILURE MODE AND EFFECT ANALYSIS
(SSDC)
DA October 12, 1990
SF FMEA
BT1 Analyses
RT Failures
RT Safety Program Reviews
RT System Safety
SO System Safety Development Center Glossary
DEF (SSDC) A basic system safety technique wherein the kinds of failures that might occur and their effect on the overall product or system are considered. Example: The effect on a system by the failure of a single component, such as a register or a hydraulic valve.

FAILURE REPORTING ANALYSIS AND CORRECTIVE ACTION SYSTEM
DA January 8, 1991
SF FRACAS
BT1 Information Systems
 BT2 Security Interests
 BT2 Systems
RT Analyses
RT Failures
RT Reports

FAILURES
DA February 12, 1991
BT1 Accidents
NT1 Catastrophic Collapses
NT1 Common Cause Failure
NT1 Containment Failure
NT1 Failed Element Monitors
NT1 Failed Fuel Element
NT1 Flow Failures
NT1 Ground Failures
 NT2 Differential Settlement
 NT2 Flow Failures
 NT2 Landslides
 NT2 Lateral Spreads
NT1 Malfunctions
NT1 No Containment Failure
NT1 PCB Transformer Rupture
NT1 Single Failure
NT1 Steam Generator Tube Rupture
NT1 Structural Collapse
RT Failure Mode and Effect Analysis
RT Failure Reporting Analysis and Corrective Action System
RT Phase Failure Relays
RT Predictive Maintenance
RT Reliability-Centered Maintenance
RT Root Cause Analysis
SO Management
SO Radiation

FALL PROTECTION DEVICE
(2660 FALL PROTECT)
DA January 3, 1991
BT1 Devices

BT1 Personal Protective Equipment
 BT2 Equipment/Parts - Personal Protective (DOE FRASE Vocabulary)
 BT3 Equipment
NT1 Safety Belt
NT1 Safety Line
NT1 Safety Net
SO DOE FRASE VOCABULARY

FAR
DA October 12, 1990
SEE Federal Aviation Regulation
SO Acronyms

FARM TANKS
(SWDA; RCRA; CFR)
DA October 12, 1990
BT1 Tanks
 BT2 Facility Components
SO Wastes
DEF (CFR) Tanks located on a tract of land devoted to the production of crops or raising animals, including fish, and associated residences and improvements. Farm tanks must be located on the farm property. "Farm" includes fish hatcheries, rangeland and nurseries with growing operations.

FAST FLUX TEST FACILITY
DA January 8, 1991
SF FFTF
BT1 Facilities
BT1 Test Area
 BT2 Area
 BT3 Sites/Areas

FAST RESPONSE GAMMA THERMOMETER
DA January 8, 1991
SF FRGT
BT1 Equipment/Parts - Instrumentation/Measuring (DOE FRASE Voc.)
 BT2 Equipment

FAULT TREE
(SSDC)
DA October 12, 1990
BT1 Analytical (Logic) Trees
 BT2 Diagrams
NT1 Management Oversight and Risk Tree
SO System Safety Development Center Glossary
DEF (SSDC) An analytical tree used to determine fault. These may be used in accident/incident investigation or to determine accident potential before one has occurred.

FAULT TREE ANALYSIS
DA January 8, 1991
SF FTA
BT1 Analyses

FAULTS
(SDWA; CFR; SSDC)
DA October 12, 1990
NT1 Capable Fault
NT1 Horizontal Displacements
NT1 Surface Faulting
 NT2 Dip-Slip
 NT2 Oblique Slip
 NT2 Strike Slip
NT1 Transmissive Faults
NT1 Transmissive Fractures
RT Control Width
RT Ground Cracks
RT Ground Failures
RT Subsidence
RT Tectonic Deformations
RT Uplifts
SO Environmental Management
SO System Safety Development Center
    Glossary
SO Water Pollution
DEF Surfaces or zones of rock fracture
    along which there has been
    displacement.

FCs
DA October 12, 1990
SEE Fluorocarbons
SO Acronyms

FDA
DA October 12, 1990
SEE Fuel Distribution Analyzer
SO Acronyms

Fe (Periodic Element)
DA October 12, 1990
SEE Iron
SO Acronyms

FEASIBILITY STUDIES
(CERCLA; CFR)
DA October 12, 1990
BT1 Studies
RT Remedial Investigations
SO Compensation and Liability
SO Environmental Management
SO Environmental Protection Agency
    Glossary
DEF (CFR) Processes undertaken by
    the lead agency (or responsible
    party if the responsible party will
    be developing a cleanup proposal)
    for developing, evaluating, and
    selecting remedial actions which
    emphasizes data analysis. See 40
    CFR 300.6 for additional
    explanation. (EPA) (1) Analyses of
    the practicability of a proposal;
    e.g., a description and analysis of
    the potential cleanup alternatives
    for a site or alternatives for a site
    on the National Priorities List.
    Feasibility studies usually
    recommend selection of a
    cost-effective alternative. They
    usually start as soon as the
    remedial investigation is
    underway; together, they are

commonly referred to as the
"RI/FS". The term can apply to a
variety of proposed corrective or
regulatory actions. (2) In research,
small-scale investigations of a
problem to ascertain whether or
not a proposed research approach
is likely to provide useful data.

FECAL COLIFORM BACTERIA
(EPA)
DA October 12, 1990
BT1 Coliform Organisms
 BT2 Bacteria
 BT3 Microorganisms
  BT4 Organisms
RT Coliform Index
RT Viruses
SO Environmental Protection Agency
    Glossary
DEF (EPA) Bacteria found in the
    intestinal tracts of mammals. Their
    presence in water or sludge is an
    indicator of pollution and possible
    contamination by pathogens.

FEDERAL AGENCIES
(SWDA; RCRA; SDWA; ESA; CFR)
DA October 12, 1990
BT1 Agencies
 BT2 Administrative Organizations
 BT3 Organizations
NT1 Atomic Energy Commission
NT1 Cognizant Federal Agencies
NT1 DOE Design Agency
NT1 DOE Production Agency
NT1 Federal Aviation Administration
NT1 Lead Agencies
NT1 National Aeronautic and Space
    Administration
NT1 Nuclear Regulatory Commission
 NT2 NRC Office of Nuclear Reactor
     Regulation
 NT2 NRC Office of Nuclear Regulatory
     Research
NT1 Occupational Safety and Health
    Administration
NT1 Office of Management and Budget
NT1 Procuring Agencies
NT1 U.S. Department of Energy
 NT2 Headquarters Operations
 NT2 Naval Reactors
 NT2 Operations Offices
  NT3 Albuquerque Operations Office
  NT3 Chicago Operations Office
  NT3 Idaho Operations Office
  NT3 Nevada Operations Office
  NT3 Oak Ridge Operations Office
  NT3 Richland Operations Office
  NT3 San Francisco Operations Office
  NT3 Savannah River Operations
      Office
 NT2 Program Offices
  NT3 Office of the Assistant Secretary
      for Nuclear Energy
  NT3 Office of the Assistant Secretary
      for Fossil Energy
  NT3 Office of the Assistant Secretary
      for Defense Programs

  NT3 Office of the Assistant Secretary
      for Conservation et.al.
  NT3 Office of the Assistant Secretary
      for Environment, et. al.
   NT4 Office of Special Projects
  NT3 Office of Environmental
      Restoration and Waste
      Management
  NT3 Office of New Production
      Reactors
  NT3 Office of Civilian Radioactive
      Waste Management
  NT3 Office of Energy Research
   NT4 Office of Basic Energy Sciences
   NT4 Office of Health and
       Environmental Research
NT1 U.S. Department of Justice
NT1 U.S. Department of Labor
NT1 U.S. Department of Transportation
 NT2 Research and Special Programs
     Administration
NT1 U.S. Environmental Protection
    Agency
 NT2 Regional Office
NT1 User Agencies
RT Agency Actions
RT Application for Federal Assistance
RT Authorized Officers
RT Natural Resource Trustees
RT Program Organizations
RT Regulatory Organizations
RT State Agencies
SO Endangered Species
SO Environmental Management
SO Wastes
SO Water Pollution
DEF Departments, agencies, or
    instrumentalities of the United
    States.

FEDERAL AVIATION ADMINISTRATION
DA January 8, 1991
SF *FAA*
BT1 Federal Agencies
 BT2 Agencies
 BT3 Administrative Organizations
 BT4 Organizations

FEDERAL AVIATION REGULATION
DA January 8, 1991
SF *FAR*
BT1 Code of Federal Regulations
 BT2 Codes, Standards, and
     Regulations
 BT2 Statutes and Regulations

FEDERAL CLASS I AREAS
(CFR)
DA November 15, 1990
BT1 Air Quality Control Regions
 BT2 Sites/Areas
BT1 Federal Facilities
 BT2 Facilities
NT1 Mandatory Class I Federal Areas
RT Federal Land Managers
DEF (CFR) Federal lands that are
    classified or reclassified "Class I".

FEDERAL COORDINATING OFFICERS
(EMER)
DA   February 1, 1991
BT1 Personnel
SO   Emergency Preparedness
DEF (EMER) The individuals designated
     by the president or his/her
     representative to coordinate
     overall federal response activities
     under the Disaster Relief Act,
     Public Law 93-288.

FEDERAL EMPLOYEE OCCUPATIONAL SAFETY AND HEALTH PROGRAM
(Doe Orders 5480.1B and 5482.1B)
DA   October 12, 1990
BT1 Programs
RT   National Institute for Occupational
     Safety and Health
RT   Occupational Safety and Health
     Administration
RT   Occupational Medical Program
SO   Construction
SO   Industrial Hygiene
DEF That program mandated by
     Executive Order 12196 and
     implemented by 29 CFR 1960,
     DOE 3790.1A, and DOE 3790.2.

FEDERAL FACILITIES
DA   January 24, 1991
BT1 Facilities
NT1 Federal Class I areas
NT2 Mandatory Class I Federal Areas
NT1 Government-Owned
     Contractor-Operated Facilities
NT2 Argonne National Laboratory
NT3 Argonne National
      Laboratory-East (Chicago)
NT3 Argonne National
      Laboratory-West (At INEL)
NT2 Energy Technology Engineering
     Center (Canoga Park)
NT2 Feed Materials Production Center
NT2 Idaho Chemical Processing Plant
NT2 Idaho National Engineering
     Laboratory
NT3 Power Burst Facility
NT2 Inhalation Toxicology Research
     Institute
NT2 Kansas City Plant
NT2 Lawrence Livermore National
     Laboratory
NT2 Lawrence Berkeley Laboratory
NT2 Los Alamos National Laboratory
NT2 Nevada Test Site
NT2 Oak Ridge Gaseous Diffusion
     Plant
NT2 Oak Ridge National Laboratory
NT2 Pacific Northwest Laboratory
NT2 Paducah Gaseous Diffusion Plant
     (Paducah)
NT2 Portsmouth Gaseous Diffusion
     Plant
NT2 Rocky Flats Plant
NT2 Sandia National Laboratories
NT3 Sandia National
      Laboratories-Alburquerque

NT3 Sandia National
      Laboratories-Livermore
NT2 Savannah River Site
NT2 Stanford Linear Accelerator Center
NT2 Strategic Petroleum Reserve
NT2 Waste Isolation Pilot Plant
NT2 Y-12 Plant
RT   Federal Facilities Agreement
RT   General Purpose Facilities Projects
SO   Environmental Management
DEF (CFR) Any building, installation,
     structure, land or public work
     owned by or leased to the federal
     government. Ships at sea, aircraft
     in the air, land forces on
     maneuvers, and other mobile
     facilities are not considered
     federal facilities. United States
     Government installations located
     on foreign soil or on land outside
     the jurisdiction of the United
     States Government are not
     considered federal facilities.

FEDERAL FACILITIES AGREEMENT
DA   January 8, 1991
SF   FFA
BT1 Agreements
NT1 Federal Facilities Compliance
     Agreement
RT   Federal Facilities

FEDERAL FACILITIES COMPLIANCE AGREEMENT
DA   January 8, 1991
SF   FFCA
BT1 Federal Facilities Agreement
BT2 Agreements
BT1 Interagency Agreement
BT2 Agreements
DEF An interagency agreement usually
     entered into between the United
     States Environmental Protection
     Agency and the Head of an
     Executive Agency. The content of
     the Agreement generally pertains
     to environmental cleanup,
     ensuring that any environmental
     impacts associated with past
     and/or present activities at a
     federal facility are thoroughly and
     adequately investigated so that
     appropriate remedial response
     actions can be formulated,
     assessed, and implemented as
     each individual situation warrants.

FEDERAL FIELD EXERCISES
(EMER)
DA   February 1, 1991
BT1 Exercises
RT   Full-Scale Exercises
SO   Emergency Preparedness
DEF (EMER) Exercises of the Federal
     Radiological Emergency
     Response Plan involving federal
     agencies in support of the
     cognizant federal agency and
     state and local governments.

FEDERAL LAND MANAGERS
(SWDA; RCRA; CAA; ARPA; CFR)
DA   April 24, 1991
BT1 Personnel
RT   Archaeological Resources
     Protection Act of 1979
RT   Archaeological Resources
RT   Federal Class I areas
SO   Air Pollution
SO   Environmental Management
SO   Wastes
DEF (ARPA) With respect to any public
     lands, the Secretary of the
     department, or the head of any
     other agency or instrumentality of
     the United States, having primary
     management authority over such
     lands. In the case of any public
     lands or Indian lands with respect
     to which no department, agency,
     or instrumentality has primary
     management authority, such term
     means the Secretary of the
     Interior. If the Secretary of the
     Interior consents, the
     responsibilities (in whole or in part)
     under ARPA of the Secretary of
     any department (other than the
     Department of the Interior) or the
     head of any other agency or
     instrumentality may be delegated
     to the Secretary of the Interior with
     respect to any land managed by
     such other Secretary or agency
     head, and in any such case, the
     term "Federal land manager"
     means the Secretary of the
     Interior.

FEDERAL RADIOLOGICAL EMERGENCY RESPONSE PLANS
(EMER)
DA   February 1, 1991
BT1 Plans
NT1 Federal Radiological Monitoring
     and Assessment Plans
RT   Procedures
SO   Emergency Preparedness
DEF (EMER) Comprehensive,
     coordinated plans broadly
     describing the entire federal
     government response to
     radiological emergencies in
     support of federal, state, and local
     government agencies.

FEDERAL RADIOLOGICAL MONITORING AND ASSESSMENT CENTERS
(EMER)
DA   February 1, 1991
BT1 Centers
RT   Assessments
SO   Emergency Preparedness
DEF (EMER) Centers at or near the
     scene of a radiological incident
     which coordinates all off-site
     federal radiological monitoring and
     assessment activities. Information
     related to the off-site radiological

impact of the accident is transmitted from the FRMAC to the cognizant federal agency and to the state(s).

FEDERAL RADIOLOGICAL MONITORING AND ASSESSMENT PLANS

(EMER)
DA February 1, 1991
BT1 Federal Radiological Emergency Response Plans
 BT2 Plans
RT Assessments
SO Emergency Preparedness
DEF (EMER) Plans contained in the Federal Radiological Emergency Response Plan for coordinating federal off-site radiological monitoring and assessment with that of the affected state(s).

FEDERAL RESPONSE CENTERS

(EMER)
DA February 1, 1991
BT1 Centers
SO Emergency Preparedness
DEF (EMER) Centers established by the Federal Emergency Management Agency at locations identified in conjunction with the state that serves as a focal point for federal response team interactions with the state.

FEDERALLY PERMITTED RELEASES

(CERCLA; CFR)
DA October 12, 1990
BT1 Releases
SO Compensation and Liability
SO Environmental Management
DEF (CFR) As defined by section 101(10) of CERCLA, means discharges in compliance with a permit under section 402 of the Federal Water Pollution Control Act; discharges resulting from circumstances identified and reviewed and made part of the public record with respect to a permit issued or modified under section 402 of the Federal Water Pollution Control Act and subject to a condition of such permit; continuous or anticipated intermittent discharges from a point source, identified in a permit or permit application under section 402 of the Federal Water Pollution Control Act, which are caused by events occurring within the scope of relevant operating or treatment systems; discharges in compliance with a legally enforceable permit under section 404 of the Federal Water Pollution Control Act;...(See 40 CFR 300.6 for continuation of definition).

FEED MATERIALS PRODUCTION CENTER

DA January 8, 1991
SF *FMPC*
BT1 Centers
BT1 Government-Owned Contractor-Operated Facilities
 BT2 Federal Facilities
 BT3 Facilities
RT Westinghouse Materials Company of Ohio
DEF A DOE-owned manufacturing facility for the production of uranium metal used in U.S. defense programs. At the present time (May 1991), production at the site has been temporarily suspended. The site is located on 1050-acres of property in a rural area about 20 miles northwest of downtown Cincinnati, Ohio. The production facilities occupy approximately 136 acres near the center of the site.

FEED WATER HEATER(S)

(2482 FEED WATER H)
DA January 3, 1991
BT1 Heater(s)
 BT2 Equipment/Parts - Heating (DOE FRASE Vocabulary)
 BT3 Equipment
SO DOE FRASE VOCABULARY

FEEDBACK

(SSDC)
DA October 12, 1990
RT Feedback Loop
RT Knowledge of Results
SO System Safety Development Center Glossary
DEF (SSDC) The two-way flow of information between personnel or organizations. In system safety we are especially concerned with good feedback between management and subordinate organizations and also between staff organizations; e.g., safety and line organizations.

FEEDBACK LOOP

(SSDC)
DA October 12, 1990
RT Feedback
SO System Safety Development Center Glossary
DEF (SSDC) The continuous communication network used to achieve feedback.

FEEDLOTS

(EPA)
DA October 12, 1990
BT1 Sites/Areas
SO Environmental Protection Agency Glossary
DEF (EPA) Relatively small, confined areas for the controlled feeding of animals that tend to concentrate large amounts of animal wastes that cannot be absorbed by the soil and, hence, may be carried to nearby streams or lakes by rainfall runoff.

FEEDWATER

DA January 8, 1991
SF *FW*
NT1 Auxiliary Feedwater
NT1 Main Feedwater
RT Emergency Feedwater System
DEF (NRC Glossary of Terms: Nuclear Power and Radiation) Water supplied to the reactor pressure vessel (in a Boiling Water Reactor) or the steam generator (in a Pressurized Water Reactor) that removes heat from the reactor fuel rods by boiling and becoming steam. The steam becomes the driving force for the plant turbine generator.

FEFGC
DA October 12, 1990
SEE Fuel Element Failure Gas Chromatograph
SO Acronyms

FENS

(EPA)
DA October 12, 1990
BT1 Wetlands
 BT2 Sites/Areas
 BT2 Surface Water Resources
 BT3 Natural Resources
RT Muck Soils
RT Swamps
SO Environmental Protection Agency Glossary
DEF (EPA) Types of wetlands that accumulate peat deposits. Fens are less acidic than bogs, deriving most of their water from groundwater rich in calcium and magnesium.

FERMENTATION

(EPA)
DA October 12, 1990
BT1 Chemical Processes
 BT2 Processes
SO Environmental Protection Agency Glossary
DEF (EPA) Chemical reactions accompanied by living microbes that are supplied with nutrients and other critical conditions such as heat, pressure, and light that are specific to the reaction at hand.

FERRY VESSELS

(SWDA; RCRA; ESH)
DA October 12, 1990
BT1 Vessels
RT Inland Waters

BT-Broader Term   NT-Narrower Term   RT Related Term

SO Hazardous Materials
DEF Vessels which are limited in their use to the carriage of deck passengers or vehicles or both, operate on a short run on a frequent schedule between two points over the most direct water route, other than in ocean or coastwise service, and are offered as a public service of a type normally attributed to a bridge or tunnel.

FERTILE
(IAEA)
DA October 12, 1990
RT Nuclides
SO Radiation
DEF (IAEA) Of a nuclide, capable of being transformed, directly or indirectly, into a fissile nuclide by neutron capture.

FERTILIZERS
(EPA)
DA October 12, 1990
RT Eutrophication
RT Nutrients
RT Plants
SO Environmental Protection Agency Glossary
DEF (EPA) Materials such as nitrogen and phosphorus that provide nutrients for plants. Commercially sold fertilizers may contain other chemicals or may be in the form of processed sewage sludge.

FFA
DA October 12, 1990
SEE Federal Facilities Agreement
SO Acronyms

FFCA
DA October 12, 1990
SEE Federal Facilities Compliance Agreement
SO Acronyms

FFE
DA October 12, 1990
SEE Failed Fuel Element
SO Acronyms

FFTF
DA October 12, 1990
SEE Fast Flux Test Facility
SO Acronyms

FHAR
DA October 12, 1990
SEE Fire Hazard Analysis Report
SO Acronyms

FICI
DA October 12, 1990

SEE Failed Instrument Component Inspection
SO Acronyms

FIDUCIARIES
(SSDC)
DA October 12, 1990
SO System Safety Development Center Glossary
DEF (SSDC) Persons or entities entrusted to act for another. For example, the fiduciary of an estate is the executor or administrator.

FIELD AREA
(1607 FIELD AREA)
DA December 10, 1990
BT1 Area
BT2 Sites/Areas
SO DOE FRASE VOCABULARY

FIELD AUDITS
(ESH)
DA October 12, 1990
BT1 Audits (SSDC)
BT2 Administrative Processes
BT3 Processes
SO Quality Assurance
DEF (ESH) Emphasis during field audits is placed on: verifying that operational aspects and procedures are in accordance with the protocols and QA/QC plan; verifying the collection of all samples including duplicates and field blanks; verifying that documentation is in order and sufficient to establish the collection location of any sample collected; determining discrepancies that exist and initiating corrective action as appropriate; and collecting independent samples.

FIELD COMMAND POSTS
(EMER)
DA February 1, 1991
BT1 Sites/Areas
RT Emergency Control Centers
SO Emergency Preparedness
DEF (EMER) Designated areas at or near the scene of an emergency which are the collection centers for the on-scene commander and which are used for response, assessment, and communications to the emergency control center.

FIELD ELEMENTS
(DOE Orders 5480.16, 5484.1, and 6430.1A)
DA October 12, 1990
SY Field Organizations
RT Program Offices
SO Construction
SO Environmental Management
SO Firearms
SO Industrial Safety
DEF A general term for any officially

established Departmental components (excluding individual duty stations) located outside the Washington, DC, metropolitan area.

FIELD FACILITY/BUILDING EMERGENCY PLANS
(EMER)
DA February 1, 1991
BT1 Emergency Plans
BT2 Plans
RT Field Site Emergency Plans
RT Procedures
SO Emergency Preparedness
DEF (EMER) Plans prepared by headquarters, field elements, field contractors, or offices under field element jurisdiction to guide response to specific buildings or facilities (e.g., office buildings, Forrestal and Germantown buildings, process buildings, laboratories).

FIELD GASES
(CAA; ESH)
DA October 12, 1990
BT1 Natural Gas
BT2 Fossil Fuels
BT3 Fuels
BT3 Geologic Resources
BT4 Natural Resources
BT2 Gases
RT Natural Gas Processing Plants
SO Air Pollution
DEF (ESH) Feedstock gases entering the gas processing plant.

FIELD LEVEL EXEMPTIONS
(Doe Order 5480.4)
DA October 12, 1990
BT1 Exemptions
RT Mandatory Standards
SO Environmental Management
SO Standards
DEF (DOE Order 5480.4) Interim releases from a mandatory standard, granted by the field organization after a request for a temporary or permanent exemption. Such exemptions shall not exceed 180 days and are not renewable.

FIELD MONITORING
(EMER)
DA February 1, 1991
BT1 Monitoring
BT2 Activities
SO Emergency Preparedness
DEF (EMER) The use of sensitive detection equipment by trained personnel to perform measurements to determine the presence and levels of radioactive or other hazardous substance contamination at selected

geographic locations in the off-site environment.

FIELD OPERATIONS PLANS

(ESH)
DA   October 12, 1990
SY   Sampling Plans
BT1 Plans
RT   Maintenance Plans
SO   Quality Assurance
DEF (ESH) Documents that are prepared for either continuous or site specific data collection activities (air, water, pesticides, hazardous waste, etc.). Plans should describe project organization and responsibilities, project description (objectives, scope, schedule of tasks and milestones, data usage, monitoring network/sampling and analysis design and rationale), data quality objectives, sampling procedures, calibration, analytical methods, documentation/data reduction/reporting, data assessment, audits, corrective action, reports and safety.

FIELD ORGANIZATIONS

(DOE Order 5483.1A; ESH)
DA   October 12, 1990
SY   Field Elements
BT1 Departmental Elements
RT   DOE Contractors
RT   Heads of Field Operations
RT   Line Organizations
RT   Program Offices
SO   Environmental Management
SO   Industrial Hygiene
SO   Industrial Safety
SO   Management
SO   Standards
DEF (DOE Order 5483.1A) Departmental Elements located outside the Washington, D.C. geographic area. (ESH) It is the first line DOE field elements that carry the organizational responsibility for (1) managing and executing assigned programs, (2) directing contractors who conduct the programs, and (3) assuring that environment, safety, and health are integral parts of each program.

FIELD SAMPLING PLANS

(EPA)
DA   October 12, 1990
BT1 Sampling Plans
BT2 Plans
BT1 Sampling and Analysis Plans
BT2 Plans
RT   Raw Data
SO   Environmental Protection Agency Glossary
DEF (EPA) Provide guidance for all fieldwork by defining in detail the sampling and data-gathering methods to be used on a project.

FIELD SITE EMERGENCY PLANS

(EMER)
DA   February 1, 1991
BT1 Emergency Plans
BT2 Plans
RT   Field Facility/Building Emergency Plans
RT   Procedures
SO   Emergency Preparedness
DEF (EMER) Plans prepared by field element contractors or other entities under field element jurisdiction to guide their responses for identified credible emergencies.

FILE(S)

(3032 FILE)
DA   January 3, 1991
BT1 Tools - Manual
BT2 Tools (DOE FRASE Vocabulary)
BT3 Equipment
SO   DOE FRASE VOCABULARY

FILL MATERIALS

(CWA; CFR)
DA   October 12, 1990
BT1 Materials
RT   Contaminant Carriers
RT   Discharge of Fill Material
SO   Water Pollution
DEF (CFR) Materials used for the primary purpose of replacing an aquatic area with dry land or of changing the bottom elevation of an waterbody. The term does not include any pollutant discharged into the water primarily to dispose of waste, as that activity is regulated under section 402 of the Clean Water Act.

FILLING

(EPA)
DA   October 12, 1990
RT   Aquatic Environment
SO   Environmental Protection Agency Glossary
DEF (EPA) Depositing dirt and mud or other materials into aquatic areas to create more dry land, usually for agricultural or commercial development purposes. Such activities often damage the ecology of the area.

FILM BADGES

(NIH)
DA   October 12, 1990
RT   Exposure
RT   Radiation Safety
SO   Radiation
DEF (NIH) Packets of photographic film used for the approximate measurement of radiation exposure for personnel monitoring purposes. Badges may contain two or more films of differing sensitivity, and they may contain

filters which shield parts of the film from certain types of radiation.

FILTERED VENTED CONTAINMENT

DA   January 8, 1991
SF   FVC
BT1 Containment
BT2 Processes

FILTERS

DA   February 26, 1991
BT1 Devices
NT1 Baghouse Filters
NT1 Fabric Filters
NT1 Filters (Radiology)
NT1 Hepa Filter
NT1 High Efficiency Particulate Air Filters
NT1 Sand Filters
NT1 Trickling Filters

FILTERS (RADIOLOGY)

(NIH)
DA   October 12, 1990
BT1 Filters
BT2 Devices
RT   Radiation Safety
SO   Radiation
DEF (NIH) Primary: Sheets of material, usually metal, placed in a beam of radiation to remove, as far as possible, the less penetrating components of the beam. Secondary: Sheets of material of lower atomic number, relative to that of the primary filter, placed in the filtered beam of radiation to remove characteristic radiation produced by the primary filter.

FILTRATION

(SDWA; CFR)
DA   October 12, 1990
BT1 Pollution Recovery Processes
BT2 Processes
NT1 Diatomaceous Earth Filtration
NT1 Direct Filtration
NT1 Flow Sand Filtration
NT1 Slow Sand Filtration
RT   Conventional Filtration Treatment
RT   Flocculation
SO   Environmental Protection Agency Glossary
SO   Water Pollution
DEF (EPA) A treatment process, under the control of qualified operators, for removing solid (particulate) matter from water by passing the water through porous media such as sand or a man-made filter. The process is often used to remove particles that contain pathogenic organisms.

FINAL AUTHORIZATIONS

(SDWA; RCRA)
DA   October 12, 1990
BT1 Administrative Processes
BT2 Processes

BT1 Permits
RT Permit (RCRA)
RT Test Authorizations
SO Environmental Management
SO Wastes
DEF (CFR) Approvals by EPA of a State
    program which has met the
    requirements of section 3006(b) of
    RCRA (Resource Conservation
    and Recovery Act) and the
    applicable requirements of 40
    CFR 271, Subpart A.

FINAL CLOSURE
(SWDA; RCRA)
DA October 12, 1990
BT1 Closure
 BT2 Administrative Processes
  BT3 Processes
RT Hazardous Waste Management
    Units
RT Site Closure and Stabilization
SO Environmental Management
SO Wastes
DEF (CFR) The closure of all hazardous
    waste management units at the
    facility in accordance with all
    applicable closure requirements so
    that hazardous waste
    management activities under 40
    CFR 264 and 265 are no longer
    conducted at the facility unless
    subject to the provisions in 40
    CFR 262.34.

FINAL COVER
DA January 24, 1991
BT1 Cover Materials
 BT2 Materials
RT Daily Cover
SO Environmental Management
DEF (CFR) "Final cover" means cover
    material that serves the same
    functions as daily cover but, in
    addition, may be permanently
    exposed on the surface.

FINAL SAFETY ANALYSIS REPORT
DA January 8, 1991
SF FSAR
BT1 Safety Analysis Reports
 BT2 Reports
RT Preliminary Safety Analysis Report

FINDING OF NO SIGNIFICANT IMPACT
(DOE Order 5440.1D; ESH)
DA October 12, 1990
SF FNSI
SF FONSI
BT1 Findings
RT Environmental Impact Statements
RT National Environmental Policy Act
    Documents
RT Programmatic NEPA Documents
RT Site-Wide NEPA Documents
SO Environmental Management
SO Environmental Protection Agency
    Glossary

SO Management
DEF (DOE Order) A document, defined
    at 40 CFR 1508.13, prepared to
    record a Departmental decision
    that the environmental impacts of
    an action considered in an
    environmental assessment will not
    have a significant effect on the
    human environment and that an
    environmental impact statement is
    not required for a proposed action.

FINDINGS
(DOE Order 5482.1B; ESH; TTGM)
DA October 12, 1990
NT1 Best Management Practice
    Findings
NT1 Compliance Findings
NT1 Finding of No Significant Impact
NT1 Noteworthy Practices
RT Computer Assisted Tracking
    System
RT Concerns
RT Environment Assessments
RT Investigations
RT Issue Identification
RT Management Appraisals
RT Performance Objectives
RT Root Causes
RT Tiger Team Assessments
RT Tiger Teams
RT Tiger Team Assessment Reports
SO Construction
SO Management
DEF Statements of fact concerning a
    condition in the ES&H program
    that was investigated during an
    appraisal. They may be a simple
    statement of proficiency, or a
    description of a deficiency (i.e., a
    variance from procedures or
    criteria). Both severity and
    potential consequences should be
    addressed in describing a deficient
    condition. (SSDC) The results of
    an investigation; the salient,
    factual, and analytical highlights of
    the accident/incident. (TTGM)
    Tiger Team Assessments include
    findings in Environmental and
    Management areas based on: 1)
    observation of routine operations,
    emergency exercises, and the
    physical condition of the site and
    facilities; 2) interviews with
    management, staff, operators, and
    craft personnel; and 3) review of
    policy statements, records,
    procedures, and other relevant
    documents. Findings support
    Concerns as indicated in Tiger
    Team Assessments.

FINGER(S)
(1104 FINGER)
DA November 28, 1990
BT1 Hand(s)
 BT2 Human Body Parts
RT Fingernail(s)

RT Thumb(s)
SO DOE FRASE VOCABULARY

FINGERNAIL(S)
(1105 FINGERNAIL)
DA November 28, 1990
BT1 Skin
 BT2 Human Body Parts
RT Finger(s)
SO DOE FRASE VOCABULARY

FIRE ALARM SYSTEM
(2033 FIRE ALARM S)
DA December 10, 1990
BT1 Alarms
 BT2 Devices
BT1 Fire Protection Systems
 BT2 Emergency Systems
  BT3 Systems
BT1 Systems (DOE FRASE Vocabulary)
 BT2 Systems
RT Fire Escape
RT Fire Station
RT Fire Suppression System
RT Firefighter
SO DOE FRASE VOCABULARY

FIRE DAMAGE
(1530 FIRE DAMAGE)
DA November 29, 1990
BT1 Damage
 BT2 Nature of Property Damage
RT Heat Damage
RT Smoke Damage
SO DOE FRASE VOCABULARY

FIRE ESCAPE
(1697 FIRE ESCAPE)
DA December 10, 1990
BT1 Site (DOE FRASE Vocabulary)
 BT2 Sites/Areas
RT Fire Alarm System
RT Fire Suppression System
SO DOE FRASE VOCABULARY

FIRE HAZARD ANALYSIS REPORT
DA January 8, 1991
SF FHAR
BT1 Reports
RT Analyses

FIRE HAZARDS
(CERCLA)
DA October 12, 1990
BT1 Hazard Categories
BT1 Hazards
 BT2 Conditions
RT Pyrophoric Materials
DEF (CFR) Including flammables,
    combustible liquids, pyrophorics,
    and oxidizers (as defined by 29
    CFR 1910.1200). (See 40 CFR
    370.2).

FIRE PROTECTION
(Doe Order 5480.7)
DA October 12, 1990
RT DOE Fire Protection Authorities

SY-Synonymous Terms   SO-Source/Subject Category   SF-See From

147

RT Halons
SO Fires
DEF (DOE Order 5480.7) Protection
       from a broad range of fire risks
       normally included in the analysis
       conducted by fire protection
       engineers. These include some
       aspects of related perils such as
       explosion, windstorm, earthquake,
       lightning, and water damage. Fire
       prevention programs are a
       necessary part of fire protection
       programs.

FIRE PROTECTION SYSTEMS
(Doe Order 5480.7)
DA October 12, 1990
BT1 Emergency Systems
 BT2 Systems
NT1 Dedicated Fire Water Systems
 NT2 Sprinkler System
NT1 Fire Alarm System
NT1 Fire Suppression System
RT Design Basis Fires
RT Fires
RT Response Time
SO Fires
DEF (DOE Order 5480.7) Systems
       designed to control or extinguish
       fires or to limit the extent of fire
       damage. These include: (1)
       Automatic suppression systems
       such as sprinklers, Halon, or
       carbon dioxide systems. (2)
       Watchmen or automatic detection
       systems, water supplies, plus a
       fire department. (3) Walls and
       doors. (4) Building separation with
       credit for water supplies plus a fire
       department.

FIRE PROTECTION TRACKING
SYSTEM
DA January 8, 1991
SF *FPTS*
BT1 Information Systems
 BT2 Security Interests
 BT2 Systems

FIRE STATION
(1779 FIRE STATION)
DA December 10, 1990
BT1 Building (DOE FRASE Vocabulary)
 BT2 Facilities and Buildings (DOE
       FRASE Vocabulary)
  BT3 Facilities
BT1 Station
 BT2 Site (DOE FRASE Vocabulary)
  BT3 Sites/Areas
RT Fire Alarm System
RT Firefighter
SO DOE FRASE VOCABULARY

FIRE SUPPRESSION SYSTEM
(2058 FS SYSTEM)
DA December 10, 1990
BT1 Fire Protection Systems
 BT2 Emergency Systems
  BT3 Systems

BT1 Systems (DOE FRASE Vocabulary)
 BT2 Systems
RT Fire Escape
RT Fire Alarm System
RT Foam System
RT Sprinkler System
SO DOE FRASE VOCABULARY

FIREARMS
(Doe Order 5480.16)
DA October 12, 1990
NT1 Authorized Weapons
NT1 Defective Firearm
NT1 Grenade Launchers
NT1 Light Antitank Weapon
NT1 Shadow Force Weapons
NT1 Small Arms
 NT2 Handguns
  NT3 Pistols
   NT4 Double Action Semiautomatic
        Pistols
   NT4 Machine Pistols
   NT4 Single Action Semiautomatic
        Pistols
  NT3 Revolvers
 NT2 Machine Guns
  NT3 Submachine Guns, Closed Bolt
  NT3 Submachine Guns, Open Bolt
 NT2 Rifles
  NT3 Automatic Rifles
 NT2 Semiautomatic Firearms
  NT3 Double Action Semiautomatic
        Pistols
  NT3 Shotgun, Semiautomatic
  NT3 Single Action Semiautomatic
        Pistols
 NT2 Shotgun, Pump
RT Armorers
RT Blank Ammunition
RT Blank Fire Adapter (BFA)
RT Central Training Academy
RT Dry Firing
RT Firearms Ranges
RT Live Round Excluders
RT Misfires
RT Munitions
RT Special Weapons
RT Tagging
RT Unauthorized Discharges
RT Weapon Simulators
DEF (DOE Order 5480.16) Include all
       weapons capable of propelling a
       missile by means of an explosive
       charge, as well as all explosive
       ordnance, ESS equipment,
       chemical weapons, and
       pyrotechnic devices.

FIREARMS RANGES
(Doe Order 5480.16)
DA October 12, 1990
SY Ranges (Firearms)
BT1 Sites/Areas
RT Central Training Academy
RT Dry Firing
RT Firearms
RT Range Masters
RT Range Safety Officers
RT Unauthorized Discharges

SO Firearms
DEF (DOE Order 5480.16) Areas
       designated for firearms training;
       they do not include training areas
       where blank ammunition or ESS
       weapons are used exclusively.

FIREFIGHTER
(0512 FIREFIGHTER)
DA November 28, 1990
BT1 Admin. Support/Clerical Employee
 BT2 Occupations
 BT2 Personnel
RT Fire Station
RT Fire Alarm System
SO DOE FRASE VOCABULARY

FIRES
DA October 19, 1990
BT1 Natural Disasters
 BT2 Natural Phenomenon
RT Accidents
RT Anticipated Operational
       Occurrences
RT Combustion
RT Deflagration
RT Design Basis Fires
RT Fire Protection Systems

FIRING RANGE
(1608 FIRING RANGE)
DA December 10, 1990
BT1 Site (DOE FRASE Vocabulary)
 BT2 Sites/Areas
SO DOE FRASE VOCABULARY

FIRST AID
(SSDC)
DA October 12, 1990
RT Medical Treatment
SO System Safety Development Center
       Glossary
DEF (SSDC) One-time treatment and
       subsequent observation of minor
       scratches, cuts, burns, splinters,
       and so forth, which do not
       ordinarily require professional
       medical care even though it may
       be provided by a physician or
       registered professional personnel.

FIRST ATTEMPT AT REPAIR
(CAA; CFR; ESH)
DA October 12, 1990
SO Air Pollution
DEF (CFR) To take rapid action for the
       purpose of stopping or reducing
       leakage of organic material to
       atmosphere using best practices.

FIRST DRAW
(EPA)
DA October 12, 1990
RT Public Water Systems
SO Environmental Protection Agency
       Glossary
DEF (EPA) The water that immediately
       comes out when a tap is first
       opened. This water is likely to

BT-Broader Term   NT-Narrower Term   RT Related Term

have the highest level of lead contamination from plumbing materials.

FIRST FEDERAL OFFICIALS
(CERCLA; CFR; EMER)
DA   October 12, 1990
BT1 Personnel
RT National Response Teams
SO Compensation and Liability
SO Emergency Preparedness
SO Environmental Management
DEF (CFR) The first Federal representatives of a participating agency of the National Response Team to arrive at the scene of a discharge or a release. These officials coordinate activities under this Plan and may initiate, in consultation with the OSC, any necessary actions until the arrival of the predesignated OSC. A State with primary jurisdiction over a site covered by a cooperative agreement will act in the stead of the First Federal official for any incident at the site.

FIRST LINE SUPERVISION
(MORT)
DA   April 3, 1991
RT Controls
RT Higher Supervision
RT Inspections
RT Maintenance
RT Operability
RT Technical Information
RT Training
DEF (MORT) MORT analysis asks: was worksite supervision adequate? were the necessary supportive services adequate? Related issues/topics include: supervisory training; time; performance errors; task assignments; task briefings; task procedures; safety analysis; and worker problems (e.g. aberrant behavior, selection, training, motivation).

FISCAL YEAR
DA   January 8, 1991
SF   *FY*
BT1 Time Designations
RT Maintenance Backlog
SO Management

FISH
(ESA)
DA   October 19, 1990
BT1 Animals
NT1 Anadromous Fish
NT1 Catanadramous Fish
NT1 Game Fish
NT1 Rough Fish
RT Artificial Reefs
RT Zooplankton
SO Endangered Species
DEF (WEBSTER) Any of numerous

cold-blooded strictly aquatic craniate vertebrates that have typically an elongated somewhat spindle-shaped body terminating in a broad caudal fin, limbs in the form of fins when present at all, and a 2-chambered heart by which blood is sent through thoracic gills to be oxygenated.

FISSILE
(IAEA; EMER)
DA   October 12, 1990
RT Nuclides
SO Emergency Preparedness
SO Radiation
DEF (IAEA) Of a nuclide, capable of undergoing fission as a result of interaction with slow neutrons.

FISSILE CLASS I
(DOE Order 5480.3;ESH)
DA   October 12, 1990
BT1 Fissile Classification
SO Hazardous Materials
DEF (DOE Order 5480.3) Packages that may be transported in unlimited numbers and in any arrangement and that require no nuclear criticality safety controls during transportation. For purposes of nuclear criticality safety control, a transport index is not assigned to Fissile Class I packages. However, the external radiation levels may require a transport index number.

FISSILE CLASS II
(ESH)
DA   October 12, 1990
BT1 Fissile Classification
SO Hazardous Materials
DEF (DOE Order 5480.3) Packages that may be transported in any arrangement but in numbers that do not exceed a transport index of 50. For purposes of nuclear criticality safety control, individual packages may have a transport index of not less than 0.1 and not more than 10. However, the external radiation levels may require a higher transport index number but not to exceed 10. Such shipments require no nuclear criticality safety control by the shipper during transportation.

FISSILE CLASS III
(ESH)
DA   October 12, 1990
BT1 Fissile Classification
SO Hazardous Materials
DEF (DOE Order 5480.3) Shipments of packages that do not meet the requirements of Fissile Class I and II and that are controlled in transportation by special arrangements between the shipper

and the carrier to provide nuclear criticality safety.

FISSILE CLASSIFICATION
(DOE Order 5480.3; ESH)
DA   October 12, 1990
NT1 Fissile Class I
NT1 Fissile Class II
NT1 Fissile Class III
RT Fissile Materials
SO Environmental Management
SO Hazardous Materials
DEF (DOE Order 5480.3) Classification of a package or shipment of fissile materials according to the controls needed to provide nuclear criticality safety during transportation.

FISSILE MATERIALS
(DOE Order 6430.1A; CFR)
DA   October 12, 1990
SY   Fissionable Materials
BT1 Hazardous Materials
 BT2 Materials
RT Fissile Classification
SO Construction
SO Environmental Management
SO Hazardous Materials
DEF (ESH) The term "fissile" means easily undergoing nuclear fission (the splitting apart of the nuclei of atoms). Fissile materials consist of or contain one or more fissile radionuclides. Fissile radionuclides are plutonium-238, plutonium-239, plutonium-241, uranium-233, uranium-235, neptunium-237, and curium-244. Fissile materials are classified according to the controls needed to provide nuclear criticality safety during transportation. (DOE Order 6430.1A) Nuclides capable of undergoing fission by interaction with slow neutrons provided the effective thermal neutron production cross section, $\overline{\nu\sigma_f}$, exceed the effective thermal neutron absorption cross section, $\overline{\sigma_a}$.

FISSION PRODUCTS
(IAEA; EMER)
DA   October 12, 1990
SF   *FP*
BT1 Nuclides
RT Radioactive Source Terms
RT Reprocessing
SO Emergency Preparedness
SO Radiation
DEF (IAEA) Nuclides produced either by fission or by the radioactive decay of nuclides formed by fission.

FISSIONABLE MATERIALS
(Doe Order 5480.5)
DA   October 12, 1990
SY   Fissile Materials

BT1 Materials
SO Environmental Management
SO Standards
DEF (DOE Order 5480.5) Nuclides capable of sustaining a neutron induced fission chain reaction (e.g., uranium-233, uranium-235, plutonium-239, plutonium-238, plutonium-241, neptunium-237, americium-241, and curium-244).

FISSIONABLE MATERIALS HANDLERS
(Doe Order 5480.5)
DA October 12, 1990
SO Environmental Management
SO Standards
DEF (DOE Order 5480.5) Individuals officially designated by management to manipulate or handle significant quantities of fissionable materials, or manipulate the controls of equipment used to produce, process, transfer, store, or package significant quantities of such materials.

FIXED CAPITAL COSTS
(CAA; CFR)
DA October 12, 1990
BT1 Costs
RT Capital Expenditure
SO Air Pollution
DEF (CFR) The capital needed to provide all of the depreciable components.

FIXED NUCLEAR FACILITIES
(EMER)
DA February 1, 1991
BT1 Nuclear Facilities
BT2 Facilities
SO Emergency Preparedness
DEF (EMER) Stationary nuclear installations that use or produce radioactive materials in their normal operations. These facilities include commercial nuclear power plants and other fixed facilities.

FIXED SOURCES
DA January 24, 1991
SY Stationary Sources
BT1 Facilities
SO Environmental Management
DEF (CFR) "Fixed source" means, for the purpose of these guidelines, a stationary facility that converts fossil fuel into energy, such as steam, hot water, electricity, etc.

FLAME RETARDANT CLOTHING
(2661 FLAME RETARD)
DA January 3, 1991
BT1 Clothing
BT1 Personal Protective Equipment
BT2 Equipment/Parts - Personal Protective (DOE FRASE Vocabulary)

BT3 Equipment
SO DOE FRASE VOCABULARY

FLAMMABLE COMPRESSED GASES
(CFR)
DA October 12, 1990
BT1 Compressed Gases
BT2 Gases
RT Hazards
SO Hazardous Materials
DEF (CFR) There are a number of technical specifications for identifying flammable compressed gases. Generally, any flammable material having in the container a pressure which exceeds 40 psi at 100°F, is considered a flammable compressed gas. Acetylene, cyclopropane, and liquefied and non-liquefied hydrocarbon gas are examples of flammable compressed gases.

FLAMMABLE LIQUIDS
(CFR)
DA October 12, 1990
BT1 Liquids
BT2 Fluids
RT Combustible Liquids
RT Flash Point
RT Hazards
RT Pyrophoric Liquids
SO Hazardous Materials
DEF (CFR) According to the hazardous materials regulations, flammable liquids are liquids having the flash point below 100°F, except those liquids which meet the definition of a compressed gas, or a mixture having one or more components with a flash point of 100°F or higher that makes up at least 99% of the total volume of the mixture. Liquid paint driers, petroleum distillates, pine oils, and liquid furniture polishes are examples of flammable liquids.

FLAMMABLE SOLIDS
(CFR)
DA October 12, 1990
BT1 Solids
BT2 Materials
RT Flash Point
RT Hazards
RT Water Reactive Materials
SO Hazardous Materials
DEF (CFR) Flammable solids are defined in the regulations as any solid material, other than an explosive, which under conditions incident to transportation is liable to cause fires through friction, retained heat from manufacturing or processing, or which can be ignited readily, and when ignited, such material burns so vigorously and persistently as to create a serious transportation hazard. Included within this class are

spontaneously combustible and water reactive materials. Wet hafnium metal, burnt fibers, and garbage tankage (containing less than 8% water) are examples of flammable solids.

FLARES
(Doe Order 5480.16)
DA October 12, 1990
BT1 Devices
SO Firearms
DEF (DOE Order 5480.16) Mechanical devices that use pyrotechnic materials to produce light for signaling, illuminating, or attracting attention.

FLASH BURN(S)
(1321 FLASH BURN)
DA November 28, 1990
BT1 Burn(s)
BT2 Injuries
SO DOE FRASE VOCABULARY

FLASH GRENADES
(Doe Order 5480.16)
DA October 12, 1990
BT1 Diversionary Devices
RT Grenade Launchers
SO Firearms
DEF (DOE Order 5480.16) Devices that produce a brilliant flash (of about 2 million candle power) and a loud report (200 decibels at a distance of about 5 feet (I 1/2 meters)) without producing lethal fragmentation.

FLASH POINT
(SWDA; RCRA; ESH)
DA October 12, 1990
BT1 Measurements
RT Combustible Liquids
RT Flammable Liquids
RT Flammable Solids
SO Hazardous Materials
DEF The minimum temperature at which a substance gives off flammable vapor which will ignite if in contact with spark or flame.

FLEXIBLE ELEMENTS
(SEA)
DA October 12, 1990
RT Structural Members
SO Construction
DEF (SEA) Flexible elements or systems are those whose deformation under lateral load is significantly larger than adjoining parts of the system.

FLEXURAL STRENGTH
(DOE Order 6430.1A)
DA October 12, 1990
BT1 Strength
SO Construction
DEF (DOE Order 6430.1A)The strength

of a material in bending, that is, resistance to fracture.

FLIGHT CREWMEMBERS
(Doe Order 5480.13)
DA   October 12, 1990
BT1 Personnel
RT   Aviation Operations
SO   Aviation Safety
DEF (DOE Order 5480.13) Persons assigned to perform flight duties or DOE mission duties in an aircraft during flight time.

FLOCCULATION
(SDWA; CFR)
DA   October 12, 1990
BT1 Wastewater Treatment Processes
 BT2 Treatment
  BT3 Waste Management Processes
   BT4 Processes
RT   Agglomeration
RT   Coagulation
RT   Conventional Filtration Treatment
RT   Filtration
RT   Flocs
RT   Precipitates
RT   Sedimentation
SO   Environmental Protection Agency Glossary
SO   Water Pollution
DEF (EPA) The process by which clumps of solids in water or sewage are made to increase in size by biological or chemical action so that they can be separated from the water. (CFR) A process to enhance agglomeration or collection of smaller floc particles into larger, more easily settleable particles through gentle stirring by hydraulic or mechanical means.

FLOCHK
(Acronyms and Abbreviations)
DA   October 12, 1990
BT1 Computer Codes
SO   Nuclear Facilities Incident Database
DEF (NFI) Computer code on the Control Computer that checks for fuel element failure

FLOCS
(EPA)
DA   October 12, 1990
RT   Flocculation
SO   Environmental Protection Agency Glossary
DEF (EPA) Clumps of solids formed in sewage by biological or chemical action.

FLOODPLAINS
(CFR)
DA   May 15, 1991
BT1 Sites/Areas
BT1 Surface Water Resources
 BT2 Natural Resources

RT   Notice of Involvement
SO   Environmental Management
DEF (CFR) Lowlands adjoining inland and coastal waters and relatively flat areas and flood-prone areas of offshore islands including, at a minimum, that area inundated by a 1 percent or greater chance flood in any given year. The base floodplain is defined as the 100-year (1.0 percent) floodplain. The critical action floodplain is defined as the 500-year (.02 percent) floodplain.

FLOOR OPENING
(1698 FLOOR OPENIN)
DA   December 10, 1990
BT1 Site (DOE FRASE Vocabulary)
 BT2 Sites/Areas
SO   DOE FRASE VOCABULARY

FLOOR SWEEPS
(EPA)
DA   October 12, 1990
BT1 Vapor Capture Systems
 BT2 Control Devices
  BT3 Devices
 BT2 Systems
SO   Environmental Protection Agency Glossary
DEF (EPA) Vapor collectors designed to capture vapors which are heavier than air and which collect along the floor.

FLOW CHARTS
(SSDC)
DA   October 12, 1990
BT1 Diagrams
SO   System Safety Development Center Glossary
DEF (SSDC) Schematic diagrams or expository outlines showing the succession of operations in an activity or system.

FLOW FAILURES
(USGS)
DA   October 12, 1990.
BT1 Failures
 BT2 Accidents
BT1 Ground Failures
 BT2 Failures
  BT3 Accidents
 BT2 Liquefaction
  BT3 Processes
RT   Soil Mechanics
RT   Soils
SO   Natural Phenomenon
DEF (USGS) Consist of liquefied soil or blocks if intact material riding on a layer of liquefied soil. Flow failures can form in loose saturated sands or soil on slopes greater than 3 degrees. These flows typically move several tens of feet and if conditions permit, can travel tens

of miles at velocities as great as many tens of miles per hour.

FLOW METER(S)
(2752 FLOW METER)
DA   January 3, 1991
BT1 Indicator(s)
 BT2 Instrument(s)
  BT3 Equipment/Parts - Instrumentation/Measuring (DOE FRASE Voc.)
   BT4 Equipment
SO   DOE FRASE VOCABULARY

FLOW MONITOR
DA   January 8, 1991
SF   DFM
BT1 Monitors
 BT2 Equipment

FLOW RATES
(SDWA; CFR)
DA   October 12, 1990
NT1 Million Gallons Per Day
RT   Fluids
RT   Subcritical Flows
RT   Supercritical Flows
SO   Water Pollution
DEF (CFR) The volume per time unit given to the flow of gases or other fluid substances which emerge from an orifice, pump, turbine or passes along a conduit or channel.

FLOW SAND FILTRATION
(EPA)
DA   October 12, 1990
BT1 Filtration
 BT2 Pollution Recovery Processes
  BT3 Processes
SO   Environmental Protection Agency Glossary

FLOW-THROUGH PROCESS TANKS
(SWDA; RCRA; CFR)
DA   October 12, 1990
BT1 Tanks
 BT2 Facility Components
SO   Wastes
DEF (CFR) Tanks that form an integral part of a production process through which there is a steady, variable, recurring, or intermittent flow of materials during the operation of the process. Flow-through process tanks do not include tanks used for the storage of materials prior to their introduction into the production process or for the storage of finished products or by-products from the production process.

FLOW WEIGHTED AVERAGE (DELTA T)
DA   January 8, 1991
SF   FWA
BT1 Measurements

FLOW ZONE
DA   January 8, 1991
SF   *FZ*
BT1 Zones
 BT2 Sites/Areas

FLUE GAS DESULFURIZATION
(EPA)
DA   October 12, 1990
RT   Desulfurization
RT   Flue Gases
RT   Scrubbers
SO   Environmental Protection Agency
      Glossary
DEF (EPA) A technology which uses a
      sorbent, usually lime or limestone,
      to remove sulfur dioxide from the
      gases produced by burning fossil
      fuels. Flue gas desulfurization is
      current the state-of-the art
      technology in use by major SO_2
      emitters, e.g., power plants.

FLUE GASES
(EPA)
DA   October 12, 1990
SY   Stack Gases
BT1 Air Pollutants
BT1 Gases
NT1 Condenser Stack Gases
RT   Acid Deposition
RT   Combustion Products
RT   Flue Gas Desulfurization
RT   Stacks
RT   Vapor Plumes
SO   Environmental Protection Agency
      Glossary
DEF (EPA) The gases coming out of a
      chimney after combustion in the
      burner it is venting. They can
      include nitrogen oxides, carbon
      oxides, water vapor, sulfur oxides,
      particles and many chemical
      pollutants.

FLUIDS
(SWDA; CFR)
DA   October 12, 1990
NT1 Formation Fluid
NT1 Liquids
 NT2 Combustible Liquids
 NT2 Cryogenic Liquids
 NT2 Flammable Liquids
 NT2 Free Liquids
 NT2 Free Products
 NT2 Influents
 NT2 Natural Gas Liquids
 NT2 Pyrophoric Liquids
 NT2 Viscous Liquids
NT1 Raffinate
RT   Flow Rates
SO   Water Pollution
DEF (CFR) Materials or substances
      which flow or move whether in a
      semisolid, liquid, gas, or any other
      form or state.

FLUMES
(EPA)

DA   October 12, 1990
RT   Earth-Lined Channels
SO   Environmental Protection Agency
      Glossary
DEF (EPA) Natural or man-made
      channels that divert water.

FLUORESCENT LIGHT BALLASTS
(TSCA; CFR)
DA   October 19, 1990
RT   PCB Equipment
SO   Hazardous Materials
DEF (CFR) Devices that electrically
      control fluorescent light fixtures
      and that include a capacitor
      containing 0.1 kg or less of
      dielectric.

FLUORIDES
(EPA)
DA   October 12, 1990
RT   Fluorosis
SO   Environmental Protection Agency
      Glossary
DEF (EPA) Gaseous, solid, or dissolved
      compounds containing fluorine
      that result from industrial
      processes. Excessive amounts in
      food can lead to fluorosis.

FLUORINE
DA   January 8, 1991
SF   *F₂*

FLUOROCARBONS
(EPA)
DA   October 12, 1990
SF   *FCs*
BT1 Halogenated Organic Compounds
 BT2 Halogenated
 BT2 Organic Chemicals
  BT3 Chemical Substances
NT1 Chlorofluorocarbons (CFCs)
SO   Environmental Protection Agency
      Glossary
DEF (EPA) Any of a number of organic
      compounds analogous to
      hydrocarbons in which one or
      more hydrogen atoms are
      replaced by fluorine. Once used in
      the United States as a propellant
      in aerosols, they are now primarily
      used in coolants and some
      industrial processes. FCs
      containing chlorine are called
      chlorofluorocarbons (CFCs). They
      are believed to be modifying the
      ozone layer in the stratosphere,
      thereby allowing more harmful
      solar radiation to reach the Earth's
      surface.

FLUOROSIS
(EPA)
DA   October 12, 1990
BT1 Illnesses
RT   Fluorides

SO   Environmental Protection Agency
      Glossary
DEF (EPA) An abnormal condition
      caused by excessive intake of
      fluorine, characterized chiefly by
      mottling of the teeth.

FLUSH
(EPA's Information Resources Directory:
Environmental Glossary)
DA   October 12, 1990
SO   Environmental Protection Agency
      Glossary
DEF (EPA) (1) To open a cold-water tap
      to clear out all the water which
      may have been sitting for a long
      time in the pipes. In new homes,
      to flush a system means to send
      large volumes of water gushing
      through the unused pipes to
      remove loose particles of solder
      and flux. (2) To force large
      amounts of water through liquid to
      clean out piping or tubing, storage
      or process tanks.

FLUX CHANNEL(S)
(2561 FLUX CHANNEL)
DA   January 3, 1991
BT1 Equipment/Parts - Nuclear (DOE
      FRASE Vocabulary)
 BT2 Equipment
 BT2 Reactor Components
SO   DOE FRASE VOCABULARY

FLY ASH
(SWDA; RCRA; ESH)
DA   October 12, 1990
BT1 Air Pollutants
BT1 Ashes
 BT2 Wastes
RT   Baffle Chambers
RT   Dust
RT   Particulates
RT   Soot
SO   Environmental Management
SO   Environmental Protection Agency
      Glossary
SO   Wastes
DEF (EPA) Non-combustible residual
      particles from the combustion
      process, carried by flue gas.
      (ESH) Suspended particles,
      charred paper, dust, soot, and
      other partially oxidized matter
      carried in the products of
      combustion.

FM
DA   October 12, 1990
SEE Factory Mutual
SO   Acronyms

FMEA
DA   October 12, 1990
SEE Failure Mode and Effect Analysis
SO   Acronyms

BT-Broader Term   NT-Narrower Term   RT Related Term

FMPC
  DA October 12, 1990
  SEE Feed Materials Production Center
  SO Acronyms

FNSI
  DA October 12, 1990
  SEE Finding of No Significant Impact
  SO Acronyms

FOAM SYSTEM
  (2034 FOAM SYSTEM)
  DA December 10, 1990
  BT1 Systems (DOE FRASE Vocabulary)
   BT2 Systems
  RT Fire Suppression System
  SO DOE FRASE VOCABULARY

FOGGING
  (EPA)
  DA October 12, 1990
  BT1 Processes
  RT Pesticides
  SO Environmental Protection Agency
       Glossary
  DEF (EPA) Applying a pesticide by
       rapidly heating the liquid chemical
       so that it forms very fine droplets
       that resemble smoke or fog. It may
       be used to destroy mosquitoes,
       black flies, and similar pests.

FONSI
  DA May 14, 1991
  SEE Finding of No Significant Impact
  SO Acronyms

FOOD-CHAIN CROPS
  (SWDA; RCRA)
  DA October 12, 1990
  RT Food Chains
  SO Wastes
  DEF Tobacco, crops grown for human
       consumption, and crops grown for
       feed for animals whose products
       are consumed by humans.

FOOD CHAINS
  (EPA)
  DA October 12, 1990
  RT Common Exposure Routes
  RT Exposure Pathways
  RT Food-chain Crops
  RT Posing an Exposure Risk to Food
       or Feed
  SO Environmental Protection Agency
       Glossary
  DEF (EPA) Sequences of organisms,
       each of which uses the next, lower
       member of the sequence as a
       food source.

FOOD SERVICE ACTIVITY
  (1232 FS ACTIVITY)
  DA November 28, 1990
  BT1 Activity Types (DOE FRASE
       Vocabulary)
   BT2 Activities

  RT Eating Area
  SO DOE FRASE VOCABULARY

FOOD SERVICE EMPLOYEE
  (0521 FOOD SERVICE)
  DA November 28, 1990
  BT1 Admin. Support/Clerical Employee
   BT2 Occupations
   BT2 Personnel
  RT Kitchen
  SO DOE FRASE VOCABULARY

FOOT PROTECTION
  (2662 FOOT PROTECT)
  DA January 3, 1991
  BT1 Personal Protective Equipment
   BT2 Equipment/Parts - Personal
        Protective (DOE FRASE
        Vocabulary)
    BT3 Equipment
  NT1 Ankle Protection
  NT1 Metatarsal Protection
  NT1 Safety Boots
  NT1 Safety Shoe(s)
  NT1 Shoe Cover(s)
   NT2 Metal Shoe Cover
  RT Foot/Feet
  RT Heel(s)
  RT Sole(s)
  RT Toe(s)
  SO DOE FRASE VOCABULARY

FOOT/FEET
  (1127 FOOTP)
  DA November 28, 1990
  BT1 Human Body Parts
  NT1 Heel(s)
  NT1 Sole(s)
  NT1 Toe(s)
  RT Ankle(s)
  RT Foot Protection
  RT Leg(s)
  RT Safety Boots
  SO DOE FRASE VOCABULARY

FORCE MAINS
  (DOE Order 6430.1A)
  DA October 12, 1990
  RT Storm Water Collection Systems
  SO Construction
  DEF (DOE Order 6430.1A) Discharge
       lines from a sewage or stormwater
       lift station.

FORCE-ON-FORCE EXERCISES
  (Doe Order 5480.16)
  DA October 12, 1990
  BT1 Exercises
  RT Full-Scale Exercises
  SO Firearms
  DEF (DOE Order 5480.16) Simulates the
       actual intrusion of an enemy force
       and the appropriate response to
       such an intrusion. This may
       include limited scope performance
       tests and similar exercises.

FOREARM(S)
  (1098 FOREARM)

  DA November 28, 1990
  BT1 Arm(s)
   BT2 Human Body Parts
  RT Forearm Protection
  SO DOE FRASE VOCABULARY

FOREARM PROTECTION
  (2663 FOREARM PROT)
  DA January 3, 1991
  RT Arm(s)
  RT Elbow(s)
  RT Forearm(s)
  SO DOE FRASE VOCABULARY

FOREHEAD
  (1080 FOREHEAD)
  DA November 28, 1990
  BT1 Head
   BT2 Human Body Parts
  RT Faceshield
  RT Head Protection
  SO DOE FRASE VOCABULARY

FOREIGN COMMERCE
  (ESA; CFR)
  DA October 12, 1990
  RT Commercial Activities
  SO Endangered Species
  DEF (CFR) Includes, among other
       things, any transaction (A)
       between persons within one
       foreign country; (B) between
       persons in two or more foreign
       countries; (C) between a person
       within the United States and a
       person in a foreign country; or (D)
       between persons within the United
       States, where the fish and wildlife
       in question are moving in any
       country or countries outside the
       United States.

FOREST WORKER
  (0570 FOREST WORKE)
  DA November 28, 1990
  BT1 Agriculture Personnel
   BT2 Occupations
   BT2 Personnel
  SO DOE FRASE VOCABULARY

FORESTS (REACTOR)
  (NFI)
  DA October 12, 1990
  BT1 Reactor Components
  SO Nuclear Facilities Incident Database
  SO Radiation
  DEF (NFI) Equipment located near top
       of reactor that checks for fuel
       element failure.

FORKLIFT(S)
  (2321 FORKLIFT)
  DA December 10, 1990
  BT1 Hoisting Apparatus
   BT2 Material Handling Device
    BT3 Devices
    BT3 Equipment/Parts - Material
         Handling (DOE FRASE
         Vocabulary)

BT4 Equipment
SO DOE FRASE VOCABULARY

FORMAL CONSULTATION
(ESA; CFR)
DA October 12, 1990
BT1 Administrative Processes
 BT2 Processes
RT Biological Opinion
RT Informal Consultation
RT Reasonable and Prudent
     Alternatives
SO Endangered Species
DEF (CFR) A process between the U.S.
    Fish and Wildlife Service and the
    Federal agency that commences
    with the Federal agency's written
    request for consultation under
    section 7(a)(2) of the Endangered
    Species Act and concludes with
    the Service's issuance of the
    biological opinion under section
    7(b)(3) of the Act.

FORMALDEHYDE
(EPA)
DA October 12, 1990
BT1 CERCLA Hazardous Substances
 BT2 Hazardous Substances
BT1 Hazardous Constituents
BT1 Organic Chemicals
 BT2 Chemical Substances
SO Environmental Protection Agency
     Glossary
DEF (EPA) A colorless, pungent,
    irritating gas, CH_2O, used chiefly
    as a disinfectant and preservative
    and in synthesizing other
    compounds and resins.

FORMATION FLUID
(SDWA; CFR)
DA October 12, 1990
BT1 Fluids
SO Water Pollution
DEF (CFR) Fluid present in a formation
    under natural conditions, as
    opposed to introduced fluids, such
    as drilling mud.

FORMATIONS
(SDWA; CFR)
DA October 12, 1990
NT1 Aquifers
 NT2 Confined Aquifers
 NT2 Exempted Aquifers
 NT2 Principal Source Aquifers
 NT2 Semi-Confined Aquifers
 NT2 Significant Source of Groundwater
 NT2 Sole Source Aquifer
 NT2 Underground Sources of Drinking
     Water
 NT2 Uppermost Aquifers
NT1 Confining Zones
NT1 Injection Zones
 NT2 Injection Interval
RT Lithology
RT Strata
RT Subsidence

SO Water Pollution
DEF Bodies of rock characterized by a
    degree of lithologic homogeneity
    which is prevailingly, but not
    necessarily, tabular and is
    mappable on the earth's surface
    or traceable in the subsurface.

FORMERLY UTILIZED SITES
REMEDIAL ACTIONS PROGRAM
DA January 8, 1991
SF *FUSRAP*
BT1 Remedial Action Program
 BT2 Programs

FORMULA QUANTITIES
DA January 24, 1991
BT1 Quantities
RT Special Nuclear Materials
SO Environmental Management
DEF (CFR) Strategic special nuclear
    material in any combination in a
    quantity of 5000 grams or more
    computed by the formula, grams =
    (grams contained U235) + 2.5
    (grams U235 + grams plutonium).

FORMULATION
(EPA)
DA October 12, 1990
NT1 Manufacturers Formulation
SO Environmental Protection Agency
     Glossary
DEF (EPA) The substance or mixture of
    substances which is comprised of
    all active and inert ingredients in a
    pesticide.

FOSSIL FUEL AND WOOD
RESIDUE-FIRED STEAM
GENERATING UNIT
(CAA; CFR)
DA October 12, 1990
BT1 Steam Generator
 BT2 Equipment
RT Fossil Fuels
RT Wood Residues
SO Air Pollution
DEF (CFR) A furnace or boiler used in
    the process of burning fossil fuel
    and wood residue for the purpose
    of producing steam by heat
    transfer.

FOSSIL FUELS
(CAA; CFR)
DA October 12, 1990
BT1 Fuels
BT1 Geologic Resources
 BT2 Natural Resources
NT1 Coal
 NT2 Ultra Clean Coal
NT1 Natural Gas
 NT2 Field Gases
 NT2 Natural Gas Liquids
NT1 Petroleum
RT Acid Deposition
RT Fossil Fuel and Wood

Residue-Fired Steam Generating
Unit
SO Air Pollution
DEF (CFR) Natural gas, petroleum, coal,
    and any form of solid, liquid, or
    gaseous fuel derived from such
    materials for the purpose of
    creating useful heat.

FOUNDATIONS
DA February 1, 1991
BT1 Research and Development
    Organizations
 BT2 Organizations
NT1 Hanford Environmental Health
    Foundation
NT1 Lovelace Medical Foundation
NT1 University of Georgia Research
    Foundation, Inc.
DEF (WEBSTER) An organization or
    institution established by
    endowment with provision for
    future maintenance.

FOUNDRIES
(CAA; CFR)
DA October 12, 1990
BT1 Facilities
RT Beryllium Alloys
SO Air Pollution
DEF (CFR) Facilities engaged in the
    melting or casting of beryllium
    metal or alloy.

FP
DA October 12, 1990
SEE Fission Products
SO Acronyms

FPTS
DA October 12, 1990
SEE Fire Protection Tracking System
SO Acronyms

FRACAS
DA October 12, 1990
SEE Failure Reporting Analysis and
     Corrective Action System
SO Acronyms

FRACTURE
(1322 FR)
DA November 28, 1990
BT1 Injuries
RT Bone(s)
SO DOE FRASE VOCABULARY

FRANTIC
(Acronyms and Abbreviations)
DA October 12, 1990
BT1 Computer Codes
DEF (NFI) A Computer code to analyze
    time-dependent reliability.

FRASE VOCABULARY
DA February 28, 1991
SY DOE FRASE VOCABULARY

BT-Broader Term   NT-Narrower Term   RT Related Term

SY Factor Relationship and Sequence
of Events Vocabulary
BT1 Vocabulary
RT DOE FRASE Categories
RT Key Word
RT Safety Performance Measurement
System
DEF A controlled vocabulary designed
for use in searching the narrative
description fields of records on the
Safety Performance Measurement
System (SPMS).

FREE LIQUIDS
(SWDA; RCRA)
DA October 12, 1990
BT1 Liquids
BT2 Fluids
SO Environmental Management
SO Wastes
DEF (CFR) Liquids which readily
separate from the solid portion of
a waste under ambient
temperature and pressure.

FREE PRODUCTS
(SDWA; RCRA)
DA October 12, 1990
BT1 Liquids
BT2 Fluids
RT Regulated Substances
SO Wastes
DEF (CFR) Refers to regulated
substances that are present as
non-aqueous phase liquids (e.g.,
liquids not dissolved in water).

FREEBOARD
(SWDA; RCRA)
DA October 12, 1990
RT Hydraulic Structures
SO Construction
SO Environmental Management
SO Wastes
DEF (CFR) The height between the
normal water surface elevation
and the top of a hydraulic
structure.

FREIGHT CONTAINERS
(SWDA; RCRA)
DA October 12, 1990
BT1 Containers
BT1 Unit Load Devices
NT1 Unit Load Devices
RT Bulk Packaging
RT Containerships
RT Rail Freight Cars
SO Hazardous Materials
DEF Reusable containers having a
volume of 64 cubic feet or more,
designed and constructed to
permit being lifted with its contents
intact and intended primarily for
containment of package during
transportation.

FREON-114
DA January 8, 1991
SF $C_2C_{12}F_4$

FREQUENCY DISTRIBUTION
(SSDC)
DA October 12, 1990
BT1 Distribution
NT1 Joint Frequency Distribution
SO System Safety Development Center
Glossary
DEF (SSDC) The relative frequency with
which a variable quantity or variate
assumes particular values.

FREQUENCY/SEVERITY
(SSDC)
DA October 12, 1990
SF *F/S*
RT Line of Balance
RT Risk Assessment
SO System Safety Development Center
Glossary
DEF (SSDC) Risk projection using all
experience data from events of
trivial consequences to the most
serious events experienced by an
organization or system under
study.

FRESH WATER
(EPA; CFR)
DA October 12, 1990
BT1 Water
RT Limnology
RT Underground Sources of Drinking
Water
SO Environmental Protection Agency
Glossary
DEF (EPA) Water that generally contains
less than 1,000 milligrams-per-liter
of dissolved solids.

FRGT
DA October 12, 1990
SEE Fast Response Gamma
Thermometer
SO Acronyms

FRIABLE ASBESTOS
(CAA)
DA October 12, 1990
BT1 Asbestos
BT2 Carcinogens
BT3 Hazardous Substances
BT2 CERCLA Hazardous Substances
BT3 Hazardous Substances
RT Asbestos Waste from Control
Devices
RT Emergency Projects
RT Friable Asbestos-Containing
Materials
RT Strip
SO Air Pollution
DEF (CFR) Asbestos insulation that is
loose and capable of becoming
airborne.

**FRIABLE ASBESTOS-CONTAINING
MATERIALS**
(TSCA; CFR; USC)
DA October 12, 1990
BT1 Asbestos-containing Materials
BT2 Asbestos Materials
BT3 Materials
RT Friable Asbestos
SO Hazardous Materials
DEF (CFR) Asbestos-containing
materials applied on ceilings,
walls, structural members, piping,
duct work, or any other part of a
building which when dry may be
crumbled, pulverized, or reduced
to powder by hand pressure. Term
includes non-friable
asbestos-containing materials after
such previously non-friable
asbestos-containing materials
become damaged to the extent
that when dry they may be
crumbled, pulverized, or reduced
to powder by hand pressure.

FRIABLE ASBESTOS MATERIALS
(CFR)
DA November 15, 1990
BT1 Asbestos Materials
BT2 Materials
RT Particulate Asbestos Materials
SO Environmental Management
SO Hazardous Materials
DEF (CFR) Materials containing more
than 1 percent asbestos by weight
that hand pressure can crumble,
pulverize, or reduce to powder
when dry.

FRMAC DIRECTOR
(EMER)
DA February 1, 1991
BT1 Personnel
SO Emergency Preparedness
DEF (EMER) The manager of the
Federal Radiological Monitoring
and Assessment Center (FRMAC).
Designated by Department of
Energy in an emergency and by
Environmental Protection Agency
(EPA) in the recovery.

**FROSTBITE/OTHER LOW TEMP
EFFECTS**
(1386 FROSTBITE)
DA November 28, 1990
BT1 Effects
BT1 Illnesses
SO DOE FRASE VOCABULARY

FSAR
DA October 12, 1990
SEE Final Safety Analysis Report
SO Acronyms

FSH
DA October 12, 1990
SEE Fuel Sleeve Housing
SO Acronyms

SY-Synonymous Terms   SO-Source/Subject Category   SF-See From

FTA
DA   October 12, 1990
SEE Fault Tree Analysis
SO Acronyms

FUEL CANISTER(S)
(2562 FUEL CANISTE)
DA   January 3, 1991
BT1 Fuel Handling Equipment
 BT2 Equipment/Parts - Nuclear (DOE
      FRASE Vocabulary)
  BT3 Equipment
  BT3 Reactor Components
RT   Fuel Element(s)
SO   DOE FRASE VOCABULARY

FUEL DISTRIBUTION ANALYZER
DA   January 8, 1991
SF   *FDA*
BT1 Equipment

FUEL DRAWER(S)
(2563 FUEL DRAWER)
DA   January 3, 1991
BT1 Fuel Handling Equipment
 BT2 Equipment/Parts - Nuclear (DOE
      FRASE Vocabulary)
  BT3 Equipment
  BT3 Reactor Components
SO   DOE FRASE VOCABULARY

FUEL ECONOMY STANDARD
(EPA)
DA   October 12, 1990
BT1 Standards
 BT2 Codes, Standards, and
      Regulations
SO   Environmental Protection Agency
       Glossary
DEF (EPA) The Corporate Average Fuel
      Economy Standard (CAFE) which
      went into effect in 1978. It was
      meant to enhance the national fuel
      conservation effort by slowing fuel
      consumption through a
      miles-per-gallon requirement for
      motor vehicles.

FUEL ELEMENT(S)
(2564 FUEL ELEMENT)
DA   January 3, 1991
BT1 Equipment/Parts - Nuclear (DOE
      FRASE Vocabulary)
 BT2 Equipment
 BT2 Reactor Components
NT1 Fuel Plate
 NT2 Cladding
 NT2 Plutonium Fuel Plate(s)
NT1 Fuel Rod(s)
NT1 Piles (Nuclear)
RT   Core(s)
RT   Fuel Canister(s)
RT   Fuel Handling Equipment
RT   Reactor Fuel
SO   DOE FRASE VOCABULARY
DEF (NRC Glossary of Terms: Nuclear
      Power and Radiation) A cluster of
      fuel rods (or plates). Also called a
      fuel assembly. Many fuel

assemblies make up a reactor
core.

**FUEL ELEMENT FAILURE GAS
CHROMATOGRAPH**
DA   January 8, 1991
SF   *FEFGC*

FUEL FABRICATION FACILITY
(1797 FUEL FABRICA)
DA   December 10, 1990
BT1 Facility (DOE FRASE Vocabulary)
 BT2 Facilities and Buildings (DOE
      FRASE Vocabulary)
  BT3 Facilities
SO   DOE FRASE VOCABULARY

FUEL HANDLING ACTIVITY
(1233 FH ACTIVITY)
DA   November 28, 1990
BT1 Activity Types (DOE FRASE
      Vocabulary)
 BT2 Activities
SO   DOE FRASE VOCABULARY

FUEL HANDLING EQUIPMENT
(2565 FH EQUIPMENT)
DA   January 3, 1991
BT1 Equipment/Parts - Nuclear (DOE
      FRASE Vocabulary)
 BT2 Equipment
 BT2 Reactor Components
NT1 Fuel Canister(s)
NT1 Fuel Drawer(s)
RT   Fuel Element(s)
SO   DOE FRASE VOCABULARY

FUEL PLATE
(2566 FUEL PLATE)
DA   January 3, 1991
BT1 Fuel Element(s)
 BT2 Equipment/Parts - Nuclear (DOE
      FRASE Vocabulary)
  BT3 Equipment
  BT3 Reactor Components
NT1 Cladding
NT1 Plutonium Fuel Plate(s)
SO   DOE FRASE VOCABULARY

FUEL ROD(S)
(2567 FUEL ROD)
DA   January 3, 1991
BT1 Fuel Element(s)
 BT2 Equipment/Parts - Nuclear (DOE
      FRASE Vocabulary)
  BT3 Equipment
  BT3 Reactor Components
SO   DOE FRASE VOCABULARY

FUEL SLEEVE HOUSING
DA   January 8, 1991
SF   *FSH*

FUEL TANKS
(SWDA; RCRA)
DA   October 12, 1990
BT1 Tanks
 BT2 Facility Components

SO   Hazardous Materials
DEF Tanks other than cargo tanks, used
      to transport flammable or
      combustible liquid, or compressed
      gas for the purpose of supplying
      fuel for propulsion of the transport
      vehicle to which it is attached, or
      for the operation of other
      equipment on the transport
      vehicle.

FUEL TOOL(S)
(2568 FUEL TOOL)
DA   January 3, 1991
BT1 Equipment/Parts - Nuclear (DOE
      FRASE Vocabulary)
 BT2 Equipment
 BT2 Reactor Components
SO   DOE FRASE VOCABULARY

FUELS
DA   February 25, 1991
NT1 Fossil Fuels
 NT2 Coal
  NT3 Ultra Clean Coal
 NT2 Natural Gas
  NT3 Field Gases
  NT3 Natural Gas Liquids
 NT2 Petroleum
NT1 Motor Fuels
NT1 Propellants
 NT2 Beryllium Propellants
NT1 Reactor Fuel
NT1 Solid-Waste-Derived Fuel
NT1 Spent Fuel

FUGITIVE EMISSIONS
(CAA; EPA)
DA   October 12, 1990
BT1 Emissions
 BT2 Air Pollutants
NT1 Uncontrolled Total Arsenic
      Emissions
RT   Particulate Matter Emissions
RT   Secondary Emissions
SO   Air Pollution
SO   Environmental Management
SO   Environmental Protection Agency
       Glossary
DEF (EPA) Emissions not caught by a
      capture system.

FULL-SCALE EXERCISES
(EMER)
DA   February 1, 1991
BT1 Exercises
RT   Federal Field Exercises
RT   Force-on-Force Exercises
SO   Emergency Preparedness
DEF (EMER) Exercises designed
      primarily for the purpose of
      validating the integrated
      emergency preparedness
      capability of a facility and state
      and local jurisdictions in an
      operational environment.

FULL-TIME EMPLOYEES
(CERCLA; CFR)

DA October 12, 1990
BT1 Personnel
RT Contractor Employees
SO Compensation and Liability
DEF (CFR) 2,000 hours per year of full-time equivalent employment. A facility would calculate the number of full-time employees by totaling the hours worked during the calendar year by all employees, including contract employees, and dividing that total by 2,000 hours.

FUMES
(EPA)
DA October 12, 1990
BT1 Emissions
BT2 Air Pollutants
RT Irritating Materials
SO Environmental Protection Agency Glossary
DEF (EPA) Tiny particles trapped in vapor in a gas stream.

FUMIGANTS
(EPA)
DA October 12, 1990
BT1 Pesticides
BT2 Hazardous Substances
RT Ethylene Dibromide
SO Environmental Protection Agency Glossary
DEF (EPA) Pesticides that are vaporized to kill pests. Used in buildings and greenhouses.

FUNCTIONAL APPRAISALS
(DOE Order 5482.1B; ESH)
DA October 12, 1990
BT1 Management Appraisals
BT2 Appraisals
RT Functional Equivalent
RT Management Appraisals
SO Construction
SO Environmental Management
SO Management
SO System Safety Development Center Glossary
DEF (DOE Order 5482.1B) Documented reviews of ES&H specialty disciplines performed in accordance with written guidance and criteria to verify, by examination and evaluation of objective evidence at the facility and/or operation, that applicable elements of the program have been developed, documented, and effectively implemented in accordance with specific ES&H requirements and needs.

FUNCTIONAL EQUIVALENT
(EPA)
DA October 12, 1990
RT Equivalent Methods
RT Functional Appraisals

SO Environmental Protection Agency Glossary
DEF (EPA) Term used to describe EPA's decision-making process and its relationship to the environmental review conducted under the National Environmental Policy Act (NEPA). A review is considered functionally equivalent when it addresses the substantive components of a NEPA review.

FUNCTIONAL UNITS
(DOE Order 4330.4A)
DA June 4, 1991
RT Audits (SSDC)
RT Facilities
RT Laboratories
RT Life Cycle Plans
RT Plants
RT Projects
SO Management
DEF (DOE Order 4330.4A) Logical and systematic groups of property that are necessary to support the site mission. A functional unit must be described in a breakdown structure for each site in order that it be properly identified and managed. Functional units will vary in size and scope within sites and from site to site, depending on the type or types of activities being carried out. A functional unit will often comprise a total facility (e.g., laboratory, production plant, or utility) but may also be a portion of a facility (e.g., production line, shop, clean room, or tooling). A functional unit is the basic entity for justifying individual projects and must be auditable in terms of mission requirements or performance standards.

FUNCTIONALLY EQUIVALENT COMPONENTS
(SWDA; RCRA)
DA October 12, 1990
BT1 Components
RT Acquisition of the Equivalent
SO Wastes
DEF (SWDA; RCRA) Components which perform the same function or measurement and which meet or exceed the performance specifications of another component.

FUNGI
(EPA; USC)
DA October 12, 1990
BT1 Plants
SO Environmental Protection Agency Glossary
DEF (EPA) (Singular, Fungus) Molds, mildews, yeasts, mushrooms, and puffballs, a group of organisms that lack chlorophyll (i.e., are not photosynthetic) and which are

usually non-mobile, filamentous, and multicellular. Some grow in the ground, others attach themselves to decaying trees and other plants, getting their nutrition from decomposing organic matter. Some cause disease, others stabilize sewage and break down solid wastes in composting. Any non-chlorophyll-bearing thallophyte (that is, any non-chlorophyll-bearing plant of a lower order than mosses and liverworts), as for example, rust, smut, mildew, mold, yeast, and bacteria, except those on or in living man or other animals and those on or in processed food, beverages, or pharmaceuticals.

FUNGICIDES
(EPA)
DA October 12, 1990
BT1 Pesticides
BT2 Hazardous Substances
NT1 Dinocap
SO Environmental Management
SO Environmental Protection Agency Glossary
DEF (EPA) Pesticides which are used to control, prevent, or destroy fungi.

FURNACE(S)
(2483 FURNACE)
DA January 3, 1991
BT1 Heater(s)
BT2 Equipment/Parts - Heating (DOE FRASE Vocabulary)
BT3 Equipment
NT1 Boiler/Industrial Furnaces
SO DOE FRASE VOCABULARY

FUSE(S)
(2394 FUSE)
DA January 3, 1991
BT1 Equipment/Parts - Electrical (DOE FRASE Vocabulary)
BT2 Equipment
SO DOE FRASE VOCABULARY

FUSION FACILITY
(1798 FUSION FACIL)
DA December 10, 1990
BT1 Facility (DOE FRASE Vocabulary)
BT2 Facilities and Buildings (DOE FRASE Vocabulary)
BT3 Facilities
SO DOE FRASE VOCABULARY

FUSRAP
DA October 12, 1990
SEE Formerly Utilized Sites Remedial Actions Program
SO Acronyms

FUSSELL-VESELY
DA January 8, 1991
SF F-V

SY-Synonymous Terms   SO-Source/Subject Category   SF-See From

FVC
DA   October 12, 1990
SEE Filtered Vented Containment
SO Acronyms

FW
DA   October 12, 1990
SEE Feedwater
SO Acronyms

FWA
DA   October 12, 1990
SEE Flow Weighted Average (Delta T)
SO Acronyms

FY
DA   October 12, 1990
SEE Fiscal Year
SO Acronyms

FZ
DA   October 12, 1990
SEE Flow Zone
SO Acronyms

F$_2$
DA   October 12, 1990
SEE Fluorine
SO Acronyms

G-M
DA   October 12, 1990
SEE Geiger-Mueller Counters
SO Acronyms

G-O-S
(NFI)
DA   October 12, 1990
SO Nuclear Facilities Incident Database
DEF (NFI) Gang control - off - Single
     control.

G/T
DA   October 12, 1990
SEE Activity per Ton
SO Acronyms

gal
DA   October 12, 1990
SEE gallon
SO Acronyms

GALLERY
(1700 GALLERY)
DA   December 10, 1990
BT1 Site (DOE FRASE Vocabulary)
 BT2 Sites/Areas
SO DOE FRASE VOCABULARY

GALLON
DA   January 8, 1991
SF *gal*
BT1 Units of Measure

GAME FISH
(EPA)
DA   October 12, 1990

BT1 Fish
 BT2 Animals
SO Environmental Protection Agency
     Glossary
DEF (EPA) Species like trout, salmon, or
     bass, caught for sport. Many of
     them show more sensitivity to
     environmental change than
     "rough" fish.

GAMMA RADIATION
(EPA)
DA   October 12, 1990
SY Gamma Rays
BT1 Ionizing Radiation
 BT2 Radiation
RT Radioactive Decay
RT X-rays
SO Environmental Protection Agency
     Glossary
DEF (EPA) Gamma rays are true rays of
     energy in contrast to alpha and
     beta radiation. The properties are
     similar to x rays and other
     electromagnetic waves. They are
     the most penetrating waves of
     radiant nuclear energy but can be
     blocked by dense materials such
     as lead.

GAMMA RAYS
(EPA; NIH; NCRP; EMER)
DA   October 12, 1990
SY Gamma Radiation
RT Half Value Layer (Half Thickness)
SO Emergency Preparedness
SO Radiation
DEF (NIH) Very penetrating
     electromagnetic radiation of
     nuclear origin. Except for origin,
     identical to x-rays. (NCRP)
     Electromagnetic radiation
     frequently accompanying alpha
     and beta emissions as radioactive
     materials decay.

GANG TEMPERATURE MONITOR
(NFI)
DA   October 12, 1990
SF *GTM*
BT1 Monitors
 BT2 Equipment
RT Gangs
SO Radiation
DEF (NFI) Activates explosive valves on
     the SSS.

GANGS
(NFI)
DA   October 12, 1990
BT1 Control Rod(s)
 BT2 Equipment/Parts - Nuclear (DOE
     FRASE Vocabulary)
  BT3 Equipment
  BT3 Reactor Components
RT Gang Temperature Monitor
SO Nuclear Facilities Incident Database
SO Radiation
DEF (NFI) Associated groups of control

rods symmetrically positioned
throughout the reactor lattice.

GARAGE
(1780 GARAGE)
DA   December 10, 1990
BT1 Building (DOE FRASE Vocabulary)
 BT2 Facilities and Buildings (DOE
     FRASE Vocabulary)
  BT3 Facilities
RT Maintenance Area
RT Maintenance Shop
RT Mechanic/Repairer
RT Parking Space
RT Parking Lot
RT Vehicle Service Area
RT Vehicle Maint/Repair Activity
SO DOE FRASE VOCABULARY

**GAS CENTRIFUGE ENRICHMENT
PLANT**
DA   January 8, 1991
SF *GCEP*
BT1 Facilities

GAS VOLUME RATIO
(NFI)
DA   October 12, 1990
SF *GVR*
BT1 Ratios
SO Radiation
DEF (NFI) Limiting value for lithium
     target irradiation.

GASEOUS DIFFUSION PLANT
DA   January 8, 1991
SF *GDP*
BT1 Facilities
DEF (NRC Glossary of Terms: Nuclear
     Power and Radiation) A nuclear
     power plant using a method of
     isotopic separation based on the
     fact that gas atoms or molecules
     with different masses will diffuse
     through a porous barrier (or
     membrane) at different rates. This
     method is used to separate
     uranium-235 from uranium-238; it
     requires large gaseous diffusion
     plants and enormous amounts of
     electric power.

GASES
DA   February 25, 1991
NT1 Atmosphere Gases
NT1 Carbon Dioxide
NT1 Carbon Monoxide
NT1 Compressed Gases
 NT2 Compressed Gases in Solution
 NT2 Flammable Compressed Gases
 NT2 Liquefied Compressed Gases
 NT2 Non-Flammable Compressed
     Gases
 NT2 Non-Liquefied Compressed Gases
NT1 Exhaust Gases
NT1 Flue Gases
 NT2 Condenser Stack Gases
NT1 Methane
NT1 Natural Gas

BT-Broader Term   NT-Narrower Term   RT Related Term

NT2 Field Gases
NT2 Natural Gas Liquids
NT1 Nitrogen Oxides (NO$_x$)
NT2 Nitric Oxides
NT2 Nitrogen Dioxide (NO$_2$)
NT1 Process Gas
NT1 Soil Gas
NT1 Stack Gases

GASIFICATION
(EPA)
DA October 12, 1990
BT1 Processes
RT Coal
SO Environmental Protection Agency
      Glossary
DEF (EPA) Conversion of solid material
      such as coal into a gas for use as
      a fuel.

GAT
DA October 12, 1990
SEE Goodyear Atomic Corporation
SO Acronyms

GATHERING LINES
(SWDA; RCRA; CFR)
DA October 12, 1990
RT Oils
RT Pipeline Facilities
SO Wastes
DEF (CFR) Pipelines, equipment,
      facilities, or buildings used in the
      transportation of oil or gas during
      oil or gas production or gathering
      operations.

GAUGE(S)
(2753 GAUGE)
DA January 3, 1991
BT1 instrument(s)
 BT2 Equipment/Parts -
      Instrumentation/Measuring (DOE
      FRASE Voc.)
  BT3 Equipment
NT1 Scale(s)
NT1 Straight Edge Ruler
NT1 Thermostat(s)
NT1 Time Clock
SO DOE FRASE VOCABULARY

GCEP
DA October 12, 1990
SEE Gas Centrifuge Enrichment Plant
SO Acronyms

GDC
DA June 4, 1991
SEE General Design Criteria
SO Acronyms

GDC PLANNING BOARD
(DOE Order 6430.1A)
DA October 12, 1990
BT1 Boards
 BT2 Administrative Organizations
  BT3 Organizations
RT Construction Projects

SO Construction
DEF (DOE Order 6430.1A) The DOE
      advisory group of major
      Headquarters and field
      organizations involved in the
      construction of facility acquisitions,
      which includes those organizations
      having planning, design,
      construction, environmental, safety
      and health, research, operations,
      and maintenance functions.

GDP
DA October 12, 1990
SEE Gaseous Diffusion Plant
SO Acronyms

GE
DA October 12, 1990
SEE General Electric
SO Acronyms

GEIGER COUNTERS
(EPA)
DA October 12, 1990
SY Geiger-Mueller Counters
BT1 Radiation Detectors
 BT2 Equipment
SO Environmental Protection Agency
      Glossary
DEF (EPA) Electrical devices that detect
      the presence of certain types of
      radioactivity.

GEIGER-MUELLER COUNTERS
(NIH)
DA October 12, 1990
SY Geiger Counters
SF *G-M*
BT1 Radiation Detectors
 BT2 Equipment
SO Radiation
DEF (NIH) Highly sensitive gas-filled
      detectors and associated circuitry
      used for radiation detection and
      measurement.

GENE LIBRARY
(EPA)
DA October 12, 1990
RT Genes
SO Environmental Protection Agency
      Glossary
DEF (EPA) A collection of DNA
      fragments from cells or organisms.
      So far, no simple way for sorting
      the contents of gene libraries has
      been devised. However, DNA
      pieces can be moved into bacterial
      cells where sorting according to
      gene function becomes feasible.

GENERAL DESIGN CRITERIA
(DOE Order 6430.1A)
DA June 4, 1991
SF *GDC*
BT1 Requirements
RT Project Design Criteria
RT Specifications

SO Construction
SO Management
DEF These criteria ensure that the
      planning, design and construction
      of the Department's [DOE]
      facilities will be performed in a
      manner that will satisfy all
      applicable executive orders,
      federal laws, and regulations.

GENERAL DESIGN PROCESS
(MORT)
DA April 3, 1991
BT1 Administrative Processes
 BT2 Processes
RT Design and Development Plans
RT Engineered Safety Features
RT Human Factors
RT Human Factors Engineering
RT Inspection Plans
RT Maintenance Plans
RT Operational Specifications
RT System Safety
DEF (MORT) MORT analysis asks: are
      commonly recognized good
      engineering practices, including
      safety, reliability, and quality
      engineering practices, adequately
      incorporated into the general
      design process? Related
      issues/topics include: code
      compliance procedures;
      engineering studies;
      standardization of parts; design
      description; acceptance criteria;
      development and qualification
      testing; change review
      procedures; and reliability and
      quality assurance.

GENERAL ELECTRIC
DA January 8, 1991
SF *GE*
BT1 Companies
 BT2 Commercial Organizations
  BT3 Organizations
BT1 DOE Contractors
 BT2 Potentially Responsible Parties
RT Albuquerque Operations Office
RT Naval Reactors

GENERAL EMERGENCIES
(EMER)
DA February 1, 1991
BT1 Emergencies
 BT2 Reportable Occurrences
  BT3 Occurrences
SO Emergency Preparedness
DEF (EMER) Emergency response
      levels which represent an event in
      progress or having occurred that
      involves actual or imminent
      substantial reduction of facility
      safety systems in which off-site
      releases of radioactive or other
      hazardous substances are
      occurring or are expected to occur
      which exceed protective action
      guidelines.

SY-Synonymous Terms   SO-Source/Subject Category   SF-See From

GENERAL ENVIRONMENT
(ANL; CFR)
DA May 24, 1991
BT1 Environment
RT Radioactive Waste Management
RT Radioactive Wastes
RT Spent Fuel
SO Environmental Management
SO Wastes
DEF (CFR) The total terrestrial,
    atmospheric, and aquatic
    environments outside sites within
    which any activity, operation, or
    process associated with the
    management and storage of spent
    nuclear fuel or radioactive waste is
    conducted.

GENERAL PERMITS
(SWDA; RCRA; CWA; CFR)
DA October 12, 1990
BT1 Permits
SO Environmental Protection Agency
    Glossary
SO Wastes
SO Water Pollution
DEF Permits applicable to a class or
    category of dischargers.

GENERAL PLANT PROJECTS
(DOE Order 4700.1)
DA January 8, 1991
SF GPP
BT1 Projects
SO Environmental Management
DEF (DOE Order 4700.1) The means by
    which Congress annually provides
    funding for miscellaneous
    construction items which are
    required during the fiscal year and
    which cannot be specifically
    identified beforehand.

GENERAL PURPOSE FACILITIES
PROJECTS
DA January 24, 1991
BT1 Projects
RT Construction
RT Federal Facilities
SO Environmental Management
DEF (DOE 4700.1) Line item
    construction estimated to cost
    greater than 1.2 million and which
    are required to support long-term
    administrative and technical needs
    of DOE-operated laboratories and
    facilities. Examples of general
    purpose facilities projects are light
    or heavy laboratories,
    administrative offices, machine
    shops, steam plants, electrical
    utilities, roads, railroads, and
    warehouses. Multiprogram general
    purpose facilities are restricted to
    general purpose facilities projects
    at the multiprogram
    laboratories/site where no one
    program will use more than
    approximately 60 percent of the
    planned facility.

GENERATORS (POLLUTION)
(SWDA; RCRA; CFR)
DA October 12, 1990
BT1 Potentially Responsible Parties
NT1 PCB Waste Generator
NT1 Small Quantity Generators
RT Hazardous Waste Generation
SO Environmental Management
SO Environmental Protection Agency
    Glossary
SO Wastes
SO Water Pollution
DEF (CFR) Facilities or mobile sources
    that emit pollutants into the air or
    release hazardous wastes into
    water or soil. Any person, by site
    location, whose act or process
    produces hazardous waste
    identified or listed in 40 CFR 261.

GENERIC EXEMPTIONS
(Doe Order 5480.1B)
DA October 12, 1990
BT1 Exemptions
SO Environmental Management
SO Industrial Hygiene
DEF (DOE Order 5480.1B) Temporary or
    permanent release from the
    requirements of this Order or other
    Orders in the DOE 5480 series,
    which extends beyond specific
    facilities and projects or applies to
    a category of facilities or activities.

GENERIC SIGNIFICANCE
(DOE Order 5000.3; ESH)
DA October 12, 1990
RT Unusual Occurrences
SO Environmental Management
SO Management
DEF (DOE Order 5000.3) Those unusual
    occurrences which by their nature
    are capable of occurring at more
    than one specific DOE facility,
    location, or site.

GENES
(EPA)
DA October 12, 1990
RT Gene Library
RT Genetic Engineering
RT Plasmids
SO Environmental Protection Agency
    Glossary
DEF (EPA) Lengths of DNA that directs
    the synthesis of a protein.

GENETIC ENGINEERING
(EPA)
DA October 12, 1990
BT1 Biological Processes
BT2 Processes
RT Genes
SO Environmental Protection Agency
    Glossary
DEF (EPA) A process of inserting new
    genetic information into existing
    cells in order to modify any

organism for the purpose of
changing one of its characteristics.

GENETIC RADIATION EFFECTS
(IAEA)
DA October 12, 1990
SY Radiation Induced Hereditary
    Effects
SY Radiation Induced Genetic Effects
BT1 Effects
RT Ionizing Radiation
RT Irradiation
DEF (IAEA) Inheritable changes, chiefly
    mutations, produced by the
    absorption of ionizing radiations.
    On the basis of present knowledge
    these effects are purely additive,
    and there is no recovery.

GENITALS
(1118 GENITALS)
DA November 28, 1990
BT1 Human Body Parts
RT Groin
SO DOE FRASE VOCABULARY

GEOLOGIC REPOSITORIES
DA January 24, 1991
BT1 Sites/Areas
NT1 Geologic Repository Operations
    Areas
RT Disposal Facilities
RT Radioactive Waste Management
SO Environmental Management
DEF (CFR) A system which is intended
    to be used for, or may be used for,
    the disposal of radioactive wastes
    in excavated geologic media. A
    geologic repository includes: (1)
    The geologic repository operations
    area and (2) the portion of the
    geologic setting that provides
    isolation of the radioactive waste.

GEOLOGIC REPOSITORY
OPERATIONS AREAS
DA January 31, 1991
BT1 Geologic Repositories
BT2 Sites/Areas
RT Disposal Facilities
RT High-Level Radioactive Wastes
SO Environmental Management

GEOLOGIC RESOURCES
(CFR)
DA November 15, 1990
BT1 Natural Resources
NT1 Fossil Fuels
NT2 Coal
NT3 Ultra Clean Coal
NT2 Natural Gas
NT3 Field Gases
NT3 Natural Gas Liquids
NT2 Petroleum
NT1 Uranium
NT2 Depleted Uranium
NT2 Enriched Uranium
NT3 Unirradiated Enriched Uranium
DEF (CFR) Those elements of the

Earth's crust such as soils, sediments, rocks, and minerals, including petroleum and natural gas, that are not included in the definitions of ground and surface water resources.

GEOMETRY
(IAEA)
DA   October 12, 1990
RT   Irradiation
SO   Radiation
DEF (IAEA) A term used to designate the arrangement in space of the various components of an irradiation or measuring system. This designation includes positions and relevant parameters of source, detector and any intervening absorber.

GERMICIDES
(EPA)
DA   October 12, 1990
NT1 Disinfectants
RT   Sterilization
SO   Environmental Protection Agency Glossary
DEF (EPA) Compounds that kill disease-causing microorganisms.

GGNS
DA   October 12, 1990
SEE Grand Gulf Nuclear Station
SO   Acronyms

GIDEP
DA   October 12, 1990
SEE Government Industry Data Exchange Program
SO   Acronyms

GJPO
DA   October 12, 1990
SEE Grand Junction Projects Office
SO   Acronyms

GLASS MELTING FURNACES
(CAA; CFR)
DA   October 12, 1990
BT1 Industrial Furnaces
 BT2 Facility Components
NT1 Pot Furnaces
RT   Cullet
RT   Rebricking
RT   Uncontrolled Total Arsenic Emissions
SO   Air Pollution
DEF (CFR) Units comprising a refractory vessel in which raw materials are charged, melted at high temperature, refined, and conditioned to produce molten glass. These units include foundations, superstructure and retaining walls, raw material charger systems, heat exchangers, melter cooling systems, exhaust systems,

refractory brick work, fuel supply and electrical boosting equipment, integral control systems and instrumentation, and appendages for conditioning and distributing molten glass to forming apparatuses. The forming apparatuses, including the float bath used in flat glass manufacturing, are not considered part of the glass melting furnace.

GLOVE(S)
(2664 GLOVE)
DA   January 3, 1991
BT1 Hand Protection
 BT2 Personal Protective Equipment
  BT3 Equipment/Parts - Personal Protective (DOE FRASE Vocabulary)
  BT4 Equipment
RT   Hand(s)
RT   Knuckle(s)
RT   Thumb(s)
SO   DOE FRASE VOCABULARY

GLOVE BOX
(2569 GLOVE BOX)
DA   January 3, 1991
BT1 Equipment/Parts - Nuclear (DOE FRASE Vocabulary)
 BT2 Equipment
 BT2 Reactor Components
SO   DOE FRASE VOCABULARY

GOALS
(MORT)
DA   April 3, 1991
RT   Hazard Analysis
RT   Risk Assessment
RT   Safety Program Reviews
RT   Technical Information Systems
SO   Management
DEF (WEBSTER) The ends toward which an effort is directed. (MORT) MORT analysis asks: are there high goals for policy and implementation criteria as well as specific goals for projects? are the goals nonconflicting, sufficiently challenging, and consistent with policy and the customer's goals?

GOCO
DA   June 12, 1991
SEE Government-Owned Contractor-Operated Facilities
SO   Acronyms

GOGGLES
(2665 GOGGLES)
DA   January 3, 1991
BT1 Eye Protection
 BT2 Personal Protective Equipment
  BT3 Equipment/Parts - Personal Protective (DOE FRASE Vocabulary)
  BT4 Equipment

RT   Eye(s)
SO   DOE FRASE VOCABULARY

GOLF BAGS
(NFI)
DA   October 12, 1990
BT1 Devices
SO   Nuclear Facilities Incident Database
SO   Radiation
DEF (NFI) Holding devices for fuel assemblies, mounted on D&E conveyors.

GOODYEAR ATOMIC CORPORATION
DA   January 8, 1991
SF   *GAT*
BT1 Companies
 BT2 Commercial Organizations
  BT3 Organizations

GOVERNMENT INDUSTRY DATA EXCHANGE PROGRAM
DA   January 8, 1991
SF   *GIDEP*
BT1 Programs

GOVERNMENT-OWNED CONTRACTOR-OPERATED FACILITIES
(ESH)
DA   October 12, 1990
SY   Atomic Energy Act Facilities
SF   *GOCO*
BT1 Federal Facilities
 BT2 Facilities
NT1 Argonne National Laboratory
 NT2 Argonne National Laboratory-East (Chicago)
 NT2 Argonne National Laboratory-West (At INEL)
NT1 Energy Technology Engineering Center (Canoga Park)
NT1 Feed Materials Production Center
NT1 Idaho Chemical Processing Plant
NT1 Idaho National Engineering Laboratory
 NT2 Power Burst Facility
NT1 Inhalation Toxicology Research Institute
NT1 Kansas City Plant
NT1 Lawrence Livermore National Laboratory
NT1 Lawrence Berkeley Laboratory
NT1 Los Alamos National Laboratory
NT1 Nevada Test Site
NT1 Oak Ridge Gaseous Diffusion Plant
NT1 Oak Ridge National Laboratory
NT1 Pacific Northwest Laboratory
NT1 Paducah Gaseous Diffusion Plant (Paducah)
NT1 Portsmouth Gaseous Diffusion Plant
NT1 Rocky Flats Plant
NT1 Sandia National Laboratories
 NT2 Sandia National Laboratories-Alburquerque
 NT2 Sandia National Laboratories-Livermore
NT1 Savannah River Site

NT1 Stanford Linear Accelerator Center
NT1 Strategic Petroleum Reserve
NT1 Waste Isolation Pilot Plant
NT1 Y-12 Plant
RT User Agencies
SO Industrial Safety
SO Management
DEF (DOE Order 5483.1A) For the purposes of this Order, facilities owned or leased by DOE or a contractor for the account of DOE in connection with which DOE prescribes and enforces through contractual provisions, occupational safety and health standards pursuant to the authority in the Atomic Energy Act of 1954, as amended, the Energy Reorganization Act of 1974, and the Department of Energy Organization Act of 1977, for contractor employees working therein. A listing of these GOCO facilities is maintained by the Office of Operational Safety (EP-32).

GOVERNMENT STANDARDS
(DOE Order 1300.2; ESH)
DA October 12, 1990
BT1 Standards
 BT2 Codes, Standards, and Regulations
RT Interagency Committee on Standards Policy
SO Management
SO Standards
DEF (DOE Order 1300.2) Federal agency standards and specifications including proposed or recommended standards developed by Federal agency personnel, outside groups under agency regulations, or by organizations or committees made up solely of Government agency representatives.

GP
DA October 12, 1990
SEE Graphic Panel
SO Acronyms

GPP
DA October 12, 1990
SEE General Plant Projects
SO Acronyms

GPU
DA October 12, 1990
SEE Grapper Pick Up
SO Acronyms

GRADE BEAMS
(DOE Order 6430.1A)
DA October 12, 1990
RT Reinforcement Ratio
SO Construction
DEF (DOE Order 6430.1A) Reinforced

concrete beams placed directly on the ground to provide the foundation for the superstructure.

GRADED APPROACH
(DOE order 4330.4A)
DA June 4, 1991
RT Activities
RT Processes
RT Requirements
RT Specifications
SO Management
DEF (DOE Order 4330.4A) By graded approach, DOE intends that the depth of detail required and the magnitude of resources expended for a particular maintenance management element be tailored to be commensurate with the element's relative importance to safety, environmental compliance, safeguards and security, programmatic importance, an/or other facility-specific requirements.

GRAIN LOADING
(EPA)
DA October 12, 1990
BT1 Loading
 BT2 Ratios
SO Environmental Protection Agency Glossary
DEF (EPA) The rate at which particles are emitted from a pollution source. Measurement is made by the number of grains per cubic foot of gas emitted.

GRAND GULF NUCLEAR STATION
DA January 8, 1991
SF *GGNS*
BT1 Facilities

GRAND JUNCTION PROJECTS OFFICE
DA January 8, 1991
SF *GJPO*
BT1 Offices
 BT2 Administrative Organizations
 BT3 Organizations
RT Projects

GRANULAR ACTIVATED CARBON TREATMENT
(EPA)
DA October 12, 1990
BT1 Treatment
 BT2 Waste Management Processes
 BT3 Processes
RT Activated Carbon
SO Environmental Protection Agency Glossary
DEF (EPA) A filtering system often used in small water systems and individual homes to remove organics. Granular Activated Carbon (GAC) can be highly effective in removing elevated levels of radon from water.

GRAPHIC PANEL
DA January 8, 1991
SF *GP*
BT1 Devices

GRAPHITAR(S)
(2754 GRAPHITAR)
DA January 3, 1991
BT1 Instrument(s)
 BT2 Equipment/Parts - Instrumentation/Measuring (DOE FRASE Voc.)
 BT3 Equipment
SO DOE FRASE VOCABULARY

GRAPPER PICK UP
DA January 8, 1991
SF *GPU*
BT1 Devices

GRATE SIFTINGS
DA January 24, 1991
SO Environmental Management
DEF (CFR) "Grate siftings" means the materials that fall from the solid waste fuel bed through the grate openings.

GRAY
(NCRP)
DA October 12, 1990
BT1 Radiation Units
 BT2 Units of Measure
RT Absorbed Dose
RT Kerma
SO Radiation
DEF (NCRP) The unit of absorbed dose. 1 Gy = 1 Joule/kg.

GRAY WATER
(EPA)
DA October 12, 1990
BT1 Wastewater
 BT2 Wastes
 BT2 Water
SO Environmental Protection Agency Glossary
DEF (EPA) The term given to domestic wastewater composed of washwater from sinks, kitchen sinks, bathroom sinks and tubs, and laundry tubs.

GREATER-THAN-CLASS-C
DA January 8, 1991
SF *GTCC*

GREEN SLUGS
(NFI)
DA October 12, 1990
SO Nuclear Facilities Incident Database
SO Radiation
DEF (NFI) Target slugs for which sufficient delay time has not been allowed after irradiation.

GREENHOUSE EFFECT
(EPA)

DA October 12, 1990
BT1 Effects
SO Environmental Protection Agency
Glossary
DEF (EPA) The warming of the Earth's
atmosphere caused by a build-up
of carbon dioxide or other trace
gases; it is believed by many
scientists that this build-up allows
light from the sun's rays to heat
the Earth but prevents a
counterbalancing loss of heat.

GRENADE LAUNCHERS
(Doe Order 5480.16)
DA October 12, 1990
BT1 Firearms
RT Flash Grenades
RT Smoke Grenades
SO Firearms
DEF (DOE Order 5480.16) Devices that
by means of gas pressure propel
a grenade.

GRENADES
(HMTA; CFR)
DA May 20, 1991
BT1 Munitions
NT1 Smoke Grenades
SO Hazardous Materials
DEF (CFR) Grenades, hand or rifle, are
small metal or other containers
designed to be thrown by hand or
projected from a rifle. They are
filled with an explosive or a liquid,
gas, or solid material such as a
toxic or tear gas or an incendiary
or smoke producing material and a
bursting charge.

GRINDER PUMPS
(EPA)
DA October 12, 1990
BT1 Devices
SO Environmental Protection Agency
Glossary
DEF (EPA) Mechanical devices that
shred solids and raise the fluid to
a higher elevation through
pressure sewers.

GRINDING MACHINE
(2123 GRINDING MAC)
DA December 10, 1990
BT1 Machines (DOE FRASE
Vocabulary)
BT2 Equipment
SO DOE FRASE VOCABULARY

GROIN
(1117 GROIN)
DA November 28, 1990
BT1 Trunk
BT2 Human Body Parts
RT Abdomen
RT Genitals
RT Thigh(s)
SO DOE FRASE VOCABULARY

GROSS ALPHA PARTICLE ACTIVITY
(SDWA; CFR; EPA)
DA October 12, 1990
BT1 Activity (Nuclear)
BT2 Measurements
RT Alpha Particles
RT Radioactivity
SO Environmental Protection Agency
Glossary
SO Water Pollution
DEF Total activity due to emission of
alpha particles. Used as the
screening measurement for
radioactivity generally due to
naturally occurring radionuclides.
Activity is commonly measured in
picocuries./ The total radioactivity
due to alpha particle emission as
inferred from measurements on a
dry sample.

GROSS BETA PARTICLE ACTIVITY
(SDWA; CFR; EPA)
DA October 12, 1990
BT1 Activity (Nuclear)
BT2 Measurements
RT Beta Particles
RT Radioactivity
SO Environmental Protection Agency
Glossary
SO Water Pollution
DEF Total activity due to emission of
beta particles. Used as the
screening measurement for
radioactivity from man-made
radionuclides since the decay
products of fission are beta
particle and gamma ray emitters.
Activity is commonly measured in
picocuries./ The total radioactivity
due to beta particle emission as
inferred from measurements on a
dry sample.

GROSS WEIGHT
(SWDA; RCRA; ESH)
DA October 12, 1990
BT1 Measurements
RT Net Weight
SO Hazardous Materials
DEF Weight of a packaging plus the
weight of its contents.

GROUND
(1609 GROUND;N)
DA December 10, 1990
BT1 Site (DOE FRASE Vocabulary)
BT2 Sites/Areas
SO DOE FRASE VOCABULARY

GROUND CRACKS
(USGS)
DA October 12, 1990
RT Faults
RT Ground Failures
RT Tectonic Deformations
SO Natural Phenomenon
DEF (USGS) Induced by earthquake
ground shaking and are a principal

cause of damage and casualties.
A partial or incomplete fracture.

GROUND CREWS
(Doe Order 5480.13)
DA October 12, 1990
BT1 Personnel
RT Aviation Operations
SO Aviation Safety
DEF (DOE Order 5480.13) All personnel
assigned to aviation operations
other than flight crewmembers and
administrative personnel.

GROUND FAILURES
(USGS)
DA October 12, 1990
BT1 Failures
BT2 Accidents
BT1 Liquefaction
BT2 Processes
NT1 Differential Settlement
NT1 Flow Failures
NT1 Landslides
NT1 Lateral Spreads
RT Faults
RT Ground Cracks
RT Tectonic Deformations
SO Natural Phenomenon
DEF (USGS) Include landslides, lateral
spreads, differential settlements,
and ground cracks.

GROUND FAULT INTERRUPTER(S)
(2396 GFI)
DA January 3, 1991
BT1 Equipment/Parts - Electrical (DOE
FRASE Vocabulary)
BT2 Equipment
SO DOE FRASE VOCABULARY

GROUNDCOVERS
(EPA)
DA October 12, 1990
RT Erosion
RT Plants
SO Environmental Protection Agency
Glossary
DEF (EPA) Plants grown to keep soil
from eroding.

GROUNDS MAINTENANCE ACTIVITY
(1234 GM ACTIVITY)
DA November 28, 1990
BT1 Activity Types (DOE FRASE
Vocabulary)
BT2 Activities
BT1 Maintenance
BT2 Activities
RT Groundskeeper
SO DOE FRASE VOCABULARY

GROUNDSKEEPER
(0562 GROUNDSKEEPE)
DA November 28, 1990
BT1 Agriculture Personnel
BT2 Occupations
BT2 Personnel

SY-Synonymous Terms SO-Source/Subject Category SF-See From

RT Grounds Maintenance Activity
SO DOE FRASE VOCABULARY

GROUNDWATER
(SWDA; CERCLA; SDWA; CFR; USC)
DA October 12, 1990
BT1 Water
NT1 Capillary Water
NT1 Groundwater Resources
RT Aquifers
RT Hydrogeologic Units
RT Leachates
RT Run-off
RT Saturated Zones
RT Significant Source of Groundwater
RT Soils
RT Special Source of Ground Water
RT Surface Waters
RT Underground Sources of Drinking
     Water
RT Water Table
SO Compensation and Liability
SO Environmental Management
SO Environmental Protection Agency
     Glossary
SO Wastes
SO Water Pollution
DEF (CFR) The supply of fresh water
     found beneath the Earth's surface,
     usually in aquifers, which is often
     used for supplying wells and
     springs. Because groundwater is a
     major source of drinking water
     there is growing concern over
     areas where leaching agricultural
     or industrial pollutants or
     substances from leaking
     underground storage tanks are
     contaminating groundwater.

GROUNDWATER RESOURCES
(CFR)
DA November 15, 1990
BT1 Groundwater
 BT2 Water
BT1 Natural Resources
DEF (CFR) Water in saturated zones or
     strata beneath the surface of land
     or water and the rocks or
     sediments through which ground
     water moves. Includes ground
     water resources that meet the
     definition of drinking water
     supplies.

GROUPS
DA February 1, 1991
BT1 Administrative Organizations
 BT2 Organizations
NT1 Accident Response Groups
NT1 Carcinogen Risk Assessment
     Verification Endeavor Workgroup
NT1 Emergency Management
     Coordination Committee
     Secretariat
NT1 Interagency Group on Energy
     Vulnerability
NT1 Purchasing Group
NT1 Reactor Materials Control Group
NT1 Risk Retention Group

NT1 Standards-developing Groups
NT1 State and Tribal Government
     Working Group
RT Boards
RT Centers
RT Committees
RT Offices
SO Management
DEF (WEBSTER) Consisting of
     individuals assembled together
     through some unifying relationship
     or purpose.

GROUT TREATMENT FACILITY
DA January 8, 1991
SF GTF
BT1 Treatment Facilities
 BT2 Facilities

GTCC
DA October 12, 1990
SEE Greater-Than-Class-C
SO Acronyms

GTF
DA October 12, 1990
SEE Grout Treatment Facility
SO Acronyms

GTM
DA October 12, 1990
SEE Gang Temperature Monitor
SO Acronyms

GUARANTORS
(CERCLA; USC)
DA October 12, 1990
RT Operators
RT Owners
SO Compensation and Liability
DEF Persons, other than the owner or
     operator, who provide evidence of
     financial responsibility for an
     owner or operator under CERCLA.

GUARD STATION
(1781 GUARD STATIO)
DA December 10, 1990
BT1 Building (DOE FRASE Vocabulary)
 BT2 Facilities and Buildings (DOE
      FRASE Vocabulary)
  BT3 Facilities
BT1 Station
 BT2 Site (DOE FRASE Vocabulary)
  BT3 Sites/Areas
RT Plant Protection System
RT Security Guard
SO DOE FRASE VOCABULARY

GUARDS
(Doe Order 5480.16)
DA October 12, 1990
BT1 Protective Force Personnel
 BT2 Personnel
RT Security
SO Firearms
DEF (DOE Order 5480.16) For the
     purpose of DOE Order 5480.16,

unarmed Departmental contractor
individuals who are employed for,
and charged with, the protection of
classified matter or Government
property.

GUIDE SPECIFICATIONS
DA January 24, 1991
BT1 Specifications
 BT2 Requirements
RT Contract Specifications
RT Project Design Criteria
SO Environmental Management
DEF (CFR) "Guide specification" means
     a general specification—often
     referred to as a design standard or
     design guideline— which is a
     model standard and is suggested
     or required for use in the design of
     all of the construction projects of
     an agency.

GUIDELINES
(SSDC)
DA October 12, 1990
NT1 Control Technique Guidelines
NT1 Effluent Limitation Guidelines
NT1 NEPA Compliance Guides
NT1 Protective Action Guides
NT1 Safety Guides
RT Codes, Standards, and Regulations
RT Recommendations
SO System Safety Development Center
     Glossary
DEF (SSDC) DOE or other government
     agency outlines or directions on
     how to conduct or control some
     program. As an example, DOE
     issues guidelines on keeping
     personnel radiation exposures "As
     Low as Reasonably Achievable."
     Guidelines normally do not have
     the legal impact of regulations,
     orders, etc.

GUST FRONTS
(SMRP)
DA October 12, 1990
BT1 Divergent Windstorms
 BT2 Act of Nature
RT Inflow
SO Natural Phenomenon
DEF (SMRP) Lines of gusty wind moving
     out from a squall line or an area of
     severe thunderstorms. A gust front
     extends tens of miles parallel to
     the line of thunderstorms. At the
     leading age of a gust front, wind
     direction shifts abruptly from
     toward the storm (inflow) to away
     from the storm (outflow). The
     gustiness factor, the ratio of the
     total range of windspeed between
     gusts and lulls divided by the
     mean windspeed, is very large in
     gust front, often reaching 0.5 to
     0.7. Peak gusts are usually less
     than F2 (113 mph).

BT-Broader Term   NT-Narrower Term   RT Related Term

GVR
DA   October 12, 1990
SEE Gas Volume Ratio
SO   Acronyms

H
DA   October 12, 1990
SEE Dose Equivalents
SO   Acronyms

H&SM
DA   October 12, 1990
SEE Health and Safety Manual,
        PUB-3000
SO   Acronyms

HABITATS
(EPA)
DA   October 12, 1990
NT1 Critical Habitats
 NT2 Proposed Critical Habitats
NT1 Riparian Habitats
NT1 Wildlife Refuges
 NT2 Manatee Refuges
RT   Environment
SO   Environmental Protection Agency
        Glossary
DEF (EPA) The places where a
        population (e.g., human, animal,
        plant, microorganism) lives and its
        surroundings, both living and
        nonliving.

HACK SAW
(3033 HACK SAW)
DA   January 3, 1991
BT1 Saws
BT1 Tools - Manual
 BT2 Tools (DOE FRASE Vocabulary)
 BT3 Equipment
SO   DOE FRASE VOCABULARY

HAIR
(1076 HAIR)
DA   November 28, 1990
BT1 Skin
 BT2 Human Body Parts
RT   Scalp
SO   DOE FRASE VOCABULARY

HALF-LIFE
(EPA; EMER)
DA   October 12, 1990
BT1 Time Designations
NT1 Biological Half-Life
NT1 Effective Half-Life
NT1 Radioactive Half-Life
SO   Emergency Preparedness
SO   Environmental Protection Agency
        Glossary
SO   Radiation
DEF (EPA) (1) The time required for a
        pollutant to lose half its affect on
        the environment. For example, the
        half-life of DDT in the environment
        is 15 years, of radium, 1,580
        years. (2) The time required for
        half of the atoms of a radioactive
        element to undergo decay. (3) The

time required for the elimination of
one half a total dose from the
body.

**HALF VALUE LAYER (HALF
THICKNESS)**
(NIH)
DA   October 12, 1990
BT1 Measurements
RT   Gamma Rays
RT   X-rays
SO   Radiation
DEF (NIH) The thickness of any
        specified material necessary to
        reduce the intensity of an x-ray or
        gamma-ray beam to one-half its
        original value.

HALOCARBONS
(CAA)
DA   November 15, 1990
BT1 Halogenated
RT   Stratosphere
SO   Environmental Management
DEF (USC) The chemical compounds
        CF_2Cl_2 and $CFCl_3$ and such other
        halogenated compounds as the
        Administrator determines may
        reasonably be anticipated to
        contribute to reductions in the
        concentration of ozone in the
        stratosphere (42 USC 7452).

HALOGENATED
(DOE Order 6430.1A)
DA   October 12, 1990
NT1 Halocarbons
NT1 Halogenated Organic Compounds
 NT2 Chlorinated Hydrocarbons
  NT3 Chlorofluorocarbons (CFCs)
  NT3 DDT
  NT3 Heptachlor
  NT3 Polychlorinated Biphenyls
   NT4 High Concentration PCBs
   NT4 Low Concentration PCBs
   NT4 Recycled PCBs
  NT3 Polyvinyl Chloride
  NT3 Trichloroethylene (TCE)
  NT3 Vinyl Chloride
 NT2 Fluorocarbons
  NT3 Chlorofluorocarbons (CFCs)
 NT2 Trihalomethane
SO   Construction
DEF (DOE Order 6430.1A) Compounds
        that contain a halogen element
        (i.e., fluorine, chlorine, bromine or
        iodine).

**HALOGENATED ORGANIC
COMPOUNDS**
(SWDA; RCRA)
DA   October 12, 1990
SF   HOCs
BT1 Halogenated
BT1 Organic Chemicals
 BT2 Chemical Substances
NT1 Chlorinated Hydrocarbons
 NT2 Chlorofluorocarbons (CFCs)
NT2 DDT

 NT2 Heptachlor
 NT2 Polychlorinated Biphenyls
  NT3 High Concentration PCBs
  NT3 Low Concentration PCBs
  NT3 Recycled PCBs
 NT2 Polyvinyl Chloride
 NT2 Trichloroethylene (TCE)
 NT2 Vinyl Chloride
NT1 Fluorocarbons
 NT2 Chlorofluorocarbons (CFCs)
NT1 Trihalomethane
RT   Agent Orange
RT   Ethylene Dibromide
SO   Environmental Management
SO   Wastes
DEF (CFR) Those compounds having a
        carbon-halogen bond.

HALOGENS
(SDWA; CFR)
DA   October 12, 1990
SO   Environmental Protection Agency
        Glossary
SO   Water Pollution
DEF (EPA) Any of a group of 5
        chemically-related nonmetallic
        elements that includes bromine,
        fluorine, chlorine, iodine, and
        astatine.

HALON SYSTEM
(2035 HALON SYSTEM)
DA   December 10, 1990
BT1 Systems (DOE FRASE Vocabulary)
 BT2 Systems
SO   DOE FRASE VOCABULARY

HALONS
(EPA)
DA   October 12, 1990
RT   Fire Protection
RT   Ozone Depletion
SO   Environmental Protection Agency
        Glossary
DEF (EPA) Bromine-containing
        compounds with long atmospheric
        lifetimes whose breakdown in the
        stratosphere cause depletion of
        ozone. Halons are used in fire
        fighting.

HAMMER
(3034 HAMMER)
DA   January 3, 1991
BT1 Tools - Manual
 BT2 Tools (DOE FRASE Vocabulary)
 BT3 Equipment
NT1 Sledgehammer(s)
SO   DOE FRASE VOCABULARY

HAMMERMILLS
(EPA)
DA   October 12, 1990
BT1 Machines (DOE FRASE
        Vocabulary)
 BT2 Equipment
SO   Environmental Protection Agency
        Glossary
DEF (EPA) High-speed machines that

SY-Synonymous Terms   SO-Source/Subject Category   SF-See From

165

hammers and cutters use to crush,
grind, chip, or shred solid wastes.

HAND(S)
(1100 HAND)
DA   November 28, 1990
BT1 Human Body Parts
NT1 Finger(s)
NT1 Palm(s)
NT1 Thumb(s)
RT   Arm(s)
RT   Glove(s)
RT   Hand Protection
RT   Knuckle(s)
RT   Wrist(s)
SO   DOE FRASE VOCABULARY

HAND IRON(S)
(3035 HAND IRON)
DA   January 3, 1991
BT1 Tools - Manual
 BT2 Tools (DOE FRASE Vocabulary)
  BT3 Equipment
SO   DOE FRASE VOCABULARY

HAND PROTECTION
(2666 HAND PROTECT)
DA   January 3, 1991
BT1 Personal Protective Equipment
 BT2 Equipment/Parts - Personal
     Protective (DOE FRASE
     Vocabulary)
  BT3 Equipment
NT1 Glove(s)
NT1 Wrist Band
RT   Hand(s)
RT   Knuckle(s)
RT   Palm(s)
RT   Thumb(s)
SO   DOE FRASE VOCABULARY

HAND SAW
(3036 HAND SAW)
DA   January 3, 1991
BT1 Saws
BT1 Tools - Manual
 BT2 Tools (DOE FRASE Vocabulary)
  BT3 Equipment
SO   DOE FRASE VOCABULARY

HAND STAPLER
(3037 HAND STAPLER)
DA   January 3, 1991
BT1 Tools - Manual
 BT2 Tools (DOE FRASE Vocabulary)
  BT3 Equipment
SO   DOE FRASE VOCABULARY

HAND TRUCK(S)
(2323 HAND TRUCK)
DA   December 10, 1990
BT1 Material Handling Device
 BT2 Devices
 BT2 Equipment/Parts - Material
     Handling (DOE FRASE
     Vocabulary)
  BT3 Equipment
NT1 Cart(s)

NT1 Dolly
SO   DOE FRASE VOCABULARY

HANDGUNS
(Doe Order 5480.16)
DA   October 12, 1990
BT1 Small Arms
 BT2 Firearms
NT1 Pistols
 NT2 Double Action Semiautomatic
      Pistols
 NT2 Machine Pistols
 NT2 Single Action Semiautomatic
      Pistols
NT1 Revolvers
SO   Firearms
DEF (DOE Order 5480.16) Hand-fired
     weapons normally carried as side
     arms. Handguns include all
     revolvers and pistols and may
     include submachine guns under
     some circumstances.

HANDLER/LABORER/HELPER
(0850 LABORER)
DA   November 28, 1990
BT1 Occupations
BT1 Personnel
SO   DOE FRASE VOCABULARY

**HANFORD ENVIRONMENTAL HEALTH
FOUNDATION**
DA   January 11, 1991
BT1 DOE Contractors
 BT2 Potentially Responsible Parties
BT1 Foundations
 BT2 Research and Development
     Organizations
  BT3 Organizations
RT   Richland Operations Office

HANFORD WASTE VITRIFICATION
DA   January 8, 1991
SF   *HWVP*

HANGFIRES
(Doe Order 5480.16)
DA   October 12, 1990
BT1 Misfires
RT   Light Antitank Weapon
SO   Firearms
DEF (DOE Order 5480.16) Missiles or
     rockets that have been fired but
     have not left the launching device.

HARD CONVERSION
(Doe Order 5900.2)
DA   October 12, 1990
BT1 Metrication
SO   Quality Assurance
DEF (DOE Order 5900.2) The process
     of changing measurement
     language to nonequivalent metric
     units necessitating changes in the
     actual physical size and
     configuration of the part, product,
     or process which exceed those
     permitted by established
     measurement tolerance. Hard

conversion allows for simplification
and rationalization of size
sequence.

HARD HAT
(2667 HARD HAT)
DA   January 3, 1991
BT1 Head Protection
 BT2 Personal Protective Equipment
  BT3 Equipment/Parts - Personal
      Protective (DOE FRASE
      Vocabulary)
   BT4 Equipment
RT   Helmet
SO   DOE FRASE VOCABULARY

HARD WATER
(EPA)
DA   October 12, 1990
BT1 Water
RT   Salts
SO   Environmental Protection Agency
     Glossary
DEF (EPA) Alkaline water containing
     dissolved salts that interfere with
     some industrial processes and
     prevent soap from lathering.

HARP
(NFI)
DA   October 12, 1990
RT   Failed Element Monitors
SO   Nuclear Facilities Incident Database
SO   Radiation
DEF (NFI) Failed fuel element container,
     cooling tubes along the side
     resemble the strings of a harp.

HATCH
(1701 HATCH)
DA   December 10, 1990
BT1 Site (DOE FRASE Vocabulary)
 BT2 Sites/Areas
SO   DOE FRASE VOCABULARY

HATCHET(S)
(3038 HATCHET)
DA   January 3, 1991
BT1 Tools - Manual
 BT2 Tools (DOE FRASE Vocabulary)
  BT3 Equipment
RT   Ax(s)
SO   DOE FRASE VOCABULARY

HATS
DA   October 12, 1990
SEE Hazard Abatement Tracking
     System
SO   Acronyms

**HAZARD ABATEMENT TRACKING
SYSTEM**
DA   January 8, 1991
SF   *HATS*
BT1 Information Systems
 BT2 Security Interests
 BT2 Systems

BT-Broader Term   NT-Narrower Term   RT Related Term

HAZARD ANALYSIS
(EPA; SSDC; EMER; MORT)
DA   October 12, 1990
BT1 Analyses
RT   Concepts and Requirements
RT   Consequence Assessments
RT   Corrective Action Triggers
RT   Critical Incident Technique
RT   Design and Development Plans
RT   Goals
RT   Hazards
RT   Hazard and Operability Study
RT   Safety Analysis
RT   Safety (Hazard) Analysis
RT   Safety Analysis Process
RT   Safety Program Reviews
RT   Technical Information Systems
SO   Emergency Preparedness
SO   Environmental Protection Agency
        Glossary
SO   System Safety Development Center
        Glossary
DEF (SSDC) The functions, steps, and
        criteria for design and plan of
        work, which identify hazards,
        provide measures to reduce the
        probability and severity potentials,
        identify residual risks, and provide
        alternative methods of further
        control. (EPA) The procedures
        involved in (1) identifying potential
        sources of release of hazardous
        materials from fixed facilities or
        transportation accidents; (2)
        determining the vulnerability of a
        geographical area to a release of
        hazardous materials; and (3)
        comparing hazards to determine
        which present greater or lesser
        risks to a community.

HAZARD AND OPERABILITY STUDY
(EMER)
DA   February 1, 1991
BT1 Studies
RT   Hazard Analysis
SO   Emergency Preparedness
DEF (EMER) A systematic technique for
        identifying hazards or operability
        problems throughout an entire
        facility. One examines each
        segment of a process and lists all
        possible deviations for normal
        operating conditions and how they
        might occur. The consequences
        on the process are assessed, and
        the means available to detect and
        correct the deviations are
        examined.

HAZARD CATEGORIES
(CERCLA; CFR)
DA   October 12, 1990
NT1 Delayed (Chronic) Health Hazards
NT1 Fire Hazards
NT1 Immediate (Acute) Health Hazards
RT   Sudden Releases of Pressure
SO   Compensation and Liability
DEF Any of the following: (1) "Immediate
        (acute) health hazard," including

highly toxic, toxic, irritant,
sensitizer, corrosive, (as defined
under 29 CFR 1910.1200) and
other hazardous chemicals that
cause an adverse effect to a
target organ and which effect
usually occurs rapidly as a result
of short term exposure and is of
short duration; (2) "Delayed
(chronic) health hazard," including
carcinogens (as defined under 29
CFR 1910.1200) and other
hazardous chemicals that cause
an adverse effect to a target organ
and which effect generally occurs
as a result of long term exposure
and is of long duration; (3) "Fire
hazard," including flammable,
combustible liquid, pyrophoric, and
oxidizer (as defined under 29 CFR
1910.12000; (4) "Sudden release
of pressure," including explosive
and compressed gas (as defined
under 29 CFR 1910.1200).

HAZARD CLASSIFICATIONS
(SSDC)
DA   October 12, 1990
SO   System Safety Development Center
        Glossary
DEF (SSDC) Alphabetical designations
        of relative potential of loss severity
        that would probably occur if a
        loss-producing event resulted from
        a hazard. Class A – potential for
        permanent disability, loss of life or
        body part, or extensive loss of
        structure, equipment or material.
        Class B – potential of serious
        injury or illness resulting in
        temporary disability or property
        damage that is disruptive. Class C
        – potential for minor injury or
        illness or nondisruptive property
        damage.

HAZARD QUOTIENT
(EPA)
DA   October 12, 1990
BT1 Ratios
RT   Hazards
SO   Environmental Protection Agency
        Glossary
DEF (EPA) The ratio of a single
        substance exposure level over a
        specified time period (e.g.,
        subchronic) to a reference dose
        for that substance derived from a
        similar exposure period.

HAZARD WASTES CHARACTERISTICS
(EPA)
DA   October 12, 1990
RT   Hazardous Wastes
SO   Environmental Protection Agency
        Glossary
DEF (EPA) The four categories used in
        defining hazardous waste:
        ignitability, corrosivity, reactivity,
        and toxicity.

HAZARDOUS AIR POLLUTANTS
(EPA)
DA   October 12, 1990
BT1 Air Pollutants
NT1 Volatile Hazardous Air Pollutant
        (VHAP)
NT2 Vinyl Chloride
SO   Environmental Management
SO   Environmental Protection Agency
        Glossary
DEF (EPA) Air pollutants which are not
        covered by ambient air quality
        standards but which, as defined in
        the Clean Air Act, may reasonably
        be expected to cause or contribute
        to irreversible illness or death.
        Such pollutants include asbestos,
        beryllium, mercury, benzene, coke
        oven emissions, radionuclides,
        and vinyl chloride. (ESH) Air
        contaminants to which no ambient
        air quality standard is applicable
        and which causes, or contributes
        to, air pollution which may
        reasonably be anticipated to result
        in an increase in mortality or an
        increase in serious irreversible or
        incapacitating reversible, illness.
        Hazardous air pollutants are
        regulated by 40 CFR 61
        (Regulations on National Emission
        Standards for Hazardous Air
        Pollutants).

HAZARDOUS AIR POLLUTION
(CAA; ESH; CFR; USC)
DA   October 12, 1990
BT1 Air Pollution
BT2 Pollution
SO   Air Pollution
DEF (ESH) Air pollution which may
        reasonably be anticipated to result
        in an increase in mortality or an
        increase in serious irreversible or
        incapacitating reversible, illness.

**HAZARDOUS AND SOLID WASTE
AMENDMENTS**
DA   January 8, 1991
SF   HSWA

HAZARDOUS CHEMICALS
(CERCLA; CFR; EMER)
DA   October 12, 1990
BT1 Hazardous Substances
NT1 Asphyxiants
NT1 Extremely Hazardous Substances
NT1 Toxic Chemicals
RT   Acceptable Daily Intake
RT   Acceptable Intake for Chronic
        Exposure
RT   Acceptable Intake for Subchronic
        Exposure
RT   Chemicals of Potential Concern
RT   Chemical Hazards Emergency
        Management System
RT   Lethal Concentration Low
RT   Lethal Dose Low
RT   Threshold Planning Quantities
SO   Compensation and Liability

SO Emergency Preparedness
DEF (CFR) Hazardous chemicals as
    defined in 29 CFR 1910.1200 (c)
    except that this term does not
    include the following substances:
    (1) Any food, food additive, color
    additive, drug, or cosmetic
    regulated by the Food and Drug
    Administration. (2) Any substance
    present as a solid in any
    manufactured item to the extent
    exposure to the substance does
    not occur under normal conditions
    of use. (3) Any substance to the
    extent it is used for personal,
    family, or household purposes, or
    is present in the same form and
    concentration as a product
    packaged for distribution and use
    by the general public. (4) Any
    substance to the extent it is used
    in a research laboratory or a
    hospital or other medical facility
    under the direct supervision of a
    technically qualified individual. (5)
    Any substance to the extent it is
    used in routine agricultural
    operations or is a fertilizer held for
    sale by a retailer to the ultimate
    customer.

HAZARDOUS CONSTITUENTS
(SWDA; RCRA; SWDA)
DA November 9, 1990
NT1 Aldicarb
NT1 DDT
NT1 Ethylene Dibromide
NT1 Formaldehyde
NT1 Heptachlor
NT1 Lead
NT1 Mercury
NT1 Trichloroethylene (TCE)
NT1 Vinyl Chloride
RT Chlorofluorocarbons (CFCs)
RT Polychlorinated Biphenyls
SO Environmental Management
SO Wastes
DEF Those constituents listed in
    Appendix VIII to 40 CFR 261.

HAZARDOUS MATERIALS
(SWDA; RCRA; EMER)
DA October 19, 1990
BT1 Materials
NT1 Corrosive Materials
NT1 Fissile Materials
NT1 Hazardous Wastes
 NT2 Extremely Hazardous Wastes
 NT2 Incompatible Wastes
 NT2 Listed Wastes
 NT2 Piles (Wastes)
  NT3 Disposal Areas
 NT2 State Hazardous Wastes
NT1 Moderators (Nuclear)
NT1 Residues
 NT2 Wood Residues
RT Chronic Effects
RT Community Awareness and
    Emergency Response Programs
RT Exempt or Limited Quantities

RT Hazardous Materials Response
    Teams
RT On-Shore Facilities
RT Process for Commercial Purposes
RT Reportable Quantities
RT Tanks
SO Construction
SO Emergency Preparedness
SO Environmental Management
DEF (CFR) Substances or materials,
    including a hazardous substance,
    that have been determined by the
    Secretary of Transportation to be
    capable of posing an
    unreasonable risk to health, safety,
    and property when transported in
    commerce, and which has been
    so designated. (40 CFR - Part
    260.10) A hazardous waste as
    defined in 40 CFR - Part 261.3.

**HAZARDOUS MATERIALS
EMERGENCIES**
(EMER)
DA February 1, 1991
BT1 Emergencies
 BT2 Reportable Occurrences
 BT3 Occurrences
SO Emergency Preparedness
DEF (EMER) Conditions or potential
    conditions that could result in the
    accidental release or loss of
    control of radioactive or toxic
    material.

**HAZARDOUS MATERIALS RESPONSE
TEAMS**
(CFR; EMER)
DA January 24, 1991
BT1 Response Teams
 BT2 Teams
  BT3 Administrative Organizations
  BT4 Organizations
RT Emergency Response Teams
RT Hazardous Materials
RT Spills
SO Environmental Management
DEF (CFR) An organized group of
    employees, designated by the
    employer, who are expected to
    perform work to handle and
    control actual or potential leaks or
    spills of hazardous substances
    requiring possible close approach
    to the substance. The team
    members perform responses to
    releases or potential releases of
    hazardous substances for the
    purpose of control or stabilization
    of the incident. A HAZMAT team is
    not a fire brigade nor is a typical
    fire brigade a HAZMAT team. A
    HAZMAT team, however, may be
    a separate component of a fire
    brigade or fire department.

HAZARDOUS RANKING SYSTEM
(EPA)
DA October 12, 1990
BT1 Systems

RT Modified Hazard Ranking System
SO Environmental Protection Agency
    Glossary
DEF (EPA) The principle screening tool
    used by EPA to evaluate risks to
    public health and the environment
    associated with abandoned or
    uncontrolled hazardous waste
    sites. The HRS calculates a score
    based on the potential of
    hazardous substances spreading
    from the site through the air,
    surface water, or ground water
    and on other factors such as
    nearby population. This score is
    the primary factor in deciding if the
    site should be on the National
    Priorities List and, if so, what
    ranking it should have compared
    to other sites on the list.

HAZARDOUS SPILL
(1531 HAZARDOUS SP)
DA November 29, 1990
BT1 Nature of Property Damage
RT Chemical Contamination
RT Environmental Release
SO DOE FRASE VOCABULARY

**HAZARDOUS SUBSTANCE UST
SYSTEM**
(SWDA; RCRA; ESA; CFR)
DA October 12, 1990
BT1 Systems
BT1 UST Systems
 BT2 Tank Systems
SO Wastes
DEF An underground storage tank (UST)
    system that contains a hazardous
    substance defined in section
    101(14) of the Comprehensive
    Environmental Response,
    Compensation and Liability Act of
    1980 (but not including any
    substance regulated as a
    hazardous waste under subtitle C)
    or any mixture of such substances
    and petroleum UST system.

HAZARDOUS SUBSTANCES
(SWDA; RCRA; CERCLA; CFR; USC;
ESH; EMER)
DA October 12, 1990
NT1 Carcinogens
 NT2 Asbestos
  NT3 Commercial Asbestos
  NT3 Friable Asbestos
 NT2 Polychlorinated Biphenyls
  NT3 High Concentration PCBs
  NT3 Low Concentration PCBs
  NT3 Recycled PCBs
NT1 CERCLA Hazardous Substances
 NT2 Agent Orange
 NT2 Aldicarb
 NT2 Asbestos
  NT3 Commercial Asbestos
  NT3 Friable Asbestos
 NT2 Beryllium
  NT3 Beryllium 7
 NT2 Cadmium

BT-Broader Term   NT-Narrower Term   RT Related Term

NT2 Chlorinated Hydrocarbons
NT3 Chlorofluorocarbons (CFCs)
NT3 DDT
NT3 Heptachlor
NT3 Polychlorinated Biphenyls
NT4 High Concentration PCBs
NT4 Low Concentration PCBs
NT4 Recycled PCBs
NT3 Polyvinyl Chloride
NT3 Trichloroethylene (TCE)
NT3 Vinyl Chloride
NT2 Chromium
NT2 DES
NT2 Diazinon
NT2 Dinoseb
NT2 Ethylene Dibromide
NT2 Formaldehyde
NT2 Lead
NT2 Nitrogen Oxides (NO$_x$)
NT3 Nitric Oxides
NT3 Nitrogen Dioxide (NO$_2$)
NT2 Phenols
NT2 Phosphorus
NT2 Radionuclides
NT3 Alpha Decay Radioisotopes
NT4 Radium 226
NT4 Radon 222
NT4 Thorium 230
NT4 Uranium 238
NT3 Beta Decay Radioisotopes
NT4 Beta-Minus Decay
  Radioisotopes
NT5 Antimony 125
NT5 Cerium 144
NT5 Cerium 141
NT5 Cesium 137
NT5 Cesium 134
NT5 Cobalt 60
NT5 Cobalt 58
NT5 Iodine 131
NT5 Radium 228
NT5 Ruthenium 103
NT5 Strontium 90
NT5 Tellurium 132
NT5 Thorium 232
NT5 Tritium
NT5 Zinc 65
NT5 Zirconium 95
NT3 Bone Seekers
NT3 Days Living Radioisotopes
NT4 Beryllium 7
NT4 Cerium 144
NT4 Cerium 141
NT4 Cesium 137
NT4 Cobalt 58
NT4 Iodine 131
NT4 Manganese 54
NT4 Radon 222
NT4 Selenium 75
NT4 Zinc 65
NT4 Zirconium 95
NT3 Delayed Neutron Precursors
NT3 Delayed Proton Precursors
NT3 Element 104 Isotopes
NT3 Element 105 Isotopes
NT3 Element 106 Isotopes
NT3 Element 107 Isotopes
NT3 Element 108 Isotopes
NT3 Element 109 Isotopes
NT3 Heavy Ion Decay Radioisotopes
NT3 Hours Living Radioisotopes

NT4 Cesium 134
NT4 Cobalt 58
NT3 Internal Conversion
  Radioisotopes
NT4 Cesium 134
NT4 Cobalt 60
NT4 Cobalt 58
NT3 Isomeric Transition Isotopes
NT4 Cesium 134
NT4 Cobalt 60
NT4 Cobalt 58
NT3 Microsec Living Radioisotopes
NT3 Millisec Living Radioisotopes
NT3 Minutes Living Radioisotopes
NT4 Cobalt 60
NT4 Thorium 232
NT3 Nanosec Living Radioisotopes
NT3 Neutron-Deficient Isotopes
NT3 Neutron-Rich Isotopes
NT3 Proton Decay Radioisotopes
NT3 Radon Progeny
NT4 Radon 222
NT3 Seconds Living Radioisotopes
NT3 Transuranic Radionuclides
NT3 Years Living Radioisotopes
NT4 Antimony 125
NT4 Cesium 134
NT4 Cobalt 60
NT4 Radium 226
NT4 Radium 228
NT4 Ruthenium 103
NT4 Strontium 90
NT4 Thorium 230
NT4 Tritium
NT4 Uranium 238
NT1 Extremely Hazardous Substances
NT1 Hazardous Chemicals
NT2 Asphyxiants
NT2 Extremely Hazardous Substances
NT2 Toxic Chemicals
NT1 Listed Hazardous Substances
NT1 Mutagens
NT1 Pesticides
NT2 Botanical Pesticides
NT2 Contact Pesticides
NT2 Fumigants
NT2 Fungicides
NT3 Dinocap
NT2 Herbicides
NT3 Alachlor
NT3 Defoliants
NT4 Agent Orange
NT3 Dinoseb
NT3 Paraquat
NT2 Insecticides
NT3 Aldicarb
NT3 DDT
NT3 Diazinon
NT3 Dicofol
NT3 Heptachlor
NT2 Microbial Pesticides
NT2 Nematocides
NT2 Organophosphates
NT2 Persistent Pesticides
NT2 Restricted-Use Pesticide
NT2 Rodenticides
NT2 Selective Pesticides
NT2 Systemic Pesticides
NT1 Teratogens
NT1 Toxic Substances
NT1 Unlisted Hazardous Substances

RT Action Levels
RT Assessment Area
RT Hazards
RT Wastes
SO Compensation and Liability
SO Emergency Preparedness
SO Environmental Management
SO Environmental Protection Agency
    Glossary
SO Hazardous Materials
DEF (DOE Order 5000.3A) (1) Materials
    that pose a threat to human health
    and/or the environment. Typical
    hazardous substances are toxic,
    corrosive, ignitable, explosive, or
    chemically reactive. (2) Any
    substance designated by EPA to
    be reported if a designated
    quantity of the substance is spilled
    in the waters of the United States
    or if otherwise emitted to the
    environment. (ESH) Materials,
    including their mixtures and
    solutions, that (1) are listed in the
    Appendix to 40 CFR 172.101; (2)
    are in a quantity, in one package,
    which equals or exceeds the
    reportable quantity (RQ) listed in
    the Appendix to 40 CFR 172.101;
    and (3) when a mixture or solution
    are in a concentration by weight
    which equals or exceeds the
    concentration corresponding to the
    RQ of the material. (ESH) Any
    substance designated under 40
    CFR 116 pursuant to §311 of the
    Clean Water Act.

HAZARDOUS WASTE DISCHARGE
(SWDA; RCRA)
DA October 19, 1990
BT1 Discharges
RT Hazardous Wastes
SO Wastes
DEF (CFR) The accidental or intentional
    spilling, leaking, pumping, pouring,
    emitting, emptying, or dumping of
    hazardous waste into or on any
    land or water.

HAZARDOUS WASTE FACILITIES
(Doe Order 5480.2)
DA October 12, 1990
BT1 Facilities
NT1 Hazardous Waste Management
    Facilities
NT2 Disposal Facilities
NT3 Land Disposal Facilities
NT4 Near Surface Disposal Facilities
NT3 Repositories
NT3 Test and Evalution Facilities
NT2 Existing Hazardous Waste
    Management (HWM) Facilities
NT2 New Hazardous Waste
    Management (HWM) Facilities
NT2 Off-Site Facilities
NT2 On-Site Facilities
NT1 Totally Enclosed Treatment
    Facilities
RT Active Life

RT Designated Facilities
RT Uncontrolled Hazardous Waste
    Sites
SO Industrial Hygiene
DEF (DOE Order 5480.2) All DOE
    owned or controlled contiguous
    land, structures, other
    appurtenances, and improvements
    on the land used for treating,
    storing, or disposing of hazardous
    waste. A facility may consist of
    several treatment, storage, or
    disposal operational units (e.g.,
    one or more landfills, surface
    impoundments, or combinations).

HAZARDOUS WASTE GENERATION
(SWDA; RCRA; USC)
DA   October 12, 1990
BT1 Processes
RT Generators (Pollution)
RT Hazardous Wastes
RT Individual Generation Sites
SO Environmental Management
SO Wastes
DEF The act or process of producing
    hazardous waste.

HAZARDOUS WASTE MANAGEMENT
(SWDA; RCRA; USC)
DA   October 12, 1990
BT1 Waste Management
 BT2 Processes
RT Hazardous Waste Management
    Units
SO Environmental Management
SO Wastes
DEF (USC) The systematic control of the
    collection, source separation,
    storage, transportation,
    processing, treatment, recovery,
    and disposal of hazardous wastes.

HAZARDOUS WASTE MANAGEMENT
FACILITIES
(SWDA; RCRA; CFR)
DA   October 19, 1990
SY   HWM Facilities
BT1 Hazardous Waste Facilities
 BT2 Facilities
NT1 Disposal Facilities
 NT2 Land Disposal Facilities
  NT3 Near Surface Disposal Facilities
 NT2 Repositories
 NT2 Test and Evalution Facilities
NT1 Existing Hazardous Waste
    Management (HWM) Facilities
NT1 New Hazardous Waste
    Management (HWM) Facilities
NT1 Off-Site Facilities
NT1 On-Site Facilities
RT Closure
RT Physical Construction
SO Environmental Management
SO Wastes
SO Water Pollution   .
DEF All contiguous land and structures,
    other appurtenances, and
    improvements on the land, used
    for treating, storing, or disposing of

hazardous waste. A facility may
consist of several treatment,
storage, or disposal operational
units (for example, one or more
landfills, surface impoundments, or
combinations of them).

HAZARDOUS WASTE MANAGEMENT
UNITS
(RCRA; ESH)
DA   October 12, 1990
BT1 Corrective Action Management
    Units
 BT2 Sites/Areas
NT1 Miscellaneous Units
RT Final Closure
RT Hazardous Waste Management
RT Land Treatment Facilities
RT Landfill Cells
RT Partial Closure
RT Surface Impoundment
RT Tanks
SO Environmental Management
SO Wastes
DEF (CFR) Contiguous areas of land on
    or in which hazardous waste is
    placed, or the largest areas in
    which there is significant likelihood
    of mixing hazardous waste
    constituents in the same area.
    Examples of hazardous waste
    management units include surface
    impoundments, waste piles, land
    treatment areas, landfill cells,
    incinerators, tanks and their
    associated piping and underlying
    containment systems, and
    container storage areas. A
    container alone does not
    constitute a unit; the unit includes
    containers and the land or pad
    upon which they are placed.

HAZARDOUS WASTE REMEDIAL
ACTIONS PROGRAM
DA   January 8, 1991
SF   *HAZRAP*
BT1 Programs

HAZARDOUS WASTES
(DOE Orders 5400.3 and 5480.2;
  SWDA; CERCLA; CFR; ESH; EMER)
DA   October 12, 1990
BT1 Hazardous Materials
 BT2 Materials
BT1 Wastes
NT1 Extremely Hazardous Wastes
NT1 Incompatible Wastes
NT1 Listed Wastes
NT1 Piles (Wastes)
 NT2 Disposal Areas
NT1 State Hazardous Wastes
RT Engineering Control Zones
RT EPA Hazardous Waste Numbers
RT Hazardous Waste Generation
RT Hazard Wastes Characteristics
RT Hazardous Waste Discharge
RT Manifests
RT Radioactive Mixed Wastes
RT Treatment Technologies

SO Compensation and Liability
SO Emergency Preparedness
SO Environmental Management
SO Environmental Protection Agency
    Glossary
SO Hazardous Materials
SO Water Pollution
DEF (EMER) (1) Wastes that are
    identified or listed in 40 CFR
    261.31 and 261.32. Source,
    special nuclear material, and by-
    product material as defined by the
    Atomic Energy Act of 1954, as
    amended, are specifically
    excluded from the term hazardous
    wastes. (2) As defined in RCRA, a
    solid waste, or combination of
    wastes, that because of its
    quantity, concentration, or
    physical, chemical, or infectious
    characteristics, may cause or
    significantly contribute to an
    increase in mortality or serious,
    irreversible, or incapacitating
    reversible illness or pose a
    substantial present or potential
    hazard to human health or the
    environment when improperly
    treated, stored, transported, or
    disposed of, or otherwise
    managed. Hazardous wastes may
    be listed or characteristic. (EPA)
    By-products of society that can
    pose a substantial or potential
    hazard to human health or the
    environment when improperly
    managed. Possesses at least one
    of four characteristics (ignitability,
    corrosivity, reactivity, or toxicity).

HAZARDS
(SSDC; MORT)
DA   October 12, 1990
BT1 Conditions
NT1 Fire Hazards
NT1 Health Hazards
 NT2 Delayed (Chronic) Health Hazards
 NT2 Immediate (Acute) Health Hazards
NT1 Imminent Hazard
NT1 Potential Hazards
 NT2 Level 1 Potential Hazards
 NT2 Level 2 Potential Hazards
 NT2 Level 3 Potential Hazards
RT Accidents
RT Activity (Nuclear)
RT Barriers
RT Charged Particles
RT Contaminants
RT Controls
RT Detonations
RT Diseases
RT Etiologic Agents
RT Explosives
RT Flammable Compressed Gases
RT Flammable Liquids
RT Flammable Solids
RT Hazardous Substances
RT Hazard Analysis
RT Hazard Quotient
RT Hazards Identification

BT-Broader Term   NT-Narrower Term   RT Related Term

RT   Heavy Metals
RT   Inadvertently Opened Relief Valve
RT   Irritating Materials
RT   Leaks
RT   Misfires
RT   Open Burning
RT   Pesticides
RT   Poisons
RT   Pollutants
RT   Radiation
RT   Radioactive Materials
RT   Releases
RT   Safety
RT   Smoke
RT   Solvents
RT   Stuck-Open Relief Valve
RT   Targets
RT   Wastes
SO   Environmental Management
SO   System Safety Development Center
      Glossary
DEF (SSDC) The potential for an energy
      flow(s) to result in an accident or
      otherwise adverse consequence.
      (MORT) MORT analysis asks:
      what was the energy flow or
      environmental condition that
      resulted in the accident?

HAZARDS IDENTIFICATION
(EPA)
DA   October 12, 1990
BT1 Administrative Processes
 BT2 Processes
RT   Hazards
SO   Environmental Protection Agency
      Glossary
DEF (EPA) (1) Providing information on
      which facilities have extremely
      hazardous substances, what those
      chemicals are, and how much
      there is at each facility. The
      process also provides information
      on how the chemicals are stored
      and whether they are used at high
      temperatures. (2) The process of
      determining whether exposure to
      an agent can cause an increase in
      the incidence of a particular
      adverse health effect (e.g., cancer,
      birth defect) and whether the
      adverse health effect is likely to
      occur in humans.

HAZRAP
DA   October 12, 1990
SEE Hazardous Waste Remedial
      Actions Program
SO   Acronyms

HBCU
DA   October 12, 1990
SEE Historically Black Colleges and
      Universities
SO   Acronyms

HC
DA   October 12, 1990

SEE Hydrocarbons
SO   Acronyms

HCF
DA   October 12, 1990
SEE Hot Channel Factor
SO   Acronyms

HDP
DA   October 12, 1990
SEE High Delta Pressure
SO   Acronyms

HE (Effective Dose)
DA   October 12, 1990
SEE Effective Dose
SO   Acronyms

HE (High Explosives)
DA   October 12, 1990
SEE High Explosives
SO   Acronyms

HE (Human Error)
DA   October 12, 1990
SEE Human Error
SO   Acronyms

HEAD
(1075 HEAD)
DA   November 28, 1990
BT1 Human Body Parts
NT1 Chin
NT1 Ear(s)
NT1 Face
 NT2 Area Around Eye
  NT3 Eyelid(s)
 NT2 Eye(s)
 NT2 Lip(s)
 NT2 Nose
NT1 Forehead
NT1 Skull (Frase)
 NT2 Jaw
RT   Brain
RT   Concussion
RT   Contusion(S)
RT   Mouth
RT   Scalp
SO   DOE FRASE VOCABULARY

HEAD PROTECTION
(2668 HEAD PROTECT)
DA   January 3, 1991
BT1 Personal Protective Equipment
 BT2 Equipment/Parts - Personal
       Protective (DOE FRASE
       Vocabulary)
  BT3 Equipment
NT1 Bump Cap
NT1 Faceshield
NT1 Hard Hat
NT1 Helmet
NT1 Sand Blaster's Hood
NT1 Welder's Hood
RT   Concussion
RT   Contusion(S)
RT   Forehead

RT   Skull (Frase)
SO   DOE FRASE VOCABULARY

**HEADQUARTERS COORDINATING
TEAMS**
(Doe Order 5480.2; EMER)
DA   October 12, 1990
BT1 Management Teams
 BT2 Teams
  BT3 Administrative Organizations
   BT4 Organizations
SO   Industrial Hygiene
DEF (DOE Order 5480.2) The group of
      representatives designated by
      program Secretarial Officers and
      other Headquarters officials
      created to assist the Assistant
      Secretary, Environmental
      Protection, Safety, and Emergency
      Preparedness (EP-1) in
      developing policy and guidance for
      the hazardous waste management
      program.

HEADQUARTERS OPERATIONS
DA   January 11, 1991
BT1 Operations
 BT2 Activities
BT1 U.S. Department of Energy
 BT2 Federal Agencies
  BT3 Agencies
   BT4 Administrative Organizations
    BT5 Organizations
RT   Bechtel Petroleum Operation, Inc.
RT   John Brown, E&C, Inc.
RT   Lawrence-Allison & Associates
RT   TRW

HEADS OF FIELD OPERATIONS
(ESH)
DA   October 12, 1990
BT1 Personnel
RT   Field Organizations
RT   Senior Management Officials
SO   Management
DEF The top management officials of the
      OPS offices and other applicable
      field organizations. They are the
      senior officials who manage the
      day-to-day operations of DOE's
      facilities under their jurisdiction.

**HEADS OF HEADQUARTERS
ELEMENTS**
(DOE Order 4330.4A)
DA   June 5, 1991
SY   Program Senior Officials
BT1 Personnel
NT1 Landlords
RT   Line Organizations
SO   Management
DEF (DOE Order 4330.4A) Senior
      program managers within a line
      organizational structure. For
      purpose of this Order, these
      positions include the Assistant
      Secretaries for Conservation and
      Renewable Energy, Defense
      Programs, Fossil Energy, and

Nuclear Energy and the Directors of Energy Research, Civilian Radioactive Waste Management, and Environmental Restoration and Waste Management. Also included are the Administrators of the Bonneville and Western Area Power Administrations.

HEADWATERS
(CWA; CFR)
DA   October 12, 1990
SO   Water Pollution
DEF (CFR) The point on a nontidal stream above which the average annual flow is less than five cubic feet per second. The district engineer may estimate this point from available data by using the mean annual area precipitation, area drainage basin maps, and the average runoff coefficient, or by similar means. For streams that are dry for long periods of the year, district engineers may establish the "headwaters" as that point on the stream where a flow of five cubic feet per second is equaled or exceeded 50 percent of the time.

HEALTH, SAFETY, AND ENVIRONMENT DIVISION
DA   January 8, 1991
SF   HSE
BT1 Divisions
 BT2 Administrative Organizations
  BT3 Organizations

HEALTH AND SAFETY MANUAL, PUB-3000
DA   January 8, 1991
SF   H&SM

HEALTH AND SAFETY STUDIES
(TSCA; CFR; USC)
DA   October 12, 1990
BT1 Studies
SO   Environmental Management
SO   Hazardous Materials
DEF (CFR) Studies of any effect of a chemical substance or mixture on health or the environment or on both, including underlying data and epidemiological studies, studies of occupational exposure to a chemical substance or mixture, toxicological, clinical, and ecological studies of a chemical substance or mixture, and any test performed pursuant to this Toxic Substances Control Act.

HEALTH EXAMINATIONS
(Doe Order 5480.6)
DA   October 12, 1990
SO   Industrial Safety
DEF (DOE Order 5480.6) Examinations performed by a licensed medical

physician to determine the physical condition and general health for duty.

HEALTH HAZARDS
DA   January 24, 1991
BT1 Hazards
 BT2 Conditions
NT1 Delayed (Chronic) Health Hazards
NT1 Immediate (Acute) Health Hazards
RT   Maximum Credible Accident
RT   Occupational Injuries
RT   Occupational Illnesses
RT   Occupational Exposure
RT   Radiation Protection
RT   Safety
SO   Environmental Management
DEF (CFR) A chemical, mixture of chemicals or a pathogen for which there is statistically significant evidence based on at least one study conducted in accordance with established scientific principles that acute or chronic health effects may occur in exposed employees. The term "health hazard" includes chemicals which are carcinogens, toxic or highly toxic agents, reproductive toxins, irritants, corrosives, sensitizers, hepatotoxins, nephrotoxins, neurotoxins, agents which act on the hematopoietic system, and agents which damage the lungs, skin, eyes, or muccous membranes. It also includes stress due to temperature extremes. Further definition of the terms used above can be found in Appendix A to 20 CFR 1910.1200.

HEALTH MONITORING
DA   January 8, 1991
SF   HM
BT1 Monitoring
 BT2 Activities

HEALTH PHYSICIST
(0184 HEALTH PHYS)
DA   November 28, 1990
BT1 Scientist
 BT2 Professional Personnel
  BT3 Occupations
  BT3 Personnel
SO   DOE FRASE VOCABULARY
DEF An individual engaged in the study of science concerned with recognition, evaluation and control of health hazards from ionizing radiation.

HEALTH PHYSICS
(NIH; EMER)
DA   October 12, 1990
SF   HP
RT   Environment, Safety, and Health (ES&H) Program
RT   Industrial Hygiene
RT   Radiation Protection

SO   Emergency Preparedness
SO   Radiation
DEF (NFI) A term in common use for that branch of radiological science dealing with the protection of personnel from harmful effects of ionizing radiation.

HEALTH TECHNICIAN
(0360 HEALTH TECH)
DA   November 28, 1990
BT1 Technicians
 BT2 Professional Personnel
  BT3 Occupations
  BT3 Personnel
SO   DOE FRASE VOCABULARY

HEAP-LEACH EXTRACTION
(NCRP)
DA   October 12, 1990
RT   In-Situ Extraction
SO   Radiation
DEF (NCRP) The application of chemical agents to ore stockpiles or mine walls for the extraction of the mineral content.

HEARING IMPAIRMENT
(1387 HEARING IMPA)
DA   November 28, 1990
BT1 Illnesses
RT   Ear Muffs
RT   Ear Plug(s)
RT   Ear(s)
SO   DOE FRASE VOCABULARY

HEARING PROTECTION
(2669 HEARING PROT)
DA   January 3, 1991
BT1 Personal Protective Equipment
 BT2 Equipment/Parts - Personal Protective (DOE FRASE Vocabulary)
  BT3 Equipment
NT1 Ear Plug(s)
RT   Ear Muffs
RT   Ear(s)
SO   DOE FRASE VOCABULARY

HEART ATTACK
(1389 HEART ATTACK)
DA   November 28, 1990
BT1 Illnesses
SO   DOE FRASE VOCABULARY

HEAT COIL
(2397 HEAT COIL)
DA   January 3, 1991
BT1 Equipment/Parts - Electrical (DOE FRASE Vocabulary)
 BT2 Equipment
SO   DOE FRASE VOCABULARY

HEAT DAMAGE
(1532 HEAT DAMAGE)
DA   November 29, 1990
BT1 Damage
 BT2 Nature of Property Damage

RT Fire Damage
RT Water Damage
SO DOE FRASE VOCABULARY

HEAT DETECTOR
(2755 HEAT DETECTO)
DA   January 3, 1991
BT1 Testing Equipment
 BT2 Instrument(s)
  BT3 Equipment/Parts -
       Instrumentation/Measuring
       (DOE FRASE Voc.)
   BT4 Equipment
SO DOE FRASE VOCABULARY

HEAT EXCHANGER
(2570 HEAT EXCHANG)
DA   January 3, 1991
BT1 Equipment/Parts - Nuclear (DOE
      FRASE Vocabulary)
 BT2 Equipment
 BT2 Reactor Components
NT1 Low Pressure Recirculation System
      Heat Exchanger
NT1 Shutdown Heat Exchanger
SO DOE FRASE VOCABULARY
DEF (NRC Glossary of Terms: Nuclear
      Power and Radiation) Any device
      that transfers heat from one fluid
      (liquid or gas) to another fluid or to
      the environment.

HEAT EXCHANGERS
DA   January 8, 1991
SF   *HX*

HEAT INPUT
(CAA; CFR)
DA   October 12, 1990
BT1 Measurements
SO Air Pollution
DEF The total gross calorific value
      (where gross calorific value is
      measured by ASTM Method
      D2015-66, D240-64, or D1826-64)
      of all fuels burned.

HEAT ISLAND EFFECTS
(EPA)
DA   October 12, 1990
BT1 Effects
SO Environmental Protection Agency
      Glossary
DEF (EPA) Domes of elevated
      temperatures over urban areas
      caused by structural and
      pavement heat fluxes, and
      pollutant emissions from the areas
      below the domes.

**HEAT STROKE/OTHER HIGH TEMP
EFFECT**
(1323 HEAT STROKE)
DA   November 28, 1990
BT1 Effects
BT1 Injuries
RT Sunburn
SO DOE FRASE VOCABULARY

HEATER(S)
(2485 HEATER)
DA   January 3, 1991
BT1 Equipment/Parts - Heating (DOE
      FRASE Vocabulary)
 BT2 Equipment
NT1 Autoclave
NT1 Boiler(s)
 NT2 Boiler Part(s)
NT1 Feed Water Heater(s)
NT1 Furnace(s)
 NT2 Boiler/Industrial Furnaces
NT1 Oven(s)
NT1 Stove(s)
RT Boiler/Industrial Furnaces
SO DOE FRASE VOCABULARY

**HEATING, VENTILATION, AIR
CONDITIONING**
DA   January 8, 1991
SY HVAC System

HEATING APPLIANCE(S)
DA   January 3, 1991
BT1 Appliance(s)
 BT2 Equipment/Parts - Electrical (DOE
       FRASE Vocabulary)
  BT3 Equipment
RT Equipment/Parts - Heating (DOE
       FRASE Vocabulary)
RT Heating Equipment
SO DOE FRASE VOCABULARY

HEATING EQUIPMENT
(2486 HEATING EQUI)
DA   January 3, 1991
SY Equipment/Parts - Heating (DOE
       FRASE Vocabulary)
NT1 Burner(s)
NT1 Evaporator(s)
NT1 Incinerators
 NT2 Catalytic Incinerators
 NT2 Qualified Incinerators
NT1 Pilot Light(s)
NT1 Retort
RT Heating Appliance(s)
SO DOE FRASE VOCABULARY

HEATING OILS
(SWDA; RCRA; CFR)
DA   October 12, 1990
BT1 Oils
RT Consumptive Use
SO Wastes
DEF (CFR) Petroleum that is No. 1, No.
      2, No. 4 – light, No. 4 – heavy,
      No. 5 – light, No. 5 – heavy, and
      No. 6 technical grades of fuel oil;
      other residual fuel oils (including
      navy Special Fuel Oil and Bunker
      C); and other fuels when used as
      substitutes for one of these fuel
      oils. Heating oil is typically used in
      the operation of heating
      equipment, boilers, or furnaces.

HEAVY ION DECAY RADIOISOTOPES
(EDB)
DA   March 29, 1991

BT1 Radionuclides
 BT2 CERCLA Hazardous Substances
  BT3 Hazardous Substances
 BT2 Nuclides
SO Radiation

HEAVY METALS
(EPA)
DA   October 12, 1990
NT1 Cadmium
NT1 Cesium (Cs)
NT1 Chromium
NT1 Lead
NT1 Mercury
NT1 Precious Metals
NT1 Thorium
NT1 Uranium
 NT2 Depleted Uranium
 NT2 Enriched Uranium
  NT3 Unirradiated Enriched Uranium
RT Hazards
SO Environmental Management
SO Environmental Protection Agency
      Glossary
DEF (EPA) Metallic elements with high
      atomic weights, e.g., mercury,
      chromium, cadmium, arsenic, and
      lead. They can damage living
      things at low concentrations and
      tend to accumulate in the food
      chain.

HEAVY MOBILE EQUIPMENT
(2324 HM EQUIPMENT)
DA   December 10, 1990
BT1 Equipment/Parts - Material
      Handling (DOE FRASE
      Vocabulary)
 BT2 Equipment
SO DOE FRASE VOCABULARY

HEEL(S)
(1128 HEEL)
DA   November 28, 1990
BT1 Foot/Feet
 BT2 Human Body Parts
RT Foot Protection
SO DOE FRASE VOCABULARY

HELIPADS
(DOE Order 5480.13)
DA   October 12, 1990
RT Heliports
SO Aviation Safety
DEF (DOE Order 5480.13) Minimum
      facility heliports without auxiliary
      facilities such as waiting rooms,
      hangars, parking, fueling, and
      maintenance.

HELIPORTS
(Doe Order 5480.13)
DA   October 12, 1990
BT1 Airports
RT Aviation Operations
RT Helipads
SO Aviation Safety
DEF (DOE Order 5480.13) Areas, either
      at ground level or elevated on a

SY-Synonymous Terms   SO-Source/Subject Category   SF-See From

structure, that are used for the
landing and takeoff of helicopters.

HELMET
(2670 HELMET)
DA   January 3, 1991
BT1 Head Protection
 BT2 Personal Protective Equipment
  BT3 Equipment/Parts - Personal
      Protective (DOE FRASE
      Vocabulary)
   BT4 Equipment
RT Hard Hat
SO DOE FRASE VOCABULARY

HEMORRHOIDS
(1390 HEMORRHOIDS)
DA November 28, 1990
BT1 Illnesses
SO DOE FRASE VOCABULARY

HEPA
DA   October 12, 1990
SEE High Efficiency Particulate Air
    Filters
SO Acronyms

HEPA FILTER
(2571 HEPA FILTER)
DA   January 3, 1991
BT1 Equipment/Parts - Nuclear (DOE
     FRASE Vocabulary)
 BT2 Equipment
 BT2 Reactor Components
BT1 Filters
 BT2 Devices
SO DOE FRASE VOCABULARY

HEPATITIS
(1392 INFECTIOUS H)
DA November 28, 1990
BT1 Diseases of Blood
 BT2 Diseases
 BT2 Illnesses
SO DOE FRASE VOCABULARY

HEPTACHLOR
(EPA)
DA   October 12, 1990
BT1 Chlorinated Hydrocarbons
 BT2 CERCLA Hazardous Substances
  BT3 Hazardous Substances
 BT2 Halogenated Organic Compounds
  BT3 Halogenated
  BT3 Organic Chemicals
   BT4 Chemical Substances
BT1 Hazardous Constituents
BT1 Insecticides
 BT2 Pesticides
  BT3 Hazardous Substances
SO Environmental Protection Agency
    Glossary
DEF (EPA) An insecticide that was
    banned on some food products in
    1975 and all of them 1978. It was
    allowed for use in seed treatment
    until in 1983. More recently it was
    found in milk and other dairy
    products in Arkansas and

Missouri, as a result of illegally
feeding treated seed to dairy
cattle.

HERBICIDES
(EPA)
DA   October 12, 1990
BT1 Pesticides
 BT2 Hazardous Substances
NT1 Alachlor
NT1 Defoliants
 NT2 Agent Orange
NT1 Dinoseb
NT1 Paraquat
RT Phytotoxic
SO Environmental Management
SO Environmental Protection Agency
    Glossary
DEF (EPA) Chemical pesticides
    designed to control or destroy
    plants, weeds, or grasses.

HERBIVORES
(EPA)
DA   October 12, 1990
BT1 Animals
SO Environmental Protection Agency
    Glossary
DEF (EPA) Animals that feed on plants.

HERMETICALLY SEALED
(SWDA; RCRA; ESH)
DA   October 12, 1990
SO Hazardous Materials
DEF (CFR) Closed by fusion, gasketing,
    crimping, or equivalent means so
    that no gas or vapor can enter or
    escape.

HERNIA
(1324 HERNIA)
DA November 28, 1990
BT1 Injuries
SO DOE FRASE VOCABULARY

HETEROTROPHIC ORGANISMS
(EPA)
DA   October 12, 1990
BT1 Organisms
SO Environmental Protection Agency
    Glossary
DEF (EPA) Consumers such as humans
    and animals, and decomposers –
    chiefly bacteria and fungi – that
    are dependent on organic matter
    for food.

HF
DA   October 12, 1990
SEE Hydrogen Fluoride
SO Acronyms

Hg
DA   June 5, 1991
SEE Mercury
SO Acronyms

HIGH CONCENTRATION PCBS
(TSCA; CFR)
DA   October 12, 1990
BT1 Polychlorinated Biphenyls
 BT2 Carcinogens
  BT3 Hazardous Substances
 BT2 Chlorinated Hydrocarbons
  BT3 CERCLA Hazardous Substances
   BT4 Hazardous Substances
  BT3 Halogenated Organic
       Compounds
   BT4 Halogenated
   BT4 Organic Chemicals
    BT5 Chemical Substances
RT Standard Wipe Test
RT Toxic Substances
SO Hazardous Materials
DEF PCBs that contain 500 ppm or
    greater PCBs, or those materials
    which EPA requires to be assumed
    to contain 500 ppm or greater
    PCBs in the absence of testing.

**HIGH-CONTACT COMMERCIAL
SURFACES**
(CFR)
DA   November 9, 1990
BT1 Sites/Areas
RT High-Contact Residential Surfaces
RT High-Contact Industrial Surfaces
RT Residential Areas
DEF (CFR) Surfaces in a
    residential/commercial area which
    are repeatedly touched, often for
    relatively long periods of time.
    Doors, wall areas below 6 feet in
    height, uncovered flooring,
    windowsills, fencing, banisters,
    stairs, automobiles, and children's
    play areas such as outdoor patios
    and sidewalks are examples of
    high-contact
    residential/commercial surfaces.
    Examples of low-contact
    residential/commercial surfaces
    include interior ceilings, interior
    wall areas above 6 feet in height,
    roofs, asphalt roadways, concrete
    roadways, wooden utility poles,
    unmanned machinery, concrete
    pads beneath electrical equipment,
    curbing, exterior structural building
    components (e.g., aluminum/vinyl
    siding, cinder block, asphalt tiles),
    and pipes.

**HIGH-CONTACT INDUSTRIAL
SURFACES**
(TSCA)
DA   October 19, 1990
BT1 Sites/Areas
RT High-Contact Residential Surfaces
RT High-Contact Commercial Surfaces
SO Hazardous Materials
DEF (CFR) Surfaces in an industrial
    setting which are repeatedly
    touched, often for relatively long
    periods of time. Manned
    machinery and control panels are
    examples of high-contact industrial

surfaces. High-contact industrial surfaces are generally of impervious solid material. Examples of low-contact industrial surfaces include ceilings, walls, floors, roofs, roadways and sidewalks in the industrial area, utility poles, unmanned machinery, concrete pads beneath electrical equipment, curbing, exterior structural building components, indoor vaults, and pipes.

HIGH-CONTACT RESIDENTIAL SURFACES
(CFR)
DA November 9, 1990
BT1 Sites/Areas
RT High-Contact Commercial Surfaces
RT High-Contact Industrial Surfaces
RT Residential Areas
DEF A surface in a residential/commercial area which is repeatedly touched, often for relatively long periods of time. Doors, wall areas below 6 feet in height, uncovered flooring, windowsills, fencing, banisters, stairs, automobiles, and children's play areas such as outdoor patios and sidewalks are examples of high-contact residential/commerical surfaces. Examples of low-contact residential/commercial surfaces include interior ceilings, interior wall areas above 6 feet in height, roofs, asphalt roadways, concrete roadways, wooden utility poles, unmanned machinery, concrete pads beneath electrical equipment, curbing, exterior structural building components (e.g., aluminum/vinyl siding, cinder block, asphalt tiles), and pipes.

HIGH DELTA PRESSURE
DA January 8, 1991
SF *HDP*
BT1 Pressure

HIGH-DENSITY POLYETHYLENE
(EPA)
DA October 12, 1990
BT1 Organic Chemicals
 BT2 Chemical Substances
SO Environmental Protection Agency Glossary
DEF (EPA) A material that produces toxic fumes when burned. Used to make plastic bottles and other products.

HIGH EFFICIENCY PARTICULATE AIR FILTERS
(DOE Order 6430.1A)
DA October 12, 1990
SF *HEPA*
BT1 Filters

 BT2 Devices
SO Construction
SO Environmental Management
DEF (DOE Order 6430.1A) High-efficiency particulate air filters having a fibrous medium that produces a particle removal efficiency of at least 99.97% for 0.3-micrometer particles of dioctylphthalate (DOP) when tested in accordance with MIL-STD- 282.

HIGH EXPLOSIVES
(DOE Order 6430.1A)
DA October 12, 1990
SF *HE (High Explosives)*
BT1 Explosives
NT1 Insensitive High Explosives
SO Construction
DEF (DOE Order 6430.1A) Explosive substances capable of mass detonation, and for which there is a significant probability of accidental initiation or transition from burning to detonation.

HIGH-GRADE PAPER
DA January 24, 1991
SO Environmental Management
DEF (CFR) "High-grade paper" means letterhead, dry copy papers, miscellaneous business forms, stationery, typing paper, tablet sheets, and computer printout paper and cards commonly sold as "whiteledge", "computer printout" and "tab card" grade by the wastepaper industry.

HIGH LEVEL CAVES
DA January 8, 1991
SF *HLC*

HIGH LEVEL FLUX MONITOR
DA January 8, 1991
SF *HLFM*
BT1 Monitors
 BT2 Equipment

HIGH-LEVEL LIQUID WASTE
DA January 8, 1991
SF *HLLW*
BT1 High-Level Wastes

HIGH-LEVEL RADIOACTIVE WASTES
(EPA)
DA October 12, 1990
SY High-Level Wastes
BT1 Radioactive Wastes
 BT2 Wastes
RT Geologic Repository Operations Areas
RT Test and Evalution Facilities
RT Transuranic Radioactive Wastes
RT Waste Isolation Pilot Plant
SO Environmental Protection Agency Glossary
DEF (EPA) Wastes generated in the fuel

of a nuclear reactor, found at nuclear reactors or nuclear fuel reprocessing plants. They are a serious threat to anyone who comes near the wastes without shielding.

HIGH-LEVEL WASTES
(DOE Order 6430.1A)
DA October 12, 1990
SY High-Level Radioactive Wastes
SF *HLW*
NT1 High-Level Liquid Waste
SO Construction
SO Environmental Management
DEF (DOE Order 6430.1A) The highly radioactive waste material that results from the reprocessing of spent nuclear fuel, including liquid waste produced directly in reprocessing and any solid waste derived from the liquid, that contains a combination of transuranic waste and fission products in concentrations requiring permanent isolation.

HIGH PERFORMANCE LIQUID CHROMATOGRAPHY
DA January 8, 1991
SF *HPLC*

HIGH POTENTIAL INCIDENTS
(SSDC)
DA October 12, 1990
SF *HIPO*
BT1 Incidents
SO System Safety Development Center Glossary
DEF (SSDC) Incidents with a large potential of significant loss.

HIGH PRESSURE COOLANT INJECTION
DA January 8, 1991
SF *HPCI*

HIGH PRESSURE CORE SPRAY
DA January 8, 1991
SF *HPCS*

HIGH PRESSURE INJECTION SYSTEM
DA January 8, 1991
SF *HPIS*
BT1 Systems

HIGH PRESSURE PUMP PAD
DA January 8, 1991
SF *HPPP*

HIGH PRESSURE RECIRCULATION SYSTEM
DA January 8, 1991
SF *HPRS*
BT1 Systems

SY-Synonymous Terms   SO-Source/Subject Category   SF-See From

HIGH PRESSURE SERVICE WATER
DA January 8, 1991
SF *HPSW*

HIGH PRESSURE SPRAY
(POST-ACCIDENT INJECTION PHASE)
DA January 8, 1991
SF *HPSI*

HIGH PRESSURE SPRAY
(POST-ACCIDENT RECIRCULATION
PHASE)
DA January 8, 1991
SF *HPSR*

HIGH RADIATION AREAS
DA January 24, 1991
BT1 Radiation Areas
 BT2 Sites/Areas
SO Environmental Management
DEF (CFR) "High radiation area" means
 any area, accessible to personnel,
 in which there exists radiation
 originating in whole or in part
 within licensed material at such
 levels that a major portion of the
 body could receive in any one
 hour a dose in excess of 100
 millirem.

HIGH TEMPERATURE ALARM
DA January 8, 1991
SF *HTA*
BT1 Alarms
 BT2 Devices

HIGH TIDE LINE
(CWA)
DA October 12, 1990
RT Maximum Probable Flood
RT Ordinary High Water Mark
RT Tidal Waters
SO Water Pollution
DEF The line of intersection of the land
 with the water's surface at the
 maximum height reached by a
 rising tide. The high tide line may
 be determined, in the absence of
 actual data, by a line of oil or
 scum along shore objects, a more
 or less continuous deposit of fine
 shell or debris on the foreshore or
 berm, other physical markings or
 characteristics, vegetation lines,
 tidal gages, or other suitable
 means that delineate the general
 height reached by a rising tide.
 The line encompasses spring high
 tides and other high tides that
 occur with periodic frequency but
 does not include storm surges in
 which there is a departure from the
 normal or predicted reach of the
 tide due to the piling up of water
 against a coast by strong winds
 such as those accompanying a
 hurricane or other intense storm.

HIGHBAY
(1703 HIGHBAY)
DA December 10, 1990
BT1 Site (DOE FRASE Vocabulary)
 BT2 Sites/Areas
SO DOE FRASE VOCABULARY

HIGHER SUPERVISION
(MORT)
DA April 3, 1991
BT1 Personnel
RT Controls
RT First Line Supervision
RT Inspections
RT Maintenance
RT Operability
RT Technical Information
DEF (MORT) MORT analysis asks: did
 upper level management provide
 the type of supportive services
 and guidance needed at lower
 organization levels for adequate
 control of unwanted work process
 energy flow? Related issues
 include: research and fact finding,
 information exchange between
 different management levels,
 appropriate standards and
 directives, adequate resources
 e.g. training, technical assistance,
 program aids, etc.

HIGHWAY
(1854 HIGHWAY)
DA December 10, 1990
BT1 Routes (Transportation)
BT1 Structures (DOE FRASE
 Vocabulary)
RT Intersection
SO DOE FRASE VOCABULARY

HIGHWAY CONSTRUCTION
EQUIPMENT
(2325 HIGHWAY CONS)
DA December 10, 1990
BT1 Equipment/Parts - Material
 Handling (DOE FRASE
 Vocabulary)
 BT2 Equipment
SO DOE FRASE VOCABULARY

HIGHWAY ROUTE CONTROLLED
QUANTITY
(CFR)
DA October 12, 1990
BT1 Quantities
SO Hazardous Materials
DEF (CFR) A quantity within a single
 package which exceeds: (1) 3000
 times the A_1 value of the
 radionuclides for special form
 radioactive materials. (2) 3000
 times the A_2 value of the
 radionuclides for normal form
 radioactive materials; or (3)
 30,000 curies; whichever is least.
 A_1 means the maximum activity of
 special form radioactive materials
 permitted in a Type A package. A_2

means the maximum activity of
radioactive materials, other than
special form or low specific activity
radioactive materials, permitted in
a Type A package.

HILL
(1610 HILL)
DA December 10, 1990
BT1 Site (DOE FRASE Vocabulary)
 BT2 Sites/Areas
SO DOE FRASE VOCABULARY

HIP(S)
(1122 HIP)
DA November 28, 1990
BT1 Joint(s)
 BT2 Human Body Parts
RT Buttock(s)
SO DOE FRASE VOCABULARY

HIPO
DA October 12, 1990
SEE High Potential Incidents
SO Acronyms

HISTOGRAMS
(SSDC)
DA October 12, 1990
BT1 Diagrams
RT Distribution
SO System Safety Development Center
 Glossary
DEF (SSDC) Pictorial representations of
 a distribution; bargraphs.

HISTORICALLY BLACK COLLEGES
AND UNIVERSITIES
DA January 8, 1991
SF *HBCU*
RT Universities

HLC
DA October 12, 1990
SEE High Level Caves
SO Acronyms

HLFM
DA October 12, 1990
SEE High Level Flux Monitor
SO Acronyms

HLLW
DA October 12, 1990
SEE High-Level Liquid Waste
SO Acronyms

HLW
DA October 12, 1990
SEE High-Level Wastes
SO Acronyms

HM
DA October 12, 1990
SEE Health Monitoring
SO Acronyms

BT-Broader Term

NT-Narrower Term

RT Related Term

HOCs
DA   October 19, 1990
SEE Halogenated Organic Compounds
SO   Acronyms

HOE(S)
(3039 HOE)
DA   January 3, 1991
BT1 Tools - Manual
  BT2 Tools (DOE FRASE Vocabulary)
  BT3 Equipment
SO   DOE FRASE VOCABULARY

HOIST(S)
(2327 HOIST)
DA   December 10, 1990
BT1 Hoisting Apparatus
  BT2 Material Handling Device
  BT3 Devices
  BT3 Equipment/Parts - Material
      Handling (DOE FRASE
      Vocabulary)
  BT4 Equipment
NT1 Air Hoist
SO   DOE FRASE VOCABULARY

HOISTING AND RIGGING MANUAL
DA   January 8, 1991
SF   HRM

HOISTING APPARATUS
(2328 HOIST APPAR)
DA   December 10, 1990
BT1 Material Handling Device
  BT2 Devices
  BT2 Equipment/Parts - Material
      Handling (DOE FRASE
      Vocabulary)
  BT3 Equipment
NT1 Derrick
NT1 Elevator
NT1 Forklift(s)
NT1 Hoist(s)
  NT2 Air Hoist
NT1 Lift Bucket
NT1 Manlift(s)
NT1 Scissor Lift
SO   DOE FRASE VOCABULARY

HOLDING PONDS
(EPA)
DA   October 12, 1990
RT   Evaporation Ponds
RT   Impoundment
RT   Oxidation Ponds
RT   PAR Ponds
RT   Stabilization Ponds
SO   Environmental Protection Agency
      Glossary
DEF (EPA) Ponds or reservoirs, usually
      made of earth, built to store
      polluted runoff.

HOLDUP (NUCLEAR MATERIAL)
(DOE Order 6430.1A)
DA   October 12, 1990
BT1 Materials
SO   Construction
DEF (DOE Order 6430.1A) Holdup is the

nuclear material that is retained in
process equipment at inventory
time.

HOOD CAPTURE EFFICIENCY
(EPA)
DA   October 12, 1990
BT1 Measurements
RT   Auxiliary Air Units
SO   Environmental Protection Agency
      Glossary
DEF (EPA) The emissions from a
      process which are captured by
      hood and directed into the control
      device, expressed as a percent of
      all emissions.

HOOK(S)
(2330 HOOK;N)
DA   December 10, 1990
BT1 Equipment/Parts - Material
      Handling (DOE FRASE
      Vocabulary)
  BT2 Equipment
SO   DOE FRASE VOCABULARY

HORIZONTAL BRACING SYSTEMS
(SEA)
DA   October 12, 1990
BT1 Building Frame Systems
  BT2 Space Frames
  BT3 Structures (DOE FRASE
      Vocabulary)
  BT2 Systems
BT1 Diaphragms
SO   Construction
DEF (SEA) Horizontal truss systems that
      serve the same function as a
      diaphragm.

HORIZONTAL DISPLACEMENTS
(USGS)
DA   October 12, 1990
SY   Strike Slip
BT1 Faults
SO   Natural Phenomenon
DEF (USGS) Displacements with no dip;
      strike slips. Faults with no vertical
      displacement.

HOSPITAL
(1782 HOSPITAL)
DA   December 10, 1990
BT1 Building (DOE FRASE Vocabulary)
  BT2 Facilities and Buildings (DOE
      FRASE Vocabulary)
  BT3 Facilities
RT   Doctor/Nurse
RT   Illnesses
RT   Injuries
RT   Mental Disorders
SO   DOE FRASE VOCABULARY

HOSTAGE NEGOTIATION TEAMS
(EMER)
DA   February 1, 1991
BT1 Response Teams
  BT2 Teams
  BT3 Administrative Organizations

  BT4 Organizations
SO   Emergency Preparedness
DEF (EMER) Teams designated to open
      and continue dialogue with
      hostage takers. The team normally
      consists of psychological, tactical,
      and analytical personnel.

HOSTAGE THROW PHONE
(EMER)
DA   February 1, 1991
SO   Emergency Preparedness
DEF (EMER) A phone designed for
      hostage negotiation emergencies.
      The phone has capabilities of
      being thrown or transported to the
      hostage taker. It is generally used
      after the hostage taker has been
      separated from exterior
      communications.

HOSTS
(EPA)
DA   October 12, 1990
BT1 Organisms
SO   Environmental Management
SO   Environmental Protection Agency
      Glossary
DEF (EPA) (1) In genetics, the organism,
      typically a bacterium, into which a
      gene from another organism is
      transplanted. (2) In medicine, an
      animal infected by or parasitized
      by another organism.

HOT CELL FACILITY
(1799 HOT CELL)
DA   December 10, 1990
BT1 Facility (DOE FRASE Vocabulary)
  BT2 Facilities and Buildings (DOE
      FRASE Vocabulary)
  BT3 Facilities
SO   DOE FRASE VOCABULARY

HOT CHANNEL FACTOR
DA   January 8, 1991
SF   HCF

HOT LINES
(DOE Order 6430.1A)
DA   October 12, 1990
SO   Construction
DEF (DOE Order 6430.1A) Phone
      numbers of local service
      companies factory-authorized to
      replace system components or
      appurtenances or value repairs to
      same. Direct customer service
      phone numbers of manufacturers
      shall also be considered as "hot
      lines."

HOT SHOP
(1783 HOT SHOP)
DA   December 10, 1990
BT1 Building (DOE FRASE Vocabulary)
  BT2 Facilities and Buildings (DOE
      FRASE Vocabulary)

BT3 Facilities
SO DOE FRASE VOCABULARY

HOURS LIVING RADIOISOTOPES
(EDB)
DA March 29, 1991
BT1 Radionuclides
BT2 CERCLA Hazardous Substances
BT3 Hazardous Substances
BT2 Nuclides
NT1 Cesium 134
NT1 Cobalt 58
SO Radiation

HP
DA October 12, 1990
SEE Health Physics
SO Acronyms

HPCI
DA October 12, 1990
SEE High Pressure Coolant Injection
SO Acronyms

HPCS
DA October 12, 1990
SEE High Pressure Core Spray
SO Acronyms

HPIS
DA October 12, 1990
SEE High Pressure Injection System
SO Acronyms

HPLC
DA October 12, 1990
SEE High Performance Liquid
     Chromatography
SO Acronyms

HPPP
DA October 12, 1990
SEE High Pressure Pump Pad
SO Acronyms

HPRS
DA October 12, 1990
SEE High Pressure Recirculation
     System
SO Acronyms

HPSI
DA October 12, 1990
SEE High Pressure Spray (Post-accident
     Injection Phase)
SO Acronyms

HPSR
DA October 12, 1990
SEE High Pressure Spray (Post-accident
     Recirculation Phase)
SO Acronyms

HPSW
DA October 12, 1990
SEE High Pressure Service Water
SO Acronyms

HQ
DA October 12, 1990
SEE DOE Headquarters
SO Acronyms

HRM
DA October 12, 1990
SEE Hoisting and Rigging Manual
SO Acronyms

HS
DA October 12, 1990
SEE Hydrogen Sulfide
SO Acronyms

HSE
DA October 12, 1990
SEE Health, Safety, and Environment
     Division
SO Acronyms

HSWA
DA October 12, 1990
SEE Hazardous and Solid Waste
     Amendments
SO Acronyms

HTA
DA October 12, 1990
SEE High Temperature Alarm
SO Acronyms

HTO
DA October 12, 1990
SEE Tritium Oxides
SO Acronyms

HUMAN BODY PARTS
(DOE FRASE Vocabulary Numeric Keys
1075-1174)
DA November 29, 1990
NT1 Arm(s)
NT2 Forearm(s)
NT2 Upper Arm
NT1 Body System(s)
NT2 Circulatory System
NT3 Blood
NT2 Digestive System
NT3 Mouth
NT4 Tooth/Teeth
NT3 Rectum
NT2 Excretory System
NT2 Nervous System
NT3 Brain
NT3 Spinal Cord
NT2 Respiratory System
NT3 Bronchial Epithelium
NT3 Nose
NT3 Throat
NT1 Bone(s)
NT2 Rib(s)
NT2 Spine
NT3 Coccyx
NT1 Breast(s)
NT1 Extremities
NT1 Foot/Feet
NT2 Heel(s)
NT2 Sole(s)

NT2 Toe(s)
NT1 Genitals
NT1 Hand(s)
NT2 Finger(s)
NT2 Palm(s)
NT2 Thumb(s)
NT1 Head
NT2 Chin
NT2 Ear(s)
NT2 Face
NT3 Area Around Eye
NT4 Eyelid(s)
NT3 Eye(s)
NT3 Lip(s)
NT3 Nose
NT2 Forehead
NT2 Skull (Frase)
NT3 Jaw
NT1 Joint(s)
NT2 Ankle(s)
NT2 Elbow(s)
NT2 Hip(s)
NT2 Knee(s)
NT2 Knuckle(s)
NT2 Shoulder(s)
NT2 Wrist(s)
NT1 Leg(s)
NT2 Lower Leg
NT2 Thigh(s)
NT1 Multiple Body Parts
NT1 Muscle/Tendon(s)
NT1 Neck
NT1 Skin
NT2 Fingernail(s)
NT2 Hair
NT2 Scalp
NT1 Trunk
NT2 Abdomen
NT2 Back
NT3 Lower Back
NT3 Upper Back
NT2 Buttock(s)
NT2 Chest
NT2 Groin
NT2 Rib(s)
RT DOE FRASE Categories
RT Targets
RT Whole Body
SO DOE FRASE VOCABULARY

HUMAN ENGINEERING
(SSDC)
DA October 12, 1990
NT1 Human Factors Engineering
RT Human Error
SO System Safety Development Center
     Glossary
DEF (SSDC) Designing hardware and
     equipment to effectively fit a wide
     range of human physical
     characteristics.

HUMAN ENVIRONMENT
(DOE 5440.1D; NEPA)
DA January 24, 1991
BT1 Environment
SO Environmental Management
DEF (CFR) Shall be interpreted
     comprehensively to include the
     natural and physical environment

BT-Broader Term | NT-Narrower Term | RT Related Term

and the relationship of people with that environment. This means that economic or social effects are not intended by themselves to require preparation of an environmental impact statement. When an environmental impact statement is prepared and economic or social and natural or physical environmental effects are interrelated, then the environmental impact statement will discuss all of these effects on the human environment.

HUMAN ERROR
DA January 8, 1991
SF *HE (Human Error)*
RT Error Sampling
RT Human Engineering

HUMAN FACTORS
(DOE Order 6430.1A; MORT)
DA October 12, 1990
SY Ergonomics
RT Design and Development Plans
RT Empirical Correlates
RT General Design Process
RT Human Performance
RT Human Factors Engineering
RT Inspection Plans
RT Maintenance Plans
RT Negligence
RT Operational Specifications
RT Reflex Arc Responses
RT Safety
RT Technique for Human Error Rate Prediction
SO Construction
SO System Safety Development Center Glossary
DEF (DOE Order 6430.1A) The biomedical, psychosocial, work place environment, and engineering considerations pertaining to people in human-machine system. Some of these considerations are allocation of functions, task analysis, human reliability, training requirements, job performance aiding, personnel qualification and selection, staffing requirements, procedures, organizational effectiveness, and workplace environmental conditions. (SSDC) The application of the human biological and psychological sciences in conjunction with the engineering sciences to achieve the optimum mutual adjustment of man and his work, the benefits being measured in terms of human efficiency and well-being. The principle disciplines involved are anthropometry, physiology, and engineering. (MORT) MORT analysis asks: has consideration been given in design, plan, and procedures to human

characteristics as they compete and interface with machine and environmental characteristics?

HUMAN FACTORS ENGINEERING
(DOE Order 6430.1A)
DA October 12, 1990
SY Ergonomics
BT1 Human Engineering
RT Accidents
RT General Design Process
RT Human Factors
RT Interfaces
RT Safety
RT Work Environment
SO Construction
SO System Safety Development Center Glossary
DEF (SSDC) The application of knowledge about human performance capabilities and behavioral principles to the design, operation, and maintenance of human-machine systems so that personnel can function at their optimum level of performance.

HUMAN PERFORMANCE
(SSDC)
DA October 12, 1990
RT Behavioral Stereotypes
RT Human Reliability
RT Human Factors
RT Stimulus-Mediation-Response
RT Technique for Human Error Rate Prediction
SO System Safety Development Center Glossary
DEF (SSDC) How a person functions, including both failure (errors) and success (reliability).

HUMAN RELIABILITY
(SSDC)
DA October 12, 1990
RT Human Performance
SO System Safety Development Center Glossary
DEF The dependability of the human within a system. In MORT, reliability for humans is determined and entered in the same analytical sense as hardware reliability to determine the overall reliability of a system.

HUMUS
(EPA)
DA October 12, 1990
RT Bogs
RT Compost
RT Soil Conditioner
RT Soils
SO Environmental Protection Agency Glossary
DEF (EPA) Decomposed organic material.

HVAC SYSTEM
(2036 HVAC SYSTEM)
DA December 10, 1990
SY Heating, Ventilation, Air Conditioning
BT1 Systems (DOE FRASE Vocabulary)
BT2 Systems
SO DOE FRASE VOCABULARY

HWM FACILITIES
(SWDA; RCRA; CFR)
DA October 19, 1990
SY Hazardous Waste Management Facilities
SO Environmental Management
SO Wastes
SO Water Pollution

HWVP
DA October 12, 1990
SEE Hanford Waste Vitrification
SO Acronyms

HX
DA October 12, 1990
SEE Heat Exchangers
SO Acronyms

HYBRID
(EPA)
DA October 12, 1990
RT Hybridoma
RT Organisms
SO Environmental Protection Agency Glossary
DEF (EPA) A cell or organism resulting from a cross between two unlike plant or animal cells or organisms.

HYBRIDOMA
(EPA)
DA October 12, 1990
RT Hybrid
SO Environmental Protection Agency Glossary
DEF (EPA) A hybrid cell that produces monoclonal antibodies in large quantities.

HYDRAULIC LIFT TANKS
(SWDA; RCRA; CFR)
DA October 12, 1990
BT1 Tanks
BT2 Facility Components
RT Lifting Station
SO Air Pollution
SO Wastes
DEF Tanks holding hydraulic fluid for a closed-loop mechanical system that uses compressed air or hydraulic fluid to operate lifts, elevators, and other similar devices.

HYDRAULIC STRUCTURES
(DOE Order 6430.1A)
DA October 12, 1990

SY-Synonymous Terms   SO-Source/Subject Category   SF-See From

BT1 Structures (DOE FRASE
    Vocabulary)
NT1 Dams
NT1 Dikes
 NT2 Terracing
RT Freeboard
SO Construction
DEF (DOE Order 6430.1A) Structures
    for the conveyance and/or control
    of water under nonpressure
    open-channel flow.

HYDROCARBONS
(EPA)
DA October 12, 1990
SF HC
BT1 Organic Chemicals
 BT2 Chemical Substances
NT1 Methane
SO Environmental Protection Agency
    Glossary
DEF (EPA) Chemical compounds that
    consist entirely of carbon and
    hydrogen.

HYDROGEN FLUORIDE
DA January 8, 1991
SF HF

HYDROGEN GAS STREAMS
(CAA; CFR)
DA October 12, 1990
RT Denuders
RT Mercury Chlor-alkali Cells
SO Air Pollution
DEF Hydrogen streams formed in the
    chlor-alkali cell denuder.

HYDROGEN SULFIDE
(EPA)
DA October 12, 1990
SF HS
BT1 Inorganic Chemicals
 BT2 Chemical Substances
SO Environmental Protection Agency
    Glossary
DEF (EPA) Gas emitted during organic
    decomposition. Also a by-product
    of oil refining and burning. It smells
    like rotten eggs and, in heavy
    concentration, can cause illness.

HYDROGEOLOGIC UNITS
DA May 24, 1991
BT1 Zones
 BT2 Sites/Areas
RT Groundwater
SO Wastes
SO Water Pollution
DEF (CFR) Any soil or rock units or
    zones which by virtue of their
    porosity or permeability, or lack
    thereof, a distinct influence on the
    storage or movement of
    groundwater.

HYDROGEOLOGY
(EPA)
DA October 12, 1990

SO Environmental Protection Agency
    Glossary
DEF (EPA) The geology of groundwater,
    with particular emphasis on the
    chemistry and movement of water.

HYDROLOGY
(EPA)
DA October 12, 1990
RT Site Characterization
RT Water
SO Environmental Protection Agency
    Glossary
DEF (EPA) The science dealing with the
    properties, distribution, and
    circulation of water.

HYDROLYSIS
DA May 20, 1991
BT1 Chemical Processes
 BT2 Processes
SO Environmental Management
SO Water Pollution
DEF (DSTT) Decomposition or alteration
    of a chemical substance by water.
    In aqueous solutions of
    electrolytes, the reactions of
    cations with water to produce a
    weak base or of anions to produce
    a weak acid.

I&C
DA October 12, 1990
SEE Instrumentation and Control
SO Acronyms

I/M
DA October 12, 1990
SEE Inspection and Maintenance
SO Acronyms

I/P
DA October 12, 1990
SEE Circuit to Pressure Converter
SO Acronyms

IA (Incident Actions)
DA October 12, 1990
SEE Incident Actions
SO Acronyms

IA (Instrument Air)
DA October 12, 1990
SEE Instrument Air
SO Acronyms

IAG
DA October 12, 1990
SEE Interagency Agreement
SO Acronyms

IBM
DA October 12, 1990
SEE International Business Machines
SO Acronyms

IC (Incident Control)
DA October 12, 1990
SEE Incident Control
SO Acronyms

IC (Isolation Condenser)
DA October 12, 1990
SEE Isolation Condenser
SO Acronyms

ICP/MS
DA October 12, 1990
SEE Inductively Coupled Plasma/Mass
    Spectrometer
SO Acronyms

ICPP
DA October 12, 1990
SEE Idaho Chemical Processing Plant
SO Acronyms

ICS
DA November 15, 1990
SEE Intermittent Control System
SO Acronyms

ICU
DA October 12, 1990
SEE Interface Control Unit
SO Acronyms

ID
DA October 12, 1990
SEE Idaho Operations Office
SO Acronyms

**IDAHO CHEMICAL PROCESSING
PLANT**
DA January 8, 1991
SF ICPP
BT1 Government-Owned
    Contractor-Operated Facilities
 BT2 Federal Facilities
  BT3 Facilities
RT Westinghouse Idaho Nuclear Co.,
    Inc.

**IDAHO NATIONAL ENGINEERING
LABORATORY**
DA January 8, 1991
SF INEL
BT1 Government-Owned
    Contractor-Operated Facilities
 BT2 Federal Facilities
  BT3 Facilities
BT1 Laboratories
 BT2 Research and Development
     Organizations
  BT3 Organizations
NT1 Power Burst Facility
RT EG&G Idaho, Inc.
RT MK-Ferguson of Idaho Co.
RT Protection Technology Idaho
DEF (Capsule Review of DOE Research
    and Development and Field
    Facilities, 1986) INEL was
    established in 1949 for the
    building and testing of nuclear

BT-Broader Term

NT-Narrower Term

RT Related Term

reactors and support equipment. INEL now reprocesses and recovers spent nuclear fuel from several test reactors, the nuclear naval fleet and other nuclear noncommercial reactors and processes liquid waste into the calcine form for intermediate storage. INEL operates the Advanced Test Reactor and a radioactive waste management complex for storage and disposal of low-level waste and conducts associated programs in materials testing, isotope production, irradiation services and training and test support. In addition, INEL functions as the lead laboratory for the multi-megawatt space reactor program and fusion reactor safety research.

IDAHO OPERATIONS OFFICE
DA   January 8, 1991
SF   *ID*
BT1 Operations Offices
 BT2 Offices
  BT3 Administrative Organizations
   BT4 Organizations
 BT2 U.S. Department of Energy
  BT3 Federal Agencies
   BT4 Agencies
    BT5 Administrative Organizations
     BT6 Organizations
RT   EG&G Idaho, Inc.
RT   MK-Ferguson
RT   MK-Ferguson of Idaho Co.
RT   Protection Technology Idaho
RT   Rockwell International Corp.
RT   West Valley Nuclear Services Co., Inc.
RT   Westinghouse Idaho Nuclear Co., Inc.
DEF (Capsule Review of DOE Research and Development and Field Facilities, 1986) The Idaho Operations Office, established in 1949, administers the Idaho National Engineering Laboratory (INEL) and associated support facilities. Major program and project assignments include: nuclear materials production; defense waste and transportation management; energy conservation; alternate energy; remedial action; the Three Mile Island Program; the Multi-Megawatt Space Reactor, etc.

IDB
DA   October 12, 1990
SEE Integrated Data Base
SO   Acronyms

IDCOR
DA   October 12, 1990

SEE Industry Degraded Core Rulemaking Program
SO   Acronyms

IDLH
DA   October 12, 1990
SEE Immediately Dangerous to Life and Health
SO   Acronyms

IE (Industrial Engineering)
DA   October 12, 1990
SEE Industrial Engineering
SO   Acronyms

IE (Office of Inspection and Enforcement)
DA   October 12, 1990
SEE NRS Office of Inspection and Enforcement
SO   Acronyms

IEEE
DA   October 12, 1990
SEE Institute of Electrical and Electronic Engineers
SO   Acronyms

IFC
DA   October 12, 1990
SEE Internal Fission Counter
SO   Acronyms

IFI
DA   October 12, 1990
SEE Inspector Follow-up Items
SO   Acronyms

IGFM
DA   October 12, 1990
SEE Internal Gamma Flux Monitor
SO   Acronyms

IGNITABILITY
(EPA)
DA   October 12, 1990
RT   Explosives
SO   Environmental Protection Agency Glossary
DEF (EPA) Capable of burning or causing a fire.

IH (Industrial Hygiene)
DA   October 12, 1990
SEE Industrial Hygiene
SO   Acronyms

IH (Inner Housing)
DA   October 12, 1990
SEE Inner Housing
SO   Acronyms

IHE
DA   October 12, 1990
SEE Insensitive High Explosives
SO   Acronyms

IHE SUBASSEMBLIES
(DOE Order 6430.1A)
DA   October 12, 1990
RT   IHE Weapons
SO   Construction
DEF (DOE Order 6430.1A) Insensitive High Explosive (IHE) hemispheres or spheres with booster charges, with or without detonators, that pass the DOE qualification tests listed in Table IX-2 of DOE/EV 06194.

IHE WEAPONS
(DOE Order 6430.1A)
DA   October 12, 1990
RT   IHE Subassemblies
RT   Insensitive High Explosives
SO   Construction
DEF (DOE Order 6430.1A) Weapons listed in DOE/DNA TP 20-7 as exempt from storage and transportation limits are classified as IHE weapons when stored or transported alone or in combination with each other. This classification is valid only by storage/shipping containers or, if out of containers, by the spacing specified in TP 20-7.

ILLNESS
(1391 ILLNESS)
DA   November 28, 1990
RT   Injury
SO   DOE FRASE VOCABULARY

ILLNESSES
(DOE FRASE Vocabulary Numeric Keys 1375-1449)
DA   January 17, 1991
NT1 Asbestosis
NT1 Benign and Unspecified Neoplasm
NT1 Chemical Reaction
NT1 Complications Peculiar to Med. Care
NT1 Cond of Respiratory Sys. Non-Toxic
 NT2 Upper Respiratory Condition
 NT2 Upper Respiratory Disease
  NT3 Tuberculosis
NT1 Conjunctivitis
NT1 Contagious or Infectious Disease
 NT2 Conjunctivitis
NT1 Dermatitis
NT1 Disease of Central Nervous System
NT1 Diseases of Blood
 NT2 Hepatitis
NT1 Fluorosis
NT1 Frostbite/Other Low Temp Effects
NT1 Hearing Impairment
NT1 Heart Attack
NT1 Hemorrhoids
NT1 Insulin Reaction
NT1 Ionizing Radiation Effects
NT1 Malignant Neoplasm
NT1 Mental Disorders
NT1 Occupational Illnesses
NT1 Other Complications
NT1 Other Skin Conditions
NT1 Physical Harm

SY-Synonymous Terms   SO-Source/Subject Category   SF-See From

NT1 Poisoning
 NT2 Systemic Poisoning
NT1 Sunburn
NT1 Synovitis
NT1 Systemic Effects
NT1 Tendonitis
NT1 Tetanus
NT1 Toxic Effects to Single System
NT1 Ulcer(s)
RT Diseases
RT DOE FRASE Categories
RT Effects
RT Hospital
RT Injuries
SO DOE FRASE VOCABULARY
DEF Equivalent to the DOE FRASE
      Vocabulary Category "Nature of
      Illness".

ILRT
DA October 12, 1990
SEE Integrated Leak Rate Test
SO Acronyms

ILS
DA October 12, 1990
SEE Integrated Logistics Support
SO Acronyms

IMM
(NFI)
DA October 12, 1990
SO Nuclear Facilities Incident Database
SO Radiation
DEF (NFI) Procedure designation for
      Instrument Department.

**IMMEDIATE (ACUTE) HEALTH
HAZARDS**
(CERCLA, CFR)
DA October 12, 1990
BT1 Hazard Categories
BT1 Health Hazards
 BT2 Hazards
  BT3 Conditions
DEF (CFR) Includes "highly toxic,"
      "toxic," "irritant," "sensitizer," and
      "corrosive," and other hazardous
      chemicals that cause an adverse
      effect to a target organ and which
      effect generally occurs rapidly as a
      result of short term exposure and
      is of short duration.

IMMEDIATE CAUSES
(SSDC)
DA October 12, 1990
BT1 Causes
SO System Safety Development Center
      Glossary
DEF (SSDC) Practices or conditions
      which physically cause an
      accident or incident at a specific
      time and place.

**IMMEDIATELY DANGEROUS TO LIFE
AND HEALTH**
(EPA; EMER)
DA October 12, 1990

SF IDLH
SO Environmental Management
SO Environmental Protection Agency
      Glossary
DEF (EPA) The maximum level to which
      a healthy individual can be
      exposed to a chemical for 30
      minutes and escape without
      suffering irreversible health effects
      or impairing symptoms. Used as a
      "level of concern."

**IMMEDIATELY DANGEROUS TO LIFE
OR HEALTH VALUES**
(EMER)
DA February 1, 1991
BT1 Conditions
SO Emergency Preparedness
DEF As defined in 29 Code of Federal
      Regulations Part 1926.103, a
      condition that either poses an
      immediate threat to life and health
      or an immediate threat of severe
      exposure to contaminants (for
      example, radioactive materials
      which are likely to have adverse
      delayed effects on health).

**IMMINENT AND SUBSTANTIAL
ENDANGERMENT**
DA January 24, 1991
BT1 Conditions
RT Imminent Danger
SO Environmental Management
DEF (USC) A qualifying condition given
      to any solid or hazardous waste
      which, when applied, allows EPA
      the authority to take necessary
      actions to protect public health
      and the environment. This action
      may include but is not limited to
      restraining any person from
      handling, storage, treatment,
      transportation, or disposal of solid
      or hazardous wastes which meet
      this condition. The test for
      "imminent and substantial
      endangerment" places a heavy
      burden of proof on the EPA and
      therefore, substantially limits
      EPA's authority to address
      disposal site problems. (Reference
      "Imminent Danger", Doe 5483.1A
      and "Substantial", DOE 5484.1)

IMMINENT DANGER
(Doe Order 5483.1A)
DA October 12, 1990
BT1 Danger
 BT2 Conditions
RT Imminent and Substantial
      Endangerment
RT Industrial Safety
SO Environmental Management
SO Industrial Safety
SO System Safety Development Center
      Glossary
DEF (DOE Order 5483.1A) Any
      condition or practice which is such
      that a hazard exists that could

reasonably be expected to cause
death or serious physical harm to
employees (permanent or
prolonged impairment of the body
or temporary disablement
requiring hospitalization), unless
immediate actions are taken to
mitigate the effects of the hazard
and/or remove employees from
the hazard.

IMMINENT HAZARD
(USC)
DA November 15, 1990
BT1 Hazards
 BT2 Conditions
RT Emergency Conditions
SO Environmental Management
DEF (USC) Situations which exist when
      the continued use of a pesticide
      during the time required for
      cancellation proceeding would be
      likely to result in unreasonable
      adverse effects on the
      environment or will involve
      unreasonable hazard to the
      survival of a species declared
      endangered or threatened by the
      Secretary pursuant to the
      Endangered Species Act of 1973
      [16 USCS sections 1531 et seq.].

IMPACT WRENCH
(3040 IMPACT WRENC)
DA January 3, 1991
BT1 Wrench(s)
 BT2 Tools (DOE FRASE Vocabulary)
  BT3 Equipment
SO DOE FRASE VOCABULARY

IMPLEMENTATION
(MORT)
DA April 3, 1991
RT Implementation Plans
RT Management System Factors
RT Policies
RT Risk Assessment
DEF (WEBSTER) To give practical effect
      to and ensure actual fulfillment by
      the establishment of concrete
      measures. (MORT) MORT
      analysis asks: does the overall
      program represent the intended
      fulfillment of the policy statement?
      if there are problems encountered
      in carrying out the policy, are
      these relayed back to the policy
      makers? is the implementation a
      continuous, balanced effort
      designed to correct systemic
      failures, and generally pre-active
      rather than re-active? Related
      issues/topics include: line and
      staff responsibility; information
      flow; directives; management
      services; budget; delays;
      accountability; and vigor and
      example.

IMPLEMENTATION PLANS
(DOE Orders 5440.1D, 5480.1B, and 5700.6B)
DA   October 12, 1990
BT1 Plans
NT1 EIS Implementation Plans
RT   Delayed Compliance Orders
RT   Environmental Impact Statements
RT   Implementation
SO   Environmental Management
SO   Management
SO   Quality Assurance
DEF Concise descriptions of the approach, resources, and time period planned for implementing Orders that require such plans on a site-wide basis. The plan includes a description of the execution of environmental protection, safety, and health responsibilities and authorities by the field organization, and any proposed generic exemptions to parts of such DOE Orders. (ESH) Written plans that record the results of the scoping process and outline the procedures by which an environmental impact statement is to be prepared. The implementation plans should be prepared in accordance with the Department's guidelines (52 FR 49662).

IMPLEMENTING AGENCIES
(SWDA; RCRA; CFR)
DA   October 19, 1990
BT1 Local Agencies
BT2 Agencies
BT3 Administrative Organizations
BT4 Organizations
BT1 State Agencies
BT2 Agencies
BT3 Administrative Organizations
BT4 Organizations
RT   Nuclear Regulatory Commission
SO   Wastes
DEF EPA, or, in the case of a state with a program approved under section 9004 (or pursuant to a memorandum of agreement with EPA), the designated state or local agency responsible for carrying out an approved UST (Underground Storage Tank) program.

IMPOUNDMENT
(SWDA; EPA; ESH)
DA   October 12, 1990
BT1 Sites/Areas
NT1 New Tailings Impoundment
NT1 PAR Ponds
NT1 Reservoirs
NT1 Surface Impoundment
RT   Evaporation Ponds
RT   Holding Ponds
RT   Sanitary Engineering Structures
RT   Sewage Lagoons

SO   Environmental Protection Agency Glossary
SO   Wastes
DEF (EPA) A body of water or sludge confined by a dam, dike, floodgate, or other barrier.

IMPROVED RISK
(DOE Orders 5480.7 and 6430.1A)
DA   October 12, 1990
BT1 Risks
SO   Construction
SO   Fires
DEF The term involves the use and application of judgment and thus does not lend itself to a precise, fixed definition applicable in all locations and situations. It has the same meaning and intent as is commonly understood when this or the term, "Highly Protected Risk," is used in the insurance industry. Generally, an improved risk property is one that would qualify for complete insurance coverage by the Factory Mutual System, the Industrial Risk Insurers, and other industrial insurance companies that limit their insurance underwriting to the best protected class of industrial risk. Essential elements of a program complying with the improved risk concept are included in this directive. Improved risk protection requires compliance with the fire protection and loss prevention standards detailed in DOE 5480.4, ENVIRONMENTAL PROTECTION, SAFETY, AND HEALTH PROTECTION STANDARDS, of 5-15-84. See DOE 5480.7 for continued definition.

IMPURITIES
(TSCA; CFR)
DA   October 12, 1990
BT1 Chemical Substances
RT   Adulterants
RT   Contamination
RT   Ethylene Dichloride Purification
RT   Vinyl Chloride Purification
SO   Hazardous Materials
DEF Chemical substances which are unintentionally present with another chemical substance.

IN GAS SERVICE
(CAA; CFR)
DA   October 12, 1990
NT1 In VHAP Service
DEF (CFR) That a piece of equipment contains process fluid that is in the gaseous state at operating conditions.

IN HEAVY LIQUID SERVICE
(CAA; CFR)
DA   October 12, 1990

BT1 In Liquid Service
SO   Air Pollution
DEF (ESH) Equipment that is not in gas/vapor service or in light liquid service. For Subpart KKK ONLY contains a liquid for which the weight percent evaporated is 10 percent or less at 150degC as determined by ASTM method D86.

IN LIGHT LIQUID SERVICE
(CAA; CFR)
DA   October 12, 1990
BT1 In Liquid Service
SO   Air Pollution
DEF (ESH) Equipment that contains a fluid for which: (1) The vapor pressure of one or more of the components is greater than 0.3 kilopascals (kPa) at 20degC; (2)the total concentration of the pure components having a vapor pressure greater than 0.3 KPa at 20degC is equal to or greater than 20 percent by weight; and (3) the fluid is a liquid at operating conditions. (For Subpart KKK ONLY contains a liquid for which the weight percent-evaporated is greater than 10 percent at 50degC as determined by ASTM method D86).

IN LIQUID SERVICE
(CAA; CFR)
DA   October 12, 1990
NT1 In Heavy Liquid Service
NT1 In Light Liquid Service
NT1 In VOC Service
SO   Air Pollution
DEF (CFR) A piece of equipment that is not in gas/vapor service.

IN-PROCESS MATERIALS
DA   October 19, 1990
SY   In-Use Materials
BT1 Materials
RT   Staging Bays (in-process)
SO   Construction
DEF (DOE Order 6430.1A) Materials that are integral to the manufacturing or production processes and are needed to maintain continuity of operations. Other material that requires temporary location near the pertinent process areas in readiness for near-term use or for movement to other process areas may also be considered "in-process". For material involved in laboratory operations, analogous definitions shall be applied to determine eligibility for the "in-process" or "in-use" category and consequent exclusion from storage requirements of these criteria.

IN-PROCESS WASTEWATER
(CAA; CFR)
DA November 15, 1990
BT1 Wastewater
 BT2 Wastes
 BT2 Water
RT Polyvinyl Chloride
RT Vinyl Chloride
SO Air Pollution
DEF (CFR) Any water which, during
    manufacturing or processing,
    comes into direct contact with vinyl
    chloride or polyvinyl chloride or
    results from the production or use
    of any raw material, intermediate
    product, finished product,
    by-product, or waste product
    containing vinyl chloride or
    polyvinyl chloride but which has
    not been discharged to a
    wastewater treatment process or
    discharged untreated as
    wastewater. Gasholder seal water
    is not inprocess wastewater until it
    is removed from the gasholder.

IN-SITU
(DOE Order 6430.1A)
DA October 12, 1990
SO Construction
DEF (DOE Order 6430.1A) In the
    existing or original location.

IN-SITU EXTRACTION
(NCRP)
DA October 12, 1990
BT1 Chemical Processes
 BT2 Processes
RT Heap-leach Extraction
SO Radiation
DEF (NCRP) Extraction of a mineral
    using chemical solutions, without
    removing the ore from its natural
    location.

IN-SITU SAMPLING SYSTEMS
(CAA; CFR)
DA October 12, 1990
BT1 Systems
RT Sampling
SO Air Pollution
DEF (CFR) Nonextractive samplers or
    in-line samplers.

IN-SITU TREATMENT
DA January 8, 1991
SF *IST*
BT1 Treatment
 BT2 Waste Management Processes
 BT3 Processes

IN-SITU VITRIFICATION
DA January 8, 1991
SF *ISV*
BT1 Processes

IN SITU VOLATIZATION
(CERCLA)
DA May 20, 1991

BT1 Chemical Processes
 BT2 Processes
RT Comprehensive Environmental
    Response, Compensation, etc.
SO Environmental Management
DEF A process presently performed at a
    number of CERCLA sites (e.g.
    Rocky Mountain Arsenal-DOD).
    (DSTT) The conversion of a
    chemical substance from a liquid
    or solid state to a gaseous or
    vapor state by the application of
    heat, by reducing pressure, or by
    a combination of these processes.

IN-USE MATERIALS
DA October 19, 1990
SY In-Process Materials
BT1 Materials
SO Construction
DEF (DOE Order 6430.1A) Materials that
    are integral to the manufacturing
    or production processes and are
    needed to maintain continuity of
    operations. Other materials that
    require temporary location near
    the pertinent process areas in
    readiness for near-term use or for
    movement to other process areas
    may also be considered
    "in-process". For materials
    involved in laboratory operations,
    analogous definitions shall be
    applied to determine eligibility for
    the "in-process" or "in-use"
    category and consequent
    exclusion from storage
    requirements of these criteria.

IN VACUUM SERVICE
(CAA; CFR)
DA October 12, 1990
RT Pressure (Surface)
SO Air Pollution
DEF (CFR) Equipment that is operating
    at an internal pressure which is at
    least 5 kilopascals (kPa) below
    ambient pressure.

IN VHAP SERVICE
(CAA; CFR)
DA October 12, 1990
BT1 In Gas Service
RT In Vinyl Chloride Service
RT In VOC Service
RT Product Accumulator Vessels
RT Volatile Hazardous Air Pollutant
    (VHAP)
SO Air Pollution
DEF (CFR) A piece of equipment that
    either contains or contacts a fluid
    (liquid or gas) that is at least 10
    percent by weight a volatile
    hazardous air pollutant (VHAP) as
    determined according to the
    provisions of 40 CFR 61.245(d).
    The provisions of section
    61.245(d) also specify how to
    determine that a piece of
    equipment is not in VHAP service.

IN VINYL CHLORIDE SERVICE
(CAA; CFR)
DA October 12, 1990
RT In VHAP Service
RT Vinyl Chloride
SO Air Pollution
DEF (CFR) A piece of equipment that
    either contains or contacts a liquid
    that is at least 10 percent vinyl
    chloride by weight or a gas that is
    at least 10 percent by volume vinyl
    chloride as determined according
    to the provisions of 40 CFR
    61.67(h). The provisions of section
    61.67(h) also specify how to
    determine that a piece of
    equipment is not in vinyl chloride
    service.

IN VITRO
(EPA)
DA October 12, 1990
SO Environmental Protection Agency
    Glossary
DEF (EPA) (1) "In glass"; a test-tube
    culture. (2) Any laboratory test
    using living cells taken from an
    organism.

IN VIVO
(EPA)
DA October 12, 1990
SO Environmental Protection Agency
    Glossary
DEF (EPA) In the living body of a plant
    or animal. In vivo tests are those
    laboratory experiments carried out
    on whole animals or human
    volunteers.

IN VOC SERVICE
(CAA; CFR)
DA October 12, 1990
BT1 In Liquid Service
RT In VHAP Service
RT Volatile Organic Compounds
SO Air Pollution
DEF (CFR) The piece of equipment that
    contains or contacts a process
    fluid that is at least 10 percent
    VOC by weight (see 40 CFR 60.2
    for the definition of volatile organic
    compound and 40 CFR 60.485(d)
    to determine whether a piece of
    equipment is not in VOC service)
    and (b) the piece of equipment is
    not in heavy liquid service as
    defined in 40 CFR 60.481. (ESH)
    Equipment that contains or
    contacts a process fluid that is at
    least 10 percent volatile organic
    compounds by weight.

INACTIVE FACILITIES
DA January 24, 1991
BT1 Facilities
RT Inactive Portions
SO Environmental Management
DEF (DOE 5480.14) An area where a

hazardous substance has been deposited, stored, disposed of, or placed or otherwise come to be located. It can be any building, structure, installation, equipment, pipe or pipeline (including any pipe into a sewer or publicly owned treatment works, well, pit, pond, lagoon, impoundment, ditch, landfill, storage container, motor vehicle, rolling stock, or aircraft. Excluded are areas that have a permit issued, or have been accorded interim status under subtitle C of the Solid Waste Disposal Act of the Memorandum of Understanding between the DOE and the EPA for hazardous waste and radioactive mixed waste management, or operated under the provisions of DOE 5480.2 and DOE 5820.2.

INACTIVE HAZARDOUS WASTE DISPOSAL SITES
DA January 24, 1991
BT1 Inactive Waste Disposal Sites
 BT2 Disposal Sites
  BT3 Sites/Areas
SO Environmental Management

INACTIVE MINES
(CAA; CFR)
DA October 12, 1990
BT1 Conventional Mines
 BT2 Sites/Areas
RT Abandoned Areas
SO Air Pollution
DEF (CFR) Mines from which uranium ore has been previously removed but which are not active mines as of the effective date of the standard. Inactive mines which become active mines after the effective date of the standard are considered new sources under the provisions of subparts A and B of 40 CFR 61.21.

INACTIVE PORTIONS
(SWDA; RCRA)
DA October 12, 1990
BT1 Sites/Areas
RT Active Portions
RT Closed Portions
RT Inactive Facilities
RT Treatment, Storage, and Disposal Facilities
SO Environmental Management
SO Wastes
DEF (CFR) Portions of a facility which are not operated after the effective date of 40 CFR 261. See also "active portions" and "closed portions."

INACTIVE WASTE DISPOSAL SITES
(CAA; CFR)
DA October 12, 1990

BT1 Disposal Sites
 BT2 Sites/Areas
NT1 Inactive Hazardous Waste Disposal Sites
RT Asbestos-Containing Waste Materials
SO Air Pollution
DEF (CFR) Disposal sites or portions of them where additional asbestos-containing waste material will not be deposited and where the surface is not disturbed by vehicular traffic.

INADVERTENTLY OPENED RELIEF VALVE
DA January 8, 1991
SF *IORV*
BT1 Relief Valves
 BT2 Valves
  BT3 Devices
RT Hazards

INCH-POUND SYSTEM OF UNITS
(Doe Order 5900.2)
DA October 12, 1990
SO Quality Assurance
DEF (DOE Order 5900.2) The system of measurement units (inch, pound, second, degree Fahrenheit, and units derived from those) most commonly used now in the United States. Synonyms: "English System," "U.S. System," "Customary System." The inch-pound system is not to be confused with "Imperial System," which describes a related but not completely identical system currently in use in Great Britain and some other English speaking countries.

INCIDENCE RATE, LOST WORKDAY CASES (LWC)
(SSDC)
DA October 12, 1990
BT1 Rates
RT Incidence Rate, Total Recordable Cases (TRC)
RT Incidence Rate, WDL
RT Incidence Rate, LWD (Lost Work Days)
RT Lost Workday Cases
SO System Safety Development Center Glossary
DEF (SSDC) The number of cases recorded in columns 2 or 9 of the OSHA 200 log shall be used to calculate this rate. Therefore, uniform calculation for the rate shall be as follows: LWC = No. of Lost Workday Cases × 200,000/Total Hours Worked.

INCIDENCE RATE, LWD (LOST WORK DAYS)
(SSDC)
DA October 12, 1990

BT1 Rates
RT Incidence Rate, Total Recordable Cases (TRC)
RT Incidence Rate, Lost Workday Cases (LWC)
RT Incidence Rate, WDLR
SO System Safety Development Center Glossary
DEF (SSDC) The total lost work days, due to days away and days restricted. LWD = (WDL + WLDR) × 200,000/Total Hours Worked.

INCIDENCE RATE, TOTAL RECORDABLE CASES (TRC)
(SSDC)
DA October 12, 1990
SY Total Recordable Cases
BT1 Rates
RT Incidence Rate, Lost Workday Cases (LWC)
RT Incidence Rate, LWD (Lost Work Days)
RT Occupational Injuries
RT Occupational Illnesses
SO System Safety Development Center Glossary
DEF (SSDC) The number of recordable injuries and illnesses per 200,000 total hours worked by all employees during the period covered. The 200,000 hours worked are equivalent to 100 full-time workers at 40 hours per week for 50 weeks. TRC = No. of Recordable Injuries & Illnesses × 200,000/Total Hours Worked. Rates for special purposes may be derived for workdays lost, workdays lost restriction.

INCIDENCE RATE, WDL
(SSDC)
DA October 12, 1990
BT1 Rates
RT Incidence Rate, Lost Workday Cases (LWC)
RT Incidence Rate, WDLR
SO System Safety Development Center Glossary
DEF (SSDC) The total number of days (away) entered in columns 4 or 11 of the OSHA 200 log shall be used to calculate this rate. WDL = No. Workdays Lost × 200,000/Total Hours Worked.

INCIDENCE RATE, WDLR
(SSDC)
DA October 12, 1990
BT1 Rates
RT Incidence Rate, WDL
RT Incidence Rate, LWD (Lost Work Days)
SO System Safety Development Center Glossary
DEF (SSDC) The total number of days lost, due to restrictions that are entered in columns 5 or 12 of the OSHA 200 log, shall be used to

calculate this rate. WDLR = No. of Workdays Restricted × 200,000/Total Hours Worked.

INCIDENT ACTIONS
DA   January 8, 1991
SF   *IA (Incident Actions)*
BT1 Actions
 BT2 Responses

INCIDENT CONTROL
DA   January 8, 1991
SF   *IC (Incident Control)*
BT1 Controls

INCIDENTAL TAKE (TAKING)
(ESA)
DA   October 12, 1990
RT   Reasonable and Prudent Measures
SO   Endangered Species
DEF (CFR) Takings that result from, but are not the purpose of, carrying out an otherwise lawful activity conducted by a Federal agency or applicant.

INCIDENTS
(SSDC; EMER)
DA   October 12, 1990
SY   Accidents
NT1 Air Pollution Episodes
NT1 Aircraft Incidents
NT1 Bomb Incidents
NT1 Bomb Threats
NT1 Criticality Incidents
NT1 Episodes (Pollution)
NT1 High Potential Incidents
NT1 Nuclear Threat Incidents
NT1 Radioactive Material Transportation Incidents
NT1 Radiological Transportation Incidents
NT1 Significant Incidents
NT1 Transportation Incidents
RT   Accidents
RT   Accident Sites
RT   Amelioration
RT   Computerized Accident/Incident Reporting System
RT   Events
RT   Occurrences
SO   Emergency Preparedness
SO   System Safety Development Center Glossary
DEF (DSTT) An occurrence of an action or situation that is a separate unit of experience; happening; something dependent on or subordinate to something else of greater or principal importance; an action likely to lead to grave consequences. (SSDC) A special definition on incidents is used in MORT analysis. They are a failure of control (barrier) without consequences.

INCINERATION
(EPA)

DA   October 12, 1990
BT1 Treatment
 BT2 Waste Management Processes
  BT3 Processes
NT1 Incineration at Sea
RT   Incinerators
RT   Open Burning
RT   Thermal Treatment
RT   Trash-To-Energy Plans
SO   Environmental Protection Agency Glossary
DEF (EPA) A treatment technology using combustion to destroy organic constituents and reduce the volume of wastes. (EPA) (1) Burning of certain types of solid, liquid or gaseous materials. (2) A treatment technology involving destruction of waste by controlled burning at high temperatures, e.g., burning sludge to remove the water and reduce the remaining residues to a safe, non-burnable ash which can be disposed of safely on land, in some waters or in underground locations.

INCINERATION AT SEA
(EPA)
DA   October 12, 1990
BT1 Incineration
 BT2 Treatment
  BT3 Waste Management Processes
   BT4 Processes
SO   Environmental Protection Agency Glossary
DEF (EPA) Disposal of waste by burning at sea on specially-designed incinerator ships.

INCINERATION VESSELS
(CERCLA; USC)
DA   October 12, 1990
BT1 Vessels
SO   Compensation and Liability
DEF (USC) Vessels which carry hazardous substances for the purpose of incineration of such substances, so long as such substances or residues of such substances are on board.

INCINERATORS
(SWDA; RCRA; CAA; TSCA; CFR; ESH)
DA   October 12, 1990
BT1 Equipment/Parts - Heating (DOE FRASE Vocabulary)
 BT2 Equipment
BT1 Heating Equipment
NT1 Catalytic Incinerators
NT1 Qualified Incinerators
RT   Afterburners
RT   Baffle Chambers
RT   Combustion
RT   Incineration
RT   Overfire Air
SO   Air Pollution
SO   Environmental Protection Agency Glossary
SO   Hazardous Materials

SO   Wastes
DEF (CFR) Furnaces used in the process of burning waste for the primary purpose of reducing the volume of the waste by removing combustible matter. (ESH) Engineered devices using controlled flame combustion to thermally degrade PCBs and PCB items. Examples of devices used for incineration include rotary kilns, liquid injection incinerators, cement kilns, and high-temperature boilers.

INCOMPATIBLE WASTES
(SWDA; RCRA)
DA   October 19, 1990
BT1 Hazardous Wastes
 BT2 Hazardous Materials
  BT3 Materials
 BT2 Wastes
SO   Environmental Management
SO   Wastes
DEF (CFR) Hazardous wastes which are unsuitable for (1) placement in a particular device or facility because it may cause corrosion or decay of containment materials (e.g., container inner liners or tank walls); or (2) commingling with another waste or material under uncontrolled conditions because the commingling might produce heat or pressure, fire or explosion, violent reaction, toxic dusts, mists fumes, or gases or flammable fumes or gases. (See 40 CFR 265 Appendix V for examples.)

INCONSISTENCIES
(DOE Order 5400.3)
DA   May 24, 1991
RT   Atomic Energy Act
SO   Wastes
DEF (DOE 5400.3) Inconsistencies between RCRA and the AEA occurs if the requirements of both laws are incompatible. RCRA applies to hazardous or radioactive requirements of the AEA.

INCREMENTS OF PROGRESS
(CAA; CFR)
DA   October 12, 1990
SY   Reasonable Further Progress
RT   Compliance Schedules
RT   Schedule of Compliance
SO   Air Pollution
DEF (CFR) Steps toward compliance which will be taken by a specific source, including (1) date of submittal of the source's final control plan to the appropriate air pollution control agency; (2) date by which contracts for emission control systems or process modifications will be awarded; or date by which orders will be issued for the purchase of component

BT-Broader Term   NT-Narrower Term   RT Related Term

parts to accomplish emission control or process modification; (3) date of initiation of on-site construction or installation of emission control equipment or process change; (4) date by which on-site construction or installation of emission control equipment or process modification is to be completed; and (5) date by which final compliance is to be achieved.

INCURRED LOSSES
(SSDC)
DA   October 12, 1990
BT1 Losses
RT   Loss Ratio
SO   System Safety Development Center Glossary
DEF (SSDC) Losses that have happened; includes amounts paid and reserved for future payments.

INDEMNIFY
(SSDC)
DA   October 12, 1990
RT   Compensation and Liability
RT   Insurance
SO   System Safety Development Center Glossary
DEF (SSDC) To reimburse an insured for loss.

INDEPENDENT (SAFETY) REVIEWS
(SSDC)
DA   October 12, 1990
BT1 Safety Reviews
 BT2 Reviews
  BT3 Administrative Processes
   BT4 Processes
SO   System Safety Development Center Glossary
DEF (SSDC) Reviews performed by personnel organizationally independent of the operating group and groups performing primary safety analysis. Other criteria for good independent review would include being financially independent of the operating group, no personal interest in review decisions, using different analytical methods from those doing the primary safety analysis, etc.

INDIAN GOVERNING BODIES
(SWDA; RCRA)
DA   October 19, 1990
BT1 Administrative Organizations
 BT2 Organizations
RT   Indian Tribes
SO   Wastes
DEF (CFR) The governing bodies of tribes, bands, or groups of Indians subject to the jurisdiction of the United States and recognized by the United States as possessing power of self government.

INDIAN LANDS
(ARPA)
DA   April 24, 1991
BT1 Natural Resources
BT1 Sites/Areas
RT   Archaeological Resources Protection Act of 1979
RT   Archaeological Resources
RT   Indian Tribes
RT   Public Lands
RT   Targets
DEF (ARPA) Lands of Indian tribes, or Indian individuals, which are either held in trust by the United States or subject to a restriction against alienation imposed by the United States, except for any subsurface interests in lands not owned or controlled by an Indian tribe or an Indian individual.

INDIAN TRIBES
(SWDA; CERCLA; SDWA; ARPA; CFR)
DA   October 12, 1990
BT1 States
RT   Alaskan Natives
RT   Archaeological Resources
RT   Authorized Officials
RT   Indian Governing Bodies
RT   Indian Lands
RT   Natural Resource Trustees
SO   Compensation and Liability
SO   Wastes
SO   Water Pollution
DEF (CFR) Except in the case of RCRA, Indian tribes, bands, nations, or other organized groups or communities, including any Alaska Native village but not including any Alaska Native regional or village corporation, which are recognized as eligible for the special programs and services provided by the United States to Indians because of their status as Indians and having a Federally recognized governing body carrying out substantial governmental duties and powers over a defined area. (ARPA) Any Indian tribe, band, nation, or other organized group or community, including any Alaska Native village or regional or village corporation as defined in or established pursuant to the Alaska Native Claims Settlement Act (85 Stat.688).

INDICATOR(S)
(2756 INDICATOR)
DA   January 3, 1991
BT1 Instrument(s)
 BT2 Equipment/Parts - Instrumentation/Measuring (DOE FRASE Voc.)
  BT3 Equipment
NT1 Ammeter(s)
NT1 Flow Meter(s)
NT1 Ohmmeter(s)
NT1 Potentiometer

NT1 Spectrometer(s)
NT1 Tachometer(s)
NT1 Thermoluminescent Dosimeters
NT1 Thermometer(s)
NT1 Volt Meter
RT   Trouble Light
SO   DOE FRASE VOCABULARY

INDIRECT DISCHARGES
(EPA)
DA   October 12, 1990
BT1 Discharges
SO   Environmental Management
SO   Environmental Protection Agency Glossary
DEF (EPA) Introductions of pollutants from non-domestic sources into publicly owned waste treatment systems. Indirect dischargers can be commercial or industrial facilities whose wastes go into the local sewers.

INDIRECTLY IONIZING PARTICLES
(IAEA)
DA   October 12, 1990
RT   Directly Ionizing Particles
RT   Ionization
SO   Radiation
DEF (IAEA) Uncharged particles (neutrons, photons, etc.) which can liberate directly ionizing particles.

INDIVIDUAL GENERATION SITES
(SWDA; RCRA)
DA   October 12, 1990
BT1 Sites/Areas
RT   Hazardous Waste Generation
SO   Environmental Management
SO   Wastes
DEF (CFR) Contiguous sites at or on which one or more hazardous wastes are generated. Individual generation sites, such as large manufacturing plants, may have one or more sources of hazardous waste but are considered single or individual generation sites if the sites or properties are contiguous.

INDIVIDUAL PERMITS
(CAA; RHA; CFR; ESA)
DA   October 12, 1990
BT1 Permits
SO   Water Pollution
DEF (CFR) DA authorizations that are issued following a case-by-case evaluation of a specific structure or work in accordance with the procedures of this regulation and 33 CFR Part 325, and a determination that the proposed structure or work is in the public interest pursuant to 33 CFR Part 320.

INDIVIDUAL PLANT EXAMINATION
DA   January 8, 1991

SF   *IPE*
BT1 Administrative Procedures Act
 BT2 Acts
  BT3 Statutes and Regulations

INDOOR AIR
(EPA)
DA   October 12, 1990
BT1 Air
RT   Indoor Air Pollution
SO   Environmental Protection Agency
     Glossary
DEF (EPA) The breathing air inside a
     habitable structure or conveyance.

INDOOR AIR POLLUTION
(EPA)
DA   October 12, 1990
BT1 Air Pollution
 BT2 Pollution
RT   Indoor Air
RT   Indoor Climate
SO   Environmental Protection Agency
     Glossary
DEF (EPA) Chemical, physical, or
     biological contaminants in indoor
     air.

INDOOR CLIMATE
(EPA)
DA   October 12, 1990
BT1 Environment
RT   Indoor Air Pollution
SO   Environmental Protection Agency
     Glossary
DEF (EPA) Temperature, humidity,
     lighting and noise levels in a
     habitable structure or conveyance.
     Indoor climate can affect indoor air
     pollution.

INDUCED RADIOACTIVITY
(IAEA)
DA   October 12, 1990
BT1 Radioactivity
RT   Activation (Nuclear)
SO   Radiation
DEF (IAEA) Radioactivity produced
     within materials by nuclear
     reactions. (NRC Glossary of
     Terms: Nuclear Power and
     Radiation) Radioactivity that is
     created when stable substances
     are bombarded by ionizing
     radiation. For example, the stable
     isotope cobalt-59 becomes the
     radioactive isotope cobalt-60
     under neutron bombardment.

INDUCTIVELY COUPLED
PLASMA/MASS SPECTROMETER
DA   January 8, 1991
SF   *ICP/MS*

INDUSTRIAL BUILDINGS
(TSCA; CFR)
DA   October 12, 1990
BT1 Facilities

SO   Hazardous Materials
DEF (CFR) Buildings directly used in
     manufacturing or technically
     productive enterprises. Industrial
     buildings are not generally or
     typically accessible to other than
     workers. Industrial buildings
     include buildings used directly in
     the production of power, the
     manufacture of products, the
     mining of raw materials, and the
     storage of textiles, petroleum
     products, wood and paper
     products, chemicals, plastics, and
     metals.

INDUSTRIAL ENGINEERING
DA   January 8, 1991
SF   *IE (Industrial Engineering)*

INDUSTRIAL FURNACES
(SWDA; RCRA)
DA   October 12, 1990
BT1 Facility Components
NT1 Glass Melting Furnaces
 NT2 Pot Furnaces
RT   Roasting
SO   Environmental Management
SO   Wastes
DEF (CFR) Any of the following
     enclosed devices that are integral
     components of manufacturing
     processes and that use controlled
     flame devices to accomplish
     recovery of materials or energy:
     cement kilns, lime kilns, aggregate
     kilns, phosphate kilns, coke ovens,
     blast furnaces, smelting, melting
     and refining furnaces (including
     pyrometallurgical devices such as
     cupolas, reverberator furnaces,
     sintering machine, roasters, and
     foundry furnaces), titanium dioxide
     chloride process oxidation
     reactors, methane reforming
     furnaces, pulping liquor recovery
     furnaces, combustion devices
     used in the recovery of sulfur
     values from sulfuric acid, and such
     other devices as the Administrator
     may, after notice and comment,
     add to this list.

INDUSTRIAL HYGIENE
(DOE Order 5480.10)
DA   October 16, 1990
SF   *IH (Industrial Hygiene)*
RT   Environment, Safety, and Health
     (ES&H) Program
RT   Health Physics
RT   Industrial Safety
RT   Maintenance Management
DEF (DOE Order 5480.10) The science
     and art devoted to the recognition,
     evaluation, and control of
     environmental factors or stresses
     arising in or from the workplace
     that may cause sickness, impaired
     health and well-being, or
     significant discomfort and

inefficiency among workers or
those with whom they come into
contact.

INDUSTRIAL RISK INSURERS
DA   January 8, 1991
SF   *IRI*

INDUSTRIAL SAFETY
DA   October 19, 1990
BT1 Safety
RT   Buddy System
RT   Critical Incident Technique
RT   Critical Jobs (task)
RT   Imminent Danger
RT   Industrial Hygiene
RT   Maintenance Management
RT   Reported Significant Observations

INDUSTRY DEGRADED CORE
RULEMAKING PROGRAM
DA   January 8, 1991
SF   *IDCOR*
BT1 Programs

INEL
DA   October 12, 1990
SEE Idaho National Engineering
     Laboratory
SO   Acronyms

INERT ATMOSPHERE
(NFI)
DA   October 12, 1990
SY   Blanket Gas
SO   Nuclear Facilities Incident Database
SO   Radiation
DEF (NFI) Use for blanket gas.

INERT INGREDIENTS
(USC; EPA)
DA   October 12, 1990
RT   Pesticides
RT   Solvents
RT   Surfactants
SO   Environmental Management
SO   Environmental Protection Agency
     Glossary
DEF (EPA) Pesticide components such
     as solvents, carriers, and
     surfactants that are not active
     against target pests. Not all inert
     ingredients are innocuous.

INERTIAL SEPARATORS
(EPA)
DA   October 12, 1990
BT1 Devices
NT1 Cyclone Collectors
SO   Environmental Protection Agency
     Glossary
DEF (EPA) Devices that use centrifugal
     force to separate waste particles.

INFECTION
(1325 INFECTION)
DA   November 28, 1990
BT1 Injuries

RT Inflammation
RT Irritation
SO DOE FRASE VOCABULARY

INFECTIOUS WASTES
(SWDA; RCRA; ESH)
DA October 12, 1990
BT1 Wastes
RT Medical Wastes
RT Pathogens
RT Viruses
SO Environmental Management
SO Wastes
DEF (ESH) (1) Equipment, instruments, utensils, and fomites of a disposal nature from the rooms of patients who are suspected to have or who have been diagnosed as having a communicable disease and must therefore be isolated as required by public health agencies. (2) Laboratory wastes, such as pathological specimens (for example, all tissues, specimens of blood elements, excreta, and secretions obtained from patients or laboratory animals) and disposable fomites (any substance that may harbor or transmit pathogenic organisms) attendant thereto. (3) Surgical operating room pathological specimens and disposable fomites attendant thereto and similar disposal materials from outpatient areas and emergency rooms.

INFILTRATION
(EPA)
DA October 12, 1990
BT1 Biological Processes
BT2 Processes
RT Percolation
SO Environmental Protection Agency Glossary
DEF (EPA) (1) The penetration of water through the ground surface into sub-surface soil or the penetration of water from the soil into sewer or other pipes through defective joints, connections, or manhole walls. (2) A land application technique where large volumes of waste water are applied to land, allowed to penetrate the surface and percolate through the underlying soil.

INFLAMMATION
(1326 INFLAMMATION)
DA November 28, 1990
BT1 Injuries
RT Infection
RT Irritation
SO DOE FRASE VOCABULARY

INFLOW
(EPA)
DA October 12, 1990

RT Gust Fronts
SO Environmental Protection Agency Glossary
DEF (EPA) Entry of extraneous rain water into a sewer system from sources other than infiltration, such as basement drains, manholes, storm drains, and street washing. (WINDS) Wind direction shifting abruptly toward the storm (inflow) as opposed to wind moving away from the storm (outflow).

INFLUENTS
(EPA)
DA October 12, 1990
BT1 Liquids
BT2 Fluids
SO Environmental Protection Agency Glossary
DEF (EPA) Water, wastewater, or other liquid flowing into a reservoir, basin, or treatment plant.

INFORMAL CONSULTATION
(ESA; CFR)
DA October 12, 1990
RT Conferences
RT Designated Non-Federal Representatives
RT Formal Consultation
RT Public Hearings
SO Endangered Species
DEF (CFR) An optional process that includes all discussions, correspondence, etc., between the Fish and Wildlife Service and the Federal agency or the designated non-Federal representative prior to formal consultation, if required.

INFORMATION FILE
(EPA)
DA October 12, 1990
SO Environmental Protection Agency Glossary
DEF (EPA) In the Superfund program, a file that contains accurate, up-to-date documents on a Superfund site. The file is usually located in a public building such as a school, library, or city hall that is convenient for local residents.

INFORMATION SYSTEMS
DA February 12, 1991
BT1 Security Interests
BT1 Systems
NT1 Emergency Management Information System
NT1 Failure Reporting Analysis and Corrective Action System
NT1 Fire Protection Tracking System
NT1 Hazard Abatement Tracking System
NT1 Integrated Risk Information System
NT1 Nuclear Facilities Incident Database
NT1 Nuclear Materials Management and Standards System

NT1 Nuclear Plant Reliability Data System
NT1 Plant Risk Status Information Management System
NT1 Radiation Records Repositories
NT1 Real Property Inventory System
NT1 Safety Performance Measurement System
NT2 Chemical Hazards Emergency Management System
NT2 Computerized Accident/Incident Reporting System
NT2 Computer Assisted Tracking System
NT2 Occurrence Reporting and Processing System
NT2 Personnel Expertise and Resource Listing
NT2 Radiation Exposure Module
NT2 Standards Information Management System
NT2 Unusual Occurrence Reporting System
NT1 Secure Automatic Communications Network
NT1 Technical Information Systems
NT1 Ultrasonic Ranging and Data System
NT1 Waste Information Network
RT Atmospheric Release Advisory Capability
RT Control Systems
RT Emergency Systems
RT Software
SO Management

INGESTION
(IAEA)
DA October 12, 1990
BT1 Intake
BT2 Measurements
RT Digestion
SO Radiation
DEF (IAEA) Intake of material by way of the gastrointestinal system.

INGESTION EXPOSURE PATHWAYS
(EMER)
DA February 1, 1991
BT1 Exposure Pathways
BT2 Routes (Exposure)
RT Exposure
SO Emergency Preparedness
DEF (EMER) The pathways in which exposure occurs after ingestion of contaminated water or foods such as milk, fresh vegetables, or aquatic foodstuffs.

INGRESS
(DOE Order 6430.1A)
DA October 12, 1990
RT Entry Control Points
SO Construction
DEF (DOE Order 6430.1A) The act of entering a structure or area through a point of access.

SY-Synonymous Terms   SO-Source/Subject Category   SF-See From

INGROUND TANKS
(SWDA; RCRA)
DA   October 12, 1990
BT1 Tanks
 BT2 Facility Components
SO Environmental Management
SO Wastes
DEF (CFR) Devices meeting the
     definition of a "tank" whereby a
     portion of the tank wall is situated
     to any degree within the ground,
     thereby preventing visual
     inspection of that external surface
     area of the tank that is in the
     ground.

INHABITED BUILDING DISTANCE
(DOE Orders 5480.16 and 6430.1A)
DA   October 12, 1990
BT1 Measurements
RT Explosives Buildings
RT Munitions
RT Occupied Area (Explosives)
RT Transients (re: Explosives
     Facilities)
SO Construction
SO Firearms
DEF (DOE Order 5480.16) The minimum
     distance permitted between
     locations containing munitions and
     inhabited buildings, administrative
     areas, site boundaries, main
     power stations, and other facilities
     of vital or strategic nature.

INHALATION
(IAEA)
DA   October 12, 1990
BT1 Intake
 BT2 Measurements
RT Aerosols
RT Air
RT Black Lung
RT Bronchial Epithelium
SO Radiation
DEF (IAEA) Intake of material by way of
     the respiratory system (including
     the material which will eventually
     go to the intestinal system).

INHALATION LC 50
DA   January 24, 1991
SO Environmental Management
DEF (CFR) A concentration of a
     substance, expressed as
     milligrams per liter of air or parts
     per million parts of air, that is
     lethal to 50% of the test population
     of animals under test conditions as
     specified in the Registration
     Guidelines.

INHALATION TOXICOLOGY
RESEARCH INSTITUTE
DA   January 8, 1991
SF   *ITRI*
BT1 Government-Owned
     Contractor-Operated Facilities
 BT2 Federal Facilities

 BT3 Facilities
BT1 Institutes
 BT2 Research and Development
     Organizations
 BT3 Organizations
RT Lovelace Medical Foundation
DEF (Capsule Review of DOE Research
     and Development and Field
     Facilities, 1986) Established in
     1960, the primary mission of the
     ITRI is to assess adverse human
     health effects associated with
     commercial energy technologies.
     The Institute investigates the
     degree to which the inhalation of
     by-products of energy
     technologies such as fugitive or
     operating emissions may harm the
     health of operators or the general
     public.

INITIAL STARTUP
(Doe Order 5480.6)
DA   October 12, 1990
BT1 Startup
 BT2 Processes
SO Industrial Safety
DEF (DOE Order 5480.6) Includes those
     activities subsequent to
     preoperational testing, starting
     with the initial loading of fuel and
     involving all actions taken,
     including tests to assure a safe,
     orderly, incremental approach to
     predefined conditions of reactor
     operation.

INITIATION STIMULI
(DOE Order 6430.1A)
DA   October 12, 1990
RT Detonations
RT Explosives
RT Pyrophoric-Igniting Spontaneously
SO Construction
DEF (DOE Order 6430.1A) Energy input
     to an explosive in a form
     potentially capable of initiating a
     rapid decomposition reaction.
     Typical initiation stimuli are heat,
     friction, impact, electrical
     discharge, and shock. An initiator
     is a device that provides initiation
     stimuli (e.g., detonators, squibs,
     etc.).

INJECTION
(CFR)
DA   October 19, 1990
BT1 Processes
NT1 Underground Injection
RT Plugging
RT Wells
SO Water Pollution
DEF (CFR) The forcing, under abnormal
     pressure, of sedimentary material
     (downward, upward, or laterally)
     into a pre-existing deposit or rock,
     either along some plane of
     weakness or into a crack or
     fissure (e.g., the transformation of

wet sands and silts to a fluid state
and their emplacement in adjacent
sediments, producing structures,
such as sandstone dikes or sand
volcanoes). Also a sedimentary
structure or rock formed by
injection. (Glossary of Geology)

INJECTION INTERVAL
(SWDA; RCRA; CFR)
DA   October 12, 1990
BT1 Injection Zones
 BT2 Formations
 BT2 Zones
 BT3 Sites/Areas
SO Wastes
SO Water Pollution
DEF (CFR) That part of the injection
     zone in which the well is
     screened, or in which the waste is
     otherwise directly emplaced.

INJECTION WELLS
(SWDA; RCRA; SDWA; CFR; EPA;
 ESH)
DA   October 12, 1990
BT1 Wells
NT1 Class II Wells
 NT2 Existing Class II Wells
 NT2 New Class II Wells
NT1 Existing Injection Wells
RT Area of Review
RT Injection Zones
RT Well Injection
RT Well Stimulation
RT Well Workovers
SO Environmental Management
SO Environmental Protection Agency
     Glossary
SO Wastes
SO Water Pollution
DEF (CFR) Wells into which fluids are
     injected for purposes such as
     waste disposal, improving the
     recovery of crude oil, or solution
     mining.

INJECTION ZONES
(SDWA; CFR; ESH)
DA   October 12, 1990
BT1 Formations
BT1 Zones
 BT2 Sites/Areas
NT1 Injection Interval
RT Injection Wells
SO Environmental Protection Agency
     Glossary
SO Water Pollution
DEF (CFR) Geological formations,
     groups of formations, or parts of a
     formation receiving fluids through
     a well.

INJURIES
(DOE FRASE Vocabulary Numeric Keys
 1300-1374)
DA   January 17, 1991
NT1 Abrasion
NT1 Amputation

NT1 Animal Bite
NT2 Snake Bite
NT1 Asphyxia
NT1 Avulsion
NT1 Blister(s)
NT1 Blood Clot
NT1 Bodily Injuries
NT1 Brain Damage
NT2 Concussion
NT2 Contusion(S)
NT1 Burn(s)
NT2 Chemical Burn(s)
NT2 Electrical Burn(s)
NT2 Flash Burn(s)
NT1 Bursitis
NT1 Damage to Prosthetic Device
NT1 Death
NT1 Dental Injury
NT1 Disabling Injuries
NT1 Dislocation
NT1 Drowning
NT1 Electric Shock
NT1 Electrocution
NT1 Fracture
NT1 Heat Stroke/Other High Temp
     Effect
NT1 Hernia
NT1 Infection
NT1 Inflammation
NT1 Insect Sting
NT1 Internal Deposition
NT1 Irritation
NT1 Laceration
NT1 Loss of Consciousness
NT1 Loss of Condenser Vacuum
NT1 Multiple Injuries
NT1 No Personnel Injury
NT1 Nosebleed
NT1 Occupational Injuries
NT1 Physical Harm
NT1 Pinched Nerve
NT1 Puncture
NT1 Radiation Exposure
NT1 Ruptured Disk
NT1 Sprain
NT1 Strain
NT1 Strangulation
NT1 Torn Cartilage
RT Diseases
RT DOE FRASE Categories
RT Effects
RT Hospital
RT Illnesses
SO DOE FRASE VOCABULARY
DEF Equivalent to the DOE FRASE
     Vocabulary Category "Nature of
     Injury".

INJURY
(1327 INJURY)
DA November 28, 1990
RT Illness
SO DOE FRASE VOCABULARY

INK
(NFI)
DA October 12, 1990
RT Differential in Hours
SO Nuclear Facilities Incident Database

SO Radiation
DEF (NFI) Gadolinium nitrate, neutron
     absorber.

INLAND WATERS
(CERCLA; CFR)
DA October 12, 1990
NT1 Lakes
NT2 Dystrophic Lakes
NT2 Eutrophic Lakes
NT2 Oligotrophic Lakes
RT Ferry Vessels
RT Inland Zones
RT Specified Ports and Harbors
SO Compensation and Liability
SO Environmental Management
DEF (CFR) For the purposes of
     classifying the size of discharges,
     means those waters of the U.S. in
     the inland zone, waters of the
     Great Lakes, and specified ports
     and harbors on inland rivers.

INLAND ZONES
(CERCLA; CFR)
DA October 12, 1990
BT1 Zones
BT2 Sites/Areas
RT Inland Waters
SO Compensation and Liability
SO Environmental Management
DEF The environment inland of the
     coastal zone excluding the Great
     Lakes and specified ports and
     harbors of inland rivers. The term
     inland zone delineates the area of
     Federal responsibility for response
     action. Precise boundaries are
     determined by EPA/USCG
     (Environmental Protection
     Agency/U.S. Coast Guard)
     agreement and identified in
     Federal regional contingency
     plans.

INNER HOUSING
DA January 8, 1991
SF *IH (Inner Housing)*

INNER LINERS
(SWDA; RCRA)
DA October 12, 1990
RT Containers
RT Liners
RT Tanks
SO Environmental Management
SO Wastes
DEF (CFR) Continuous layers of
     material placed inside tanks or
     containers which protect the
     construction materials of the tanks
     or containers from the contained
     waste or reagents used to treat
     the waste.

INNER TARGETS
DA January 8, 1991
SF *IT*

INNOVATION DIFFUSION
(SSDC)
DA October 12, 1990
BT1 Administrative Processes
BT2 Processes
SO System Safety Development Center
     Glossary
DEF (SSDC) A structured,
     research-based, step-by-step
     process for developing acceptance
     of a new mode of behavior.

INOCULA
(EPA)
DA October 12, 1990
RT Bacteria
SO Environmental Protection Agency
     Glossary
DEF (EPA) (1) Bacteria placed in
     compost to start biological action.
     (2) A medium containing
     organisms which is introduced into
     cultures or living organisms.

INORGANIC ARSENIC
(CAA; CFR)
DA October 12, 1990
RT Arsenic Kitchens
RT Commercial Arsenic
RT Theoretical Arsenic Emissions
     Factor
RT Uncontrolled Total Arsenic
     Emissions
SO Air Pollution
DEF (CFR) The oxides and other
     noncarbon compounds of the
     element arsenic included in
     particulate matter, vapors, and
     aerosols.

INORGANIC CHEMICALS
(EPA)
DA October 12, 1990
BT1 Chemical Substances
NT1 Carbon Dioxide
NT1 Carbon Monoxide
NT1 Hydrogen Sulfide
NT1 Nitrogen Oxides (NO_x)
NT2 Nitric Oxides
NT2 Nitrogen Dioxide (NO_2)
SO Environmental Protection Agency
     Glossary
DEF (EPA) Chemical substances of
     mineral origin, not of basically
     carbon structure.

INPO
DA October 12, 1990
SEE Institute for Nuclear Power
     Operations
SO Acronyms

INSECT STING
(1328 INSECT STING)
DA November 28, 1990
BT1 Injuries
SO DOE FRASE VOCABULARY

INSECTICIDES
(EPA)
DA   October 12, 1990
BT1 Pesticides
 BT2 Hazardous Substances
NT1 Aldicarb
NT1 DDT
NT1 Diazinon
NT1 Dicofol
NT1 Heptachlor
RT   Insects
SO   Environmental Management
SO   Environmental Protection Agency
     Glossary
DEF (EPA) Pesticide compounds
     specifically used to kill or control
     the growth of insects.

INSECTS
(USC)
DA   November 15, 1990
BT1 Animals
RT   Attractants
RT   Insecticides
DEF (USC) Any of the numerous small
     invertebrate animals generally
     having the body more or less
     obviously segmented, for the most
     part belonging to the class insecta,
     comprising six-legged, usually
     winged forms, as for example,
     beetles, bugs, bees, flies, and to
     other allied classes of arthropods
     whose members are wingless and
     usually have more than six legs,
     as for example, spiders, mites,
     ticks, centipedes, and wood lice.

INSENSITIVE HIGH EXPLOSIVES
(DOE Order 6430.1A)
DA   October 12, 1990
SF   IHE
BT1 High Explosives
 BT2 Explosives
RT   IHE Weapons
SO   Construction
DEF (DOE Order 6430.1A) Explosive
     substances that, although mass
     detonating, are so insensitive that
     there is negligible probability of
     accidental initiation or transition
     from burning to detonation. The
     materials passing the DOE
     qualification tests in Table IX-1 of
     DOE/EV-06194 are classified as
     IHE, and are listed in Table IX-2 of
     the same document.

INSPECTION AND EVALUATION
(EMER)
DA   February 1, 1991
BT1 Administrative Processes
 BT2 Processes
RT   Assessments
RT   Reviews
RT   Safety Program Reviews
SO   Emergency Preparedness
DEF A documented review of the
     safeguards and security program
     activities, performed in accordance

with written guidance and criteria,
to verify by examination and
evaluation of objective evidence
(including site visits) that
applicable protection measures
have been developed,
documented, and effectively
implemented in accordance with
current U.S. Department of Energy
safeguards and security policies,
standards, and procedures, as
modified by agreements contained
in the Master Safeguards and
Security Agreement.

INSPECTION AND MAINTENANCE
(EPA)
DA   October 12, 1990
SF   I/M
NT1 Surveillance and Maintenance
RT   Secondary Environmental Monitor
SO   Environmental Protection Agency
     Glossary
DEF (EPA) (1) Activities to assure
     proper emissions related operation
     of mobile sources of air pollutants,
     particularly automobile emissions
     controls. (2) Also applies to
     wastewater treatment plants and
     other anti-pollution facilities and
     processes.

INSPECTION PLANS
(MORT)
DA   April 3, 1991
RT   Design and Development Plans
RT   General Design Process
RT   Human Factors
RT   Inspections
RT   Maintenance Plans
RT   Operational Specifications
DEF (MORT) MORT analysis asks: is
     inspection of an operation or
     facility given consideration during
     the conceptual phase and on
     throughout the rest of the life
     cycle? is there an adequate
     inspection plan? was the plan
     scope broad enough to include all
     the areas that should be
     inspected? was management
     aware of those areas not included
     in the plan? was there adequate
     execution of the inspection plan?

INSPECTION/MONITORING ACTIVITY
(1235 IM ACTIVITY)
DA   November 28, 1990
BT1 Activity Types (DOE FRASE
     Vocabulary)
 BT2 Activities
SO   DOE FRASE VOCABULARY

INSPECTIONS
(RCRA; TSCA; CFR; SSDC; MORT)
DA   January 15, 1991
BT1 Administrative Processes
 BT2 Processes
NT1 Compliance Inspections

NT1 Condition Assessment Surveys
NT1 Failed Instrument Component
     Inspection
NT1 Maintenance Team Inspection
NT1 Safety System Functional
     Inspection
NT1 Safety System Outage Modification
     Inspection
NT1 Site Inspections
RT   Accuracy
RT   Authorized Inspectors
RT   Calibration
RT   Controls
RT   First Line Supervision
RT   Higher Supervision
RT   Inspection Plans
RT   Installation Inspectors
RT   Maintenance
RT   Maintenance Management
RT   Operability
RT   Predictive Maintenance
RT   Sampling
RT   Security Inspectors
RT   Technical Information
RT   Technical Support
SO   System Safety Development Center
     Glossary
DEF (SSDC) Deliberate, systematic
     scrutiny or examinations of
     activities or projects; thorough,
     close critical examinations,
     checking or testing against
     established standards. (MORT)
     MORT analysis asks: was there
     adequate inspection of equipment,
     processes, utilities, operations,
     etc?

INSPECTIONS (NUCLEAR)
(DOE Order 5480.6; TSCA; CFR; SSDC)
DA   October 12, 1990
BT1 Administrative Processes
 BT2 Processes
RT   Nuclear Criticality Safety
RT   Nuclear Facilities
RT   Nuclear Criticality
RT   Safeguards
SO   Industrial Safety
SO   System Safety Development Center
     Glossary
DEF (DOE Order 5480.6) Deliberate and
     systematic examinations at the
     reactors including, but not limited
     to, physical inspection of reactor
     systems, operating and
     maintenance procedures, logs,
     records, and reactor operations.

INSPECTOR FOLLOW-UP ITEMS
DA   January 8, 1991
SF   IFI

INSTALLATION INSPECTORS
(SWDA; RCRA)
DA   October 19, 1990
BT1 Technically Qualified Individuals
 BT2 Personnel
RT   Inspections
RT   Tank Systems
SO   Environmental Management

BT-Broader Term   NT-Narrower Term   RT Related Term

SO Wastes
DEF (CFR) People, who, by reason of their knowledge of the physical sciences and the principles of engineering, acquired by a professional education and related practical experience, are qualified to supervise the installation of tank systems.

INSTALLATIONS
(CAA; CFR)
DA October 12, 1990
RT Equipment
SO Air Pollution
DEF (CFR) Identifiable pieces of process equipment.

INSTALLED EQUIPMENT
(DOE Order 4330.4A)
DA June 4, 1991
BT1 Equipment
SO Construction
SO Management
DEF (DOE Order 4330.4A) Includes the mechanical and electrical systems that are installed as part of basic building construction and are essential to the normal functioning of the facility and its intended use. Examples are heating, ventilating, and air conditioning (HVAC) systems; elevators; and communications systems.

INSTITUTE FOR NUCLEAR POWER OPERATIONS
DA January 8, 1991
SF INPO
BT1 Institutes
BT2 Research and Development Organizations
BT3 Organizations

INSTITUTE OF ELECTRICAL AND ELECTRONIC ENGINEERS
DA January 8, 1991
SF IEEE
BT1 Institutes
BT2 Research and Development Organizations
BT3 Organizations

INSTITUTES
DA February 1, 1991
BT1 Research and Development Organizations
BT2 Organizations
NT1 American National Standards Institute
NT1 American Petroleum Institute
NT1 Battelle Memorial Institute
NT1 Electric Power Research Institute
NT2 Nuclear Safety Analysis Center
NT1 Inhalation Toxicology Research Institute
NT1 Institute of Electrical and Electronic Engineers

NT1 Institute for Nuclear Power Operations
NT1 Midwest Research Institute
NT1 National Institute for Occupational Safety and Health
NT1 National Institute of Standards and Testing
SO Management
DEF (WEBSTER) An organization existing for the promotion of a cause or discipline.

INSTITUTION CONTROL
DA January 24, 1991
BT1 Controls
RT Waste Management
SO Environmental Management
DEF (DOE 5820.2A) A period of time, assumed to be about 100 years, during which human institutions continue to control waste management facilities.

INSTITUTIONAL SOLID WASTES
DA January 29, 1991
BT1 Solid Wastes
BT2 Wastes
SO Environmental Management
DEF (CFR) Solid Wastes generated by educational, health care, correctional, and other institutional facilities.

INSTREAM USE
(EPA)
DA October 12, 1990
BT1 Uses
SO Environmental Protection Agency Glossary
DEF (EPA) Water use taking place within a stream channel, e.g., hydro-electric power generation, navigation, water quality.

INSTRUMENT(S)
(2757 INSTRUMENT)
DA January 3, 1991
BT1 Equipment/Parts - Instrumentation/Measuring (DOE FRASE Voc.)
BT2 Equipment
NT1 Gauge(s)
NT2 Scale(s)
NT2 Straight Edge Ruler
NT2 Thermostat(s)
NT2 Time Clock
NT1 Graphitar(s)
NT1 Indicator(s)
NT2 Ammeter(s)
NT2 Flow Meter(s)
NT2 Ohmmeter(s)
NT2 Potentiometer
NT2 Spectrometer(s)
NT2 Tachometer(s)
NT2 Thermoluminescent Dosimeters
NT2 Thermometer(s)
NT2 Volt Meter
NT1 Level Alarm
NT1 Log N Recorder

NT1 Ram(s)
NT1 Recorder
NT1 Testing Equipment
NT2 Heat Detector
NT2 Leak Detector
NT2 Oscilloscope(s)
NT2 Scintilliation Probe(s)
NT2 Sensor(s)
SO DOE FRASE VOCABULARY

INSTRUMENT AIR
DA January 8, 1991
SF IA (Instrument Air)

INSTRUMENT AIR SYSTEM
(2037 IAS)
DA December 10, 1990
BT1 Systems (DOE FRASE Vocabulary)
BT2 Systems
SO DOE FRASE VOCABULARY

INSTRUMENTATION AND CONTROL
DA January 8, 1991
SF I&C
BT1 Controls

INSULATED URANIUM OXIDE
DA January 8, 1991
SF IRO

INSULATOR
(2436 INSULATOR)
DA January 3, 1991
BT1 Equipment/Parts - Electrical (DOE FRASE Vocabulary)
BT2 Equipment
SO DOE FRASE VOCABULARY

INSULIN REACTION
(1393 INSULIN REAC)
DA November 28, 1990
BT1 Illnesses
RT Chemical Reaction
SO DOE FRASE VOCABULARY

INSURANCE
(CERCLA; USC)
DA October 12, 1990
RT Factory Mutual
RT Indemnify
RT Liabilities
RT Risks
SO Compensation and Liability
DEF (USC) Primary insurance, excess insurance, reinsurance, surplus lines insurance, and any other arrangement for shifting and distributing risk which is determined to be insurance under applicable State or Federal law.

INTAKE
(EPA; IAEA)
DA October 12, 1990
BT1 Measurements
NT1 Acceptable Daily Intake
NT1 Annual Limit on Intake
NT1 Ingestion

NT1 Inhalation
NT1 Radioactive Nuclide Intake
RT Acceptable Intake for Chronic
    Exposure
RT Acceptable Intake for Subchronic
    Exposure
RT Uptake
SO Environmental Protection Agency
    Glossary
SO Radiation
DEF (EPA) (1) A measure of exposure
    expressed as the mass of
    substance in contact with the
    exchange boundary per unity body
    weight per unit time (e.g.,
    mg/kg-day). Also termed the
    normalized exposure rate. (2) The
    amount of radioactive material
    taken into the body by inhalation,
    absorption through the skin,
    ingestion, or through wounds.

INTANGIBLE RISKS
(SSDC)
DA October 12, 1990
BT1 Risks
SO System Safety Development Center
    Glossary
DEF (SSDC) Those risks for which it is
    usually not possible to determine
    dollar values, such as public
    opinion, employee morale, etc.

INTEGRAL VISTA
(CAA; CFR)
DA November 15, 1990
NT1 Visibility in any Mandatory Class I
    Federal Area
RT Mandatory Class I Federal Areas
RT Visibility Impairments
SO Air Pollution
DEF (CFR) Views perceived from within
    the mandatory Class I Federal
    area of specific landmarks or
    panoramas located outside the
    boundary of the mandatory Class I
    Federal area.

INTEGRATED DATA BASE
DA January 8, 1991
SF IDB

INTEGRATED LEAK RATE TEST
DA January 8, 1991
SF ILRT

INTEGRATED LOGISTICS SUPPORT
DA January 8, 1991
SF ILS

INTEGRATED PEST MANAGEMENT
(EPA)
DA October 12, 1990
SF IPM
SO Environmental Protection Agency
    Glossary
DEF (EPA) A mixture of pesticide and
    non-pesticide methods to control
    pests.

**INTEGRATED RISK INFORMATION
SYSTEM**
(EPA)
DA October 12, 1990
SF IRIS
BT1 Information Systems
 BT2 Security Interests
 BT2 Systems
RT Risk Assessment
SO Environmental Protection Agency
    Glossary
DEF (EPA) IRIS is an EPA data base
    containing verified RfDs
    (Reference Doses) and slope
    factors and up-to-date health risk
    and EPA regulatory information for
    numerous chemicals. IRIS is
    EPA's preferred source for toxicity
    information for Superfund.

INTERAGENCY AGREEMENT
DA January 8, 1991
SF IAG
BT1 Agreements
NT1 Federal Facilities Compliance
    Agreement

**INTERAGENCY COMMITTEE ON
STANDARDS POLICY**
(Doe Order 1300.2)
DA October 12, 1990
BT1 Committees
 BT2 Administrative Organizations
 BT3 Organizations
RT Government Standards
SO Management
SO Standards
DEF (DOE Order 1300.2) A committee
    established under the auspices of
    the Department of Commerce to
    coordinate and provide policy
    guidance to the heads of Federal
    agencies on standards. It is
    comprised of representatives from
    the major Federal departments
    and agencies which have an
    interest in standards. The
    Committee is chaired by the
    Deputy Assistant Secretary for
    Product Standards, Office of the
    Assistant Secretary for
    Productivity, Technology and
    Innovation, U.S. Department of
    Commerce.

**INTERAGENCY GROUP ON ENERGY
VULNERABILITY**
DA January 24, 1991
BT1 Groups
 BT2 Administrative Organizations
 BT3 Organizations
RT Emergency Preparedness
SO Environmental Management
DEF (DOE 5500 5.1) The Interagency
    Group on Energy Vulnerability
    (IG-EV) is a forum chartered under
    the Senior Interagency Group for
    National Security Decision
    Directive (NSDD) 188,
    "Government Coordination for

National Security Emergency
Preparedness." It consists of
senior representatives with
national security emergency
preparedness responsibilities from
departments and agencies. The
IG-EV facilitates government-wide
coordination of national policy
issues relating to the vulnerability
of US energy systems in advance
of crises and coordinates crisis
assessments and response
recommendations in an
emergency.

**INTERAGENCY RADIOLOGICAL
ASSISTANCE PLANS**
(EMER)
DA February 1, 1991
BT1 Plans
SO Emergency Preparedness
DEF (EMER) This has been superseded
    by the Federal Radiological
    Monitoring and Assessment Plan
    (FRMAP).

INTERCEPTOR SEWERS
(EPA)
DA October 12, 1990
BT1 Sewers
SO Environmental Protection Agency
    Glossary
DEF (EPA) Large sewer lines that, in a
    combined system, control the flow
    of the sewage to the treatment
    plant. In a storm, they allow some
    of the sewage to flow directly into
    a receiving stream, thus
    preventing an overload by a
    sudden surge of water into the
    sewers. They are also used in
    separate systems to collect the
    flows from main and trunk sewers
    and carry them to treatment points.

INTERFACE CONTROL UNIT
DA January 8, 1991
SF ICU
BT1 Controls

INTERFACES
(DOE Order 6430.1A)
DA October 12, 1990
RT Human Factors Engineering
RT Work Environment
SO Construction
DEF (DOE Order 6430.1A) The
    relationships between two or more
    system components, or between
    the work environment and one or
    more system components. Human
    performance is a function of the
    physical interfaces between
    people and equipment; the
    environments within which people
    or equipment work; the type and
    amount of training people receive;
    the accuracy and ease of use of
    the procedures people are given

for guidance; and the
effectiveness of the organizations
in which people work.

INTERFERENCES
DA   January 24, 1991
BT1 Discharges
SO   Environmental Management
DEF (CFR) The term "interference"
     means a discharge which, alone
     or in conjunction with a discharge
     or discharges from other sources,
     both: (1) Inhibits or disrupts the
     POTW, its treatment processes or
     operations, or its sludge
     processes, use or disposal; and
     (2) Therefore is a cause of a
     violation of any requirement of the
     POTW's NPDES permit (including
     an increase in the magnitude or
     duration of a violation) or of the
     prevention of sewage sludge use
     or disposal in compliance with the
     following statutory provisions and
     regulations or permits issued
     thereunder (or more stringent state
     or local regulations): Section 405
     of the Clean Water Act, the Solid
     Waste Disposal Act (including
     state regulations contained in any
     state sludge management plan
     prepared pursuant to Subtitle D of
     the SWDA), the Clean Air Act, the
     Toxic Substances Control Act, and
     the Marine Protection, Research
     and Sanctuaries Act.

INTERIM (PERMIT) STATUS
(EPA)
DA   October 12, 1990
SF   *Part B Permits*
SF   *Part A Permits*
BT1 Permits
RT   Permit (RCRA)
RT   Treatment, Storage, and Disposal
     Facilities
SO   Environmental Management
SO   Environmental Protection Agency
     Glossary
DEF (EPA) Period during which
     treatment, storage and disposal
     facilities coming under RCRA in
     1980 are temporarily permitted to
     operate while awaiting denial or
     issuance of a permanent permit.
     Permits issued under these
     circumstances are usually called
     "Part A" or "Part B" permits.

INTERIM ACTIONS
(DOE Order 5440.1D; NEPA)
DA   May 14, 1991
BT1 Actions
 BT2 Responses
RT   Environmental Impact Statements
SO   Environmental Management
DEF (DOE Order) Actions that are within
     the scope of an ongoing
     Environmental Impact Statement
     and that DOE propose to take

before issuing a record of decision
(ROD).

INTERIM AUTHORIZATIONS
(SWDA; RCRA)
DA   October 12, 1990
BT1 Permits
RT   State/EPA Agreements
RT   Test Authorizations
SO   Environmental Management
SO   Wastes
DEF (CFR) Approval by EPA of a State
     hazardous waste program which
     has met the requirements of
     section 3006(c) of RCRA and
     applicable requirements of 40
     CFR Part 271, Subpart B.

INTERIM RELIABILITY EVALUTION PROGRAM
DA   January 8, 1991
SF   *IREP*
BT1 Programs

INTERLOCK SYSTEM
(2038 INTERLOCK SY)
DA   December 10, 1990
BT1 Systems (DOE FRASE Vocabulary)
 BT2 Systems
SO   DOE FRASE VOCABULARY

INTERMEDIATE COVER
DA   January 24, 1991
BT1 Cover Materials
 BT2 Materials
RT   Daily Cover
SO   Environmental Management
DEF (CFR) Cover material that serves
     the same functions as daily cover
     but must resist erosion for a
     longer period of time, because it is
     applied on areas where additional
     cells are not to be constructed for
     extended period of time.

INTERMEDIATE MOMENT RESISTING SPACE FRAME
(SEA)
DA   October 12, 1990
BT1 Moment Resisting Space Frame
 BT2 Space Frames
  BT3 Structures (DOE FRASE
       Vocabulary)
RT   Dual Systems
SO   Construction
DEF (SEA) A concrete space frame
     design.

INTERMEDIATE PRE-MANUFACTURE NOTICES
DA   January 24, 1991
BT1 Pre-Manufacture Notices
 BT2 Section 5 Notices
SO   Environmental Management
DEF (CFR) Any PMN submitted to EPA
     for a chemical substance which is
     an intermediate (as "intermediate"
     is defined in P.L. 720.3 of this
     chapter) in the production of a final

product, provided that the PMN for
the intermediate is submitted to
EPA at the same time.

INTERMEDIATES
DA   January 29, 1991
BT1 Chemical Substances
NT1 Nitrites
NT1 Non-Isolated Intermediates
NT1 Site-Limited Intermediates
SO   Environmental Management
DEF (CFR) Chemical substances that
     are consumed, in whole or in part,
     in chemical reactions used for the
     intentional manufacture of other
     chemical substances or mixtures
     or that is intentionally present for
     the purpose of altering the rates of
     such chemical reactions.

INTERMITTENT CONTROL SYSTEM
(CAA; CFR)
DA   October 12, 1990
SF   *ICS*
BT1 Control Systems
 BT2 Controls
 BT2 Systems
BT1 Dispersion Techniques
SO   Air Pollution
DEF (CFR) A dispersion technique which
     varies the rate at which pollutants
     are emitted to the atmosphere
     according to meteorological
     conditions and/or ambient
     concentrations of the pollutant, in
     order to prevent ground-level
     concentrations in excess of
     applicable ambient air quality
     standards. Such a dispersion
     technique is an ICS whether used
     alone, used with other dispersion
     techniques, or used as a
     supplement to continuous
     emission controls (i.e., used as a
     supplemental control system).

INTERMODAL PORTABLE TANKS
DA   June 3, 1991
BT1 Portable Tanks
 BT2 Tanks
  BT3 Facility Components
RT   International Shipments
SO   Hazardous Materials
DEF Specific class of portable tanks
     designed primarily for internation
     intermodal use.

INTERMUNICIPAL AGENCIES
(SWDA; RCRA; USC)
DA   October 19, 1990
BT1 Local Agencies
 BT2 Agencies
  BT3 Administrative Organizations
   BT4 Organizations
RT   Solid Waste Management
SO   Wastes
DEF (USC) Agencies established by two
     or more municipalities with

responsibility for planning or
administration of solid waste.

INTERNAL APPRAISALS
(DOE Order 5482.1B)
DA   October 12, 1990
BT1 Management Appraisals
 BT2 Appraisals
RT   Management Appraisals
SO   Construction
SO   Environmental Management
SO   Management
DEF (DOE Order 5482.1B) Examinations
    and evaluations by the operating
    level (either Federal or contractor)
    of those portions of its internal
    ES&H program, program plan
    implementation, and operations
    retained under its direct control.

INTERNAL AUDITS
(SSDC)
DA   October 12, 1990
SY   Audits (SSDC)
BT1 Audits (SSDC)
 BT2 Administrative Processes
 BT3 Processes
SO   System Safety Development Center
     Glossary
DEF (SSDC) See "Audits".

INTERNAL CONVERSION
RADIOISOTOPES
DA   March 29, 1991
BT1 Radionuclides
 BT2 CERCLA Hazardous Substances
 BT3 Hazardous Substances
 BT2 Nuclides
NT1 Cesium 134
NT1 Cobalt 60
NT1 Cobalt 58

INTERNAL DEPOSITION
(1344 INT DEPOSITI)
DA   November 28, 1990
BT1 Injuries
SO   DOE FRASE VOCABULARY

INTERNAL FISSION COUNTER
DA   January 8, 1991
SF   IFC
BT1 Radiation Detectors
 BT2 Equipment

INTERNAL GAMMA FLUX MONITOR
DA   January 8, 1991
SF   IGFM
BT1 Monitors
 BT2 Equipment
BT1 Radiation Detectors
 BT2 Equipment

INTERNAL SECURITY REPORTS
(EMER)
DA   February 1, 1991
BT1 Reports
RT   Threats

SO   Emergency Preparedness
DEF (EMER) Field office periodic reports
    submitted to the Safeguards and
    Security Division director
    concerning known or suspected
    potential threats to U.S.
    Department of Energy (DOE) and
    DOE contractors.

INTERNATIONAL BUSINESS
MACHINES
DA   January 8, 1991
SF   IBM
BT1 Companies
 BT2 Commercial Organizations
 BT3 Organizations

INTERNATIONAL SHIPMENTS
(SWDA; RCRA)
DA   October 12, 1990
RT   Intermodal Portable Tanks
RT   Transport (Hazardous Substances)
SO   Environmental Management
SO   Wastes
DEF (CFR) The transportation of
    hazardous waste into or out of the
    jurisdiction of the United States.

INTERNATIONAL SYSTEM OF UNITS
(DOE Order 5900.2)
DA   October 16, 1990
SY   Metric System
SO   Quality Assurance
DEF (DOE Order 5900.2) The system
    popularly known as the
    modernized metric system is a
    coherent system of units based
    upon and including the meter
    (length), kilogram (mass), second
    (time), kelvin (temperature),
    ampere (electric current), candela
    (luminous intensity) and mole
    (amount of a substance). The
    radian (plane angle) and the
    steradian (solid angle) are
    supplemental units of the system.

INTERSECTION
(1611 INTERSECTION)
DA   December 10, 1990
BT1 Site (DOE FRASE Vocabulary)
 BT2 Sites/Areas
RT   Highway
SO   DOE FRASE VOCABULARY

INTERSTATE AGENCIES
(SWDA; RCRA; CFR; USC)
DA   October 12, 1990
BT1 State Agencies
 BT2 Agencies
 BT3 Administrative Organizations
 BT4 Organizations
NT1 Interstate Air Pollution Control
     Agencies
RT   Solid Waste Management
SO   Wastes
SO   Water Pollution
DEF (CFR) Any agencies of two or more
    States established by or under an

agreement or compact approved
by the Congress, or any other
agencies of two or more States or
Indian tribes having substantial
powers or duties pertaining to the
control of pollution as determined
and approved by the Administrator
under the "appropriate Act and
regulations." (USC) Agencies of
two or more municipalities in
different States, or agencies
established by two or more States,
with authority to provide for the
management of solid wastes and
serving two or more municipalities
located in different States.

INTERSTATE AIR POLLUTION
CONTROL AGENCIES
(CAA; USC)
DA   October 12, 1990
BT1 Air Pollution Control Agencies
 BT2 State Agencies
 BT3 Agencies
 BT4 Administrative Organizations
 BT5 Organizations
BT1 Interstate Agencies
 BT2 State Agencies
 BT3 Agencies
 BT4 Administrative Organizations
 BT5 Organizations
SO   Air Pollution
DEF (CAA) (1) Air pollution control
    agencies established by two or
    more States, or (2) air pollution
    control agencies of two or more
    municipalities located in different
    States.

INTERSTATE CARRIER WATER
SUPPLIES
(EPA)
DA   October 12, 1990
SO   Environmental Protection Agency
     Glossary
DEF (EPA) Sources of water for drinking
    and sanitary use on planes,
    buses, trains, and ships operating
    in more than one state. These
    sources are federally regulated.

INTERSTATE WATERS
(EPA)
DA   October 12, 1990
BT1 Water
SO   Environmental Protection Agency
     Glossary
DEF (EPA) Waters that flow across or
    form part of state or international
    boundaries, e.g., the Great Lakes,
    the Mississippi River, or coastal
    waters.

INTERSTITIAL MONITORING
(EPA)
DA   October 12, 1990
BT1 Monitoring
 BT2 Activities

BT-Broader Term   NT-Narrower Term   RT Related Term

SO Environmental Protection Agency
Glossary
DEF (EPA) The continuous surveillance
of the space between the walls of
an underground storage tank.

INTERVENTION LEVELS
(IAEA)
DA November 20, 1990
SY Emergency Reference Levels
SY Protective Action Guides
BT1 Reference Levels
RT Positive Data
SO Radiation
DEF (IAEA) Levels usually specified in
advance by the competent
authority or management for use in
abnormal situations; if the value of
the quantity of interest exceeds or
is predicted to exceed a particular
level, the appropriate remedial
action may have to be taken. Also
referred to as Protective Actions
Guides (PAG) or Emergency
Reference Levels (ERL).

INTRALINE SEPARATION
(BARRICADED)
(DOE Order 6430.1A)
DA October 12, 1990
BT1 Quantity Distances
BT2 Measurements
SO Construction
DEF (DOE Order 6430.1A) The
minimum quantity-distance
separation allowed between
buildings as described in the
paragraph below when an
effective barricade (as defined in
DOD 6055.9) is interposed
between buildings. The distance is
one-half the unbarricaded intraline
separation. This distance
(corresponding to approximately
82.7Kpa (12 psi) peak
overpressure for Class 1.1
explosives) shall be determined
based on the maximum explosives
weight, using the tables in DOD
6055.9.

INTRALINE SEPARATION
(UNBARRICADED)
(DOE Order 6430.1A)
DA October 12, 1990
BT1 Quantity Distances
BT2 Measurements
SO Construction
DEF (DOE Order 6430.1A) The
minimum quantity-distance
separation allowed between
explosives buildings on a plant site
unless equivalent protection to
personnel and property is provided
by building design and
construction, or a barricade. This
distance (corresponding to
approximately 24kPa (3.5 psi)
peak overpressure for Class 1.1
explosives) shall be determined

based on the maximum explosives
weight, using the tables in DOD
6055.9.

INTRUDER BARRIERS
(ANL; CFR)
DA May 24, 1991
BT1 Barriers
SO Industrial Safety
SO Wastes
DEF (CFR) A sufficient depth of cover
over the waste that inhibits contact
with waste and helps to ensure
that radiation exposures to an
inadvertent intruder will meet the
performance objectives set forth in
this part, or engineered structures
that provide equivalent protection
to the inadvertent intruder.

INTRUSION ALARM SYSTEM
(PERIMETER OR INTERIOR)
(DOE Order 6430.1A)
DA October 12, 1990
BT1 Alarms
BT2 Devices
BT1 Detection Equipment
BT2 Equipment
BT1 Emergency Systems
BT2 Systems
SO Construction
DEF (DOE Order 6430.1A) Detection
hardware and/or software
composed of sensors, alarm
assessment systems, and alarm
reporting systems (including alarm
communications and information
display equipment).

INVENTORY OF OPEN DUMPS
DA January 24, 1991
RT Open Dumps
SO Environmental Management
DEF (CFR) The inventory required under
section 4005(b) and is defined as
the list published by EPA of those
disposal facilities which do not
meet the criteria.

INVERSE SQUARE LAW
(NIH)
DA October 12, 1990
SO Radiation
DEF (NIH) The intensity of radiation at
any distance from a point source
varies inversely as the square of
that distance. For example: If the
radiation exposure is 100 R/hr at
1 inch from a source, the exposure
will be 0.01 R/hr at 100 inches.

INVERSION
(EPA)
DA October 12, 1990
RT Air Mass
SO Environmental Protection Agency
Glossary
DEF (EPA) An atmospheric condition
caused by a layer of warm air

preventing the rise of cooling air
trapped beneath it. This prevents
the rise of pollutants that might
otherwise be dispersed and can
cause an air pollution episode.

INVERTED SIPHONS
(DOE Order 6430.1A)
DA October 12, 1990
SO Construction
DEF (DOE Order 6430.1A) Pressure
pipelines crossing under a
highway or other obstruction.

INVESTIGATION LEVELS
(IAEA)
DA November 20, 1990
BT1 Reference Levels
SO Radiation
DEF (IAEA) Levels for the values of
quantities (such as dose
equivalent, intake, contamination
per unit area, etc.) above which
further investigations are
considered to be justified.

INVESTIGATION REPORT
(Doe Order 5484.1)
DA October 12, 1990
BT1 Reports
RT Investigations
SO Management
DEF (DOE Order 5484.1) A clear and
concise written account of the
results of an investigation.

INVESTIGATIONS
(Doe Order 5484.1)
DA October 12, 1990
BT1 Administrative Processes
BT2 Processes
NT1 Remedial Investigations
NT2 RCRA Remedial Investigation
NT1 Special Health Supervision
RT Findings
RT Investigation Report
RT Occurrences
RT Trained Investigators
SO Management
SO System Safety Development Center
Glossary
DEF (DOE Order 5484.1) Detailed,
systematic searches to uncover
the "who, what, when, where, why,
and how" of occurrences and to
determine what corrective actions
are needed in order to prevent a
recurrence.

IODINE 131
(EDB)
DA March 29, 1991
BT1 Beta-Minus Decay Radioisotopes
BT2 Beta Decay Radioisotopes
BT3 Radionuclides
BT4 CERCLA Hazardous
Substances
BT5 Hazardous Substances
BT4 Nuclides

SY-Synonymous Terms   SO-Source/Subject Category   SF-See From

BT1 Days Living Radioisotopes
 BT2 Radionuclides
  BT3 CERCLA Hazardous Substances
   BT4 Hazardous Substances
  BT3 Nuclides
 SO Radiation

ION EXCHANGE
(DOE Order 6430.1A)
DA October 12, 1990
SF IX
RT Desalinization
RT Ion Exchange Treatment
SO Construction
DEF (DOE Order 6430.1A) A chemical
     reaction used in water or
     wastewater treatment processes in
     which mobile dehydrated ions of a
     solid are exchanged (with ions of
     like charge in solution).

ION EXCHANGE COLUMN
(2574 ION EXCHANGE)
DA January 3, 1991
BT1 Equipment/Parts - Nuclear (DOE
     FRASE Vocabulary)
 BT2 Equipment
 BT2 Reactor Components
SO DOE FRASE VOCABULARY

ION EXCHANGE TREATMENT
(EPA)
DA October 12, 1990
BT1 Treatment
 BT2 Waste Management Processes
  BT3 Processes
RT Ion Exchange
SO Environmental Protection Agency
     Glossary
DEF (EPA) A common water-softening
     method often found on a large
     scale at water purification plants
     that remove some organics and
     radium by adding calcium oxide or
     calcium hydroxide to increase the
     ph to a level where the metals will
     precipitate out.

ION PUMP
(2575 ION PUMP)
DA January 3, 1991
BT1 Equipment/Parts - Nuclear (DOE
     FRASE Vocabulary)
 BT2 Equipment
 BT2 Reactor Components
BT1 Pump(s)
 BT2 Machines (DOE FRASE
      Vocabulary)
  BT3 Equipment
SO DOE FRASE VOCABULARY

IONIZATION
(NIH)
DA October 12, 1990
BT1 Processes
RT Directly Ionizing Particles
RT Indirectly Ionizing Particles
RT Ionization Chambers
RT Ionizing Radiation

RT Ions
RT Kerma
RT Linear Energy Transfer
RT Specific Ionization
SO Radiation
DEF (NIH) The process by which a
     neutral atom or molecule acquires
     either a positive or negative
     charge.

IONIZATION CHAMBERS
(EPA; NIH; EMER)
DA October 12, 1990
BT1 Radiation Detectors
 BT2 Equipment
RT Ionization
RT Ionizing Radiation
SO Emergency Preparedness
SO Environmental Protection Agency
     Glossary
SO Radiation
DEF (EPA) Devices that measure the
     intensity of ionizing radiation.
     (NIH) Instruments designed to
     measure the quantity of ionizing
     radiation in terms of the charge of
     electricity associated with ions
     produced within a defined volume.

IONIZING RADIATION
(EPA; NIH; IAEA; EMER)
DA October 12, 1990
BT1 Radiation
NT1 Background Radiation
NT1 Gamma Radiation
NT1 Lethal Dose of Radiation
NT1 X-rays
RT Absorbed Dose
RT Absorbed Dose Rate
RT Alpha Particles
RT Alpha Rays
RT Beta Particles
RT Beta Rays
RT Environmental Exposure
RT Exposure
RT Genetic Radiation Effects
RT Ionization
RT Ionization Chambers
RT Occupational Doses
RT Radiation Shields
SO Emergency Preparedness
SO Environmental Protection Agency
     Glossary
SO Radiation
DEF (EPA) Radiations that can remove
     electrons from atoms, i.e., alpha,
     beta, and gamma radiation. (NIH)
     Any electromagnetic or particulate
     radiations capable of producing
     ions, directly or indirectly, in their
     passage through matter. (IAEA)
     For the purposes of radiation
     protection, radiations capable of
     producing ion pairs in biological
     material(s).

IONIZING RADIATION EFFECTS
(1394 IONIZ RAD EF)
DA November 28, 1990
BT1 Effects

BT1 Illnesses
SO DOE FRASE VOCABULARY

IONS
(EPA; NIH)
DA October 12, 1990
BT1 Charged Particles
RT Ionization
SO Environmental Protection Agency
     Glossary
SO Radiation
DEF (EPA) Electrically charged atoms or
     groups of atoms which can be
     drawn from waste water during the
     electrodialysis process. (NIH)
     Atomic particles, atoms, or
     chemical radicals bearing an
     electrical charge, either negative
     or positive.

IORV
DA October 12, 1990
SEE Inadvertently Opened Relief Valve
SO Acronyms

IOWA STATE UNIVERSITY
DA January 11, 1991
BT1 Universities
 BT2 Educational Organizations
  BT3 Organizations
RT Chicago Operations Office

IP
DA May 14, 1991
SEE EIS Implementation Plans
SO Acronyms

IPE
DA October 12, 1990
SEE Individual Plant Examination
SO Acronyms

IPM
DA October 12, 1990
SEE Integrated Pest Management
SO Acronyms

IR CIRCUIT
(NFI)
DA October 12, 1990
SO Nuclear Facilities Incident Database
SO Radiation
DEF (NFI) Powers all the scram relays.

IREP
DA October 12, 1990
SEE Interim Reliability Evalution
     Program
SO Acronyms

IRI
DA October 12, 1990
SEE Industrial Risk Insurers
SO Acronyms

IRIS
DA October 12, 1990

SEE Integrated Risk Information System
SO   Acronyms

IRO
DA   October 12, 1990
SEE Insulated Uranium Oxide
SO   Acronyms

IRON
DA   January 8, 1991
SF   *Fe (Periodic Element)*

IRRADIATED FOOD
(EPA)
DA   October 12, 1990
RT   Irradiation
SO   Environmental Protection Agency
     Glossary
DEF (EPA) Food that has been subject
     to brief radioactivity, usually by
     gamma rays, to kill insects,
     bacteria, and mold, and preserve it
     without refrigeration or freezing.

**IRRADIATED FUEL PROCESS
FACILITY**
(1800 IFP FACILITY)
DA   December 10, 1990
BT1 Facility (DOE FRASE Vocabulary)
 BT2 Facilities and Buildings (DOE
      FRASE Vocabulary)
  BT3 Facilities
SO   DOE FRASE VOCABULARY

**IRRADIATED FUEL STORAGE
FACILITY**
(1801 IRS FACILITY)
DA   December 10, 1990
BT1 Facility (DOE FRASE Vocabulary)
 BT2 Facilities and Buildings (DOE
      FRASE Vocabulary)
  BT3 Facilities
SO   DOE FRASE VOCABULARY

IRRADIATION
(EPA)
DA   October 12, 1990
BT1 Exposure
RT   Absorbed Dose
RT   Critical Organs
RT   Genetic Radiation Effects
RT   Geometry
RT   Irradiated Food
RT   Non-Stochastic Radiation Effects
RT   Radiation
RT   Radiation Safety
RT   Somatic Radiation Effects
RT   Stochastic Radiation Effects
RT   Weighting Factors
SO   Environmental Protection Agency
     Glossary
DEF (EPA) Exposure to radiation of
     wavelengths shorter than those of
     visible light (gamma, x-ray, or
     ultraviolet), for medical purposes,
     the destruction of bacteria in milk
     or other foodstuffs, or for inducing
     polymerization of monomers or
     vulcanization of rubber.

IRRAS
(Acronyms and Abbreviations)
DA   October 12, 1990
BT1 Computer Codes
DEF A probabilistic risk assessment
     computer code developed at the
     Idaho National Engineering
     Laboratory.

**IRREVERSIBLE COMMITMENT OF
RESOURCES**
DA   October 19, 1990
RT   Committed Uses
RT   Reasonable and Prudent
     Alternatives
SO   Endangered Species
DEF (CFR) Any commitment of
     resources which has the effect of
     foreclosing the formulation or
     implementation of any reasonable
     or prudent alternatives which
     would not violate section 7(a)(2) of
     the Endangered Species Act.

IRRIGATION
(EPA)
DA   October 12, 1990
BT1 Processes
RT   Cultivating
SO   Environmental Protection Agency
     Glossary
DEF (EPA) Technique for applying water
     or wastewater to land areas to
     supply the water and nutrient
     needs of plants.

IRRITATING MATERIALS
(CFR)
DA   October 12, 1990
BT1 Materials
RT   Fumes
RT   Hazards
SO   Hazardous Materials
DEF (CFR) Irritating materials are liquid
     or solid substances which upon
     contact with fire or when exposed
     to air give off dangerous or
     intensely irritating fumes, but do
     not include any material classed
     as Poison Class A. Tear gas
     grenades, brombenzylcyanide, gas
     identification kits, and
     diphenylchlorarsine are examples
     of irritating materials.

IRRITATION
(1329 IRRITATION)
DA   November 28, 1990
BT1 Injuries
RT   Infection
RT   Inflammation
SO   DOE FRASE VOCABULARY

ISOLATION CONDENSER
DA   January 8, 1991
SF   *IC (Isolation Condenser)*
BT1 Equipment

ISOLATION ZONES
(DOE Order 6430.1A)
DA   October 12, 1990
BT1 Zones
 BT2 Sites/Areas
RT   Security Facilities
SO   Construction
DEF (DOE Order 6430.1A) Areas
     surrounding a protected facility
     that have been cleared of any
     objects that could conceal vehicles
     or individuals, and that afford
     unobstructed observation of, or
     other means of detection of, entry
     into the area.

ISOMERIC TRANSITION ISOTOPES
(EDB)
DA   March 29, 1991
BT1 Radionuclides
 BT2 CERCLA Hazardous Substances
  BT3 Hazardous Substances
 BT2 Nuclides
NT1 Cesium 134
NT1 Cobalt 60
NT1 Cobalt 58
SO   Radiation

ISOSTATIC PRESS
(2124 ISOSTATIC PR)
DA   December 10, 1990
BT1 Press
 BT2 Machines (DOE FRASE
      Vocabulary)
  BT3 Equipment
SO   DOE FRASE VOCABULARY

ISOTOPES
(EPA; NIH; EMER)
DA   October 12, 1990
BT1 Nuclides
NT1 Isotopic Tracers
SO   Emergency Preparedness
SO   Environmental Protection Agency
     Glossary
SO   Radiation
DEF (EPA) Variations of an element that
     have the same atomic number but
     a different weight because of its
     neutrons. Various isotopes of the
     same element may have different
     radioactive behaviors. (NIH)
     Nuclides having the same number
     of protons in their nuclei, and
     hence having the same atomic
     number, but differing in the
     number of neutrons, and therefore
     also differing in the mass number.
     Almost identical chemical
     properties exist between isotopes
     of a particular element.

ISOTOPIC TRACERS
(NIH)
DA   October 12, 1990
BT1 Isotopes
 BT2 Nuclides
RT   Labeled Compounds

SY-Synonymous Terms   SO-Source/Subject Category   SF-See From

SO Radiation
DEF (NIH) Isotopes or nonnatural mixtures of isotopes of an element which may be incorporated into a sample to make possible observation of the course of that element, alone or in combination, through a chemical, biological, or physical process. The observations may be made by measurement of radioactivity or of isotopic abundance.

ISSUE IDENTIFICATION
(TTGM)
DA May 6, 1991
BT1 Administrative Processes
 BT2 Processes
RT Concerns
RT Findings
RT Tiger Team Assessments
SO Management
DEF (TTGM) In general, the first step in the development of a Tiger Team Assessment finding is the identification of a specific condition or practice. The following types of information should be obtained including: 1) the specific nature of the problem, issues, condition or practice; 2) a detailed location, if appropriate; 3) the framework or perspective within which the problem or practice exists; 4) the regulatory or performance standards being violated, met or exceeded; 5) supporting information describing the problem or practice, or events leading to the problem; 6) information on actions being taken with respect to the problem or practice; and 7) information regarding how the assessment team member learned of the problem or practice.

IST
 DA October 12, 1990
 SEE In-Situ Treatment
 SO Acronyms

ISV
 DA October 12, 1990
 SEE In-Situ Vitrification
 SO Acronyms

IT
 DA October 12, 1990
 SEE Inner Targets
 SO Acronyms

ITRI
 DA October 12, 1990
 SEE Inhalation Toxicology Research Institute
 SO Acronyms

IX
 DA October 12, 1990

SEE Ion Exchange
SO Acronyms

JACKHAMMER
(2953 JACKHAMMER)
DA January 3, 1991
BT1 Tools - Powered
 BT2 Tools (DOE FRASE Vocabulary)
 BT3 Equipment
RT Construction Activity
SO DOE FRASE VOCABULARY

JANITOR
(0524 JANITOR)
DA November 28, 1990
BT1 Admin. Support/Clerical Employee
 BT2 Occupations
 BT2 Personnel
RT Janitorial/Housekeeping Activity
SO DOE FRASE VOCABULARY

JANITORIAL/HOUSEKEEPING ACTIVITY
(1236 JH ACTIVITY)
DA November 28, 1990
BT1 Activity Types (DOE FRASE Vocabulary)
 BT2 Activities
RT Janitor
SO DOE FRASE VOCABULARY

JAW
(1090 JAW)
DA November 28, 1990
BT1 Skull (Frase)
 BT2 Head
 BT3 Human Body Parts
RT Chin
SO DOE FRASE VOCABULARY

JET THRUST UNITS
(HMTA; CFR)
DA May 20, 1991
RT Aircraft
SO Hazardous Materials
DEF (CFR) Class B explosives, are metal cylinders containing a mixture of chemicals capable of burning rapidly and producing considerable pressure. Jet thrust units are designed to be ignited by an electric igniter. They are used to assist airplanes in take-off.

JOB SAFETY ANALYSIS
(SSDC)
DA October 12, 1990
SF JSA
BT1 Safety Analysis
 BT2 Analyses
RT Potential Hazards
SO System Safety Development Center Glossary
DEF (SSDC) A systematic technique for the safety review of a job used to uncover inherent or potential hazards. A JSA includes five steps: (1) select a job; (2) break the job down into steps; (3)

identify the hazards and determine the necessary hazard controls; (4) apply the controls; and (5) evaluate the controls.

JOHN BROWN, E&C, INC.
DA January 11, 1991
BT1 Companies
 BT2 Commercial Organizations
 BT3 Organizations
BT1 DOE Contractors
 BT2 Potentially Responsible Parties
RT Headquarters Operations

JOINT(S)
(1137 JOINT)
DA November 28, 1990
BT1 Human Body Parts
NT1 Ankle(s)
NT1 Elbow(s)
NT1 Hip(s)
NT1 Knee(s)
NT1 Knuckle(s)
NT1 Shoulder(s)
NT1 Wrist(s)
SO DOE FRASE VOCABULARY

JOINT FREQUENCY DISTRIBUTION
(DOE Order 6430.1A)
DA October 12, 1990
BT1 Frequency Distribution
 BT2 Distribution
SO Construction
DEF (DOE Order 6430.1A) The result of a frequency analysis of the probability of the occurrence of two or more random events (e.g., hydrologic or meteorological parameters).

JOINT INFORMATION CENTER
(EMER)
DA February 1, 1991
BT1 Centers
SO Emergency Preparedness
DEF (EMER) A center established to coordinate the federal public information activities on scene. It may be colocated with the press activities of an affected facility or state and local authorities. Also known as Joint Public Information Center. May also be referred to as Joint Information Bureau.

JOINT NUCLEAR ACCIDENT COORDINATION CENTER
(EMER)
DA February 1, 1991
BT1 Centers
SO Emergency Preparedness
DEF A joint U.S. Department of Energy (DOE) and Department of Defense (DOD) capability responsible for maintaining current information on the location of specialized DOE and DOD teams or organizations capable of providing nuclear weapons accidents assistance.

BT-Broader Term

NT-Narrower Term

RT Related Term

The DOE and DOD elements of JNACC are also responsible for initiating actions to deploy response in the event of a nuclear weapon accident or significant incident. The DOE component is located at Kirtland Air Force Base, New Mexico, and the DOD component is located at Springfield, Virginia, Defense Nuclear Agency (DNA) Headquarters.

JSA
DA   October 12, 1990
SEE Job Safety Analysis
SO Acronyms

JUDICIAL OFFICERS
(SWDA; RCRA)
DA   October 12, 1990
BT1 Personnel
DEF (CFR) Permanent or temporary employees of the Agency appointed as a Judicial Officer (JO) by the Administrator [EPA] under these regulations and subject to the following conditions: (a) A JO shall be a licensed attorney; shall not be employed in the Office of Enforcement or the Office of Water and Waste Management; and shall not participate in the consideration or decision of any case in which he or she performed investigative or prosecutorial functions, or which is factually related to such a case.The Administrator may delegate any authority to act in an appeal of a given case under 40 CFR 124 to a JO who, in addition may perform other duties for EPA, provided such delegations shall not preclude a JO from referring any motion or case to the Administrator, when the JO decides such action would be appropriate. (See Part 124.72 for continued definition).

JUMPER
(2399 JUMPER)
DA   January 3, 1991
BT1 Equipment/Parts - Electrical (DOE FRASE Vocabulary)
 BT2 Equipment
SO DOE FRASE VOCABULARY

KAISER ENGINEERS HANFORD, CO.
DA   January 11, 1991
BT1 Companies
 BT2 Commercial Organizations
  BT3 Organizations
BT1 DOE Contractors
 BT2 Potentially Responsible Parties
RT Richland Operations Office

KANNE ALARM
(NFI)
DA   October 12, 1990
BT1 Alarms
 BT2 Devices
BT1 Radiation Detectors
 BT2 Equipment
SO Nuclear Facilities Incident Database
SO Radiation
DEF (NFI) Radioactive gas detector.

KANSAS CITY PLANT
DA   January 8, 1991
SF   *KCP*
BT1 Government-Owned Contractor-Operated Facilities
 BT2 Federal Facilities
  BT3 Facilities
RT Allied-Signal Inc.
DEF (Capsule Review of DOE Research and Development and Field Facilities, 1986) The KCP was established in 1949. The plant is a highly diversified, technically oriented operation that embraces the full spectrum of work on non-nuclear products—from research on new materials to production of complex and reliable weapons components. Production activities are directed toward three basic areas: electrical and electronics work, including microelectronics, mechanical products and plastic products.

KARST TERRAIN
(DOE Order 6430.1A)
DA   October 12, 1990
SO Construction
DEF (DOE Order 6430.1A) An irregular limestone region with sinks, underground streams and caverns.

KCP
DA   October 12, 1990
SEE Kansas City Plant
SO Acronyms

KCR
DA   October 12, 1990
SEE Crane Control Room
SO Acronyms

KERMA
(IAEA)
DA   October 12, 1990
RT Gray
RT Ionization
SO Radiation
DEF (IAEA) The sum of the initial kinetic energies of all charged ionizing particles liberated by uncharged ionizing particles divided by the mass of the material.

KEY WORD
(SSDC)
DA   October 12, 1990

RT FRASE Vocabulary
SO System Safety Development Center Glossary
DEF (SSDC) A computer term meaning a word affording a means of access to a particular part of the total data bank.

Kg/yr
DA   October 12, 1990
SEE Kilograms/year
SO Acronyms

KILOGRAMS/YEAR
DA   January 8, 1991
SF   *Kg/yr*
BT1 Units of Measure

KILOMETER
DA   January 8, 1991
SF   *Km*
BT1 Units of Measure

KILOVOLT
DA   January 8, 1991
SF   *KV*
BT1 Units of Measure

KINETIC RATE COEFFICIENT
(EPA)
DA   October 12, 1990
BT1 Measurements
RT Biochemical Oxygen Demand
RT Dissolved Oxygen
SO Environmental Protection Agency Glossary
DEF (EPA) A number that describes the rate at which a water constituent such as a biochemical oxygen demand or dissolved oxygen increases or decreases.

KITCHEN
(1705 KITCHEN)
DA   December 10, 1990
BT1 Site (DOE FRASE Vocabulary)
 BT2 Sites/Areas
RT Cafeteria
RT Eating Area
RT Food Service Employee
SO DOE FRASE VOCABULARY

Km
DA   October 12, 1990
SEE Kilometer
SO Acronyms

Km^2
DA   October 12, 1990
SEE square kilometers
SO Acronyms

KNEE(S)
(1124 KNEE)
DA   November 28, 1990
BT1 Joint(s)
 BT2 Human Body Parts
RT Leg(s)

RT Leggings
SO DOE FRASE VOCABULARY

KNIFE(S)
(3041 KNIFE)
DA January 3, 1991
BT1 Tools - Manual
 BT2 Tools (DOE FRASE Vocabulary)
  BT3 Equipment
NT1 Pocket Knife(s)
RT Laceration
SO DOE FRASE VOCABULARY

KNOWLEDGE
(MORT)
DA April 3, 1991
RT Analyses
RT Communication
RT Corrective Action Triggers
RT Monitoring
RT Technical Information Systems
RT Trending
DEF (WEBSTER) The fact or condition
    of knowing something with
    familiarity gained through
    experience or association.
    (MORT) MORT analysis asks:
    was knowledge of the work flow
    process adequate? based upon
    known precedent; was application
    of knowledge obtainable from
    codes and manuals adequate?
    was the list of experts to contact
    for knowledge adequate? was any
    existing but unwritten precedent
    relevant to the work flow process
    known to the appropriate person?
    were there studies directed to the
    solution of known work flow
    process problems? were there
    investigations and analyses of
    prior similar accidents/incidents?
    was there research directed to the
    obtaining of knowledge about the
    work flow process? etc.

KNOWLEDGE OF RESULTS
(SSDC)
DA October 12, 1990
SF *KOR*
RT Feedback
SO System Safety Development Center
    Glossary
DEF (SSDC) Having useful feedback on
    the operation so that the
    information is adequate and in the
    right form for proper interpretation
    to make decisions.

KNOWN HUMAN EFFECTS
(TSCA; CFR)
DA October 12, 1990
BT1 Effects
RT Critical Organs
RT Effects on Welfare
RT Toxicity
SO Hazardous Materials
DEF (CFR) Commonly recognized
    human health effects of a

particular substance or mixture as
described either in (1) scientific
articles or publications abstracted
in standard reference sources or
(2) the firm's product labeling of
material safety data sheets
(MSDS). However, an effect is not
a "known human effect" if it (1)
was a significantly more severe
toxic effect than previously
described; (2) was a manifestation
of a toxic effect after a significantly
shorter exposure period or lower
exposure level than described;
and (3) was a manifestation of a
toxic effect by an exposure route
different from that described.

KNOWN PRECEDENTS
(SSDC)
DA October 12, 1990
RT Actuarial
SO System Safety Development Center
    Glossary
DEF (SSDC) The use of technical
    information and knowledge from
    past experience and existing data
    to assist in hazard or accident
    analysis. This knowledge could
    come from such sources as:
    codes, manuals and
    recommendations; written reports
    and case histories; lists of
    expertise (experts); and studies
    directed toward solutions of known
    problems.

KNUCKLE(S)
(1102 KNUCKLE)
DA November 28, 1990
BT1 Joint(s)
 BT2 Human Body Parts
RT Glove(s)
RT Hand Protection
RT Hand(s)
SO DOE FRASE VOCABULARY

KOR
DA October 12, 1990
SEE Knowledge of Results
SO Acronyms
SO System Safety Development Center
    Glossary

KV
DA October 12, 1990
SEE Kilovolt
SO Acronyms

LAB COAT
(2671 LAB COAT)
DA January 3, 1991
BT1 Anticontamination Clothing
 BT2 Clothing
 BT2 Personal Protective Equipment
  BT3 Equipment/Parts - Personal
      Protective (DOE FRASE
      Vocabulary)

 BT4 Equipment
SO DOE FRASE VOCABULARY

LABELED COMPOUNDS
(NIH)
DA October 12, 1990
RT Isotopic Tracers
SO Radiation
DEF (NIH) Compounds consisting, in
    part, of labeled molecules. By
    observations of radioactivity or
    isotopic composition these
    compounds or their fragments
    may be followed through physical,
    chemical, or biological processes.

LABELING
(USC)
DA November 15, 1990
BT1 Labels
NT1 Marks
RT Pesticides
RT Signal Words
DEF (USC) All labels and all other
    written, printed, or graphic matter
    (A) accompanying the pesticide or
    device at any time; or (B) to which
    reference is made on the label or
    in literature accompanying the
    pesticide or device, except to
    current official publications of the
    Environmental Protection Agency,
    the United States Departments of
    Agriculture and Interior, the
    Department of Health and Human
    Services, State experiment
    stations, State agricultural
    colleges, and other similar Federal
    or State institutions or agencies
    authorized by law to conduct
    research in the field of pesticides.

LABELS
(USC)
DA November 15, 1990
NT1 Labeling
 NT2 Marks
RT Pesticides
DEF (USC) The written, printed, or
    graphic matter on, or attached to,
    the pesticide or device or any of
    its containers or wrappers.

LABORATORIES
DA February 1, 1991
BT1 Research and Development
    Organizations
 BT2 Organizations
NT1 Argonne National Laboratory
 NT2 Argonne National Laboratory-East
     (Chicago)
 NT2 Argonne National Laboratory-West
     (At INEL)
NT1 Argonne National Laboratory-East
    (Chicago)
NT1 Argonne National Laboratory-West
    (At INEL)
NT1 Battelle Columbus Laboratories
NT1 Brookhaven National Laboratory

NT1 Idaho National Engineering
Laboratory
NT2 Power Burst Facility
NT1 Laboratory for Energy-Related
Health Research
NT1 Lawrence Livermore National
Laboratory
NT1 Lawrence Berkeley Laboratory
NT1 Los Alamos National Laboratory
NT1 Nationally Recognized Testing
Laboratories
NT1 Oak Ridge National Laboratory
NT1 Pacific Northwest Laboratory
NT1 Performance Testing Laboratories
NT1 Research & Development
Laboratory
NT1 Sandia National Laboratories
NT2 Sandia National
Laboratories-Alburquerque
NT2 Sandia National
Laboratories-Livermore
RT Functional Units
SO Environmental Management
SO Industrial Hygiene
DEF (WEBSTER) A place equipped for
experimental study in a science or
for testing and analysis.

LABORATORY
(1706 LABORATORY)
DA December 10, 1990
BT1 Site (DOE FRASE Vocabulary)
BT2 Sites/Areas
SO DOE FRASE VOCABULARY

LABORATORY AUDITS
(ESH)
DA October 12, 1990
RT Audits (SSDC)
SO Quality Assurance
DEF (ESH) Evaluations to assure that all
the necessary quality control is
being applied by a laboratory to
deliver a quality product. Should
allow the evaluators to determine
that the organization and
personnel are qualified to perform
assigned tasks; adequate facilities
and equipment are available;
complete documentation, including
chain-of-custody of sample is
being implemented; proper
analytical methodology is being
used; adequate analytical quality
control, including reference
samples, control charts, and
documented corrective action
measures, is being provided; and
acceptable data handling and
documentation techniques are
being used.

**LABORATORY FOR
ENERGY-RELATED HEALTH
RESEARCH**
DA January 8, 1991
SF *LEHR*
BT1 Laboratories

BT2 Research and Development
Organizations
BT3 Organizations

LACERATION
(1330 LA)
DA November 28, 1990
BT1 Injuries
RT Knife(s)
RT Puncture
SO DOE FRASE VOCABULARY

LAGOONS
(EPA)
DA October 12, 1990
SY Stabilization Ponds
RT Sewage Lagoons
SO Environmental Protection Agency
Glossary
DEF (EPA) (1) Shallow ponds where
sunlight, bacterial action, and
oxygen work to purify wastewater;
also used to store wastewaters or
spent nuclear fuel rods. (2)
Shallow bodies of water, often
separated from the sea by coral
reefs or sandbars.

LAKES
(CWA; RHA; CFR)
DA October 12, 1990
BT1 Inland Waters
BT1 Surface Waters
BT2 Water
BT1 Surface Water Resources
BT2 Natural Resources
NT1 Dystrophic Lakes
NT1 Eutrophic Lakes
NT1 Oligotrophic Lakes
RT Eutrophication
RT Limnology
SO Water Pollution
DEF (CFR) Standing bodies of open
water that occur in a natural
depression fed by one or more
streams from which a stream may
flow, that occur due to the
widening or natural blockage or
cutoff of a river or stream, or that
occur in an isolated natural
depression that is not a part of a
surface river or stream. The term
also includes standing bodies of
open water created by artificially
blocking or restricting the flow of a
river, stream, or tidal area. As
used in this regulation, the term
does not include artificial lakes or
ponds created by excavating
and/or diking dry land to collect
and retain water for such purposes
as stock watering, irrigation,
settling basins, cooling, or rice
growing.

LAND APPLICATION
(EPA)
DA October 12, 1990
NT1 Overland Flow

SO Construction
SO Environmental Protection Agency
Glossary
DEF (EPA) Discharge of wastewater
onto the ground for treatment or
reuse.

LAND DISPOSAL
(SWDA; RCRA; CFR; ESH)
DA October 12, 1990
BT1 Disposal
BT2 Waste Management Processes
BT3 Processes
NT1 Near Surface Disposal
NT2 Shallow Land Burial
RT Land Disposal Units
RT Land Reclamation
RT Land Disposal Restrictions
RT Piles (Wastes)
RT Reclamation
RT Sanitary Landfills
RT Surface Impoundment
RT Treatment Technologies
SO Environmental Management
SO Wastes
DEF (CFR) Placement in or on the land
and includes, but is not limited to,
placement in a landfill, surface
impoundment, waste pile, injection
well, land treatment facility, salt
dome formation, salt bed
formation, underground mine or
cave, or placement in a concrete
vault or bunker intended for
disposal purposes.

LAND DISPOSAL FACILITIES
(ANL; CFR)
DA May 24, 1991
BT1 Disposal Facilities
BT2 Hazardous Waste Management
Facilities
BT3 Hazardous Waste Facilities
BT4 Facilities
NT1 Near Surface Disposal Facilities
SO Environmental Management
SO Wastes
DEF (CFR) The land, buildings, and
equipment which are intended to
be used for the disposal of
radioactive wastes into the
subsurface of the land. For
purposes of this chapter, a
geologic repository as defined in
40 CFR 60 is not considered a
land disposal facility.

LAND DISPOSAL RESTRICTIONS
DA January 8, 1991
SF *LDR*
BT1 Requirements
RT Land Disposal

LAND DISPOSAL UNITS
DA January 24, 1991
BT1 Disposal Units
BT2 Corrective Action Management
Units
BT3 Sites/Areas

SY-Synonymous Terms   SO-Source/Subject Category   SF-See From

NT1 Landfill Cells
NT1 Landfills
NT2 Chemical Waste Landfills
NT2 Sanitary Landfills
NT2 Specially Designated Landfills
RT Land Disposal
RT Land Treatment Facilities
SO Environmental Management
DEF (CFR) Includes, but is not limited to, placement in a landfill, surface impoundment, waste pile, injection well, land treatment facility, salt dome formation, salt bed formation, underground mine or cave, or placement in a concrete vault or bunker intended for disposal purposes.

LAND FARMING (OF WASTE)
(EPA)
DA October 12, 1990
BT1 Disposal
 BT2 Waste Management Processes
 BT3 Processes
SO Environmental Protection Agency Glossary
DEF (EPA) A disposal process in which hazardous waste deposited on or in the soil is naturally degraded by microbes.

LAND RECLAMATION
(EDB)
DA February 1, 1991
BT1 Reclamation
 BT2 Resource Recovery
 BT3 Pollution Recovery Processes
 BT4 Processes
RT Abandoned Areas
RT Conservation
RT Land Disposal
SO Environmental Management
DEF (DSTT) The recovery of land previously abandoned due to some form of natural resource damage.

LAND TREATMENT FACILITIES
(SWDA; RCRA, CFR)
DA October 12, 1990
BT1 Facilities
BT1 Treatment Facilities
 BT2 Facilities
RT Hazardous Waste Management Units
RT Land Disposal Units
RT Landfills
RT Reclamation
RT Treatment Zones
SO Environmental Management
SO Wastes
DEF (CFR) Facilities or parts of facilities at which hazardous wastes are applied onto or incorporated into the soil surface; such facilities are disposal facilities if the waste will remain after closure.

LANDFILL CELLS
(SWDA; RCRA)
DA October 12, 1990
BT1 Cells
BT1 Land Disposal Units
 BT2 Disposal Units
 BT3 Corrective Action Management Units
 BT4 Sites/Areas
RT Hazardous Waste Management Units
RT Landfills
RT Liners
RT Solid Waste Disposal
SO Environmental Management
DEF (CFR) Sites for disposal of solid waste in which compacted layers are covered with soil. (CFR) Discrete volumes of a hazardous waste landfill that use a liner to provide isolation of wastes from adjacent cells or wastes. Examples of landfill cells are trenches and pits.

LANDFILLS
(EPA)
DA October 12, 1990
BT1 Land Disposal Units
 BT2 Disposal Units
 BT3 Corrective Action Management Units
 BT4 Sites/Areas
NT1 Chemical Waste Landfills
NT1 Sanitary Landfills
NT1 Specially Designated Landfills
RT Caps
RT Land Treatment Facilities
RT Landfill Cells
RT Liners
SO Construction
SO Environmental Management
SO Environmental Protection Agency Glossary
SO Wastes
DEF (EPA) (1) Sanitary landfills are land disposal sites for nonhazardous solid wastes at which the waste is spread in layers, compacted to the smallest practical volume, and cover material applied at the end of each operating day. (2) Secure chemical landfills are disposal sites for hazardous waste. They are selected and designed to minimize the chance of release of hazardous substances into the environment.

LANDLORDS
(DOE Order 4330.4A)
DA June 5, 1991
BT1 Heads of Headquarters Elements
 BT2 Personnel
RT Line Organizations
RT Program Senior Officials
SO Management
DEF (DOE Order 4330.4A) Heads of Headquarters Elements with overall capital improvement and

common support responsibility for a site; also represents the various Headquarters interests at the site.

LANDSLIDES
(USGS)
DA October 12, 1990
BT1 Ground Failures
 BT2 Failures
 BT3 Accidents
 BT2 Liquefaction
 BT3 Processes
BT1 Natural Disasters
 BT2 Natural Phenomenon
SO Natural Phenomenon
DEF (USGS) Rock falls, avalanches, or slides as a result of an earthquake.

LANL
DA October 12, 1990
SEE Los Alamos National Laboratory
SO Acronyms

LARGE HIGH VOLTAGE CAPACITORS
(TSCA; CFR)
DA October 19, 1990
BT1 Capacitors
 BT2 Electrical Equipment
 BT3 Equipment
SO Hazardous Materials
DEF (ESH) Capacitors that contain 1.36 kg (3 lbs.) or more of dielectric fluid and that operate at 2,000 volts (a.c. or d.c.) or above.

LARGE LOW VOLTAGE CAPACITORS
(TSCA; CFR)
DA October 19, 1990
BT1 Capacitors
 BT2 Electrical Equipment
 BT3 Equipment
SO Hazardous Materials
DEF (ESH) Capacitors that contain 1.36 kg (3 lbs.) or more of dielectric fluid and that operate below 2,000 volts (a.c. or d.c.).

LARGE QUANTITIES
(EMER)
DA February 1, 1991
BT1 Quantities
RT Radioactive Materials
RT Transportation
SO Emergency Preparedness
DEF (EMER) In transportation of radioactive materials, a quantity which exceeds the Type B quantity limits [49 Code of Federal Regulations Part 173.389(b)]. Large quantities are subject to requirements for being carried on highways designated as preferred routes.

LASER
(2400 LASER)
DA January 3, 1991

BT1 Equipment/Parts - Electrical (DOE
      FRASE Vocabulary)
 BT2 Equipment
 SO DOE FRASE VOCABULARY

LASER EYE SAFETY DISTANCE
(Doe Order 5480.16)
DA October 12, 1990
RT Multiple Integrated Laser
      Engagement System
SO Firearms
DEF (DOE Order 5480.16) The minimum
      distance required to protect the
      eye from corneal or retinal damage
      caused by a specific laser beam.

LATERAL FORCE RESISTING SYSTEM
(SEA)
DA October 12, 1990
BT1 Building Frame Systems
 BT2 Space Frames
  BT3 Structures (DOE FRASE
        Vocabulary)
 BT2 Systems
RT Collectors
SO Construction
DEF (SEA) That part of the structural
      system assigned to resist lateral
      forces.

LATERAL SEWERS
(EPA)
DA October 12, 1990
BT1 Sewers
SO Environmental Protection Agency
      Glossary
DEF (EPA) Pipes that run under city
      streets and receive the sewage
      from homes and businesses.

LATERAL SPREADS
(USGS)
DA October 12, 1990
BT1 Ground Failures
 BT2 Failures
  BT3 Accidents
 BT2 Liquefaction
  BT3 Processes
SO Natural Phenomenon
DEF (USGS) Lateral movement of large
      blocks of soil on top of a liquified
      subsurface layer. These lateral
      spreads, which break up in
      numerous fissures and scarps,
      generally develop on gentle
      slopes, most commonly on those
      between 0.3 and 3 degrees.
      Horizontal movements on lateral
      spreads commonly are as much
      as 10 to 15 feet, but, where slopes
      are favorable and the duration of
      ground shaking is long, lateral
      movement may be as much as
      100 to 150 feet.

LATEX RESINS
(CAA; CFR)
DA October 12, 1990
BT1 Types of Resin

 BT2 Resin Grade
 SO Air Pollution
DEF (CFR) Resins that are produced by
      a polymerization process that
      initiates from free radical catalyst
      sites and are sold undried.

LATHE
(2125 LATHE)
DA December 10, 1990
BT1 Machines (DOE FRASE
      Vocabulary)
 BT2 Equipment
 SO DOE FRASE VOCABULARY

LAW
DA October 12, 1990
SEE Light Antitank Weapon
SO Acronyms

LAW HAZARD ZONE
(Doe Order 5480.16)
DA October 12, 1990
BT1 Zones
 BT2 Sites/Areas
RT Light Antitank Weapon
SO Firearms
DEF (DOE Order 5480.16) The zone at
      the rear of a light antitank weapon
      (LAW) or LAW simulator where
      flame, hot gases, or fragments
      may be present during discharge
      of the weapon. The hazard zone is
      defined as a 30 degree cone
      truncated at 10 feet wide by 30
      feet deep at the rear of the LAW
      tube.

LAW SIMULATORS
(Doe Order 5480.16)
DA October 12, 1990
BT1 Simulators
BT1 Weapon Simulators
RT Light Antitank Weapon
SO Firearms
DEF (DOE Order 540.16) Weapons that
      simulate the firing of a light
      antitank weapon (LAW) and emit a
      coded laser beam in the direction
      aimed. The simulators do not fire
      a projectile but do expel fragments
      and a hot flash from the rear of
      the launch tube.

LAWRENCE-ALLISON & ASSOCIATES
DA January 11, 1991
BT1 Companies
 BT2 Commercial Organizations
  BT3 Organizations
BT1 DOE Contractors
 BT2 Potentially Responsible Parties
RT Headquarters Operations

LAWRENCE BERKELEY LABORATORY
DA January 8, 1991
SF *LBL*
BT1 Government-Owned
      Contractor-Operated Facilities
 BT2 Federal Facilities

 BT3 Facilities
BT1 Laboratories
 BT2 Research and Development
      Organizations
  BT3 Organizations
RT University of California
DEF (Capsule Review of DOE Research
      and Development and Field
      Facilities, 1986) LBL was founded
      in 1931 to advance the
      development of the cyclotron
      invented by Ernest Lawrence.
      Currently, the major roles of LBL
      are to perform multidisciplinary
      research in the general and energy
      sciences; develop and operate
      unique national experimental
      facilities; educate and train the
      next generation of scientists and
      engineers; and foster productive
      relationships between LBL
      research programs and industry.

LAWRENCE LIVERMORE NATIONAL
LABORATORY
DA January 11, 1991
BT1 Government-Owned
      Contractor-Operated Facilities
 BT2 Federal Facilities
  BT3 Facilities
BT1 Laboratories
 BT2 Research and Development
      Organizations
  BT3 Organizations
RT University of California
DEF (Capsule Review of DOE Research
      and Development and Field
      Facilities, 1986) LLNL, established
      in 1952, is a scientific and
      technical resource for the nation's
      nuclear weapons program and
      other programs of national interest.
      LLNL primary role is to perform
      the research, development and
      testing associated with the nuclear
      design aspects of all phases of the
      nuclear weapon life cycle and
      associated national security
      activities. LLNL has developed
      expertise in inertial fusion,
      magnetic fusion, biomedical and
      environmental research, isotope
      separation and applied energy
      technology.

LBL
DA October 12, 1990
SEE Lawrence Berkeley Laboratory
SO Acronyms

lbs/year
DA October 12, 1990
SEE Pounds/year
SO Acronyms

LC
DA October 12, 1990
SEE Locked Closed
SO Acronyms

LC 50
(EPA)
DA October 12, 1990
SY Lethal Concentration
NT1 Subacute Dietary LC 50
SO Environmental Protection Agency
    Glossary
DEF (EPA) Median level concentration, a
    standard measure of toxicity. It
    tells how much of a substance is
    needed to kill half of a group of
    experimental organisms at a
    specific time of observation.

LCO
DA October 12, 1990
SEE Limiting Condition for Operation
SO Acronyms

LD 0
(EPA)
DA October 12, 1990
BT1 Doses
SO Environmental Protection Agency
    Glossary
DEF (EPA) The highest concentration of
    a toxic substance at which none of
    the test organisms die.

LD 50
(EPA)
DA October 12, 1990
SY Lethal Dose
BT1 Doses
NT1 Acute LD 50
NT1 Acute Oral LD 50
SO Environmental Protection Agency
    Glossary
DEF (EPA) The dose of a toxicant that
    will kill 50 percent of the test
    organisms within a designated
    period of time. The lower the LD
    50, the more toxic the compound.

LD L0
(EPA)
DA October 12, 1990
BT1 Doses
BT1 Lethal Dose
SO Environmental Protection Agency
    Glossary
DEF (EPA) The lowest concentration
    and dosage of a toxic substance
    that kills test organisms.

LDR
DA October 12, 1990
SEE Land Disposal Restrictions
SO Acronyms

LEACHATE COLLECTION SYSTEMS
(EPA)
DA October 12, 1990
BT1 Systems
RT Leachates
SO Environmental Protection Agency
    Glossary
DEF (EPA) System that gather leachate

and pump it to the surface for
treatment.

LEACHATES
(DOE Order 6430.1A; SWDA; RCRA;
ESH)
DA October 12, 1990
NT1 Dissolved Solids
NT1 Suspended Solids
NT2 Settleable Solids
RT Groundwater
RT Leachate Collection Systems
RT Leaching
RT Mixtures
RT Percolation
RT Solutions
SO Construction
SO Environmental Management
SO Environmental Protection Agency
    Glossary
SO Wastes
DEF (EPA) Liquids that result from water
    collecting contaminants as it
    trickles through wastes,
    agricultural pesticides, or
    fertilizers. Leaching may occur in
    farming areas, feedlots, and
    landfills and may result in
    hazardous substances entering
    surface water, groundwater, or
    soil. (ESH) Liquids that have
    percolated through solid waste
    and have extracted dissolved or
    suspended materials from the
    wastes.

LEACHING
(EPA)
DA October 12, 1990
BT1 Processes
RT Leachates
RT Soil Column
SO Environmental Protection Agency
    Glossary
DEF (EPA) The process by which
    soluble constituents are dissolved
    and carried down through the soil
    by a percolating fluid. Leaching
    may occur in farming areas,
    feedlots, and landfills and may
    result in hazardous substances
    entering surface water,
    groundwater, or soil.

LEAD
(EPA)
DA October 12, 1990
SF Pb (Periodic Element)
BT1 CERCLA Hazardous Substances
BT2 Hazardous Substances
BT1 Criteria Pollutants
BT2 Pollutants
BT1 Hazardous Constituents
BT1 Heavy Metals
SO Environmental Protection Agency
    Glossary
DEF (EPA) A heavy metal that is
    hazardous to health if breathed or
    swallowed. Its use in gasoline,
    paints, and plumbing compounds

has been sharply restricted or
eliminated by federal laws and
regulations.

LEAD AGENCIES
(CERCLA; CFR; EMER)
DA October 12, 1990
BT1 Federal Agencies
BT2 Agencies
BT3 Administrative Organizations
BT4 Organizations
RT State Agencies
SO Compensation and Liability
SO Environmental Management
DEF (CFR) The Federal agency (or
    State agency operating pursuant
    to a contract or cooperative
    agreement executed pursuant to
    section 104(d)(1) of CERCLA) that
    has primary responsibility for
    coordinating response action
    under this Plan. A Federal lead
    agency is the agency that provides
    the OSC or RPM as specified
    elsewhere in this Plan. In the case
    of a State as lead agency, the
    State shall carry out the same
    responsibilities delineated for
    OSCs/RPMs in this Plan (except
    coordinating and directing Federal
    agency response actions).

LEAD AUTHORIZED OFFICIALS
(CFR)
DA November 15, 1990
BT1 Authorized Officials
BT2 Personnel
DEF (CFR) A Federal or State official
    authorized to act on behalf of all
    affected Federal or State agencies
    acting as trustees where there are
    multiple agencies, or an official
    designated by multiple tribes
    where there are multiple tribes,
    affected because of coexisting or
    contiguous natural resources or
    concurrent jurisdiction.

LEAD FREE
(SDWA; CFR; USC)
DA October 12, 1990
RT Drinking Water Coolers
SO Water Pollution
DEF (USC) That each part of a drinking
    water cooler or component of the
    cooler which may come in contact
    with drinking water contains not
    more than 8 percent lead, except
    that no drinking water cooler that
    contains any solder, flux, or
    storage tank interior surface that
    may come in contact with drinking
    water shall be considered lead
    free if the solder, flux, or storage
    tank interior surface contains more
    than 0.2 percent lead. More
    stringent requirements for treating
    any part or component of a
    drinking water cooler as lead free
    may be established whenever it is

BT-Broader Term   NT-Narrower Term   RT Related Term

determined that any such part may constitute an important source of lead in drinking water.

LEAD MATTES
(CAA; CFR)
DA   October 12, 1990
RT   Copper Converters
RT   Copper Mattes
SO   Air Pollution
DEF (CFR) Any molten solutions of copper and other metal sulfides produced by reduction of sinter product from the oxidation of lead sulfide ore concentrates.

LEADER SEATER
(NFI)
DA   October 12, 1990
BT1 Tools (DOE FRASE Vocabulary)
 BT2 Equipment
SO   Nuclear Facilities Incident Database
DEF (NFI) Tool used to check position of instrument rods in the reactor lattice; contains device to confirm seating.

LEAK DETECTION SYSTEMS
(SWDA; RCRA; CFR; ESH)
DA   October 12, 1990
BT1 Systems
NT1 Beetles
NT1 Leak Detector
RT   Containment
RT   Detection
RT   Leaks
RT   Releases
SO   Environmental Management
SO   Wastes
DEF (ESH) Systems that can detect the failure of either the primary or secondary containment structure or the presence of a release of hazardous waste or accumulated liquid in the secondary containment structure. Such systems must use operational controls (for example, daily visual inspections for releases into the secondary containment system or aboveground tanks) or consist of an interstitial monitoring device designed to detect continuously and automatically the failure of the primary or secondary containment structure or the presence of a release of hazardous waste into the secondary containment structure.

LEAK DETECTOR
(2759 LEAK DETECTO)
DA   January 3, 1991
BT1 Leak Detection Systems
 BT2 Systems
BT1 Testing Equipment
 BT2 Instrument(s)
  BT3 Equipment/Parts -

Instrumentation/Measuring (DOE FRASE Voc.)
 BT4 Equipment
RT   Environmental Release
SO   DOE FRASE VOCABULARY

LEAKS
(CAA; TSCA; CFR; ESH; ASMT)
DA   October 12, 1990
SY   Spills
RT   Concrete Encasement
RT   Hazards
RT   Leak Detection Systems
RT   Management of Migrations
RT   Releases
RT   Spills
SO   Air Pollution
SO   Hazardous Materials
DEF (ASTM) Holes or voids in the wall of an enclosure, capable of passing liquid or gas from one side of the wall to the other under action of pressure or concentration differential existing across the wall, independent of the quantity of fluid flowing.

LEG(S)
(1121 LEGP)
DA   November 28, 1990
BT1 Human Body Parts
NT1 Lower Leg
NT1 Thigh(s)
RT   Ankle(s)
RT   Foot/Feet
RT   Knee(s)
RT   Leggings
SO   DOE FRASE VOCABULARY

LEGAL DEFENSE COSTS
(SWDA; RCRA; CFR)
DA   October 19, 1990
BT1 Costs
RT   Liabilities
SO   Wastes
DEF (CFR) Any expenses that an insurer incurs in defending against claims of third parties brought under the terms and conditions of an insurance policy.

LEGGINGS
(2672 LEGGINGS)
DA   January 3, 1991
BT1 Personal Protective Equipment
 BT2 Equipment/Parts - Personal Protective (DOE FRASE Vocabulary)
  BT3 Equipment
RT   Knee(s)
RT   Leg(s)
RT   Lower Leg
RT   Thigh(s)
SO   DOE FRASE VOCABULARY

LEGIONELLA
(SDWA; CFR)
DA   October 12, 1990
BT1 Bacteria

BT2 Microorganisms
 BT3 Organisms
SO   Water Pollution
DEF (CFR) A genus of bacteria, some species of which have caused a type of pneumonia called Legionnaires Disease.

LEGM
DA   October 12, 1990
SEE Low Energy Gamma Monitor
SO   Acronyms

LEHR
DA   October 12, 1990
SEE Laboratory for Energy-Related Health Research
SO   Acronyms

LEL
DA   October 12, 1990
SEE Lower Explosive Limit
SO   Acronyms

LENS OF EYE DOSE EQUIVALENTS
(DOE Order 5480.11)
DA   October 16, 1990
BT1 Dose Equivalents
 BT2 Radiation Units
  BT3 Units of Measure
SO   Industrial Hygiene
DEF (DOE Order 5480.11) The dose equivalent at the respective depths of 0.007 cm, 1.0 cm, and 0.3 cm in tissue.

LEPC
DA   October 12, 1990
SEE Local Emergency Planning Committees
SO   Acronyms

LER
DA   October 12, 1990
SEE Licensee Event Report
SO   Acronyms

LESS THAN ADEQUATE
(SSDC)
DA   October 12, 1990
SF   *LTA*
RT   Analytical (Logic) Trees
SO   System Safety Development Center Glossary
DEF (SSDC) Does not meet minimum requirements. Used in MORT and other analytical tree logic to indicate areas of increased risk.

LET
DA   October 12, 1990
SEE Linear Energy Transfer
SO   Acronyms

LETHAL CONCENTRATION
(EPA)
DA   January 10, 1991
SY   LC 50

NT1 Subacute Dietary LC 50
RT Lethal Dose
RT Toxicity
SO Environmental Protection Agency
   Glossary
DEF (EPA) Median Level concentration,
   a standard measure of toxicity. It
   tells how much of a substance is
   needed to kill half of a group of
   experimental organisms at a
   specific time of observation. (See
   LD 50)

LETHAL CONCENTRATION LOW
(EMER)
DA February 1, 1991
RT Exposure
RT Hazardous Chemicals
SO Emergency Preparedness
DEF (EMER) The lowest concentration
   of a chemical at which some test
   animals die following inhalation
   exposure.

LETHAL DOSE
(EPA)
DA January 10, 1991
SY LD 50
NT1 Acute Oral LD 50
NT1 LD L0
RT Lethal Concentration
SO Environmental Protection Agency
   Glossary
DEF The dose of a toxicant that will kill
   50 percent of the test organisms
   within a designated period of time.
   The lower the LD 50, the more
   toxic the compound.

LETHAL DOSE LOW
(EMER)
DA February 1, 1991
BT1 Doses
RT Exposure
RT Hazardous Chemicals
SO Emergency Preparedness
DEF (EMER) The lowest dose of
   chemical at which some test
   animals die following exposure.

LETHAL DOSE OF RADIATION
(EMER)
DA February 1, 1991
BT1 Ionizing Radiation
 BT2 Radiation
RT Exposure
SO Emergency Preparedness
DEF (EMER) The amount of ionizing
   radiation exposure required to
   cause death. A brief (within four
   days) whole body gamma
   exposure of 600 roentgens would
   be a lethal dose for most people.

LETTER OF PERMISSION
(CWA; RHA; CFR)
DA October 12, 1990
BT1 Permits

SO Water Pollution
DEF (CFR) A type of individual permit
   issued in accordance with the
   abbreviated procedures of 33 CFR
   325.2(e).

LEVEL 1 COMPLIANCE
(ESH)
DA November 19, 1990
BT1 Compliance Considerations
RT Level 1 Potential Hazards
SO Standards
DEF (ESH) Does not comply with
   mandatory DOE requirement
   (DOE Orders), prescribed policies
   and standards, and documented
   accepted practice (the latter is a
   professional judgement based on
   the acceptance and applicability of
   national concensus standards not
   prescribed by DOE requirements).

LEVEL 1 POTENTIAL HAZARDS
(ESH)
DA October 12, 1990
BT1 Potential Hazards
 BT2 Hazards
  BT3 Conditions
RT Level 1 Compliance
DEF (ESH) Have the potential for
   causing a severe injury or fatality,
   potentially fatal occupational
   illness, or loss of the facility.

LEVEL 2 COMPLIANCE
(ESH)
DA November 19, 1990
BT1 Compliance Considerations
RT Level 2 Potential Hazards
SO Standards
DEF (ESH) Does not comply with
   recommended DOE references,
   standards, guidance, or with good
   practice (as derived from industry
   experience, but not based on
   national concensus standards).

LEVEL 2 POTENTIAL HAZARDS
(ESH)
DA October 12, 1990
BT1 Potential Hazards
 BT2 Hazards
  BT3 Conditions
RT Level 2 Compliance
DEF (ESH) Has the potential for causing
   minor injury, minor occupational
   illness, major property damage, or
   has the potential for resulting in or
   contributing to unnecessary
   exposure to radiation toxic
   substances.

LEVEL 3 COMPLIANCE
(ESH)
DA November 19, 1990
BT1 Compliance Considerations
RT Level 3 Potential Hazards
SO Standards
DEF (ESH) Has little or no compliance

consideration; these concerns are
based on professional judgement
in pursuit of excellence in design
or practice (i.e., these are
improvements for their own sake,
not deficiency-driven).

LEVEL 3 POTENTIAL HAZARDS
(ESH)
DA October 12, 1990
BT1 Potential Hazards
 BT2 Hazards
  BT3 Conditions
RT Level 3 Compliance
DEF (ESH) Has little potential for
   threatening safety, health, or
   property.

LEVEL ALARM
(2760 LEVEL ALARM)
DA January 3, 1991
BT1 Alarms
 BT2 Devices
BT1 Instrument(s)
 BT2 Equipment/Parts -
    Instrumentation/Measuring (DOE
    FRASE Voc.)
  BT3 Equipment
SO DOE FRASE VOCABULARY

LEVEL OF CONCERN
(EPA; EMER)
DA January 8, 1991
SF *LOC*
RT Exposure
SO Emergency Preparedness
DEF (EMER) The concentration of an
   extremely hazardous substance in
   the air above which there may be
   serious irreversible health effects
   or death as a result of a single
   exposure for a relatively short
   period of time.

LIABILITIES
(SWDA; RCRA; CFR)
DA October 12, 1990
NT1 Contingent Liability
NT1 Contractual Liability
NT1 Current Liabilities
NT1 Pollution Liability
NT1 Product Liability
RT Assets
RT Compensation and Liability
RT Insurance
RT Legal Defense Costs
RT Net Worth
SO Wastes
SO Water Pollution
DEF (CFR) Liabilities are probable future
   sacrifices of economic benefits
   arising from present obligations to
   transfer assets or provide services
   to other entities in the future as a
   result of past transactions or
   events.

LIAISON OFFICERS
(EMER)

BT-Broader Term

NT-Narrower Term

RT Related Term

DA February 1, 1991
BT1 Personnel
SO Emergency Preparedness
DEF (EMER) Federal agency officials
sent to another agency or another
emergency response facility to
facilitate interagency
communications and coordination.

LICENSE APPLICANTS
(USC)
DA October 19, 1990
SY Permit Applicants
SO Endangered Species
DEF (USC) When used with respect to
an action of a Federal agency for
which exemption is sought under
section 7 [16 USCS 1536], any
person whose application to such
agency for a permit or license has
been denied primarily because of
the application of section 7(a) [16
USCS 1536(a)] to such agency
action.

LICENSED MATERIALS
(AEA, CFR)
DA January 29, 1991
BT1 Materials
NT1 Source Materials
NT1 Special Nuclear Materials
NT2 Special Nuclear Material Scrap
NT2 Special Nuclear Material of Low
Strategic Significance
NT2 Strategic Special Nuclear
Materials
DEF (CFR) Source material, special
nuclear material or by-product
material received, possessed,
used, or transferred under a
general or specific license issued
by the Atomic Energy Commission
pursuant to regulations of the
Atomic Energy Act.

LICENSED SITES
(CAA; CFR)
DA November 15, 1990
BT1 Sites/Areas
RT Nuclear Regulatory Commission
RT Uranium By-Product Materials
DEF (CFR) Areas contained within the
boundary of a location under the
control of persons generating or
storing uranium byproduct
materials under a license issued
by the Commission. These include
such areas licensed by Agreement
States, i.e., those States that have
entered into an effective
agreement under section 274(b) of
the Atomic Energy Act of 1954, as
amended.

LICENSEE EVENT REPORT
DA January 8, 1991
SF *LER*
BT1 Reports
RT Events

LICENSES
(CFR, USC, et. al.; EMER)
DA January 29, 1991
NT1 Approvals Necessary to Begin
Physical Construction
NT1 Nuclear By-product Material
License
RT Requirements
SO Emergency Preparedness
SO Environmental Management
DEF (AEA) Licenses issued under
appropriate regulations, including
licenses to operate production or
utilization facilities, licenses to
possess power reactor spent fuel
in an independent spent fuel
storage installation, etc.

LIFE CYCLE COSTS
(DOE Order 6430.1A)
DA October 12, 1990
BT1 Costs
SO Construction
DEF (DOE Order 6430.1A) All costs
except the cost of personnel
occupying the facility incurred from
the time that space requirement is
defined until that facility passes
out of the government's hands.

LIFE CYCLE PLANS
(DOE Order 4330.4A)
DA June 5, 1991
BT1 Plans
RT Functional Units
DEF (DOE Order 4330.4A) Present an
analysis and description of the
major events and activities in the
life of a functional unit from
planning through decommissioning
and site restoration. The plan
documents the history of the
functional unit and forecasts future
activities, including major line item
and expense projects and their
duration, relationships, and impact
on life expectancy. The plan also
describes maintenance practices
and costs.

LIFT BUCKET
(2331 LIFT BUCKET)
DA December 10, 1990
BT1 Hoisting Apparatus
BT2 Material Handling Device
BT3 Devices
BT3 Equipment/Parts - Material
Handling (DOE FRASE
Vocabulary)
BT4 Equipment
SO DOE FRASE VOCABULARY

LIFTING STATION
(EPA)
DA October 12, 1990
BT1 Devices
RT Hydraulic Lift Tanks

SO Environmental Protection Agency
Glossary
DEF (EPA) Mechanical device installed
in sewer or water system or other
liquid-carrying pipeline that moves
the liquid to a higher level.

LIFTS
(EPA)
DA October 12, 1990
RT Sanitary Landfills
SO Environmental Protection Agency
Glossary
DEF (EPA) In a sanitary landfill, a
compacted layer of solid waste
and the top layer of cover material.

LIGHT(S)
(2401 LIGHT;N)
DA January 3, 1991
BT1 Equipment/Parts - Electrical (DOE
FRASE Vocabulary)
BT2 Equipment
SO DOE FRASE VOCABULARY

LIGHT ANTITANK WEAPON
(Doe Order 5480.16)
DA October 12, 1990
SF *LAW*
BT1 Firearms
RT Hangfires
RT Law Hazard Zone
RT Law Simulators
SO Firearms
DEF (DOE Order 5480.16) A portable,
shoulder-fired, recoilless weapon
capable of launching explosive
projectiles.

LIGHT WATER REACTOR
DA January 8, 1991
SF *LWR*
BT1 Reactors
DEF (NRC Glossary of Terms: Nuclear
Power and Radiation) A term used
to designate reactors using
ordinary water as coolant,
including boiling water reactors
and pressurized water reactors,
the most common types used in
the United States.

LIMESTONE SCRUBBING
(EPA)
DA October 12, 1990
BT1 Pollution Recovery Processes
BT2 Processes
RT Scrubbers
SO Environmental Protection Agency
Glossary
DEF (EPA) Process in which sulfur
gases moving toward a
smokestack are passed through a
limestone and water solution to
remove sulfur before it reaches
the atmosphere.

LIMITED AREAS
(DOE Order 6430.1A)

SY-Synonymous Terms   SO-Source/Subject Category   SF-See From

209

DA   October 12, 1990
BT1 Security Areas
 BT2 Sites/Areas
SO   Construction
DEF (DOE Order 6430.1A) Security
     areas for the protection of
     classified matter where guards,
     security inspectors, or other
     internal controls can prevent
     access by unauthorized persons
     to classified matter.

LIMITED QUANTITY
(SWDA; RCRA; ESH)
DA   October 12, 1990
BT1 Quantities
RT   Exceptions
SO   Hazardous Materials
DEF (CFR) The maximum amount of
     hazardous material for which there
     is a specific labeling and
     packaging exception, when
     specified as such in a section
     applicable to a particular material,
     with the exception of Poison B
     materials.

LIMITED RESPONSES
(EMER)
DA   February 1, 1991
BT1 Responses
SO   Emergency Preparedness
DEF (EMER) Responses to a request for
     radiological assistance that
     involves limited U.S. Department
     of Energy or other agency
     resources and does not require the
     formal field management structure.

LIMITING CONDITION FOR
OPERATION
DA   January 8, 1991
SF   LCO
BT1 Conditions

LIMITING FACTORS
(EPA)
DA   October 12, 1990
BT1 Conditions
SO   Environmental Protection Agency
     Glossary
DEF (EPA) A condition, whose absence,
     or excessive concentration, is
     incompatible with the needs or
     tolerance of a species or
     population and which may have a
     negative influence on their ability
     to grow or even survive.

LIMITS
(IAEA)
DA   October 12, 1990
NT1 Alternate Concentration Limits
NT1 Authorized Limits
NT1 Categorical Pretreatment Standards
NT1 Channel Effluent Limit
NT1 Computer Inoperative Limits
NT1 Confinement Protection Limits

NT1 Contract-Required Quantitation
     Limits
NT1 Derived Limits
NT1 Detection Limits
 NT2 Environmental Detection Limits
 NT2 Lower Limit of Detection
NT1 Dose Limit
 NT2 Annual Dose Equivalent Limit
NT1 Effluent Limitations
NT1 Emission Limitation
NT1 Lower Explosive Limit
NT1 Maximum Concentration Limits
NT1 Method Quantification Limits
NT1 MQL
NT1 Operational (Radiation) Limits
NT1 Operating Limit
NT1 Permissible Exposure Limits
NT1 Prescribed Limits
NT1 Primary Limits
NT1 Published Exposure Levels
NT1 Quantitation Limit
NT1 Safety Limits
NT1 Secondary Limits
NT1 Shaft Break Limit
NT1 Short-Term Exposure Limit
NT1 Tolerance Limits
NT1 Total Indicated Runout
NT1 Transient Protection Limit
RT   Radiation Protection
SO   Radiation
DEF (IAEA) The values of a quantity that
     must not be exceeded. Limits in
     radiation protection are as follows:
     primary, secondary, derived,
     authorized, and operational.

LIMNOLOGY
(EPA)
DA   October 12, 1990
RT   Aquatic Environment
RT   Fresh Water
RT   Lakes
SO   Environmental Protection Agency
     Glossary
DEF (EPA) The study of the physical,
     chemical, meteorological, and
     biological aspects of fresh water.

LINE MANAGEMENT
(SSDC)
DA   October 12, 1990
BT1 Management
RT   Line Organizations
SO   System Safety Development Center
     Glossary
DEF (SSDC) Those management
     positions whose responsibility is
     the accomplishment of the
     organization's primary mission(s),
     as distinguished from staff
     organization which supports the
     organization's primary mission(s).

LINE OF BALANCE
(SSDC)
DA   October 12, 1990
RT   Frequency/Severity
SO   System Safety Development Center
     Glossary
DEF (SSDC) A 45° line indicating the

norm in accident
frequency/severity plotting on
log-log paper. Deviations from the
line of balance will provide
important clues to future risk.

LINE ORGANIZATIONS
(DOE Orders 5480.1B, 5481.1B and
 5482.1B; ESH)
DA   October 12, 1990
BT1 Organizations
RT   Field Organizations
RT   Heads of Headquarters Elements
RT   Landlords
RT   Line Management
RT   Program Secretarial Officers
RT   Program Senior Officials
RT   Second Line Organization Level
RT   Staff Organization
SO   Construction
SO   Industrial Hygiene
SO   Management
SO   Standards
SO   System Safety Development Center
     Glossary
DEF (DOE Orders 5480.1b; 5481.1b;
     5482.1b) That unbroken chain of
     command which extends from the
     Secretary through the Under
     Secretary, to the Program Senior
     Officials (PSO) who set program
     policy and plans and develop
     assigned programs, to the field
     organization managers who are
     responsible to the PSO for
     execution of these programs, to
     the contractors who conduct the
     programs. Environment, safety,
     and health are integral parts of
     each program. Accordingly, line
     management responsibility for
     ES&H functions flows from the
     Secretary through the Under
     Secretary, to the PSO, to the field
     organization managers, to the
     contractors. (SSDC) The
     organization within a company or
     project responsible for
     accomplishing the primary goals.

LINEAR ENERGY TRANSFER
(IAEA)
DA   October 12, 1990
SF   LET
BT1 Measurements
RT   Energy Flow
RT   Ionization
SO   Radiation
DEF (IAEA) Of charged particles in a
     medium, the quotient of dE by dl,
     where dE is the energy lost by a
     charged particle in transversing a
     distance dl as a result of those
     collisions with electrons in which
     the energy loss is less than some
     specified value.

LINERS
(SWDA; RCRA; CFR; ESH)
DA   October 12, 1990

BT1 Barriers
RT Containment
RT Inner Liners
RT Landfill Cells
RT Landfills
RT Surface Impoundment
SO Environmental Management
SO Environmental Protection Agency
  Glossary
SO Wastes
DEF (CFR) (1) Relatively impermeable
  barriers designed to prevent
  leachate from leaking from a
  landfill. Liner materials include
  plastic and dense clay. (2) Inserts
  or sleeves for sewer pipes to
  prevent leakage or infiltration.
  (ESH) Continuous layer of natural
  or man-made materials, beneath
  or on the sides of a surface
  impoundment, landfill, or landfill
  cell, that restricts the downward or
  lateral escape of hazardous
  waste, hazardous waste
  constituents, or leachate.

LIP(S)
DA November 28, 1990
BT1 Face
BT2 Head
BT3 Human Body Parts
RT Mouth
SO DOE FRASE VOCABULARY

LIPID SOLUBILITY
(EPA)
DA October 12, 1990
SO Environmental Protection Agency
  Glossary
DEF (EPA) The maximum concentration
  of a chemical that will dissolve in
  fatty substances; lipid soluble
  substances are insoluble in water.
  If a substance is lipid soluble it will
  very selectively disperse through
  the environment via living tissue.

LIQUEFACTION
(USGS)
DA October 12, 1990
BT1 Processes
NT1 Ground Failures
NT2 Differential Settlement
NT2 Flow Failures
NT2 Landslides
NT2 Lateral Spreads
RT Catastrophic Collapses
SO Environmental Protection Agency
  Glossary
SO Natural Phenomenon
DEF (USGS) When seismic shear waves
  pass through a saturated granular
  soil layer, distorting its granular
  structure, this distortion causes
  some of the void spaces to
  collapse. Disruptions to the soil
  generated by these collapses
  cause transfer of the ground
  shaking load from grain-to-grain
  contacts to the pore water. This

transfer of load increases pressure
  in the pore water, either causing
  drainage or, if drainage is
  restricted, a sudden buildup of
  pore-water pressures. When
  pore-water pressures reach a
  critical level (grain-to-grain
  stresses approach zero), the
  granular material suddenly
  behaves as a liquid rather than as
  a solid.

LIQUEFIED COMPRESSED GASES
DA June 3, 1991
BT1 Compressed Gases
BT2 Gases
SO Hazardous Materials
DEF Gases which, under the charged
  pressure, are partially liquid at a
  temperature of 70 degrees F.

LIQUID TRAPS
(SWDA; RCRA; ESA; CFR)
DA October 12, 1990
NT1 Sumps
SO Wastes
DEF (CFR) Sumps, well cellars, and
  other traps used in association
  with oil and gas production,
  gathering, and extraction
  operations (including gas
  production plants), for the purpose
  of collecting oil, water, and other
  liquids. These liquid traps may
  temporarily collect liquids for
  subsequent disposition or
  reinjection into a production or
  pipeline stream, or may collect
  and separate liquids from a gas
  stream.

LIQUIDS
(SWDA; RCRA; CFR; ESH)
DA October 12, 1990
BT1 Fluids
NT1 Combustible Liquids
NT1 Cryogenic Liquids
NT1 Flammable Liquids
NT1 Free Liquids
NT1 Free Products
NT1 Influents
NT1 Natural Gas Liquids
NT1 Pyrophoric Liquids
NT1 Viscous Liquids
SO Hazardous Materials
DEF (CFR) Material that has a vertical
  flow of over 2 inches (50 mm)
  within a three-minute period, or a
  material having one gram (1 g) or
  more liquid separation, when
  determined in accordance with the
  procedures specified in ASTM D
  4359-84, Standard Test Method for
  Determining Whether a Material is
  a Liquid or Solid, 1984 edition.

LIST
(EPA; ESA)
DA October 12, 1990

SO Endangered Species
SO Environmental Protection Agency
  Glossary
DEF (EPA) Shorthand term for EPA list
  of violating facilities or lists of firms
  debarred from obtaining
  government contracts because
  they violated certain sections of
  the Clean Air or Clean Water Acts.
  The list is maintained by the Office
  of Enforcement and Compliance
  Monitoring. (ESA) The list of
  Endangered and Threatened
  Wildlife and Plants as found in
  federal regulations.

LISTED HAZARDOUS SUBSTANCES
(CERCLA)
DA October 12, 1990
SY CERCLA Hazardous Substances
BT1 Hazardous Substances
RT Chemical Substances
RT Extremely Hazardous Substances
RT Regulated Substances
RT Toxic Substances
SO Compensation and Liability
DEF (CFR) The elements and
  compounds and hazardous wastes
  appearing in Table 302.4 (40 CFR
  302.4) are designated as
  hazardous substances under
  section 102(a) of the act.

LISTED SPECIES
(ESA; CFR)
DA October 12, 1990
BT1 Endangered Species
BT2 Species
RT Biological Assessments
RT Biological Opinion
RT Recovery
SO Endangered Species
DEF (CFR) Any species of fish, wildlife,
  or plant which has been
  determined to be endangered or
  threatened under section 4 of the
  Act. Listed species are found in 50
  CFR 17.11-17.12.

LISTED WASTES
(EPA)
DA October 12, 1990
BT1 Hazardous Wastes
BT2 Hazardous Materials
BT3 Materials
BT2 Wastes
SO Environmental Protection Agency
  Glossary
DEF (EPA) Wastes listed as hazardous
  under RCRA but which have not
  been subjected to the Toxic
  Characteristics Listing Process
  because the dangers they present
  are considered self-evident.

LITHOLOGY
(SDWA; CFR)
DA October 12, 1990
RT Formations

SO Water Pollution
DEF (CFR) The study of rocks on the
basis of their physical and
chemical characteristics.

LITHOSPHERE
(ANL; CFR)
DA May 24, 1991
RT Atmosphere
RT Environment
SO Environmental Management
SO Wastes
DEF (CFR) The solid part of the Earth
below the surface, including any
ground water contained within it.

LIVE LOADS
(DOE Order 6430.1A)
DA October 12, 1990
RT Dead Loads
SO Construction
DEF (DOE Order 6430.1A) A moving
load or a load of variable force
acting on a structure, in addition to
its own weight.

LIVE ROUND EXCLUDERS
(Doe Order 5480.16Definitions)
DA October 12, 1990
BT1 Devices
RT Blank Ammunition
RT Firearms
SO Firearms
DEF (DOE Order 5480.16) Obstructive
devices mounted in the cylinder of
an Engagement Simulation
System (ESS) revolver or the
breech of other ESS weapons,
permitting chambering and firing of
blank ammunition but preventing
chambering of a live round.

LLD
DA October 12, 1990
SEE Lower Limit of Detection
SO Acronyms

LLRW
DA October 12, 1990
SEE Low Level Radioactive Wastes
SO Acronyms

LLW
DA October 12, 1990
SEE Low Level Wastes
SO Acronyms
SO Construction
SO Environmental Protection Agency
Glossary

LLWDDD
DA October 12, 1990
SEE Low Level Waste Disposal
Development Demonstration
SO Acronyms

LNP
DA October 12, 1990

SEE Loss of Normal Power
SO Acronyms

LO
DA October 12, 1990
SEE Locked Open
SO Acronyms

LOAD FACTOR
(DOE Order 6430.1A)
DA October 12, 1990
NT1 Capacity Factor
RT Structural Members
SO Construction
DEF (DOE Order 6430.1A) The
strength-to-service-load ratio.

LOAD SECURING DEVICE
(2332 LOAD SECURIN)
DA December 11, 1990
BT1 Devices
BT1 Equipment/Parts - Material
Handling (DOE FRASE
Vocabulary)
BT2 Equipment
RT Barriers
RT Controls
SO DOE FRASE VOCABULARY

LOADING
DA January 24, 1991
BT1 Ratios
NT1 Grain Loading
NT1 Particulate Loading
RT Biomass
DEF (CFR) The ratio of the biomass of
gammarids (grams, wet weight) to
the volume (liters) of test solution
in either a test chamber or passing
through it in a 24-hour period.

LOADING DOCK
(1855 LOADING DOCK)
DA December 10, 1990
BT1 Structures (DOE FRASE
Vocabulary)
SO DOE FRASE VOCABULARY

LOBBY
(1707 LOBBY)
DA December 10, 1990
BT1 Site (DOE FRASE Vocabulary)
BT2 Sites/Areas
SO DOE FRASE VOCABULARY

LOC
DA October 12, 1990
SEE Level of Concern
SO Acronyms

LOCA
DA October 12, 1990
SEE Loss of Coolant Accident
SO Acronyms

LOCAL AGENCIES
(CAA; CFR)
DA October 12, 1990

BT1 Agencies
BT2 Administrative Organizations
BT3 Organizations
NT1 Implementing Agencies
NT1 Intermunicipal Agencies
NT1 Local Governments
NT1 Local Educational Agencies
SO Air Pollution
DEF (CFR) Any local government
agencies other than the State
agencies, which are charged with
responsibility for carrying out a
portion of a plan or program.

LOCAL EDUCATIONAL AGENCIES
(TSCA; USC)
DA October 19, 1990
BT1 Local Agencies
BT2 Agencies
BT3 Administrative Organizations
BT4 Organizations
SO Hazardous Materials
SO Water Pollution
DEF (USC) (A) Any local educational
agency as defined in section 198
of the Elementary and Secondary
Education Act of 1965 (20 U.S.C.
3381), (B) the owner of any
private, nonprofit elementary or
secondary school building, and (C)
the governing authority of any
school operated under the defense
dependents' education system
provided for under the Defense
Dependents' Education Act of
1978 (20 U.S.C. 921 et seq.).

**LOCAL EMERGENCY PLANNING
COMMITTEES**
(EPA; EMER)
DA October 12, 1990
SF *LEPC*
BT1 Committees
BT2 Administrative Organizations
BT3 Organizations
SO Emergency Preparedness
SO Environmental Protection Agency
Glossary
DEF (EPA) A committee appointed by
the state emergency response
commission, as required by SARA
(Superfund Amendments and
Reauthorization Act) Title III to
formulate a comprehensive
emergency plan for its jurisdiction.

LOCAL GOVERNMENTS
(EMER)
DA February 1, 1991
BT1 Local Agencies
BT2 Agencies
BT3 Administrative Organizations
BT4 Organizations
SO Emergency Preparedness
DEF (EMER) Any county, city, village,
town, district, or political
subdivisions of any state, Indian
tribe or authorized tribal
organization, or Alaska native
village or organization, including

any rural community or
unincorporated town or village or
any other public entity.

LOCAL MAGNITUDE (M$_L$)
(USGS)
DA October 12, 1990
SY Body Wave Magnitude
SY Richter Magnitude (M$_L$)
SF M$_L$
BT1 Measurements
RT Earthquake Magnitude
SO Natural Phenomenon
DEF (USGS) The logarithm, to the base
   10, of the amplitude in
   micrometers of the maximum
   amplitude of seismic waves that
   would be observed on a standard
   torsion seismograph at a distance
   of about 60 miles from the
   epicenter.

LOCKED CLOSED
DA January 8, 1991
SF LC

LOCKED OPEN
DA January 8, 1991
SF LO

LOCKER/SHOWER ROOM
(1708 LOCKER ROOM)
DA December 10, 1990
BT1 Room
 BT2 Sites/Areas
SO DOE FRASE VOCABULARY

LOCV
DA October 12, 1990
SEE Loss of Condenser Vacuum
SO Acronyms

LOG N CHANNEL
(2577 LOG N CHANNE)
DA January 3, 1991
BT1 Equipment/Parts - Nuclear (DOE
   FRASE Vocabulary)
 BT2 Equipment
 BT2 Reactor Components
SO DOE FRASE VOCABULARY

LOG N RECORDER
(2761 LOG N RECORD)
DA January 3, 1991
BT1 Instrument(s)
 BT2 Equipment/Parts -
   Instrumentation/Measuring (DOE
   FRASE Voc.)
 BT3 Equipment
SO DOE FRASE VOCABULARY

LOGIC GATES
(SSDC)
DA October 12, 1990
NT1 Conditional AND gate
NT1 Conditional OR Gate
NT1 Summation Gates
RT Analytical (Logic) Trees

RT Constraints
SO System Safety Development Center
   Glossary
DEF (SSDC) Using analytical trees, logic
   gates are symbols connecting an
   event with the next lower tier or
   level on the tree. Gates indicate
   what contribution is required from
   lower events to cause the top
   event.

LOGIC GATES (BOOLEAN)
(SSDC)
DA October 12, 1990
NT1 AND gate
NT1 Conditional AND gate
NT1 Conditional OR Gate
NT1 OR gate
NT1 Summation Gates
RT Constraints
SO System Safety Development Center
   Glossary

LOGIC TREES
(SSDC)
DA October 12, 1990
SY Analytical (Logic) Trees
BT1 Diagrams
SO System Safety Development Center
   Glossary
DEF (SSDC) Diagrams, in the shape of
   a tree, using different geometrical
   symbols to aid a user in
   systematically portraying
   information in a logical sequence
   and showing relationships
   between elements of the tree.
   Trees may be positive or negative
   (fault tree).

LOGICAL CORRELATES
(SSDC)
DA October 12, 1990
RT Empirical Correlates
SO System Safety Development Center
   Glossary
DEF (SSDC) Those performance indices
   and factors which correlate with
   ES&H performance in a logical
   manner. (See Empirical
   Correlates).

LONG PLENUM PLUGS
DA January 8, 1991
SF LPP

LONG TERM CONTRACTS
(SWDA; RCRA; USC)
DA October 12, 1990
RT Solid Waste Management
RT Substantial Business Relationships
SO Wastes
DEF (USC) When used in relation to
   solid waste supply, contracts of
   sufficient duration to assure the
   viability of a resource recovery
   facility (to the extent that such
   viability depends upon solid waste
   supply).

LOOP
DA October 12, 1990
SEE Loss of Offsite Power
SO Acronyms

LOP
DA October 12, 1990
SEE Loss of Offsite Power
SO Acronyms

**LOS ALAMOS NATIONAL
LABORATORY**
DA January 8, 1991
SF LANL
BT1 Government-Owned
   Contractor-Operated Facilities
 BT2 Federal Facilities
 BT3 Facilities
BT1 Laboratories
 BT2 Research and Development
   Organizations
 BT3 Organizations
RT University of California
DEF (Capsule Review of DOE Research
   and Development and Field
   Facilities, 1986) LANL was
   established in 1943 as part of the
   Manhattan Engineer District during
   World War II to develop the
   world's first nuclear weapons.
   Currently, LANL's primary mission
   is the application of science and
   technology to problems of national
   security, including the
   maintenance of a strong defense,
   the fulfillment of arms control
   commitments and the guarantee
   of a secure energy supply for the
   future. LANL also undertakes
   multidisciplinary fundamental and
   applied research.

LOSP
DA October 12, 1990
SEE Loss of Offsite Power
SO Acronyms

LOSS CONTROL MANAGEMENT
(SSDC)
DA October 12, 1990
BT1 Management
SO System Safety Development Center
   Glossary
DEF (SSDC) The application of
   professional management
   techniques and skills to those
   program activities – risk
   avoidance, loss prevention and
   loss reduction – specifically
   intended to minimize losses
   involved with undesired events
   resulting from the pure
   (nonspeculative) risks of business.

LOSS OF CONDENSER VACUUM
DA January 8, 1991
SF LOCV
BT1 Injuries
BT1 Losses

SY-Synonymous Terms   SO-Source/Subject Category   SF-See From

213

LOSS OF CONSCIOUSNESS
(1331 LOSS OF CONS)
DA November 28, 1990
BT1 Injuries
BT1 Losses
RT Concussion
RT Contusion(S)
SO DOE FRASE VOCABULARY

LOSS OF COOLANT ACCIDENT
DA January 8, 1991
SF LOCA
BT1 Accidents
BT1 Losses
NT1 Small Break LOCA

LOSS OF EXPERIMENT
(1547 LOSS OF EXPE)
DA November 29, 1990
BT1 Losses
BT1 Nature of Property Damage
SO DOE FRASE VOCABULARY

LOSS OF INSTRUMENT AIR
(1548 LOSS OF INST)
DA November 29, 1990
BT1 Losses
BT1 Nature of Property Damage
SO DOE FRASE VOCABULARY

LOSS OF MATERIAL
(1533 LOSS OF MATE)
DA November 29, 1990
BT1 Losses
BT1 Nature of Property Damage
SO DOE FRASE VOCABULARY

LOSS OF NORMAL POWER
DA January 8, 1991
SF LNP
BT1 Losses

LOSS OF OFFSITE POWER
DA January 8, 1991
SF LOOP
SF LOP
SF LOSP
BT1 Losses

LOSS OF OPERATING TIME
(1578 LOSS OF OPER)
DA November 28, 1990
BT1 Losses
BT1 Nature of Programmatic Impact
BT1 Time Designations
SO DOE FRASE VOCABULARY

LOSS OF PRODUCTION
(1579 LOSS OF PROD)
DA November 28, 1990
BT1 Losses
BT1 Nature of Programmatic Impact
SO DOE FRASE VOCABULARY

LOSS OF SPECIMEN
(1551 LOSS OF SPEC)
DA November 29, 1990
BT1 Losses

BT1 Nature of Property Damage
SO DOE FRASE VOCABULARY

LOSS OF TARGET ACCIDENT
DA January 8, 1991
SF LOTA
BT1 Accidents
BT1 Losses

LOSS OF TRANSMISSION
(1552 LOSS OF TRAN)
DA November 29, 1990
BT1 Losses
BT1 Nature of Property Damage
SO DOE FRASE VOCABULARY

LOSS RATIO
(SSDC)
DA October 12, 1990
BT1 Ratios
RT Consequential Losses
RT Incurred Losses
SO System Safety Development Center
     Glossary
DEF (SSDC) Used in insurance. A ratio
     calculated by dividing the amount
     of loss(es) by the amount of the
     premium(s). Normally expressed
     as a percentage of the premiums.

LOSSES
(CFR)
DA November 15, 1990
NT1 Consequential Losses
NT1 Incurred Losses
NT1 Loss of Transmission
NT1 Loss of Consciousness
NT1 Loss of Operating Time
NT1 Loss of Production
NT1 Loss of Material
NT1 Loss of Specimen
NT1 Loss of Instrument Air
NT1 Loss of Experiment
NT1 Loss of Normal Power
NT1 Loss of Coolant Accident
 NT2 Small Break LOCA
NT1 Loss of Condenser Vacuum
NT1 Loss of Offsite Power
NT1 Loss of Target Accident
NT1 Lost Workday Cases
NT1 Maximum Credible Loss
NT1 Maximum Foreseeable Loss
NT1 Maximum Probable Loss
NT1 Maximum Possible Loss
 NT2 Maximum Possible Fire Loss
NT1 Mean Annual Loss
NT1 Power Losses
NT1 Property Loss
NT1 Reactor Opening Loss
NT1 Significant Economic Loss
DEF (CFR) Measurable adverse
     reductions of chemical or physical
     qualities or viabilities of natural
     resources.

LOST WORKDAY CASES
DA January 8, 1991
SF LWC

BT1 Losses
RT Incidence Rate, Lost Workday
     Cases (LWC)

LOTA
DA October 12, 1990
SEE Loss of Target Accident
SO Acronyms

LOVE SEISMIC WAVES
(USGS)
DA October 12, 1990
RT Earthquake Magnitude
RT Rayleigh Seismic Waves
RT Surface Wave Magnitude
SO Natural Phenomenon
DEF (USGS) Types of surface waves
     having a horizontal motion that is
     shear or transverse to the
     direction of propagation. Its
     velocity depends only on density
     and rigidity modulus, and not on
     bulk modulus.

LOVELACE MEDICAL FOUNDATION
DA January 11, 1991
BT1 DOE Contractors
 BT2 Potentially Responsible Parties
BT1 Foundations
 BT2 Research and Development
      Organizations
  BT3 Organizations
RT Albuquerque Operations Office
RT Inhalation Toxicology Research
     Institute

LOW CONCENTRATION PCBS
(TSCA; CFR)
DA October 12, 1990
BT1 Polychlorinated Biphenyls
 BT2 Carcinogens
  BT3 Hazardous Substances
 BT2 Chlorinated Hydrocarbons
  BT3 CERCLA Hazardous Substances
   BT4 Hazardous Substances
  BT3 Halogenated Organic
       Compounds
   BT4 Halogenated
   BT4 Organic Chemicals
    BT5 Chemical Substances
SO Hazardous Materials
DEF (CFR) PCBs that are tested and
     found to contain less than 500
     ppm PCBs, or those
     PCB-containing materials which
     EPA requires to be assumed to be
     at concentrations below 500 ppm
     (i.e., untested mineral oil dielectric
     fluid).

LOW ENERGY GAMMA MONITOR
DA January 8, 1991
SF LEGM
BT1 Monitors
 BT2 Equipment
BT1 Radiation Detectors
 BT2 Equipment

LOW LEVEL RADIOACTIVE WASTES
(EPA)
DA October 12, 1990
SY Low Level Wastes
SF *LLRW*
BT1 Radioactive Wastes
 BT2 Wastes
SO Environmental Protection Agency
    Glossary
DEF (EPA) Wastes less hazardous than
    most of those generated by a
    nuclear reactor. Usually generated
    by hospitals, research
    laboratories, and certain
    industries. The Department of
    Energy, Nuclear Regulatory
    Commission, and EPA share
    responsibilities for managing them.

**LOW LEVEL WASTE DISPOSAL
DEVELOPMENT DEMONSTRATION**
DA January 8, 1991
SF *LLWDDD*

LOW LEVEL WASTES
(DOE Order 6430.1A; EMER)
DA October 12, 1990
SY Low Level Radioactive Wastes
SF *LLW*
BT1 Radioactive Wastes
 BT2 Wastes
RT Below Regulatory Concern
SO Construction
SO Emergency Preparedness
DEF (DOE Order 6430.1A) Radioactive
    wastes not classified as high-level
    waste, transuranic waste, spent
    nuclear fuel, or by- product
    material.

LOW POWER REACTOR FACILITY
(1802 LOW POWER RE)
DA December 10, 1990
BT1 Facility (DOE FRASE Vocabulary)
 BT2 Facilities and Buildings (DOE
     FRASE Vocabulary)
  BT3 Facilities
BT1 Reactor Facilities
 BT2 Nuclear Facilities
 BT3 Facilities
SO DOE FRASE VOCABULARY

**LOW PRESSURE COOLANT
INJECTION**
DA January 8, 1991
SF *LPCI*

**LOW PRESSURE COOLING
RECIRCULATION PHASE**
DA January 8, 1991
SF *LPCR*

LOW PRESSURE CORE SPRAY
DA January 8, 1991
SF *LPCS*

LOW PRESSURE INJECTION
DA January 8, 1991
SF *LPI*

LOW PRESSURE PUMP PAD
DA January 8, 1991
SF *LPPP*

**LOW PRESSURE RECIRCULATION
SYSTEM**
DA January 8, 1991
SF *LPRS*
BT1 Systems

**LOW PRESSURE RECIRCULATION
SYSTEM HEAT EXCHANGER**
DA January 8, 1991
SF *LPRSX*
BT1 Heat Exchanger
 BT2 Equipment/Parts - Nuclear (DOE
     FRASE Vocabulary)
  BT3 Equipment
  BT3 Reactor Components

LOW PRESSURE SERVICE WATER
DA January 8, 1991
SF *LPSW*

LOW SPECIFIC ACTIVITY
(DOE Order 5480.3; ESH)
DA October 12, 1990
BT1 Specific Activity
 BT2 Activity (Nuclear)
  BT3 Measurements
RT Low Specific Activity Materials
SO Hazardous Materials
DEF (DOE Order 5480.3) Material of low
    radioactivity level such as ores and
    chemical concentrations of those
    ores. The low specific activity
    definition is in 49 CFR 173.403.

LOW SPECIFIC ACTIVITY MATERIALS
(CFR; EMER)
DA October 12, 1990
BT1 Radioactive Materials
 BT2 Materials
RT Low Specific Activity
SO Emergency Preparedness
SO Hazardous Materials
DEF (CFR) Any of the materials such as
    uranium or thorium ores and
    physical or chemical concentrates
    of those ores, unirradiated natural
    or depleted uranium or
    unirradiated natural thorium, etc.

LOWER BACK
(1114 LOWER BACK)
DA November 28, 1990
BT1 Back
 BT2 Trunk
  BT3 Human Body Parts
SO DOE FRASE VOCABULARY

LOWER EXPLOSIVE LIMIT
(EPA)
DA October 12, 1990

SF *LEL*
BT1 Limits
SO Environmental Protection Agency
    Glossary
DEF The concentration of a compound
    in air below which a flame will not
    propagate if the mixture is ignited.

LOWER LEG
(1125 LOWER LEG)
DA November 28, 1990
BT1 Leg(s)
 BT2 Human Body Parts
RT Leggings
SO DOE FRASE VOCABULARY

LOWER LIMIT OF DETECTION
(ESH)
DA October 12, 1990
SF *LLD*
BT1 Detection Limits
 BT2 Limits
SO Management
DEF (ESH) The smallest amount of a
    contaminant that can be
    distinguished in a sample by a
    given measurement procedure at
    a given confidence level.

**LOWEST ACHIEVABLE EMISSION
RATE**
(CAA; CFR; USC; ESH)
DA October 12, 1990
BT1 Rates
RT National Emissions Standards for
    Hazardous Air Pollutants
SO Air Pollution
SO Environmental Protection Agency
    Glossary
DEF (EPA) Under the Clean Air Act, this
    is the rate of emissions that
    reflects (a) the most stringent
    emission limitation that is
    contained in the implementation
    plan of any state for such source
    unless the owner or operator of
    the proposed source demonstrates
    such limitations are not
    achievable; or (b) the most
    stringent emissions limitation
    achieved in practice, which ever is
    more stringent. Application of this
    term does not permit a proposed
    new or modified source to emit
    pollutants in excess of existing
    new source standards./ (ESH)
    The use of the "lowest achievable
    emission rate" is required of new
    or modified sources locating in
    nonattainment areas (40 CFR
    51.165).

**LOWEST-OBSERVED-ADVERSE-
EFFECT LEVEL**
(EPA)
DA October 12, 1990
RT Dose-Response Evaluation

SO Environmental Protection Agency
    Glossary
DEF (EPA) (LOAEL) In dose-response
    experiments, the experimental
    exposure level representing the
    lowest level tested at which
    adverse effects were
    demonstrated.

LPCI
DA   October 12, 1990
SEE Low Pressure Coolant Injection
SO Acronyms

LPCR
DA   October 12, 1990
SEE Low Pressure Cooling Recirculation
    Phase
SO Acronyms

LPCS
DA   October 12, 1990
SEE Low Pressure Core Spray
SO Acronyms

LPI
DA   October 12, 1990
SEE Low Pressure Injection
SO Acronyms

LPP
DA   October 12, 1990
SEE Long Plenum Plugs
SO Acronyms

LPPP
DA   October 12, 1990
SEE Low Pressure Pump Pad
SO Acronyms

LPRS
DA   October 12, 1990
SEE Low Pressure Recirculation System
SO Acronyms

LPRSX
DA   October 12, 1990
SEE Low Pressure Recirculation System
    Heat Exchanger
SO Acronyms

LPSW
DA   October 12, 1990
SEE Low Pressure Service Water
SO Acronyms

LTA
DA   October 12, 1990
SEE Less than Adequate
SO Acronyms

LUBRICATING OILS
(SWDA; RCRA; USC)
DA   October 12, 1990
BT1 Oils
RT Re-refined Oils

SO Wastes
DEF (USC) The fraction of crude oil that
    is sold for purposes of reducing
    friction in any industrial or
    mechanical device. Such term
    includes re-refined oil.

LUNG CLASSES
(IAEA)
DA   October 12, 1990
RT Biological Clearance Rate
SO Radiation
DEF (IAEA) A classification scheme
    used by the ICRP to designate the
    clearance of inhaled radioactive
    materials from the lung. Materials
    are classified on the basis of their
    period of retention in the
    pulmonary region. D (Day)
    indicates a biological half- life of
    less than 10 days. W (Week) a
    half-life of 10-100 days. Y (Year) a
    half-life greater than 100 days.

LWC
DA   October 12, 1990
SEE Lost Workday Cases
SO Acronyms

LWR
DA   October 12, 1990
SEE Light Water Reactor
SO Acronyms

M
DA   October 12, 1990
SEE Moment Magnitude
SO Acronyms

M&O
DA   October 12, 1990
SEE Management and Operating
    Contractor for DOE Facility
SO Acronyms

M&T
DA   October 12, 1990
SEE Main and Trim
SO Acronyms

MACCS
(Acronyms and Abbreviations)
DA   October 12, 1990
BT1 Computer Codes
DEF A computer code used in accident
    consequence analysis.

MACHINE BASIN
DA   January 8, 1991
SF   *MB*

MACHINE GUNS
(Doe Order 5480.16)
DA   October 12, 1990
BT1 Small Arms
BT2 Firearms
NT1 Submachine Guns, Closed Bolt
NT1 Submachine Guns, Open Bolt

SO Firearms
DEF (DOE Order 5480.16) A fully
    automatic weapon capable of
    firing multiple rounds with a single
    pull of the trigger; it is belt fed and
    is usually mounted on a bipod,
    tripod, or another fixture.

MACHINE PISTOLS
(Doe Order 5480.16)
DA   October 12, 1990
BT1 Pistols
BT2 Handguns
BT3 Small Arms
BT4 Firearms
RT Magazines
SO Firearms
DEF (DOE Order 5480.16) Capable of
    being fired in the fully automatic
    mode.

MACHINE SETUP/OPERATOR
(0710 MACHINE OPER)
DA   November 28, 1990
BT1 Precision/Production Personnel
BT2 Occupations
BT2 Personnel
SO DOE FRASE VOCABULARY

MACHINE SHOP
(1784 MACHINE SHOP)
DA   December 10, 1990
BT1 Building (DOE FRASE Vocabulary)
BT2 Facilities and Buildings (DOE
    FRASE Vocabulary)
BT3 Facilities
SO DOE FRASE VOCABULARY

**MACHINES (DOE FRASE
VOCABULARY)**
(DOE FRASE Vocabulary Numeric Keys
    2100-2199)
DA   December 10, 1990
BT1 Equipment
NT1 Agitator
NT1 Agricultural Machine
NT1 Air Dryer
NT1 Band Saw
NT1 Biostabilizers
NT1 Boring Machine
NT1 Centrifuge
NT1 Comminuters
NT1 Compactor
NT1 Compressor
NT2 Air Compressor
NT1 Concrete Saw
NT1 Crushing Machine
NT1 Discharge Machine
NT1 Drilling Machine
NT1 Drying Machine
NT1 Electrical Office Machine
NT1 Emergency Diesel Generator
NT1 Grinding Machine
NT1 Hammermills
NT1 Lathe
NT1 Milling Machine
NT2 Ball Mill
NT2 Roll Mill
NT1 Mixing Machine

BT-Broader Term   NT-Narrower Term   RT Related Term

NT1 Polishing Machine
NT1 Press
 NT2 Isostatic Press
NT1 Printing Machine
NT1 Pump(s)
 NT2 Coolant Return Pump
 NT2 Ion Pump
 NT2 Reactor Coolant Pump
 NT2 Sump Pump
NT1 Recirculator
NT1 Sanding Machine
NT1 Slicing Machine
NT1 Stitching/Sewing Machine
NT1 Table Saw
NT1 Turbine
NT1 Vacuum Cleaner
 RT DOE FRASE Categories
 SO DOE FRASE VOCABULARY
DEF A subject category used with the
    DOE FRASE Vocabulary.

MACHINIST
(0681 MACHINIST)
DA November 28, 1990
BT1 Precision/Production Personnel
 BT2 Occupations
 BT2 Personnel
 SO DOE FRASE VOCABULARY

MAGAZINE SEPARATION
(DOE Order 6430.1A)
DA October 12, 1990
BT1 Quantity Distances
 BT2 Measurements
 RT Magazines (Buildings)
 SO Construction
DEF (DOE Order 6430.1A) The
    minimum quantity-distance
    separation between magazines
    (not including service magazines)
    within a storage area. Siting of
    magazines within a storage area
    with respect to one another and
    location of facilities such as guard
    shelters and loading docks in
    storage areas are covered in DOD
    6055.9. Maximum explosives
    weight shall be used in
    determining separation distances.

MAGAZINE VESSELS
(SWDA; RCRA; ESH)
DA October 12, 1990
 RT Explosives Activities
 SO Hazardous Materials
DEF (CFR) Vessels used for receiving,
    storing, or dispensing of
    explosives.

MAGAZINES
(DOE Order 5480.16)
DA October 12, 1990
 RT Automatic Rifles
 RT Double Action Semiautomatic
    Pistols
 RT Machine Pistols
 RT Semiautomatic Firearms
 RT Single Action Semiautomatic Pistols

 SO Firearms
DEF (DOE Order 5480.16) Mechanical
    devices used to hold a
    predetermined number of
    cartridges in position for feeding
    into a weapon.

MAGAZINES (BUILDINGS)
(DOE Order 6430.1A)
DA October 16, 1990
NT1 Service Magazine
 RT Magazine Separation
 RT Quantity Distances
 SO Construction
DEF (DOE Order 6430.1A) Buildings or
    structures, except an operating
    building, used for the storage of
    ammunition or explosives.

MAGNEFORMING
(NFI)
DA October 12, 1990
BT1 Manufacturing Processes
 BT2 Processes
 SO Nuclear Facilities Incident Database
 SO Radiation
DEF (NFI) Part of fabrication process of
    fuel subassemblies.

MAGNESIUM FLUORIDE
DA January 8, 1991
 SF MgF_2

**MAGNETIC/ELECTROLYTIC
APPARATUS**
(2402 MEA)
DA January 3, 1991
BT1 Equipment/Parts - Electrical (DOE
    FRASE Vocabulary)
 BT2 Equipment
 SO DOE FRASE VOCABULARY

MAIN AND TRIM
(NFI)
DA October 12, 1990
 SF M&T
 SO Radiation
DEF (NFI) Switches on control rod drive
    amplifiers.

MAIN FEEDWATER
DA January 8, 1991
 SF MFW
BT1 Feedwater

**MAIN FEEDWATER AND
CONDENSATE SYSTEM**
DA January 8, 1991
 SF MFWCS
BT1 Systems

MAIN STEAM ISOLATION VALVE
DA January 8, 1991
 SF MSIV
BT1 Valves
 BT2 Devices

MAINTENANCE
(DOE Order 4330.4A; SWDA; RCRA;
  ESA; CFR; MORT)
DA October 12, 1990
BT1 Activities
NT1 Active Maintenance
NT1 Building/Equip Maint/Repair
    Activity
NT1 Corrective Repair Maintenance
NT1 Grounds Maintenance Activity
NT1 Predictive Maintenance
NT1 Preventive Maintenance
NT1 Reliability-Centered Maintenance
NT1 Source Control Maintenance
    Measures
NT1 Surveillance and Maintenance
NT1 Vehicle Maint/Repair Activity
NT1 Wastewater Operations and
    Maintenance
 RT Condition Assessment Surveys
 RT Controls
 RT First Line Supervision
 RT Higher Supervision
 RT Inspections
 RT Maintenance Plans
 RT Maintenance Backlog
 RT Operability
 RT Self-Assessment
 RT Technical Information Systems
 RT Technical Information
 RT Underground Storage Tanks
 SO Wastes
DEF (DOE Order 4330.4A) Day-to-day
    work that is required to maintain
    and preserve plant and capital
    equipment in a condition suitable
    for it to be used for its designated
    purpose. This includes preventive,
    predictive, and corrective repair
    maintenance. (DSTT) The
    required upkeep of industrial
    facilities and equipment. (MORT)
    MORT analysis asks: was there
    adequate maintenance of
    equipment, processes, utilities,
    operations, etc.?

MAINTENANCE AREA
(1612 MAINT AREA)
DA December 10, 1990
BT1 Area
 BT2 Sites/Areas
 RT Garage
 RT Mechanic/Repairer
 SO DOE FRASE VOCABULARY

MAINTENANCE BACKLOG
(DOE Order 4330.4A)
DA June 5, 1991
 RT Fiscal Year
 RT Maintenance
 SO Management
DEF (DOE Order 4330.4A) The amount
    of maintenance and repair work
    not accomplished at the end of the
    fiscal year that is needed or
    planned to sustain the assigned
    mission.

SY-Synonymous Terms   SO-Source/Subject Category   SF-See From

MAINTENANCE INFORMATION AND CONTROL
DA January 8, 1991
SF *MIAC*
BT1 Controls

MAINTENANCE MANAGEMENT
(DOE Order 4330.4A)
DA June 5, 1991
BT1 Management
RT Compliance Considerations
RT Industrial Hygiene
RT Industrial Safety
RT Inspections
RT Plans
RT Procedures
RT Quality Control
RT Self-Assessment
SO Management
DEF (DOE Order 4330.4A) The administration of a program utilizing such concepts as organization, plans, procedures, schedules, cost control, periodic evaluation, and feedback for the effective performance and control of maintenance with adequate provisions for interface with other concerned disciplines such as health, safety, environmental compliance, quality control, and security. All work done in conjunction with existing property is either maintenance (preserving), repair (restoring), service (cleaning and making usable), or improvements. The work to be considered under the DOE maintenance management program is only that for maintenance and repair.

MAINTENANCE PLANS
(DOE Order 4330.4A; MORT)
DA April 3, 1991
BT1 Plans
RT Design and Development Plans
RT Field Operations Plans
RT General Design Process
RT Human Factors
RT Inspection Plans
RT Maintenance
RT Operational Safety Requirements
RT Operational Specifications
DEF (DOE Order 4330.4A) A narrative description of a site's maintenance program. The plan should be a real time document which is updated at least annually and which addresses all elements of a successful Maintenance Program. The plan should describe the backlog and strategies to reduce the backlog, as well as the maintenance funding required to sustain the assigned mission. The maintenance plan should integrate individual maintenance activities addressed under each functional unit life cycle plan. (MORT) MORT

analysis asks: is maintenance of an operation or facility given consideration during the conceptual phase and on through the rest of the life cycle? is there an adequate maintenance plan? was the maintenance plan broad enough to include all the areas that should be maintained? was management aware of those areas not included in the plan? was there adequate execution of the maintenance plan?

MAINTENANCE REQUIREMENT CARD
DA January 8, 1991
SF *MRC*

MAINTENANCE SHOP
(1785 MAINT SHOP)
DA December 10, 1990
BT1 Building (DOE FRASE Vocabulary)
BT2 Facilities and Buildings (DOE FRASE Vocabulary)
BT3 Facilities
RT Garage
RT Mechanic/Repairer
SO DOE FRASE VOCABULARY

MAINTENANCE TEAM INSPECTION
DA January 8, 1991
SF *MTI*
BT1 Inspections
BT2 Administrative Processes
BT3 Processes

MAINTENANCE WORK ORDER
DA January 8, 1991
SF *MWO*

MAJOR CONSTRUCTION ACTIVITIES
(ESA; CFR)
DA October 12, 1990
BT1 Activities
BT1 Construction Projects
BT2 Projects
SO Endangered Species
DEF (CFR) Construction projects (or other undertakings having similar physical impacts) that are major Federal actions significantly affecting the quality of the human environment as referred to in the National environmental Policy Act [NEPA, 42 U.S.C. 4332(2)(C)].

MAJOR DISCHARGES
(CERCLA; CFR)
DA October 12, 1990
BT1 Discharges
BT1 Oil Spills
BT1 Size Classes of Discharges
DEF (CFR) Discharges of more than 10,000 gallons of oil to the inland waters or more than 100,000 gallons of oil to the coastal waters.

MAJOR EMITTING FACILITIES
(CAA; USC)
DA October 12, 1990
BT1 Major Stationary Source
BT2 Stationary Sources
BT3 Facilities
SO Air Pollution
DEF (USC) Any stationary facilities or sources of air pollutants which directly emits, or has the potential to emit, one hundred tons per year or more of any air pollutant (including any major emitting facility or source of fugitive emissions of any such pollutant, as determined by rule by the administrator).

MAJOR FACILITIES
(SWDA; RCRA; CFR)
DA October 12, 1990
BT1 Facilities
SO Wastes
DEF (CFR) Any RCRA, UIC, NPDES, or 404 "facility or activity" classified as such by the Regional Administrator, or, in the case of "approved State programs," the Regional Administrator in conjunction with the State Director.

MAJOR MODIFICATION
(EPA)
DA October 12, 1990
SY Major PSD Modification
BT1 Modification
BT2 Administrative Processes
BT3 Processes
RT Best Available Retrofit Technology
RT Major Stationary Source
SO Environmental Protection Agency Glossary
DEF (EPA) This term is used to define modifications with respect to Prevention of Significant Deterioration and New Source Review under the Clean Air Act and refers to modifications to major stationary sources of emissions and provides significant pollutant increase levels below which a modification is not considered major.

MAJOR PSD MODIFICATION
(CFR)
DA October 26, 1990
SY Major Modification
BT1 Modification
BT2 Administrative Processes
BT3 Processes
RT Major Stationary Source
RT Regulated Activities
DEF (CFR) A "major modification" as defined in 40 CFR 52.21.

MAJOR PSD STATIONARY SOURCE
(SWDA; RCRA; CFR)
DA October 19, 1990

BT1 Major Stationary Source
 BT2 Stationary Sources
  BT3 Facilities
RT Regulated Activities
SO Wastes
DEF (CFR) A "major stationary source"
    as defined in 40 CFR 52.21(b)(1).

MAJOR RELEASE
(CERCLA; CFR)
DA October 12, 1990
BT1 Releases
BT1 Size Classes of Releases
DEF (CFR) A release of any quantity of
    hazardous substance(s),
    pollutant(s), or contaminant(s) that
    poses a substantial threat to public
    health or welfare or the
    environment or results in
    significant public concern.

MAJOR STATIONARY SOURCE
(CAA; USC; CFR; ESH)
DA November 15, 1990
BT1 Stationary Sources
 BT2 Facilities
NT1 Major Emitting Facilities
NT1 Major PSD Stationary Source
RT Major Modification
RT Major PSD Modification
SO Air Pollution
SO Environmental Protection Agency
    Glossary
DEF (EPA) Term used to determine
    applicability of Prevention of
    Significant Deterioration and new
    source regulations. In a
    nonattainment area, any stationary
    pollutant source that has a
    potential to emit more than 100
    tons per year is considered a
    major stationary source. In PSD
    areas the cutoff level may be
    either 100 or 250 tons, depending
    upon the type of source. (See 42
    USC 7491 for continuation of
    definition.)

MAJOR SYSTEM ACQUISITION
DA January 8, 1991
SF *MSA*
RT Procurement Items
RT Procuring Agencies

MALFUNCTIONS
(CAA; CFR)
DA October 12, 1990
BT1 Failures
 BT2 Accidents
SO Air Pollution
DEF (CFR) Any sudden failures of air
    pollution control equipment or
    process equipment or of a process
    to operate in a normal or usual
    manner so that emissions of
    arsenic are increased.

MALIGNANT NEOPLASM
(1395 MALIGNANT NE)

DA November 28, 1990
BT1 Illnesses
SO DOE FRASE VOCABULARY

MAN-MADE AIR POLLUTION
(CAA)
DA October 12, 1990
BT1 Air Pollution
 BT2 Pollution
SO Air Pollution
DEF (USC) Air pollution that results
    directly or indirectly from human
    activities.

MAN-MADE BETA PARTICLE AND PHOTON EMITTERS
(SDWA; CFR)
DA October 12, 1990
RT Beta Particles
SO Water Pollution
DEF (CFR) All radionuclides emitting
    beta particles and/or photons
    listed in Maximum Permissible
    Body Burdens and Maximum
    Permissible Concentration of
    Radionuclides in Air or Water for
    Occupational Exposure, NBS
    Handbook 69, except the daughter
    products of thorium-232,
    uranium-235 and uranium-238.

MANAGEMENT
(SWDA; RCRA; WTID)
DA October 19, 1990
NT1 Best Management Practices
NT1 Environmental Management
NT1 Line Management
NT1 Loss Control Management
NT1 Maintenance Management
NT1 Management by Objective
NT1 Risk Management
NT1 Subordinate Management
RT Chain of Command
RT Concept of Operations
RT Emergency Management
    Coordination Committee
RT Program Offices
RT Program Secretarial Officers
RT Program Senior Officials
RT Program Managers
RT Senior Management Officials
RT Top Management
RT Water Management Division
    Director
SO System Safety Development Center
    Glossary
SO Wastes
DEF (SSDC) All personnel above the
    level of job and task supervisors
    serving in a command role within
    the organizational structure.
    (WTID) The act or art of
    managing. This includes the
    executive function of planning,
    organizing, coordinating, directing,
    controlling and supervising any
    industrial or business projects or
    activities with responsibility for the
    results.

MANAGEMENT AND OPERATING CONTRACTOR FOR DOE FACILITY
DA January 8, 1991
SF *M&O*

MANAGEMENT AND ORGANIZATION ASSESSMENT
(TTGM)
DA May 6, 1991
BT1 Tiger Team Assessments
 BT2 Assessments
  BT3 Administrative Processes
   BT4 Processes
RT Environment Assessments
RT Safety and Health Assessments
SO Management
DEF (TTGM) An assessment to evaluate
    the effectiveness and identify the
    strengths and weaknesses in DOE
    and site contractor management
    and administration of ES&H
    programs. A pragmatic "bottom
    up" assessment which is
    integrated with, and somewhat
    driven by, the findings identified by
    the concurrently performed
    Environment, and Safety and
    Health Assessments.
    Information/documents gathered
    in support of this assessment
    would include: organization charts
    and functions, facility/site layout,
    description of operations, prior
    TSAs, Environmental Survey,
    Operations Office and other
    appraisals, MOUs, contract
    provisions, legislated federal and
    state requirements, past
    appraisal/inspection
    results/corrective actions, internal
    self-assessments/corrective
    actions, selected DOE and
    contractor policies, Orders, and
    correspondence, incident reports,
    UORs, budgetary requests for
    corrective actions, etc.

MANAGEMENT APPRAISALS
(DOE Orders 5482.1B and 5700.6B;
    ESH)
DA October 12, 1990
BT1 Appraisals
NT1 Functional Appraisals
NT1 Internal Appraisals
RT Criteria
RT Findings
RT Functional Appraisals
RT Internal Appraisals
RT Quality Assurance
SO Construction
SO Management
SO System Safety Development Center
    Glossary
DEF (SSDC) Documented
    determinations of managerial
    effectiveness in establishing and
    implementing ES&H program
    plans that conform to DOE policy
    requirements. These are based on
    analyses of functional appraisals,

internal appraisals, and other information, and on the application of appropriate criteria. The appraisals are reviews and evaluations of management performance covering all ES&H disciplines and management responsibilities to assure proper program balance.

MANAGEMENT BY OBJECTIVE
(SSDC)
DA   October 12, 1990
SF   *MBO*
BT1   Management
RT   Management Systems
SO   System Safety Development Center Glossary
DEF (SSDC) A management system wherein each manager establishes objectives (goals) consistent with the overall organizational objectives. The four basic steps in MBO include: (1) define the job (key responsibilities and duties); (2) define the expected results (objectives); (3) measure the results; and (4) appraisal (providing feedback on results and establishing necessary modification to objectives to achieve expected results during the next performance period).

MANAGEMENT OF MIGRATIONS
(CERCLA; CFR)
DA   October 12, 1990
RT   Discharges
RT   Leaks
RT   Plumes
RT   Releases
RT   Spills
SO   Compensation and Liability
DEF (CFR) Actions that are taken to minimize and mitigate the migration of hazardous substances or pollutants or contaminants and the effects of such migration. Management of migration actions may be appropriate where the hazardous substances or pollutants or contaminants are no longer at or near the area where they were originally located or situations where a source cannot be adequately identified or characterized. Measures may include, but are not limited to, provision of alternative water supplies, management of a plume of contamination, or treatment of a drinking water aquifer.

MANAGEMENT OVERSIGHT AND RISK TREE
(SSDC)
DA   October 12, 1990
SF   *MORT*
BT1   Fault Tree
 BT2   Analytical (Logic) Trees

BT3   Diagrams
RT   Oversights and Omissions
RT   Root Causes
SO   System Safety Development Center Glossary
DEF (SSDC) A formal, disciplined logic or decision tree to relate and integrate a wide variety of safety concepts systematically. As an accident analysis technique, it focuses on three main concerns: specific oversights and omissions, assumed risks, and general management system weaknesses.

MANAGEMENT SYSTEM FACTORS
(MORT)
DA   April 2, 1991
RT   Implementation
RT   Policies
RT   Risk Assessment
RT   Root Causes
RT   Root Cause Analysis
RT   Specific Control Factors
SO   Management
DEF (MORT) MORT analysis ask this question: Are all the factors of the management system necessary, sufficient, and organized in such a manner as to assure that the overall program will be "as advertised" to the customer, to the public, to the organization itself, and to other groups as appropriate?

MANAGEMENT SYSTEMS
(SSDC)
DA   October 12, 1990
BT1   Systems
RT   Management by Objective
RT   Priority Problem List
RT   Program Evaluation and Review Technique
SO   System Safety Development Center Glossary
DEF (SSDC) The organizational structures and operating philosophies of companies, projects, or organizations. Management may follow a textbook model, such as PERT or MBO, in their management system but the model is usually modified in philosophy and even more in actual operation.

MANAGEMENT TEAMS
DA   February 14, 1991
BT1   Teams
 BT2   Administrative Organizations
 BT3   Organizations
NT1   Accreditation Review Teams
NT1   Continuity of Government Emergency Management Teams
NT1   Crisis Management Teams
NT1   Emergency Management Teams
 NT2   Energy Emergency Management Teams

NT2   Operational Emergency Management Teams
NT1   Headquarters Coordinating Teams

MANAGER/ADMINISTRATOR
(0110 MANAGER)
DA   November 28, 1990
SO   DOE FRASE VOCABULARY

MANATEE PROTECTION AREAS
(ESA; CFR)
DA   October 12, 1990
BT1   Sites/Areas
NT1   Manatee Refuges
NT1   Manatee Sanctuaries
SO   Endangered Species
DEF (CFR) Manatee refuges or a manatee sanctuaries.

MANATEE REFUGES
(ESA; CFR)
DA   October 12, 1990
BT1   Manatee Protection Areas
 BT2   Sites/Areas
BT1   Wildlife Refuges
 BT2   Habitats
SO   Endangered Species
DEF Areas in which the Director has determined that certain waterborne activity would result in the taking of one or more manatees, or that certain waterborne activity must be restricted to prevent the taking of one or more manatees, including but not limited to a taking by harassment.

MANATEE SANCTUARIES
(ESA; CFR)
DA   October 12, 1990
BT1   Manatee Protection Areas
 BT2   Sites/Areas
SO   Endangered Species
DEF Areas in which the Director has determined that any waterborne activity would result in a taking of one or more manatees, including but not limited to a taking by harassment.

MANDATORY CLASS I FEDERAL AREAS
(CAA; CFR)
DA   November 15, 1990
BT1   Federal Class I areas
 BT2   Air Quality Control Regions
 BT3   Sites/Areas
 BT2   Federal Facilities
 BT3   Facilities
RT   Integral Vista
RT   Significant Impairment
SO   Air Pollution
DEF (CFR) Areas identified in 40 CFR 81, Subpart D. (CAA) Federal areas which may not be designated as other an class I under 42 USC 7491 Sec. 169A.

MANDATORY STANDARDS
(Doe Order 5480.4)
DA   October 12, 1990
BT1 Standards
 BT2 Codes, Standards, and
     Regulations
RT Field Level Exemptions
RT Permanent Exemptions
RT Temporary Exemptions
SO Management
SO Standards
DEF (DOE Order 5480.4) Those
    standards of this Order adopted by
    DOE that define the minimum
    requirements that DOE and its
    contractors must comply with to
    the extent they apply to the
    activities being conducted.

MANGANESE 54
(EDB)
DA   March 29, 1991
BT1 Days Living Radioisotopes
 BT2 Radionuclides
  BT3 CERCLA Hazardous Substances
   BT4 Hazardous Substances
  BT3 Nuclides
SO Radiation

MANIFEST DOCUMENT NUMBER
(SWDA; RCRA)
DA October 12, 1990
RT Manifests
RT Reports
SO Wastes
DEF The U.S. EPA twelve-digit
    identification number assigned to
    the generator plus a unique
    five-digit document number
    assigned to the Manifest by the
    generator for recording and
    reporting purposes.

MANIFESTS
(RCRA; ANL)
DA May 24, 1991
RT Hazardous Wastes
RT Manifest Document Number
RT Transportation
SO Management
SO Wastes
DEF (RCRA) The forms used for
    identifying the quantity,
    composition, and the origin,
    routing, and destination of
    hazardous waste during
    transportation from the point of
    generation to the point of disposal,
    treatment, or storage.

MANIPULATOR
(2578 MANIPULATOR)
DA January 3, 1991
BT1 Equipment/Parts - Nuclear (DOE
    FRASE Vocabulary)
 BT2 Equipment
 BT2 Reactor Components
NT1 Electromechanical Manipulator(s)

RT Manipulator Tape
SO DOE FRASE VOCABULARY

MANIPULATOR TAPE
(2333 MANIP TAPE)
DA December 11, 1990
BT1 Equipment/Parts - Material
    Handling (DOE FRASE
    Vocabulary)
 BT2 Equipment
RT Manipulator
SO DOE FRASE VOCABULARY

MANLIFT(S)
(2334 MANLIFT)
DA December 10, 1990
BT1 Hoisting Apparatus
 BT2 Material Handling Device
  BT3 Devices
  BT3 Equipment/Parts - Material
      Handling (DOE FRASE
      Vocabulary)
   BT4 Equipment
SO DOE FRASE VOCABULARY

MANNED CONTROL CENTERS
(TSCA; CFR)
DA   October 19, 1990
BT1 Centers
SO Hazardous Materials
DEF (CFR) Electrical power distribution
    control rooms where the operating
    conditions of a PCB Transformer
    are continuously monitored during
    the normal hours of operation (of
    the facility), and, where the duty
    engineers, electricians, or other
    trained personnel have the
    capability to deenergize a PCB
    Transformer completely within 1
    minute of the receipt of a signal
    indicating abnormal operating
    conditions such as an
    overtemperature condition or
    overpressure condition in a PCB
    Transformer.

MANUAL REACTOR SCRAM
(1553 MANUAL REACT)
DA November 29, 1990
SY Action Charlie
BT1 Nature of Property Damage
BT1 Scram
 BT2 Reactor Shutdown
RT Automatic Reactor Scram
RT Nuclear Facility
RT Unscheduled Shutdown
RT Violations
SO DOE FRASE VOCABULARY

MANUFACTURE
(CERCLA; TSCA; SDWA; CFR; USC;
 ESH)
DA October 12, 1990
SO Compensation and Liability
SO Environmental Management
SO Hazardous Materials
DEF (CFR) Activity associated with
    manufacturing for commercial

purposes. (ES&H Audit Manual)
Manufacture means to produce,
manufacture, or import into the
customs territory of the United
States for commercial purposes.
[Regarding Toxic Chemicals] To
produce, prepare, import, or
compound a toxic chemical.
Manufacture also applies to a toxic
chemical that is produced
coincidentally during the
manufacture, processing, use, or
disposal of another chemical or
mixture of chemicals, including a
toxic chemical that is separated
from that other chemical or mixture
of chemicals as a byproduct, and
a toxic chemical that remains in
that other chemical or mixture of
chemicals as an impurity.

MANUFACTURERS FORMULATION
(EPA)
DA October 12, 1990
BT1 Formulation
SO Environmental Protection Agency
    Glossary
DEF (EPA) A list of substances or
    component parts as described by
    the maker of a coating, pesticide.

MANUFACTURING PROCESSES
DA February 26, 1991
BT1 Processes
NT1 Fabricating
NT1 Magneforming
NT1 Pouring
NT1 Producing (Special Nuclear
    Materials)
NT1 Roasting
NT1 Side-Stream Extraction
RT Acid Deposition
RT Acid Rain

MAP
DA May 15, 1991
SEE Mitigation Action Plans
SO Acronyms

MAPS
(SSDC)
DA October 12, 1990
RT Bench Marks
RT Diagrams
SO System Safety Development Center
    Glossary
DEF (SSDC) Drawings used to illustrate
    the physical relationships of
    elements of people, equipment,
    materials and environmental
    structures associated with an
    accident or incident.

MARGIN OF CONTROL (CRITICALITY)
DA January 8, 1991
SF *MOC*

MARINE SANITATION DEVICES
(EPA)

DA   October 12, 1990
BT1 Devices
SO   Environmental Protection Agency
     Glossary
DEF (EPA) Any equipment installed on
     board a vessel to receive, retain,
     treat, or discharge sewage and
     any process to treat such sewage.

MARKET/MARKETERS
(TSCA; CFR)
DA   October 12, 1990
SO   Hazardous Materials
DEF (CFR) The processing or
     distributing in commerce, or the
     persons who process or distribute
     in commerce, used oil fuels to
     burners or other marketers, and
     may include the generators of the
     fuel if they market the fuel directly
     to the burner.

MARKS
(TSCA; CFR; ESH)
DA   October 12, 1990
BT1 Labeling
 BT2 Labels
RT   Name of Contents
RT   Polychlorinated Biphenyls
RT   Proper Shipping Name
SO   Hazardous Materials
DEF (CFR) The descriptive name,
     instructions, cautions, or other
     information applied to PCBs and
     PCB Items, or other objects
     subject to these regulations.

MARSHES
(EPA)
DA   October 12, 1990
BT1 Wetlands
 BT2 Sites/Areas
 BT2 Surface Water Resources
  BT3 Natural Resources
NT1 Tidal Marshes
RT   Channelization
RT   Surface Waters
RT   Swamps
SO   Environmental Protection Agency
     Glossary
DEF (EPA) A type of wetland that does
     not accumulate appreciable peat
     deposits and is dominated by
     herbaceous vegetation. Marshes
     may be either freshwater or
     saltwater and tidal or nontidal.

**MARTIN MARIETTA ENERGY
SYSTEMS**
DA   January 8, 1991
SF   MMES
BT1 Companies
 BT2 Commercial Organizations
  BT3 Organizations
BT1 DOE Contractors
 BT2 Potentially Responsible Parties
RT   Oak Ridge Operations Office
RT   Oak Ridge Gaseous Diffusion Plant
RT   Oak Ridge National Laboratory

RT   Portsmouth Gaseous Diffusion
     Plant
RT   Waste Information Network
RT   Y-12 Plant

MASON
(0641 MASON)
DA   November 28, 1990
BT1 Repair/Construction Personnel
 BT2 Occupations
 BT2 Personnel
SO   DOE FRASE VOCABULARY

**MASON AND HANGER-SILAS MASON
CO.**
DA   January 11, 1991
BT1 Companies
 BT2 Commercial Organizations
  BT3 Organizations
BT1 DOE Contractors
 BT2 Potentially Responsible Parties
RT   Albuquerque Operations Office

MASS ATTENUATION COEFFICIENT
(IAEA)
DA   October 12, 1990
BT1 Measurements
SO   Radiation
DEF (IAEA) For a material for uncharged
     ionizing particles, the quotient of
     dN/N by pdl, where dN/N is the
     fraction of particles that
     experience interactions in
     traversing a distance dl in a
     material of density p.

MASS CONCRETE
(DOE Order 6430.1A)
DA   October 12, 1990
SO   Construction
DEF (DOE Order 6430.1A) A large
     volume of cast-in-place concrete
     with dimensions large enough to
     require that measures be taken to
     cope with the generation of heat
     and attendant volume change and
     to minimize cracking.

**MASS ENERGY ABSORPTION
COEFFICIENT**
(IAEA)
DA   October 12, 1990
BT1 Measurements
RT   Bremsstrahlung
RT   Mass Energy Transfer Coefficient
SO   Radiation
DEF (IAEA) For a material for uncharged
     ionizing particles the Mass Energy
     Absorption Coefficient is the
     product of the mass energy
     transfer coefficient and $(1 - g)$,
     where g is the fraction of the
     energy of secondary charged
     particles that is lost to
     bremsstrahlung in the material.

**MASS ENERGY TRANSFER
COEFFICIENT**
(IAEA)

DA   October 12, 1990
BT1 Measurements
RT   Mass Energy Absorption Coefficient
SO   Radiation
DEF (IAEA) For a material for uncharged
     ionizing particles, the quotient of
     dE_{tr}/EN by pdl, where E is the
     energy of each particle (excluding
     rest energy), N is the number of
     particles, and dE_{tr}/EN is the
     fraction of incident particle energy
     that is transferred to kinetic energy
     of charged particles by interactions
     in traversing a distance dl in the
     material of density p.

**MASSACHUSETTS INSTITUTE OF
TECHNOLOGY**
DA   January 8, 1991
SF   MIT
BT1 Universities
 BT2 Educational Organizations
  BT3 Organizations

MATERIAL ACCESS AREAS
(DOE Order 6430.1A)
DA   October 12, 1990
BT1 Sites/Areas
RT   Protected Areas
SO   Construction
DEF (DOE Order 6430.1A) Areas that
     contain a Category I quantity of
     special nuclear material and are
     specifically defined by physical
     barriers, located within a protected
     area, and subject to specific
     access controls.

MATERIAL BALANCE AREA
(DOE Order 6430.1A)
DA   October 12, 1990
SO   Construction
DEF (DOE Order 6430.1A) A subsidiary
     account of a facility designed to
     establish accountability and to
     localize inventory differences.

MATERIAL HANDLING ACTIVITY
(1237 MH ACTIVITY)
DA   November 28, 1990
BT1 Activity Types (DOE FRASE
     Vocabulary)
 BT2 Activities
SO   DOE FRASE VOCABULARY

MATERIAL HANDLING DEVICE
(2335 MH DEVICE)
DA   December 10, 1990
BT1 Devices
BT1 Equipment/Parts - Material
     Handling (DOE FRASE
     Vocabulary)
 BT2 Equipment
NT1 Conveyor(s)
NT1 Crane(s)
 NT2 Boom Crane
 NT2 Bridge Crane
 NT2 Mobile Crane
 NT2 Overhead Crane

BT-Broader Term   NT-Narrower Term   RT Related Term

NT1 Earth Moving Equipment
NT2 Dredge(s)
NT1 Hand Truck(s)
NT2 Cart(s)
NT2 Dolly
NT1 Hoisting Apparatus
NT2 Derrick
NT2 Elevator
NT2 Forklift(s)
NT2 Hoist(s)
NT3 Air Hoist
NT2 Lift Bucket
NT2 Manlift(s)
NT2 Scissor Lift
SO DOE FRASE VOCABULARY

MATERIAL NONCONFORMANCE REPORT
DA January 8, 1991
SF *MNCR*
BT1 Reports

MATERIAL SAFETY DATA SHEETS
(CERCLA; CFR; USC; EPA; EMER)
DA October 12, 1990
SF *MSDS*
RT Occupational Safety and Health Administration
RT Safety Guides
SO Compensation and Liability
SO Emergency Preparedness
SO Environmental Management
SO Environmental Protection Agency Glossary
DEF (CFR) Compilation of information required under the OSHA Communication Standard on the identity of hazardous chemicals, health, and physical hazards, exposure limits, and precautions. Section 311 of SARA (Superfund Amendments and Reauthorization Act) required facilities to submit MSDSs under certain circumstances. The sheet required to be developed is under 29 CFR 1910.1200(g).

MATERIALS
DA February 4, 1991
NT1 Asbestos Materials
NT2 Asbestos-containing Materials
NT3 Friable Asbestos-Containing Materials
NT2 Asbestos-Containing Waste Materials
NT3 Asbestos Tailings
NT3 Asbestos Waste from Control Devices
NT2 Friable Asbestos Materials
NT2 Particulate Asbestos Materials
NT1 By-products
NT2 By-Product Materials
NT3 Uranium By-Product Materials
NT1 Cover Materials
NT2 Daily Cover
NT2 Final Cover
NT2 Intermediate Cover
NT1 Dielectric Materials
NT1 Discarded Materials

NT1 Dredged Materials
NT1 Fill Materials
NT1 Fissionable Materials
NT1 Hazardous Materials
NT2 Corrosive Materials
NT2 Fissile Materials
NT2 Hazardous Wastes
NT3 Extremely Hazardous Wastes
NT3 Incompatible Wastes
NT3 Listed Wastes
NT3 Piles (Wastes)
NT4 Disposal Areas
NT3 State Hazardous Wastes
NT2 Moderators (Nuclear)
NT2 Residues
NT3 Wood Residues
NT1 Holdup (Nuclear Material)
NT1 In-Process Materials
NT1 In-Use Materials
NT1 Irritating Materials
NT1 Licensed Materials
NT2 Source Materials
NT2 Special Nuclear Materials
NT3 Special Nuclear Material Scrap
NT3 Special Nuclear Material of Low Strategic Significance
NT3 Strategic Special Nuclear Materials
NT1 Other Regulated Material
NT2 Other Regulated Material - A
NT2 Other Regulated Material - B
NT2 Other Regulated Material - C
NT2 Other Regulated Material - D
NT2 Other Regulated Material - E
NT1 Oxidizing Materials
NT1 Pyrophoric Materials
NT2 Pyrophoric Liquids
NT1 Radioactive Materials
NT2 Airborne Radioactive Materials
NT2 Low Specific Activity Materials
NT2 Naturally Occurring Radioactive Material
NT2 Normal Form Radioactive Materials
NT2 Residual Radioactive Materials
NT2 Special Form Radioactive Materials
NT1 Radioactive Substances
NT1 Reactor Materials
NT2 Nuclear Poisons
NT1 Recovered Materials
NT1 Recycled Materials
NT1 Shielding Materials
NT1 Solids
NT2 Dissolved Solids
NT2 Flammable Solids
NT2 Precipitates
NT2 Settleable Solids
NT2 Total Dissolved Solids
NT2 Total Suspended Solids
NT1 Special Form
NT1 Spent Materials
NT1 Spontaneously Combustible Materials
NT1 Transuranic (TRU) Contaminated Materials
NT1 Virgin Materials
NT1 Water Reactive Materials

MATERIALS BALANCE AREA
DA January 8, 1991
SF *MBA*

MATERIALS TEST REACTOR
DA January 8, 1991
SF *MTR*
BT1 Reactors

MATRICES
(SSDC)
DA October 12, 1990
SO System Safety Development Center Glossary
DEF (SSDC) Any multi-dimensional classification or chart.

MATRIX ORGANIZATIONS
(SSDC)
DA October 12, 1990
BT1 Organizations
RT Project Managers
SO System Safety Development Center Glossary
DEF (SSDC) Organizations wherein a project manager temporarily borrows the talent he needs on the project from a functional department. The project manager supervises the people borrowed during the loan period, not the functional department head.

MATRIX/SPIKE-DUPLICATE ANALYSIS
(ESH)
DA October 12, 1990
BT1 Analyses
SO Quality Assurance
DEF (ESH) In matrix/spike duplicate analysis, predetermined quantities of stock solutions of certain analytes are added to a sample matrix prior to sample extraction/digestion and analysis. Samples are split into duplicate, spike and analyzed. Percent recoveries are calculated for each of the analytes detected. The relative percent difference between the samples is calculated and used to assess analytical precision. The concentration of the spike should be at the regulatory standard level or the estimated or actual method quantification limit.

MAXIMAL EFFECTIVE PRESSURE
(DOE Order 6430.1A)
DA October 12, 1990
BT1 Pressure
RT Peak Positive Incident Pressure
SO Construction
DEF (DOE Order 6430.1A) The highest of: (1) the peak incident pressure, (2) the incident plus dynamic pressure, or (3) the reflected pressure.

MAXIMUM CONCENTRATION LIMITS
DA January 8, 1991
SF *MCLs*
BT1 Limits

MAXIMUM CONTAMINANT LEVEL GOAL
(SDWA; CFR; ESH)
DA October 12, 1990
SF *MCLG*
RT Contaminants
RT Maximum Contaminant Levels
SO Water Pollution
DEF (CFR) The maximum level of a contaminant in drinking water at which no known or anticipated adverse effect on the health of persons would occur, and which allows an adequate margin of safety. Maximum contaminant level goals are nonenforceable health goals.

MAXIMUM CONTAMINANT LEVELS
(SDWA; CFR; USC; ESH)
DA October 12, 1990
SF *MCL*
NT1 Secondary Maximum Contaminant Levels
NT1 Secure Maximum Contaminant Levels
RT Contaminants
RT Maximum Contaminant Level Goal
RT Recommended Maximum Contaminant Level
RT Sanitary Surveys
SO Environmental Management
SO Environmental Protection Agency Glossary
SO Water Pollution
DEF (CFR) The maximum permissible levels of a contaminant in water which is delivered to the free flowing outlet of the ultimate user of a public water system, except in the case of turbidity where the maximum permissible level is measured at the point of entry to the distribution system. Contaminants added to the water under circumstances controlled by the user, except those resulting from corrosion of piping and plumbing caused by water quality, are excluded from this definition.

MAXIMUM CREDIBLE ACCIDENT
DA January 8, 1991
SF *MCA*
BT1 Credible Accidents
BT2 Accidents
RT Health Hazards

MAXIMUM CREDIBLE LOSS
(Doe Order 5480.7)
DA October 12, 1990
BT1 Losses
RT Maximum Possible Fire Loss

SO Fires
DEF (DOE Order 5480.7) The maximum loss that could occur from a combination of events resulting from a single fire. Considerable judgment is required to evaluate the full range of potential losses, but in general, readily conceivable fires in sensitive areas are considered. Examples are power wiring failures in cable trays, flammable liquid spills, and high-value parts storage areas or combustible exposures to sensitive machines. Any installed fire protection systems are assumed to function as designed. Due to the uncertainties of predicting human action, the effect of emergency response is generally omitted except for post-fire actions such as salvage work, shutting down water systems, and restoring production.

MAXIMUM FORESEEABLE LOSS
(SSDC)
DA October 12, 1990
SY Maximum Possible Loss
SF *MFL*
BT1 Losses
RT Maximum Probable Loss
SO System Safety Development Center Glossary
DEF (SSDC) The largest loss that could possibly happen under the worst circumstances.

MAXIMUM NORMAL OPERATING PRESSURE
(DOE Order 5480.3; ESH)
DA October 12, 1990
BT1 Pressure
RT Containment Vessels
SO Environmental Management
SO Hazardous Materials
DEF (DOE Order 5480.3) The maximum gauge pressure that is expected to develop in the containment vessel under the normal conditions of transport.

MAXIMUM PERMISSIBLE CONCENTRATION
(NIH; NCRP)
DA October 12, 1990
SF *MPC*
RT Critical Organs
SO Radiation
DEF (NIH; NCRP) The concentration in air or water that would lead to an amount of radionuclide in the critical organ that would just deliver the maximum permissible dose rate to that organ. A phrase used in the conventional system of units.

MAXIMUM PERMISSIBLE DOSE
(NIH)
DA October 12, 1990
SF *MPD*
BT1 Doses
RT Dose Limit
SO Radiation
DEF (NIH) Maximum dose of radiation that may be received by persons working with ionizing radiation, that will produce no detectable damage over the normal life span.

MAXIMUM POSSIBLE FIRE LOSS
(Doe Order 5480.7)
DA October 12, 1990
BT1 Maximum Possible Loss
BT2 Losses
RT Maximum Credible Loss
SO Fires
DEF (DOE Order 5480.7) The maximum possible loss that could occur in a single fire area assuming the failure of both automatic and manual fire extinguishing actions.

MAXIMUM POSSIBLE LOSS
(SSDC)
DA October 12, 1990
SY Maximum Foreseeable Loss
SF *MPL*
BT1 Losses
NT1 Maximum Possible Fire Loss
SO System Safety Development Center Glossary
DEF (SSDC) The largest loss that could possibly happen under the worst circumstances.

MAXIMUM PROBABLE FLOOD
(DOE Order 6430.1A)
DA October 12, 1990
RT Design Basis Floods
RT High Tide Line
RT Watershed
SO Construction
DEF (DOE Order 6430.1A) A hypothetical flood (peak discharge, volume, and hydrograph shape) that is considered to be the most severe reasonably possible, based on comprehensive hydro-meteorological application of probable maximum precipitation and other hydrological factors favorable for maximum flood runoff such as sequential storms and snowmelts.

MAXIMUM PROBABLE LOSS
(SSDC)
DA October 12, 1990
BT1 Losses
RT Maximum Foreseeable Loss
SO System Safety Development Center Glossary

BT-Broader Term

NT-Narrower Term

RT Related Term

MAXIMUM TOTAL TRIHALOMETHANE POTENTIAL (MTP)
(SDWA; CFR; ESH)
DA October 19, 1990
RT Trihalomethane
SO Water Pollution
DEF (CFR) The maximum concentration of total trihalomethanes produced in a given water containing a disinfectant residual after 7 days at a temperature of 25 deg. C or above.

MB
DA October 12, 1990
SEE Machine Basin
SO Acronyms

MBA
DA October 12, 1990
SEE Materials Balance Area
SO Acronyms

MBO
DA October 12, 1990
SEE Management by Objective
SO Acronyms

MCA
DA October 12, 1990
SEE Maximum Credible Accident
SO Acronyms

MCC
DA October 12, 1990
SEE Motor Control Center
SO Acronyms

MCL
DA October 12, 1990
SEE Maximum Contaminant Levels
SO Acronyms

MCLG
DA October 12, 1990
SEE Maximum Contaminant Level Goal
SO Acronyms

MCLs
DA October 12, 1990
SEE Maximum Concentration Limits
SO Acronyms

10 MDLM
DA October 12, 1990
SEE 10-minute Delay Line Gamma Monitor
SO Acronyms
SO Nuclear Facilities Incident Database

MDNR
DA October 12, 1990
SEE Missouri Department of Natural Resources
SO Acronyms

MEAN
(SSDC)
DA November 16, 1990
RT Mean Annual Loss
RT Standard Deviation
SO System Safety Development Center Glossary
DEF (SSDC) The arithmetic average of all observations or values.

MEAN ANNUAL LOSS
(SSDC)
DA October 12, 1990
BT1 Losses
RT Mean
SO System Safety Development Center Glossary
DEF (SSDC) The average loss per year over a period of years. Sum of losses during a period, divided by years in the period.

MEAN TIME BETWEEN FAILURES
DA January 8, 1991
SF MTBF
BT1 Time Designations

MEAN TIME TO REPAIR
DA January 8, 1991
SF MTTR
BT1 Time Designations

MEASUREMENT AND TEST EQUIPMENT
DA January 8, 1991
SF MTE
BT1 Equipment

MEASUREMENTS
DA February 27, 1991
NT1 Accuracy
NT1 Activity (Nuclear)
 NT2 Gross Alpha Particle Activity
 NT2 Gross Beta Particle Activity
 NT2 Specific Activity
  NT3 Low Specific Activity
NT1 Activity-Median Aerodynamic Diameter
NT1 Air Changes Per Hour
NT1 Biochemical Oxygen Demand
NT1 Biological Oxygen Demand
NT1 Body Content
NT1 Carrying Capacity
NT1 Cation Exchange Capacity
NT1 CBOD5
NT1 Chemical Oxygen Demand
NT1 Coefficient of Haze
NT1 Control Width
NT1 Count (Radiation Measurement)
NT1 Critical Mass
NT1 Cumulative Working Level Months
NT1 Deposition
NT1 Derived Concentration Guide
NT1 Difference Deviation
NT1 Dose Commitments
NT1 Energy Fluence
NT1 Energy Flux
NT1 Exposure Point Concentration
NT1 Extreme Value

NT1 Extreme Value Projection
NT1 Flash Point
NT1 Flow Weighted Average (Delta T)
NT1 Gross Weight
NT1 Half Value Layer (Half Thickness)
NT1 Heat Input
NT1 Hood Capture Efficiency
NT1 Inhabited Building Distance
NT1 Intake
 NT2 Acceptable Daily Intake
 NT2 Annual Limit on Intake
 NT2 Ingestion
 NT2 Inhalation
 NT2 Radioactive Nuclide Intake
NT1 Kinetic Rate Coefficient
NT1 Linear Energy Transfer
NT1 Local Magnitude (M_L)
NT1 Mass Attenuation Coefficient
NT1 Mass Energy Absorption Coefficient
NT1 Mass Energy Transfer Coefficient
NT1 Method Quantification Limits
NT1 Moment Magnitude
NT1 MQL
NT1 Net Explosive Weight
NT1 Net Weight
NT1 Particle Fluence
NT1 Particle Flux
NT1 pH
 NT2 Background Soil Ph
NT1 Precision
NT1 Process Weight
NT1 Process Waste Assessment
NT1 Quantity Distances
 NT2 Intraline Separation (Barricaded)
 NT2 Intraline Separation (Unbarricaded)
 NT2 Magazine Separation
NT1 Radioactivity Decay Constant
NT1 Relative Importance Measure
NT1 Specific Ionization
NT1 Standard Temperature
NT1 Total Suspended Particulates
NT1 Total Dissolved Solids
NT1 Total Trihalomethanes (TTHM)
NT1 Total Mass Stopping Power
NT1 Total Indicated Runout
NT1 Total Exposure Points
NT1 Total Suspended Solids
NT1 Total Recordable Cases
NT1 Ullage
NT1 Vulnerable Zones Radius
RT Calibration
RT Doses
RT Pressure
RT Rates
RT Ratios
RT Standard Curve
RT Standard Deviation
RT Time Designations
DEF (WTID) Areas, quantities, degrees, or capacities obtained by measuring.

MECHANIC/REPAIRER
(0610 MECHANIC)
DA November 28, 1990
BT1 Repair/Construction Personnel
 BT2 Occupations
 BT2 Personnel
RT Garage

RT Maintenance Area
RT Maintenance Shop
SO DOE FRASE VOCABULARY

MECHANICAL AERATION
(EPA)
DA October 12, 1990
BT1 Aeration
 BT2 Biological Processes
  BT3 Processes
SO Environmental Protection Agency
    Glossary
DEF (EPA) Use of mechanical energy to
    inject air into water to cause a
    waste stream to absorb oxygen.

MECHANICAL DAMAGE
(1534 MECH DAMAGE)
DA November 29, 1990
BT1 Damage
 BT2 Nature of Property Damage
SO DOE FRASE VOCABULARY

MECHANICAL TURBULENCE
(EPA)
DA October 12, 1990
SO Environmental Protection Agency
    Glossary
DEF (EPA) Random irregularities of fluid
    motion in air caused by buildings
    or mechanical, nonthermal,
    processes.

MEDIA
(EPA)
DA October 12, 1990
BT1 Environment
SO Environmental Protection Agency
    Glossary
DEF (EPA) Specific environments – air,
    water, soil – which are the subject
    of regulatory concern and
    activities.

MEDIAN LETHAL CONCENTRATION
(EMER)
DA February 1, 1991
RT Exposure
SO Emergency Preparedness
DEF (EMER) Concentration level at
    which 50 percent of the test
    animals die when exposed by
    inhalation for a specified time
    period.

MEDIAN LETHAL DOSE
(EMER)
DA February 1, 1991
BT1 Doses
SO Emergency Preparedness
DEF (EMER) Dose at which 50 percent
    of test animals die following
    exposure. Dose is usually given in
    milligrams per kilogram of body
    weight of the test animal.

MEDICAL EXPOSURE
(IAEA)

DA October 12, 1990
BT1 Exposure
RT Medical Treatment
SO Radiation
DEF (IAEA) Exposure of individuals
    resulting from their medical
    examination or treatment involving
    radiation.

MEDICAL TREATMENT
(SSDC)
DA October 12, 1990
RT Approved Medical Practitioners
RT First Aid
RT Medical Exposure
RT Triage
SO System Safety Development Center
    Glossary
DEF (SSDC) Medical treatment includes
    treatment (other than first aid)
    administered by a physician or by
    registered professional personnel
    under the standing orders of a
    physician. Medical treatment does
    not include first aid treatment
    (one- time treatment and
    subsequent observation of minor
    scratches, cuts, burns, splinters,
    etc., which do not ordinarily
    require professional care) even
    though such care is provided by a
    physician or registered
    professional personnel.

MEDICAL WASTES
(SWDA; RCRA; USC)
DA October 12, 1990
BT1 Solid Wastes
 BT2 Wastes
RT Infectious Wastes
SO Wastes
DEF (USC) Solid wastes that are
    generated in the diagnosis,
    treatment, or immunization of
    human beings or animals, in
    research pertaining thereto, or in
    the production or testing of
    biologicals. This term does not
    include any hazardous waste
    identified or listed under subtitle C
    [42 USCS 6921 et seq.] or any
    household waste as defined in
    regulations under subtitle C [42
    USCS 6921 et seq.].

MEDIUM DISCHARGES
(CERCLA; CFR)
DA October 12, 1990
BT1 Discharges
BT1 Oil Spills
BT1 Size Classes of Discharges
DEF (CFR) Discharge of 1,000 to 10,000
    gallons of oil to the inland waters
    or discharges of 10,000 to
    100,000 gallons of oil to the
    coastal waters. Any oil discharge
    that poses a substantial threat to
    the public health or welfare or
    results in critical public concern
    shall be classified as a major

discharge regardless of this
quantitative measure.

MEGAWATT THERMAL
DA January 8, 1991
SF *MWt*
BT1 Units of Measure

MEGGER TESTING
(NFI)
DA October 12, 1990
RT Electrical Equipment
SO Nuclear Facilities Incident Database
SO Radiation
DEF (NFI) Testing for electrical faults.

MEMBERS OF THE PUBLIC
(DOE Order 5400.5)
DA October 16, 1990
RT Comment Periods
RT Critical Group
RT Public Doses
DEF (DOE Order 5400.5) Persons who
    are not occupationally associated
    with the DOE facility or operations,
    i.e., persons whose assigned
    occupational duties do not require
    them to enter the DOE site.

MEMORANDUM OF UNDERSTANDING
DA January 8, 1991
SF *MOU*

MENTAL DISORDERS
(1396 MENTAL DISOR)
DA November 28, 1990
BT1 Illnesses
RT Hospital
SO DOE FRASE VOCABULARY

MEPS
DA October 12, 1990
SEE Multimedia Environmental Pollutant
    Assessment System
SO Acronyms

MERCURY
(CAA; CFR)
DA October 12, 1990
SF *Hg*
BT1 Hazardous Constituents
BT1 Heavy Metals
RT Mercury Ore
SO Air Pollution
SO Environmental Protection Agency
    Glossary
DEF (CFR) The element mercury, a
    heavy metal, that can accumulate
    in the environment and is highly
    toxic if breathed or swallowed.
    Includes mercury in particulates,
    vapors, aerosols, and compounds.

MERCURY CHLOR-ALKALI CELLS
(CAA; CFR)
DA October 12, 1990
RT Cell Rooms
RT Denuders

BT-Broader Term

NT-Narrower Term

RT Related Term

RT   Hydrogen Gas Streams
RT   Mercury Chlor-alkali Electrolyzers
SO   Air Pollution
DEF (CFR) Devices that are basically
     composed of an electrolyzer
     section and a denuder
     (decomposer) section and utilize
     mercury to produce chlorine gas,
     hydrogen gas, and alkali metal
     hydroxide.

MERCURY CHLOR-ALKALI ELECTROLYZERS
(CAA; CFR)
DA   October 12, 1990
RT   End Boxes
RT   Mercury Chlor-alkali Cells
SO   Air Pollution
DEF (CFR) Electrolytic devices that are
     part of a mercury chlor-alkali cell
     and utilize a flowing mercury
     cathode to produce chlorine gas
     and alkali metal amalgam.

MERCURY ORE
(CAA; CFR)
DA   October 12, 1990
RT   Mercury
RT   Mercury Ore Processing Facilities
SO   Air Pollution
DEF (CFR) A mineral mined specifically
     for its mercury content.

MERCURY ORE PROCESSING FACILITIES
(CAA; CFR)
DA   October 12, 1990
BT1 Facilities
RT   Condenser Stack Gases
RT   Mercury Ore
SO   Air Pollution
DEF (CFR) Facilities processing mercury
     ore to obtain mercury.

METABOLITES
(EPA)
DA   October 12, 1990
SO   Environmental Management
SO   Environmental Protection Agency
     Glossary
DEF (EPA) Substances produced in or
     by biological processes and
     derived from a pesticide.

METAL SHOE COVER
(2673 METAL SHOE C)
DA   January 3, 1991
BT1 Shoe Cover(s)
 BT2 Foot Protection
  BT3 Personal Protective Equipment
   BT4 Equipment/Parts - Personal
        Protective (DOE FRASE
        Vocabulary)
    BT5 Equipment
SO   DOE FRASE VOCABULARY

METATARSAL PROTECTION
(2674 METATARSAL P)
DA   January 3, 1991

BT1 Foot Protection
 BT2 Personal Protective Equipment
  BT3 Equipment/Parts - Personal
       Protective (DOE FRASE
       Vocabulary)
   BT4 Equipment
RT   Toe(s)
SO   DOE FRASE VOCABULARY

METHANE
(EPA)
DA   October 12, 1990
BT1 Gases
BT1 Hydrocarbons
 BT2 Organic Chemicals
  BT3 Chemical Substances
RT   Agricultural Pollution
SO   Environmental Protection Agency
     Glossary
DEF (EPA) A colorless, nonpoisonous,
     flammable gas created by
     anaerobic decomposition of
     organic compounds.

METHOD 18
(EPA)
DA   October 12, 1990
SO   Environmental Protection Agency
     Glossary
DEF (EPA) An EPA test method which
     uses gas chromatographic
     techniques to measure the
     concentration of individual volatile
     organic compounds in a gas
     stream.

METHOD 24
(EPA)
DA   October 12, 1990
BT1 Reference Methods
SO   Environmental Protection Agency
     Glossary
DEF (EPA) An EPA reference method to
     determine density, water content
     and total volatile content (water
     and VOC) of coatings.

METHOD 25
(EPA)
DA   October 12, 1990
BT1 Reference Methods
SO   Environmental Protection Agency
     Glossary
DEF (EPA) An EPA reference method to
     determine the VOC concentration
     in gas stream.

METHOD QUANTIFICATION LIMITS
(ESH)
DA   February 25, 1991
SY   MQL
BT1 Limits
BT1 Measurements
SO   Quality Assurance
DEF (ESH) The minimum concentrations
     of substances that can be
     measured and reported.

METRIC COORDINATOR
(DOE Order 5900.2)
DA   October 16, 1990
BT1 Personnel
RT   Metric Transition Committee
DEF (DOE Order 5900.2) A person
     designated by the Assistant
     Secretary for Environment to act
     as the Department's central point
     of contact for metrication matters.

METRIC SYSTEM
(Doe Order 5900.2)
DA   October 12, 1990
SY   International System of Units
BT1 Systems
RT   American National Metric Council
RT   Units of Measure
SO   Quality Assurance
DEF (DOE Order 5900.2) The
     International System of Units (SI)
     as established by the General
     Conference on weights and
     Measures in 1960 and as
     interpreted or modified for the
     United States by the Secretary of
     Commerce.

METRIC TRANSITION COMMITTEE
(Doe Order 5900.2)
DA   October 12, 1990
BT1 Committees
 BT2 Administrative Organizations
  BT3 Organizations
RT   Metric Coordinator
SO   Quality Assurance
DEF (DOE Order 5900.2) A committee
     with representatives from each
     major Headquarters element,
     established to develop a
     Department-wide policy. The
     representatives also serve as the
     metric coordinators for their
     organizations.

METRIC TRANSITION PLAN
(Doe Order 5900.2)
DA   October 12, 1990
BT1 Plans
RT   Programs
SO   Quality Assurance
DEF (DOE Order 5900.2) A summary
     description of the metrication
     objectives, the actions required to
     attain the stated objectives, and
     the actions underway or planned
     for this purpose consistent with
     national policy, interagency metric
     policy, Department policy, and
     organizational element policy.

METRICATION
(Doe Order 5900.2)
DA   October 12, 1990
NT1 Hard Conversion
NT1 Soft Conversion
RT   American National Metric Council
SO   Quality Assurance
DEF (DOE Order 5900.2) An activity

tending to increase the use of the International System of Units (SI). It may include metric training and the initiation or conversion to metric of new or existing measurement sensitive processes, software or hardware systems, and engineering standards.

METRICATION OPERATING COMMITTEE
(Doe Order 5900.2)
DA   October 12, 1990
BT1 Committees
 BT2 Administrative Organizations
 BT3 Organizations
SO Quality Assurance
DEF (DOE Order 5900.2) A committee of the Interagency Committee on Metric Policy, which serves as the vehicle for coordination of Federal interagency metrication activities, and recommends policy guidance to the parent committee. The Metrication Operating Committee is comprised of representatives from the major Federal departments and agencies who serve as their agencies' metric coordinators.

MFL
DA   October 12, 1990
SEE Maximum Foreseeable Loss
SO Acronyms

MFW
DA   October 12, 1990
SEE Main Feedwater
SO Acronyms

MFWCS
DA   October 12, 1990
SEE Main Feedwater and Condensate System
SO Acronyms

MG
DA   October 12, 1990
SEE Motor Generator
SO Acronyms

MG-MA
DA   October 12, 1990
SEE Motor Generator-Motor Alternator
SO Acronyms

mg/kg
DA   October 12, 1990
SEE milligram/killigram
SO Acronyms

mg/l
DA   October 12, 1990
SEE milligram/liter
SO Acronyms

mgallon
DA   October 12, 1990
SEE million gallons
SO Acronyms

MGD
DA   October 12, 1990
SEE Million Gallons Per Day
SO Acronyms

MgF₂
DA   October 12, 1990
SEE Magnesium Fluoride
SO Acronyms

MIAC
DA   October 12, 1990
SEE Maintenance Information and Control
SO Acronyms

MICROBES
(EPA)
DA   October 12, 1990
BT1 Microorganisms
 BT2 Organisms
RT Pathogens
SO Environmental Protection Agency Glossary
DEF (EPA) Microscopic organisms such as algae, animals, viruses, bacteria, fungus, and protozoa, some of which cause diseases.

MICROBIAL PESTICIDES
(EPA)
DA   October 12, 1990
BT1 Pesticides
 BT2 Hazardous Substances
SO Environmental Protection Agency Glossary
DEF (EPA) Microorganisms that are used to control a pest. They are of low toxicity to man.

MICROBURSTS
(SMRP)
DA   October 12, 1990
BT1 Downbursts
 BT2 Divergent Windstorms
 BT3 Act of Nature
SO Natural Phenomenon
DEF (SMRP) Small downbursts with horizontal scale of less than the square root of 10 miles (3.16 miles=5.1 km). The mature life of microbursts is only a few to 10 minutes, making their detection and warning extremely difficult. The maximum windspeed of microbursts is less than mid-F1 (90 mph). Damage caused by microbursts is limited mostly to outbuildings and mobile homes.

MICROCURIES
DA   January 8, 1991

SF   μCi
BT1 Units of Measure

MICROGRAMS/LITER
DA   January 8, 1991
BT1 Units of Measure

MICROORGANISMS
(EPA)
DA   October 12, 1990
BT1 Organisms
NT1 Bacteria
 NT2 Coliform Organisms
  NT3 Fecal Coliform Bacteria
 NT2 Legionella
 NT2 Recombinant Bacteria
NT1 Microbes
NT1 Pathogens
NT1 Viruses
RT Aerobic Treatment
RT Anaerobic Digestion
RT Biological Additives
RT Etiologic Agents
SO Environmental Protection Agency Glossary
DEF (EPA) Living organisms so small that individually they can usually only be seen through a microscope.

MICROROENTGEN/HOUR
DA   January 8, 1991
SF   mR/h
BT1 Radiation Units
 BT2 Units of Measure

MICROSEC LIVING RADIOISOTOPES
(EDB)
DA   March 29, 1991
BT1 Radionuclides
 BT2 CERCLA Hazardous Substances
 BT3 Hazardous Substances
 BT2 Nuclides
SO Radiation

MIDWEST RESEARCH INSTITUTE
DA   January 11, 1991
BT1 Institutes
 BT2 Research and Development Organizations
 BT3 Organizations
RT Chicago Operations Office

MIGRATION
(DOE Order 5480.14; ANL)
DA   May 24, 1991
BT1 Processes
RT Disposal Sites
SO Hazardous Materials
SO Management
DEF (DOE Order 5480.14) The movement of hazardous substances from the disposal site by means of air, surface water, or groundwater.

MILES
DA   October 12, 1990

BT-Broader Term   NT-Narrower Term   RT Related Term

SEE Multiple Integrated Laser
Engagement System
SO Acronyms

MILESTONES
(DOE Order 4700.1; ANL)
DA May 24, 1991
RT Project Managers
RT Projects
RT Project Planning and Control
SO Management
DEF (DOE 4700.1) Important or critical
events and/or activities that must
occur in the project cycle in order
to achieve the project objective(s).

MILITARY PERSONNEL
(0910 MILITARY)
DA November 28, 1990
BT1 Occupations
BT1 Personnel
SO DOE FRASE VOCABULARY

MILLIGRAM/KILLIGRAM
DA January 8, 1991
SF *mg/kg*
BT1 Units of Measure

MILLIGRAM/LITER
DA January 8, 1991
SF *mg/l*
BT1 Units of Measure

MILLING MACHINE
(2128 MILLING MACH)
DA December 10, 1990
BT1 Machines (DOE FRASE
Vocabulary)
BT2 Equipment
NT1 Ball Mill
NT1 Roll Mill
SO DOE FRASE VOCABULARY

MILLION GALLONS
DA January 8, 1991
SF *mgallon*

MILLION GALLONS PER DAY
(EPA)
DA October 12, 1990
SF *MGD*
BT1 Flow Rates
BT1 Units of Measure
SO Environmental Protection Agency
Glossary
DEF (EPA) A measure of water flow.

MILLIROENTGEN
(NIH)
DA October 12, 1990
SF *mR*
BT1 Units of Measure
RT Roentgen
SO Radiation
DEF (NIH) A submultiple of the roentgen
equal to one one-thousandth
(1/1000th) of a roentgen.

MILLISEC LIVING RADIOISOTOPES
(EDB)
DA March 29, 1991
BT1 Radionuclides
BT2 CERCLA Hazardous Substances
BT3 Hazardous Substances
BT2 Nuclides
SO Radiation

MINE
(1856 MINE)
DA December 10, 1990
BT1 Structures (DOE FRASE
Vocabulary)
SO DOE FRASE VOCABULARY

MINER/DRILLER
(0650 MINER;DRILLE)
DA November 28, 1990
BT1 Repair/Construction Personnel
BT2 Occupations
BT2 Personnel
RT Mining/Drilling Activity
SO DOE FRASE VOCABULARY

MINERAL OIL PCB TRANSFORMERS
(TSCA)
DA October 19, 1990
BT1 PCB Transformers
BT2 Electrical Equipment
BT3 Equipment
SO Hazardous Materials
DEF (CFR) Any transformers originally
designed to contain mineral oil as
the dielectric fluid and which have
been tested and found to contain
500 ppm or greater PCBs.

MINIMUM FRAGMENT DISTANCE
(DOE Order 5480.16)
DA October 16, 1990
RT Munitions
DEF (DOE Order 5480.16) The minimum
distance required for the protection
of personnel in the open, inhabited
buildings, and public traffic routes
from hazardous fragments.

**MINIMUM REQUIREMENTS AND
STANDARDS**
(Doe Order 5480.8)
DA October 12, 1990
BT1 Requirements
RT Occupational Medical Program
RT Policies
RT Standards
SO Industrial Hygiene
DEF (DOE Order 5480.8) The program
content necessary to satisfy the
policies and objectives of this
directive.

MINING WASTES
(SWDA; RCRA; CFR; ESH)
DA October 12, 1990
BT1 Wastes
SO Environmental Management
SO Wastes
DEF Residues resulting from the

extraction of raw materials from
the earth.

MINING/DRILLING ACTIVITY
(1251 MD ACTIVITY)
DA November 28, 1990
BT1 Activity Types (DOE FRASE
Vocabulary)
BT2 Activities
NT1 Strip Mining
RT Miner/Driller
SO DOE FRASE VOCABULARY

MINOR ACCIDENTS
(SSDC)
DA October 12, 1990
BT1 Accidents
SO System Safety Development Center
Glossary
DEF (SSDC) Any accidents which are
not serious accidents, or do not
have high potential for being
serious accidents.

MINOR DISCHARGES
(CERCLA; CFR)
DA October 12, 1990
BT1 Discharges
BT1 Oil Spills
BT1 Size Classes of Discharges
DEF (CFR) Discharges to the inland
waters of less than 1,000 gallons
of oil or discharges to the coastal
waters of less than 10,000 gallons
of oil. Any oil discharge that poses
a substantial threat to the public
health or welfare or results in
critical public concern shall be
classified as a major discharge
regardless of this quantitative
measure.

MINOR DRAINAGE
(CAA; RHA)
DA October 12, 1990
RT Watershed
SO Water Pollution
DEF The discharge of dredged or fill
material incidental to connecting
upland drainage facilities to waters
of the United States, adequate to
effect the removal of excess soil
moisture from upland croplands.
Construction and maintenance of
upland (dryland) facilities, such as
ditching and tiling, incidental to the
planting, cultivating, protecting, or
harvesting of crops.

MINOR RELEASES
(CERCLA; CFR)
DA October 12, 1990
BT1 Releases
BT1 Size Classes of Releases
DEF (CFR) Releases of a quantity of
hazardous substance(s),
pollutant(s), or contaminant(s) that
pose minimal threat to public

health or welfare or the
environment.

MINORITY INSTITUTIONS
DA   January 8, 1991
SF   *MIs*
BT1 Organizations

MINUTES LIVING RADIOISOTOPES
(EDB)
DA   March 29, 1991
BT1 Radionuclides
 BT2 CERCLA Hazardous Substances
  BT3 Hazardous Substances
 BT2 Nuclides
NT1 Cobalt 60
NT1 Thorium 232
SO   Radiation

MIs
DA   October 12, 1990
SEE Minority Institutions
SO   Acronyms

MISC AGRICULTURE EMPLOYEE
(0580 MISC AGRICUL)
DA   November 28, 1990
BT1 Agriculture Personnel
 BT2 Occupations
 BT2 Personnel
BT1 Misc Employee
 BT2 Occupations
 BT2 Personnel
SO   DOE FRASE VOCABULARY

MISC EMPLOYEE
(0990 MISC EMPLOYE)
DA   November 28, 1990
BT1 Occupations
BT1 Personnel
NT1 Misc Service Employee
NT1 Misc Agriculture Employee
NT1 Misc Repair/Construction
      Employee
NT1 Misc Precision/Production
      Employee
NT1 Misc Transport Employee
SO   DOE FRASE VOCABULARY

MISC PRECISION/PRODUCTION
EMPLOYEE
(0780 MISC PRECIS)
DA   November 28, 1990
BT1 Misc Employee
 BT2 Occupations
 BT2 Personnel
BT1 Precision/Production Personnel
 BT2 Occupations
 BT2 Personnel
SO   DOE FRASE VOCABULARY

MISC PROFESSIONAL
(0200 MISC PROF)
DA   November 28, 1990
BT1 Professional Personnel
 BT2 Occupations
 BT2 Personnel
SO   DOE FRASE VOCABULARY

MISC REPAIR/CONSTRUCTION
EMPLOYEE
(0660 MISC REPAIR)
DA   November 28, 1990
BT1 Misc Employee
 BT2 Occupations
 BT2 Personnel
BT1 Repair/Construction Personnel
 BT2 Occupations
 BT2 Personnel
RT   Construction Activity
SO   DOE FRASE VOCABULARY

MISC SERVICE EMPLOYEE
(0525 MISC SERVICE)
DA   November 28, 1990
BT1 Admin. Support/Clerical Employee
 BT2 Occupations
 BT2 Personnel
BT1 Misc Employee
 BT2 Occupations
 BT2 Personnel
SO   DOE FRASE VOCABULARY

MISC TRANSPORT EMPLOYEE
(0840 MISC TRANSPO)
DA   November 28, 1990
BT1 Misc Employee
 BT2 Occupations
 BT2 Personnel
BT1 Transport Personnel
 BT2 Occupations
 BT2 Personnel
RT   Transportation Activity
SO   DOE FRASE VOCABULARY

MISCELLANEOUS TECHNICIAN
(0390 MISC TECH)
DA   November 28, 1990
BT1 Technicians
 BT2 Professional Personnel
  BT3 Occupations
  BT3 Personnel
SO   DOE FRASE VOCABULARY

MISCELLANEOUS UNITS
DA   January 24, 1991
BT1 Hazardous Waste Management
      Units
 BT2 Corrective Action Management
       Units
  BT3 Sites/Areas
SO   Environmental Management
DEF (CFR) Hazardous waste
      management units where
      hazardous waste are treated,
      stored or disposed of and that is
      not a container, tank, surface
      impoundment, pile, land treatment
      unit, landfill, incinerator, boiler,
      industrial furnace, underground
      injection well with appropriate
      technical standards under 40 CFR
      Part 146, or unit eligible for a
      research, development, and
      demonstration permit under
      270.65.

MISFIRES
(Doe Order 5480.16)
DA   October 12, 1990
NT1 Hangfires
RT   Armorers
RT   Defective Firearm
RT   Firearms
RT   Hazards
SO   Firearms
DEF (DOE Order 5480.16) Any
      cartridge, missile, or rocket that
      does not properly fire when
      triggered.

MISHAPS
(SSDC)
DA   October 12, 1990
SY   Accidents
SO   System Safety Development Center
      Glossary
DEF (SSDC) A synonym for accident.
      Used by some government
      organizations, including NASA and
      DOD.

MISSISSIPPI POWER AND LIGHT
DA   January 8, 1991
SF   *MP&L*
BT1 Companies
 BT2 Commercial Organizations
  BT3 Organizations

MISSOURI DEPARTMENT OF
NATURAL RESOURCES
DA   January 8, 1991
SF   *MDNR*
BT1 State Agencies
 BT2 Agencies
  BT3 Administrative Organizations
   BT4 Organizations

MISTS
(EPA)
DA   October 12, 1990
SO   Environmental Protection Agency
      Glossary
DEF (EPA) Liquid particles measuring
      500 to 40 microns, that are formed
      by condensation of vapor. By
      comparison, "fog" particles are
      smaller than 40 microns.

MIT
DA   October 12, 1990
SEE Massachusetts Institute of
      Technology
SO   Acronyms

MITIGATION
(EPA)
DA   October 12, 1990
BT1 Amelioration
RT   Mitigation Action Plans
SO   Environmental Management
SO   Environmental Protection Agency
      Glossary
DEF (EPA) Measures taken to reduce
      adverse impacts on the
      environment.

MITIGATION ACTION PLANS
(DOE Order 5440.1D)
DA   May 14, 1991
SF   *MAP*
BT1 Plans
RT   Environmental Assessments
RT   Environmental Impact Statements
RT   Mitigation
RT   Monitoring
RT   Record of Decision (NEPA)
SO   Environmental Management
DEF (DOE Order) Documents that
        describe the plan for implementing
        commitments made in a DOE
        Environmental Impact Statement
        (EIS) and its associated record of
        decision (ROD) or in an
        Environmental
        Assessment/Finding of No
        Significant Impact (EA/FONSI)
        (where a FONSI is based, in
        significant part, on such a
        commitment) to mitigate adverse
        environmental impacts associated
        with an action.

MIXED LIQUOR
(EPA)
DA   October 12, 1990
BT1 Mixtures
RT   Activated Sludges
SO   Environmental Protection Agency
        Glossary
DEF (EPA) A mixture of activated sludge
        and water containing organic
        matter undergoing activated
        sludge treatment in an aeration
        tank.

**MIXED WASTE MANAGEMENT
FACILITY**
DA   January 8, 1991
SF   *MWMF*
BT1 Waste Receiving and Processing
        Plant
 BT2 Waste Treatment Plants
  BT3 Facilities

MIXED WASTES
(SWDA; RCRA; ESH)
DA   October 12, 1990
SY   Radioactive Mixed Wastes
SF   *MW*
BT1 Wastes
SO   Environmental Management
SO   Wastes
DEF Contain both radioactive and
        hazardous components as defined
        by the Atomic Energy Act and the
        Resource Conservation and
        Recovery Act, respectively. (ESH)
        Any wastes that meet the
        definition of a hazardous waste
        and contain radioactive waste.

MIXER(S)
(2954 MIXER)
DA   January 3, 1991
BT1 Tools - Powered

 BT2 Tools (DOE FRASE Vocabulary)
  BT3 Equipment
SO   DOE FRASE VOCABULARY

MIXING MACHINE
(2130 MIXING MACHI)
DA   December 10, 1990
BT1 Machines (DOE FRASE
        Vocabulary)
 BT2 Equipment
SO   DOE FRASE VOCABULARY

MIXING ZONES
(CAA; CFR)
DA   October 12, 1990
BT1 Zones
 BT2 Sites/Areas
RT   Discharges
SO   Water Pollution
DEF (CFR) Limited volumes of water
        serving as zones of initial dilution
        in the immediate vicinity of a
        discharge point where receiving
        water quality may not meet quality
        standards or other requirements
        otherwise applicable to the
        receiving water. The mixing zone
        should be considered as a place
        where wastes and water mix and
        not as a place where effluents are
        treated.

MIXTURES
(SWDA; CERCLA; TSCA; CFR; USC)
DA   October 12, 1990
NT1 Mixed Liquor
NT1 Slurries
NT1 Solutions
 NT2 Copper Mattes
NT1 Test Mixtures
RT   Leachates
SO   Compensation and Liability
SO   Environmental Management
SO   Hazardous Materials
DEF (CFR) Any combinations of two or
        more chemical substances if the
        combination does not occur in
        nature and is not, in whole or in
        part, the result of a chemical
        reaction. Heterogeneous
        associations of substances where
        the various individual substances
        retain their identities and can
        usually be separated by
        mechanical means. Include
        solutions or compounds but do not
        include alloys or amalgams.

MK-FERGUSON
DA   January 11, 1991
BT1 Companies
 BT2 Commercial Organizations
  BT3 Organizations
BT1 DOE Contractors
 BT2 Potentially Responsible Parties
RT   Idaho Operations Office
RT   Oak Ridge Operations Office
RT   Paducah Gaseous Diffusion Plant
        (Paducah)

MK-FERGUSON OF IDAHO CO.
DA   January 11, 1991
BT1 Companies
 BT2 Commercial Organizations
  BT3 Organizations
BT1 DOE Contractors
 BT2 Potentially Responsible Parties
RT   Idaho Operations Office
RT   Idaho National Engineering
        Laboratory

MMES
DA   October 12, 1990
SEE Martin Marietta Energy Systems
SO   Acronyms

MNCR
DA   October 12, 1990
SEE Material Nonconformance Report
SO   Acronyms

MOBILE COMMAND VEHICLES
(EMER)
DA   February 1, 1991
BT1 Motor Vehicles
 BT2 Transport Vehicles
SO   Emergency Preparedness
DEF (EMER) Vehicles designed
        specifically to support the
        on-scene commander with
        communications to the emergency
        operations center.

MOBILE CRANE
(2337 MOBILE CRANE)
DA   December 10, 1990
BT1 Crane(s)
 BT2 Material Handling Device
  BT3 Devices
  BT3 Equipment/Parts - Material
        Handling (DOE FRASE
        Vocabulary)
   BT4 Equipment
SO   DOE FRASE VOCABULARY

MOBILE SOURCES
(EPA)
DA   October 12, 1990
RT   Area Sources
RT   Pollution
RT   Releases
RT   Stationary Sources
SO   Environmental Protection Agency
        Glossary
DEF (EPA) Moving producers of air
        pollution, mainly forms of
        transportation such as cars,
        trucks, motorcycles, airplanes.

MOC
DA   October 12, 1990
SEE Margin of Control (Criticality)
SO   Acronyms

MODE
(SWDA; RCRA)
DA   October 12, 1990
RT   Transportation

SY-Synonymous Terms   SO-Source/Subject Category   SF-See From

SO Hazardous Materials
SO System Safety Development Center
    Glossary
DEF (SSDC) In statistics, the most
    common or frequent observation
    or value. (ESH) Any of the
    following transportation methods:
    rail, highway, air, or water.

MODEL PLANTS
(EPA)
DA October 12, 1990
BT1 Facilities
SO Environmental Protection Agency
    Glossary
DEF (EPA) Descriptions of typical but
    theoretical plants used for
    developing economic,
    environmental impact, and energy
    impact analyses as support for
    regulations or regulatory
    guidelines. They are imaginary
    plants, with features of existing or
    future plants used to estimate the
    cost of incorporating air pollution
    control technology as the first step
    in exploring the economic impact
    of a potential New Source
    Performance Standard (NSPS).

MODELING
(EPA)
DA October 12, 1990
RT ABEL
RT Phantoms
SO Environmental Protection Agency
    Glossary
DEF (EPA) An investigative technique
    using a mathematical or physical
    representation of a system or
    theory that accounts for all or
    some its known properties. Models
    are often used to test the effect of
    changes of system components
    on the overall performance of the
    system.

MODERATOR RECOVERY SYSTEM
DA January 8, 1991
SF *MRS*
BT1 Systems

MODERATORS (NUCLEAR)
(DOE Order 5480.3; ESH)
DA October 12, 1990
BT1 Hazardous Materials
 BT2 Materials
SO Hazardous Materials
DEF (DOE Order 5480.3) Materials used
    to reduce the kinetic energy of
    neutrons by scattering collisions
    without appreciable neutron
    capture.

MODERN AIRCRAFT
(Doe Order 5480.13)
DA October 12, 1990
BT1 Aircraft

SO Aviation Safety
DEF (DOE Order 5480.13) Aircraft
    whose performance capabilities,
    age, and ease of maintenance
    meet current state-of-the-art and
    technology for the type aircraft
    involved.

MODIFICATION
(DOE Order 5480.6; CAA; CFR)
DA October 12, 1990
BT1 Administrative Processes
 BT2 Processes
NT1 Major Modification
NT1 Major PSD Modification
RT Stationary Sources
SO Air Pollution
SO Industrial Safety
DEF (DOE Order 5480.6) Any change
    made to structures, systems,
    components, or procedures during
    any phase of the life of the reactor
    project. As applied to an active
    underground uranium mine any
    major change in the method of
    operation or mining procedure
    which will result in an increase in
    the amount of radon-222 emitted
    to air. The normal development or
    operation of an active mine, even
    though it results in an increase in
    emissions, is not considered a
    modification for the purposes of
    this 40 CFR 61.21.

MODIFIED HAZARD RANKING SYSTEM
DA January 24, 1991
BT1 Systems
RT Hazardous Ranking System
SO Environmental Management
DEF (DOE 5480.14) The methodology
    developed by DOE to rank sites
    containing hazardous substances
    and/or radionuclides.

MODIFIED SOURCES
(CFR)
DA November 15, 1990
BT1 Stationary Sources
 BT2 Facilities
DEF (CFR) Any physical changes in, or
    changes in the method of
    operation of, stationary sources
    which increase the emission rate
    of any pollutant for which a
    national standard has been
    promulgated under Part 50 of this
    chapter or which result in the
    emission of any such pollutant not
    previously emitted, except that: (1)
    Routine maintenance, repair, and
    replacement shall not be
    considered a physical change, and
    (2) The following shall not be
    considered a change in the
    method of operation: (i) An
    increase in the production rate, if
    such increase does not exceed
    the operating design capacity of

the source; (ii) An increase in the
hours of operation; (iii) Use of an
alternative fuel or raw material, if
prior to the effective date of a
paragraph in this part which
imposes conditions on or limits
modifications, the source is
designed to accommodate such
alternative use.

MOMENT MAGNITUDE
(USGS)
DA October 12, 1990
SF *M*
BT1 Measurements
BT1 Units of Measure
RT Earthquake Magnitude
RT Surface Faulting
SO Natural Phenomenon
DEF (USGS) Derived from seismic
    moment, M_0, the product of the
    surface area of the fault, the
    average displacement on the fault
    plane, and the rigidity of the
    material of the fault. After certain
    corrections, M_0 can be calculated
    from measurements of long-period
    waves (200-300 seconds) that
    typically accompany great
    earthquakes.

MOMENT RESISTING SPACE FRAME
(SEA)
DA October 12, 1990
BT1 Space Frames
 BT2 Structures (DOE FRASE
    Vocabulary)
NT1 Intermediate Moment Resisting
    Space Frame
NT1 Ordinary Moment Resisting Space
    Frame
NT1 Special Moment Resisting Space
    Frame
SO Construction
DEF (SEA) A space frame in which the
    members and joints are capable of
    resisting forces primarily by
    flexure.

MONITORED PERSONNEL LOCATOR FILE
(Doe Order 5484.1)
DA October 12, 1990
RT Personnel
SO Industrial Safety
DEF (DOE Order 5484.1) A DOE
    centralized file maintained at the
    System Safety Development
    Center, EG&G Idaho, that
    contains all monitored DOE and
    DOE contractor personnel
    employed, and visitors who have
    positive exposures. The file
    consists of identification
    information only, e.g., name,
    social security number, birth year,
    and employer organization (or
    organization visited). The file is
    used to identify personnel work
    locations so that inquiries can be

BT-Broader Term   NT-Narrower Term   RT Related Term

made to the reporting organization for official dose records.

MONITORED VISITORS
(Doe Order Definitions)
DA   October 12, 1990
RT   Personnel
SO   Management
DEF (DOE Order 5484.1) Nonemployees, including subcontractors, not classified as a "nonemployee radiation workers" visiting a facility that is operated by DOE or a DOE contractor under circumstances requiring that they be monitored for radiation exposure.

MONITORED WORKERS
(Doe Order 5484.1)
DA   October 12, 1990
BT1   Personnel
SO   Industrial Safety
DEF (DOE 5484.1) Employees of the reporting organization who work with, or are in the proximity of, ionizing radiation or radioactive material and who are monitored in accordance with DOE 5480.11.

MONITORING
(NCRP; IAEA; EMER; MORT)
DA   October 12, 1990
BT1   Activities
NT1   Air Monitoring
NT1   Biomonitoring
NT1   Environmental Monitoring
 NT2   Effluent Monitoring
 NT2   Environmental Surveillance
NT1   Field Monitoring
NT1   Health Monitoring
NT1   Interstitial Monitoring
NT1   Radiological Monitoring
 NT2   Smear
NT1   Vibration and Acoustic Monitoring
NT1   Well Monitoring
RT   Analyses
RT   Communication
RT   Corrective Action Triggers
RT   Knowledge
RT   Mitigation Action Plans
RT   Monitoring Wells
RT   Predictive Maintenance
RT   Sanitary Surveys
RT   Secondary Environmental Monitor
RT   Sensors
RT   Slip Gauges
RT   Standard Sample
RT   Technical Information
RT   Trending
SO   Emergency Preparedness
SO   Environmental Management
SO   Environmental Protection Agency Glossary
SO   Industrial Hygiene
SO   Radiation
SO   System Safety Development Center Glossary
DEF Actions intended to detect and evaluate radiological conditions.

(EPA) Periodic or continuous surveillance or testing to determine the level of compliance with statutory requirements and/or pollutant levels in various media or in humans, animals, and other living things. (SSDC) A set of observation and data collection methods to detect and measure deviations in current operations. (IAEA) The measurement of radiation or activity for reasons related to the assessment or control of exposure to radiation or radioactive material, and the interpretation of such measurements. (MORT) MORT analysis asks: was the monitoring system adequate? were the principal elements of a good monitoring system present? Also included are safety observation plans, safety inspections, error sampling techniques, work site inspections, general health monitoring, a system for Reported Significant Observations, etc.

MONITORING SYSTEMS
(CAA; CFR)
DA   October 12, 1990
BT1   Systems
NT1   Energy Monitoring and Control Systems
RT   Air Monitoring
SO   Air Pollution
DEF (CFR) Any systems, required under the monitoring sections in applicable subparts, used to sample and condition (if applicable), to analyze, and to provide a record of emissions or process parameters.

MONITORING WELLS
(EPA)
DA   October 12, 1990
BT1   Wells
RT   Monitoring
SO   Environmental Protection Agency Glossary
DEF (EPA) Wells drilled at a hazardous waste management facility or Superfund site to collect groundwater samples for the purpose of physical, chemical, or biological analysis to determine the amounts, types, and distribution of contaminants in the ground water beneath the site.

MONITORS
DA   February 12, 1991
BT1   Equipment
NT1   10-minute Delay Line Gamma Monitor
NT1   Axial Power Monitors
NT1   Constant Air Monitor
NT1   Continuous Air Monitors
NT1   Cooling Water Gamma Monitor

NT1   Crane Drip Pan Monitor
NT1   Essential Equipment Monitor
NT1   Failed Element Monitors
NT1   Flow Monitor
NT1   Gang Temperature Monitor
NT1   High Level Flux Monitor
NT1   Internal Gamma Flux Monitor
NT1   Low Energy Gamma Monitor
NT1   Nuclear Incident Monitor
NT1   Power Density Monitor
NT1   Primary Environmental Monitors
NT1   Process Water Gamma Monitor
NT1   Radial Power Monitor
NT1   Secondary Environmental Monitor
NT1   Secondary Environmental Monitors
NT1   Selection and Monitoring Chassis
NT1   Stack Tritium Monitor
NT1   Temperature Alarm Monitor
NT1   Temperature Scram Circuit Monitor
NT1   Traveling Wire Flux Monitor
RT   Controls

MONOCLONAL ANTIBODIES
(EPA)
DA   October 12, 1990
BT1   Antibodies
 BT2   Proteins
  BT3   Organic Chemicals
   BT4   Chemical Substances
SO   Environmental Protection Agency Glossary
DEF (EPA) (Also called MABs and MCAs) Molecules of living organisms that selectively find and attach to other molecules to which their structure conforms exactly. This could also apply to equivalent activity by chemical molecules.

MONTHLY NEPA REPORTS
(DOE Order 5440.1D)
DA   May 14, 1991
BT1   Reports
SO   Environmental Management
DEF (DOE Order) Documents submitted monthly to the Secretary that identify Environmental Assessments and Environmental Impact Statements that Secretarial Officers expect to forward for approval during the subsequent 3 months.

MONUMENTATION
(DOE Order 6430.1A)
DA   October 12, 1990
RT   Bench Marks
RT   Construction
RT   Datum
SO   Construction
DEF (DOE Order 6430.1A) The act of setting a permanent survey control point.

MOP(S)
(3043 MOP)
DA   January 3, 1991
BT1   Tools - Manual
 BT2   Tools (DOE FRASE Vocabulary)

BT3 Equipment
SO DOE FRASE VOCABULARY

MORT
DA October 12, 1990
SEE Management Oversight and Risk
     Tree
SO Acronyms
SO System Safety Development Center
     Glossary

MOTEL/HOTEL
(1786 MOTEL)
DA December 10, 1990
BT1 Building (DOE FRASE Vocabulary)
 BT2 Facilities and Buildings (DOE
     FRASE Vocabulary)
  BT3 Facilities
SO DOE FRASE VOCABULARY

MOTOR(S)
(2403 MOTOR)
DA January 3, 1991
BT1 Equipment/Parts - Electrical (DOE
     FRASE Vocabulary)
 BT2 Equipment
SO DOE FRASE VOCABULARY

MOTOR CONTROL CENTER
DA January 8, 1991
SF MCC
BT1 Centers

MOTOR FUELS
(SWDA; RCRA; CFR)
DA October 19, 1990
BT1 Fuels
RT Petroleum
SO Hazardous Materials
SO Wastes
DEF (CFR) Petroleum or a
     petroleum-based substances that
     are motor gasoline, aviation
     gasoline, No. 1 or No. 2 diesel
     fuel, or any grade of gasohol, and
     are typically used in the operation
     of a motor engine.

MOTOR GENERATOR
DA January 8, 1991
SF MG
BT1 Equipment

**MOTOR GENERATOR-MOTOR
ALTERNATOR**
DA January 8, 1991
SF MG-MA
BT1 Equipment

MOTOR OPERATED VALVES
DA January 8, 1991
SF MOV
BT1 Valves
 BT2 Devices

MOTOR VEHICLES
(SWDA; CAA; USC; ESH)
DA October 12, 1990

BT1 Transport Vehicles
NT1 COMVAN
NT1 Mobile Command Vehicles
RT Cargo Tanks
RT Catalytic Converters
RT Transportation Control Measures
SO Air Pollution
SO Hazardous Materials
DEF (USC) Self-propelled vehicles
     designed for transporting persons
     or property on a street or highway.
     (ESH) Vehicles, machines,
     tractors, trailers, or semitrailers, or
     any combinations thereof,
     propelled or drawn by mechanical
     power and used upon the
     highways in the transportation of
     passengers or property. They do
     not include a vehicle, locomotive,
     or car operated exclusively on a
     rail or rails, or a trolley bus
     operated by electric power derived
     from a fixed overhead wire,
     furnishing local passenger
     transportation similar to
     street-railway service.

MOU
DA October 12, 1990
SEE Memorandum of Understanding
SO Acronyms

MOUTH
(1088 MOUTH)
DA November 28, 1990
BT1 Digestive System
 BT2 Body System(s)
  BT3 Human Body Parts
NT1 Tooth/Teeth
RT Dental Injury
RT Face
RT Head
RT Lip(s)
SO DOE FRASE VOCABULARY

MOV
DA October 12, 1990
SEE Motor Operated Valves
SO Acronyms

MOVEMENT
(SWDA; RCRA)
DA October 12, 1990
RT Transport Vehicles
RT Transportation
SO Environmental Management
SO Wastes
DEF Means that hazardous waste
     transported to a facility in an
     individual vehicle.

MP&L
DA October 12, 1990
SEE Mississippi Power and Light
SO Acronyms

MPC
DA October 12, 1990

SEE Maximum Permissible
     Concentration
SO Acronyms

MPD
DA October 12, 1990
SEE Maximum Permissible Dose
SO Acronyms

MPL
DA October 12, 1990
SEE Maximum Possible Loss
SO Acronyms

MQL
(ESH)
DA October 12, 1990
SY Method Quantification Limits
BT1 Limits
BT1 Measurements
SO Quality Assurance
DEF The method quantification limit
     (MQL) is the minimum
     concentration of a substance that
     can be measured and reported.

mR
DA October 12, 1990
SEE Milliroentgen
SO Acronyms

mR/h
DA October 12, 1990
SEE Microroentgen/hour
SO Acronyms

MRC
DA October 12, 1990
SEE Maintenance Requirement Card
SO Acronyms

MRS
DA October 12, 1990
SEE Moderator Recovery System
SO Acronyms

MSA
DA October 12, 1990
SEE Major System Acquisition
SO Acronyms

MSCANS
(Acronyms and Abbreviations)
DA October 12, 1990
BT1 Computer Codes
SO Nuclear Facilities Incident Database
DEF A computer code for miscellaneous
     computer scans.

MSDS
DA October 12, 1990
SEE Material Safety Data Sheets
SO Acronyms

MSIV
DA October 12, 1990

BT-Broader Term   NT-Narrower Term   RT Related Term

SEE Main Steam Isolation Valve
SO Acronyms

MTBF
DA   October 12, 1990
SEE Mean Time Between Failures
SO Acronyms

MTE
DA   October 12, 1990
SEE Measurement and Test Equipment
SO Acronyms

MTI
DA   October 12, 1990
SEE Maintenance Team Inspection
SO Acronyms

MTR
DA   October 12, 1990
SEE Materials Test Reactor
SO Acronyms

MTSAA
DA   October 12, 1990
SEE Multidiscipline Technical Safety
      Assurance Appraisal
SO Acronyms

MTTR
DA   October 12, 1990
SEE Mean Time to Repair
SO Acronyms

MUCK SOILS
(EPA)
DA   October 12, 1990
BT1 Soils
RT Dystrophic Lakes
RT Fens
RT Swamps
RT Wetlands
SO Environmental Protection Agency
      Glossary
DEF (EPA) Earth made from decaying
      plant materials.

MULCHES
(EPA)
DA   October 12, 1990
SO Environmental Protection Agency
      Glossary
DEF (EPA) Layers of material (wood
      chips, straw, leaves, etc.) placed
      around plants to hold moisture,
      prevent weed growth, protect the
      plants, and enrich the soil.

**MULTIDISCIPLINE TECHNICAL
SAFETY ASSURANCE APPRAISAL**
DA   January 8, 1991
SF   *MTSAA*
BT1 Appraisals

**MULTIMEDIA ENVIRONMENTAL
POLLUTANT ASSESSMENT SYSTEM**
DA   January 8, 1991

SF   *MEPS*
BT1 Systems
RT Assessments

MULTIPLE BODY PARTS
(1134 MULTIPLE BP)
DA   November 28, 1990
BT1 Human Body Parts
SO DOE FRASE VOCABULARY

MULTIPLE INJURIES
(1332 MULTIPLE INJ)
DA   November 28, 1990
BT1 Injuries
SO DOE FRASE VOCABULARY

**MULTIPLE INTEGRATED LASER
ENGAGEMENT SYSTEM**
(Doe Order 5480.16)
DA   October 12, 1990
SF   *MILES*
BT1 Engagement Simulation Systems
 BT2 Simulators
BT1 Systems
RT Laser Eye Safety Distance
SO Firearms
DEF (DOE Order 5480.16) See
      Engagement Simulation Systems.

MULTIPLE USES
(EPA)
DA   October 12, 1990
BT1 Uses
SO Environmental Protection Agency
      Glossary
DEF (EPA) Uses of land for more than
      one purpose, i.e., grazing of
      livestock, wildlife production,
      recreation, watershed, and timber
      production. Can also apply to use
      of bodies of water for recreational
      purposes, fishing, and water
      supply.

MULTIPLEXOR
DA   January 8, 1991
SF   *MUX*
BT1 Equipment
NT1 TRAnsient Multiplexor

MUNICIPAL SOLID WASTES
(TSCA; CFR)
DA   October 19, 1990
BT1 Solid Wastes
 BT2 Wastes
NT1 Sewage
 NT2 Raw Sewage
RT Municipalities
RT Publicly Owned Treatment Works
RT Septic Tanks
RT Sludge
SO Environmental Management
SO Hazardous Materials
DEF (CFR) Garbage, refuse, sludges,
      wastes, and other discarded
      materials resulting from residential
      and nonindustrial operations and
      activities, such as household

activities, office functions, and
commercial housekeeping wastes.

MUNICIPALITIES
(SWDA; RCRA; CFR; USC)
DA   October 12, 1990
RT Air Pollution Control Agencies
RT Bay Area Air Quality Management
      District
RT Municipal Solid Wastes
RT Urban Runoff
SO Environmental Management
SO Wastes
SO Water Pollution
DEF Cities, towns, boroughs, counties,
      parishes, districts, or other public
      bodies created by or pursuant to
      State law, with responsibility for
      the planning or administration of
      solid waste management, or
      Indian tribes or authorized tribal
      organizations or Alaska Native
      villages or organizations, and
      include any rural communities or
      unincorporated towns or villages
      or any other public entities for
      which an application for
      assistance is made by a State or
      political subdivision thereof.

MUNITIONS
(Doe Order 5480.16)
DA   October 12, 1990
NT1 Ammunition
 NT2 Chemical Ammunition
  NT3 Smoke Grenades
 NT2 Rocket Ammunition
 NT2 Small Arms Ammunition
NT1 Duds
NT1 Explosive Projectiles
NT1 Explosive Bombs
NT1 Explosive Mines
NT1 Explosive Torpedoes
NT1 Grenades
 NT2 Smoke Grenades
RT Armorers
RT Blank Ammunition
RT Explosives
RT Firearms
RT Inhabited Building Distance
RT Minimum Fragment Distance
RT Net Explosive Weight
RT Public Traffic Route Distance
RT Quantity Distances
RT Special Weapons
SO Firearms
DEF (DOE Order 5480.16) Ammunition
      and explosives used by protective
      force personnel, Transportation
      Safeguards Division (AL) couriers,
      and Departmental safeguards and
      security staff personnel. An
      explosive is any chemical
      compound or mechanical mixture
      that, when subjected to such
      stimuli as heat, impact, friction, or
      shock undergoes a very rapid
      chemical change that releases
      large volumes of highly heated
      gases that exert pressure in the

surrounding medium. The term applies to materials that either detonate or deflagrate.

MUSCLE/TENDON(S)
(1138 MUSCLE)
DA November 28, 1990
BT1 Human Body Parts
RT Tendonitis
SO DOE FRASE VOCABULARY

MUTAGENS
(EPA)
DA October 12, 1990
BT1 Hazardous Substances
RT Mutate
RT Pesticides
SO Environmental Protection Agency Glossary
DEF (EPA) Substances that can cause a change in genetic material.

MUTATE
(EPA)
DA October 12, 1990
RT Mutagens
SO Environmental Protection Agency Glossary
DEF (EPA) To bring about a change in the genetic constitution of a cell by altering its DNA. In turn, "mutagenesis" is any process by which cells are mutated.

MUTUAL ASSISTANCE AGREEMENTS
(EMER)
DA February 1, 1991
BT1 Agreements
SO Emergency Preparedness
DEF (EMER) Agreements between contractors and U.S. Department of Energy (DOE) and/or DOE and other government agencies or municipalities to share emergency response and techniques. Also known as Mutual Aid Agreements.

MUX
DA October 12, 1990
SEE Multiplexor
SO Acronyms

μCi
DA October 12, 1990
SEE Microcuries
SO Acronyms

MW
DA October 12, 1990
SEE Mixed Wastes
SO Acronyms

MWMF
DA October 12, 1990
SEE Mixed Waste Management Facility
SO Acronyms

MWO
DA October 12, 1990
SEE Maintenance Work Order
SO Acronyms

MWt
DA October 12, 1990
SEE Megawatt thermal
SO Acronyms

M_b
DA October 12, 1990
SEE Body Wave Magnitude
SO Acronyms

M_L
DA October 12, 1990
SEE Local Magnitude (M_L)
SO Acronyms

M_s
DA October 12, 1990
SEE Surface Wave Magnitude
SO Acronyms

N-UNIT
(NFI)
DA October 12, 1990
SO Nuclear Facilities Incident Database
SO Radiation
DEF Control rod drive for full rods.

NAAQS
DA October 12, 1990
SEE National Ambient Air Quality Standards
SO Acronyms

NAIL GUN
(2955 NAIL GUN)
DA January 3, 1991
BT1 Tools - Powered
 BT2 Tools (DOE FRASE Vocabulary)
  BT3 Equipment
SO DOE FRASE VOCABULARY

NAME OF CONTENTS
(SWDA; RCRA; ESH)
DA October 12, 1990
SY Proper Shipping Name
RT Marks
RT Technical Names
SO Hazardous Materials
DEF The proper shipping name (PSN).

NANOSEC LIVING RADIOISOTOPES
(EDB)
DA March 29, 1991
BT1 Radionuclides
 BT2 CERCLA Hazardous Substances
  BT3 Hazardous Substances
 BT2 Nuclides
SO Radiation

NAS
DA October 12, 1990

SEE National Academy of Sciences
SO Acronyms

NASA
DA October 12, 1990
SEE National Aeronautic and Space Administration
SO Acronyms

NATIONAL ACADEMY OF SCIENCES
DA January 8, 1991
SF NAS
BT1 Research and Development Organizations
 BT2 Organizations

NATIONAL AERONAUTIC AND SPACE ADMINISTRATION
DA January 8, 1991
SF NASA
BT1 Federal Agencies
 BT2 Agencies
  BT3 Administrative Organizations
   BT4 Organizations
DEF (U.S. Government Manual, 1990-1991) The NASA conducts research for the solution of problems of flight within and outside the Earth's atmosphere and develops, constructs, tests, and operates aeronautical and space vehicles. It conducts activities required for the exploration of space with manned and unmanned vehicles and arranges for the most effective utilization of the scientific and engineering resources of the United States with other nations engaged in aeronautical and space activities for peaceful purposes.

NATIONAL AMBIENT AIR QUALITY STANDARDS
(EPA)
DA October 12, 1990
SF NAAQS
BT1 Ambient Air Quality Standards
 BT2 Air Quality Standards
  BT3 Standards
   BT4 Codes, Standards, and Regulations
NT1 Primary Standards
NT1 Secondary Standards
RT Attainment Areas
RT Non-Attainment Areas
RT Primary Standard Attainment Date
RT Reasonable Further Progress
SO Environmental Protection Agency Glossary
DEF (EPA) Air quality standards established by EPA that apply to outside air.

NATIONAL CAPACITY VARIANCES
DA January 24, 1991
BT1 Variances

BT-Broader Term   NT-Narrower Term   RT Related Term

SO Environmental Management
DEF (CFR) Nationwide variances based on inadequate treatment capacity. EPA included provisions within regulations for specific wastes which allows a variance from the land disposal restrictions. For example, the regulation on solvent wastes in 40 CFR 268.30 provides a two-year variance on dilute mixtures of solvents and other wastes.

NATIONAL CONTINGENCY PLAN
(CERCLA; EPA; CFR; EMER)
DA November 15, 1990
SF *NCP*
BT1 Plans
RT Oil Spills
RT Pollutants
RT Replacement
RT Superfund Amendments and Reauthorization Act of 1986
RT Support Agencies
RT Support Agency Coordinators
RT United States
SO Emergency Preparedness
DEF (CFR) The National Oil and Hazardous Substances Contingency Plan and revisions promulgated by EPA, pursuant to section 105 of CERCLA and codified in 40 CFR Part 300. The scope plan includes: (1) Discharges or substantial threats of discharges of oil to or upon the navigable waters of the United States and adjoining shorelines; (2) Releases or substantial threats of releases of hazardous substances into the environment, and releases or substantial threats of releases of pollutants or contaminants which may present imminent and substantial danger to public health and welfare. (3) Establishing requirements for Federal regional and Federal local contingency plans; (4) Procedures for removal operations pursuant to §311 of the Clean Water Act (CWA); (5) Procedures for response operations pursuant to CERCLA: (6) Designating trustees for natural resources (CERCLA); and (7) National policies and procedures for using dispersants and other chemicals in removal and response actions.

NATIONAL DEFENSE AREAS
(EMER)
DA February 1, 1991
BT1 Sites/Areas
SO Emergency Preparedness
DEF (EMER) Areas established on nonfederal or public accessible lands located within the United States, its possessions, or its territories for the purpose of safeguarding classified defense information or protecting Department of Defense equipment and/or material. Establishment of an NDA temporarily places such nonfederal lands under the effective control of the Department of Defense and results only from an emergency event. The landowner's consent and cooperation will be obtained whenever possible; however, military necessity will dictate the final decision regarding location, shape, and size of the NDA.

NATIONAL ELECTRICAL CODE
DA January 8, 1991
SF *NEC*

NATIONAL EMERGENCIES
(EMER)
DA February 1, 1991
BT1 Emergencies
 BT2 Reportable Occurrences
  BT3 Occurrences
RT Vital Records
SO Emergency Preparedness
DEF (EMER) Conditions proclaimed by the President of the United States or declared by Congress, including probable, immediate, or actual attack upon the United States.

NATIONAL EMISSIONS STANDARDS FOR HAZARDOUS AIR POLLUTANTS
(EPA)
DA October 12, 1990
SF *NESHAPS*
BT1 Emission Standards
 BT2 Requirements
 BT2 Standards
  BT3 Codes, Standards, and Regulations
RT Emission Limitation
RT Lowest Achievable Emission Rate
RT Particulate Matter
SO Environmental Protection Agency Glossary
DEF (EPA) Also known as NESHAPS, these emissions standards set by EPA for an air pollutant not covered by National Ambient Air Quality Standards Emissions NAAQS that may cause an increase in deaths or in serious, irreversible, or incapacitating illness. Primary standards are designed to protect human health, secondary standards to protect public welfare.

NATIONAL ENVIRONMENTAL POLICY ACT
(DOE Order 5440.1D; NEPA)
DA January 8, 1991
SF *NEPA*
BT1 Acts
 BT2 Statutes and Regulations
RT Biosphere
RT Council on Environmental Quality
RT Natural Resources
RT Quarterly Environmental Compliance Reports
RT Significance
SO Environmental Management
DEF (NEPA) To declare a national policy which will encourage productive and enjoyable harmony between man and his environment; to promote efforts which will prevent or eliminate damage to the environment and biosphere and stimulate the health and welfare of man; to enrich the understanding of the ecological systems and natural resources important to the Nation; and to establish a Council on Environmental Quality. (DOE Order) Also see the DOE NEPA Guidelines and relevant references from the Council on Environmental Quality Regulations (40 CFR Parts 1500-1508, as amended 7-1-86).

NATIONAL ENVIRONMENTAL POLICY ACT DOCUMENTS
(Doe Order 5440.1D; NEPA)
DA October 12, 1990
SF *NEPA Documents*
BT1 Reports
NT1 Programmatic NEPA Documents
NT1 Site-Wide NEPA Documents
RT Categorical Exclusions
RT Environmental Assessments
RT Environmental Impact Statements
RT Finding of No Significant Impact
RT Notice of Intent
RT Record of Decision (NEPA)
SO Environmental Management
SO Management
DEF (Doe Order 5440.1D)
        Environmental assessments, findings of no significant impact, notices of intent to prepare an environmental impact statement, record of decision, categorical exclusion determination, or any other documents prepared pursuant to a requirement of NEPA or the Council on Environmental Quality Regulations (40 CFR parts 1500-1508, as amended 7-1- 86).

NATIONAL FIRE PROTECTION ASSOCIATION
DA January 8, 1991
SF *NFPA*
BT1 Professional Bodies
 BT2 Organizations

NATIONAL INSTITUTE FOR OCCUPATIONAL SAFETY AND HEALTH
(Doe Order 5483.1A)
DA October 12, 1990
BT1 Institutes
 BT2 Research and Development Organizations

SY-Synonymous Terms   SO-Source/Subject Category   SF-See From

BT3 Organizations
RT Federal Employee Occupational
    Safety and Health Program
RT Occupational Safety and Health
    Administration
SO Industrial Safety
DEF (DOE Order 5483.1A) An Agency
    of the U.S. Department of Health
    and Human Services, established
    under Public Law 91.596 with
    major responsibility to undertake
    National occupational safety and
    health research and development
    activities.

**NATIONAL INSTITUTE OF
STANDARDS AND TESTING**
DA January 8, 1991
SF *NIST*
BT1 Institutes
 BT2 Research and Development
     Organizations
  BT3 Organizations
RT Standards

**NATIONAL OIL AND HAZARDOUS
SUBSTANCES CONTINGENCY PLAN**
(EPA)
DA October 12, 1990
SF *NOHSCP*
BT1 Plans
SO Environmental Protection Agency
    Glossary
DEF (EPA) The federal regulation that
    guides determination of the sites
    to be corrected under the
    Superfund program and the
    program to prevent or control spills
    into surface waters or other
    portions of the environment. [Also
    known as NOHSCP/NCP].

**NATIONAL POLLUTANT DISCHARGE
ELIMINATION SYSTEM**
(SWDA; RCRA; EPA)
DA October 12, 1990
SF *NPDES*
BT1 Systems
NT1 Pollutant Discharge Elimination
    System
SO Environmental Management
SO Environmental Protection Agency
    Glossary
SO Wastes
DEF Section 402 of the Federal Water
    Pollution Act (the Clean Water
    Act) that establishes a permit for
    discharges to water and provides
    standards by which such permits
    may be granted.

NATIONAL PRIORITIES LIST
(EPA)
DA October 12, 1990
SF *NPL*
SO Environmental Protection Agency
    Glossary
DEF (EPA) Formal Listing of the nation's
    worst hazardous waste sites, as

established by the Comprehensive
Environmental Response,
Compensation, and Liability Act
(CERCLA). (EPA) EPA's list of the
most serious uncontrolled or
abandoned hazardous waste sites
identified by possible long-term
remedial action under Superfund.
A site must be on the NPL to
receive money from the Trust
Fund for remedial action. The list
is based primarily on the score a
site receives from the Hazard
Ranking System. EPA is required
to update the NPL at least once a
year.

NATIONAL RESPONSE CENTERS
(EPA; EMER)
DA October 12, 1990
BT1 Centers
SO Emergency Preparedness
SO Environmental Protection Agency
    Glossary
DEF (EPA) The federal operations
    center that receives notifications of
    all releases of oil and hazardous
    substances into the environment.
    The Center, open 24 hours a day,
    is operated by the U.S. Coast
    Guard, which evaluates all reports
    and notifies the appropriate
    agency.

NATIONAL RESPONSE TEAMS
(EPA; EMER)
DA October 12, 1990
SF *NRT*
BT1 Response Teams
 BT2 Teams
  BT3 Administrative Organizations
  BT4 Organizations
RT Activation (Emergency)
RT First Federal Officials
SO Emergency Preparedness
SO Environmental Protection Agency
    Glossary
DEF (EPA) Consist of representatives of
    13 federal agencies that, as
    teams, coordinate federal
    responses to nationally significant
    incidents of pollution and provide
    advice and technical assistance to
    the responding agency(ies) before
    and during a response action.

NATIONAL SECURITY
(Doe Order 5480.7)
DA October 12, 1990
RT Classified Information
RT Classified Interests
SO Environmental Management
SO Fires
DEF (DOE 5480.7) Those aspects of
    national security as outlined in the
    Atomic Energy Act that could be
    affected adversely by fire,
    explosion, or other catastrophes.

NATIONAL SECURITY AREAS
(EMER)
DA February 1, 1991
BT1 Sites/Areas
SO Emergency Preparedness
DEF (EMER) Areas established on
    nonfederal or public accessible
    lands located within the United
    States, its possessions, or its
    territories for the purpose of
    safeguarding classified information
    or protecting DOE equipment
    and/or material. Establishment of
    an NSA temporarily places such
    nonfederal lands under the
    effective control of DOE and
    results only from an emergency
    event. The senior DOE
    representative having custody of
    the material at the scene will
    define the boundary, mark it with a
    physical barrier, and post warning
    signs. The landowner's consent
    and cooperation will be obtained
    whenever possible; however,
    operational necessity will dictate
    the final decision regarding
    location, shape, and size of NSA.

NATIONAL SECURITY INFORMATION
(EMER)
DA February 1, 1991
BT1 Classified Information
 BT2 Classified Matter
  BT3 Security Interests
SO Emergency Preparedness
DEF (EMER) information that has been
    determined pursuant to Executive
    Order 12356, National Security
    Information (NSI), or any
    predecessor order to require
    protection against unauthorized
    disclosure and that is so
    designated.

**NATIONAL STANDARDS (AIR
POLLUTION)**
(CAA; CFR)
DA October 12, 1990
BT1 Standards
 BT2 Codes, Standards, and
     Regulations
RT Air Quality Standards
SO Air Pollution
DEF (CFR) Either a primary or
    secondary standard.

NATIONAL WARNING SYSTEM
(EMER)
DA February 1, 1991
BT1 Emergency Systems
 BT2 Systems
SO Emergency Preparedness
DEF (EMER) The federal portion of the
    Civil Defense Warning System,
    used for the dissemination of
    warning and other emergency
    information from the warning
    centers or regions to warning
    points in each state.

NATIONALLY RECOGNIZED TESTING LABORATORIES
(DOE Order 6430.1A)
DA   October 12, 1990
BT1 Laboratories
 BT2 Research and Development
       Organizations
  BT3 Organizations
SO Construction
DEF (DOE Order 6430.1A)
      Organizations that are recognized
      by OSHA in accordance with
      Appendix A of 29 CFR 1910.7 and
      that test for safety, and list, label
      or accept equipment or material.
      (Example: Underwriters
      Laboratory [UL]).

NATURAL BACKGROUND RADIATION
(NCRP)
DA   October 12, 1990
SY   Background Radiation
BT1 Radiation
SO Radiation
DEF (NCRP) The radiation in man's
      natural environment, including
      radiation originating outside the
      earth's atmosphere and radiation
      from the naturally occurring
      radioactive elements on earth.
      These elements may be found
      both in the environment and inside
      the bodies of men and animals.

NATURAL BARRIERS
DA   January 24, 1991
BT1 Barriers
RT Disposal Systems
SO Environmental Management
DEF (DOE 5820.2A) Physical, chemical,
      and hydrological characteristics of
      the geological environment at the
      disposal site that, individually and
      collectively, act to retard or
      preclude waste migration.

NATURAL CONDITIONS
(CAA; CFR)
DA   October 12, 1990
BT1 Conditions
RT Visibility Impairments
SO Air Pollution
DEF (CFR) Include naturally occurring
      phenomena that reduce visibility
      as measured in terms of visual
      range, contrast, or coloration.

NATURAL DISASTERS
(EMER)
DA   February 1, 1991
BT1 Natural Phenomenon
NT1 Fires
NT1 Landslides
NT1 Tornadoes
NT1 Tsunamis
SO Emergency Preparedness
DEF (EMER) Any flood, high water,
      wind-driven water, drought, fire,
      hurricane, tornado, storm,

earthquake, tidal wave, volcano,
or other natural occurrence
causing significant damage.

NATURAL GAS
(EPA)
DA   October 12, 1990
BT1 Fossil Fuels
 BT2 Fuels
 BT2 Geologic Resources
  BT3 Natural Resources
BT1 Gases
NT1 Field Gases
NT1 Natural Gas Liquids
RT Natural Gas Processing Plants
SO Environmental Protection Agency
      Glossary
DEF (EPA) A natural fuel containing
      primarily methane and ethane that
      occurs in certain geologic
      formations.

NATURAL GAS LIQUIDS
(CAA; CFR; USC; ESH)
DA   October 12, 1990
BT1 Liquids
 BT2 Fluids
BT1 Natural Gas
 BT2 Fossil Fuels
  BT3 Fuels
  BT3 Geologic Resources
   BT4 Natural Resources
 BT2 Gases
SO Air Pollution
DEF (ESH) Hydrocarbons, such ethane,
      propane, butane, and pentane,
      that are extracted from field gas.

NATURAL GAS PROCESSING PLANTS
(CAA; CFR; USC; ESH)
DA   October 12, 1990
BT1 Facilities
RT Field Gases
RT Natural Gas
SO Air Pollution
DEF (ESH) Processing sites engaged in
      the extraction of natural gas
      liquids from field gas, fractionation
      of mixed natural gas liquids to
      natural gas products, or both.

NATURAL PHENOMENA EMERGENCIES
DA   January 24, 1991
BT1 Emergencies
 BT2 Reportable Occurrences
  BT3 Occurrences
RT Natural Phenomenon
SO Environmental Management
DEF (DOE 5500.1A) Conditions caused
      by flood, earthquake, fire, storm,
      or other natural occurrences.

NATURAL PHENOMENON
DA   October 19, 1990
SY   Act of Nature
NT1 Algae Bloom
NT1 Anticipated Processes and Events
NT1 Natural Disasters

NT2 Fires
NT2 Landslides
NT2 Tornadoes
NT2 Tsunamis
NT1 Red Tide
NT1 Water Bloom
RT Natural Phenomena Emergencies
RT Safe Shutdown Earthquake 135
DEF (WEBSTER) Exceptional, unusual,
      or abnormal natural occurrences.

NATURAL RESOURCE DAMAGE ASSESSMENT
(DOE Order 5400.4; CFR)
DA   October 16, 1990
BT1 Damage Assessments
 BT2 Assessments
  BT3 Administrative Processes
   BT4 Processes
RT Compensation and Liability
DEF (DOE Order 5400.4) An
      assessment (conducted under 43
      CFR Part 11), based on the
      results of a Natural Resource
      Damage Preassessment Screen
      of a release, that (1) establishes
      whether a natural resource injury
      has occurred and resulted from
      the release, (2) quantifies the
      effects of the release in injury, and
      (3) determines the financial
      compensation appropriate for the
      injury. / The process of collecting,
      compiling, and analyzing
      information, statistics, or data
      through prescribed methodologies
      to determine damages for injuries
      to natural resources as set forth in
      this part.

NATURAL RESOURCE DAMAGE PREASSESSMENT SCREENS
(DOE Order 5400.4)
DA   October 16, 1990
BT1 Damage Assessments
 BT2 Assessments
  BT3 Administrative Processes
   BT4 Processes
SO Environmental Management
DEF (DOE Order 5400.4) Desk-top
      reviews of existing data
      (conducted under 43 CFR Part 11)
      that are triggered when DOE is
      notified by an on-scene
      coordinator or lead agency of a
      potential injury due to a release to
      a natural resource for which DOE
      is a trustee. Such a review is to be
      completed as expeditiously as
      possible, with a minimal amount of
      field work, and provide a
      preliminary identification of the
      substance released and its
      source, initial estimates of the
      pathways for the purposes of
      identifying resources that may be
      impacted, and further identification
      of important resources that may
      justify further assessment.

SY-Synonymous Terms   SO-Source/Subject Category   SF-See From

NATURAL RESOURCE DEFENSE COUNCIL
DA   January 8, 1991
SF   *NRDC*
BT1 Professional Bodies
 BT2 Organizations

NATURAL RESOURCE TRUSTEES
(CFR)
DA   November 15, 1990
SY   Trustees
RT   Federal Agencies
RT   Indian Tribes
RT   State Agencies
DEF (CFR) Any Federal natural
     resources management agencies
     designated in the National Oil and
     Hazardous Substances
     Contingency Plan (NCP) and any
     State agencies designated by the
     Governor of each State, pursuant
     to section 107(f)(2)(B) of
     CERCLA, that may prosecute
     claims for damages under section
     107(f) or 111(b) of CERCLA; or an
     Indian tribes, that may commence
     an action under section 126(d) of
     CERCLA.

NATURAL RESOURCES
(CERCLA; CFR; USC)
DA   October 12, 1990
SY   Resource
NT1 Air Resources
NT1 Biological Resources
NT1 Geologic Resources
 NT2 Fossil Fuels
  NT3 Coal
   NT4 Ultra Clean Coal
  NT3 Natural Gas
   NT4 Field Gases
   NT4 Natural Gas Liquids
  NT3 Petroleum
 NT2 Uranium
  NT3 Depleted Uranium
  NT3 Enriched Uranium
   NT4 Unirradiated Enriched Uranium
NT1 Groundwater Resources
NT1 Indian Lands
NT1 Public Lands
NT1 Surface Water Resources
 NT2 Brackish Waters
 NT2 Estuaries
 NT2 Floodplains
 NT2 Lakes
  NT3 Dystrophic Lakes
  NT3 Eutrophic Lakes
  NT3 Oligotrophic Lakes
 NT2 River/Creek
 NT2 Wetlands
  NT3 Bogs
  NT3 Bottomland Hardwoods
  NT3 Fens
  NT3 Marshes
   NT4 Tidal Marshes
  NT3 Swamps
RT   Archaeological Resources
     Protection Act of 1979
RT   Archaeological Resources
RT   Assessment Area

RT   Assessments
RT   National Environmental Policy Act
SO   Compensation and Liability
SO   Environmental Management
DEF (CFR) As defined by section
     101(16) of CERCLA, land, fish,
     wildlife, biota, air, water, ground
     water, drinking water supplies, and
     other such resources belonging to,
     managed by, held in trust by,
     appertaining to, or otherwise
     controlled by the United States
     (including the resources of the
     fishery conservation zone
     established by the Magnuson
     Fishery Conservation and
     Management Act of 1976), any
     State or local government, any
     foreign government, any Indian
     tribe, or, if such resources are
     subject to a trust restriction on
     alienation, any member of an
     Indian tribe. These natural
     resources have been categorized
     into the following five groups:
     surface water resources, ground
     water resources, air resources,
     geologic resources, and biological
     resources.

NATURAL SELECTION
(EPA)
DA   October 12, 1990
BT1 Biological Processes
 BT2 Processes
RT   Adaptations
SO   Environmental Protection Agency
     Glossary
DEF (EPA) The process of survival of
     the fittest, by which organisms that
     adapt to their environment survive
     and those that do not disappear.

NATURALLY OCCURRING BACKGROUND LEVELS
(EPA)
DA   October 12, 1990
SO   Environmental Protection Agency
     Glossary
DEF (EPA) Ambient concentrations of
     chemicals that are present in the
     environment and have not been
     influenced by humans (e.g.,
     aluminum, manganese).

NATURALLY OCCURRING RADIOACTIVE MATERIAL
(DOE Order 5820.2A; ANL)
DA   May 24, 1991
BT1 Radioactive Materials
 BT2 Materials
RT   Accelerator Produced Radioactive
     Materials
SO   Radiation

NATURE OF PROGRAMMATIC IMPACT
(DOE FRASE Vocabulary Numeric Keys
1575-1599)
DA   January 17, 1991

NT1 Delay of Experiment
NT1 Delay of Response
NT1 Delay of Sampling
NT1 Loss of Operating Time
NT1 Loss of Production
NT1 No Programmatic Impact
NT1 Unscheduled Shutdown
NT1 Unscheduled Evacuation
RT   Damage Assessments
RT   Delays
RT   DOE FRASE Categories
SO   DOE FRASE VOCABULARY
DEF A subject category used with the
     DOE FRASE Vocabulary.

NATURE OF PROPERTY DAMAGE
(DOE FRASE Vocabulary Numeric Keys
1525-1574)
DA   January 17, 1991
NT1 Automatic Reactor Scram
NT1 Damage
 NT2 Electronic Damage
 NT2 Fire Damage
 NT2 Heat Damage
 NT2 Mechanical Damage
 NT2 Smoke Damage
 NT2 Structural Damage
 NT2 Water Damage
 NT2 Wind Damage
NT1 Dent
NT1 Electrical Short
NT1 Environmental Release
NT1 Hazardous Spill
NT1 Loss of Transmission
NT1 Loss of Material
NT1 Loss of Specimen
NT1 Loss of Instrument Air
NT1 Loss of Experiment
NT1 Manual Reactor Scram
NT1 Power Outage
NT1 Power Shutdown
NT1 Radioactive Contamination
 NT2 Non-fixed Radioactive
      Contamination
NT1 Vehicle Accident
NT1 Violation of Working Limit
NT1 Violation of Operational Safety Req
NT1 Violation of Tech Specification
NT1 Violation of Occup Safety
     Regulation
NT1 Violation of Criticality Specifictn
RT   Damage Assessments
RT   DOE FRASE Categories
RT   Violations
SO   DOE FRASE VOCABULARY
DEF A subject category used with the
     DOE FRASE Vocabulary.

NAVAL REACTORS
DA   January 11, 1991
BT1 Reactors
BT1 U.S. Department of Energy
 BT2 Federal Agencies
  BT3 Agencies
   BT4 Administrative Organizations
    BT5 Organizations
RT   General Electric
RT   Westinghouse Electric Corp.

NAVIGABLE WATERS
(SWDA; RCRA; CERCLA; CFR; USC; ESH)
DA   October 12, 1990
RT   Transportation
SO   Compensation and Liability
SO   Environmental Management
SO   Environmental Protection Agency
        Glossary
SO   Hazardous Materials
DEF Traditionally, waters sufficiently
        deep and wide for navigation by
        all, or specified sizes of vessels;
        such waters in the United States
        come under federal jurisdiction
        and are included in certain
        provisions of the Clean Water Act.
        (ESH) Limited to waters of the
        United States, including the
        territorial seas.

NC
DA   October 12, 1990
SEE Normally Closed
SO   Acronyms

NCF
DA   October 12, 1990
SEE No Containment Failure
SO   Acronyms

NCO
DA   May 14, 1991
SEE NEPA Compliance Officers
SO   Acronyms

NCP
(CERCLA; EPA; CFR)
DA   November 15, 1990
SEE National Contingency Plan
SO   Acronyms
DEF The National Oil and Hazardous
        Substances Contingency Plan and
        revisions promulgated by EPA,
        pursuant to section 105 of
        CERCLA and codified in 40 CFR
        Part 300.

NCR
DA   October 12, 1990
SEE Non-Conformance Report
SO   Acronyms

NCSTATE
DA   October 12, 1990
SEE North Carolina State University
SO   Acronyms

NDE
DA   October 12, 1990
SEE Non-Destructive Evaluation
SO   Acronyms

NDT
DA   October 12, 1990
SEE Non-Destructive Test
SO   Acronyms

NE
DA   October 12, 1990
SEE Nuclear Energy
SO   Acronyms

NEAR SURFACE DISPOSAL
DA   January 24, 1991
BT1 Land Disposal
 BT2 Disposal
  BT3 Waste Management Processes
  BT4 Processes
NT1 Shallow Land Burial
SO   Environmental Management
DEF (DOE 5820.2A) Disposal in the
        upper 30 meters of the earth's
        surface (e.g., shallow land burial).

**NEAR SURFACE DISPOSAL
FACILITIES**
(ANL; CFR)
DA   May 24, 1991
BT1 Land Disposal Facilities
 BT2 Disposal Facilities
  BT3 Hazardous Waste Management
          Facilities
   BT4 Hazardous Waste Facilities
   BT5 Facilities
RT   Radioactive Wastes
SO   Wastes
DEF (CFR) Land disposal facilities in
        which radioactive waste is
        disposed of in or within the upper
        30 meters of the earth's surface.

NEC
DA   October 12, 1990
SEE National Electrical Code
SO   Acronyms

NECK
(1093 NECK)
DA   November 28, 1990
BT1 Human Body Parts
RT   Spine
RT   Throat
SO   DOE FRASE VOCABULARY

NECROSIS
(EPA)
DA   October 12, 1990
RT   Plants
SO   Environmental Protection Agency
        Glossary
DEF (EPA) Death of plant or animal
        cells. In plants, necrosis can
        discolor areas on the plant or kill it
        entirely.

NEGLIGENCE
(SSDC)
DA   October 12, 1990
RT   Human Factors
SO   System Safety Development Center
        Glossary
DEF (SSDC) Failure to use such care as
        a reasonably prudent and careful
        person would under similar
        circumstances.

NEMATOCIDES
(EPA)
DA   October 12, 1990
BT1 Pesticides
 BT2 Hazardous Substances
RT   Nematodes
SO   Environmental Protection Agency
        Glossary
DEF (EPA) A chemical agent which is
        destructive to nematodes (round
        worms or threadworms).

NEMATODES
(USC)
DA   November 15, 1990
BT1 Animals
RT   Nematocides
DEF (USC) Invertebrate animals of the
        phylum nemathelminthes and
        class nematoda, that is,
        unsegmented round worms with
        elongated, fusiform, or saclike
        bodies covered with cuticle, and
        inhabiting soil, water, plants, or
        plant parts; may also be called
        nemas or eelworms.

NEPA
(Acronyms and Abbreviations)
DA   October 12, 1990
SEE National Environmental Policy Act
SO   Acronyms
DEF (NEPA) To declare a national policy
        which will encourage productive
        and enjoyable harmony between
        man and his environment; to
        promote efforts which will prevent
        or eliminate damage to the
        environment and biosphere and
        stimulate the health and welfare of
        man; to enrich the understanding
        of the ecological systems and
        natural resources important to the
        Nation; and to establish a Council
        on Environmental Quality.

NEPA COMPLIANCE GUIDES
(DOE Order 5440.1D)
DA   May 14, 1991
BT1 Guidelines
SO   Environmental Management
DEF (DOE Order) A DOE collection of
        written guidance and reference
        material to assist DOE staff in
        both planning for and achieving
        compliance with NEPA and
        various related environmental
        statutes. The Guide, which is
        updated periodically by EH,
        provides information on the NEPA
        process, the content of the NEPA
        documents, the substantive and
        timing relationships between
        NEPA reviews and review
        requirements of other
        environmental statutes, and timing
        relationships between the NEPA
        process and the development of
        DOE actions.

NEPA COMPLIANCE OFFICERS
(DOE Order 5440.1D)
DA   May 14, 1991
SF   *NCO*
BT1 Personnel
SO   Environmental Management
DEF (DOE Order) DOE employees at
    Program Offices and Operations
    Offices, and optionally at other
    offices, designated to coordinate,
    assist, and generally oversee the
    NEPA compliance activities in that
    office.

NEPA Documents
(DOE Order 5440.1D)
DA   May 15, 1991
SEE National Environmental Policy Act
    Documents
SO   Environmental Management
DEF (DOE Order) Environmental
    assessments, findings of no
    significant impact, notices of intent
    to prepare an environmental
    impact statement, record of
    decision, categorical exclusion
    determination, or any other
    documents prepared pursuant to a
    requirement of NEPA or the
    Council on Environmental Quality
    Regulations (40 CFR Parts
    1500-=1508, as amended 7-1-86).

NEPA STATUS REPORTS
(DOE Order 5440.1D)
DA   May 15, 1991
BT1 Reports
RT   Activity Data Sheets
RT   Quarterly Environmental
    Compliance Reports
SO   Environmental Management
DEF (DOE Order) Reports on the status
    of existing or planned NEPA
    compliance activities, which are
    included in internal budget review
    documents (i.e., project data
    sheets or activity data sheets)
    prepared pursuant to DOE Order
    5100.3.

NERVOUS SYSTEM
(1144 NERVOUS SYST)
DA   November 28, 1990
BT1 Body System(s)
 BT2 Human Body Parts
NT1 Brain
NT1 Spinal Cord
SO   DOE FRASE VOCABULARY

NESHAPS
DA   October 12, 1990
SEE National Emissions Standards for
    Hazardous Air Pollutants
SO   Acronyms

NET EXPLOSIVE WEIGHT
(Doe Order 5480.16)
DA   October 12, 1990
BT1 Measurements

RT   Munitions
SO   Firearms
DEF (DOE Order 5480.16) The weight of
    the energy-producing material in
    munitions.

NET POSITIVE SUCTION HEAD
DA   January 8, 1991
SF   *NPSH*

NET WEIGHT
(ESH)
DA   October 19, 1990
BT1 Measurements
RT   Gross Weight
SO   Hazardous Materials
DEF (ESH) A measure of weight
    referring only to the contents of a
    package, and does not include the
    weight of any packaging material.

NET WORKING CAPITAL
(SWDA; RCRA; CFR)
DA   October 12, 1990
RT   Current Assets
RT   Current Liabilities
RT   Current Plugging and
    Abandonment Cost Estimates
SO   Wastes
SO   Water Pollution
DEF Current assets minus current
    liabilities.

NET WORTH
(SWDA; RCRA; CFR; ESH)
DA   October 12, 1990
NT1 Tangible Net Worth
RT   Liabilities
SO   Wastes
SO   Water Pollution
DEF (CFR) Total assets minus total
    liabilities and is equivalent to
    owner's equity. Used in the
    specifications for the financial
    tests for closure, post-closure
    care, and liability coverage. The
    definition is intended to assist in
    the understanding of these
    regulations and is not intended to
    limit the meanings of terms in a
    way that conflicts with generally
    accepted accounting practices.

NEUTRALIZATION
(EPA)
DA   October 12, 1990
BT1 Chemical Processes
 BT2 Processes
RT   Antagonism
RT   Chemical Treatments
RT   Elementary Neutralization Units
RT   pH
SO   Environmental Protection Agency
    Glossary
DEF (EPA) Decreasing the acidity or
    alkalinity of a substance by adding
    to it alkaline or acidic materials
    respectively.

NEUTRON(S)
(2581 NEUTRON)
DA   January 3, 1991
BT1 Equipment/Parts - Nuclear (DOE
    FRASE Vocabulary)
 BT2 Equipment
 BT2 Reactor Components
RT   Neutron Source
SO   DOE FRASE VOCABULARY
DEF (NRC Glossary of Terms: Nuclear
    Power and Radiation) Uncharged
    elementary particles with a mass
    slightly greater than that of the
    proton, and found in the nucleus of
    every atom heavier than hydrogen.

NEUTRON-DEFICIENT ISOTOPES
(EDB)
DA   March 29, 1991
BT1 Radionuclides
 BT2 CERCLA Hazardous Substances
 BT3 Hazardous Substances
 BT2 Nuclides
SO   Radiation

NEUTRON POISONS
(SSDC)
DA   May 24, 1991
SY   Nuclear Poisons
SO   Radiation
DEF (SSDC) Substances with large
    neutron cross-sections. (EDB)
    Neutron absorbers in a reactor.

NEUTRON-RICH ISOTOPES
(EDB)
DA   March 29, 1991
BT1 Radionuclides
 BT2 CERCLA Hazardous Substances
 BT3 Hazardous Substances
 BT2 Nuclides
SO   Radiation

NEUTRON SOURCE
(2579 NEUTRON SOUR)
DA   January 3, 1991
BT1 Equipment/Parts - Nuclear (DOE
    FRASE Vocabulary)
 BT2 Equipment
 BT2 Reactor Components
RT   Neutron(s)
SO   DOE FRASE VOCABULARY
DEF (NRC Glossary of Terms: Nuclear
    Power and Radiation) A
    radioactive material (decays by
    neutron emission) that can be
    inserted into a reactor to ensure
    that a sufficient quantity of
    neutrons is available to start a
    chain reaction and register on
    neutron detection equipment.

NEVADA OPERATIONS OFFICE
DA   January 8, 1991
SF   *NV*
BT1 Operations Offices
 BT2 Offices
 BT3 Administrative Organizations
 BT4 Organizations

BT-Broader Term   NT-Narrower Term   RT Related Term

BT2 U.S. Department of Energy
BT3 Federal Agencies
BT4 Agencies
BT5 Administrative Organizations
BT6 Organizations
RT EG&G Energy Measurements, Inc.
RT Raytheon Services Nevada
RT Reynolds Electrical and
Engineering Company
RT Wackenhut Services Inc.
DEF (Capsule Review of DOE Research
and Development and Field
Facilities, 1986) The Nevada
Operations Office, established by
the Atomic Energy Commission in
1962, is responsible for programs
at the Nevada Test Site and
conducts all U.S. nuclear tests,
including weapons development
and weapons effects. The Site is
an integral part of the Office and
does not operate as a separate
entity. Located 65 miles northwest
of Las Vegas, the Test Site was
chosen as a continental proving
ground in 1950 to reduce the
expense and logistical problems of
testing in the Pacific. Through Site
operations, Office contractors
have developed unique expertise
in fields such as big hole drilling,
mining and downhole diagnostics
in support of the weapons
development laboratories and the
Department of Defense. Other
programs and projects include:
waste management; geologic,
hydrologic and seismic
investigations; biological and
medical experiments, and the
effects of liquid gaseous fuel spills.

NEVADA TEST SITE
DA January 8, 1991
SF *NTS*
BT1 Government-Owned
Contractor-Operated Facilities
BT2 Federal Facilities
BT3 Facilities
BT1 Test Area
BT2 Area
BT3 Sites/Areas
RT Atomic Weapons
RT EG&G Energy Measurements, Inc.
RT Raytheon Services Nevada
RT Reynolds Electrical and
Engineering Company
RT Wackenhut Services Inc.

NEW CHEMICAL SUBSTANCES
(TSCA; USC)
DA October 19, 1990
BT1 Chemical Substances
SO Environmental Management
SO Hazardous Materials
DEF (USC) Chemical substances that
are not included in the chemical
substance list compiled and
published under section 8(b) [15
USCS 2607(b).

NEW CLASS II WELLS
(SDWA; CFR)
DA October 12, 1990
BT1 Class II Wells
BT2 Injection Wells
BT3 Wells
SO Water Pollution
DEF (CFR) Wells constructed or
converted after the effective date
of this program, or which are
under construction on the effective
date of this program.

NEW FACILITIES
(SDWA; RCRA)
DA October 19, 1990
SY New Hazardous Waste
Management (HWM) Facilities
BT1 Facilities
SO Wastes
DEF (CFR) Facilities which began
operation, or for which
construction began commenced
after October 21, 1976. (See also
"Existing hazardous waste
management facilities").

**NEW HAZARDOUS WASTE
MANAGEMENT (HWM) FACILITIES**
(SWDA; RCRA)
DA October 12, 1990
SY New Facilities
BT1 Hazardous Waste Management
Facilities
BT2 Hazardous Waste Facilities
BT3 Facilities
SO Environmental Management
SO Wastes
DEF (CFR) Facilities which began
operation, or for which
construction commenced after
October 21, 1976. (See also
Existing Hazardous Waste
Management Facilities).

**NEW MEXICO ENVIRONMENTAL
IMPROVEMENT DIVISION**
DA January 8, 1991
SF *NMEID*
BT1 Divisions
BT2 Administrative Organizations
BT3 Organizations
BT1 State Agencies
BT2 Agencies
BT3 Administrative Organizations
BT4 Organizations

**NEW MEXICO WATER QUALITY
CONTROL COMMISSION
REGULATIONS**
DA January 8, 1991
SF *NMWQCCR*

NEW PRODUCTION REACTOR
DA January 8, 1991
SF *NPR*
BT1 Reactors

**NEW SOURCE PERFORMANCE
STANDARDS**
(CAA; CFR; USC; EPA; ESH)
DA October 12, 1990
SF *NSPS*
BT1 Emission Standards
BT2 Requirements
BT2 Standards
BT3 Codes, Standards, and
Regulations
SO Air Pollution
SO Environmental Protection Agency
Glossary
DEF (EPA) Uniform national EPA air
emission and water effluent
standards which limit the amount
of pollution allowed from new
sources or from existing sources
that have been modified. (ESH)
Emission standards for new
stationary sources or for major
modifications of existing sources.
New Source Performance
Standards are found in 40 CFR
60. These standards reflect the
best system of emission reduction
which (taking into account the cost
of achieving such reduction) the
EPA Administrator determines has
been adequately demonstrated.

NEW SOURCES
(CAA; CFR)
DA October 12, 1990
BT1 Stationary Sources
BT2 Facilities
SO Air Pollution
SO Environmental Management
SO Environmental Protection Agency
Glossary
SO Hazardous Materials
DEF (CFR) Any stationary sources that
are built or modified after
publication of final or proposed
regulations that prescribe a
standard of performance which is
intended to apply to that type of
emission source. (ESH) Any
building, structure, facility, or
installation from which there is or
may be a discharge of pollutants,
the construction of which is
commenced after the publication
of proposed regulations
prescribing a standard of
performance under §306 of the
Clean Air Act.

NEW STORAGE FACILITIES
(DOE Order 6430.1A)
DA October 12, 1990
BT1 Nuclear Facilities
BT2 Facilities
RT Unirradiated Enriched Uranium
SO Construction
DEF (DOE Order 6430.1A) Newly
constructed facilities, or the
conversion of existing facilities, or
portions of existing facilities, for

use as unirradiated enriched
uranium storage facilities.

NEW TAILINGS
(CAA; CFR)
DA   October 12, 1990
BT1 Tailings
 BT2 Solid Wastes
  BT3 Wastes
RT   Existing Tailings Piles
RT   New Tailings Impoundment
RT   Uranium By-Product Materials
SO   Air Pollution
DEF Uranium tailings produced after the
     effective date of 40 CFR 61.251.

NEW TAILINGS IMPOUNDMENT
(CAA; CFR)
DA   October 12, 1990
BT1 Impoundment
 BT2 Sites/Areas
RT   New Tailings
RT   Tailings
SO   Air Pollution
DEF (CFR) Any location or structure at
     which uranium mill tailings are
     temporarily or permanently stored
     and which is placed in operation
     after the promulgation of 40 CFR
     61.251.

NEW TANK COMPONENTS
(SWDA; RCRA)
DA   October 19, 1990
SY   New Tank Systems
BT1 Facility Components
SO   Wastes
DEF (USC) Systems or components that
     will be used for the storage or
     treatment of hazardous waste and
     for which installation has
     commenced after July 14, 1986;
     except, however, for purposes of
     40 CFR 264.193(g)(2) and 40
     CFR 265(g)(2), new tank systems
     are ones for which construction
     commences after July 14, 1986.
     (See also Existing Tank Systems).

NEW TANK SYSTEMS
(SWDA; RCRA; CFR)
DA   October 19, 1990
SY   New Tank Components
BT1 Systems
BT1 Tank Systems
SO   Environmental Management
SO   Wastes
DEF (CFR) A tank system that will be
     used to contain an accumulation
     of regulated substances and for
     which installation has commenced
     after December 22, 1988.

NFA
DA   October 12, 1990
SEE No Flow Assemblies
SO   Acronyms

NFPA
DA   October 12, 1990
SEE National Fire Protection Association
SO   Acronyms

NIGHT SIMULATION GLASSES
(DOE Order 5480.16)
DA   October 16, 1990
RT Night Simulations
DEF (DOE Order 5480.16) Goggles for
     day use that simulate night
     conditions.

NIGHT SIMULATIONS
(Doe Order 5480.16)
DA   October 12, 1990
RT Night Simulation Glasses
SO   Firearms

NIGHT VISION GOGGLES
(Doe Order 5480.16)
DA   October 12, 1990
SO   Firearms
DEF (DOE Order 5480.16) The principal
     component of a night vision
     imaging system.

NIM
DA   October 12, 1990
SEE Nuclear Incident Monitor
SO   Acronyms

NIST
DA   October 12, 1990
SEE National Institute of Standards and
     Testing
SO   Acronyms

NITRATES
(EPA)
DA   October 12, 1990
RT Denitrification
SO   Environmental Protection Agency
     Glossary
DEF (EPA) Compounds containing
     nitrogen which can exist in the
     atmosphere or as a dissolved gas
     in water and which can have
     harmful effects on humans and
     animals. Nitrates in water can
     cause severe illness in infants and
     cows.

NITRIC ACID PLANTS
(CAA; CFR)
DA   October 12, 1990
BT1 Facilities
RT Acid Deposition
SO   Air Pollution
DEF (CFR) Facilities producing nitric
     acid 30 to 70 percent in strength
     by either the pressure or
     atmospheric pressure process.

NITRIC OXIDES
(EPA)
DA   October 12, 1990
SF   *NO (Nitric Oxide)*

BT1 Nitrogen Oxides (NO$_x$)
 BT2 CERCLA Hazardous Substances
  BT3 Hazardous Substances
 BT2 Criteria Pollutants
  BT3 Pollutants
 BT2 Gases
 BT2 Inorganic Chemicals
  BT3 Chemical Substances
SO   Environmental Protection Agency
     Glossary
DEF (EPA) Gases formed by combustion
     under high temperature and high
     pressure in an internal combustion
     engine, change into nitrogen
     dioxide in the ambient air and
     contribute to photochemical smog.

NITRIFICATION
(EPA)
DA   October 12, 1990
BT1 Chemical Processes
 BT2 Processes
RT Denitrification
RT Nitrites
SO   Environmental Protection Agency
     Glossary
DEF (EPA) The process whereby
     ammonia in wastewater is oxidized
     to nitrite and then to nitrate by
     bacterial or chemical reactions.

NITRILOTRIACETIC ACID
(EPA)
DA   October 12, 1990
SF   *NTA*
BT1 Organic Chemicals
 BT2 Chemical Substances
SO   Environmental Protection Agency
     Glossary
DEF (EPA) A compound being used to
     replace phosphates in detergents.

NITRITES
(EPA)
DA   October 12, 1990
BT1 Intermediates
 BT2 Chemical Substances
RT Nitrification
SO   Environmental Protection Agency
     Glossary
DEF (EPA) (1) An intermediate in the
     process of nitrification. (2) Nitrous
     oxide salts used in food
     preservation.

NITROGEN DIOXIDE (NO$_2$)
(EPA)
DA   October 12, 1990
SF   *NO$_2$*
BT1 Nitrogen Oxides (NO$_x$)
 BT2 CERCLA Hazardous Substances
  BT3 Hazardous Substances
 BT2 Criteria Pollutants
  BT3 Pollutants
 BT2 Gases
 BT2 Inorganic Chemicals
  BT3 Chemical Substances

BT-Broader Term   NT-Narrower Term   RT Related Term

SO Environmental Protection Agency
  Glossary
DEF (EPA) The result of nitric oxide
  combining with oxygen in the
  atmosphere. A major component
  of photochemical smog.

NITROGEN OXIDES (NO$_x$)
(EPA)
DA   October 12, 1990
BT1 CERCLA Hazardous Substances
 BT2 Hazardous Substances
BT1 Criteria Pollutants
 BT2 Pollutants
BT1 Gases
BT1 Inorganic Chemicals
 BT2 Chemical Substances
NT1 Nitric Oxides
NT1 Nitrogen Dioxide (NO$_2$)
RT Acid Deposition
SO Environmental Protection Agency
  Glossary
DEF (EPA) Product of combustion from
  transportation and stationary
  sources and a major contributor to
  the formation of ozone in the
  troposphere and acid deposition.

NITROGENOUS WASTES
(EPA)
DA   October 12, 1990
BT1 Wastes
SO Environmental Protection Agency
  Glossary
DEF (EPA) Animal or vegetable residues
  that contain significant amounts of
  nitrogen.

NLU
DA   October 12, 1990
SEE Normal Latch Up
SO Acronyms

NMEID
DA   October 12, 1990
SEE New Mexico Environmental
  Improvement Division
SO Acronyms

NMWQCCR
DA   October 12, 1990
SEE New Mexico Water Quality Control
  Commission Regulations
SO Acronyms

NO (Nitric Oxide)
DA   October 12, 1990
SEE Nitric Oxides
SO Acronyms

NO (normally open)
DA   October 12, 1990
SEE Normally Open
SO Acronyms

NO ACTIVITY
(1254 NO ACTIVITY)
DA   November 28, 1990

BT1 Activity Types (DOE FRASE
  Vocabulary)
 BT2 Activities
SO DOE FRASE VOCABULARY

NO CONTAINMENT FAILURE
DA   January 8, 1991
SF   *NCF*
BT1 Failures
 BT2 Accidents

NO FLOW ASSEMBLIES
DA   January 8, 1991
SF   *NFA*

NO-OBSERVED-EFFECT-LEVEL
(EPA)
DA   October 12, 1990
RT Dose-Response Evaluation
SO Environmental Protection Agency
  Glossary
DEF (EPA) In dose-response
  experiments, the experimental
  exposure level representing the
  highest level tested at which no
  effects at all were demonstrated.

NO PERSONNEL INJURY
(1333 NO PERSONNEL)
DA   November 28, 1990
BT1 Injuries
SO DOE FRASE VOCABULARY

NO PROGRAMMATIC IMPACT
(1582 NO PROG IMPA)
DA   November 28, 1990
BT1 Nature of Programmatic Impact
SO DOE FRASE VOCABULARY

NO PROPERTY DAMAGE OR INJURY
(1535 NO PROP DAM)
DA   November 29, 1990
SO DOE FRASE VOCABULARY

NOA
DA   May 16, 1991
SEE Notice of Availability
SO Acronyms

NOAEL
DA   October 12, 1990
SEE Non-Observed Adverse Effect Level
SO Acronyms

NOD
DA   October 12, 1990
SEE Notice of Deficiency
SO Acronyms

NODULIZING KILNS
(CAA)
DA   October 12, 1990
RT Calciners
RT Pyrolysis
RT Roasting
SO Air Pollution
DEF Units in which phosphate rock is

heated to high temperatures to
remove organic material and/or to
convert it to a nodular form.

NOHSCP
DA   January 11, 1991
SEE National Oil and Hazardous
  Substances Contingency Plan
SO Acronyms

NOI
DA   October 12, 1990
SEE Notice of Intent
SO Acronyms

NOISE POLLUTION ABATEMENT
(NFI)
DA   October 12, 1990
BT1 Abatement
 BT2 Pollution Recovery Processes
  BT3 Processes
RT A-Scale Sound Levels
RT Pollution
SO Nuclear Facilities Incident Database
SO Radiation
DEF Used for noise suppression circuits.

**NOMINAL AUTOMATIC BURNOUT
SAFETY FACTOR**
DA   January 8, 1991
SF   *ABOSFn*
BT1 Burnout Safety Factor

NOMINAL BOSF
DA   January 8, 1991
SF   *BOSFn*
BT1 Burnout Safety Factor

NON
DA   October 12, 1990
SEE Notification of Non-Compliance
SO Acronyms

NON-ATTAINMENT AREAS
(EPA; CAA; CFR; USC; ESH)
DA   October 12, 1990
BT1 Sites/Areas
RT National Ambient Air Quality
  Standards
RT Reasonably Available Control
  Technology
SO Air Pollution
SO Environmental Protection Agency
  Glossary
DEF (USC) For any air pollutant, any
  area which is shown by monitored
  data or which is calculated by air
  quality modeling (or other methods
  determined by the administrator to
  be reliable) to exceed any national
  ambient air quality standard for
  such pollutant. (EPA) Geographic
  areas that do not meet one or
  more of the National Ambient Air
  Quality Standards for the criteria
  pollutants designed in the Clean
  Air Act.

NON-BULK PACKAGING
DA   June 3, 1991
BT1 Packaging
 BT2 Packages
RT   Bulk Packaging
SO   Hazardous Materials
DEF Packaging which has (1) an internal
     volume of 450 liters (118.9
     gallons) or less as a receptacle for
     a liquid; (2) a capacity of 400
     kilograms (881.8 pounds) or less
     as a receptacle for a solid; or (3) a
     water capacity of 1000 pounds
     (453.6 kilograms) or less as a
     receptacle for a gas as defined in
     @173.300.

NON-COMMUNITY WATER SYSTEMS
(SDWA; CFR; EPA; ESH)
DA   October 12, 1990
BT1 Public Water Systems
 BT2 Water Systems
  BT3 Publicly Owned Treatment Works
   BT4 Treatment Facilities
    BT5 Facilities
  BT3 Systems
SO   Environmental Protection Agency
     Glossary
SO   Water Pollution
DEF Public water systems that are not a
     community water system, e.g., the
     water supplies at camp sites or
     national parks.

NON-CONFORMANCE REPORT
DA   January 8, 1991
SF   NCR
BT1 Reports

NON-CONTACT COOLING WATER
DA   January 29, 1991
BT1 Water
RT   Cooling Water Gamma Monitor
SO   Environmental Management
DEF (CFR) Water used for cooling which
     does not come into direct contact
     with any raw material,
     intermediate product, waste
     product or finished product.

NON-CONTACT COOLING WATER POLLUTANTS
DA   January 24, 1991
BT1 Pollutants
SO   Environmental Management
DEF (CFR) Pollutants present in
     non-contact cooling waters.

NON-CONVENTIONAL POLLUTANTS
(EPA)
DA   October 12, 1990
BT1 Pollutants
SO   Environmental Protection Agency
     Glossary
DEF (EPA) Any pollutants that are not a
     statutorily listed or that are poorly
     understood by the scientific
     community.

NON-DESTRUCTIVE EVALUATION
DA   January 8, 1991
SF   NDE

NON-DESTRUCTIVE TEST
DA   January 8, 1991
SF   NDT

NON-DETECTS
(EPA)
DA   October 12, 1990
BT1 Chemical Substances
SO   Environmental Protection Agency
     Glossary
DEF (EPA) Chemicals that are not
     detected in a particular sample
     above a certain limit. This limit
     usually will be the quantitation limit
     for the chemical in that sample.
     (Note, however, that it is possible
     to detect and estimate
     concentrations of chemicals below
     the quantitation limit but above the
     detection limit.)

NON-EMPLOYEE RADIATION WORKERS
(Doe Order 5484.1)
DA   October 12, 1990
BT1 Radiation Workers
 BT2 Occupational Workers
  BT3 Personnel
SO   Environmental Management
SO   Industrial Safety
DEF (DOE Order 5484.1) Individuals
     who are either a subcontractor to
     a DOE contractor or who visits a
     DOE site to perform work for or in
     conjunction with DOE or utilizes
     DOE facilities and who are
     monitored for occupational
     exposure as required in DOE
     5480.11.

NON-FIXED RADIOACTIVE CONTAMINATION
(HMTA; CFR)
DA   May 20, 1991
BT1 Radioactive Contamination
 BT2 Contamination
 BT2 Nature of Property Damage
SO   Hazardous Materials
DEF (CFR) Non-fixed radioactive
     contamination means radioactive
     contamination that can be readily
     removed from a surface by wiping
     with an absorbent material.

NON-FLAMMABLE COMPRESSED GASES
(CFR)
DA   October 12, 1990
BT1 Compressed Gases
 BT2 Gases
SO   Hazardous Materials
DEF (CFR) Any compressed gases that
     do not meet the definition as
     flammable compressed gases are
     considered as nonflammable

compressed gases for the purpose
of the hazardous materials
regulations.

NON-FRACTIONATING PLANTS
(CAA; CFR; USC; ESH)
DA   October 12, 1990
BT1 Facilities
SO   Air Pollution
DEF (ESH) Any gas plants that do not
     fractionate mixed natural gas
     liquids into natural gas products.

NON-IMPERVIOUS SOLID SURFACES
(TSCA; CFR)
DA   October 12, 1990
SO   Hazardous Materials
DEF (CFR) Solid surfaces that are
     porous and are more likely to
     absorb spilled PCBs prior to
     completion of the cleanup
     requirements prescribed in this
     policy. Nonimpervious solid
     surfaces include, but are not
     limited to, wood, concrete, asphalt,
     and plasterboard.

NON-IONIZING ELECTROMAGNETIC RADIATION
(EPA)
DA   October 12, 1990
SY   Radio Frequency Radiation
BT1 Radiation
SO   Environmental Protection Agency
     Glossary
DEF (EPA) (1) Radiation that does not
     change the structure of atoms but
     does heat tissue and may cause
     harmful biological effects. (2)
     Microwaves, radio waves, and
     low-frequency electromagnetic
     fields from high-voltage
     transmission lines.

NON-ISOLATED INTERMEDIATES
DA   January 29, 1991
BT1 Intermediates
 BT2 Chemical Substances
SO   Environmental Management
DEF (CFR) Intermediates that are not
     intentionally removed from the
     equipment in which they are
     manufactured, including the
     reaction vessel in which they are
     manufactured, equipment which is
     ancillary to the reaction vessel,
     and any equipment through which
     the chemical substances pass
     during a continuous flow process,
     but not including tanks or other
     vessels in which the substances
     are stored after their manufacture.

NON-LIQUEFIED COMPRESSED GASES
DA   June 3, 1991
BT1 Compressed Gases
 BT2 Gases
RT   Compressed Gases in Solution

SO Hazardous Materials
DEF Gases, other than in solution, which
    under the charged pressure are
    entirely gaseous at a temperature
    of 70 degrees F.

NON-NUCLEAR FACILITIES
(EMER)
DA February 1, 1991
BT1 Facilities
RT Nuclear Facilities
SO Emergency Preparedness
DEF (EMER) Facilities which meet
    neither the nuclear nor reactor
    facility definitions. Office buildings
    and nonradioactive hazardous
    waste storage facilities that are
    staffed by persons whose work
    does not involve hazardous
    quantities of radioactive material
    are examples of non-nuclear
    facilities.

NON-OBSERVED ADVERSE EFFECT
LEVEL
(EPA)
DA December 5, 1990
SF NOAEL
RT Dose-Response Evaluation
SO Environmental Protection Agency
    Glossary
DEF (EPA) In dose-response
    experiments, the experimental
    exposure level representing the
    highest level tested at which no
    adverse effects were
    demonstrated.

NON-OPERATIONAL STORAGE TANKS
(SWDA; RCRA; USC)
DA October 12, 1990
BT1 Underground Storage Tanks
 BT2 Underground Tanks
  BT3 Tanks
   BT4 Facility Components
SO Wastes
DEF Any underground storage tanks in
    which regulated substances will
    not be deposited or from which
    regulated substances will not be
    dispensed after the date of the
    enactment of the Hazardous and
    Solid Waste Amendments of 1984
    [enacted Nov. 8, 1984].

NON-PCB TRANSFORMERS
(TSCA; CFR)
DA October 12, 1990
RT PCB Transformers
RT Retrofill
SO Hazardous Materials
DEF (CFR) Any transformers that
    contain less that 50 ppm PCB;
    except that any transformers that
    have been converted from a PCB
    Transformer or a
    PCB-Contaminated transformer
    cannot be classified as a non-PCB
    Transformer until reclassification

has occurred, in accordance with
the requirements of [40 CFR - Part
761.30(a)(2)(v)].

NON-PENETRATING RADIATION
(EMER)
DA February 1, 1991
BT1 Radiation
NT1 Alpha Rays
NT1 Beta Rays
SO Emergency Preparedness
DEF (EMER) A general term used to
    describe external radiations of
    such low penetrating power that
    the absorbed dose from exposures
    to humans is principally in the skin
    and does not reach deeper organs
    to any significant extent. It refers
    to alpha, beta, and very low
    energy gamma or x-ray radiations.

NON-POINT SOURCES
(EPA)
DA October 12, 1990
SO Environmental Protection Agency
    Glossary
DEF (EPA) Pollution sources that are
    diffuse and do not have a single
    point of origin or are not
    introduced into a receiving stream
    from a specific outlet. The
    pollutants are generally carried off
    the land by stormwater runoff. The
    commonly used categories for
    non-point sources are: agriculture,
    forestry, urban, mining,
    construction, dams and channels,
    land disposal, and saltwater
    intrusion.

NON-POWERED HANDTOOL(S)
(3044 NP HANDTOOL)
DA January 3, 1991
BT1 Tools - Manual
 BT2 Tools (DOE FRASE Vocabulary)
  BT3 Equipment
SO DOE FRASE VOCABULARY

NON-RESTRICTED ACCESS AREAS
(TSCA; CFR)
DA October 12, 1990
BT1 Sites/Areas
RT Controlled Areas
RT Residential Areas
SO Hazardous Materials
DEF (CFR) Any areas other than
    restricted access, outdoor
    electrical substations, and other
    restricted access locations, as
    defined in this section. In addition
    to residential/commercial areas,
    these areas include unrestricted
    access rural areas (areas of
    low-density development and
    population where access is
    uncontrolled by either man-made
    barriers or naturally occurring
    barriers, such as rough terrain,
    mountains, or cliffs).

NON-REUSEABLE CONTAINERS
(SWDA; RCRA)
DA October 12, 1990
BT1 Containers
DEF (ESH) Containers whose reuse is
    restricted.

NON-STOCHASTIC EFFECTS
(Doe Order 5480.11)
DA October 12, 1990
BT1 Effects
NT1 Non-Stochastic Radiation Effects
SO Environmental Management
SO Industrial Hygiene
DEF (DOE Order 5480.11) Effects such
    as the opacity of the lens of the
    eye for which the severity of the
    effect varies with the dose, and for
    which a threshold may exist.

NON-STOCHASTIC RADIATION
EFFECTS
(IAEA)
DA October 12, 1990
BT1 Non-Stochastic Effects
 BT2 Effects
RT Irradiation
SO Radiation
DEF (IAEA) Radiation effects for which a
    threshold exists above which the
    severity of the effect varies with
    the dose.

NON-SUDDEN ACCIDENTAL
OCCURRENCES
(SWDA; RCRA; CFR)
DA October 12, 1990
BT1 Accidental Occurrences
 BT2 Occurrences
SO Wastes
DEF (CFR) Occurrences that take place
    over time and involve continuous
    or repeated exposure.

NON-TARGET ORGANISMS
DA January 24, 1991
BT1 Organisms
SO Environmental Management
DEF (CFR) Those flora and fauna
    (including man) are not intended
    to be controlled, injured, killed or
    detrimentally affected in any way
    by a pesticide.

NON-TRANSIENT NON-COMMUNITY
WATER SYSTEM
(SWDA; CFR; EPA; ESH)
DA October 19, 1990
SF NTNCWS
BT1 Public Water Systems
 BT2 Water Systems
  BT3 Publicly Owned Treatment Works
   BT4 Treatment Facilities
    BT5 Facilities
  BT3 Systems
SO Water Pollution
DEF (CFR) A public water system that is
    not a community water system
    and that regularly serves at least

25 of the same persons over 6 months per year.

NORMAL FORM
(EMER)
DA   February 1, 1991
RT   Contamination
RT   Radioactive Materials
RT   Special Form
SO   Emergency Preparedness
DEF   (EMER) Those materials which by nature of their physical form or encapsulation if released from a package, might present some possibility of contamination as well as direct radiation. For example, materials in the form of liquids or powdery-like substances are more likely to be dispersible. (See special form.)

NORMAL FORM RADIOACTIVE MATERIALS
(CFR)
DA   October 12, 1990
BT1   Radioactive Materials
 BT2 Materials
SO   Environmental Management
SO   Hazardous Materials
DEF   (CFR) Radioactive materials that have not been demonstrated to qualify as special form radioactive materials.

NORMAL LATCH UP
DA   January 8, 1991
SF   NLU

NORMALLY CLOSED
DA   January 8, 1991
SF   NC

NORMALLY OPEN
DA   January 8, 1991
SF   NO (normally open)

NORTH CAROLINA STATE UNIVERSITY
DA   January 8, 1991
SF   NCSTATE
BT1   Universities
 BT2 Educational Organizations
  BT3 Organizations

NOS
DA   October 12, 1990
SEE   Not Otherwise Specified
SO   Acronyms
SO   Hazardous Materials

NOSE
(1086 NOSE)
DA   November 28, 1990
BT1   Face
 BT2 Head
  BT3 Human Body Parts
BT1   Respiratory System
 BT2 Body System(s)

 BT3 Human Body Parts
RT   Air Mask
RT   Dust Mask
SO   DOE FRASE VOCABULARY

NOSEBLEED
(1334 NOSEBLEED)
DA   November 28, 1990
BT1   Injuries
SO   DOE FRASE VOCABULARY

NOT OTHERWISE SPECIFIED
(SWDA; RCRA; ESH)
DA   October 12, 1990
SF   NOS
SO   Hazardous Materials

NOTEWORTHY PRACTICES
(TTGM)
DA   April 26, 1991
BT1   Findings
RT   Best Management Practice Findings
RT   Compliance Findings
RT   Procedures
RT   Tiger Team Assessments
RT   Tiger Teams
SO   Management
DEF   (TTGM) In addition to Compliance and Best Management Practice Findings, a Tiger Team Assessment may identify practices, which in the judgement of the Tiger Team, may be noteworthy and have general application to DOE facilities and should be documented for the purposes of information transfer.

NOTICE OF AVAILABILITY
(EPA)
DA   May 16, 1991
SF   NOA
RT   Environmental Impact Statements
SO   Environmental Management
SO   Environmental Protection Agency Glossary
DEF   (CFR) A formal notice, published in the Federal Register, that announces the issuance and public availability of a draft or final Environmental Impact Statement. The Environmental Protection Agency Notice of Availability is the official public notification of an Environmental Impact Statement; a DOE Notice of Availability is an optional notice used to provide information to the public.

NOTICE OF DEFICIENCY
DA   January 8, 1991
SF   NOD

NOTICE OF INTENT
(NEPA; DOE Order 5440.1D)
DA   January 8, 1991
SF   NOI
RT   Environmental Impact Statements

RT   National Environmental Policy Act Documents
RT   Programmatic NEPA Documents
RT   Site-Wide NEPA Documents
SO   Environmental Management
DEF   (NEPA) A notice that an environmental impact statement will be prepared and considered. The notice shall briefly: (a) describe the proposed action and possible alternatives (b) describe the agency's proposed scoping process including whether, when, and where any scoping meeting will be held (c) state the name and address of a person within the agency who can answer questions about the proposed action and the environmental impact statement. (DOE Order) A document, defined at 40 CFR 1508.22, that announces the intent to prepare an Environmental Impact Statement for a proposed action.

NOTICE OF INVOLVEMENT
(CFR)
DA   May 15, 1991
SY   Public Notices
RT   Floodplains
RT   Wetlands
SO   Management
DEF   (CFR) A brief notice published in the Federal Register, and circulated to affected and interested persons and agencies, which describes a proposed floodplain/wetlands action and affords the opportunity for public review.

NOTICE OF VIOLATION
DA   May 20, 1991
SF   NOV
SO   Environmental Management
SO   Management

NOTIFICATION
(DOE Order 5484.1; EMER)
DA   October 12, 1990
NT1   Notification of Non-Compliance
NT1   Off-site Notification/Warning
RT   Occurrences
RT   Officially Reportable Events
SO   Emergency Preparedness
SO   Management
DEF   (DOE Order 5484.1) The actions taken to notify cognizant Department of Energy officials of an occurrence, and the subsequent actions taken at successive levels within the Department of Energy to notify the Secretary of an occurrence.

NOTIFICATION OF NON-COMPLIANCE
DA   January 8, 1991
SF   NON
BT1   Notification

NOV
DA May 20, 1991
SEE Notice of Violation
SO Acronyms

NO$_2$
DA October 12, 1990
SEE Nitrogen Dioxide (NO$_2$)
SO Acronyms

NPAR
DA October 12, 1990
SEE Nuclear Plant Aging Research
       Program
SO Acronyms

NPDES
DA October 12, 1990
SEE National Pollutant Discharge
       Elimination System
SO Acronyms

NPL
DA October 12, 1990
SEE National Priorities List
SO Acronyms

NPP
DA October 12, 1990
SEE Nuclear Power Plants
SO Acronyms

NPR
DA October 12, 1990
SEE New Production Reactor
SO Acronyms

NPRDS
DA October 12, 1990
SEE Nuclear Plant Reliability Data
       System
SO Acronyms

NPSH
DA October 12, 1990
SEE Net Positive Suction Head
SO Acronyms

NQA-1
DA October 12, 1990
SEE Nuclear Quality Assurance-1
SO Acronyms

NRC
DA October 12, 1990
SEE Nuclear Regulatory Commission
SO Acronyms

NRC (NON-REUSEABLE CONTAINER)
(SWDA; RCRA; ESH)
DA October 19, 1990
SY Non-Reuseable Containers
BT1 Containers
SO Acronyms
SO Hazardous Materials
DEF (ESH) A container whose reuse is
       restricted.

NRC-LICENSED FACILITIES
(CAA; CFR)
DA October 12, 1990
BT1 Facilities
RT Nuclear Regulatory Commission
SO Air Pollution
DEF (CFR) Any facilities licensed by the
       Nuclear Regulatory Commission
       or any Agreement State to receive
       title to, receive, possess, use,
       transfer, or deliver any source,
       byproduct, or special nuclear
       material, except facilities regulated
       by 40 CFR Parts 190, 191, or 192.

**NRC OFFICE OF NUCLEAR REACTOR
REGULATION**
DA January 8, 1991
SF *NRR*
BT1 Nuclear Regulatory Commission
  BT2 Federal Agencies
    BT3 Agencies
      BT4 Administrative Organizations
      BT5 Organizations
RT Reactors

**NRC OFFICE OF NUCLEAR
REGULATORY RESEARCH**
DA January 8, 1991
SF *RES*
BT1 Nuclear Regulatory Commission
  BT2 Federal Agencies
    BT3 Agencies
      BT4 Administrative Organizations
      BT5 Organizations

NRDC
DA October 12, 1990
SEE Natural Resource Defense Council
SO Acronyms

NRR
DA October 12, 1990
SEE NRC Office of Nuclear Reactor
       Regulation
SO Acronyms

**NRS OFFICE FOR ANALYSIS AND
EVALUATION OF OPERATIONAL
DATA**
DA January 8, 1991
SF *AEOD*
BT1 Offices
  BT2 Administrative Organizations
    BT3 Organizations
RT Analyses

**NRS OFFICE OF INSPECTION AND
ENFORCEMENT**
DA January 8, 1991
SF *IE (Office of Inspection and
       Enforcement)*
BT1 Offices
  BT2 Administrative Organizations
    BT3 Organizations

NRT
DA October 12, 1990

SEE National Response Teams
SO Acronyms

NSAC
DA October 12, 1990
SEE Nuclear Safety Analysis Center
SO Acronyms

NSPKTR
(Acronyms and Abbreviations)
DA October 12, 1990
BT1 Computer Codes
DEF A Computer code to analyze PRA
       information; developed at the
       Brookhaven National Laboratory.

NSPS
DA October 12, 1990
SEE New Source Performance
       Standards
SO Acronyms

NTA
DA October 12, 1990
SEE Nitrilotriacetic Acid
SO Acronyms

NTG
DA October 12, 1990
SEE Nuclear Test Gage
SO Acronyms

NTNCWS
DA October 12, 1990
SEE Non-Transient Non-Community
       Water System
SO Acronyms

NTS
DA October 12, 1990
SEE Nevada Test Site
SO Acronyms

NUCLEAR ACCIDENT DOSIMETER
(EMER)
DA February 1, 1991
BT1 Dosimeters
  BT2 Radiation Detectors
    BT3 Equipment
SO Emergency Preparedness
DEF (EMER) A device containing
       several materials responsive to
       different amounts and energies of
       neutron and gamma radiation. See
       also dosimeter.

**NUCLEAR BY-PRODUCT MATERIAL
LICENSE**
(EMER)
DA February 1, 1991
BT1 Licenses
SO Emergency Preparedness
DEF (EMER) A license issued to a
       facility owner or operator by a
       federal agency pursuant to the
       conditions of the Atomic Energy
       Act of 1954 (as amended) or

issued by an agreement state
pursuant to appropriate state laws.

NUCLEAR CRITICALITY
(DOE Orders 5480.4 and 5480.5)
DA   October 16, 1990
SY   Criticality
RT   Inspections (Nuclear)
SO   Standards
DEF (DOE Order 5480.5) A
    self-sustaining chain reaction, i.e.,
    the state in which the effective
    neutron multiplication constant of
    a system of fissionable material
    equals or exceeds unity.

NUCLEAR CRITICALITY SAFETY
(DOE Order 5480.5; EMER)
DA   October 12, 1990
BT1 Safety
RT   Criticality Accidents
RT   Inspections (Nuclear)
SO   Emergency Preparedness
SO   Standards
DEF (DOE Order 5480.5) The prevention
    or termination of inadvertent
    nuclear criticality, mitigation of
    consequences, and protection
    against injury or damage due to
    an accidental nuclear criticality.

NUCLEAR EMERGENCY SEARCH TEAMS
(EMER)
DA   February 1, 1991
BT1 Response Teams
 BT2 Teams
  BT3 Administrative Organizations
  BT4 Organizations
SO   Emergency Preparedness
DEF (EMER) Groups of DOE and DOE
    contractor experts assigned
    responsibility to provide assistance
    without geographical limitations
    following a nuclear threat.
    Resources include radiation
    detection systems and personnel
    for searching and identifying lost or
    stolen nuclear weapons or special
    nuclear materials, responding to
    radiation dispersal threats, and
    providing assistance in disabling
    devices or mitigating their effects.

NUCLEAR ENERGY
DA   January 8, 1991
SF   NE
BT1 Energy
DEF (DSTT) Energy released by nuclear
    fission or nuclear fusion.

NUCLEAR FACILITIES
(DOE Order 5480.18; EMER)
DA   October 12, 1990
BT1 Facilities
NT1 Fixed Nuclear Facilities
NT1 New Storage Facilities
NT1 Nuclear Power Plants

NT1 Plutonium Processing and Handling
    Facilities
NT1 Plutonium Storage Facilities
NT1 Reactor Facilities
 NT2 Low Power Reactor Facility
 NT2 Production Reactor Facility
 NT2 Test Reactor Facility
NT1 Waste Isolation Pilot Plant
RT   Decommissioning
RT   Inspections (Nuclear)
RT   Non-Nuclear Facilities
RT   Operating Organizations
RT   Reactor Facilities
RT   Technical Safety Appraisals
SO   Construction
SO   Emergency Preparedness
SO   Standards
DEF (DOE Order 5480.18) Facilities
    whose operations involve
    radioactive materials in such form
    and quantity that a significant
    nuclear hazard potentially exists to
    the employees or the general
    public. Included are facilities that
    (1) produce, process, or store
    radioactive liquid or solid waste,
    fissionable materials, or tritium; (2)
    conduct separations operations;
    (3) conduct irradiated materials
    inspection, fuel fabrication,
    decontamination, or recovery
    operations; or (4) conduct fuel
    enrichment operations. Incidental
    use of radioactive materials in a
    facility operation (e.g., check
    sources, radioactive sources, and
    X- ray machines) does not
    necessarily require the facility to
    be included in this definition.
    Accelerators and reactors and
    their operations are not included.

NUCLEAR FACILITIES INCIDENT DATABASE
DA   December 13, 1990
BT1 Information Systems
 BT2 Security Interests
 BT2 Systems
DEF The database is sponsored by
    DOE's Offices of Environment,
    Safety and Health (EH) and
    Defense Programs (DP) for the
    storage, analysis, and retrieval of
    reactor incident/event reports
    from the Savannah River Site
    (SRS) near Aiken, South Carolina.

NUCLEAR FACILITY
(1787 NUCLEAR FAC)
DA   December 10, 1990
BT1 Facility (DOE FRASE Vocabulary)
 BT2 Facilities and Buildings (DOE
    FRASE Vocabulary)
  BT3 Facilities
RT   Automatic Reactor Scram
RT   Decommissioning Activity
RT   Decontamination Activity
RT   Manual Reactor Scram
RT   Particle Accelerator Facility

RT   Production Reactor Facility
SO   DOE FRASE VOCABULARY

NUCLEAR FUEL CYCLE
(EMER)
DA   February 1, 1991
SO   Emergency Preparedness
DEF (EMER) The complete cycle of
    nuclear activities which includes
    mining, milling, conversion,
    enrichment, fuel fabrication,
    nuclear power plant operation,
    spent fuel storage, reprocessing (if
    applicable), and waste
    management operations.

NUCLEAR INCIDENT MONITOR
DA   January 8, 1991
SF   NIM
BT1 Monitors
 BT2 Equipment

NUCLEAR INCIDENTS
(AEA; ANL)
DA   May 24, 1991
BT1 Occurrences
NT1 Extraordinary Nuclear Occurrences
DEF (AEA) Any occurrences, including
    an extraordinary nuclear
    occurrence, within the United
    States causing, within or outside
    the United States, bodily injury,
    sickness, disease, or death, or
    loss of or damage to property, or
    loss of use of property, arising out
    of or resulting from the radioactive,
    toxic, explosive, or other
    hazardous properties of source,
    special nuclear, or byproduct
    material: Provided, however, That
    as the term is used in subsection
    170 l., it shall include any such
    occurrence outside the United
    States: And provided further, That
    as the term is used in subsection
    170 d., it shall include any such
    occurence outside the United
    States if such occurrence involves
    source, special nuclear, or
    byproduct material owned by, and
    used by or under contract with, the
    United States: And provided
    further, That as the term is used in
    subsection 170 c., it shall include
    any such occurrence outside both
    the United States and any other
    nation is such occurrence arises
    out of or etc.

NUCLEAR MATERIALS MANAGEMENT AND STANDARDS SYSTEM
(EMER)
DA   February 1, 1991
BT1 Information Systems
 BT2 Security Interests
 BT2 Systems
SO   Emergency Preparedness
DEF (EMER) A reporting and analytical

BT-Broader Term   NT-Narrower Term   RT Related Term

system used in safeguarding and
managing nuclear materials.

**NUCLEAR PLANT AGING RESEARCH
PROGRAM**
DA   January 8, 1991
SF   *NPAR*
BT1 Programs

**NUCLEAR PLANT RELIABILITY DATA
SYSTEM**
DA   January 8, 1991
SF   *NPRDS*
BT1 Information Systems
 BT2 Security Interests
 BT2 Systems

NUCLEAR POISONS
(SSDC; EDB)
DA   May 24, 1991
SY   Neutron Poisons
BT1 Reactor Materials
 BT2 Materials
SO   Radiation
DEF (SSDC) Substances with large
     neutron cross-sections. (EDB)
     Neutron absorbers in a reactor.

NUCLEAR POWER PLANTS
(EPA)
DA   October 12, 1990
SF   *NPP*
BT1 Nuclear Facilities
 BT2 Facilities
SO   Environmental Protection Agency
     Glossary
DEF (EPA) Facilities that convert atomic
     energy into usable power; heat
     produced by a reactor makes
     steam to drive turbines that
     produce electricity.

NUCLEAR QUALITY ASSURANCE-1
DA   January 8, 1991
SF   *NQA-1*
BT1 Quality Assurance

**NUCLEAR REGULATORY
COMMISSION**
(EMER)
DA   January 8, 1991
SF   *NRC*
BT1 Federal Agencies
 BT2 Agencies
  BT3 Administrative Organizations
   BT4 Organizations
NT1 NRC Office of Nuclear Reactor
     Regulation
NT1 NRC Office of Nuclear Regulatory
     Research
RT   Agreement States
RT   Implementing Agencies
RT   Licensed Sites
RT   NRC-Licensed Facilities
RT   U.S. Department of Energy
SO   Emergency Preparedness
SO   Environmental Management
SO   Radiation
DEF (EMER) The federal agency

responsible for regulating
commercial nuclear power plants
and other commercial nuclear
operations pursuant to the Atomic
Energy Act of 1954, as amended,
and covered by provisions under
section 170(a) of that Act.

**NUCLEAR REGULATORY
COMMISSION LICENSED ACTIVITIES**
(EMER)
DA   February 1, 1991
BT1 Activities
SO   Emergency Preparedness
DEF (EMER) Activities licensed pursuant
     to the Atomic Energy Act of 1954,
     as amended, and covered by
     provisions under Section 170(a) of
     that Act.

**NUCLEAR SAFETY ANALYSIS
CENTER**
DA   January 8, 1991
SF   *NSAC*
BT1 Centers
BT1 Electric Power Research Institute
 BT2 Institutes
  BT3 Research and Development
       Organizations
   BT4 Organizations
RT   Analyses

NUCLEAR TEST GAGE
DA   January 8, 1991
SF   *NTG*

NUCLEAR THREAT INCIDENTS
(EMER)
DA   February 1, 1991
BT1 Incidents
BT1 Threats
 BT2 Emergencies
  BT3 Reportable Occurrences
   BT4 Occurrences
SO   Emergency Preparedness
DEF (EMER) Situations involving the
     threatened, attempted, or actual
     theft, loss, unauthorized
     possession of source or special
     nuclear material, radioactive
     by-products, nuclear explosive
     devices, improvised devices
     (either separately or in
     combination with explosives), or
     radioactive dispersal devices, or
     the threatened use of said items.

NUCLEAR WASTE POLICY ACT
DA   January 8, 1991
SF   *NWPA*
BT1 Acts
 BT2 Statutes and Regulations

NUCLEAR WEAPONS ACCIDENTS
(EMER)
DA   February 1, 1991
BT1 Accidents
SO   Emergency Preparedness
DEF (EMER) Unexpected events

involving nuclear weapons or
radiological nuclear weapon
components which result in any of
the following: 1) accidental or
unauthorized launching, firing, or
use of a nuclear explosive 2)
nuclear detonation 3) nonnuclear
detonation/burning of a nuclear
weapon 4) radioactive
contamination 5) seizure, theft, or
loss of a nuclear weapon or an
actual component of a nuclear
weapon 6) public hazard, actual or
implied

NUCLEAR WINTER
(EPA)
DA   October 12, 1990
RT   Atomic Weapons
RT   Chilling Effect
SO   Environmental Protection Agency
     Glossary
DEF (EPA) Prediction by some scientists
     that smoke and debris rising from
     massive fires resulting from a
     nuclear war could enter the
     atmosphere and block out sunlight
     for weeks or months. The
     scientists making this prediction
     project a cooling of the earth's
     surface, and changes in climate
     which could, for example,
     negatively affect world agricultural
     and weather patterns.

NUCLIDES
(NCRP; EMER)
DA   October 12, 1990
NT1 Fission Products
NT1 Isotopes
 NT2 Isotopic Tracers
NT1 Radionuclides
 NT2 Alpha Decay Radioisotopes
  NT3 Radium 226
  NT3 Radon 222
  NT3 Thorium 230
  NT3 Uranium 238
 NT2 Beta Decay Radioisotopes
  NT3 Beta-Minus Decay Radioisotopes
   NT4 Antimony 125
   NT4 Cerium 144
   NT4 Cerium 141
   NT4 Cesium 137
   NT4 Cesium 134
   NT4 Cobalt 60
   NT4 Cobalt 58
   NT4 Iodine 131
   NT4 Radium 228
   NT4 Ruthenium 103
   NT4 Strontium 90
   NT4 Tellurium 132
   NT4 Thorium 232
   NT4 Tritium
   NT4 Zinc 65
   NT4 Zirconium 95
 NT2 Bone Seekers
 NT2 Days Living Radioisotopes
  NT3 Beryllium 7
  NT3 Cerium 144
  NT3 Cerium 141

SY-Synonymous Terms   SO-Source/Subject Category   SF-See From

NT3 Cesium 137
NT3 Cobalt 58
NT3 Iodine 131
NT3 Manganese 54
NT3 Radon 222
NT3 Selenium 75
NT3 Zinc 65
NT3 Zirconium 95
NT2 Delayed Neutron Precursors
NT2 Delayed Proton Precursors
NT2 Element 104 Isotopes
NT2 Element 105 Isotopes
NT2 Element 106 Isotopes
NT2 Element 107 Isotopes
NT2 Element 108 Isotopes
NT2 Element 109 Isotopes
NT2 Heavy Ion Decay Radioisotopes
NT2 Hours Living Radioisotopes
NT3 Cesium 134
NT3 Cobalt 58
NT2 Internal Conversion Radioisotopes
NT3 Cesium 134
NT3 Cobalt 60
NT3 Cobalt 58
NT2 Isomeric Transition Isotopes
NT3 Cesium 134
NT3 Cobalt 60
NT3 Cobalt 58
NT2 Microsec Living Radioisotopes
NT2 Millisec Living Radioisotopes
NT2 Minutes Living Radioisotopes
NT3 Cobalt 60
NT3 Thorium 232
NT2 Nanosec Living Radioisotopes
NT2 Neutron-Deficient Isotopes
NT2 Neutron-Rich Isotopes
NT2 Proton Decay Radioisotopes
NT2 Radon Progeny
NT3 Radon 222
NT2 Seconds Living Radioisotopes
NT2 Transuranic Radionuclides
NT2 Years Living Radioisotopes
NT3 Antimony 125
NT3 Cesium 134
NT3 Cobalt 60
NT3 Radium 226
NT3 Radium 228
NT3 Ruthenium 103
NT3 Strontium 90
NT3 Thorium 230
NT3 Tritium
NT3 Uranium 238
NT1 Transuranic Elements
NT1 Transuranic Nuclides
RT Fertile
RT Fissile
SO Emergency Preparedness
SO Radiation
DEF (NCRP) (1) A species of atom
    characterized by its mass number,
    atomic number, and energy state
    of its nucleus, provided that the
    atom is capable of existing for a
    measurable time. (2) A species of
    atom characterized by the
    constitution of its nucleus.

NUTRIENTS
(EPA)
DA October 12, 1990

RT Eutrophication
RT Fertilizers
SO Environmental Protection Agency
    Glossary
DEF (EPA) Any substances assimilated
    by living things that promote
    growth. The term is generally
    applied to nitrogen and
    phosphorus in wastewater, but is
    also applied to other essential and
    trace elements.

NV
DA October 12, 1990
SEE Nevada Operations Office
SO Acronyms

NWC
DA October 12, 1990
SEE DOE Nuclear Weapons Complex
SO Acronyms

NWPA
DA October 12, 1990
SEE Nuclear Waste Policy Act
SO Acronyms

O&M
DA October 12, 1990
SEE Operations and Maintenance
SO Acronyms

O-UNIT
(NFI)
DA October 12, 1990
SO Nuclear Facilities Incident Database
SO Radiation
DEF (NFI) Control rod drive for partial
    rods.

OAC
DA October 12, 1990
SEE Ohio Administrative Code
SO Acronyms

**OAK RIDGE ASSOCIATED
UNIVERSITIES**
DA January 8, 1991
SF *ORAU*
BT1 DOE Contractors
 BT2 Potentially Responsible Parties
BT1 Research and Development
    Organizations
 BT2 Organizations
RT Oak Ridge Operations Office
RT Universities
DEF (Capsule Review of DOE Research
    and Development and Field
    Facilities, 1986) A private,
    not-for-profit association
    comprised of 55 colleges and
    universities. ORAU conducts
    research and educational
    programs in energy, health and
    the environment for DOE, member
    institutions of ORAU, other
    colleges and universities and

private and governmental
organizations.

**OAK RIDGE GASEOUS DIFFUSION
PLANT**
DA January 8, 1991
SF *ORGDP*
BT1 Government-Owned
    Contractor-Operated Facilities
 BT2 Federal Facilities
  BT3 Facilities
RT Martin Marietta Energy Systems

OAK RIDGE NATIONAL LABORATORY
DA January 8, 1991
SF *ORNL*
BT1 Government-Owned
    Contractor-Operated Facilities
 BT2 Federal Facilities
  BT3 Facilities
BT1 Laboratories
 BT2 Research and Development
    Organizations
  BT3 Organizations
RT Martin Marietta Energy Systems
RT Oak Ridge Operations Office
DEF (Capsule Review of DOE Research
    and Development and Field
    Facilities, 1986) Founded in 1943,
    ORNL primarily supports the
    fission nuclear fuel cycle and
    development of magnetic fusion
    energy through scientific research
    and technology. In addition, ORNL
    identifies and solves generic
    research problems in energy
    technologies such as materials,
    separation techniques, chemical
    processes and biotechnology and
    is the major national source of
    stable and radioactive isotopes.

OAK RIDGE OPERATIONS OFFICE
DA January 8, 1991
SF *OR*
SF *ORO*
BT1 Operations Offices
 BT2 Offices
  BT3 Administrative Organizations
   BT4 Organizations
 BT2 U.S. Department of Energy
  BT3 Federal Agencies
   BT4 Agencies
    BT5 Administrative Organizations
     BT6 Organizations
RT Boeing Petroleum Services
RT Martin Marietta Energy Systems
RT MK-Ferguson
RT Oak Ridge Associated Universities
RT Oak Ridge National Laboratory
RT Southeastern Universities Research
    Association, Inc.
RT Westinghouse Materials Company
    of Ohio
RT Y-12 Plant
DEF (Capsule Review of DOE Research
    and Development and Field
    Facilities, 1986) The Oak Ridge
    Operations Office was established
    under the Atomic Energy

Commission in 1942. The principal programs of the Office are the production of nuclear weapon components in support of national defense programs, production of enriched uranium for defense requirements and for fueling nuclear power plants in the U.S. and abroad, processing of uranium feed materials and fuel cores for DOE's plutonium production reactors and extensive energy research and development in all DOE program areas. In addition, the Office provides assistance to local, state and Federal government agencies and the private sector, and administers DOE contracts with universities in the Southeast. The Office reservation serves as a national environmental research site.

OAR
DA   October 12, 1990
SEE Operation Assessment and
       Readiness
SO   Acronyms

OBA
DA   October 12, 1990
SEE Operating Basis Accident
SO   Acronyms

OBE
DA   October 12, 1990
SEE Operating Basis Earthquake
SO   Acronyms

OBES
DA   October 12, 1990
SEE Office of Basic Energy Sciences
SO   Acronyms

OBLIQUE SLIP
(USGS)
DA   October 12, 1990
BT1 Surface Faulting
 BT2 Faults
SO   Natural Phenomenon
DEF (USGS) A combination of strike-slip
       and dip-slip movement; movement
       or slip that is intermediate in
       orientation between the dip slip
       and the strike slip. Also called a
       diagonal slip fault.

OC
DA   October 12, 1990
SEE Outer Chuck
SO   Acronyms

OCAW
DA   October 12, 1990
SEE Oil Chemical and Atomic Workers
SO   Acronyms

OCB
DA   October 12, 1990
SEE Oil Circuit Breaker
SO   Acronyms

OCCUPATION (UNK)
(0001 OCCUP;UNK)
DA   November 28, 1990
SO   DOE FRASE VOCABULARY

OCCUPATIONAL DOSES
DA   January 24, 1991
BT1 Doses
RT   Ionizing Radiation
RT   Occupational Exposure
SO   Environmental Management
DEF (CFR) Exposure of individuals to
       radiation: (a) In a restrictive area
       or (b) In the course of employment
       in which the individual's duties
       involve exposure to radiation,
       provided that occupational dose
       shall not be deemed to include
       any exposure of an individual to
       radiation for the purpose of
       medical diagnosis or medical
       therapy of such individual.

OCCUPATIONAL EXPOSURE
(NCRP; IAEA)
DA   October 12, 1990
BT1 Exposure
RT   Annual Limit on Intake
RT   Asbestosis
RT   Health Hazards
RT   Occupational Workers
RT   Occupational Doses
RT   Occupational Injuries
RT   Occupational Illnesses
RT   Special Health Supervision
RT   Work Environment
RT   Working Conditions
SO   Nuclear Facilities Incident Database
SO   Radiation
DEF (NCRP; IAEA) Exposure of a
       worker during a period of work.

OCCUPATIONAL HEALTH DIVISION
DA   January 8, 1991
SF   *OHD*
BT1 Divisions
 BT2 Administrative Organizations
  BT3 Organizations

OCCUPATIONAL ILLNESSES
(SSDC)
DA   October 12, 1990
BT1 Illnesses
RT   Asbestosis
RT   Health Hazards
RT   Incidence Rate, Total Recordable
       Cases (TRC)
RT   Occupational Injuries
RT   Occupational Exposure
RT   Total Recordable Cases
SO   System Safety Development Center
       Glossary
DEF (SSDC) Abnormal physical
       conditions or disorders of an

employee, other than one resulting from an occupational injury, caused by exposure to environmental factors associated with the employment. It includes (but may not be limited to) acute and chronic illnesses or diseases that may be caused by inhalation, absorption, ingestion, direct contact, or radiation.

OCCUPATIONAL INJURIES
(SSDC)
DA   October 12, 1990
BT1 Injuries
RT   Disabling Injuries
RT   Health Hazards
RT   Incidence Rate, Total Recordable
       Cases (TRC)
RT   Occupational Illnesses
RT   Occupational Exposure
RT   Total Recordable Cases
RT   Work Environment
SO   System Safety Development Center
       Glossary
DEF (SSDC) Any injuries, such as cuts,
       fractures, sprains, or amputations,
       which result from a work-related
       accident or from exposure to the
       work environment. In some cases,
       an employer may classify an
       exposure to chemicals or toxic
       agents as an injury when there is
       a single traumatic event
       associated with the exposure.

OCCUPATIONAL MEDICAL PROGRAM
(DOE Order 5480.8)
DA   October 16, 1990
BT1 Programs
RT   Federal Employee Occupational
       Safety and Health Program
RT   Minimum Requirements and
       Standards
RT   Occupational Medicine
SO   Industrial Hygiene
DEF (DOE 5480.8) A program to (1)
       assure the health and safety of
       employees in their work
       environments through the
       application of occupational
       medical principles; (2) determine
       the physical and mental fitness of
       employees to perform job
       assignments without undue hazard
       to themselves, fellow employees,
       or the public at large; (3) assure
       the early detection and treatment
       of employee illness or injuries by
       means of scheduled periodic
       health evaluations and
       unscheduled employee health
       visits; and (4) contribute to the
       maintenance of good employee
       health through the application of
       preventive medical measures,
       such as immunizations, alcohol
       and drug abuse programs, and
       health counseling.

SY-Synonymous Terms   SO-Source/Subject Category   SF-See From

OCCUPATIONAL MEDICINE

(Doe Order 5480.8)
DA October 12, 1990
RT Approved Medical Practitioners
RT Environment, Safety, and Health
  (ES&H) Program
RT Occupational Medical Program
RT Special Health Supervision
SO Industrial Hygiene
DEF (DOE Order 5480.8) A specialty
  branch of the profession of
  medicine that deals with the health
  protection and health maintenance
  of employees, with special
  reference to job hazards, job
  stresses, and work environment
  hazards.

OCCUPATIONAL SAFETY AND HEALTH ADMINISTRATION

(Doe Orders 5480.4 and 5483.1A)
DA October 12, 1990
SF *OSHA*
BT1 Federal Agencies
 BT2 Agencies
  BT3 Administrative Organizations
   BT4 Organizations
RT Federal Employee Occupational
  Safety and Health Program
RT Material Safety Data Sheets
RT National Institute for Occupational
  Safety and Health
SO Industrial Safety
SO Standards
DEF (DOE Order 5483.1A) An agency of
  the U.S. Department of Labor,
  established under Public Law
  91-596 with major responsibilities
  to promulgate, prescribe, and
  enforce occupational safety and
  health standards.

OCCUPATIONAL WORKERS

(Doe Order 5480.11)
DA October 12, 1990
BT1 Personnel
NT1 Radiation Workers
 NT2 Non-Employee Radiation Workers
RT Occupational Exposure
SO Environmental Management
SO Industrial Hygiene
DEF (DOE Order 5480.11) Individuals
  who are either DOE or DOE
  contractor employees; employees
  of a subcontractor to a DOE
  contractor; or individuals who visit
  to perform work for or in
  conjunction with DOE or utilizes
  DOE facilities.

OCCUPATIONS

(DOE FRASE Vocabulary Numeric Keys
0001-0999)
DA November 29, 1990
NT1 Admin. Support/Clerical Employee
NT2 Firefighter
NT2 Food Service Employee
NT2 Janitor
NT2 Misc Service Employee
NT2 Security Guard

NT1 Agriculture Personnel
NT2 Forest Worker
NT2 Groundskeeper
NT2 Misc Agriculture Employee
NT1 Handler/Laborer/Helper
NT1 Military Personnel
NT1 Misc Employee
NT2 Misc Service Employee
NT2 Misc Agriculture Employee
NT2 Misc Repair/Construction
  Employee
NT2 Misc Precision/Production
  Employee
NT2 Misc Transport Employee
NT1 Precision/Production Personnel
NT2 Machinist
NT2 Machine Setup/Operator
NT2 Misc Precision/Production
  Employee
NT2 Operator, Plant/System/Utility
NT2 Sheet Metal Worker
NT2 Welder/Solderer
NT1 Professional Personnel
NT2 Doctor/Nurse
NT2 Engineer
NT2 Misc Professional
NT2 Scientist
 NT3 Health Physicist
NT2 Technicians
 NT3 Engineering Technician
 NT3 Health Technician
 NT3 Miscellaneous Technician
 NT3 Radiation Monitor/Technician
 NT3 Science Technician
NT1 Repair/Construction Personnel
NT2 Carpenter
NT2 Electrician
NT2 Mason
NT2 Mechanic/Repairer
NT2 Miner/Driller
NT2 Misc Repair/Construction
  Employee
NT2 Painter
NT2 Pipe Fitter
NT1 Sales Person
NT1 Transport Personnel
NT2 Aircraft Pilot
NT2 Bus Driver
NT2 Equipment Operator
NT2 Misc Transport Employee
NT2 Truck Driver
RT DOE FRASE Categories
SO DOE FRASE VOCABULARY
DEF A subject category used with the
  DOE FRASE Vocabulary.

OCCUPIABLE AREA

(DOE Order 6430.1A)
DA October 12, 1990
BT1 Sites/Areas
SO Construction
DEF That portion of the gross area that
  is available for use by an
  occupant's personnel or
  furnishings, including space that is
  available jointly to the various
  occupants of the buildings, such
  as auditoriums, health units, and
  snack bars. Occupiable area does
  not include space in the building

that is devoted to its operations
and maintenance, including craft
shops, gear rooms, and building
supply storage and issue rooms.
Ceiling-high corridors solely
serving a single space assignment
are "occupiable." Occupiable area
is computed by measuring from
the occupant's side of ceiling-high
corridor partitions or partitions
enclosing mechanical, toilet,
and/or custodial space to the
inside finish of permanent exterior
building walls or the face of the
convector if the convector
occupies at least 50% of the
length of exterior wall. When
computing occupiable area
separated by partitions,
measurements are taken from the
center line of the partitions.

OCCUPIED AREA (EXPLOSIVES)

(DOE Order 6430.1A)
DA October 12, 1990
BT1 Sites/Areas
RT Explosives Buildings
RT Inhabited Building Distance
SO Construction
DEF (DOE Order 6430.1A) Any work
  area to which personnel are
  assigned or any non-work area
  where persons regularly
  congregate. In the context of Class
  II bays for explosives facilities,
  access ramps and plant roads are
  not considered occupied areas.

OCCURRENCE REPORTING AND PROCESSING SYSTEM

(SSDC; DOE Order 5000.3A)
DA May 20, 1991
SF *ORPS*
BT1 Safety Performance Measurement
  System
 BT2 Information Systems
  BT3 Security Interests
  BT3 Systems
RT Computerized Accident/Incident
  Reporting System
RT Occurrence Reports
RT Unusual Occurrence Reporting
  System
SO Environmental Management
DEF (SSDC) One of the actuarial
  modules available on the Safety
  Performance Measurement
  System (SPMS) maintained by
  EG&G Idaho's System Safety
  Development Center (SSDC) at
  Idaho Falls, Idaho. The system
  provides a departmental (DOE)
  mechanism for submission,
  collection, transmission, and
  analysis of occurrence reports as
  required by DOE Order 5000.3A.

OCCURRENCE REPORTS

(DOE Order 5000.3A)
DA December 14, 1990

BT1 Reports
NT1 Unusual Occurrence Reports
RT Occurrence Reporting and
    Processing System
SO Management
DEF (DOE Order 5000.3A) Written
    evaluations of an event or
    condition that are prepared in
    sufficient detail to enable the
    reader to assess their significance,
    consequences, or implications and
    to evaluate the actions being
    proposed or employed to correct
    the condition or to avoid
    recurrence.

OCCURRENCES
(DOE Order 5484.1; SSDC)
DA October 12, 1990
SY Accidents
NT1 Accidental Occurrences
 NT2 Non-Sudden Accidental
     Occurrences
 NT2 Sudden Accidental Occurrences
  NT3 Pressure Releases
   NT4 Sudden Releases of Pressure
NT1 Anticipated Operational
    Occurrences
NT1 Environmental Occurrences
NT1 Epidemics
 NT2 Waterborne Disease Outbreaks
NT1 Nuclear Incidents
 NT2 Extraordinary Nuclear
     Occurrences
NT1 Reportable Occurrences
 NT2 Emergencies
  NT3 Abnormal Exposure Conditions
  NT3 Bomb Incidents
  NT3 Bomb Threats
  NT3 Emergency Situations
  NT3 Energy Emergencies
  NT3 General Emergencies
  NT3 Hazardous Materials
      Emergencies
  NT3 National Emergencies
  NT3 Natural Phenomena Emergencies
  NT3 Operational Emergencies
  NT3 Safeguards and Security
      Emergencies
   NT4 Safeguards and Security Alerts
    NT5 Alpha
    NT5 Bravo
    NT5 Charlie
    NT5 Safeguards and Security Alert I
        (Code Designator: Alpha)
    NT5 Safeguards and Security Alert
        II (Code Designator: Bravo)
    NT5 Safeguards and Security Alert
        III (Code Designator: Charlie)
   NT4 Special Nuclear Materials
       Emergencies
  NT3 Site Emergencies
  NT3 Threats
   NT4 Bomb Threats
   NT4 Covert Threats
   NT4 Nuclear Threat Incidents
   NT4 Terrorist Threats
 NT2 Off Normal Occurrences
 NT2 Officially Reportable Events
 NT2 Unusual Occurrences

RT Accidents
RT Accident Sites
RT Amelioration
RT Emergency Conditions
RT Events
RT Incidents
RT Investigations
RT Notification
SO Environmental Management
SO Management
SO System Safety Development Center
   Glossary
DEF (SSDC) Deviations from the
    planned or expected behavior or
    course of events in connection
    with any Department of Energy or
    Department of Energy-controlled
    operation if the deviation has
    environmental protection, safety,
    or health protection significance.

OCM
DA October 12, 1990
SEE Operational Change Memos
SO Acronyms

OCRWM
DA October 12, 1990
SEE Office of Civilian Radioactive Waste
    Management
SO Acronyms

ODC
DA October 12, 1990
SEE Ohio Department of Commerce
SO Acronyms

ODH
DA October 12, 1990
SEE Ohio Department of Health
SO Acronyms

OEPA
DA October 12, 1990
SEE Ohio Environmental Protection
    Agency
SO Acronyms

OER
DA October 12, 1990
SEE Office of Energy Research
SO Acronyms

OFF GAS SYSTEM
(2039 OFF GAS SYST)
DA December 10, 1990
BT1 Systems (DOE FRASE Vocabulary)
 BT2 Systems
SO DOE FRASE VOCABULARY

OFF NORMAL OCCURRENCES
(DOE Order 5000.3A)
DA December 14, 1990
BT1 Reportable Occurrences
 BT2 Occurrences
DEF (DOE Order 5000.3A) Off normal
    occurrences are abnormal or
    unplanned events or conditions

that adversely affect, potentially
affect, or are indicative of
degradation in the safety, security,
environmental or health protection
performance or operation of a
facility. The types of occurrences
that are to be categorized as off
normal occurrences are specified
in DOE Order 5000.3A.

OFF-SITE
(SWDA; RCRA; EMER)
DA October 19, 1990
BT1 Sites/Areas
RT On-Site
RT Site Boundaries
SO Emergency Preparedness
SO Wastes
DEF (CFR) Any site which is not on-site.

OFF-SITE FACILITIES
(EPA)
DA October 12, 1990
BT1 Hazardous Waste Management
    Facilities
 BT2 Hazardous Waste Facilities
  BT3 Facilities
SO Environmental Protection Agency
   Glossary
DEF (EPA) A hazardous waste
    treatment, storage or disposal
    area that is located at a place
    away from the generating site.

OFF-SITE FEDERAL SUPPORT
(EMER)
DA February 1, 1991
BT1 Amelioration
SO Emergency Preparedness
DEF (EMER) Federal assistance in
    mitigating the off-site
    consequences of an emergency
    and protecting the public health
    and safety, including the
    assistance with determining and
    implementing public protective
    action measures.

OFF-SITE NOTIFICATION/WARNING
(EMER)
DA February 1, 1991
BT1 Notification
SO Emergency Preparedness
DEF (EMER) An emergency notification
    and/or warning message issued
    to the state/local government
    and/or the public.

OFFICE ACTIVITY
(1226 O ACTIVITY)
DA November 28, 1990
BT1 Activity Types (DOE FRASE
    Vocabulary)
 BT2 Activities
RT Admin. Support/Clerical Employee
RT Office Area
SO DOE FRASE VOCABULARY

OFFICE AREA
(1709 OFFICE AREA)
DA   December 10, 1990
BT1 Area
 BT2 Sites/Areas
RT   Office Activity
SO   DOE FRASE VOCABULARY

OFFICE OF BASIC ENERGY SCIENCES
DA   January 8, 1991
SF   *OBES*
BT1 Office of Energy Research
 BT2 Program Offices
  BT3 Offices
   BT4 Administrative Organizations
    BT5 Organizations
  BT3 U.S. Department of Energy
   BT4 Federal Agencies
    BT5 Agencies
     BT6 Administrative Organizations
      BT7 Organizations
SO   Management

OFFICE OF CIVILIAN RADIOACTIVE WASTE MANAGEMENT
DA   January 8, 1991
SF   *OCRWM*
SF   *RW (DOE Program Office)*
BT1 Program Offices
 BT2 Offices
  BT3 Administrative Organizations
   BT4 Organizations
 BT2 U.S. Department of Energy
  BT3 Federal Agencies
   BT4 Agencies
    BT5 Administrative Organizations
     BT6 Organizations
SO   Management
DEF (U.S. Government Manual) The Office of Civilian Radioactive Waste Management was established by the Nuclear Waste Policy Act of 1982 (42 U.S.C. 10224). The Office has responsibility for the Nuclear Waste Fund and for the management of Federal programs for recommending, constructing, and operating repositories for disposal of high-level radioactive waste and spent nuclear fuel; interim storage of spent nuclear fuel; monitored retrievable storage; and research, development, and demonstration regarding disposal of high-level radioactive waste and spent nuclear fuel.

OFFICE OF DEFENSE WASTE AND TRANSPORTATION MANAGEMENT
DA   January 8, 1991
SF   *DWTM*
BT1 Offices
 BT2 Administrative Organizations
  BT3 Organizations
SO   Management

OFFICE OF ENERGY RESEARCH
DA   January 8, 1991

SF   *ER (Office of Energy Research)*
SF   *OER*
BT1 Program Offices
 BT2 Offices
  BT3 Administrative Organizations
   BT4 Organizations
 BT2 U.S. Department of Energy
  BT3 Federal Agencies
   BT4 Agencies
    BT5 Administrative Organizations
     BT6 Organizations
NT1 Office of Basic Energy Sciences
NT1 Office of Health and Environmental · Research
SO   Management
DEF (U.S. Government Manual) The Office of Energy Research advises the Secretary on the physical and energy research and development programs of the Department, the use of multipurpose laboratories, education, and training for basic and applied research, and financial assistance and budgetary priorities for these activities. The Office manages the basic energy sciences, high energy physics, and fusion energy research programs; administers DOE programs supporting university researchers, funds research in mathematical and computational sciences critical to the use and development of supercomputers; and administers a financial support program for research and development projects not funded elsewhere in the department.

OFFICE OF ENVIRONMENTAL RESTORATION AND WASTE MANAGEMENT
DA   February 4, 1991
SF   *ER (Environmental Restoration)*
BT1 Program Offices
 BT2 Offices
  BT3 Administrative Organizations
   BT4 Organizations
 BT2 U.S. Department of Energy
  BT3 Federal Agencies
   BT4 Agencies
    BT5 Administrative Organizations
     BT6 Organizations
RT   Waste Information Network
SO   Management
DEF (U.S. Government Manual) The Director of the Office of Environmental Restoration and Waste Management provides program policy guidance and manages the assessment and cleanup of inactive waste sites and facilities, continues safe and effective waste management operations, and develops and implements an aggressively applied waste research and development program to provide innovative environmental technologies that yield permanent

disposal solutions at reduced costs. The Director provides centralized management for the Department for waste management operations, environmental restoration, and applied research and development programs and activities, including environmental restoration and waste management program policy and guidance to DOE Operations Offices in these areas.

OFFICE OF HEALTH AND ENVIRONMENTAL RESEARCH
DA   January 8, 1991
SF   *OHER*
BT1 Office of Energy Research
 BT2 Program Offices
  BT3 Offices
   BT4 Administrative Organizations
    BT5 Organizations
  BT3 U.S. Department of Energy
   BT4 Federal Agencies
    BT5 Agencies
     BT6 Administrative Organizations
      BT7 Organizations
SO   Management

OFFICE OF MANAGEMENT AND BUDGET
DA   January 8, 1991
SF   *OMB*
BT1 Federal Agencies
 BT2 Agencies
  BT3 Administrative Organizations
   BT4 Organizations
BT1 Offices
 BT2 Administrative Organizations
  BT3 Organizations
SO   Management
DEF (U.S. Government Manual) The OMB evaluates, formulates, and coordinates management procedures and program objectives within and among Federal departments and agencies. It also controls the administration of the Federal budget, while routinely providing the President with recommendations regarding budget proposals and relevant legislative enactments.

OFFICE OF NEW PRODUCTION REACTORS
DA   February 4, 1991
BT1 Program Offices
 BT2 Offices
  BT3 Administrative Organizations
   BT4 Organizations
 BT2 U.S. Department of Energy
  BT3 Federal Agencies
   BT4 Agencies
    BT5 Administrative Organizations
     BT6 Organizations
SO   Management
DEF (U.S. Government Manual) The Office of New Production Reactors

manages and directs a program for the acquisition and construction of new production reactor capacity to meet national security requirements.

OFFICE OF SPECIAL PROJECTS
(TTGM)
DA   May 6, 1991
SF   *OSP (Office of Special Projects)*
BT1 Office of the Assistant Secretary for Environment, et. al.
 BT2 Program Offices
 BT3 Offices
  BT4 Administrative Organizations
  BT5 Organizations
  BT3 U.S. Department of Energy
  BT4 Federal Agencies
  BT5 Agencies
  BT6 Administrative Organizations
  BT7 Organizations
RT   Tiger Teams
RT   Tiger Team Assessment Program
SO   Environmental Management
SO   Management
DEF (TTGM) The Tiger Team Assessment Program is implemented by the Office of Special Projects (OSP), which is within the Office of Environment, Safety and Health. The OSP manages the staffing and conduct of the Tiger Team Assessments; ensures that the resulting assessment reports are completed, reviewed, revised and approved as necessary; provides liaison with other involved DOE and external entities; provides guidance to the Tiger Team Assessment process; maintains a data base of Tiger Team roster, candidates, activities and findings; and prepares summaries and trend analyses of Tiger Team findings.

OFFICE OF SPECIAL PROJECTS COORDINATORS
(TTGM)
DA   May 7, 1991
SF   *OSP Coordinators*
BT1 Personnel
RT   Tiger Team Assessment Program
RT   Tiger Team Leaders
RT   Tiger Team Administrators
DEF (TTGM) The OSP Coordinators, designated by the Director of the Office of Special Projects, will have primary oversight of the assessment process and schedules. The Coordinators work directly with the Team Leaders throughout the assessment. Major responsibilities include: 1) assisting in the assembly of Tiger Team members; 2) providing guidance to Tiger Team Leaders/members concerning the overall assessment process; 3)

participating/attending major meetings; 4) establishing, maintaining, and participating in daily telecommunications between Tiger Team Leaders and the OSP; 5) serving as the point of contact at Headquarters for requests regarding Tiger Team activities and reports; 6) informing the Tiger Team Leaders of any DOE/HQ policy and/or program changes; 7) assisting the Team leaders in the review and concurrence process for the Report and Action Plan; and 8) ensuring that the final Assessment Report and Action plan are incorporated into appropriate systems for analysis, etc.

OFFICE OF TECHNOLOGY DEVELOPMENT
DA   January 8, 1991
SF   *OTD*
BT1 Offices
 BT2 Administrative Organizations
 BT3 Organizations
SO   Management

OFFICE OF THE ASSISTANT SECRETARY FOR CONSERVATION ET. AL.
(Office of the Assistant Secretary for Conservation and Renewable Energy)
DA   February 4, 1991
BT1 Program Offices
 BT2 Offices
 BT3 Administrative Organizations
 BT4 Organizations
 BT2 U.S. Department of Energy
 BT3 Federal Agencies
 BT4 Agencies
  BT5 Administrative Organizations
  BT6 Organizations
SO   Management
DEF (Capsule Review of DOE Research and Development and Field Facilities) The Assistant Secretary for Conservation Renewable Energy manages programs designed to increase the production of renewable energy (including solar heat and electric energy, geothermal energy, biofuels energy and municipal waste energy) and to improve the energy efficiency of transportation, buildings, industrial and community systems and related processes. These programs include support of high-risk, high-payoff research and development that would not otherwise be carried out by the private sector and dissemination of the results to a broad spectrum of private and public sector interests. The Assistant Secretary also has responsibility for administering statutorily mandated programs that

provide assistance to conservation programs at the state level.

OFFICE OF THE ASSISTANT SECRETARY FOR DEFENSE PROGRAMS
DA   February 4, 1991
SY   Defense Programs
BT1 Program Offices
 BT2 Offices
 BT3 Administrative Organizations
 BT4 Organizations
 BT2 U.S. Department of Energy
 BT3 Federal Agencies
 BT4 Agencies
  BT5 Administrative Organizations
  BT6 Organizations
SO   Management
DEF (U.S. Government Manual) The Assistant Secretary for Defense Programs directs the Nation's nuclear weapons research, development, testing, production, and surveillance program, as well as the production of the special nuclear materials used by the weapons program within the Department, and management of defense nuclear waste and byproducts. The Office also manages research in inertial fusion; the safeguards and security program; classification, declassification, and reclassification of documents; unclassified controlled nuclear information; departmental intelligence; export control; test ban treaty verification and monitoring technology; defense- and energy-related intelligence activities; and coordinates the Department's emergency management activities.

OFFICE OF THE ASSISTANT SECRETARY FOR ENVIRONMENT, ET. AL.
DA   February 4, 1991
SY   Environment, Safety, and Health
SF   *EH*
BT1 Program Offices
 BT2 Offices
 BT3 Administrative Organizations
 BT4 Organizations
 BT2 U.S. Department of Energy
 BT3 Federal Agencies
 BT4 Agencies
  BT5 Administrative Organizations
  BT6 Organizations
NT1 Office of Special Projects
SO   Management
DEF (U.S. Government Manual) The Office of the Assistant Secretary for Environment, Safety and Health ensures that departmental programs are in compliance with environmental safety and health regulations and that environmental and safety impacts of Department

programs receive management
review.

OFFICE OF THE ASSISTANT SECRETARY FOR FOSSIL ENERGY
DA   February 4, 1991
BT1 Program Offices
 BT2 Offices
  BT3 Administrative Organizations
   BT4 Organizations
 BT2 U.S. Department of Energy
  BT3 Federal Agencies
   BT4 Agencies
    BT5 Administrative Organizations
     BT6 Organizations
SO   Management
DEF (U.S. Government Manual) The
     Assistant Secretary for Fossil
     Energy is responsible for research
     and development programs
     involving fossil fuels - coal,
     petroleum, and gas. The fossil
     energy program involves applied
     research, exploratory
     development, and limited
     proof-of-concept testing targeted
     to high-risk and high-payoff
     endeavors. The objective of the
     program is to provide the general
     technology and knowledge base
     that the private sector can use to
     complete development and initiate
     commercialization of advanced
     processes and energy systems.
     The program is principally
     executed through two Energy
     Technology Centers located in the
     field.

OFFICE OF THE ASSISTANT SECRETARY FOR NUCLEAR ENERGY
DA   February 4, 1991
BT1 Program Offices
 BT2 Offices
  BT3 Administrative Organizations
   BT4 Organizations
 BT2 U.S. Department of Energy
  BT3 Federal Agencies
   BT4 Agencies
    BT5 Administrative Organizations
     BT6 Organizations
SO   Management
DEF (U.S. Government Manual) The
     Assistant Secretary for Nuclear
     Energy administers the
     Department's research and
     development programs associated
     with fission energy. This includes
     programs relating to nuclear
     reactor development, both civilian
     and naval; nuclear fuel cycle;
     space nuclear applications; and
     uranium enrichment. The Assistant
     Secretary also manages the
     Department's Remedial Action
     Program to treat or stabilize
     radioactive wastes and perform
     decontamination and
     decommissioning at DOE surplus
     sites. The Assistant Secretary

conducts technical analyses and
provides advice concerning
nonproliferation; assesses
alternative nuclear systems and
new reactor and fuel cycle
concepts; and evaluates proposed
advanced nuclear fission energy
concepts and technical
improvements for possible
application to nuclear powerplant
systems.

OFFICE OF WEAPONS PRODUCTION
DA   January 8, 1991
SF   *OWP*
BT1 Offices
 BT2 Administrative Organizations
  BT3 Organizations
SO   Management

OFFICES
DA   February 1, 1991
BT1 Administrative Organizations
 BT2 Organizations
NT1 Grand Junction Projects Office
NT1 NRS Office for Analysis and
     Evaluation of Operational Data
NT1 NRS Office of Inspection and
     Enforcement
NT1 Office of Defense Waste and
     Transportation Management
NT1 Office of Management and Budget
NT1 Office of Technology Development
NT1 Office of Weapons Production
NT1 Operations Offices
 NT2 Albuquerque Operations Office
 NT2 Chicago Operations Office
 NT2 Idaho Operations Office
 NT2 Nevada Operations Office
 NT2 Oak Ridge Operations Office
 NT2 Richland Operations Office
 NT2 San Francisco Operations Office
 NT2 Savannah River Operations Office
NT1 Program Offices
 NT2 Office of the Assistant Secretary
      for Nuclear Energy
 NT2 Office of the Assistant Secretary
      for Fossil Energy
 NT2 Office of the Assistant Secretary
      for Defense Programs
 NT2 Office of the Assistant Secretary
      for Conservation et.al.
 NT2 Office of the Assistant Secretary
      for Environment, et. al.
  NT3 Office of Special Projects
 NT2 Office of Environmental
      Restoration and Waste
      Management
 NT2 Office of New Production Reactors
 NT2 Office of Civilian Radioactive
      Waste Management
 NT2 Office of Energy Research
  NT3 Office of Basic Energy Sciences
  NT3 Office of Health and
       Environmental Research
NT1 Program Enrichment Office
NT1 Regional Coordinating Offices
NT1 Sample Management Offices
RT   Boards
RT   Centers

RT   Committees
RT   Groups
SO   Management
DEF (WEBSTER) The directing
     headquarters of an enterprise or
     organization. A place where a
     particular kind of business is
     transacted or service is supplied.

OFFICIAL USE ONLY
(EMER)
DA   February 1, 1991
BT1 Uses
SO   Emergency Preparedness
DEF (EMER) A designation identifying
     unclassified information that may
     be exempt from mandatory
     disclosure under the Freedom of
     Information Act (FOIA).

OFFICIALLY REPORTABLE EVENTS
(EMER)
DA   February 1, 1991
BT1 Events
BT1 Reportable Occurrences
 BT2 Occurrences
RT   Notification
RT   Reports
SO   Emergency Preparedness
DEF (EMER) Events that require official
     notification to state and federal
     governments as outlined in
     Section 103(c) of the
     Comprehensive Environmental
     Response, Compensation and
     Liability Act of 1980; Title 40 Code
     of Federal Regulations Part 117,
     Determination of Reportable
     Quantities for Hazardous
     Substances; Title 33 Code of
     Federal Regulations Part 153.201,
     Notice of Discharge of Oil or
     Hazardous Substances; and Title
     40 Code of Federal Regulations
     Parts 110 and 112.

OFFSHORE FACILITIES
(CERCLA; CFR; USC; ESH)
DA   October 12, 1990
BT1 Facilities
SO   Compensation and Liability
SO   Environmental Management
SO   Hazardous Materials
DEF As defined by section 101(17) of
     CERCLA. (ESH) Any facility of any
     kind located in, on, or under any
     of the navigable waters of the
     United States, and any facility of
     any kind that is subject to the
     jurisdiction of the United States
     and is located in, on, or under any
     other waters, other than a vessel
     or a public vessel.

OH
DA   October 12, 1990
SEE Outer Housing
SO   Acronyms

BT-Broader Term   NT-Narrower Term   RT Related Term

OHD
DA   October 12, 1990
SEE Occupational Health Division
SO   Acronyms

OHER
DA   October 12, 1990
SEE Office of Health and Environmental
      Research
SO   Acronyms

OHIO ADMINISTRATIVE CODE
DA   January 8, 1991
SF   *OAC*
BT1   Codes, Standards, and Regulations

**OHIO BUREAU OF UNDERGROUND
STORAGE TANK REGULATION**
DA   January 8, 1991
SF   *BUSTR*
BT1   State Agencies
 BT2   Agencies
  BT3   Administrative Organizations
  BT4   Organizations
RT   Underground Storage Tanks

OHIO DEPARTMENT OF COMMERCE
DA   January 8, 1991
SF   *ODC*
BT1   State Agencies
 BT2   Agencies
  BT3   Administrative Organizations
  BT4   Organizations

OHIO DEPARTMENT OF HEALTH
DA   January 8, 1991
SF   *ODH*
BT1   State Agencies
 BT2   Agencies
  BT3   Administrative Organizations
  BT4   Organizations

**OHIO ENVIRONMENTAL PROTECTION
AGENCY**
DA   January 8, 1991
SF   *OEPA*
BT1   State Agencies
 BT2   Agencies
  BT3   Administrative Organizations
  BT4   Organizations

OHIO REVISED CODE
DA   January 8, 1991
SF   *ORC*
BT1   Codes, Standards, and Regulations

OHIO VALLEY ELECTRIC COMPANY
DA   January 8, 1991
SF   *OVEC*
BT1   Companies
 BT2   Commercial Organizations
  BT3   Organizations

OHMMETER(S)
(2762 OHMMETER)
DA   January 3, 1991
BT1   Indicator(s)
 BT2   Instrument(s)

 BT3   Equipment/Parts -
       Instrumentation/Measuring
       (DOE FRASE Voc.)
 BT4   Equipment
SO   DOE FRASE VOCABULARY

**OIL CHEMICAL AND ATOMIC
WORKERS**
DA   January 8, 1991
SF   *OCAW*
BT1   Personnel

OIL CIRCUIT BREAKER
DA   January 8, 1991
SF   *OCB*

OIL FINGERPRINTING
(EPA)
DA   October 12, 1990
BT1   Chemical Processes
 BT2   Processes
RT   Oil Spills
SO   Environmental Protection Agency
       Glossary
DEF (EPA) A method that identifies
       sources of oil and allows spills to
       be traced back to their source.

OIL POLLUTION FUND
(CERCLA; CFR)
DA   October 12, 1990
RT   Oil Spills
SO   Compensation and Liability
SO   Environmental Management
DEF (CFR) The fund established by
       section 311(k) of the Clean Water
       Act.

OIL SPILLS
(EPA)
DA   October 12, 1990
NT1   Major Discharges
NT1   Medium Discharges
NT1   Minor Discharges
RT   Air Curtains
RT   Assessment Area
RT   Biological Additives
RT   Booms
RT   National Contingency Plan
RT   Oil Fingerprinting
RT   Oil Pollution Fund
RT   Sinking Agents
RT   Sinking
RT   Skimming
RT   Surface Collecting Agents
SO   Environmental Protection Agency
       Glossary
DEF (EPA) An accidental or intentional
       discharge of oil which reaches
       bodies of water. Can be controlled
       by chemical dispersion,
       combustion, mechanical
       containment, and/or adsorption.

OILS
(DOE ORDER 5000.3A; CERCLA; CFR;
CWA; ESH)
DA   October 12, 1990
NT1   Heating Oils

NT1   Lubricating Oils
NT1   Petroleum
NT1   Re-refined Oil
NT1   Used Oils
 NT2   Recycled Oils
  NT3   Re-refined Oils
NT1   Waste Oils
RT   Gathering Lines
SO   Compensation and Liability
SO   Hazardous Materials
DEF As defined by section 311(a)(1) of
       the Clean Water Act, oils of any
       kind or in any forms, including, but
       not limited to, petroleum, fuel oils,
       sludges, oil refuses, and oils
       mixed with wastes other than
       dredged spoil.

OL
DA   October 12, 1990
SEE Operating Limit
SO   Acronyms

OLC
DA   October 12, 1990
SEE On-Line Computer
SO   Acronyms

OLIGOTROPHIC LAKES
(EPA)
DA   October 12, 1990
BT1   Lakes
 BT2   Inland Waters
 BT2   Surface Waters
  BT3   Water
 BT2   Surface Water Resources
  BT3   Natural Resources
RT   Eutrophic Lakes
SO   Environmental Protection Agency
       Glossary
DEF (EPA) Deep clear lakes with low
       nutrient supplies. They contain
       little organic matter and have a
       high dissolved-oxygen level.

OM
DA   October 12, 1990
SEE Operating Methods
SO   Acronyms

OMB
DA   October 12, 1990
SEE Office of Management and Budget
SO   Acronyms

OMEGA WEST REACTOR
DA   January 8, 1991
SF   *OWR*
BT1   Reactors

ON-GOING PROJECTS
(ESA; CFR)
DA   October 12, 1990
BT1   Projects
RT   Project Segment
SO   Endangered Species
DEF (CFR) Activities for scientific
       purposes or to enhance the

SY-Synonymous Terms   SO-Source/Subject Category   SF-See From

propagation or survival of threatened species which are not conducted in the course of a commercial activity initiated before the listing of the effected species.

ON GROUND TANKS
(SWDA; RCRA)
DA   October 12, 1990
BT1 Tanks
 BT2 Facility Components
SO   Environmental Management
SO   Wastes
DEF (CFR) Stationary devices designed to contain an accumulation of regulated substances and constructed of non-earthen materials (e.g., concrete, steel, plastic) that provide structural support and that are situated in such ways that the bottom of the tanks are on the same levels as the adjacent surrounding surfaces so that the external tank bottoms cannot be visually inspected.

ON-LINE COMPUTER
DA   January 8, 1991
SF   OLC

ON-SCENE
(EMER)
DA   February 1, 1991
BT1 Sites/Areas
SO   Emergency Preparedness
DEF (EMER) The area surrounding an accident or incident site that is, or potentially could be, affected by the accident or incident. This area includes both the on-site and off-site areas.

ON-SCENE COMMANDERS
(EMER)
DA   February 1, 1991
BT1 Personnel
NT1 Containment Commanders
SO   Emergency Preparedness
DEF (EMER) Officers or senior officials who command DOD and/or DOE operations at the scene of a DOD or DOE nuclear weapon accident or significant incident. For a security event, the officers or senior officials who command operations at the scene of the event.

ON-SCENE COORDINATORS
(CERCLA; CFR; EMER)
DA   October 12, 1990
SF   OSC
BT1 Personnel
RT   Support Agencies
SO   Compensation and Liability
SO   Emergency Preparedness
SO   Environmental Management

SO   Environmental Protection Agency Glossary
DEF (CFR) Predesignated EPA, U.S. Coast Guard, or Department of Defense officials who coordinate and direct Superfund removal actions or Clean Water Act oil- or hazardous-spill corrective actions. Federal officials predesignated by the EPA or USCG to coordinate and direct Federal responses; or DOD officials designated to coordinate and direct the removal actions from releases of hazardous substances, pollutants, or contaminants from DOD vessels and facilities.

ON-SHORE FACILITIES
(CERCLA; CFR; USC; ESH)
DA   October 12, 1990
BT1 Facilities
RT   Hazardous Materials
RT   Storage
SO   Compensation and Liability
SO   Environmental Management
SO   Hazardous Materials
DEF (ESH) Any facility, including , but not limited to, motor vehicles and rolling stock, of any kind located in, on, or under any land within the United States other than submerged land.

ON-SITE
(SWDA; RCRA; TSCA; CFR; EMER)
DA   October 19, 1990
BT1 Sites/Areas
RT   Off-site
RT   Site Boundaries
SO   Emergency Preparedness
SO   Environmental Management
SO   Hazardous Materials
SO   Wastes
DEF (CFR) On the same or geographically contiguous property that may be divided by public or private right(s)-of-way, provided the entrance and exit between the properties is at a cross-roads intersection, and access is by crossing as opposed to going along, the right(s)-of-way. Noncontiguous properties owned by the same person but connected by a right-of-way which the person controls and to which the public does not have access, are also considered on-site property.

ON-SITE DISCHARGES
(DOE Order 5484.1)
DA   October 12, 1990
BT1 Discharges
SO   Environmental Management
SO   Management
DEF (DOE Order 5484.1) Airborne and liquid wastes discharged to onsite treatment or disposal systems, e.g., sewage lagoons, retention

ponds, and cribs, for retention, settling, decay, or storage onsite.

ON-SITE FACILITIES
(EPA)
DA   October 12, 1990
BT1 Hazardous Waste Management Facilities
 BT2 Hazardous Waste Facilities
 BT3 Facilities
SO   Environmental Protection Agency Glossary
DEF (EPA) Hazardous waste treatment, storage or disposal areas that are located on the generating site.

ON-SITE FEDERAL SUPPORT
(EMER)
DA   February 1, 1991
BT1 Amelioration
SO   Emergency Preparedness
DEF (EMER) Federal assistance that is the primary responsibility of the federal agency that owns, authorizes, regulates, or is otherwise deemed responsible for the radiological facility or material being transported (i.e., the cognizant federal agency). This response supports state and local efforts by supporting the owner or operator's efforts to bring the incident under control and thereby preventing or minimizing off-site consequences.

ON-SITE TECHNICAL DIRECTORS
(EMER)
DA   February 1, 1991
BT1 Personnel
SO   Emergency Preparedness
DEF (EMER) Officials, selected by the U.S. Department of Energy (DOE) team leader or on-scene commander, who are responsible for directing the on-site operations for the DOE team leader or on-scene commander for a nuclear weapons accident response.

ONCOGENIC
(EPA)
DA   October 12, 1990
SO   Environmental Management
SO   Environmental Protection Agency Glossary
DEF (EPA) A substance that causes tumors, whether benign or malignant.

ONE-INCH COMPONENTS
(NFI)
DA   October 12, 1990
BT1 Components
SO   Nuclear Facilities Incident Database
SO   Radiation
DEF (NFI) Instrument rods, tie rods, or scram rods located at 1-inch positions of lattice.

BT-Broader Term   NT-Narrower Term   RT Related Term

OP
DA October 12, 1990
SEE Operating Procedures
SO Acronyms

OPACITY
(CAA; CFR)
DA October 12, 1990
RT Visibility Impairments
SO Air Pollution
SO Environmental Protection Agency
Glossary
DEF (EPA) The amount of light obscured
by particulate pollution in the air;
clear window glass has a zero
opacity, a brick wall has 100
percent opacity. Opacity is used
as an indicator of changes in
performance of particulate matter
pollution control systems. (CFR)
The degree to which emissions
reduce the transmission of light.

OPEN BURNING
(SWDA; RCRA)
DA October 12, 1990
RT Hazards
RT Incineration
RT Open Dumps
RT Thermal Treatment
SO Environmental Management
SO Environmental Protection Agency
Glossary
SO Wastes
DEF (EPA) Uncontrolled fires in an open
dump.

OPEN DUMPS
(SWDA; RCRA; USC)
DA October 12, 1990
BT1 Dumps
BT2 Sites/Areas
RT Inventory of Open Dumps
RT Open Burning
RT Solid Waste Disposal
SO Environmental Management
SO Environmental Protection Agency
Glossary
SO Wastes
DEF Any facilities or sites where solid
waste is disposed of which are not
sanitary landfills which meet the
criteria promulgated under section
4004 [42 USCS 6944] and which
are not a facility for disposal of
hazardous waste.

OPEN-ENDED LINE
(CAA; CFR)
DA October 19, 1990
RT Open Ended Valve
SO Air Pollution
DEF Any line, except pressure release
lines, having one end in contact
with process fluid and one end
open to atmosphere, either directly
or through an open valve.

OPEN ENDED VALVE
(CAA; CFR)
DA October 19, 1990
BT1 Valves
BT2 Devices
RT Open-Ended Line
RT Relief Valves
SO Air Pollution
DEF (CFR) Any valve, except pressure
relief valves, having one side of
the valve seat in contact with
process fluid and one side open to
atmosphere, either directly or
through open piping.

OPERABILITY
(MORT)
DA April 3, 1991
RT Controls
RT First Line Supervision
RT Higher Supervision
RT Inspections
RT Maintenance
RT Technical Information
DEF (MORT) MORT analysis asks: was
the facility and process
operationally ready? were the
necessary supplementary
operations supportive to the main
process ready? Operability refers
to the status of "upstream
processes" e.g. design, training,
etc. which support the ingredients
of the work process (hardware,
procedures, and people). Two
major upstream processes are
examined: (1) the original design,
construction, test, and qualification
plus documents defining operating
limits and performance
specification and (2) modification
projects to the facility. Each
upstream process can be
analyzed as to hardware,
procedures, and personnel.

OPERABLE
(DOE Order 5480.6)
DA October 12, 1990
BT1 Conditions
SO Industrial Safety
DEF (DOE Order 5480.6) When the
reactor is being operated or has
the potential for being operated. A
reactor that cannot be operated on
a day-to-day basis because of
refueling, extensive modifications,
or technical problems is still
considered to be operable.

OPERABLE UNITS
(CERCLA; CFR)
DA October 12, 1990
BT1 Response Actions
BT2 Actions
BT3 Responses
RT Superfund
SO Compensation and Liability
SO Environmental Management

SO Environmental Protection Agency
Glossary
DEF Term for each of a number of
separate activities undertaken as
part of a Superfund site cleanup.
A typical operable unit would be
removing drums and tanks from
the surface of a site. A discrete
part of the entire response action
that decreases a release, threat of
release, or pathway of exposure.

OPERATING AREA COMPARTMENT
(DOE Order 6430.1A)
DA October 12, 1990
BT1 Sites/Areas
RT Confinement Areas
SO Construction
DEF (DOE Order 6430.1A) An area or
series of areas that contain
process enclosures, and/or their
attendant equipment located within
that area or series of areas.

OPERATING BASIS ACCIDENT
(DOE Order 6430.1A)
DA October 12, 1990
SF *OBA*
BT1 Accidents
SO Construction
DEF (DOE Order 6430.1A) Maximum
severity accident under which the
plant structure, systems, and
components are designed to either
remain operable or be readily
restored to operating condition.
This is the highest severity event
that the operating contractor may
recover from without DOE
approval.

OPERATING BASIS EARTHQUAKE
(CFR)
DA January 8, 1991
SF *OBE*
SO Environmental Management
DEF (CFR) The operating basis
earthquake which, considering the
regional and local geology and
seismology and specific
characteristics of local subsurface
material, could reasonably be
expected to affect the plant site
during the operating life of the
plant; it is that earthquake which
produces the vibratory growth
motion for which those features of
the nuclear power plant necessary
for continued operation, without
undue risk to the health and safety
of the public, are designed to
remain functional.

OPERATING LEVEL
(DOE Orders 5480.2 and 5482.1B; ESH)
DA October 12, 1990
RT Contractors
RT Second Line Organization Level
SO Construction

SY-Synonymous Terms   SO-Source/Subject Category   SF-See From

SO Industrial Hygiene
SO Management
DEF (ESH) The organization performing the actual work or job related tasks. It may be a contractor performing work for DOE or it may be a particular DOE element, such as an energy technology center or a power administration.

OPERATING LIMIT
DA January 8, 1991
SF *OL*
BT1 Limits

OPERATING METHODS
DA January 8, 1991
SF *OM*

OPERATING ORGANIZATIONS
(IAEA)
DA October 12, 1990
BT1 Administrative Organizations
 BT2 Organizations
RT Nuclear Facilities
SO Radiation
DEF (IAEA) The organization authorized by the regulatory body to operate a nuclear facility.

OPERATING PROCEDURES
DA January 8, 1991
SF *OP*
BT1 Procedures
NT1 Detailed Operating Procedures
NT1 Emergency Operating Procedures
NT1 Standard Operating Procedures
SO Management

OPERATION ASSESSMENT AND READINESS
DA January 8, 1991
SF *OAR*
BT1 Assessments
 BT2 Administrative Processes
 BT3 Processes

OPERATIONAL (RADIATION) LIMITS
(IAEA)
DA November 20, 1990
BT1 Limits
SO Radiation
DEF (IAEA) Limits of any quantity specified by the management for a given radiation practice or source. These are equal to or lower than the authorized limits.

OPERATIONAL ACCIDENTS
(EMER)
DA February 1, 1991
BT1 Accidents
SO Emergency Preparedness
DEF (EMER) Events stemming from technological and man-made hazards which present a potential risk to life, health, property, or the environment.

OPERATIONAL CHANGE MEMOS
DA January 8, 1991
SF *OCM*

OPERATIONAL DBA
(DOE Order 6430.1A)
DA October 12, 1990
BT1 Design Basis Accidents
 BT2 Accidents
SO Construction
DEF (DOE Order 6430.1A) Any design basis accident caused by an internal event. Direct causes are usually poor design or procedures, operator errors, equipment failures, or inadequate technical development (unknowns) that lead to the accident. The major accident categories are explosion, fire, nuclear criticality, leaks to the atmosphere, and leaks to the aquatic environment.

OPERATIONAL EMERGENCIES
(EMER)
DA January 24, 1991
BT1 Emergencies
 BT2 Reportable Occurrences
 BT3 Occurrences
BT1 Emergency Response Levels
RT DOE Operations
SO Emergency Preparedness
SO Environmental Management
DEF (DOE 5500.1A) Significant events involving DOE operations and activities that present a potential risk to life, health, property, or the environment.

OPERATIONAL EMERGENCY MANAGEMENT TEAMS
(EMER)
DA February 1, 1991
BT1 Emergency Management Teams
 BT2 Management Teams
 BT3 Teams
 BT4 Administrative Organizations
 BT5 Organizations
SO Emergency Preparedness
DEF (EMER) U.S. Department of Energy (DOE) teams predesignated to manage activities during operational emergencies involving DOE or requiring DOE assistance.

OPERATIONAL EMERGENCY PREPAREDNESS MANAGEMENT PLANS
(EMER)
DA February 1, 1991
BT1 Emergency Plans
 BT2 Plans
SO Emergency Preparedness
DEF (EMER) Site-specific plans required by U.S. Department of Energy Order which identify emergency planning, preparedness, and response capabilities, personnel staffing, and equipment resources

necessary to minimize emergency consequences to people, property, and the environment.

OPERATIONAL EMERGENCY RESPONSE LEVELS
(EMER)
DA February 1, 1991
BT1 Emergency Response Levels
NT1 Alert
SO Emergency Preparedness
DEF (EMER) Response levels to hazardous materials incidents, natural phenomena occurring at nuclear and nonnuclear facilities, safeguards and security incidents, Radiological Assistance Program requests for assistances, and nuclear weapon accidents or significant incidents. For hazardous material emergencies the response levels are unusual event, alert, site emergency, and general emergency. For safeguards and security emergencies, response levels are Alert III (Charlie), Alert II (Bravo), and Alert I (Alpha). For Radiological Assistance Program emergencies, the response levels are radiological Assistance Program alert and Radiological Assistance Program emergency. Nuclear weapons accident or significant incident emergencies include all responses by the Accident Response Group, and there are no specific response levels. May also be referred to as emergency action levels.

OPERATIONAL LIFE
(SWDA; RCRA; CFR)
DA October 12, 1990
BT1 Time Designations
RT Tank Systems
SO Wastes
DEF Refers to the period beginning when installation of the tank system has commenced until the time the tank system is properly closed.

OPERATIONAL READINESS
(SSDC)
DA October 12, 1990
BT1 Conditions
RT Operational Readiness Reviews
RT Upstream Processes
SO System Safety Development Center Glossary
DEF (SSDC) A state attained (1) by verification of every significant detail of preparation for operations to assure safe start up and operation and (2) by correction of hazardous omissions in design and plans and the initiation of retrofit prior to accidents.

BT-Broader Term   NT-Narrower Term   RT Related Term

OPERATIONAL READINESS REVIEWS
(DOE Orders 5480.4, 5480.5, and
5480.6)
DA October 16, 1990
BT1 Reviews
 BT2 Administrative Processes
 BT3 Processes
RT Operational Readiness
RT Safety Program Reviews
SO Environmental Management
SO Standards
DEF Structured methods for determining
that a project, process, or facility
are ready to operate and occupy
and include, as a minimum, a
review of the readiness of the
plant and hardware, personnel,
and procedures. Reviews include
a determination of compliance with
Environment, Safety, and Health
(ES&H) Orders.

**OPERATIONAL SAFETY
PROCEDURES**
DA January 8, 1991
SF *OSP (Operational Safety
 Procedures)*
BT1 Procedures

**OPERATIONAL SAFETY
REQUIREMENTS**
(Doe Order 5480.5)
DA October 12, 1990
SF *OSR*
BT1 Requirements
RT Maintenance Plans
SO Construction
SO Standards
DEF (DOE 5480.5) Those requirements
that define the conditions, safe
boundaries, and bases thereof,
and management or administrative
controls required to assure the
safe operation of a nuclear facility.

OPERATIONAL SPECIFICATIONS
(MORT)
DA April 3, 1991
BT1 Specifications
 BT2 Requirements
RT Design and Development Plans
RT General Design Process
RT Human Factors
RT Inspection Plans
RT Maintenance Plans
RT Procedures
RT Training
DEF (MORT) MORT analysis asks: are
there adequate operational
specifications for all phases of the
system operation? These phases
include: proper testing and
qualification; adequate
supervision; proper task
procedures; motivation; training;
and adequate monitoring points.

OPERATIONS
(DOE Order 5480.4; ESH)

DA October 12, 1990
BT1 Activities
NT1 Aviation Operations
 NT2 Charter Operations
NT1 Burial Operations
 NT2 Shallow Land Burial
NT1 Cleanup Operations
NT1 Department of Energy Operations
NT1 DOE Operations
NT1 Emergency Renovation Operations
NT1 Headquarters Operations
NT1 Planned Renovation Operations
NT1 Reactor Operations
NT1 Wastewater Operations and
 Maintenance
NT1 Waste Management Operations
SO Environmental Management
SO Management
SO Standards
DEF (WTID) Denoting specific
responsibility for principal planning
and functional activities of
subordinate organizational units.
(ESH) Those activities funded by
DOE for which DOE has
responsibility for environmental
protection, safety, and health
protection.

OPERATIONS AND MAINTENANCE
(EPA)
DA October 12, 1990
SF *O&M*
BT1 Activities
SO Environmental Protection Agency
 Glossary
DEF (EPA) (1) Activities conducted at a
site after a Superfund site action is
completed to ensure that the
action is effective and operating
properly. (2) Actions taken after
construction to assure that
facilities constructed to treat
wastewater will be properly
operated, maintained, and
managed to achieve efficiency
levels and prescribed effluent
limitations in an optimum manner.

OPERATIONS OFFICE MANAGERS
(DOE Order 5440.1D)
DA May 15, 1991
BT1 Personnel
SO Environmental Management
DEF (DOE Order) For the purposes of
DOE Order 5440.1D, these are
people who are responsible for
managing an Operations Office or
similar field organization, who
report directly to the Under
Secretary or a Secretarial Officer.
The Operations Offices are
Albuquerque, Chicago, Idaho,
Nevada, Oak Ridge, Richland,
San Francisco, and Savannah
River. For the purposes of DOE
Order 5440.1D, Operations Office
Managers include but may not be
limited to: the Manager of the
Rocky Flats Office; the

Administrators of the Alaska,
Southeastern, and Southwestern
Power Administrations; and the
Directors of the Morgantown and
Pittsburgh Energy Technology
Centers.

OPERATIONS OFFICES
DA February 4, 1991
BT1 Offices
 BT2 Administrative Organizations
 BT3 Organizations
BT1 U.S. Department of Energy
 BT2 Federal Agencies
 BT3 Agencies
 BT4 Administrative Organizations
 BT5 Organizations
NT1 Albuquerque Operations Office
NT1 Chicago Operations Office
NT1 Idaho Operations Office
NT1 Nevada Operations Office
NT1 Oak Ridge Operations Office
NT1 Richland Operations Office
NT1 San Francisco Operations Office
NT1 Savannah River Operations Office
RT Program Offices

OPERATOR, PLANT/SYSTEM/UTILITY
(0690 OP;UTIL)
DA November 28, 1990
BT1 Precision/Production Personnel
 BT2 Occupations
 BT2 Personnel
SO DOE FRASE VOCABULARY

OPERATORS
(DOE Order 5480.5; SWDA; RCRA;
 CFR; USC; ESH)
DA October 12, 1990
SY Owners
BT1 Potentially Responsible Parties
BT1 Responsible Parties
RT Authorized Representatives
RT Guarantors
RT Water Suppliers
SO Air Pollution
SO Compensation and Liability
SO Environmental Management
SO Hazardous Materials
SO Industrial Safety
SO Management
SO Wastes
SO Water Pollution
DEF Owners or operators of any "facility
or activity" subject to regulation
under the Resource Conservation
and Recovery Act (RCRA),
Underground Injection Control
(UIC), National Pollutant
Discharge Elimination System
(NPDES), or 404 programs.(ESH)
Individuals designated by
management to perform
operations or conduct activities
with radioactive materials at a
nuclear facility. (ESH) Persons
who control the use of an aircraft,
vessel, or vehicle.

SY-Synonymous Terms SO-Source/Subject Category SF-See From

OPTIMUM INTERSPERSED HYDROGENOUS MODERATION
(DOE Order 5480.3; ESH)
DA   October 12, 1990
RT   Containment Vessels
SO   Hazardous Materials
DEF (ESH) The occurrence of
     hydrogenous material between
     containment vessels to such an
     extent that the maximum nuclear
     reactivity results.

OR
DA   October 12, 1990
SEE Oak Ridge Operations Office
SO   Acronyms

OR GATE
(SSDC)
DA   October 12, 1990
BT1 Logic Gates (Boolean)
SO   System Safety Development Center
     Glossary
DEF (SSDC) A logic gate that produces
     an output when one or more of the
     input events occur. May contain
     the identifying word "OR."

ORAU
DA   October 12, 1990
SEE Oak Ridge Associated Universities
SO   Acronyms

ORC
DA   October 12, 1990
SEE Ohio Revised Code
SO   Acronyms

ORDINARY HIGH WATER MARK
(CWA; RHA; CFR)
DA   October 12, 1990
RT   High Tide Line
RT   Tidal Waters
SO   Water Pollution
DEF (CFR) That line on the shore
     established by the fluctuations of
     water and indicated by physical
     characteristics such as clear,
     natural line impressed on the
     bank, shelving, changes in the
     character of soil, destruction of
     terrestrial vegetation, the presence
     of litter and debris, or other
     appropriate means that consider
     the characteristics of the
     surrounding areas.

ORDINARY MOMENT RESISTING SPACE FRAME
(SEA)
DA   October 12, 1990
BT1 Moment Resisting Space Frame
BT2 Space Frames
BT3 Structures (DOE FRASE
       Vocabulary)
SO   Construction
DEF (SEA) A moment resisting space
     frame not meeting special detailing
     requirements for ductile behavior.

ORGANIC
(EPA)
DA   October 12, 1990
SO   Environmental Protection Agency
     Glossary
DEF (EPA) (1) Referring to or derived
     from living organisms. (2) In
     chemistry, any compound
     containing carbon.

ORGANIC CHEMICALS
(EPA)
DA   October 12, 1990
BT1 Chemical Substances
NT1 Acetylcholine
NT1 Alar
NT1 Dioxins
NT1 DNA
NT2 Plasmids
NT2 Recombinant DNA
NT1 Formaldehyde
NT1 Halogenated Organic Compounds
NT2 Chlorinated Hydrocarbons
NT3 Chlorofluorocarbons (CFCs)
NT3 DDT
NT3 Heptachlor
NT3 Polychlorinated Biphenyls
NT4 High Concentration PCBs
NT4 Low Concentration PCBs
NT4 Recycled PCBs
NT3 Polyvinyl Chloride
NT3 Trichloroethylene (TCE)
NT3 Vinyl Chloride
NT2 Fluorocarbons
NT3 Chlorofluorocarbons (CFCs)
NT2 Trihalomethane
NT1 High-Density Polyethylene
NT1 Hydrocarbons
NT2 Methane
NT1 Nitrilotriacetic Acid
NT1 Organic Peroxides
NT1 Phenols
NT1 Proteins
NT2 Antibodies
NT3 Monoclonal Antibodies
NT2 Carboxyhemoglobin
NT2 Restriction Enzymes
NT1 Ribonucleic Acid
NT1 Synthetic Organic Chemicals
NT2 Polyelectrolytes
NT2 Volatile Synthetic Organic
       Chemicals
NT1 Volatile Organic Compounds
NT2 Volatile Synthetic Organic
       Chemicals
RT   Common Laboratory Contaminants
RT   Technical Names
SO   Environmental Protection Agency
     Glossary
DEF (EPA) Animal or plant-produced
     substances containing mainly
     carbon, hydrogen, and oxygen.

ORGANIC MATTER
(EPA)
DA   October 12, 1990
NT1 Soil Conditioner
NT2 Compost

SO   Environmental Protection Agency
     Glossary
DEF (EPA) Carbonaceous waste
     contained in plant or animal matter
     and originating from domestic or
     industrial sources.

ORGANIC PEROXIDES
(CFR)
DA   October 12, 1990
BT1 Organic Chemicals
BT2 Chemical Substances
SO   Hazardous Materials
DEF An organic peroxide is defined as
     an organic compound containing
     the bivalent –O–O– structure and
     which may be considered a
     derivative of hydrogen peroxide
     where one or more of the
     hydrogen atoms have been
     replaced by organic radicals.
     Lauroyl peroxide and
     paramenthane hydroperoxide are
     examples of these types of
     materials.

ORGANISMS
(EPA)
DA   October 12, 1990
NT1 Beneficial Organisms
NT1 Benthic Organisms (Benthos)
NT1 Heterotrophic Organisms
NT1 Hosts
NT1 Microorganisms
NT2 Bacteria
NT3 Coliform Organisms
NT4 Fecal Coliform Bacteria
NT3 Legionella
NT3 Recombinant Bacteria
NT2 Microbes
NT2 Pathogens
NT2 Viruses
NT1 Non-Target Organisms
RT   Acclimatization
RT   Adaptations
RT   Aerobic
RT   Autotrophic
RT   Hybrid
RT   Targets
SO   Environmental Protection Agency
     Glossary
DEF (CFR) Any living things.

ORGANIZATIONS
DA   February 1, 1991
NT1 Administrative Organizations
NT2 Agencies
NT3 Federal Agencies
NT4 Atomic Energy Commission
NT4 Cognizant Federal Agencies
NT4 DOE Design Agency
NT4 DOE Production Agency
NT4 Federal Aviation Administration
NT4 Lead Agencies
NT4 National Aeronautic and Space
       Administration
NT4 Nuclear Regulatory Commission
NT5 NRC Office of Nuclear Reactor
       Regulation

BT-Broader Term   NT-Narrower Term   RT Related Term

NT5 NRC Office of Nuclear
Regulatory Research
NT4 Occupational Safety and Health
Administration
NT4 Office of Management and
Budget
NT4 Procuring Agencies
NT4 U.S. Department of Energy
NT5 Headquarters Operations
NT5 Naval Reactors
NT5 Operations Offices
NT6 Albuquerque Operations Office
NT6 Chicago Operations Office
NT6 Idaho Operations Office
NT6 Nevada Operations Office
NT6 Oak Ridge Operations Office
NT6 Richland Operations Office
NT6 San Francisco Operations
Office
NT6 Savannah River Operations
Office
NT5 Program Offices
NT6 Office of the Assistant
Secretary for Nuclear Energy
NT6 Office of the Assistant
Secretary for Fossil Energy
NT6 Office of the Assistant
Secretary for Defense
Programs
NT6 Office of the Assistant
Secretary for Conservation
et.al.
NT6 Office of the Assistant
Secretary for Environment,
et. al.
NT7 Office of Special Projects
NT6 Office of Environmental
Restoration and Waste
Management
NT6 Office of New Production
Reactors
NT6 Office of Civilian Radioactive
Waste Management
NT6 Office of Energy Research
NT7 Office of Basic Energy
Sciences
NT7 Office of Health and
Environmental Research
NT4 U.S. Department of Justice
NT4 U.S. Department of Labor
NT4 U.S. Department of
Transportation
NT5 Research and Special
Programs Administration
NT4 U.S. Environmental Protection
Agency
NT5 Regional Office
NT4 User Agencies
NT3 Local Agencies
NT4 Implementing Agencies
NT4 Intermunicipal Agencies
NT4 Local Governments
NT4 Local Educational Agencies
NT3 State Agencies
NT4 Air Pollution Control Agencies
NT5 Interstate Air Pollution Control
Agencies
NT4 Colorado Department of Health
NT4 Division of the State Fire
Marshall
NT4 Implementing Agencies

NT4 Interstate Agencies
NT5 Interstate Air Pollution Control
Agencies
NT4 Missouri Department of Natural
Resources
NT4 New Mexico Environmental
Improvement Division
NT4 Ohio Bureau of Underground
Storage Tank Regulation
NT4 Ohio Department of Commerce
NT4 Ohio Department of Health
NT4 Ohio Environmental Protection
Agency
NT4 Procuring Agencies
NT4 State Routing Agencies
NT4 State Emergency Response
Commission
NT4 State Authority
NT4 Tennessee Department of
Commerce
NT5 Bureau of Environment
NT4 Utah Department of Health
NT3 Support Agencies
NT2 Boards
NT3 Accrediting Boards
NT3 Energy Research Advisory Board
NT3 GDC Planning Board
NT3 Regional Water Quality Control
Board
NT2 Committees
NT3 Accident Response Capabilities
Coordinating Committee
NT3 Advisory Committee on Reactor
Safety
NT3 Emergency Management
Coordination Committee
NT3 Interagency Committee on
Standards Policy
NT3 Local Emergency Planning
Committees
NT3 Metrication Operating Committee
NT3 Metric Transition Committee
NT3 Plant Oversight Review
Committee
NT3 Regional Assistance Committees
NT3 Safety Review Committee
NT3 Subcommittee on Federal
Response
NT3 Waste Acceptance Criteria
Committee
NT2 Divisions
NT3 Computer Operations Division
NT3 Health, Safety, and Environment
Division
NT3 New Mexico Environmental
Improvement Division
NT3 Occupational Health Division
NT2 Groups
NT3 Accident Response Groups
NT3 Carcinogen Risk Assessment
Verification Endeavor Workgroup
NT3 Emergency Management
Coordination Committee
Secretariat
NT3 Interagency Group on Energy
Vulnerability
NT3 Purchasing Group
NT3 Reactor Materials Control Group
NT3 Risk Retention Group
NT3 Standards-developing Groups

NT3 State and Tribal Government
Working Group
NT2 Indian Governing Bodies
NT2 Offices
NT3 Grand Junction Projects Office
NT3 NRS Office for Analysis and
Evaluation of Operational Data
NT3 NRS Office of Inspection and
Enforcement
NT3 Office of Defense Waste and
Transportation Management
NT3 Office of Management and
Budget
NT3 Office of Technology
Development
NT3 Office of Weapons Production
NT3 Operations Offices
NT4 Albuquerque Operations Office
NT4 Chicago Operations Office
NT4 Idaho Operations Office
NT4 Nevada Operations Office
NT4 Oak Ridge Operations Office
NT4 Richland Operations Office
NT4 San Francisco Operations Office
NT4 Savannah River Operations
Office
NT3 Program Offices
NT4 Office of the Assistant Secretary
for Nuclear Energy
NT4 Office of the Assistant Secretary
for Fossil Energy
NT4 Office of the Assistant Secretary
for Defense Programs
NT4 Office of the Assistant Secretary
for Conservation et.al.
NT4 Office of the Assistant Secretary
for Environment, et. al.
NT5 Office of Special Projects
NT4 Office of Environmental
Restoration and Waste
Management
NT4 Office of New Production
Reactors
NT4 Office of Civilian Radioactive
Waste Management
NT4 Office of Energy Research
NT5 Office of Basic Energy
Sciences
NT5 Office of Health and
Environmental Research
NT3 Program Enrichment Office
NT3 Regional Coordinating Offices
NT3 Sample Management Offices
NT2 Operating Organizations
NT2 Program Organizations
NT2 Teams
NT3 Management Teams
NT4 Accreditation Review Teams
NT4 Continuity of Government
Emergency Management
Teams
NT4 Crisis Management Teams
NT4 Emergency Management Teams
NT5 Energy Emergency
Management Teams
NT5 Operational Emergency
Management Teams
NT4 Headquarters Coordinating
Teams
NT3 Response Teams
NT4 Airborne Response Teams

SY-Synonymous Terms SO-Source/Subject Category SF-See From

NT4 Emergency Support Teams
NT4 Emergency Response Teams
NT4 Energy Emergency Teams
NT4 Environmental Response Teams
NT4 Explosive Ordnance Disposal
    Teams
NT4 Hazardous Materials Response
    Teams
NT4 Hostage Negotiation Teams
NT4 National Response Teams
NT4 Nuclear Emergency Search
    Teams
NT4 Radiological Assistance Teams
NT4 Regional Response Teams
NT4 Special Response Teams
NT4 Tactical Response Teams
NT3 Tiger Teams
NT1 Commercial Organizations
NT2 Companies
NT3 Allied-Signal Inc.
NT3 Associated Universities, Inc.
NT3 Bechtel Petroleum Operation, Inc.
NT3 Boeing Petroleum Services
NT3 EG&G Energy Measurements,
    Inc.
NT3 EG&G Mound Applied
    Technologies, Inc.
NT3 EG&G Idaho, Inc.
NT4 System Safety Development
    Center
NT3 General Electric
NT3 Goodyear Atomic Corporation
NT3 International Business Machines
NT3 John Brown, E&C, Inc.
NT3 Kaiser Engineers Hanford, Co.
NT3 Lawrence-Allison & Associates
NT3 Martin Marietta Energy Systems
NT3 Mason and Hanger-Silas Mason
    Co.
NT3 Mississippi Power and Light
NT3 MK-Ferguson
NT3 MK-Ferguson of Idaho Co.
NT3 Ohio Valley Electric Company
NT3 Pan American World Services,
    INC.
NT3 Raytheon Services Nevada
NT3 Reactive Metals, Inc.
NT3 Reynolds Electrical and
    Engineering Company
NT3 Rockwell International Corp.
NT3 Ross Aviation, Inc.
NT3 Sandia Corporation AT&T
    Technologies, Inc.
NT3 Science Applications International
    Corporation
NT3 United Electric
NT3 Universities Research
    Association, Inc.
NT3 Wackenhut Services Inc.
NT3 West Valley Nuclear Services
    Co., Inc.
NT3 Westinghouse Electric Corp.
NT3 Westinghouse Idaho Nuclear Co.,
    Inc.
NT3 Westinghouse Savannah River
    Company
NT3 Westinghouse Hanford Company
NT3 Westinghouse Materials
    Company of Ohio
NT2 Retailers
NT1 Educational Organizations

NT2 Central Training Academy
NT2 Universities
NT3 Iowa State University
NT3 Massachusetts Institute of
    Technology
NT3 North Carolina State University
NT3 Stanford University
NT3 University of Chicago
NT3 University of California
NT3 University of California-Irvine
NT1 Line Organizations
NT1 Matrix Organizations
NT1 Minority Institutions
NT1 Professional Bodies
NT2 American Society of Mechanical
    Engineers
NT2 Atomic Industrial Forum
NT2 Conference of Federal
    Environmental Engineers
NT2 Council on Environmental Quality
NT2 National Fire Protection
    Association
NT2 Natural Resource Defense Council
NT1 Regulatory Organizations
NT1 Research and Development
    Organizations
NT2 American National Metric Council
NT2 American Nuclear Society
NT2 Foundations
NT3 Hanford Environmental Health
    Foundation
NT3 Lovelace Medical Foundation
NT3 University of Georgia Research
    Foundation, Inc.
NT2 Institutes
NT3 American National Standards
    Institute
NT3 American Petroleum Institute
NT3 Battelle Memorial Institute
NT3 Electric Power Research Institute
NT4 Nuclear Safety Analysis Center
NT3 Inhalation Toxicology Research
    Institute
NT3 Institute of Electrical and
    Electronic Engineers
NT3 Institute for Nuclear Power
    Operations
NT3 Midwest Research Institute
NT3 National Institute for Occupational
    Safety and Health
NT3 National Institute of Standards
    and Testing
NT2 Laboratories
NT3 Argonne National Laboratory
NT4 Argonne National
    Laboratory-East (Chicago)
NT4 Argonne National
    Laboratory-West (At INEL)
NT3 Argonne National
    Laboratory-East (Chicago)
NT3 Argonne National
    Laboratory-West (At INEL)
NT3 Battelle Columbus Laboratories
NT3 Brookhaven National Laboratory
NT3 Idaho National Engineering
    Laboratory
NT4 Power Burst Facility
NT3 Laboratory for Energy-Related
    Health Research
NT3 Lawrence Livermore National
    Laboratory

NT3 Lawrence Berkeley Laboratory
NT3 Los Alamos National Laboratory
NT3 Nationally Recognized Testing
    Laboratories
NT3 Oak Ridge National Laboratory
NT3 Pacific Northwest Laboratory
NT3 Performance Testing Laboratories
NT3 Research & Development
    Laboratory
NT3 Sandia National Laboratories
NT4 Sandia National
    Laboratories-Alburquerque
NT4 Sandia National
    Laboratories-Livermore
NT2 National Academy of Sciences
NT2 Oak Ridge Associated Universities
NT2 Southeastern Universities
    Research Association, Inc.
NT1 Staff Organization
NT1 Voluntary Standards Bodies
RT Personnel
RT Procedures
SO Management
DEF (WTID) A group of people with a
    more or less constant
    membership, a body of officers, a
    purpose, and usually, a set of
    regulations. The group will also
    exhibit an administrative and
    functional structure including
    established relationships of
    personnel through lines of
    authority and responsibility with
    delegated and assigned duties.

ORGANOPHOSPHATES
(EPA)
DA October 12, 1990
BT1 Pesticides
BT2 Hazardous Substances
SO Environmental Protection Agency
    Glossary
DEF (EPA) Pesticide chemicals that
    contain phosphorus; used to
    control insects. They are
    short-lived, but some can be toxic
    when first applied.

ORGANOTINS
(EPA)
DA October 12, 1990
RT TBT Paints (Trybutilin)
SO Environmental Protection Agency
    Glossary
DEF (EPA) Chemical compounds used
    in antifoulant paints to protect the
    hulls of boats and ships, buoys,
    and dock pilings from marine
    organisms such as barnacles.

ORGDP
DA October 12, 1990
SEE Oak Ridge Gaseous Diffusion Plant
SO Acronyms

ORM
DA October 12, 1990
SEE Other Regulated Material
SO Acronyms

ORNL
DA   October 12, 1990
SEE Oak Ridge National Laboratory
SO   Acronyms

ORO
DA   October 12, 1990
SEE Oak Ridge Operations Office
SO   Acronyms

ORPS
DA   May 20, 1991
SEE Occurrence Reporting and
     Processing System
SO   Acronyms

ORTHOGONAL EFFECTS
(SEA)
DA   October 12, 1990
BT1   Effects
SO   Construction
DEF (SEA) The effects on the structure
     due to earthquake motions acting
     in directions other than parallel to
     the direction of resistance under
     consideration.

OSC
DA   October 12, 1990
SEE On-Scene Coordinators
SO   Acronyms
SO   Environmental Management

OSCILLOSCOPE(S)
(2763 OSCILLOSCOPE)
DA   January 3, 1991
BT1   Testing Equipment
 BT2 Instrument(s)
  BT3 Equipment/Parts -
        Instrumentation/Measuring
        (DOE FRASE Voc.)
   BT4 Equipment
SO   DOE FRASE VOCABULARY

OSHA
DA   October 12, 1990
SEE Occupational Safety and Health
     Administration
SO   Acronyms

OSMOSIS
(EPA)
DA   October 12, 1990
BT1   Biological Processes
 BT2 Processes
NT1 Reverse Osmosis
RT   Permeability
SO   Environmental Protection Agency
     Glossary
DEF (EPA) The tendency of a fluid to
     pass through a permeable
     membrane such as the wall of a
     living cell into a less concentrated
     solution so as to equalize the
     concentrations on both sides of
     the membrane.

OSP (Office of Special Projects)
DA   May 7, 1991
SEE Office of Special Projects
SO   Acronyms

OSP (Operational Safety Procedures)
DA   October 12, 1990
SEE Operational Safety Procedures
SO   Acronyms

OSP Coordinators
DA   May 7, 1991
SEE Office of Special Projects
     Coordinators
SO   Acronyms

OSR
DA   October 12, 1990
SEE Operational Safety Requirements
SO   Acronyms

OTD
DA   October 12, 1990
SEE Office of Technology Development
SO   Acronyms

OTHER COMPLICATIONS
(1397 OTHER COMPLI)
DA   November 28, 1990
BT1   Illnesses
SO   DOE FRASE VOCABULARY

OTHER EQUIPMENT
(DOE Order 4330.4A)
DA   June 4, 1991
BT1   Equipment
SO   Management
DEF (DOE Order 4330.4A) Some
     examples in this category of
     equipment are office machines,
     vehicles and mobile equipment,
     helicopters, airplanes, and
     computers and other automated
     data-processing equipment.

OTHER NON-TASK ACTIVITY
(1253 ONT ACTIVITY)
DA   November 28, 1990
BT1   Activity Types (DOE FRASE
     Vocabulary)
 BT2 Activities
SO   DOE FRASE VOCABULARY

OTHER REGULATED MATERIAL
(SWDA: RCRA; CFR)
DA   October 12, 1990
SF   *ORM*
BT1   Materials
NT1 Other Regulated Material - A
NT1 Other Regulated Material - B
NT1 Other Regulated Material - C
NT1 Other Regulated Material - D
NT1 Other Regulated Material - E
SO   Hazardous Materials
DEF (CFR) Any material that does not
     meet the definition of a hazardous
     material, other than a combustible
     liquid in packaging having a

capacity of 110 gallons or less,
and is specifically listed in the
Table of Hazardous Materials (40
CFR 172.101) as an ORM. A
material not listed in the Table of
Hazardous Materials may also be
considered an ORM.

OTHER REGULATED MATERIAL - A
(CFR)
DA   October 12, 1990
BT1   Other Regulated Material
 BT2 Materials
DEF (CFR) An ORM-A material has an
     anesthetic, irritating, noxious, toxic
     or other similar property and can
     cause extreme annoyance or
     discomfort to passengers and
     crews in the event of leakage
     during transportation. Phencapton,
     thiram, chloroform, and carbaryl
     are examples of Other Regulated
     Materials-A.

OTHER REGULATED MATERIAL - B
(CFR)
DA   October 12, 1990
BT1   Other Regulated Material
 BT2 Materials
DEF (CFR) An ORM-B material is
     capable of causing significant
     damage to a transport vehicle or
     vessel from leakage during
     transportation (includes a solid
     when wet with water). An ORM-B
     material will be so designated in
     the Table of Hazardous Materials
     (172.101), or be a liquid substance
     that has a corrosion rate
     exceeding 0.250 inch per year on
     aluminum at a test temperature of
     $130°F(173.500(a)(2))$. Ammonium
     hydrogen fluoride (solid),
     chloroplatinic acid (solid), and
     barium oxide are examples of
     materials classed as ORM-B.

OTHER REGULATED MATERIAL - C
(CFR)
DA   October 12, 1990
BT1   Other Regulated Material
 BT2 Materials
DEF (CFR) Every material classed as an
     ORM-C will be so listed and
     identified in the Table of
     Hazardous Materials (172.101).
     An ORM-C material has other
     inherent characteristics not
     included within the ORM-A or
     ORM-B classes, but which render
     the material unsuitable for
     shipment unless properly identified
     and prepared for transportation.
     One material considered in the
     Table of Hazardous Materials as
     an ORM-C is a magnetized
     material. A substance is
     considered to be a magnetized
     material if, when packaged for
     transport by air, it has a magnetic

field strength of 0.002 gauss or more at a distance of seven feet from any point of the surface of the package, or is of such a mass that it could affect the aircraft instruments, particularly the compasses. Petroleum coke, sawdust (when dry, clean, and free from oil), copra, castor beans and feed (wet, mixed) are examples of materials classed as ORM-C.

OTHER REGULATED MATERIAL - D
(CFR)
DA   October 12, 1990
BT1 Other Regulated Material
 BT2 Materials
DEF (CFR) Materials, such as consumer commodities that would otherwise be subject to the hazardous materials regulations, that present a limited hazard during transportation because of their form, quantity, and packaging are classed as ORM-D; and ORM-D classification is only allowed when the material is granted an exception in a reference from the Table of Hazardous Materials. For example, a consumer commodity that contains a flammable solid and would ordinarily be classed as Flammable Solid, could be reclassified as an ORM-D material, provided the flammable solid is inside containers each having a net weight of one pound or less, packed in strong outside packagings each having a net weight of 25 pounds or less.

OTHER REGULATED MATERIAL - E
(CFR)
DA   October 12, 1990
BT1 Other Regulated Material
 BT2 Materials
DEF Materials that are not included in any other hazard class but that are hazardous wastes or hazardous substances and so designated by the Environmental Protection Agency's regulations (Title 40, Code of Federal Regulations).

OTHER RESTRICTED ACCESS (NONSUBSTATION) LOCATIONS
(TSCA; CFR)
DA   October 19, 1990
BT1 Sites/Areas
RT   Outdoor Electrical Substations
SO   Hazardous Materials
DEF (CFR) Areas other than electrical substations that are at least 0.1 kilometer (km) from a residential/commercial area and limited by man-made barriers (e.g., fences and walls) to substantially limited by naturally occurring barriers such as

mountains, cliffs, or rough terrain. These areas generally include industrial facilities and extremely remote rural locations. (Areas where access is restricted but are less than 0.1 km from a residential/commercial area are considered to be residential/commercial areas.)

OTHER SKIN CONDITIONS
(1398 OTHER SKIN C)
DA   November 28, 1990
BT1 Illnesses
RT   Dermatitis
RT   Skin
SO   DOE FRASE VOCABULARY

OTHERWISE USE
(CERCLA; CFR)
DA   October 12, 1990
BT1 Uses
RT   Toxic Chemicals
SO   Compensation and Liability
DEF (CFR) Any use of a toxic chemical that is not covered by the terms "manufacture" or "process" and includes use of a toxic chemical contained in a mixture or trade name product. Relabeling or redistributing a container of a toxic chemical where no repackaging of the toxic chemical occurs does not constitute use or processing of the toxic chemical.

OUTAGE
(ESH)
DA   October 19, 1990
SY   Ullage
SO   Hazardous Materials
DEF (ESH) The amount by which a packaging falls short of being liquid full, usually expressed in percent by volume.

OUTDOOR ELECTRICAL SUBSTATIONS
(TSCA; CFR)
DA   October 12, 1990
BT1 Sites/Areas
RT   Other Restricted Access (nonsubstation) Locations
SO   Hazardous Materials
DEF (CFR) Outdoor, fenced-off, and restricted access areas used in the transmission and/or distribution of electrical power. Outdoor electrical substations restrict public access by being fenced or walled off as defined in 40 CFR 761.30(l)(1)(ii). For purposes of this Toxic Substances Control Act (TSCA) policy, outdoor electrical substations are defined as being located at least 0.1 km from a residential/commercial area. Outdoor fenced-off and restricted access areas used in

the transmission and/or distribution of electrical power which are located less than 0.1.km from a residential/commercial area are considered to be residential/commercial areas.

OUTER CHUCK
DA   January 8, 1991
SF   OC
BT1 D Machines
DEF (NFI) Part of the reactor discharging machine.

OUTER HOUSING
DA   January 8, 1991
SF   OH
DEF (NFI) Denotes the outer housing of a fuel assembly.

OUTFALL
(EPA)
DA   October 12, 1990
BT1 Sites/Areas
SO   Environmental Protection Agency Glossary
DEF (EPA) The place where an effluent is discharged into receiving waters.

OUTLET/RECEPTACLE(S)
(2404 OUTLET)
DA   January 3, 1991
BT1 Equipment/Parts - Electrical (DOE FRASE Vocabulary)
 BT2 Equipment
SO   DOE FRASE VOCABULARY

OUTLIER
(ESH)
DA   October 12, 1990
RT   Distribution
SO   Management
DEF (ESH) An extreme value in a data set so far removed from the other values with which it is associated that the chance probability of its being a valid member of the group is very small. Such a questionable value may be eliminated from the group on the basis of further statistical investigations of the data set.

OUTSIDE AIR
(CAA; CFR)
DA   October 12, 1990
SY   Ambient Air
BT1 Air
SO   Air Pollution
DEF (CFR) The air outside buildings and structures.

OUTSIDE CONTAINER
(SWDA; RCRA; ESH)
DA   October 12, 1990
BT1 Containers
NT1 Strong Outside Containers

SO   Hazardous Materials
DEF The outermost enclosure used in transporting a hazardous material other than a freight container.

OVEC
DA   October 12, 1990
SEE Ohio Valley Electric Company
SO   Acronyms

OVEN(S)
(2488 OVEN)
DA   January 3, 1991
BT1 Heater(s)
 BT2 Equipment/Parts - Heating (DOE FRASE Vocabulary)
  BT3 Equipment
SO   DOE FRASE VOCABULARY

OVERBURDEN
(EPA)
DA   October 12, 1990
SO   Environmental Protection Agency Glossary
DEF (EPA) The rock and soil cleared away before mining.

OVERFILL RELEASES
(SWDA; RCRA; CFR)
DA   October 12, 1990
BT1 Releases
SO   Wastes
DEF (CFR) Releases that occur when a tank is filled beyond its capacity, resulting in a discharge of the regulated substance to the environment.

OVERFIRE AIR
DA   November 8, 1990
BT1 Air
RT   Incinerators
SO   Environmental Management
DEF Air forced into the top of an incinerator or boiler to fan the flames.

OVERHEAD CRANE
(2338 OVERHEAD CRA)
DA   December 10, 1990
BT1 Crane(s)
 BT2 Material Handling Device
  BT3 Devices
  BT3 Equipment/Parts - Material Handling (DOE FRASE Vocabulary)
   BT4 Equipment
SO   DOE FRASE VOCABULARY

OVERLAND FLOW
(EPA)
DA   October 12, 1990
BT1 Land Application
RT   Surface Waters
SO   Environmental Protection Agency Glossary
DEF (EPA) A land application technique that cleanses wastewater by allowing it to flow over a sloped surface. As the water flows over the surface, the contaminants are removed and the water is collected at the bottom of the slope for reuse.

OVERPACK
(SWDA; RCRA; ESH)
DA   October 12, 1990
BT1 Packages
SO   Hazardous Materials
DEF Enclosure that is used by a single consignor to provide protection or convenience in handling of a package or to consolidate two or more packages. Overpack does not include a freight container.

OVERPRESSURE
(DOE Order 6430.1A)
DA   October 12, 1990
BT1 Pressure
RT   PCB Transformer Rupture
RT   Peak Positive Incident Pressure
RT   Relief Valve Discharge
SO   Construction
DEF (DOE Order 6430.1A) The maximal effective pressure is the highest of (1) the peak incident pressure, (2) the incident plus dynamic pressure, or (3) the reflected pressure (ref. TM 5-1300).

OVERSIGHTS AND OMISSIONS
(SSDC; MORT)
DA   October 12, 1990
RT   Assumed Risks
RT   Events
RT   Management Oversight and Risk Tree
RT   Protection of the Public Health and Welfare
RT   Root Cause Analysis
SO   System Safety Development Center Glossary
DEF (SSDC) Those things which are overlooked or left out of an organization's plans due to inadvertence, which cause delays, problems, or failures in achieving stated goals. (MORT) In MORT analysis the tree structure depicts two fundamental causes of the adverse consequences of an Event: (1) Management Oversights and Omissions or (2) Assumed Risks. All contributing factors in an accident sequence are seen as Specific Oversights and Omissions until such time as they are transferred to Assumed Risks.

OVERTURN
(EPA)
DA   October 12, 1990
SO   Environmental Protection Agency Glossary
DEF (EPA) The period of mixing (turnover), by top to bottom circulation, of previously stratified water masses. This phenomenon may occur in spring and/or fall, or after storms. It results in a uniformity of chemical and physical properties of the water at all depths.

OWNERS
(SWDA; CAA; CFR; ESH; EMER)
DA   October 12, 1990
SY   Operators
BT1 Potentially Responsible Parties
BT1 Responsible Parties
RT   Guarantors
RT   Small Manufacturers
RT   Water Suppliers
SO   Air Pollution
SO   Compensation and Liability
SO   Emergency Preparedness
SO   Environmental Management
SO   Hazardous Materials
SO   Wastes
SO   Water Pollution
DEF (CFR) Owners or operators of any "facility or activity" subject to regulation under the Resource Conservation and Recovery Act (RCRA), Underground Injection Control (UIC), National Pollutant Discharge Elimination System (NPDES), or 404 programs. (a) In the case of an underground storage tank (UST) system in use on November 8, 1984, or brought into use after that date, persons who own an UST system used for storage, use, or dispensing of regulated substances; and (b) in the case of any UST system in use before November 8, 1984, but no longer in use on that date, persons who owned such UST immediately before the discontinuation of its use. Persons who own or operate a uranium mill or an existing tailings pile or a new impoundment.

OWP
DA   October 12, 1990
SEE Office of Weapons Production
SO   Acronyms

OWR
DA   October 12, 1990
SEE Omega West Reactor
SO   Acronyms

OX2 ASSEMBLY
(NFI)
DA   October 12, 1990
SO   Nuclear Facilities Incident Database

SO Radiation
DEF (NFI) Oxide assembly vs. metal
   alloy type.

OXIDANTS
(EPA)
DA   October 12, 1990
BT1 Photochemical Smog
 BT2 Air Pollutants
 BT2 Air Pollution
  BT3 Pollution
RT Oxidation
RT Oxidizing Materials
RT Photochemical Smog
SO Environmental Protection Agency
   Glossary
DEF (EPA) Substances containing
   oxygen that react chemically in air
   to produce a new substance. The
   primary ingredient of
   photochemical smog.

OXIDATION
(EPA)
DA   October 12, 1990
BT1 Chemical Processes
 BT2 Processes
NT1 Combustion
RT Corrosion
RT Disinfection
RT Oxidants
RT Oxidizing Materials
SO Environmental Protection Agency
   Glossary
DEF (EPA) (1) The addition of oxygen ,
   which breaks down organic waste
   or chemicals such as cyanides,
   phenols, and organic sulfur
   compounds in sewage by bacterial
   and chemical means. (2) Oxygen
   combining with other elements. (3)
   The process in chemistry whereby
   electrons are removed from a
   molecule.

OXIDATION PONDS
(EPA)
DA   October 12, 1990
SY   Sewage Lagoons
RT Holding Ponds
RT Wastewater Treatment Processes
SO Environmental Protection Agency
   Glossary
DEF (EPA) Man-made lakes or bodies of
   water in which waste is consumed
   by bacteria. It is used most
   frequently with other waste-
   treatment processes. An oxidation
   pond is basically the same as a
   sewage lagoon.

OXIDIZING MATERIALS
(CFR)
DA   October 12, 1990
BT1 Materials
RT Oxidants
RT Oxidation
SO Hazardous Materials
DEF (CFR) Substances that yields

oxygen readily to stimulate the
combustion of organic matter are
considered as an oxidizing
material for the purposes of the
Hazardous Materials Regulations.
Peracetic Acid Solution, nitric acid
(over 40%), and sodium peroxide
are all examples of oxidizing
material.

OXYGEN DEFICIENCIES
DA   January 24, 1991
SO Environmental Management
DEF (CFR) Concentration of oxygen by
   volume below which atmosphere
   supplying respiratory protection
   must be provided. It exists in
   atmospheres where the
   percentage of oxygen by volume
   is less than 19.5 percent oxygen.

OXYGENATED SOLVENTS
(EPA)
DA   October 12, 1990
BT1 Solvents
SO Environmental Protection Agency
   Glossary
DEF (EPA) Organic solvents containing
   oxygen as part of their molecular
   structure. Alcohols and ketones
   are oxygenated compounds often
   used as paint solvents.

OZONATORS
(EPA)
DA   October 12, 1990
BT1 Devices
SO Environmental Protection Agency
   Glossary
DEF (EPA) Devices that add ozone to
   water.

OZONE
(EPA)
DA   October 12, 1990
SF O_3
BT1 Criteria Pollutants
 BT2 Pollutants
RT Photochemical Smog
RT Stratosphere
RT Troposphere
SO Environmental Protection Agency
   Glossary
DEF (EPA) Found in two layers of the
   atmosphere, the stratosphere and
   the troposphere. In the
   stratosphere (the atmospheric
   layer beginning 7 to 10 miles
   above the earth's surface), ozone
   is a form of oxygen found naturally
   which provides a protective layer
   shielding the earth from ultraviolet
   radiation's harmful health effects
   on humans and the environment.
   In the troposphere (the layer
   extending up 7 to 10 miles from
   the earth's surface), ozone is a
   chemical oxidant and major
   component of photochemical

smog. Ozone can seriously affect
the human respiratory system and
is one of the most prevalent and
widespread of all the criteria
pollutants for which the Clean Air
Act required EPA to set standards.
Ozone in the troposphere is
produced through complex
chemical reactions of nitrogen
oxides, which are among the
primary pollutants emitted by
combustion sources;
hydrocarbons, released into the
atmosphere through the
combustion, handling and
processing of petroleum products;
and sunlight.

OZONE DEPLETION
(EPA)
DA   October 12, 1990
NT1 Antarctic "Ozone Hole"
RT Antarctic "Ozone Hole"
RT Halons
SO Environmental Protection Agency
   Glossary
DEF Destruction of the stratospheric
   ozone layer which shields the
   earth from ultraviolet radiation
   harmful to biological life. This
   destruction of ozone is caused by
   the breakdown of certain chlorine
   and/or bromine containing
   compounds (chlorofluorocarbons
   or halons) which break down when
   they reach the stratosphere and
   catalytically destroy ozone
   molecules.

O_3
DA   October 12, 1990
SEE Ozone
SO Acronyms

P
DA   October 12, 1990
SEE Primary Body Waves
SO Acronyms

P&ID
DA   October 12, 1990
SEE Piping and Instrumentation Drawing
SO Acronyms

P-DELTA EFFECT
(SEA)
DA   October 12, 1990
BT1 Effects
RT Building Frame Systems
SO Construction
DEF The secondary effect on shears
   and moments of frame members
   induced by the vertical loads
   acting on the laterally displaced
   building frame.

PA
DA   October 12, 1990

BT-Broader Term

NT-Narrower Term

RT Related Term

SEE DOE Production Agency
SO Acronyms

PA (Preliminary Assessment)
DA   October 12, 1990
SEE Preliminary Assessment
SO Acronyms

PACIFIC NORTHWEST LABORATORY
DA   January 8, 1991
SF   *PNL*
BT1 Government-Owned
      Contractor-Operated Facilities
 BT2 Federal Facilities
  BT3 Facilities
BT1 Laboratories
 BT2 Research and Development
       Organizations
  BT3 Organizations
RT Battelle Memorial Institute
DEF (Capsule Review of DOE Research
      and Development and Field
      Facilities, 1986) Established in
      1965, PNL has two principal
      missions. First, as a DOE
      multiprogram laboratory, PNL
      develops and deploys technology
      for energy security and national
      defense and transfers technology
      to enhance the international
      competitiveness of the U.S.
      Second, as the research and
      development laboratory for the
      Hanford site, PNL provides
      advanced technology and
      environmental surveillance to
      support Hanford Operations.

PACKAGES
DA   January 29, 1991
NT1 Disposal Packages
NT1 Overpack
NT1 Packaging
 NT2 Break-bulk
 NT2 Bulk Packaging
 NT2 DOE Specification Packaging
 NT2 Non-bulk Packaging
 NT2 Type A Packaging
 NT2 Type B Packaging
RT Waste Packages
SO Environmental Management
DEF (CFR) The packaging together with
      its radioactive contents as
      presented for transport.

PACKAGING
(DOE Order 5480.3; SWDA; RCRA;
  CFR; ESH; IAEA; EMER)
DA   October 12, 1990
BT1 Packages
NT1 Break-bulk
NT1 Bulk Packaging
NT1 DOE Specification Packaging
NT1 Non-bulk Packaging
NT1 Type A Packaging
NT1 Type B Packaging
RT Containers
RT Transport (Hazardous Substances)
SO Emergency Preparedness

SO Environmental Management
SO Hazardous Materials
SO Radiation
DEF (EMER) One or more receptacles
      and wrappers and their contents
      excluding fissile material and other
      radioactive material, but including
      absorbent material, spacing
      structures, thermal insulation,
      radiation shielding, devices for
      cooling and for absorbing
      mechanical shock, external
      fittings, neutron moderators,
      nonfissile neutron absorbers, and
      other supplementary equipment.
      (ESH) The assembly of one or
      more containers and any other
      components necessary to assure
      compliance with the minimum
      packaging and includes containers
      (other than freight containers or
      overpacks), portable tanks, cargo
      tanks, tank cars, and multi-unit
      tank car tanks.

PACKED TOWERS
(EPA)
DA   October 12, 1990
BT1 Control Devices
 BT2 Devices
SO Environmental Protection Agency
       Glossary
DEF (EPA) Pollution control devices that
      force dirty air through a tower
      packed with crushed rock or wood
      chips while liquid is sprayed over
      the packing material. The
      pollutants in the air stream either
      dissolve or chemically react with
      the liquid.

PACKERS
(SWDA; CFR)
DA   October 12, 1990
BT1 Devices
RT Plugging
RT Surface Casings
RT Well Plugs
SO Water Pollution
DEF (CFR) Devices lowered into a well
      to produce a fluid-tight seal.

PAD
(NFI)
DA   October 12, 1990
SO Nuclear Facilities Incident Database
SO Radiation
DEF (NFI) Difference between operating
      value and operating limit.

PADDING
(2675 PADDING)
DA   January 3, 1991
BT1 Personal Protective Equipment
 BT2 Equipment/Parts - Personal
       Protective (DOE FRASE
       Vocabulary)
  BT3 Equipment
SO DOE FRASE VOCABULARY

**PADUCAH GASEOUS DIFFUSION
PLANT (PADUCAH)**
DA   January 8, 1991
SF   *PGDP*
BT1 Government-Owned
      Contractor-Operated Facilities
 BT2 Federal Facilities
  BT3 Facilities
RT MK-Ferguson

PAG
DA   October 12, 1990
SEE Protective Action Guides
SO Acronyms

PAINTER
(0644 PAINTER)
DA   November 28, 1990
BT1 Repair/Construction Personnel
 BT2 Occupations
 BT2 Personnel
SO DOE FRASE VOCABULARY

PALM(S)
(1101 PALM)
DA   November 28, 1990
BT1 Hand(s)
 BT2 Human Body Parts
RT Hand Protection
RT Skin
SO DOE FRASE VOCABULARY

Pan Am
DA   October 12, 1990
SEE Pan American World Services, INC.
SO Acronyms

**PAN AMERICAN WORLD SERVICES,
INC.**
DA   January 8, 1991
SF   *Pan Am*
BT1 Companies
 BT2 Commercial Organizations
  BT3 Organizations
BT1 DOE Contractors
 BT2 Potentially Responsible Parties

PANDEMIC
(EPA)
DA   October 12, 1990
SO Environmental Protection Agency
       Glossary
DEF (EPA) Widespread throughout an
      area, nation or the world.

PAR PONDS
(NFI)
DA   October 12, 1990
BT1 Impoundment
 BT2 Sites/Areas
RT Holding Ponds
SO Nuclear Facilities Incident Database
SO Radiation
DEF (NFI) Cooling water impoundments
      for secondary coolant circuits and
      auxiliary water systems.

SY-Synonymous Terms   SO-Source/Subject Category   SF-See From

PARAQUAT
(EPA)
DA October 12, 1990
BT1 Herbicides
 BT2 Pesticides
  BT3 Hazardous Substances
SO Environmental Protection Agency
     Glossary
DEF (EPA) A standard herbicide used to
    kill various types of crops,
    including marijuana.

PARD
DA June 13, 1991
SEE Protect as Restricted Data
SO Acronyms

PARENT CORPORATIONS
(SWDA; RCRA; CFR)
DA October 19, 1990
RT Companies
SO Wastes
SO Water Pollution
DEF Corporations which directly own at
    least 50 percent of the voting
    stock of the corporations which
    are the facility owner or operator;
    the latter corporations are deemed
    "subsidiaries" of the parent
    corporations.

PARKING LOT
(1857 PARKING LOT)
DA December 10, 1990
BT1 Structures (DOE FRASE
    Vocabulary)
RT Garage
RT Parking Space
SO DOE FRASE VOCABULARY

PARKING SPACE
(1613 PARKING SPAC)
DA December 10, 1990
BT1 Site (DOE FRASE Vocabulary)
 BT2 Sites/Areas
RT Garage
RT Parking Lot
SO DOE FRASE VOCABULARY

Part A Permits
(RCRA; EPA)
DA November 19, 1990
SEE Interim (Permit) Status
SO Environmental Management
DEF (EPA) Interim (permit) status is the
    period during which treatment,
    storage, and disposal facilities
    coming under RCRA in 1980 are
    temporarily permitted to operate
    while awaiting denial or issuance
    of a permanent permit. Permits
    issued under these circumstances
    are usually called "Part A" or "Part
    B" permits. During the interim
    status period the facility shall not
    (1) Treat, store, or dispose of
    hazardous waste not specified in
    Part A of the permit application;
    (2) Employ processes not

specified in Part A of the
application; or (3) Exceed the
design capacities specified in Part
A of the permit application.

Part B Permits
(RCRA; EPA)
DA November 8, 1990
SEE Interim (Permit) Status
SO Environmental Management
DEF (EPA) (Interim Permit Status)
    Period during which treatment,
    storage, and disposal facilities
    coming under RCRA in 1980 are
    temporarily permitted to operate
    while awaiting denial or issuance
    of a permanent permit. Permits
    issued under these circumstances
    are usually called "Part A" or "Part
    B" permits.

PARTIAL CLOSURE
(SWDA; RCRA; ESH)
DA October 12, 1990
BT1 Closure
 BT2 Administrative Processes
  BT3 Processes
RT Hazardous Waste Management
     Units
SO Environmental Management
SO Wastes
DEF The closure of a hazardous waste
    management unit in accordance
    with the applicable closure
    requirements of 40 CFR - Parts
    264 and 265 at a facility that
    contains other active hazardous
    waste management units. For
    example, partial closure may
    include the closure of a tank
    (including its associated piping
    and underlying containment
    systems), landfill cell, surface
    impoundment, waste pile, or other
    hazardous waste management
    unit, while other units of the same
    facility continue to operate.

PARTICLE ACCELERATOR FACILITY
(1803 PARTICLE ACC)
DA December 10, 1990
BT1 Facility (DOE FRASE Vocabulary)
 BT2 Facilities and Buildings (DOE
     FRASE Vocabulary)
  BT3 Facilities
RT Nuclear Facility
SO DOE FRASE VOCABULARY

PARTICLE FLUENCE
(IAEA)
DA October 12, 1990
BT1 Measurements
RT Particle Flux
SO Radiation
DEF (IAEA) The quotient of dN by da,
    where dN is the number of
    particles incident on an elementary
    sphere of cross-sectional area da.

PARTICLE FLUX
(IAEA)
DA October 12, 1990
BT1 Measurements
RT Particle Fluence
SO Radiation
DEF (IAEA) For any defined area the
    quotient of dN by dt, where dN is
    the increment of particle number in
    the time interval dt.

PARTICULATE ASBESTOS MATERIALS
(CAA; CFR)
DA October 12, 1990
BT1 Asbestos Materials
 BT2 Materials
BT1 Particulate Matter
 BT2 Airborne Particulates
  BT3 Particulates
   BT4 Air Pollutants
RT Friable Asbestos Materials
RT Visible Emissions
SO Air Pollution
DEF Finely divided particles of asbestos
    material.

PARTICULATE LOADING
(EPA)
DA October 12, 1990
BT1 Loading
 BT2 Ratios
SO Environmental Protection Agency
     Glossary
DEF (EPA) The mass of particulates per
    unit volume of air or water.

PARTICULATE MATTER
(CAA; CFR)
DA October 12, 1990
BT1 Airborne Particulates
 BT2 Particulates
  BT3 Air Pollutants
NT1 Particulate Asbestos Materials
NT1 PM$_{10}$
RT Baghouse Filters
RT National Emissions Standards for
     Hazardous Air Pollutants
RT Particulate Matter Emissions
RT Total Suspended Particulates
SO Air Pollution
DEF (CFR) Any airborne finely divided
    solid or liquid material with an
    aerodynamic diameter smaller
    than 100 micrometers. (40 CFR
    61) Any finely divided solid or
    liquid material, other than
    uncombined water, as measured
    by the specified reference method.

PARTICULATE MATTER EMISSIONS
(CAA; CFR)
DA October 12, 1990
BT1 Emissions
 BT2 Air Pollutants
NT1 Allowable Emissions
NT1 Excess Emissions
NT1 PM$_{10}$ Emissions
NT1 Visible Emissions
RT Fugitive Emissions

RT Particulate Matter
RT Point Sources
RT Primary Emission Control Systems
RT Stacks
SO Air Pollution
DEF (CFR) All finely divided solid or liquid material, other than uncombined water, emitted to the ambient air as measured by applicable reference methods, or an equivalent or alternative method, specified in this chapter, or by a test method specified in an approved State implementation plan.

PARTICULATES
(EPA)
DA October 12, 1990
BT1 Air Pollutants
NT1 Airborne Particulates
 NT2 Particulate Matter
  NT3 Particulate Asbestos Materials
  NT3 PM_{10}
RT Aerosols
RT Air Contaminants
RT Air Pollution
RT Atomize
RT Conventional Filtration Treatment
RT Dust
RT Fly Ash
SO Environmental Protection Agency Glossary
DEF (EPA) Fine liquid or solid particles such as dust, smoke, mist, fumes, or smog, found in air or emissions.

PARTS PER MILLION
DA January 8, 1991
SY PPM/PPB
BT1 Units of Measure

PASSENGER-CARRYING AIRCRAFT
(SWDA; RCRA)
DA October 12, 1990
BT1 Aircraft
SO Hazardous Materials
DEF An aircraft that carries any person other than a crew member, company employee, an authorized representative of the United States, or a person accompanying the shipment.

PASSENGER VESSELS
(SWDA; RCRA; ESH)
DA October 12, 1990
BT1 Vessels
SO Hazardous Materials
DEF (1) A vessel subject to any of the requirements of the International Convention for the Safety of Life at Sea, 1960, which carries more than 12 passengers; (2) a cargo vessel documented under the laws of the United States and not subject to the Convention, which carries more than 16 passengers; (3) a cargo vessel of any foreign

nation that extends reciprocal privileges and is not subject to the Convention and which carries more than 16 passengers; and (4) a vessel engaged in a ferry operation and which carries passengers.

PASSENGERS (AIRCRAFT)
(Doe Order 5480.13; SWDA; RCRA; ESH)
DA October 12, 1990
RT Aircraft
SO Aviation Safety
DEF (DOE Order 5480.13) Occupants of aircraft who do not have assigned flight duties or other duties related to the mission to which the aircraft is assigned.

PATHOGENS
(EPA)
DA October 12, 1990
BT1 Microorganisms
 BT2 Organisms
RT Disinfection
RT Infectious Wastes
RT Microbes
RT Vectors
RT Waterborne Disease Outbreaks
SO Environmental Protection Agency Glossary
DEF (EPA) Microorganisms that can cause disease in other organisms or in humans, animals and plants. They may be bacteria, viruses, or parasites and are found in sewage, in runoff from animal farms or rural areas populated with domestic and/or wild animals, and in water used for swimming. Fish and shellfish contaminated by pathogens, or the contaminated water itself, can cause serious illnesses.

Pb (Periodic Element)
DA October 12, 1990
SEE Lead
SO Acronyms

PBF
DA May 24, 1991
SEE Power Burst Facility
SO Acronyms

PCB ARTICLES
(TSCA; CFR; ESH)
DA October 12, 1990
BT1 Articles
RT PCB Containers
RT PCB Items
SO Hazardous Materials
DEF (CFR) Any manufactured articles, other than a PCB Container, that contains PCBs and whose surface(s) has been in direct contact with PCBs. PCB articles include capacitors, transformers,

electric motors, pumps, pipes and any other manufactured items (1) which are formed to a specific shape or design during manufacture, (2) which have end use function(s) dependent in whole or in part upon their shape or design during end use, and (3) which have either no change of chemical composition during their end use or only those changes of composition which have no commercial purpose separate from that of the PCB articles.

PCB CONTAINERS
(TSCA; CFR)
DA October 12, 1990
BT1 Containers
RT PCB Articles
RT PCB Items
SO Hazardous Materials
DEF (CFR) Any packages, cans, bottles, bags, barrels, drums, tanks, or other devices that contain PCBs or PCB articles and whose surfaces have been in direct contact with PCBs.

PCB EQUIPMENT
(TSCA; CFR; ESH)
DA October 12, 1990
BT1 Equipment
RT Fluorescent Light Ballasts
RT PCB Items
RT Responsible Parties
SO Hazardous Materials
DEF (CFR) Any manufactured item, other than a PCB Container or a PCB Article Container, which contains a PCB Article or other PCB Equipment, and includes microwave ovens, electronic equipment, and fluorescent light ballasts and fixtures.

PCB ITEMS
(TSCA; CFR; ESH)
DA October 12, 1990
RT PCB Articles
RT PCB Containers
RT PCB Equipment
RT Posing an Exposure Risk to Food or Feed
SO Hazardous Materials
DEF Any PCB Article, PCB Article Container, PCB Container, or PCB Equipment that deliberately or unintentionally contains or has as part of it any PCB or PCBs.

PCB TRANSFORMER RUPTURE
(TSCA; CFR)
DA October 12, 1990
BT1 Failures
 BT2 Accidents
RT Overpressure
RT PCB Transformers
RT Releases

SO Hazardous Materials
DEF (CFR) A violent or non-violent
      break in the integrity of a PCB
      Transformer caused by an
      overtemperature and/or
      overpressure condition that results
      in the release of PCBs.

PCB TRANSFORMERS
(TSCA; CFR; ESH)
DA October 12, 1990
BT1 Electrical Equipment
 BT2 Equipment
NT1 Mineral Oil PCB Transformers
RT Non-PCB Transformers
RT PCB Transformer Rupture
RT Retrofill
SO Hazardous Materials
DEF (ESH) Any transformers with
      dielectric fluid that contain 500
      ppm PCB or greater concentration.

PCB WASTE GENERATOR
(TSCA)
DA October 19, 1990
BT1 Generators (Pollution)
 BT2 Potentially Responsible Parties
RT PCB Wastes
SO Hazardous Materials
DEF (CFR) Any person whose act or
      process produces PCBs that are
      regulated for disposal under 40
      CFR 761 Subpart D, or whose act
      first causes PCBs or PCB Items to
      become subject to the disposal
      requirements of subpart D, or who
      has physical control over the
      PCBs when a decision is made
      that the use of the PCBs has been
      terminated and therefore is subject
      to the disposal requirements of
      subpart D. Unless another
      provision of this part specifically
      requires a site-specific meaning,
      "generator of PCB waste" includes
      all of the sites of PCB waste
      generation owned or operated by
      the person who generates PCB
      waste.

PCB WASTES
(TSCA)
DA October 19, 1990
BT1 Wastes
RT Annual Document Logs
RT Commercial Storers of PCB Waste
RT Disposer of PCB Waste
RT PCB Waste Generator
RT Transfer Facilities
SO Hazardous Materials
DEF (TSCA) Those PCBs and PCB
      items that are subject to the
      disposal requirements of 40 CFR
      761.60. PCBs at concentrations of
      500 ppm or greater must be
      disposed of in an incinerator that
      complies with 40 CFR 761.70.

PCBs
DA October 19, 1990
SEE Polychlorinated Biphenyls
SO Acronyms
SO Environmental Protection Agency
      Glossary
SO Hazardous Materials
SO Wastes

PCC
DA October 12, 1990
SEE Primary Component Cooling
SO Acronyms

pCi/g
DA October 12, 1990
SEE Picocuries/gram
SO Acronyms

pCi/L
DA October 12, 1990
SEE Picocuries Per Liter
SO Acronyms

PCS
DA October 12, 1990
SEE Power Conversion System
SO Acronyms

PDB
DA October 12, 1990
SEE Plant Damage Bin
SO Acronyms

PDC
DA October 12, 1990
SEE Program Development Computer
SO Acronyms

PDM
DA October 12, 1990
SEE Power Density Monitor
SO Acronyms

PEAK POSITIVE INCIDENT PRESSURE
(DOE Order 6430.1A)
DA October 12, 1990
BT1 Pressure
RT Maximal Effective Pressure
RT Overpressure
SO Construction
DEF (DOE Order 6430.1A) The almost
      instantaneous rise from the
      ambient pressure caused by a
      blast wave's pressure disturbance.

PEAK-TO-PEAK RATIO
(NFI)
DA October 12, 1990
SF *PTP*
BT1 Ratios
SO Radiation
DEF (NFI) The ratio of the two axial flux
      shape peaks when axial shapes
      are saddled, used to position
      partial control rods correctly
      (K-1858). PTP.

PEARL
DA October 12, 1990
SEE Personnel Expertise and Resource
      Listing
SO Acronyms

PECSS
DA October 12, 1990
SEE Poisoned Emergency Cooling
      System Sampling
SO Acronyms

PEL
DA October 12, 1990
SEE Permissible Exposure Limits
SO Acronyms
SO Emergency Preparedness
SO Industrial Hygiene
SO Industrial Safety

PENETRATING RADIATION
(EMER)
DA February 1, 1991
BT1 Radiation
SO Emergency Preparedness
DEF (EMER) A general term used to
      describe external radiations with
      sufficient penetrating power that
      the absorbed dose from
      exposures to humans are
      delivered in significant quantities
      to tissues and organs other than
      the skin. It refers to most gamma
      radiation, x radiation (excluding
      those with very low energy), and
      neutron radiation.

PEO
DA October 12, 1990
SEE Program Enrichment Office
SO Acronyms

PERCEIVED RISKS
(SSDC)
DA October 12, 1990
BT1 Risks
SO System Safety Development Center
      Glossary
DEF (SSDC) Risks as a person believes
      or understands them to be,
      whether actual or not. Perceived
      risks may be what the actual risks
      are or something far different. As
      an example, the actual risk from
      smoking is thousands of times
      greater than the risk from nuclear
      power generation and yet, a
      smoker may perceive his risk from
      a nuclear power plant to be far
      greater.

PERCOLATION
(EPA)
DA October 12, 1990
RT Infiltration
RT Leachates
RT Recharge

SO Environmental Protection Agency
   Glossary
DEF (EPA) The movement of water
   downward and radially through the
   subsurface soil layers, usually
   continuing downward to the
   groundwater.

PERFORMANCE ASSESSMENTS
DA January 24, 1991
BT1 Assessments
 BT2 Administrative Processes
  BT3 Processes
RT Risk Analysis
SO Environmental Management
DEF (DOE 5820.2A) Systematic analysis
   of the potential risks posed by
   waste management systems to
   the public and environment and a
   comparison of those risks to
   established performance
   objectives.

PERFORMANCE-BASED TRAINING
(Doe Order 5480.18)
DA October 12, 1990
BT1 Training Programs
 BT2 Programs
RT Accreditation
RT Proper Job Analysis
RT Proper Job Instruction
SO Standards
DEF (DOE Order 5480.18) A systematic
   approach to training that is based
   on tasks and the related
   knowledges and skills required for
   competent job performance.

PERFORMANCE EVALUATION
SAMPLES
(SDWA; CFR)
DA October 12, 1990
BT1 Samples
SO Water Pollution
DEF (CFR) Reference samples provided
   to a laboratory for the purpose of
   demonstrating that the laboratory
   can successfully analyze the
   sample within limits of
   performance specified by the
   Agency. The true value of the
   concentration of the reference
   material is unknown to the
   laboratory at the time of the
   analysis.

PERFORMANCE OBJECTIVES
(TTGM)
DA April 26, 1991
RT Concerns
RT Findings
RT Tiger Team Assessments
RT Tiger Teams
SO Management
DEF (TTGM) Findings and concerns as
   stated within a Tiger Team
   Assessment are prefaced by a
   statement of the Performance
   Objective in each discipline area.

Performance Objectives for
Compliance Findings are derived
from promulgated regulations and
final DOE Orders, consent orders,
agreements, and permit
conditions. Performance
Objectives for Best Management
Practice Findings are derived from
regulatory agency guidance,
accepted industry practices, and
professional judgement.

PERFORMANCE TESTING
LABORATORIES
(Doe Order 5480.15)
DA October 12, 1990
BT1 Laboratories
 BT2 Research and Development
    Organizations
  BT3 Organizations
RT Personnel Dosimetry Programs
RT Personnel Dosimeters
SO Standards
DEF (DOE Order 5480.15) Calibration
   laboratories designated by the
   DOE Laboratory Accreditation
   Administrator to test dosimeters.

PERMAFROST
(DOE Order 6430.1A)
DA October 12, 1990
RT Soils
RT Tundra
SO Construction
DEF (DOE Order 6430.1A) A
   permanently frozen layer of
   variable depth below the earth's
   surface in frigid regions.

PERMANENT EXEMPTIONS
(DOE Order 5480.4; ESH)
DA October 12, 1990
BT1 Exemptions
RT Mandatory Standards
SO Environmental Management
SO Management
SO Standards
DEF (ESH) Release from a mandatory
   standard. Such exemptions are
   not time-specified.

PERMANENT PARTIAL DISABILITIES
(SSDC)
DA October 12, 1990
BT1 Disabilities
SO System Safety Development Center
   Glossary
DEF (SSDC) Injuries other than death or
   permanent total disability which
   result in the loss, or complete loss
   of use, of any member or part of a
   member of the body or any
   permanent impairment of functions
   of the body or body part. Loss
   disregards any pre-existing
   disability of the injured member or
   impaired body function.

PERMANENT TOTAL DISABILITIES
(SSDC)
DA October 12, 1990
BT1 Disabilities
SO System Safety Development Center
   Glossary
DEF (SSDC) Injuries other than death
   which permanently and totally
   incapacitate an employee, which
   deny gainful occupation or result
   in the loss or complete loss of use
   (in one accident) of: (a) both eyes;
   (b) one eye and one hand, arm,
   leg or foot; or (c) any two of the
   following not on the same limb:
   hand, arm, foot, leg.

PERMANENT VARIANCES
(Doe Order 5483.1A)
DA October 12, 1990
BT1 Variances
RT Temporary Variances
SO Environmental Management
SO Industrial Safety
DEF (DOE Order 5483.1A) Releases
   from a DOE-prescribed OSHA
   standard. Such variances are not
   time-specified.

PERMEABILITY
(EPA)
DA October 12, 1990
RT Confining Beds
RT Osmosis
RT Pervious
RT Transmissive Faults
RT Water Resistant
SO Environmental Protection Agency
   Glossary
DEF (EPA) The rate at which liquids
   pass through soil or other
   materials in a specified direction.
   (Glossary of Geology) It is a
   measure of the relative ease of
   fluid flow under unequal pressure.

PERMISSIBLE EXPOSURE LIMITS
(EMER)
DA January 8, 1991
SF PEL
BT1 Limits
SO Emergency Preparedness
SO Environmental Management
SO Industrial Hygiene
SO Industrial Safety
DEF (OSHA) The exposure, inhalation
   or dermal, permissible exposure
   limit specified in 20 CFR Part
   1910, Subparts G and Z.

PERMIT (RCRA)
(RCRA)
DA June 13, 1991
BT1 Permits
RT Emergency Permits
RT Final Authorizations
RT Interim (Permit) Status
RT Permit-by-Rule

SO Environmental Management
DEF (RCRA) Means an authorization, license, or equivalent control document issued by EPA or an approved State to implement the requirements of 40 CFR Parts 124, 270, and 274. Permit includes Permit-by-Rule, and Emergency Permit. Permit does not include RCRA interim status, or any permit which has not yet been the subject of final agency action, such as a draft permit or a proposed permit.

PERMIT APPLICANTS
(ESA)
DA October 19, 1990
SY License Applicants
SO Endangered Species
DEF (ESA) Any persons whose application to an agency for a permit or license has been denied primarily because of the application of section 7(a)(2) of the Act, 16 U.S.C. 1563(a)(2).

PERMIT-BY-RULE
(SWDA; RCRA)
DA October 12, 1990
BT1 Permits
RT Permit (RCRA)
SO Wastes
DEF A provision of these regulations stating that a facility or activity is deemed to have a RCRA permit if it meets the requirements of the provision.

PERMITS
(SWDA; RCRA; SDWA; CFR; ESA; CFR; EPA)
DA October 12, 1990
NT1 Approvals Necessary to Begin Physical Construction
NT1 Draft Permits
NT1 Emergency Permits
NT1 Final Authorizations
NT1 General Permits
NT1 Individual Permits
NT1 Interim Authorizations
NT1 Interim (Permit) Status
NT1 Letter of Permission
NT1 Permit-by-Rule
NT1 Permit (RCRA)
NT1 Permits Necessary to Begin Physical Construction
NT1 PSD Permits
NT1 Safe Work Permit
NT1 Safety Work Permit
NT1 Test Authorizations
SO Endangered Species
SO Environmental Management
SO Environmental Protection Agency Glossary
SO Hazardous Materials
SO Wastes
SO Water Pollution
DEF Authorization, license, or equivalent control documents issued by EPA or an "approved state" to

implement the requirements of this part and 40 CFR 122, 123, 144, 145, 233, 270, and 271. "Permits" include RCRA "permit by rule" (section 270.60), UIC (Underground Injection Control) area permit (section 144.33), NPDES or 404 "general permit" (sections 270.61, 144.34, and 233.38). Permits do not include RCRA interim status (section 270.70), UIC authorization by rule (section 144.21), or any permit which has not yet been the subject of final agency action, such as a "draft permit" or a "proposed permit." (EPA) Authorization, license, or equivalent control documents issued by EPA or an approved state agency to implement the requirements of an environmental regulation; e.g., permits to operate a wastewater treatment plant or to operate a facility that may generate harmful emissions.

PERMITS NECESSARY TO BEGIN PHYSICAL CONSTRUCTION
(SWDA; RCRA)
DA October 19, 1990
BT1 Permits
SO Wastes
DEF Permits and approvals required under Federal, State or local hazardous waste control statutes, regulations or ordinances.

PERMITTING AUTHORITIES
(CWA; RHA; CFR)
DA October 12, 1990
SO Water Pollution
DEF The District Engineer of the U.S. Army Corps of Engineers or such other individuals as may be designated by the Secretary of the Army to issue or deny permits under section 404 of the Act; or the State Director of a permit program approved by EPA under section 404(g) and section 404(h) or his delegated representative.

PERSISTENT PESTICIDES
(EPA)
DA October 12, 1990
BT1 Pesticides
 BT2 Hazardous Substances
SO Environmental Protection Agency Glossary
DEF (EPA) Pesticides that do not break down chemically or break down very slowly and that remain in the environment after a growing season.

PERSONAL PROPERTY
(DOE Order 4330.4A)
DA June 5, 1991

BT1 Property
NT1 Related Personal Property
RT Equipment
RT Real Property
SO Management
DEF (DOE Order 4330.4A) Generally capitablizable property that can be moved, that is not permanently affixed to and part of real estate. Generally, items remain personal property if they can be removed without seriously damaging or diminishing the functional value of either the property or real estate. Examples of personal property are shop equipment and automated data- processing and peripheral equipment.

PERSONAL PROTECTIVE EQUIPMENT
(2676 PPE)
DA January 3, 1991
BT1 Equipment/Parts - Personal Protective (DOE FRASE Vocabulary)
 BT2 Equipment
NT1 Air Mask
NT1 Anticontamination Clothing
 NT2 Acid Suit
 NT2 Lab Coat
 NT2 Radiation Suit
NT1 Chin Strap
NT1 Dust Mask
NT1 Eye Protection
 NT2 Goggles
 NT2 Safety Glasses
  NT3 Safety Glasses W Side Shields
  NT3 Tinted Safety Glasses
 NT2 Side Shields
NT1 Fall Protection Device
 NT2 Safety Belt
 NT2 Safety Line
 NT2 Safety Net
NT1 Flame Retardant Clothing
NT1 Foot Protection
 NT2 Ankle Protection
 NT2 Metatarsal Protection
 NT2 Safety Boots
 NT2 Safety Shoe(s)
 NT2 Shoe Cover(s)
  NT3 Metal Shoe Cover
NT1 Hand Protection
 NT2 Glove(s)
 NT2 Wrist Band
NT1 Head Protection
 NT2 Bump Cap
 NT2 Faceshield
 NT2 Hard Hat
 NT2 Helmet
 NT2 Sand Blaster's Hood
 NT2 Welder's Hood
NT1 Hearing Protection
 NT2 Ear Plug(s)
NT1 Leggings
NT1 Padding
NT1 Respirator(s)
NT1 Seat Belt(s)
SO DOE FRASE VOCABULARY

PERSONNEL
(SWDA: RCRA)
DA   October 19, 1990
NT1 Accreditation Coordinators
NT1 Administrators
 NT2 Assistant Administrator for
       Fisheries
 NT2 Regional Administrator
NT1 Admin. Support/Clerical Employee
 NT2 Firefighter
 NT2 Food Service Employee
 NT2 Janitor
 NT2 Misc Service Employee
 NT2 Security Guard
NT1 Agency Lead Officials
NT1 Agriculture Personnel
 NT2 Forest Worker
 NT2 Groundskeeper
 NT2 Misc Agriculture Employee
NT1 Architect Engineer
NT1 Armorers
NT1 Assessment Team Leaders
NT1 Authorized Representatives
NT1 Authorized Officers
NT1 Authorized Officials
 NT2 Lead Authorized Officials
NT1 Authorized Inspectors
NT1 BPS Technical Representative
NT1 Captain of the Port
NT1 Cathodic Protection Testers
NT1 Certifying Officials
NT1 Cognizant Federal Agency Officials
NT1 Communicators
NT1 Controllers
 NT2 Senior Controllers
NT1 Contracting Officers
NT1 Contracting Officer's
       Representative
NT1 Contractor Employees
NT1 Couriers
NT1 Crisis Managers
NT1 Department's Metric Coordinator
NT1 Designated Non-Federal
       Representatives
NT1 Directors
NT1 Director of Emergency Operations
NT1 District Commander
NT1 DOE Accident Response Group
       Team Leaders
NT1 DOE Energy Management
       Coordinators
NT1 DOE Facility Representatives
NT1 DOE Fire Protection Authorities
NT1 DOE Representatives
NT1 DOE Safeguards and Security
       Coordinators
NT1 Duty Officers
NT1 EEMT Member
NT1 Emergency Management Team
       Directors
NT1 Emergency Medical Technician
NT1 Employees' Representative
NT1 Facility Managers
NT1 Facility Authority
NT1 Federal Coordinating Officers
NT1 Federal Land Managers
NT1 First Federal Officials
NT1 Flight Crewmembers
NT1 FRMAC Director
NT1 Full-time Employees
NT1 Ground Crews

NT1 Handler/Laborer/Helper
NT1 Heads of Field Operations
NT1 Heads of Headquarters Elements
 NT2 Landlords
NT1 Higher Supervision
NT1 Judicial Officers
NT1 Liaison Officers
NT1 Metric Coordinator
NT1 Military Personnel
NT1 Misc Employee
 NT2 Misc Service Employee
 NT2 Misc Agriculture Employee
 NT2 Misc Repair/Construction
       Employee
 NT2 Misc Precision/Production
       Employee
 NT2 Misc Transport Employee
NT1 Monitored Workers
NT1 NEPA Compliance Officers
NT1 Occupational Workers
 NT2 Radiation Workers
  NT3 Non-Employee Radiation Workers
NT1 Office of Special Projects
       Coordinators
NT1 Oil Chemical and Atomic Workers
NT1 On-Scene Coordinators
NT1 On-Scene Commanders
 NT2 Containment Commanders
NT1 On-Site Technical Directors
NT1 Operations Office Managers
NT1 Precision/Production Personnel
 NT2 Machinist
 NT2 Machine Setup/Operator
 NT2 Misc Precision/Production
       Employee
 NT2 Operator, Plant/System/Utility
 NT2 Sheet Metal Worker
 NT2 Welder/Solderer
NT1 Professional Personnel
 NT2 Doctor/Nurse
 NT2 Engineer
 NT2 Misc Professional
 NT2 Scientist
  NT3 Health Physicist
 NT2 Technicians
  NT3 Engineering Technician
  NT3 Health Technician
  NT3 Miscellaneous Technician
  NT3 Radiation Monitor/Technician
  NT3 Science Technician
NT1 Program Secretarial Officers
NT1 Program Senior Officials
 NT2 Secretarial Officers
NT1 Program Managers
NT1 Project Managers
 NT2 Remedial Project Managers
NT1 Protective Force Personnel
 NT2 Guards
 NT2 Security Inspectors
 NT2 Shadow Forces
 NT2 Tactical Response Forces
NT1 Public Affairs (Personnel)
 NT2 Public Information Officers
NT1 Radiation Protection Officers
NT1 Range Masters
NT1 Range Safety Officers
NT1 Reactor Operator
 NT2 Senior Reactor Operator
NT1 Regional Administrator
NT1 Remedial Project Managers
NT1 Remedial Investigator

NT1 Repair/Construction Personnel
 NT2 Carpenter
 NT2 Electrician
 NT2 Mason
 NT2 Mechanic/Repairer
 NT2 Miner/Driller
 NT2 Misc Repair/Construction
       Employee
 NT2 Painter
 NT2 Pipe Fitter
NT1 Safety and Health Directors
NT1 Sales Person
NT1 Senior Management Officials
NT1 Senior Reactor Operator
NT1 Senior Federal Emergency
       Management Agency Officials
NT1 Senior Scientific Advisers
NT1 Senior Security Supervisors
NT1 Shift Supervisor
NT1 Shift Technical Advisor
NT1 State Director
NT1 State Coordinating Officers
NT1 Study Directors
NT1 Supervisors
NT1 Support Agency Coordinators
NT1 Technically Qualified Individuals
 NT2 Approved Medical Practitioners
 NT2 Certified Applicators
 NT2 Competent Authorities
 NT2 Corrosion Experts
 NT2 Installation Inspectors
 NT2 Reactor Supervisor
 NT2 Subject Matter Experts
 NT2 Technical Experts
 NT2 Trained Investigators
NT1 Terminated Employees
NT1 Tiger Team Leaders
NT1 Tiger Team Administrators
NT1 Training Accreditation Program
       Staff
NT1 Transport Personnel
 NT2 Aircraft Pilot
 NT2 Bus Driver
 NT2 Equipment Operator
 NT2 Misc Transport Employee
 NT2 Truck Driver
NT1 Water Management Division
       Director
RT   Employers
RT   Monitored Visitors
RT   Monitored Personnel Locator File
RT   Organizations
RT   Personnel Dosimetry Programs
RT   Personnel Expertise and Resource
       Listing
RT   Targets
SO   Wastes
DEF (WEBSTER) Persons employed (as
       in a factory, office or organization).

PERSONNEL DOSIMETERS
(DOE Order 5480.15)
DA   October 12, 1990
BT1 Dosimeters
 BT2 Radiation Detectors
  BT3 Equipment
BT1 Personnel Monitoring Equipment
 BT2 Equipment
RT   Dosimetry Processors
RT   Performance Testing Laboratories

SY-Synonymous Terms   SO-Source/Subject Category   SF-See From

RT Personnel Dosimetry Programs
SO Standards
DEF (DOE Order 5480.15) Devices containing one or more radiation-responsive elements (e.g., film, thermoluminescent, nuclear track detector) and possibly one or more absorbers. For the purposes of this Order, personnel dosimeter means the type of dosimeter worn to assess "whole body" dose equivalents. Specifically excluded are dosimeters expressly designed for extremities such as finger ring or wrist dosimeters.

PERSONNEL DOSIMETRY PROGRAMS
(DOE Order 5480.15)
DA October 12, 1990
BT1 Programs
RT Accredited
RT Performance Testing Laboratories
RT Personnel Dosimeters
RT Personnel
RT Technical Experts
SO Standards
DEF (DOE Order 5480.15) Programs using personnel dosimeters to determine, record, report, and archive the dose equivalents received by personnel.

PERSONNEL EXPERTISE AND RESOURCE LISTING
(SSDC)
DA January 8, 1991
SF *PEARL*
BT1 Safety Performance Measurement System
 BT2 Information Systems
  BT3 Security Interests
  BT3 Systems
RT Personnel
SO Environmental Management
DEF (SSDC) This database consists of an environmental, safety and health (ES&H) personnel directory which is continually updated.

PERSONNEL MONITORING EQUIPMENT
DA January 24, 1991
BT1 Equipment
NT1 Personnel Dosimeters
RT Controls
DEF (CFR) A device designed to be worn or carried by an individual for the purpose of estimating the dose received by the individual [e.g., film badges, pocket dosimeters, and thermoluminescent dosimeters (TLD)].

PERT
 DA October 12, 1990
 SEE Program Evaluation and Review Technique

SO Acronyms
SO System Safety Development Center Glossary

PERVIOUS
(DOE Order 6430.1A)
DA October 12, 1990
RT Permeability
SO Construction
DEF (DOE Order 6430.1A) That property of a surface that allows water or other fluids to pass through.

PESTICIDE RELATED WASTES
DA January 24, 1991
BT1 Wastes
RT Pesticides
SO Environmental Management
DEF (CFR) All pesticide-containing wastes or by-products which are produced in the manufacturing or processing of a pesticide and which are to be discarded but which, pursuant to acceptable pesticide manufacturing or processing operations, are not ordinarily a part of or contained within an industrial waste stream discharged into a sewer or the waters of a state.

PESTICIDE TOLERANCE
(EPA)
DA October 12, 1990
RT Pesticides
RT Tolerance Limits
RT Tolerances
SO Environmental Protection Agency Glossary
DEF (EPA) The amount of pesticide residue allowed by law to remain in or on a harvested crop. By using various safety factors, EPA sets these levels well below the point where the chemicals might be harmful to consumers.

PESTICIDES
(EPA)
DA October 12, 1990
BT1 Hazardous Substances
NT1 Botanical Pesticides
NT1 Contact Pesticides
NT1 Fumigants
NT1 Fungicides
 NT2 Dinocap
NT1 Herbicides
 NT2 Alachlor
 NT2 Defoliants
  NT3 Agent Orange
 NT2 Dinoseb
 NT2 Paraquat
NT1 Insecticides
 NT2 Aldicarb
 NT2 DDT
 NT2 Diazinon
 NT2 Dicofol
 NT2 Heptachlor
NT1 Microbial Pesticides

NT1 Nematocides
NT1 Organophosphates
NT1 Persistent Pesticides
NT1 Restricted-Use Pesticide
NT1 Rodenticides
NT1 Selective Pesticides
NT1 Systemic Pesticides
RT Action Levels
RT Active Ingredients
RT Agricultural Pollution
RT Band Application
RT Basal Application
RT Certification
RT Changed Use Pattern
RT Degradation Products
RT Disinfectants
RT Drift
RT Economic Poisons
RT Ecosystems
RT Emergency Conditions
RT Fogging
RT Hazards
RT Inert Ingredients
RT Labels
RT Labeling
RT Mutagens
RT Pesticide Related Wastes
RT Pesticide Tolerance
RT Pests
RT Pollution
RT Pollutants
RT Reregistrations
RT Soil Injection
RT Suspension
RT Triple rinse
RT Water Dumping
RT Xenobiotic
SO Environmental Management
SO Environmental Protection Agency Glossary
DEF (EPA) Substances or mixtures of substances intended for preventing, destroying, repelling, or mitigating any pest. Also, any substances or mixtures of substances intended for use as a plant regulator, defoliant, or desiccant. Pesticides can accumulate in the food chain and/or contaminate the environment if misused. (ESH) Any substances or mixtures of substances intended for preventing, destroying, repelling, or mitigating any pest, or any substances or mixtures of substances intended for use as a plant regulator, defoliant, or desiccant.

PESTS
(EPA)
DA October 12, 1990
RT Attractants
RT Biological Controls
RT Pesticides
SO Environmental Protection Agency Glossary
DEF (EPA) Insects, rodents, nematodes, fungi, weeds or other forms of

terrestrial or aquatic plant or
animal life or virus, bacterial or
microorganism that is injurious to
health or the environment.

PETROLEUM
(SWDA; RCRA; USC)
DA   October 19, 1990
BT1 Fossil Fuels
 BT2 Fuels
 BT2 Geologic Resources
  BT3 Natural Resources
BT1 Oils
RT   Motor Fuels
RT   Petroleum UST Systems
SO   Wastes
DEF (USC) Petroleum, including crude
      oil or any fraction thereof, which is
      liquid at standard conditions of
      temperature and pressure (60
      degrees Fahrenheit and 14.7
      pounds per square inch absolute).

PETROLEUM UST SYSTEMS
(SWDA; RCRA; CFR)
DA   October 12, 1990
BT1 Systems
BT1 UST Systems
 BT2 Tank Systems
RT   Petroleum
SO   Wastes
DEF (CFR) Underground storage tank
      systems that contain petroleum or
      a mixture of petroleum with de
      minimis quantities of other
      regulated substances. Such
      systems include those containing
      motor fuels, jet fuels, distillate fuel
      oils, residual fuel oils, lubricants,
      petroleum solvents, and used oils.

PETS
DA   October 12, 1990
SEE Procedures for Evaluating
      Technical Specifications Program
SO   Acronyms

PFR
DA   October 12, 1990
SEE Phase Failure Relays
SO   Acronyms

PG
DA   October 12, 1990
SEE Process Gas
SO   Acronyms

PGDP
DA   October 12, 1990
SEE Paducah Gaseous Diffusion Plant
      (Paducah)
SO   Acronyms

PH
(EPA)
DA   October 12, 1990
BT1 Measurements
NT1 Background Soil Ph

RT   Neutralization
SO   Construction
SO   Environmental Protection Agency
      Glossary
DEF (EPA) A measure of the acidity or
      alkalinity of a liquid or solid
      material.

PHANTOMS
(IAEA)
DA   October 12, 1990
RT   Modeling
RT   Tissue Equivalents
SO   Radiation
DEF (IAEA) Mathematical or physical
      models used to simulate the
      radiation interaction characteristics
      of a human or animal body.

PHASE FAILURE RELAYS
DA   January 8, 1991
SF   *PFR*
BT1 Relay(s)
 BT2 Equipment/Parts - Electrical (DOE
      FRASE Vocabulary)
 BT3 Equipment
RT   Failures

PHASED DISPOSAL
(CFR)
DA   October 12, 1990
BT1 Solid Waste Disposal
 BT2 Disposal
  BT3 Waste Management Processes
   BT4 Processes
RT   Tailings
SO   Air Pollution
DEF (CFR) A method of tailings
      management and disposal which
      uses lined impoundments meeting
      the requirements of 40 CFR
      192.32, no greater than 40 acres
      in area, which immediately filled,
      upon becoming dried, and covered
      to Federal standards.

PHENOLS
(EPA)
DA   October 12, 1990
BT1 CERCLA Hazardous Substances
 BT2 Hazardous Substances
BT1 Organic Chemicals
 BT2 Chemical Substances
SO   Environmental Protection Agency
      Glossary
DEF (EPA) Organic compounds that are
      by-products of petroleum refining,
      tanning, and textile, dye, and resin
      manufacturing. Low concentrations
      cause taste and odor problems in
      water; higher concentrations can
      kill aquatic life and humans.

PHEROMONES
(EPA)
DA   October 12, 1990
SO   Environmental Protection Agency
      Glossary
DEF A hormonal chemical produced by

members of a species that
influences the behavior of other
members of the same species.
(EPA) Hormonal chemicals
produced by females of a species
to attract a mate.

PHOSPHATE ROCKS
(NCRP)
DA   October 12, 1990
RT   Elemental Phosphorus Plants
RT   Phosphorus
SO   Radiation
DEF (NCRP) Ores from which
      phosphorus is extracted and which
      often contain low concentrations of
      uranium.

PHOSPHATES
(EPA)
DA   October 12, 1990
RT   Phosphorus
SO   Environmental Protection Agency
      Glossary
DEF (EPA) Certain chemical compounds
      containing phosphorus.

PHOSPHORUS
(EPA)
DA   October 12, 1990
BT1 CERCLA Hazardous Substances
 BT2 Hazardous Substances
RT   Elemental Phosphorus Plants
RT   Phosphate Rocks
RT   Phosphates
SO   Environmental Protection Agency
      Glossary
DEF (EPA) An essential chemical food
      element that can contribute to the
      eutrophication of lakes and other
      water bodies. Increased
      phosphorus levels result from
      discharge of
      phosphorus-containing materials
      into surface waters.

PHOTOCHEMICAL SMOG
(EPA)
DA   October 12, 1990
SY   Smog
BT1 Air Pollutants
BT1 Air Pollution
 BT2 Pollution
NT1 Oxidants
RT   Oxidants
RT   Ozone
SO   Environmental Protection Agency
      Glossary
DEF (EPA) Air pollution caused by
      chemical reactions.

PHOTOSYNTHESIS
(EPA)
DA   October 12, 1990
BT1 Biological Processes
 BT2 Processes
RT   Plants

SY-Synonymous Terms   SO-Source/Subject Category   SF-See From

SO Environmental Protection Agency
    Glossary
DEF (EPA) The manufacture by plants of
    carbohydrates and oxygen from
    carbon dioxide and water in the
    presence of chlorophyll, using
    sunlight as an energy source.

PHYSICAL AND CHEMICAL TREATMENTS
(EPA)
DA October 12, 1990
BT1 Wastewater Treatment Processes
 BT2 Treatment
  BT3 Waste Management Processes
   BT4 Processes
SO Environmental Protection Agency
    Glossary
DEF (EPA) Processes generally used in
    large-scale wastewater treatment
    facilities. Physical processes may
    involve air-stripping or filtration.
    Chemical treatment includes
    coagulation, chlorination, or ozone
    addition. The term can also refer
    to treatment processes, treatment
    of toxic materials in surface waters
    and groundwaters, oil spills, and
    some methods of dealing with
    hazardous materials on or in the
    ground.

PHYSICAL CONSTRUCTION
(SWDA; RCRA)
DA October 12, 1990
BT1 Construction
 BT2 Processes
RT Hazardous Waste Management
    Facilities
SO Environmental Management
SO Wastes
DEF Excavation, movement of earth,
    erection of forms or structures, or
    similar activity to prepare a
    hazardous waste management
    (HWM) facility to accept
    hazardous waste.

PHYSICAL FITNESS TRAINING ACTIVITY
(1238 PFT ACTIVITY)
DA November 28, 1990
BT1 Activity Types (DOE FRASE
    Vocabulary)
 BT2 Activities
SO DOE FRASE VOCABULARY

PHYSICAL HARM
(SSDC)
DA October 12, 1990
BT1 Illnesses
BT1 Injuries
SO System Safety Development Center
    Glossary
DEF (SSDC) Injury and/or illness as
    well as adverse mental,
    neurological, or systemic effects
    resulting from an exposure or from

circumstances encountered in the
course of employment.

PHYSICAL PROTECTION (PHYSICAL SECURITY)
(DOE Order 6430.1A)
DA October 12, 1990
BT1 Security
RT Detection
RT Safeguards
SO Construction
DEF (DOE Order 6430.1A) The
    application of methods for
    preventing diversion of nuclear
    material or for detecting such
    diversion as it occurs.

PHYSICALLY SEPARATED
(DOE Order 6430.1A)
DA October 12, 1990
SO Construction
DEF (DOE Order 6430.1A) Set apart by
    distance, fences, walls or similar
    obstructions.

PHYTOPLANKTON
(EPA)
DA October 12, 1990
BT1 Plankton
NT1 Algae
SO Environmental Protection Agency
    Glossary
DEF (EPA) That portion of the plankton
    community comprised of tiny
    plants, e.g., algae, diatoms.

PHYTOTOXIC
(EPA)
DA November 8, 1990
RT Herbicides
RT Plants
DEF (EPA) Something that harms plants.

PICK(S)
(3045 PICK;N)
DA January 3, 1991
BT1 Tools - Manual
 BT2 Tools (DOE FRASE Vocabulary)
  BT3 Equipment
SO DOE FRASE VOCABULARY

PICOCURIE
(SWDA; EPA; NPDWR)
DA October 12, 1990
BT1 Radiation Units
 BT2 Units of Measure
RT Radioactive Materials
SO Environmental Protection Agency
    Glossary
SO Water Pollution
DEF The quantity of radioactive material
    producing 2.22 nuclear
    transformations per minute.

PICOCURIES PER LITER
(EPA)
DA October 12, 1990
SF pCi/L

BT1 Radiation Units
 BT2 Units of Measure
SO Environmental Protection Agency
    Glossary
DEF (EPA) A unit of measure used for
    expressing levels of radon gas.
    (See: Picocurie)

PICOCURIES/GRAM
DA January 8, 1991
SF pCi/g
BT1 Radiation Units
 BT2 Units of Measure

PIER
(1858 PIER)
DA December 10, 1990
BT1 Structures (DOE FRASE
    Vocabulary)
SO DOE FRASE VOCABULARY

PIGS
(EPA)
DA October 12, 1990
SY Casks
SY Coffins
BT1 Containers
SO Environmental Protection Agency
    Glossary
DEF (EPA) Containers, usually lead,
    used to ship or store radioactive
    materials.

PILES (NUCLEAR)
DA January 29, 1991
BT1 Fuel Element(s)
 BT2 Equipment/Parts - Nuclear (DOE
    FRASE Vocabulary)
  BT3 Equipment
  BT3 Reactor Components
SO Radiation
DEF (DSTT) Fuel elements in a nuclear
    reactor. (NRC Glossary of Terms:
    Nuclear Power and Radiation) A
    nuclear reactor; called a pile
    because the earliest reactors were
    "piles" of graphite and uranium
    blocks.

PILES (WASTES)
(SWDA; RCRA)
DA October 12, 1990
BT1 Hazardous Wastes
 BT2 Hazardous Materials
  BT3 Materials
 BT2 Wastes
NT1 Disposal Areas
RT Existing Tailings Piles
RT Land Disposal
RT Solid Wastes
SO Environmental Management
SO Environmental Protection Agency
    Glossary
SO Wastes
DEF (1) Heaps of waste. (2) Any
    non-containerized accumulations
    of solid, nonflowing hazardous
    waste that are used for treatment
    or storage.

PILOT
(Acronyms and Abbreviations)
DA   October 12, 1990
BT1 Computer Codes
SO   Nuclear Facilities Incident Database
DEF (NFI) A computer code that
        updates control rod information.

PILOT LIGHT(S)
(2489 PILOT LIGHT)
DA   January 3, 1991
BT1 Equipment/Parts - Heating (DOE
        FRASE Vocabulary)
 BT2 Equipment
BT1 Heating Equipment
SO   DOE FRASE VOCABULARY

PINCHED NERVE
(1335 PINCH NERV)
DA   November 28, 1990
BT1 Injuries
RT   Spinal Cord
RT   Upper Back
SO   DOE FRASE VOCABULARY

PIPE FITTER
(0645 PIPE FITTER)
DA   November 28, 1990
BT1 Repair/Construction Personnel
 BT2 Occupations
 BT2 Personnel
SO   DOE FRASE VOCABULARY

PIPELINE FACILITIES
(SWDA; RCRA)
DA   October 12, 1990
BT1 Facilities
RT   Gathering Lines
RT   Piping
SO   Wastes
DEF Are new and existing pipe
        rights-of-way and any associated
        equipment, facilities, or buildings.

PIPING
(SWDA; RCRA)
DA   October 19, 1990
BT1 Facility Components
NT1 Connected Piping
NT1 Vitrified Clay Pipe
RT   Connectors
RT   Pipeline Facilities
RT   Service Connectors
SO   Wastes
DEF (CFR) A hollow cylinder or tubular
        conduit that is constructed of
        non-earthen materials.

**PIPING AND INSTRUMENTATION
DRAWING**
DA   January 8, 1991
SF   P&ID
BT1 Diagrams

PISTOL GRIP SWITCHES
(NFI)
DA   October 12, 1990
BT1 Control Rod Drive

 BT2 Control Rod Drive Control System
  BT3 Control Systems
   BT4 Controls
   BT4 Systems
  BT3 Reactor Components
SO   Nuclear Facilities Incident Database
SO   Radiation
DEF Control rod drive switches.

PISTOLS
(Doe Order 5480.16)
DA   October 12, 1990
BT1 Handguns
 BT2 Small Arms
  BT3 Firearms
NT1 Double Action Semiautomatic
        Pistols
NT1 Machine Pistols
NT1 Single Action Semiautomatic Pistols
SO   Firearms
DEF (NFI) Short firearms that can be
        armed and fired with one hand.

PIT AREA
(1614 PIT AREA)
DA   December 10, 1990
BT1 Area
 BT2 Sites/Areas
SO   DOE FRASE VOCABULARY

PJA
DA   October 12, 1990
SEE Proper Job Analysis
SO   Acronyms

PJI
DA   October 12, 1990
SEE Proper Job Instruction
SO   Acronyms

PLACARDED CARS
(SWDA; RCRA; ESH)
DA   October 12, 1990
RT   Rail Freight Cars
SO   Hazardous Materials
DEF Rail cars that are placarded in
        accordance with requirements
        except those cars displaying only
        the Fumigation placards.

PLANE(S)
(3046 PLANE)
DA   January 3, 1991
BT1 Tools - Manual
 BT2 Tools (DOE FRASE Vocabulary)
  BT3 Equipment
SO   DOE FRASE VOCABULARY

PLANKTON
(EPA)
DA   October 12, 1990
NT1 Phytoplankton
 NT2 Algae
NT1 Zooplankton
SO   Environmental Protection Agency
        Glossary
DEF (EPA) Tiny plants and animals that
        live in water.

PLANNED RENOVATION OPERATIONS
(CAA; CFR)
DA   October 12, 1990
BT1 Operations
 BT2 Activities
RT   Removal
SO   Air Pollution
DEF (CFR) Renovation operations in
        which the amount of friable
        asbestos material that will be
        removed or stripped within a given
        period of time can be predicted.
        Individual nonscheduled
        operations are included if a
        number of such operations can be
        predicted to occur during a given
        period of time based on operating
        experience.

PLANNED SPECIAL EXPOSURE
(IAEA)
DA   October 12, 1990
BT1 Exposure
RT   Dose Limit
SO   Radiation
DEF (IAEA) An exposure in excess of
        recommended dose limits,
        authorized only infrequently under
        special circumstances during
        normal operations, when
        alternative procedures not
        involving such exposures cannot
        be used.

PLANNING ZONES
(EMER)
DA   February 1, 1991
BT1 Zones
 BT2 Sites/Areas
SO   Emergency Preparedness
DEF (EMER) Areas for which planning is
        done to ensure that prompt and
        effective actions can be taken to
        protect emergency personnel, the
        public health and safety, and the
        environment in the event of a
        major emergency.

PLANS
DA   January 29, 1991
NT1 Action Plans
NT1 Assessment Plans
NT1 Closure Plans
NT1 Defense Waste Management Plan
NT1 Design and Development Plans
NT1 Emergency Plans
 NT2 Contingency Plans
 NT2 Field Facility/Building Emergency
        Plans
 NT2 Field Site Emergency Plans
 NT2 Operational Emergency
        Preparedness Management
        Plans
 NT2 Primary Emergency Plans
 NT2 Spill Prevention Control and
        Countermeasures Plan
 NT2 Terrorist Response Plans
NT1 Federal Radiological Emergency
        Response Plans

SY-Synonymous Terms   SO-Source/Subject Category   SF-See From

NT2 Federal Radiological Monitoring
   and Assessment Plans
NT1 Field Operations Plans
NT1 Implementation Plans
NT2 EIS Implementation Plans
NT1 Interagency Radiological
   Assistance Plans
NT1 Life Cycle Plans
NT1 Maintenance Plans
NT1 Metric Transition Plan
NT1 Mitigation Action Plans
NT1 National Oil and Hazardous
   Substances Contingency Plan
NT1 National Contingency Plan
NT1 Plugging and Abandonment Plans
NT1 Post-Closure Plan
NT1 Quality Assurance Plans
 NT2 Site Quality Assurance Plan
NT1 Sampling Plans
 NT2 Field Sampling Plans
NT1 Sampling and Analysis Plans
 NT2 Field Sampling Plans
 NT2 Quality Assurance Project Plans
NT1 Site Development and Facility
   Utilization Plans
NT1 Site-Specific Safeguards and
   Security Plan
NT1 State Implementation Plans
NT1 Training Program Accreditation
   Plan
NT1 Trash-To-Energy Plans
RT Maintenance Management
RT Technical Support
DEF (DSTT) Documents containing
   comprehensive procedures,
   regulations, requirements, etc. for
   the purpose of achieving specific
   objectives.

PLANT DAMAGE BIN
DA January 8, 1991
SF *PDB*

PLANT OVERSIGHT REVIEW
COMMITTEE
DA January 8, 1991
SF *PORC*
BT1 Committees
 BT2 Administrative Organizations
  BT3 Organizations

PLANT PROTECTION SYSTEM
(2040 PPS)
DA December 10, 1990
BT1 Emergency Systems
 BT2 Systems
BT1 Systems (DOE FRASE Vocabulary)
 BT2 Systems
RT Guard Station
RT Security Guard
RT Security Activity
SO DOE FRASE VOCABULARY

PLANT RISK STATUS INFORMATION
MANAGEMENT SYSTEM
DA January 8, 1991
SF *PRISIM*
BT1 Information Systems

BT2 Security Interests
BT2 Systems

PLANTS
(ESA; CFR)
DA October 12, 1990
NT1 Algae
NT1 Fungi
NT1 Weeds
RT Biomass
RT Biota
RT Buffer Strips
RT Chlorosis
RT Evapotranspiration
RT Fertilizers
RT Functional Units
RT Groundcovers
RT Necrosis
RT Photosynthesis
RT Phytotoxic
RT Pollen
RT Threatened Species
SO Endangered Species
DEF Members of the plant kingdom,
   including seeds, roots, and other
   parts thereof.

PLASMIDS
(EPA)
DA October 12, 1990
BT1 DNA
 BT2 Organic Chemicals
  BT3 Chemical Substances
RT Genes
RT Vectors
SO Environmental Protection Agency
   Glossary
DEF (EPA) Circular pieces of DNA that
   exists apart from the chromosome
   and replicates independently of it.
   Bacterial plasmids carry
   information that renders the
   bacteria resistant to antibiotics.
   Plasmids are often used in genetic
   engineering to carry desired genes
   into organisms.

PLASTIC YIELDING
(DOE Order 6430.1A)
DA October 12, 1990
SO Construction
DEF (DOE Order 6430.1A) The point at
   which permanent deformation
   occurs when tensile stress is
   imposed on a material.

PLASTICS
(EPA)
DA October 12, 1990
NT1 Polyvinyl Chloride
RT Polymers
SO Environmental Protection Agency
   Glossary
DEF (EPA) Nonmetallic compounds that
   result from a chemical reaction
   and are molded or formed into
   rigid or pliable construction
   materials or fabrics.

PLASTRONS
(ESA; CFR)
DA October 12, 1990
RT Sea Turtles
SO Endangered Species
DEF (CFR) The ventral part of the shell
   of a sea turtle consisting typically
   of nine symmetrically placed
   bones overlaid by horny plates.

PLATFORMS
(SEA)
DA October 12, 1990
BT1 Structures (DOE FRASE
   Vocabulary)
SO Construction
DEF (SEA) Lower rigid portions of
   structures having a vertical
   combination of structural systems.

PLENUM PRESSURE
DA January 8, 1991
SF *PP*
BT1 Pressure

PLIERS
(3047 PLIERS)
DA January 3, 1991
BT1 Tools - Manual
 BT2 Tools (DOE FRASE Vocabulary)
  BT3 Equipment
SO DOE FRASE VOCABULARY

PLUGGING
(NPDWR; EPA; CFR)
DA October 12, 1990
BT1 Processes
RT Cementing
RT Current Plugging and
   Abandonment Cost Estimates
RT Injection
RT Packers
RT Plugging Records
RT Plugging and Abandonment Plans
RT Surface Casings
RT Well Plugs
RT Wells
SO Environmental Protection Agency
   Glossary
SO Nuclear Facilities Incident Database
SO Radiation
SO Water Pollution
DEF (1) The act or process of stopping
   the flow of water, oil, or gas into or
   out of a formation through a
   borehole or well penetrating that
   formation. (2) Stopping a leak or
   sealing off a pipe or hose.

PLUGGING AND ABANDONMENT
PLANS
(SDWA; NPDWR; CFR)
DA October 12, 1990
BT1 Plans
RT Plugging
SO Water Pollution
DEF Plans for the plugging and
   abandonment of wells; prepared in

accordance with applicable
statutory and legal requirements.

PLUGGING RECORDS
(SDWA; NPDWR; CFR)
DA   October 12, 1990
BT1 Reports
RT   Plugging
SO   Water Pollution
DEF Systematic listings of permanent or
temporary abandonment of water,
oil, gas, test, exploration and
waste injection wells, and may
contain a well log, description of
amounts and types of plugging
material used, the method
employed for plugging, a
description of formations which
are sealed and a graphic log of
the well showing formation
location, formation thickness, and
location of plugging structures.

PLUME EXPOSURE PATHWAYS
(EMER)
DA   February 1, 1991
BT1 Exposure Pathways
 BT2 Routes (Exposure)
SO   Emergency Preparedness
DEF (EMER) Principal exposure sources
for pathways are: 1) whole body
external exposure (gamma
radiation) and/or contact exposure
to skin or eyes (hazardous
substances) from contact with
materials from the plume and from
deposited material 2) inhalation
and absorption of constituents in
the passing plume

PLUMES
(EPA; EMER)
DA   October 12, 1990
BT1 Air Pollutants
BT1 Discharges
NT1 Vapor Plumes
RT   Air Pollution
RT   Dispersion Techniques
RT   Management of Migrations
RT   Smoke
RT   Thermal Pollution
RT   Water Pollution
SO   Emergency Preparedness
SO   Environmental Protection Agency
      Glossary
DEF (EPA) (1) Visible or measurable
discharges of a contaminant from
a given point of origin. Can be
visible or thermal in water, or
visible in the air as, for example,
plumes of smoke. (2) Areas of
measurable and potentially
harmful radiation leaking from a
damaged reactor. (3) The
distances from a toxic release
considered dangerous for those
exposed to the leaking fumes.

PLUTONIUM
(EPA)
DA   October 12, 1990
SO   Environmental Protection Agency
      Glossary
DEF (EPA) A radioactive metallic
element similar chemically to
uranium.

PLUTONIUM FUEL PLATE(S)
(2582 PU FUEL PLAT)
DA   January 3, 1991
BT1 Fuel Plate
 BT2 Fuel Element(s)
  BT3 Equipment/Parts - Nuclear (DOE
       FRASE Vocabulary)
   BT4 Equipment
   BT4 Reactor Components
SO   DOE FRASE VOCABULARY

PLUTONIUM PROCESS HOOD
(2583 PU PROC HOOD)
DA   January 3, 1991
BT1 Equipment/Parts - Nuclear (DOE
     FRASE Vocabulary)
 BT2 Equipment
 BT2 Reactor Components
RT   Barriers
RT   Plutonium Recovery Process
SO   DOE FRASE VOCABULARY

**PLUTONIUM PROCESSING AND
HANDLING FACILITIES**
(DOE Order 6430.1A)
DA   October 12, 1990
BT1 Nuclear Facilities
 BT2 Facilities
SO   Construction
DEF (DOE Order 6430.1A) Facilities
constructed primarily to process
plutonium (including Pu 238) and
handle substantial quantities of
in-process plutonium where there
is a possibility of a release of
plutonium to the environs under
normal operations or design basis
accident conditions in excess of
limits set forth in the directive on
Radiation Protection of the Public
and the Environment in the DOE
Orders, 5400 series.

PLUTONIUM RECOVERY PROCESS
(2584 PLUT PROC HOOD)
DA   January 3, 1991
BT1 Equipment/Parts - Nuclear (DOE
     FRASE Vocabulary)
 BT2 Equipment
 BT2 Reactor Components
RT   Plutonium Process Hood
SO   DOE FRASE VOCABULARY

PLUTONIUM STORAGE FACILITIES
(DOE Order 6430.1A)
DA   October 12, 1990
BT1 Nuclear Facilities
 BT2 Facilities
SO   Construction
DEF (DOE Order 6430.1A) Facilities

constructed to store strategic
(category I) quantities of
plutonium.

PLWA
DA   October 12, 1990
SEE Primary Light Water Addition
SO   Acronyms

PM
DA   October 12, 1990
SEE Preventive Maintenance
SO   Acronyms

PMA
DA   May 15, 1991
SEE Power Marketing Administrations
SO   Acronyms

PMF
DA   October 12, 1990
SEE Probable Maximum Flood
SO   Acronyms

PM$_{10}$
(CAA; CFR; ESH)
DA   October 12, 1990
BT1 Particulate Matter
 BT2 Airborne Particulates
  BT3 Particulates
   BT4 Air Pollutants
SO   Air Pollution
DEF (ESH) Particulate matter with an
aerodynamic diameter less than or
equal to a nominal 10 micrometers
as measured by a reference
method based on Appendix J of
40 CFR 50 and designated in
accordance with or by an
equivalent method designated in
accordance with 40 CFR 53.

PM$_{10}$ EMISSIONS
(CAA)
DA   November 15, 1990
BT1 Particulate Matter Emissions
 BT2 Emissions
  BT3 Air Pollutants
SO   Air Pollution
DEF (CAA) Finely divided solid or liquid
materials, with an aerodynamic
diameter less than or equal to a
nominal 10 micrometers emitted to
the ambient air as measured by
an applicable reference method,
or an equivalent or alternative
method, or by a test method
specified in an approved State
implementation plan.

PNL
DA   October 12, 1990
SEE Pacific Northwest Laboratory
SO   Acronyms

POCKET KNIFE(S)
(3048 POCKET KNIFE)
DA   January 3, 1991

BT1 Knife(s)
 BT2 Tools - Manual
  BT3 Tools (DOE FRASE Vocabulary)
   BT4 Equipment
 SO DOE FRASE VOCABULARY

POINT OF COMPLIANCE
DA January 24, 1991
BT1 Sites/Areas
SO Environmental Management
DEF (CFR) A vertical surface located at the hydraulically down gradient limit of the waste management area that extends down into the uppermost aquifer underlying the regulated units.

POINT OF DISINFECTANT APPLICATION
(SDWA; CFR)
DA October 12, 1990
BT1 Sites/Areas
RT Disinfectant Contact Time
RT Disinfection
SO Water Pollution
DEF The point where the disinfectant is applied and water downstream of that point is not subject to recontamination by surface water runoff.

POINT-OF-ENTRY TREATMENT DEVICES
(SDWA; NPDWR; CFR)
DA October 12, 1990
BT1 Devices
RT Point-of-use Treatment Devices
SO Water Pollution
DEF Treatment devices applied to the drinking water entering a house or building for the purpose of reducing contaminants in the drinking water distributed throughout the house or building.

POINT OF NEAREST PUBLIC ACCESS
(DOE Order 6430.1A)
DA October 12, 1990
BT1 Sites/Areas
RT Public Travel Routes
SO Construction
DEF (DOE Order 6430.1A) Location inside or outside the site boundary where a member of the public could legally be (e.g., visitor center or public highway) without the specific knowledge of the owner or operator.

POINT-OF-USE TREATMENT DEVICES
(SDWA; NPDWR; CFR)
DA October 12, 1990
BT1 Devices
RT Point-of-entry Treatment Devices
SO Water Pollution
DEF Treatment devices applied to a single tap used for the purpose of reducing contaminants in drinking water at that one tap.

POINT SOURCES
(SWDA; CAA; ESH; CFR)
DA October 12, 1990
BT1 Stationary Sources
 BT2 Facilities
RT Area Sources
RT Direct Dischargers
RT Particulate Matter Emissions
RT Stacks
SO Air Pollution
SO Environmental Management
SO Environmental Protection Agency Glossary
SO Hazardous Materials
SO Wastes
DEF Stationary locations or fixed facilities from which pollutants are discharged or emitted. Also, any single identifiable sources of pollution, e.g., a pipe, ditch, ship, ore pit, factory smokestack. (ESH) Any discernable, confined and discrete conveyances, including but not limited to pipe, ditch, channel, tunnel, conduit, well, landfill leachate, discrete fissure, container, rolling stock, concentrated animal feeding operation, or vessel or other floating craft from which pollutants are or may be discharged. The term does not include return flows from irrigated agriculture.

POISON A
(CFR)
DA October 12, 1990
BT1 Poisons
DEF (CFR) Extremely dangerous poisons, Class A are poisonous gases or liquids of such a nature that a very small amount of the gas, vapor of the liquid, mixed with air would be dangerous to life. Some of Class A poisons include liquid bromoacetone, cyanogen gas, liquid nitrogen peroxide and phosphine.

POISON B
(CFR)
DA October 12, 1990
BT1 Poisons
DEF (CFR) Class B Poisons are not considered as dangerous as Class A Poisons. This class includes those substances, liquid or solid, (including pastes and semisolids), other than Class A or Irritating Material which are known to be so toxic to man as to provide a hazard to health during transportation, or which in the absence of adequate data on human toxicity, are presumed to be toxic to man based on results with test animals (e.g., oral toxicity testing). Arsenic acid (solid), liquid chloropicrin, and solid cocculus

are examples of Poison B materials.

POISONED EMERGENCY COOLING SYSTEM SAMPLING
DA January 8, 1991
SF *PECSS*
BT1 Sampling
 BT2 Sampling and Analysis
BT1 Sampling and Analysis

POISONING
(1399 POISONING)
DA November 28, 1990
BT1 Illnesses
NT1 Systemic Poisoning
RT Toxic Effects to Single System
SO DOE FRASE VOCABULARY

POISONS
(CFR)
DA October 12, 1990
NT1 Commercial Arsenic
NT1 Economic Poisons
NT1 Poison A
NT1 Poison B
NT1 Toxicants
RT Hazards
SO Hazardous Materials
DEF (DSTT) Substances that in relatively small doses have an action that either destroys life or impairs seriously the functions of organs or tissues.

POLICIES
(SSDC; MORT)
DA October 12, 1990
NT1 Bubble Policy
NT1 Emissions Trading
RT Implementation
RT Management System Factors
RT Minimum Requirements and Standards
RT Protection of the Public Health and Welfare
RT Risk Assessment
SO System Safety Development Center Glossary
DEF (SSDC) Written statements that express the wisdom, philosophy, experience, and belief of an organization's senior managers for future guidance toward attainment of stated goals. Lesser categories of guidance include practices or directives: Standard methods of performing work or communicating; procedures: step-by-step methods of performing a task. (MORT) MORT analysis asks: are there a written, up-to-date policies with a broad enough scope to address major problems likely to be encountered? Are they also sufficiently comprehensive to include the major motivations (e.g., humane, cost, efficiency,

---

legal compliance)? Can they be implemented without conflict?

POLISHING MACHINE
(2133 POLISHING MA)
DA   December 10, 1990
BT1 Machines (DOE FRASE Vocabulary)
 BT2 Equipment
SO   DOE FRASE VOCABULARY

POLLEN
(EPA)
DA   October 12, 1990
RT   Plants
SO   Environmental Protection Agency Glossary
DEF (EPA) (1) A fine dust produced by plants. (2) The fertilizing element of flowering plants. (3) A natural or background air pollutant.

POLLUTANT DISCHARGE ELIMINATION SYSTEM
(SWDA; RCRA)
DA   October 19, 1990
BT1 National Pollutant Discharge Elimination System
 BT2 Systems
RT   Abatement
RT   Discharge of Pollutants
SO   Wastes
DEF Section 402 of the Federal Water Pollution Control Act (a.k.a. Clean Water Act) that establishes a permit for discharges to water and provides standards by which such permits may be granted.

POLLUTANT OR CONTAMINANT
(CERCLA; ESH)
DA   October 12, 1990
RT   Contaminants
RT   Pollutants
DEF (ESH) Pollutant or contaminant with certain exceptions, is any element, substance, compound, or mixture, including disease- causing agents, which after release into the environment and upon exposure, ingestion, inhalation, or assimilation into any organism, either directly from the environment or indirectly by ingestion through food chains, will or may reasonably be anticipated to cause death, disease, behavioral abnormalities, cancer, genetic mutation, physiological malfunctions (including malfunctions in reproduction), or physical deformations, in such organisms or their offspring.

POLLUTANT STANDARD INDEX
(EPA)
DA   October 12, 1990
SF   *PSI*
RT   Air Pollution

SO   Environmental Protection Agency Glossary
DEF (EPA) Measure of adverse health effects of air pollution levels in major cities.

POLLUTANTS
(CERCLA; CFR; EPA; ESH)
DA   October 12, 1990
NT1 Conventional Pollutants
NT1 Criteria Pollutants
NT2 Carbon Monoxide
NT2 Lead
NT2 Nitrogen Oxides (NO$_x$)
 NT3 Nitric Oxides
 NT3 Nitrogen Dioxide (NO$_2$)
NT2 Ozone
NT2 Sulfur Dioxide
NT2 Total Suspended Particulates
NT1 Non-Contact Cooling Water Pollutants
NT1 Non-Conventional Pollutants
NT1 Process Waste Water Pollutants
NT1 Toxic Pollutants
RT   Assimilation
RT   Baseline Concentration
RT   Contaminants
RT   Contamination
RT   Discharge of Pollutants
RT   Effluents
RT   Hazards
RT   National Contingency Plan
RT   Pesticides
RT   Pollution
RT   Pollutant or Contaminant
RT   Wastes
SO   Compensation and Liability
SO   Environmental Management
SO   Environmental Protection Agency Glossary
SO   Hazardous Materials
SO   Water Pollution
DEF Generally, any substance introduced into the environment that adversely affects the usefulness of a resource. (ESH) Dredge spoil, solid waste, incinerator residue, filter backwash, sewage, garbage, sewage sludge, munitions, chemical wastes, biological materials , radioactive materials (except those regulated under the Atomic Energy Act of 1954, as amended). Examples of material not covered include radium and accelerator-produced isotopes, heat, wrecked or discarded equipment, rock, sand, cellar dirt, and industrial, municipal, and agricultural waste discharged into water. It does not mean (1) sewage from vessels or (2) water, gas, or other material which is injected into a well to facilitate production of oil or gas, or disposal.

POLLUTION
(CWA; RHA; CFR; EPA; ESH)

DA   October 12, 1990
NT1 Agricultural Pollution
NT1 Air Pollution
NT2 Hazardous Air Pollution
NT2 Indoor Air Pollution
NT2 Man-Made Air Pollution
NT2 Photochemical Smog
 NT3 Oxidants
NT2 Smog
NT2 Wood-Burning Stove Pollution
NT1 Thermal Pollution
NT1 Water Pollution
RT   Anti-Degradation Clause
RT   Bottle Bill
RT   Effluents
RT   Environment
RT   Mobile Sources
RT   Noise Pollution Abatement
RT   Pesticides
RT   Pollutants
RT   Spills
RT   Stationary Sources
SO   Environmental Management
SO   Environmental Protection Agency Glossary
SO   Hazardous Materials
SO   Water Pollution
DEF Generally, the presence of matter or energy whose nature, location or quantity produces undesired environmental effects. Under the Clean Water Act, for example, the term is defined as the man-made or man-induced alteration of the physical, biological, and radiological integrity of water. (ESH) The man-made or man-induced alteration of the chemical, physical, biological, and radiological integrity of water.

POLLUTION LIABILITY
(CERCLA)
DA   October 12, 1990
BT1 Liabilities
RT   Purchasing Group
RT   Risk Retention Group
SO   Compensation and Liability
DEF (CERCLA) Liability for injuries arising from the release of hazardous substances or pollutants or contaminants.

POLLUTION RECOVERY PROCESSES
DA   February 26, 1991
BT1 Processes
NT1 Abatement
NT2 Noise Pollution Abatement
NT1 Assimilation
NT1 Banking (Air Pollution)
NT1 Desulfurization
NT1 Filtration
NT2 Diatomaceous Earth Filtration
NT2 Direct Filtration
NT2 Flow Sand Filtration
NT2 Slow Sand Filtration
NT1 Limestone Scrubbing
NT1 Resource Recovery
NT2 Reclamation
NT3 Land Reclamation

NT3 Refuse Reclamation
NT2 Solid Waste Planning
NT1 Screening
RT Sorption

POLYBOR
(NFI)
DA   October 12, 1990
SO Nuclear Facilities Incident Database
SO Radiation
DEF (NFI) Sodium polyborate, neutron
    absorber mixed with Emergency
    Core Cooling Systems (ECCS)
    light water.

POLYCHLORINATED BIPHENYLS
(SWDA; RCRA)
DA   October 19, 1990
SF   *PCBs*
BT1 Carcinogens
 BT2 Hazardous Substances
BT1 Chlorinated Hydrocarbons
 BT2 CERCLA Hazardous Substances
  BT3 Hazardous Substances
 BT2 Halogenated Organic Compounds
  BT3 Halogenated
  BT3 Organic Chemicals
   BT4 Chemical Substances
NT1 High Concentration PCBs
NT1 Low Concentration PCBs
NT1 Recycled PCBs
RT Changed Use Pattern
RT Hazardous Constituents
RT Marks
RT Responsible Parties
RT Totally Enclosed Manner
SO Environmental Management
SO Wastes
DEF (CFR) Chemical substances that
    are limited to the biphenyl
    molecule that have been
    chlorinated to varying degrees or
    any combination of substances
    which contains such substance.
    (EPA) A group of toxic, persistent
    chemicals used in transformers
    and capicitors for insulating
    purposes and in gas pipeline
    systems as a lubricant. Further
    sale of new use was banned by
    law in 1979.

POLYELECTROLYTES
(EPA)
DA   October 12, 1990
BT1 Synthetic Organic Chemicals
 BT2 Organic Chemicals
  BT3 Chemical Substances
SO Environmental Protection Agency
    Glossary
DEF (EPA) Synthetic chemicals that help
    solids to clump during sewage
    treatment.

POLYMERS
(EPA)
DA   October 12, 1990
RT Plastics
RT Types of Resin

SO Environmental Protection Agency
    Glossary
DEF (EPA) Basic molecular ingredients
    in plastic.

POLYVINYL CHLORIDE
(EPA)
DA   October 12, 1990
SF   *PVC*
BT1 Chlorinated Hydrocarbons
 BT2 CERCLA Hazardous Substances
  BT3 Hazardous Substances
 BT2 Halogenated Organic Compounds
  BT3 Halogenated
  BT3 Organic Chemicals
   BT4 Chemical Substances
BT1 Plastics
RT In-Process Wastewater
RT Reactors (Chemical)
SO Environmental Protection Agency
    Glossary
DEF (EPA) A tough, environmentally
    indestructible plastic that releases
    hydrochloric acid when burned.

POLYVINYL CHLORIDE PLANTS
(CAA; CFR)
DA   October 12, 1990
BT1 Facilities
NT1 Vinyl Chloride Plants
RT Vinyl Chloride
SO Air Pollution
DEF (CFR) Any plants where vinyl
    chloride alone or in combination
    with other materials is
    polymerized.

POND
(1859 POND)
DA   December 10, 1990
BT1 Structures (DOE FRASE
    Vocabulary)
RT Body of Water
RT Dam
SO DOE FRASE VOCABULARY

POOL
(1727 POOL)
DA   December 10, 1990
BT1 Site (DOE FRASE Vocabulary)
 BT2 Sites/Areas
RT Drowning
SO DOE FRASE VOCABULARY

POPULATION DOSE PROJECTIONS
(EMER)
DA   February 1, 1991
SO Emergency Preparedness
DEF (EMER) Estimates of total radiation
    dose to which the population may
    be exposed.

POPULATIONS
(ESA; CFR; EPA)
DA   October 12, 1990
NT1 Experimental Populations
 NT2 Essential Experimental
     Populations
NT1 Special Populations

RT Biotic Communities
RT Ecosystems
RT Epidemiology
SO Endangered Species
SO Environmental Protection Agency
    Glossary
DEF Groups of interbreeding organisms
    of the same kind occupying a
    particular space. Generically, the
    number of humans or other living
    creatures in a designated area.

PORC
DA   October 12, 1990
SEE Plant Oversight Review Committee
SO Acronyms

PORTABLE TANKS
(SWDA; RCRA; ESH)
DA   October 12, 1990
BT1 Tanks
 BT2 Facility Components
NT1 Intermodal Portable Tanks
RT Containerships
SO Hazardous Materials
DEF Bulk packagings (except a cylinder
    having a water capacity of 1000
    pounds or less) designed primarily
    to be loaded into, or on, or
    temporarily attached to a transport
    vehicle or ship and equipped with
    skids, mountings, or accessories
    to facilitate handling of the tank by
    mechanical means.

PORTLAND CEMENT
(DOE Order 6430.1A)
DA   October 12, 1990
RT Cementing
SO Construction
DEF (DOE Order 6430.1A) A mixture of
    lime- and clay-bearing materials
    that are calcined to form a clinker,
    which is then pulverized to form a
    fine powder for mortar and
    concrete mixtures.

PORTS
DA   October 12, 1990
SEE Portsmouth Gaseous Diffusion
    Plant
SO Acronyms

PORTSMOUTH GASEOUS DIFFUSION
PLANT
DA   January 8, 1991
SF   *PORTS*
BT1 Government-Owned
    Contractor-Operated Facilities
 BT2 Federal Facilities
  BT3 Facilities
RT Martin Marietta Energy Systems

PORTSMOUTH URANIUM
ENRICHMENT COMPLEX
DA   January 8, 1991
SF   *PUEC*
BT1 Facilities

PORV
DA   October 12, 1990
SEE Power Operated Relief Valve
SO   Acronyms

POSING AN EXPOSURE RISK TO FOOD OR FEED
(TSCA; CFR)
DA   October 12, 1990
RT   Food Chains
RT   PCB Items
SO   Hazardous Materials
DEF (CFR) Being in any location where human food or animal feed products could be exposed to PCBs released from a PCB Item. A PCB item poses an exposure risk to food or feed if PCBs released in any way from the PCB item have a potential pathway to human food or animal feed. EPA considers human food or animal feed to include items regulated by the U.S. Department of Agriculture or the Food and Drug Administration as human food or animal feed; this includes direct additives. Food or feed is excluded from this definition if it is used or stored in private homes.

POSITIVE (OBJECTIVE) TREES
(SSDC)
DA   October 12, 1990
BT1 Analytical (Logic) Trees
 BT2 Diagrams
SO   System Safety Development Center Glossary
DEF (SSDC) Analytical trees that are positive in nature; the top event is a goal or desired happening.

POSITIVE DATA
(EPA)
DA   October 12, 1990
RT   Intervention Levels
SO   Environmental Protection Agency Glossary
DEF (EPA) Analytical results for which measurable concentrations (i.e., above a quantitation limit) are reported. May have data qualifiers attached.

POSITIVE EXPOSURE
(Doe Order 5484.1)
DA   October 12, 1990
BT1 Exposure
SO   Environmental Management
SO   Management
DEF (DOE Order 5484.1) Any recorded exposure, corrected for background, greater than the established minimum detection limit of the monitoring device or the measuring technique employed.

POST-CLOSURE
(EPA)
DA   October 12, 1990
BT1 Time Designations
RT   Post-Closure Plan
RT   Site Closure and Stabilization
SO   Environmental Protection Agency Glossary
DEF (EPA) The time period following the shutdown of a waste management or manufacturing facility. For monitoring purposes, this is often considered to be 30 years.

POST-CLOSURE PLAN
(SWDA; RCRA)
DA   October 12, 1990
BT1 Plans
RT   Post-Closure
DEF The plan for post-closure care prepared in accordance with the requirements of 40 CFR 264.117 through 264.120.

POST CONSUMER WASTES
DA   January 24, 1991
BT1 Wastes
RT   Recycled Materials
SO   Environmental Management
DEF (CFR) A material or product that has served its intended use and has been discarded for disposal or recovery after passing through the hands of a final user. PCW are a part of the broader category "Recycled Materials."

POST DRIVER(S)
(2956 POST DRIVER)
DA   January 3, 1991
BT1 Tools - Powered
 BT2 Tools (DOE FRASE Vocabulary)
 BT3 Equipment
SO   DOE FRASE VOCABULARY

POST EMERGENCY RESPONSES
DA   January 24, 1991
BT1 Responses
RT   Emergency Response
SO   Environmental Management
DEF (CFR) That portion of an emergency response performed after the immediate threat of a release has been stabilized or eliminated and cleanup of the site has begun. If post emergency response is performed by an employer's own employees who were part of the initial emergency response, it is considered to be part of the initial response and not post emergency response. However, if a group of an employer's own employees, separate from the group providing initial response, performs the cleanup operation, then the separate group of employees would be considered to be

performing post-emergency response and subject to paragraph (q)(11) of this section.

POST-INCIDENT ACTIVITIES
(EMER)
DA   February 1, 1991
BT1 Activities
SO   Emergency Preparedness
DEF (EMER) Those activities occurring after the cessation of an emergency where immediate health and/or safety hazards no longer exist; however, long-term recovery actions or monitoring functions may be required.

POST REMOVAL SITE CONTROL
(ANL; CFR)
DA   May 24, 1991
BT1 Activities
BT1 Controls
RT   Comprehensive Environmental Response, Compensation, etc.
SO   Management
SO   Wastes
DEF (CFR) Those activities that are necessary to sustain the integrity of a Fund-financed removal action following its conclusion. Post-removal site control may be a removal or remedial action under CERCLA. The term includes, but is not limited to, activities such as relighting gas flares, replacing filters, and collecting leachate.

POT FURNACES
(CAA; CFR)
DA   October 12, 1990
BT1 Glass Melting Furnaces
 BT2 Industrial Furnaces
 BT3 Facility Components
RT   Refractories
SO   Air Pollution
DEF (CFR) Glass melting furnaces that contains one or more refractory vessels in which glass is melted by indirect heating. The openings of the vessels are in the outside wall of the furnace and are covered with refractory stoppers during melting.

POTABLE WATER
(EPA)
DA   October 12, 1990
BT1 Water
RT   Drinking Water Coolers
RT   Water Supply System
SO   Environmental Protection Agency Glossary
DEF (EPA) Water that is safe for drinking and cooking.

POTASSIUM IODIDE
(EMER)
DA   February 1, 1991

SO Emergency Preparedness
DEF (EMER) Thyroid blocking agent that
may be used in radiological events
involving releases of radioiodine.

POTENTIAL HAZARDS
(ESH; TTGM)
DA October 12, 1990
BT1 Hazards
 BT2 Conditions
NT1 Level 1 Potential Hazards
NT1 Level 2 Potential Hazards
NT1 Level 3 Potential Hazards
RT Concerns
RT Environment Assessments
RT Job Safety Analysis
RT Technical Safety Appraisals
SO Management
SO Standards
DEF (TTGM) Regarding Concerns, as
identified within Technical Safety
Appraisals and/or the Safety and
Health Assessment of Tiger Team
Assessments, potential hazards
are one of three characteristics
(also including Seriousness and
Compliance Considerations) used
to indicate a specific concern's
position relative to other concerns.

POTENTIAL TO EMIT
(CAA; CFR)
DA October 12, 1990
BT1 Conditions
RT Baseline Concentration
RT Stationary Sources
SO Air Pollution
DEF (CFR) The maximum capacity of a
stationary source to emit a
pollutant under its physical and
operational design. Any physical
or operational limitation on the
capacity of the source to emit a
pollutant, including air pollution
control equipment and restrictions
on hours of operation or on the
type or amount of material
combusted, stored, or processed,
shall be treated as part of its
design if the limitation or the effect
it would have on emissions is
federally enforceable. Secondary
emissions do not count in
determining the potential to emit of
a stationary source.

POTENTIALLY RESPONSIBLE
PARTIES
(CFR; EPA)
DA October 12, 1990
SF PRP
NT1 DOE Contractors
 NT2 Allied-Signal Inc.
 NT2 Battelle Memorial Institute
 NT2 Battelle Columbus Laboratories
 NT2 Bechtel Petroleum Operation, Inc.
 NT2 Department of Energy Resident
 Construction Contractors
 NT2 Department of Energy Project
 Construction Contractors

 NT2 EG&G Energy Measurements, Inc.
 NT2 EG&G Mound Applied
 Technologies, Inc.
 NT2 EG&G Idaho, Inc.
 NT3 System Safety Development
 Center
 NT2 General Electric
 NT2 Hanford Environmental Health
 Foundation
 NT2 John Brown, E&C, Inc.
 NT2 Kaiser Engineers Hanford, Co.
 NT2 Lawrence-Allison & Associates
 NT2 Lovelace Medical Foundation
 NT2 Martin Marietta Energy Systems
 NT2 Mason and Hanger-Silas Mason
 Co.
 NT2 MK-Ferguson
 NT2 MK-Ferguson of Idaho Co.
 NT2 Oak Ridge Associated Universities
 NT2 Pan American World Services,
 INC.
 NT2 Raytheon Services Nevada
 NT2 Reynolds Electrical and
 Engineering Company
 NT2 Rockwell International Corp.
 NT2 Ross Aviation, Inc.
 NT2 Sandia Corporation AT&T
 Technologies, Inc.
 NT2 Science Applications International
 Corporation
 NT2 Southeastern Universities
 Research Association, Inc.
 NT2 Stanford University
 NT2 Trustees of Princeton University
 NT2 TRW
 NT2 Universities Research Association,
 Inc.
 NT2 University of Georgia Research
 Foundation, Inc.
 NT2 University of California
 NT2 Wackenhut Services Inc.
 NT2 West Valley Nuclear Services Co.,
 Inc.
 NT2 Westinghouse Electric Corp.
 NT2 Westinghouse Idaho Nuclear Co.,
 Inc.
 NT2 Westinghouse Savannah River
 Company
 NT2 Westinghouse Hanford Company
 NT2 Westinghouse Materials Company
 of Ohio
NT1 Generators (Pollution)
 NT2 PCB Waste Generator
 NT2 Small Quantity Generators
NT1 Operators
NT1 Owners
NT1 Transporters
 NT2 Transporter of PCB Waste
RT Responsible Parties
SO Environmental Protection Agency
Glossary
DEF (EPA) Any individuals or companies
– including owners, operators,
transporters or generators –
potentially responsible for, or
contributing to, the contamination
problems at a Superfund site.
Whenever possible, EPA requires
PRPs, through administrative and
legal actions, to clean up

hazardous waste sites they have
contaminated.

POTENTIOMETER
(2764 POTENTIOMETE)
DA January 3, 1991
SF pots
BT1 Indicator(s)
 BT2 Instrument(s)
 BT3 Equipment/Parts -
 Instrumentation/Measuring
 (DOE FRASE Voc.)
 BT4 Equipment
SO DOE FRASE VOCABULARY

pots
DA October 12, 1990
SEE Potentiometer
SO Acronyms

POTW
DA October 12, 1990
SEE Publicly Owned Treatment Works
SO Acronyms
SO Environmental Management

POUNDS PER SQUARE INCH
(SWDA; RCRA; ESH)
DA October 12, 1990
SF PSI (Pounds per Square Inch)
RT Pressure (Surface)
SO Hazardous Materials
DEF (Engineering Encyclopedia) The
pressure of steam, air, water, or
any gases or fluids is given
ordinarily in relation to the square
inch. One pound per square inch =
144 pounds per square foot =
0.0703 kilogram per square
centimeter = 2.31 feet of water at
62 degrees Fahrenheit = 27.7
inches of water at 62 degrees
Fahrenheit = 2.042 inches of
mercury at 62 degrees Fahrenheit
= 0.08 atmosphere.

POUNDS PER SQUARE INCH
ABSOLUTE
DA January 8, 1991
SF PSIA

POUNDS PER SQUARE INCH GAUGE
(ESH)
DA October 19, 1990
SF PSIG
RT Pressure (Surface)
SO Hazardous Materials

POUNDS/YEAR
DA January 8, 1991
SF lbs/year
BT1 Units of Measure

POURING
(CAA; CFR)
DA October 12, 1990
BT1 Manufacturing Processes
 BT2 Processes

RT Copper Converters
SO Air Pollution
DEF (CFR) The removal of blister copper
  from the copper converter bath.

POWER ACTUATED TOOL(S)
(2957 POWER ACTUA)
DA January 3, 1991
BT1 Tools - Powered
 BT2 Tools (DOE FRASE Vocabulary)
 BT3 Equipment
SO DOE FRASE VOCABULARY

POWER BUFFER(S)
(2958 POWER BUFFER)
DA January 3, 1991
BT1 Tools - Powered
 BT2 Tools (DOE FRASE Vocabulary)
 BT3 Equipment
SO DOE FRASE VOCABULARY

POWER BURST FACILITY
(SSDC)
DA May 24, 1991
SF *PBF*
BT1 Idaho National Engineering
  Laboratory
 BT2 Government-Owned
   Contractor-Operated Facilities
  BT3 Federal Facilities
   BT4 Facilities
 BT2 Laboratories
  BT3 Research and Development
    Organizations
   BT4 Organizations
SO Radiation
DEF (User's Guide to DOE Facilities,
  1984) A facility centered around a
  light water moderated reactor
  designed for up to 28 MW steady
  state or 270 GW burst operation in
  fuels research work. The high
  neutron flux region is a right
  circular cylinder about 0.2 meters
  in diameter and 1 meter long.
  Experiments are conducted within
  a high pressure central in-pile tube
  connected to an external loop.
  Heated, high pressure water is
  circulated through the loop and
  external equipment allows the
  generation of the specific
  environment to meet a wide
  variety of test conditions. A wide
  assortment of instrumentation is
  used to record the experiment
  conditions and results with a high
  capacity data collection and
  display system. The primary
  sponsor, the U.S. Nuclear
  Regulatory Commission, has
  conducted a wide variety of
  nuclear fuel behavior and fission
  product release experiments.

POWER CHISEL(S)
(2959 POWER CHISEL)
DA January 3, 1991
BT1 Tools - Powered

BT2 Tools (DOE FRASE Vocabulary)
 BT3 Equipment
SO DOE FRASE VOCABULARY

POWER CONVERSION SYSTEM
DA January 8, 1991
SF *PCS*
BT1 Systems

POWER CORD(S)
(2405 POWER CORD)
DA January 3, 1991
BT1 Equipment/Parts - Electrical (DOE
   FRASE Vocabulary)
 BT2 Equipment
SO DOE FRASE VOCABULARY

POWER CUTTER(S)
(2960 POWER CUTTER)
DA January 3, 1991
BT1 Tools - Powered
 BT2 Tools (DOE FRASE Vocabulary)
 BT3 Equipment
SO DOE FRASE VOCABULARY

POWER DENSITY MONITOR
DA January 8, 1991
SF *PDM*
BT1 Monitors
 BT2 Equipment

POWER DRILL(S)
(2961 POWER DRILL)
DA January 3, 1991
BT1 Tools - Powered
 BT2 Tools (DOE FRASE Vocabulary)
 BT3 Equipment
SO DOE FRASE VOCABULARY

POWER FILE(S)
(2962 POWER FILE)
DA January 3, 1991
BT1 Tools - Powered
 BT2 Tools (DOE FRASE Vocabulary)
 BT3 Equipment
SO DOE FRASE VOCABULARY

POWER GRINDER(S)
(2963 POWER GRINDE)
DA January 3, 1991
BT1 Tools - Powered
 BT2 Tools (DOE FRASE Vocabulary)
 BT3 Equipment
SO DOE FRASE VOCABULARY

POWER HAMMER(S)
(2964 POWER HAMMER)
DA January 3, 1991
BT1 Tools - Powered
 BT2 Tools (DOE FRASE Vocabulary)
 BT3 Equipment
SO DOE FRASE VOCABULARY

POWER IMPACT WRENCH(S)
(2965 POWER IMPACT)
DA January 3, 1991
BT1 Tools - Powered
 BT2 Tools (DOE FRASE Vocabulary)

 BT3 Equipment
BT1 Wrench(s)
 BT2 Tools (DOE FRASE Vocabulary)
 BT3 Equipment
SO DOE FRASE VOCABULARY

POWER KEY(S)
(2406 POWER KEY)
DA January 3, 1991
BT1 Equipment/Parts - Electrical (DOE
   FRASE Vocabulary)
 BT2 Equipment
SO DOE FRASE VOCABULARY

POWER LEAD
(2407 POWER LEAD)
DA January 3, 1991
BT1 Equipment/Parts - Electrical (DOE
   FRASE Vocabulary)
 BT2 Equipment
SO DOE FRASE VOCABULARY

POWER LINE(S)
(2408 POWER LINE)
DA January 3, 1991
BT1 Equipment/Parts - Electrical (DOE
   FRASE Vocabulary)
 BT2 Equipment
SO DOE FRASE VOCABULARY

POWER LOSSES
(NFI)
DA October 12, 1990
BT1 Losses
RT Electrical Equipment
SO Nuclear Facilities Incident Database
SO Radiation
DEF (NFI) Use for onsite electrical
  power losses.

**POWER MARKETING
ADMINISTRATIONS**
DA May 15, 1991
SF *PMA*
NT1 Alaska Power Administration
NT1 Bonneville Power Administration
NT1 Southwestern Power Administration
NT1 Southeastern Power Administration
NT1 Western Area Power Administration
SO Management
DEF (The Secretary's Annual Report to
  Congress, 1989-90) DOE markets
  energy in three forms: electricity
  produced by Federal hydropower
  projects and acquired from thermal
  resources under the five power
  marketing administrations (PMA);
  enriched uranium produced for
  domestic and international
  customers; and crude oil,
  petroleum products, and natural
  gas produced from the Naval
  Petroleum Reserves. The five
  Power Marketing Administrations
  market the power generated at all
  federal multipurpose water
  projects except those under the
  jurisdiction of the Tennessee
  Valley Authority. To carry out their

responsibilities, the administrations contract for the sale and purchase of power and wheeling; develop rates; construct, operate, and maintain transmission lines, substations, switchyards, and attendant facilities; conduct appropriate energy conservation programs; and oversee environmental reviews.

POWER OPERATED RELIEF VALVE
DA January 8, 1991
SF *PORV*
BT1 Relief Valves
 BT2 Valves
  BT3 Devices

POWER OUTAGE
(1562 POWER OUTAGE)
DA November 29, 1990
BT1 Nature of Property Damage
RT Emergency Situations
RT Power Shutdown
SO DOE FRASE VOCABULARY

POWER POLE(S)
(2409 POWER POLE)
DA January 3, 1991
BT1 Equipment/Parts - Electrical (DOE
   FRASE Vocabulary)
 BT2 Equipment
SO DOE FRASE VOCABULARY

POWER POLISHER(S)
(2967 POWER POLISH)
DA January 3, 1991
BT1 Tools - Powered
 BT2 Tools (DOE FRASE Vocabulary)
  BT3 Equipment
SO DOE FRASE VOCABULARY

POWER RIVETER(S)
(2968 POWER RIVETE)
DA January 3, 1991
BT1 Tools - Powered
 BT2 Tools (DOE FRASE Vocabulary)
  BT3 Equipment
SO DOE FRASE VOCABULARY

POWER SANDBLASTER(S)
(2969 POWER SANDBL)
DA January 3, 1991
BT1 Tools - Powered
 BT2 Tools (DOE FRASE Vocabulary)
  BT3 Equipment
SO DOE FRASE VOCABULARY

POWER SANDER(S)
(2970 POWER SANDER)
DA January 3, 1991
SO DOE FRASE VOCABULARY

POWER SCREWDRIVER(S)
(2972 POWER SCREWD)
DA January 3, 1991
BT1 Tools - Powered
 BT2 Tools (DOE FRASE Vocabulary)

 BT3 Equipment
SO DOE FRASE VOCABULARY

POWER SHUTDOWN
(1554 POWER SHUTDO)
DA November 29, 1990
BT1 Nature of Property Damage
RT Power Outage
SO DOE FRASE VOCABULARY

POWER SPRAY GUN(S)
(2973 POWER SPRAY)
DA January 3, 1991
BT1 Tools - Powered
 BT2 Tools (DOE FRASE Vocabulary)
  BT3 Equipment
SO DOE FRASE VOCABULARY

POWER STAPLER(S)
(2975 POWER STAPLE)
DA January 3, 1991
BT1 Tools - Powered
 BT2 Tools (DOE FRASE Vocabulary)
  BT3 Equipment
SO DOE FRASE VOCABULARY

POWER SUPPLY
(2410 POWER SUPPLY)
DA January 3, 1991
BT1 Equipment/Parts - Electrical (DOE
   FRASE Vocabulary)
 BT2 Equipment
NT1 Uninterruptible Power Supplies
SO DOE FRASE VOCABULARY

POWER WAXER(S)
(2976 POWER WAXER)
DA January 3, 1991
BT1 Tools - Powered
 BT2 Tools (DOE FRASE Vocabulary)
  BT3 Equipment
SO DOE FRASE VOCABULARY

POWERED HANDTOOL(S)
(2977 POWERED HT)
DA January 3, 1991
BT1 Tools - Powered
 BT2 Tools (DOE FRASE Vocabulary)
  BT3 Equipment
SO DOE FRASE VOCABULARY

PP
DA October 12, 1990
SEE Plenum Pressure
SO Acronyms

PPC
DA October 12, 1990
SEE Project Planning and Control
SO Acronyms

PPL
DA October 12, 1990
SEE Priority Problem List
SO Acronyms

PPM/PPB
(EPA)
DA October 12, 1990
SY parts per million
BT1 Units of Measure
SO Environmental Protection Agency
   Glossary
DEF (EPA) Parts per million/parts per
   billion, a way of expressing tiny
   concentrations of pollutants in air,
   water, soil, human tissue, food, or
   other products.

PRA
DA October 12, 1990
SEE Probabilistic Risk Assessment
SO Acronyms

PRACTICABLE
(CAA; RHA; CFR)
DA November 15, 1990
SO Water Pollution
DEF (CFR) Available and capable of
   being done after taking into
   consideration cost, existing
   technology, and logistics in light of
   overall project purposes.

PRE-MANUFACTURE NOTICES
DA January 24, 1991
BT1 Section 5 Notices
NT1 Consolidated Pre-Manufacture
   Notices
NT1 Intermediate Pre-Manufacture
   Notices
RT Process for Commercial Purposes
SO Environmental Management
DEF (CFR) A notice submitted by
   persons who intend to
   manufacture a new chemical
   substance in the United States for
   commercial purposes must submit
   a notice unless the substance is
   excluded under 720.30. If a person
   contracts with a manufacturer to
   manufacture or produce a new
   chemical substance, and (i) the
   manufacturer manufactures or
   produces the substance
   exclusively for that person, and (ii)
   that person specifies the identity of
   the substance, and controls the
   total amount produced and the
   basic technology for the plant
   process; that person must submit
   the notice. If it is unclear who must
   report, EPA should be contacted
   to determine who must submit the
   notice. Only manufacturers that
   are incorporated, licensed, or
   doing business in the United
   States may submit a notice.

**PRE START-UP/ CALIBRATION
ACTIVITY**
(1239 PSC ACTIVITY)
DA November 28, 1990
BT1 Activity Types (DOE FRASE
   Vocabulary)

BT-Broader Term   NT-Narrower Term   RT Related Term

BT2 Activities
RT Start-Up Procedure
SO DOE FRASE VOCABULARY

PREAMPLIFIER(S)
(2411 PREAMPLIFIER)
DA January 3, 1991
BT1 Equipment/Parts - Electrical (DOE
    FRASE Vocabulary)
BT2 Equipment
SO DOE FRASE VOCABULARY

PRECIOUS METALS
DA January 24, 1991
BT1 Heavy Metals
SO Environmental Management
DEF (DOE 1540.1 5.L) Uncommon and
    highly valuable metals
    characterized by their superior
    resistance to corrosion and
    oxidation. Included are gold, silver,
    and the platinum group metals -
    platinum, palladium, rhodium,
    iridium, ruthenium and osmium.

PRECIPITATES
(EPA)
DA October 12, 1990
BT1 Solids
BT2 Materials
RT Coagulation
RT Flocculation
SO Environmental Protection Agency
    Glossary
DEF (EPA) Solids that separate from a
    solution because of some
    chemical or physical change.

PRECIPITATION
(EPA)
DA November 8, 1990
BT1 Chemical Processes
BT2 Processes
BT1 Waste Management Processes
BT2 Processes
DEF (EPA) Removal of solids from liquid
    waste so that the hazardous solid
    portion can be disposed of safely;
    removal of particles from airborne
    emissions.

PRECIPITATORS
(EPA)
DA October 12, 1990
BT1 Control Devices
BT2 Devices
NT1 Electrostatic Precipitators
SO Environmental Protection Agency
    Glossary
DEF Air pollution control devices that
    collect particles from an emission.

PRECISION
(ESH)
DA October 12, 1990
BT1 Measurements
SO Quality Assurance
DEF (ESH) The measurement of
    agreement of a set of replicate

results among themselves without
assumption of any prior
information as to the true result.
Precision is assessed by means of
duplicate/replicate sample
analysis.

PRECISION/PRODUCTION
PERSONNEL
DA November 29, 1990
BT1 Occupations
BT1 Personnel
NT1 Machinist
NT1 Machine Setup/Operator
NT1 Misc Precision/Production
    Employee
NT1 Operator, Plant/System/Utility
NT1 Sheet Metal Worker
NT1 Welder/Solderer
SO DOE FRASE VOCABULARY

PRECURSORS
(EPA)
DA October 12, 1990
SO Environmental Protection Agency
    Glossary
DEF (EPA) In photochemical
    terminology, compounds such as
    a volatile organic compound
    (VOC) that "precede" an oxidant.
    Precursors react in sunlight to
    form ozone or other
    photochemical oxidants.

PREDICTIVE MAINTENANCE
(DOE Order 4330.4A)
DA June 5, 1991
BT1 Maintenance
BT2 Activities
RT Analyses
RT Failures
RT Inspections
RT Monitoring
SO Management
DEF (DOE Order 4330.4A) Actions
    necessary to monitor, find trends,
    and analyze the parameters,
    performance characteristics,
    properties, and signatures
    associated with equipment,
    systems, or facilities that are
    indicative of decreasing
    performance or impending failure.

PREFERRED HIGHWAY
(ESH)
DA October 19, 1990
SY Preferred Routes
BT1 Routes (Transportation)
RT Roadways
SO Hazardous Materials
DEF (ESH) A highway for shipment of
    highway route controlled quantities
    of radioactive materials so
    designated by a State routing
    agency, and any Interstate System
    highway for which an alternative
    highway has not been designated
    by such State Agency.

PREFERRED ROUTES
(ESH)
DA October 19, 1990
SY Preferred Highway
BT1 Routes (Transportation)
NT1 State-Designated Routes
RT Roadways
RT State Routing Agencies
SO Hazardous Materials
DEF (ESH) A highway for shipment of
    highway route controlled quantities
    of radioactive materials so
    designated by a State routing
    agency, and any Interstate System
    highway for which an alternative
    highway has not been designated
    by such State Agency.

PRELIMINARY ASSESSMENT
(EPA)
DA October 12, 1990
SF PA (Preliminary Assessment)
BT1 Assessments
BT2 Administrative Processes
BT3 Processes
SO Environmental Protection Agency
    Glossary
DEF (EPA) The process of collecting
    and reviewing available
    information about a known or
    suspected waste site or release.

PRELIMINARY BIOLOGICAL OPINION
(ESA; CFR)
DA October 12, 1990
BT1 Biological Opinion
BT2 Reports
SO Endangered Species
DEF (CFR) An opinion issued as a result
    of early consultation.

PRELIMINARY DESIGNS
DA January 24, 1991
RT Conceptual Designs
RT Project Design Criteria
RT Technical Support
SO Environmental Management
DEF (DOE 4700.1) Continues the design
    effort utilizing the conceptual
    design and the project design
    criteria as a basis for project
    development. Title I design
    develops topographical and
    subsurface data and determines
    the requirements and criteria
    which will govern the definitive
    design. Tasks include preparation
    of preliminary planning and
    engineering studies preliminary
    drawings and outline
    specifications, life-cycle cost
    analysis, preliminary costs
    estimates, and scheduling for
    project completion. Preliminary
    design provides identification of
    long lead procurement items and
    detailed description of the services
    provided during preliminary
    design.

SY-Synonymous Terms   SO-Source/Subject Category   SF-See From

PRELIMINARY SAFETY ANALYSIS
(SSDC)
DA   October 12, 1990
SF   *PSA*
BT1 Safety Analysis
 BT2 Analyses
RT   Preliminary Safety Analysis Report
SO   System Safety Development Center
       Glossary
DEF (SSDC) Any safety analysis that is
       not based on final design
       information.

**PRELIMINARY SAFETY ANALYSIS
REPORT**
(DOE Order 6430.1A)
DA   October 12, 1990
BT1 Safety Analysis Reports
 BT2 Reports
RT   Final Safety Analysis Report
RT   Preliminary Safety Analysis
SO   Construction
DEF (DOE Order 6430.1A) Any safety
       analysis that is not based on final
       design information.

PREPAREDNESS
(EMER)
DA   February 1, 1991
SO   Emergency Preparedness
DEF (EMER) The training of personnel,
       acquisition of resources and
       facilities, and testing of emergency
       plans and procedures to ensure
       an effective response.

PREPP
DA   October 12, 1990
SEE Process Experimental Pilot Plant
SO   Acronyms

PRESCRIBED LIMITS
(IAEA)
DA   October 12, 1990
SY   Authorized Limits
BT1 Limits
SO   Radiation
DEF (IAEA) Limits of any quantity
       specified by the competent
       authority for a given radiation
       practice or source. These are
       generally lower than the primary,
       secondary, or derived limits.

PRESS
(2134 PRESS;N)
DA   December 10, 1990
BT1 Machines (DOE FRASE
       Vocabulary)
 BT2 Equipment
NT1 Isostatic Press
RT   Printing Machine
SO   DOE FRASE VOCABULARY

PRESSURE
DA   February 26, 1991
NT1 Differential Pressure
NT1 End Fitting Delta Pressure
NT1 High Delta Pressure

NT1 Maximum Normal Operating
       Pressure
NT1 Maximal Effective Pressure
NT1 Overpressure
NT1 Peak Positive Incident Pressure
NT1 Plenum Pressure
NT1 Pressure (Surface)
 NT2 Standard Pressure
RT   Measurements
RT   Units of Measure
SO   Environmental Management
SO   Industrial Safety
DEF (WEBSTER) A measure of the
       application of force brought to
       bear on an object.

PRESSURE (SURFACE)
(SDWA; CFR)
DA   October 12, 1990
BT1 Pressure
NT1 Standard Pressure
RT   Atmosphere
RT   In Vacuum Service
RT   Pounds per Square Inch
RT   Pounds per Square Inch Gauge
SO   Water Pollution
DEF (CFR) The total load or force per
       unit area acting on a surface.

PRESSURE REGULATOR VALVE
DA   January 8, 1991
SF   *PRV*
BT1 Valves
 BT2 Devices

PRESSURE RELEASES
(CAA; CFR)
DA   October 12, 1990
BT1 Sudden Accidental Occurrences
 BT2 Accidental Occurrences
  BT3 Occurrences
NT1 Sudden Releases of Pressure
RT   Relief Valves
SO   Air Pollution
DEF (CFR) The emission of materials
       resulting from the system pressure
       being greater than the set
       pressure of the pressure relief
       device.

PRESSURE RELIEF SYSTEM
(2041 PRESSURE REL)
DA   December 10, 1990
BT1 Systems (DOE FRASE Vocabulary)
 BT2 Systems
SO   DOE FRASE VOCABULARY

PRESSURE SEWERS
(EPA)
DA   October 12, 1990
BT1 Sewers
SO   Environmental Protection Agency
       Glossary
DEF (EPA) A system of pipes in which
       water, wastewater, or other liquid
       is transported to a higher elevation
       by use of pumping force.

PRESSURE VESSEL(S)
(2586 PRESSURE VES)
DA   January 3, 1991
BT1 Equipment/Parts - Nuclear (DOE
       FRASE Vocabulary)
 BT2 Equipment
 BT2 Reactor Components
RT   Pressure Vessel Part
RT   Reactor(s)
SO   DOE FRASE VOCABULARY
DEF (NRC Glossary of Terms: Nuclear
       Power and Radiation) A
       strong-walled container housing
       the core of most types of power
       reactors; it usually also contains
       the moderator, neutron reflector,
       thermal shield and control rods.

PRESSURE VESSEL PART
(2585 PRESSURE VP)
DA   January 3, 1991
BT1 Equipment/Parts - Nuclear (DOE
       FRASE Vocabulary)
 BT2 Equipment
 BT2 Reactor Components
RT   Pressure Vessel(s)
SO   DOE FRASE VOCABULARY

PRESSURIZED WATER REACTOR
DA   January 8, 1991
SF   *PWR*
BT1 Reactors
DEF (NRC Glossary of Terms: Nuclear
       Power and Radiation) A power
       reactor in which heat is transferred
       from the core to a heat exchanger
       by high-temperature water kept
       under high pressure in the primary
       system. Steam is generated in a
       secondary circuit. Many reactors
       producing electric power are
       pressurized water reactors.

PRETREATMENT
(EPA)
DA   October 12, 1990
BT1 Wastewater Treatment Processes
 BT2 Treatment
  BT3 Waste Management Processes
   BT4 Processes
SO   Environmental Management
SO   Environmental Protection Agency
       Glossary
DEF (EPA) Processes used to reduce,
       eliminate, or alter the nature of
       wastewater pollutants from
       non-domestic sources before they
       are discharged into publicly owned
       treatment works.

PRETREATMENT FACILITIES
(EPA)
DA   May 20, 1991
BT1 Facilities
RT   Publicly Owned Treatment Works
RT   Wastewater
SO   Water Pollution
DEF (EPA) Water treatment facilities
       which reduce, eliminate, or alter

BT-Broader Term   NT-Narrower Term   RT-Related Term the nature of wastewater pollutants from non-domestic sources before they are discharged into publicly owned treatment works.

PREVENTION OF SIGNIFICANT DETERIORATION
(EPA)
DA   October 12, 1990
SY   PSD Permits
SF   *PSD*
RT   Significant Deterioration
SO   Environmental Protection Agency Glossary
DEF (EPA) EPA program in which state and/or federal permits are required that are intended to restrict emissions for new or modified sources in places where air quality is already better than required to meet primary and secondary ambient air quality standards.

PREVENTION OF SIGNIFICANT DETERIORATION REGULATIONS
(CAA; ESH)
DA   October 12, 1990
BT1 Code of Federal Regulations
BT2 Codes, Standards, and Regulations
BT2 Statutes and Regulations
SO   Air Pollution
DEF (ESH) Regulations promulgated by EPA governing source construction and modification in attainment areas to satisfy the nondegradation philosophy of the Clean Air Act (Section 160-169).

PREVENTIVE MAINTENANCE
DA   January 8, 1991
SF   *PM*
BT1 Maintenance
BT2 Activities

PRIMARY BODY WAVES
(USGS)
DA   October 12, 1990
SY   Compressional Waves
SF   *P*
RT   Earthquake Magnitude
RT   Secondary Body Waves
SO   Natural Phenomenon
DEF (USGS) Waves that alternately push (compress) and pull (dilate) the material through which they travel. P waves, or compressional waves, are the first waves to cause vibration in a building.

PRIMARY COMPONENT COOLING
DA   January 8, 1991
SF   *PCC*

PRIMARY CONFINEMENT SYSTEMS
(DOE Order 6430.1A)
DA   October 12, 1990
BT1 Confinement Systems

BT2 Systems
NT1 Enclosures
SO   Construction
DEF (DOE Order 6430.1A) Areas having structures or systems from which releases of hazardous materials are controlled. The primary confinement systems are the process enclosures (glove boxes, conveyors, transfer boxes, other spaces normally containing hazardous materials), which are surrounded by one or more secondary confinement areas (operating area compartments).

PRIMARY COOLANTS
(DOE Order 5480.3; ESH)
DA   October 12, 1990
BT1 Coolants
SO   Hazardous Materials
DEF (ESH) Gases, liquids, or solids, or combinations of them, in contact with radioactive material, or, if the material is in special form, in contact with its capsule, and used to remove decay heat.

PRIMARY COPPER SMELTER
(CAA; CFR)
DA   October 12, 1990
BT1 Smelters
BT2 Facilities
RT   Copper Converters
RT   Process Emissions
SO   Air Pollution
DEF (CFR) Any installation or intermediate process engaged in the production of copper from copper-bearing materials through the use of pyrometallurgical techniques.

PRIMARY DOCUMENTS
(Doe Order 5400.4)
DA   October 12, 1990
BT1 Reports
NT1 Secondary Documents
RT   Remedial Investigations
RT   Remedial Actions
RT   Remedial Design
RT   Remedies
SO   Compensation and Liability
DEF (DOE Order 5400.4) Those reports that are major, discrete portions of a remedial investigation/feasibility study or remedial design/remedial action.

PRIMARY DRINKING WATER REGULATIONS
(SDWA; CFR; USC)
DA   October 12, 1990
BT1 Code of Federal Regulations
BT2 Codes, Standards, and Regulations
BT2 Statutes and Regulations
RT   Approved State Primacy Program
RT   Primary Enforcement Responsibility

RT   Public Water Systems
RT   Secondary Drinking Water Regulations
RT   State Primary Drinking Water Regulation
RT   Treatment Technique Requirements
SO   Environmental Management
SO   Environmental Protection Agency Glossary
SO   Water Pollution
DEF Applies to public water systems and specifies a contaminant level, which, in the judgement of the EPA Administrator, will have no adverse effect on human health. See 42 USCS 300(f) for complete definition.

PRIMARY EMERGENCY PLANS
(EMER)
DA   February 1, 1991
BT1 Emergency Plans
BT2 Plans
SO   Emergency Preparedness
DEF (EMER) Plans prepared by field offices to guide the response of field elements and contractors for major emergencies having departmental, national, or international implications.

PRIMARY EMISSION CONTROL SYSTEMS
(CAA; CFR)
DA   October 12, 1990
BT1 Control Devices
BT2 Devices
BT1 Control Systems
BT2 Controls
BT2 Systems
RT   Particulate Matter Emissions
SO   Air Pollution
DEF The hoods, ducts, and control devices used to capture, convey, and collect process emissions.

PRIMARY ENFORCEMENT RESPONSIBILITY
(CFR)
DA   October 19, 1990
RT   Primary Drinking Water Regulations
SO   Water Pollution
DEF (CFR) The primary responsibility for administration and enforcement of primary drinking water regulations and related requirements applicable to public water systems within a State.

PRIMARY ENVIRONMENTAL MONITORS
(DOE Order 5000.3A)
DA   December 14, 1990
BT1 Monitors
BT2 Equipment
RT   Class A Equipment
RT   Secondary Environmental Monitor
DEF (DOE Order 5000.3A) Monitoring equipment legally required to

monitor ongoing discharges. In general, this term applies to monitors closest to the point of discharge which are used to determine if discharges are within specified limits. It also includes any equipment that actuates automatically in response to set level signals from such a monitor. It does not include equipment in general area, remediation or compliance monitoring programs. Significant equipment in such programs will fit in the Class B Equipment definition.

PRIMARY LIGHT WATER ADDITION
DA January 8, 1991
SF *PLWA*

PRIMARY LIMITS
(IAEA)
DA November 20, 1990
BT1 Limits
SO Radiation
DEF (IAEA) Values of dose equivalent and/or effective dose equivalent applying to an individual. In the case of a member of the public the limit is taken to apply to the average dose in the critical group.

PRIMARY STANDARD ATTAINMENT DATE
(CAA; USC)
DA October 12, 1990
BT1 Time Designations
RT Effective Date
RT National Ambient Air Quality Standards
RT Schedule of Compliance
SO Air Pollution
DEF The date specified in the applicable implementation plan for the attainment of a national primary ambient air quality standard for any air pollutant.

PRIMARY STANDARDS
(CAA; CFR)
DA October 12, 1990
BT1 National Ambient Air Quality Standards
BT2 Ambient Air Quality Standards
BT3 Air Quality Standards
BT4 Standards
BT5 Codes, Standards, and Regulations
SO Air Pollution
DEF (CFR) National primary ambient air quality standards promulgated pursuant to section 109 of the Clean Air Act.

PRIMARY SYSTEM RELIEF
DA January 8, 1991
SF *PSR*

PRIMARY WASTE TREATMENT
(EPA)
DA October 12, 1990
BT1 Treatment
BT2 Waste Management Processes
BT3 Processes
BT1 Wastewater Treatment Processes
BT2 Treatment
BT3 Waste Management Processes
BT4 Processes
RT Sedimentation
SO Environmental Protection Agency Glossary
DEF (EPA) First steps in wastewater treatment; screens and sedimentation tanks are used to remove most material that floats or will settle. Primary treatment results in the removal of about 30 percent of carbonaceous biochemical oxygen demand from domestic sewage.

PRINCIPAL SOURCE AQUIFERS
(SDWA; CFR)
DA October 19, 1990
BT1 Aquifers
BT2 Formations
SO Water Pollution

PRINTING MACHINE
(2135 PRINTING MAC)
DA December 10, 1990
BT1 Machines (DOE FRASE Vocabulary)
BT2 Equipment
RT Press
SO DOE FRASE VOCABULARY

PRIORITY PROBLEM LIST
(SSDC)
DA October 12, 1990
SF *PPL*
RT Management Systems
SO System Safety Development Center Glossary
DEF (SSDC) A management system wherein problems and deficiencies within an organization are categorized as to need for attention and importance (based on input from the organizations line and staff managers). After categorization, the most money and time is spent on these items given the highest priority.

PRISIM
DA October 12, 1990
SEE Plant Risk Status Information Management System
SO Acronyms

PRIVATE APPLICATORS
(USC)
DA November 15, 1990
RT Certified Applicators
RT Commercial Applicators

SO Environmental Management
DEF (USC) Certified applicators who use or supervise the use of any pesticide that is classified for restricted use for purposes of producing any agricultural commodity on property owned or rented by them or their employer(s) or (if applied without compensation other than trading of personal services between producers of agricultural commodities) on the property of another person.

PRIVATE SIDING
(ESH)
DA October 19, 1990
SY Private Track
SO Hazardous Materials
DEF (ESH) Track located outside of a carrier's right-of-way, yard, or terminals where the carrier does not own the rails, ties, roadbed, or right-of-way and includes track or portion of track which is devoted to the purpose of its user either by lease or written agreement, in which case the lease or written agreement is considered equivalent to ownership.

PRIVATE TRACK
(ESH)
DA October 19, 1990
SY Private Siding
SO Hazardous Materials
DEF (ESH) Track located outside of a carrier's right-of-way, yard, or terminals where the carrier does not own the rails, ties, roadbed, or right-of-way and includes track or portion of track that is devoted to the purpose of its user either by lease or written agreement, in which case the lease or written agreement is considered equivalent to ownership.

PROBABILISTIC RISK ASSESSMENT
DA January 8, 1991
SF *PRA*
BT1 Risk Assessment
BT2 Assessments
BT3 Administrative Processes
BT4 Processes

PROBABILISTIC SAFETY STUDY
DA January 8, 1991
SF *PSS*
BT1 Studies

PROBABILITY
(SSDC)
DA October 12, 1990
NT1 Conditional Probability
RT Confidence
RT Cumulative Frequency (probability)
RT Probable Causes

RT Risk Assessment
RT Risks
SO System Safety Development Center
 Glossary
DEF (SSDC) A ratio of the number of
 times an event occurs divided by
 the number of times it is possible
 to occur. A probability of 1 means
 the event is certain. A probability
 of 0 means the event is
 impossible. A flipped coin has a
 probability of 0.5 of being heads
 (or tails).

PROBABLE CAUSES
(SSDC)
DA October 12, 1990
BT1 Causes
RT Probability
SO System Safety Development Center
 Glossary
DEF (SSDC) Causes attributed to an
 accident or incident as most likely
 in the absence of positive proof.

PROBABLE MAXIMUM FLOOD
DA January 8, 1991
SF *PMF*

PROCEDURES
(SSDC; MORT)
DA October 12, 1990
NT1 Bioassay Procedures
NT1 Emergency Procedures
NT1 Extraction Procedure Toxic
NT1 Operating Procedures
 NT2 Detailed Operating Procedures
 NT2 Emergency Operating Procedures
 NT2 Standard Operating Procedures
NT1 Operational Safety Procedures
NT1 Process Unit Shutdown
NT1 Standard Job Procedures
NT1 Standard Practice Procedures
NT1 Tagging
NT1 Toxicity Characteristic Leaching
 Procedure
RT Administrative Procedures Act
RT Contingency Plans
RT Daily Instruction Logs
RT Emergency Plans
RT EPA Guidance Document
RT Federal Radiological Emergency
 Response Plans
RT Field Facility/Building Emergency
 Plans
RT Field Site Emergency Plans
RT Maintenance Management
RT Noteworthy Practices
RT Operational Specifications
RT Organizations
RT Protection of the Public Health and
 Welfare
RT Regulations and Procedures
 Manual, PUB-201
RT Requirements
RT Sampling Plans
RT Spill Prevention Control and
 Countermeasures Plan
RT Temporary Procedure Change
RT Training

SO System Safety Development Center
 Glossary
DEF (SSDC) The written or established
 methods by which a company or
 organization operates to
 accomplish its objectives; a
 sequence of actions that
 collectively accomplish some
 desired task. (MORT) MORT
 analysis asks: do the procedures
 for each task meet selection and
 training criteria and applicable
 operating criteria? are the
 procedures responsive to
 supervisory problems that can be
 addressed in written procedures?
 Related issues/topics include:
 clarity and adequacy; accuracy;
 emergency provisions; cautions
 and warnings; event sequences;
 lockouts; and communication
 interfaces. Also, do work task
 completion procedures, as
 directed by oral or written
 instruction, agree with the actual
 requirements of the work task?

PROCEDURES FOR EVALUATING TECHNICAL SPECIFICATIONS PROGRAM
DA January 8, 1991
SF *PETS*
BT1 Programs

PROCESS (TOXIC CHEMICAL)
(CERCLA; CFR)
DA October 12, 1990
BT1 Chemical Processes
 BT2 Processes
NT1 Process for Commercial Purposes
RT By-products
RT Toxic Chemicals
SO Compensation and Liability
SO Environmental Management
SO Hazardous Materials
DEF (CFR) Preparation of a toxic
 chemical, after its manufacture, for
 distribution in commerce: (1) In
 the same form or physical state
 as, or in a different form of
 physical state from, that in which it
 was received by the person so
 preparing such substance; or (2)
 as part of an article containing the
 toxic chemical. Process also
 applies to the processing of a toxic
 chemical contained in a mixture or
 trade name product.

PROCESS EMISSIONS
(CAA;CFR)
DA October 12, 1990
BT1 Emissions
 BT2 Air Pollutants
RT Primary Copper Smelter
SO Air Pollution
DEF (CFR) Inorganic arsenic emissions
 from copper converters that are
 captured directly at the source of
 generation.

PROCESS EXPERIMENTAL PILOT PLANT
DA January 8, 1991
SF *PREPP*
BT1 Facilities

PROCESS FOR COMMERCIAL PURPOSES
(TSCA; CFR; ESH)
DA October 19, 1990
BT1 Process (Toxic Chemical)
 BT2 Chemical Processes
 BT3 Processes
RT Hazardous Materials
RT Pre-Manufacture Notices
RT Toxic Chemicals
SO Hazardous Materials
DEF The preparation of a chemical
 substance or mixture, after its
 manufacture, for distribution in
 commerce with the purpose of
 obtaining an immediate or
 eventual commercial advantage
 for the processor. Processing of
 any amount of a chemical
 substance or mixture is included. If
 a chemical substance or mixture
 containing impurities is processed
 or commercial purposes, then
 those impurities are also
 processed for commercial
 purposes.

PROCESS GAS
DA January 8, 1991
SF *PG*
BT1 Gases

PROCESS UNIT SHUTDOWN
(CAA; CFR)
DA October 12, 1990
BT1 Procedures
BT1 Shutdown (Emissions)
RT Process Units
RT Reactor Shutdown
SO Air Pollution
DEF (CFR) A work practice or
 operational procedure that stops
 production from a process unit or
 part of a process unit. An
 unscheduled work practice or
 operational procedure that stops
 production from a process unit or
 part of a process unit for less than
 24 hours is not a process unit
 shutdown. The use of spare
 equipment and technically feasible
 bypassing of equipment without
 stopping production are not
 process unit shutdowns.

PROCESS UNITS
(CAA; CFR)
DA October 12, 1990
BT1 Equipment
RT Process Unit Shutdown
RT Volatile Hazardous Air Pollutant
 (VHAP)

SY-Synonymous Terms   SO-Source/Subject Category   SF-See From

SO Air Pollution
DEF (CFR) Equipment assembled to produce a volatile hazardous air pollutant (VHAP) or its derivatives as intermediates or final products, or equipment assembled to use a VHAP in the production of a product. Process units can operate independently if supplied with sufficient feed or raw materials and sufficient product storage facilities.

PROCESS WASTE ASSESSMENT
DA   January 8, 1991
SF   *PWA*
BT1 Assessments
 BT2 Administrative Processes
  BT3 Processes
BT1 Measurements

PROCESS WASTE WATER
DA   January 24, 1991
BT1 Wastewater
 BT2 Wastes
 BT2 Water
RT Process Waste Water Pollutants
SO Environmental Management
DEF (CFR) Any water which, during manufacturing or processing, comes into direct contact with or results from the production or use of any raw material, intermediate product, finished product, by-product, or waste product.

PROCESS WASTE WATER POLLUTANTS
DA   January 24, 1991
BT1 Pollutants
RT Process Waste Water
SO Environmental Management
DEF (CFR) Pollutants present in process wastewater.

PROCESS WATER
DA   January 8, 1991
SF   *PW*
RT Automatic Incident Actions

PROCESS WATER GAMMA MONITOR
DA   January 8, 1991
SF   *PWGM*
BT1 Monitors
 BT2 Equipment
BT1 Radiation Detectors
 BT2 Equipment

PROCESS WEIGHT
(EPA)
DA   October 12, 1990
BT1 Measurements
SO Environmental Protection Agency Glossary
DEF (EPA) Total weight of all materials, including fuel, used in a manufacturing process. It is used to calculate the allowable particulate emission rate from the process.

PROCESSES
DA   February 25, 1991
NT1 Acclimatization
NT1 Adhesion
NT1 Administrative Processes
 NT2 Accreditation
 NT2 Activation (Emergency)
 NT2 Arbitration
 NT2 Assessments
  NT3 Biological Assessments
  NT3 Consequence Assessments
  NT3 Credibility Assessments
  NT3 Damage Assessments
   NT4 Natural Resource Damage Assessment
   NT4 Natural Resource Damage Preassessment Screens
  NT3 Data Quality Assessments
  NT3 Endangered Assessment
  NT3 Environmental Assessments
  NT3 Exposure Assessment
  NT3 Operation Assessment and Readiness
  NT3 Performance Assessments
  NT3 Preliminary Assessment
  NT3 Process Waste Assessment
  NT3 RCRA Facility Assessment
  NT3 Risk Assessment
   NT4 Probabilistic Risk Assessment
  NT3 Seabrook Station Probabilistic Safety Assessment
  NT3 Self-Assessment
  NT3 Systematic Assessment of Licensee Performance
  NT3 Tiger Team Assessments
   NT4 Environment Assessments
   NT4 Management and Organization Assessment
   NT4 Safety and Health Assessments
  NT3 Type A Assessments
  NT3 Type B Assessments
  NT3 Unified Dose Assessments
 NT2 Audits (SSDC)
  NT3 Environmental Audits
  NT3 Field Audits
  NT3 Internal Audits
  NT3 System Audits
 NT2 Classification
 NT2 Closure
  NT3 Current Closure Cost Estimate
  NT3 Current Post-Closure Cost Estimate
  NT3 Final Closure
  NT3 Partial Closure
  NT3 Site Closure and Stabilization
 NT2 Comprehensive Planning
 NT2 Construction Project Planning
 NT2 Coordination Processes
 NT2 Cost Recovery
 NT2 Critical Incident Technique
 NT2 Decommissioning
 NT2 Final Authorizations
 NT2 Formal Consultation
 NT2 General Design Process
 NT2 Hazards Identification
 NT2 Innovation Diffusion
 NT2 Inspections
  NT3 Compliance Inspections
  NT3 Condition Assessment Surveys
  NT3 Failed Instrument Component Inspection

  NT3 Maintenance Team Inspection
  NT3 Safety System Functional Inspection
  NT3 Safety System Outage Modification Inspection
  NT3 Site Inspections
 NT2 Inspections (Nuclear)
 NT2 Inspection and Evaluation
 NT2 Investigations
  NT3 Remedial Investigations
   NT4 RCRA Remedial Investigation
  NT3 Special Health Supervision
 NT2 Issue Identification
 NT2 Modification
  NT3 Major Modification
  NT3 Major PSD Modification
 NT2 Program Evaluation and Review Technique
 NT2 Project Planning and Control
 NT2 Reconstruction
 NT2 Release of Property
 NT2 Renovation
  NT3 Emergency Renovation Operations
 NT2 Reviews
  NT3 Environmental Reviews
  NT3 Operational Readiness Reviews
  NT3 Safety Reviews
   NT4 Backfit Reviews
   NT4 Independent (Safety) Reviews
  NT3 Safety Program Reviews
  NT3 Special Reviews
 NT2 Safety Analysis Process
 NT2 Siting
 NT2 State Coordination
 NT2 State Notification
 NT2 Surveillance
  NT3 Environmental Surveillance
 NT2 Test Authorizations
 NT2 Tiering
 NT2 Upstream Processes
 NT2 Validation
 NT2 Verification
  NT3 Calibration Checks
  NT3 Verification of Training and Retraining
 NT2 Work Processes
NT1 Agglomeration
NT1 Agglutination
NT1 Anticipated Processes and Events
NT1 Application (Pesticide)
 NT2 Band Application
 NT2 Basal Application
 NT2 Broadcast Application
NT1 Atomize
NT1 Attenuation
NT1 Attrition
NT1 Benefaction
NT1 Biological Processes
 NT2 Adaptations
 NT2 Aeration
  NT3 Diffused Air Process
  NT3 Mechanical Aeration
 NT2 Bioassays
  NT3 Direct Bioassays
 NT2 Biological Magnification
 NT2 Biological Oxidation
 NT2 Cloning
 NT2 Composting
 NT2 Decomposition
 NT2 Denitrification

NT2 Destruction
NT2 Differentiation
NT2 Evapotranspiration
NT2 Genetic Engineering
NT2 Infiltration
NT2 Natural Selection
NT2 Osmosis
  NT3 Reverse Osmosis
NT2 Photosynthesis
NT2 Regeneration
NT2 Sterilization
NT2 Transpiration
NT1 Blowing
NT1 Cathodic Protection
NT1 Cementing
NT1 Channelization
NT1 Charging
NT1 Chemical Processes
NT2 Absorption (Chemical)
NT2 Corrosion
  NT3 External Corrosion
NT2 Dechlorination
NT2 Degradation
NT2 Desalinization
NT2 Fermentation
NT2 Hydrolysis
NT2 In Situ Volatization
NT2 In-Situ Extraction
NT2 Neutralization
NT2 Nitrification
NT2 Oil Fingerprinting
NT2 Oxidation
  NT3 Combustion
NT2 Precipitation
NT2 Process (Toxic Chemical)
  NT3 Process for Commercial
      Purposes
NT2 Pyrolysis
NT2 Reduction
NT2 Sorption
NT1 Chlorination
NT1 Construction
NT2 Physical Construction
NT2 Substantial Construction
NT1 Containment
NT2 Filtered Vented Containment
NT1 Criticality Excursions
NT1 Decontamination
NT1 Digestion
NT2 Anaerobic Digestion
NT1 DNA Hybridization
NT1 Dose-Response Evaluation
NT1 Electrodialysis
NT1 Enclosed Processes
NT1 Enrichment
NT1 Erosion
NT1 Eutrophication
NT2 Cultural Eutrophication
NT2 Senescence
NT1 Fogging
NT1 Gasification
NT1 Hazardous Waste Generation
NT1 In-Situ Vitrification
NT1 Injection
NT2 Underground Injection
NT1 Ionization
NT1 Irrigation
NT1 Leaching
NT1 Liquefaction
NT2 Ground Failures
  NT3 Differential Settlement

NT3 Flow Failures
NT3 Landslides
NT3 Lateral Spreads
NT1 Manufacturing Processes
NT2 Fabricating
NT2 Magneforming
NT2 Pouring
NT2 Producing (Special Nuclear
      Materials)
NT2 Roasting
NT2 Side-Stream Extraction
NT1 Migration
NT1 Plugging
NT1 Pollution Recovery Processes
NT2 Abatement
  NT3 Noise Pollution Abatement
NT2 Assimilation
NT2 Banking (Air Pollution)
NT2 Desulfurization
NT2 Filtration
  NT3 Diatomaceous Earth Filtration
  NT3 Direct Filtration
  NT3 Flow Sand Filtration
  NT3 Slow Sand Filtration
NT2 Limestone Scrubbing
NT2 Resource Recovery
  NT3 Reclamation
    NT4 Land Reclamation
    NT4 Refuse Reclamation
  NT3 Solid Waste Planning
NT2 Screening
NT1 Purification Processes
NT2 Disinfection
NT2 Distillation
NT2 Ethylene Dichloride Purification
NT2 Vinyl Chloride Purification
NT1 Reprocessing
NT1 Salt Water Intrusion
NT1 Sanitation
NT1 Sinking
NT1 Skimming
NT1 Smear
NT1 Startup
NT2 Initial Startup
NT1 Strip
NT2 Removal
  NT3 Dredging
NT1 Suppression Pool Cooling
NT1 Sweeping the Gas Plenum
NT1 Transformation
NT1 Triage
NT1 Triple rinse
NT1 Ventilation
NT1 Waste Management
NT2 Cleanup Operations
NT2 Hazardous Waste Management
NT2 Radioactive Waste Management
NT2 Salvage
NT2 Solid Waste Management
  NT3 Compaction
  NT3 Refuse Reclamation
NT2 Waste Minimization
NT2 Waste Management Operations
NT1 Waste Management Processes
NT2 Coagulation
NT2 Comminution
NT2 Deposition (Process)
  NT3 Sedimentation
NT2 Disposal
  NT3 Continuous Disposal
  NT3 Land Disposal

    NT4 Near Surface Disposal
      NT5 Shallow Land Burial
  NT3 Land Farming (of waste)
  NT3 Solid Waste Disposal
    NT4 Phased Disposal
  NT3 Water Dumping
NT2 Precipitation
NT2 Recycling
  NT3 Closed-Loop Recycling
NT2 Removal
  NT3 Dredging
NT2 Retrofill
NT2 Screening
NT2 Storage
  NT3 Burial Operations
    NT4 Shallow Land Burial
  NT3 Contaminated Water Storage
  NT3 Vertical Tube Storage
NT2 Treatment
  NT3 Biological Treatment
    NT4 Aerobic Treatment
  NT3 Chemical Treatments
  NT3 Conventional Filtration Treatment
  NT3 Granular Activated Carbon
      Treatment
  NT3 In-Situ Treatment
  NT3 Incineration
    NT4 Incineration at Sea
  NT3 Ion Exchange Treatment
  NT3 Primary Waste Treatment
  NT3 Secondary Treatment
  NT3 Stabilization
  NT3 Thermal Treatment
  NT3 Wastewater Treatment Processes
    NT4 Absorption (Waste)
    NT4 Advanced Waste Water
        Treatment
    NT4 Clarification
    NT4 Diatomaceous Earth Filtration
    NT4 Flocculation
    NT4 Physical and Chemical
        Treatments
    NT4 Pretreatment
    NT4 Primary Waste Treatment
    NT4 Secondary Treatment
    NT4 Solidification and Stabilization
    NT4 Tertiary Treatment
NT2 Treatment, Storage, and Disposal
NT2 Trenching
NT2 Waste Minimization
NT1 Well Stimulation
RT Activities
RT Graded Approach

PROCUREMENT ITEMS
(SWDA; RCRA; USC)
DA October 12, 1990
RT Major System Acquisition
RT Procuring Agencies
SO Environmental Management
SO Wastes
DEF Any devices, goods, substances,
    materials, products, or other items
    whether real or personal property
    that are the subject of any
    purchase, barter, or other
    exchange made to procure such
    items.

SY-Synonymous Terms   SO-Source/Subject Category   SF-See From

PROCURING AGENCIES
(SWDA; RCRA; USC)
DA   October 12, 1990
BT1 Federal Agencies
 BT2 Agencies
  BT3 Administrative Organizations
  BT4 Organizations
BT1 State Agencies
 BT2 Agencies
  BT3 Administrative Organizations
  BT4 Organizations
RT   Major System Acquisition
RT   Procurement Items
SO   Environmental Management
SO   Wastes
DEF Any Federal agencies, or any State
    agencies or agencies of a political
    subdivision of a State that are
    using appropriated Federal funds
    for such procurement, or any
    people contracting with any such
    agencies with respect to work
    performed under such contracts.

PRODUCING (SPECIAL NUCLEAR MATERIALS)
(AEA; ANL)
DA   May 24, 1991
BT1 Manufacturing Processes
 BT2 Processes
RT   Special Nuclear Materials
SO   Radiation
DEF (AEA) When used in relation to
    special nuclear material, (1)to
    manufacture, make, produce, or
    refine special nuclear material; (2)
    to separate special nuclear
    material from other substances in
    which such material may be
    contained; or (3) to make or to
    produce new special nuclear
    material.

PRODUCT ACCUMULATOR VESSELS
(CAA; CFR)
DA   October 12, 1990
RT   Distillation
RT   In VHAP Service
SO   Air Pollution
DEF (CFR) Any distillate receivers,
    bottoms receivers, surge control
    vessels, or product separators in
    volatile hazardous air pollutant
    (VHAP) service that are vented to
    the atmosphere either directly or
    through a vacuum-producing
    system. Product accumulator
    vessels are in VHAP service if the
    liquids or the vapors in the vessel
    are at least 10 percent by weight
    VHAP.

PRODUCT LIABILITY
(SSDC)
DA   October 12, 1990
BT1 Liabilities
RT   Compensation and Liability
SO   System Safety Development Center
     Glossary
DEF (SSDC) Liability arising out of

manufactured goods after they
leave the premises. Applies to the
manufacturer and those who
handle or distribute the goods.

PRODUCTION FACILITIES
(AEA; ANL)
DA   May 24, 1991
BT1 Facilities
RT   Devices
RT   Equipment
RT   Special Nuclear Materials
SO   Management
SO   Radiation
DEF (AEA) (1) Any equipment or
    devices determined by rule of the
    Commission to be capable of the
    production of special nuclear
    material in such quantity as to be
    of significance to the common
    defense and security, or in such
    manner as to affect the health and
    safety of the public; or (2) any
    important component part
    especially designed for such
    equipment or device as
    determined by the Commission.

PRODUCTION REACTOR FACILITY
(1804 PROD REACTOR)
DA   December 10, 1990
BT1 Facility (DOE FRASE Vocabulary)
 BT2 Facilities and Buildings (DOE
      FRASE Vocabulary)
  BT3 Facilities
BT1 Reactor Facilities
 BT2 Nuclear Facilities
  BT3 Facilities
RT   Nuclear Facility
SO   DOE FRASE VOCABULARY

PRODUCTION/OPERATION ACTIVITY
(1240 PO ACTIVITY)
DA   November 28, 1990
BT1 Activity Types (DOE FRASE
     Vocabulary)
 BT2 Activities
RT   Production/Operations Area
SO   DOE FRASE VOCABULARY

PRODUCTION/OPERATIONS AREA
(1710 P;O AREA)
DA   December 10, 1990
BT1 Area
 BT2 Sites/Areas
RT   Production/Operation Activity
SO   DOE FRASE VOCABULARY

PROFESSIONAL BODIES
DA   February 25, 1991
BT1 Organizations
NT1 American Society of Mechanical
    Engineers
NT1 Atomic Industrial Forum
NT1 Conference of Federal
    Environmental Engineers
NT1 Council on Environmental Quality
NT1 National Fire Protection Association
NT1 Natural Resource Defense Council

PROFESSIONAL PERSONNEL
DA   November 29, 1990
BT1 Occupations
BT1 Personnel
NT1 Doctor/Nurse
NT1 Engineer
NT1 Misc Professional
NT1 Scientist
 NT2 Health Physicist
NT1 Technicians
 NT2 Engineering Technician
 NT2 Health Technician
 NT2 Miscellaneous Technician
 NT2 Radiation Monitor/Technician
 NT2 Science Technician
SO   DOE FRASE VOCABULARY

PROGRAM
(ESA; CFR)
DA   October 12, 1990
RT   Programs
SO   Endangered Species
SO   Management
DEF (CFR) A State-developed plan for
    the conservation and management
    of all resident species which are
    deemed by the Secretary to be
    endangered or threatened and
    those which are deemed by the
    State to be endangered or
    threatened, which includes goals,
    priorities, strategies, actions, and
    funding necessary to accomplish
    the objectives on an individual
    species basis.

PROGRAM DEVELOPMENT COMPUTER
DA   January 8, 1991
SF   PDC
RT   Programs

PROGRAM ENRICHMENT OFFICE
DA   January 8, 1991
SF   PEO
BT1 Offices
 BT2 Administrative Organizations
  BT3 Organizations
RT   Programs

PROGRAM EVALUATION AND REVIEW TECHNIQUE
(SSDC)
DA   October 12, 1990
SF   PERT
BT1 Administrative Processes
 BT2 Processes
RT   Management Systems
RT   Programs
SO   System Safety Development Center
     Glossary
DEF (SSDC) A management system,
    originally developed on the U.S.
    Navy's Polaris Weapons Systems,
    used to analyze and control the
    timing aspects of a major project.
    In PERT, events leading to the
    accomplishment of a task or
    project are indicated by circles on

a chart, with arrows showing the sequence and time between events. The critical path –the longest sequence– may be color-indicated on the chart.

PROGRAM MANAGERS
(DOE Order 5000.3A)
DA   December 14, 1990
BT1 Personnel
RT   Management
RT   Program Senior Officials
RT   Programs
DEF (DOE Order 5000.3A) The DOE Headquarters (HQ) individual, or his or her designee designated by and under the direction of a Program Senior Official, who is directly involved in the operation of facilities under his or her cognizance, and with signature authority to provide technical direction through DOE Field Organizations to contractors for these facilities.

PROGRAM OFFICES
(DOE Orders 5440.1D, 5480.4 and 5483.1A; ESH)
DA   October 12, 1990
BT1 Offices
 BT2 Administrative Organizations
 BT3 Organizations
BT1 U.S. Department of Energy
 BT2 Federal Agencies
 BT3 Agencies
 BT4 Administrative Organizations
 BT5 Organizations
NT1 Office of the Assistant Secretary for Nuclear Energy
NT1 Office of the Assistant Secretary for Fossil Energy
NT1 Office of the Assistant Secretary for Defense Programs
NT1 Office of the Assistant Secretary for Conservation et.al.
NT1 Office of the Assistant Secretary for Environment, et. al.
 NT2 Office of Special Projects
NT1 Office of Environmental Restoration and Waste Management
NT1 Office of New Production Reactors
NT1 Office of Civilian Radioactive Waste Management
NT1 Office of Energy Research
 NT2 Office of Basic Energy Sciences
 NT2 Office of Health and Environmental Research
RT   Field Organizations
RT   Field Elements
RT   Management
RT   Operations Offices
RT   Programs
SO   Industrial Safety
SO   Management
SO   Standards
DEF A Headquarters organization that is responsible for assisting and supporting field organizations in safety and health, administrative,

management, and technical areas and reports to the cognizant program Secretarial Officer.

PROGRAM ORGANIZATIONS
(DOE Order 5440.1C; ESH)
DA   October 19, 1990
SY   Regulatory Organizations
BT1 Administrative Organizations
 BT2 Organizations
RT   Federal Agencies
RT   Programs
SO   Management
DEF Organizations (Assistant Secretary, Administrator or Director level) responsible for the decision making and implementation of the Department's programmatic or regulatory action requiring a National Environmental Policy Act review.

PROGRAM SECRETARIAL OFFICERS
(DOE Orders 5000.3, 5480.2, and 5480.4; ESH)
DA   October 12, 1990
SY   Program Senior Officials
SF   *PSO (Program Secretarial Officer)*
BT1 Personnel
RT   Line Organizations
RT   Management
SO   Environmental Management
SO   Industrial Safety
SO   Management
DEF Outlay program managers who include: the Assistant Secretaries for Conservation and Renewable Energy; Fossil Energy; Nuclear Energy; Policy, Safety, and Environment; Defense Programs; and the Directors of Energy Research and Civilian Radioactive Waste Management. For purposes of this Order, this definition also includes the Administrators of the Bonneville Power and Western Area Power Administrations.

PROGRAM SENIOR OFFICIALS
(DOE Orders 5400.1, 5480.1, 5481.1B, 5482.1B, and 5700.6B; ESH; EMER)
DA   October 12, 1990
SY   Heads of Headquarters Elements
SY   Program Secretarial Officers
SF   *PSO (Program Senior Official)*
BT1 Personnel
NT1 Secretarial Officers
RT   Landlords
RT   Line Organizations
RT   Management
RT   Program Managers
SO   Construction
SO   Emergency Preparedness
SO   Environmental Management
SO   Industrial Hygiene
SO   Management
DEF (EMER) Senior outlay program managers; includes the Assistant Secretaries for Conservation and Renewable Energy; Fossil Energy;

Nuclear Energy: Defense Programs; and the Directors of Energy Research, Civilian Radioactive Waste Management, New Production Reactors, and Environmental Restoration and Waste Management. May also include the Administrators of the Bonneville and Western Area Power Administrations.

PROGRAM SIGNIFICANT COST
(DOE Order 5000.3A)
DA   December 14, 1990
BT1 Costs
RT   Programs
DEF (DOE Order 5000.3A) Meets the criteria of DOE Order 5000.3A's Category 7, Value Basis Reporting: Any event specifying cost as a basis for reporting, unless otherwise stated, will be classified by the monetary values necessary to repair, replace, or otherwise restore a facility/system/component to acceptable operation. Costs used for reporting should be reasonable initial estimates.

PROGRAM SIGNIFICANT DELAY
(DOE Order 5000.3A)
DA   December 14, 1990
BT1 Delays
RT   Programs
RT   Unusual Occurrences
DEF (DOE Order 5000.3) Meets the criteria of DOE Order 5000.3A's Unusual Occurrence Category 8, Facility Status: any unplanned occurrence in any portion of a program conducted in accordance with approved requirements and procedures that results in facility, process, etc. securing or significantly curtailing operations.

PROGRAMMATIC EQUIPMENT
(DOE Order 4330.4A)
DA   June 4, 1991
BT1 Equipment
SO   Management
DEF (DOE Order 4330.4A) Equipment (both real and personal) dedicated for a specific programmatic use. Examples are accelerators, microscopes, radiation detection equipment, gloveboxes, and hotcells.

PROGRAMMATIC IMPACT
(1583 PROGRAMMATIC)
DA   November 28, 1990
SO   DOE FRASE VOCABULARY

PROGRAMMATIC NEPA DOCUMENTS
(DOE Order 5440.1D)
DA   May 15, 1991

BT1 National Environmental Policy Act
Documents
BT2 Reports
RT Environmental Assessments
RT Environmental Impact Statements
RT Finding of No Significant Impact
RT Notice of Intent
RT Record of Decision (NEPA)
SO Environmental Management
DEF (DOE Order) Broad-scope
Environmental Impact Statements
or Environmental Assessments
that identify and assess the
environmental impacts of a DOE
program; they may also refer to an
associated NEPA document such
as a Notice of Intent, record of
decision, or findings of no
significant impact.

PROGRAMS
DA February 4, 1991
NT1 Approved Programs
NT1 Approved State Primacy Program
NT2 State Program Revisions
NT1 Community Awareness and
Emergency Response Programs
NT1 Comprehensive Environmental
Assessment and Response
Program
NT1 Consultant Fire Protection Survey
Program
NT1 Contract Laboratory Program
NT1 Defense Facility Decommissioning
Program
NT1 DOE Programs
NT2 Defense Programs
NT2 Environment, Safety, and Health
(ES&H) Program
NT1 Emergency Planning,
Preparedness, and Response
Program
NT1 Federal Employee Occupational
Safety and Health Program
NT1 Government Industry Data
Exchange Program
NT1 Hazardous Waste Remedial
Actions Program
NT1 Industry Degraded Core
Rulemaking Program
NT1 Interim Reliability Evalution
Program
NT1 Nuclear Plant Aging Research
Program
NT1 Occupational Medical Program
NT1 Personnel Dosimetry Programs
NT1 Procedures for Evaluating
Technical Specifications Program
NT1 Radiological Assistance Programs
NT1 Reactor Safety Study Methodology
Application Program
NT1 Remedial Action Program
NT2 Formerly Utilized Sites Remedial
Actions Program
NT2 Uranium Mill Tailings Remedial
Action Program
NT1 Site Characterization
NT1 Superfund
NT1 Surplus Facilities Managment
Program

NT1 Systematic Evaluation Program
NT1 Tiger Team Assessment Program
NT1 Training Programs
NT2 Performance-Based Training
NT1 Underground Injection Control
NT1 Vital Programs
RT Metric Transition Plan
RT Program Offices
RT Program
RT Program Evaluation and Review
Technique
RT Program Managers
RT Program Significant Delay
RT Program Significant Cost
RT Program Development Computer
RT Program Enrichment Office
RT Program Organizations
RT Research and Special Programs
Administration
RT Training Accreditation Program
Staff
RT Training Program Accreditation
Plan

PROHIBIT SPECIFICATION
(CWA; RHA; CFR)
DA October 12, 1990
RT Deny
RT Disposal Sites
SO Water Pollution
DEF To prevent the designation of an
area as a present or future
disposal site.

PROJECT DESIGN CRITERIA
(DOE Order 6430.1A)
DA October 12, 1990
BT1 Specifications
BT2 Requirements
RT Conceptual Designs
RT Design and Development Plans
RT General Design Criteria
RT Guide Specifications
RT Preliminary Designs
SO Construction
DEF (DOE Order 6430.1A) Those
technical data and other project
information developed during the
project identification, conceptual
design and/or preliminary design
phases. They define the project
scope, construction features and
requirements, design parameters,
applicable design codes,
standards, and regulations;
applicable health, safety, fire
protection, safeguards, security,
energy conservation, and quality
assurance requirements; and
other requirements. The project
design criteria are normally
consolidated into a document that
provides the technical base for
any further design performed after
the criteria are developed.

PROJECT MANAGERS
(SSDC)
DA October 12, 1990
BT1 Personnel

NT1 Remedial Project Managers
RT Authorized Representatives
RT Matrix Organizations
RT Milestones
RT Projects
SO System Safety Development Center
Glossary
DEF (SSDC) The managers responsible
for a specific project assignment,
as distinguished from line
managers responsible for a
general functional area.

PROJECT PLANNING AND CONTROL
DA January 8, 1991
SF PPC
BT1 Administrative Processes
BT2 Processes
RT Milestones
RT Projects

PROJECT SEGMENT
(ESA; CFR)
DA October 12, 1990
RT On-Going Projects
RT Projects
SO Endangered Species
DEF (CFR) An essential part or a
division of a project, usually
separated as a period of time,
occasionally as a unit of work.

PROJECTED DOSE
(EMER)
DA February 1, 1991
BT1 Doses
SO Emergency Preparedness
DEF (EMER) An estimate of the
radiation dose that affected
individuals could receive.

PROJECTS
DA February 4, 1991
NT1 Capital Improvement Project
NT1 Construction Projects
NT2 Major Construction Activities
NT1 Emergency Projects
NT1 General Purpose Facilities Projects
NT1 General Plant Projects
NT1 On-Going Projects
NT1 Reactor Projects
RT Construction Project Planning
RT Department of Energy Project
Construction Contractors
RT Functional Units
RT Grand Junction Projects Office
RT Milestones
RT Project Segment
RT Project Managers
RT Project Planning and Control
RT Quality Assurance Project Plans

PROPELLANT PLANTS
(CAA;CFR)
DA October 12, 1990
BT1 Facilities
RT Propellants
SO Air Pollution
DEF (CFR) Any facilities engaged in the

mixing, casting, or machining of propellant.

PROPELLANTS
(CAA:CFR)
DA October 12, 1990
BT1 Fuels
NT1 Beryllium Propellants
RT Propellant Plants
SO Air Pollution
SO Environmental Management
DEF (CFR) Fuels and oxidizers physically or chemically combined which undergo combustion to provide rocket propulsion.

PROPER JOB ANALYSIS
(SSDC)
DA October 12, 1990
SF PJA
BT1 Analyses
RT Performance-Based Training
RT Standard Job Procedures
SO System Safety Development Center Glossary
DEF (SSDC) The breaking down of a job (task) into its component steps and the determination of downgrading incidents or problems and their controls associated with each step of that job. This is a tool to provide assurance that all important aspects of a job have been considered and evaluated in order to determine one unified procedure for doing the job the proper way.

PROPER JOB INSTRUCTION
(SSDC)
DA October 12, 1990
SF PJI
RT Performance-Based Training
SO System Safety Development Center Glossary
DEF (SSDC) An instructional technique employed by the supervisor when teaching a worker to do a new or different task for the first time, or when reviewing a task with an experienced worker. It involves telling, showing, testing and checking a worker to make sure he will perform the job properly.

PROPER SHIPPING NAME
(SWDA; RCRA; ESH)
DA October 12, 1990
SY Name of Contents
RT Marks
RT Technical Names
SO Hazardous Materials
DEF The name of the hazardous material.

PROPERTY
(Doe Order 5480.7)
DA October 12, 1990
NT1 Personal Property

NT2 Related Personal Property
NT1 Real Property
RT DOE Property
SO Fires
DEF (DOE Order 5480.7) All Government-owned or -leased property for which the Department has responsibility, except: (1) Property furnished under contract requiring contractor assumption of the risk of loss or damage to Government-furnished property. (2) Property covered by a private insurance policy specifying the Department of Energy as the beneficiary.

PROPERTY LOSS
(Doe Order 5480.7)
DA October 12, 1990
BT1 Losses
RT Actual Cash Value
SO Fires
DEF The dollar cost of restoring damaged facilities or equipment to their original condition, whether or not such restoration actually occurs. In determining loss, the estimated damage to the building and contents shall include replacement cost, less salvage value, plus the cost of decontamination and cleanup. Effects upon program continuity, auxiliary costs of fire extinguishment, and consequent effects on related areas should be included if the effects can be determined.

PROPERTY PROTECTION AREA
(DOE Order 6430.1A)
DA October 12, 1990
BT1 Sites/Areas
SO Construction
DEF (DOE Order 6430.1A) An area set aside for the protection of property as required by these criteria.

PROPOSED ACTIONS
(ESA; CFR)
DA October 12, 1990
BT1 Actions
BT2 Responses
RT Record of Decision (NEPA)
SO Endangered Species
DEF (CFR) Actions proposed by the Federal agency or by a permit or license applicant, for which exemption is sought.

PROPOSED CRITICAL HABITATS
(ESA; CFR)
DA October 12, 1990
BT1 Critical Habitats
BT2 Habitats
SO Endangered Species
DEF (CFR) Habitats proposed in the FEDERAL REGISTER to be

designated or revised as critical habitat under section 4 of the Endangered Species Act for any listed or proposed species.

PROPOSED SPECIES
(ESA; CFR)
DA October 12, 1990
RT Biological Assessments
RT Endangered Species
SO Endangered Species
DEF (CFR) Any species of fish, wildlife, or plant that is proposed in the FEDERAL REGISTER to be listed under section 4 of the Endangered Species Act.

PROTECT AS RESTRICTED DATA
(EMER)
DA February 1, 1991
SF PARD
RT Classified Information
SO Emergency Preparedness
DEF (EMER) The PARD designation is assigned to computer generated numerical data or related information for which it is not operationally feasible to establish a security classification because detailed knowledge of weapon design or other significant information is essential for determination or because of the high volume of output and low density of potentially classified data.

PROTECTED AREAS
(DOE Order 6430.1A)
DA October 12, 1990
BT1 Sites/Areas
RT Material Access Areas
RT Restricted Areas
RT Security Areas
SO Construction
DEF (DOE Order 6430.1A) Areas encompassed by physical barriers (e.g., walls or fences), subject to access controls, surrounding a material access area or containing Category II special nuclear material.

PROTECTION OF THE PUBLIC HEALTH AND WELFARE
(DOE Order 5480.7; ANL)
DA May 24, 1991
BT1 Activities
RT Amelioration
RT Barriers
RT Controls
RT Emergency Management Systems
RT Oversights and Omissions
RT Policies
RT Procedures
DEF (DOE 5480.7) Control of fire, explosion, or effects of hazards to minimize potential injury to the

public and damage to property not owned by the Department.

PROTECTION TECHNOLOGY IDAHO
DA   January 11, 1991
RT   Idaho Operations Office
RT   Idaho National Engineering Laboratory

PROTECTIVE ACTION GUIDES
(DOE Order 5400.5; EMER)
DA   October 12, 1990
SY   Emergency Reference Levels
SY   Intervention Levels
SF   PAG
BT1 Guidelines
RT   Doses
SO   Emergency Preparedness
SO   Wastes
DEF (DOE Order 5400.5) Projected numerical dose values established by EPA, DOE, or States for individuals in the population. These values may trigger protective actions that would reduce or avoid the projected dose.

PROTECTIVE ACTION RECOMMENDATIONS
(EMER)
DA   February 1, 1991
BT1 Recommendations
SO   Emergency Preparedness
DEF (EMER) Advice to states on emergency measures they should consider in determining action for the public to take to avoid or reduce their exposure to radiation or other hazardous materials.

PROTECTIVE ACTIONS
(EMER)
DA   February 1, 1991
BT1 Actions
 BT2 Responses
BT1 Amelioration
NT1 Denials
SO   Emergency Preparedness
DEF (EMER) Actions taken during an emergency for the purpose of preventing or minimizing hazards.

PROTECTIVE BARRIERS
(NIH)
DA   October 12, 1990
BT1 Barriers
RT   Radiation Safety
RT   Radiation Shields
RT   Shielding Materials
SO   Radiation
DEF (NIH) Barriers of radiation-absorbing material, such as lead, concrete, plaster, and plastic, that are used to reduce radiation exposure. Primary: Barriers sufficient to attenuate the useful beam to the required degree. Secondary: Barriers

sufficient to attenuate stray or scattered radiation to the required degree.

PROTECTIVE FORCE PERSONNEL
(Doe Order 5480.16)
DA   October 12, 1990
BT1 Personnel
NT1 Guards
NT1 Security Inspectors
NT1 Shadow Forces
NT1 Tactical Response Forces
RT   Armorers
RT   Authorized Weapons
RT   Central Training Academy
RT   Security Facilities
RT   Special Weapons
RT   Unauthorized Discharges
SO   Firearms
DEF (DOE Order 5480.16) Guards, security inspectors, couriers, authorized escorts, and personnel assigned to protective details, who are employed to protect the security interests of the Department.

PROTECTIVE MEASURES
(EMER)
DA   February 1, 1991
BT1 Amelioration
SO   Emergency Preparedness
DEF (EMER) Measures taken during an emergency for the purpose of preventing or minimizing hazards which are likely to develop if the actions were not taken.

PROTECTIVE RELAY SYSTEM
(2042 PROTECTIVE R)
DA   December 10, 1990
BT1 Systems (DOE FRASE Vocabulary)
 BT2 Systems
SO   DOE FRASE VOCABULARY

PROTEINS
(EPA)
DA   October 12, 1990
BT1 Organic Chemicals
 BT2 Chemical Substances
NT1 Antibodies
 NT2 Monoclonal Antibodies
NT1 Carboxyhemoglobin
NT1 Restriction Enzymes
SO   Environmental Protection Agency Glossary
DEF (EPA) Complex nitrogenous organic compounds of high molecular weight that contain amino acids as their basic unit and are essential for growth and repair of animal tissue. Many proteins are enzymes.

PROTON DECAY RADIOISOTOPES
(EDB)
DA   March 29, 1991
BT1 Radionuclides
 BT2 CERCLA Hazardous Substances

BT3 Hazardous Substances
BT2 Nuclides
SO   Radiation

PROTONS
(EMER)
DA   February 1, 1991
BT1 Charged Particles
SO   Emergency Preparedness
DEF (EMER) Elementary particles with a single positive electrical charge and a mass approximately 1837 times that of the electron. The nucleus of an ordinary or light hydrogen atom. Protons are constituents of all nuclei. The atomic number (Z) of an atom is equal to the number of protons in its nucleus.

PROTOPLASTS
(EPA)
DA   October 12, 1990
SO   Environmental Protection Agency Glossary
DEF (EPA) Membrane-bound cells from which the outer cell wall has been partially or completely removed. The term often is applied to plant cells.

PROXIMATE CAUSES
(SSDC)
DA   October 12, 1990
BT1 Causes
SO   System Safety Development Center Glossary
DEF (SSDC) The cause factors that directly produce the effect without the intervention of any other cause. The causes nearest to the effect in time and space.

PRP
DA   October 12, 1990
SEE Potentially Responsible Parties
SO   Acronyms

PRV
DA   October 12, 1990
SEE Pressure Regulator Valve
SO   Acronyms

PSA
DA   October 12, 1990
SEE Preliminary Safety Analysis
SO   Acronyms

PSB
DA   October 12, 1990
SEE Pump Shaft Break
SO   Acronyms

PSD
DA   October 12, 1990
SEE Prevention of Significant Deterioration
SO   Acronyms

PSD PERMITS
(SWDA; RCRA; CFR)
DA   October 19, 1990
SY   Prevention of Significant
     Deterioration
BT1 Permits
RT   Approved Programs
RT   State Director
SO   Wastes
DEF Permits issued under 40 CFR
     52.21 or by an approved State.

PSI
DA   October 12, 1990
SEE Pollutant Standard Index
SO   Acronyms

PSI (Pounds per Square Inch)
DA   January 16, 1991
SEE Pounds per Square Inch
SO   Acronyms

PSIA
DA   October 12, 1990
SEE Pounds per Square Inch Absolute
SO   Acronyms

PSIG
DA   October 12, 1990
SEE Pounds per Square Inch Gauge
SO   Acronyms

PSO (Program Secretarial Officer)
DA   October 12, 1990
SEE Program Secretarial Officers
SO   Acronyms

PSO (Program Senior Official)
DA   October 12, 1990
SEE Program Senior Officials
SO   Acronyms
SO   Environmental Management

PSR
DA   October 12, 1990
SEE Primary System Relief
SO   Acronyms

PSS
DA   October 12, 1990
SEE Probabilistic Safety Study
SO   Acronyms

PTP
DA   October 12, 1990
SEE Peak-To-Peak Ratio
SO   Acronyms

PUBLIC ADDRESS SYSTEM
(2043 PA SYSTEM)
DA   December 10, 1990
BT1 Systems (DOE FRASE Vocabulary)
 BT2 Systems
SO   DOE FRASE VOCABULARY

PUBLIC AFFAIRS (PERSONNEL)
(EMER)

DA   February 1, 1991
BT1 Personnel
NT1 Public Information Officers
RT   Risk Communication
SO   Emergency Preparedness
DEF (EMER) Personnel involved in
     summarizing, coordinating, and
     providing the release of
     information to the news media in
     an emergency situation.

PUBLIC AIRCRAFT
(DOE Order 5480.13)
DA   October 12, 1990
BT1 Aircraft
SO   Aviation Safety
DEF (DOE Order 5480.13) Aircraft used
     only in the service of a
     Government or political
     subdivision. This does not include
     any Government-owned aircraft
     engaged in carrying persons or
     property for commercial purposes.

PUBLIC DOSES
(DOE Order 5400.5)
DA   October 12, 1990
BT1 Doses
RT   Critical Group
RT   Members of the Public
SO   Industrial Hygiene
DEF (DOE Order 5400.5) The doses
     received by member(s) of the
     public from exposure to radiation
     and to radioactive material
     released by a DOE facility or
     operation, whether the exposure is
     within a DOE site boundary or
     off-site. It does not include dose
     received from occupational
     exposures, doses received from
     naturally occurring "background"
     radiation, doses received as a
     patient from medical practices, or
     doses received from consumer
     products.

PUBLIC HEARINGS
(ESA; CFR)
DA   October 12, 1990
RT   Informal Consultation
RT   Risk Communication
SO   Endangered Species
DEF (CFR) Informal hearings to provide
     the public with the opportunity to
     give comments and to permit an
     exchange of information and
     opinion on a proposed rule.

PUBLIC INFORMATION OFFICERS
(EMER)
DA   February 1, 1991
BT1 Public Affairs (Personnel)
 BT2 Personnel
RT   Relations
RT   Risk Communication
SO   Emergency Preparedness
DEF (EMER) Department of Energy
     officers at headquarters and in the

field responsible for preparing and
coordinating the dissemination of
public information in cooperation
with other responding federal,
state, and local agencies.

PUBLIC LANDS
(ARPA)
DA   April 24, 1991
BT1 Natural Resources
BT1 Sites/Areas
RT   Archaeological Resources
     Protection Act of 1979
RT   Archaeological Resources
RT   Indian Lands
SO   Environmental Management
DEF (ARPA) Lands which are: (A)
     owned and administered by the
     United States as part of 1) the
     national park system, 2) the
     national wildlife refuge system or
     3) the national forest system; and
     (B) all other lands the fee title to
     which is held by the United States,
     other than lands on the Outer
     Continental Shelf and lands which
     are under the jurisdiction of the
     Smithsonian Institution.

PUBLIC NOTICES
(CFR)
DA   May 15, 1991
SY   Notice of Involvement
SO   Management

PUBLIC TRAFFIC ROUTE DISTANCE
(DOE Order 5480.16)
DA   October 12, 1990
RT   Munitions
SO   Firearms
DEF (DOE Order 5480.16) The minimum
     distance permitted between
     locations containing munitions and
     any public street, road, highway,
     or passenger railroad (including
     roads on DOE-controlled land
     open to public travel).

PUBLIC TRAVEL ROUTES
(DOE Order 6430.1A)
DA   October 12, 1990
BT1 Routes (Transportation)
RT   Point of Nearest Public Access
RT   Roadways
SO   Construction
DEF (DOE Order 6430.1A) Any public
     streets, roads, highways, or
     passenger railroads (including
     roads on DOE-controlled land
     open to public travel).

PUBLIC VESSELS
(SWDA; RCRA; ESH)
DA   October 12, 1990
BT1 Vessels
SO   Hazardous Materials
DEF Vessels owned by and being used
     in the public service of the United
     States. This does not include

vessels owned by the United
States and engaged in a trade or
commercial service or vessels
under contract or charter to the
United States. (ESH) Vessels
owned or bareboat-chartered and
operated by the United States, or
by a State or political subdivision
thereof, or by a foreign nation,
except when such vessels are
engaged in commerce.

PUBLIC WATER SYSTEMS
(SDWA; CFR; ESH; EPA)·
DA October 12, 1990
BT1 Water Systems
 BT2 Publicly Owned Treatment Works
  BT3 Treatment Facilities
   BT4 Facilities
 BT2 Systems
NT1 Community Water Systems
NT1 Drinking Water System
NT1 Non-Community Water Systems
NT1 Non-Transient Non-Community
 Water System
NT1 Water Supply System
RT CTcalc
RT Drinking Water Supplies
RT First Draw
RT Primary Drinking Water Regulations
RT Safe Drinking Water Act
RT Sanitary Surveys
RT Secondary Drinking Water
 Regulations
RT Secondary Maximum Contaminant
 Levels
RT Waterborne Disease Outbreaks
SO Environmental Management
SO Environmental Protection Agency
 Glossary
SO Water Pollution
DEF (EPA) Systems that provide piped
 water for human consumption to at
 least 15 service connections or
 regularly serves 25 individuals.
 Public water systems are either
 "community water systems" or
 "noncommunity water systems."
 (ESH) Public water systems are
 defined in 40 CFR 141.2 as
 systems that provide piped water
 for human consumption and have
 at least 15 service connections or
 regularly serve at least 25
 individuals daily for a total of at
 least 60 days per year.

PUBLICLY OWNED TREATMENT
WORKS
(SWDA; SDWA; NPDWR; EPA; ESH)
DA October 12, 1990
SF POTW
BT1 Treatment Facilities
 BT2 Facilities
NT1 Water Systems
 NT2 Public Water Systems
  NT3 Community Water Systems
  NT3 Drinking Water System
  NT3 Non-Community Water Systems

  NT3 Non-Transient Non-Community
   Water System
  NT3 Water Supply System
 NT2 Storm Water Collection Systems
 NT2 Waste Water System
  NT3 Waste Water Collection Systems
RT Municipal Solid Wastes
RT Pretreatment Facilities
SO Environmental Management
SO Environmental Protection Agency
 Glossary
SO Wastes
SO Water Pollution
DEF Devices or systems used in the
 treatment (including recycling and
 reclamation) of municipal sewage
 or industrial wastes of a liquid
 nature which is owned by a State
 or municipality. This definition
 includes sewers, pipes, or other
 conveyances only if they convey
 wastewater to a POTW providing
 treatment.

PUBLISHED EXPOSURE LEVELS
DA January 24, 1991
BT1 Limits
RT Threshold Limit Values
SO Environmental Management
DEF (CFR) Exposure limits published in
 "NIOSH Recommendations for
 Occupational Health Standards"
 dated 1986, incorporated by
 reference, or, if none is specified,
 the exposure limits published in
 the standards specified by the
 American Conference of
 Governmental Industrial Hygienists
 in their publication "Threshold Limit
 Values and Biological Exposure
 Indices for 1987-88" dated 1987,
 incorporated by reference.

PUEC
DA October 12, 1990
SEE Portsmouth Uranium Enrichment
 Complex
SO Acronyms

PULL BOX
(2412 PULL BOX)
DA January 3, 1991
BT1 Equipment/Parts - Electrical (DOE
 FRASE Vocabulary)
 BT2 Equipment
SO DOE FRASE VOCABULARY

PUMP(S)
(2137 PUMP;N)
DA December 10, 1990
BT1 Machines (DOE FRASE
 Vocabulary)
 BT2 Equipment
NT1 Coolant Return Pump
NT1 Ion Pump
NT1 Reactor Coolant Pump
NT1 Sump Pump
SO DOE FRASE VOCABULARY

PUMP AREA
(1615 PUMP AREA)
DA December 10, 1990
BT1 Area
 BT2 Sites/Areas
SO DOE FRASE VOCABULARY

PUMP SHAFT BREAK
DA January 8, 1991
SF PSB

PUMPING STATIONS
(EPA)
DA October 12, 1990
BT1 Devices
RT Sewerage
SO Environmental Protection Agency
 Glossary
DEF (EPA) Mechanical devices installed
 in sewer or water systems or other
 liquid-carrying pipelines that move
 the liquids to a higher level.

PUNCH
(3049 PUNCH)
DA January 3, 1991
BT1 Tools - Manual
 BT2 Tools (DOE FRASE Vocabulary)
  BT3 Equipment
SO DOE FRASE VOCABULARY

PUNCTURE
(1336 PUNCTURE)
DA November 28, 1990
BT1 Injuries
RT Laceration
SO DOE FRASE VOCABULARY

PUNCTURE PROTECTION
(2677 PUNCTURE PRO)
DA January 3, 1991
SO DOE FRASE VOCABULARY

PURCHASING GROUP
(CERCLA; USC)
DA October 12, 1990
BT1 Groups
 BT2 Administrative Organizations
  BT3 Organizations
RT Pollution Liability
SO Compensation and Liability
DEF (USC) Any group of persons that
 has as one of its purposes the
 purchase of pollution liability
 insurance on a group basis.

PURGE SYSTEM
(2044 PURGE SYSTEM)
DA December 10, 1990
BT1 Systems (DOE FRASE Vocabulary)
 BT2 Systems
SO DOE FRASE VOCABULARY

PURIFICATION PROCESSES
DA February 26, 1991
BT1 Processes
NT1 Disinfection
NT1 Distillation

NT1 Ethylene Dichloride Purification
NT1 Vinyl Chloride Purification

PUTRESCIBLE
(EPA)
DA   October 12, 1990
RT   Decomposition
SO   Environmental Protection Agency
         Glossary
DEF (EPA) Able to rot quickly enough to
         cause odors and attract flies.

PVC
DA   October 12, 1990
SEE Polyvinyl Chloride
SO   Acronyms
SO   Environmental Protection Agency
         Glossary

PW
DA   October 12, 1990
SEE Process Water
SO   Acronyms

PWA
DA   October 12, 1990
SEE Process Waste Assessment
SO   Acronyms

PWGM
DA   October 12, 1990
SEE Process Water Gamma Monitor
SO   Acronyms

PWR
DA   October 12, 1990
SEE Pressurized Water Reactor
SO   Acronyms

PYROLYSIS
(EPA)
DA   October 12, 1990
BT1 Chemical Processes
 BT2 Processes
RT   Calciners
RT   Nodulizing Kilns
SO   Environmental Protection Agency
         Glossary
DEF (EPA) Decomposition of a chemical
         by extreme heat.

**PYROPHORIC-IGNITING
SPONTANEOUSLY**
(DOE Order 6430.1A)
DA   October 12, 1990
RT   Initiation Stimuli
SO   Construction
DEF (DOE Order 6430.1A) Emitting
         sparks when scratched or struck
         especially with steel.

PYROPHORIC LIQUIDS
(HMTA; CFR)
DA   May 20, 1991
BT1 Liquids
 BT2 Fluids
BT1 Pyrophoric Materials
 BT2 Materials

RT   Flammable Liquids
SO   Hazardous Materials
DEF (CFR) Pyrophoric liquids are any
         liquids that ignite spontaneously in
         dry or moist air at or below 130
         degrees Fahrenheit. A pyrophoric
         solid is any solid material, other
         than one classed as an explosive,
         which under normal conditions is
         liable to cause fires through
         friction, retained heat from
         manufacturing or processing, or
         which can be ignited readily and
         when ignited burns so vigorously
         and persistently as to create a
         serious transportation, handling, or
         disposal hazard. Included are
         spontaneously combustible and
         water-reactive materials.

PYROPHORIC MATERIALS
DA   January 24, 1991
BT1 Materials
NT1 Pyrophoric Liquids
RT   Fire Hazards
SO   Environmental Management
DEF (DOE 5820.2A) Materials which
         under normal conditions are liable
         to cause fires through friction,
         retained heat from manufacturing
         or processing, persistently as to
         create a serious transportation,
         handling or disposal hazard.

Q
DA   October 12, 1990
SEE Quality Factors
SO   Acronyms

QA
DA   October 12, 1990
SEE Quality Assurance
SO   Acronyms

QAOK
DA   October 12, 1990
SEE Quality Assurance Acceptance
SO   Acronyms

QC
DA   October 12, 1990
SEE Quality Control
SO   Acronyms

QIC
DA   October 12, 1990
SEE Quality Inspection Control
SO   Acronyms

QUALIFIED INCINERATORS
(TSCA; CFR)
DA   October 12, 1990
BT1 Incinerators
 BT2 Equipment/Parts - Heating (DOE
         FRASE Vocabulary)
  BT3 Equipment
 BT2 Heating Equipment

SO   Hazardous Materials
DEF (CFR) One of the following:
         (1)Incinerators approved under the
         provisions of Section 761.70. Any
         level of PCB concentration can be
         destroyed in an incinerator
         approved under Section 761.70.
         (2) High-efficiency boilers which
         comply with the criteria of Section
         761.60(a)(2)(iii)(A), and for which
         the operator has given written
         notice to the appropriate EPA
         Regional Administrator in
         accordance with the notification
         requirements for the burning of
         mineral oil dielectric fluid under
         Section 761.60(a)(2)(iii)(B). (3)
         Incinerators approved under
         Section 3005(c) of the Resource
         Conservation and Recovery Act
         (42 U.S.C. 6925(c)) (RCRA). (4)
         Industrial furnaces and boilers
         which are identified in 40 CFR
         260.10 and 40 CFR 266.41(b)
         when operating at their normal
         operating temperatures (this
         prohibits feeding fluids, above the
         level of detection, during either
         startup or shutdown operations).

QUALITY ASSURANCE
(DOE Order 5700.6B; ESH)
DA   October 12, 1990
SY   Reliability and Quality Assurance
SF   QA
NT1 Nuclear Quality Assurance-1
RT   Appraisals
RT   Data Quality Assessments
RT   Implementation Plans
RT   Management Appraisals
RT   Quality Assurance Records
RT   Redundance
SO   Construction
SO   Environmental Management
SO   System Safety Development Center
         Glossary
DEF (DOE Order 5700.6B) Involves all
         those planned and systematic
         actions necessary to provide
         adequate confidence that a facility,
         structure, system, or component
         will perform satisfactorily and
         safely in service. The goal of
         quality assurance is to assure that:
         research, development,
         demonstration, scientific
         investigations, and production
         activities are performed in a
         controlled manner; that
         components, systems, and
         processes are designed,
         developed, constructed, tested,
         operated, and maintained
         according to engineering
         standards, quality practices, and
         Technical Specifications
         Operational Safety Requirements;
         and that resulting technology data
         are valid and retrievable. Quality
         assurance includes quality control,

which comprises all those actions
necessary to Control and verify
the features and characteristics of
a material, process, product, or
service to specified requirements.

QUALITY ASSURANCE ACCEPTANCE
DA   January 8, 1991
SF   *QAOK*

QUALITY ASSURANCE OVERVIEWS
(DOE Order 5700.6B)
DA   October 16, 1990
BT1 Quality Control
 BT2 Controls
NT1 Quality Assurance Plans
 NT2 Site Quality Assurance Plan
RT   Verification
SO   Quality Assurance
DEF (DOE Order 5700.6B) An organized
    set of activities performed as
    independent functions. Their
    purpose is to assure that all
    aspects of quality related activities
    at the program, project, and
    contractor level of management
    are adequately addressed. Such
    activities include: (1) Periodic and
    timely reviews of program/project
    documents, activities, actions and
    plans; (2) review of new major
    procurements and management
    and operating contracts; (3) review
    of extend/compete packages for
    management and operating
    contracts; and (4) review of DOE
    Orders with relevance to the
    incorporation of the DOE quality
    assurance policy, where
    necessary.

QUALITY ASSURANCE PLANS
(DOE Order 5700.6B; ESH)
DA   October 12, 1990
BT1 Plans
BT1 Quality Assurance Overviews
 BT2 Quality Control
  BT3 Controls
NT1 Site Quality Assurance Plan
SO   Quality Assurance
DEF (DOE Order 5700.6B) Documents
    that contain or reference the
    quality assurance elements
    established for an activity, group
    of activities, a scientific
    investigation or a project and
    describes how conformance with
    such requirements is to be
    assured for structures, systems,
    computer software, components,
    and their operation commensurate
    with (1) the scope, complexity,
    duration, and importance to
    satisfactory performance, (2) the
    potential impact on environment,
    safety and health, and (3)
    requirements for reliability and
    continuity of operation.

**QUALITY ASSURANCE PROJECT
PLANS**
(EPA)
DA   October 12, 1990
SY   Site Quality Assurance Plan
BT1 Sampling and Analysis Plans
 BT2 Plans
RT   Projects
SO   Environmental Protection Agency
     Glossary
DEF (EPA) Describe the necessary
    policy, organization, functional
    activities, and quality assurance
    and quality control protocols.

QUALITY ASSURANCE RECORDS
(DOE Order 6430.1A)
DA   October 12, 1990
RT   Quality Assurance
SO   Construction
DEF (DOE Order 6430.1A) Includes
    results of reviews, inspections,
    audits, and material analyses;
    monitoring of work performance;
    qualification of personnel,
    procedures, and equipment; and
    other documentation such as
    drawings, special reports, and
    corrective action reports.

QUALITY ASSURANCE UNITS
(TSCA; CFR)
DA   October 19, 1990
RT   Studies
SO   Environmental Management
SO   Hazardous Materials
DEF (CFR) Persons or organizational
    elements, except the study
    director, designated by testing
    facility management to perform the
    duties relating to quality assurance
    of the studies.

QUALITY CONTROL
(EPA)
DA   October 12, 1990
SF   *QC*
BT1 Controls
NT1 Quality Assurance Overviews
 NT2 Quality Assurance Plans
  NT3 Site Quality Assurance Plan
NT1 Quality Inspection Control
NT1 Quality Verification
RT   Maintenance Management
RT   Sampling
RT   Specifications
RT   Total Indicated Runout
SO   Environmental Protection Agency
     Glossary
DEF (EPA) A system of procedures,
    checks, audits, and corrective
    actions to insure that all research
    design and performance,
    environmental monitoring and
    sampling, and other technical and
    reporting activities are of the
    highest achievable quality.

QUALITY FACTORS
(DOE Orders 5400.5 and 5480.11;
EMER)
DA   October 12, 1990
SF   *Q*
RT   Dose Equivalents
RT   Relative Biological Effectiveness
SO   Emergency Preparedness
SO   Environmental Management
SO   Industrial Hygiene
SO   Radiation
DEF (EMER) Principal modifying factors
    used to calculate the dose
    equivalent from the absorbed
    dose. For the purposes of this
    Order, the following quality factors,
    which are taken from DOE
    5480.11, are to be used.
    [Radiation Type Quality Factor.
    X-rays, gamma rays, positrons,
    and electrons (including tritium)]
    [Neutrons, <10 keV 3] [Neutrons,
    >10 keV 10, Protons and single
    charged, particles of unknown
    energy with rest mass > one
    atomic mass unit] [Alpha particles
    20, Multiple-charged particles,
    (and particles of unknown charge)
    of unknown energy]. For neutrons
    of known energies, the more
    detailed quality factors given in
    DOE 5480.11 may be used.
    (IAEA) A factor that weights the
    absorbed dose, defined as a
    function of the collision-stopping
    power in water at the point of
    interest. Values of Q are specified
    by the International Commission
    on Radiological Protection (ICRP).

QUALITY INSPECTION CONTROL
DA   January 8, 1991
SF   *QIC*
BT1 Quality Control
 BT2 Controls

QUALITY VERIFICATION
DA   January 8, 1991
SF   *QV*
BT1 Quality Control
 BT2 Controls

QUANTITATION LIMIT
(EPA)
DA   October 12, 1990
BT1 Limits
SO   Environmental Protection Agency
     Glossary
DEF (EPA) The lowest level at which a
    chemical may be accurately and
    reproducibly quantitated. Usually
    equal to the detection limit
    multiplied by a factor of 3 to 5, but
    varies between chemicals and
    between samples.

QUANTITIES
DA   February 25, 1991
NT1 Exempt or Limited Quantities

NT1 Formula Quantities
NT1 Highway Route Controlled Quantity
NT1 Large Quantities
NT1 Limited Quantity
NT1 Reportable Quantities
NT1 Significant Quantities
NT1 Small Quantities for Research and Development
NT1 Threshold Planning Quantities
NT1 Type A Quantities
NT1 Type B Quantities

QUANTITY DISTANCES
(Doe Order 5480.16)
DA   October 12, 1990
BT1 Measurements
NT1 Intraline Separation (Barricaded)
NT1 Intraline Separation (Unbarricaded)
NT1 Magazine Separation
RT   Magazines (Buildings)
RT   Munitions
SO   Construction
SO   Firearms
DEF (DOE Order 5480.16) Distances required for a specific level of protection for a particular hazard class/division of ammunition and explosives.

QUARTERLY ENVIRONMENTAL COMPLIANCE REPORTS
(DOE Order 5440.1D)
DA   May 15, 1991
BT1 Reports
RT   National Environmental Policy Act
RT   NEPA Status Reports
SO   Environmental Management
DEF (DOE Order) Quarterly reports prepared by DOE facilities and sent to the appropriate Secretarial Officer(s) and to the Assistant Secretary for Environment, Safety and Health (EH-1) in response to the initiatives in SEN-7A-90. Quarterly environmental compliance reports include the status of the line organization's NEPA compliance activities.

QUENCH TANKS
(EPA)
DA   October 12, 1990
BT1 Tanks
BT2 Facility Components
SO   Environmental Protection Agency Glossary
DEF (EPA) Water-filled tanks used to cool incinerator residues or hot materials during industrial processes.

QUICK CLAYS
(USGS)
DA   October 12, 1990
BT1 Soils
SO   Natural Phenomenon
DEF (USGS) Clays that have lost their shear strength. Sometimes also called "sensitive" clays.

OV
DA   October 12, 1990
SEE Quality Verification
SO   Acronyms

R
DA   October 12, 1990
SEE Roentgen
SO   Acronyms

R&D
DA   October 12, 1990
SEE Research and Development
SO   Acronyms

R&PM
DA   October 12, 1990
SEE Regulations and Procedures Manual, PUB-201
SO   Acronyms

R&QA
DA   October 12, 1990
SEE Reliability and Quality Assurance
SO   Acronyms

RA
DA   October 12, 1990
SEE Remedial Actions
SO   Acronyms

RAAS
DA   October 12, 1990
SEE Remedial Action Assessment System
SO   Acronyms

RABBITS
(NFI)
DA   October 12, 1990
RT   Reactor Components
SO   Nuclear Facilities Incident Database
SO   Radiation
DEF (NFI) Slugs irradiated to obtain reaction products (Irradiation Capsules).

RACT
DA   October 12, 1990
SEE Reasonably Available Control Technology
SO   Acronyms

RAD
(EPA; EMER)
DA   October 12, 1990
SEE Radiation Absorbed Dose
SO   Acronyms
DEF (EPA) A unit of absorbed dose replaced by the gray.

RAD WASTE TREATMENT/STORAGE FACILITY
(1806 RWT FACILITY)
DA   December 10, 1990
BT1 Facility (DOE FRASE Vocabulary)

BT2 Facilities and Buildings (DOE FRASE Vocabulary)
BT3 Facilities
SO   DOE FRASE VOCABULARY

RADIAL POWER MONITOR
DA   January 8, 1991
SF   RPM (Radial Power Monitor)
BT1 Monitors
BT2 Equipment

RADIANT ENERGY
(IAEA)
DA   October 12, 1990
BT1 Energy
RT   Energy Fluence
RT   Energy Flux
SO   Radiation
DEF (IAEA) The energy (excluding rest energy) emitted, transferred, or received in the form of radiation.

RADIATION
(NIH; IAEA; NFI)
DA   October 12, 1990
NT1 Bremsstrahlung
NT1 Ionizing Radiation
NT2 Background Radiation
NT2 Gamma Radiation
NT2 Lethal Dose of Radiation
NT2 X-rays
NT1 Natural Background Radiation
NT1 Non-Ionizing Electromagnetic Radiation
NT1 Non-Penetrating Radiation
NT2 Alpha Rays
NT2 Beta Rays
NT1 Penetrating Radiation
NT1 Radio Frequency Radiation
NT1 Ultraviolet Rays
RT   Accelerators
RT   Buildup Factor
RT   Hazards
RT   Irradiation
SO   Environmental Protection Agency Glossary
DEF Any form of energy propagated as rays, waves, or streams of energetic particles. The term is frequently used in relation to the emission of rays from the nucleus of an atom. (NIH) (1) The emission and propagation of energy through space or through a material medium in the form of waves; for instance, the emission and propagation of electromagnetic waves, or of sound and elastic waves. (2) The energy propagated through a material medium as waves; for example, energy in the form of electromagnetic waves or of elastic waves. The term "radiation" or "radiant energy," when unqualified, usually refers to electromagnetic radiation. Such radiation commonly is classified according to frequency as Hertzian, infrared, visible (light), ultraviolet, x ray, and gamma ray.

(3) By extension, corpuscular emissions such as alpha and beta radiation, or rays of mixed or unknown type, as cosmic radiation.

RADIATION ABSORBED DOSE
(EPA)
DA   October 12, 1990
SF   *RAD*
BT1 Radiation Units
 BT2 Units of Measure
SO   Environmental Protection Agency Glossary
DEF (EPA) A unit of absorbed dose of radiation. One RAD of absorbed dose is equal to .01 joules per kilogram.

RADIATION ALARM SYSTEM
(2045 RADIATION AS)
DA   December 10, 1990
BT1 Alarms
 BT2 Devices
BT1 Emergency Systems
 BT2 Systems
BT1 Systems (DOE FRASE Vocabulary)
 BT2 Systems
SO   DOE FRASE VOCABULARY

RADIATION AREAS
DA   January 24, 1991
BT1 Sites/Areas
NT1 High Radiation Areas
RT   Exposure
SO   Environmental Management
DEF (CFR) Any area, accessible to individuals, in which there exists radiation at such levels that a major portion of the body could receive in any one (1) hour a dose in excess of five (5) millirems or in any five (5) consecutive days a dose in excess of 100 millirems.

RADIATION DETECTORS
DA   January 30, 1991
BT1 Equipment
NT1 Cooling Water Gamma Monitor
NT1 Cutie Pies
NT1 Dose Rate Meters
NT1 Dosimeters
 NT2 Nuclear Accident Dosimeter
 NT2 Personnel Dosimeters
 NT2 Thermoluminescent Dosimeters
NT1 External Fission Counter
NT1 Geiger-Mueller Counters
NT1 Geiger Counters
NT1 Internal Fission Counter
NT1 Internal Gamma Flux Monitor
NT1 Ionization Chambers
NT1 Kanne Alarm
NT1 Low Energy Gamma Monitor
NT1 Process Water Gamma Monitor
NT1 Scintillation Counters
NT1 Vibrating Reed Electrometers
NT1 Whole Body Counter
RT   Controls

RT   Count Rate Meter
RT   Dosimeters

RADIATION EMERGENCY ASSISTANCE CENTER/TRAINING SITES
(EMER)
DA   February 1, 1991
BT1 Sites/Areas
SO   Emergency Preparedness
DEF (EMER) Multipurpose medical facilities located in Oak Ridge, Tennessee, prepared to deal with all types of radiation exposure emergencies and provide medical and health physics advice and assistance in radiological emergencies.

RADIATION EXPOSURE
(1343 RAD EXPOSURE)
DA   November 28, 1990
BT1 Injuries
RT   Absorption (Radiation)
RT   Radiation Exposure Module
SO   DOE FRASE VOCABULARY

RADIATION EXPOSURE MODULE
(SSDC)
DA   January 8, 1991
SF   *REM (Radiation Exposure Module)*
BT1 Safety Performance Measurement System
 BT2 Information Systems
 BT3 Security Interests
 BT3 Systems
RT   Radiation Exposure
SO   Management
SO   Radiation
DEF (SSDC) This module of the Safety Performance Measurement System is divided into two databases; a locator file and a database containing the annual dose of all monitored DOE and DOE contractor personnel.

RADIATION INDUCED GENETIC EFFECTS
(IAEA)
DA   October 12, 1990
SY   Genetic Radiation Effects
SY   Radiation Induced Hereditary Effects
BT1 Stochastic Effects
 BT2 Effects
SO   Radiation
DEF (IAEA) Changes induced by radiation in the genetic material of both somatic and germinal cells. Loosely used in radiation protection as a synonym for radiation induced hereditary diseases.

RADIATION INDUCED HEREDITARY EFFECTS
(IAEA)
DA   November 20, 1990

SY   Genetic Radiation Effects
SY   Radiation Induced Genetic Effects
BT1 Stochastic Effects
 BT2 Effects
SO   Radiation
DEF (IAEA) Stochastic effects that occur in the progeny of the exposed 'individual'.

RADIATION MONITOR/TECHNICIAN
(0383 RAD TECH)
DA   November 28, 1990
BT1 Technicians
 BT2 Professional Personnel
 BT3 Occupations
 BT3 Personnel
SO   DOE FRASE VOCABULARY

RADIATION PROTECTION
DA   November 7, 1990
RT   Absorption (Radiation)
RT   Accidents
RT   As Low As Reasonably Achievable
RT   Confinement Systems
RT   Containment
RT   Decontamination
RT   Differential Cost Benefit Analysis
RT   Enclosures
RT   Health Hazards
RT   Health Physics
RT   Limits
RT   Radiation Protection Officers
RT   Radiation Safety
RT   Radioactive Contamination
RT   Radiological Monitoring
RT   Reference Man
RT   Remedial Measures
RT   Restricted Areas
SO   Industrial Safety
SO   Radiation

RADIATION PROTECTION OFFICERS
(IAEA)
DA   October 12, 1990
BT1 Personnel
RT   Radiation Protection
SO   Radiation
DEF (IAEA) Technically competent persons designated by management to supervise the application of radiation protection regulations and to provide advice on all relevant aspects of radiation protection.

RADIATION RECORDS REPOSITORIES
(Doe Order 5484.1)
DA   October 12, 1990
BT1 Information Systems
 BT2 Security Interests
 BT2 Systems
RT   System Safety Development Center
SO   Management
DEF (DOE Order 5484.1) The DOE centralized data base located at the System Safety Development Center, EG&G Idaho, Inc., which contains statistical summaries of occupational radiation exposure

information for activities associated with DOE operations. Individual occupational exposure records are maintained by DOE sites. The Radiation Records Repository also contains summary data submitted for the DOE predecessor agencies, the Atomic Energy Commission, and the Energy Research and Development Administration activities.

RADIATION SAFETY
(NCRP)
DA   October 12, 1990
BT1 Safety
RT Annual Dose Equivalent Limit
RT Annual Limit on Intake
RT As Low As Reasonably Achievable
RT Cognizant Federal Agencies
RT Competent Authorities
RT Controlled Areas
RT Decontamination
RT Dose Limit
RT Dose Upper Bounds
RT Doses
RT Film Badges
RT Filters (Radiology)
RT Irradiation
RT Protective Barriers
RT Radiation Protection
RT Radiation Shields
RT Radioactive Contamination
RT Radiological Surveys
RT Sealed Sources
RT Supervised Areas
RT Whole Body
RT Working Conditions
SO Radiation
DEF (NCRP) Concerned with
     recognition, evaluation, and control
     of risks due to radiation exposure.

RADIATION SHIELDS
(IAEA)
DA   October 12, 1990
RT Absorption (Radiation)
RT Ionizing Radiation
RT Protective Barriers
RT Radiation Safety
RT Shielding Materials
SO Radiation
DEF (IAEA) Materials interposed
     between a source of radiation and
     persons, or equipment or other
     objects, in order to attenuate the
     radiation.

RADIATION STANDARDS
(EPA)
DA   October 12, 1990
BT1 Standards
 BT2 Codes, Standards, and
      Regulations
NT1 Annual Limit on Intake
NT1 As Low As Reasonably Achievable
NT1 As Low As Practicable

SO Environmental Protection Agency
     Glossary
DEF (EPA) Regulations that set
     maximum exposure limits for
     protection of the public from
     radioactive materials.

RADIATION SUIT
(2678 RADIATION SU)
DA   January 3, 1991
BT1 Anticontamination Clothing
 BT2 Clothing
 BT2 Personal Protective Equipment
  BT3 Equipment/Parts - Personal
       Protective (DOE FRASE
       Vocabulary)
   BT4 Equipment
SO DOE FRASE VOCABULARY

RADIATION UNITS
DA   January 30, 1991
BT1 Units of Measure
NT1 Absorbed Dose
NT1 Becquerel
NT1 Curie
NT1 Dose Equivalents
 NT2 Annual Dose Equivalent
  NT3 Annual Effective Dose Equivalent
   NT4 Cumulative Annual Effective
        Dose Equivalent
 NT2 Collective Dose Equivalent
  NT3 Collective Effective Dose
       Equivalent
 NT2 Committed Dose Equivalent
  NT3 Committed Effective Dose
       Equivalent
 NT2 Deep Dose Equivalent
  NT3 Deep Eye Dose Equivalent
 NT2 Effective Dose Equivalent
  NT3 Annual Effective Dose Equivalent
   NT4 Cumulative Annual Effective
        Dose Equivalent
  NT3 Collective Effective Dose
       Equivalent
  NT3 Committed Effective Dose
       Equivalent
 NT2 Lens of Eye Dose Equivalents
 NT2 Shallow Eye Dose Equivalents
 NT2 Shallow Dose Equivalents
NT1 Gray
NT1 Microroentgen/hour
NT1 Picocurie
NT1 Picocuries Per Liter
NT1 Picocuries/gram
NT1 Radiation Absorbed Dose
NT1 Relative Biological Effectiveness
NT1 Roentgen Equivalent Man
NT1 Roentgen
NT1 Sievert
NT1 Working Level
NT1 Working Level Month
 NT2 Cumulative Working Level Months
RT Dose Limit
DEF Units of measure of radiations,
     radiation doses, and radioactivity.

RADIATION WORKERS
(DOE Order 5480.11)
DA   October 12, 1990
BT1 Occupational Workers

 BT2 Personnel
NT1 Non-Employee Radiation Workers
SO Industrial Hygiene
DEF (DOE Order 5480.11) Occupational
     workers whose job assignments
     require work on, with, or in the
     proximity of radiation- producing
     machines or radioactive materials,
     and/or who have the potential of
     being routinely exposed above 0.1
     rem (0.001 sievert) per year,
     which is the sum of the annual
     effective dose equivalent from
     external irradiation and the
     committed effective dose
     equivalent from internal irradiation.

RADIATION ZONE
DA   January 8, 1991
SF   *RZ*
BT1 Zones
 BT2 Sites/Areas

RADIO
(2437 RADIO)
DA   January 3, 1991
BT1 Equipment/Parts - Electrical (DOE
     FRASE Vocabulary)
 BT2 Equipment
SO DOE FRASE VOCABULARY

RADIO FREQUENCY RADIATION
(EPA)
DA   November 9, 1990
SY Non-Ionizing Electromagnetic
     Radiation
BT1 Radiation
DEF (EPA) (1) Radiation that does not
     change the structure of atoms but
     does heat tissue and may cause
     harmful biological effects. (2)
     Microwaves, radio waves, and
     low-frequency electromagnetic
     fields from high-voltage
     transmission lines.

RADIO REPEATER STATIONS
(DOE Order 6430.1A)
DA   October 12, 1990
BT1 Facilities
SO Construction
DEF (DOE Order 6430.1A) Unmanned
     radio transmission facilities,
     usually located in remote areas.

RADIOACTIVE ARTICLES
(HMTA; CFR)
DA   May 20, 1991
BT1 Articles
SO Hazardous Materials
DEF (CFR) Radioactive articles are any
     manufactured instruments and
     articles such as an instrument,
     clock, electronic tube or
     apparatus, or similar instruments
     and articles having radioactive
     material as a component part.

RADIOACTIVE CONTAMINATION
(NIH; IAEA)
DA October 12, 1990
SY Contamination
BT1 Contamination
BT1 Nature of Property Damage
NT1 Non-fixed Radioactive
    Contamination
RT Body Content
RT Radiation Protection
RT Radiation Safety
RT Radiological Areas
RT Radioactivity
RT Radiological Monitoring
RT Smear
SO Radiation
DEF (NIH) Deposition of radioactive
    material in any place where it is
    not desired, and particularly in any
    place where its presence may be
    harmful. The harm may be vitiating
    the validity of an experiment or a
    procedure, or in actually being a
    source of excessive exposure to
    personnel.

RADIOACTIVE CONTENTS
(HMTA; CFR)
DA May 20, 1991
RT Disposal Packages
SO Hazardous Materials
DEF (CFR) Radioactive contents are the
    radioactive material, together with
    any contaminated liquids or gases,
    within the package.

RADIOACTIVE DECAY
(NIH)
DA October 12, 1990
NT1 Beta Decay
 NT2 Electron Capture
RT Alpha Particles
RT Gamma Radiation
RT Radioactivity
RT Radioactivity Decay Constant
RT Radioactive Half-Life
RT Radioactive Series
SO Radiation
DEF (NIH) Disintegration of the nucleus
    of an unstable nuclide by the
    spontaneous emission of charged
    particles and/or photons.

RADIOACTIVE EFFLUENTS
(IAEA)
DA October 12, 1990
BT1 Effluents
 BT2 Discharges
 BT2 Emissions
  BT3 Air Pollutants
BT1 Radioactive Wastes
 BT2 Wastes
SO Radiation
DEF (IAEA) Airborne or liquid radioactive
    materials that are discharged into
    the environment.

RADIOACTIVE HALF-LIFE
(NIH)

DA October 12, 1990
BT1 Half-Life
 BT2 Time Designations
RT Radioactive Decay
SO Radiation
DEF (NIH) Time required for a
    radioactive substance to lose 50
    percent of its activity by decay.
    Each radionuclide has a unique
    half-life.

RADIOACTIVE MATERIAL TRANSPORTATION ACCIDENTS
(EMER)
DA February 1, 1991
BT1 Accidents
SO Emergency Preparedness
DEF (EMER) Incidents in which the
    conveyance transporting a
    radioactive material package is
    involved in an accident. Accidents
    are subsets of radioactive material
    transportation incidents and
    include a wide range of severities
    ranging from minor mishaps to the
    very severe. This definition is not
    limited by financial costs, injuries,
    or fatalities.

RADIOACTIVE MATERIAL TRANSPORTATION INCIDENTS
(EMER)
DA February 1, 1991
BT1 Incidents
SO Emergency Preparedness
DEF (EMER) Events occurring during
    the course of transportation of
    radioactive materials (including
    loading, transport, unloading, and
    temporary storage) which result in
    actual or suspected release of
    radioactive material. The events
    may include fire, breakage,
    spillage, release or suspected
    release of radioactive material,
    excessive or suspected radiation,
    loss of possession, or accident
    conditions.

RADIOACTIVE MATERIALS
(SWDA; RCRA; ESH)
DA October 12, 1990
SY Radioactive Substances
BT1 Materials
NT1 Airborne Radioactive Materials
NT1 Low Specific Activity Materials
NT1 Naturally Occurring Radioactive
    Material
NT1 Normal Form Radioactive Materials
NT1 Residual Radioactive Materials
NT1 Special Form Radioactive Materials
RT Containment
RT Disposal Packages
RT Hazards
RT Large Quantities
RT Normal Form
RT Picocurie
RT Radioactive Wastes
RT Radioactive Source Terms
RT Release of Property

RT Tailings
RT Type A Quantities
RT Type B Quantities
SO Environmental Management
SO Hazardous Materials
DEF (ESH) Any material or combination
    of materials that spontaneously
    emit ionizing radiation. (ESH)
    There are sundry definitions
    provided within the regulations
    pertaining to the various aspects
    of radioactive materials. The term
    radioactive materials is defined as
    any material having a specific
    activity greater than 0.002
    microcuries per gram.

RADIOACTIVE MIXED WASTES
(DOE Order 5400.3 and 5480.2)
DA October 12, 1990
SY Mixed Wastes
BT1 Radioactive Wastes
 BT2 Wastes
RT Hazardous Wastes
RT Radioactive Wastes
SO Environmental Management
SO Industrial Hygiene
SO Wastes
DEF Radioactive waste that also
    contains hazardous waste
    constituents.

RADIOACTIVE NUCLIDE INTAKE
(IAEA)
DA October 12, 1990
BT1 Intake
 BT2 Measurements
SO Radiation
DEF (IAEA) Amount of radioactive
    material introduced into the body
    by inhalation or ingestion, or
    through the skin. Also, used to
    denote the process.

RADIOACTIVE SERIES
(NCRP)
DA October 12, 1990
RT Equilibrium
RT Radioactive Decay
SO Radiation
DEF (NCRP) A succession of nuclides,
    each of which transforms by
    radioactive decay into the next
    until a stable nuclide results. The
    first member is called the parent
    and the subsequent members are
    called progeny, daughters, or
    decay products.

RADIOACTIVE SOURCE TERMS
(IAEA)
DA October 12, 1990
BT1 Vocabulary
RT Fission Products
RT Radioactive Materials
RT Risk Assessment
SO Radiation
DEF (IAEA) Expressions used to denote
    information about the actual or

potential release of radioactive
material from a given source,
which may include a specification
of the composition, the amount,
the rate, and the mode of the
release.

RADIOACTIVE SUBSTANCES
(EPA)
DA   October 12, 1990
SY   Radioactive Materials
BT1 Materials
SO   Environmental Protection Agency
      Glossary
DEF (EPA) Substances that emit
      radiation.

RADIOACTIVE WASTE MANAGEMENT
DA   November 7, 1990
BT1 Waste Management
 BT2 Processes
RT   Burial Grounds
RT   General Environment
RT   Geologic Repositories
RT   Radioactive Wastes
SO   Radiation
SO   Wastes
DEF (USC) The systematic
      administration of activities that
      provide for the collection, storage,
      transportation, transfer,
      processing, treatment, and
      disposal of radioactive wastes.

RADIOACTIVE WASTES
(DOE Order 5480.2; SDWA; SWDA;
 CFR; ESH)
DA   October 12, 1990
BT1 Wastes
NT1 High-Level Radioactive Wastes
NT1 Low Level Radioactive Wastes
NT1 Low Level Wastes
NT1 Radioactive Mixed Wastes
NT1 Radioactive Effluents
NT1 Transuranic Wastes (TRU Waste)
 NT2 Contact-Handled Transuranic
      Wastes
 NT2 Remote-Handled Transuranic
      Wastes
NT1 Transuranic Radioactive Wastes
NT1 TRU Wastes
RT   General Environment
RT   Near Surface Disposal Facilities
RT   Radioactive Waste Management
RT   Radioactive Materials
RT   Radioactive Mixed Wastes
SO   Environmental Management
SO   Industrial Hygiene
SO   Wastes
SO   Water Pollution
DEF Solid or fluid materials of no value
      containing radioactivity; discarded
      items such as clothing, containers,
      equipment, rubble, residues, or
      soils contaminated with
      radioactivity; or soils, rubble,
      equipment, or other items
      containing induced radioactivity
      such that the levels exceed safe
      limits for unconditional release.

Any waste that contains
radioactive material in
concentrations that exceed those
listed in 10 CFR Part 20, Appendix
B, Table II Column 2. (ESH) Solid,
liquid, or gaseous material that
contains radionuclides regulated
under the Atomic Energy Act of
1954, as amended, and of
negligible economic value
considering costs of recovery.

RADIOACTIVITY
(DOE Order 5400.5; EMER)
DA   October 12, 1990
SY   Activity (Nuclear)
NT1 Induced Radioactivity
RT   Becquerel
RT   Curie
RT   Gross Alpha Particle Activity
RT   Gross Beta Particle Activity
RT   Radioactive Contamination
RT   Radioactive Decay
RT   Specific Activity
SO   Emergency Preparedness
SO   Environmental Management
SO   Radiation
SO   Wastes
DEF (DOE Order 5400.5) The property
      or characteristic of radioactive
      material to spontaneously
      "disintegrate" with the emission of
      energy in the form of radiation.
      The unit of radioactivity is the
      curie (or becquerel).

RADIOACTIVITY DECAY CONSTANT
(IAEA)
DA   October 12, 1990
BT1 Measurements
RT   Radionuclides
RT   Radioactive Decay
SO   Radiation
DEF (IAEA) For a radioactive nuclide in
      a particular energy state, the
      quotient of dP by dt, where dP is
      the probability of a given nucleus
      undergoing a spontaneous nuclear
      transition from that energy state in
      the time interval dt.

RADIOBIOLOGY
(EPA)
DA   October 12, 1990
SO   Environmental Protection Agency
      Glossary
DEF (EPA) The study radiation effects
      on living things.

RADIOLOGICAL ACCIDENTS
(EMER)
DA   February 1, 1991
BT1 Accidents
SO   Emergency Preparedness
DEF (EMER) Loss of control of
      radioactive materials which
      present a potential hazard to
      personnel, public health, property,
      or environment, or the exceeding

of the established limit for
exposure to ionizing radiation.

RADIOLOGICAL AREAS
(Doe Order 5480.11)
DA   October 12, 1990
BT1 Sites/Areas
RT   Radioactive Contamination
SO   Environmental Management
SO   Industrial Hygiene
DEF (DOE Order 5480.11) Any areas
      within a controlled area where an
      individual can receive a dose
      equivalent greater than 5 mrem
      (50 microsieverts) in 1 hour at 30
      cm from the radiation source or
      any surface through which the
      radiation penetrates, or where
      airborne radioactive
      concentrations greater than 1/10
      of the derived air concentrations
      are present (or are likely to be), or
      where surface contamination
      levels greater than those specified
      in Attachment 2 of this Order are
      present.

RADIOLOGICAL ASSISTANCE
PROGRAMS
(EMER)
DA   February 1, 1991
BT1 Programs
SO   Emergency Preparedness
DEF (EMER) Department of Energy
      programs which provide for
      radiological assistance to federal
      and state agencies and private
      entities in the event of an incident
      involving radioactive materials.

RADIOLOGICAL ASSISTANCE TEAMS
(EMER)
DA   February 1, 1991
BT1 Response Teams
 BT2 Teams
  BT3 Administrative Organizations
   BT4 Organizations
SO   Emergency Preparedness
DEF (EMER) Teams dispatched to the
      site of a radiological incident by
      the U.S. Department of Energy
      (DOE) regional office responding
      to a radiological incident.
      Radiological Assistance Teams
      are located at DOE operations
      offices and national laboratories
      and most area offices and
      associated contractor sites.

RADIOLOGICAL MONITORING
(NIH)
DA   October 12, 1990
BT1 Monitoring
 BT2 Activities
NT1 Smear
RT   Aerial Measuring Systems
RT   Autoradiographs
RT   Radiation Protection
RT   Radioactive Contamination

SO Radiation
DEF (NIH) Periodic or continuous
    determination of the amount of
    ionizing radiation or radioactive
    contamination present in an
    occupied region as a safety
    measure for purposes of health
    protection. Area Monitoring:
    Routine monitoring of the level of
    radiation or of radioactive
    contamination of any particular
    area, building, room, or
    equipment. Personnel Monitoring:
    Monitoring any part of an
    individual, his breath, excretions,
    or any part of his clothing.

RADIOLOGICAL RELEASES
(EMER)
DA February 1, 1991
BT1 Releases
SO Emergency Preparedness
DEF (EMER) Incidents in which
    radiological material (gas, liquid,
    and/or solid) are discharged into
    the biosphere in an unplanned
    manner.

RADIOLOGICAL SURVEYS
(NIH)
DA October 12, 1990
BT1 Surveys
RT Radiation Safety
SO Radiation
DEF (NIH) Evaluation of the radiation
    hazards incident to the production,
    use of, existence of radioactive
    materials or other sources of
    radiation under a specific set of
    conditions. Such evaluation
    customarily includes a physical
    survey of the disposition of
    materials and equipment,
    measurements or estimates of the
    levels of radiation that may be
    involved, and a sufficient
    knowledge of processes using or
    affecting these materials to predict
    hazards resulting from expected or
    possible changes in materials or
    equipment.

RADIOLOGICAL TRANSPORTATION
INCIDENTS
(EMER)
DA February 1, 1991
BT1 Incidents
SO Emergency Preparedness
DEF (EMER) Incidents that involve a
    transportation vehicle or shipment
    containing radioactive materials.
    (See Transportation Incidents.)

RADIONUCLIDE BARRIERS (NATURAL
OR ENGINEERED)
(IAEA)
DA October 12, 1990
BT1 Barriers
NT1 Bulkheads

RT Radionuclides
SO Radiation
DEF (IAEA) Structures that delay or
    prevent radionuclide migration
    from the source material.

RADIONUCLIDES
(CAA; CFR: NIH; EMER)
DA October 12, 1990
BT1 CERCLA Hazardous Substances
BT2 Hazardous Substances
BT1 Nuclides
NT1 Alpha Decay Radioisotopes
NT2 Radium 226
NT2 Radon 222
NT2 Thorium 230
NT2 Uranium 238
NT1 Beta Decay Radioisotopes
NT2 Beta-Minus Decay Radioisotopes
NT3 Antimony 125
NT3 Cerium 144
NT3 Cerium 141
NT3 Cesium 137
NT3 Cesium 134
NT3 Cobalt 60
NT3 Cobalt 58
NT3 Iodine 131
NT3 Radium 228
NT3 Ruthenium 103
NT3 Strontium 90
NT3 Tellurium 132
NT3 Thorium 232
NT3 Tritium
NT3 Zinc 65
NT3 Zirconium 95
NT1 Bone Seekers
NT1 Days Living Radioisotopes
NT2 Beryllium 7
NT2 Cerium 144
NT2 Cerium 141
NT2 Cesium 137
NT2 Cobalt 58
NT2 Iodine 131
NT2 Manganese 54
NT2 Radon 222
NT2 Selenium 75
NT2 Zinc 65
NT2 Zirconium 95
NT1 Delayed Neutron Precursors
NT1 Delayed Proton Precursors
NT1 Element 104 Isotopes
NT1 Element 105 Isotopes
NT1 Element 106 Isotopes
NT1 Element 107 Isotopes
NT1 Element 108 Isotopes
NT1 Element 109 Isotopes
NT1 Heavy Ion Decay Radioisotopes
NT1 Hours Living Radioisotopes
NT2 Cesium 134
NT2 Cobalt 58
NT1 Internal Conversion Radioisotopes
NT2 Cesium 134
NT2 Cobalt 60
NT2 Cobalt 58
NT1 Isomeric Transition Isotopes
NT2 Cesium 134
NT2 Cobalt 60
NT2 Cobalt 58
NT1 Microsec Living Radioisotopes
NT1 Millisec Living Radioisotopes

NT1 Minutes Living Radioisotopes
NT2 Cobalt 60
NT2 Thorium 232
NT1 Nanosec Living Radioisotopes
NT1 Neutron-Deficient Isotopes
NT1 Neutron-Rich Isotopes
NT1 Proton Decay Radioisotopes
NT1 Radon Progeny
NT2 Radon 222
NT1 Seconds Living Radioisotopes
NT1 Transuranic Radionuclides
NT1 Years Living Radioisotopes
NT2 Antimony 125
NT2 Cesium 134
NT2 Cobalt 60
NT2 Radium 226
NT2 Radium 228
NT2 Ruthenium 103
NT2 Strontium 90
NT2 Thorium 230
NT2 Tritium
NT2 Uranium 238
RT Body Content
RT Carrier Free
RT Environmental Detection Limits
RT Radionuclide Barriers (Natural or
    Engineered)
RT Radioactivity Decay Constant
RT Transuranic Elements
RT Transuranic Nuclides
SO Air Pollution
SO Emergency Preparedness
SO Environmental Management
SO Environmental Protection Agency
    Glossary
SO Radiation
DEF (EMER) Radioactive elements
    characterized according to their
    atomic mass and atomic number
    which can be man-made or
    naturally occurring. Radioisotopes
    can have a long life as soil or
    water pollutants, and are believed
    to have potentially mutagenic
    effects on the human body. (NIH)
    Nuclides with unstable ratios of
    neutrons to protons placing the
    nucleus in a state of stress. In an
    attempt to reorganize to a more
    stable state, they may undergo
    various types of rearrangement
    that involve the release of
    radiation.

RADIOTOXICITY
(NIH)
DA October 12, 1990
BT1 Toxicity
SO Radiation
DEF (NIH) Term referring to the potential
    of an isotope to cause damage to
    living tissue by absorption of
    energy from the disintegration of
    the radioactive material introduced
    into the body.

RADIUM 226
(EDB)
DA March 29, 1991
BT1 Alpha Decay Radioisotopes

BT2 Radionuclides
 BT3 CERCLA Hazardous Substances
  BT4 Hazardous Substances
 BT3 Nuclides
BT1 Years Living Radioisotopes
 BT2 Radionuclides
  BT3 CERCLA Hazardous Substances
   BT4 Hazardous Substances
  BT3 Nuclides
RT Uranium Mill Tailings
SO Radiation

RADIUM 228
(EDB)
DA March 29, 1991
BT1 Beta-Minus Decay Radioisotopes
 BT2 Beta Decay Radioisotopes
  BT3 Radionuclides
   BT4 CERCLA Hazardous
    Substances
    BT5 Hazardous Substances
   BT4 Nuclides
BT1 Years Living Radioisotopes
 BT2 Radionuclides
  BT3 CERCLA Hazardous Substances
   BT4 Hazardous Substances
  BT3 Nuclides
RT Uranium Mill Tailings
SO Radiation

RADON
(TSCA; EPA)
DA October 12, 1990
RT Radon Flux
RT Radon Progeny
RT Radon Decay Products
RT Uranium Mill Tailings
SO Environmental Protection Agency
   Glossary
SO Hazardous Materials
SO Radiation
DEF (EPA) A colorless naturally
   occurring, radioactive, inert
   gaseous element formed by
   radioactive decay of radium atoms
   in soil or rocks. The radioactive
   gaseous element and its
   short-lived decay products
   produced by the disintegration of
   the element radium occurring in
   air, water, soil, or other media.

RADON 222
(EDB)
DA March 29, 1991
BT1 Alpha Decay Radioisotopes
 BT2 Radionuclides
  BT3 CERCLA Hazardous Substances
   BT4 Hazardous Substances
  BT3 Nuclides
BT1 Days Living Radioisotopes
 BT2 Radionuclides
  BT3 CERCLA Hazardous Substances
   BT4 Hazardous Substances
  BT3 Nuclides
BT1 Radon Progeny
 BT2 Radionuclides
  BT3 CERCLA Hazardous Substances
   BT4 Hazardous Substances

BT3 Nuclides
SO Radiation

RADON DECAY PRODUCTS
(EPA)
DA October 12, 1990
SY Radon Progeny
RT Radon
SO Environmental Protection Agency
   Glossary
DEF (EPA) A term used to refer
   collectively to the immediate
   products of the radon decay chain.
   These include Po-218, Pb-214,
   Bi-214, and Po-214, which have
   an average combined half life of
   about 30 minutes.

RADON FLUX
(NCRP)
DA October 12, 1990
RT Radon
SO Radiation
DEF (NCRP) The number of radon
   atoms passing through a unit
   cross-sectional area per unit time.

RADON PROGENY
(NCRP)
DA October 12, 1990
SY Radon Decay Products
BT1 Radionuclides
 BT2 CERCLA Hazardous Substances
  BT3 Hazardous Substances
 BT2 Nuclides
NT1 Radon 222
RT Radon
RT Unattached Fractions
RT Working Level
RT Working Level Month
SO Radiation
DEF (NCRP) The short-lived
   radionuclides formed as a result of
   decay of radon. For radon-222,
   they consist of polonium-218
   (RaA), Lead-214 (RaB),
   bismuth-214 (RaC) and
   polonium-214 (RaC'). Combined
   numbers of these radionuclides
   are reduced by one- half
   approximately every thirty minutes.

RAFFINATE
(NCRP)
DA October 12, 1990
BT1 Fluids
SO Radiation
DEF (NCRP) Fluid from the purification
   step in a mill, depleted in the
   mineral of interest relative to the
   fluid entering the purification
   process. The raffinate may be
   reused in the process stream or
   discarded.

RAIL FREIGHT CARS
DA June 3, 1991
RT Freight Containers
RT Placarded Cars

SO Hazardous Materials
DEF Cars designed to carry freight or
   non-passenger personnel by rail,
   and include a box car, flat car,
   gondola car, tank car, and
   occupied car.

RAINCOATS
DA October 12, 1990
BT1 Control Rod(s)
 BT2 Equipment/Parts - Nuclear (DOE
   FRASE Vocabulary)
  BT3 Equipment
  BT3 Reactor Components
SO Nuclear Facilities Incident Database
SO Radiation
DEF (NFI) Control rod sheaths, broken
   off during disassembly, located
   between rods and cans.

RAKE(S)
(3050 RAKE)
DA January 3, 1991
BT1 Tools - Manual
 BT2 Tools (DOE FRASE Vocabulary)
  BT3 Equipment
SO DOE FRASE VOCABULARY

RAM
DA October 12, 1990
SEE Reliability, Availability, and
   Maintainability
SO Acronyms

RAM(S)
(2765 RAM)
DA January 3, 1991
BT1 Instrument(s)
 BT2 Equipment/Parts -
   Instrumentation/Measuring (DOE
   FRASE Voc.)
  BT3 Equipment
SO DOE FRASE VOCABULARY

RANGE (STATISTICAL)
(SSDC)
DA October 12, 1990
RT Distribution
SO System Safety Development Center
   Glossary
DEF (SSDC) A measurement of the
   difference between two
   observations. The entire range is
   the difference between the
   smallest and largest value. We
   also speak of the inner two quartile
   range which includes 50% of all
   values (excluding the smallest
   25% and the largest 25%).

RANGE MASTERS
(Doe Order 5480.16)
DA October 12, 1990
BT1 Personnel
RT Firearms Ranges
RT Range Safety Officers
SO Firearms
DEF (DOE Order 5480.16) Individuals
   responsible for daily range

BT3 Nuclides
SO Radiation

operations; the range master
ensures that the range is always
safe and that only qualified
firearms instructors conduct
training activities.

RANGE SAFETY OFFICERS
(Doe Order 5480.16)
DA   October 12, 1990
BT1 Personnel
RT   Firearms Ranges
RT   Range Masters
SO   Firearms
DEF (DOE Order 5480.16) The
     designated and specifically trained
     individuals responsible for safety
     at a live firing range.

RANGES (FIREARMS)
DA   April 19, 1991
SY   Firearms Ranges
BT1 Sites/Areas
SO   Firearms

RAP
DA   October 12, 1990
SEE Remedial Action Program
SO   Acronyms

RASPS
(EPA)
DA   October 12, 1990
BT1 Equipment
SO   Environmental Protection Agency
     Glossary
DEF (EPA) A machine that grinds waste
     into a manageable material and
     helps prevent odor.

RATCHET WRENCH(S)
(3051 RATCHET WREN)
DA   January 3, 1991
BT1 Wrench(s)
 BT2 Tools (DOE FRASE Vocabulary)
  BT3 Equipment
SO   DOE FRASE VOCABULARY

RATES
DA   February 25, 1991
NT1 Absorbed Dose Rate
NT1 Biological Clearance Rate
NT1 Collective Effective Dose
     Equivalent Rate
NT1 Contact Rate
NT1 Converter Arsenic Charging Rate
NT1 Energy Fluence Rate
NT1 Exposure Rate
NT1 Incidence Rate, Total Recordable
     Cases (TRC)
NT1 Incidence Rate, Lost Workday
     Cases (LWC)
NT1 Incidence Rate, WDL
NT1 Incidence Rate, WDLR
NT1 Incidence Rate, LWD (Lost Work
     Days)
NT1 Lowest Achievable Emission Rate
RT   Measurements
RT   Units of Measure

RATIONAL METHOD
(DOE Order 6430.1A)
DA   October 12, 1990
SO   Construction
DEF (DOE Order 6430.1A) As applied to
     drainage design, the expression of
     peak discharge as equal to the
     product of rainfall intensity,
     drainage area and a runoff
     coefficient depending on drainage
     basin characteristics.

RATIOS
DA   February 25, 1991
NT1 Buildup Factor
NT1 Capacity Factor
NT1 Capture Efficiency
NT1 Critical Temperature Ratio
NT1 Decontamination Factor
NT1 Dilution Ratio
NT1 Emission Factor
 NT2 Theoretical Arsenic Emissions
      Factor
NT1 Gas Volume Ratio
NT1 Hazard Quotient
NT1 Loading
 NT2 Grain Loading
 NT2 Particulate Loading
NT1 Loss Ratio
NT1 Peak-To-Peak Ratio
NT1 Reinforcement Ratio
NT1 Roof-Top-Ratio
NT1 Story Drift Ratio
RT   Measurements
RT   Units of Measure

RAW DATA
(TSCA;CFR)
DA   October 19, 1990
RT   Field Sampling Plans
SO   Environmental Management
SO   Hazardous Materials
DEF Any laboratory worksheets,
     records, memoranda, notes, or
     exact copies thereof, that are the
     result of original observations and
     activities of a study and are
     necessary for the reconstruction
     and evaluation of the report of that
     study. In the event that exact
     transcripts of raw data have been
     prepared (e.g., tapes that have
     been transcribed verbatim, dated,
     and verified accurate by
     signature), the exact copy or exact
     transcript may be substituted for
     the original source as raw data.
     "Raw data" may include
     photographs, microfilm or
     microfiche copies, computer
     printouts, magnetic media,
     including dictated observations,
     and recorded data from automated
     instruments.

RAW SEWAGE
(EPA)
DA   October 12, 1990
BT1 Sewage
 BT2 Municipal Solid Wastes

 BT3 Solid Wastes
  BT4 Wastes
SO   Environmental Protection Agency
     Glossary
DEF (EPA) Untreated wastewater.

RAW WATER INTAKE STRUCTURE
DA   January 8, 1991
SF   *RWIS*
BT1 Structures (DOE FRASE
     Vocabulary)

RAW WATER SYSTEM
(2046 RAW WATER SY)
DA   December 10, 1990
BT1 Systems (DOE FRASE Vocabulary)
 BT2 Systems
RT   Drinking Water System
SO   DOE FRASE VOCABULARY

RAYLEIGH SEISMIC WAVES
(USGS)
DA   October 12, 1990
RT   Earthquake Magnitude
RT   Love Seismic Waves
RT   Surface Wave Magnitude
SO   Natural Phenomenon
DEF (USGS) Types of surface waves
     having a retrograde, elliptical
     motion at the free surface. They,
     along with Love waves, mainly
     cause low-frequency vibrations,
     which are more likely to make tall
     buildings vibrate.

RAYTHEON SERVICES NEVADA
DA   January 11, 1991
BT1 Companies
 BT2 Commercial Organizations
  BT3 Organizations
BT1 DOE Contractors
 BT2 Potentially Responsible Parties
RT   Nevada Test Site
RT   Nevada Operations Office

RAZOR(S)
(3052 RAZOR)
DA   January 3, 1991
BT1 Tools - Manual
 BT2 Tools (DOE FRASE Vocabulary)
  BT3 Equipment
RT   Abrasion
SO   DOE FRASE VOCABULARY

RBC
DA   October 12, 1990
SEE Reactor Building Cooling
SO   Acronyms

RBCCW
DA   October 12, 1990
SEE Reactor Building Closed Cooling
     Water
SO   Acronyms

RBCLCW
DA   October 12, 1990

SEE Reactor Building Closed Loop
    Cooling Water
SO Acronyms

RBCS
DA   October 12, 1990
SEE Reactor Building Fan Coolers
SO Acronyms

RBE
DA   October 12, 1990
SEE Relative Biological Effectiveness
SO Acronyms

RBS
DA   October 12, 1990
SEE Reactor Building Spray
SO Acronyms

RBSI
DA   October 12, 1990
SEE Reactor Building Spray Injection
SO Acronyms

RBSR
DA   October 12, 1990
SEE Reactor Building Spray
    Recirculation
SO Acronyms

RBSVS
DA   October 12, 1990
SEE Reactor Building Standby
    Ventilation System
SO Acronyms

RCB
DA   October 12, 1990
SEE Retrieval Containment Building
SO Acronyms

RCIC
DA   October 12, 1990
SEE Reactor Core Isolation Cooling
SO Acronyms

RCM
DA   October 12, 1990
SEE Reliability-Centered Maintenance
SO Acronyms

RCP
DA   October 12, 1990
SEE Reactor Coolant Pump
SO Acronyms

RCRA
(SDWA; RCRA; CFR)
DA   October 12, 1990
SEE Resource Conservation and
    Recovery Act
SO Acronyms
DEF (RCRA) To promote the protection
    of health and the environment and
    to conserve valuable material and
    energy resources.

**RCRA CORRECTIVE MEASURES
STUDY**
DA   January 8, 1991
SF   *CMS*
BT1 Studies

RCRA FACILITY ASSESSMENT
DA   January 8, 1991
SF   *RFA*
BT1 Assessments
  BT2 Administrative Processes
  BT3 Processes

RCRA REMEDIAL INVESTIGATION
DA   January 8, 1991
SF   *RRI*
BT1 Remedial Investigations
  BT2 Investigations
   BT3 Administrative Processes
    BT4 Processes

RCS
DA   October 12, 1990
SEE Reactor Coolant System
SO Acronyms

RCSI
DA   October 12, 1990
SEE Reactor Coolant System Integrity
SO Acronyms

RCW
DA   October 12, 1990
SEE Recirculated Cooling Water
SO Acronyms

RD
DA   October 12, 1990
SEE Remedial Design
SO Acronyms

RDDT&E
DA   October 12, 1990
SEE Research, Development,
    Demonstration, Testing and
    Evaluation
SO Acronyms

rDNA
DA   October 12, 1990
SEE Recombinant DNA
SO Acronyms

RE-REFINED OIL
(RCRA; ANL)
DA   May 24, 1991
BT1 Oils
SO Wastes
DEF (RCRA Sec. 1004(39)) Used oil
    from which the physical and
    chemical contaminants acquired
    through previous use have been
    removed through a refining
    process.

RE-REFINED OILS
(SWDA; RCRA)

DA   October 12, 1990
BT1 Recycled Oils
 BT2 Used Oils
  BT3 Oils
RT   Lubricating Oils
SO Wastes
DEF Used oils from which the physical
    and chemical contaminants
    acquired through previous use
    have been removed through a
    refining process.

REACTIVE METALS, INC.
DA   January 8, 1991
SF   *RMI*
BT1 Companies
  BT2 Commercial Organizations
   BT3 Organizations

REACTOR(S)
(2589 REACTOR)
DA   January 3, 1991
BT1 Equipment/Parts - Nuclear (DOE
    FRASE Vocabulary)
  BT2 Equipment
  BT2 Reactor Components
RT   Core(s)
RT   Pressure Vessel(s)
SO DOE FRASE VOCABULARY

**REACTOR AND REACTOR MATERIAL
TECHNOLOGY**
DA   January 8, 1991
SF   *RRMT*

**REACTOR BUILDING CLOSED
COOLING WATER**
DA   January 8, 1991
SF   *RBCCW*
RT   Reactor Facilities

**REACTOR BUILDING CLOSED LOOP
COOLING WATER**
DA   January 8, 1991
SF   *RBCLCW*
RT   Reactor Facilities

REACTOR BUILDING COOLING
DA   January 8, 1991
SF   *RBC*
RT   Reactor Facilities

REACTOR BUILDING FAN COOLERS
DA   January 8, 1991
SF   *RBCS*
RT   Reactor Facilities

REACTOR BUILDING SPRAY
DA   January 8, 1991
SF   *RBS*
RT   Reactor Facilities

**REACTOR BUILDING SPRAY
INJECTION**
DA   January 8, 1991
SF   *RBSI*
RT   Reactor Facilities

SY-Synonymous Terms   SO-Source/Subject Category   SF-See From

REACTOR BUILDING SPRAY RECIRCULATION
DA   January 8, 1991
SF   *RBSR*
RT   Reactor Facilities

REACTOR BUILDING STANDBY VENTILATION SYSTEM
DA   January 8, 1991
SF   *RBSVS*
BT1 Ventilation System
  BT2 Systems (DOE FRASE Vocabulary)
   BT3 Systems
RT   Reactor Facilities

REACTOR COMPONENTS
DA   February 12, 1991
NT1 Control Rod Drive Control System
NT2 Control Rod Drive Assembly
NT2 Control Rod Drive
  NT3 Pistol Grip Switches
NT2 Control Rod Drive Mechanism
NT1 Equipment/Parts - Nuclear (DOE FRASE Vocabulary)
NT2 Birdcage
NT2 Burial Box(s)
NT2 Cold Trap
NT2 Control Panel(s)
NT2 Control Rod Drive Assembly
NT2 Control Rod(s)
  NT3 Gangs
  NT3 Raincoats
NT2 Coolant Return Pump
NT2 Core(s)
NT2 Exhaust Stack
NT2 Flux Channel(s)
NT2 Fuel Element(s)
  NT3 Fuel Plate
   NT4 Cladding
   NT4 Plutonium Fuel Plate(s)
  NT3 Fuel Rod(s)
  NT3 Piles (Nuclear)
NT2 Fuel Handling Equipment
  NT3 Fuel Canister(s)
  NT3 Fuel Drawer(s)
NT2 Fuel Tool(s)
NT2 Glove Box
NT2 Heat Exchanger
  NT3 Low Pressure Recirculation System Heat Exchanger
  NT3 Shutdown Heat Exchanger
NT2 Hepa Filter
NT2 Ion Exchange Column
NT2 Ion Pump
NT2 Log N Channel
NT2 Manipulator
  NT3 Electromechanical Manipulator(s)
NT2 Neutron Source
NT2 Neutron(s)
NT2 Plutonium Process Hood
NT2 Plutonium Recovery Process
NT2 Pressure Vessel Part
NT2 Pressure Vessel(s)
NT2 Reactor Fuel
NT2 Reactor Waste
NT2 Reactor(s)
NT2 Rod Drive
NT2 Shield Plug(s)
NT2 Sodium Scrubber(s)

NT2 Test Train
NT2 Transient Operation
NT2 Transient Test
NT1 Forests (Reactor)
NT1 Reactor Fuel
NT1 Reactor Coolant Pump
NT1 Reactor Coolant System
 NT2 Emergency Spray Water System
NT1 Reactor Pressure Vessel
NT1 Reactor Vessel
NT1 Shaft Guides
NT1 Special Isotope Separator
NT1 Spiders
RT   Rabbits
RT   Reactor Refueling Activity
RT   Reactor Protection System
RT   Reactors
RT   Targets

REACTOR COOLANT PUMP
DA   January 8, 1991
SF   *RCP*
BT1 Pump(s)
  BT2 Machines (DOE FRASE Vocabulary)
   BT3 Equipment
BT1 Reactor Components

REACTOR COOLANT SYSTEM
DA   January 8, 1991
SF   *RCS*
BT1 Reactor Components
BT1 Systems
NT1 Emergency Spray Water System
RT   Reactor Core Isolation Cooling
RT   Reactor Coolant System Integrity
RT   Reactor Heat Removal
DEF (NRC Glossary of Terms: Nuclear Power Radiation) The cooling system used to remove energy from the reactor core and transfer that energy either directly or indirectly to the steam turbine.

REACTOR COOLANT SYSTEM INTEGRITY
DA   January 8, 1991
SF   *RCSI*
RT   Reactor Coolant System

REACTOR CORE ISOLATION COOLING
DA   January 8, 1991
SF   *RCIC*
RT   Reactor Coolant System

REACTOR ELECTRIC DISTRIBUTION SYSTEM
DA   January 8, 1991
SF   *REDS*
BT1 Systems

REACTOR FACILITIES
(DOE Order 5480.6; EMER)
DA   October 12, 1990
BT1 Nuclear Facilities
 BT2 Facilities
NT1 Low Power Reactor Facility
NT1 Production Reactor Facility
NT1 Test Reactor Facility

RT   Nuclear Facilities
RT   Reactor Building Cooling
RT   Reactor Building Closed Cooling Water
RT   Reactor Building Closed Loop Cooling Water
RT   Reactor Building Fan Coolers
RT   Reactor Building Spray
RT   Reactor Building Spray Injection
RT   Reactor Building Spray Recirculation
RT   Reactor Building Standby Ventilation System
RT   Reactor Works Engineering
RT   Reactors
RT   Test Reactor Area
SO   Emergency Preparedness
SO   Industrial Safety
DEF (DOE Order 5480.6) Unless modified by words, such as containment, vessel, or core, means the entire reactor facility including housing, equipment, and associated areas devoted to operation and maintenance of one or more reactor cores.

REACTOR FUEL
(2587 REACTOR FUEL)
DA   January 3, 1991
BT1 Equipment/Parts - Nuclear (DOE FRASE Vocabulary)
 BT2 Equipment
 BT2 Reactor Components
BT1 Fuels
BT1 Reactor Components
RT   Buckled Zones
RT   Core Melt Accidents
RT   Fuel Element(s)
SO   DOE FRASE VOCABULARY

REACTOR HEAT REMOVAL
DA   January 8, 1991
SF   *REHR*
RT   Reactor Coolant System

REACTOR MATERIALS
DA   May 24, 1991
BT1 Materials
NT1 Nuclear Poisons
SO   Radiation

REACTOR MATERIALS CONTROL GROUP
DA   January 8, 1991
SF   *RMCG*
BT1 Groups
 BT2 Administrative Organizations
  BT3 Organizations

REACTOR OPENING LOSS
(CAA; CFR)
DA   October 12, 1990
BT1 Emissions
 BT2 Air Pollutants
BT1 Losses
RT   Vinyl Chloride
SO   Air Pollution
DEF The emissions of vinyl chloride

occurring when a reactor is vented to the atmosphere for any purpose other than an emergency relief discharge as defined in 40 CFR 61.65(a).

REACTOR OPERATIONS
(Doe Order 5480.6)
DA   October 12, 1990
BT1 Operations
  BT2 Activities
RT   Alternate Removal Systems
RT   Backup Systems
RT   Computer Inoperative Limits
RT   Reactor Operator
RT   Reactors
SO   Industrial Safety
DEF (DOE Order 5480.6) All those activities or functions involved in operating and using a reactor which, for purposes of this Order, begin with the initial loading of fuel in the reactor vessel and end with the removal of fuel to officially decommission or place the reactor in a standby status.

REACTOR OPERATOR
(Doe Order 5480.6)
DA   October 12, 1990
SF   *RO (Reactor Operator)*
BT1 Personnel
NT1 Senior Reactor Operator
RT   Reactor Operations
SO   Industrial Safety
DEF (DOE Order 5480.6) An individual certified by contractor management to operate (manipulate the controls of) a DOE-owned reactor.

REACTOR PRESSURE VESSEL
DA   January 8, 1991
SF   *RPV*
BT1 Reactor Components

REACTOR PROJECTS
(Doe Order 5480.6)
DA   October 12, 1990
BT1 Projects
SO   Industrial Safety
DEF (DOE Order 5480.6) Those activities that contribute to siting, designing, constructing, operating, or decommissioning a reactor, and those activities involving the operation or maintenance of operable and standby reactors, including shutdown reactors containing fuel.

REACTOR PROTECTION SYSTEM
DA   January 8, 1991
SF   *RPS*
BT1 Emergency Systems
  BT2 Systems
NT1 Emergency Core Cooling System
RT   Reactor Components

REACTOR REFUELING ACTIVITY
(1248 RR ACTIVITY)
DA   November 28, 1990
BT1 Activity Types (DOE FRASE Vocabulary)
  BT2 Activities
RT   Reactor Components
SO   DOE FRASE VOCABULARY

REACTOR SAFETY STUDY
DA   January 8, 1991
SF   *RSS*
BT1 Studies

REACTOR SAFETY STUDY METHODOLOGY APPLICATION PROGRAM
DA   January 8, 1991
SF   *RSSMAP*
BT1 Programs
RT   Studies

REACTOR SHUTDOWN
(NFI)
DA   October 12, 1990
NT1 Scram
  NT2 Automatic Reactor Scram
  NT2 Manual Reactor Scram
RT   Anticipated Transient Without Scram
RT   Automatic Backup Shutdown of the Safety Computer
RT   Core Melt Accidents
RT   Engineered Safety Features (Shutdown)
RT   Process Unit Shutdown
RT   Reactors
RT   Safe Shutdown System
RT   Shutdown Cooling
RT   Shutdown Heat Exchanger
RT   Shutdown Sequencer
RT   Very Low Flow Constant
SO   Nuclear Facilities Incident Database
SO   Radiation
DEF (NFI) Used as extra descriptor or as qualifier to reactor when the fact that reactor was in shutdown mode is significant to the severity of the incident.

REACTOR SUBCRITICALITY
DA   January 8, 1991
SF   *RESC*
BT1 Criticality

REACTOR SUPERVISOR
(Doe Order 5480.6)
DA   October 12, 1990
BT1 Technically Qualified Individuals
  BT2 Personnel
SO   Industrial Safety
DEF (DOE Order 5480.6) An individual certified by contractor management to operate or to direct the operation of a DOE-owned Category B reactor.

REACTOR TECHNOLOGY DEPARTMENT
DA   January 8, 1991
SF   *RTD*

REACTOR VESSEL
DA   January 8, 1991
SF   *RV*
BT1 Reactor Components

REACTOR VOLUME CONTROL
DA   January 8, 1991
SF   *RVC*
BT1 Controls
DEF (NFI) Related to inventory control.

REACTOR WASTE
(2588 REACTOR WAST)
DA   January 3, 1991
BT1 Equipment/Parts - Nuclear (DOE FRASE Vocabulary)
  BT2 Equipment
  BT2 Reactor Components
SO   DOE FRASE VOCABULARY

REACTOR WATER LEVEL
DA   January 8, 1991
SF   *RWL*

REACTOR WORKS ENGINEERING
DA   January 8, 1991
SF   *RWE*
RT   Reactor Facilities

REACTORS
DA   January 8, 1991
SF   *RX*
NT1 Advanced Test Reactor
NT1 Annular Core Research Reactor
NT1 Arkansas Nuclear One-1
NT1 Boiling Water Reactor
NT1 Category A Reactors
NT1 Experimental Boiling Water Reactor
NT1 Light Water Reactor
NT1 Materials Test Reactor
NT1 Naval Reactors
NT1 New Production Reactor
NT1 Omega West Reactor
NT1 Pressurized Water Reactor
NT1 Sandia Pulse Reactor III
RT   Advisory Committee on Reactor Safety
RT   NRC Office of Nuclear Reactor Regulation
RT   Reactor Components
RT   Reactor Facilities
RT   Reactor Operations
RT   Reactor Shutdown
DEF (DOE Order 5480.6) Any apparatus that is designed or used to sustain nuclear chain reactions in a controlled manner, including critical and pulsed assemblies and research, test, and power reactors, is defined as a reactor. All assemblies designed to perform subcritical experiments which could potentially reach criticality are also to be considered

reactors. Critical assemblies are special nuclear devices designed and used to sustain nuclear reactions. Critical assemblies may be subject to frequent core and lattice configuration changes and may be used often as mockups of reactor configurations. Therefore, requirements for modifications do not apply unless the overall assembly room is modified, a new assembly room proposed, or a new configuration not covered in previous safety evaluations.

REACTORS (CHEMICAL)
(CAA; CFR)
DA   October 12, 1990
BT1 Facility Components
RT   Polyvinyl Chloride
RT   Vinyl Chloride
SO   Air Pollution
DEF Includes any vessels in which vinyl chloride is partially or totally polymerized into polyvinyl chloride.

READINESS ASSURANCE
(EMER)
DA   February 1, 1991
SO   Emergency Preparedness
DEF (EMER) The actions taken to provide assurance that headquarters, field elements, and facility contractors implement appropriate aspects of U.S. Department of Energy (DOE) emergency management program policies and requirements as established by DOE orders.

REAGENT GRADE
(ESH)
DA   October 12, 1990
SO   Quality Assurance
DEF (ESH) Analytical reagent (AR) grade, ACS reagent grade, and reagent grade are synonymous terms for reagents that conform to the current specifications of the Committee on Analytical Reagents of the American Chemical Society.

REAL PROPERTY
(DOE Order 4330.4A)
DA   June 5, 1991
BT1 Property
RT   Equipment
RT   Facilities
RT   Personal Property
RT   Sites/Areas
SO   Management
DEF (DOE Order 4330.4A) Includes land, improvements on the land, or both, including interests therein. The chief characteristics of real property (real estate) are its immobility and tangibility. It comprises land and all things of a permanent and substantial nature

affixed thereto, whether by nature or by the hand of man. By "nature" is meant trees, the products of the land, natural resources; by "the hand of man", those objects, buildings, fences, bridges, etc. that are erected upon the land. All equipment or fixtures (such as plumbing, electrical, heating, built-in cabinets, and elevators) that are installed in a building in a more or less permanent manner or which are essential to its primary purpose, usually are held to part of the real property.

REAL PROPERTY INVENTORY SYSTEM
(DOE Order 6430.1A)
DA   October 12, 1990
SF   RPIS
BT1 Information Systems
BT2 Security Interests
BT2 Systems
SO   Construction
DEF (DOE Order 6430.1A) The Department of Energy's automated real property reporting system.

REASONABLE AND PRUDENT ALTERNATIVES
(ESA; CFR)
DA   October 12, 1990
SY   Alternative Courses of Action
BT1 Actions
BT2 Responses
RT   Formal Consultation
RT   Irreversible Commitment of Resources
RT   Reasonable and Prudent Measures
SO   Endangered Species
DEF Alternative actions identified during formal consultation that can be implemented in a manner consistent with the intended purpose of the action, that can be implemented consistent with the scope of the Federal agency's legal authority and jurisdiction, that are economically and technologically feasible, and that the Director believes would avoid the likelihood of jeopardizing the continued existence of listed species or resulting in the destruction or adverse modification of critical habitat.

REASONABLE AND PRUDENT MEASURES
(ESA; CFR)
DA   October 12, 1990
BT1 Actions
BT2 Responses
RT   Incidental Take (Taking)
RT   Reasonable and Prudent Alternatives
SO   Endangered Species
DEF Those actions the Director believes necessary or appropriate to

minimize the impacts (i.e., amount or extent) of incidental take.

REASONABLE COSTS
(CERCLA)
DA   November 15, 1990
BT1 Costs
SO   Compensation and Liability
DEF (CERCLA) Amounts that may be recovered for the cost of performing a damage assessment. Costs are reasonable when: the Injury Determination, Quantification, and Damage Determination phases have a well-defined relationship to one another and are coordinated; the anticipated increment of extra benefits in terms of the precision or accuracy of estimates obtained by using a more costly injury, quantification, or damage determination methodology are greater than the anticipated increment of extra costs of that methodology; and the anticipated cost of the assessment is expected to be less than the anticipated damage amount determined in the Injury, Quantification, and Damage Determination phases.

REASONABLE FURTHER PROGRESS
(CAA)
DA   October 12, 1990
SY   Increments of Progress
RT   National Ambient Air Quality Standards
RT   Schedule of Compliance
SO   Air Pollution
DEF (CAA) Annual incremental reductions in emissions of the applicable air pollutant (including substantial reductions in the early years following approval or promulgation of plan provisions under this part [42 USCS 7501 et seq.] and section 110(a)(2)(I) [42 USCS 7410(a)(2)(I)] and regular reductions thereafter) which are sufficient in the judgment of the Administrator to provide for attainment of the applicable national ambient air quality standard by the date required in section 172(a) [42 USCS @ 7502(a)].

REASONABLY ATTRIBUTABLE
(CAA; CFR)
DA   October 12, 1990
RT   Causes
SO   Air Pollution
DEF (CFR) Attributable by visual observation or any other technique the State deems appropriate.

REASONABLY AVAILABLE CONTROL TECHNOLOGY
(CAA; CFR)
DA   October 12, 1990
SF   *RACT*
RT   Best Available Control Technology
RT   Existing Sources
RT   Non-Attainment Areas
SO   Air Pollution
SO   Environmental Management
SO   Environmental Protection Agency Glossary
DEF (CFR) The lowest emissions limit that a particular source is capable of meeting by the application of control technology that is both reasonably available, as well as technologically and economically feasible. RACT is usually applied to existing sources in nonattainment areas and in most cases is less stringent than new source performance standards. Devices, systems process modifications, or other apparatus or techniques that are reasonably available taking into account (1) the necessity of imposing such controls in order to attain and maintain a national ambient air quality standard; (2) the social, environmental, and economic impact of such controls; and (3) alternative means of providing for attainment and maintenance of such standard. (This provision defines RACT for the purposes of 40 CFR 51.110(c)(2) and 51.341(b) only.)

REBRICKING
(CAA; CFR)
DA   October 12, 1990
RT   Glass Melting Furnaces
RT   Refractories
SO   Air Pollution
DEF Cold replacement of damaged or worn refractory parts of the glass melting furnace. Rebricking includes replacement of the refractories comprising the bottom, sidewalls, or roof of the melting vessel; replacement of refractory work in the heat exchanger; and replacement of refractory portions of the glass conditioning and distribution system.

RECEIVING STREAMS
(DOE Order 6430.1A)
DA   October 12, 1990
BT1 Receiving Waters
SO   Construction
DEF (DOE Order 6430.1A) Streams that receive outfall discharge of wastewater effluents.

RECEIVING WATERS
(EPA)
DA   October 12, 1990

NT1 Receiving Streams
RT   Wastewater
RT   Watershed
SO   Environmental Protection Agency Glossary
DEF (EPA) A river, lake, ocean, stream or other watercourse into which wastewater or treated effluent is discharged.

RECHARGE
DA   November 9, 1990
RT   Percolation
RT   Saturated Zones
DEF The process by which water is added to a zone of saturation, usually by percolation from the soil surface, e.g., the recharge of an aquifer.

RECHARGE AREAS
(EPA)
DA   October 12, 1990
BT1 Sites/Areas
SO   Environmental Protection Agency Glossary
DEF (EPA) Land areas in which water reaches to the zone of saturation from surface infiltration, e.g., areas where rainwater soaks through the earth to reach an aquifer.

RECIRCULATED COOLING WATER
DA   January 8, 1991
SF   *RCW*

RECIRCULATION PUMP TRIP
DA   January 8, 1991
SF   *RPT*

RECIRCULATOR
(2138 RECIRCULATOR)
DA   December 10, 1990
BT1 Machines (DOE FRASE Vocabulary)
BT2 Equipment
SO   DOE FRASE VOCABULARY

RECLAMATION
(EDB)
DA   February 1, 1991
BT1 Resource Recovery
BT2 Pollution Recovery Processes
BT3 Processes
NT1 Land Reclamation
NT1 Refuse Reclamation
RT   Abatement
RT   Air Pollution
RT   Land Disposal
RT   Land Treatment Facilities
RT   Waters of the United States
RT   Water Pollution
SO   Environmental Management
DEF (DSTT) The recovery of natural resources previously abandoned due to some form of damage.

RECOMBINANT BACTERIA
(EPA)
DA   October 12, 1990
BT1 Bacteria
BT2 Microorganisms
BT3 Organisms
SO   Environmental Protection Agency Glossary
DEF (EPA) A type of microorganism whose genetic makeup has been altered by deliberate introduction of new genetic elements. The offspring of these altered bacteria also contain these new genetic elements.

RECOMBINANT DNA
(EPA)
DA   October 12, 1990
SF   *rDNA*
BT1 DNA
BT2 Organic Chemicals
BT3 Chemical Substances
SO   Environmental Protection Agency Glossary
DEF (EPA) The new DNA that is formed by combining pieces of DNA from different organisms or cells.

RECOMMENDATIONS
(SSDC)
DA   October 12, 1990
NT1 Conservation Recommendations
NT1 Corrective Actions
NT1 Protective Action Recommendations
RT   Guidelines
RT   Responses
SO   System Safety Development Center Glossary
DEF (SSDC) Specific methods and corrective actions believed feasible, logical, practical, and sufficient to fulfill the judgments of needs. In general, each need is expressed in two kinds of recommendations: (1) for fixing the specific problems involved in the occurrence; and (2) for fixing systemic problems revealed during the investigation.

RECOMMENDED MAXIMUM CONTAMINANT LEVEL
(EPA)
DA   October 12, 1990
SF   *RMCL*
RT   Maximum Contaminant Levels
SO   Environmental Protection Agency Glossary
DEF (EPA) The maximum level of a contaminant in drinking water at which no known or anticipated adverse affect on human health would occur, and which includes an adequate margin of safety. Recommended levels are nonenforceable health goals.

SY-Synonymous Terms   SO-Source/Subject Category   SF-See From

RECONSTRUCTED SOURCES
(EPA)
DA   October 12, 1990
BT1 Existing Facilities
BT2 Facilities
RT   Reconstruction
SO   Environmental Protection Agency
     Glossary
DEF (EPA) Existing facilities in which
     components are replaced to such
     an extent that the fixed capital
     cost of the new components
     exceed 50 percent of the capital
     cost that would be required to
     construct a comparable entirely
     new facility. New source
     performance standards may be
     applied to sources that are
     reconstructed after the proposal of
     the standard if it is technologically
     and economically feasible to meet
     the standard.

RECONSTRUCTION
(CAA; CFR; ESH)
DA   October 12, 1990
BT1 Administrative Processes
BT2 Processes
RT   Reconstructed Sources
SO   Air Pollution
DEF Will be presumed to have taken
     place where the fixed capital cost
     of the new component exceeds 50
     percent of the fixed capital cost of
     a comparable entirely new source.
     Any final decision as to whether
     reconstruction has occurred must
     be made in accordance with the
     provisions of 40 CFR 60.15(f) (1)
     through (3).

RECORD OF DECISION (CERCLA)
(EPA; CERCLA)
DA   May 16, 1991
SF   ROD (CERCLA Record of Decision)
SO   Environmental Management
SO   Environmental Protection Agency
     Glossary
DEF (EPA) A public document prepared
     by the Environmental Protection
     Agency that explains which
     cleanup alternatives(s) will be
     used at National Priorities List
     sites where, under CERCLA, the
     Trust Fund pays for the cleanup.
     (DOE) The Record of Decision,
     under CERCLA, is a legally
     binding decision document
     required for all remedial actions. It:
     1) summarizes problems posed by
     a site and analyzes methods to
     address them; 2) documents the
     decision process; 3) demonstrates
     consistency with CERCLA, SARA,
     and the NCP; 4) supports future
     cost recovery actions; and 5)
     serves as the centerpiece for the
     administrative record.

RECORD OF DECISION (NEPA)
(DOE Order 5440.1D; NEPA; ESH)
DA   October 12, 1990
SY   Enforcement Decision Documents
SF   ROD (NEPA Record of Decision)
RT   Alternative Courses of Action
RT   Mitigation Action Plans
RT   National Environmental Policy Act
     Documents
RT   Programmatic NEPA Documents
RT   Proposed Actions
RT   Site-Wide NEPA Documents
RT   Supplement Analyses
SO   Environmental Management
SO   Environmental Protection Agency
     Glossary
SO   Management
DEF (ESH) A concise public record of
     the Department's decision on a
     proposed action for which an
     environmental impact statement
     was prepared which includes the
     alternatives considered, the
     environmentally preferable
     alternative, factors balanced in the
     decision, and mitigation measures
     and monitoring to minimize harm.
     (DOE Order) A document
     prepared in accordance with the
     requirements of 40 CFR 1505.2,
     that provides a concise public
     record of the Department's
     decision on a proposed action for
     which an Environmental Impact
     Statement was prepared, and
     identifies the alternatives
     considered in reaching the
     decision, the environmentally
     preferable alternative(s), factors
     balanced by the Department in
     making the decision, whether all
     practicable means to avoid or
     minimize environmental harm
     have been adopted, and if not,
     why they were not.

RECORDED DOSES
(Doe Order 5484.1)
DA   October 12, 1990
BT1 Doses
SO   Management
DEF (DOE Order 5484.1) Those
     numbers (corrected for
     background), zero (minimal or
     negligible) and above, which are
     recorded as representing
     individuals' doses from external
     radiation sources or internally
     deposited radioactive materials
     determined in accordance with
     DOE 5480.1B, Chapter XI,
     requirements.

RECORDER
(2766 RECORDER)
DA   January 3, 1991
BT1 Instrument(s)
BT2 Equipment/Parts -
     Instrumentation/Measuring (DOE
     FRASE Voc.)

BT3 Equipment
SO   DOE FRASE VOCABULARY

RECORDING LEVELS
(IAEA)
DA   November 20, 1990
BT1 Reference Levels
SO   Radiation
DEF (IAEA) Levels defined by the
     competent authority or the
     management for any of the
     quantities determined in the
     practice of radiation protection,
     above which recording of the
     information is taken to be
     necessary.

RECOVERABLE
(SWDA; RCRA)
DA   October 12, 1990
RT   Recovered Materials
SO   Wastes
DEF Refers to the capability and
     likelihood of being recovered from
     solid waste for a commercial or
     industrial use.

RECOVERED MATERIALS
(SWDA; RCRA)
DA   October 12, 1990
BT1 Materials
RT   Recoverable
RT   Solid Wastes
SO   Wastes
DEF Waste materials and by-products
     that have been recovered or
     diverted from solid waste. This
     term does not include those
     materials and by-products
     generated from, and commonly
     reused within, an original
     manufacturing process.

RECOVERED RESOURCES
(RCRA; ANL)
DA   May 24, 1991
BT1 Resource
RT   Resource Recovery
RT   Solid Wastes
SO   Environmental Management
SO   Wastes
DEF (RCRA Sec. 1004(20)) Material or
     energy recovered from solid
     waste.

RECOVERY
(ESA; CFR)
DA   October 12, 1990
RT   Listed Species
RT   Resource Conservation and
     Recovery Act
SO   Endangered Species
DEF Improvement in the status of listed
     species to the point at which listing
     is no longer appropriate under the
     criteria set out in section 4(a)(1) of
     the Endangered Species Act.

RECOVERY ACTIONS
(EMER; MORT)
DA   February 1, 1991
BT1 Actions
 BT2 Responses
BT1 Recovery Plans
 BT2 Amelioration
RT   Corrective Actions
RT   Emergency Actions
RT   Relations
SO   Emergency Preparedness
DEF (EMER) Those actions taken after
     an emergency to restore the
     affected areas as nearly as
     possible to the preemergency
     condition. (MORT) MORT analysis
     asks: were recovery actions taken
     with respect to persons and
     objects after the accident? if an
     injury was disabling, could it's
     overall disabling effect have been
     reduced and/or the individual
     made more functional? was
     damaged equipment, buildings, or
     other property expeditiously
     repaired, salvaged, or replaced?

RECOVERY PLANS
(EMER)
DA   February 1, 1991
BT1 Amelioration
NT1 Recovery Actions
SO   Emergency Preparedness
DEF (EMER) Plans developed to restore
     the affected area with federal
     assistance if needed.

RECREATION AREA
(1616 RECREATION A)
DA   December 10, 1990
BT1 Area
 BT2 Sites/Areas
RT   Recreation/Break Activity
SO   DOE FRASE VOCABULARY

RECREATION/BREAK ACTIVITY
(1241 RB ACTIVITY)
DA   November 28, 1990
BT1 Activity Types (DOE FRASE
     Vocabulary)
 BT2 Activities
RT   Recreation Area
SO   DOE FRASE VOCABULARY

RECTIFIER(S)
(2413 RECTIFIER)
DA   January 3, 1991
BT1 Equipment/Parts - Electrical (DOE
     FRASE Vocabulary)
 BT2 Equipment
SO   DOE FRASE VOCABULARY

RECTUM
(1120 RECTUM)
DA   November 28, 1990
BT1 Digestive System
 BT2 Body System(s)
 BT3 Human Body Parts
SO   DOE FRASE VOCABULARY

RECYCLED MATERIALS
DA   January 24, 1991
BT1 Materials
RT   Post Consumer Wastes
RT   Recycling
RT   Scrap
DEF (CFR) Materials that can be utilized
     in place of raw or virgin materials
     in manufacturing a product and
     consist of materials derived from
     post consumer waste, industrial
     scrap, material derived from
     agricultural wastes and other
     items, all of which can be used in
     the manufacture of new products.

RECYCLED OILS
(SWDA; RCRA)
DA   October 12, 1990
BT1 Used Oils
 BT2 Oils
NT1 Re-refined Oils
SO   Wastes
DEF Any used oils that are reused,
     following their original use, for any
     purpose (including the purpose for
     which the oil was originally used).
     Such terms includes oils that are
     re-refined, reclaimed, burned, or
     reprocessed.

RECYCLED PCBS
(TSCA; CFR)
DA   October 12, 1990
BT1 Polychlorinated Biphenyls
 BT2 Carcinogens
 BT3 Hazardous Substances
 BT2 Chlorinated Hydrocarbons
 BT3 CERCLA Hazardous Substances
 BT4 Hazardous Substances
 BT3 Halogenated Organic
     Compounds
 BT4 Halogenated
 BT4 Organic Chemicals
 BT5 Chemical Substances
SO   Hazardous Materials
DEF Defined as those intentionally
     manufactured PCBs that appear in
     the processing of paper products
     or asphalt roofing materials as
     PCB-contaminated raw materials
     and that meet these requirements:
     (1) The concentration of Aroclor
     PCBs in paper products leaving
     any manufacturing site or imported
     into the United States must have
     an annual average of less than 25
     ppm with a 50 ppm maximum. (2)
     There are no detectable
     concentrations of Aroclor PCBs in
     asphalt roofing materials. (3) The
     release of Aroclor PCBs at the
     point at which emissions are
     vented to ambient air must be less
     than 10 ppm. (4) The amount of
     Aroclor PCBs added to water
     discharged from a processing site
     must at all times be less than 3
     micrograms per liter (mg/1) for
     total Aroclors (roughly 3 parts per

billion (3 ppb)). (5) Disposal of any
other process wastes above
concentrations of 50 ppm PCB
must be in accordance with
Subpart D of 40 CFR 761.

RECYCLING
(EPA)
DA   October 12, 1990
SY   Reuse
BT1 Waste Management Processes
 BT2 Processes
NT1 Closed-Loop Recycling
RT   Cullet
RT   Discarded Materials
RT   Recycled Materials
RT   Refuse Reclamation
RT   Scrap
SO   Environmental Protection Agency
     Glossary
DEF (EPA) The process of minimizing
     the generation of waste by
     recovering usable products that
     might otherwise become waste.
     Examples are the recycling of
     aluminum cans, wastepaper, and
     bottles.

RED BORDERS
(EPA)
DA   October 12, 1990
BT1 Reports
SO   Environmental Protection Agency
     Glossary
DEF (EPA) EPA documents undergoing
     final review before being submitted
     for final management decision.

RED TIDE
(EPA)
DA   October 12, 1990
SY   Algae Bloom
SY   Water Bloom
BT1 Natural Phenomenon
SO   Environmental Protection Agency
     Glossary
DEF (EPA) A proliferation of a marine
     plankton that is toxic and often
     fatal to fish. This natural
     phenomenon may be stimulated
     by the addition of nutrients. A tide
     can be called red, green, or brown
     depending on the coloration of the
     plankton.

REDAC
DA   October 12, 1990
SEE Remote Detection and Control
SO   Acronyms

REDS
DA   October 12, 1990
SEE Reactor Electric Distribution System
SO   Acronyms

REDUCTION
DA   May 20, 1991
BT1 Chemical Processes
 BT2 Processes

SY-Synonymous Terms   SO-Source/Subject Category   SF-See From

SO Radiation
DEF (DSTT) Reaction of hydrogen with
another substance. Chemical
reaction in which an element gains
an electron (has a decrease in
positive valence).

REDUNDANCE
(SSDC)
DA   October 12, 1990
RT Controls
RT Quality Assurance
SO System Safety Development Center
Glossary
DEF (SSDC) A planned duplication of
controls to increase assurance of
reliability.

REECo
DA   October 12, 1990
SEE Reynolds Electrical and
Engineering Company
SO Acronyms

REEL(S)
(2339 REEL)
DA   December 10, 1990
BT1 Equipment/Parts - Material
Handling (DOE FRASE
Vocabulary)
BT2 Equipment
SO DOE FRASE VOCABULARY

REENTRY INTERVALS
(EPA)
DA   October 12, 1990
BT1 Time Designations
SO Environmental Protection Agency
Glossary
DEF (EPA) Periods of time immediately
following the application of a
pesticide during which unprotected
workers should not enter a field.

REFERENCE DOSE (RFD)
(EPA)
DA   October 12, 1990
SY RFD
BT1 Doses
NT1 Chronic RfD
NT1 Developmental RfD
NT1 Subchronic RfD (RfD)
SO Environmental Protection Agency
Glossary
DEF (EPA) Toxicity value used most
often in evaluating
noncarcinogenic effects resulting
from exposures at Superfund
sites. See specific entries for
chronic RfDs, subchronic RfDs,
and developmental RfDs.

REFERENCE LEVELS
(IAEA)
DA   February 27, 1991
NT1 Emergency Reference Levels
NT1 Intervention Levels
NT1 Investigation Levels
NT1 Recording Levels

NT1 Waste Load Allocations
NT1 Water Quality Criteria
SO Emergency Preparedness
SO Radiation
DEF (IAEA) The values of quantities
which govern a particular course
of action. Such levels may be
established for any of the
quantities determined in the
practice of radiation protection;
when they are reached or
exceeded, all relevant information
is considered and the appropriate
action may be taken. Reference
levels are not to be confused with
the limits. Reference levels in the
IAEA Basic Safety Standards are:
recording levels, investigation
levels, and intervention levels.

REFERENCE MAN
(DOE Order 5400.5; IAEA; NCRP)
DA   October 12, 1990
RT Annual Limit on Intake
RT Derived Air Concentration
RT Derived Concentration Guide
RT Radiation Protection
SO Radiation
SO Wastes
DEF A hypothetical aggregation of
human (male and female) physical
and physiological characteristics
arrived at by international
consensus (ICRP Publication 23).
These characteristics may be used
by researchers and public health
workers to standardize results of
experiments and to relate
biological insult from ionizing
radiation to a common base. The
"reference man" is assumed to
inhale 8400 cubic meters of air in
a year and to ingest 730 liters of
water in a year. (IAEA) A model of
a hypothetical adult with the
anatomical and physiological
characteristics defined in the
report of the ICRP Task Group on
Reference Man, used in dosimetry
for radiation protection purposes.

REFERENCE METHODS
(CAA; ESH)
DA   October 12, 1990
NT1 Method 24
NT1 Method 25
RT Air Monitoring
RT Alternate Methods
RT Equivalent Methods
RT Sampling
SO Air Pollution
SO Environmental Management
DEF (CFR) Methods of sampling and
analyzing the ambient air for an air
pollutant that are specified as
reference methods in an appendix
to 40 CFR 50, or a method that
has been designated as a
reference method in accordance
with 40 CFR 53; it does not

include a method for which a
reference method designated has
been cancelled in accordance with
40 CFR 53.11 or 53.16.

REFERENCE STANDARDS
(DOE Order 5480.4; ESH)
DA   October 12, 1990
BT1 Standards
BT2 Codes, Standards, and
Regulations
SO Environmental Management
SO Management
SO Standards
DEF Those guides or standards of DOE
Order 5480.4 adopted by DOE
that DOE and its contractors
should consider for guidance, as
applicable, in addition to the
mandatory standards.

REFERENCE SUBSTANCES
(TSCA)
DA   October 19, 1990
BT1 Samples
SO Hazardous Materials
DEF (TSCA) Any chemical substances
or mixtures, or analytical
standards, or materials other than
a test substance, feed, or water,
that are administered to or used in
analyzing the test system in the
course of a study for the purposes
of establishing a basis for
comparison with the test
substance for known chemical or
biological measurements.

REFLEX ARC RESPONSES
(SSDC)
DA   October 12, 1990
BT1 Stimulus-Response
RT Human Factors
SO System Safety Development Center
Glossary
DEF (SSDC) Those involuntary
responses to physical stimulus,
such as the leg jerking when the
knee is tapped in the right spot.

REFRACTORIES
(DOE Order 6430.1A)
DA   October 12, 1990
RT Pot Furnaces
RT Rebricking
SO Construction
DEF (DOE Order 6430.1A) Refractories
include nonmetallic materials
having those chemical and
physical properties that make
them applicable for structures, or
as components of systems, that
are exposed to environments
above 1,000°F.

REFRIGERATION ROOM
(1711 REFRIG ROOM)
DA   December 10, 1990
BT1 Room

BT2 Sites/Areas
SO DOE FRASE VOCABULARY

REFUELING WATER STORAGE TANK
DA January 8, 1991
SF *RWST*
BT1 Tanks
BT2 Facility Components

REFUSE
(EPA)
DA November 9, 1990
SY Solid Wastes
NT1 Coal Refuse
DEF (EPA) Synonymous to Solid Waste: Any garbage, refuse, sludge from a waste treatment plant, water supply treatment plant, or air pollution control facility and other discarded material, including solid, liquid, semiliquid, or contained gaseous material resulting from industrial, commercial, mining, and agricultural operations, and from community activities, but does not include solid or dissolved materials in irrigation return flows or industrial discharges that are point sources subject to permits under section 402 of the Federal Water Pollution Control Act, as amended (86 Stat. 880)[33 USCS 1342], or source, special nuclear, or by-product material as defined by the Atomic Energy Act of 1954, as amended(68 Stat. 923)[42 USCS 2011 et seq.].

REFUSE RECLAMATION
(EPA)
DA October 12, 1990
BT1 Reclamation
BT2 Resource Recovery
BT3 Pollution Recovery Processes
BT4 Processes
BT1 Solid Waste Management
BT2 Waste Management
BT3 Processes
RT Recycling
SO Environmental Protection Agency Glossary
DEF (EPA) Conversion of solid waste into useful products, e.g., composting organic wastes to make soil conditioners or separating aluminum and other metals for melting and recycling.

REGENERATION
(EPA)
DA October 12, 1990
BT1 Biological Processes
BT2 Processes
SO Environmental Protection Agency Glossary
DEF (EPA) Manipulation of individual cells or masses of cells to cause them to develop into whole plants.

REGIONAL ADMINISTRATOR
(SWDA; RCRA; CFR)
DA October 12, 1990
BT1 Administrators
BT2 Personnel
BT1 Personnel
RT Directors
SO Wastes
SO Water Pollution
DEF The Regional Administrator of the appropriate Regional Office of the Environmental Protection Agency or the authorized representative of the Regional Administrator.

REGIONAL ASSISTANCE COMMITTEES
(EMER)
DA February 1, 1991
BT1 Committees
BT2 Administrative Organizations
BT3 Organizations
SO Emergency Preparedness
DEF (EMER) Regional committees chaired by a regional Federal Emergency Management Agency representative with members from numerous federal departments and agencies including the U.S. Department of Energy. The purpose of regional assistance committees are to assist state and local government officials in the development of their radiological emergency response plans including reviewing plans and observing exercises.

REGIONAL AUTHORITY
(SWDA; RCRA)
DA October 12, 1990
SO Wastes
DEF The authority established or designated under section 4006 [42 USCS 6946].

REGIONAL COORDINATING OFFICES
(EMER)
DA February 1, 1991
BT1 Offices
BT2 Administrative Organizations
BT3 Organizations
SO Emergency Preparedness
DEF (EMER) Operations offices located at Oak Ridge, Savannah River, Albuquerque, Chicago, Idaho, Richland, and San Francisco which provide radiological assistance coordination, national contingency planning, and regional preparedness committee coordination.

REGIONAL FREQUENCY ANALYSIS
(DOE Order 6430.1A)
DA October 12, 1990
BT1 Analyses
SO Construction
DEF (DOE Order 6430.1A) An analysis

that addresses the probability of the occurrence of two or more random hydrologic events.

REGIONAL OFFICE
(CAA; CFR)
DA October 12, 1990
BT1 U.S. Environmental Protection Agency
BT2 Federal Agencies
BT3 Agencies
BT4 Administrative Organizations
BT5 Organizations
RT Water Management Division Director
SO Air Pollution
DEF (CFR) One of the ten (10) EPA Regional Offices.

REGIONAL RESPONSE TEAMS
(EPA; EMER)
DA October 12, 1990
SF *RRT*
BT1 Response Teams
BT2 Teams
BT3 Administrative Organizations
BT4 Organizations
RT Activation (Emergency)
SO Emergency Preparedness
SO Environmental Protection Agency Glossary
DEF (EPA) Representatives of federal, local, and state agencies who may assist in coordination of activities at the request of the On-Scene Coordinator before and during a Superfund response action.

REGIONAL WATER QUALITY CONTROL BOARD
DA January 8, 1991
SF *RWQCB*
BT1 Boards
BT2 Administrative Organizations
BT3 Organizations

REGIONS
(CAA; CFR)
DA October 12, 1990
BT1 Air Quality Control Regions
BT2 Sites/Areas
SO Air Pollution
DEF (CFR) Area designated as an air quality control region (AQCR).

REGISTRANTS
(EPA)
DA October 12, 1990
SO Environmental Protection Agency Glossary
DEF (EPA) Any manufacturers or formulators who obtain registration for a pesticide active ingredient or product.

REGISTRATION
(EPA)
DA October 12, 1990
NT1 Conditional Registration

NT1 Reregistrations
RT Cancellations
RT Changed Use Pattern
SO Environmental Protection Agency
Glossary
DEF (EPA) Formal listing with EPA of a
new pesticide before it can be sold
or distributed in intra- or inter-state
commerce. The product must be
registered under the Federal
Insecticide, Fungicide, and
Rodenticide Act. EPA is
responsible for registration
(pre-market licensing) of
pesticides on the basis of data
demonstrating that they will not
cause unreasonable adverse
effects on human health or the
environment when used according
to approved label directions.
Includes reregistration.

REGISTRATION STANDARDS
(EPA)
DA October 12, 1990
BT1 Reports
RT Data Call-In
SO Environmental Protection Agency
Glossary
DEF (EPA) Published reviews of all the
data available on pesticide active
ingredients.

REGULATED ACTIVITIES
(SWDA; RCRA; CFR)
DA October 19, 1990
BT1 Activities
RT Major PSD Stationary Source
RT Major PSD Modification
RT Regulated Substances
SO Wastes
DEF "Major Prevention of Significant
Deterioration (PSD) stationary
sources" or "major PSD
modifications."

REGULATED SUBSTANCES
(SWDA; RCRA; CFR)
DA October 12, 1990
RT CERCLA Hazardous Substances
RT Free Products
RT Listed Hazardous Substances
RT Regulated Activities
SO Wastes
DEF Chemicals, compounds, or
materials the manufacture,
generation, transportation,
alteration, or disposition of which
are regulated under any of the
Federal or State statutes.

REGULATIONS AND PROCEDURES
MANUAL, PUB-201
DA January 8, 1991
SF R&PM
RT Procedures

REGULATOR
(2414 REGULATOR)

DA January 3, 1991
BT1 Equipment/Parts - Electrical (DOE
FRASE Vocabulary)
BT2 Equipment
SO DOE FRASE VOCABULARY

REGULATORY GUIDE
DA January 8, 1991
SF RG

REGULATORY ORGANIZATIONS
(DOE Order 5440.1C; ESH)
DA October 19, 1990
SY Program Organizations
BT1 Organizations
RT Federal Agencies
SO Management
DEF Organizations (Assistant Secretary,
Administrator, or Director level)
responsible for the decision
making and implementation of the
Department's programmatic or
regulatory action requiring a
National Environmental Policy Act
review.

REHR
DA October 12, 1990
SEE Reactor Heat Removal
SO Acronyms

REIMBURSEMENT PERIOD
DA January 24, 1991
BT1 Time Designations
SO Environmental Management
DEF (CFR) Refers to a period that
begins when the data from the last
non-duplicative test to be
completed under a test rule are
submitted to EPA and ends after
an amount of time equal to that
which had been required to
develop data or after five years.

REINFORCEMENT RATIO
(DOE Order 6430.1A)
DA October 12, 1990
BT1 Ratios
RT Grade Beams
SO Construction
DEF (DOE Order 6430.1A) The
percentage of tension
reinforcement in a reinforced
concrete beam.

RELATED PERSONAL PROPERTY
(DOE Order 4330.4A)
DA June 5, 1991
BT1 Personal Property
BT2 Property
RT Equipment
SO Management
DEF (DOE Order 4330.4A) Means any
personal property that, once
installed, becomes an integral part
of the real property in which it is
installed or is related to, designed
for, or specially adapted to the
functional or productive capacity of

the real property. The removal of
related personal property would
significantly diminish the economic
value of the real property or the
related personal property.
Examples of related personal
property are communications and
telephone systems.

RELATIONS
(MORT)
DA April 3, 1991
RT Corrective Actions
RT Emergency Actions
RT Public Information Officers
RT Recovery Actions
DEF (MORT) MORT analysis asks: Was
there a management plan outlining
the protocol to be followed and
steps to be taken subsequent to a
significant accident? was the
accident news disseminated to all
concerned parties in a proper and
timely manner?

RELATIVE BIOLOGICAL
EFFECTIVENESS
(IAEA, NIH; EMER)
DA October 12, 1990
SF RBE
BT1 Radiation Units
BT2 Units of Measure
RT Absorbed Dose
RT Quality Factors
SO Emergency Preparedness
SO Radiation
DEF (NIH) For a particular living
organism or part of an organism,
the ratio of the absorbed dose of a
reference radiation that produces
a specified biological effect to the
absorbed dose of the radiation of
interest that produces the same
biological effect.

RELATIVE IMPORTANCE MEASURE
DA January 8, 1991
SF RIM
BT1 Measurements

RELAY(S)
(2415 RELAY)
DA January 3, 1991
BT1 Equipment/Parts - Electrical (DOE
FRASE Vocabulary)
BT2 Equipment
NT1 Building Power Low Voltage Relay
NT1 Building Time Delay Relay
NT1 Phase Failure Relays
NT1 Scram Relay
SO DOE FRASE VOCABULARY

RELEASE DETECTION
(SWDA; RCRA; CFR)
DA October 12, 1990
BT1 Detection
RT Releases
RT UST Systems

SO Wastes

DEF Determining whether a release of a regulated substance has occurred from the underground storage tank (UST) system into the environment or into the interstitial space between the UST system and its secondary barrier or secondary containment around it.

RELEASE OF PROPERTY

(Doe Order 5400.5)

DA October 12, 1990

BT1 Administrative Processes

 BT2 Processes

RT Radioactive Materials

SO Wastes

DEF (DOE Order 5400.5) As used in this Order, means the exercising of DOE's authority to release property from its control after confirming that residual radioactive material (over which DOE has authority) on the property has been determined to meet the guidelines for residual radioactive material in Chapter IV or any other applicable radio- logical requirements. There may be instances in which DOE or other authority will impose restrictions on the management and/or use of the property if the residual radioactive material guidelines of Chapter IV are not met or if other applicable Federal, State, or local requirements cause the imposition of such restrictions.

RELEASES

(SWDA; CERCLA; CFR; ESA; ESH; EMER)

DA October 12, 1990

NT1 Aboveground Releases

NT1 Airborne Releases

NT1 Belowground Releases

NT1 Environmental Release

NT1 Federally Permitted Releases

NT1 Major Release

NT1 Minor Releases

NT1 Overfill Releases

NT1 Radiological Releases

NT1 Underground Releases

RT Community Awareness and Emergency Response Programs

RT Discharges

RT Emissions

RT Enclosed Processes

RT Hazards

RT Leak Detection Systems

RT Leaks

RT Management of Migrations

RT Mobile Sources

RT PCB Transformer Rupture

RT Release Detection

RT Reportable Quantities

RT Size Classes of Releases

RT Spills

RT Stationary Sources

RT Water Pollution

SO Compensation and Liability

SO Emergency Preparedness

SO Environmental Management

SO Wastes

DEF (DOE Order 5000.3A) Any spilling, leaking, pumping, pouring, emitting, emptying, discharging, injecting, escaping, leaching, dumping, or otherwise disposing of substances into the environment. This includes abandoning/discarding any type of receptacle containing substances, or the stockpiling of a reportable quantity of a hazardous substance in an enclosed containment structure. (EMER) Spills, leaks, emissions, discharges, escapes, leachings or disposings from an underground storage tank (UST) into groundwater, surface water, or subsurface soils. (ESH) With certain exceptions, any spilling, leaking, pumping, pouring, emitting, emptying, discharging, injecting, escaping, leaching, dumping, or disposing into the environment (including the abandonment or discarding of barrels, containers or other closed receptacles) of any hazardous substance, or CERCLA hazardous substance.

RELEVANT AND APPROPRIATE REQUIREMENTS

(CERCLA; CFR)

DA October 12, 1990

BT1 Requirements

SO Compensation and Liability

SO Environmental Management

DEF Those Federal requirements that, while not "applicable," are designed to apply to problems sufficiently similar to those encountered at CERCLA sites that their application is appropriate. Requirements may be relevant and appropriate if they would be "applicable" but for jurisdictional restrictions associated with the requirement.

RELIABILITY, AVAILABILITY, AND MAINTAINABILITY

DA January 8, 1991

SF *RAM*

RELIABILITY AND QUALITY ASSURANCE

(SSDC)

DA October 12, 1990

SY Quality Assurance

SF *R&QA*

SO System Safety Development Center Glossary

DEF (SSDC) See Quality Assurance. Involves all those planned and systematic actions necessary to provide adequate confidence that

a facility, structure, system, or component will perform satisfactorily and safely in service.

RELIABILITY-CENTERED MAINTENANCE

(DOE Order 4330.4A)

DA June 5, 1991

SF *RCM*

BT1 Maintenance

 BT2 Activities

RT Failures

RT Safety

SO Management

DEF (DOE Order 4330.4A) A maintenance system that determines the most effective maintenance activity, based on an analysis of an items failure modes, failure rates, and the importance of the item to the safety operation of the facility.

RELIEF VALVE DISCHARGE

(CAA; CFR)

DA October 12, 1990

BT1 Discharges

RT Overpressure

RT Relief Valves

SO Air Pollution

DEF Any nonleak discharge through a relief valve. Relief valve discharge does not include discharges ducted to a control system from which the concentration of vinyl chloride in the exhaust gases does not exceed 10 ppm (average for 3-hour period), or equivalent as provided in 40 CFR 61.66.

RELIEF VALVES

(CAA; CFR)

DA October 12, 1990

BT1 Valves

 BT2 Devices

NT1 Emergency Relief Valve

NT1 Inadvertently Opened Relief Valve

NT1 Power Operated Relief Valve

NT1 Safety/Relief Valve

NT1 Stuck-Open Relief Valve

RT Open Ended Valve

RT Pressure Releases

RT Relief Valve Discharge

RT Sudden Releases of Pressure

SO Air Pollution

DEF Pressure relief devices including pressure relief valves, rupture disks, and other pressure relief systems used to protect process components from overpressure conditions. "Relief valve" does not include polymerization shortstop systems, refrigerated water systems or control valves, or other devices used to control flow to an incinerator or other air pollution control device.

REM (Radiation Exposure Module)
DA   October 12, 1990
SEE Radiation Exposure Module
SO   Acronyms
SO   Radiation

REM (Roentgen Equivalent Man)
DA   October 12, 1990
SEE Roentgen Equivalent Man
SO   Acronyms
SO   Construction
SO   Emergency Preparedness
SO   Environmental Protection Agency
       Glossary
SO   Radiation
SO   Water Pollution

REMAINS OPEN
DA   January 8, 1991
SF   RO (Remains Open)

**REMEDIAL ACTION ASSESSMENT
SYSTEM**
DA   January 8, 1991
SF   RAAS
BT1 Systems
RT   Assessments

REMEDIAL ACTION PROGRAM
DA   January 8, 1991
SF   RAP
BT1 Programs
NT1 Formerly Utilized Sites Remedial
       Actions Program
NT1 Uranium Mill Tailings Remedial
       Action Program

REMEDIAL ACTIONS
(DOE Order 5400.5; CFR; EPA)
DA   October 12, 1990
SY   Cleanup
SY   Remedies
SY   Removal Actions
SY   Response Actions
SY   Respond
SF   RA
BT1 Actions
 BT2 Responses
NT1 Remedial Design
NT1 Source Control Remedial Actions
NT1 Uranium Mill Tailings Remedial
       Action
RT   Cleanup Operations
RT   Consent Decrees
RT   Corrective Actions
RT   Decommissioning
RT   Decontamination
RT   Primary Documents
RT   Remedial Project Managers
RT   Remedial Measures
RT   Remedial Responses
RT   Schedule of Compliance
RT   Superfund
RT   Superfund State Contracts
SO   Compensation and Liability
SO   Environmental Management
SO   Environmental Protection Agency
       Glossary

SO   Wastes
DEF Those actions consistent with
       permanent remedy taken instead
       of, or in addition to, removal action
       in the event of a release or
       threatened release of a hazardous
       substance into the environment, to
       prevent or minimize the release of
       hazardous substances so that
       they do not migrate to cause
       substantial danger to present or
       future public health or welfare or
       the environment.

REMEDIAL DESIGN
(EPA)
DA   October 12, 1990
SF   RD
BT1 Remedial Actions
 BT2 Actions
  BT3 Responses
RT   Primary Documents
SO   Environmental Protection Agency
       Glossary
DEF (EPA) A phase of remedial action
       that follows the remedial
       investigation/feasibility study and
       includes development of
       engineering drawings and
       specifications for a site cleanup.

REMEDIAL INVESTIGATIONS
(CERCLA; CFR)
DA   October 12, 1990
SF   RI
BT1 Investigations
 BT2 Administrative Processes
  BT3 Processes
NT1 RCRA Remedial Investigation
RT   Endangered Assessment
RT   Feasibility Studies
RT   Primary Documents
SO   Compensation and Liability
SO   Environmental Management
SO   Environmental Protection Agency
       Glossary
DEF In-depth studies designed to gather
       the data necessary to determine
       the nature and extent of
       contamination at a Superfund site;
       establish criteria for cleaning up
       the site; identify preliminary
       alternatives for remedial actions;
       and support the technical and cost
       analyses of the alternatives. The
       remedial investigation is usually
       done with the feasibility study.
       Together, they are usually referred
       to as the "RI/FS."

REMEDIAL INVESTIGATOR
DA   January 8, 1991
BT1 Personnel

REMEDIAL MEASURES
(IAEA)
DA   October 12, 1990
BT1 Actions
 BT2 Responses

RT   Radiation Protection
RT   Remedial Actions
SO   Radiation
DEF (IAEA) Actions taken to reduce
       radiation doses that might
       otherwise be received in abnormal
       exposure conditions. Remedial
       measures are sometimes called
       protective actions or
       countermeasures.

REMEDIAL PROJECT MANAGERS
(CERCLA; CFR; EPA; EMER)
DA   October 12, 1990
SF   RPM (Remedial Project Manager)
BT1 Personnel
BT1 Project Managers
 BT2 Personnel
RT   Remedial Actions
RT   Support Agencies
SO   Compensation and Liability
SO   Emergency Preparedness
SO   Environmental Management
SO   Environmental Protection Agency
       Glossary
DEF (EMER) EPA or state officials
       responsible for overseeing
       remedial action at a site.

REMEDIAL RESPONSES
(EPA)
DA   October 12, 1990
BT1 Responses
RT   Remedial Actions
SO   Environmental Protection Agency
       Glossary
DEF (EPA) Long-term actions that stop
       or substantially reduce a release
       or threat of a release of hazardous
       substances that is serious but not
       an immediate threat to public
       health.

REMEDIES
(CERCLA; CFR; USC)
DA   October 19, 1990
SY   Remedial Actions
SY   Respond
BT1 Actions
 BT2 Responses
RT   Primary Documents
SO   Compensation and Liability
DEF As defined by section 101(24) of
       CERCLA, means those actions
       consistent with permanent remedy
       taken instead of, or in addition to,
       removal action in the event of a
       release or threatened release of a
       hazardous substance into the
       environment, to prevent or
       minimize the release of hazardous
       substances so that they do not
       migrate to cause substantial
       danger to present or future public
       health or welfare or the
       environment.

REMOTE DETECTION AND CONTROL
DA   January 8, 1991

SF   *REDAC*
BT1 Controls
RT   Remote Detection and Control
        Systems

**REMOTE DETECTION AND CONTROL
SYSTEMS**
(NFI)
DA   October 12, 1990
BT1 Control Systems
 BT2 Controls
 BT2 Systems
RT   Remote Detection and Control
SO   Radiation
DEF (NFI) Systems for remote
        monitoring and operation process
        variables.

REMOTE HANDLED
DA   January 8, 1991
SF   *RH*
RT   Remote-Handled Transuranic
        Wastes

**REMOTE-HANDLED TRANSURANIC
WASTES**
DA   January 24, 1991
BT1 Transuranic Wastes (TRU Waste)
 BT2 Radioactive Wastes
  BT3 Wastes
RT   Remote Handled
SO   Environmental Management
DEF (DOE 5820.2A) Packaged
        transuranic waste whose external
        surface dose rate exceeds 200
        mrem per hour. Test specimens of
        fissionable material irradiated for
        research and development
        purposes only and not for the
        production of power or plutonium
        may be classified as
        remote-handled transuranic waste.

REMOTE INTERROGATION POINTS
(DOE Order 6430.1A)
DA   October 12, 1990
BT1 Sites/Areas
SO   Construction
DEF (DOE Order 6430.1A) Locations for
        receiving information (e.g.,
        printouts) transmitted by automatic
        data processing centers.

REMOVAL
(EPA; USC; CFR; ESH)
DA   January 16, 1991
SY   Respond
BT1 Strip
 BT2 Processes
BT1 Waste Management Processes
 BT2 Processes
NT1 Dredging
RT   Cleanup
RT   Planned Renovation Operations
RT   Spills
SO   Compensation and Liability
SO   Environmental Management
SO   Hazardous Materials
DEF The cleanup or removal of released

hazardous substances from the
environment; such actions as may
be necessary taken in the event of
the threat of release of hazardous
substances into the environment;
such actions may be necessary to
monitor, assess, and evaluate the
release or threat of release of
hazardous substances, the
disposal of removed material, or
the taking of such other actions as
may be necessary to prevent,
minimize, or mitigate damage to
the public health or welfare or to
the environment, which may
otherwise result from a release or
threat of release.

REMOVAL ACTIONS
(EPA)
DA   October 12, 1990
SY   Cleanup
SY   Remedial Actions
SY   Response Actions
SY   Respond
BT1 Actions
 BT2 Responses
RT   Response Actions
SO   Environmental Protection Agency
        Glossary
DEF (EPA) Short-term immediate actions
        taken to address releases of
        hazardous substances that require
        expedited response.

REMOVAL SITE EVALUATIONS
(EPA)
DA   May 16, 1991
SF   *RSE*

RENOVATION
(CAA; CFR)
DA   October 12, 1990
BT1 Administrative Processes
 BT2 Processes
NT1 Emergency Renovation Operations
RT   Asbestos-Containing Waste
        Materials
RT   Demolition
RT   Strip
SO   Air Pollution
SO   Environmental Management
DEF Altering in any way one or more
        facility components. Operations in
        which load-supporting structural
        members are wrecked or taken
        out are excluded.

**REPAIR/CONSTRUCTION
PERSONNEL**
DA   November 29, 1990
BT1 Occupations
BT1 Personnel
NT1 Carpenter
NT1 Electrician
NT1 Mason
NT1 Mechanic/Repairer
NT1 Miner/Driller

NT1 Misc Repair/Construction
        Employee
NT1 Painter
NT1 Pipe Fitter
SO   DOE FRASE VOCABULARY

REPLACEMENT
(CERCLA)
DA   November 15, 1990
SY   Acquisition of the Equivalent
BT1 Actions
 BT2 Responses
RT   National Contingency Plan
DEF (CERCLA) The substitution for an
        injured resource with a resource
        that provides the same or
        substantially similar services,
        when such substitutions are in
        addition to any substitutions made
        or anticipated as part of response
        actions and when such
        substitutions exceed the level of
        response actions determined
        appropriate to the site pursuant to
        the National Contingency Plan
        (NCP).

REPLACEMENT VALUES
(SSDC)
DA   October 12, 1990
BT1 Costs
RT   Actual Cash Value
SO   System Safety Development Center
        Glossary
DEF (SSDC) The costs to replace
        something with like kind, quality,
        and capacity.

REPLICATE SAMPLES
(ESH)
DA   October 12, 1990
BT1 Samples
RT   Aliquots
SO   Quality Assurance
DEF (ESH) Samples prepared by
        dividing a sample into two or more
        separate aliquots. Laboratory
        duplicate sample are considered
        to be two replicates.

REPORT OF DISCREPANCY
DA   January 8, 1991
SF   *ROD (Report of Discrepancy)*
BT1 Reports

REPORTABLE OCCURRENCES
(DOE Order 5000.3A)
DA   December 14, 1990
BT1 Occurrences
NT1 Emergencies
 NT2 Abnormal Exposure Conditions
 NT2 Bomb Incidents
 NT2 Bomb Threats
 NT2 Emergency Situations
 NT2 Energy Emergencies
 NT2 General Emergencies
 NT2 Hazardous Materials Emergencies
 NT2 National Emergencies
 NT2 Natural Phenomena Emergencies

SY-Synonymous Terms   SO-Source/Subject Category   SF-See From

NT2 Operational Emergencies
NT2 Safeguards and Security
　Emergencies
　NT3 Safeguards and Security Alerts
　NT4 Alpha
　NT4 Bravo
　NT4 Charlie
　NT4 Safeguards and Security Alert I
　　(Code Designator: Alpha)
　NT4 Safeguards and Security Alert II
　　(Code Designator: Bravo)
　NT4 Safeguards and Security Alert III
　　(Code Designator: Charlie)
　NT3 Special Nuclear Materials
　　Emergencies
NT2 Site Emergencies
NT2 Threats
　NT3 Bomb Threats
　NT3 Covert Threats
　NT3 Nuclear Threat Incidents
　NT3 Terrorist Threats
NT1 Off Normal Occurrences
NT1 Officially Reportable Events
NT1 Unusual Occurrences
RT Reports
DEF (DOE Order 5000.3A) Events or
　conditions to be reported in
　accordance with the criteria
　defined in DOE Order 5000.3A.

REPORTABLE QUANTITIES
(DOE Order 5000.3A; SWDA; RCRA;
　CERCLA; ESH; EMER)
DA　October 12, 1990
SF　*RQ*
BT1　Quantities
RT　CERCLA Hazardous Substances
RT　Hazardous Materials
RT　Releases
RT　Reports
RT　Significant Quantities
RT　Spills
RT　Threshold Planning Quantities
SO　Compensation and Liability
SO　Emergency Preparedness
SO　Environmental Management
SO　Environmental Protection Agency
　　Glossary
SO　Hazardous Materials
DEF (DOE Order 5000.3A) For any
　CERCLA hazardous substance,
　the quantity established in Table
　302.4 and Appendix B of 40 CFR
　Part 302, the release of which
　requires notification unless
　federally permitted. (EMER)
　Quantities of a hazardous
　substance that trigger reports
　under CERCLA. If a substance is
　released in amounts exceeding its
　reportable quantity (RQ), the
　release must be reported to the
　National Response Center, the
　State Emergency Response
　Commission (SERC), and
　community emergency
　coordinators for areas likely to be
　affected.

**REPORTED SIGNIFICANT
OBSERVATIONS**
(SSDC)
DA　October 12, 1990
SF　*RSO (Reported Significant
　　Observations)*
RT　Critical Incident Technique
RT　Industrial Safety
RT　Reports
SO　System Safety Development Center
　　Glossary
DEF (SSDC) A means of obtaining
　feedback from the work level on
　things that could contribute to an
　accident situation. Questionnaires
　are used to obtain reports on
　these observations. Similar to
　Critical Incident Technique.

REPORTS
DA　February 4, 1991
NT1 Accreditation Maintenance Reports
NT1 Activity Data Sheets
NT1 Advisory
NT1 Annual Document Logs
NT1 Annual Reports
NT1 Annual Site Environmental Reports
NT1 Biological Opinion
　NT2 Preliminary Biological Opinion
NT1 Conceptual Design Report
NT1 Contractor Self-Evaluation Reports
NT1 Emergency and Hazardous
　　Chemical Inventory Forms
NT1 Enforcement Decision Documents
NT1 Environmental Information
　　Documents
NT1 EPA Guidance Document
NT1 Fire Hazard Analysis Report
NT1 Internal Security Reports
NT1 Investigation Report
NT1 Licensee Event Report
NT1 Material Nonconformance Report
NT1 Monthly NEPA Reports
NT1 National Environmental Policy Act
　　Documents
　NT2 Programmatic NEPA Documents
　NT2 Site-Wide NEPA Documents
NT1 NEPA Status Reports
NT1 Non-Conformance Report
NT1 Occurrence Reports
　NT2 Unusual Occurrence Reports
NT1 Plugging Records
NT1 Primary Documents
　NT2 Secondary Documents
NT1 Quarterly Environmental
　　Compliance Reports
NT1 Red Borders
NT1 Registration Standards
NT1 Report of Discrepancy
NT1 Safety Analysis Reports
　NT2 Final Safety Analysis Report
　NT2 Preliminary Safety Analysis Report
　NT2 Updated Final Safety Analysis
　　Report
NT1 Safety Documents
NT1 Safety Evaluation Report
NT1 Tiger Team Assessment Reports
NT1 Toxic Chemical Release Forms
RT　Annual Document Logs

RT　Computerized Accident/Incident
　　Reporting System
RT　Failure Reporting Analysis and
　　Corrective Action System
RT　Manifest Document Number
RT　Officially Reportable Events
RT　Reportable Quantities
RT　Reported Significant Observations
RT　Reportable Occurrences

REPOSITORIES
(DOE Order 5820.2A; ANL;)
DA　May 24, 1991
BT1　Disposal Facilities
　BT2　Hazardous Waste Management
　　Facilities
　BT3　Hazardous Waste Facilities
　BT4　Facilities
RT　Transuranic Wastes (TRU Waste)
DEF (DOE 5820.2A) Facilities for the
　permanent deep geological
　disposal of High Level or
　Transuranic Waste.

REPRESENTATIVE SAMPLES
(SWDA; RCRA)
DA　October 12, 1990
BT1　Samples
SO　Environmental Management
SO　Wastes
DEF Samples of a universe or whole
　(e.g., waste pile, lagoon,
　groundwater) which can be
　expected to exhibit the average
　properties of the universe or
　whole.

REPROCESSING
(IAEA)
DA　October 12, 1990
BT1　Processes
RT　Fission Products
RT　Transuranic Elements
SO　Radiation
DEF The dissolution of spent reactor fuel
　and separation of uranium,
　transuranic elements, and fission
　products. (IAEA) The processing
　of nuclear fuel (material), after its
　use in a reactor, to recover
　valuable material and to remove
　fission products.

REQUIRED STRENGTH
(DOE Order 6430.1A)
DA　October 12, 1990
BT1　Strength
RT　Structural Members
SO　Construction
DEF (DOE Order 6430.1A) Required
　strength to resist factored loads or
　related internal moments and
　forces.

REQUIREMENTS
(TSCA; CFR)
DA　October 19, 1990
NT1 Anti-Degradation Clause
NT1 Constraints

NT1 Emission Limitation
NT1 Emission Standards
NT2 National Emissions Standards for
Hazardous Air Pollutants
NT2 New Source Performance
Standards
NT1 General Design Criteria
NT1 Land Disposal Restrictions
NT1 Minimum Requirements and
Standards
NT1 Operational Safety Requirements
NT1 Relevant and Appropriate
Requirements
NT1 Specifications
NT2 Contract Specifications
NT2 Guide Specifications
NT2 Operational Specifications
NT2 Project Design Criteria
NT2 Technical Specifications
NT1 Standard of Performance
NT1 Testing Requirements Rule
NT1 Testing Requirements
NT1 Treatment Technique Requirements
RT Approvals Necessary to Begin
Physical Construction
RT Concepts and Requirements
RT Graded Approach
RT Licenses
RT Procedures
SO Hazardous Materials
DEF (WTID) Something called for or
demanded: a requisite or essential
condition.

RER
DA October 12, 1990
SEE Rod Equipment Room
SO Acronyms

REREGISTRATIONS
(EPA)
DA October 12, 1990
BT1 Registration
RT Pesticides
SO Environmental Protection Agency
Glossary
DEF (EPA) The reevaluation and
relicensing of existing pesticides
originally registered prior to
current scientific and regulatory
standards. EPA reregisters
pesticides through its Registration
Standards Program.

RES
DA October 12, 1990
SEE NRC Office of Nuclear Regulatory
Research
SO Acronyms

RESC
DA October 12, 1990
SEE Reactor Subcriticality
SO Acronyms

**RESEARCH, DEVELOPMENT,
DEMONSTRATION, TESTING AND
EVALUATION**
DA January 10, 1991

SF *RDDT&E*
BT1 Activities

**RESEARCH & DEVELOPMENT
LABORATORY**
(1805 R&D LABORATO)
DA December 10, 1990
BT1 Building (DOE FRASE Vocabulary)
BT2 Facilities and Buildings (DOE
FRASE Vocabulary)
BT3 Facilities
BT1 Laboratories
BT2 Research and Development
Organizations
BT3 Organizations
SO DOE FRASE VOCABULARY

RESEARCH AND DEVELOPMENT
(CFR)
DA January 8, 1991
SF *R&D*
BT1 Activities
NT1 Applied Research
NT1 Basic Research
RT Small Quantities for Research and
Development
SO Environmental Management
DEF (CFR) (a) Theoretical analysis,
exploration, or experimentation; or
(b) the extension of investigative
findings and theories of a scientific
or technical nature into practical
application for experimental and
demonstration purposes, including
the experimental production and
testing of models, devices,
equipment, materials, and
processes. Research and
development does not include the
internal or external administration
of radiation of radioactive material
to human beings.

**RESEARCH AND DEVELOPMENT
ORGANIZATIONS**
DA February 5, 1991
BT1 Organizations
NT1 American National Metric Council
NT1 American Nuclear Society
NT1 Foundations
NT2 Hanford Environmental Health
Foundation
NT2 Lovelace Medical Foundation
NT2 University of Georgia Research
Foundation, Inc.
NT1 Institutes
NT2 American National Standards
Institute
NT2 American Petroleum Institute
NT2 Battelle Memorial Institute
NT2 Electric Power Research Institute
NT3 Nuclear Safety Analysis Center
NT2 Inhalation Toxicology Research
Institute
NT2 Institute of Electrical and
Electronic Engineers
NT2 Institute for Nuclear Power
Operations
NT2 Midwest Research Institute

NT2 National Institute for Occupational
Safety and Health
NT2 National Institute of Standards and
Testing
NT1 Laboratories
NT2 Argonne National Laboratory
NT3 Argonne National
Laboratory-East (Chicago)
NT3 Argonne National
Laboratory-West (At INEL)
NT2 Argonne National Laboratory-East
(Chicago)
NT2 Argonne National Laboratory-West
(At INEL)
NT2 Battelle Columbus Laboratories
NT2 Brookhaven National Laboratory
NT2 Idaho National Engineering
Laboratory
NT3 Power Burst Facility
NT2 Laboratory for Energy-Related
Health Research
NT2 Lawrence Livermore National
Laboratory
NT2 Lawrence Berkeley Laboratory
NT2 Los Alamos National Laboratory
NT2 Nationally Recognized Testing
Laboratories
NT2 Oak Ridge National Laboratory
NT2 Pacific Northwest Laboratory
NT2 Performance Testing Laboratories
NT2 Research & Development
Laboratory
NT2 Sandia National Laboratories
NT3 Sandia National
Laboratories-Alburquerque
NT3 Sandia National
Laboratories-Livermore
NT1 National Academy of Sciences
NT1 Oak Ridge Associated Universities
NT1 Southeastern Universities Research
Association, Inc.
RT Administrative Organizations
RT Commercial Organizations
RT Educational Organizations

**RESEARCH AND SPECIAL
PROGRAMS ADMINISTRATION**
(SWDA; RCRA; ESH)
DA October 12, 1990
SF *RSPA*
BT1 U.S. Department of Transportation
BT2 Federal Agencies
BT3 Agencies
BT4 Administrative Organizations
BT5 Organizations
RT Programs
SO Hazardous Materials

RESEARCH/TESTING ACTIVITY
(1242 RT ACTIVITY)
DA November 28, 1990
BT1 Activity Types (DOE FRASE
Vocabulary)
BT2 Activities
RT Scientist
RT Test Area
RT Test Reactor Facility
RT Testing Equipment
RT Transient Test
SO DOE FRASE VOCABULARY

RESERVOIR ROUTING
(DOE Order 6430.1A)
DA   October 12, 1990
RT   Reservoirs
SO   Construction
DEF (DOE Order 6430.1A) A technique
     used in hydrology to compute the
     effect of reservoir inflow on
     reservoir outflow.

RESERVOIRS
(EPA)
DA   October 12, 1990
BT1 Impoundment
 BT2 Sites/Areas
RT   Reservoir Routing
RT   Water
SO   Environmental Protection Agency
     Glossary
DEF (EPA) Natural or artificial holding
     areas used to store, regulate, or
     control water.

RESET DEVICE
(2416 RESET DEVICE)
DA   January 3, 1991
BT1 Devices
BT1 Equipment/Parts - Electrical (DOE
     FRASE Vocabulary)
 BT2 Equipment
SO   DOE FRASE VOCABULARY

RESIDENT SPECIES
(ESA; CFR)
DA   October 12, 1990
BT1 Species
SO   Endangered Species
DEF For purposes of these regulations,
     with respect to a State, a species
     that exists in the wild in that State
     during any part of its life.

RESIDENTIAL AREAS
(CFR)
DA   November 9, 1990
BT1 Sites/Areas
RT   Commercial Areas
RT   High-Contact Residential Surfaces
RT   High-Contact Commercial Surfaces
RT   Non-Restricted Access Areas
DEF (CFR) Those areas where people
     live or reside. Residential areas
     include housing and the property
     on which housing is located, as
     well as playgrounds, roadways,
     sidewalks, parks, and other similar
     areas within a residential
     community.

RESIDENTIAL SOLID WASTES
DA   January 29, 1991
BT1 Solid Wastes
 BT2 Wastes
SO   Environmental Management
DEF (CFR) Garbage, rubbish, trash, and
     other solid waste resulting from the
     normal activities of households.

RESIDENTIAL TANKS
(SWDA; RCRA)
DA   October 12, 1990
BT1 Tanks
 BT2 Facility Components
SO   Wastes
DEF A tank located on property used
     primarily for dwelling purposes.

RESIDUAL
(EPA)
DA   October 12, 1990
RT   Treatment
SO   Environmental Protection Agency
     Glossary
DEF (EPA) Amount of a pollutant
     remaining in the environment after
     a natural or technological process
     has taken place, e.g., the sludge
     remaining after initial wastewater
     treatment, or particulates
     remaining in air after the air
     passes through a scrubbing or
     process.

RESIDUAL DISINFECTANT CONCENTRATION
(SWDA; CFR)
DA   October 12, 1990
RT   CT (Public Water Disinfection
     Formula)
RT   CTcalc
RT   Disinfectant Contact Time
RT   Disinfection
RT   Water Quality Criteria
SO   Water Pollution
DEF The concentration of disinfectant
     measured in mg/l in a
     representative sample of water.

RESIDUAL HEAT REMOVAL
DA   January 8, 1991
SF   RHR

RESIDUAL RADIOACTIVE MATERIALS
(DOE Order 5400.5)
DA   October 12, 1990
BT1 Radioactive Materials
 BT2 Materials
SO   Wastes
DEF (DOE Order 5400.5) Radioactive
     materials that are in or on soil, air,
     equipment, or structures as a
     consequence of past operations or
     activities.

RESIDUAL RISKS
(SSDC)
DA   October 12, 1990
SY   Acceptable Risks
BT1 Risks
SO   System Safety Development Center
     Glossary
DEF (SSDC) Risks remaining after the
     application of resources for
     prevention or mitigation.

RESIDUES
(SWDA; RCRA; ESH)

DA   October 12, 1990
BT1 Hazardous Materials
 BT2 Materials
NT1 Wood Residues
SO   Hazardous Materials
DEF Hazardous material remaining in a
     packaging, including a tank car,
     after its contents have been
     unloaded to the maximum extent
     practicable and before the
     packaging is either refilled or
     cleaned of hazardous material and
     purged to remove any hazardous
     vapors.

RESIN GRADE
(CAA; CFR)
DA   October 12, 1990
NT1 Types of Resin
 NT2 Bulk Resins
 NT2 Dispersion Resins
 NT2 Latex Resins
RT   Types of Resin
SO   Air Pollution
DEF (CFR) The subdivision of resin
     classification which describes it as
     a unique resin, i.e., the most exact
     description of a resin with no
     further subdivision.

RESISTOR(S)
(2417 RESISTOR)
DA   January 3, 1991
BT1 Equipment/Parts - Electrical (DOE
     FRASE Vocabulary)
 BT2 Equipment
SO   DOE FRASE VOCABULARY

RESOURCE
(CFR; EPA)
DA   October 12, 1990
SY   Natural Resources
NT1 Recovered Resources
RT   Resource Conservation and
     Recovery Act
SO   Environmental Protection Agency
     Glossary
DEF (EPA) A person, thing, or action
     needed for living or to improve the
     quality of life.

RESOURCE CONSERVATION
(SWDA; RCRA)
DA   October 12, 1990
BT1 Conservation
RT   Comprehensive Planning
RT   Solid Waste Planning
SO   Environmental Management
SO   Wastes
DEF Reduction of the amounts of solid
     waste that are generated,
     reduction of overall resource
     consumption, and utilization of
     recovered resources.

RESOURCE CONSERVATION AND RECOVERY ACT
(RCRA; EMER)
DA   January 10, 1991

SF *RCRA*
BT1 Acts
 BT2 Statutes and Regulations
BT1 Appropriate Act and Regulations
RT Conservation
RT Recovery
RT Resource
SO Emergency Preparedness
DEF (RCRA) To promote the protection
   of health and the environment and
   to conserve valuable material and
   energy resources.

RESOURCE RECOVERY
(SWDA; RCRA)
DA October 12, 1990
BT1 Pollution Recovery Processes
 BT2 Processes
NT1 Reclamation
 NT2 Land Reclamation
 NT2 Refuse Reclamation
NT1 Solid Waste Planning
RT Comprehensive Planning
RT Recovered Resources
RT Resource Recovery Systems
RT Resource Recovery Facilities
SO Environmental Management
SO Environmental Protection Agency
   Glossary
SO Wastes
DEF The process of obtaining matter or
   energy from materials formerly
   discarded. The recovery of
   material or energy from solid
   waste.

RESOURCE RECOVERY FACILITIES
(SWDA; RCRA)
DA October 12, 1990
BT1 Facilities
RT Resource Recovery
RT Resource Recovery Systems
RT Solid Waste Management Facilities
SO Environmental Management
SO Wastes
DEF Any facilities at which solid waste is
   processed for the purpose of
   extracting, converting to energy, or
   otherwise separating and
   preparing solid waste for reuse.

RESOURCE RECOVERY SYSTEMS
(SWDA; RCRA)
DA October 12, 1990
BT1 Systems
RT Resource Recovery
RT Resource Recovery Facilities
RT Solid Waste Management Facilities
SO Environmental Management
SO Wastes
DEF Solid waste management systems
   which provide collection,
   separation, recycling, and recovery
   of solid wastes, including disposal
   of nonrecoverable waste residues.

RESPIRATOR(S)
(2679 RESPIRATOR)
DA January 3, 1991

BT1 Personal Protective Equipment
 BT2 Equipment/Parts - Personal
   Protective (DOE FRASE
   Vocabulary)
  BT3 Equipment
SO DOE FRASE VOCABULARY

RESPIRATORY SYSTEM
(1145 RESP SYSTEM)
DA November 28, 1990
BT1 Body System(s)
 BT2 Human Body Parts
NT1 Bronchial Epithelium
NT1 Nose
NT1 Throat
RT Chest
RT Cond of Respiratory Sys. Non-Toxic
RT Upper Respiratory Condition
RT Upper Respiratory Disease
SO DOE FRASE VOCABULARY

RESPOND
(CERCLA; CFR)
DA November 15, 1990
SY Remedial Actions
SY Remedies
SY Removal
SY Removal Actions
SY Response Actions
SY Responses
DEF As defined by section 101(25) of
   CERCLA, remove, removal,
   remedy, or remedial action.

RESPONSE ACTIONS
(TSCA; CFR; EPA; EMER)
DA October 12, 1990
SY Cleanup
SY Remedial Actions
SY Removal Actions
SY Respond
BT1 Actions
 BT2 Responses
NT1 Operable Units
RT Action Levels
RT Removal Actions
SO Environmental Protection Agency
   Glossary
SO Hazardous Materials
DEF (EMER) CERCLA-authorized
   actions involving either a
   short-term removal action or a
   long-term removal response that
   may include but is not limited to:
   removing hazardous materials
   from a site to an EPA-approved
   hazardous waste facility for
   treatment, containment, or
   destruction; containing the waste
   safely on-site; destroying or
   treating the waste on-site; and
   identifying and removing the
   source of groundwater
   contamination and halting further
   migration of contaminants. (See:
   Cleanup). Methods that protect
   human health and the environment
   from asbestos-containing material.
   Such methods include methods
   described in chapters 3 and 5 of

the Environmental Protection
Agency's "Guidance for Controlling
Asbestos-Containing Materials in
Buildings."

RESPONSE SPECTRUM
DA January 24, 1991
BT1 Time Designations
SO Environmental Management
DEF (CFR) The time interval from a step
   change in the input concentration
   at the instrument inlet to a reading
   of 90 percent (nominally
   equivalent to 2.2 time constants)
   of the ultimate record output.

RESPONSE TEAMS
DA February 14, 1991
BT1 Teams
 BT2 Administrative Organizations
  BT3 Organizations
NT1 Airborne Response Teams
NT1 Emergency Support Teams
NT1 Emergency Response Teams
NT1 Energy Emergency Teams
NT1 Environmental Response Teams
NT1 Explosive Ordnance Disposal
   Teams
NT1 Hazardous Materials Response
   Teams
NT1 Hostage Negotiation Teams
NT1 National Response Teams
NT1 Nuclear Emergency Search Teams
NT1 Radiological Assistance Teams
NT1 Regional Response Teams
NT1 Special Response Teams
NT1 Tactical Response Teams

RESPONSE TIME
(DOE Order 6430.1A)
DA October 12, 1990
BT1 Time Designations
RT Fire Protection Systems
SO Construction
DEF (DOE Order 6430.1A) This term
   when used to specify performance
   of a rapid action deluge fire
   protection system represents the
   elapsed time between the initiation
   of the incident and water
   application to the material being
   protected.

RESPONSES
(CERCLA; CFR; EMER)
DA November 15, 1990
SY Respond
NT1 Actions
 NT2 Agency Actions
 NT2 Alternative Courses of Action
 NT2 Assessment Actions
 NT2 Automatic Incident Actions
 NT2 Categorical Exclusions
 NT2 Corrective Actions
 NT2 Emergency Actions
 NT2 Incident Actions
 NT2 Interim Actions
 NT2 Proposed Actions
 NT2 Protective Actions

NT3 Denials
NT2 Reasonable and Prudent
Alternatives
NT2 Reasonable and Prudent
Measures
NT2 Recovery Actions
NT2 Remedial Actions
NT3 Remedial Design
NT3 Source Control Remedial Actions
NT3 Uranium Mill Tailings Remedial
Action
NT2 Remedial Measures
NT2 Remedies
NT2 Removal Actions
NT2 Replacement
NT2 Response Actions
NT3 Operable Units
NT2 Restoration
NT1 Emergency Response
NT1 Limited Responses
NT1 Post Emergency Responses
NT1 Remedial Responses
RT Compensation and Liability
RT Recommendations
SO Emergency Preparedness
DEF (EMER) As defined by section
101(25) of CERCLA, removing,
removals, remedies, or remedial
actions. (ESH) Any actions taken
to remove, remedy, or clean up
releases of hazardous substances.

RESPONSIBLE PARTIES
(TSCA; CFR)
DA   October 12, 1990
NT1 Operators
NT1 Owners
RT PCB Equipment
RT Polychlorinated Biphenyls
RT Potentially Responsible Parties
SO Hazardous Materials
DEF Owners of the PCB equipment,
facility, or other source of PCBs or
his/her designated agent (e.g., a
facility manager or foreman).

REST ROOM
(1712 REST ROOM)
DA   December 10, 1990
BT1 Room
BT2 Sites/Areas
SO DOE FRASE VOCABULARY

RESTAURANT
(1789 RESTAURANT)
DA   December 10, 1990
BT1 Building (DOE FRASE Vocabulary)
BT2 Facilities and Buildings (DOE
FRASE Vocabulary)
BT3 Facilities
RT Cafeteria
RT Eating Area
SO DOE FRASE VOCABULARY

RESTORATION
(CFR; EPA)
DA   October 12, 1990
BT1 Actions
BT2 Responses

SO Environmental Protection Agency
Glossary
DEF (EPA) Actions undertaken to return
an injured resource to its baseline
condition, as measured in terms of
the injured resource's physical,
chemical, or biological properties
or the services it previously
provided, when such actions are in
addition to response actions
completed or anticipated, and
when such actions exceed the
level of response actions
determined appropriate to the site
pursuant to the National
Contingency Plan (NCP).

RESTRICTED AREAS
DA   January 24, 1991
BT1 Sites/Areas
RT Protected Areas
RT Radiation Protection
SO Environmental Management
DEF (CFR) Any area access to which is
controlled by the licensee for
purposes of protection of
individuals from exposure to
radiation and radioactive
materials. "Restricted area" shall
not include any areas used as
residential quarters, although a
separate room or rooms in a
residential building may be set
apart as a restricted area.

RESTRICTED USE
(EPA)
DA   October 12, 1990
BT1 Uses
SO Environmental Protection Agency
Glossary
DEF (EPA) When a pesticide is
registered, some or all of its uses
may be classified under the
Federal Insecticide Fungicide
Rodenticide Act (FIFRA)
regulations for restricted use if the
pesticide requires special handling
because of its toxicity.
Restricted-use pesticides may be
applied only by trained, certified
applicators or those under their
direct supervision.

RESTRICTED-USE PESTICIDE
(ESH)
DA   November 20, 1990
BT1 Pesticides
BT2 Hazardous Substances
DEF (ESH) A pesticide that is classified
for restricted use under the
provisions of Section 3(d)(1)(C) of
the Act which states: "If the
Administrator determines that the
pesticide, when applied in
accordance with its directions for
use, warnings and cautions and
for the uses for which it is
registered, or for one or more of
such uses, or in accordance with a

widespread and commonly
recognized practice, may generally
cause, without additional
regulatory restrictions,
unreasonable adverse effects on
the environment, including injury
to the applicator, he shall classify
the pesticide, or the particular use
or uses to which the determination
applies, for restricted use."

RESTRICTION ENZYMES
(EPA)
DA   October 12, 1990
BT1 Proteins
BT2 Organic Chemicals
BT3 Chemical Substances
SO Environmental Protection Agency
Glossary
DEF (EPA) Enzymes that recognize
certain specific regions of a long
DNA molecule and then cut the
DNA into smaller pieces.

RETAILERS
(TSCA; CFR)
DA   October 12, 1990
BT1 Commercial Organizations
BT2 Organizations
RT Chemical Substances
SO Hazardous Materials
DEF A person who distributes in
commerce a chemical substance,
mixture, or article to ultimate
purchasers who are not
commercial entities.

RETAINING WALLS
(DOE Order 6430.1A)
DA   October 12, 1990
SO Construction
DEF (DOE Order 6430.1A) Walls
designed to maintain differences in
ground elevations by holding back
a bank of material.

RETENTION
(SSDC; IAEA)
DA   October 12, 1990
RT Body Content
RT Deposition
RT Uptake
SO Radiation
SO System Safety Development Center
Glossary
DEF (IAEA) The fraction of deposited
material remaining in the body or
in some organ of interest at any
given time after deposition.

RETORT
(2491 RETORT)
DA   January 3, 1991
BT1 Equipment/Parts - Heating (DOE
FRASE Vocabulary)
BT2 Equipment
BT1 Heating Equipment
SO DOE FRASE VOCABULARY

BT-Broader Term   NT-Narrower Term   RT Related Term

RETRIEVAL CONTAINMENT BUILDING
DA   January 8, 1991
SF   *RCB*
BT1 Facilities

RETROFILL
(TSCA; CFR)
DA   October 12, 1990
BT1 Waste Management Processes
 BT2 Processes
RT   Non-PCB Transformers
RT   PCB Transformers
SO   Hazardous Materials
DEF To remove PCB or
     PCB-contaminated dielectric fluid
     and to replace it with either PCB,
     PCB-contaminated, or non-PCB
     dielectric fluid.

RETURN PERIOD
(SSDC)
DA   October 12, 1990
BT1 Time Designations
SO   Construction
SO   System Safety Development Center
     Glossary
DEF (SSDC) The average number of
     years within which a given event
     will be equaled or exceeded.

REUSE
(EPA)
DA   November 19, 1990
SY   Recycling
BT1 Uses
NT1 Closed-Loop Recycling
SO   Environmental Protection Agency
     Glossary
DEF (EPA) The process of minimizing
     the generation of waste by
     recovering usable products that
     might otherwise become waste.
     Examples are the recycling of
     aluminum cans, waste paper, and
     bottles.

REVERSE OSMOSIS
(EPA)
DA   October 12, 1990
BT1 Osmosis
 BT2 Biological Processes
 BT3 Processes
SO   Environmental Protection Agency
     Glossary
DEF (EPA) A water treatment process
     used in small water systems by
     adding pressure to force water
     through a semipermeable
     membrane. Reverse osmosis
     removes most drinking water
     contaminants. Also used in
     wastewater treatment. Large-scale
     reverse osmosis plants are now
     being developed.

REVIEWS
(SSDC)
DA   October 12, 1990
BT1 Administrative Processes

 BT2 Processes
NT1 Environmental Reviews
NT1 Operational Readiness Reviews
NT1 Safety Reviews
 NT2 Backfit Reviews
 NT2 Independent (Safety) Reviews
NT1 Safety Program Reviews
NT1 Special Reviews
RT   Inspection and Evaluation
RT   Sanitary Surveys
SO   System Safety Development Center
     Glossary
DEF (SSDC) Inspecting or examining;
     critical evaluations using stated
     criteria; "reviews" of an analysis
     done by others.

REVOLVERS
(Doe Order 5480.16)
DA   October 12, 1990
BT1 Handguns
 BT2 Small Arms
 BT3 Firearms
DEF (DOE Order 5480.16) Firearms
     (most commonly pistols) with a
     cylinder of several chambers so
     arranged as to revolve on an axis
     and be discharged in succession
     by the same lock.

REYNOLDS ELECTRICAL AND ENGINEERING COMPANY
DA   January 8, 1991
SF   *REECo*
BT1 Companies
 BT2 Commercial Organizations
 BT3 Organizations
BT1 DOE Contractors
 BT2 Potentially Responsible Parties
RT   Nevada Test Site
RT   Nevada Operations Office

RFA
DA   October 12, 1990
SEE RCRA Facility Assessment
SO   Acronyms

RFD
(EPA)
DA   January 9, 1991
SY   Reference Dose (RfD)
BT1 Doses
RT   Risk Assessment
SO   Environmental Protection Agency
     Glossary

RFP
DA   October 12, 1990
SEE Rocky Flats Plant
SO   Acronyms

RG
DA   October 12, 1990
SEE Regulatory Guide
SO   Acronyms

RH
DA   October 12, 1990

SEE Remote Handled
SO   Acronyms

RHEOSTAT(S)
(2418 RHEOSTAT)
DA   January 3, 1991
BT1 Equipment/Parts - Electrical (DOE
     FRASE Vocabulary)
 BT2 Equipment
SO   DOE FRASE VOCABULARY

RHR
DA   October 12, 1990
SEE Residual Heat Removal
SO   Acronyms

RI
DA   October 12, 1990
SEE Remedial Investigations
SO   Acronyms

RIB(S)
(1111 RIB)
DA   November 28, 1990
BT1 Bone(s)
 BT2 Human Body Parts
BT1 Trunk
 BT2 Human Body Parts
RT   Chest
SO   DOE FRASE VOCABULARY

RIBONUCLEIC ACID
DA   November 9, 1990
SF   *RNA*
BT1 Organic Chemicals
 BT2 Chemical Substances
RT   DNA
DEF A molecule that carries the genetic
     message from DNA to a cell's
     protein producing mechanisms;
     similar to, but chemically different
     from, DNA.

RICHLAND OPERATIONS OFFICE
DA   January 11, 1991
BT1 Operations Offices
 BT2 Offices
 BT3 Administrative Organizations
 BT4 Organizations
 BT2 U.S. Department of Energy
 BT3 Federal Agencies
 BT4 Agencies
 BT5 Administrative Organizations
 BT6 Organizations
RT   Battelle Memorial Institute
RT   Hanford Environmental Health
     Foundation
RT   Kaiser Engineers Hanford, Co.
RT   Westinghouse Hanford Company
DEF (Capsule Review of DOE Research
     and Development and Field
     Facilities, 1986) The Richland
     Operations Office, established in
     1943 as the Manhattan Project
     Plutonium Production Facility, has
     several current missions. The
     Office manages the 562-square
     mile Hanford Site where a broad
     range of nuclear and non-nuclear

energy programs are conducted under all DOE Secretarial Offices. The Office provides continuing operation of nuclear production, chemical processing and waste management programs, including interim storage and ultimate disposal of high-level radioactive wastes, and manages advanced reactor development programs. The Office manages programs for facilities and site services and provide administrative and policy guidance to several operating and support service contractors.

RICHTER MAGNITUDE (M_L)
(USGS)
DA October 12, 1990
SY Local Magnitude (M_L)
RT Earthquake Magnitude
SO Natural Phenomenon
DEF (USGS) The logarithm, to the base 10, of the amplitude in micrometers of the maximum amplitude of seismic waves that would be observed on a standard torsion seismograph at a distance of about 60 miles from the epicenter. Also called the local magnitude.

RIFLES
(DOE Order 5480.16)
DA October 12, 1990
BT1 Small Arms
 BT2 Firearms
NT1 Automatic Rifles
SO Firearms
DEF Firearms fired from the shoulder, as distinct from artillery and pistols.

RIG
DA October 12, 1990
SEE Risk-Based Inspection Guide
SO Acronyms

RIM
DA October 12, 1990
SEE Relative Importance Measure
SO Acronyms

RINGLEMANN CHART
(EPA)
DA October 12, 1990
BT1 Diagrams
SO Environmental Protection Agency Glossary
DEF (EPA) A series of shaded illustrations used to measure the opacity of air pollution emissions. The chart ranges from light grey through black and is used to set and enforce emissions standards.

RIPARIAN HABITATS
(EPA)
DA October 12, 1990
BT1 Habitats

SO Environmental Protection Agency Glossary
DEF (EPA) Areas adjacent to rivers and streams that have a high density, diversity, and productivity of plant and animal species relative to nearby uplands.

RIPARIAN RIGHTS
(EPA)
DA October 12, 1990
SO Environmental Protection Agency Glossary
DEF (EPA) Entitlement of a land owner to the water on or bordering his property, including the right to prevent diversion or misuse of upstream waters. Generally, a matter of state law.

RISK ANALYSIS
(SSDC)
DA October 12, 1990
SY Risk Assessment
BT1 Analyses
RT Performance Assessments
RT Risk Assessment
RT Risk Management
RT Risks
SO System Safety Development Center Glossary
DEF (SSDC) The quantification of the degree of risk.

RISK ASSESSMENT
(SSDC)
DA October 12, 1990
SY Risk Analysis
BT1 Assessments
 BT2 Administrative Processes
 BT3 Processes
NT1 Probabilistic Risk Assessment
RT Acute Effects
RT Carcinogen Risk Assessment Verification Endeavor Workgroup
RT Corrective Action Triggers
RT Derived Projection
RT Extreme Value Projection
RT Frequency/Severity
RT Goals
RT Implementation
RT Integrated Risk Information System
RT Management System Factors
RT Policies
RT Probability
RT Radioactive Source Terms
RT RFD
RT Risk Analysis
RT Risk Management
RT Risks
RT Safety Program Reviews
RT Special Reviews
RT Testing Requirements
RT Vulnerability Analyses
SO Environmental Protection Agency Glossary
SO System Safety Development Center Glossary
DEF (SSDC) The qualitative and quantitative evaluation performed

in an effort to define the risk posed to human health and/or the environment by the presence or potential presence and/or use of specific pollutants.

RISK-BASED INSPECTION GUIDE
DA January 8, 1991
SF *RIG*

RISK COMMUNICATION
(EPA)
DA October 12, 1990
RT Public Hearings
RT Public Affairs (Personnel)
RT Public Information Officers
SO Environmental Protection Agency Glossary
DEF (EPA) The exchange of information about health or environmental risks between risk assessors, risk managers, the general public, news media, interest groups, etc.

RISK MANAGEMENT
(SSDC)
DA October 12, 1990
BT1 Management
RT Advisory
RT Risk Analysis
RT Risk Assessment
RT Risks
SO Environmental Protection Agency Glossary
SO System Safety Development Center Glossary
DEF (SSDC) The process of evaluating alternative regulatory and nonregulatory responses to risk and selecting among them. The selection process necessarily requires the consideration of legal, economic, and social factors.

RISK RETENTION GROUP
(CERCLA)
DA October 12, 1990
BT1 Groups
 BT2 Administrative Organizations
 BT3 Organizations
RT Pollution Liability
SO Compensation and Liability
DEF (CERCLA) Any corporation or other limited liability association taxable as a corporation, or as an insurance company, formed under the laws of any State- (A) whose primary activity consists of assuming and spreading all, or any portion, of the pollution liability of its group members; (B) which is organized for the primary purpose of conducting the activity described under subparagraph (A); (C) which is chartered or licensed as an insurance company and authorized to engage in the business of insurance under the laws of any State; and (D) which

does not exclude any person from membership in the group solely to provide for members of such a group a competitive advantage over such a person.

RISKS
(DOE Orders 5480.5, 5480.6, and 5481.1B; ESH)
DA   October 12, 1990
NT1 Acceptable Risks
NT1 Assigned Risks
NT1 Assumed Risks
NT1 Detriment
NT1 Improved Risk
NT1 Intangible Risks
NT1 Perceived Risks
NT1 Residual Risks
NT1 Undue Risks
RT   Exposure
RT   Insurance
RT   Probability
RT   Risk Analysis
RT   Risk Assessment
RT   Risk Management
RT   Safe
RT   Safety
RT   Weighting Factors
SO   Environmental Management
SO   Industrial Safety
SO   Management
SO   Radiation
SO   System Safety Development Center Glossary
DEF Quantitative or qualitative expressions of possible loss which consider both the probability that a hazard will cause harm and the consequences of that event.

RIVER BASINS
DA   November 9, 1990
BT1 Sites/Areas
RT   Watershed
DEF Land areas drained by a river and its tributaries.

RIVER WATER
DA   January 8, 1991
SF   *RW (River Water)*
BT1 Surface Waters
 BT2 Water

RIVER/CREEK
(1617 RIVER)
DA   December 10, 1990
BT1 Surface Water Resources
 BT2 Natural Resources
RT   Body of Water
SO   DOE FRASE VOCABULARY

RL
DA   October 12, 1990
SO   Acronyms

RMCG
DA   October 12, 1990
SEE Reactor Materials Control Group
SO   Acronyms

RMCL
DA   October 12, 1990
SEE Recommended Maximum Contaminant Level
SO   Acronyms

RMI
DA   October 12, 1990
SEE Reactive Metals, Inc.
SO   Acronyms

RNA
DA   October 12, 1990
SEE Ribonucleic Acid
SO   Acronyms

RO (Reactor Operator)
DA   October 12, 1990
SEE Reactor Operator
SO   Acronyms

RO (Remains Open)
DA   October 12, 1990
SEE Remains Open
SO   Acronyms

ROAD
(1861 ROAD)
DA   December 10, 1990
BT1 Structures (DOE FRASE Vocabulary)
SO   DOE FRASE VOCABULARY

ROADWAY CROWN
(DOE Order 6430.1A)
DA   October 12, 1990
BT1 Roadways
 BT2 Routes (Transportation)
RT   Roadways
SO   Construction
DEF (DOE Order 6430.1A) The high point of a roadway cross-section (usually at the centerline).

ROADWAYS
(CAA; CFR)
DA   October 12, 1990
BT1 Routes (Transportation)
NT1 Roadway Crown
RT   Base Course
RT   Preferred Routes
RT   Preferred Highway
RT   Public Travel Routes
RT   Roadway Crown
RT   Subbase
RT   Superelevation
RT   Transportation
SO   Air Pollution
DEF Surfaces on which motor vehicles travel. This term includes highways, roads, streets, parking areas, and driveways.

ROASTING
(CAA; CFR)
DA   October 12, 1990
BT1 Manufacturing Processes
 BT2 Processes

RT   Industrial Furnaces
RT   Nodulizing Kilns
SO   Air Pollution
DEF The use of a furnace to heat arsenic plant feed material for the purpose of eliminating a significant portion of the volatile materials contained in the feed.

ROCKET AMMUNITION
(HMTA; CFR)
DA   May 20, 1991
BT1 Ammunition
 BT2 Munitions
SO   Hazardous Materials
DEF (CFR) Rocket ammunition (including guided missiles) is ammunition designed for launching from a tube, launcher, rails, trough, or other launching device, in which the propellant explosive. It consists of an igniter, rocket motor, and projectile (warhead) either fused or unfused, containing high explosives or chemicals. Rocket ammunition may be shipped completely assembled or may be shipped unassembled in one outside container.

ROCKET MOTOR TEST SITES
(CAA; CFR)
DA   October 12, 1990
BT1 Test Area
 BT2 Area
  BT3 Sites/Areas
SO   Air Pollution
DEF Buildings, structures, facilities, or installations where the static test firing of a beryllium rocket motor and/or the disposal of beryllium propellant is conducted.

ROCKWELL INTERNATIONAL CORP.
DA   January 11, 1991
BT1 Companies
 BT2 Commercial Organizations
  BT3 Organizations
BT1 DOE Contractors
 BT2 Potentially Responsible Parties
RT   Energy Technology Engineering Center (Canoga Park)
RT   Idaho Operations Office
RT   San Francisco Operations Office

ROCKY FLATS PLANT
DA   January 8, 1991
SF   *RFP*
BT1 Government-Owned Contractor-Operated Facilities
 BT2 Federal Facilities
  BT3 Facilities

ROD (CERCLA Record of Decision)
DA   June 6, 1991
SEE Record of Decision (CERCLA)
SO   Acronyms

ROD (NEPA Record of Decision)
DA October 12, 1990
SEE Record of Decision (NEPA)
SO Acronyms
SO Environmental Protection Agency
　　Glossary
SO Management

ROD (Report of Discrepancy)
DA October 12, 1990
SEE Report of Discrepancy
SO Acronyms

ROD DRIVE
(2590 ROD DRIVE)
DA January 3, 1991
BT1 Equipment/Parts - Nuclear (DOE
　　FRASE Vocabulary)
BT2 Equipment
BT2 Reactor Components
RT Control Rod Drive Assembly
SO DOE FRASE VOCABULARY

ROD EQUIPMENT ROOM
DA January 8, 1991
SF *RER*

RODENTICIDES
(EPA)
DA October 12, 1990
BT1 Pesticides
BT2 Hazardous Substances
SO Environmental Protection Agency
　　Glossary
DEF (EPA) A chemical or agent used to
　　destroy rats or other rodent pests,
　　or to prevent them from damaging
　　food, crops, etc.

ROENTGEN
DA January 8, 1991
SF *R*
BT1 Radiation Units
BT2 Units of Measure
RT Milliroentgen
RT Roentgen Equivalent Man
RT Sievert
SO Environmental Management
SO Radiation
DEF (DSTT) An exposure dose of x- or
　　y- radiation such that the electrons
　　and positrons liberated by this
　　radiation produce, in air, when
　　stopped completely, ions carrying
　　positive and negative charges of
　　2.58 x 10 to the negative 4
　　coulomb per kilogram of air.
　　Abbreviated R (formerly r).

ROENTGEN EQUIVALENT MAN
(EPA)
DA October 12, 1990
SY Sievert
SF *REM (Roentgen Equivalent Man)*
BT1 Radiation Units
BT2 Units of Measure
RT Roentgen

SO Environmental Protection Agency
　　Glossary
DEF (DSTT) A unit of ionizing radiation,
　　equal to the amount that produces
　　the same damage to humans as 1
　　roentgen of high- voltage x-rays.

ROLL MILL
(2139 ROLL MILL)
DA December 10, 1990
BT1 Milling Machine
BT2 Machines (DOE FRASE
　　Vocabulary)
BT3 Equipment
SO DOE FRASE VOCABULARY

ROOF-TOP-RATIO
(NFI)
DA October 12, 1990
SF *RTR*
BT1 Ratios
SO Radiation
DEF (NFI) Ratio of the flux at points 1/4
　　and 3/4 of the distance from the
　　top to the bottom of the reactor
　　core, used to position partial
　　control rods correctly.

ROOM
(1713 Room)
DA January 18, 1991
BT1 Sites/Areas
NT1 Air Lock Room
NT1 Central Control Room
NT1 Clean Rooms
NT1 Control Room
NT1 Equipment Room
NT1 Locker/Shower Room
NT1 Refrigeration Room
NT1 Rest Room
NT1 Shipping Room
NT1 Stock Room
RT Chamber
SO DOE FRASE VOCABULARY

ROOT CAUSE ANALYSIS
(DOE Order 4330.4A)
DA June 5, 1991
BT1 Analyses
RT Failures
RT Management System Factors
RT Oversights and Omissions
RT Specific Control Factors
SO Management
DEF (DOE Order 4330.4A) An analysis
　　performed to determine the cause
　　of part, system, and component
　　failures.

ROOT CAUSES
(TTGM; SSDC)
DA October 12, 1990
SY Basic Causes
BT1 Causes
RT Concerns
RT Contributing Causes
RT Events and Causal Factors
RT Findings

RT Management Oversight and Risk
　　Tree
RT Management System Factors
RT Specific Control Factors
SO System Safety Development Center
　　Glossary
DEF (TTGM) Within a Tiger Team
　　Assessment, a) those items which
　　if corrected could prevent
　　occurrence or reoccurrence of the
　　situations and conditions that were
　　found and b) those
　　specific/systemic factors that
　　could cause or create conditions
　　that may be less than adequate or
　　could result in accidents or
　　incidents. (SSDC) A lack of
　　adequate management control
　　contributing to substandard
　　practices or conditions (immediate
　　causes) which result in accidents
　　or incidents. Management control
　　failures can result from the lack of
　　adequate policy, inadequate policy
　　implementation or from the lack of
　　or insufficient risk evaluation.

ROSS AVIATION, INC.
DA January 11, 1991
BT1 Companies
BT2 Commercial Organizations
BT3 Organizations
BT1 DOE Contractors
BT2 Potentially Responsible Parties
RT Albuquerque Operations Office

ROTATIONAL WINDSTORMS
(SMRP)
DA October 12, 1990
BT1 Act of Nature
NT1 Dust Devils
NT1 Tornadoes
NT1 Waterspouts
RT Divergent Windstorms
RT Subvortices
SO Natural Phenomenon
DEF (SMRP) Tornadoes, waterspouts,
　　and dust devils; characterized by a
　　rotational flow field.

ROUGH FISH
(EPA)
DA October 12, 1990
BT1 Fish
BT2 Animals
SO Environmental Protection Agency
　　Glossary
DEF (EPA) Those fish, not prized for
　　eating, such as gar and suckers.
　　Most are more tolerant of
　　changing environmental conditions
　　than game species.

ROUTER(S)
(2978 ROUTER)
DA January 3, 1991
BT1 Tools - Powered
BT2 Tools (DOE FRASE Vocabulary)

BT-Broader Term　　　　　　　　　NT-Narrower Term　　　　　　　　　RT Related Term

BT3 Equipment
SO DOE FRASE VOCABULARY

ROUTES (EXPOSURE)
DA February 25, 1991
NT1 Common Exposure Routes
NT1 Critical Pathways
NT1 Exposure Pathways
 NT2 Ingestion Exposure Pathways
 NT2 Plume Exposure Pathways
NT1 Exposure Route
RT Exposure

ROUTES (TRANSPORTATION)
DA February 25, 1991
NT1 Highway
NT1 Preferred Routes
 NT2 State-Designated Routes
NT1 Preferred Highway
NT1 Public Travel Routes
NT1 Roadways
 NT2 Roadway Crown

ROUTINE ANALYTICAL SERVICES
(EPA)
DA October 12, 1990
RT Analyses
SO Environmental Protection Agency
 Glossary
DEF (EPA) The set of Contract
 Laboratory Program (CLP)
 analytical protocols that are used
 to analyze most Superfund site
 samples. These protocols are
 provided in the EPA Statements of
 Work (SOW) for the CLP (SOW
 for Inorganics; SOW for Organics)
 and must be followed by every
 CLP laboratory.

ROUTINE WASTES
(DOE Order 6430.1A)
DA October 12, 1990
BT1 Wastes
SO Construction
DEF (DOE Order 6430.1A) Wastes
 generated due to normal
 operations and anticipated
 abnormal events.

RPIS
DA October 12, 1990
SEE Real Property Inventory System
SO Acronyms

RPM (Radial Power Monitor)
DA October 12, 1990
SEE Radial Power Monitor
SO Acronyms

RPM (Remedial Project Manager)
DA October 12, 1990
SEE Remedial Project Managers
SO Acronyms

RPS
DA October 12, 1990

SEE Reactor Protection System
SO Acronyms

RPT
DA October 12, 1990
SEE Recirculation Pump Trip
SO Acronyms

RPV
DA October 12, 1990
SEE Reactor Pressure Vessel
SO Acronyms

RQ
DA October 12, 1990
SEE Reportable Quantities
SO Acronyms

RRI
DA October 12, 1990
SEE RCRA Remedial Investigation
SO Acronyms

RRMT
DA October 12, 1990
SEE Reactor and Reactor Material
 Technology
SO Acronyms

RRT
DA October 12, 1990
SEE Regional Response Teams
SO Acronyms

RSE
DA May 16, 1991
SEE Removal Site Evaluations
SO Acronyms

RSO (Reported Significant Observations)
DA October 12, 1990
SEE Reported Significant Observations
SO Acronyms

RSPA
DA October 12, 1990
SEE Research and Special Programs
 Administration
SO Acronyms

RSS
DA October 12, 1990
SEE Reactor Safety Study
SO Acronyms

RSSMAP
DA October 12, 1990
SEE Reactor Safety Study Methodology
 Application Program
SO Acronyms

RTD
DA October 12, 1990
SEE Reactor Technology Department
SO Acronyms

RTR
DA October 12, 1990
SEE Roof-Top-Ratio
SO Acronyms

RUBBISH
(EPA)
DA October 12, 1990
BT1 Solid Wastes
 BT2 Wastes
SO Environmental Protection Agency
 Glossary
DEF (EPA) Solid waste, excluding food
 waste and ashes, from homes,
 institutions, and work-places.

RUN
(CAA; CFR)
DA October 12, 1990
BT1 Time Designations
RT Air Monitoring
RT Sampling
SO Air Pollution
DEF The net period of time during which
 an emission sample is collected.
 Unless otherwise specified, a run
 may be either intermittent or
 continuous within the limits of
 good engineering practice.

RUN-OFF
(SWDA; RCRA; EPA)
DA October 12, 1990
NT1 Urban Runoff
RT Contour Plowing
RT Erosion
RT Groundwater
RT Storm Water
SO Environmental Management
SO Environmental Protection Agency
 Glossary
SO Wastes
DEF That part of precipitation, snow
 melt, or irrigation water that runs
 off the land into streams or other
 surface water. It can carry
 pollutants from the air and land
 into the receiving waters.

RUN-ON
(SWDA; RCRA)
DA October 12, 1990
SO Environmental Management
SO Wastes
DEF Any rainwater, leachate, or other
 liquid that drains over land onto
 any part of a facility.

RUPTURED DISK
(1337 RUPTURED DIS)
DA November 28, 1990
BT1 Injuries
RT Back
RT Spinal Cord
RT Upper Back
SO DOE FRASE VOCABULARY

RUTHENIUM 103
(EDB)

SY-Synonymous Terms   SO-Source/Subject Category   SF-See From

DA   March 29, 1991
BT1 Beta-Minus Decay Radioisotopes
 BT2 Beta Decay Radioisotopes
  BT3 Radionuclides
   BT4 CERCLA Hazardous
        Substances
    BT5 Hazardous Substances
   BT4 Nuclides
BT1 Years Living Radioisotopes
 BT2 Radionuclides
  BT3 CERCLA Hazardous Substances
   BT4 Hazardous Substances
  BT3 Nuclides
SO   Radiation

RV
DA   October 12, 1990
SEE Reactor Vessel
SO   Acronyms

RVC
DA   October 12, 1990
SEE Reactor Volume Control
SO   Acronyms

RW (DOE Program Office)
DA   October 12, 1990
SEE Office of Civilian Radioactive Waste
       Management
SO   Acronyms

RW (River Water)
DA   October 12, 1990
SEE River Water
SO   Acronyms

RWE
DA   October 12, 1990
SEE Reactor Works Engineering
SO   Acronyms

RWIS
DA   October 12, 1990
SEE Raw Water Intake Structure
SO   Acronyms

RWL
DA   October 12, 1990
SEE Reactor Water Level
SO   Acronyms

RWQCB
DA   October 12, 1990
SEE Regional Water Quality Control
       Board
SO   Acronyms

RWST
DA   October 12, 1990
SEE Refueling Water Storage Tank
SO   Acronyms

RX
DA   October 12, 1990
SEE Reactors
SO   Acronyms

RZ
DA   October 12, 1990
SEE Radiation Zone
SO   Acronyms

S&A
DA   October 12, 1990
SEE Sampling and Analysis
SO   Acronyms

S&M
DA   October 12, 1990
SEE Surveillance and Maintenance
SO   Acronyms

S Waves
DA   October 12, 1990
SEE Secondary Body Waves
SO   Acronyms

S/R
DA   October 12, 1990
SEE Safety/Relief
SO   Acronyms

SAFE
(SSDC)
DA   October 12, 1990
BT1 Conditions
RT   Risks
SO   System Safety Development Center
      Glossary
DEF (SSDC) A condition wherein risks
      are as low as practicable and
      present no significant residual risk.

SAFE DRINKING WATER ACT
DA   June 6, 1991
SF   *SDWA*
BT1 Acts
 BT2 Statutes and Regulations
BT1 Appropriate Act and Regulations
RT   Drinking Water Supplies
RT   Public Water Systems
RT   Water Quality Criteria
SO   Water Pollution

SAFE MASS
(Doe Order 5480.5)
DA   October 12, 1990
SO   Environmental Management
SO   Industrial Safety
DEF (DOE Order 5480.5) That mass of
      fissionable materials which is
      subcritical for all conditions to
      which it could reasonably be
      expected to be exposed, including
      processing, handling, storing, and
      procedural uncertainties.

SAFE SHUTDOWN EARTHQUAKE 135
DA   January 24, 1991
BT1 Design Basis Earthquakes
 BT2 Design Basis Accidents
  BT3 Accidents
RT   Natural Phenomenon
SO   Environmental Management
DEF (CFR) That earthquake which is

based upon an evaluation of the
maximum earthquake potential
considering the regional and local
geology and seismology and
specific characteristics of local
subsurface material. It is that
earthquake which produces the
maximum vibratory ground motion
for which certain structures,
systems, and components are
designed to remain functional.

SAFE SHUTDOWN SYSTEM
DA   January 8, 1991
SF   *SSS (Safe Shutdown System)*
BT1 Emergency Systems
 BT2 Systems
NT1 Sparger Assemblies
RT   Reactor Shutdown

SAFE WORK PERMIT
(SSDC)
DA   October 12, 1990
BT1 Permits
RT   Safeguards
SO   System Safety Development Center
      Glossary
DEF (SSDC) A permit allowing
      employees to perform potential
      hazardous work, which outlines
      necessary safeguards,
      procedures, etc. An SWP usually
      requires several approval
      procedures before work can begin.
      The form may have other names
      in different companies.

SAFEGUARDS
(DOE Order 6430.1A; EMER)
DA   October 12, 1990
RT   Departmental-Approved Equipment
RT   DOE Safeguards and Security
      Coordinators
RT   Inspections (Nuclear)
RT   Physical Protection (physical
      security)
RT   Safe Work Permit
RT   Site-Specific Safeguards and
      Security Plan
SO   Construction
SO   Emergency Preparedness
DEF (DOE Order 6430.1A) An
      integrated system of physical
      protection, material accounting,
      and material control measures
      designed to date, prevent, detect,
      and respond to unauthorized
      possession, use, or sabotage of
      special nuclear materials. In
      practice, safeguards involve the
      development and application of
      techniques and procedures
      dealing with the establishment and
      continued maintenance of a
      system of activities including
      physical protection, quantitative
      knowledge of the location and use
      of special nuclear materials, and
      administrative controls and
      surveillance to assure that

BT-Broader Term   NT-Narrower Term   RT Related Term

procedures and techniques of the system are effective and are being carried out. Safeguards include the timely indication of possible diversion or credible assurances by audits and inventory verification that no diversion has occurred.

SAFEGUARDS AND SECURITY ALERT I (CODE DESIGNATOR: ALPHA)
(EMER)
DA   February 1, 1991
SY   Alpha
BT1 Safeguards and Security Alerts
 BT2 Safeguards and Security
      Emergencies
  BT3 Emergencies
   BT4 Reportable Occurrences
    BT5 Occurrences
SO   Emergency Preparedness
DEF (EMER) A maximum alert action which shall be effected when the head of a field office determines that conditions warrant maximum security measures at U.S. Department of Energy (DOE) or DOE contractor facilities.

SAFEGUARDS AND SECURITY ALERT II (CODE DESIGNATOR: BRAVO)
(EMER)
DA   February 1, 1991
SY   Bravo
BT1 Safeguards and Security Alerts
 BT2 Safeguards and Security
      Emergencies
  BT3 Emergencies
   BT4 Reportable Occurrences
    BT5 Occurrences
SO   Emergency Preparedness
DEF (EMER) A substantial alert action which shall be put into effect when the head of a field office determines that conditions or information received warrants more than the preparatory safeguards actions under Safeguards and Security Alert III.

SAFEGUARDS AND SECURITY ALERT III (CODE DESIGNATOR: CHARLIE)
(EMER)
DA   February 1, 1991
SY   Charlie
BT1 Safeguards and Security Alerts
 BT2 Safeguards and Security
      Emergencies
  BT3 Emergencies
   BT4 Reportable Occurrences
    BT5 Occurrences
SO   Emergency Preparedness
DEF (EMER) A preparatory alert action which shall be effected when the head of a field office determines that existing preemergency conditions warrant increased safeguards and security measures at facilities under his/her jurisdiction; however, the Office of the Assistant Secretary for

Defense Programs may establish a general U.S. Department of Energy-wide or a locally confined Safeguards and Security Alert III without prior consultation with the head of field offices.

SAFEGUARDS AND SECURITY ALERTS
(EMER)
DA   February 1, 1991
BT1 Safeguards and Security
      Emergencies
 BT2 Emergencies
  BT3 Reportable Occurrences
   BT4 Occurrences
NT1 Alpha
NT1 Bravo
NT1 Charlie
NT1 Safeguards and Security Alert I
      (Code Designator: Alpha)
NT1 Safeguards and Security Alert II
      (Code Designator: Bravo)
NT1 Safeguards and Security Alert III
      (Code Designator: Charlie)
SO   Emergency Preparedness
DEF (EMER) Conditions effected when additional safeguards and security measures may be required at U.S. Department of Energy (DOE) or DOE contractor or subcontractor facilities.

SAFEGUARDS AND SECURITY EMERGENCIES
(EMER)
DA   January 24, 1991
BT1 Emergencies
 BT2 Reportable Occurrences
  BT3 Occurrences
NT1 Safeguards and Security Alerts
NT2 Alpha
NT2 Bravo
NT2 Charlie
NT2 Safeguards and Security Alert I
      (Code Designator: Alpha)
NT2 Safeguards and Security Alert II
      (Code Designator: Bravo)
NT2 Safeguards and Security Alert III
      (Code Designator: Charlie)
NT1 Special Nuclear Materials
      Emergencies
RT   DOE Safeguards and Security
      Coordinators
RT   Site-Specific Safeguards and
      Security Plan
SO   Environmental Management
DEF (DOE Order 5500.1A) Conditions in which there are potential or actual malevolent acts that create or appear likely to create a condition resulting in sabotage, bodily harm, unlawful access to DOE or DOE contractor facilities, or the interruption or loss of vital services. Also included are special nuclear materials emergencies in which there is a situation involving stolen, lost, or unauthorized possession of source material,

special nuclear material, by-product material of U.S. and/or foreign manufacture, improvised nuclear

SAFETY
(SSDC)
DA   October 12, 1990
NT1 Aviation Safety
NT1 Industrial Safety
NT1 Nuclear Criticality Safety
NT1 Radiation Safety
NT1 System Safety
RT   Accidents
RT   Hazards
RT   Health Hazards
RT   Human Factors
RT   Human Factors Engineering
RT   Reliability-Centered Maintenance
RT   Risks
RT   Uncertainty
SO   System Safety Development Center
      Glossary
DEF (SSDC) The control of accidental loss or injury.

SAFETY (HAZARD) ANALYSIS
(SSDC)
DA   October 12, 1990
SY   Safety Analysis
BT1 Analyses
RT   Hazard Analysis
SO   System Safety Development Center
      Glossary
DEF (SSDC) The entire complex of safety (hazard) analysis methods and techniques ranging from relatively informal job and task safety analyses to large complex safety analysis studies and reports.

SAFETY ANALYSIS
(Doe Orders 5480.5, 5481.1B, and 6430.1A)
DA   October 12, 1990
SY   Safety (Hazard) Analysis
BT1 Analyses
NT1 Job Safety Analysis
NT1 Preliminary Safety Analysis
NT1 System (Safety) Analyses
RT   Hazard Analysis
RT   Safety Analysis Reports
RT   Safety Appraisals
RT   Safety Analysis Process
SO   Construction
SO   Industrial Safety
DEF A documented process to systematically identify the hazards of a DOE operation; to describe and analyze the adequacy of the measures taken to eliminate, control, or mitigate identified hazards; and to analyze and evaluate potential accidents and their associated risks. (MORT) MORT analysis asks: Was the task safety analysis adequate? Was the task safety analysis scaled properly for the hazards

SY-Synonymous Terms   SO-Source/Subject Category   SF-See From

involved? Was the preparation (and content) of the task safety analysis adequate? Were safety hazards associated with the work task adequately identified and selected? Was the knowledge input to the task safety analysis adequate? Was there adequate employee participation? Technical information? Was the development of the specific task safety analysis by the first line supervisor adequate? Were adequate worksite controls placed on the work process, facility, equipment, and personnel?

SAFETY ANALYSIS PROCESS
(SSDC)
DA October 12, 1990
BT1 Administrative Processes
 BT2 Processes
RT Analyses
RT Hazard Analysis
RT Safety Analysis Reports
RT Safety Analysis
SO System Safety Development Center
    Glossary
DEF (SSDC) The identification of hazards, analysis of hazards control, and analysis of residual risk.

SAFETY ANALYSIS REPORTS
(Doe Order 5480.6, 5480.16; SSDC)
DA October 12, 1990
SF SAR
BT1 Reports
NT1 Final Safety Analysis Report
NT1 Preliminary Safety Analysis Report
NT1 Updated Final Safety Analysis Report
RT Analyses
RT Safety Analysis
RT Safety Analysis Process
RT Security Facilities
SO Construction
SO Environmental Management
SO Firearms
SO Industrial Safety
SO System Safety Development Center
    Glossary
DEF Prepared in accordance with DOE Order 5481.1B; safety documents providing a concise but complete description and safety evaluation of the site, the design, normal and emergency operation, potential accidents, and predicted consequences of such accidents, and the means proposed to prevent such accidents or mitigate the consequences of such accidents. Safety Analysis Reports are designated as final when they are based on final design information. Otherwise, they are designated as preliminary.

SAFETY AND HEALTH ASSESSMENTS
(TTGM)
DA May 6, 1991
BT1 Tiger Team Assessments
 BT2 Assessments
  BT3 Administrative Processes
   BT4 Processes
RT Concerns
RT Environment Assessments
RT Management and Organization
    Assessment
SO Management
DEF (TTGM) One of the three major assessments conducted as part of a Tiger Team Assessment. The Technical Safety Appraisal (TSA) process, which is a comprehensive, periodic review of field office operations conducted by DOE's Office of Safety Appraisals, has been modified and expanded, on a site-specific basis, to fulfill the needs of the Safety and Health component of the Tiger Team Assessments. It is expected that TSAs will be performed as a part of all future Tiger Team Assessments. Major disciplines included in this assessment are: organization and administration, operations, maintenance, training and certification, auxiliary systems, emergency preparedness, technical support, security/safety interfaces, experimental activities, site/facility safety review, nuclear criticality safety, radiological protection, fire protection, packaging and transportation, quality verification, personnel protection, aviation safety, and medical services.

SAFETY AND HEALTH DIRECTORS
(Doe Order 5483.1A)
DA October 12, 1990
BT1 Personnel
SO Industrial Safety
DEF (DOE Order 5483.1A) The primary field organization staff managers responsible for the overview and coordination of the occupational safety and health program administered by a field organization for its contractor operations.

SAFETY APPRAISALS
(SSDC)
DA October 12, 1990
BT1 Appraisals
NT1 Technical Safety Appraisals
RT Safety Analysis
RT Safety Assurance System
    Summary
RT Safety Program Reviews
SO System Safety Development Center
    Glossary
DEF (SSDC) Safety appraisals are determinations by external,

independent reviewers of safety/loss control program effectiveness. They may be conducted as a management appraisal, a functional appraisal, or a comprehensive appraisal, which combines the previous two. A management appraisal is a review and evaluation of management performance covering all safety disciplines and management responsibilities to assure proper safety/loss control program balance. A functional appraisal is a review of a safety specialty discipline (industrial safety, industrial hygiene, fire protection, etc.) to verify that applicable elements of the safety/loss control program have been developed, documented, and effectively implemented in accordance with specific safety/loss control requirements and needs.

SAFETY ASSURANCE SYSTEM SUMMARY
(SSDC)
DA October 12, 1990
SF SASS
BT1 Diagrams
RT Safety Appraisals
SO System Safety Development Center
    Glossary
DEF (SSDC) A safety appraisal tree containing nine basic ES&H appraisal factors. Eight of the factors are each branched into 11 major program elements. The ninth, documentation, is considered in each of the others, as well as in the overall appraisal process. (See SSDC-24.)

SAFETY BARRIERS
(SSDC)
DA October 12, 1990
BT1 Barriers
SO System Safety Development Center
    Glossary
DEF (SSDC) Those barriers used to prevent, control, or minimize the impact of unwanted energy flows, such as a guardrail.

SAFETY BELT
(2680 SAFETY BELT)
DA January 3, 1991
BT1 Fall Protection Device
 BT2 Devices
 BT2 Personal Protective Equipment
  BT3 Equipment/Parts - Personal
      Protective (DOE FRASE
      Vocabulary)
   BT4 Equipment
SO DOE FRASE VOCABULARY

SAFETY BOOTS
(2681 SAFETY BOOTS)

DA   January 3, 1991
BT1 Foot Protection
 BT2 Personal Protective Equipment
  BT3 Equipment/Parts - Personal
     Protective (DOE FRASE
     Vocabulary)
   BT4 Equipment
RT Foot/Feet
SO DOE FRASE VOCABULARY

SAFETY CLASS ITEMS
(DOE Order 6430.1A)
DA   October 12, 1990
RT   Critical Facilities
RT   Design Basis Accidents
RT   Engineered Safety Features
RT   Seismic Category I
SO   Construction
DEF (DOE Order 6430.1A) Systems,
     components, and structures,
     including portions of process
     systems, whose failure could
     adversely affect the environment
     or safety and health of the public.
     Determination of classification is
     based on analysis of the potential
     abnormal and accidental scenario
     consequences as presented in the
     Safety Analysis Report (SAR) (as
     required by 5481.1B).

SAFETY COMPUTER
DA   January 8, 1991
SF   *SC*

SAFETY DEVICE/SYSTEM
(DOE Order 5000.3A)
DA   December 14, 1990
BT1 Devices
RT   Class A Equipment
RT   Class B Equipment
DEF (DOE Order 5000.3A) This term is
     generally intended to mean all
     permanently installed
     safety-related equipment which
     relates to processes, other major
     equipment, major personnel
     hazards, etc. It is not intended to
     include boundary ropes, chains,
     goggles, handrails, and any other
     of a host of minor items which
     could be included under literal
     compliance. Problems with minor
     items fall under this reporting
     system when they result in
     consequences of a level with
     reportable criteria.

SAFETY DOCUMENTS
(Doe Order 5480.6)
DA   October 12, 1990
BT1 Reports
SO   Industrial Safety
DEF (DOE Order 5480.6) Documents
     prepared specifically to assure
     that the safety aspects of part or
     all of the activities conducted at a
     reactor are formally and
     thoroughly analyzed, evaluated,

and recorded (e.g., Technical
Specifications, Safety Analysis
Reports and addenda, and
documented reports of special
safety reviews and studies).

SAFETY EVALUATION REPORT
DA   January 8, 1991
SF   *SER*
BT1 Reports

SAFETY GLASSES
(2682 SAF GLASS)
DA   January 3, 1991
BT1 Eye Protection
 BT2 Personal Protective Equipment
  BT3 Equipment/Parts - Personal
     Protective (DOE FRASE
     Vocabulary)
   BT4 Equipment
NT1 Safety Glasses W Side Shields
NT1 Tinted Safety Glasses
RT   Eye(s)
SO   DOE FRASE VOCABULARY

SAFETY GLASSES W SIDE SHIELDS
(2683 SAF GLASS SS)
DA   January 3, 1991
BT1 Safety Glasses
 BT2 Eye Protection
  BT3 Personal Protective Equipment
   BT4 Equipment/Parts - Personal
     Protective (DOE FRASE
     Vocabulary)
    BT5 Equipment
RT   Eye(s)
SO   DOE FRASE VOCABULARY

SAFETY GUIDES
(Doe Order 5480.5)
DA   October 12, 1990
BT1 Guidelines
RT   Material Safety Data Sheets
SO   Industrial Safety
DEF (DOE Order 5480.5) Documents
     designated or recognized as an
     acceptable basis for nuclear
     criticality safety evaluations. These
     guides may be used as aids by
     DOE field organizations in
     establishing acceptable safety
     practices, and include material
     developed by DOE contractors,
     professional societies, industrial
     organizations, and foreign atomic
     energy industries.

SAFETY INJECTION CONTROL SYSTEM
DA   January 8, 1991
SF   *SICS*
BT1 Control Systems
 BT2 Controls
 BT2 Systems

SAFETY INJECTION TANK
DA   January 8, 1991
SF   *SIT*

BT1 Tanks
 BT2 Facility Components

SAFETY LIMITS
(DOE Order 6430.1A)
DA   October 12, 1990
BT1 Limits
RT   Barriers
SO   Construction
DEF (DOE Order 6430.1A) Limits on
     important process variables that
     are necessary to provide
     reasonable protection to the
     integrity of certain physical barriers
     that guard against the uncontrolled
     release of radioactivity or an
     accidental criticality.

SAFETY LINE
(2684 SAFETY LINE)
DA   January 3, 1991
BT1 Fall Protection Device
 BT2 Devices
 BT2 Personal Protective Equipment
  BT3 Equipment/Parts - Personal
     Protective (DOE FRASE
     Vocabulary)
   BT4 Equipment
SO   DOE FRASE VOCABULARY

SAFETY NET
(2685 SAFETY NET)
DA   January 3, 1991
BT1 Fall Protection Device
 BT2 Devices
 BT2 Personal Protective Equipment
  BT3 Equipment/Parts - Personal
     Protective (DOE FRASE
     Vocabulary)
   BT4 Equipment
SO   DOE FRASE VOCABULARY

SAFETY PERFORMANCE MEASUREMENT SYSTEM
(SSDC)
DA   January 8, 1991
SF   *SPMS*
BT1 Information Systems
 BT2 Security Interests
 BT2 Systems
NT1 Chemical Hazards Emergency
     Management System
NT1 Computerized Accident/Incident
     Reporting System
NT1 Computer Assisted Tracking
     System
NT1 Occurrence Reporting and
     Processing System
NT1 Personnel Expertise and Resource
     Listing
NT1 Radiation Exposure Module
NT1 Standards Information Management
     System
NT1 Unusual Occurrence Reporting
     System
RT   Factor Relationship and Sequence
     of Events Vocabulary
RT   FRASE Vocabulary

RT System Safety Development Center
DEF (SSDC) The Safety Performance
Measurement System (SPMS) is a
collection of automated
environmental, safety, and health
(ES&H) information modules.
SPMS is maintained for DOE by
the System Safety Development
Center (SSDC) at the Idaho
National Engineering Laboratory
(INEL) Super Computing Center.

SAFETY PRECEDENCE SEQUENCE
(SSDC)
DA   October 12, 1990
SF   *SPS*
RT   Safety Program Reviews
RT   System (Safety) Analyses
SO   System Safety Development Center
Glossary
DEF (SSDC) A ranking of safety
processes by their effectiveness.
This system safety process is part
of the Hazard Analysis Process.
The elements of SPS in MORT
are: (1) Design; (2) Safety
Devices; (3) Warning Devices; (4)
Human Factors Review; (5)
Procedures, including emergency
procedures; (6) Personnel
including supervision, selection,
training, motivation, and residual
risks. NASA's SPS includes
elements 1, 2, 3, 5 and 7, but
covers items 4 and 6 in other
documents.

SAFETY PROGRAM REVIEWS
(MORT)
DA   April 3, 1991
BT1 Reviews
 BT2 Administrative Processes
  BT3 Processes
RT   Failure Mode and Effect Analysis
RT   Goals
RT   Hazard Analysis
RT   Inspection and Evaluation
RT   Operational Readiness Reviews
RT   Risk Assessment
RT   Safety Appraisals
RT   Safety Precedence Sequence
RT   Technical Information Systems
DEF (MORT) MORT analysis asks: does
the safety program review assure
a planned and measured program,
with low cost/high volume
services, professional growth, and
use of modern methods? Related
issues/topics include: definition of
ideals and policy; description and
schematics; monitoring, audit, and
comparison techniques;
professional staff, management
peer committees, organization for
improvement; block function and
work schematics; scope and
integration of the review; and
safety program services.

SAFETY REVIEW COMMITTEE
DA   January 8, 1991
SF   *SRC*
BT1 Committees
 BT2 Administrative Organizations
  BT3 Organizations

SAFETY REVIEWS
(Doe Order 5480.6)
DA   October 12, 1990
BT1 Reviews
 BT2 Administrative Processes
  BT3 Processes
NT1 Backfit Reviews
NT1 Independent (Safety) Reviews
RT   Unreviewed Safety Questions
SO   Industrial Safety
DEF (DOE Order 5480.6) Deliberate and
critical examinations of the safety
impact of proposed activities or an
ongoing activities during the siting,
designing, constructing, operating,
maintaining, modifying, or
decommissioning of a reactor,
which could affect health and
safety. Documentation of the
safety reviews serves to provide
management with adequate
identification of the safety issues
and their possible implications,
and also to allow others not
directly involved in the program or
review process to independently
evaluate the completeness or
adequacy of the review.

SAFETY SHOE(S)
(2686 SAFETY SHOE)
DA   January 3, 1991
BT1 Foot Protection
 BT2 Personal Protective Equipment
  BT3 Equipment/Parts - Personal
Protective (DOE FRASE
Vocabulary)
   BT4 Equipment
SO   DOE FRASE VOCABULARY

**SAFETY SYSTEM FUNCTIONAL
INSPECTION**
DA   January 8, 1991
SF   *SSFI*
BT1 Inspections
 BT2 Administrative Processes
  BT3 Processes

**SAFETY SYSTEM OUTAGE
MODIFICATION INSPECTION**
DA   January 8, 1991
SF   *SSOMI*
BT1 Inspections
 BT2 Administrative Processes
  BT3 Processes

SAFETY WORK PERMIT
DA   January 8, 1991
SF   *SWP*
BT1 Permits

SAFETY/RELIEF
DA   January 8, 1991
SF   *S/R*

SAFETY/RELIEF VALVE
DA   January 8, 1991
SF   *SRV*
BT1 Relief Valves
 BT2 Valves
  BT3 Devices

SAFETY/RELIEF VALVE OPEN
DA   January 8, 1991
SF   *SR/VO*

SAIC
DA   October 12, 1990
SEE Science Applications International
Corporation
SO   Acronyms

SALES PERSON
(0400 SALES PERSON)
DA   November 28, 1990
BT1 Occupations
BT1 Personnel
SO   DOE FRASE VOCABULARY

SALP
DA   October 12, 1990
SEE Systematic Assessment of
Licensee Performance
SO   Acronyms

SALT WATER INTRUSION
(EPA)
DA   October 12, 1990
BT1 Processes
SO   Environmental Protection Agency
Glossary
DEF (EPA) The invasion of fresh surface
water or groundwater by salt
water. If the salt water comes from
the ocean it may be called sea
water intrusion.

SALT WATER SYSTEM
DA   January 8, 1991
SF   *SWS (Salt Water System)*
BT1 Systems

SALTS
(EPA)
DA   October 12, 1990
RT   Hard Water
SO   Environmental Protection Agency
Glossary
DEF (EPA) Minerals that water picks up
as it passes through the air, over
and under the ground, and as it is
used by households and industry.
(DSTT) The reaction product when
a metal displaces the hydrogen of
an acid.

SALVAGE
(EPA)
DA   October 12, 1990

BT-Broader Term   NT-Narrower Term   RT Related Term

BT1 Waste Management
BT2 Processes
SO Environmental Protection Agency
     Glossary
DEF (EPA) The utilization of waste
     materials.

SAM
DA October 12, 1990
SEE Site Availability Mode
SO Acronyms

SAMPLE MANAGEMENT OFFICES
(EPA)
DA October 12, 1990
BT1 Offices
BT2 Administrative Organizations
BT3 Organizations
RT Contract Laboratory Program
RT U.S. Environmental Protection
     Agency
SO Environmental Protection Agency
     Glossary
DEF (EPA) EPA contractors providing
     management, operational, and
     administrative support to the
     Contract Laboratory Program
     (CLP) to facilitate optimal use of
     the program.

SAMPLES
DA February 25, 1991
NT1 Aliquots
NT1 Batches
NT2 Analytical Batches
NT1 Blanks
NT1 Check Samples
NT1 Performance Evaluation Samples
NT1 Reference Substances
NT1 Replicate Samples
NT1 Representative Samples
NT1 Specimens
NT1 Spiked Samples
NT1 Standard Sample
RT Sampling and Analysis
RT Studies

SAMPLING
(NCRP)
DA October 12, 1990
BT1 Sampling and Analysis
NT1 Error Sampling
NT1 Poisoned Emergency Cooling
     System Sampling
RT Analytes
RT Borings
RT Check Samples
RT Contract-Required Quantitation
     Limits
RT Equivalent Methods
RT In-Situ Sampling Systems
RT Inspections
RT Quality Control
RT Reference Methods
RT Run
SO Radiation
DEF (NCRP) The process of taking a
     representative small portion or

quantity of something for testing or
analysis.

SAMPLING AND ANALYSIS
DA January 8, 1991
SF *S&A*
NT1 Poisoned Emergency Cooling
     System Sampling
NT1 Sampling
NT2 Error Sampling
NT2 Poisoned Emergency Cooling
     System Sampling
RT Analyses
RT Sampling and Analysis Plans
RT Samples

SAMPLING AND ANALYSIS PLANS
(EPA)
DA October 12, 1990
BT1 Plans
NT1 Field Sampling Plans
NT1 Quality Assurance Project Plans
RT Analyses
RT Sampling and Analysis
SO Environmental Protection Agency
     Glossary
DEF (EPA) Consist of a Quality
     Assurance Project Plan (QAPP)
     and a Field Sampling Plan (FSP).

SAMPLING PLANS
(ESH)
DA October 12, 1990
SY Field Operations Plans
BT1 Plans
NT1 Field Sampling Plans
RT Procedures
RT Site Quality Assurance Plan
SO Quality Assurance
DEF (ESH) Documents that are
     prepared for either continuous or
     site specific data collection
     activities (air, water, pesticides,
     hazardous waste, etc.). The plan
     should describe project
     organization and responsibilities,
     project description (objectives,
     scope, schedule of tasks and
     milestones, data usage,
     monitoring network/sampling and
     analysis, design rationale), data
     quality objectives, sampling
     procedures, calibration, analytical
     methods, documentation/data
     reduction/reporting, data
     assessment, audits, corrective
     action, reports and safety.

SAN
DA October 12, 1990
SEE San Francisco Operations Office
SO Acronyms

**SAN FRANCISCO OPERATIONS
OFFICE**
DA January 8, 1991
SF *SAN*
BT1 Operations Offices
BT2 Offices

BT3 Administrative Organizations
BT4 Organizations
BT2 U.S. Department of Energy
BT3 Federal Agencies
BT4 Agencies
BT5 Administrative Organizations
BT6 Organizations
RT Rockwell International Corp.
RT Stanford University
RT University of California
DEF (Capsule Review of DOE Research
     and Development and Field
     Facilities, 1986) Established in
     1952, this Office oversees defense
     and energy-related programs. It is
     responsible for the management,
     coordination and support of
     programs and projects involving
     weapons research and
     development, basic research and
     energy technologies. The Office
     administers contracts for the
     operation of several laboratory
     and engineering facilities with a
     work force of approximately
     20,000, including two major
     national laboratories. The Office
     also manages field construction of
     major technical facilities.

SAND BLASTER'S HOOD
(2687 SAND BL HOOD)
DA January 3, 1991
BT1 Head Protection
BT2 Personal Protective Equipment
BT3 Equipment/Parts - Personal
     Protective (DOE FRASE
     Vocabulary)
BT4 Equipment
RT Sand Blasting Area
SO DOE FRASE VOCABULARY

SAND BLASTING AREA
(1714 SB AREA)
DA December 10, 1990
BT1 Area
BT2 Sites/Areas
RT Sand Blaster's Hood
SO DOE FRASE VOCABULARY

SAND FILTERS
(EPA)
DA October 12, 1990
BT1 Filters
BT2 Devices
RT Trickling Filters
SO Environmental Protection Agency
     Glossary
DEF (EPA) Devices that remove some
     suspended solids from sewage.
     Air and bacteria decompose
     additional wastes filtering through
     the sand so that cleaner water
     drains from the bed.

**SANDIA CORPORATION AT&T
TECHNOLOGIES, INC.**
DA January 11, 1991
BT1 Companies

BT2 Commercial Organizations
BT3 Organizations
BT1 DOE Contractors
BT2 Potentially Responsible Parties
RT Albuquerque Operations Office
RT Sandia National Laboratories

SANDIA LABORATORY INSTRUCTION
DA January 8, 1991
SF *SLI*

SANDIA NATIONAL LABORATORIES
DA January 8, 1991
SF *SNL*
BT1 Government-Owned
    Contractor-Operated Facilities
BT2 Federal Facilities
BT3 Facilities
BT1 Laboratories
BT2 Research and Development
    Organizations
BT3 Organizations
NT1 Sandia National
    Laboratories-Alburquerque
NT1 Sandia National
    Laboratories-Livermore
RT Sandia Corporation AT&T
    Technologies, Inc.
DEF (Capsule Review of DOE Research
    and Development and Field
    Facilities, 1986) The principal
    mission of Sandia National
    Laboratories, established in 1949,
    is research, development and
    engineering of nuclear weapon
    systems, except for the nuclear
    explosive. The result of this effort
    is the existence of a national
    stockpile of operational nuclear
    weapons that are safe, secure,
    reliable and strictly controlled.
    Sandia also conducts energy
    programs in fossil, solar, fission
    and basic energy sciences.

SANDIA NATIONAL LABORATORIES-ALBURQUERQUE
DA January 8, 1991
SF *SNLA*
BT1 Sandia National Laboratories
BT2 Government-Owned
    Contractor-Operated Facilities
BT3 Federal Facilities
BT4 Facilities
BT2 Laboratories
BT3 Research and Development
    Organizations
BT4 Organizations

SANDIA NATIONAL LABORATORIES-LIVERMORE
DA January 8, 1991
SF *SNLL*
BT1 Sandia National Laboratories
BT2 Government-Owned
    Contractor-Operated Facilities
BT3 Federal Facilities
BT4 Facilities
BT2 Laboratories

BT3 Research and Development
    Organizations
BT4 Organizations

SANDIA PULSE REACTOR III
DA January 8, 1991
SF *SPR III*
BT1 Reactors

SANDING MACHINE
(2140 SANDING MACH)
DA December 10, 1990
BT1 Machines (DOE FRASE
    Vocabulary)
BT2 Equipment
SO DOE FRASE VOCABULARY

SANITARY ENGINEERING STRUCTURES
(DOE Order 6430.1A)
DA October 12, 1990
RT Impoundment
RT Tanks
SO Construction
DEF (DOE Order 6430.1A) Tanks,
    reservoirs, and other structures
    commonly used in water and
    waste treatment works, where
    dense, impermeable concrete with
    high resistance to chemical attack
    is required.

SANITARY LANDFILLS
(DOE Order 6430.1A; SWDA; RCRA;
EPA)
DA October 12, 1990
BT1 Landfills
BT2 Land Disposal Units
BT3 Disposal Units
BT4 Corrective Action Management
    Units
BT5 Sites/Areas
RT Cover Materials
RT Daily Cover
RT Disposal Facilities
RT Land Disposal
RT Lifts
RT Solid Waste Disposal
SO Construction
SO Environmental Management
SO Wastes
DEF A facility for the disposal of solid
    waste which meets the criteria
    published under section 4004 (42
    USCS 6944). (ESH) A disposal
    facility employing an engineered
    method of disposing of solid
    wastes on land in a manner which
    minimizes environmental hazards
    by spreading the solid wastes in
    thin layers, compacting the solid
    wastes to the smallest practical
    volume, and applying cover
    material at the end of each
    working day. Such facilities must
    comply with the Environmental
    Protection Agency Guidelines for
    the Land Disposal of Solid Wastes
    as prescribed in 40 CFR 241.

SANITARY SEWERS
(EPA)
DA October 12, 1990
BT1 Sewers
SO Environmental Protection Agency
    Glossary
DEF (EPA) Underground pipes that carry
    off only domestic or industrial
    waste, not storm water.

SANITARY SURVEYS
(SDWA; CFR)
DA October 12, 1990
BT1 Surveys
RT Confluent Growth
RT Maximum Contaminant Levels
RT Monitoring
RT Public Water Systems
RT Reviews
RT Secondary Maximum Contaminant
    Levels
RT Total Trihalomethanes (TTHM)
SO Environmental Protection Agency
    Glossary
SO Water Pollution
DEF (CFR) On-site reviews of the water
    sources, facilities, equipment,
    operation and maintenance of a
    public water system to evaluate
    the adequacy of those elements
    for producing and distributing safe
    drinking water.

SANITARY TREATMENT SYSTEM
(2048 SANITARY TS)
DA December 10, 1990
BT1 Systems (DOE FRASE Vocabulary)
BT2 Systems
RT Sewage Plant
SO DOE FRASE VOCABULARY

SANITATION
(EPA)
DA October 12, 1990
BT1 Processes
SO Environmental Protection Agency
    Glossary
DEF (EPA) Control of physical factors in
    the human environment that could
    harm development, health, or
    survival.

SANITIZED
(CFR)
DA November 15, 1990
SO Management
DEF (CFR) A version of a document
    from which information claimed as
    trade secret or confidential has
    been omitted or withheld.

SAR
DA October 12, 1990
SEE Safety Analysis Reports
SO Acronyms

SARA
(CFR)
DA October 12, 1990

SEE Superfund Amendments and
    Reauthorization Act of 1986
SO Acronyms

SAS
DA October 12, 1990
SEE Semi-annual soils
SO Acronyms

SASS
DA October 12, 1990
SEE Safety Assurance System
    Summary
SO Acronyms

SAT
DA October 12, 1990
SEE Semi-Automatic Trim
SO Acronyms

SATURATED ZONES
(DOE Order 6430.1A; SWDA; RCRA;
  EPA)
DA October 12, 1990
BT1 Zones
 BT2 Sites/Areas
RT Aeration Zones
RT Groundwater
RT Recharge
RT Unsaturated Zones
SO Construction
SO Environmental Management
SO Environmental Protection Agency
    Glossary
SO Wastes
DEF Subsurface areas in which all pores
    and cracks are filled with water
    under pressure equal to or greater
    than that of the atmosphere.

SAV
DA October 12, 1990
SEE Semi-annual vegetation
SO Acronyms

**SAVANNAH RIVER OPERATIONS
OFFICE**
DA January 8, 1991
SF *SR*
BT1 Operations Offices
 BT2 Offices
  BT3 Administrative Organizations
   BT4 Organizations
 BT2 U.S. Department of Energy
  BT3 Federal Agencies
   BT4 Agencies
    BT5 Administrative Organizations
     BT6 Organizations
RT University of Georgia Research
    Foundation, Inc.
RT Wackenhut Services Inc.
RT Westinghouse Savannah River
    Company
DEF (Capsule Review of DOE Research
    and Development and Field
    Facilities, 1986) The Savannah
    River Operations Office,
    established in 1950, manages the
    Savannah River Site, a key DOE

installation for nuclear materials
production and research. The
mission of the Site is to produce
tritium and plutonium for
fabrication into weapons
components for the nation's
defense program. In addition,
some of the Site's nuclear
products have peacetime
applications. Plutonium-238 is
produced as a long-lasting fuel
source to generate electrical
power and warm instruments
during space exploration.
Research in nuclear energy and
ecology are conducted at the
Savannah River Laboratory and
Savannah River Ecology
Laboratory. In 1972, SRS was
designated the nation's first
National Environmental Research
Park, where scientists from other
Government agencies, private
foundations and universities could
use SRS land as a protected
outdoor laboratory for long-term
studies.

SAVANNAH RIVER SITE
DA January 8, 1991
SF *SRS*
BT1 Government-Owned
    Contractor-Operated Facilities
 BT2 Federal Facilities
  BT3 Facilities
RT Wackenhut Services Inc.
RT Westinghouse Savannah River
    Company
DEF (Capsule Review of DOE Research
    and Development and Field
    Facilities, 1986) Conducts fuel and
    target fabrication, isotope
    production in nuclear reactors,
    chemical separations, waste
    management and heavy-water
    extraction. The plant was
    established in 1950. Although
    activities at the facility are oriented
    primarily to defense, many of the
    programs and products
    (radioisotopes) have peacetime
    applications. Major facilities of the
    Savannah River Site include
    nuclear production reactors, two
    chemical separation plants, a fuel
    fabrication plant, waste
    management facilities and
    extensive research and
    development facilities.

SAVANNAH RIVER WATER
DA January 8, 1991
SF *SRW*

SAWS
DA January 30, 1991
NT1 Band Saw
NT1 Chain Saw
NT1 Concrete Saw
NT1 Hack Saw

NT1 Hand Saw
NT1 Table Saw
SO DOE FRASE VOCABULARY

SBLOCA
DA October 12, 1990
SEE Small Break LOCA
SO Acronyms

SBO
DA October 12, 1990
SEE Station Blackout
SO Acronyms

SC
DA October 12, 1990
SEE Safety Computer
SO Acronyms

SCALE(S)
(2767 SCALE)
DA January 3, 1991
BT1 Gauge(s)
 BT2 Instrument(s)
  BT3 Equipment/Parts -
    Instrumentation/Measuring
    (DOE FRASE Voc.)
   BT4 Equipment
SO DOE FRASE VOCABULARY

SCALP
(1077 SCALP)
DA November 28, 1990
BT1 Skin
 BT2 Human Body Parts
RT Hair
RT Head
SO DOE FRASE VOCABULARY

SCALPEL(S)
(3053 SCALPEL)
DA January 3, 1991
BT1 Tools - Manual
 BT2 Tools (DOE FRASE Vocabulary)
  BT3 Equipment
SO DOE FRASE VOCABULARY

SCF
DA October 12, 1990
SEE Standard Cubic Foot
SO Acronyms

SCG
DA October 12, 1990
SEE Standby Diesel Generator
SO Acronyms

SCHEDULE OF COMPLIANCE
(CAA)
DA October 12, 1990
SY Compliance Schedules
RT Control Strategies
RT Increments of Progress
RT Primary Standard Attainment Date
RT Reasonable Further Progress
RT Remedial Actions
RT Transportation Control Measures
SO Air Pollution

SO Environmental Management
SO Wastes
SO Water Pollution
DEF (CAA) A schedule of required measures including an enforceable sequence of actions or operations leading to compliance with an emission limitation, other limitation, prohibition, or standard.

SCHEMATICS
(SSDC)
DA October 12, 1990
BT1 Diagrams
SO System Safety Development Center Glossary
DEF (SSDC) Block functional diagrams showing the secession of functions or processes required to attain a desired output.

SCHOOLS
(SDWA; TSCA; CFR; ESH)
DA October 12, 1990
SO Hazardous Materials
SO Water Pollution
DEF Any elementary or secondary school as defined in section 198 of the Elementary and Secondary Education Act of 1965 (20 U.S.C. 2854).

SCIENCE APPLICATIONS INTERNATIONAL CORPORATION
DA January 8, 1991
SF SAIC
BT1 Companies
BT2 Commercial Organizations
BT3 Organizations
BT1 DOE Contractors
BT2 Potentially Responsible Parties

SCIENCE TECHNICIAN
(0380 SCIENCE TECH)
DA November 28, 1990
BT1 Technicians
BT2 Professional Personnel
BT3 Occupations
BT3 Personnel
SO DOE FRASE VOCABULARY

SCIENTIST
(0170 SCIENTIST)
DA November 28, 1990
BT1 Professional Personnel
BT2 Occupations
BT2 Personnel
NT1 Health Physicist
RT Research/Testing Activity
SO DOE FRASE VOCABULARY

SCINTILLATION COUNTERS
(NIH)
DA October 12, 1990
BT1 Radiation Detectors
BT2 Equipment
RT Count (Radiation Measurement)
RT Cutie Pies
RT Dosimeters

RT Efficiency (Counters)
SO Radiation
DEF (NIH) Counters in which light flashes produced in a scintillator by ionizing radiation are converted into electrical pulses by a photomultiplier tube.

SCINTILLLATION PROBE(S)
(2768 SCINTIL PROB)
DA January 3, 1991
BT1 Testing Equipment
BT2 Instrument(s)
BT3 Equipment/Parts - Instrumentation/Measuring (DOE FRASE Voc.)
BT4 Equipment
SO DOE FRASE VOCABULARY

SCISSOR LIFT
(2340 SCISSOR LIFT)
DA December 10, 1990
BT1 Hoisting Apparatus
BT2 Material Handling Device
BT3 Devices
BT3 Equipment/Parts - Material Handling (DOE FRASE Vocabulary)
BT4 Equipment
SO DOE FRASE VOCABULARY

SCISSORS
(3054 SCISSORS)
DA January 3, 1991
BT1 Tools - Manual
BT2 Tools (DOE FRASE Vocabulary)
BT3 Equipment
RT Cutter(s)
RT Snips
SO DOE FRASE VOCABULARY

SCRAM
DA January 8, 1991
SF ABLE
BT1 Reactor Shutdown
NT1 Automatic Reactor Scram
NT1 Manual Reactor Scram
RT Temperature Scram Circuit Monitor
DEF (NRC Glossary of Terms: Nuclear Power and Radiation) Sudden shutting down of a nuclear reactor, usually by rapid insertion of control rods, either automatically or manually by the reactor operator. (WEBSTER) A sudden or emergency shutting down of a nuclear reactor.

SCRAM RELAY
DA January 8, 1991
SF CSR
BT1 Relay(s)
BT2 Equipment/Parts - Electrical (DOE FRASE Vocabulary)
BT3 Equipment

SCRAP
(EPA)
DA October 12, 1990

NT1 Special Nuclear Material Scrap
NT1 Spent Materials
RT Recycled Materials
RT Recycling
RT Solid Wastes
SO Environmental Protection Agency Glossary
DEF (EPA) Materials discarded from manufacturing operations that may be suitable for reprocessing.

SCRAPER
(2341 SCRAPER)
DA December 10, 1990
BT1 Equipment/Parts - Material Handling (DOE FRASE Vocabulary)
BT2 Equipment
SO DOE FRASE VOCABULARY

SCREENING
(EPA)
DA October 12, 1990
BT1 Pollution Recovery Processes
BT2 Processes
BT1 Waste Management Processes
BT2 Processes
SO Environmental Protection Agency Glossary
DEF (EPA) Use of screens to remove coarse floating and suspended solids from sewage.

SCREWDRIVER(S)
(3055 SCREWDRIVER)
DA January 3, 1991
BT1 Tools - Manual
BT2 Tools (DOE FRASE Vocabulary)
BT3 Equipment
SO DOE FRASE VOCABULARY

SCRUBBERS
(EPA)
DA October 12, 1990
BT1 Devices
RT Cyclone Collectors
RT Flue Gas Desulfurization
RT Limestone Scrubbing
SO Environmental Protection Agency Glossary
DEF (EPA) Air pollution devices that use a spray of water or reactant or a dry process to trap pollutants in emissions.

SCTI
DA October 12, 1990
SEE Sodium Components Test Installation
SO Acronyms

SCW
DA October 12, 1990
SEE Service Clarified Water
SO Acronyms

SDC
DA October 12, 1990

SEE Shutdown Cooling
SO Acronyms

SDHX
DA October 12, 1990
SEE Shutdown Heat Exchanger
SO Acronyms

SDS
DA October 12, 1990
SEE Shutdown Sequencer
SO Acronyms

SDWA
DA June 6, 1991
SEE Safe Drinking Water Act
SO Acronyms

SEA TURTLES
(ESA; CFR)
DA October 12, 1990
BT1 Animals
RT Plastrons
SO Endangered Species
DEF (CFR) Those sea turtle species enumerated in 50 CFR 227.4 and any part(s), product(s), egg(s) or offspring thereof, or the dead body or part(s) thereof.

SEABROOK STATION PROBABILISTIC SAFETY ASSESSMENT
DA January 8, 1991
SF *SSPSA*
BT1 Assessments
 BT2 Administrative Processes
  BT3 Processes

SEAL HEAD TANK
DA January 8, 1991
SF *SHT*
DEF Refers specifically in this context to a seal head tank found in a blanket gas system.

SEALED SOURCES
(IAEA)
DA October 12, 1990
RT Radiation Safety
SO Environmental Management
SO Radiation
DEF (IAEA) Radiation sources whose structures are such as to prevent, under normal conditions of use, any dispersion of the radioactive material into the environment.

SEAT BELT(S)
(2688 SEAT BELT)
DA January 3, 1991
BT1 Personal Protective Equipment
 BT2 Equipment/Parts - Personal Protective (DOE FRASE Vocabulary)
  BT3 Equipment
RT Vehicle Accident
SO DOE FRASE VOCABULARY

SECOND LINE ORGANIZATION LEVEL
(DOE Order 5482.1B; ESH)
DA October 12, 1990
RT Line Organizations
RT Operating Level
RT U.S. Department of Energy
SO Construction
SO Management
DEF The DOE element that is contractually or organizationally responsible for the work or job tasks being performed by an operating level. It may be an operations office, or an Assistant Secretary directly responsible for an energy technology center or a power administration.

SECONDARY BODY WAVES
(USGS)
DA October 12, 1990
SF *S Waves*
RT Body Wave Magnitude
RT Primary Body Waves
SO Natural Phenomenon
DEF (USGS) Earthquake waves that shear the rock sideways at right angles to the direction of travel, producing an up-and-down and side-to-side oscillation like the snapping of a rope. P and S waves are the most damaging waves because buildings are more susceptible to damage for horizontal motion than from vertical motion.

SECONDARY DOCUMENTS
(Doe Order 5400.4)
DA October 12, 1990
BT1 Primary Documents
 BT2 Reports
SO Compensation and Liability
DEF (DOE Order 5400.4) Those reports that are discrete portions of primary documents and are typically input or feeder documents within the remedial investigation/feasibility study or remedial design/remedial action process.

SECONDARY DRINKING WATER REGULATIONS
(SDWA; EPA)
DA October 19, 1990
BT1 Code of Federal Regulations
 BT2 Codes, Standards, and Regulations
 BT2 Statutes and Regulations
RT Primary Drinking Water Regulations
RT Public Water Systems
SO Environmental Management
SO Environmental Protection Agency Glossary
SO Water Pollution
DEF (USC) Regulations that apply to public water systems and that specify the maximum contaminant levels which, in the judgment of

the Administrator, are requisite to protect the public welfare. Such regulations may apply to any contaminant in drinking water (A) which may adversely affect the odor or appearance of such water and consequently may cause a substantial number of the persons served by the public water system providing such water to discontinue its use, or (B) which may otherwise adversely affect the public welfare. Such regulations may vary according to geographic and other circumstances.

SECONDARY EMISSIONS
(CAA; CFR)
DA October 12, 1990
BT1 Emissions
 BT2 Air Pollutants
RT Fugitive Emissions
RT Transportation Control Measures
RT Uncontrolled Total Arsenic Emissions
SO Air Pollution
DEF (CFR) Emissions that occur as a result of the construction or operation of an existing stationary facility but do not come from the existing stationary facility. Secondary emissions may include, but are not limited to, emissions from ships or trains coming to or from the existing stationary facility. (CFR) Inorganic arsenic emissions that escape capture by a primary emission control system.

SECONDARY ENVIRONMENTAL MONITOR
(DOE Order 5000.3A)
DA June 7, 1991
BT1 Monitors
 BT2 Equipment
RT Inspection and Maintenance
RT Monitoring
RT Primary Environmental Monitors
SO Environmental Management
DEF (DOE Order 5000.3A) Environmental monitoring equipment or activities which, if degraded, will produce a more than minor disruption of a monitoring program. An example of a minor effect would be the failure of a unit whose place in the program is effectively duplicated by overlap between one or more other components. An example of a more than minor effect would be failure of sufficient units to preclude continued coverage, or the failure of a unit which provides the only coverage for large areas, such as a groundwater monitoring well.

SECONDARY ENVIRONMENTAL MONITORS
(DOE Order 5000.3A)
DA   December 14, 1990
BT1 Monitors
 BT2 Equipment
RT   Class B Equipment
DEF (DOE Order 5000.3A)
     Environmental monitoring
     equipment or activities which, if
     degraded, will produce a more
     than minor disruption of a
     monitoring program. An example
     of a minor effect would be the
     failure of a unit whose place in the
     program is effectively duplicated
     by overlap between one or more
     other components. An example of
     a more than minor effect would be
     the failure of sufficient units to
     preclude continued coverage, or
     the failure of a unit which provides
     the only coverage for large areas,
     such as a groundwater monitoring
     well.

SECONDARY HOOD SYSTEMS
(CAA; CFR)
DA   October 12, 1990
BT1 Control Devices
 BT2 Devices
BT1 Systems
RT   Theoretical Arsenic Emissions
     Factor
SO   Air Pollution
DEF (CFR) The equipment (including
     hoods, ducts, fans, and dampers)
     used to capture and transport
     secondary inorganic arsenic
     emissions.

SECONDARY LIMITS
(IAEA)
DA   November 20, 1990
BT1 Limits
SO   Radiation
DEF (IAEA) Values of the dose
     equivalent indices (deep and
     shallow), in the case of external
     exposure, or of annual limits on
     intake, in the case of internal
     exposure, which can be used to
     obtain an an indirect assessment
     of compliance with primary limits.

SECONDARY MAXIMUM CONTAMINANT LEVELS
(SWDA; CFR; ESH)
DA   October 19, 1990
SF   SMCL
BT1 Maximum Contaminant Levels
RT   Contaminants
RT   Public Water Systems
RT   Sanitary Surveys
SO   Water Pollution
DEF (CFR) Secondary maximum
     contaminant levels, or SMCLs,
     apply to public water systems and
     which, in the judgement of the
     Administrator, are requisite to

protect the public welfare. (ESH)
The SMCLs mean the maximum
permissible levels of a
contaminant in water which are
delivered to the free flowing outlet
of the ultimate user of public water
system. Contaminants added to
the water under circumstances
controlled by the user, except
those resulting from corrosion of
piping and plumbing caused by
water quality, are excluded from
this definition.

SECONDARY STANDARDS
(CAA; CFR)
DA   October 12, 1990
BT1 National Ambient Air Quality
     Standards
 BT2 Ambient Air Quality Standards
 BT3 Air Quality Standards
 BT4 Standards
 BT5 Codes, Standards, and
     Regulations
SO   Air Pollution
DEF (CFR) A national secondary
     ambient air quality standard
     promulgated pursuant to section
     109 of the Act.

SECONDARY SYSTEM RELIEF
DA   January 8, 1991
SF   SSR

SECONDARY TREATMENT
(EPA)
DA   October 12, 1990
BT1 Treatment
 BT2 Waste Management Processes
 BT3 Processes
BT1 Wastewater Treatment Processes
 BT2 Treatment
 BT3 Waste Management Processes
 BT4 Processes
SO   Environmental Protection Agency
     Glossary
DEF (CFR) The second step in most
     publicly owned waste treatment
     systems in which bacteria
     consume the organic parts of the
     waste. It is accomplished by
     bringing together waste, bacteria,
     and oxygen in trickling filters or in
     the activated sludge process. This
     treatment removes floating and
     settleable solids and about 90
     percent of the oxygen demanding
     substances and suspended solids.
     Disinfection is the final stage of
     secondary treatment. (See:
     Primary, Tertiary Treatment).

SECONDS LIVING RADIOISOTOPES
(EDB)
DA   March 29, 1991
BT1 Radionuclides
 BT2 CERCLA Hazardous Substances
 BT3 Hazardous Substances

 BT2 Nuclides
SO   Radiation

SECRETARIAL OFFICERS
(DOE Order 5440.1D)
DA   May 15, 1991
BT1 Program Senior Officials
 BT2 Personnel
SO   Environmental Management
DEF (DOE Order) For purposes of DOE
     Order 5440.1D, Secretarial
     Officers are Assistant Secretaries;
     the Deputy Under Secretary for
     Policy, Planning, and Analysis; the
     Directors of the Offices of
     Environmental Restoration and
     Waste Management, Energy
     Research, New Production
     Reactors, Nuclear Safety. Civilian
     Radioactive Waste Management,
     and Administration and Human
     Resource Management; and the
     Administrators of the Western
     Area Power Administration and the
     Bonneville Power Administration.

SECTION 5 NOTICES
DA   January 24, 1991
NT1 Pre-Manufacture Notices
 NT2 Consolidated Pre-Manufacture
     Notices
 NT2 Intermediate Pre-Manufacture
     Notices
SO   Environmental Management
DEF (CFR) Any PMN, consolidated
     PMN, intermediate PMN,
     significant new use notice,
     exemption notice, or exemption
     application.

SECURE AUTOMATIC COMMUNICATIONS NETWORK
(EMER)
DA   February 1, 1991
BT1 Information Systems
 BT2 Security Interests
 BT2 Systems
SO   Emergency Preparedness
DEF (EMER) A data system designed to
     handle U.S. Department of
     Energy's normal requirements for
     secure message and data traffic
     within the continental United
     States and to exchange such
     traffic via the Department of
     Defense automatic digital network
     and which allows access to the
     Department of State Diplomatic
     Telecommunications System and
     the General Services
     Administration Advanced Record
     System.

SECURE COMMUNICATIONS CENTERS
(DOE Order 6430.1A; EMER)
DA   October 12, 1990
BT1 Centers
BT1 Security Interests

BT-Broader Term   NT-Narrower Term   RT Related Term

SO Construction
SO Emergency Preparedness
DEF (DOE Order 6430.1A) Security
     areas devoted in whole or in part
     to the encryption and decryption of
     sensitive and/or classified
     information.

SECURE MAXIMUM CONTAMINANT LEVELS
(EPA)
DA October 12, 1990
BT1 Maximum Contaminant Levels
SO Environmental Protection Agency
     Glossary
DEF (EPA) Maximum permissible levels
     of a contaminant in water that is
     delivered to the free flowing outlet
     of the ultimate user of a water
     supply, the consumer, or of
     contamination resulting from
     corrosion of piping and plumbing
     caused by water quality.

SECURITY
(DOE Order 6430.1A)
DA October 12, 1990
NT1 Physical Protection (physical
     security)
RT DOE Safeguards and Security
     Coordinators
RT Guards
SO Construction
DEF (DOE Order 6430.1A) Activities
     through which DOE defines,
     develops, and implements its
     responsibilities, under the Atomic
     Energy Act of 1954, as amended,
     Federal statutes, Executive
     Orders, and other directives, for
     the protection of Restricted Data
     and other classified information or
     matter, nuclear weapons and
     nuclear weapon components, and
     for the protection of Department
     and Departmental contractor
     facilities, property, and equipment.
     Security is also applied to special
     nuclear materials. When physical,
     personnel, and technical security
     are combined with material control
     and material accountability, the
     protection is referred to as
     safeguards.

SECURITY ACTIVITY
(1243 SEC ACTIVITY)
DA November 28, 1990
BT1 Activity Types (DOE FRASE
     Vocabulary)
BT2 Activities
RT Plant Protection System
RT Security Guard
SO DOE FRASE VOCABULARY

SECURITY AREAS
(DOE Order 6430.1A)
DA October 12, 1990
BT1 Sites/Areas

NT1 Exclusion Areas
NT1 Limited Areas
NT1 SNM Vault
NT1 Vault-type Rooms
NT1 Vital Areas
RT Entry Control Points
RT Protected Areas
SO Construction
DEF (DOE Order 6430.1A) A physically
     defined space containing a
     Departmental security interest and
     subject to physical protection and
     access controls.

SECURITY COMMUNICATIONS CONTROL CENTER
(EMER)
DA February 1, 1991
BT1 Centers
SO Emergency Preparedness
DEF (EMER) A continuously staffed
     operation located in Albuquerque,
     New Mexico which is staffed,
     equipped, and operated by the
     Albuquerque Operations Office to
     provide necessary
     communications and actions to
     initiate immediate response to
     U.S. Department of Energy
     transportation safeguards system
     emergencies involving nuclear
     weapons, components and
     devices, and strategic quantities of
     government-owned special nuclear
     materials.

SECURITY EVENTS
(EMER)
DA February 1, 1991
BT1 Events
SO Emergency Preparedness
DEF (EMER) Emergency conditions
     which threaten the security of
     personnel, property, or special
     nuclear material.

SECURITY FACILITIES
(Doe Order 5480.16)
DA October 12, 1990
BT1 Facilities
RT Isolation Zones
RT Protective Force Personnel
RT Safety Analysis Reports
RT Security Inspectors
RT Shadow Forces
SO Firearms
DEF (DOE Order 5480.16) Any facilities
     that have been approved by the
     Department for generating,
     receiving, using, processing,
     storing, reproducing, transmitting,
     destroying, or handling special
     nuclear material or classified
     matter.

SECURITY GUARD
(0513 SECURITY GUA)
DA November 28, 1990
BT1 Admin. Support/Clerical Employee

BT2 Occupations
BT2 Personnel
RT Guard Station
RT Plant Protection System
RT Security Activity
SO DOE FRASE VOCABULARY

SECURITY INSPECTORS
(Doe Order 5480.16)
DA October 12, 1990
BT1 Protective Force Personnel
BT2 Personnel
RT Inspections
RT Security Facilities
SO Firearms
DEF (DOE Order 5480.16) A uniformed
     Departmental contractor employed
     for, and charged with, the
     protection of classified matter,
     special nuclear material, or other
     Government property and is
     authorized under section 161k of
     the Atomic Energy Act of 1954, as
     amended, or other statutory
     authority, to carry firearms and to
     make arrests without warrant.
     (See DOE 5632.4.)

SECURITY INTERESTS
(DOE Order 6430.1A)
DA October 12, 1990
NT1 Classified Interests
NT1 Classified Matter
NT2 Classified Information
NT3 National Security Information
NT3 Sensitive Compartmented
     Information
NT3 Sensitive Nuclear Material
     Production Information
NT1 DOE Property
NT1 Information Systems
NT2 Emergency Management
     Information System
NT2 Failure Reporting Analysis and
     Corrective Action System
NT2 Fire Protection Tracking System
NT2 Hazard Abatement Tracking
     System
NT2 Integrated Risk Information
     System
NT2 Nuclear Facilities Incident
     Database
NT2 Nuclear Materials Management
     and Standards System
NT2 Nuclear Plant Reliability Data
     System
NT2 Plant Risk Status Information
     Management System
NT2 Radiation Records Repositories
NT2 Real Property Inventory System
NT2 Safety Performance Measurement
     System
NT3 Chemical Hazards Emergency
     Management System
NT3 Computerized Accident/Incident
     Reporting System
NT3 Computer Assisted Tracking
     System
NT3 Occurrence Reporting and
     Processing System

NT3 Personnel Expertise and
    Resource Listing
NT3 Radiation Exposure Module
NT3 Standards Information
    Management System
NT3 Unusual Occurrence Reporting
    System
NT2 Secure Automatic
    Communications Network
NT2 Technical Information Systems
NT2 Ultrasonic Ranging and Data
    System
NT2 Waste Information Network
NT1 Secure Communications Centers
NT1 Special Nuclear Materials
NT2 Special Nuclear Material Scrap
NT2 Special Nuclear Material of Low
    Strategic Significance
NT2 Strategic Special Nuclear
    Materials
NT1 Suspicious Devices
RT Confidential
SO Construction
DEF (DOE Order 6430.1A) Any of the
    following that requires special
    protection: classified matter,
    special nuclear material, security
    shipments, secure
    communications centers, sensitive
    compartmented information
    facilities, automatic data
    processing centers, or other
    systems including classified
    information, or Departmental
    property.

SEDIMENTATION
(SDWA; CFR; EPA)
DA October 12, 1990
BT1 Deposition (Process)
 BT2 Waste Management Processes
  BT3 Processes
RT Agglomeration
RT Clarification
RT Coagulation
RT Conventional Filtration Treatment
RT Flocculation
RT Primary Waste Treatment
RT Settleable Solids
RT Silt
SO Environmental Protection Agency
    Glossary
SO Water Pollution
DEF Letting solids settle out of
    wastewater by gravity during
    wastewater treatment. Process of
    deposition.

SEDIMENTATION TANKS
(EPA)
DA October 12, 1990
BT1 Tanks
 BT2 Facility Components
RT Wastewater Treatment Tanks
SO Environmental Protection Agency
    Glossary
DEF (EPA) Holding areas for wastewater
    where floating wastes are
    skimmed off and settled solids are
    removed for disposal.

SEDIMENTS
(EPA)
DA October 12, 1990
SO Environmental Protection Agency
    Glossary
DEF (EPA) Soil, sand, and minerals
    washed from land into water
    usually after rain. They pile up in
    reservoirs, rivers, and harbors
    destroying fish-nesting areas and
    holes of water animals, and
    clouding the water so that needed
    sunlight might not reach aquatic
    plants. Careless farming, mining,
    and building activities will expose
    sediment materials, allowing them
    to be washed off the land after
    rainfalls.

SEISMIC CATEGORY I
(DOE Order 6430.1A)
DA October 12, 1990
RT Safety Class Items
SO Construction
DEF (DOE Order 6430.1A) A level and
    method of seismic qualification
    that provides documented
    assurance that an item,
    component, or system can
    continue to perform its required
    function. Qualification includes all
    SC-1 and selected SC-2 and SC-3
    items, components, or systems.

**SELECTION AND MONITORING
CHASSIS**
(NFI)
DA October 12, 1990
SF *SMC*
BT1 Monitors
 BT2 Equipment
SO Radiation
DEF (NFI) Monitors pump and motor
    temperature when computer is
    inoperative.

SELECTIVE PESTICIDES
(EPA)
DA October 12, 1990
BT1 Pesticides
 BT2 Hazardous Substances
SO Environmental Protection Agency
    Glossary
DEF (EPA) Chemicals designed to affect
    only certain types of pests, leaving
    other plant and animals unharmed.

SELENIUM 75
(EDB)
DA March 29, 1991
BT1 Days Living Radioisotopes
 BT2 Radionuclides
  BT3 CERCLA Hazardous Substances
   BT4 Hazardous Substances
  BT3 Nuclides
SO Radiation

SELF-ASSESSMENT
(DOE Order 4330.4A)

DA June 5, 1991
BT1 Assessments
 BT2 Administrative Processes
  BT3 Processes
RT Facilities
RT Maintenance
RT Maintenance Management
SO Management
DEF (DOE Order 4330.4A) A systematic
    evaluation of a facility
    maintenance program, including
    the activities and practices,
    utilizing the performance
    objectives and criteria from each
    element of the Maintenance
    Management Program as defined
    in this Order.

SELF-EVALUATION
(Doe Order 5480.18)
DA October 12, 1990
NT1 Contractor Self-Evaluation Reports
RT Training Programs
RT Training Program Accreditation
    Plan
SO Standards
DEF (DOE Order 5480.18) A critical
    evaluation of a facility training
    program measured against the
    accreditation objectives and
    criteria. This evaluation is
    conducted by the contractor.

SELF-INSURANCE
(SSDC)
DA October 12, 1990
SO System Safety Development Center
    Glossary
DEF (SSDC) Retention of risk. Generally
    refers to planned program for
    financing or otherwise recognizing
    losses. A poor term because it is
    not insurance.

SEMI-ANNUAL SOILS
DA January 8, 1991
SF *SAS*
BT1 Soils

SEMI-ANNUAL VEGETATION
DA January 8, 1991
SF *SAV*

SEMI-AUTOMATIC TRIM
DA January 8, 1991
SF *SAT*

SEMI-CONFINED AQUIFERS
(EPA)
DA October 12, 1990
BT1 Aquifers
 BT2 Formations
SO Environmental Protection Agency
    Glossary
DEF (EPA) An aquifer that is partially
    confined by a soil layer (or layers)
    of low permeability through which
    recharge and discharge can occur.

SEMIANNUAL
(CAA; CFR)
DA   October 12, 1990
BT1 Time Designations
SO   Air Pollution
DEF A 6-month period; the first
     semiannual period concludes on
     the last day of the last month
     during the 180 days following initial
     startup for new sources; and the
     first semiannual period concludes
     on the last day of the last full
     month during the 180 days after
     June 6, 1984, for existing sources.

SEMIAUTOMATIC FIREARMS
(Doe Order 5480.16)
DA   October 12, 1990
BT1 Small Arms
 BT2 Firearms
NT1 Double Action Semiautomatic
     Pistols
NT1 Shotgun, Semiautomatic
NT1 Single Action Semiautomatic Pistols
RT   Magazines
SO   Firearms
DEF (DOE Order 5480.16) A type of
     firearm that employs either gas
     pressure or recoil force and
     mechanical spring action in
     ejecting the empty cartridge case
     after the first shot and loading the
     next cartridge from the magazine
     but that requires release and
     another pressure of the trigger for
     firing each successive shot.

SENESCENCE
(EPA)
DA   October 12, 1990
BT1 Eutrophication
 BT2 Processes
RT   Eutrophic Lakes
SO   Environmental Protection Agency
     Glossary
DEF (EPA) Term for the aging process.
     Sometimes used to describe lakes
     or other bodies of water in
     advanced stages of eutrophication.

SENIOR CONTROLLERS
(DOE Order 5480.16; EMER)
DA   October 12, 1990
BT1 Controllers
 BT2 Personnel
SO   Emergency Preparedness
SO   Firearms
SO   Management
DEF (DOE Order 5480.16) Those
     responsible for assigning tasks
     and coordinating the efforts of all
     controllers during advanced
     firearms training and
     force-on-force exercises.

**SENIOR FEDERAL EMERGENCY
MANAGEMENT AGENCY OFFICIALS**
(EMER)
DA   February 1, 1991

BT1 Personnel
SO   Emergency Preparedness
DEF (EMER) The officials appointed by
     the director of the Federal
     Emergency Management Agency
     (FEMA), or his/her representative,
     to direct FEMA response at the
     scene of a radiological
     emergency. The senior FEMA
     officials serve as the focal point for
     promoting the coordination of the
     federal response activities at the
     scene of an emergency.

SENIOR MANAGEMENT OFFICIALS
(CERCLA; CFR)
DA   October 12, 1990
BT1 Personnel
RT   Heads of Field Operations
RT   Management
SO   Compensation and Liability
SO   Management
DEF Officials with management
     responsibility for the person or
     persons completing the report, or
     the manager of environmental
     programs for the facility or
     establishments, or for the
     corporation owning or operating
     the facility or establishments
     responsible for certifying similar
     reports under other environmental
     regulatory requirements.

SENIOR REACTOR OPERATOR
(Doe Order 5480.6; NFI)
DA   October 12, 1990
SF   SRO
BT1 Personnel
BT1 Reactor Operator
 BT2 Personnel
SO   Industrial Safety
DEF An individual certified by contractor
     management to operate or to
     direct the operation of a
     DOE-owned Category A reactor.

SENIOR SCIENTIFIC ADVISERS
(EMER)
DA   February 1, 1991
BT1 Personnel
SO   Emergency Preparedness
DEF (EMER) Senior scientists selected
     by the U.S. Department of Energy
     team leader, Federal Radiological
     Monitoring and Assessment
     Center director, or on-scene
     commander to serve as his/her
     primary scientific/technical
     advisor. The person will have
     special technical expertise related
     to the radioactive source produced
     by the accident.

SENIOR SECURITY SUPERVISORS
(EMER)
DA   February 1, 1991
BT1 Personnel

SO   Emergency Preparedness
DEF (EMER) Security contractor officials
     at the incident site who assume
     tactical control of the incident
     response.

**SENSITIVE COMPARTMENTED
INFORMATION**
(EMER)
DA   February 1, 1991
BT1 Classified Information
 BT2 Classified Matter
 BT3 Security Interests
SO   Emergency Preparedness
DEF (EMER) All classified information
     and materials bearing intelligence
     community special access controls
     formally limiting access and
     dissemination. SCI does not
     include restricted data as defined
     in the Atomic Energy Act of 1954,
     as amended.

**SENSITIVE NUCLEAR MATERIAL
PRODUCTION INFORMATION**
(EMER)
DA   February 1, 1991
BT1 Classified Information
 BT2 Classified Matter
 BT3 Security Interests
SO   Emergency Preparedness
DEF (EMER) (1) Classified production
     rate or stockpile quantity
     information relating to plutonium,
     tritium, enriched lithium-6, and
     uranium-235 and -233. (2) Laser
     separation technology.

SENSOR(S)
(2769 SENSOR)
DA   January 3, 1991
BT1 Testing Equipment
 BT2 Instrument(s)
 BT3 Equipment/Parts -
     Instrumentation/Measuring
     (DOE FRASE Voc.)
 BT4 Equipment
SO   DOE FRASE VOCABULARY

SENSORS
(CAA; CFR; SSDC)
DA   October 12, 1990
BT1 Devices
RT   Monitoring
SO   Air Pollution
SO   System Safety Development Center
     Glossary
DEF Devices that measure a physical
     quantity or the change in a
     physical quantity, such as
     temperature, pressure, flow rate,
     pH, or liquid level.

SEP
DA   October 12, 1990
SEE Systematic Evaluation Program
SO   Acronyms

SEPARATOR
(2142 SEPARATOR)
DA December 10, 1990
SO DOE FRASE VOCABULARY

SEPTIC TANKS
(SWDA; RCRA; CFR; EPA)
DA October 12, 1990
BT1 Underground Storage Tanks
 BT2 Underground Tanks
  BT3 Tanks
   BT4 Facility Components
RT Municipal Solid Wastes
SO Environmental Protection Agency
    Glossary
SO Wastes
DEF Underground storage tanks for
    wastes from homes having no
    sewer line to a treatment plant.
    The waste goes directly from the
    homes to the tanks, where the
    organic waste is decomposed by
    bacteria and the sludge settles to
    the bottom. The effluent flows out
    of the tanks into the ground
    through drains; the sludge is
    pumped out periodically.

SEPTIFOILS
(NFI)
DA October 12, 1990
SO Nuclear Facilities Incident Database
SO Radiation
DEF (NFI) S-foils. These guide control
    rod movement and provide cooling
    (JACKETS).

SER
DA October 12, 1990
SEE Safety Evaluation Report
SO Acronyms

SERC
DA October 12, 1990
SEE State Emergency Response
    Commission
SO Acronyms

SERIOUS VIOLATIONS
(SSDC)
DA October 12, 1990
BT1 Violations
SO System Safety Development Center
    Glossary
DEF To determine if a violation is
    serious, two questions must be
    answered "Yes": (1) Is there a
    substantial probability that death
    or serious physical harm could
    result? And if so, (2) Did the
    employer know, or with the
    exercise of reasonable diligence,
    should have known of the hazard?
    (OSHA).

SERIOUSNESS
(TSA; TTGM)
DA October 12, 1990
BT1 Conditions

NT1 Category I Seriousness
NT1 Category II Seriousness
NT1 Category III Seriousness
RT Concerns
RT Environment Assessments
RT Technical Safety Appraisals
SO Standards
DEF As applies to any situation for
    which a clear and present danger
    exists to workers or member of the
    public. Three categories of
    "seriousness" have been defined
    for DOE within the area of
    environment, safety and health.

SERVICE
(SWDA; ESA; CFR)
DA October 12, 1990
SO Endangered Species
DEF The U.S. Fish and Wildlife Service
    or the National Marine Fisheries
    Service, as appropriate.

SERVICE CLARIFIED WATER
DA January 8, 1991
SF *SCW*
BT1 Water

SERVICE CONNECTORS
(EPA)
DA October 12, 1990
RT Piping
SO Environmental Protection Agency
    Glossary
DEF (EPA) Pipes that carry tap water
    from the public water main to a
    building.

SERVICE MAGAZINE
(DOE Order 6430.1A)
DA October 12, 1990
BT1 Magazines (Buildings)
SO Construction
DEF (DOE Order 6430.1A) An auxiliary
    building of an operating line used
    for the intermediate storage of
    explosives within the operational
    plant area. The amount of
    explosives is normally limited to a
    maximum consistent with intraline
    separation from other explosives
    buildings based on the quantity of
    explosives in the service
    magazine.

SERVICE WATER SYSTEM
DA January 8, 1991
SF *SWS (Service Water System)*
BT1 Systems

SETBACK
(DOE Order 6430.1A)
DA October 12, 1990
SO Construction
DEF (DOE Order 6430.1A) Building
    offset from a property line,
    sidewalk, or street right-of-way.

SETS
(Acronyms and Abbreviations)
DA October 12, 1990
BT1 Computer Codes
DEF A computer code to simplify
    Boolean expressions.

SETTLEABLE SOLIDS
(DOE Order 5400.5; EPA)
DA October 16, 1990
BT1 Solids
 BT2 Materials
BT1 Suspended Solids
 BT2 Leachates
RT Sedimentation
RT Wastewater
SO Environmental Protection Agency
    Glossary
SO Wastes
DEF Those solids suspended in waste
    water that are determined to be
    settleable using Method 209 E,
    Settleable Solids pp 98 and 99,
    16th edition, Standard Methods for
    Examination of Water and Waste
    Water. (EPA) Material heavy
    enough to sink to the bottom of a
    wastewater treatment tank.

SETTLING CHAMBER
(EPA)
DA October 12, 1990
SO Environmental Protection Agency
    Glossary
DEF (EPA) A series of screens placed in
    the way of flue gases to slow the
    stream of air, thus helping gravity
    to pull particles out of the emission
    into a collection area.

SETTLING TANKS
(EPA)
DA October 12, 1990
BT1 Tanks
 BT2 Facility Components
RT Wastewater Treatment Tanks
SO Environmental Protection Agency
    Glossary
DEF (EPA) A holding area for
    wastewater, where heavier
    particles sink to the bottom for
    removal and disposal.

SEWAGE
(Doe Order 5400.5; EPA)
DA October 12, 1990
BT1 Municipal Solid Wastes
 BT2 Solid Wastes
  BT3 Wastes
NT1 Raw Sewage
RT Sewage Sludge
RT Sewers
RT Sewerage
RT Significant Municipal Facilities
SO Environmental Protection Agency
    Glossary
DEF The waste matter that passes
    through sewers. (EPA) The waste
    and wastewater produced by

BT-Broader Term   NT-Narrower Term   RT Related Term

residential and commercial establishments and discharged into sewers.

SEWAGE LAGOONS
(EPA)
DA   November 9, 1990
SY   Oxidation Ponds
RT   Impoundment
RT   Lagoons
DEF (EPA) A man-made lake or body of water in which waste is consumed by bacteria. It is used most frequently with other waste treatment processes.

SEWAGE PLANT
(1790 SEWAGE PLANT)
DA   December 10, 1990
BT1 Building (DOE FRASE Vocabulary)
 BT2 Facilities and Buildings (DOE FRASE Vocabulary)
  BT3 Facilities
RT   Sanitary Treatment System
RT   Sewer System
SO   DOE FRASE VOCABULARY

SEWAGE SLUDGE
(EPA)
DA   October 12, 1990
BT1 Sludge
 BT2 Wastes
RT   Evaporation Ponds
RT   Sewage
SO   Environmental Protection Agency Glossary
DEF (EPA) Sludge produced at a Publicly Owned Treatment Works, the disposal of which is regulated under the Clean Water Act.

SEWER SYSTEM
(2050 SEWER SYSTEM)
DA   December 10, 1990
BT1 Systems (DOE FRASE Vocabulary)
 BT2 Systems
RT   Sewage Plant
SO   DOE FRASE VOCABULARY

SEWERAGE
(Doe Order 5400.5; EPA)
DA   October 12, 1990
RT   Pumping Stations
RT   Sewage
RT   Sewers
SO   Environmental Protection Agency Glossary
SO   Wastes
DEF The system of sewers. (EPA) The entire system of sewage collection, treatment, and disposal.

SEWERS
(Doe Order 5400.5; EPA)
DA   October 12, 1990
NT1 Combined Sewers
NT1 Interceptor Sewers
NT1 Lateral Sewers
NT1 Pressure Sewers
NT1 Sanitary Sewers
NT1 Storm Sewers
RT   Concrete Encasement
RT   Curb Inlets
RT   Sewage
RT   Sewerage
RT   Urban Runoff
SO   Environmental Protection Agency Glossary
SO   Wastes
DEF The artificial conduit, usually underground, for carrying off wastewater and refuse. (EPA) A channel or conduit that carries wastewater and stormwater runoff from the source to a treatment plant or receiving stream. Sanitary sewers carry household, industrial, and commercial waste. Storm sewers carry runoff from rain or snow. Combined sewers are used for both purposes.

SFMP
DA   October 12, 1990
SEE Surplus Facilities Managment Program
SO   Acronyms

SG
DA   October 12, 1990
SEE Steam Generator
SO   Acronyms

SGTR
DA   October 12, 1990
SEE Steam Generator Tube Rupture
SO   Acronyms

SGTS
DA   October 12, 1990
SEE Standby Gas Treatment System
SO   Acronyms

SH
DA   October 12, 1990
SEE Sleeve Housings
SO   Acronyms

SHADOW FORCE WEAPONS
(Doe Order 5480.16)
DA   October 12, 1990
BT1 Firearms
RT   Shadow Forces
SO   Firearms
DEF (DOE Order 8480.16) Live fire weapons used by shadow forces to respond to an actual security alarm that may occur during a security exercise.

SHADOW FORCES
(Doe Order 5480.16)
DA   October 12, 1990
BT1 Protective Force Personnel
 BT2 Personnel
RT   Security Facilities
RT   Shadow Force Weapons

SO   Firearms
DEF (DOE Order 5480.16) Armed security forces stationed away from an exercise area and under the continuous supervision of a controller, preferably a security officer.

SHAFT BREAK LIMIT
(NFI)
DA   October 12, 1990
BT1 Limits
SO   Nuclear Facilities Incident Database
SO   Radiation
DEF (NFI) Average assembly effluent temperature limit.

SHAFT GUIDES
(NFI)
DA   October 12, 1990
BT1 Reactor Components
SO   Nuclear Facilities Incident Database
SO   Radiation
DEF (NFI) Use for machine part guides, e.g., control rods.

SHALLOW DOSE EQUIVALENTS
DA   October 19, 1990
BT1 Dose Equivalents
 BT2 Radiation Units
  BT3 Units of Measure
SO   Industrial Hygiene
DEF The maximum dose equivalent within the spherical shell extending form a depth of 0.07 to a depth of 1 cm from the surface of a 30 cm diameter sphere centered at this point and consisting of material equivalent to soft tissue with a density of $1 \text{ g} \cdot \text{cm}^{-3}$.

SHALLOW EYE DOSE EQUIVALENTS
(DOE Order 5480.11)
DA   October 16, 1990
BT1 Dose Equivalents
 BT2 Radiation Units
  BT3 Units of Measure
DEF The dose equivalent at the respective depths of 0.007 cm, 1.0 cm, and 0.3 cm in tissue.

SHALLOW LAND BURIAL
DA   January 8, 1991
SF   SLB
BT1 Burial Operations
 BT2 Operations
  BT3 Activities
 BT2 Storage
  BT3 Waste Management Processes
   BT4 Processes
BT1 Near Surface Disposal
 BT2 Land Disposal
  BT3 Disposal
   BT4 Waste Management Processes
    BT5 Processes

SHEAR(S)
(3056 SHEAR)
DA   January 3, 1991

BT1 Tools - Manual
 BT2 Tools (DOE FRASE Vocabulary)
 BT3 Equipment
SO DOE FRASE VOCABULARY

SHEAR WALLS
(SEA)
DA   October 12, 1990
RT   Boundary Elements
RT   Dual Systems
SO   Construction
DEF (SEA) A wall designed to resist
     lateral forces parallel to the plane
     of the wall (sometimes referred to
     as a vertical diaphragm or a
     structural wall).

SHEATHING
(SWDA; RCRA; ESH)
DA   October 12, 1990
SO   Hazardous Materials
DEF A covering consisting of a smooth
     layer of wood placed over metal
     and secured to prevent any
     movement.

SHEET METAL WORKER
(0682 METAL WORKER)
DA   November 28, 1990
BT1 Precision/Production Personnel
 BT2 Occupations
 BT2 Personnel
SO DOE FRASE VOCABULARY

SHEET PILING
(DOE Order 6430.1A)
DA   October 12, 1990
SO   Construction
DEF (DOE Order 6430.1A) Closely
     spaced piles of wood, steel, or
     concrete driven vertically into the
     ground to obstruct lateral
     movement of earth or water.

SHELTERS
(EMER)
DA   February 1, 1991
BT1 Facilities
SO   Emergency Preparedness
DEF (EMER) Facilities used to protect,
     house, and supply the essential
     needs of designated individuals
     during the period of an
     emergency. A shelter may or may
     not be specifically constructed for
     such use, depending on the type
     of emergency and the specific
     programmatic requirements.

SHIELD PLUG(S)
(2591 SHIELD PLUG)
DA   January 3, 1991
BT1 Equipment/Parts - Nuclear (DOE
     FRASE Vocabulary)
 BT2 Equipment
 BT2 Reactor Components
SO DOE FRASE VOCABULARY

SHIELDING MATERIALS
(NIH)
DA   October 12, 1990
BT1 Materials
RT   Protective Barriers
RT   Radiation Shields
SO   Radiation
DEF (NIH) Any material that is used to
     absorb radiation and thus
     effectively reduce the intensity of
     radiation, and in some cases
     eliminate it. Lead, concrete,
     aluminum, water, and plastic are
     examples of commonly used
     shielding material.

SHIFT SUPERVISOR
DA   January 8, 1991
SF   SS
BT1 Personnel

SHIFT TECHNICAL ADVISOR
DA   January 8, 1991
SF   STA
BT1 Personnel

SHIPPING PAPERS
(SWDA; RCRA; ESH)
DA   October 12, 1990
SO   Hazardous Materials
DEF A shipping order, bill of lading,
     manifest, or other shipping
     document servicing a similar
     purpose and containing
     information.

SHIPPING ROOM
(1715 SHIPPING ROO)
DA   December 10, 1990
BT1 Room
 BT2 Sites/Areas
SO DOE FRASE VOCABULARY

SHOE COVER(S)
(2690 SHOE COVER)
DA   January 3, 1991
BT1 Foot Protection
 BT2 Personal Protective Equipment
  BT3 Equipment/Parts - Personal
      Protective (DOE FRASE
      Vocabulary)
   BT4 Equipment
NT1 Metal Shoe Cover
SO DOE FRASE VOCABULARY

SHORING
(DOE Order 6430.1A)
DA   October 12, 1990
BT1 Building Frame Systems
 BT2 Space Frames
  BT3 Structures (DOE FRASE
      Vocabulary)
 BT2 Systems
SO   Construction
DEF (DOE Order 6430.1A) Temporary
     bracing of an existing building
     foundation to provide support
     during adjacent excavations. Also

applies to supporting construction
of above grade floors.

SHORT-TERM EXPOSURE LIMIT
(EMER)
DA   February 1, 1991
BT1 Limits
SO   Emergency Preparedness
DEF (EMER) The maximum
     concentration allowed for a
     continuous 15-minute exposure
     period. There may be no more
     than four such exposures each
     day with at least one hour between
     exposures. The daily threshold
     limit value - time weighted average
     (TLV-TWA) may not be exceeded.

SHOTGUN
(EPA)
DA   October 12, 1990
RT   DNA
SO   Environmental Protection Agency
     Glossary
DEF (EPA) Non-scientific term for the
     process of breaking up the DNA
     derived from an organism and
     then moving each separate and
     unidentified DNA fragment into a
     bacterium.

SHOTGUN, PUMP
(Doe Order 5480.16)
DA   October 12, 1990
BT1 Small Arms
 BT2 Firearms
SO   Firearms
DEF (DOE Order 5480.16) A shotgun
     that uses a pumping or sliding
     action to eject the fired round and
     load the next round from the
     magazine into the chamber.

SHOTGUN, SEMIAUTOMATIC
(Doe Order 5480.16)
DA   October 12, 1990
BT1 Semiautomatic Firearms
 BT2 Small Arms
  BT3 Firearms
SO   Firearms
DEF (DOE Order 5480.16) A shotgun
     that by means of a gas tube, or
     recoil, automatically ejects the
     round fired and loads the next
     round from the magazine
     automatically into the chamber.

SHOULDER(S)
(1094 SHOULDER)
DA   November 28, 1990
BT1 Joint(s)
 BT2 Human Body Parts
RT   Upper Arm
SO   DOE FRASE VOCABULARY

SHOVEL(S)
(3057 SHOVEL)
DA   January 3, 1991
BT1 Tools - Manual

BT-Broader Term   NT-Narrower Term   RT Related Term

BT2 Tools (DOE FRASE Vocabulary)
  BT3 Equipment
SO DOE FRASE VOCABULARY

SHT
DA October 12, 1990
SEE Seal Head Tank
SO Acronyms

SHUT-DOWN CIRCUIT(S)
(2419 SHUT-DOWN CI)
DA January 3, 1991
BT1 Equipment/Parts - Electrical (DOE
    FRASE Vocabulary)
  BT2 Equipment
SO DOE FRASE VOCABULARY

SHUTDOWN (EMISSIONS)
(CAA; CFR; ESH)
DA October 12, 1990
NT1 Process Unit Shutdown
RT Closure
RT Curtail
SO Air Pollution
DEF (CFR) The cessation of operation
    of an affected source for any
    purpose.

SHUTDOWN COOLING
DA January 8, 1991
SF *SDC*
RT Reactor Shutdown
SO Radiation

SHUTDOWN HEAT EXCHANGER
DA January 8, 1991
SF *SDHX*
BT1 Heat Exchanger
  BT2 Equipment/Parts - Nuclear (DOE
      FRASE Vocabulary)
    BT3 Equipment
    BT3 Reactor Components
RT Reactor Shutdown
SO Radiation

SHUTDOWN SEQUENCER
DA January 8, 1991
SF *SDS*
RT Reactor Shutdown
SO Radiation

SI
DA October 12, 1990
SEE Site Inspections
SO Acronyms

Si (Periodic Element)
DA October 12, 1990
SEE Silicon
SO Acronyms

SICS
DA October 12, 1990
SEE Safety Injection Control System
SO Acronyms

SIDE SHIELDS
(2691 SIDE SHIELDS)
DA January 3, 1991
BT1 Eye Protection
  BT2 Personal Protective Equipment
    BT3 Equipment/Parts - Personal
        Protective (DOE FRASE
        Vocabulary)
      BT4 Equipment
RT Eye(s)
SO DOE FRASE VOCABULARY

SIDE-STREAM EXTRACTION
(NCRP)
DA October 12, 1990
BT1 Manufacturing Processes
  BT2 Processes
SO Radiation
DEF (NCRP) The extraction of a mineral
    that is a by-product of the principal
    mineral being extracted.

SIEVERT
(IAEA, NCRP)
DA October 12, 1990
SY Roentgen Equivalent Man
BT1 Radiation Units
  BT2 Units of Measure
RT Dose Equivalents
RT Effective Dose Equivalent
    Commitment
RT Roentgen
SO Radiation
DEF (DSTT) A unit of ionizing radiation,
    equal to the amount that produces
    the same damage to humans as 1
    roentgen of high- voltage x-rays.

SIGNAL WORDS
(EPA)
DA October 12, 1990
RT Labeling
SO Environmental Protection Agency
    Glossary
DEF (EPA) The words used on a
    pesticide label – Danger, Warning,
    Caution – to indicate the level of
    toxicity of the chemicals.

SIGNIFICANCE
(NEPA)
DA May 16, 1991
RT National Environmental Policy Act
SO Environmental Management

SIGNIFICANT ADVERSE REACTIONS
(TSCA; CFR)
DA October 12, 1990
RT Environment
SO Hazardous Materials
DEF Reactions that may indicate a
    substantial impairment of normal
    activities, or long-lasting or
    irreversible damage to health or
    the environment.

SIGNIFICANT DETERIORATION
(EPA)
DA October 12, 1990

RT Prevention of Significant
    Deterioration
SO Environmental Protection Agency
    Glossary
DEF (EPA) Pollution resulting from a
    new source in previously "clean"
    areas. (See: Prevention of
    Significant Deterioration.)

SIGNIFICANT ECONOMIC LOSS
(ESH)
DA November 20, 1990
BT1 Losses
DEF (ESH)Under emergency
    conditions,for a productive
    activity,profitability would be
    greatly below expected profitability
    for that activity; or, for other types
    of activities, where profits can't be
    calculated, the value of public or
    private fixed assets would be
    greatly below the expected value
    for those assets. See ES&H
    Environmental Audit Manual vol. 1
    (DOE/EH-0125 vol 1) for
    continued definition.

**SIGNIFICANT ENVIRONMENTAL
COMPLIANCE ISSUES**
(DOE Order 5400.2A; ESH)
DA October 12, 1990
RT Coordination Processes
SO Environmental Management
SO Management
DEF Issues that are or have the
    potential of being precedent
    setting or controversial, and/or
    involve Headquarters notification,
    concurrence, or approval. See
    DOE Order 5400.2A for Examples
    of Environmental Compliance
    issues.

SIGNIFICANT IMPAIRMENT
(CAA; CFR)
DA October 12, 1990
BT1 Visibility Impairments
RT Mandatory Class I Federal Areas
SO Air Pollution
DEF (CFR) Visibility impairment which,
    in the judgement of the
    Administrator, interferes with the
    management, protection,
    preservation, or enjoyment of the
    visitor's visual experience of the
    mandatory Class I Federal area.
    This determination must be made
    on a case-by-case basis taking
    into account the geographic
    extent, intensity, duration,
    frequency, and time of the visibility
    impairment, and how these factors
    correlate with: (1) times of visitor
    use of the mandatory Class I
    Federal area, and (2) the
    frequency and timing of natural
    conditions that reduce visibility.

SY-Synonymous Terms   SO-Source/Subject Category   SF-See From

SIGNIFICANT INCIDENTS
(EMER)
DA February 1, 1991
BT1 Incidents
SO Emergency Preparedness
DEF (EMER) Unexpected events
involving nuclear weapons or
radiological nuclear weapon
components which results in any
of the following: 1) accidental or
unauthorized launching, firing, or
use of a nuclear explosive 2)
nuclear detonation 3) nonnuclear
detonation/burning of a nuclear
weapon 4) radioactive
contamination 5) seizure, theft, or
loss of a nuclear weapon or an
actual component of a nuclear
weapon 6) public hazard, actual or
implied

SIGNIFICANT MODIFICATION
(Doe Orders 6430.1A; 5480.5, and
5481.1B)
DA October 12, 1990
RT Construction
SO Construction
SO Environmental Management
SO Industrial Safety
DEF A change to a nuclear facility that
involves an unreviewed safety
question.

SIGNIFICANT MUNICIPAL FACILITIES
(EPA)
DA October 12, 1990
BT1 Facilities
RT Sewage
SO Environmental Protection Agency
Glossary
DEF (EPA) Those publicly owned
sewage treatment plants that
discharge a million gallons per day
or more and are therefore
considered by states to have the
potential for substantial effect on
the quality of receiving waters.

SIGNIFICANT QUANTITIES
(Doe Order 5480.5)
DA October 12, 1990
BT1 Quantities
RT Reportable Quantities
SO Environmental Management
SO Industrial Safety
DEF (DOE Order 5480.5) Masses of
fissionable materials greater than
a safe mass.

SIGNIFICANT SOURCE OF
GROUNDWATER
(ANL; CFR)
DA May 24, 1991
BT1 Aquifers
BT2 Formations
RT Groundwater
DEF (CFR) (1) An aquifer that: (i) is
saturated with water having less
than 10,000 milligrams per liter of

total dissolved solids; (ii) is within
2,500 feet of the land surface; (iii)
has a transmissivity greater than
200 gallons per day per foot,
provided that any formation or part
of a formation included within the
source of groundwater has a
hydraulic conductivity greater than
2 gallons per day per square foot;
and (iv) is capable of continuously
yielding at least 10,000 gallons per
day to a pumped or flowing well
for a period of at least a year; or
(2) an aquifer that provides the
primary source of water for a
community water system.

SIGNIFICANT VIOLATIONS
(EPA)
DA October 12, 1990
BT1 Violations
SO Environmental Protection Agency
Glossary
DEF (EPA) Violations by point source
dischargers of sufficient magnitude
and/or duration to be a regulatory
priority.

SILICON
DA January 8, 1991
SF Si (Periodic Element)

SILO
(1863 SILO)
DA December 10, 1990
BT1 Structures (DOE FRASE
Vocabulary)
SO DOE FRASE VOCABULARY

SILT
(EPA)
DA October 12, 1990
BT1 Soils
RT Sedimentation
SO Environmental Protection Agency
Glossary
DEF (EPA) Fine particles of sand or rock
that can be picked up by the air or
water and deposited as sediment.

SILVICULTURE
(EPA)
DA October 12, 1990
RT Clear Cut
SO Environmental Protection Agency
Glossary
DEF (EPA) Management of forest land
for timber. Sometimes contributes
to water pollution, as in
clear-cutting.

SIMS
DA October 12, 1990
SEE Standards Information Management
System
SO Acronyms
SO Environmental Management
SO Management

SIMULATORS
(EDB)
DA January 31, 1991
NT1 Engagement Simulation Systems
NT2 Multiple Integrated Laser
Engagement System
NT1 Law Simulators
SO Firearms
DEF (DSTT) A computer or other piece
of equipment which imitates or
mimics the actions and reactions
of a system or condition, showing
the effects of various applied
changes.

SINGLE ACTION SEMIAUTOMATIC
PISTOLS
(Doe Order 5480.16)
DA October 12, 1990
BT1 Pistols
BT2 Handguns
BT3 Small Arms
BT4 Firearms
BT1 Semiautomatic Firearms
BT2 Small Arms
BT3 Firearms
RT Magazines
SO Firearms
DEF (DOE Order 5480.16)
Magazine-fed, must be fired with
the hammer in the cocked position
and will fire one shot each time
the trigger is pulled.

SINGLE FAILURE
(DOE Order 6430.1A)
DA October 12, 1990
BT1 Failures
BT2 Accidents
RT Single Failure Analysis
SO Construction
DEF (DOE Order 6430.1A) An
occurrence that results in the loss
of capability of a component to
perform its intended safety
function(s). Multiple failures, i.e.,
loss of capability of several
components, resulting from a
single occurrence are considered
to be a single failure. Systems are
considered to be designed against
an assumed single failure if
neither (1) a single failure of any
active component (assuming
passive components function
properly) nor, (2) a single failure of
any passive component (assuming
active components function
properly) results in loss of the
system's capability to perform its
safety function(s).

SINGLE FAILURE ANALYSIS
(SSDC)
DA October 12, 1990
BT1 Analyses
RT Single Failure
SO System Safety Development Center
Glossary
DEF (SSDC) Primary failure analysis to

detect where a single failure could
shut down a system or process.

SINGLE SHELL TANKS
DA   January 8, 1991
SF   *SST*
BT1 Tanks
 BT2 Facility Components

SINGLE STRIP CONTAINERS
(SWDA; RCRA; ESH)
DA   October 12, 1990
SF   *STC*
BT1 Containers
SO   Hazardous Materials

SINKING
(EPA)
DA   October 12, 1990
BT1 Processes
RT   Oil Spills
RT   Sinking Agents
SO   Environmental Protection Agency
     Glossary
DEF (EPA) Controlling oil spills by using
     an agent to trap the oil and sink it
     to the bottom of the body of water
     where the agent and the oil are
     biodegraded.

SINKING AGENTS
(CERCLA; CFR)
DA   October 12, 1990
BT1 Additives
 BT2 Chemical Substances
RT   Cleanup
RT   Oil Spills
RT   Sinking
SO   Compensation and Liability
DEF Those additives applied to oil
     discharges to sink floating
     pollutants below the water surface.

SIP
DA   October 12, 1990
SEE State Implementation Plans
SO   Acronyms

SIS
DA   October 12, 1990
SEE Special Isotope Separator
SO   Acronyms

SIT
DA   October 12, 1990
SEE Safety Injection Tank
SO   Acronyms

SITE (DOE FRASE VOCABULARY)
(1618 SITE; EMER)
DA   December 10, 1990
BT1 Sites/Areas
NT1 Alley
NT1 Balcony
NT1 Basement
NT1 Body of Water
NT1 Cafeteria
NT1 Catwalk

NT1 Cell
NT1 Chamber
NT1 Corridor/Hall
NT1 Dwelling
NT1 En Route
NT1 Escalator
NT1 Escape Hatch
NT1 Exit
NT1 Fire Escape
NT1 Firing Range
NT1 Floor Opening
NT1 Gallery
NT1 Ground
NT1 Hatch
NT1 Highbay
NT1 Hill
NT1 Intersection
NT1 Kitchen
NT1 Laboratory
NT1 Lobby
NT1 Parking Space
NT1 Pool
NT1 Stall/Booth(s)
NT1 Station
 NT2 Fire Station
 NT2 Guard Station
 NT2 Work Station
NT1 Substation
NT1 Switchyard
NT1 Tank Farm
NT1 Vault
 NT2 SNM Vault
NT1 Waste Disposal Site
NT1 Well Pad
NT1 Work Station
SO   DOE FRASE VOCABULARY
SO   Emergency Preparedness
SO   Environmental Management

SITE AVAILABILITY MODE
DA   January 8, 1991
SF   *SAM*

SITE BOUNDARIES
(DOE Order 6430.1A)
DA   October 12, 1990
BT1 Sites/Areas
RT   Off-site
RT   On-Site
SO   Construction
DEF (DOE Order 6430.1A) Well-marked
     boundaries of the property over
     which the owner or operator can
     exercise strict control without the
     aid of outside authorities.

SITE CHARACTERIZATION
DA   January 24, 1991
BT1 Programs
RT   Hydrology
RT   Site Development and Facility
     Utilization Plans
SO   Environmental Management
DEF (CFR) The program of exploration
     and research, both in the
     laboratory and in the field,
     undertaken to establish the
     geologic conditions and the
     ranges of those parameters of a
     particular site relevant to the

procedures under this part. Site
characterization includes borings,
surface excavations, excavation of
exploratory shafts, limited
subsurface lateral excavations and
borings, and in situ testing at
depth needed to determine the
suitability of the site for a geologic
repository but does not include
preliminary borings and geological
testing needed to decide whether
site characterization should be
undertaken.

SITE CLOSURE AND STABILIZATION
(ANL; CFR)
DA   May 24, 1991
BT1 Closure
 BT2 Administrative Processes
 BT3 Processes
RT   Closure Plans
RT   Final Closure
RT   Post-Closure
SO   Management
DEF (CFR) Those actions that are taken
     upon completion of operations that
     prepare the disposal site for
     custodial care and that assure that
     the disposal site will remain stable
     and will not need ongoing active
     maintenance.

SITE DEVELOPMENT AND FACILITY UTILIZATION PLANS
(DOE Orders 4320.1B; 4700.1)
DA   January 24, 1991
BT1 Plans
RT   Site Characterization
SO   Environmental Management
DEF Formal written documents
     summarizing all of the various
     data necessary to plan for the
     most effective utilization and
     orderly future development and
     disposal of facilities at an
     individual site. Such planning shall
     be in accordance with site related
     program objectives and
     requirements and shall represent
     consolidated views of site
     management, the field
     organization, and the resource
     sponsor.

SITE EMERGENCIES
(EMER)
DA   February 1, 1991
BT1 Emergencies
 BT2 Reportable Occurrences
 BT3 Occurrences
SO   Emergency Preparedness
DEF (EMER) Emergency response
     levels which represent an event in
     progress or having occurred that
     involves actual or likely major
     failures of facility functions that are
     needed for the protection of
     on-site personnel, the public
     health and safety, and the
     environment. Releases off site of

radioactive or other hazardous
substances not exceeding
protective action guidelines are
occurring or are likely to occur.

SITE INSPECTIONS
(EPA)
DA   October 12, 1990
SF   *SI*
BT1 Inspections
 BT2 Administrative Processes
  BT3 Processes
SO   Environmental Protection Agency
     Glossary
DEF (EPA) Information collected from a
     Superfund site to determine the
     extent and severity of hazards
     posed by the site. It follows and is
     more extensive than a preliminary
     assessment. The purpose is to
     gather information necessary to
     score the site, using the Hazard
     Ranking System, and to determine
     if the site presents an immediate
     threat that requires prompt
     removal action.

SITE-LIMITED INTERMEDIATES
DA   January 24, 1991
BT1 Intermediates
 BT2 Chemical Substances
SO   Environmental Management
DEF (CFR) An intermediate
     manufactured, processed, and
     used only within a site and not
     distributed in commerce other than
     as an impurity or for disposal.
     Imported intermediates cannot be
     "site-limited."

SITE QUALITY ASSURANCE PLAN
(CERCLA)
DA   October 12, 1990
SY   Quality Assurance Project Plans
BT1 Quality Assurance Plans
 BT2 Plans
 BT2 Quality Assurance Overviews
  BT3 Quality Control
   BT4 Controls
RT   Sampling Plans
SO   Compensation and Liability
DEF (CERCLA) A written document,
     associated with site sampling
     activities, which presents in
     specific terms the organization
     (where applicable), objectives,
     functional activities, and specific
     quality assurance (QA) and quality
     control (QC) activities designed to
     achieve the data quality goals of a
     specific project(s) or continuing
     operation(s). The QA Project Plan
     is prepared for each specific
     project or continuing operation (or
     group of similar projects of
     continuing operations). The QA
     Project Plan will be prepared by
     the responsible program office,
     regional office, laboratory,
     contractor, recipient of an

assistance agreement, or other
organization.

SITE-SPECIFIC SAFEGUARDS AND
SECURITY PLAN
(DOE Order 6430.1A)
DA   October 12, 1990
BT1 Plans
RT   Safeguards and Security
     Emergencies
RT   Safeguards
SO   Construction
DEF (DOE Order 6430.1A) A specific
     description of the systems and
     procedures implemented and
     planned to protect Departmental
     security interests and other
     property. The format for
     site-specific safeguards and
     security plans can be obtained
     from DP-34.

SITE-WIDE NEPA DOCUMENTS
(DOE Order 5440.1D)
DA   May 15, 1991
BT1 National Environmental Policy Act
     Documents
 BT2 Reports
RT   Environmental Assessments
RT   Environmental Impact Statements
RT   Finding of No Significant Impact
RT   Notice of Intent
RT   Record of Decision (NEPA)
SO   Environmental Management
DEF (DOE Order) Broad-scope
     Environmental Impact Statements
     or Environmental Assessments
     that identify and assess the
     individual and cumulative impact
     of the continuing and reasonably
     foreseeable future actions at a
     DOE site; they may also refer to
     an associated NEPA document
     such as a notice of intent, record
     of decision, or finding of no
     significant impact.

SITES/AREAS
(DOE FRASE Vocabulary Numeric Keys
 16000-1774)
DA   January 17, 1991
NT1 Abandoned Areas
 NT2 Temporarily Abandoned Areas
NT1 Accident Sites
NT1 Action Areas
NT1 Active Portions
NT1 Air Quality Control Regions
 NT2 Federal Class I areas
  NT3 Mandatory Class I Federal Areas
 NT2 Regions
NT1 Area
 NT2 Canal Area
 NT2 Construction Area
 NT2 Customer Service Area
 NT2 Data Processing Area
 NT2 Decontamination Areas
 NT2 Eating Area
  NT3 Cafeteria
 NT2 Field Area
 NT2 Maintenance Area

 NT2 Office Area
 NT2 Pit Area
 NT2 Production/Operations Area
 NT2 Pump Area
 NT2 Recreation Area
 NT2 Sand Blasting Area
 NT2 Storage Area
  NT3 Stock Room
  NT3 Warehouse
 NT2 Sump Area
 NT2 Test Area
  NT3 Fast Flux Test Facility
  NT3 Nevada Test Site
  NT3 Rocket Motor Test Sites
  NT3 Test Reactor Facility
  NT3 Test Stations
  NT3 Test Reactor Area
  NT3 Tonopah Test Range
 NT2 Vehicle Service Area
 NT2 Yard Area
NT1 Area of Review
NT1 Assessment Area
NT1 Attainment Areas
NT1 Burial Grounds
NT1 Central Alarm Stations
NT1 Commercial Areas
NT1 Confinement Areas
NT1 Controlled Areas
 NT2 Disturbed Zones
NT1 Conventional Mines
 NT2 Active Mines
 NT2 Inactive Mines
 NT2 Underground Uranium Mines
NT1 Corrective Action Management
    Units
 NT2 Disposal Units
  NT3 Land Disposal Units
   NT4 Landfill Cells
   NT4 Landfills
    NT5 Chemical Waste Landfills
    NT5 Sanitary Landfills
    NT5 Specially Designated Landfills
 NT2 Hazardous Waste Management
     Units
  NT3 Miscellaneous Units
 NT2 Solid Waste Management Units
NT1 Crane Maintenance Area
NT1 Crawl Spaces
NT1 Critical Areas
NT1 Cultural Resource Sites
NT1 Department of Energy Sites
NT1 Depositories
NT1 Discharge Point
NT1 Disposal Sites
 NT2 Inactive Waste Disposal Sites
  NT3 Inactive Hazardous Waste
      Disposal Sites
NT1 Disposal Areas
NT1 DOE Property
NT1 DOE Reservation
NT1 Dog Houses
NT1 Dry Caves
NT1 Dumps
 NT2 Open Dumps
NT1 Emergency Control Stations
NT1 Entry Control Points
NT1 EPA Region
NT1 Evaporation Ponds
NT1 Extraction Sites
NT1 Feedlots
NT1 Field Command Posts

NT1 Firearms Ranges
NT1 Floodplains
NT1 Geologic Repositories
 NT2 Geologic Repository Operations
     Areas
NT1 High-Contact Residential Surfaces
NT1 High-Contact Commercial Surfaces
NT1 High-Contact Industrial Surfaces
NT1 Impoundment
 NT2 New Tailings Impoundment
 NT2 PAR Ponds
 NT2 Reservoirs
 NT2 Surface Impoundment
NT1 Inactive Portions
NT1 Indian Lands
NT1 Individual Generation Sites
NT1 Licensed Sites
NT1 Manatee Protection Areas
 NT2 Manatee Refuges
 NT2 Manatee Sanctuaries
NT1 Material Access Areas
NT1 National Defense Areas
NT1 National Security Areas
NT1 Non-Attainment Areas
NT1 Non-Restricted Access Areas
NT1 Occupiable Area
NT1 Occupied Area (Explosives)
NT1 Off-site
NT1 On-Scene
NT1 On-Site
NT1 Operating Area Compartment
NT1 Other Restricted Access
     (nonsubstation) Locations
NT1 Outdoor Electrical Substations
NT1 Outfall
NT1 Point of Disinfectant Application
NT1 Point of Compliance
NT1 Point of Nearest Public Access
NT1 Property Protection Area
NT1 Protected Areas
NT1 Public Lands
NT1 Radiation Areas
 NT2 High Radiation Areas
NT1 Radiation Emergency Assistance
     Center/Training Sites
NT1 Radiological Areas
NT1 Ranges (Firearms)
NT1 Recharge Areas
NT1 Remote Interrogation Points
NT1 Residential Areas
NT1 Restricted Areas
NT1 River Basins
NT1 Room
 NT2 Air Lock Room
 NT2 Central Control Room
 NT2 Clean Rooms
 NT2 Control Room
 NT2 Equipment Room
 NT2 Locker/Shower Room
 NT2 Refrigeration Room
 NT2 Rest Room
 NT2 Shipping Room
 NT2 Stock Room
NT1 Security Areas
 NT2 Exclusion Areas
 NT2 Limited Areas
 NT2 SNM Vault
 NT2 Vault-type Rooms
 NT2 Vital Areas
NT1 Site (DOE FRASE Vocabulary)
 NT2 Alley

NT2 Balcony
NT2 Basement
NT2 Body of Water
NT2 Cafeteria
NT2 Catwalk
NT2 Cell
NT2 Chamber
NT2 Corridor/Hall
NT2 Dwelling
NT2 En Route
NT2 Escalator
NT2 Escape Hatch
NT2 Exit
NT2 Fire Escape
NT2 Firing Range
NT2 Floor Opening
NT2 Gallery
NT2 Ground
NT2 Hatch
NT2 Highbay
NT2 Hill
NT2 Intersection
NT2 Kitchen
NT2 Laboratory
NT2 Lobby
NT2 Parking Space
NT2 Pool
NT2 Stall/Booth(s)
NT2 Station
 NT3 Fire Station
 NT3 Guard Station
 NT3 Work Station
NT2 Substation
NT2 Switchyard
NT2 Tank Farm
NT2 Vault
 NT3 SNM Vault
NT2 Waste Disposal Site
NT2 Well Pad
NT2 Work Station
NT1 Site Boundaries
NT1 Soil Adsorption Field
NT1 Solid Waste Storage Area
NT1 Specified Ports and Harbors
NT1 Special Aquatic Sites
NT1 Spill Area
NT1 Spill Boundaries
NT1 Stabilization Ponds
NT1 Storage Area Compartments
NT1 Supervised Areas
NT1 Test Area North
NT1 Transfer Stations
NT1 Unattended Openings
NT1 Uncontrolled Hazardous Waste
     Sites
NT1 Underground Areas
NT1 Unpackaging Rooms
NT1 Watershed
NT1 Wetlands
 NT2 Bogs
 NT2 Bottomland Hardwoods
 NT2 Fens
 NT2 Marshes
 NT3 Tidal Marshes
 NT2 Swamps
NT1 Zones
 NT2 Aeration Zones
 NT2 Buckled Zones
 NT2 Buffer Zone
 NT2 Coastal Zones
 NT2 Confining Zones

NT2 Contiguous Zones
NT2 Contingency Planning Zones
NT2 Disturbed Zones
NT2 Emergency Planning Zones
NT2 Engineering Control Zones
NT2 Excavation Zone
NT2 Flow Zone
NT2 Hydrogeologic Units
NT2 Injection Zones
 NT3 Injection Interval
NT2 Inland Zones
NT2 Isolation Zones
NT2 Law Hazard Zone
NT2 Mixing Zones
NT2 Planning Zones
NT2 Radiation Zone
NT2 Saturated Zones
NT2 Treatment Zones
NT2 Unsaturated Zones
NT2 Vulnerable Zones
RT DOE FRASE Categories
RT Facilities
RT Real Property
RT Siting
RT Structures (DOE FRASE
     Vocabulary)
RT Targets
SO DOE FRASE VOCABULARY
DEF Designations of particular locations.
     Also, A subject category used with
     the DOE FRASE Vocabulary.

SITING
(EPA)
DA October 12, 1990
BT1 Administrative Processes
 BT2 Processes
RT Facilities
RT Sites/Areas
SO Environmental Protection Agency
     Glossary
DEF (EPA) The process of choosing a
     location for a facility.

SIZE CLASSES OF DISCHARGES
(CERCLA)
DA October 12, 1990
NT1 Major Discharges
NT1 Medium Discharges
NT1 Minor Discharges
RT Size Classes of Releases
SO Compensation and Liability
DEF (CERCLA) Size classes of oil
     discharges which are provided as
     guidance to the On-Scene
     Coordinator (OSC) and serve as
     the criteria for further actions. They
     are not meant to imply associated
     degrees of hazard to public health
     or welfare, nor are they a measure
     of environmental damage.

SIZE CLASSES OF RELEASES
(CERCLA)
DA October 12, 1990
NT1 Major Release
NT1 Minor Releases
RT Releases
RT Size Classes of Discharges

SO Compensation and Liability
DEF (CERCLA) Refers to size classifications that are provided as guidance to the On-Scene Coordinator (OSC) for meeting pollution reporting requirements.

SJP
DA October 12, 1990
SEE Standard Job Procedures
SO Acronyms

SKIMMING
(CAA; CFR; EPA)
DA October 12, 1990
BT1 Processes
RT Oil Spills
RT Smelters
SO Air Pollution
SO Environmental Protection Agency Glossary
DEF Using a machine to remove oil or scum from the surface of the water. (40 CFR 61.171) The removal of slag from the molten converter bath.

SKIN
(1132 SKIN)
DA November 28, 1990
BT1 Human Body Parts
NT1 Fingernail(s)
NT1 Hair
NT1 Scalp
RT Dermatitis
RT Other Skin Conditions
RT Palm(s)
SO DOE FRASE VOCABULARY

SKULL (FRASE)
(1078 SKULL)
DA November 28, 1990
BT1 Head
 BT2 Human Body Parts
NT1 Jaw
RT Brain
RT Brain Damage
RT Concussion
RT Contusion(S)
RT Head Protection
SO DOE FRASE VOCABULARY

SLAC
DA October 12, 1990
SEE Stanford Linear Accelerator Center
SO Acronyms

SLANTING
(DOE Order 6430.1A)
DA October 12, 1990
SO Construction
DEF (DOE Order 6430.1A) The incorporation, without appreciable extra cost or reduction in efficiency, of certain architectural and engineering features into new structures (except temporary type) or portions of the structures to improve their ability to resist the

effects of an attack and to offer protection to personnel and material.

SLB
DA October 12, 1990
SEE Shallow Land Burial
SO Acronyms

SLCS
DA October 12, 1990
SEE Standby Liquid Control System
SO Acronyms

SLEDGEHAMMER(S)
(3059 SLEDGEHAMMER)
DA January 3, 1991
BT1 Hammer
 BT2 Tools - Manual
  BT3 Tools (DOE FRASE Vocabulary)
  BT4 Equipment
SO DOE FRASE VOCABULARY

SLEEVE HOUSINGS
DA January 8, 1991
SF SH

SLEEVE TARGET
DA January 8, 1991
SF ST

SLI
DA October 12, 1990
SEE Sandia Laboratory Instruction
SO Acronyms

SLICING MACHINE
(2144 SLICING MACH)
DA December 10, 1990
BT1 Machines (DOE FRASE Vocabulary)
 BT2 Equipment
SO DOE FRASE VOCABULARY

SLING(S)
(2342 SLING)
DA December 10, 1990
BT1 Equipment/Parts - Material Handling (DOE FRASE Vocabulary)
 BT2 Equipment
SO DOE FRASE VOCABULARY

SLIP GAUGES
(CAA; CFR)
DA October 12, 1990
BT1 Equipment
RT Monitoring
SO Air Pollution
DEF Gauges that have a probe that moves through the gas/liquid interface in a storage or transfer vessel and indicates the level of vinyl chloride in the vessel by the physical state of the material the gauge discharges.

SLOPE FACTOR
(EPA)
DA October 12, 1990
RT Exposure
SO Environmental Protection Agency Glossary
DEF (EPA) A plausible upper-bound estimate of the probability of a response per unit intake of a chemical over a lifetime. The slope factor is used to estimate an upper-bound probability of an individual developing cancer as a result of a lifetime of exposure to a particular level of a potential carcinogen.

SLOW SAND FILTRATION
(SWDA; CFR)
DA October 12, 1990
BT1 Filtration
 BT2 Pollution Recovery Processes
  BT3 Processes
SO Water Pollution
DEF A process involving passage of raw water through a bed of sand at low velocity (generally less than 0.4 m/h) resulting in substantial particulate removal by physical and biological mechanisms. (EPA) Treatment process involving passage of raw water through a bed of a sand at low velocity which results in the substantial removal of chemical and biological contaminants.

SLUDGE
(SWDA; RCRA; CAA; CFR; EPA)
DA October 12, 1990
BT1 Wastes
NT1 Activated Sludges
NT1 Sewage Sludge
RT Municipal Solid Wastes
RT Sludge Dryer
RT Slurries
SO Air Pollution
SO Environmental Management
SO Environmental Protection Agency Glossary
SO Wastes
DEF (SWDA) A semisolid residue from any of a number of air or water treatment processes. Sludge can be a hazardous waste. (ESH) Any solid, semisolid, or liquid waste generated from a municipal, commercial, or industrial wastewater treatment plant, water supply treatment plant, or air pollution control facility or any other such waste having similar characteristics and effect.

SLUDGE DRYER
(CAA; CFR)
DA October 12, 1990
BT1 Devices
RT Sludge

SO Air Pollution
DEF (CAA) A device used to reduce the moisture content of sludge by heating to temperatures above 65 deg. C (ca. 150 deg. F) directly with combustion gases.

SLURRIES
(EPA)
DA   October 12, 1990
BT1 Mixtures
RT Sludge
SO Environmental Protection Agency Glossary
DEF (EPA) Watery mixtures of insoluble matter that result from some pollution control techniques.

SMALL ARMS
(Doe Order 5480.16)
DA   October 12, 1990
BT1 Firearms
NT1 Handguns
 NT2 Pistols
  NT3 Double Action Semiautomatic Pistols
  NT3 Machine Pistols
  NT3 Single Action Semiautomatic Pistols
 NT2 Revolvers
NT1 Machine Guns
 NT2 Submachine Guns, Closed Bolt
 NT2 Submachine Guns, Open Bolt
NT1 Rifles
 NT2 Automatic Rifles
NT1 Semiautomatic Firearms
 NT2 Double Action Semiautomatic Pistols
 NT2 Shotgun, Semiautomatic
 NT2 Single Action Semiautomatic Pistols
NT1 Shotgun, Pump
RT Ammunition
RT Small Arms Ammunition
SO Firearms
DEF (DOE Order 5480.16) Handcarried firearms, including revolvers, pistols, rifles, and submachine guns.

SMALL ARMS AMMUNITION
(HMTA; CFR)
DA   May 20, 1991
BT1 Ammunition
 BT2 Munitions
RT Class C Explosives
RT Small Arms
SO Hazardous Materials
DEF (CFR) Small arms ammunition is fixed ammunition consisting of a metallic, plastic composition, or paper cartridge case, a primer, and a propelling charge, with or without bullet, projectile, shot, tear gas material, tracer components, or incendiary compositions, or mixtures, and is further limited to the following: (1) Ammunition designed to be fired from a pistol, revolver, rifle, or shotgun held by the hand or to the shoulder. (2) Ammunition of caliber less than 20 millimeters with incendiary solid inert or empty projectiles (with or without tracers), designed to be fired from machine guns or cannons. (3) Blank cartridges including canopy remover cartridges, starter cartridges, and seat ejector cartridges, containing not more than 500 grains of propellant powder, provided that such cartridges shall be incapable of functioning en masse as a result of the functioning of any single cartridge in the container or as a result of exposure to external flame.

SMALL BREAK LOCA
DA   January 8, 1991
SF   *SBLOCA*
BT1 Loss of Coolant Accident
 BT2 Accidents
 BT2 Losses

SMALL CAPACITORS
(TSCA; CFR)
DA   October 12, 1990
BT1 Capacitors
 BT2 Electrical Equipment
 BT3 Equipment
SO Hazardous Materials
DEF (ESH) Capacitors that contain less than 1.36 kg (3 lbs.) of dielectric fluid. The following assumptions may be used if the actual weight of the dielectric fluid is unknown. Capacitors whose total volume is less than 1,639 cubic centimeters (100 cubic inches) may be considered less than 1.36 kgs (3 lbs.) of dielectric fluid and a capacitors whose total volume is more than 3,278 cubic centimeters (200 cubic inches) must be considered to contain more than 1.36 kg (3 lbs.) of dielectric fluid. Capacitors whose volume is between 1,639 and 3,278 cubic centimeters may be considered to contain less then 1.36 kg (3 lbs.) of dielectric fluid if the total weight of the capacitor is less than 4.08 kg (9 lbs.).

SMALL MANUFACTURERS
(CFR; ESH)
DA   October 12, 1990
RT Chemical Substances
RT Companies
RT Owners
SO Hazardous Materials
DEF (ESH) Small manufacturers should read the introductory paragraph of 704.5 and paragraph (d) of 704.5 for complete information on TSCA 8(a) small manufacturer exemption. First standard. A manufacturer of a chemical substance if its total annual sales, when combined with those of its parent company (if any) are less than $40 million. However, if annual production volume of a particular chemical substance at any individual site owned or controlled by the manufacturer is greater than 45,400 kilograms (100,000 lbs.), the manufacturer shall not qualify as small for purposes of reporting on the production of that chemical at that site, unless the manufacturer qualifies as small under "small manufacturer" paragraph (2) standard that a manufacturer is small if its total annual sales, when combined with those of its parent company are less than $4 million, regardless of the quantity of chemicals produced by that manufacturer.

SMALL QUANTITIES FOR RESEARCH AND DEVELOPMENT
(CFR)
DA   October 12, 1990
BT1 Quantities
RT Research and Development
SO Hazardous Materials
DEF (CFR) Quantities of PCBs (1) that are originally packaged in one or more hermetically sealed containers of a volume of no more than five (5.0) milliliters, and (2) that are used only for purposes of scientific experimentation or analysis, or chemical research on, or analysis of, PCBs, but not for research or analysis for the development of a PCB product.

SMALL QUANTITY GENERATORS
(SWDA; RCRA; EMER)
DA   October 19, 1990
BT1 Generators (Pollution)
 BT2 Potentially Responsible Parties
SO Emergency Preparedness
SO Environmental Management
SO Wastes
DEF (RCRA) Generators who generate less than 1000 kg of hazardous waste in a calendar month.

SMC
DA   October 12, 1990
SEE Selection and Monitoring Chassis
SO Acronyms

SMCL
DA   October 12, 1990
SEE Secondary Maximum Contaminant Levels
SO Acronyms

SMEAR
(NIH)
DA   October 12, 1990

BT1 Processes
BT1 Radiological Monitoring
 BT2 Monitoring
  BT3 Activities
RT Radioactive Contamination
SO Radiation
DEF (NIH) A procedure in which a swab,
    e.g., a circle of filter paper, is
    rubbed on a surface and its
    radioactivity measured to
    determine if the surface is
    contaminated with loose
    radioactive material.

SMELTERS
(EPA)
DA October 12, 1990
BT1 Facilities
NT1 Primary Copper Smelter
RT Skimming
SO Environmental Protection Agency
    Glossary
DEF (EPA) Facilities that melt or fuse
    ore, often with an accompanying
    chemical change, to separate the
    metal. Emissions are known to
    cause pollution. Smelting is the
    process involved.

SMOA
DA May 24, 1991
SEE Superfund Memorandum of
    Agreement
SO Acronyms

SMOG
(EPA)
DA October 12, 1990
SY Photochemical Smog
BT1 Air Pollutants
BT1 Air Pollution
 BT2 Pollution
RT Air Mass
SO Environmental Protection Agency
    Glossary
DEF (EPA) Air pollution associated with
    oxidants. (See: Photochemical
    Smog.)

SMOKE
(EPA)
DA October 12, 1990
BT1 Air Pollutants
RT Ashes
RT Dispersion Techniques
RT Hazards
RT Plumes
RT Soot
RT Visibility Impairments
SO Environmental Protection Agency
    Glossary
DEF (EPA) Particles suspended in air
    after incomplete combustion of
    materials.

SMOKE DAMAGE
(1537 SMOKE DAMAGE)
DA November 29, 1990
BT1 Damage

 BT2 Nature of Property Damage
RT Fire Damage
RT Water Damage
SO DOE FRASE VOCABULARY

SMOKE GRENADES
(Doe Order 5480.16)
DA October 12, 1990
BT1 Chemical Ammunition
 BT2 Ammunition
  BT3 Munitions
BT1 Diversionary Devices
BT1 Grenades
 BT2 Munitions
RT Grenade Launchers
SO Firearms
DEF (DOE Order 5480.16) Pyrotechnic
    devices capable of generating
    large amounts of smoke. While
    smoke grenades do not emit
    projectiles, they may emit
    fragments on actuation and may
    generate sufficient heat to cause
    fires and to injure personnel.

SMRSF
DA October 12, 1990
SEE Special Moment Resisting Space
    Frame
SO Acronyms

SNAKE BITE
(1338 SNAKE BITE)
DA November 28, 1990
BT1 Animal Bite
 BT2 Injuries
SO DOE FRASE VOCABULARY

SNIPS
(3060 SNIPS)
DA January 3, 1991
BT1 Tools - Manual
 BT2 Tools (DOE FRASE Vocabulary)
  BT3 Equipment
RT Cutter(s)
RT Scissors
SO DOE FRASE VOCABULARY

SNL
DA October 12, 1990
SEE Sandia National Laboratories
SO Acronyms

SNLA
DA October 12, 1990
SEE Sandia National
    Laboratories-Alburquerque
SO Acronyms

SNLL
DA October 12, 1990
SEE Sandia National
    Laboratories-Livermore
SO Acronyms

SNM
DA October 12, 1990

SEE Special Nuclear Materials
SO Acronyms

SNM VAULT
(DOE Order 6430.1A)
DA October 12, 1990
BT1 Security Areas
 BT2 Sites/Areas
BT1 Vault
 BT2 Site (DOE FRASE Vocabulary)
  BT3 Sites/Areas
RT Special Nuclear Materials
SO Construction
DEF (DOE Order 6430.1A) A
    penetration-resistant, windowless
    enclosure that has (a) walls, floor,
    and ceiling substantially
    constructed of materials that afford
    penetration resistance at least
    equal to that of 8-inch thick
    reinforced concrete; (b) any
    openings greater than 96 square
    inches in area and over 6 inches
    in the smallest dimension
    protected by embedded steel bars
    at least 5/8 inches in diameter on
    6-inch centers both horizontally
    and vertically; (c) a built-in
    combination locked steel door that
    in existing structures is at least
    1-inch thick exclusive of bolt work
    and locking devices and that for
    new structures at least meets the
    Class 5 standards as set forth in
    FS AA-D-6008 of the Federal
    Specifications and Standards cited
    in 41 CFR 101.

SNUBBER(S)
(3061 SNUBBER)
DA January 3, 1991
BT1 Tools - Manual
 BT2 Tools (DOE FRASE Vocabulary)
  BT3 Equipment
SO DOE FRASE VOCABULARY

SOCs
DA October 12, 1990
SEE Synthetic Organic Chemicals
SO Acronyms

**SODIUM COMPONENTS TEST
INSTALLATION**
DA January 8, 1991
SF SCTI
BT1 Facilities

SODIUM SCRUBBER(S)
(2593 SODIUM SCRUB)
DA January 3, 1991
BT1 Equipment/Parts - Nuclear (DOE
    FRASE Vocabulary)
 BT2 Equipment
 BT2 Reactor Components
SO DOE FRASE VOCABULARY

SOFT CONVERSION
(DOE Order 5900.2)
DA October 12, 1990

BT1 Metrication
SO Quality Assurance
DEF (DOE Order 5900.2) The process of changing measurement language from inch-pound measurement units to equivalent metric units within acceptable measurement tolerances without changing the actual physical size or configuration of the part, product, or process.

SOFT DETERGENTS
(EPA)
DA October 12, 1990
BT1 Chemical Agents
BT1 Detergents
SO Environmental Protection Agency Glossary
DEF (EPA) Cleaning agents that break down in nature.

SOFT STORIES
(SEA)
DA October 12, 1990
BT1 Stories
SO Construction
DEF (SEA) Stories in which the lateral stiffness is less than 70 percent of the stiffness of the story above.

SOFT WATER
(EPA)
DA October 12, 1990
BT1 Water
SO Environmental Protection Agency Glossary
DEF (EPA) Any water that is not "hard," i.e., does not contain a significant amount of dissolved minerals such as salts containing calcium or magnesium.

SOFTWARE
(SSDC)
DA October 12, 1990
RT Computer Codes
RT Information Systems
SO System Safety Development Center Glossary
DEF (SSDC) Any of the written programs, flow charts, etc., that are used to support the operation of computer equipment.

SOIL ADSORPTION FIELD
(EPA)
DA October 12, 1990
BT1 Sites/Areas
SO Environmental Protection Agency Glossary
DEF (EPA) A sub-surface area containing a trench or bed with clean stones and a system of distribution piping through which treated sewage may seep into the surrounding soil for further treatment and disposal.

SOIL COLUMN
(DOE Order 5400.5)
DA October 12, 1990
BT1 Soils
RT Leaching
SO Wastes
DEF (DOE Order 5400.5) An in situ volume of soil down through which liquid wastes percolate from ponds, cribs, seepage basins, or trenches.

SOIL CONDITIONER
(EPA)
DA October 12, 1990
BT1 Organic Matter
NT1 Compost
RT Humus
SO Environmental Protection Agency Glossary
DEF (EPA) An organic material like humus or compost that helps soil absorb water, build a bacterial community, and distribute nutrients and minerals.

SOIL GAS
(EPA)
DA October 12, 1990
BT1 Gases
SO Environmental Protection Agency Glossary
DEF (EPA) Gaseous elements and compounds that occur in the small spaces between particles of the earth and soil. Such gases can move through or leave the soil or rock, depending on changes in pressure.

SOIL INJECTION
(ESH)
DA November 20, 1990
RT Pesticides
DEF (ESH) The emplacement of pesticides by ordinary tillage practices within the plow layer of a soil.

SOIL MECHANICS
(DOE Order 6430.1A)
DA October 12, 1990
RT Flow Failures
RT Soils
RT Subsidence
SO Construction
DEF (DOE Order 6430.1A) The application of the laws of solid and fluid mechanics to soils and similar granular materials as a basis for design, construction, and maintenance of stable foundations and earth structures.

SOIL RESISTIVITY
(DOE Order 6430.1A)
DA October 12, 1990
RT Soils

SO Construction
DEF (DOE Order 6430.1A) The measured potential difference between two points in a naturally occurring soil between which a known electric current is passed.

SOIL-STRUCTURE RESONANCE
(SEA)
DA October 12, 1990
SO Construction
DEF (SEA) The coincidence of the natural period of a structure with a dominant frequency in the ground motion.

SOILS
(TSCA; CFR)
DA October 12, 1990
NT1 Bentonite Clays
NT1 Muck Soils
NT1 Quick Clays
NT1 Semi-annual soils
NT1 Silt
NT1 Soil Column
RT Background Soil Ph
RT Cation Exchange Capacity
RT Ecosystems
RT Flow Failures
RT Groundwater
RT Humus
RT Permafrost
RT Soil Resistivity
RT Soil Mechanics
RT Spill Area
SO Hazardous Materials
DEF (CFR) All vegetation, soils and other ground media, including but not limited to, sand, grass, gravel, and oyster shells. It does not include concrete and asphalt.

SOLAR COLLECTOR(S)
(2420 SOLAR COLLEC)
DA January 3, 1991
BT1 Equipment/Parts - Electrical (DOE FRASE Vocabulary)
BT2 Equipment
SO DOE FRASE VOCABULARY

SOLDER
(EPA)
DA October 12, 1990
SO Environmental Protection Agency Glossary
DEF (EPA) A metallic compound used to seal the joints between pipes. Until recently, most solder contained 50 percent lead.

SOLDERING IRON(S)
(2979 SOLDERING IR)
DA January 3, 1991
BT1 Tools - Powered
BT2 Tools (DOE FRASE Vocabulary)
BT3 Equipment
SO DOE FRASE VOCABULARY

SOLE(S)
(1130 SOLE)
DA November 28, 1990
BT1 Foot/Feet
 BT2 Human Body Parts
RT Foot Protection
SO DOE FRASE VOCABULARY

SOLE SOURCE AQUIFER
(SDWA; CFR; EPA)
DA October 12, 1990
BT1 Aquifers
 BT2 Formations
RT Underground Sources of Drinking
     Water
SO Environmental Protection Agency
     Glossary
DEF An aquifer that supplies 50 percent
     or more of the drinking water of an
     area.

SOLENOID(S)
(2421 SOLENOID)
DA January 3, 1991
BT1 Equipment/Parts - Electrical (DOE
     FRASE Vocabulary)
 BT2 Equipment
SO DOE FRASE VOCABULARY

SOLID STATE PROTECTION SYSTEM
DA January 8, 1991
SF SSPS
BT1 Systems

SOLID WASTE ASSESSMENT TEST
DA January 8, 1991
SF SWAT
RT Assessments

SOLID-WASTE-DERIVED FUEL
(CFR)
DA January 24, 1991
BT1 Fuels
RT Solid Wastes
SO Environmental Management
DEF (CFR) A fuel that is produced from
     solid waste that can be used as a
     primary or supplementary fuel in
     conjunction with or in place of
     fossil fuels. The
     solid-waste-derived fuel can be in
     the form of raw (unprocessed)
     solid waste, shredded (or pulped)
     and classified solid waste, gas or
     oil derived from pyrolyzed solid
     waste, or gas derived from the
     biodegradation of solid waste.

SOLID WASTE DISPOSAL
(EPA)
DA October 12, 1990
BT1 Disposal
 BT2 Waste Management Processes
  BT3 Processes
NT1 Phased Disposal
RT Cells
RT Landfill Cells
RT Open Dumps
RT Sanitary Landfills

RT Solid Wastes
RT Solid Waste Management
SO Environmental Protection Agency
     Glossary
DEF (EPA) The final placement of refuse
     that is not salvaged or recycled.

SOLID WASTE DISPOSAL ACT
DA January 8, 1991
SF SWDA
BT1 Acts
 BT2 Statutes and Regulations
BT1 Appropriate Act and Regulations
DEF The original Act of 1965 provided
     for a national research and
     development program into
     improved methods of disposal,
     and for a program of technical and
     financial assistance to state and
     local governments. Under the Act,
     solid waste experts were
     assembled into a new office and
     grants were made for research,
     technology demonstration,
     development of area-wide
     management systems,
     development of state and
     interstate plans, and personnel
     training.

SOLID WASTE MANAGEMENT
(SWDA; RCRA; USC; EPA)
DA October 12, 1990
BT1 Waste Management
 BT2 Processes
NT1 Compaction
NT1 Refuse Reclamation
RT Comminution
RT Interstate Agencies
RT Intermunicipal Agencies
RT Long Term Contracts
RT Solid Wastes
RT Solid Waste Management Facilities
RT Solid Waste Planning
RT Solid Waste Disposal
SO Environmental Protection Agency
     Glossary
SO Wastes
DEF Supervised handling of waste
     materials from their source through
     recovery processes to disposal.

SOLID WASTE MANAGEMENT FACILITIES
(SWDA; RCRA; CFR; USC)
DA October 12, 1990
BT1 Facilities
RT Resource Recovery Systems
RT Resource Recovery Facilities
RT Solid Waste Management
SO Environmental Management
SO Wastes
DEF (USC) (A) Any resource recovery
     systems or components thereof;
     (B) any systems, programs, or
     facilities for resource conservation;
     and (C) any facilities for the
     collection, source separation,
     storage, transportation, transfer,
     processing, treatment or disposal

of solid wastes, including
hazardous wastes, whether such
facilities are associated with
facilities generating such wastes
or otherwise.

SOLID WASTE MANAGEMENT UNITS
DA January 8, 1991
SF SWMU
BT1 Corrective Action Management
     Units
 BT2 Sites/Areas

SOLID WASTE PLANNING
(SWDA; RCRA; USC)
DA October 12, 1990
BT1 Resource Recovery
 BT2 Pollution Recovery Processes
  BT3 Processes
RT Resource Conservation
RT Solid Wastes
RT Solid Waste Management
SO Wastes
DEF (USC) Includes planning or
     management respecting resource
     recovery and resource
     conservation.

SOLID WASTE STORAGE AREA
DA January 8, 1991
SF SWSA
BT1 Sites/Areas

SOLID WASTES
(SWDA; SDWA; EPA; ESH; EMER)
DA November 13, 1990
SY Refuse
BT1 Wastes
NT1 Agricultural Solid Wastes
NT1 Bulky Wastes
NT1 Commercial Solid Wastes
NT1 Institutional Solid Wastes
NT1 Medical Wastes
NT1 Municipal Solid Wastes
 NT2 Sewage
  NT3 Raw Sewage
NT1 Residential Solid Wastes
NT1 Rubbish
NT1 Special Wastes
NT1 Tailings
 NT2 Asbestos Tailings
 NT2 New Tailings
RT Bottom Ash
RT Discarded Materials
RT Effluents
RT Piles (Wastes)
RT Recovered Materials
RT Recovered Resources
RT Scrap
RT Solid Waste Management
RT Solid Waste Planning
RT Solid-Waste-Derived Fuel
RT Solid Waste Disposal
RT Transfer Stations
SO Emergency Preparedness
SO Environmental Management
SO Environmental Protection Agency
     Glossary

SO Wastes
DEF (EPA) Nonliquid, nonsoluble
materials ranging from municipal
garbage to industrial wastes that
contain complex, and sometimes
hazardous, substances. Solid
wastes also include sewage
sludge, agricultural refuse,
demolition wastes, and mining
residues. Technically, solid waste
also refers to liquids and gases in
containers. (ESH) Garbage,
refuse, sludges, and other
discarded solid materials resulting
from industrial and commercial
operations and from community
activities. It does not include solids
or dissolved material in domestic
sewage or other significant
pollutants in water resources, such
as silt, dissolved or suspended
solids in industrial wastewater
effluents, dissolved materials in
irrigation return flows or other
common water pollutants.

SOLIDIFICATION AND STABILIZATION
(EPA)
DA October 12, 1990
BT1 Wastewater Treatment Processes
 BT2 Treatment
  BT3 Waste Management Processes
  BT4 Processes
SO Environmental Protection Agency
Glossary
DEF (EPA) Removal of wastewater from
a waste or changing it chemically
to make the waste less permeable
and susceptible to transport by
water.

SOLIDS
(SWDA; RCRA)
DA October 12, 1990
BT1 Materials
NT1 Dissolved Solids
NT1 Flammable Solids
NT1 Precipitates
NT1 Settleable Solids
NT1 Total Dissolved Solids
NT1 Total Suspended Solids
RT Bar Screens
DEF (SWDA; RCRA) Materials that have
a vertical flow of two inches (50
mm) or less within a three-minute
period, or a separation of one
gram (1 g) or less of liquid when
determined in accordance with the
procedures specified in ASTM D
4359-84 Standard Test Method for
Determining Whether a Material is
a Liquid or Solid, 1984 edition.

SOLUTIONS
(SWDA; RCRA; CFR)
DA October 12, 1990
BT1 Mixtures
NT1 Copper Mattes
RT Leachates

RT Vehicles
DEF (CFR) Any homogeneous liquid
mixture of two or more chemical
compounds or elements that will
undergo segregation under
conditions normal to
transportation.

SOLVENTS
(EPA)
DA October 12, 1990
NT1 Chlorinated Solvents
NT1 Exempt Solvents
NT1 Oxygenated Solvents
RT Double Wash/Rinse
RT Hazards
RT Inert Ingredients
SO Environmental Management
SO Environmental Protection Agency
Glossary
DEF (EPA) Substances (usually liquid)
capable of dissolving or dispersing
one or more other substances.

SOMATIC RADIATION EFFECTS
(IAEA)
DA October 12, 1990
BT1 Effects
RT Irradiation
SO Radiation
DEF (IAEA) Radiation effects occurring
in the exposed individual. (NRC
Glossary of Terms: Nuclear Power
and Radiation) Effects of radiation
limited to the exposed individual,
as distinguished from genetic
effects, which may also affect
subsequent unexposed
generations.

SOOT
(EPA)
DA October 12, 1990
BT1 Dust
 BT2 Air Pollutants
RT Air Pollution
RT Coal
RT Fly Ash
RT Smoke
SO Environmental Protection Agency
Glossary
DEF (EPA) Carbon dust formed by
incomplete combustion.

SOP (Management)
DA October 12, 1990
SEE Standard Operating Procedures
SO Acronyms
SO Management

SOP (Standard Operating Power)
DA October 12, 1990
SEE Standard Operating Power
SO Acronyms

SORPTION
(EPA)
DA October 12, 1990
BT1 Chemical Processes

 BT2 Processes
RT Absorption (Waste)
RT Absorption (Chemical)
RT Activated Carbon
RT Carbon Adsorbers
RT Pollution Recovery Processes
SO Environmental Protection Agency
Glossary
DEF (EPA) The action of soaking up or
attracting substances. A process
used in many pollution control
systems.

SORV
DA October 12, 1990
SEE Stuck-Open Relief Valve
SO Acronyms

SOURCE CONTROL MAINTENANCE MEASURES
(ANL; CFR)
DA May 24, 1991
BT1 Maintenance
 BT2 Activities
RT Source Control Remedial Actions
SO Management
SO Wastes
DEF (CFR) Those measures intended to
maintain the effectiveness of
source control actions once such
actions are operating and
functioning properly, such as the
maintenance of landfill caps and
leachate collection systems.

SOURCE CONTROL REMEDIAL ACTIONS
(CERCLA; CFR)
DA October 12, 1990
BT1 Remedial Actions
 BT2 Actions
  BT3 Responses
RT Engineering Control Zones
RT Source Control Maintenance
Measures
SO Compensation and Liability
DEF Measures that are intended to
contain the hazardous substances
or pollutants or contaminants
where they are located or
eliminate potential contamination
by transporting the hazardous
substances or pollutants or
contaminants to a new location.
Source control remedial actions
may be appropriate if a substantial
concentration or amount of
hazardous substances or
pollutants or contaminants remains
at or near the area where they are
originally located and inadequate
barriers exist to retard migration of
hazardous substances or
pollutants or contaminants into the
environment. Source control
remedial actions may not be
appropriate if most hazardous
substances or pollutants or
contaminants have migrated from
the area where originally located

or if the lead agency determines that the hazardous substances or pollutants or contaminants are adequately contained.

SOURCE MATERIALS
(NCRP; EMER)
DA   October 12, 1990
BT1 Licensed Materials
 BT2 Materials
RT   Thorium
RT   Uranium
SO   Emergency Preparedness
SO   Environmental Management
SO   Radiation
DEF (NCRP) Materials that contain 0.05 percent or more of natural uranium, thorium, or any combination of the two.

SOUTHEASTERN POWER ADMINISTRATION
DA   May 15, 1991
BT1 Power Marketing Administrations
SO   Management
DEF (U.S. Government Manual) The Administration was created by the Secretary of the Interior in 1950 to carry out functions assigned to the Secretary by the Flood Control Act of 1944, which pertain to the transmission and disposition of surplus electric power and energy generated at reservoir projects that are or may be under the control of the Department of the Army in the states of West Virginia, Virginia, North Carolina, South Carolina, Georgia, Florida, Alabama, Mississippi, Tennessee, and Kentucky.

SOUTHEASTERN UNIVERSITIES RESEARCH ASSOCIATION, INC.
DA   January 11, 1991
BT1 DOE Contractors
 BT2 Potentially Responsible Parties
BT1 Research and Development Organizations
 BT2 Organizations
RT   Oak Ridge Operations Office
RT   Universities

SOUTHWESTERN POWER ADMINISTRATION
DA   May 15, 1991
BT1 Power Marketing Administrations
SO   Management
DEF (U.S. Government Manual) The Administration carries out the functions assigned to the Secretary by the Flood Control Act of 1944 in the states of Arkansas, Kansas, Louisiana, Missouri, Oklahoma, and Texas. It transmits and disposes of the electric power and energy generated at Federal reservoir projects, supplemented by power purchased from public

and private utilities, in such a manner as to encourage the most widespread and economical use.

SO₂
SO_2
DA   October 12, 1990
SEE Sulfur Dioxide
SO   Acronyms

SO₂F₂
SO_2F_2
DA   October 12, 1990
SEE Sulfuryl Fluoride
SO   Acronyms

SP (Suppression Pool)
DA   October 12, 1990
SEE Suppression Pool
SO   Acronyms

SPACE FRAMES
(SEA)
DA   October 12, 1990
BT1 Structures (DOE FRASE Vocabulary)
NT1 Building Frame Systems
 NT2 Bearing Wall Systems
 NT2 Braced Frame
  NT3 Concentric Braced Frames
  NT3 Eccentric Braced Frames
 NT2 Horizontal Bracing Systems
 NT2 Lateral Force Resisting System
 NT2 Shoring
NT1 Moment Resisting Space Frame
 NT2 Intermediate Moment Resisting Space Frame
 NT2 Ordinary Moment Resisting Space Frame
 NT2 Special Moment Resisting Space Frame
NT1 Vertical Load Carrying Space Frames
SO   Construction
DEF (SEA) Three-dimensional structural systems without bearing walls composed of members interconnected so as to function as complete self-contained units with or without the aid of horizontal diaphragms or floor bracing systems.

SPARGER ASSEMBLIES
(NFI)
DA   October 12, 1990
BT1 Safe Shutdown System
 BT2 Emergency Systems
 BT3 Systems
SO   Nuclear Facilities Incident Database
SO   Radiation
DEF (NFI) Direct flow of neutron poison (ink) and direct stream of heavy water into bulk moderator to mix moderator and reduce temperature gradients, part of Safe Shutdown System (SSS); jets.

SPATULA(S)
(3062 SPATULA)
DA   January 3, 1991

BT1 Tools - Manual
 BT2 Tools (DOE FRASE Vocabulary)
 BT3 Equipment
SO   DOE FRASE VOCABULARY

SPC
DA   October 12, 1990
SEE Suppression Pool Cooling
SO   Acronyms

SPCC
DA   October 12, 1990
SEE Spill Prevention Control and Countermeasures Plan
SO   Acronyms

SPECIAL ANALYTICAL SERVICES
(EPA)
DA   October 12, 1990
BT1 Analyses
SO   Environmental Protection Agency Glossary
DEF (EPA) Non-standardized analyses conducted under the Contract Laboratory Program (CLP) to meet user requirements that cannot be met using RAS, such as shorter analytical turnaround time, lower detection limits, and analysis of non-standard matrices or non-TCL compounds.

SPECIAL AQUATIC SITES
(CWA; RHA; CFR)
DA   October 12, 1990
BT1 Sites/Areas
RT   Aquatic Environment
SO   Water Pollution
DEF Geographic areas, large or small, possessing special ecological characteristics of productivity, habitat, wildlife protection, or other important and easily disrupted ecological values. These areas are generally recognized as significantly influencing or positively contributing to the general overall environmental health or vitality of the entire ecosystem of a region. (See 40 CFR 230.10(a)(3).)

SPECIAL FORM
(EMER)
DA   February 1, 1991
BT1 Materials
RT   Normal Form
SO   Emergency Preparedness
DEF (EMER) Materials which are not dispersible because of their form (e.g., strongly encapsulated solids) and, therefore, present little or no possibility of contamination although they might present some direct radiation hazard [49 Code of Federal Regulations Part 173.389 (g)]. (See normal form.)

BT-Broader Term

NT-Narrower Term

RT Related Term

SPECIAL FORM RADIOACTIVE MATERIALS
(DOE Order 5480.3; CFR)
DA   October 12, 1990
BT1 Radioactive Materials
 BT2 Materials
SO   Environmental Management
SO   Hazardous Materials
DEF (CFR) To qualify as special form the radioactive material must either be in massive solid form or encapsulated. Special tests that are required of special form material are explained in 49 CFR 173.403. (ESH) Radioactive materials that satisfy the following conditions: (1) It is either a single solid piece or is contained in a sealed capsule that can be opened only by destroying the capsule; (2) the piece or capsule has at least one dimension not less than 5 millimeters (0.197); and (3) it satisfies the test requirements of Sec. 173.469. Special form encapsulations designed in accordance with the requirements of Sec. 173.389(g) in effect on June 30, 1983, and constructed prior to July 1, 1985, may continue to be used. Special form encapsulations designed or constructed after July 1, 1985, must meet the requirements of this paragraph.

SPECIAL HEALTH SUPERVISION
(IAEA)
DA   October 12, 1990
BT1 Investigations
 BT2 Administrative Processes
 BT3 Processes
RT   Occupational Medicine
RT   Occupational Exposure
RT   Working Condition A
SO   Radiation
DEF (IAEA) Special medical investigations carried out for workers exposed under Working Condition A.

SPECIAL ISOTOPE SEPARATOR
DA   January 8, 1991
SF   SIS
BT1 Reactor Components

SPECIAL MOMENT RESISTING SPACE FRAME
(SEA)
DA   October 12, 1990
SF   SMRSF
BT1 Moment Resisting Space Frame
 BT2 Space Frames
 BT3 Structures (DOE FRASE
        Vocabulary)
RT   Dual Systems
SO   Construction
DEF (SEA) A moment resisting space frame specially detailed to provide ductile behavior.

SPECIAL NUCLEAR MATERIAL OF LOW STRATEGIC SIGNIFICANCE
(CFR)
DA   January 29, 1991
BT1 Special Nuclear Materials
 BT2 Licensed Materials
 BT3 Materials
 BT2 Security Interests
SO   Environmental Management
DEF (CFR) Less than a formula quantity of strategic special nuclear material but more than 1000 grams of uranium-235 (contained in uranium enriched to 20% or more in the U235 isotope) or more than 500 grams of uranium-233 or plutonium or in a combined quantity of more than 100 grams when computed by the equation, grams = (grams contained U235) + 2(grams U235 + grams plutonium); or (2) 10,000 grams or more of uranium-235 (contained in uranium enriched to 10% or more but less than 20% in the U235 isotope.

SPECIAL NUCLEAR MATERIAL SCRAP
(CFR)
DA   January 24, 1991
BT1 Scrap
BT1 Special Nuclear Materials
 BT2 Licensed Materials
 BT3 Materials
 BT2 Security Interests
SO   Environmental Management
DEF (CFR) The various forms of special nuclear material generated during chemical and mechanical processing, other than recycle material and normal process intermediates, which are unsuitable for use in their present form, but all or part of which will be used after further processing.

SPECIAL NUCLEAR MATERIALS
(DOE Order 6430.1A; EMER)
DA   October 12, 1990
SF   SNM
BT1 Licensed Materials
 BT2 Materials
BT1 Security Interests
NT1 Special Nuclear Material Scrap
NT1 Special Nuclear Material of Low
       Strategic Significance
NT1 Strategic Special Nuclear Materials
RT   Formula Quantities
RT   Producing (Special Nuclear
       Materials)
RT   Production Facilities
RT   SNM Vault
RT   Utilization Facilities
SO   Construction
SO   Emergency Preparedness
SO   Environmental Management
DEF (DOE Order 6430.1A) Plutonium, uranium-233, uranium enriched in uranium-233 or in uranium-235, or any material artificially enriched in

any of the foregoing (but does not include source material) and any other material that, pursuant to the provisions of Section 51 of the Atomic Energy Act of 1954. as amended, has been determined to be special nuclear material.

SPECIAL NUCLEAR MATERIALS EMERGENCIES
(EMER)
DA   February 1, 1991
BT1 Safeguards and Security
       Emergencies
 BT2 Emergencies
 BT3 Reportable Occurrences
 BT4 Occurrences
SO   Emergency Preparedness
DEF (EMER) Any situation involving stolen, lost, or unauthorized possession of source material, special nuclear material, by-product material of U.S. and/or foreign manufacture, improvised nuclear explosives, or radioactive dispersal devices or the threatened use of said items.

SPECIAL POPULATIONS
(EMER)
DA   February 1, 1991
BT1 Populations
SO   Emergency Preparedness
DEF (EMER) Public groups located within the plume exposure   . emergency planning zone who, in the event the general public is instructed to evacuate, may require special transportation or protective provisions due to institutional confinement, lack of transportation, disability, or the need to staff certain industrial plants or public utilities.

SPECIAL RESPONSE TEAMS
(EMER)
DA   February 1, 1991
SY   Tactical Response Teams
BT1 Response Teams
 BT2 Teams
 BT3 Administrative Organizations
 BT4 Organizations
SO   Emergency Preparedness
DEF (EMER) Groups of one or more armed security inspectors who are specially trained and equipped to respond to security incidents. Also called Tactical Response Teams.

SPECIAL REVIEWS
(EPA)
DA   October 12, 1990
BT1 Reviews
 BT2 Administrative Processes
 BT3 Processes
RT   Risk Assessment
SO   Environmental Management

SO Environmental Protection Agency Glossary

DEF (EPA) Formerly known as Rebuttable Presumption Against Registration (RPAR), this is the regulatory process through which existing pesticides suspected of posing unreasonable risks to human health, non-target organisms, or the environment are referred for review by EPA. The review requires an intensive risk/benefit analysis with opportunity for public comment. If the risk of any use of a pesticide is found to outweigh social and economic benefits, regulatory actions — ranging from label revisions and use-restriction to cancellation or suspended registration — can be initiated.

SPECIAL SOURCE OF GROUND WATER
(ANL; CFR)
DA May 24, 1991
RT Controlled Areas
RT Drinking Water Supplies
RT Groundwater
SO Wastes
SO Water Pollution
DEF (CFR) Those Class I ground waters identified in accordance with Agency's Ground-Water Protection Strategy published in August 1984 that: (1) Are within the controlled area encompassing a disposal system or are less than five kilometers beyond the controlled area; (2) are supplying drinking water for thousands of persons as of the data that the Department chooses a location within that area for detailed characterization as a potential site for a disposal system (e.g., in accordance with Section (112(b)(1)(B) of the NWPA); and (3) are irreplaceable in that no reasonable alternative source of drinking water is available to that population.

SPECIAL WASTES
(CFR)
DA January 24, 1991
BT1 Solid Wastes
 BT2 Wastes
SO Environmental Management
DEF (CFR) Nonhazardous solid wastes requiring handling other than that normally used for municipal solid waste.

SPECIAL WEAPONS
(DOE Order 5480.16)
DA October 12, 1990
RT Firearms
RT Munitions
RT Protective Force Personnel

SO Firearms
DEF (DOE Order 5480.16) Include all weapons that are not normally issued to protective force personnel and include certain firearms, land mines, booby traps, and demolition changes.

SPECIALLY DESIGNATED LANDFILLS
(ESH)
DA November 20, 1990
BT1 Landfills
 BT2 Land Disposal Units
  BT3 Disposal Units
   BT4 Corrective Action Management Units
    BT5 Sites/Areas
DEF (ESH) Landfills at which complete long-term protection is provided for the quality of surface and subsurface waters from pesticides, pesticide containers, and pesticide-related wastes deposited therein, and against hazard to public health and the environment. Such sites should be located and engineered to avoid direct hydraulic continuity with surface and subsurface waters, and any leachate or subsurface flow into the disposal area should be contained within the site unless treatment is provided. Monitoring wells should be established and a sampling and analysis program conducted. The location of the disposal site should be permanently recorded in the appropriate local office of legal jurisdiction. Such facility complies with the Agency Guidelines for the Land Disposal of Solid Wastes as prescribed in 40 CFR Part 241.

SPECIES
(SWDA; ESA; CFR; EPA)
DA October 12, 1990
NT1 Candidates
NT1 Endangered Species
 NT2 Listed Species
NT1 Resident Species
NT1 Threatened Species
SO Endangered Species
SO Environmental Protection Agency Glossary
DEF (CFR) Includes any subspecies of fish or wildlife or plants, and any distinct population segment of any species of vertebrate fish or wildlife which interbreeds when mature. Excluded is any species of the Class Insecta determined by the Secretary to constitute a pest whose protection under the provisions of the Endangered Species Act would present an overwhelming and overriding risk to man.

SPECIFIC ACTIVITY
(NIH)
DA October 12, 1990
BT1 Activity (Nuclear)
 BT2 Measurements
NT1 Low Specific Activity
RT Radioactivity
SO Environmental Management
SO Radiation
DEF (NIH) Total radioactivity of a given nuclide per gram of a compound, element or radioactive nuclide.

SPECIFIC CONTROL FACTORS
(MORT)
DA April 2, 1991
RT Accidents
RT Assumed Risks
RT Corrective Actions
RT Management System Factors
RT Root Causes
RT Root Cause Analysis
SO Management
DEF (MORT) MORT analysis asks: What were the specific control factors of the management system that were overlooked or omitted? A detailed understanding of the incident/accident sequence leads naturally to: (1) consideration of the Management System Factors and (2) judgement whether the fault (failure potential was an assumed risk.

SPECIFIC IONIZATION
DA November 15, 1990
BT1 Measurements
RT Ionization
SO Environmental Management
SO Radiation
DEF The number of ion pairs per unit length of path of ionizing radiation in a medium (e.g., per centimeter of air or per micron of tissue).

SPECIFICATIONS
(CFR)
DA January 24, 1991
BT1 Requirements
NT1 Contract Specifications
NT1 Guide Specifications
NT1 Operational Specifications
NT1 Project Design Criteria
NT1 Technical Specifications
RT General Design Criteria
RT Graded Approach
RT Quality Control
RT Standards
SO Environmental Management
DEF (CFR) A clear and accurate description of the technical requirement for materials, products or services, which specifies the minimum requirement for quality and construction of materials and equipment necessary for an acceptable product. In general, specifications are in the form of written descriptions, drawings,

prints, commercial designations, industry standards, and other descriptive references.

SPECIFIED PORTS AND HARBORS
(CERCLA: CFR)
DA October 12, 1990
BT1 Sites/Areas
RT Inland Waters
SO Compensation and Liability
DEF Those port and harbor areas on inland rivers, and land areas immediately adjacent to those waters, where the USCG acts as predesignated on-scene coordinator. Precise locations are determined by EPA/USCG regional agreements and identified in Federal regional contingency plans.

SPECIMENS
(ESA; TSCA; CFR)
DA October 12, 1990
BT1 Samples
SO Endangered Species
SO Environmental Management
SO Hazardous Materials
DEF Materials derived from a test system for examination or analysis. (50 CFR 17.3) Any animal or plant, or any part, product, egg, seed or root of any animal or plant.

SPECTROMETER(S)
(2770 SPECTROMETER)
DA January 3, 1991
BT1 Indicator(s)
 BT2 Instrument(s)
  BT3 Equipment/Parts -
    Instrumentation/Measuring
    (DOE FRASE Voc.)
   BT4 Equipment
SO DOE FRASE VOCABULARY

SPENT FUEL
(EMER)
DA February 1, 1991
BT1 Fuels
RT General Environment
RT Test and Evalution Facilities
SO Emergency Preparedness
DEF (EMER) Nuclear reactor fuel that has been used in a nuclear reactor and that contains large amounts of highly radioactive fission products.

SPENT MATERIALS
(CFR)
DA January 24, 1991
BT1 Materials
BT1 Scrap
SO Environmental Management
DEF (CFR) Any material that has been used and as a result of contamination can no longer serve the purpose for which it was produced without processing.

SPIDERS
(NFI)
DA October 12, 1990
BT1 Reactor Components
SO Nuclear Facilities Incident Database
SO Radiation
DEF (NFI) Parts of fuel assemblies to retain orifices and provide top seating surface between assemblies and housings; full-flow spiders used for high coolant flow; blanket spiders used in low power and coolant flow in outer regions of lattice.

SPIKED SAMPLES
(ESH)
DA October 12, 1990
BT1 Samples
SO Management
DEF (ESH) Normal samples of material (gas, liquid, or solid) to which a known amount of some substance of interest is added. Spiked samples are used to check on the performance of a routine analysis or the recovery efficiency of an analytical method.

SPILL AREA
(TSCA; CFR)
DA October 12, 1990
BT1 Sites/Areas
RT Soils
RT Spills
RT Spill Boundaries
SO Hazardous Materials
DEF The area of soil on which visible traces of the spill can be observed plus a buffer zone of 1 foot beyond the visible traces. Any surface or object (e.g., concrete sidewalk or automobile) within the visible traces area or on which visible traces of the spilled material are observed is included in the spill area. This area represents the minimum area assumed to be contaminated by PCBs in the absence of precleanup sampling data and is thus the minimum area that must be cleaned.

SPILL BOUNDARIES
(TSCA; CFR)
DA October 12, 1990
BT1 Sites/Areas
RT Spills
RT Spill Area
SO Hazardous Materials
DEF (CFR) The actual area of contamination as determined by postcleanup verification sampling or by precleanup sampling to determine actual spill boundaries. EPA can require additional cleanup when necessary to decontaminate all areas within the spill boundaries to the levels required in this policy (e.g.,

additional cleanup will be required if postcleanup sampling indicates that the area decontaminated by the responsible party, such as the spill area as defined in this section, did not encompass the actual boundaries of PCB contamination).

SPILL PREVENTION CONTROL AND COUNTERMEASURES PLAN
(EPA)
DA October 12, 1990
SF SPCC
BT1 Emergency Plans
 BT2 Plans
RT Procedures
SO Environmental Protection Agency Glossary
DEF (EPA) Plan covering the release of hazardous substances as defined in the Clean Water Act.

SPILLS
(TSCA; CFR)
DA October 12, 1990
SY Leaks
SY Unauthorized Discharges
RT Anticipated Operational Occurrences
RT Discharges
RT Hazardous Materials Response Teams
RT Leaks
RT Management of Migrations
RT Pollution
RT Releases
RT Removal
RT Reportable Quantities
RT Spill Area
RT Spill Boundaries
RT Standard Wipe Test
RT Threshold Planning Quantities
SO Environmental Management
SO Hazardous Materials
DEF (CFR) Both intentional and unintentional spills, leaks, and other uncontrolled discharges where the release results in any quantity of PCBs running off or about to run off the external surface of the equipment or other PCB source, as well as the contamination resulting from those releases. This policy applies to spills of 50 ppm or greater PCBs. The concentration of PCBs spilled is determined by the PCB concentration in the material spilled as oppose to the concentration of PCBs in the material onto which the PCBs were spilled. Where a spill of untested mineral oil occurs, the oil is presumed to contain greater than 50 ppm but less than 500 ppm PCBs and is subject to the relevant requirements of this policy.

SPINAL CORD
(1113 SPINAL CORD)
DA November 28, 1990
BT1 Nervous System
 BT2 Body System(s)
  BT3 Human Body Parts
RT Back
RT Pinched Nerve
RT Ruptured Disk
RT Spine
RT Upper Back
SO DOE FRASE VOCABULARY

SPINE
(1112 SPINE)
DA November 28, 1990
BT1 Bone(s)
 BT2 Human Body Parts
NT1 Coccyx
RT Neck
RT Spinal Cord
SO DOE FRASE VOCABULARY

SPMS
DA October 12, 1990
SEE Safety Performance Measurement
    System
SO Acronyms

SPOIL
(EPA)
DA October 12, 1990
SO Environmental Protection Agency
    Glossary
DEF (EPA) Dirt or rock that has been
    removed from its original location,
    destroying the composition of the
    soil in the process, as with
    strip-mining or dredging.

**SPONTANEOUSLY COMBUSTIBLE
MATERIALS**
(SWDA; RCRA; ESH)
DA October 12, 1990
BT1 Materials
SO Hazardous Materials
DEF Solid substances (including sludges
    and pastes) which may undergo
    spontaneous heating or
    self-ignition under conditions
    normally incident to transportation
    or which may upon contact with
    the atmosphere undergo an
    increase in temperature and ignite.

SPP
DA October 12, 1990
SEE Standard Practice Procedures
SO Acronyms

SPR
DA October 12, 1990
SEE Strategic Petroleum Reserve
SO Acronyms

SPR III
DA October 12, 1990

SEE Sandia Pulse Reactor III
SO Acronyms

SPRAIN
(1339 SPRAIN)
DA November 28, 1990
BT1 Injuries
RT Strain
SO DOE FRASE VOCABULARY

SPRAWL
(EPA)
DA October 12, 1990
SO Environmental Protection Agency
    Glossary
DEF (EPA) Unplanned development of
    open land.

SPRAY GUN(S)
(2980 SPRAY GUN)
DA January 3, 1991
BT1 Tools - Powered
 BT2 Tools (DOE FRASE Vocabulary)
  BT3 Equipment
SO DOE FRASE VOCABULARY

SPRINKLER SYSTEM
(2051 SPRINKLER SY)
DA December 10, 1990
BT1 Dedicated Fire Water Systems
 BT2 Fire Protection Systems
  BT3 Emergency Systems
   BT4 Systems
BT1 Systems (DOE FRASE Vocabulary)
 BT2 Systems
RT Fire Suppression System
SO DOE FRASE VOCABULARY

SPS
DA October 12, 1990
SEE Safety Precedence Sequence
SO Acronyms

SQUARE KILOMETERS
DA January 8, 1991
SF *Km²*
BT1 Units of Measure

SQUIMP
(Acronyms and Abbreviations)
DA October 12, 1990
BT1 Computer Codes
DEF A computer code to analyze
    Probabilistic Risk Assessment
    (PRA) information; developed at
    the Idaho National Engineering
    Laboratory.

SR
DA October 12, 1990
SEE Savannah River Operations Office
SO Acronyms

SR/VO
DA October 12, 1990
SEE Safety/Relief Valve Open
SO Acronyms

SRC
DA October 12, 1990
SEE Safety Review Committee
SO Acronyms

SRO
DA October 12, 1990
SEE Senior Reactor Operator
SO Acronyms

SRS
DA October 12, 1990
SEE Savannah River Site
SO Acronyms

SRV
DA October 12, 1990
SEE Safety/Relief Valve
SO Acronyms

SRW
DA October 12, 1990
SEE Savannah River Water
SO Acronyms

SS
DA October 12, 1990
SEE Shift Supervisor
SO Acronyms

SSDC
DA January 8, 1991
SEE System Safety Development Center
SO Acronyms
DEF (SSDC) A DOE center located at
    the Idaho National Engineering
    Laboratory and operated by
    EG&G Idaho, Inc. Its functions
    include the development and
    application of new technologies,
    the implementation and analysis of
    safety information systems, the
    presentation of training seminars
    and workshops, and the
    publication and dissemination of
    SSDC documents.

SSFI
DA October 12, 1990
SEE Safety System Functional
    Inspection
SO Acronyms

SSOMI
DA October 12, 1990
SEE Safety System Outage Modification
    Inspection
SO Acronyms

SSPS
DA October 12, 1990
SEE Solid State Protection System
SO Acronyms

SSPSA
DA October 12, 1990

BT-Broader Term   NT-Narrower Term   RT Related Term

SEE Seabrook Station Probabilistic
    Safety Assessment
SO Acronyms

SSR
DA   October 12, 1990
SEE Secondary System Relief
SO Acronyms

SSS (Safe Shutdown System)
DA   October 12, 1990
SEE Safe Shutdown System
SO Acronyms

SSS (Supplementary Safety System)
DA   October 12, 1990
SEE Supplementary Safety System
SO Acronyms

SST
DA   October 12, 1990
SEE Single Shell Tanks
SO Acronyms

ST
DA   October 12, 1990
SEE Sleeve Target
SO Acronyms

STA
DA   October 12, 1990
SEE Shift Technical Advisor
SO Acronyms

STABILIZATION
(EPA)
DA   October 12, 1990
BT1 Treatment
 BT2 Waste Management Processes
  BT3 Processes
SO Environmental Protection Agency
    Glossary
DEF (EPA) Conversion of the active
    organic matter in sludge into inert,
    harmless material.

STABILIZATION PONDS
(EPA)
DA   November 13, 1990
SY   Lagoons
BT1 Sites/Areas
RT   Holding Ponds
DEF (EPA) A shallow pond where
    sunlight, bacterial action, and
    oxygen work to purify wastewater;
    also used for the storage of
    wastewaters or spent nuclear fuel
    rods.

STABLE AIR
(EPA)
DA   October 12, 1990
BT1 Air
SO Environmental Protection Agency
    Glossary
DEF (EPA) A mass of air that is not
    moving normally, so that it holds
    rather than disperses pollutants.

STACK EFFECT
(EPA)
DA   October 12, 1990
SY   Exhaust Stack
BT1 Effects
RT   Stacks
SO Environmental Protection Agency
    Glossary
DEF (EPA) Used air, as in a chimney,
    that moves upward because it is
    warmer than the surrounding
    atmosphere.

STACK GASES
(EPA)
DA   November 13, 1990
SY   Flue Gases
BT1 Gases
DEF (EPA) The air coming out of a
    chimney after combustion in the
    burner it is venting. It can include
    nitrogen oxides, carbon oxides,
    water vapor, sulfur oxides,
    particles, and many chemical
    pollutants.

STACK TRITIUM MONITOR
DA   January 8, 1991
SF   *STM*
BT1 Monitors
 BT2 Equipment

STACKS
(CAA; CFR; EPA)
DA   November 15, 1990
RT   Dispersion Techniques
RT   Exhaust Gases
RT   Flue Gases
RT   Particulate Matter Emissions
RT   Point Sources
RT   Stack Effect
SO Air Pollution
SO Environmental Protection Agency
    Glossary
DEF (CFR) Points in a source designed
    to emit solids, liquids, or gases
    into the air, including a pipe or
    duct but not including flares. (EPA)
    Chimneys or smokestacks; vertical
    pipes that discharge used air.

STAFF ORGANIZATION
(SSDC)
DA   October 12, 1990
BT1 Organizations
RT   Line Organizations
SO System Safety Development Center
    Glossary
DEF (SSDC) The portion of the total
    organization which assists the line
    organization in performance of
    required function necessary to
    accomplish the organization's
    primary mission. This includes
    both individuals (e.g., staff
    assistance, coordinators,
    specialists) and entire
    organizational elements (e.g.

safety, quality, reliability, legal
groups).

STAGING BAYS (IN-PROCESS)
(DOE Order 6430.1A)
DA   October 12, 1990
BT1 Explosives Bays
 BT2 Explosives Buildings
  BT3 Structures (DOE FRASE
      Vocabulary)
RT   In-Process Materials
SO Construction
DEF (DOE Order 6430.1A) Bays within
    an operating building used to
    stage explosives in excess of four
    hours supply. This practice is
    permissible as long as the bays
    are designed to provide Class II
    level of protection.

STAGNATION
(EPA)
DA   October 12, 1990
SO Environmental Protection Agency
    Glossary
DEF (EPA) Lack of motion in a mass of
    air or water, which tends to hold
    pollutants.

STALL/BOOTH(S)
(1716 STALL;N)
DA   December 10, 1990
BT1 Site (DOE FRASE Vocabulary)
 BT2 Sites/Areas
SO DOE FRASE VOCABULARY

STANDARD CUBIC FOOT
(SWDA; RCRA; ESH)
DA   October 12, 1990
SF   *SCF*
BT1 Units of Measure
SO Hazardous Materials

STANDARD CURVE
(ESH)
DA   October 12, 1990
RT   Measurements
RT   Units of Measure
SO Management
SO Quality Assurance
DEF (ESH) A curve that plots
    concentrations of known analyte
    standard versus the instrument
    response to the analyte.

STANDARD DEVIATION
(SSDC)
DA   October 12, 1990
RT   Mean
RT   Measurements
RT   Units of Measure
SO Management
SO System Safety Development Center
    Glossary
DEF (SSDC) The square root of the
    average of the squares of the
    deviations from the mean. In a
    normal distribution, 68, 95, and
    99.7% of all values occur between

> 1, 2, or 3 standard deviations from the means, respectively.

STANDARD JOB PROCEDURES
(SSDC)
DA October 12, 1990
SF *SJP*
BT1 Procedures
RT Critical Jobs (task)
RT Proper Job Analysis
RT Work Processes
SO System Safety Development Center Glossary
DEF (SSDC) Tools for teaching the most systematic way to do a critical job consistently with maximum efficiency. Developed by Proper Job Analysis.

STANDARD OF PERFORMANCE
(CAA; CFR; USC)
DA October 12, 1990
BT1 Requirements
RT Excess Emissions
RT Stationary Sources
SO Air Pollution
SO Environmental Management
SO Hazardous Materials
DEF (USC) A requirement of continuous emission reduction, including any requirement relating to the operation or maintenance of a source to assure continuous emission reduction. (ESH) Any restriction established by the Administrator pursuant to §306 of the Act on quantities, rates, and concentrations of chemical, physical, and biological and other constituents which are or may be discharged from new sources into navigable waters, the waters of the contiguous zone, or the ocean.

STANDARD OPERATING POWER
DA January 8, 1991
SF *SOP (Standard Operating Power)*

STANDARD OPERATING PROCEDURES
(CAA; CFR; EMER)
DA October 12, 1990
SF *SOP (Management)*
BT1 Operating Procedures
 BT2 Procedures
SO Air Pollution
SO Emergency Preparedness
SO Management
DEF (CFR) Formal written procedures officially adopted by the plant owner or operator and available on a routine basis to those persons responsible for carrying out the procedures.

STANDARD PRACTICE PROCEDURES
DA January 8, 1991
SF *SPP*

BT1 Procedures
SO Management

STANDARD PRESSURE
(CAA; CFR)
DA October 12, 1990
BT1 Pressure (Surface)
 BT2 Pressure
SO Air Pollution
DEF (CFR) A temperature of 760 mm of Hg (29.92 in. of Hg).

STANDARD SAMPLE
(SWDA; CFR)
DA October 12, 1990
BT1 Samples
RT Aliquots
RT Coliform Organisms
RT Monitoring
SO Water Pollution
DEF (CFR) The aliquot of finished drinking water that is examined for the presence of coliform bacteria.

STANDARD TEMPERATURE
(CAA; CFR)
DA October 12, 1990
BT1 Measurements
SO Air Pollution
DEF (CFR) A temperature of 20 deg. C (69 deg. F).

STANDARD WIPE TEST
(TSCA; CFR)
DA October 12, 1990
RT Cleanup
RT High Concentration PCBs
RT Spills
SO Hazardous Materials
DEF (CFR) For spills of high-concentration PCBs on solid surfaces, a cleanup to numerical surface standards and sampling by a standard wipe test to verify that the numerical standards have been met. This definition constitutes the minimum requirements for an appropriate wipe testing protocol. A standard-size template (10 centimeters (cm) x 10 cm) will be used to delineate the area of cleanup; the wiping medium will be a gauze pad or glass wool of known size which has been saturated with hexane. It is important that the wipe be performed very quickly after the hexane is exposed to air. EPA strongly recommends that the gauze (or glass wool) be prepared with hexane in the laboratory and that the wiping medium be stored in sealed glass vials until it is used for the wipe test. Further, EPA requires the collection and testing of field blanks and replicates.

STANDARDS
(DOE Orders 5480.1A, 5480.4; CFR; EPA; SSDC)
DA October 12, 1990
BT1 Codes, Standards, and Regulations
NT1 Acceptable Daily Intake
NT1 Acceptable Intake for Chronic Exposure
NT1 Acceptable Intake for Subchronic Exposure
NT1 Air Quality Standards
NT2 Ambient Air Quality Standards
NT3 National Ambient Air Quality Standards
NT4 Primary Standards
NT4 Secondary Standards
NT1 Applicable Standards and Limitations
NT1 Applicable or Relevant and Appropriate Requirements
NT1 Categorical Pretreatment Standards
NT1 Commercial Standards
NT1 Emission Standards
NT2 National Emissions Standards for Hazardous Air Pollutants
NT2 New Source Performance Standards
NT1 Environmental Protection Standards
NT1 Fuel Economy Standard
NT1 Government Standards
NT1 Mandatory Standards
NT1 National Standards (Air Pollution)
NT1 Radiation Standards
NT2 Annual Limit on Intake
NT2 As Low As Reasonably Achievable
NT2 As Low As Practicable
NT1 Reference Standards
NT1 Technology-Based Standards
NT1 Voluntary Standards
NT1 Water Quality Standards
RT American National Standards Institute
RT Minimum Requirements and Standards
RT National Institute of Standards and Testing
RT Specifications
RT Standards Information Management System
SO Environmental Management
SO Environmental Protection Agency Glossary
SO Hazardous Materials
DEF (ASTM) (1) Physical references used as a basis for comparison or calibration; (2) Concepts that has been established by authority, custom, or agreement to serve as a model or rule in the measurement of quality or the establishment of a practice or procedure.

STANDARDS-DEVELOPING GROUPS
(DOE Order 1300.2; EPA)
DA October 12, 1990
BT1 Groups
 BT2 Administrative Organizations
 BT3 Organizations
RT Commercial Standards

BT-Broader Term   NT-Narrower Term   RT Related Term

RT Voluntary Standards Bodies
SO Standards
DEF (DOE Order 1300.2) Committees,
    subcommittees, boards or other
    principal subdivisions of voluntary
    standards bodies, established by
    such bodies for the purpose of
    developing, revising, or reviewing
    standards, and which are bound
    by the procedures of those bodies.

STANDARDS INFORMATION MANAGEMENT SYSTEM
(SSDC)
DA January 8, 1991
SF *SIMS*
BT1 Safety Performance Measurement
    System
 BT2 Information Systems
 BT3 Security Interests
 BT3 Systems
RT Standards
SO Environmental Management
SO Management
DEF (SSDC) This system provides
    access to two standards
    databases: (1) a set of general
    industrial standards developed by
    Argonne National Laboratory, and
    (2) a set of safety standards
    developed by Reynolds Electric
    and Engineering Company at the
    Nevada Test Site.

STANDBY
(DOE Orders 5480.6, 5480.18; EPA)
DA October 12, 1990
BT1 Conditions
SO Industrial Safety
SO Standards
DEF That condition in which a reactor
    facility is neither operable nor
    declared excess, and
    documentary authorization exists
    to maintain the reactor for possible
    future operation.

STANDBY DIESEL GENERATOR
DA January 8, 1991
SF *SCG*
BT1 Diesel Generator
 BT2 Equipment

STANDBY GAS TREATMENT SYSTEM
DA January 8, 1991
SF *SGTS*
BT1 Systems

STANDBY LIQUID CONTROL SYSTEM
DA January 8, 1991
SF *SLCS*
BT1 Control Systems
 BT2 Controls
 BT2 Systems

STANDBY POWER
(DOE Order 6430.1A)
DA October 12, 1990
RT Emergency Power

RT Uninterruptible Power Supplies
SO Construction
DEF (DOE Order 6430.1A) A reserve
    power generation or supply with
    switching devices that will supply
    power to selected loads in the
    event of a normal power failure. It
    is not required to have redundant
    equipment or to operate through
    events greater than UBC. A
    standby power system shall not be
    classified SC-1.

STANFORD LINEAR ACCELERATOR CENTER
DA January 8, 1991
SF *SLAC*
BT1 Centers
BT1 Government-Owned
    Contractor-Operated Facilities
 BT2 Federal Facilities
 BT3 Facilities
RT Stanford University
DEF (Capsule Review of DOE Research
    and Development and Field
    Facilities, 1986) Carries out
    experimental and theoretical
    research in high energy physics
    and developmental work in new
    techniques for particle acceleration
    and experimental instrumentation.
    The Center's main research
    instrument is a two mile long linear
    electron accelerator, the largest in
    the world.

STANFORD UNIVERSITY
DA January 11, 1991
BT1 DOE Contractors
 BT2 Potentially Responsible Parties
BT1 Universities
 BT2 Educational Organizations
 BT3 Organizations
RT San Francisco Operations Office
RT Stanford Linear Accelerator Center

START-UP PROCEDURE
(1244 SU PROCEDURE)
DA November 28, 1990
RT Pre Start-Up/ Calibration Activity
SO DOE FRASE VOCABULARY

STARTER
(2422 STARTER)
DA January 3, 1991
BT1 Equipment/Parts - Electrical (DOE
    FRASE Vocabulary)
 BT2 Equipment
SO DOE FRASE VOCABULARY

STARTER CARTRIDGES
(HMTA; CFR)
DA May 20, 1991
BT1 Class B Explosives
 BT2 Explosives
RT Aircraft
SO Hazardous Materials
DEF (CFR) Jet engine, Class B
    explosives consist of plastic

and/or rubber cases, each
containing a pressed cylindrical
block of propellant explosive and
having in the top of the case a
small compartment that encloses
an electrical squib, small amounts
of black powder, and smokeless
powder, which constitutes an
igniter. The starter cartridge is
used to activate a mechanical
starter for jet engines.

STARTUP
(CAA; CFR)
DA October 12, 1990
BT1 Processes
NT1 Initial Startup
RT Stationary Sources
SO Air Pollution
DEF (CFR) The setting in operation of a
    source for any purpose. (40 CFR
    60.02) The setting in operation of
    a stationary source for any
    purpose.

STATE AGENCIES
(CAA; ESA; CFR)
DA October 12, 1990
BT1 Agencies
 BT2 Administrative Organizations
 BT3 Organizations
NT1 Air Pollution Control Agencies
 NT2 Interstate Air Pollution Control
     Agencies
NT1 Colorado Department of Health
NT1 Division of the State Fire Marshall
NT1 Implementing Agencies
NT1 Interstate Agencies
 NT2 Interstate Air Pollution Control
     Agencies
NT1 Missouri Department of Natural
    Resources
NT1 New Mexico Environmental
    Improvement Division
NT1 Ohio Bureau of Underground
    Storage Tank Regulation
NT1 Ohio Department of Commerce
NT1 Ohio Department of Health
NT1 Ohio Environmental Protection
    Agency
NT1 Procuring Agencies
NT1 State Routing Agencies
NT1 State Emergency Response
    Commission
NT1 State Authority
NT1 Tennessee Department of
    Commerce
 NT2 Bureau of Environment
NT1 Utah Department of Health
RT Air Pollution Control Agencies
RT Approved Programs
RT Approved State Primacy Program
RT Authorized Officers
RT Federal Agencies
RT Lead Agencies
RT Natural Resource Trustees
RT States
SO Air Pollution
SO Endangered Species
DEF Any State agencies, departments,

SY-Synonymous Terms

SO-Source/Subject Category

SF-See From

boards, commissions, or other governmental entities that are responsible for the oversight and management of official programs designated for their respective states

STATE AND TRIBAL GOVERNMENT WORKING GROUP
DA   January 8, 1991
SF   *STGWG*
BT1 Groups
 BT2 Administrative Organizations
 BT3 Organizations

STATE AUTHORITY
(SWDA; RCRA; CFR)
DA   October 19, 1990
BT1 State Agencies
 BT2 Agencies
 BT3 Administrative Organizations
 BT4 Organizations
RT   States
SO   Wastes
DEF The agency established or designated under section 4007 [42 USCS 6947].

STATE COORDINATING OFFICERS
(EMER)
DA   February 1, 1991
BT1 Personnel
SO   Emergency Preparedness
DEF (EMER) Officials designated by the governor of the affected state to work with the cognizant federal agency official and senior FEMA official in coordinating the response effort of federal, state, local, volunteer, and private agencies.

STATE COORDINATION
(DOE Order 5440.1D)
DA   May 15, 1991
BT1 Administrative Processes
 BT2 Processes
RT   Environmental Assessments
SO   Environmental Management
DEF (DOE Order) The process by which a host State and, as appropriate, adjacent states are provided the opportunity to review and comment on an Environmental Assessment before DOE approval.

STATE-DESIGNATED ROUTES
(SWDA; RCRA; ESH)
DA   October 12, 1990
BT1 Preferred Routes
 BT2 Routes (Transportation)
SO   Hazardous Materials
DEF Preferred routes selected in accordance with U.S. DOT Guidelines for Selecting Preferred Highway Routes for Highway Route Controlled Quantities of Radioactive Materials or an equivalent routing analysis which

adequately considers overall risk to the public.

STATE DIRECTOR
(SWDA; RCRA; CFR)
DA   October 12, 1990
BT1 Personnel
RT   Directors
RT   PSD Permits
SO   Wastes
SO   Water Pollution
DEF (CFR) The chief administrative officer of any State or interstate agency operating an "approved program," or the delegated representative of the state Director. If responsibility is divided among two or more State or interstate agencies, "State Director" means the chief administrative officer of the State or interstate agency authorized to perform the particular procedure or function to which reference is made.

STATE EMERGENCY RESPONSE COMMISSION
(EPA; EMER)
DA   October 12, 1990
SF   *SERC*
BT1 State Agencies
 BT2 Agencies
 BT3 Administrative Organizations
 BT4 Organizations
SO   Emergency Preparedness
SO   Environmental Protection Agency Glossary
DEF (EPA) Commission appointed by each state governor according to the requirements of SARA Title III. The SERC's designated emergency planning districts, appoint local emergency planning committees, and supervise and coordinate their activities.

STATE HAZARDOUS WASTES
(DOE Order 5400.3)
DA   October 12, 1990
BT1 Hazardous Wastes
 BT2 Hazardous Materials
 BT3 Materials
 BT2 Wastes
SO   Wastes
DEF (DOE Order 5400.3) Waste defined as hazardous by a State. Pursuant to RCRA Section 6001, DOE is subject to and must comply with State requirements respective to solid and hazardous waste management.

STATE IMPLEMENTATION PLANS
(CAA; EPA; ESH)
DA   October 12, 1990
SF   *SIP*
BT1 Plans
RT   Allowable Emissions

SO   Air Pollution
SO   Environmental Protection Agency Glossary
DEF EPA-approved state plans for the establishment, regulation, and enforcement of air pollution standards. (ESH) Plans required of each State under Section 110 of the Clean Air Act which provide for implementation, maintenance, and enforcement of primary and secondary ambient air quality standards within the State. Among other requirements, they contain the following key elements: (1)Attainment and Maintenance of primary ambient air quality standards within three years, and the secondary standards within a reasonable time thereafter; (2) emission standards which limit the quantity of pollutants which can be emitted from facilities or groups of facilities. These limits at the very least must be applicable to the criteria pollutants; and (3)procedures for reviewing the impact a new source will have on existing and future air quality.

STATE NOTIFICATION
(DOE Order 5440.1D)
DA   May 15, 1991
BT1 Administrative Processes
 BT2 Processes
SO   Environmental Management
DEF (DOE Order) The process by which a host State and, as appropriate, adjacent States are informed of an initial DOE determination to prepare an Environmental Assessment or Environmental Impact Statement.

STATE PRIMARY DRINKING WATER REGULATION
(CFR)
DA   October 19, 1990
BT1 Code of Federal Regulations
 BT2 Codes, Standards, and Regulations
 BT2 Statutes and Regulations
RT   Approved State Primacy Program
RT   Primary Drinking Water Regulations
SO   Water Pollution
DEF (CFR) A drinking water regulation of a State which is comparable to a national primary drinking water regulation.

STATE PROGRAM REVISIONS
(CFR)
DA   October 19, 1990
BT1 Approved State Primacy Program
 BT2 Programs
SO   Water Pollution
DEF (CFR) Changes in an approved State primacy program.

STATE ROUTING AGENCIES
(SWDA; RCRA; ESH)
DA October 12, 1990
BT1 State Agencies
 BT2 Agencies
  BT3 Administrative Organizations
  BT4 Organizations
RT Preferred Routes
SO Hazardous Materials
DEF Entities (including a common
 agency of more than one State
 such as one established by
 Interstate compact) which are
 authorized to use the State legal
 process to impose routing
 requirements, enforceable by
 State agencies, on carriers of
 radioactive materials without
 regard to intrastate jurisdictional
 boundaries. This term also
 includes Indian tribal authorities
 which have police powers to
 regulate and enforce highway
 routing requirements within their
 lands.

STATE/EPA AGREEMENTS
(SWDA; RCRA)
DA October 19, 1990
BT1 Agreements
NT1 Superfund Memorandum of
 Agreement
RT Interim Authorizations
SO Wastes
DEF Agreements between the Regional
 Administrator and the State which
 coordinate EPA and State
 activities, responsibilities and
 programs.

STATES
(SWDA; RCRA; SDWA; CERCLA; ESA;
TSCA; CFR)
DA October 12, 1990
NT1 Agreement States
NT1 Approved States
NT1 Delegated States
NT1 Indian Tribes
RT Assigned Risks
RT EPA Region
RT State Agencies
RT State Authority
RT Superfund State Contracts
SO Compensation and Liability
SO Endangered Species
SO Hazardous Materials
SO Wastes
SO Water Pollution
DEF One of the States of the United
 States, the District of Columbia,
 the Commonwealth of Puerto
 Rico, the Virgin Islands, Guam,
 American Samoa, the Trust
 Territory of the Pacific Islands
 (except in the case of RCRA), the
 Commonwealth of the Northern
 Mariana Islands, or an Indian Tribe
 treated as a State (except in the
 case of RCRA). "State Director"
 means the chief administrative

officer of any State, interstate, or
 Tribal agency operating an
 approved program, or the
 delegated representative of the
 State director. If the responsibility
 is divided among two or more
 States, interstate, or Tribal
 agencies, "State Director" means
 the chief administrative officer of
 any State, interstate, or Tribal
 agency authorized to perform the
 particular procedure or function to
 which reference is made.

STATION
(1717 STATION)
DA December 10, 1990
BT1 Site (DOE FRASE Vocabulary)
 BT2 Sites/Areas
NT1 Fire Station
NT1 Guard Station
NT1 Work Station
RT Substation
SO DOE FRASE VOCABULARY

STATION BLACKOUT
DA January 8, 1991
SF SBO
BT1 Conditions

STATIONARY SOURCES
(CAA; CFR; EPA)
DA October 12, 1990
SY Fixed Sources
BT1 Facilities
NT1 Existing Stationary Facilities
NT1 Existing Sources
NT1 Major Stationary Source
 NT2 Major Emitting Facilities
 NT2 Major PSD Stationary Source
NT1 Modified Sources
NT1 New Sources
NT1 Point Sources
RT Affected Facilities
RT Allowable Emissions
RT Construction
RT Mobile Sources
RT Modification
RT Pollution
RT Potential to Emit
RT Releases
RT Standard of Performance
RT Startup
SO Air Pollution
SO Environmental Protection Agency
 Glossary
DEF Fixed, non-moving producers of
 pollution, mainly power plants and
 other facilities using industrial
 combustion processes. (40 CFR)
 Any buildings, structures, facilities,
 or installations that emit or may
 emit an air pollution for which a
 national standard is in effect.

STATUTES AND REGULATIONS
DA January 28, 1991
NT1 Acts
 NT2 Administrative Procedures Act

NT3 Individual Plant Examination
NT2 Archaeological Resources
 Protection Act of 1979
NT2 Atomic Energy Act
NT2 Clean Air Act
NT2 Clean Water Act
NT2 Comprehensive Environmental
 Response, Compensation, etc.
NT2 Endangered Species Act
NT2 National Environmental Policy Act
NT2 Nuclear Waste Policy Act
NT2 Resource Conservation and
 Recovery Act
NT2 Safe Drinking Water Act
NT2 Solid Waste Disposal Act
NT2 Superfund Amendments and
 Reauthorization Act of 1986
NT2 Toxic Substances Control Act
NT1 Bottle Bill
NT1 Code of Federal Regulations
 NT2 Federal Aviation Regulation
 NT2 Prevention of Significant
 Deterioration Regulations
 NT2 Primary Drinking Water
 Regulations
 NT2 Secondary Drinking Water
 Regulations
 NT2 State Primary Drinking Water
 Regulation
RT Appropriate Act and Regulations
RT Codes, Standards, and Regulations

STC
DA October 12, 1990
SEE Single Strip Containers
SO Acronyms

STEAM GENERATOR
DA January 8, 1991
SF SG
BT1 Equipment
NT1 Fossil Fuel and Wood
 Residue-Fired Steam Generating
 Unit
DEF (NRC Glossary of Terms: Nuclear
 Power and Radiation) The heat
 exchanger used in some reactor
 designs to transfer heat from the
 primary (reactor coolant) system
 to the secondary (steam) system.
 This design permits heat exchange
 with little or no contamination of
 the secondary system equipment.

STEAM GENERATOR TUBE RUPTURE
DA January 8, 1991
SF SGTR
BT1 Failures
 BT2 Accidents

STEAM PLANT
(1807 STEAM PLANT)
DA December 10, 1990
BT1 Building (DOE FRASE Vocabulary)
 BT2 Facilities and Buildings (DOE
 FRASE Vocabulary)
 BT3 Facilities
SO DOE FRASE VOCABULARY

STERILIZATION
(EPA)
DA   October 12, 1990
BT1 Biological Processes
 BT2 Processes
RT Chemosterilants
RT Germicides
SO Environmental Protection Agency
     Glossary
DEF (EPA) (1) In pest control, the use of
     radiation and chemicals to
     damage body cells needed for
     reproduction. (2) The destruction
     of all living organisms in water or
     on the surface of various
     materials. In contrast, disinfection
     is the destruction of most living
     organisms in water or on surfaces.

STGWG
DA   October 12, 1990
SEE State and Tribal Government
     Working Group
SO Acronyms

STI
DA   October 12, 1990
SEE Surveillance Test Interval
SO Acronyms

STIMULUS-MEDIATION-RESPONSE
(SSDC)
DA   October 12, 1990
BT1 Stimulus-Response
RT Behavioral Stereotypes
RT Human Performance
SO System Safety Development Center
     Glossary
DEF (SSDC) The normal way, other than
     reflex arc response, that people
     respond to a stimulus. They
     receive a stimulus and mediate
     between the input and previous
     experience before responding.

STIMULUS-RESPONSE
(SSDC)
DA   October 12, 1990
NT1 Reflex Arc Responses
NT1 Stimulus-Mediation-Response
SO System Safety Development Center
     Glossary
DEF (SSDC) The type of response
     expected from most hardware
     wherein the response is
     consistently the same for a given
     input. Normally, people do not
     respond in this manner other than
     reflex arc response.

STITCHING/SEWING MACHINE
(2145 STITCHING MA)
DA   December 10, 1990
BT1 Machines (DOE FRASE
     Vocabulary)
 BT2 Equipment
SO DOE FRASE VOCABULARY

STM
DA   October 12, 1990
SEE Stack Tritium Monitor
SO Acronyms

STOCHASTIC EFFECTS
(DOE Orders 5480.11, 5400.5)
DA   October 12, 1990
BT1 Effects
NT1 Radiation Induced Hereditary
     Effects
NT1 Radiation Induced Genetic Effects
NT1 Stochastic Radiation Effects
SO Industrial Hygiene
SO Wastes
DEF Biological effects, the probability,
     rather than the severity, of which is
     a function of the magnitude of the
     radiation dose without threshold;
     i.e., stochastic effects are random
     in nature. Nonstochastic Effects
     are biological effects, the severity
     of which, in affected individuals,
     varies with the magnitude of the
     dose above a threshold value.

STOCHASTIC RADIATION EFFECTS
(IAEA)
DA   October 12, 1990
BT1 Stochastic Effects
 BT2 Effects
RT Irradiation
SO Radiation
DEF (IAEA) Radiation effects, the
     severity of which is independent of
     dose and the probability of which
     is assumed by the ICRP to be
     proportional to the dose without
     threshold at the low doses of
     interest in radiation protection.

STOCK ROOM
(1719 STOCK ROOM)
DA   December 10, 1990
BT1 Room
 BT2 Sites/Areas
BT1 Storage Area
 BT2 Area
  BT3 Sites/Areas
SO DOE FRASE VOCABULARY

STORAGE
(SWDA; RCRA; EPA; USC)
DA   October 12, 1990
BT1 Waste Management Processes
 BT2 Processes
NT1 Burial Operations
 NT2 Shallow Land Burial
NT1 Contaminated Water Storage
NT1 Vertical Tube Storage
RT Containers
RT Dry Caves
RT On-Shore Facilities
RT Tanks
RT Treatment, Storage, and Disposal
SO Environmental Protection Agency
     Glossary
SO Wastes
DEF (USC) The holding of hazardous

waste for a temporary period, at
the end of which the hazardous
waste is treated, disposed, or
stored elsewhere. (EPA)
Temporary holding of waste
pending treatment or disposal.
Storage methods include
containers, tanks, waste piles, and
surface impoundments.

STORAGE AREA
(1619 STORAGE AREA)
DA   December 10, 1990
BT1 Area
 BT2 Sites/Areas
NT1 Stock Room
NT1 Warehouse
SO DOE FRASE VOCABULARY

STORAGE AREA COMPARTMENTS
(DOE Order 6430.1A)
DA   October 12, 1990
BT1 Sites/Areas
SO Construction
DEF (DOE Order 6430.1A) An area or
     series of areas that contain
     storage enclosures.

STORIES
(SEA)
DA   October 12, 1990
NT1 Soft Stories
NT1 Weak Stories
RT Story Drifts
RT Story Shear
SO Construction
DEF (SEA) The spaces between levels.
     Story x is the story below level x.

STORM SEWERS
(EPA)
DA   October 12, 1990
BT1 Sewers
RT Storm Water
RT Urban Runoff
SO Environmental Protection Agency
     Glossary
DEF (EPA) Systems of pipes (separate
     from sanitary sewers) that carry
     only water runoff from building and
     land surfaces.

STORM WATER
(SWDA; RCRA)
DA   October 19, 1990
BT1 Water
RT Run-off
RT Storm Water Collection Systems
RT Storm Sewers
SO Wastes
DEF Surface water runoff resulting from
     precipitation.

**STORM WATER COLLECTION
SYSTEMS**
(SWDA; RCRA; CFR)
DA   October 12, 1990
SY Waste Water Collection Systems
BT1 Water Systems

BT2 Publicly Owned Treatment Works
  BT3 Treatment Facilities
    BT4 Facilities
BT2 Systems
RT Force Mains
RT Storm Water
DEF Piping, pumps, conduits, and any
  other equipment necessary to
  collect and transport the flow of
  surface water run-off resulting
  from precipitation, or domestic,
  commercial, or industrial
  wastewater to and from retention
  areas or any areas where
  treatment is designated to occur.
  The collection of storm water and
  wastewater does not include
  treatment except where incidental
  to conveyance.

STORY DRIFT RATIO
(SEA)
DA October 12, 1990
BT1 Ratios
RT Story Drifts
SO Construction
DEF (SEA) The story drift divided by the
  story height.

STORY DRIFTS
(SEA)
DA October 12, 1990
RT Stories
RT Story Drift Ratio
SO Construction
DEF (SEA) The displacement of one
  level relative to the level above or
  below.

STORY SHEAR
(SEA)
DA October 12, 1990
RT Stories
SO Construction
DEF (SEA) The summation of design
  lateral forces above the story
  under consideration.

STOVE(S)
(2492 STOVE)
DA January 3, 1991
BT1 Heater(s)
  BT2 Equipment/Parts - Heating (DOE
    FRASE Vocabulary)
  BT3 Equipment
SO DOE FRASE VOCABULARY

STOWAGE
(SWDA; RCRA; ESH)
DA October 12, 1990
RT Vessels
SO Hazardous Materials
DEF The act of placing hazardous
  materials on board a vessel.

STRAIGHT EDGE RULER
(2771 STRAIGHT EDG)
DA January 3, 1991
BT1 Gauge(s)

BT2 Instrument(s)
  BT3 Equipment/Parts -
    Instrumentation/Measuring
    (DOE FRASE Voc.)
  BT4 Equipment
SO DOE FRASE VOCABULARY

STRAIN
(1340 STRAIN)
DA November 28, 1990
BT1 Injuries
RT Sprain
SO DOE FRASE VOCABULARY

STRANGULATION
(1341 STRANGULATIO)
DA November 28, 1990
BT1 Injuries
SO DOE FRASE VOCABULARY

STRATA
(SDWA; CFR)
DA October 12, 1990
RT Confining Beds
RT Confining Zones
RT Formations
SO Water Pollution
DEF Single sedimentary beds or layers,
  regardless of thickness, that
  consist of generally the same kind
  of rock material. (Singular:
  stratum).

STRATEGIC PETROLEUM RESERVE
DA January 8, 1991
SF *SPR*
BT1 Government-Owned
    Contractor-Operated Facilities
  BT2 Federal Facilities
    BT3 Facilities
RT Boeing Petroleum Services
DEF (Capsule Review of DOE Research
  and Development and Field
  Facilities, 1986) The SPR was
  created by the Energy Policy and
  Conservation Act in December
  1975. The objective of the SPR
  program is to stockpile crude oil to
  reduce the vulnerability of the
  United States to a severe energy
  emergency involving a disruption
  in its petroleum supplies.

**STRATEGIC SPECIAL NUCLEAR
MATERIALS**
(CFR)
DA January 29, 1991
BT1 Special Nuclear Materials
  BT2 Licensed Materials
    BT3 Materials
  BT2 Security Interests
SO Environmental Management
DEF (CFR) Uranium-235 (contained in
  uranium enriched to 20% or more
  in the U235 isotope), uranium-233,
  or plutonium.

STRATOSPHERE
(EPA)

DA October 12, 1990
BT1 Atmosphere
RT Halocarbons
RT Ozone
SO Environmental Management
SO Environmental Protection Agency
    Glossary
DEF (EPA) The portion of the
  atmosphere that is 10 to 25 miles
  above the Earth's surface.

STRENGTH
(SEA)
DA October 12, 1990
NT1 Flexural Strength
NT1 Required Strength
SO Construction
DEF (SEA) The usable capacity of a
  structure or its members to resist
  load within the deformation limits.

STRIKE SLIP
(USGS)
DA October 12, 1990
SY Horizontal Displacements
BT1 Surface Faulting
  BT2 Faults
SO Natural Phenomenon
DEF (USGS) One of the three main
  types of surface faulting; in a fault,
  the component of the movement
  or slip that is parallel to the strike
  or perpendicular to the dip of the
  fault. Horizontal displacement
  parallel to the strike.

STRIP
(CAA; CFR)
DA October 12, 1990
BT1 Processes
NT1 Removal
  NT2 Dredging
RT Friable Asbestos
RT Renovation
SO Air Pollution
DEF (CFR) To take off friable asbestos
  materials from any part of a
  facility.

STRIP CROPPING
(EPA)
DA November 13, 1990
BT1 Cultivating
RT Contour Plowing
SO Environmental Protection Agency
    Glossary
DEF (EPA) Growing crops in a
  systematic arrangement of strips
  or bands which serve as barriers
  to wind and water erosion.

STRIP MINING
(EPA)
DA November 13, 1990
BT1 Mining/Drilling Activity
  BT2 Activity Types (DOE FRASE
    Vocabulary)
    BT3 Activities
RT Conventional Mines

STRIP MINING

SO Environmental Protection Agency
Glossary
DEF (EPA) A process that uses
machines to scrape soil or rock
away from mineral deposits just
under the Earth's surface.

STRIPPERS

(CAA; CFR)
DA   October 12, 1990
BT1 Equipment
RT Vinyl Chloride
SO Air Pollution
DEF (CFR) Any vessels in which
residual vinyl chloride is removed
from polyvinyl chloride resin,
except bulk resin, in the slurry
form by the use of heat and/or
vacuum. In the case of bulk resin,
strippers include any vessel that is
used to remove residual vinyl
chloride from polyvinyl chloride
resin immediately following the
polymerization step in the plant
process flow.

STRONG OUTSIDE CONTAINERS

(SWDA; RCRA; ESH)
DA   October 12, 1990
BT1 Outside Container
 BT2 Containers
SO Hazardous Materials
DEF Outermost enclosure which
provides protection against the
unintentional release of its
contents under conditions normally
incident to transportation.

STRONTIUM 90

(EDB)
DA   March 29, 1991
BT1 Beta-Minus Decay Radioisotopes
 BT2 Beta Decay Radioisotopes
  BT3 Radionuclides
   BT4 CERCLA Hazardous
     Substances
    BT5 Hazardous Substances
   BT4 Nuclides
BT1 Years Living Radioisotopes
 BT2 Radionuclides
  BT3 CERCLA Hazardous Substances
   BT4 Hazardous Substances
  BT3 Nuclides

STRUCTURAL COLLAPSE

(DOE Order 6430.1A)
DA   October 12, 1990
BT1 Failures
 BT2 Accidents
SO Construction
DEF (DOE Order 6430.1A) The failure of
a structural component as a direct
result of loss of structural integrity
of the facility being subjected to
various loadings.

STRUCTURAL DAMAGE

(1538 STRUCTURAL D)
DA   November 29, 1990

BT1 Damage
 BT2 Nature of Property Damage
RT Structures (DOE FRASE
Vocabulary)
SO DOE FRASE VOCABULARY

STRUCTURAL MEMBERS

(CAA; CFR)
DA   October 12, 1990
NT1 Bearing Wall Systems
RT Bearing Capacity
RT Construction
RT Demolition
RT Flexible Elements
RT Load Factor
RT Required Strength
SO Air Pollution
DEF (CFR) Load-supporting members of
a facility, such as beams and load
supporting walls; or any
nonload-supporting member, such
as ceilings and nonload-supporting
walls.

STRUCTURES (DOE FRASE VOCABULARY)

(DOE Order 4330.4A; DOE FRASE
Vocabulary Numeric Keys 1850-1924)
DA   January 17, 1991
NT1 Artificial Reefs
NT1 Bridge
NT1 Cooling Towers
NT1 Dam
NT1 Earth-Lined Channels
NT1 Explosives Buildings
 NT2 Explosives Bays
  NT3 Staging Bays (in-process)
NT1 Facility Boundaries
NT1 Highway
NT1 Hydraulic Structures
 NT2 Dams
 NT2 Dikes
  NT3 Terracing
NT1 Loading Dock
NT1 Mine
NT1 Parking Lot
NT1 Pier
NT1 Platforms
NT1 Pond
NT1 Raw Water Intake Structure
NT1 Road
NT1 Silo
NT1 Space Frames
 NT2 Building Frame Systems
  NT3 Bearing Wall Systems
  NT3 Braced Frame
   NT4 Concentric Braced Frames
   NT4 Eccentric Braced Frames
  NT3 Horizontal Bracing Systems
  NT3 Lateral Force Resisting System
  NT3 Shoring
 NT2 Moment Resisting Space Frame
  NT3 Intermediate Moment Resisting
    Space Frame
  NT3 Ordinary Moment Resisting
    Space Frame
  NT3 Special Moment Resisting Space
    Frame
 NT2 Vertical Load Carrying Space
   Frames

NT1 Trailer Building
NT1 Trench/Ditch
RT DOE FRASE Categories
RT Facilities
RT Sites/Areas
RT Structural Damage
RT Targets
RT Towers
SO DOE FRASE VOCABULARY
DEF (DOE Order 4330.4A) Any fixed
real property improvement
constructed on or in the land that
is not a building or utility (e.g.,
bridges, towers, and tanks).
Denotes something constructed or
built. Also, a subject category used
with the DOE FRASE Vocabulary.

STUCK-OPEN RELIEF VALVE

DA   January 8, 1991
SF   *SORV*
BT1 Relief Valves
 BT2 Valves
  BT3 Devices
RT Hazards

STUDIES

(TSCA; CFR)
DA   October 19, 1990
NT1 Feasibility Studies
NT1 Hazard and Operability Study
NT1 Health and Safety Studies
NT1 Probabilistic Safety Study
NT1 RCRA Corrective Measures Study
NT1 Reactor Safety Study
NT1 Treatability Studies
NT1 Zion Probabilistic Safety Study
RT Quality Assurance Units
RT Reactor Safety Study Methodology
Application Program
RT Samples
RT Study Completion Date
RT Study Directors
RT Study Initiation Date
SO Hazardous Materials
DEF (WTID) Consists of observations
and/or analyses in detail (of a
phenomenon, development or
question) usually within a
restricted area with a view to
some action. (CFR) Any in vivo or
in vitro experiments in which test
substances are studied
prospectively in a test system
under conditions to determine or
help predict their fate, toxicity,
metabolism, or other
characteristics in humans, or other
animals and plants. The term does
not include studies utilizing human
subjects or clinical studies. The
term does not include basic
exploratory studies carried out to
determine whether a test
substance has any potential utility.

STUDY COMPLETION DATE

(TSCA)
DA   October 19, 1990
BT1 Time Designations

BT-Broader Term   NT-Narrower Term   RT Related Term

RT Studies
SO Hazardous Materials
DEF (TSCA) The date the final report is signed by the study director.

STUDY DIRECTORS
(TSCA; CFR)
DA October 19, 1990
BT1 Personnel
RT Studies
SO Hazardous Materials
DEF (CFR) Individuals responsible for the overall conduct of a study.

STUDY INITIATION DATE
(TSCA)
DA October 19, 1990
BT1 Time Designations
RT Studies
SO Hazardous Materials
DEF (TSCA) The date the protocol is signed by the study director.

SUBACUTE DIETARY LC 50
(CFR)
DA January 24, 1991
BT1 LC 50
BT1 Lethal Concentration
SO Environmental Management
DEF (CFR) A concentration of a substance, expressed as parts per million, in food that is lethal to 50 percent of the test population of animals under test conditions as specified in the Registration Guidelines.

SUBBASE
(DOE Order 6430.1A)
DA October 12, 1990
RT Roadways
SO Construction
DEF (DOE Order 6430,1A) A layer of granular material located beneath the base course of a highway pavement.

SUBCHRONIC RFD (RFD)
(EPA)
DA October 12, 1990
BT1 Reference Dose (RfD)
BT2 Doses
RT Chronic RfD
SO Environmental Protection Agency Glossary
DEF (EPA) An estimate (with uncertainty spanning perhaps an order of magnitude or greater) of a daily exposure level for the human population, including sensitive subpopulations, that is likely to be without an appreciable risk of deleterious effects if the exposure were to occur for a period of less than 7 years.

SUBCOMMITTEE ON FEDERAL RESPONSE
(EMER)

DA February 1, 1991
BT1 Committees
BT2 Administrative Organizations
BT3 Organizations
SO Emergency Preparedness
DEF (EMER) A subcommittee of the Federal Radiological Preparedness Coordinating Committee formed to develop and test the Federal Radiological Emergency Response Plan. Most agencies that would participate in the federal radiological emergency response are represented on this subcommittee.

SUBCRITICAL FLOWS
(DOE Order 6430.1A)
DA October 12, 1990
RT Flow Rates
RT Supercritical Flows
SO Construction
DEF (DOE Order6430.1A) Open channel flows having a low velocity and a froude number less than unity (also described as tranquil or streaming flow).

SUBGRADE MODULUS
(DOE Order 6430.1A)
DA October 12, 1990
SO Construction
DEF (DOE Order 6430.1A) The slope of a load-settlement diagram constructed with data from field-loading tests on the actual subgrade.

SUBJECT MATTER EXPERTS
(DOE Order 5480.18)
DA October 12, 1990
BT1 Technically Qualified Individuals
BT2 Personnel
RT Training Accreditation Program Staff
SO Standards
DEF (DOE Order 5480.18) Individuals qualified, or previously qualified, and experienced in performing a particular task.

SUBMACHINE GUNS, CLOSED BOLT
(DOE Order 5480.16)
DA October 12, 1990
BT1 Machine Guns
BT2 Small Arms
BT3 Firearms
SO Firearms
DEF (DOE Order 5480.16) Small caliber magazine-fed weapons that must be fired in the closed bolt position and containing a spring in the bolt to thrust the firing pin forward to detonate the primer.

SUBMACHINE GUNS, OPEN BOLT
(Doe Order 5480.16)
DA October 12, 1990
BT1 Machine Guns

BT2 Small Arms
BT3 Firearms
SO Firearms
DEF (DOE Order 5480.16) A small caliber magazine-fed weapon that must be fired from the open bolt position. This is a fixed firing pin against which the bolt produces the thrust necessary to detonate the primer.

SUBORDINATE MANAGEMENT
(SSDC)
DA October 12, 1990
BT1 Management
SO System Safety Development Center Glossary
DEF (SSDC) For purposes of this document, all management subordinate to the particular management level under consideration.

SUBSIDENCE
(SDWA; CFR; USGS)
DA October 12, 1990
RT Catastrophic Collapses
RT Differential Settlement
RT Dip-Slip
RT Faults
RT Formations
RT Soil Mechanics
RT Tectonic Deformations
SO Natural Phenomenon
SO Water Pollution
DEF (CFR) The lowering of the natural land surface in response to earth movements; lowering of fluid pressure; removal of underlying supporting material by mining or solution of solids, either artificially or from natural causes; compaction due to wetting (hydrocompaction); oxidation of organic matter in soils; or added load on the land surface.

SUBSISTENCE
(ESA; CFR)
DA October 12, 1990
RT Wasteful Manner
SO Endangered Species
DEF (CFR) The use of endangered or threatened wildlife for food, clothing, shelter, heating, transportation, and other uses necessary to maintain the life of the taker of the wildlife, or those who depend upon the taker to provide them with such subsistence, and includes selling any edible portions of such wildlife in native villages and towns in Alaska for native consumption within native villages and towns.

SUBSLABS
(DOE Order 6430.1A)
DA October 12, 1990

SY-Synonymous Terms

SO-Source/Subject Category

SF-See From

RT Construction
SO Construction
DEF (DOE Order 6430.1A) Also known as a structural slab, base slab, mud slab, or wearing slab. The concrete slab below the waterproofing membrane in a double-slab configuration.

SUBSTANTIAL
(Doe Order 5484.1)
DA October 12, 1990
SO Environmental Management
SO Management
DEF (DOE Order 5480.1) Clearly outside normally accepted or experienced bounds.

SUBSTANTIAL BUSINESS RELATIONSHIPS
(SWDA; RCRA; CFR)
DA October 12, 1990
RT Contractual Relationships
RT Long Term Contracts
SO Wastes
DEF The extent of business relationships necessary under applicable State law to make a guarantee contract issued incident to that relationship valid and enforceable. A "substantial business relationship" must arise from a pattern of recent or ongoing business transactions, in addition to the guarantee itself, such that a currently existing business relationship between the guarantor and the owner or operator is demonstrated to the satisfaction of the applicable EPA Regional Administrator.

SUBSTANTIAL CONSTRUCTION
(DOE Order 6430.1A)
DA October 12, 1990
BT1 Construction
 BT2 Processes
SO Construction
DEF (DOE Order 6430.1A) If determined by the cognizant DOE security personnel, classified matter shall be stored in a building or portion thereof that provides a physical barrier of the required penetration times and resistance. NBS Technical Note 837 shall be used for a comparison of the forcible penetration time through different structural barriers.

SUBSTATION
(1620 SUBSTATION)
DA December 10, 1990
BT1 Site (DOE FRASE Vocabulary)
 BT2 Sites/Areas
RT Station
SO DOE FRASE VOCABULARY

SUBVORTICES
(SMRP)

DA October 12, 1990
RT Rotational Windstorms
RT Waterspouts
SO Natural Phenomenon
DEF (SMRP) Spinning columns of air embedded inside a tornado, waterspout, or dust devil. A large windstorm may be accompanied by several subvortices simultaneously. The travel of these subvortices is governed by the airflow in and around the parent storm.

SUDDEN ACCIDENTAL OCCURRENCES
(SWDA; RCRA; CFR)
DA October 12, 1990
BT1 Accidental Occurrences
 BT2 Occurrences
NT1 Pressure Releases
 NT2 Sudden Releases of Pressure
RT Emergency Projects
SO Wastes
DEF (CFR) Occurrences that are not continuous or repeated in nature.

SUDDEN RELEASES OF PRESSURE
(CERCLA)
DA October 12, 1990
BT1 Pressure Releases
 BT2 Sudden Accidental Occurrences
  BT3 Accidental Occurrences
   BT4 Occurrences
RT Hazard Categories
RT Relief Valves
DEF (CFR) A sudden, almost instantaneous release of pressure, gas, and heat when subjected to sudden shock or pressure or high temperature.

SULFUR DIOXIDE
(EPA)
DA October 12, 1990
SF SO_2
BT1 Criteria Pollutants
 BT2 Pollutants
RT Acid Deposition
SO Environmental Protection Agency Glossary
DEF (EPA) A heavy, pungent, colorless, gaseous air pollutant formed primarily by the combustion of fossil plants.

SULFURIC ACID PLANTS
(CAA; CFR)
DA October 12, 1990
BT1 Facilities
RT Acid Deposition
SO Air Pollution
DEF (CFR) Any facilities producing sulfuric acid by the contact process by burning elemental sulfur, alkylation acid, hydrogen sulfide, or acid sludge. Does not include facilities where conversion to sulfuric acid is utilized primarily

as a means of preventing emissions to the atmosphere of sulfur dioxide or other sulfur compounds.

SULFURYL FLUORIDE
DA January 8, 1991
SF SO_2F_2

SUMMATION GATES
(SSDC)
DA October 12, 1990
BT1 Logic Gates (Boolean)
BT1 Logic Gates
SO System Safety Development Center Glossary
DEF (SSDC) Special logic gates which require that an acceptable combination of input events be present to produce an output. Inputs can be present in varying proportions, as long as the sum of the inputs is adequate to generate an output.

SUMP AREA
(1621 SUMP AREA)
DA December 10, 1990
BT1 Area
 BT2 Sites/Areas
SO DOE FRASE VOCABULARY

SUMP PUMP
(2146 SUMP PUMP)
DA December 10, 1990
BT1 Pump(s)
 BT2 Machines (DOE FRASE Vocabulary)
  BT3 Equipment
SO DOE FRASE VOCABULARY

SUMPS
(SWDA; RCRA; EPA)
DA October 12, 1990
BT1 Liquid Traps
SO Environmental Management
SO Environmental Protection Agency Glossary
SO Wastes
DEF Containment Sumps. (EPA) Pits or tanks that catch liquid runoff for drainage or disposal.

SUNBURN
(1400 SUNBURN)
DA November 28, 1990
BT1 Illnesses
RT Heat Stroke/Other High Temp Effect
SO DOE FRASE VOCABULARY

SUPERCRITICAL FLOWS
(DOE Order 6430.1A)
DA October 12, 1990
RT Flow Rates
RT Subcritical Flows
SO Construction
DEF (DOE Order 6430.1A) Open

channel flows having a high velocity and a froude number greater than unity (also described as rapid, shooting or torrential flow).

SUPERELEVATION
(DOE Order 6430.1A)
DA   October 12, 1990
RT   Roadways
SO   Construction
DEF (DOE Order 6430.1A) The practice of elevating one side of a roadway over the other on curves in alignment.

SUPERFUND
(CERCLA; CFR; EPA)
DA   October 12, 1990
BT1 Programs
RT   Action Levels
RT   Applicable or Relevant and Appropriate Requirements
RT   Operable Units
RT   Remedial Actions
RT   Superfund Memorandum of Agreement
RT   Superfund State Contracts
RT   Superfund Amendments and Reauthorization Act of 1986
SO   Compensation and Liability
SO   Environmental Protection Agency Glossary
DEF The program operated under the legislative authority of CERCLA and the Superfund Amendments and Reauthorization Act (SARA) that funds and carries out the EPA solid waste emergency and long-term removal remedial activities. These activities include establishing the National Priorities List, investigating sites for inclusion on the list, determining their priority level on the list, and conducting and/or supervising the ultimately determined cleanup and other remedial actions.

SUPERFUND AMENDMENTS AND REAUTHORIZATION ACT OF 1986
(SARA; EMER)
DA   January 8, 1991
SF   *SARA*
BT1 Acts
BT2 Statutes and Regulations
RT   National Contingency Plan
RT   Superfund
SO   Emergency Preparedness
DEF (SARA) To extend and amend the Comprehensive Environmental Response, Compensation, and Liability Act of 1980, and for other purposes.

SUPERFUND MEMORANDUM OF AGREEMENT
(ANL; CFR)
DA   May 24, 1991

SF   *SMOA*
BT1 State/EPA Agreements
BT2 Agreements
RT   Superfund
SO   Management
SO   Wastes
DEF (CFR) A nonbinding, written document executed by an EPA Regional Administrator and the head of a state agency that may establish the nature and extent of EPA and state interaction during the removal, pre-remedial, remedial, and/or enforcement response process. The SMOA is not a site-specific document, although attachments may address specific sites. The SMOA generally defines the role and responsibilities of both the lead and support agencies.

SUPERFUND STATE CONTRACTS
(ANL; CFR)
DA   May 24, 1991
RT   Remedial Actions
RT   States
RT   Superfund
SO   Management
DEF (CFR) A joint, legally binding agreement between EPA and a state to obtain the necessary assurances before a federal lead remedial action can begin at a site. In the case of a political subdivision-lead remedial response, a three-party Superfund state contract among EPA, the state, and the political subdivision thereof, is required before a political subdivision takes the lead for any phase of remedial response to ensure state involvement pursuant to section 121(f)(1) of CERCLA. The Superfund state contract may be amended to provide the states CERCLA section 104 assurances before a political subdivision can take the lead for remedial action.

SUPERVISED AREAS
(IAEA)
DA   October 12, 1990
BT1 Sites/Areas
RT   Radiation Safety
SO   Radiation
DEF (IAEA) Areas where radiation levels are such that annual exposure is most unlikely to exceed three-tenths of the occupational dose equivalent limits but may exceed one-tenth of those limits, and where special forms of supervision (such as area monitoring) are accordingly applied.

SUPERVISORS
(DOE Order 5480.5; ESH)

DA   October 12, 1990
BT1 Personnel
SO   Industrial Safety
SO   Management
DEF (DOE Order 5480.5) Individuals officially designated by management to direct the activities of operators or fissionable materials handlers and to supervise the operation of equipment that handles, produces, processes, stores, packages, or uses radioactive material or significant quantities of fissionable materials.

SUPPLEMENT ANALYSES
(DOE Order 5440.1D)
DA   May 15, 1991
BT1 Analyses
RT   Environmental Impact Statements
RT   Record of Decision (NEPA)
RT   Supplemental EIS
SO   Environmental Management
DEF (DOE Order) DOE documents that describe any changes in a proposed action that are relevant to environmental concerns, or any significant new circumstances and information relevant to environmental concerns and bearing on the proposed action or its impacts. A supplement analysis is used to determine whether a supplemental environmental impact statement should be prepared pursuant to 40 CFR 1502.9(c), or to support a decision to prepare a new environmental impact statement or a revised record of decision.

SUPPLEMENTAL EIS
(DOE Order 5440.1D)
DA   May 15, 1991
BT1 Environmental Impact Statements
RT   Supplement Analyses
SO   Environmental Management
DEF (DOE Order) An environmental impact statement prepared to supplement a prior environmental impact statement, as provided at 40 CFR 1502.9(c).

SUPPLEMENTARY SAFETY SYSTEM
DA   January 8, 1991
SF   *SSS (Supplementary Safety System)*
BT1 Emergency Systems
BT2 Systems

SUPPORT AGENCIES
(ANL; CFR)
DA   May 24, 1991
BT1 Agencies
BT2 Administrative Organizations
BT3 Organizations
RT   Comprehensive Environmental Response, Compensation, etc.

SY-Synonymous Terms   SO-Source/Subject Category   SF-See From

RT National Contingency Plan
RT On-Scene Coordinators
RT Remedial Project Managers
SO Management
DEF (CFR) The agencies that provide the support agency coordinator to furnish necessary data and documents, and provide other assistance as requested by the On-Scene Coordinator (OSC) or the Remedial Project Manager (RPM). EPA, the USCG, another federal agency, or a state may be support agencies for a response action if operating pursuant to a contract executed under section 104(d)1 of CERCLA or designated pursuant to a Superfund Memorandum of Agreement entered into pursuant to subpart F of the NCP or other agreement. The support may also concur on decision documents.

SUPPORT AGENCY COORDINATORS
(ANL; CFR)
DA May 24, 1991
BT1 Personnel
RT National Contingency Plan
SO Management
DEF (CFR) The official designated by the support agency, as appropriate, to interact and coordinate with the lead agency in response actions under subpart E of the NCP.

SUPPORT BUILDINGS
(DOE Order 6430.1A)
DA October 12, 1990
BT1 Facilities
RT Explosives Activities
SO Construction
DEF (DOE Order 6430.1A) Any structures (including utilities) directly supporting explosives activities but containing no explosives.

SUPPRESSION POOL
DA January 8, 1991
SF *SP (Suppression Pool)*

SUPPRESSION POOL COOLING
DA January 8, 1991
SF *SPC*
BT1 Processes

SURFACE CASINGS
(SDWA; CFR)
DA October 12, 1990
BT1 Casings
RT Packers
RT Plugging
RT Wells
SO Water Pollution
DEF The first string of well casing to be installed in the well.

SURFACE COLLECTING AGENTS
(CERCLA; CFR)
DA October 12, 1990
BT1 Chemical Agents
RT Cleanup
RT Oil Spills
SO Compensation and Liability
DEF Those chemical agents that form a surface film to control the layer thickness of oil.

SURFACE FAULTING
(USGS)
DA October 12, 1990
BT1 Faults
NT1 Dip-Slip
NT1 Oblique Slip
NT1 Strike Slip
RT Catastrophic Collapses
RT Moment Magnitude
SO Natural Phenomenon
DEF (USGS) The offset or tearing of the earth's surface by differential movement across a fault.

SURFACE IMPOUNDMENT
(SWDA; RCRA; CFR; EPA; ESH)
DA October 12, 1990
BT1 Impoundment
 BT2 Sites/Areas
RT Hazardous Waste Management Units
RT Land Disposal
RT Liners
SO Environmental Management
SO Environmental Protection Agency Glossary
SO Wastes
DEF (CFR) A natural topographic depression, man-made excavation, or diked area formed primarily of earthen materials (although it may be lined with man-made materials) that is not an injection well. (EPA) Treatment, storage, or disposal of liquid hazardous wastes in ponds. (ESH) A facility or part of a facility that is a natural topographic depression, manmade excavation, or diked area, formed primarily of earthen materials (although it may be lined with man-made materials), that is designed to hold an accumulation of liquid wastes or wastes containing free liquids and is not an injection well. Examples of surface impoundments are holding, storage, settling, and aeration pits, ponds, and lagoons.

SURFACE WATER RESOURCES
(CFR)
DA November 15, 1990
SY Surface Waters
BT1 Natural Resources
NT1 Brackish Waters
NT1 Estuaries
NT1 Floodplains
NT1 Lakes

NT2 Dystrophic Lakes
NT2 Eutrophic Lakes
NT2 Oligotrophic Lakes
NT1 River/Creek
NT1 Wetlands
NT2 Bogs
NT2 Bottomland Hardwoods
NT2 Fens
NT2 Marshes
 NT3 Tidal Marshes
NT2 Swamps
RT Watershed
DEF (CFR) The waters of the United States, including the sediments suspended in water or lying on the bank, bed, or shoreline and sediments in or transported through coastal and marine areas. This term does not include groundwater or water or sediments in ponds, lakes, or reservoirs designed for waste treatment under the Resource Conservation and Recovery Act of 1976 (RCRA), 42 U.S.C. 6901-6987 or the CWA, and applicable regulations.

SURFACE WATERS
(SDWA; CFR; EPA)
DA October 12, 1990
SY Surface Water Resources
BT1 Water
NT1 Brackish Waters
NT1 Estuaries
NT1 Lakes
NT2 Dystrophic Lakes
NT2 Eutrophic Lakes
NT2 Oligotrophic Lakes
NT1 River Water
RT Accessible Environment
RT Aquatic Environment
RT Assimilation
RT Groundwater
RT Marshes
RT Overland Flow
RT Swamps
RT Waters of the United States
RT Wetlands
SO Environmental Protection Agency Glossary
SO Water Pollution
DEF (CFR) All waters that are open to the atmosphere and subject to surface runoff. (EPA) All waters naturally open to the atmosphere (rivers, lakes, reservoirs, streams, impoundments, seas, estuaries, etc.) and all springs, wells, or other collectors that are directly influenced by surface water.

SURFACE WAVE MAGNITUDE
(USGS)
DA October 12, 1990
SF *M$_s$*
BT1 Units of Measure
RT Body Wave Magnitude
RT Earthquake Magnitude
RT Love Seismic Waves

BT-Broader Term   NT-Narrower Term   RT Related Term

RT Rayleigh Seismic Waves
SO Natural Phenomenon
DEF (USGS) Measures the amplitude of surface waves with a period of 20 seconds (a wavelength of about 38 miles) and are often dominant on the seismograms.

SURFACTANTS
(DOE Order 6430.1A; EPA)
DA October 12, 1990
RT Chemical Agents
RT Detergents
RT Dispersants
RT Inert Ingredients
SO Construction
SO Environmental Protection Agency Glossary
DEF (DOE Order 6430.1A) Surface-active agents used in detergents to cause lathering.

SURPLUS FACILITIES
(DOE Order 5480.2A)
DA January 29, 1991
BT1 Facilities
SO Environmental Management
DEF (DOE Order 5820.2A) Any facilities or sites (including equipment) that have no identified or planned programmatic use and are contaminated with radioactivity to levels that require controlled access.

SURPLUS FACILITIES MANAGMENT PROGRAM
DA January 8, 1991
SF *SFMP*
BT1 Programs

SURROGATES
(ESH)
DA October 12, 1990
SO Quality Assurance
DEF (ESH) Organic compounds which are similar to analytes of interest in chemical composition, extraction, and chromatography, but which are not normally found in environmental samples. These compounds are spiked into all blanks, standards, samples, and spiked samples prior to analysis. Percent recoveries are calculated for each surrogate.

SURVEILLANCE
(DOE Order 4330.4A; IAEA)
DA October 12, 1990
BT1 Administrative Processes
 BT2 Processes
NT1 Environmental Surveillance
SO Radiation
DEF (DOE Order 4330.4A) Any periodic monitoring for performance adequacy. (IAEA) All planned activities performed to ensure compliance with operational specifications established for a particular installation.

SURVEILLANCE AND MAINTENANCE
DA January 8, 1991
SF *S&M*
BT1 Inspection and Maintenance
BT1 Maintenance
 BT2 Activities

SURVEILLANCE AND NUCLEAR DETECTION SYSTEMS
(EMER)
DA February 1, 1991
BT1 Emergency Systems
 BT2 Systems
SO Emergency Preparedness
DEF (EMER) Research and development efforts for developing gamma ray and neutron detector arrays used in searching for or mapping nuclear material contamination in support of the Nuclear Emergency Search Team.

SURVEILLANCE SYSTEMS
(EPA)
DA October 12, 1990
BT1 Emergency Systems
 BT2 Systems
SO Environmental Protection Agency Glossary
DEF (EPA) A series of monitoring devices designed to determine environmental quality.

SURVEILLANCE TEST INTERVAL
DA January 8, 1991
SF *STI*
BT1 Time Designations

SURVEYS
(SSDC)
DA October 12, 1990
NT1 Condition Assessment Surveys
NT1 Environmental Surveys
NT1 Radiological Surveys
NT1 Sanitary Surveys
SO Environmental Management
SO System Safety Development Center Glossary
DEF (SSDC) A comprehensive examination of policies, procedures, practices, facilities, and equipment, including field observation of actual conditions, within a stated broad scope. (Less methodical and detailed than an appraisal; for example, an examination by a professional, expert, consultant, or committee.)

SUSPENDED SOLIDS
(EPA)
DA October 12, 1990
BT1 Leachates
NT1 Settleable Solids
RT Turbidity

SO Environmental Protection Agency Glossary
DEF (EPA) Small particles of solid pollutants that float on the surface of, or are suspended in sewage or other liquids. They resist removal by conventional means. (See: Total Suspended Solids.)

SUSPENSION
(EPA)
DA October 12, 1990
RT Pesticides
SO Environmental Protection Agency Glossary
DEF (EPA) The act of suspending the use of a pesticide when EPA deems it necessary to do so in order to prevent an imminent hazard resulting from continued use of the pesticide. An emergency suspension takes effect immediately; under an ordinary suspension a registrant can request a hearing before the suspension goes into effect. Such a hearing process might take six months.

SUSPENSION CULTURES
(EPA)
DA October 12, 1990
SO Environmental Protection Agency Glossary
DEF (EPA) Individual cells or small clumps of cells growing in a liquid nutrient medium.

SUSPICIOUS DEVICES
(EMER)
DA February 1, 1991
BT1 Devices
BT1 Security Interests
SO Emergency Preparedness
DEF (EMER) Any devices or packages that arouse suspicion by sound, geographic location, or shape.

SWAMPS
(EPA)
DA October 12, 1990
BT1 Wetlands
 BT2 Sites/Areas
 BT2 Surface Water Resources
 BT3 Natural Resources
RT Bogs
RT Fens
RT Marshes
RT Muck Soils
RT Surface Waters
SO Environmental Protection Agency Glossary
DEF (EPA) Types of wetland dominated by woody vegetation and not accumulating appreciable peat deposits. Swamps may be fresh or salt water and tidal or non-tidal.

SY-Synonymous Terms   SO-Source/Subject Category   SF-See From

SWAT
DA   October 12, 1990
SEE Solid Waste Assessment Test
SO   Acronyms

SWDA
DA   October 12, 1990
SEE Solid Waste Disposal Act
SO   Acronyms

SWEEPING THE GAS PLENUM
(NFI)
DA   October 12, 1990
BT1 Processes
SO   Nuclear Facilities Incident Database
SO   Radiation
DEF (NFI) Removal of hydrogen,
     oxygen, and deuterium.

SWGR
DA   October 12, 1990
SEE Switchgear Room
SO   Acronyms

SWITCH BOX
(2423 SWITCH BOX)
DA   January 3, 1991
BT1 Equipment/Parts - Electrical (DOE
     FRASE Vocabulary)
 BT2 Equipment
SO   DOE FRASE VOCABULARY

SWITCHBOARD
(2424 SWITCHBOARD)
DA   January 3, 1991
BT1 Equipment/Parts - Electrical (DOE
     FRASE Vocabulary)
 BT2 Equipment
SO   DOE FRASE VOCABULARY

SWITCHGEAR
(2425 SWITCHGEAR)
DA   January 3, 1991
BT1 Equipment/Parts - Electrical (DOE
     FRASE Vocabulary)
 BT2 Equipment
SO   DOE FRASE VOCABULARY

SWITCHGEAR ROOM
DA   January 8, 1991
SF   *SWGR*

SWITCHYARD
(1622 SWITCHYARD)
DA   December 10, 1990
BT1 Site (DOE FRASE Vocabulary)
 BT2 Sites/Areas
SO   DOE FRASE VOCABULARY

SWMU
DA   October 12, 1990
SEE Solid Waste Management Units
SO   Acronyms

SWP
DA   October 12, 1990

SEE Safety Work Permit
SO   Acronyms

SWP CONDITIONS
(NFI)
DA   October 12, 1990
BT1 Conditions
SO   Radiation
DEF (NFI) Safe Work Permit conditions.

SWS (Salt Water System)
DA   October 12, 1990
SEE Salt Water System
SO   Acronyms

SWS (Service Water System)
DA   October 12, 1990
SEE Service Water System
SO   Acronyms

SWSA
DA   October 12, 1990
SEE Solid Waste Storage Area
SO   Acronyms

SYNERGISM
(EPA)
DA   October 12, 1990
SO   Environmental Protection Agency
     Glossary
DEF (EPA) The cooperative interaction
     of two or more chemicals or other
     phenomena producing a greater
     total effect than the sum of their
     individual effects.

SYNOVITIS
(1401 SYNOVITIS)
DA   November 28, 1990
BT1 Illnesses
RT   Tendonitis
SO   DOE FRASE VOCABULARY

SYNTHETIC ORGANIC CHEMICALS
(EPA)
DA   October 12, 1990
SF   *SOCs*
BT1 Organic Chemicals
 BT2 Chemical Substances
NT1 Polyelectrolytes
NT1 Volatile Synthetic Organic
     Chemicals
RT   Xenobiotic
SO   Environmental Protection Agency
     Glossary
DEF (EPA) Man-made organic
     chemicals. Some SOCs are
     volatile, others tend to stay
     dissolved in water rather than
     evaporate out of it.

SYSTEM (SAFETY) ANALYSES
(SSDC)
DA   October 12, 1990
BT1 Safety Analysis
 BT2 Analyses
RT   Safety Precedence Sequence

SO   System Safety Development Center
     Glossary
DEF (SSDC) The formal analyses of a
     system and the interrelationships
     among its various parts (including
     plant and hardware, policies and
     procedures, and personnel) to
     determine the real and potential
     hazards within the system, and
     suggest ways to reduce and
     control those hazards.

SYSTEM AUDITS
(ESH)
DA   October 12, 1990
BT1 Audits (SSDC)
 BT2 Administrative Processes
 BT3 Processes
SO   Quality Assurance
DEF (ESH) Overall evaluations of
     projects to: verify that sampling
     methodology is being performed in
     accordance with program
     requirement; check on the use of
     appropriate QA/QC measures;
     check methods of sample
     handling; identify any existing
     quality problems; check program
     documentation; initiate corrective
     action if a problem is identified;
     assess personnel experience and
     qualifications as required; provide
     on-site debriefings for sampling
     personnel; and provide written
     evaluations of the sampling
     program.

SYSTEM SAFETY
(SSDC)
DA   October 12, 1990
BT1 Safety
RT   Failure Mode and Effect Analysis
RT   General Design Process
SO   System Safety Development Center
     Glossary
DEF (SSDC) Safety analysis (usually
     specialized and sophisticated)
     applied as an adjunct to design of
     an engineered system. While
     many associate system safety
     primarily with the hardware portion
     of the system, it includes all
     aspects of configuration control.

**SYSTEM SAFETY DEVELOPMENT
CENTER**
(SSDC)
DA   October 12, 1990
SF   *SSDC*
BT1 EG&G Idaho, Inc.
 BT2 Companies
 BT3 Commercial Organizations
 BT4 Organizations
 BT2 DOE Contractors
 BT3 Potentially Responsible Parties
RT   Radiation Records Repositories
RT   Safety Performance Measurement
     System

BT-Broader Term   NT-Narrower Term   RT Related Term

SO System Safety Development Center
Glossary
DEF (SSDC) A DOE center located at
the Idaho National Engineering
Laboratory and operated by
EG&G Idaho, Inc. Its functions
include the development and
application of new technologies,
the implementation and analysis of
safety information systems, the
presentation of training seminars
and workshops, and the
publication and dissemination of
SSDC documents.

**SYSTEM SAFETY DEVELOPMENT
CENTER GLOSSARY**
(Acronyms and Abbreviations)
DA October 12, 1990
DEF A glossary of terms and definitions
as used by the System Safety
Development Center at EG&G
Idaho, Inc. in Idaho Falls, Idaho.

**SYSTEMATIC ASSESSMENT OF
LICENSEE PERFORMANCE**
DA January 8, 1991
SF *SALP*
BT1 Assessments
BT2 Administrative Processes
BT3 Processes

SYSTEMATIC EVALUATION PROGRAM
DA January 8, 1991
SF *SEP*
BT1 Programs

SYSTEMIC EFFECTS
(1402 SYSTEMIC EFF)
DA November 28, 1990
BT1 Effects
BT1 Illnesses
SO DOE FRASE VOCABULARY

SYSTEMIC PESTICIDES
(EPA)
DA October 12, 1990
BT1 Pesticides
BT2 Hazardous Substances
SO Environmental Protection Agency
Glossary
DEF (EPA) Chemicals that are taken up
from the ground or absorbed
through the surface and carried
through the system of the
organism being protected, making
the organism toxic to pests.

SYSTEMIC POISONING
(1403 SYSTEMIC POI)
DA November 28, 1990
BT1 Poisoning
BT2 Illnesses
SO DOE FRASE VOCABULARY

SYSTEMS
DA February 11, 1991
NT1 Aerial Measuring Systems

NT1 Alternate Removal Systems
NT1 Automatic Depressurization System
NT1 Buddy System
NT1 Building Frame Systems
NT2 Bearing Wall Systems
NT2 Braced Frame
NT3 Concentric Braced Frames
NT3 Eccentric Braced Frames
NT2 Horizontal Bracing Systems
NT2 Lateral Force Resisting System
NT2 Shoring
NT1 Centrifugal Collectors
NT1 Closed Vent Systems
NT2 Double Block and Bleed System
NT1 Confinement Systems
NT2 Airborne Activity Confinement
Systems
NT2 Primary Confinement Systems
NT3 Enclosures
NT1 Control Systems
NT2 Best Available Control Technology
NT2 Best Available Retrofit Technology
NT2 Chemical Volume and Control
System
NT2 Control Rod Drive Control System
NT3 Control Rod Drive Assembly
NT3 Control Rod Drive
NT4 Pistol Grip Switches
NT3 Control Rod Drive Mechanism
NT2 Energy Monitoring and Control
Systems
NT2 Intermittent Control System
NT2 Primary Emission Control Systems
NT2 Remote Detection and Control
Systems
NT2 Safety Injection Control System
NT2 Standby Liquid Control System
NT1 Conventional Systems
NT1 Core Flood System
NT1 Disposal Systems
NT1 Double Block and Bleed System
NT1 Dual Systems
NT1 Emergency Systems
NT2 Backup Systems
NT2 Containment Systems
NT3 Containment Spray System
NT4 Containment Spray System
(Post-accident Injection Phase)
NT4 Containment Spray System
(Post-accident Recirculation
Phase)
NT2 Criticality Alarm System
NT2 Diagnosis of Multiple Alarms
System
NT2 Duress Systems
NT2 Emergency Core Cooling System
NT2 Emergency Power System
NT2 Emergency Spray Water System
NT2 Emergency Broadcast System
NT2 Emergency Management
Information System
NT2 Emergency Management Systems
NT2 Emergency Cooling System
NT2 Emergency Feedwater System
NT2 Emergency Raw Cooling Water
System
NT2 Energy Emergency System
NT2 Explosion Suppression System
NT2 Fire Protection Systems
NT3 Dedicated Fire Water Systems
NT4 Sprinkler System

NT3 Fire Alarm System
NT3 Fire Suppression System
NT2 Intrusion Alarm System (Perimeter
or Interior)
NT2 National Warning System
NT2 Plant Protection System
NT2 Radiation Alarm System
NT2 Reactor Protection System
NT3 Emergency Core Cooling System
NT2 Safe Shutdown System
NT3 Sparger Assemblies
NT2 Supplementary Safety System
NT2 Surveillance Systems
NT2 Surveillance and Nuclear
Detection Systems
NT1 End Box Ventilation System
NT1 Energy Management Systems
NT1 Engineered Safety Features
Actuation System
NT1 Engineered Barrier Systems
NT2 Engineered Barriers
NT1 Hazardous Substance UST System
NT1 Hazardous Ranking System
NT1 High Pressure Injection System
NT1 High Pressure Recirculation
System
NT1 In-Situ Sampling Systems
NT1 Information Systems
NT2 Emergency Management
Information System
NT2 Failure Reporting Analysis and
Corrective Action System
NT2 Fire Protection Tracking System
NT2 Hazard Abatement Tracking
System
NT2 Integrated Risk Information
System
NT2 Nuclear Facilities Incident
Database
NT2 Nuclear Materials Management
and Standards System
NT2 Nuclear Plant Reliability Data
System
NT2 Plant Risk Status Information
Management System
NT2 Radiation Records Repositories
NT2 Real Property Inventory System
NT2 Safety Performance Measurement
System
NT3 Chemical Hazards Emergency
Management System
NT3 Computerized Accident/Incident
Reporting System
NT3 Computer Assisted Tracking
System
NT3 Occurrence Reporting and
Processing System
NT3 Personnel Expertise and
Resource Listing
NT3 Radiation Exposure Module
NT3 Standards Information
Management System
NT3 Unusual Occurrence Reporting
System
NT2 Secure Automatic
Communications Network
NT2 Technical Information Systems
NT2 Ultrasonic Ranging and Data
System
NT2 Waste Information Network
NT1 Leachate Collection Systems

NT1 Leak Detection Systems
 NT2 Beetles
 NT2 Leak Detector
NT1 Low Pressure Recirculation System
NT1 Main Feedwater and Condensate
      System
NT1 Management Systems
NT1 Metric System
NT1 Moderator Recovery System
NT1 Modified Hazard Ranking System
NT1 Monitoring Systems
 NT2 Energy Monitoring and Control
      Systems
NT1 Multiple Integrated Laser
      Engagement System
NT1 Multimedia Environmental Pollutant
      Assessment System
NT1 National Pollutant Discharge
      Elimination System
 NT2 Pollutant Discharge Elimination
      System
NT1 New Tank Systems
NT1 Petroleum UST Systems
NT1 Power Conversion System
NT1 Reactor Coolant System
 NT2 Emergency Spray Water System
NT1 Reactor Electric Distribution System
NT1 Remedial Action Assessment
      System
NT1 Resource Recovery Systems
NT1 Salt Water System
NT1 Secondary Hood Systems
NT1 Service Water System
NT1 Solid State Protection System
NT1 Standby Gas Treatment System
NT1 Systems (DOE FRASE Vocabulary)
 NT2 Cooling System
 NT2 Criticality Alarm System
 NT2 Deluge System
 NT2 Drinking Water System
 NT2 Emergency Core Cooling System
 NT2 Emergency Power System
 NT2 Explosion Suppression System
 NT2 Fire Alarm System
 NT2 Fire Suppression System
 NT2 Foam System
 NT2 Halon System
 NT2 HVAC System
 NT2 Instrument Air System
 NT2 Interlock System
 NT2 Off Gas System
 NT2 Plant Protection System
 NT2 Pressure Relief System
 NT2 Protective Relay System
 NT2 Public Address System
 NT2 Purge System
 NT2 Radiation Alarm System
 NT2 Raw Water System
 NT2 Sanitary Treatment System
 NT2 Sewer System
 NT2 Sprinkler System
 NT2 Vacuum System
 NT2 Ventilation System
  NT3 End Box Ventilation System
  NT3 Reactor Building Standby
      Ventilation System
 NT2 Waste Water System
  NT3 Waste Water Collection Systems
NT1 Test Systems
NT1 Vapor Capture Systems
 NT2 Floor Sweeps

NT1 Water Systems
 NT2 Public Water Systems
  NT3 Community Water Systems
  NT3 Drinking Water System
  NT3 Non-Community Water Systems
  NT3 Non-Transient Non-Community
      Water System
  NT3 Water Supply System
 NT2 Storm Water Collection Systems
 NT2 Waste Water System
  NT3 Waste Water Collection Systems
RT Controls
RT Targets

SYSTEMS (DOE FRASE
VOCABULARY)
(DOE FRASE Vocabulary Numeric Keys
 2025-2099)
DA   December 10, 1990
BT1 Systems
NT1 Cooling System
NT1 Criticality Alarm System
NT1 Deluge System
NT1 Drinking Water System
NT1 Emergency Core Cooling System
NT1 Emergency Power System
NT1 Explosion Suppression System
NT1 Fire Alarm System
NT1 Fire Suppression System
NT1 Foam System
NT1 Halon System
NT1 HVAC System
NT1 Instrument Air System
NT1 Interlock System
NT1 Off Gas System
NT1 Plant Protection System
NT1 Pressure Relief System
NT1 Protective Relay System
NT1 Public Address System
NT1 Purge System
NT1 Radiation Alarm System
NT1 Raw Water System
NT1 Sanitary Treatment System
NT1 Sewer System
NT1 Sprinkler System
NT1 Vacuum System
NT1 Ventilation System
 NT2 End Box Ventilation System
 NT2 Reactor Building Standby
      Ventilation System
NT1 Waste Water System
 NT2 Waste Water Collection Systems
RT Controls
RT DOE FRASE Categories
SO DOE FRASE VOCABULARY
DEF A subject category used with the
     DOE FRASE Vocabulary.

T&B
DA   October 12, 1990
SEE Top and Bottom
SO Acronyms

TA
DA   October 12, 1990
SEE Test Authorizations
SO Acronyms

TABLE SAW
(2147 TABLE SAW)
DA   December 10, 1990
BT1 Machines (DOE FRASE
     Vocabulary)
 BT2 Equipment
BT1 Saws
SO DOE FRASE VOCABULARY

TABLETOP
(EMER)
DA   February 1, 1991
BT1 Exercises
SO Emergency Preparedness
DEF (EMER) An exercise in which key
     officials and agency
     representatives are presented with
     a series of simulated problems
     based on a scenario. Participants
     are seated together and problems
     are presented through questions.
     This exercise is moderated to
     control play and can be formal or
     informal.

TACHOMETER(S)
(2773 TACHOMETER)
DA   January 3, 1991
BT1 Indicator(s)
 BT2 Instrument(s)
  BT3 Equipment/Parts -
      Instrumentation/Measuring
      (DOE FRASE Voc.)
   BT4 Equipment
SO DOE FRASE VOCABULARY

TACTICAL RESPONSE FORCES
(DOE Order 6430.1A)
DA   October 12, 1990
BT1 Protective Force Personnel
 BT2 Personnel
SO Construction
DEF (DOE Order 6430.1A) Armed
     combat forces trained in security
     protection.

TACTICAL RESPONSE TEAMS
(EMER)
DA   February 13, 1991
SY Special Response Teams
BT1 Response Teams
 BT2 Teams
  BT3 Administrative Organizations
   BT4 Organizations
SO Emergency Preparedness
DEF (EMER) Groups of one or more
     armed security inspectors who are
     specially trained and equipped to
     respond to security incidents. Also
     called Special Response Teams.

TAGGING
(Doe Order 5480.16)
DA   October 12, 1990
BT1 Procedures
RT Armorers
RT Firearms
SO Firearms
DEF (DOE Order 5480.16) A safety

procedure involving labeling a
defective weapon in order to
identify the weapon's status (e.g.,
faulty, safe, requires cleaning).

TAILINGS
(CFR; EPA; NCRP)
DA   October 12, 1990
BT1 Solid Wastes
 BT2 Wastes
NT1 Asbestos Tailings
NT1 New Tailings
RT   Coal Refuse
RT   Continuous Disposal
RT   Dewatered
RT   Existing Tailings Piles
RT   New Tailings Impoundment
RT   Phased Disposal
RT   Radioactive Materials
RT   Uranium By-Product Materials
SO   Environmental Protection Agency
     Glossary
SO   Radiation
DEF (EPA) Residues of raw materials or
     wastes separated out during the
     processing of crops or mineral
     ores. (CFR) The wastes produced
     by the extraction or concentration
     of uranium from any ore processed
     primarily for its source material
     content. Ore bodies depleted by
     uranium solution extractions and
     which remain underground do not
     constitute byproduct material.

TALM
DA   October 12, 1990
SEE Temperature Alarm Monitor
SO   Acronyms

TAN
DA   October 12, 1990
SEE Test Area North
SO   Acronyms

TANGIBLE NET WORTH
(SWDA; RCRA; CFR; ESH)
DA   October 12, 1990
BT1 Net Worth
SO   Wastes
SO   Water Pollution
DEF The tangible assets that remain
     after deducting liabilities; such
     assets would not include
     intangibles such as goodwill and
     rights to patents or royalties.

TANK FARM
(1623 TANK FARM)
DA   December 10, 1990
BT1 Site (DOE FRASE Vocabulary)
 BT2 Sites/Areas
SO   DOE FRASE VOCABULARY

TANK SYSTEMS
(SWDA; RCRA; CFR)
DA   October 12, 1990
NT1 Existing Tank Systems
NT1 New Tank Systems

NT1 UST Systems
 NT2 Hazardous Substance UST
     System
 NT2 Petroleum UST Systems
RT   Excavation Zone
RT   Installation Inspectors
RT   Operational Life
SO   Environmental Management
SO   Wastes
DEF (CFR) Underground storage tanks,
     connected underground piping,
     underground ancillary equipment,
     and containment systems, if any.

TANKS
(SWDA; RCRA; CFR; ESH)
DA   October 12, 1990
BT1 Facility Components
NT1 Aboveground Tanks
NT1 Aeration Tanks
NT1 Boron Injection Tank
NT1 Cargo Tanks
NT1 Clarifiers
NT1 Condensate Storage Tank
NT1 Containment Storage Tanks
NT1 Coolant Return Tank
NT1 Core Flood Tanks
NT1 Double-shell Tank
NT1 Evaporator Condensate Tank
NT1 Farm Tanks
NT1 Flow-Through Process Tanks
NT1 Fuel Tanks
NT1 Hydraulic Lift Tanks
NT1 Inground Tanks
NT1 On Ground Tanks
NT1 Portable Tanks
 NT2 Intermodal Portable Tanks
NT1 Quench Tanks
NT1 Refueling Water Storage Tank
NT1 Residential Tanks
NT1 Safety Injection Tank
NT1 Sedimentation Tanks
NT1 Settling Tanks
NT1 Single Shell Tanks
NT1 Underground Tanks
 NT2 Underground Storage Tanks
  NT3 Non-Operational Storage Tanks
  NT3 Septic Tanks
NT1 Wastewater Treatment Tanks
RT   Barriers
RT   Containers
RT   Hazardous Waste Management
     Units
RT   Hazardous Materials
RT   Inner Liners
RT   Sanitary Engineering Structures
RT   Storage
SO   Environmental Management
SO   Wastes
DEF (ESH) A stationary device,
     designed to contain an
     accumulation of hazardous waste,
     that is constructed primarily of
     nonearthen material (such as
     wood, concrete, steel, or plastic),
     that provides structural support.

TAP
DA   October 12, 1990

SEE Toxic Air Pollutants
SO   Acronyms

TARGET COMPOUND LIST
(EPA)
DA   October 12, 1990
SO   Environmental Protection Agency
     Glossary
DEF (EPA) Developed by EPA for
     Superfund site sample analyses.
     The TCL is a list of analytes (34
     volatile organic chemicals, 65
     semivolatile organic chemicals, 19
     pesticides, 7 polychlorinated
     biphenyls, 23 metals, and total
     cyanide) for which every
     Superfund sample must be
     analyzed using the RAS of the
     EPA Contract Laboratory Program.

TARGET SLEEVE HOUSING
DA   January 8, 1991
SF   TSH

TARGETS
(SSDC; MORT)
DA   October 12, 1990
RT   Accidents
RT   Aircraft
RT   Animals
RT   Archaeological Resources
RT   Barriers
RT   Controls
RT   Ecosystems
RT   Environment
RT   Equipment
RT   Facilities and Buildings (DOE
     FRASE Vocabulary)
RT   Facilities
RT   Hazards
RT   Human Body Parts
RT   Indian Lands
RT   Organisms
RT   Personnel
RT   Reactor Components
RT   Sites/Areas
RT   Structures (DOE FRASE
     Vocabulary)
RT   Systems
RT   Vessels
SO   System Safety Development Center
     Glossary
DEF (SSDC) Persons, objects, or
     animals upon which an unwanted
     energy flow may act to cause
     damage, injury, or death. (MORT)
     MORT analysis asks: what
     vulnerable people and/or objects
     of value were exposed to the
     harmful energy flow or
     environmental condition?

TASK ANALYSIS
(SSDC)
DA   October 12, 1990
BT1 Analyses
SO   System Safety Development Center
     Glossary
DEF (SSDC) The systematic review of a

collection of actions or behaviors necessary and sufficient to complete a given task. It includes an extremely thorough look at the individual elements and supplements comprising a task.

TAXONOMY
(SSDC)
DA October 12, 1990
SO System Safety Development Center Glossary
DEF (SSDC) The systematic classification of a subject.

TB
DA October 12, 1990
SEE Turbine Building
SO Acronyms

TBSCCW
DA October 12, 1990
SEE Turbine Building Secondary Closed Cooling Water
SO Acronyms

TBSW
DA October 12, 1990
SEE Turbine Building Service Water
SO Acronyms

TBT PAINTS (TRYBUTILIN)
(EPA)
DA November 13, 1990
RT Organotins
SO Water Pollution
DEF (EPA) Anti-foulant paints used to protect the hulls of boats and ships, buoys, and dock pilings from marine organisms such as barnacles.

TBV
DA October 12, 1990
SEE Turbine Block Valve
SO Acronyms

Tc (Periodic element)
DA October 12, 1990
SEE Technetium
SO Acronyms

TC (Thermocouple)
DA October 12, 1990
SEE Thermocouple(s)
SO Acronyms

TCE
DA October 12, 1990
SEE Trichloroethylene (TCE)
SO Acronyms

TCLP
DA October 12, 1990
SEE Toxicity Characteristic Leaching Procedure
SO Acronyms

TD
DA October 12, 1990
SEE Time Delay
SO Acronyms

TDC
DA June 4, 1991
SEE Tennessee Department of Commerce
SO Acronyms

TEAMS
DA February 13, 1991
BT1 Administrative Organizations
 BT2 Organizations
NT1 Management Teams
 NT2 Accreditation Review Teams
 NT2 Continuity of Government Emergency Management Teams
 NT2 Crisis Management Teams
 NT2 Emergency Management Teams
  NT3 Energy Emergency Management Teams
  NT3 Operational Emergency Management Teams
 NT2 Headquarters Coordinating Teams
NT1 Response Teams
 NT2 Airborne Response Teams
 NT2 Emergency Support Teams
 NT2 Emergency Response Teams
 NT2 Energy Emergency Teams
 NT2 Environmental Response Teams
 NT2 Explosive Ordnance Disposal Teams
 NT2 Hazardous Materials Response Teams
 NT2 Hostage Negotiation Teams
 NT2 National Response Teams
 NT2 Nuclear Emergency Search Teams
 NT2 Radiological Assistance Teams
 NT2 Regional Response Teams
 NT2 Special Response Teams
 NT2 Tactical Response Teams
NT1 Tiger Teams
DEF (WTID) Consist of a number of persons associated together in work or activity; a group of specialists or scientists functioning as a collaborative unit.

TECHNETIUM
DA January 8, 1991
SF *Tc (Periodic element)*

TECHNICAL EXPERTS
(DOE Order 5480.15; ESH)
DA October 12, 1990
BT1 Technically Qualified Individuals
 BT2 Personnel
RT Personnel Dosimetry Programs
SO Management
SO Standards
DEF (DOE Order 5480.15) Technically trained individuals with professional experience in personnel radiation dosimetry (programmatic or research) for a minimum of 5 years who may serve as a site assessor or as a

member of the Oversight or Appeals Boards. Technical experts are nominated by Managers of Operations Offices and selected by the DOE Laboratory Accreditation Program Administrator.

TECHNICAL INFORMATION
(MORT)
DA April 15, 1991
RT Analyses
RT Communication
RT Controls
RT Corrective Action Triggers
RT First Line Supervision
RT Higher Supervision
RT Inspections
RT Maintenance
RT Monitoring
RT Operability
RT Technical Information Systems
RT Trending
DEF (MORT) MORT analysis asks: was adequate technical information relevant to the work flow process available to the "action" person? Often relevant information exists but is not available to the "action" persons associated with the process. Possible reasons for this are: less that adequate knowledge of work flow processes; less than adequate (LTA) communication of relevant knowledge; LTA monitoring systems; LTA data collection and analysis; LTA audit and appraisal of the total safety system; and LTA initiators of the Hazard Analysis process.

TECHNICAL INFORMATION SYSTEMS
(MORT)
DA April 3, 1991
BT1 Information Systems
 BT2 Security Interests
 BT2 Systems
RT Goals
RT Hazard Analysis
RT Knowledge
RT Maintenance
RT Safety Program Reviews
RT Technical Information
DEF (MORT) MORT analysis asks: was the technical information system adequate with respect to the unwanted energy flow? Complex work flow processes must be supported by complete technical information systems. It is axiomatic that complex systems will depart from plans and procedures to some degree. Therefore, information systems need to detect deviations, determine rates and trends, initiate corrections, and assure that goals are attained. MORT conceives a technical information system as consisting of "research" persons,

"program" persons, and "action" persons obtaining, handling, and providing technical information relevant to the work flow process in a "communication" network.

TECHNICAL NAMES
(SWDA; RCRA; ESH)
DA   October 12, 1990
RT   Name of Contents
RT   Organic Chemicals
RT   Proper Shipping Name
SO   Hazardous Materials
DEF Recognized chemical names currently used in scientific and technical handbooks, journals, and texts. Generic descriptions authorized for use as technical names are: Organic phosphate compound, Organic phosphorus compound, Organic phosphate compound mixture, Organic phosphorus compound mixture, Methyl parathion, and Parathion.

TECHNICAL SAFETY APPRAISALS
(DOE Orders 5480.1B, 5480.5, 5480.6, 5482.1B and 6430.1A)
DA   October 12, 1990
SF   *TSA*
BT1 Safety Appraisals
BT2 Appraisals
RT   Compliance Considerations
RT   Environmental Surveys
RT   Nuclear Facilities
RT   Potential Hazards
RT   Seriousness
SO   Construction
SO   Environmental Management
SO   Industrial Hygiene
DEF Documented, multidiscipline appraisals of selected department reactors and nuclear facilities conducted by a team selected by the Deputy Assistant Secretary for Safety, Health, and Quality Assurance (EH-30). They assure proper Department-wide application of particular safety elements of the ES&H program, nuclear industry lessons learned, and appropriate licensed facility requirements, as described in DOE 5482.1B, paragraph 9b.

TECHNICAL SPECIFICATIONS
(DOE Order 5480.6)
DA   October 12, 1990
SF   *TS*
BT1 Specifications
BT2 Requirements
SO   Industrial Safety
DEF (DOE Order 5480.6) Safety documents approved by DOE which in a specified format defines the conditions, safety boundaries, and procedures under which activities are to be carried out at a reactor. See Code of Federal Regulations, Title 10, Part 50.36.

TECHNICAL SUPPORT
(DOE Order 4330.4A)
DA   June 5, 1991
RT   Conceptual Designs
RT   Inspections
RT   Plans
RT   Preliminary Designs
SO   Management
DEF (DOE Order 4330.4A) The engineering, design, specialized inspections, planning, or other such support of capital asset maintenance and repair.

TECHNICALLY QUALIFIED INDIVIDUALS
DA   January 24, 1991
BT1 Personnel
NT1 Approved Medical Practitioners
NT1 Certified Applicators
NT1 Competent Authorities
NT1 Corrosion Experts
NT1 Installation Inspectors
NT1 Reactor Supervisor
NT1 Subject Matter Experts
NT1 Technical Experts
NT1 Trained Investigators
SO   Environmental Management
DEF (CFR) A person or persons (1) who, because of education, training, or experience, or a combination of these factors, is capable of understanding the health and environmental risks associated with the chemical substance which is used under his or her supervision, (2) who is responsible for enforcing appropriate methods of conducting scientific experimentation, analysis, or chemical research to minimize such risks, and (3) who is responsible for the safety assessments and clearances related to the procurement, storage, use and disposal of the chemical substance as may be appropriate or required within the scope of conducting a research and development activity.

TECHNICIANS
DA   November 29, 1990
BT1 Professional Personnel
BT2 Occupations
BT2 Personnel
NT1 Engineering Technician
NT1 Health Technician
NT1 Miscellaneous Technician
NT1 Radiation Monitor/Technician
NT1 Science Technician
SO   DOE FRASE VOCABULARY

TECHNIQUE FOR HUMAN ERROR RATE PREDICTION
DA   January 8, 1991
SF   *THERP*
RT   Human Performance
RT   Human Factors

TECHNOLOGY-BASED STANDARDS
(EPA)
DA   October 12, 1990
BT1 Standards
BT2 Codes, Standards, and Regulations
SO   Environmental Protection Agency Glossary
DEF (EPA) Effluent limitations applicable to direct and indirect sources which are developed on a category-by-category basis using statutory factors, not including water-quality effects.

TECTONIC DEFORMATIONS
(USGS)
DA   October 12, 1990
BT1 Act of Nature
RT   Earthquake Magnitude
RT   Faults
RT   Ground Cracks
RT   Ground Failures
RT   Subsidence
RT   Uplifts
SO   Natural Phenomenon
DEF (USGS) Accompany surface faulting. The deformation may be local, affecting a narrow zone near a fault break, or it may involve differential vertical and horizontal movements over broad parts of the Earth's crust.

TELECOMMUNICATION
(EMER)
DA   February 1, 1991
SO   Emergency Preparedness
DEF (EMER) Any transmission, emission, or reception of signs, signals, writings, images, and sounds or intelligence of any nature by wire, radio, visual, or other electromagnetic systems.

TELLURIUM 132
(EDB)
DA   March 29, 1991
BT1 Beta-Minus Decay Radioisotopes
BT2 Beta Decay Radioisotopes
BT3 Radionuclides
BT4 CERCLA Hazardous Substances
BT5 Hazardous Substances
BT4 Nuclides
SO   Radiation

TEMPERATURE ALARM MONITOR
DA   January 8, 1991
SF   *TALM*
BT1 Monitors
BT2 Equipment

TEMPERATURE SCRAM CIRCUIT MONITOR
DA   January 8, 1991
SF   *TSCM*
BT1 Monitors

BT2 Equipment
RT Scram

TEMPORARILY ABANDONED AREAS
(CAA; CFR)
DA October 12, 1990
BT1 Abandoned Areas
BT2 Sites/Areas
SO Air Pollution
DEF (CFR) Mine areas in which further
work is not intended for at least
six months. Areas which function
as escapeways, formerly used
lunchrooms, shops, and
transformer or pumping stations
are not considered abandoned
areas. Except for designated
ventilation passageways designed
to minimize the distance to vents,
worked-out mine areas are
considered temporarily abandoned
areas for the purpose of this
subpart if work is not intended in
the area for at least six months.

TEMPORARY EXEMPTIONS
(DOE Order 5480.4; ESH)
DA October 12, 1990
BT1 Exemptions
RT Mandatory Standards
SO Environmental Management
SO Management
SO Standards
DEF (DOE Order 5480.4) Short-term
releases from a mandatory
standard of this Order. Such
exemptions shall not exceed 1
year, except that in unusual cases
a renewal may be granted, not to
exceed an additional year.

TEMPORARY PROCEDURE CHANGE
DA January 8, 1991
SF TPC
BT1 Change
RT Procedures

TEMPORARY VARIANCES
(DOE Order 5483.1A)
DA October 12, 1990
BT1 Variances
RT Permanent Variances
SO Environmental Management
SO Industrial Safety
DEF (DOE Order 5483.1A) A short-term
release from a DOE-prescribed
OSHA standard. Such variances
shall not exceed 1 year, except
that in unusual cases a renewal
may be granted, not to exceed an
additional year.

TENDONITIS
(1404 TENDONITIS)
DA November 28, 1990
BT1 Illnesses
RT Muscle/Tendon(s)
RT Synovitis
SO DOE FRASE VOCABULARY

**TENNESSEE DEPARTMENT OF
COMMERCE**
DA June 4, 1991
SF TDC
BT1 State Agencies
BT2 Agencies
BT3 Administrative Organizations
BT4 Organizations
NT1 Bureau of Environment

TENSION WIRES
(DOE Order 6430.1A)
DA October 12, 1990
RT Facility Boundaries
SO Construction
DEF (DOE Order 6430.1A) Wires placed
along the top and bottom of a
chain link fence to provide tension
and structural rigidity.

TERATOGENS
(EPA)
DA October 12, 1990
BT1 Hazardous Substances
SO Environmental Protection Agency
Glossary
DEF (EPA) Substances that causes
malformation or serious deviation
from normal development of
embryos and fetuses.

TERMINATED EMPLOYEES
(Doe Order 5484.1)
DA October 12, 1990
BT1 Personnel
SO Management
DEF (DOE Order 5484.1) For the
purpose of this Order, individuals
employed by DOE or a DOE
contractor who terminate their
employment, transfer to another
DOE or contractor facility or office,
begin a leave of absence which
results in the termination of
radiation monitoring of greater
than 12 months duration, or all
employees of a contractor whose
contracts with DOE are
terminated.

TERRACING
(EPA)
DA October 12, 1990
BT1 Dikes
BT2 Hydraulic Structures
BT3 Structures (DOE FRASE
Vocabulary)
SO Environmental Protection Agency
Glossary
DEF (EPA) Diking, built along the
contour of sloping agricultural
land, that holds runoff and
sediment to reduce erosion.

TERRITORIAL SEAS
(CWA; RHA; CFR)
DA October 12, 1990
RT Coastal Zones
RT Contiguous Zones

RT Waters of the United States
SO Water Pollution
DEF The belt of the seas measured from
the line of ordinary low water
along that portion of the coast
which is in direct contact with the
open sea and the line marking the
seaward limit of inland waters, and
extending seaward a distance of
three miles.

TERRORIST
(EMER)
DA February 1, 1991
SO Emergency Preparedness
DEF (EMER) A person, or group of
persons, who organize, conspire,
or act in a manner that threatens
or carries out violence for the
purpose of political coercion.

TERRORIST ACTS
(EMER)
DA February 1, 1991
RT Terrorist Threats
SO Emergency Preparedness
DEF (EMER) Any potential or actual
occurrence which creates or
appears likely to create a condition
resulting in sabotage, bodily harm,
unlawful access to U.S.
Department of Energy (DOE) or
DOE contractor facilities, or the
interruption or loss of vital
services.

TERRORIST RESPONSE PLANS
(EMER)
DA February 1, 1991
BT1 Emergency Plans
BT2 Plans
SO Emergency Preparedness
DEF (EMER) Plans for response to
hostile action taken against a U.S.
Department of Energy (DOE)
contractor. Response plans may
be written by contractor security
organizations, DOE security
organizations, local municipal and
state organizations, or the Federal
Bureau of Investigation (FBI).

TERRORIST THREATS
(EMER)
DA February 1, 1991
BT1 Threats
BT2 Emergencies
BT3 Reportable Occurrences
BT4 Occurrences
RT Terrorist Acts
SO Emergency Preparedness
DEF (EMER) Any potential or actual
occurrence which creates or
appears likely to create a condition
resulting in sabotage, bodily harm,
unlawful access to U.S.
Department of Energy (DOE) or
DOE contractor facilities, or the

BT-Broader Term

NT-Narrower Term

RT Related Term

interruption or loss of vital services.

TERTIARY TREATMENT
(EPA)
DA October 12, 1990
BT1 Wastewater Treatment Processes
BT2 Treatment
BT3 Waste Management Processes
BT4 Processes
SO Environmental Protection Agency Glossary
DEF (EPA) Advanced cleaning of wastewater that goes beyond the secondary or biological stage. It removes nutrients such as phosphorus and nitrogen and most BOD and suspended solids.

TEST AND EVALUTION FACILITIES
(NWPA; ANL)
DA May 24, 1991
BT1 Disposal Facilities
BT2 Hazardous Waste Management Facilities
BT3 Hazardous Waste Facilities
BT4 Facilities
RT High-Level Radioactive Wastes
RT Spent Fuel
RT Transuranic Wastes (TRU Waste)
RT Waste Management
SO Radiation
SO Wastes

TEST AREA
(1624 TEST AREA)
DA December 10, 1990
BT1 Area
BT2 Sites/Areas
NT1 Fast Flux Test Facility
NT1 Nevada Test Site
NT1 Rocket Motor Test Sites
NT1 Test Reactor Facility
NT1 Test Stations
NT1 Test Reactor Area
NT1 Tonopah Test Range
RT Research/Testing Activity
SO DOE FRASE VOCABULARY

TEST AREA NORTH
DA January 8, 1991
SF TAN
BT1 Sites/Areas
SO Management

TEST AUTHORIZATIONS
DA January 8, 1991
SF TA
BT1 Administrative Processes
BT2 Processes
BT1 Permits
RT Final Authorizations
RT Interim Authorizations
SO Management

TEST MIXTURES
(CFR)
DA November 9, 1990

BT1 Mixtures
DEF (CFR) Substances or mixtures administered or added to a test system in a study, which substance or mixture is used to develop data to meet the requirements of a TSCA (Toxic Substances Control Act) section 4(a) test rule and/or is developed under a negotiated testing agreement or section 5 rule/order to the extent the agreement or rule/order references this part.

TEST REACTOR AREA
DA January 8, 1991
SY Advanced Test Reactor
SF TRA
BT1 Test Area
BT2 Area
BT3 Sites/Areas
RT Reactor Facilities

TEST REACTOR FACILITY
(1808 TEST REACTOR)
DA December 10, 1990
BT1 Facility (DOE FRASE Vocabulary)
BT2 Facilities and Buildings (DOE FRASE Vocabulary)
BT3 Facilities
BT1 Reactor Facilities
BT2 Nuclear Facilities
BT3 Facilities
BT1 Test Area
BT2 Area
BT3 Sites/Areas
RT Research/Testing Activity
SO DOE FRASE VOCABULARY
DEF (NWPA) An at-depth, prototypic, underground cavity with subsurface lateral excavations extending from a central shaft that is used for research and development purposes, including the development of data and experience for the safe handling and disposal of solidified high-level radioactive waste, transuranic waste, or spent nuclear fuel.

TEST STATIONS
(NFI)
DA October 12, 1990
BT1 Test Area
BT2 Area
BT3 Sites/Areas
SO Nuclear Facilities Incident Database
SO Radiation
DEF (NFI) Testing of fuel assemblies; fluid flow tests.

TEST SUBSTANCES
(TSCA; CFR)
DA October 19, 1990
RT Batches
RT Control Substances
RT Test Systems
RT Vehicles
SO Environmental Management

SO Hazardous Materials
DEF (CFR) Substances or mixtures administered or added to a test system in a study, which substance or mixture is used to develop data to meet the requirements of a TSCA (Toxic Substances Control Act) section 4(a) test rule and/or is developed under a negotiated testing agreement or section 5 rule/order to the extent the agreement, rule or order references 40 CFR 792.3.

TEST SYSTEMS
(TSCA; CFR)
DA October 19, 1990
BT1 Systems
RT Biomonitoring
RT Test Substances
SO Environmental Management
SO Hazardous Materials
DEF Any animals, plants, microorganisms, or subparts thereof, to which the test or control substance is administered or added for study. Test systems also include appropriate groups or components of the system not treated with the test or control substances.

TEST TRAIN
(2594 TEST TRAIN)
DA January 3, 1991
BT1 Equipment/Parts - Nuclear (DOE FRASE Vocabulary)
BT2 Equipment
BT2 Reactor Components
SO DOE FRASE VOCABULARY

TESTING EQUIPMENT
(2774 TEST EQUIPMT)
DA January 3, 1991
BT1 Instrument(s)
BT2 Equipment/Parts - Instrumentation/Measuring (DOE FRASE Voc.)
BT3 Equipment
NT1 Heat Detector
NT1 Leak Detector
NT1 Oscilloscope(s)
NT1 Scintilliation Probe(s)
NT1 Sensor(s)
RT Research/Testing Activity
SO DOE FRASE VOCABULARY

TESTING REQUIREMENTS
(TSCA; CFR)
DA October 12, 1990
BT1 Requirements
RT Chemical Substances
RT Risk Assessment
RT Testing Requirements Rule
SO Hazardous Materials
DEF (CFR) The Administrator shall by rule require that testing be conducted on such substance or mixture as may present an

unreasonable risk of injury to health or the environment in order to develop data with respect to the health and environmental effects for which there is an insufficiency of data and experience and which are relevant to a determination that the manufacture, distribution in commerce, processing, use, or disposal of such substance or mixture, or any combination of such activities, does or does not present an unreasonable risk of injury to health or the environment. More complete discussion at 15 U.S.C. 2603, Testing of Chemical Substances and Mixtures.

TESTING REQUIREMENTS RULE
(TSCA, CFR)
DA   October 12, 1990
BT1 Requirements
RT   Chemical Substances
RT   Testing Requirements
SO   Hazardous Materials
DEF (CFR) A rule under [15 U.S.C.2603, Section 4, subsection (a)—Testing Requirements] shall include (A) identification of the chemical substance or mixture for which testing is required under the rule, (B) standards for the development of test data for such substance or mixture, and (C)with respect to chemical substances which are not new chemical substances and to mixtures, a specification of the period (which period may not be of unreasonable duration) within which the persons required to conduct the testing shall submit to the Administrator data developed in accordance with the standards referred to in subparagraph (B).

TETANUS
(1406 TETANUS)
DA   November 28, 1990
BT1 Illnesses
SO   DOE FRASE VOCABULARY

THEORETICAL ARSENIC EMISSIONS FACTOR
(CAA; CFR)
DA   October 12, 1990
BT1 Emission Factor
BT2 Ratios
RT   Arsenic Kitchens
RT   Arsenic Containing Glass Types
RT   Inorganic Arsenic
RT   Secondary Hood Systems
RT   Uncontrolled Total Arsenic Emissions
SO   Air Pollution
DEF (CFR) The amount of inorganic arsenic, expressed in grams per kilogram of glass produced, as determined based on a material balance.

THERMAL POLLUTION
(EPA)
DA   October 12, 1990
BT1 Pollution
RT   Plumes
SO   Environmental Protection Agency Glossary
DEF (EPA) Discharge of heated water from industrial processes that can affect the life processes of aquatic organisms.

THERMAL TREATMENT
(SWDA; RCRA)
DA   October 19, 1990
BT1 Treatment
BT2 Waste Management Processes
BT3 Processes
RT   Incineration
RT   Open Burning
SO   Environmental Management
SO   Wastes
DEF The treatment of hazardous waste in a device which uses elevated temperatures as the primary means to change the chemical, physical, or biological character of composition of the hazardous waste. Examples of thermal treatment processes are incineration, molten salt, pyrolysis, calcination, wet air oxidation, and microwave discharge. (See also "incinerator" and "open burning").

THERMOCOUPLE(S)
(2426 THERMOCOUPLE)
DA   January 3, 1991
SF   TC (Thermocouple)
BT1 Equipment/Parts - Electrical (DOE FRASE Vocabulary)
BT2 Equipment
SO   DOE FRASE VOCABULARY

THERMOLUMINESCENT DOSIMETERS
(NCRP; NIH)
DA   October 12, 1990
SF   TLD
BT1 Dosimeters
BT2 Radiation Detectors
BT3 Equipment
BT1 Indicator(s)
BT2 Instrument(s)
BT3 Equipment/Parts - Instrumentation/Measuring (DOE FRASE Voc.)
BT4 Equipment
SO   Radiation
DEF (NIH) Dosimeters made of certain crystalline materials which are capable of both storing a fraction of absorbed ionizing radiation and releasing this energy in the form of visible photons when heated. The amount of light released can be used as a measure of radiation exposure to these crystals.

THERMOMETER(S)
(2776 THERMOMETER)
DA   January 3, 1991
BT1 Indicator(s)
BT2 Instrument(s)
BT3 Equipment/Parts - Instrumentation/Measuring (DOE FRASE Voc.)
BT4 Equipment
SO   DOE FRASE VOCABULARY

THERMOSTAT(S)
(2777 THERMOSTAT)
DA   January 3, 1991
BT1 Gauge(s)
BT2 Instrument(s)
BT3 Equipment/Parts - Instrumentation/Measuring (DOE FRASE Voc.)
BT4 Equipment
SO   DOE FRASE VOCABULARY

THERP
DA   October 12, 1990
SEE Technique for Human Error Rate Prediction
SO   Acronyms

THIGH(S)
(1123 THIGH)
DA   November 28, 1990
BT1 Leg(s)
BT2 Human Body Parts
RT   Groin
RT   Leggings
SO   DOE FRASE VOCABULARY

THM
DA   October 12, 1990
SEE Trihalomethane
SO   Acronyms

THORIUM
(NCRP)
DA   October 12, 1990
BT1 Heavy Metals
RT   Source Materials
RT   Uranium Mill Tailings
SO   Radiation
DEF (NCRP) A naturally radioactive element. Thorium-232 is the parent of one radioactive series, and specific thorium nuclides are members of the three radionuclide series.

THORIUM 230
(EDB)
DA   March 29, 1991
BT1 Alpha Decay Radioisotopes
BT2 Radionuclides
BT3 CERCLA Hazardous Substances
BT4 Hazardous Substances
BT3 Nuclides
BT1 Years Living Radioisotopes
BT2 Radionuclides
BT3 CERCLA Hazardous Substances
BT4 Hazardous Substances

BT-Broader Term

NT-Narrower Term

RT Related Term

BT3 Nuclides
SO Radiation

THORIUM 232
(EDB)
DA March 29, 1991
BT1 Beta-Minus Decay Radioisotopes
 BT2 Beta Decay Radioisotopes
  BT3 Radionuclides
   BT4 CERCLA Hazardous
        Substances
     BT5 Hazardous Substances
   BT4 Nuclides
BT1 Minutes Living Radioisotopes
 BT2 Radionuclides
  BT3 CERCLA Hazardous Substances
   BT4 Hazardous Substances
  BT3 Nuclides

THREADER(S)
(3063 THREADER)
DA January 3, 1991
BT1 Tools - Manual
 BT2 Tools (DOE FRASE Vocabulary)
  BT3 Equipment
SO DOE FRASE VOCABULARY

THREATENED SPECIES
(ESA; CFR)
DA October 12, 1990
BT1 Species
RT Animals
RT Candidates
RT Endangered Species
RT Plants
SO Endangered Species
DEF (CFR) Any species which is likely
      to become an endangered species
      within the foreseeable future
      throughout all or a significant
      portion of its range.

THREATS
DA February 13, 1991
BT1 Emergencies
 BT2 Reportable Occurrences
  BT3 Occurrences
NT1 Bomb Threats
NT1 Covert Threats
NT1 Nuclear Threat Incidents
NT1 Terrorist Threats
RT Credibility Assessments
RT Internal Security Reports
SO Emergency Preparedness
DEF (WTID) Conditions which by their
      very nature pose harm or injury;
      something or someone intending
      to inflict evil, injury, or damage.

THRESHOLD LIMIT VALUES
(EPA; SSDC; EMER)
DA October 12, 1990
SF *TLV*
RT Published Exposure Levels
RT Time Weighted Average
SO Emergency Preparedness
SO Environmental Protection Agency
      Glossary

SO System Safety Development Center
      Glossary
DEF (EPA) Represent the air
      concentrations of chemical
      substances to which it is believed
      that workers may be daily exposed
      without adverse effect.

THRESHOLD PLANNING QUANTITIES
(CERCLA; EPA; ESH; EMER)
DA October 12, 1990
SF *TPQ*
BT1 Quantities
RT Extremely Hazardous Substances
RT Hazardous Chemicals
RT Reportable Quantities
RT Spills
SO Compensation and Liability
SO Emergency Preparedness
SO Environmental Protection Agency
      Glossary
DEF (EPA) Quantities designated for
      each chemical on the list of
      extremely hazardous substances
      that trigger notification by facilities
      to the state emergency response
      commission that such facilities are
      subject to emergency planning
      under SARA (Superfund
      Amendments and Reauthorization
      Act) Title III. (ESH) The amount of
      an extremely hazardous
      substance equal to or above which
      a facility must participate in SARA
      Title III emergency planning. The
      TPQs are provided in 40 CFR
      355, Appendices A and B.

THROAT
(1092 THROAT)
DA November 28, 1990
BT1 Respiratory System
 BT2 Body System(s)
  BT3 Human Body Parts
RT Neck
SO DOE FRASE VOCABULARY

THUMB(S)
(1103 THUMB)
DA November 28, 1990
BT1 Hand(s)
 BT2 Human Body Parts
RT Finger(s)
RT Glove(s)
RT Hand Protection
SO DOE FRASE VOCABULARY

TIDAL MARSHES
(EPA)
DA October 12, 1990
BT1 Marshes
 BT2 Wetlands
  BT3 Sites/Areas
  BT3 Surface Water Resources
   BT4 Natural Resources
RT Coastal Zones
RT Tidal Waters

SO Environmental Protection Agency
      Glossary
DEF (EPA) Low, flat marshlands
      traversed by channels and tidal
      hollows and subject to tidal
      inundation; normally, the only
      vegetation present are salt-tolerant
      bushes and grasses.

TIDAL WATERS
(CWA; RHA; CFR)
DA October 12, 1990
BT1 Water
RT Coastal Waters
RT High Tide Line
RT Ordinary High Water Mark
RT Tidal Marshes
SO Water Pollution
DEF (CFR) Those waters that rise and
      fall in a predictable and
      measurable rhythm or cycle due to
      the gravitational pulls of the moon
      and sun. Tidal waters end where
      the rise and fall of the water
      surface can no longer be
      practically measured in a
      predictable rhythm due to masking
      by hydrologic, wind, or other
      effects.

TIERING
(CFR)
DA January 24, 1991
BT1 Administrative Processes
 BT2 Processes
RT Environmental Impact Statements
SO Environmental Management
DEF (CFR) Refers to the coverage of
      general matters in broader
      environmental impact statements
      with subsequent narrower
      statements or environmental
      analyses incorporating by
      reference the general discussions
      and concentrating solely on the
      issues specific to the statement
      subsequently prepared.

TIGER TEAM ADMINISTRATORS
(TTGM)
DA May 7, 1991
BT1 Personnel
RT Assessment Team Leaders
RT Office of Special Projects
      Coordinators
RT Tiger Team Assessments
RT Tiger Team Leaders
SO Management
DEF (TTGM) The Tiger Team
      Administrators provide on-site
      logistics and administrative
      support to the Tiger Teams,
      reporting to the Tiger Team
      Leaders. The Administrators
      generally establish contacts with
      appropriate site personnel to
      ensure that the needs of the Tiger
      Teams are met.

TIGER TEAM ASSESSMENT PROGRAM
(TTGM)
DA May 6, 1991
BT1 Programs
RT Office of Special Projects
RT Office of Special Projects
    Coordinators
RT Tiger Team Assessments
RT Tiger Teams
SO Management
DEF (TTGM) A program to provide the
    Secretary of Energy with concise
    information on the: 1) current
    ES&H compliance status of each
    facility and associated
    vulnerabilities; 2) root cause(s) for
    noncompliance; 3) adequacy of
    DOE and site contractor ES&H
    management programs; 4)
    response actions to address
    identified problem areas; and 5)
    DOE-wide ES&H compliance
    trends and root causes.

TIGER TEAM ASSESSMENT REPORTS
(TTGM)
DA May 6, 1991
BT1 Reports
RT Action Plans
RT Concerns
RT Findings
RT Tiger Team Assessments
SO Management
DEF (TTGM) The findings of the Tiger
    Team Assessments are
    documented in a written report.
    The Tiger Team Assessment
    Report contains individual findings
    (e.g. compliance findings, best
    management practices, and
    noteworthy practices), root
    causes, and requisite supporting
    documentation. This report does
    not contain specific
    recommendations for correcting
    the problems identified during the
    Tiger Team assessment. Broad
    recommendations as to how the
    findings and concerns should be
    acted upon by the relevant
    Program Senior Official may be
    provided in the memorandum
    transmitting the final assessment
    report from the Assistant
    Secretary for Environment, Safety
    and Health to the Secretary.

TIGER TEAM ASSESSMENTS
DA April 22, 1991
BT1 Assessments
 BT2 Administrative Processes
  BT3 Processes
NT1 Environment Assessments
NT1 Management and Organization
    Assessment
NT1 Safety and Health Assessments
RT Action Plans
RT Assessment Team Leaders
RT Assessment Plans

RT Best Management Practice
    Findings
RT Compliance Findings
RT Concerns
RT Findings
RT Issue Identification
RT Noteworthy Practices
RT Performance Objectives
RT Tiger Teams
RT Tiger Team Assessment Program
RT Tiger Team Assessment Reports
RT Tiger Team Leaders
RT Tiger Team Administrators
SO Management
DEF The purpose of the Tiger Team
    Assessments are to provide the
    Secretary of Energy with concise
    information on: 1) current
    environment, safety and health
    (ES&H) compliance status of each
    facility and associated
    vulnerabilities; 2) root causes for
    noncompliance; 3) adequacy of
    DOE and site contractor ES&H
    management programs; 4)
    response actions to address
    identified problem areas; and 5)
    input to evaluation of DOE-wide
    ES&H compliance trends and root
    causes. Scope includes but is not
    limited to: 1) compliance with
    applicable federal, state and local
    regulations, permits, agreements,
    and enforcement actions; 2)
    compliance with DOE Order
    requirements for ES&H activities;
    3) adequacy of the facility's and
    the site contractor's ES&H
    management programs, including
    planning, organization, resources,
    training, and relationships with
    regulatory agencies; 4)
    conformance with applicable "best"
    and "accepted industry" practices;
    and 5) identification of root causes.

TIGER TEAM LEADERS
(TTGM)
DA May 7, 1991
BT1 Personnel
RT Assessment Team Leaders
RT Assessment Plans
RT Office of Special Projects
    Coordinators
RT Tiger Team Assessments
RT Tiger Teams
RT Tiger Team Administrators
SO Management
DEF (TTGM) These individuals manage
    the Tiger Team and serve as the
    primary contact point with the
    Operations Office and the site
    office during a Tiger Team
    Assessment. The principal
    responsibilities of the Tiger Team
    Leaders are to ensure that the
    Tiger Teams are organized, staffed
    and supported as necessary to
    meet the objectives of the Tiger
    Team Assessments; to ensure that

the assessment groups interact
effectively and that their respective
activities and results are
functionally integrated; and to
ensure that the Tiger Team
Assessment report is accurate,
objective and thorough. The Tiger
Team Leaders provide overall
management and policy guidance
to the assessment teams.

TIGER TEAMS
(TTGM)
DA April 22, 1991
BT1 Teams
 BT2 Administrative Organizations
  BT3 Organizations
RT Assessment Team Leaders
RT Best Management Practice
    Findings
RT Compliance Findings
RT Concerns
RT Findings
RT Noteworthy Practices
RT Office of Special Projects
RT Performance Objectives
RT Tiger Team Assessments
RT Tiger Team Assessment Program
RT Tiger Team Leaders
SO Management
DEF (TTGM) A special investigatory
    group operating under the
    designated authority of DOE's
    Tiger Team Assessment program.
    These assessments are/will be
    performed by teams at over 100
    DOE operating facilities. The
    assessments are part of a
    ten-point initiative announced on
    June 27, 1989 by the Secretary of
    Energy, Admiral James D.
    Watkins, USN (Ret.), to conduct
    independent oversight compliance
    and management assessments of
    the ES&H programs at DOE
    facilities. Tiger Teams include
    individuals from: 1) the
    Department of Energy, 2)
    Department of Energy
    Contractors, and 3) private
    consulting organizations.

TIME CLOCK
(2778 TIME CLOCK)
DA January 3, 1991
BT1 Gauge(s)
 BT2 Instrument(s)
  BT3 Equipment/Parts -
      Instrumentation/Measuring
      (DOE FRASE Voc.)
   BT4 Equipment
SO DOE FRASE VOCABULARY

TIME DELAY
DA January 8, 1991
SF TD
RT Allowable Outage Time

TIME DESIGNATIONS
DA February 25, 1991
NT1 Active Life
NT1 Allowable Outage Time
NT1 Biological Half-Time
NT1 Closure Period
NT1 Comment Periods
NT1 Compliance Schedules
NT1 Disinfectant Contact Time
NT1 Effective Date
NT1 Experimental Start Date
NT1 Experimental Termination Date
NT1 Fiscal Year
NT1 Half-Life
 NT2 Biological Half-Life
 NT2 Effective Half-Life
 NT2 Radioactive Half-Life
NT1 Loss of Operating Time
NT1 Mean Time Between Failures
NT1 Mean Time to Repair
NT1 Operational Life
NT1 Post-Closure
NT1 Primary Standard Attainment Date
NT1 Reentry Intervals
NT1 Reimbursement Period
NT1 Response Spectrum
NT1 Response Time
NT1 Return Period
NT1 Run
NT1 Semiannual
NT1 Study Completion Date
NT1 Study Initiation Date
NT1 Surveillance Test Interval
NT1 Time to Repair
NT1 Useful Life
RT Measurements
RT Units of Measure

TIME LOSS (T/L) ANALYSIS
(SSDC)
DA October 12, 1990
BT1 Analyses
SO System Safety Development Center
   Glossary
DEF (SSDC) The analysis of
   amelioration process following an
   accident and the positive and
   negative effects of intervenors on
   the extent of the accident.

TIME TO REPAIR
DA January 8, 1991
SF *TTR (Time to Repair)*
BT1 Time Designations

TIME WEIGHTED AVERAGE
(SSDC)
DA October 12, 1990
SF *TWA*
RT Threshold Limit Values
SO System Safety Development Center
   Glossary
DEF (SSDC) Used in determining
   threshold limit values; TLV-TWA is
   the time weighted average
   concentration for a normal 8-hour
   workday or 40-hour workweek, to
   which nearly all workers may be
   repeatedly exposed, day after day,
   without adverse effect.

TINE(S)
(2343 TINE)
DA December 10, 1990
BT1 Equipment/Parts - Material
   Handling (DOE FRASE
   Vocabulary)
 BT2 Equipment
SO DOE FRASE VOCABULARY

TINTED SAFETY GLASSES
(2692 TINTED SAFET)
DA January 3, 1991
BT1 Safety Glasses
 BT2 Eye Protection
  BT3 Personal Protective Equipment
   BT4 Equipment/Parts - Personal
     Protective (DOE FRASE
     Vocabulary)
    BT5 Equipment
SO DOE FRASE VOCABULARY

TIR
DA October 12, 1990
SEE Total Indicated Runout
SO Acronyms

TISSUE EQUIVALENTS
(IAEA)
DA October 12, 1990
RT Dose Equivalents
RT Phantoms
SO Radiation
DEF (IAEA) Materials whose absorbing
   and scattering properties for
   radiation of a given type and
   energy simulate those of a
   specified biological tissue.

TLD
DA October 12, 1990
SEE Thermoluminescent Dosimeters
SO Acronyms

TLV
DA October 12, 1990
SEE Threshold Limit Values
SO Acronyms

TNT EQUIVALENT
(DOE Order 6430.1A)
DA October 12, 1990
SO Construction
DEF (DOE Order 6430.1A) A measure
   of the blast effects from the
   explosion for a given quantity of
   material expressed in terms of the
   weight of TNT that would produce
   the same blast effects when
   detonated. For safety and design
   purposes, a reasonable value can
   be obtained by substituting a
   measurement of energy release of
   blast effects.

TOC
DA October 12, 1990
SEE Total Organic Carbon
SO Acronyms

TOE(S)
(1131 TOE)
DA November 28, 1990
BT1 Foot/Feet
 BT2 Human Body Parts
RT Foot Protection
RT Metatarsal Protection
SO DOE FRASE VOCABULARY

TOLERANCE LIMITS
(SSDC)
DA October 12, 1990
BT1 Limits
RT Pesticide Tolerance
SO System Safety Development Center
   Glossary
DEF (SSDC) The criteria as to what is
   deemed acceptable in
   performance, significant error, or
   harmful energy.

TOLERANCES
(EPA; NFI)
DA October 12, 1990
RT Action Levels
RT Pesticide Tolerance
SO Environmental Protection Agency
   Glossary
SO Radiation
DEF (EPA) The permissible residue
   levels for pesticides in raw
   agricultural produce and
   processed foods. Whenever a
   pesticide is registered for use on a
   food or a feed crop, tolerances (or
   exemption from the tolerance
   requirement) must be established.
   (NFI) EPA establishes the
   tolerance levels, which are
   enforced by the Food and Drug
   Administration and the Department
   of Agriculture. Also, use for
   engineering tolerances and
   clearances between parts.

TONGS
(3064 TONGS)
DA January 3, 1991
BT1 Tools - Manual
 BT2 Tools (DOE FRASE Vocabulary)
  BT3 Equipment
SO DOE FRASE VOCABULARY

TONOPAH TEST RANGE
DA January 8, 1991
SF *TTR (Tonopah Test Range)*
BT1 Test Area
 BT2 Area
  BT3 Sites/Areas

TOOLS (DOE FRASE VOCABULARY)
(3065 TOOL)
DA January 3, 1991
BT1 Equipment
NT1 Leader Seater
NT1 Tools - Manual
 NT2 Ax(s)
 NT2 Broom(s)
 NT2 Brush(s)

NT2 Chisel(s)
NT2 Crowbar(s)
NT2 Cutter(s)
NT2 File(s)
NT2 Hack Saw
NT2 Hammer
 NT3 Sledgehammer(s)
NT2 Hand Iron(s)
NT2 Hand Saw
NT2 Hand Stapler
NT2 Hatchet(s)
NT2 Hoe(s)
NT2 Knife(s)
 NT3 Pocket Knife(s)
NT2 Mop(s)
NT2 Non-Powered Handtool(s)
NT2 Pick(s)
NT2 Plane(s)
NT2 Pliers
NT2 Punch
NT2 Rake(s)
NT2 Razor(s)
NT2 Scalpel(s)
NT2 Scissors
NT2 Screwdriver(s)
NT2 Shear(s)
NT2 Shovel(s)
NT2 Snips
NT2 Snubber(s)
NT2 Spatula(s)
NT2 Threader(s)
NT2 Tongs
NT2 Torch(s)
NT2 Tweezers
NT2 Vise
NT2 Visegrips
NT1 Tools - Powered
NT2 Air Tamper
NT2 Chain Saw
NT2 Disk Sander(s)
NT2 Jackhammer
NT2 Mixer(s)
NT2 Nail Gun
NT2 Post Driver(s)
NT2 Power Actuated Tool(s)
NT2 Power Buffer(s)
NT2 Power Chisel(s)
NT2 Power Cutter(s)
NT2 Power Drill(s)
NT2 Power File(s)
NT2 Power Grinder(s)
NT2 Power Hammer(s)
NT2 Power Impact Wrench(s)
NT2 Power Polisher(s)
NT2 Power Riveter(s)
NT2 Power Sandblaster(s)
NT2 Power Screwdriver(s)
NT2 Power Spray Gun(s)
NT2 Power Stapler(s)
NT2 Power Waxer(s)
NT2 Powered Handtool(s)
NT2 Router(s)
NT2 Soldering Iron(s)
NT2 Spray Gun(s)
NT2 Weed Eater
NT1 Wrench(s)
NT2 Allen Wrench
NT2 Impact Wrench
NT2 Power Impact Wrench(s)
NT2 Ratchet Wrench(s)

NT2 Torque Wrench(s)
SO DOE FRASE VOCABULARY

TOOLS - MANUAL
(DOE FRASE Vocabulary Numeric Keys
  3025-3099)
DA January 17, 1991
BT1 Tools (DOE FRASE Vocabulary)
 BT2 Equipment
NT1 Ax(s)
NT1 Broom(s)
NT1 Brush(s)
NT1 Chisel(s)
NT1 Crowbar(s)
NT1 Cutter(s)
NT1 File(s)
NT1 Hack Saw
NT1 Hammer
 NT2 Sledgehammer(s)
NT1 Hand Iron(s)
NT1 Hand Saw
NT1 Hand Stapler
NT1 Hatchet(s)
NT1 Hoe(s)
NT1 Knife(s)
 NT2 Pocket Knife(s)
NT1 Mop(s)
NT1 Non-Powered Handtool(s)
NT1 Pick(s)
NT1 Plane(s)
NT1 Pliers
NT1 Punch
NT1 Rake(s)
NT1 Razor(s)
NT1 Scalpel(s)
NT1 Scissors
NT1 Screwdriver(s)
NT1 Shear(s)
NT1 Shovel(s)
NT1 Snips
NT1 Snubber(s)
NT1 Spatula(s)
NT1 Threader(s)
NT1 Tongs
NT1 Torch(s)
NT1 Tweezers
NT1 Vise
NT1 Visegrips
RT DOE FRASE Categories
RT Tools - Powered
SO DOE FRASE VOCABULARY
DEF A subject category used with the
     DOE FRASE Vocabulary.

TOOLS - POWERED
(DOE FRASE Vocabulary Numeric Keys
  2950-3024)
DA January 17, 1991
BT1 Tools (DOE FRASE Vocabulary)
 BT2 Equipment
NT1 Air Tamper
NT1 Chain Saw
NT1 Disk Sander(s)
NT1 Jackhammer
NT1 Mixer(s)
NT1 Nail Gun
NT1 Post Driver(s)
NT1 Power Actuated Tool(s)
NT1 Power Buffer(s)
NT1 Power Chisel(s)

NT1 Power Cutter(s)
NT1 Power Drill(s)
NT1 Power File(s)
NT1 Power Grinder(s)
NT1 Power Hammer(s)
NT1 Power Impact Wrench(s)
NT1 Power Polisher(s)
NT1 Power Riveter(s)
NT1 Power Sandblaster(s)
NT1 Power Screwdriver(s)
NT1 Power Spray Gun(s)
NT1 Power Stapler(s)
NT1 Power Waxer(s)
NT1 Powered Handtool(s)
NT1 Router(s)
NT1 Soldering Iron(s)
NT1 Spray Gun(s)
NT1 Weed Eater
RT DOE FRASE Categories
RT Tools - Manual
SO DOE FRASE VOCABULARY
DEF A subject category used with the
     DOE FRASE Vocabulary.

TOOTH/TEETH
(1089 TOOTH)
DA November 28, 1990
BT1 Mouth
 BT2 Digestive System
  BT3 Body System(s)
   BT4 Human Body Parts
RT Dental Injury
SO DOE FRASE VOCABULARY

TOP AND BOTTOM
DA January 8, 1991
SF T&B

TOP MANAGEMENT
(SSDC)
DA October 12, 1990
RT Management
SO System Safety Development Center
     Glossary
DEF (SSDC) The individual manager or
     group of managers serving in a
     policymaking capacity for the total
     organization. These individuals
     also may be, and usually are,
     involved in policy implementation.

TORCH(S)
(3066 TORCH)
DA January 3, 1991
BT1 Tools - Manual
 BT2 Tools (DOE FRASE Vocabulary)
  BT3 Equipment
SO DOE FRASE VOCABULARY

TORN CARTILAGE
(1342 TORN CARTILA)
DA November 28, 1990
BT1 Injuries
SO DOE FRASE VOCABULARY

TORNADOES
(SMRP)
DA October 12, 1990
BT1 Natural Disasters

BT2 Natural Phenomenon
BT1 Rotational Windstorms
 BT2 Act of Nature
RT Design Basis Tornadoes
RT Downbursts
RT Waterspouts
SO Natural Phenomenon
DEF (SMRP) Violently rotating columns of air, pendant from the base of a convective cloud, and often observable as funnel cloud attached to the cloud base. A tornado must be accompanied by damaging winds (F0 or stronger) at structure levels. If not, the windstorm is classified as "funnel aloft".

TORPEDOES
(NFI)
DA October 12, 1990
BT1 Devices
SO Nuclear Facilities Incident Database
SO Radiation
DEF (NFI) Weight sensing devices to determine if control rods are latched.

TORQUE WRENCH(S)
(3067 TORQUE WRENC)
DA January 3, 1991
BT1 Wrench(s)
 BT2 Tools (DOE FRASE Vocabulary)
 BT3 Equipment
SO DOE FRASE VOCABULARY

TORTS
(SSDC)
DA October 12, 1990
SO System Safety Development Center Glossary
DEF (SSDC) Legal wrongs arising from a duty owed to people generally, rather than specifically as by contract.

TOTAL DISSOLVED SOLIDS
(SDWA; CFR)
DA October 12, 1990
BT1 Measurements
BT1 Solids
 BT2 Materials
SO Water Pollution
DEF (CFR) The total dissolved (filterable) solids as determined by use of the method specified in 40 CFR 136.

TOTAL EXPOSURE POINTS
(EPA)
DA October 12, 1990
BT1 Exposure Point
BT1 Measurements
SO Environmental Protection Agency Glossary
DEF (EPA) Points of potential exposure to substances from more than one exposure pathway.

TOTAL INDICATED RUNOUT
(NFI)
DA October 12, 1990
SF *TIR*
BT1 Limits
BT1 Measurements
RT Quality Control
SO Radiation
DEF (NFI) Quality control limit on fuel assemblies, possibly alignment of components.

TOTAL MASS STOPPING POWER
(IAEA)
DA October 12, 1990
BT1 Measurements
SO Radiation
DEF (IAEA) For charged particles, the quotient of dE by pdl, where dE is the energy lost by a charged particle in traversing a distance dl in a given material of density p.

TOTAL ORGANIC CARBON
DA January 8, 1991
SF *TOC*

TOTAL RECORDABLE CASES
(SSDC)
DA January 8, 1991
SY Incidence Rate, Total Recordable Cases (TRC)
SF *TRC*
BT1 Measurements
RT Occupational Injuries
RT Occupational Illnesses
SO Industrial Hygiene
SO Industrial Safety
SO Management
SO System Safety Development Center Glossary
DEF (SSDC) The number of recordable injuries and illnesses per 200,000 total hours worked by all employees during the period covered. The 200,000 hours worked are equivalent to 100 full-time workers at 40 hours per week for 50 weeks. TRC = No. of recordable injuries and illnesses X 200,000/total hours worked.

TOTAL SUSPENDED PARTICULATES
(CAA; CFR)
DA October 12, 1990
BT1 Criteria Pollutants
 BT2 Pollutants
BT1 Measurements
RT Particulate Matter
SO Air Pollution
DEF (CFR) Particulate matter as measured by the method described in Appendix B of 40 CFR Part 50 of this chapter.

TOTAL SUSPENDED SOLIDS
(EPA)
DA October 12, 1990
SF *TSS*

BT1 Measurements
BT1 Solids
 BT2 Materials
SO Environmental Protection Agency Glossary
DEF (EPA) A measure of the suspended solids in wastewater, effluent, or water bodies, determined by using tests for "total suspended non-filterable solids."

TOTAL TOXIC ORGANICS
DA January 8, 1991
SF *TTO*

TOTAL TRIHALOMETHANES (TTHM)
(SDWA; CFR)
DA October 12, 1990
BT1 Measurements
RT Sanitary Surveys
RT Trihalomethane
SO Water Pollution
DEF The sum of the concentration in milligrams per liter of the trihalomethan per liter of the trihalomethane compounds (trichloromethane [chloroform], dibromochloromethane, bromodichloromethane and tribromomethane [bromoform]), rounded to two significant figures.

TOTALLY ENCLOSED MANNER
(TSCA; CFR; ESH)
DA October 12, 1990
RT Polychlorinated Biphenyls
SO Hazardous Materials
DEF (ESH) Any manner that will ensure no exposure of human beings or the environment to any concentration of PCBs.

TOTALLY ENCLOSED TREATMENT FACILITIES
(SWDA; RCRA; ESH)
DA October 12, 1990
BT1 Hazardous Waste Facilities
 BT2 Facilities
BT1 Treatment Facilities
 BT2 Facilities
BT1 Treatment, Storage, and Disposal Facilities
 BT2 Facilities
SO Environmental Management
SO Wastes
DEF (ESH) Hazardous waste treatment facilities that are directly connected to an industrial production process and that are constructed and operated in a manner that prevents the release of any hazardous waste or any constituent thereof into the environment during treatment. An example is a pipe in which waste acid is neutralized.

TOWERS
(SEA)

SY-Synonymous Terms

SO-Source/Subject Category

SF-See From

DA   October 12, 1990
RT   Structures (DOE FRASE
        Vocabulary)
SO   Construction
DEF (SEA) The upper flexible portions of
        a structure having a vertical
        combination of structural systems.

TOXIC AIR POLLUTANTS
DA   January 8, 1991
SF   *TAP*
BT1 Air Pollutants

TOXIC CHEMICAL RELEASE FORMS
(EPA; EMER)
DA   October 12, 1990
BT1 Reports
SO   Emergency Preparedness
SO   Environmental Protection Agency
        Glossary
DEF (EPA) Information forms required to
        be submitted by facilities that
        manufacture, process, or use (in
        quantities above a specific
        amount) chemicals listed under
        SARA (Superfund Amendments
        and Reauthorization Act) Title III.

TOXIC CHEMICALS
(CERCLA; CFR; ESH; EMER)
DA   October 12, 1990
BT1 Hazardous Chemicals
 BT2 Hazardous Substances
RT   Otherwise Use
RT   Process (Toxic Chemical)
RT   Process for Commercial Purposes
SO   Compensation and Liability
SO   Emergency Preparedness
DEF (ESH) Chemicals or chemical
        categories listed in 40 CFR
        372.65, the release of which under
        certain conditions must be
        reported in accordance with SARA
        (Superfund Amendment and
        Reauthorization Act) Title III,
        Section 313.

TOXIC CLOUDS
(EPA)
DA   October 12, 1990
BT1 Air Pollutants
SO   Environmental Protection Agency
        Glossary
DEF (EPA) Airborne masses of gases,
        vapors, fumes, or aerosols
        containing toxic materials.

TOXIC EFFECTS TO SINGLE SYSTEM
(1407 TOXIC EFFECT)
DA   November 28, 1990
BT1 Effects
BT1 Illnesses
RT   Poisoning
SO   DOE FRASE VOCABULARY

TOXIC POLLUTANTS
(EPA)
DA   October 12, 1990
BT1 Pollutants

SO   Environmental Management
SO   Environmental Protection Agency
        Glossary
SO   Hazardous Materials
DEF (EPA) Materials contaminating the
        environment that cause death,
        disease, birth defects in organisms
        that ingest or absorb them. The
        quantities and length of exposure
        necessary to cause these effects
        can vary widely. (ESH) Any
        pollutant listed as toxic under
        §307(a)(1) of the Toxic Substances
        Control Act.

TOXIC SUBSTANCES
(EPA)
DA   October 12, 1990
BT1 Hazardous Substances
RT   Background Levels
RT   CERCLA Hazardous Substances
RT   High Concentration PCBs
RT   Listed Hazardous Substances
SO   Environmental Management
SO   Environmental Protection Agency
        Glossary
DEF (EPA) Chemicals or mixtures that
        may present an unreasonable risk
        of injury to health or the
        environment.

TOXIC SUBSTANCES CONTROL ACT
DA   January 8, 1991
SF   *TSCA*
BT1 Acts
 BT2 Statutes and Regulations

TOXICANTS
(EPA)
DA   October 12, 1990
BT1 Poisons
SO   Environmental Protection Agency
        Glossary
DEF (EPA) Poisonous agents that kill or
        injure animal or plant life.

TOXICITY
(EPA)
DA   October 12, 1990
NT1 Acute Toxicity
NT1 Chronic Toxicity
NT1 Dermal Toxicity
NT1 Radiotoxicity
RT   Known Human Effects
RT   Lethal Concentration
SO   Environmental Management
SO   Environmental Protection Agency
        Glossary
DEF (EPA) The degree of danger posed
        by a substance to animal or plant
        life.

TOXICITY CHARACTERISTIC LEACHING PROCEDURE
DA   January 8, 1991
SF   *TCLP*
BT1 Procedures

TPC
DA   October 12, 1990
SEE Temporary Procedure Change
SO   Acronyms

TPL
DA   October 12, 1990
SEE Transient Protection Limit
SO   Acronyms

TPQ
DA   October 12, 1990
SEE Threshold Planning Quantities
SO   Acronyms

TR
DA   October 12, 1990
SEE Transformer Room
SO   Acronyms

TRA
DA   October 12, 1990
SEE Test Reactor Area
SO   Acronyms

TRACEABLE
(CAA; CFR)
DA   October 12, 1990
SO   Air Pollution
SO   Environmental Management
DEF (CFR) That a local standard has
        been compared and certified either
        directly or via not more than one
        intermediate standard, to a primary
        standard such as a National
        Bureau of Standards Standard
        Reference Material (NBS SRM), or
        an EPA/NBS-approved Certified
        Reference Material (CRM).

TRAILER BUILDING
(1866 TRAILER; BLD)
DA   December 10, 1990
BT1 Structures (DOE FRASE
        Vocabulary)
SO   DOE FRASE VOCABULARY

TRAILERSHIP
(SWDA; RCRA; ESH)
DA   October 12, 1990
BT1 Vessels
SO   Hazardous Materials
DEF A vessel other than a carfloat,
        specifically equipped to handle
        highway vehicles, and fitted with
        installed securing devices to tie
        down each vehicle.

TRAINED INVESTIGATORS
(DOE Order 5484.1; SSDC)
DA   October 12, 1990
BT1 Technically Qualified Individuals
 BT2 Personnel
RT   Investigations
SO   Management
SO   System Safety Development Center
        Glossary
DEF (SSDC) Individuals who have

completed the Department of
Energy Accident Investigation
Workshop.

TRAINING
(MORT)
DA   April 24, 1991
RT   First Line Supervision
RT   Operational Specifications
RT   Procedures
SO   Management
DEF (Webster) The development of a
skill or group of skills. Instruction
in an art, profession or occupation.
(MORT) MORT analysis asks:
Was the training of personnel
adequate? Was the individual
trained for the task he or she
performed? Were the criteria used
to establish the training program
adequate in scope, depth, and
detail? Were the methods used in
training adequate to the training
requirements? e.g. realistic
simulation, programmed
self-instruction, and other special
training in addition to basic
indoctrination, plant familiarization,
etc. Was the basic professional
skill of the trainers adequate to
implement the prescribed training
program? Was the verification of
the person's current trained status
adequate? Were retraining and
requalification requirements of the
task defined?

**TRAINING ACCREDITATION
PROGRAM STAFF**
(DOE Order 5480.18)
DA   October 12, 1990
BT1   Personnel
RT   Accreditation Review Teams
RT   Accreditation Coordinators
RT   Programs
RT   Subject Matter Experts
RT   Training Programs
RT   Training Program Accreditation
Plan
SO   Standards
DEF (DOE Order 5480.18) An
organization contracted by EH-1,
responsible for developing and
providing documents, training, and
assistance to those who must
comply with this Order. This staff
also manages the conduct of the
team evaluations for accreditation.

**TRAINING PROGRAM
ACCREDITATION PLAN**
(DOE Order 5480.18)
DA   October 12, 1990
BT1   Plans
RT   Accreditation
RT   Contractor Self-Evaluation Reports
RT   Programs
RT   Self-Evaluation
RT   Training Accreditation Program
Staff

RT   Training Programs
SO   Standards
DEF (DOE Order 5480.18) An action
plan developed following a
thorough contractor self-evaluation
and identification of training
programs requiring accreditation.
The Training Program
Accreditation Plan identifies scope
and resource needs for
accomplishing accreditation for all
programs at a facility.

TRAINING PROGRAMS
(DOE Order 5480.18)
DA   October 12, 1990
BT1   Programs
NT1   Performance-Based Training
RT   Accreditation
RT   Contractor Self-Evaluation Reports
RT   Self-Evaluation
RT   Training Accreditation Program
Staff
RT   Training Program Accreditation
Plan
RT   Verification of Training and
Retraining
SO   Standards
DEF (DOE Order 5480.18) Planned,
organized sequences of activities
designed to prepare individuals to
perform their jobs, to meet a
specific position or classification
need, and to maintain or improve
their performance on the job.

TRAINING/EDUCATION ACTIVITY
(1245 TE ACTIVITY)
DA   November 28, 1990
BT1   Activity Types (DOE FRASE
Vocabulary)
BT2   Activities
SO   DOE FRASE VOCABULARY

TRAINSHIPS
(SWDA; RCRA; ESH)
DA   October 12, 1990
BT1   Vessels
SO   Hazardous Materials
DEF Vessels other than a rail car ferries
or carfloats, specifically equipped
to transport railroad vehicles, and
fitted with installed securing
devices to tie down each vehicle.

TRAM
DA   October 12, 1990
SEE TRAnsient Multiplexor
SO   Acronyms

TRAM VALUES
(NIF)
DA   October 12, 1990
SO   Radiation
DEF (NFI) TRAnsient Multiplexer values,
used to calculate BOR (Burn Out
Risk).

TRANSDUCER(S)
(2427 TRANSDUCER)
DA   January 3, 1991
BT1   Equipment/Parts - Electrical (DOE
FRASE Vocabulary)
BT2   Equipment
SO   DOE FRASE VOCABULARY

TRANSFER FACILITIES
(SWDA; RCRA; TSCA; CFR)
DA   October 12, 1990
BT1   Facilities
RT   PCB Wastes
RT   Transporter of PCB Waste
SO   Environmental Management
SO   Hazardous Materials
SO   Wastes
DEF Transportation-related facilities
including loading docks, parking
areas, storage areas and other
similar areas where shipments of
hazardous waste are held during
the normal course of
transportation, excluding transport
vehicles unless they are used for
the storage of PCB waste.
Regarding the 10-day limit, see 40
CFR 263.12.

TRANSFER STATIONS
(SWDA; RCRA; ESH)
DA   October 12, 1990
BT1   Sites/Areas
RT   Solid Wastes
RT   Transport (Hazardous Substances)
SO   Wastes
DEF (ESH) Sites at which solid wastes
are concentrated for transport to a
processing facility or land disposal
site. Transfer stations may be
fixed or mobile.

TRANSFORMATION
(EPA)
DA   October 12, 1990
BT1   Processes
SO   Environmental Protection Agency
Glossary
DEF (EPA) The process of placing new
genes into a host cell, thereby
inducing the host cell to exhibit
functions encoded by the DNA.

TRANSFORMER(S)
(2428 TRANSFORMER)
DA   January 3, 1991
BT1   Equipment/Parts - Electrical (DOE
FRASE Vocabulary)
BT2   Equipment
SO   DOE FRASE VOCABULARY

TRANSFORMER ROOM
DA   January 8, 1991
SF   *TR*

TRANSIENT MULTIPLEXOR
DA   January 8, 1991
SF   *TRAM*

BT1 Multiplexor
BT2 Equipment

TRANSIENT OPERATION
(2596 TRANSIENT OP)
DA January 3, 1991
BT1 Equipment/Parts - Nuclear (DOE
    FRASE Vocabulary)
BT2 Equipment
BT2 Reactor Components
SO DOE FRASE VOCABULARY

TRANSIENT PROTECTION LIMIT
DA January 8, 1991
SF *TPL*
BT1 Limits

TRANSIENT TEST
(2597 TRANSIENT TE)
DA January 3, 1991
BT1 Equipment/Parts - Nuclear (DOE
    FRASE Vocabulary)
BT2 Equipment
BT2 Reactor Components
RT Research/Testing Activity
SO DOE FRASE VOCABULARY

TRANSIENTS (RE: EXPLOSIVES FACILITIES)
(DOE Order 6430.1A)
DA October 12, 1990
RT Inhabited Building Distance
SO Construction
DEF (DOE Order 6430.1A) Any persons
    within inhabited building distance
    but not inside an explosives bay or
    other occupied areas (offices,
    break areas, shops, etc.).

TRANSMISSIVE FAULTS
(SWDA; RCRA; ESH)
DA October 19, 1990
SY Transmissive Fractures
BT1 Faults
RT Permeability
SO Wastes
SO Water Pollution
DEF Faults or fractures that have
    sufficient permeability and vertical
    extent to allow fluids to move
    between formations.

TRANSMISSIVE FRACTURES
(SWDA; RCRA; CFR)
DA October 19, 1990
SY Transmissive Faults
BT1 Faults
SO Wastes
SO Water Pollution
DEF (CFR) Faults or fractures that have
    sufficient permeability and vertical
    extent to allow fluids to move
    between formations.

TRANSMITTER(S)
(2429 TRANSMITTER)
DA January 3, 1991

BT1 Equipment/Parts - Electrical (DOE
    FRASE Vocabulary)
BT2 Equipment
SO DOE FRASE VOCABULARY

TRANSPIRATION
(EPA)
DA October 12, 1990
BT1 Biological Processes
BT2 Processes
RT Evapotranspiration
SO Environmental Protection Agency
    Glossary
DEF (EPA) The process by which water
    vapor is lost to the atmosphere
    from living plants. The term can
    also be applied to the quantity of
    water thus dissipated.

TRANSPORT (HAZARDOUS SUBSTANCES)
(USC)
DA October 19, 1990
SY Transportation
RT Carfloats
RT Cargo Aircraft Only
RT Containers
RT Containerships
RT International Shipments
RT Packaging
RT Transfer Stations
SO Compensation and Liability
DEF (USC) The movement of a
    hazardous substance by any
    mode, including pipeline (as
    defined in the Pipeline Safety Act),
    and in the case of a hazardous
    substance which has been
    accepted for transportation by a
    common or contract carrier, the
    term "transport" or "transportation"
    shall include any stoppage in
    transit which is temporary,
    incidental to the transportation
    movement, and at the ordinary
    operating convenience of a
    common or contract carrier, and
    any such stoppage shall be
    considered as a continuity of
    movement and not as the storage
    of a hazardous substance.

TRANSPORT INDICES
(DOE Order 5480.3; ESH)
DA October 12, 1990
SO Environmental Management
SO Hazardous Materials
DEF (DOE Order 5480.3) The numbers
    placed on a package to designate
    the degree of control to be
    exercised by the carrier during
    transportation. The transport index
    to be assigned to a package of
    radioactive material shall be
    determined by either numbers (1)
    or (2) below, whichever is larger.
    The number expressing the
    transport index shall be rounded
    up to the next higher tenth; e.g.,
    1.01 becomes 1.1. (1) The highest

radiation dose rate in millirem per
hour at 1 meter from any
accessible external surface of the
package. (2) The transport index
of each Fissile Class II package is
calculated by dividing the number
50 by the number of such Fissile
Class II packages that may be
transported together as
determined under the limitations of
10 CFR 71.

TRANSPORT PERSONNEL
DA November 29, 1990
BT1 Occupations
BT1 Personnel
NT1 Aircraft Pilot
NT1 Bus Driver
NT1 Equipment Operator
NT1 Misc Transport Employee
NT1 Truck Driver
RT Transportation Activity
RT Travel Activity
SO DOE FRASE VOCABULARY

TRANSPORT VEHICLES
(SWDA; RCRA; TSCA; CFR; ESH)
DA October 12, 1990
NT1 Compartmentalized Vehicles
NT1 Motor Vehicles
NT2 COMVAN
NT2 Mobile Command Vehicles
RT Movement
RT Transportation Control Measures
SO Environmental Management
SO Hazardous Materials
SO Wastes
DEF (ESH) Motor vehicles or rail cars
    used for the transportation of
    cargo by any mode. Each
    cargo-carrying body (e.g., trailer,
    railroad freight car) is a separate
    transport vehicle.

TRANSPORTATION
(SWDA; RCRA)
DA October 19, 1990
SY Transport (Hazardous Substances)
RT Break-bulk
RT Carriers
RT Competent Authorities
RT Large Quantities
RT Manifests
RT Mode
RT Movement
RT Navigable Waters
RT Roadways
RT Water(bulk shipment)
SO Compensation and Liability
SO Environmental Management
SO Wastes
DEF (USC) Means the movement of a
    hazardous substance by any
    mode, including pipeline (as
    defined in the Pipeline Safety Act),
    and in the case of a hazardous
    substance which has been
    accepted for transportation by a
    common or contract carrier, the
    term "transport" or "transportation"

shall include any stoppage in transit which is temporary, incidental to the transportation movement, and at the ordinary operating convenience of a common or contract carrier, and any such stoppage shall be considered as a continuity of movement and not as the storage of a hazardous substance.

TRANSPORTATION ACTIVITY

(1246 TRS ACTIVITY)
DA   November 28, 1990
BT1 Activity Types (DOE FRASE
     Vocabulary)
BT2 Activities
RT   Misc Transport Employee
RT   Transport Personnel
SO   DOE FRASE VOCABULARY

TRANSPORTATION CONTROL MEASURES

(CAA; CFR)
DA   October 12, 1990
BT1 Control Strategies
RT   Catalytic Converters
RT   Motor Vehicles
RT   Schedule of Compliance
RT   Secondary Emissions
RT   Transport Vehicles
RT   Vessels
SO   Air Pollution
DEF (CFR) Measures that are directed
    toward reducing emissions of air
    pollutants from transportation
    sources. Such measures include,
    but are not limited to, those listed
    in section 108(f) of the Clean Air
    Act.

TRANSPORTATION INCIDENTS

(EMER)
DA   February 1, 1991
BT1 Incidents
SO   Emergency Preparedness
DEF (EMER) Incidents that involves a
    transportation vehicle or shipment
    containing hazardous or
    radioactive materials.

TRANSPORTER OF PCB WASTE

(TSCA)
DA   October 19, 1990
BT1 Transporters
BT2 Potentially Responsible Parties
RT   Transfer Facilities
SO   Hazardous Materials

TRANSPORTERS

(SWDA; RCRA)
DA   October 19, 1990
BT1 Potentially Responsible Parties
NT1 Transporter of PCB Waste
SO   Wastes
DEF People engaged in the off-site
    transportation of hazardous waste
    by air, rail, highway or water.

TRANSURANIC (TRU) CONTAMINATED MATERIALS

(DOE Order 5480.2)
DA   October 12, 1990
SY   TRU Wastes
BT1 Materials
RT   Transuranic Wastes (TRU Waste)
SO   Environmental Management
SO   Industrial Hygiene
DEF (DOE Order 5480.2) Materials
    which, without regard to source or
    form, contain certain
    alpha-emitting radionuclides with
    long half-lives in concentrations
    greater than 100nCi/g.

TRANSURANIC ELEMENTS

(DOE Order 6430.1A; EMER)
DA   October 12, 1990
SY   Transuranic Nuclides
BT1 Nuclides
RT   Radionuclides
RT   Reprocessing
SO   Construction
SO   Emergency Preparedness
DEF (DOE Order 6430.1A) Those
    elements having an atomic
    number greater than 92 (uranium).

TRANSURANIC NUCLIDES

DA   January 8, 1991
SY   Transuranic Elements
SF   TRU
BT1 Nuclides
RT   Radionuclides
SO   Radiation
DEF (DSTT) Elements that have atomic
    numbers greater than 92; all are
    radioactive, are products of
    artificial nuclear changes, and are
    members of the actinide group.

TRANSURANIC RADIOACTIVE WASTES

(ANL; CFR)
DA   May 24, 1991
BT1 Radioactive Wastes
BT2 Wastes
RT   High-Level Radioactive Wastes
SO   Radiation
DEF (CFR) Waste containing more than
    100 nanocuries of alpha-emitting
    transuranic isotopes, with half-lives
    greater than twenty years, per
    gram of waste, except for: (1)
    High level radioactive wastes; (2)
    wastes that the Department has
    determined, with the concurrence
    of the Administrator, do not need
    the degree of isolation required by
    this Part; or (3) wastes that the
    Commission has approved for
    disposal on a case-by-case basis
    in accordance with 20 CFR Part
    61.

TRANSURANIC RADIONUCLIDES

(DOE Order 5480.2; ANL)
DA   May 24, 1991

BT1 Radionuclides
BT2 CERCLA Hazardous Substances
BT3 Hazardous Substances
BT2 Nuclides
SO   Radiation
DEF (DOE 5820.2A) Any radionuclide
    having an atomic number greater
    than 92.

TRANSURANIC WASTE FACILITY

DA   January 8, 1991
SF   TWF
BT1 Facilities

TRANSURANIC WASTE TREATMENT AND STORAGE FACILITY

DA   January 8, 1991
SF   TWTSF
BT1 Facilities

TRANSURANIC WASTES (TRU WASTE)

(DOE Order 5480.2; EMER)
DA   October 12, 1990
SY   TRU Wastes
BT1 Radioactive Wastes
BT2 Wastes
NT1 Contact-Handled Transuranic
    Wastes
NT1 Remote-Handled Transuranic
    Wastes
RT   Repositories
RT   Test and Evalution Facilities
RT   Transuranic (TRU) Contaminated
    Materials
RT   Waste Isolation Pilot Plant
SO   Emergency Preparedness
SO   Environmental Management
SO   Industrial Hygiene
DEF (DOE Order 5480.2) TRU
    contaminated materials which
    have been declared as having no
    significant economic value or use.

TRANSVERSE

(DOE Order 6430.1A)
DA   October 12, 1990
SO   Construction
DEF (DOE Order 6430.1A) That which is
    extended or is lying across.

TRASH-TO-ENERGY PLANS

(EPA)
DA   October 12, 1990
BT1 Plans
RT   Incineration
SO   Environmental Protection Agency
    Glossary
DEF (EPA) Plans for putting waste back
    to work by burning trash to
    produce energy.

TRAVEL ACTIVITY

(1249 TRA ACTIVITY)
DA   November 28, 1990
BT1 Activity Types (DOE FRASE
    Vocabulary)
BT2 Activities
RT   Transport Personnel
SO   DOE FRASE VOCABULARY

SY-Synonymous Terms   SO-Source/Subject Category   SF-See From

TRAVELING WIRE FLUX MONITOR
DA   January 8, 1991
SF   *TWFM*
BT1 Monitors
 BT2 Equipment

TRC
DA   October 12, 1990
SEE Total Recordable Cases
SO   Acronyms

TRE
DA   October 12, 1990
SEE Turbidity Removal Evaporator
SO   Acronyms

TREATABILITY STUDIES
(SWDA; RCRA)
DA   October 12, 1990
BT1 Studies
SO   Environmental Management
SO   Wastes
DEF (CFR) Studies in which a
        hazardous waste is subjected to a
        treatment process to determine (1)
        whether the waste is amenable to
        the treatment process, (2) what
        pretreatment (if any) is required,
        (3) the optimal process conditions
        needed to achieve the desired
        treatment, (4) the efficiency of a
        treatment process for a specific
        waste or wastes, or (5) the
        characteristics and volumes of
        residuals from a particular
        treatment process. Also included
        in this definition for the purpose of
        the 40 CFR 261.4 (e) and (f)
        exemptions are liner compatibility,
        corrosion, and other material
        compatibility studies and
        toxicological and health effect
        studies. A "treatability study" is not
        a means to commercially treat or
        dispose of hazardous waste.

TREATMENT
(DOE Order 5820.2A; SWDA; RCRA)
DA   October 12, 1990
SY   Treatment Technologies
BT1 Waste Management Processes
 BT2 Processes
NT1 Biological Treatment
 NT2 Aerobic Treatment
NT1 Chemical Treatments
NT1 Conventional Filtration Treatment
NT1 Granular Activated Carbon
        Treatment
NT1 In-Situ Treatment
NT1 Incineration
 NT2 Incineration at Sea
NT1 Ion Exchange Treatment
NT1 Primary Waste Treatment
NT1 Secondary Treatment
NT1 Stabilization
NT1 Thermal Treatment
NT1 Wastewater Treatment Processes
 NT2 Absorption (Waste)
 NT2 Advanced Waste Water Treatment

NT2 Clarification
NT2 Diatomaceous Earth Filtration
NT2 Flocculation
NT2 Physical and Chemical Treatments
NT2 Pretreatment
NT2 Primary Waste Treatment
NT2 Secondary Treatment
NT2 Solidification and Stabilization
NT2 Tertiary Treatment
RT   Residual
RT   Treatment, Storage, and Disposal
SO   Environmental Management
SO   Wastes
DEF (WEBSTER) The techniques or
        actions customarily applied in a
        specified situation. (DOE Order
        5820.2A) Any method, technique,
        or process designed to change the
        physical or chemical character of
        waste to render it less hazardous,
        safer to transport, store or dispose
        of, or reduced in volume. (USC)
        When used in connection with
        hazardous waste, means any
        method, technique, or process,
        including neutralization, designed
        to change the physical, chemical,
        or biological character or
        composition of any hazardous
        waste so as to neutralize such
        waste or so as to render such
        waste nonhazardous, safer for
        transport, amenable for recovery,
        amenable for storage, or reduced
        in volume.

**TREATMENT, STORAGE, AND
DISPOSAL**
DA   January 8, 1991
SF   *TSD*
BT1 Waste Management Processes
 BT2 Processes
RT   Disposal
RT   Storage
RT   Treatment

**TREATMENT, STORAGE, AND
DISPOSAL FACILITIES**
(EPA)
DA   October 12, 1990
BT1 Facilities
NT1 Designated Facilities
NT1 Totally Enclosed Treatment
        Facilities
RT   Active Portions
RT   Closed Portions
RT   EPA Identification Number
RT   Inactive Portions
RT   Interim (Permit) Status
SO   Environmental Protection Agency
        Glossary
DEF (EPA) Sites where a hazardous
        substance [hazardous wastes] is
        treated, stored, or disposed. TSD
        facilities are regulated by EPA and
        states under RCRA (Resource
        Conservation and Recovery Act).

TREATMENT FACILITIES
(DOE Order 5820.2A)

DA   January 29, 1991
BT1 Facilities
NT1 Grout Treatment Facility
NT1 Land Treatment Facilities
NT1 Publicly Owned Treatment Works
 NT2 Water Systems
  NT3 Public Water Systems
   NT4 Community Water Systems
   NT4 Drinking Water System
   NT4 Non-Community Water Systems
   NT4 Non-Transient Non-Community
        Water System
   NT4 Water Supply System
  NT3 Storm Water Collection Systems
  NT3 Waste Water System
   NT4 Waste Water Collection Systems
NT1 Totally Enclosed Treatment
        Facilities
SO   Environmental Management
DEF (DOE Order 5820.2A) Specific
        areas of land, structures, and
        equipment dedicated to waste
        treatment and related activities.

**TREATMENT TECHNIQUE
REQUIREMENTS**
(CFR)
DA   October 19, 1990
BT1 Requirements
RT   Primary Drinking Water Regulations
SO   Environmental Management
SO   Water Pollution
DEF (CFR) Requirements of the national
        primary drinking water regulations
        which specify for a contaminant a
        specific treatment technique(s)
        known to the Administrator which
        leads to a reduction in the level of
        such contaminant sufficient to
        comply with the requirements of
        40 CFR 141.

TREATMENT TECHNOLOGIES
(ANL; CFR)
DA   May 24, 1991
SY   Treatment
RT   Hazardous Wastes
RT   Land Disposal
SO   Environmental Management
SO   Wastes
SO   Water Pollution
DEF (CFR) Any unit operations or series
        of unit operations that alter the
        composition of a hazardous
        substance or pollutant or
        contaminant through chemical,
        biological, or physical means so
        as to reduce toxicity, mobility, or
        volume of contaminated materials
        being treated. Treatment
        technologies are an alternative to
        land disposal of hazardous wastes
        without treatment.

TREATMENT ZONES
(SWDA; RCRA)
DA   October 12, 1990
BT1 Zones
 BT2 Sites/Areas
RT   Land Treatment Facilities

SO Environmental Management
SO Wastes
DEF Soil areas of the unsaturated zone of a land treatment unit within which hazardous constituents are degraded, transformed, or immobilized.

TRENCH/DITCH
(1867 TRENCH)
DA December 10, 1990
BT1 Structures (DOE FRASE Vocabulary)
SO DOE FRASE VOCABULARY

TRENCHING
(CFR)
DA January 24, 1991
SY Burial Operations
BT1 Waste Management Processes
BT2 Processes
RT Waste Management
SO Environmental Management
DEF (CFR) Any method, technique or process, including storage for radioactive decay, designed to change the physical, chemical or biological characteristics or composition of any waste in order to render the waste for transport, storage or disposal, amendable to recovery, convertible to another usable material or reduced in volume.

TRENDING
(DOE Order 4330.4A)
DA June 5, 1991
RT Analyses
RT Communication
RT Corrective Action Triggers
RT Knowledge
RT Monitoring
RT Technical Information
SO Management
DEF (DOE Order 4330.4A) An analysis of parts, systems, component surveillances, performance, and operating histories to determine such things as failure causes, operational effectiveness, cost-effectiveness, and other attributes.

TRIAC
(NFI)
DA October 12, 1990
SO Nuclear Facilities Incident Database
SO Radiation
DEF (NFI) Circuit board for computer, also undervoltage circuit for Emergency Diesel Generators.

TRIAGE
(DOE Order 5480.8)
DA October 12, 1990
BT1 Processes
RT Medical Treatment

SO Industrial Hygiene
DEF (DOE Order 5480.8) The medical screening of patients to determine their priority for treatment; the separation of a large number of casualties in military or civilian disaster medical care into three groups: (1) those who cannot be expected to survive even with treatment; (2) those who will recover without treatment; and (3) the priority group of those who need treatment in order to survive.

TRICHLOROETHYLENE (TCE)
(EPA)
DA October 12, 1990
SF *TCE*
BT1 Chlorinated Hydrocarbons
BT2 CERCLA Hazardous Substances
BT3 Hazardous Substances
BT2 Halogenated Organic Compounds
BT3 Halogenated
BT3 Organic Chemicals
BT4 Chemical Substances
BT1 Hazardous Constituents
SO Environmental Protection Agency Glossary
DEF (EPA) A stable, low boiling colorless liquid, toxic by inhalation. TCE is used as a solvent, metal degreasing agent, and in other industrial applications.

TRICKLING FILTERS
(EPA)
DA October 12, 1990
BT1 Filters
BT2 Devices
RT Sand Filters
SO Environmental Protection Agency Glossary
DEF (EPA) Coarse, biological treatment systems in which wastewater is trickled over a bed of stones or other material covered with bacterial growth.

TRIGGER
(SSDC)
DA October 12, 1990
RT Trigger Event
SO System Safety Development Center Glossary
DEF (SSDC) In MORT, a report of an event or condition that initiates a hazard analysis or correction function.

TRIGGER EVENT
(SSDC)
DA October 12, 1990
BT1 Events
RT Trigger
SO System Safety Development Center Glossary
DEF (SSDC) In MORT, a report of an event or condition which initiates a

hazard analysis or correction function.

TRIHALOMETHANE
(SDWA; CFR; ESH; EPA)
DA October 12, 1990
SF *THM*
BT1 Halogenated Organic Compounds
BT2 Halogenated
BT2 Organic Chemicals
BT3 Chemical Substances
RT Disinfectants
RT Maximum Total Trihalomethane Potential (MTP)
RT Total Trihalomethanes (TTHM)
SO Environmental Protection Agency Glossary
SO Water Pollution
DEF (EPA)One of a family of organic compounds, named as derivatives of methane. THM's are generally the by-product from chlorination of drinking water that contains organic material.(ESH) One of the family of organic compounds, named as derivatives of methane, wherein three of the four hydrogen atoms in methane are each substituted by a halogen atom in the molecular structure.

TRIPLE RINSE
(ESH)
DA November 20, 1990
BT1 Processes
RT Double Wash/Rinse
RT Pesticides
DEF (ESH) The flushing of containers three times, each time using a volume of the normal diluent equal to approximately ten percent of the container's capacity, and adding the rinse liquid to the spray mixture or disposing of it by a method prescribed for disposing of the pesticide.

TRITIUM
(EDB)
DA March 29, 1991
BT1 Beta-Minus Decay Radioisotopes
BT2 Beta Decay Radioisotopes
BT3 Radionuclides
BT4 CERCLA Hazardous Substances
BT5 Hazardous Substances
BT4 Nuclides
BT1 Years Living Radioisotopes
BT2 Radionuclides
BT3 CERCLA Hazardous Substances
BT4 Hazardous Substances
BT3 Nuclides
SO Radiation
DEF (NRC Glossary of Terms: Nuclear Power and Radiation) A radioactive isotope of hydrogen (one proton, two neutrons). Because it is chemically identical to natural hydrogen, tritium can easily be taken into the body by

SY-Synonymous Terms

SO-Source/Subject Category

SF-See From

any ingestion path. Decays by
beta emission. Its radioactive
halflife is about 12 1/2 years.

TRITIUM OXIDES
DA   January 8, 1991
SF   *HTO*

TROPOSPHERE
(EPA)
DA   October 12, 1990
BT1 Atmosphere
RT   Ozone
SO   Environmental Protection Agency
     Glossary
DEF (EPA) The lower atmosphere, the
     portion of the atmosphere
     between seven and ten miles from
     the Earth's surface where clouds
     are formed.

TROUBLE LIGHT
(2430 TROUBLE LIGH)
DA   January 3, 1991
BT1 Equipment/Parts - Electrical (DOE
     FRASE Vocabulary)
BT2 Equipment
RT   Indicator(s)
SO   DOE FRASE VOCABULARY

TRU
DA   October 12, 1990
SEE Transuranic Nuclides
SO   Acronyms

TRU WASTES
(DOE Order 6430.1A)
DA   October 12, 1990
SY   Transuranic (TRU) Contaminated
     Materials
SY   Transuranic Wastes (TRU Waste)
BT1 Radioactive Wastes
BT2 Wastes
SO   Construction
DEF (DOE Order 6430.1A) Without
     regard to source or form,
     radioactive wastes that at the end
     of institutional control periods are
     contaminated with alpha-emitting
     transuranic radionuclides with
     half-lives greater than 20 years
     and concentrations greater than
     100 nCi/g. Regarding the Waste
     Isolation Pilot Plant, high-level
     waste and spent nuclear fuel as
     defined by DOE 5820.2A are
     specifically excluded by this
     definition.

TRUCK DRIVER
(0820 TRUCK DRIVER)
DA   November 28, 1990
BT1 Transport Personnel
BT2 Occupations
BT2 Personnel
SO   DOE FRASE VOCABULARY

TRUNK
(1106 TRUNK)
DA   November 28, 1990
BT1 Human Body Parts
NT1 Abdomen
NT1 Back
NT2 Lower Back
NT2 Upper Back
NT1 Buttock(s)
NT1 Chest
NT1 Groin
NT1 Rib(s)
SO   DOE FRASE VOCABULARY

TRUST FUND (CERCLA)
(CERCLA; CFR; EPA)
DA   October 12, 1990
SO   Compensation and Liability
SO   Environmental Management
SO   Environmental Protection Agency
     Glossary
DEF (CFR) A fund set up under the
     Comprehensive Environmental
     Response, Compensation and
     Liability Act (CERCLA section
     221) to help pay for cleanup of
     hazardous waste sites and for
     legal action to force those
     responsible for the sites to clean
     them up.

TRUSTEES
(CERCLA; CFR)
DA   October 12, 1990
SY   Natural Resource Trustees
RT   Comprehensive Environmental
     Response, Compensation, etc.
SO   Compensation and Liability
DEF (CFR) Federal natural resources
     management agencies designated
     in Subpart G of this Plan, and any
     State agency which may pursue
     claims for damages under section
     107(f) of CERCLA.

TRUSTEES OF PRINCETON UNIVERSITY
DA   January 11, 1991
BT1 DOE Contractors
BT2 Potentially Responsible Parties
RT   Chicago Operations Office

TRW
DA   January 11, 1991
BT1 DOE Contractors
BT2 Potentially Responsible Parties
RT   Headquarters Operations

TS
DA   October 12, 1990
SEE Technical Specifications
SO   Acronyms
SO   Industrial Safety

TSA
DA   October 12, 1990
SEE Technical Safety Appraisals
SO   Acronyms

SO   Construction
SO   Industrial Hygiene

TSCA
DA   October 12, 1990
SEE Toxic Substances Control Act
SO   Acronyms

TSCANS
(NFI)
DA   October 12, 1990
BT1 Computer Codes
SO   Nuclear Facilities Incident Database
DEF (NFI) Computer code that
     processes temperature data for
     alarm and rod reversal purposes.

TSCAT
(NFI)
DA   October 12, 1990
DEF Saturation temperature used to
     calculate Critical Temperature
     Ratio.

TSCM
DA   October 12, 1990
SEE Temperature Scram Circuit Monitor
SO   Acronyms

TSD
DA   October 12, 1990
SEE Treatment, Storage, and Disposal
SO   Acronyms

TSH
DA   October 12, 1990
SEE Target Sleeve Housing
SO   Acronyms

TSS
DA   October 12, 1990
SEE Total Suspended Solids
SO   Acronyms

TSUNAMIS
(USGS)
DA   October 12, 1990
BT1 Natural Disasters
BT2 Natural Phenomenon
RT   Earthquake Magnitude
SO   Natural Phenomenon
DEF (USGS) Series of waves of
     extremely long length and period
     typically caused by a sudden
     vertical displacement of a large
     area of the sea floor during an
     undersea earthquake.

TTO
DA   October 12, 1990
SEE Total Toxic Organics
SO   Acronyms

TTR (Time to Repair)
DA   October 12, 1990
SEE Time to Repair
SO   Acronyms

TTR (Tonopah Test Range)
DA   October 12, 1990
SEE Tonopah Test Range
SO   Acronyms

TUBERCULOSIS
(1409 TUBERCULOSIS)
DA   November 28, 1990
BT1 Upper Respiratory Disease
 BT2 Cond of Respiratory Sys.
       Non-Toxic
  BT3 Diseases
  BT3 Illnesses
RT   Upper Respiratory Condition
SO   DOE FRASE VOCABULARY

TUNDRA
(EPA)
DA   October 12, 1990
BT1 Ecosystems
RT   Permafrost
SO   Environmental Protection Agency
       Glossary
DEF (EPA) A type of ecosystem
     dominated by lichens, mosses,
     grasses, and woody plants.
     Tundra is found at high latitudes
     (arctic tundra) and high altitudes
     (alpine tundra). Arctic tundra is
     underlain by permafrost and is
     usually very wet.

TURBIDIMETERS
(EPA)
DA   October 12, 1990
BT1 Devices
SO   Environmental Protection Agency
       Glossary
DEF (EPA) Devices that measure the
     amount of suspended solids in a
     liquid.

TURBIDITY
(EPA)
DA   October 12, 1990
RT   Coefficient of Haze
RT   Suspended Solids
SO   Environmental Protection Agency
       Glossary
DEF (EPA) (1) Haziness in air caused by
     the presence of particles and
     pollutants. (2) A similar cloudy
     condition in water due to
     suspended silt or organic matter.

TURBIDITY REMOVAL EVAPORATOR
DA   January 8, 1991
SF   *TRE*
BT1 Equipment

TURBINE
(2148 TURBINE)
DA   December 10, 1990
BT1 Facility Components
BT1 Machines (DOE FRASE
       Vocabulary)
 BT2 Equipment
SO   DOE FRASE VOCABULARY
DEF (NRC Glossary of Terms: Nuclear

Power and Radiation) A rotary
engine made with a series of
curved vanes on a rotating shaft.
Usually turned by water or steam.
Turbines are considered to be the
most economical means to turn
large electrical generators.

TURBINE BLOCK VALVE
DA   January 8, 1991
SF   *TBV*
BT1 Valves
 BT2 Devices

TURBINE BUILDING
DA   January 8, 1991
SF   *TB*
BT1 Facilities

**TURBINE BUILDING SECONDARY
CLOSED COOLING WATER**
DA   January 8, 1991
SF   *TBSCCW*

TURBINE BUILDING SERVICE WATER
DA   January 8, 1991
SF   *TBSW*

TWA
DA   October 12, 1990
SEE Time Weighted Average
SO   Acronyms

TWEEZERS
(3068 TWEEZERS)
DA   January 3, 1991
BT1 Tools - Manual
 BT2 Tools (DOE FRASE Vocabulary)
  BT3 Equipment
SO   DOE FRASE VOCABULARY

TWF
DA   October 12, 1990
SEE Transuranic Waste Facility
SO   Acronyms

TWFM
DA   October 12, 1990
SEE Traveling Wire Flux Monitor
SO   Acronyms

TWTSF
DA   October 12, 1990
SEE Transuranic Waste Treatment and
       Storage Facility
SO   Acronyms

TYPE A ASSESSMENTS
(CFR)
DA   November 15, 1990
BT1 Assessments
 BT2 Administrative Processes
  BT3 Processes
DEF (CFR) Standard procedures for
     simplified assessments requiring
     minimal field observation to

determine damages as specified in
section 301(c)(2)(A) of CERCLA.

TYPE A PACKAGING
(EMER)
DA   February 1, 1991
BT1 Packaging
 BT2 Packages
SO   Emergency Preparedness
DEF (EMER) Packaging which is
     designed in accordance with the
     general packaging requirements
     and which is adequate to prevent
     the loss or dispersal of radioactive
     contents and to retain the
     efficiency of its radiation shielding
     properties if the package is
     subjected to the prescribed tests
     [49 Code of Federal Regulations
     Part 173.398(b)] which represent
     the normal, rough handling
     conditions of transport.

TYPE A QUANTITIES
(CFR)
DA   January 24, 1991
BT1 Quantities
RT   Radioactive Materials
SO   Environmental Management
DEF (CFR) Quantities of radioactive
     material, the aggregate
     radioactivity of which does not
     exceed A1 for special normal form
     radioactive material, where A1 and
     A2 are given in Appendix A of this
     part or may be determined by
     procedures described in Appendix
     A of this part.

TYPE B ASSESSMENTS
(CFR)
DA   November 15, 1990
BT1 Assessments
 BT2 Administrative Processes
  BT3 Processes
DEF (CFR) Alternative methodologies
     for conducting assessments in
     individual cases to determine the
     type and extent of short- and
     long-term injury and damages, as
     specified in section 301(c)(2)(B) of
     CERCLA.

TYPE B PACKAGING
(EMER)
DA   February 1, 1991
BT1 Packaging
 BT2 Packages
SO   Emergency Preparedness
DEF (EMER) Packaging radioactive
     material which meets the
     standards for Type A packaging
     and, in addition, meets the
     standards for the hypothetical
     accident conditions of transport as
     prescribed in 49 Code of Federal
     Regulations Part 173.398(c).

SY-Synonymous Terms   SO-Source/Subject Category   SF-See From

TYPE B QUANTITIES

(CFR)
DA   January 24, 1991
BT1 Quantities
RT   Radioactive Materials
SO   Environmental Management
DEF (CFR) Quantities of radioactive
     material greater than Type A
     quantities.

TYPES OF RESIN

(CAA; CFR)
DA   October 12, 1990
BT1 Resin Grade
NT1 Bulk Resins
NT1 Dispersion Resins
NT1 Latex Resins
RT   Polymers
RT   Resin Grade
SO   Air Pollution
DEF Broad classifications of resin
     referring to the basic
     manufacturing processes for
     producing that resin, including, but
     not limited to, the suspension,
     dispersion, latex, bulk, and
     solution processes.

U.S. ARMY TOXIC AND HAZARDOUS MATERIALS CENTER

DA   January 8, 1991
SF   USATHAMA
BT1 Centers

U.S. DEPARTMENT OF ENERGY

DA   January 8, 1991
SF   DOE
BT1 Federal Agencies
 BT2 Agencies
  BT3 Administrative Organizations
   BT4 Organizations
NT1 Headquarters Operations
NT1 Naval Reactors
NT1 Operations Offices
 NT2 Albuquerque Operations Office
 NT2 Chicago Operations Office
 NT2 Idaho Operations Office
 NT2 Nevada Operations Office
 NT2 Oak Ridge Operations Office
 NT2 Richland Operations Office
 NT2 San Francisco Operations Office
 NT2 Savannah River Operations Office
NT1 Program Offices
 NT2 Office of the Assistant Secretary
      for Nuclear Energy
 NT2 Office of the Assistant Secretary
      for Fossil Energy
 NT2 Office of the Assistant Secretary
      for Defense Programs
 NT2 Office of the Assistant Secretary
      for Conservation et.al.
 NT2 Office of the Assistant Secretary
      for Environment, et. al.
  NT3 Office of Special Projects
 NT2 Office of Environmental
      Restoration and Waste
      Management
 NT2 Office of New Production Reactors
 NT2 Office of Civilian Radioactive
      Waste Management

 NT2 Office of Energy Research
  NT3 Office of Basic Energy Sciences
  NT3 Office of Health and
       Environmental Research
RT   Atomic Energy Commission
RT   Energy Research and Development
     Administration
RT   Nuclear Regulatory Commission
RT   Second Line Organization Level
DEF (U.S. Government Manual,
     1990-1991) Provides the
     framework for a comprehensive
     and balanced national energy plan
     through the coordination and
     administration of the energy
     functions of the Federal
     Government. The Department is
     responsible for long- term,
     high-risk research and
     development of energy
     technology; the marketing of
     Federal power; energy
     conservation; the nuclear weapons
     program; energy regulatory
     programs; and a central energy
     data collection and analysis
     program.

U.S. DEPARTMENT OF JUSTICE

DA   January 8, 1991
SF   DOJ
BT1 Federal Agencies
 BT2 Agencies
  BT3 Administrative Organizations
   BT4 Organizations
DEF (U.S. Government Manual,
     1990-1991) Serves as counsel for
     U.S. citizens. It represents them in
     enforcing the law in the public
     interest. The Department plays a
     key role in protection against
     criminals and subversion, in
     ensuring healthy competition of
     business, in safeguarding the
     consumer, and in enforcing drug,
     immigration, and naturalization
     laws. The Department plays a
     significant role in protecting
     citizens through law enforcement,
     crime prevention, crime detection,
     and prosecution and rehabilitation
     of offenders. The Department
     conducts all suits in the Supreme
     Court in which the United States is
     concerned. It represents the
     Government in legal matters
     generally, rendering legal advice
     and opinions, upon request, to the
     President and to the heads of
     executive departments.

U.S. DEPARTMENT OF LABOR

DA   January 8, 1991
SF   DOL
BT1 Federal Agencies
 BT2 Agencies
  BT3 Administrative Organizations
   BT4 Organizations
DEF (U.S. Government Manual,
     1990-1991) To foster, promote,

and develop the welfare of the
wage earners of the United States,
to improve the working conditions,
and to advance their opportunities
for profitable employment. The
Department administers a variety
of Federal labor laws guaranteeing
workers' rights to safe and
healthful working conditions, a
minimum hourly wage and
overtime pay, freedom from
employment discrimination,
unemployment insurance, and
workers' compensation. It also
protects workers' pension rights,
provides for job training programs;
helps workers find jobs; works to
strengthen free collective
bargaining; and keeps track of
changes in employment, prices
and other national economic
measurements.

U.S. DEPARTMENT OF TRANSPORTATION

DA   January 8, 1991
SF   DOT
BT1 Federal Agencies
 BT2 Agencies
  BT3 Administrative Organizations
   BT4 Organizations
NT1 Research and Special Programs
     Administration
DEF (U.S. Government Manual,
     1990-1991) Establishes the
     Nation's overall transportation
     policy. Under its umbrella there
     are nine administrations whose
     jurisdictions include highway
     planning, development, and
     construction; urban mass transit;
     railroads, aviation; and the safety
     of waterways, ports, highways,
     and oil and gas pipelines.
     Decisions made by the
     Department in conjunction with the
     appropriate State and local
     officials strongly affect other
     programs such as land planning,
     energy conservation, scarce
     resource utilization, and
     technological change.

U.S. ENVIRONMENTAL PROTECTION AGENCY

(SWDA; RCRA; CFR)
DA   October 19, 1990
SF   EPA
BT1 Federal Agencies
 BT2 Agencies
  BT3 Administrative Organizations
   BT4 Organizations
NT1 Regional Office
RT   Cooperative Agreements
RT   Sample Management Offices
SO   Wastes
SO   Water Pollution
DEF (U.S. Government Manual,
     1990-1991) The EPA Protects and
     enhances our environment today

and for future generations to the
fullest extent possible under the
laws enacted by Congress. The
Agency's mission is to control and
abate pollution in the areas of air,
water, solid wastes, pesticides,
radiation, and toxic substances. Its
mandate is to mount an
integrated, coordinated attack on
environmental pollution in
cooperation with State and local
governments.

UBIQUITOUS BACKGROUND LEVELS
(EPA)
DA   October 12, 1990
BT1 Background Levels
SO   Environmental Protection Agency
     Glossary
DEF (EPA) Concentrations of chemicals
     that are present in the environment
     due to anthropogenic sources
     (e.g., industry, automobiles).

UC
DA   October 12, 1990
SEE University of California
SO   Acronyms

UCC
DA   October 12, 1990
SEE Ultra Clean Coal
SO   Acronyms

UCI
DA   October 12, 1990
SEE University of California-Irvine
SO   Acronyms

UDH
DA   October 12, 1990
SEE Utah Department of Health
SO   Acronyms

UE
DA   October 12, 1990
SEE United Electric
SO   Acronyms

UFSAR
DA   October 12, 1990
SEE Updated Final Safety Analysis
     Report
SO   Acronyms

UF$_4$
DA   October 12, 1990
SEE Uranium Tetrafluoride
SO   Acronyms

UF$_6$
DA   October 12, 1990
SEE Uranium Hexafluoride
SO   Acronyms

UIC
DA   October 12, 1990

SEE Underground Injection Control
SO   Acronyms
SO   Water Pollution

ULCER(S)
(1410 ULCER)
DA   November 28, 1990
BT1 Illnesses
SO   DOE FRASE VOCABULARY

ULLAGE
(ESH)
DA   October 19, 1990
SY   Outage
BT1 Measurements
SO   Hazardous Materials
DEF (ESH) The amount by which a
     packaging falls short of being
     liquid full, usually expressed in
     percent by volume.

ULTRA CLEAN COAL
(EPA)
DA   October 12, 1990
SF   *UCC*
BT1 Coal
  BT2 Fossil Fuels
    BT3 Fuels
    BT3 Geologic Resources
    BT4 Natural Resources
SO   Environmental Protection Agency
     Glossary
DEF (EPA) Coal that has been washed,
     ground into fine particles, then
     chemically treated to remove
     sulfur, ash, silicone, and other
     substances; usually briquetted and
     coated with a sealant made from
     coal.

**ULTRASONIC RANGING AND DATA
SYSTEM**
DA   January 8, 1991
SF   *USRADS*
BT1 Information Systems
  BT2 Security Interests
  BT2 Systems

ULTRAVIOLET EQUIPMENT
(2431 ULTRAVIOLET)
DA   January 3, 1991
BT1 Equipment/Parts - Electrical (DOE
     FRASE Vocabulary)
  BT2 Equipment
SO   DOE FRASE VOCABULARY

ULTRAVIOLET RAYS
(EPA)
DA   October 12, 1990
BT1 Radiation
SO   Environmental Protection Agency
     Glossary
DEF (EPA) Radiation from the sun that
     can be useful or potentially
     harmful. UV rays from one part of
     the spectrum enhance plant life
     and are useful in some medical
     and dental procedures; UV rays
     from other parts of the spectrum to

which humans are exposed (e.g.,
while getting a suntan) can cause
skin cancer or other tissue
damage. The ozone layer in the
atmosphere provides a protective
shield that limits the amount of
ultraviolet rays that reach the
Earth's surface.

UMT
DA   October 12, 1990
SEE Uranium Mill Tailings
SO   Acronyms

UMTRA
DA   October 12, 1990
SEE Uranium Mill Tailings Remedial
     Action
SO   Acronyms

UMTRAP
DA   October 12, 1990
SEE Uranium Mill Tailings Remedial
     Action Program
SO   Acronyms

UNACCEPTABLE ADVERSE EFFECTS
(CAA; RHA; CFR)
DA   October 12, 1990
BT1 Effects
RT   Aquatic Environment
RT   Wetlands
SO   Water Pollution
DEF (CFR) Impacts on an aquatic or
     wetland ecosystem which are
     likely to result in significant
     degradation of municipal water
     supplies (including surface or
     ground water) or significant loss of
     or damage to fisheries,
     shell-fishing, or wildlife habitat or
     recreation areas.

UNATTACHED FRACTIONS
(NCRP)
DA   October 12, 1990
RT   Radon Progeny
SO   Radiation
DEF (NCRP) The fractions of any
     short-lived daughter products of
     radon that are not attached to the
     ambient aerosol.

UNATTENDED OPENINGS
(DOE Order 6430.1A)
DA   October 12, 1990
BT1 Sites/Areas
SO   Construction
DEF (DOE Order 6439.1A) Doors,
     operable windows, hatches,
     louvered openings, etc., that are
     not attended by security guards or
     guarded by safety devices.

UNAUTHORIZED DISCHARGES
(DOE Order 5480.16)
DA   October 12, 1990
SY   Spills

BT1 Discharges
RT Firearms
RT Firearms Ranges
RT Protective Force Personnel
SO Firearms
DEF (DOE Order 5480.16) Discharges
of a firearm under circumstances
other than during firearms training
with the firearm properly pointed
downrange (or toward a target) or
the intentional firing at hostile
parties when deadly force is
unauthorized.

UNCERTAINTY
(SSDC)
DA October 12, 1990
BT1 Conditions
RT Safety
SO System Safety Development Center
Glossary
DEF (SSDC) Nonquantifiable increase in
accident probability resulting from
the lack of named safety program
features (e.g., human factors
review, hazard analysis, training);
oversight and omissions in a
management system. This may be
termed descriptive uncertainty. A
"degree of uncertainty" results
from a quantitative inadequacy of
estimation of a stated known
factor.

UNCONTROLLED HAZARDOUS
WASTE SITES
(CFR)
DA January 24, 1991
BT1 Sites/Areas
RT Hazardous Waste Facilities
SO Environmental Management
DEF (CFR) Areas where an
accumulation of hazardous waste
create a threat to the health and
safety of individuals or the
environment or both. Some sites
are found on public lands, such as
those created by former municipal,
county or state landfills where
illegal or poorly managed waste
disposal has taken place. Other
sites are found on private property,
often belonging to generators or
former generators of hazardous
waste. Examples of such sites
include, but are not limited,
surface impoundments, landfills,
dumps, and tank or drum farms.
Normal operations at TSD sites
are not covered by this definition.

UNCONTROLLED TOTAL ARSENIC
EMISSIONS
(CAA; CFR)
DA October 12, 1990
BT1 Fugitive Emissions
BT2 Emissions
BT3 Air Pollutants
RT Glass Melting Furnaces
RT Inorganic Arsenic

RT Secondary Emissions
RT Theoretical Arsenic Emissions
Factor
SO Air Pollution
DEF (CFR) The total inorganic arsenic in
the glass melting furnace exhaust
gas preceding any add-on
emission control device.

UNDER CONSTRUCTION
(DOE Order 5480.6)
DA October 12, 1990
SO Industrial Safety
DEF (DOE Order 5480.6) When the
authorization for construction has
been issued and authorization for
operation has not yet been issued.

UNDERFIRE AIR
(CFR)
DA January 24, 1991
BT1 Air
SO Environmental Management
DEF (CFR) Any forced or induced air,
under control as to quantity and
direction, that is supplied from
beneath and which passes
through the solid wastes fuel bed.

UNDERGROUND AREAS
(SWDA; RCRA; CFR)
DA October 12, 1990
BT1 Sites/Areas
SO Wastes
DEF (CFR) Underground rooms, such
as basements, cellars, shafts or
vaults, providing enough space for
physical inspection of the exterior
of a tank situated on or above the
surface of the floor.

UNDERGROUND INJECTION
(SWDA; RCRA; CFR)
DA October 12, 1990
SY Well Injection
BT1 Injection
BT2 Processes
RT Underground Injection Control
SO Environmental Management
SO Wastes
SO Water Pollution
DEF (CFR) The subsurface
emplacement of fluids through a
bored, drilled or driven well; or
through a dug well, where the
depth of the dug well is greater
than the surface dimension. A well
injection.

UNDERGROUND INJECTION
CONTROL
(SDWA; CFR; ESH)
DA October 12, 1990
SF UIC
BT1 Controls
BT1 Programs
RT Underground Injection
SO Environmental Management

SO Water Pollution
DEF (SDWA) Effective programs to
prevent underground injection that
endangers drinking water sources.
Authorization of underground
injection by permit issued by the
State; includes inspection,
monitoring, recordkeeping, and
reporting requirements.

UNDERGROUND RELEASES
(SWDA; RCRA; CFR)
DA October 12, 1990
SY Belowground Releases
BT1 Releases
SO Wastes
DEF Any below ground releases.

UNDERGROUND SOURCES OF
DRINKING WATER
(SWDA; RCRA; SDWA; CFR; EPA;
ESH)
DA October 19, 1990
SF USDW
BT1 Aquifers
BT2 Formations
RT Exempted Aquifers
RT Fresh Water
RT Groundwater
RT Sole Source Aquifer
RT Wells
SO Environmental Management
SO Environmental Protection Agency
Glossary
SO Wastes
SO Water Pollution
DEF Aquifers or a portion of one (a)(1)
which supplies any public water
system; or (2) which contains a
sufficient quantity of ground water
to supply a public water system;
and (i) currently supplies drinking
water for human consumption; or
(ii) contains fewer than 10,000
mg/l total dissolved solids; and (b)
which is not an exempted aquifer.
(EPA) As defined in the UIC
(Underground Injection Control)
program, this term refers to
aquifers that are currently being
used as a source of drinking
water, and those that are capable
of supplying a public water
system. They have a total
dissolved solids content of 10,000
milligrams per liter or less, and are
not "exempted aquifers."

UNDERGROUND STORAGE TANKS
(SWDA; RCRA; CFR; EPA; ESH)
DA October 12, 1990
SF UST
BT1 Underground Tanks
BT2 Tanks
BT3 Facility Components
NT1 Non-Operational Storage Tanks
NT1 Septic Tanks
RT Ancillary Equipment
RT Maintenance

RT Ohio Bureau of Underground
    Storage Tank Regulation
SO Environmental Protection Agency
    Glossary
SO Wastes
DEF Any one or combination of tanks
    (including underground pipes
    connected thereto) that is used to
    contain an accumulation of
    regulated substances, and the
    volume of which (including the
    volume of underground pipes
    connected thereto) is 10% or more
    beneath the surface of the ground.
    (EPA) Tanks located all or partially
    under ground that is designed to
    hold gasoline or other petroleum
    products or chemical solutions.

UNDERGROUND TANKS
(SWDA; RCRA)
DA October 12, 1990
BT1 Tanks
 BT2 Facility Components
NT1 Underground Storage Tanks
 NT2 Non-Operational Storage Tanks
 NT2 Septic Tanks
SO Environmental Management
SO Wastes
DEF Devices meeting the definition of
    "tanks" in 40 CFR 260.10 whose
    entire surface areas are totally
    below the surface of and covered
    by the ground.

UNDERGROUND URANIUM MINES
(CAA; CFR)
DA October 12, 1990
BT1 Conventional Mines
 BT2 Sites/Areas
SO Air Pollution
DEF (CFR) Man-made underground
    excavations made for the purpose
    of removing material containing
    uranium for the principal purpose
    of recovering uranium.

UNDERPINNINGS
(DOE Order 6430.1A)
DA October 12, 1990
SO Construction
DEF (DOE Order 6430.1A) Permanent
    supports replacing or reinforcing
    the older supports beneath a wall
    or column.

UNDERVOLTAGE
DA January 8, 1991
SF UV

UNDUE RISKS
(DOE Order 5480.5; ESH)
DA October 12, 1990
BT1 Risks
SO Industrial Safety
SO Management
DEF (DOE Order 5480.5) Levels of
    identifiable risks which are
    unacceptable to DOE.

UNIFIED DOSE ASSESSMENTS
(EMER)
DA February 1, 1991
BT1 Assessments
 BT2 Administrative Processes
  BT3 Processes
SO Emergency Preparedness
DEF (EMER) A functional capability to
    coordinate monitoring teams,
    collection of monitoring data,
    calculation of off-site radiation
    dose projections and used for the
    recommendation of protective
    actions for the plume and
    ingestion exposure emergency
    planning zones.

**UNINTERRUPTIBLE POWER
SUPPLIES**
(DOE Order 6430.1A)
DA October 12, 1990
SF UPS
BT1 Power Supply
 BT2 Equipment/Parts - Electrical (DOE
    FRASE Vocabulary)
  BT3 Equipment
RT Emergency Power
RT Standby Power
SO Construction
DEF (DOE Order 6430.1A) Power
    supplies that provide automatic,
    instantaneous power, without
    delay or transients, on failure of
    normal power. They can consist of
    batteries or full-time operating
    generators. They can be
    designated as standby or
    emergency power depending on
    the application. Emergency
    installations must meet the
    requirements specified for
    emergency power.

UNIRRADIATED ENRICHED URANIUM
(DOE Order 6430.1A)
DA October 12, 1990
BT1 Enriched Uranium
 BT2 Uranium
  BT3 Geologic Resources
   BT4 Natural Resources
  BT3 Heavy Metals
RT New Storage Facilities
SO Construction
DEF (DOE Order 6430.1A) Naturally
    occurring uranium enriched with
    U-235 above its natural
    abundance of approximately
    0.72% (weight percent) that has
    not been exposed to a neutron
    flux.

UNIT LOAD DEVICES
(SWDA; RCRA; ESH)
DA October 12, 1990
BT1 Freight Containers
 BT2 Containers
 BT2 Unit Load Devices
NT1 Freight Containers
SO Hazardous Materials
DEF Types of freight containers, aircraft

containers, aircraft pallets with
nets, or aircraft pallets with nets
over igloos.

UNIT MASONRY
(DOE Order 6430.1A)
DA October 12, 1990
SO Construction
DEF (DOE Order 6430.1A) Includes
    brick made of clay or shale, sand
    lime, and concrete; structural clay,
    concrete masonry units, solid load
    bearings, tile, load-bearing, and
    non-load-bearing, hollow
    load-bearing, and hollow
    non-load-bearing; natural stone
    and cast stone; ceramic glazed
    clay masonry, solid units, and
    hollow units; and prefaced
    concrete masonry units.

UNITED ELECTRIC
DA January 8, 1991
SF UE
BT1 Companies
 BT2 Commercial Organizations
  BT3 Organizations

UNITED STATES
(SWDA; RCRA; CERCLA; CFR; ESH)
DA October 12, 1990
RT EPA Region
RT National Contingency Plan
SO Compensation and Liability
SO Hazardous Materials
SO Wastes
DEF (USC) Includes the several States
    of the United States, the District of
    Columbia, the Commonwealth of
    Puerto Rico, Guam, American
    Samoa, the United States Virgin
    Islands, the Commonwealth of the
    Northern Marianas, and any other
    territory or possession over which
    the United States has jurisdiction.

UNITS OF MEASURE
DA February 27, 1991
NT1 Activity per Ton
NT1 Body Wave Magnitude
NT1 British Thermal Unit
NT1 Centimeters per second
NT1 Cubic Feet Per Minute
NT1 Cubic Yards
NT1 Disintegrations per minute per gram
NT1 Drips per Minute
NT1 gallon
NT1 Kilograms/year
NT1 Kilometer
NT1 Kilovolt
NT1 Megawatt thermal
NT1 Microcuries
NT1 micrograms/liter
NT1 Milliroentgen
NT1 Million Gallons Per Day
NT1 milligram/killigram
NT1 milligram/liter
NT1 Moment Magnitude
NT1 parts per million

NT1 Pounds/year
NT1 PPM/PPB
NT1 Radiation Units
  NT2 Absorbed Dose
  NT2 Becquerel
  NT2 Curie
  NT2 Dose Equivalents
    NT3 Annual Dose Equivalent
      NT4 Annual Effective Dose
          Equivalent
        NT5 Cumulative Annual Effective
            Dose Equivalent
    NT3 Collective Dose Equivalent
      NT4 Collective Effective Dose
          Equivalent
    NT3 Committed Dose Equivalent
      NT4 Committed Effective Dose
          Equivalent
    NT3 Deep Dose Equivalent
      NT4 Deep Eye Dose Equivalent
    NT3 Effective Dose Equivalent
      NT4 Annual Effective Dose
          Equivalent
        NT5 Cumulative Annual Effective
            Dose Equivalent
      NT4 Collective Effective Dose
          Equivalent
      NT4 Committed Effective Dose
          Equivalent
    NT3 Lens of Eye Dose Equivalents
    NT3 Shallow Eye Dose Equivalents
    NT3 Shallow Dose Equivalents
  NT2 Gray
  NT2 Microroentgen/hour
  NT2 Picocurie
  NT2 Picocuries Per Liter
  NT2 Picocuries/gram
  NT2 Radiation Absorbed Dose
  NT2 Relative Biological Effectiveness
  NT2 Roentgen Equivalent Man
  NT2 Roentgen
  NT2 Sievert
  NT2 Working Level
  NT2 Working Level Month
    NT3 Cumulative Working Level
        Months
NT1 square kilometers
NT1 Standard Cubic Foot
NT1 Surface Wave Magnitude
RT Doses
RT Metric System
RT Pressure
RT Rates
RT Ratios
RT Standard Curve
RT Standard Deviation
RT Time Designations
DEF (DSTT) Quantities adopted as
    standards of measurement.

UNIVERSAL SLEEVE HOUSING
DA January 8, 1991
SF *USH*

UNIVERSITIES
DA February 1, 1991
BT1 Educational Organizations
  BT2 Organizations
NT1 Iowa State University

NT1 Massachusetts Institute of
    Technology
NT1 North Carolina State University
NT1 Stanford University
NT1 University of Chicago
NT1 University of California
NT1 University of California-Irvine
RT *Historically Black Colleges and*
    Universities
RT Oak Ridge Associated Universities
RT Southeastern Universities Research
    Association, Inc.
RT Universities Research Association,
    Inc.
RT University of Georgia Research
    Foundation, Inc.
RT University of Tennessee Center for
    Biotechnology
SO Management
DEF (WEBSTER) An institution of higher
    learning providing facilities for
    teaching and research and
    authorized to grant academic
    degrees.

**UNIVERSITIES RESEARCH
ASSOCIATION, INC.**
DA January 11, 1991
BT1 Companies
  BT2 Commercial Organizations
    BT3 Organizations
BT1 DOE Contractors
  BT2 Potentially Responsible Parties
RT Chicago Operations Office
RT Universities

UNIVERSITY OF CALIFORNIA
DA January 8, 1991
SF *UC*
BT1 DOE Contractors
  BT2 Potentially Responsible Parties
BT1 Universities
  BT2 Educational Organizations
    BT3 Organizations
RT Albuquerque Operations Office
RT Lawrence Livermore National
    Laboratory
RT Lawrence Berkeley Laboratory
RT Los Alamos National Laboratory
RT San Francisco Operations Office

UNIVERSITY OF CALIFORNIA-IRVINE
DA January 8, 1991
SF *UCI*
BT1 Universities
  BT2 Educational Organizations
    BT3 Organizations

UNIVERSITY OF CHICAGO
DA January 11, 1991
BT1 Universities
  BT2 Educational Organizations
    BT3 Organizations
RT Argonne National Laboratory-East
    (Chicago)
RT Argonne National Laboratory-West
    (At INEL)
RT Chicago Operations Office

**UNIVERSITY OF GEORGIA RESEARCH
FOUNDATION, INC.**
DA January 11, 1991
BT1 DOE Contractors
  BT2 Potentially Responsible Parties
BT1 Foundations
  BT2 Research and Development
      Organizations
    BT3 Organizations
RT Savannah River Operations Office
RT Universities

**UNIVERSITY OF TENNESSEE CENTER
FOR BIOTECHNOLOGY**
DA January 8, 1991
SF *UTENN*
BT1 Centers
RT Universities

**UNKNOWN/UNDETERMINED
ACTIVITY**
(1252 UU ACTIVITY)
DA November 28, 1990
BT1 Activity Types (DOE FRASE
    Vocabulary)
  BT2 Activities
SO DOE FRASE VOCABULARY

**UNLISTED HAZARDOUS
SUBSTANCES**
(CERCLA)
DA October 12, 1990
BT1 Hazardous Substances
SO Compensation and Liability
DEF (CERCLA) Solid wastes, as defined
    in 40 CFR 261.2, which are not
    excluded from regulation as a
    hazardous waste under 40 CFR
    261.4(b), is a hazardous
    substance under section 101(14)
    of the Act if it exhibits any of the
    characteristics identified in 40
    CFR 261.20 through 261.24.

UNPACKAGING ROOMS
(DOE Order 6430.1A)
DA October 12, 1990
BT1 Sites/Areas
SO Construction
DEF (DOE Order 6430.1A) The spaces
    in which receiving containers are
    opened and unpackaged and
    repackaged for storage or
    shipment and are surrounded by
    one or more secondary
    confinement areas.

UNREVIEWED SAFETY QUESTIONS
(DOE Orders 5480.1B, 5480.5 and
5480.6)
DA October 12, 1990
RT Safety Reviews
SO Industrial Hygiene
SO Industrial Safety
DEF Proposed changes, tests, or
    experiments shall be deemed to
    involve an unreviewed safety
    question if: (1) the probability of
    occurrence or the consequences

BT-Broader Term   NT-Narrower Term   RT Related Term

of an accident or malfunction of
equipment important to safety
evaluated previously by safety
analyses will be significantly
increased, or (2) a possibility for
an accident or malfunction of a
different type than any evaluated
previously by safety analyses will
be created which could result in
significant safety consequences.

UNSATURATED ZONES
(SWDA; RCRA; EPA)
DA October 12, 1990
SY Aeration Zones
BT1 Zones
 BT2 Sites/Areas
RT Saturated Zones
SO Environmental Management
SO Environmental Protection Agency
     Glossary
SO Wastes
DEF Areas above the water table where
     the soil pores are not fully
     saturated, although some water
     may be present.

UNSCHEDULED EVACUATION
(1581 UNSCHED EVAC)
DA November 28, 1990
BT1 Nature of Programmatic Impact
SO DOE FRASE VOCABULARY

UNSCHEDULED SHUTDOWN
(1580 UNSCH SHUTDN)
DA November 28, 1990
BT1 Nature of Programmatic Impact
RT Automatic Reactor Scram
RT Manual Reactor Scram
RT Violations
SO DOE FRASE VOCABULARY

UNUSUAL EVENTS
(EMER)
DA February 1, 1991
BT1 Emergency Response Levels
BT1 Events
SO Emergency Preparedness
DEF (EMER) Emergency response
     levels which represent an event in
     progress or having occurred that
     normally would not constitute an
     emergency but which indicates a
     potential reduction of safety of a
     facility in which no potential exists
     for significant off- site release of
     radioactive or other hazardous
     substances.

**UNUSUAL OCCURRENCE REPORTING
SYSTEM**
(SSDC)
DA February 12, 1991
SF UORS (System)
BT1 Safety Performance Measurement
     System
 BT2 Information Systems
 BT3 Security Interests
 BT3 Systems

RT Occurrence Reporting and
     Processing System
RT Unusual Occurrences
RT Unusual Occurrence Reports
DEF (SSDC) A database of unusual
     occurrence reports submitted
     through the DOE reporting system.

UNUSUAL OCCURRENCE REPORTS
(DOE Order 5000.3; ESH)
DA October 12, 1990
SF UORS (Reports)
BT1 Occurrence Reports
 BT2 Reports
RT Unusual Occurrences
RT Unusual Occurrence Reporting
     System
SO Management
DEF (DOE Order 5000.3) Written
     evaluations of unusual
     occurrences that are prepared in
     sufficient detail to enable the
     reviewer to assess their
     significance, consequences, or
     implications and to determine the
     means of avoiding a recurrence
     with minimal additional inquiry.

UNUSUAL OCCURRENCES
(DOE Order 5000.3; ESH)
DA October 12, 1990
BT1 Reportable Occurrences
 BT2 Occurrences
RT Environmental Occurrences
RT Generic Significance
RT Program Significant Delay
RT Unusual Occurrence Reports
RT Unusual Occurrence Reporting
     System
SO Environmental Management
SO Management
DEF (DOE Order 5000.3) Unusual or
     unplanned events having
     programmatic significance such
     that they adversely affect or
     potentially affect the performance,
     reliability, or safety of a facility.
     (DOE Order 5000.3A) Unusual
     occurrences are non- emergency
     occurrences that have significant
     impact or potential for impact on
     safety, environment, health,
     security, or operations. The types
     of occurrences that are to be
     categorized as unusual
     occurrences are specified in DOE
     Order 5000.3A.

UORS (Reports)
DA February 12, 1991
SEE Unusual Occurrence Reports
SO Acronyms

UORS (System)
(SSDC)
DA March 1, 1991
SEE Unusual Occurrence Reporting
     System

SO Acronyms
DEF (SSDC) A database of unusual
     occurrence reports submitted
     through the DOE reporting system.

**UPDATED FINAL SAFETY ANALYSIS
REPORT**
DA January 8, 1991
SF UFSAR
BT1 Safety Analysis Reports
 BT2 Reports

UPGRADES
(SWDA; RCRA; CFR)
DA October 12, 1990
RT UST Systems
SO Wastes
DEF Additions or retrofitting of some
     systems such as cathodic
     protection, lining, or spill and
     overfill controls to improve the
     ability of an underground storage
     tank system to prevent the release
     of product.

UPLIFTS
(USGS)
DA October 12, 1990
RT Faults
RT Tectonic Deformations
SO Natural Phenomenon
DEF (USGS) Structurally high areas in
     the crust of the earth, produced by
     positive movements that raise or
     upthrust the rocks, as in a dome
     or arch. They are the elevation of
     any extensive part of the earth's
     surface relative to some other
     parts. Also, as an engineering
     term, they are any force that tends
     to raise an engineering structure
     and its foundation relative to its
     surroundings. They may be
     caused by pressure of subjacent
     ground, surface water, expansive
     soil under the base of the
     structure, or lateral forces such as
     wind.

UPPER ARM
(1096 UPPER ARM)
DA November 28, 1990
BT1 Arm(s)
 BT2 Human Body Parts
RT Shoulder(s)
SO DOE FRASE VOCABULARY

UPPER BACK
(1108 UPPER BACK)
DA November 28, 1990
BT1 Back
 BT2 Trunk
 BT3 Human Body Parts
RT Pinched Nerve
RT Ruptured Disk
RT Spinal Cord
SO DOE FRASE VOCABULARY

SY-Synonymous Terms   SO-Source/Subject Category   SF-See From

UPPER RESPIRATORY CONDITION
(1411 UPPER RC)
DA November 28, 1990
BT1 Cond of Respiratory Sys. Non-Toxic
 BT2 Diseases
 BT2 Illnesses
RT Respiratory System
RT Tuberculosis
SO DOE FRASE VOCABULARY

UPPER RESPIRATORY DISEASE
(1412 UPPER RD)
DA November 28, 1990
BT1 Cond of Respiratory Sys. Non-Toxic
 BT2 Diseases
 BT2 Illnesses
NT1 Tuberculosis
RT Respiratory System
SO DOE FRASE VOCABULARY

UPPERMOST AQUIFERS
(SWDA; RCRA)
DA October 12, 1990
BT1 Aquifers
 BT2 Formations
SO Environmental Management
SO Wastes
DEF Geologic formations nearest the
    natural ground surface that are
    acquifers, as well as lower
    aquifers that are hydraulically
    interconnected with these aquifers
    within a facility's property
    boundary.

UPS
DA October 12, 1990
SEE Uninterruptible Power Supplies
SO Acronyms

UPSTREAM PROCESSES
(SSDC)
DA October 12, 1990
BT1 Administrative Processes
 BT2 Processes
RT Operational Readiness
SO System Safety Development Center
    Glossary
DEF (SSDC) Methods and steps used to
    prepare designs and plans,
    hardware, people, procedures, or
    managerial controls to attain
    operational readiness.

UPTAKE
(IAEA; NCRP)
DA October 12, 1990
RT Deposition
RT Intake
RT Retention
SO Radiation
DEF (NCRP) Quantity of radionuclide
    taken up by the systemic
    circulation, e.g., by absorption
    from compartments in the
    respiratory or gastrointestinal
    tracts. (IAEA) Amount of
    radioactive material absorbed into

the extracellular fluids. Also used
to denote the process.

URANIUM
(EPA)
DA October 12, 1990
BT1 Geologic Resources
 BT2 Natural Resources
BT1 Heavy Metals
NT1 Depleted Uranium
NT1 Enriched Uranium
 NT2 Unirradiated Enriched Uranium
RT Source Materials
RT Uranium Mill Tailings
RT Yellowcake
SO Environmental Protection Agency
    Glossary
SO Radiation
DEF (EPA) A radioactive heavy metal
    element used in nuclear reactors
    and the production of nuclear
    weapons. Term refers usually to
    U-238, the most abundant radium
    isotope, although a small
    percentage of naturally occurring
    uranium is U-235.

URANIUM 238
(EDB)
DA March 29, 1991
BT1 Alpha Decay Radioisotopes
 BT2 Radionuclides
  BT3 CERCLA Hazardous Substances
   BT4 Hazardous Substances
  BT3 Nuclides
BT1 Years Living Radioisotopes
 BT2 Radionuclides
  BT3 CERCLA Hazardous Substances
   BT4 Hazardous Substances
  BT3 Nuclides
SO Radiation

URANIUM BY-PRODUCT MATERIALS
(CAA; CFR)
DA October 12, 1990
BT1 By-Product Materials
 BT2 By-products
  BT3 Materials
RT Licensed Sites
RT New Tailings
RT Tailings
SO Air Pollution
DEF (CFR) Tailings are the wastes
    produced by the extraction or
    concentration of uranium from any
    ore processed primarily for its
    source material content. Ore
    bodies depleted by uranium
    solution extractions and which
    remain underground do not
    constitute byproduct material for
    the purposes of this subpart.

URANIUM HEXAFLUORIDE
(EMER)
DA January 8, 1991
SF *UF₆*
SO Emergency Preparedness

URANIUM MILL TAILINGS
DA January 8, 1991
SF *UMT*
RT Radium 226
RT Radium 228
RT Radon
RT Thorium
RT Uranium
SO Radiation
DEF (NRC Glossary of Terms: Nuclear
    Power and Radiation) Naturally
    radioactive residue from the
    processing of uranium ore into
    yellowcake in a mill. Although the
    milling process recovers about 93
    percent of the uranium, the
    residues, or tailings, contain
    several radioactive elements,
    including uranium, thorium,
    radium, polonium and radon.

**URANIUM MILL TAILINGS REMEDIAL
ACTION**
DA January 8, 1991
SF *UMTRA*
BT1 Remedial Actions
 BT2 Actions
  BT3 Responses

**URANIUM MILL TAILINGS REMEDIAL
ACTION PROGRAM**
DA January 8, 1991
SF *UMTRAP*
BT1 Remedial Action Program
 BT2 Programs

URANIUM TETRAFLUORIDE
DA January 8, 1991
SF *UF₄*

URBAN RUNOFF
(EPA)
DA October 12, 1990
BT1 Run-off
RT Municipalities
RT Sewers
RT Storm Sewers
SO Environmental Protection Agency
    Glossary
DEF (EPA) Stormwater from city streets
    and adjacent domestic or
    commercial properties that may
    carry pollutants of various kinds
    into the sewer systems and/or
    receiving waters.

USATHAMA
DA October 12, 1990
SEE U.S. Army Toxic and Hazardous
    Materials Center
SO Acronyms

USDW
DA October 12, 1990
SEE Underground Sources of Drinking
    Water
SO Acronyms

BT-Broader Term   NT-Narrower Term   RT Related Term

USED OILS
(SWDA; RCRA)
DA   October 12, 1990
SY   Waste Oils
BT1 Oils
NT1 Recycled Oils
 NT2 Re-refined Oils
SO   Wastes
DEF (USC) Any oils which have been
     (A) refined from crude oil, (B)
     used, and (C) as a result of such
     use, contaminated by physical or
     chemical impurities.

USEFUL LIFE
(DOE Order 6430.1A)
DA   October 12, 1990
BT1 Time Designations
SO   Construction
DEF (DOE Order 6430.1A) The time
     period in which a building element
     can be expected to perform
     effectively with proper
     maintenance.

USER AGENCIES
(ESH)
DA   October 12, 1990
BT1 Federal Agencies
 BT2 Agencies
  BT3 Administrative Organizations
  BT4 Organizations
RT   Government-Owned
     Contractor-Operated Facilities
SO   Management
DEF (ESH) Government agencies or
     contractors thereof other than
     DOE who utilize facilities and
     equipment at DOE Government-
     owned, contractor-operated sites.

USES
DA   February 25, 1991
NT1 Committed Uses
NT1 Consumptive Use
NT1 Designated Uses
NT1 Instream Use
NT1 Multiple Uses
NT1 Official Use Only
NT1 Otherwise Use
NT1 Restricted Use
NT1 Reuse
 NT2 Closed-Loop Recycling

USH
DA   October 12, 1990
SEE Universal Sleeve Housing
SO   Acronyms

USRADS
DA   October 12, 1990
SEE Ultrasonic Ranging and Data
     System
SO   Acronyms

UST
DA   October 19, 1990
SEE Underground Storage Tanks
SO   Acronyms

UST SYSTEMS
(SWDA; RCRA; CFR)
DA   October 19, 1990
BT1 Tank Systems
NT1 Hazardous Substance UST System
NT1 Petroleum UST Systems
RT   Release Detection
RT   Upgrades
SO   Wastes
DEF (CFR) Underground storage tanks,
     connected underground piping,
     underground ancillary equipment,
     and containment system, if any.

UTAH DEPARTMENT OF HEALTH
DA   January 8, 1991
SF   *UDH*
BT1 State Agencies
 BT2 Agencies
  BT3 Administrative Organizations
   BT4 Organizations

UTENN
DA   October 12, 1990
SEE University of Tennessee Center for
     Biotechnology
SO   Acronyms

UTILIZATION FACILITIES
(AEA; ANL)
DA   May 24, 1991
BT1 Facilities
RT   Special Nuclear Materials
SO   Management
SO   Radiation
DEF (AEA) (1) Any equipment or device,
     except an atomic weapon,
     determined by rule of the
     Commission to be capable of
     making use of special nuclear
     material in such quantity as to be
     of significance to the common
     defense and security, or in such
     manner as to affect the health and
     safety of the public, or peculiarly
     adapted for making use of atomic
     energy in such quantity as to be of
     significance to the common
     defense and security, or in such
     manner as to affect the health and
     safety of the public; or (2) any
     important component part
     especially designed for such
     equipment or device as
     determined by the Commission.

UV
DA   October 12, 1990
SEE Undervoltage
SO   Acronyms

VAC
DA   October 12, 1990
SEE Volts Alternating Current
SO   Acronyms

VACCINES
(EPA)
DA   October 12, 1990

RT   Antigens
SO   Environmental Protection Agency
     Glossary
DEF (EPA) Dead or partial or modified
     antigens used to induce immunity
     to certain infectious diseases.

VACUUM CLEANER
(2150 VACUUM CLEAN)
DA   December 10, 1990
BT1 Machines (DOE FRASE
     Vocabulary)
 BT2 Equipment
RT   Vacuum System
SO   DOE FRASE VOCABULARY

VACUUM SYSTEM
(2054 VAC SYSTEM)
DA   December 10, 1990
BT1 Systems (DOE FRASE Vocabulary)
 BT2 Systems
RT   Vacuum Cleaner
SO   DOE FRASE VOCABULARY

VALIDATION
(DOE Order 4700.1)
DA   January 24, 1991
BT1 Administrative Processes
 BT2 Processes
RT   Verification
SO   Environmental Management
DEF (DOE Order 4700.1) The process
     of evaluating project planning,
     development, baselines and
     proposed funding prior to inclusion
     of new projects or system
     acquisition in the DOE budget or
     seeking increased funding for a
     prior project or system. It requires
     a review of project planning and
     conceptual development
     documentation, as well as
     discussion with the program or
     field element and principle
     contributing contractors to
     determine the source basis,
     procedures, and validity of
     proposed requirements, scope,
     cost, schedule, finding, and so
     forth. Findings and
     recommendations resulting from
     the validation process will be
     provided for use in the annual
     budget formulation.

VALUES
(SSDC)
DA   October 12, 1990
SO   System Safety Development Center
     Glossary
DEF (SSDC) The consequences of risks
     or accidents which are
     nonmonetary and may be
     nonquantifiable or intangible, such
     as pain, humane consideration,
     pride, or employee or public
     opinion.

VALVES
DA February 26, 1991
BT1 Devices
NT1 Back Pressure Valve
NT1 Check Valves
NT1 Main Steam Isolation Valve
NT1 Motor Operated Valves
NT1 Open Ended Valve
NT1 Pressure Regulator Valve
NT1 Relief Valves
NT2 Emergency Relief Valve
NT2 Inadvertently Opened Relief Valve
NT2 Power Operated Relief Valve
NT2 Safety/Relief Valve
NT2 Stuck-Open Relief Valve
NT1 Turbine Block Valve
RT Controls
DEF (WEBSTER) Mechanical devices by which the flow of material may be started, stopped, or regulated by a movable part that opens, shuts, or partially obstructs one or more ports or passageways.

VAM
DA October 12, 1990
SEE Vibration and Acoustic Monitoring
SO Acronyms

VAPOR CAPTURE SYSTEMS
(EPA)
DA October 12, 1990
BT1 Control Devices
BT2 Devices
BT1 Systems
NT1 Floor Sweeps
SO Environmental Protection Agency Glossary
DEF (EPA) Any combination of hoods and ventilation systems that capture or contain organic vapors in order that they may be directed to an abatement or recovery device.

VAPOR DISPERSION
(EPA)
DA October 12, 1990
SO Environmental Protection Agency Glossary
DEF (EPA) The movement of vapor clouds in air due to wind, gravity spreading, and mixing.

VAPOR PLUMES
(EPA)
DA October 12, 1990
BT1 Plumes
BT2 Air Pollutants
BT2 Discharges
RT Flue Gases
SO Environmental Protection Agency Glossary
DEF (EPA) Flue gases that are visible because they contain water droplets.

VAPORIZATION
(EPA)

DA October 12, 1990
RT Distillation
RT Vapors
SO Environmental Protection Agency Glossary
DEF (EPA) The change of a substance from a liquid to a gas.

VAPORS
(EPA)
DA October 12, 1990
RT Vaporization
SO Environmental Protection Agency Glossary
DEF (EPA) The gaseous phase of substances that are liquid or solid at atmospheric temperature and pressure, e.g., steam.

VARIANCES
(DOE Orders 5480.1B and 5483.1A; CAA; CFR; EPA; SSDC)
DA October 12, 1990
SY Exceptions
NT1 DOE Alternative
NT1 National Capacity Variances
NT1 Permanent Variances
NT1 Temporary Variances
SO Air Pollution
SO Environmental Management
SO Environmental Protection Agency Glossary
SO Industrial Hygiene
SO System Safety Development Center Glossary
DEF Releases from a standard of the type specified under the Occupational Safety and Health Act which is processed in accordance with DOE 5483.1A. (EPA) Government permission for a delay or exception in the application of a given law, ordinance, or regulation. (CFR) The temporary deferral of a final compliance date for an individual source subject to an approved regulation, or a temporary change to an approved regulation as it applies to an individual source.

VAULT
(1723 VAULT)
DA December 10, 1990
BT1 Site (DOE FRASE Vocabulary)
BT2 Sites/Areas
NT1 SNM Vault
SO DOE FRASE VOCABULARY

VAULT-TYPE ROOMS
(DOE Order 6430.1A)
DA October 12, 1990
BT1 Security Areas
BT2 Sites/Areas
SO Construction
DEF (DOE Order 6430.1A) DOE-approved rooms having combination-locked door(s) and protected by a

Departmental-approved intrusion alarm system activated by any penetration of walls, floor, ceiling, or openings or by motion with the room.

VCAPCD
DA October 12, 1990
SEE Ventura County Air Pollution Control Districty
SO Acronyms

VCP
DA October 12, 1990
SEE Vitrified Clay Pipe
SO Acronyms

VDC
DA October 12, 1990
SEE Volts Direct Current
SO Acronyms

VECTORS
(DOE Order 6430.1A; EPA)
DA October 12, 1990
NT1 Viruses
RT Bacteria
RT Pathogens
RT Plasmids
SO Construction
SO Environmental Management
SO Environmental Protection Agency Glossary
DEF (DOE Order 6430.1A) (1) Organisms, often insects or rodents, that carry disease. (2) Objects that are used to transport genes into a host cell (vectors can be plasmids, viruses, or other bacteria). A gene is placed in the vector, the vector then "infects" the bacterium.

VEEDER UNITS
DA January 8, 1991
SF *VU*

VEHICLE ACCIDENT
(1539 VEHICLE ACC)
DA November 29, 1990
BT1 Nature of Property Damage
RT Seat Belt(s)
SO DOE FRASE VOCABULARY

VEHICLE MAINT/REPAIR ACTIVITY
(1247 VMR ACTIVITY)
DA November 28, 1990
BT1 Activity Types (DOE FRASE Vocabulary)
BT2 Activities
BT1 Maintenance
BT2 Activities
RT Garage
SO DOE FRASE VOCABULARY

VEHICLE SERVICE AREA
(1625 VS AREA)
DA December 10, 1990

BT-Broader Term   NT-Narrower Term   RT Related Term

BT1 Area
 BT2 Sites/Areas
RT Garage
SO DOE FRASE VOCABULARY

VEHICLES
(TSCA)
DA October 19, 1990
RT Solutions
RT Test Substances
SO Hazardous Materials
DEF (TSCA) Agents that facilitate the mixture, dispersion, or solubilization of a test substance with a carrier.

VENTILATION
(EPA)
DA October 12, 1990
BT1 Processes
RT Air
RT Asphyxiants
RT Ventilation System
SO Environmental Protection Agency Glossary
DEF (EPA) The act of admitting fresh air into a space in order to replace stale or contaminated air; achieved by blowing air into the space.

VENTILATION SYSTEM
(2055 VENTIL SYST)
DA December 10, 1990
BT1 Systems (DOE FRASE Vocabulary)
 BT2 Systems
NT1 End Box Ventilation System
NT1 Reactor Building Standby Ventilation System
RT Ventilation
SO DOE FRASE VOCABULARY

VENTURA COUNTY AIR POLLUTION CONTROL DISTRICTY
DA January 8, 1991
SF VCAPCD

VERIFICATION
(ESH)
DA October 12, 1990
BT1 Administrative Processes
 BT2 Processes
NT1 Calibration Checks
NT1 Verification of Training and Retraining
RT Quality Assurance Overviews
RT Validation
SO Quality Assurance
DEF (ESH) The act of reviewing, inspecting, testing, checking, auditing, or otherwise determining and documenting whether items, processes, services or documents conform to specified requirements.

VERIFICATION OF TRAINING AND RETRAINING
(DOE Order 5480.5)
DA October 12, 1990
BT1 Verification

 BT2 Administrative Processes
 BT3 Processes
RT Training Programs
SO Industrial Safety
DEF (DOE Order 5480.5) The confirmation by an auditable record of the experience, education, medical conditions, training, and testing pertinent to a candidate's specific job assignment and responsibilities.

VERTICAL LOAD CARRYING SPACE FRAMES
(SEA)
DA October 12, 1990
BT1 Space Frames
 BT2 Structures (DOE FRASE Vocabulary)
SO Construction
DEF (SEA) Space frames designed to carry all vertical (gravity) loads.

VERTICAL TUBE STORAGE
DA January 8, 1991
SF VTS
BT1 Storage
 BT2 Waste Management Processes
 BT3 Processes

VERY HIGH TEMPERATURE
DA January 8, 1991
SF VHT

VERY LOW FLOW
DA January 8, 1991
SF VLF

VERY LOW FLOW CONSTANT
(NFI)
DA October 12, 1990
SF VLFC
RT Reactor Shutdown
SO Radiation
DEF (NFI) Reduced assembly flow at which the safety computer must initiate a scram.

VESSELS
(SWDA; RCRA; CERCLA; CFR; ESH)
DA October 12, 1990
SY Water Vehicles
NT1 Barges
NT1 Carfloats
NT1 Cargo Vessels
 NT2 Containerships
NT1 Ferry Vessels
NT1 Incineration Vessels
NT1 Passenger Vessels
NT1 Public Vessels
NT1 Trailership
NT1 Trainships
RT Stowage
RT Targets
RT Transportation Control Measures
SO Compensation and Liability
SO Hazardous Materials
SO Wastes
DEF Every description of watercraft or

other artificial contrivance used, or capable of being used, as a means of transportation on water.

VHAP
(CFR)
DA November 15, 1990
SEE Volatile Hazardous Air Pollutant (VHAP)
SO Acronyms
DEF (CFR) A substance regulated under this part for which a standard for equipment leaks of the substance has been proposed and promulgated. Benzene is a VHAP. Vinyl chloride is a VHAP.

VHT
DA October 12, 1990
SEE Very High Temperature
SO Acronyms

VIBRATING REED ELECTROMETERS
(NFI)
DA October 12, 1990
SF VRE
BT1 Radiation Detectors
 BT2 Equipment
SO Radiation
DEF (NFI) Radiation monitors for air sampling for tritium.

VIBRATION AND ACOUSTIC MONITORING
DA January 8, 1991
SF VAM
BT1 Monitoring
 BT2 Activities

VINYL CHLORIDE
(EPA)
DA October 12, 1990
BT1 Chlorinated Hydrocarbons
 BT2 CERCLA Hazardous Substances
 BT3 Hazardous Substances
 BT2 Halogenated Organic Compounds
 BT3 Halogenated
 BT3 Organic Chemicals
 BT4 Chemical Substances
BT1 Hazardous Constituents
BT1 Volatile Hazardous Air Pollutant (VHAP)
 BT2 Hazardous Air Pollutants
 BT3 Air Pollutants
RT In Vinyl Chloride Service
RT In-Process Wastewater
RT Polyvinyl Chloride Plants
RT Reactors (Chemical)
RT Reactor Opening Loss
RT Strippers
RT Vinyl Chloride Plants
RT Vinyl Chloride Purification
SO Environmental Protection Agency Glossary
DEF (EPA) A chemical compound, used in producing some plastics, that is believed to be carcinogenic.

VINYL CHLORIDE PLANTS
(CAA; CFR)
DA October 12, 1990
BT1 Polyvinyl Chloride Plants
 BT2 Facilities
RT Vinyl Chloride
RT Vinyl Chloride Purification
SO Air Pollution
DEF Includes any plants which produce
    vinyl chloride by any process.

VINYL CHLORIDE PURIFICATION
(CFR)
DA November 15, 1990
BT1 Purification Processes
 BT2 Processes
RT Impurities
RT Vinyl Chloride Plants
RT Vinyl Chloride
DEF (CFR) Includes any part of the
    process of vinyl chloride
    production which follows vinyl
    chloride formation.

VIOLATION OF CRITICALITY
SPECIFICTN
(1555 VIOLATION CS)
DA November 29, 1990
BT1 Nature of Property Damage
BT1 Violations
RT Criticality Alarm System
SO DOE FRASE VOCABULARY

VIOLATION OF OCCUP SAFETY
REGULATION
(1556 VIOLATION SR)
DA November 29, 1990
BT1 Nature of Property Damage
BT1 Violations
SO DOE FRASE VOCABULARY

VIOLATION OF OPERATIONAL
SAFETY REQ
(1558 VIOLATIO OSR)
DA November 29, 1990
BT1 Nature of Property Damage
BT1 Violations
SO DOE FRASE VOCABULARY

VIOLATION OF TECH SPECIFICATION
(1557 VIOLATION TS)
DA November 29, 1990
BT1 Nature of Property Damage
BT1 Violations
SO DOE FRASE VOCABULARY

VIOLATION OF WORKING LIMIT
(1559 VI WRK LIMIT)
DA November 29, 1990
BT1 Nature of Property Damage
BT1 Violations
SO DOE FRASE VOCABULARY

VIOLATIONS
DA January 30, 1991
NT1 De Minimus Violations
NT1 Serious Violations
NT1 Significant Violations

NT1 Violation of Working Limit
NT1 Violation of Operational Safety Req
NT1 Violation of Tech Specification
NT1 Violation of Occup Safety
    Regulation
NT1 Violation of Criticality Specifictn
RT Automatic Reactor Scram
RT Manual Reactor Scram
RT Nature of Property Damage
RT Unscheduled Shutdown
SO DOE FRASE VOCABULARY

VIRGIN MATERIALS
(SWDA; RCRA; ESH)
DA October 12, 1990
BT1 Materials
SO Environmental Management
SO Hazardous Materials
SO Wastes
DEF Raw materials, including previously
    unused copper, aluminum, lead,
    zinc, iron, or other metals or metal
    ores, any undeveloped resources
    that are, or with new technology
    will become, a source of raw
    materials.

VIRUSES
(SDWA; CFR; EPA)
DA October 12, 1990
BT1 Microorganisms
 BT2 Organisms
BT1 Vectors
RT Fecal Coliform Bacteria
RT Infectious Wastes
RT Waterborne Disease Outbreaks
SO Environmental Protection Agency
    Glossary
SO Water Pollution
DEF Viruses of fecal origin which are
    infectious to humans by
    waterborne transmission. (EPA)
    The smallest form of
    microorganisms capable of
    causing disease.

VISCOUS LIQUIDS
(SWDA; RCRA; ESH)
DA October 12, 1990
BT1 Liquids
 BT2 Fluids
SO Hazardous Materials
DEF Liquid material which has a
    measured by viscosity in excess of
    2500 centistokes at 25°C (77°F)
    when determined in accordance
    with the procedures specified in
    ASTM Method D 445-72.

VISE
(3069 VISE)
DA January 3, 1991
BT1 Tools - Manual
 BT2 Tools (DOE FRASE Vocabulary)
  BT3 Equipment
SO DOE FRASE VOCABULARY

VISEGRIPS
(3070 VISEGRIPS)

DA January 3, 1991
BT1 Tools - Manual
 BT2 Tools (DOE FRASE Vocabulary)
  BT3 Equipment
SO DOE FRASE VOCABULARY

VISIBILITY IMPAIRMENTS
(CAA; CFR; ESH)
DA October 12, 1990
NT1 Significant Impairment
RT Adverse Impact on Visibility
RT Coefficient of Haze
RT Integral Vista
RT Natural Conditions
RT Opacity
RT Smoke
SO Air Pollution
DEF (ESH) Any humanly perceptible
    changes in visibility (visual range,
    contrast, coloration) from that
    which would have existed under
    natural conditions. (CAA) "Visibility
    impairment" and "impairment of
    visibility" shall include reduction in
    visual range and atmospheric
    discoloration.

VISIBILITY IN ANY MANDATORY
CLASS I FEDERAL AREA
(CFR)
DA November 15, 1990
BT1 Integral Vista
DEF (CFR) Includes any integral vista
    associated with that area.

VISIBLE EMISSIONS
(CAA; CFR)
DA October 12, 1990
BT1 Particulate Matter Emissions
 BT2 Emissions
  BT3 Air Pollutants
RT Particulate Asbestos Materials
SO Air Pollution
DEF Any emissions containing
    particulate asbestos material that
    are visually detectable without the
    aid of instruments. This does not
    include condensed uncombined
    water vapor.

VITAL ACTIVITIES
(DOE Order 6430.1A)
DA October 12, 1990
BT1 Activities
RT Vital Areas
RT Vital Facilities
RT Vital Programs
SO Construction
DEF (DOE Order 6430.1A) Relating to
    integrity of a national security
    program or a public health and
    safety function.

VITAL AREAS
(DOE Order 6430.1A)
DA October 12, 1990
BT1 Security Areas
 BT2 Sites/Areas
RT Vital Activities

RT Vital Equipment
SO Construction
DEF (DOE Order 6430.1A) Security
  areas for the protection of vital
  equipment.

VITAL EQUIPMENT
(DOE Order 6430.1A)
DA October 12, 1990
BT1 Equipment
RT Fail-Safe
RT Vital Areas
SO Construction
DEF (DOE Order 6430.1A) Equipment,
  systems, or components whose
  failure or destruction would cause
  unacceptable interruption to a
  national security program or harm
  to the health and safety of the
  public.

VITAL FACILITIES
(DOE Order 6430.1A)
DA October 12, 1990
BT1 Facilities
RT Vital Activities
SO Construction
DEF (DOE Order 6430.1A) Facilities
  where vital activities occur.

VITAL PROGRAMS
(DOE Order 6430.1A)
DA October 12, 1990
BT1 Programs
RT Vital Activities
SO Construction
DEF (DOE Order 6430.1A) Programs
  designated vital by the program
  senior official.

VITAL RECORDS
(EMER)
DA February 1, 1991
RT Continuity of Government
RT National Emergencies
SO Emergency Preparedness
DEF (EMER) Records essential for
  maintaining the continuity of
  government activities during a
  national emergency.

VITRIFIED CLAY PIPE
DA January 8, 1991
SF VCP
BT1 Piping
  BT2 Facility Components

VLF
DA October 12, 1990
SEE Very Low Flow
SO Acronyms

VLFC
DA October 12, 1990
SEE Very Low Flow Constant
SO Acronyms

VM
DA October 12, 1990
SEE Volt Meter
SO Acronyms

VOCABULARY
(SSDC)
DA October 12, 1990
NT1 DOE FRASE VOCABULARY
NT1 Factor Relationship and Sequence
  of Events Vocabulary
NT1 FRASE Vocabulary
NT1 Radioactive Source Terms
RT Catalogs
SO System Safety Development Center
  Glossary
DEF (SSDC) The set of words employed
  by a language, group, individual,
  or work, or in relation to a subject.

VOCs
DA October 12, 1990
SEE Volatile Organic Compounds
SO Acronyms

VOLATILE
(EPA)
DA October 12, 1990
SO Environmental Protection Agency
  Glossary
DEF (EPA) Description of any substance
  that evaporates readily.

**VOLATILE HAZARDOUS AIR
POLLUTANT (VHAP)**
(CAA; CFR)
DA October 12, 1990
SF VHAP
BT1 Hazardous Air Pollutants
  BT2 Air Pollutants
NT1 Vinyl Chloride
RT In VHAP Service
RT Process Units
SO Air Pollution
DEF Substances regulated and for
  which a standard for equipment
  leaks of the substance has been
  proposed and promulgated.
  Examples of VHAPs are benzene
  and vinyl chloride.

VOLATILE ORGANIC COMPOUNDS
(EPA)
DA October 12, 1990
SF VOCs
BT1 Organic Chemicals
  BT2 Chemical Substances
NT1 Volatile Synthetic Organic
  Chemicals
RT Catalytic Incinerators
RT In VOC Service
SO Environmental Management
SO Environmental Protection Agency
  Glossary
DEF (EPA) Any organic compounds
  which participate in atmospheric
  photochemical reactions except for
  those designated by the EPA

Administrator as having negligible
photochemical reactivity.

**VOLATILE SYNTHETIC ORGANIC
CHEMICALS**
(EPA)
DA October 12, 1990
BT1 Synthetic Organic Chemicals
  BT2 Organic Chemicals
  BT3 Chemical Substances
BT1 Volatile Organic Compounds
  BT2 Organic Chemicals
  BT3 Chemical Substances
SO Environmental Protection Agency
  Glossary
DEF (EPA) Chemicals that tend to
  volatilize or evaporate from water.

VOLATILITY
(SWDA; RCRA; ESH)
DA October 12, 1990
RT Distillation
SO Hazardous Materials
DEF The relative rate of evaporation of
  materials to assume the vapor
  state.

VOLT METER
DA January 8, 1991
SF VM
BT1 Indicator(s)
  BT2 Instrument(s)
  BT3 Equipment/Parts -
    Instrumentation/Measuring
    (DOE FRASE Voc.)
  BT4 Equipment

VOLT METER(S)
(2779 VOLT METER)
DA January 3, 1991
SO DOE FRASE VOCABULARY

VOLTAGE
(2432 VOLTAGE)
DA January 3, 1991
BT1 Equipment/Parts - Electrical (DOE
  FRASE Vocabulary)
  BT2 Equipment
SO DOE FRASE VOCABULARY

VOLTAGE REGULATOR
(2433 VOLTAGE REG)
DA January 3, 1991
BT1 Equipment/Parts - Electrical (DOE
  FRASE Vocabulary)
  BT2 Equipment
SO DOE FRASE VOCABULARY

VOLTS ALTERNATING CURRENT
DA January 8, 1991
SF VAC
BT1 Alternating Current
RT Electron Volt

VOLTS DIRECT CURRENT
DA January 8, 1991
SF VDC

BT1 Direct Current
RT Electron Volt

VOLUNTARY STANDARDS
(DOE Order 1300.2; ESH)
DA October 12, 1990
SY Commercial Standards
BT1 Standards
 BT2 Codes, Standards, and
   Regulations
RT Voluntary Standards Bodies
SO Management
SO Standards
DEF (DOE Order 1300.2) Those
   standards that are established
   generally by private sector bodies
   and are available for use by any
   person or organization, private or
   governmental. Voluntary standards
   are also referred to as "industry
   standards" as well as "consensus
   standards" (standards developed
   under due process procedures)
   but do not include professional
   standards of personal conduct,
   private standards of individual
   firms, standards mandated by law,
   or standards of individual
   organizations for their internal use.

VOLUNTARY STANDARDS BODIES
(DOE Order 1300.2; ESH)
DA October 12, 1990
BT1 Organizations
RT Commercial Standards
RT Standards-developing Groups
RT Voluntary Standards
SO Management
SO Standards
DEF (DOE Order 1300.2)
   Nongovernmental bodies which
   are broadly based,
   multimembered, domestic and
   multinational organizations,
   industry associations, and
   professional or technical societies
   which develop, establish, or
   coordinate voluntary standards
   activities.

VRE
DA October 12, 1990
SEE Vibrating Reed Electrometers
SO Acronyms

VTS
DA October 12, 1990
SEE Vertical Tube Storage
SO Acronyms

VU
DA October 12, 1990
SEE Veeder Units
SO Acronyms

VULNERABILITY ANALYSES
(EPA)
DA October 12, 1990
BT1 Analyses

RT Risk Assessment
SO Environmental Protection Agency
   Glossary
DEF (EPA) Assessment of elements in
   the community that are susceptible
   to damage should a release of
   hazardous materials occur.

VULNERABLE ZONES
(EPA; EMER)
DA October 12, 1990
SY Vulnerable Zones Radius
BT1 Zones
 BT2 Sites/Areas
SO Emergency Preparedness
SO Environmental Protection Agency
   Glossary
DEF (EPA) Areas over which the
   airborne concentration of a
   chemical involved in an accidental
   release could reach the level of
   concern.

VULNERABLE ZONES RADIUS
(EPA)
DA October 12, 1990
SY Vulnerable Zones
BT1 Measurements
SO Environmental Protection Agency
   Glossary
DEF (EPA) The maximum distance from
   the point of release of a
   hazardous substance in which the
   airborne concentration could reach
   the level of concern under
   specified weather conditions.

WACC
DA October 12, 1990
SEE Waste Acceptance Criteria
   Committee
SO Acronyms

WACKENHUT SERVICES INC.
DA January 11, 1991
BT1 Companies
 BT2 Commercial Organizations
  BT3 Organizations
BT1 DOE Contractors
 BT2 Potentially Responsible Parties
RT Nevada Test Site
RT Nevada Operations Office
RT Savannah River Operations Office
RT Savannah River Site

WAG
DA October 12, 1990
SEE Waste Area Grouping
SO Acronyms

WAREHOUSE
(1791 WAREHOUSE)
DA December 10, 1990
BT1 Storage Area
 BT2 Area
  BT3 Sites/Areas
SO DOE FRASE VOCABULARY

WASTE ACCEPTANCE CRITERIA
COMMITTEE
DA January 8, 1991
SF WACC
BT1 Committees
 BT2 Administrative Organizations
  BT3 Organizations

WASTE AREA GROUPING
DA January 8, 1991
SF WAG

WASTE CONTAINERS
(DOE Order 5820.2A)
DA January 24, 1991
BT1 Containers
RT Disposal
RT Waste Packages
SO Environmental Management
DEF (DOE Order 5820.2A) Receptacles
   for waste, including any liner or
   shielding material that are
   intended to accompany the waste
   in disposal.

WASTE DISPOSAL SITE
(1626 WD SITE)
DA December 10, 1990
BT1 Site (DOE FRASE Vocabulary)
 BT2 Sites/Areas
RT Depositories
SO DOE FRASE VOCABULARY

WASTE EXPERIMENTAL REDUCTION
FACILITY
DA January 8, 1991
SF WERF
BT1 Waste Treatment Plants
 BT2 Facilities

WASTE HANDLING AND PACKAGING
PLANT
DA January 8, 1991
SF WHPP
BT1 Waste Treatment Plants
 BT2 Facilities
RT Waste Receiving and Processing
   Plant

WASTE INFORMATION NETWORK
DA January 8, 1991
SF WIN
BT1 Information Systems
 BT2 Security Interests
 BT2 Systems
RT Martin Marietta Energy Systems
RT Office of Environmental Restoration
   and Waste Management
SO Environmental Management
DEF An information network developed
   through the efforts of the national
   Hazardous Waste Remedial
   Actions Program (HAZRAP) in
   support of DOE's Office of
   Environmental Restoration and
   Waste Management. WIN is
   designed to promote information
   exchange among the DOE
   installations and waste

management community through
communications and data base
applications.

WASTE ISOLATION PILOT PLANT
DA   November 5, 1990
SF   *WIPP*
BT1 Government-Owned
      Contractor-Operated Facilities
 BT2 Federal Facilities
  BT3 Facilities
BT1 Nuclear Facilities
 BT2 Facilities
RT High-Level Radioactive Wastes
RT Transuranic Wastes (TRU Waste)
RT Westinghouse Electric Corp.
SO Wastes
DEF Research and demonstration facility
      located at Carlsbad, New Mexico,
      intended to demonstrate safe
      disposal of radioactive waste in a
      deep geologic environment. A
      decision on whether to convert the
      Waste Isolation Pilot Plant to a
      disposal facility for transuranic
      waste will be made after
      successful testing is
      demonstrated.

WASTE LOAD ALLOCATIONS
(EPA)
DA   October 12, 1990
BT1 Reference Levels
RT Discharge of Pollutants
RT Water Pollution
SO Environmental Protection Agency
      Glossary
DEF (EPA) The maximum loads of
      pollutants each discharger of
      waste is allowed to release into a
      particular waterway. Discharge
      limits are usually required for each
      specific water quality criterion
      being, or expected to be, violated.

WASTE MANAGEMENT
DA   November 7, 1990
BT1 Processes
NT1 Cleanup Operations
NT1 Hazardous Waste Management
NT1 Radioactive Waste Management
NT1 Salvage
NT1 Solid Waste Management
 NT2 Compaction
 NT2 Refuse Reclamation
NT1 Waste Minimization
NT1 Waste Management Operations
RT Burial Operations
RT Closure Period
RT Corrective Action Management
      Units
RT Delisting Petitions
RT Disposal Systems
RT Environmental Management
RT Institution Control
RT Test and Evalution Facilities
RT Trenching
RT Waste Management Processes
SO Environmental Management

SO Wastes
DEF (USC) The systematic
      administration of activities that
      provide for the collection, storage,
      transportation, transfer,
      processing, treatment, and
      disposal of waste.

WASTE MANAGEMENT OPERATIONS
DA   January 8, 1991
SF   *WMO*
BT1 Operations
 BT2 Activities
BT1 Waste Management
 BT2 Processes
SO Environmental Management
SO Wastes

WASTE MANAGEMENT PROCESSES
DA   February 26, 1991
BT1 Processes
NT1 Coagulation
NT1 Comminution
NT1 Deposition (Process)
 NT2 Sedimentation
NT1 Disposal
 NT2 Continuous Disposal
 NT2 Land Disposal
  NT3 Near Surface Disposal
   NT4 Shallow Land Burial
 NT2 Land Farming (of waste)
 NT2 Solid Waste Disposal
  NT3 Phased Disposal
 NT2 Water Dumping
NT1 Precipitation
NT1 Recycling
 NT2 Closed-Loop Recycling
NT1 Removal
 NT2 Dredging
NT1 Retrofill
NT1 Screening
NT1 Storage
 NT2 Burial Operations
  NT3 Shallow Land Burial
 NT2 Contaminated Water Storage
 NT2 Vertical Tube Storage
NT1 Treatment
 NT2 Biological Treatment
  NT3 Aerobic Treatment
 NT2 Chemical Treatments
 NT2 Conventional Filtration Treatment
 NT2 Granular Activated Carbon
      Treatment
 NT2 In-Situ Treatment
 NT2 Incineration
  NT3 Incineration at Sea
 NT2 Ion Exchange Treatment
 NT2 Primary Waste Treatment
 NT2 Secondary Treatment
 NT2 Stabilization
 NT2 Thermal Treatment
 NT2 Wastewater Treatment Processes
  NT3 Absorption (Waste)
  NT3 Advanced Waste Water
      Treatment
  NT3 Clarification
  NT3 Diatomaceous Earth Filtration
  NT3 Flocculation
  NT3 Physical and Chemical
      Treatments

  NT3 Pretreatment
  NT3 Primary Waste Treatment
  NT3 Secondary Treatment
  NT3 Solidification and Stabilization
  NT3 Tertiary Treatment
NT1 Treatment, Storage, and Disposal
NT1 Trenching
NT1 Waste Minimization
RT Waste Management

WASTE MINIMIZATION
DA   January 8, 1991
SF   *WMin*
BT1 Waste Management
 BT2 Processes
BT1 Waste Management Processes
 BT2 Processes
SO Environmental Management
SO Wastes

WASTE MONITORING FACILITY
(1792 WM FACILITY)
DA   December 10, 1990
BT1 Facility (DOE FRASE Vocabulary)
 BT2 Facilities and Buildings (DOE
      FRASE Vocabulary)
  BT3 Facilities
SO DOE FRASE VOCABULARY

WASTE OILS
(TSCA; CFR; ESH)
DA   October 12, 1990
SY   Used Oils
BT1 Oils
SO Hazardous Materials
DEF Used products primarily derived
      from petroleum, which include, but
      are not limited to, fuel oils, motor
      oils, gear oils, cutting oils,
      transmission fluids, hydraulic
      fluids, and dielectric fluids.

WASTE PACKAGES
(DOE Order 5820.2A)
DA   January 24, 1991
RT Disposal
RT Packages
RT Waste Containers
SO Environmental Management
DEF (DOE Order 5820.2A) The waste,
      waste container, and any
      absorbent that are intended for
      disposal as a unit. In the case of
      surface contaminated, damaged,
      leaking, or breached waste
      packages, any overpack shall be
      considered the waste container,
      and the original container shall be
      considered part of the waste.

**WASTE RECEIVING AND
PROCESSING PLANT**
DA   January 8, 1991
SF   *WRAP*
BT1 Waste Treatment Plants
 BT2 Facilities
NT1 Mixed Waste Management Facility
RT Waste Handling and Packaging
      Plant

SY-Synonymous Terms   SO-Source/Subject Category   SF-See From

WASTE TREATMENT PLANTS
(EPA)
DA October 12, 1990
BT1 Facilities
NT1 Waste Experimental Reduction
     Facility
NT1 Waste Handling and Packaging
     Plant
NT1 Waste Receiving and Processing
     Plant
 NT2 Mixed Waste Management Facility
RT Ballistic Separators
RT Waste Treatment Stream
SO Environmental Protection Agency
     Glossary
DEF (EPA) Facilities containing a series
     of tanks, screens, filters and other
     processes by which pollutants are
     removed from water.

WASTE TREATMENT STREAM
(EPA)
DA October 12, 1990
RT Waste Treatment Plants
SO Environmental Protection Agency
     Glossary
DEF (EPA) The continuous movement of
     waste from generator to treater
     and disposer.

WASTE WATER COLLECTION SYSTEMS
(SWDA; RCRA)
DA October 19, 1990
SY Storm Water Collection Systems
BT1 Waste Water System
 BT2 Systems (DOE FRASE
       Vocabulary)
  BT3 Systems
 BT2 Water Systems
  BT3 Publicly Owned Treatment Works
   BT4 Treatment Facilities
    BT5 Facilities
  BT3 Systems
RT Conventional Systems
SO Wastes
DEF Piping, pumps, conduits, and any
     other equipment necessary to
     collect and transport the flow of
     surface water run-off resulting
     from precipitation, or domestic,
     commercial, or industrial
     wastewater to and from retention
     areas or any areas where
     treatment is designated to occur.
     The collection of storm water and
     wastewater does not include
     treatment except where incidental
     to conveyance.

WASTE WATER SYSTEM
(2057 WW SYSTEM)
DA December 10, 1990
BT1 Systems (DOE FRASE Vocabulary)
 BT2 Systems
BT1 Water Systems
 BT2 Publicly Owned Treatment Works
  BT3 Treatment Facilities
   BT4 Facilities
 BT2 Systems

NT1 Waste Water Collection Systems
RT Drinking Water System
SO DOE FRASE VOCABULARY

WASTEFUL MANNER
(ESA; CFR)
DA October 12, 1990
RT Subsistence
SO Endangered Species
DEF (CFR) Any taking or method of
     taking which is likely to result in
     the killing or injury of endangered
     or threatened wildlife beyond those
     needed for subsistence purposes,
     or which results in the waste of a
     substantial portion of the wildlife,
     and includes without limitation the
     employment of a method of taking
     which is not likely to assure the
     capture or killing of the wildlife, or
     which is not immediately followed
     by a reasonable effort to retrieve
     the wildlife.

WASTES
(EPA)
DA October 12, 1990
NT1 Agricultural Pollution
NT1 Asbestos-Containing Waste
     Materials
 NT2 Asbestos Tailings
 NT2 Asbestos Waste from Control
       Devices
NT1 Ashes
 NT2 Bottom Ash
 NT2 Fly Ash
NT1 Beryllium-Containing Wastes
NT1 Certified Wastes
NT1 Class C Wastes
NT1 Classified Wastes
NT1 Construction and Demolition
     Wastes
NT1 Constituents
NT1 Department of Energy Wastes
NT1 Hazardous Wastes
 NT2 Extremely Hazardous Wastes
 NT2 Incompatible Wastes
 NT2 Listed Wastes
 NT2 Piles (Wastes)
  NT3 Disposal Areas
 NT2 State Hazardous Wastes
NT1 Infectious Wastes
NT1 Mining Wastes
NT1 Mixed Wastes
NT1 Nitrogenous Wastes
NT1 PCB Wastes
NT1 Pesticide Related Wastes
NT1 Post Consumer Wastes
NT1 Radioactive Wastes
 NT2 High-Level Radioactive Wastes
 NT2 Low Level Radioactive Wastes
 NT2 Low Level Wastes
 NT2 Radioactive Mixed Wastes
 NT2 Radioactive Effluents
 NT2 Transuranic Wastes (TRU Waste)
  NT3 Contact-Handled Transuranic
        Wastes
  NT3 Remote-Handled Transuranic
        Wastes
 NT2 Transuranic Radioactive Wastes

 NT2 TRU Wastes
NT1 Routine Wastes
NT1 Sludge
 NT2 Activated Sludges
 NT2 Sewage Sludge
NT1 Solid Wastes
 NT2 Agricultural Solid Wastes
 NT2 Bulky Wastes
 NT2 Commercial Solid Wastes
 NT2 Institutional Solid Wastes
 NT2 Medical Wastes
 NT2 Municipal Solid Wastes
  NT3 Sewage
   NT4 Raw Sewage
 NT2 Residential Solid Wastes
 NT2 Rubbish
 NT2 Special Wastes
 NT2 Tailings
  NT3 Asbestos Tailings
  NT3 New Tailings
NT1 Wastewater
 NT2 Blackwater
 NT2 Gray Water
 NT2 In-Process Wastewater
 NT2 Process Waste Water
RT Hazardous Substances
RT Hazards
RT Pollutants
SO Environmental Management
SO Environmental Protection Agency
     Glossary
DEF (EPA) (1) Unwanted materials left
     over from a manufacturing
     process. (2) Refuse from places of
     human or animal habitation.

WASTEWATER
(EPA)
DA October 12, 1990
BT1 Wastes
BT1 Water
NT1 Blackwater
NT1 Gray Water
NT1 In-Process Wastewater
NT1 Process Waste Water
RT Pretreatment Facilities
RT Receiving Waters
RT Settleable Solids
RT Wastewater Treatment Processes
SO Environmental Protection Agency
     Glossary
DEF (EPA) The spent or used water from
     individual homes, a community, a
     farm, or an industry that contains
     dissolved or suspended matter.

WASTEWATER OPERATIONS AND MAINTENANCE
(EPA)
DA October 12, 1990
BT1 Maintenance
 BT2 Activities
BT1 Operations
 BT2 Activities
RT Wastewater Treatment Processes
SO Environmental Protection Agency
     Glossary
DEF (EPA) Actions taken after
     construction to assure that
     facilities constructed to treat

wastewater will be properly operated, maintained, and managed to achieve efficiency levels and prescribed effluent levels in an optimum manner.

WASTEWATER TREATMENT PROCESSES
(CAA; CFR)
DA   October 12, 1990
BT1 Treatment
 BT2 Waste Management Processes
  BT3 Processes
NT1 Absorption (Waste)
NT1 Advanced Waste Water Treatment
NT1 Clarification
NT1 Diatomaceous Earth Filtration
NT1 Flocculation
NT1 Physical and Chemical Treatments
NT1 Pretreatment
NT1 Primary Waste Treatment
NT1 Secondary Treatment
NT1 Solidification and Stabilization
NT1 Tertiary Treatment
RT   Activated Carbon
RT   Activated Sludges
RT   Bar Screens
RT   Comminution
RT   Conventional Filtration Treatment
RT   Oxidation Ponds
RT   Wastewater Treatment Units
RT   Wastewater
RT   Wastewater Operations and
     Maintenance
SO   Air Pollution
DEF Any processes which modify
     characteristics such as BOD,
     COD, TSS, and pH, usually for the
     purpose of meeting effluent
     guidelines and standards; They do
     not include any process the
     purpose of which is to remove
     vinyl chloride from water to meet
     requirements of 40 CFR 61.61.

WASTEWATER TREATMENT TANKS
(SWDA; RCRA; CFR)
DA   October 19, 1990
BT1 Tanks
 BT2 Facility Components
BT1 Wastewater Treatment Units
 BT2 Devices
RT   Sedimentation Tanks
RT   Settling Tanks
SO   Wastes
DEF Tanks that are designed to receive
     and treat influent wastewater
     through physical, chemical, or
     biological methods.

WASTEWATER TREATMENT UNITS
(SWDA; RCRA; ESH)
DA   October 12, 1990
BT1 Devices
NT1 Digesters
NT1 Wastewater Treatment Tanks
RT   Eligible Costs
RT   Wastewater Treatment Processes
SO   Environmental Management

SO   Wastes
DEF Devices which (a) are part of a
     wastewater treatment facility which
     is subject to regulation under
     either section 402 or 307(b) of the
     Clean Water Act; and (b) receive
     and treat or store an influent
     wastewater which is a hazardous
     waste as defined in 40 CFR 261.3,
     or generates and accumulates a
     wastewater treatment sludge
     which is a hazardous waste as
     defined in 40 CFR 261.3, or treat
     or store a wastewater treatment
     sludge which is a hazardous
     waste as defined in 40 CFR 261.3;
     and (c) meet the definition of tank
     or tank system in 40 CFR 260.10.

WATER
(ESH)
DA   October 12, 1990
NT1 Coastal Waters
NT1 Deionized Water
NT1 Fresh Water
NT1 Groundwater
 NT2 Capillary Water
 NT2 Groundwater Resources
NT1 Hard Water
NT1 Interstate Waters
NT1 Non-Contact Cooling Water
NT1 Potable Water
NT1 Service Clarified Water
NT1 Soft Water
NT1 Storm Water
NT1 Surface Waters
 NT2 Brackish Waters
 NT2 Estuaries
 NT2 Lakes
  NT3 Dystrophic Lakes
  NT3 Eutrophic Lakes
  NT3 Oligotrophic Lakes
 NT2 River Water
NT1 Tidal Waters
NT1 Wastewater
 NT2 Blackwater
 NT2 Gray Water
 NT2 In-Process Wastewater
 NT2 Process Waste Water
RT   Hydrology
RT   Reservoirs
SO   Quality Assurance
DEF (ESH) Clear, odorless, tasteless
     liquid that is essential for most
     plant and animal life; descends
     from clouds as rain and forms
     streams, lakes, and seas.

WATER(BULK SHIPMENT)
(SWDA; RCRA)
DA   October 12, 1990
RT   Transportation
SO   Wastes
DEF (40 CFR 260.10) The bulk
     transportation of hazardous waste
     that is loaded or carried on board
     a vessel without containers or
     labels.

WATER BLOOM
(SSDC)
DA   May 24, 1991
SY   Algae Bloom
SY   Red Tide
BT1 Natural Phenomenon
SO   Water Pollution

WATER DAMAGE
(1540 WATER DAMAGE)
DA   November 29, 1990
BT1 Damage
 BT2 Nature of Property Damage
RT   Heat Damage
RT   Smoke Damage
SO   DOE FRASE VOCABULARY

WATER DUMPING
(ESH)
DA   November 20, 1990
BT1 Disposal
 BT2 Waste Management Processes
  BT3 Processes
RT   Pesticides
RT   Water Pollution
SO   Environmental Management
DEF (ES&H Audit Manual) The disposal
     of pesticides in or on lakes, ponds,
     rivers, sewers, or other water
     systems as defined in P.L. 92-500.

WATER HAMMERS
(DOE Order 6430.1A)
DA   October 12, 1990
SO   Construction
DEF (DOE Order 6430.1A) Pressure
     rises in a pipeline caused by a
     sudden change in the rate of flow
     or stoppage of flow in the line.

WATER MANAGEMENT DIVISION DIRECTOR
(CFR)
DA   January 24, 1991
BT1 Personnel
RT   Management
RT   Regional Office
SO   Environmental Management
DEF (CFR) One of the Directors of the
     Water Management Divisions
     within the regional offices of the
     Environmental Protection Agency
     or this person's delegated
     representative.

WATER POLLUTION
(EPA)
DA   October 12, 1990
BT1 Pollution
RT   Agricultural Pollution
RT   Algal Blooms
RT   Discharge of Pollutants
RT   Eutrophication
RT   Plumes
RT   Reclamation
RT   Releases
RT   Waste Load Allocations
RT   Waters of the United States
RT   Water Dumping

SY-Synonymous Terms   SO-Source/Subject Category   SF-See From

RT Wells
SO Environmental Protection Agency
　　Glossary
DEF (EPA) The presence in water of
　　enough harmful or objectionable
　　material to damage the water's
　　quality.

WATER QUALITY CRITERIA
(EPA)
DA October 12, 1990
BT1 Reference Levels
RT Anti-Degradation Clause
RT Residual Disinfectant Concentration
RT Safe Drinking Water Act
RT Waterborne Disease Outbreaks
SO Environmental Protection Agency
　　Glossary
DEF (EPA) Specific levels of water
　　quality which, if reached, are
　　expected to render a body of
　　water suitable for its designated
　　use. The criteria are based on
　　specific levels of pollutants that
　　would make the water harmful if
　　used for drinking, swimming,
　　farming, fish production, or
　　industrial processes.

WATER QUALITY STANDARDS
(EPA)
DA October 12, 1990
BT1 Standards
BT2 Codes, Standards, and
　　Regulations
SO Environmental Protection Agency
　　Glossary
DEF (EPA) State-adopted and
　　EPA-approved ambient standards
　　for water bodies. The standards
　　cover the use of the water body
　　and the water quality criteria which
　　must be met to protect the
　　designated use or uses.

WATER REACTIVE MATERIALS
(SWDA; RCRA; ESH)
DA October 12, 1990
BT1 Materials
RT Flammable Solids
SO Hazardous Materials
DEF Solid substances (including sludges
　　and pastes) which, by interaction
　　with water, are likely to become
　　spontaneously flammable or to
　　give off flammable or toxic gases
　　in dangerous quantities.

WATER RESISTANT
(SWDA; RCRA; ESH)
DA October 12, 1990
RT Permeability
SO Hazardous Materials
DEF Having a degree of resistance to
　　permeability by and damage
　　caused by water in liquid form.

WATER SOLUBILITY
(EPA)

DA October 12, 1990
SO Environmental Protection Agency
　　Glossary
DEF (EPA) The maximum concentration
　　of a chemical compound which
　　can result when it is dissolved in
　　water. If a substance is water
　　soluble it can very readily disperse
　　through the environment.

WATER SUPPLIERS
(EPA)
DA October 12, 1990
RT Operators
RT Owners
SO Environmental Protection Agency
　　Glossary
DEF (EPA) Persons who own or operate
　　a public water system.

WATER SUPPLY SYSTEM
(EPA)
DA October 12, 1990
BT1 Public Water Systems
BT2 Water Systems
BT3 Publicly Owned Treatment Works
BT4 Treatment Facilities
BT5 Facilities
BT3 Systems
RT Chlorine-Contact Chambers
RT Potable Water
SO Environmental Protection Agency
　　Glossary
DEF (EPA) The collection, treatment,
　　storage, and distribution of potable
　　water from source to consumer.

WATER SYSTEMS
DA February 13, 1991
BT1 Publicly Owned Treatment Works
BT2 Treatment Facilities
BT3 Facilities
BT1 Systems
NT1 Public Water Systems
NT2 Community Water Systems
NT2 Drinking Water System
NT2 Non-Community Water Systems
NT2 Non-Transient Non-Community
　　Water System
NT2 Water Supply System
NT1 Storm Water Collection Systems
NT1 Waste Water System
NT2 Waste Water Collection Systems
RT Control Systems
RT Emergency Systems
SO Environmental Management
SO Water Pollution

WATER TABLE
(EPA)
DA October 12, 1990
RT Groundwater
SO Environmental Management
SO Environmental Protection Agency
　　Glossary
DEF (EPA) The level of ground water.

WATER TIGHT
(SWDA; RCRA)

DA October 12, 1990
SF *WT*
RT Construction
DEF (WEBSTER) Of such tight
　　construction or fit as to be
　　impermeable to water except
　　when under sufficient pressure to
　　produce structural discontinuity.

WATER VEHICLES
(ESA; CFR)
DA October 12, 1990
SY Vessels
RT Waterborne Activities
SO Endangered Species
DEF (CFR) Includes, but is not limited
　　to, boats (whether powered by
　　engine, wind, or other means),
　　ships (whether powered by
　　engine, wind, or other means),
　　barges, surfboards, water skis, or
　　any other device or mechanism
　　the primary or an incidental
　　purpose of which is locomotion on,
　　across, or underneath the surface
　　of the water.

WATERBORNE ACTIVITIES
(ESA; CFR)
DA October 12, 1990
BT1 Activities
RT Water Vehicles
SO Endangered Species
DEF (CFR) Include, but are not limited
　　to, swimming, diving (including
　　skin and scuba diving), snorkeling,
　　water skiing, surfing, fishing, the
　　use of water vehicles, and
　　dredging and filling operations.

WATERBORNE DISEASE OUTBREAKS
(SDWA; CFR)
DA October 12, 1990
BT1 Epidemics
BT2 Occurrences
RT Pathogens
RT Public Water Systems
RT Viruses
RT Water Quality Criteria
SO Water Pollution
DEF The significant occurrences of
　　acute infections illness,
　　epidemiologically associated with
　　the ingestion of water from a
　　public water system which is
　　deficient in treatment, as
　　determined by the appropriate
　　local or State agency.

WATERS OF THE UNITED STATES
(CWA; RHA; CFR)
DA October 12, 1990
RT Reclamation
RT Surface Waters
RT Territorial Seas
RT Water Pollution
SO Hazardous Materials
SO Water Pollution
DEF (1) All waters currently used,used

BT-Broader Term　　　　　　NT-Narrower Term　　　　　　RT Related Term

in the past, or susceptible to use in interstate or foreign commerce, including all waters which are subject to the ebb and flow of the tide... (see 40 CFR 230.3 for continued definition).

WATERSHED
(EPA)
DA   October 12, 1990
BT1 Sites/Areas
RT   Maximum Probable Flood
RT   Minor Drainage
RT   Receiving Waters
RT   River Basins
RT   Surface Water Resources
SO   Environmental Protection Agency
       Glossary
DEF (EPA) Land areas that drain into a stream.

WATERSPOUTS
(SMRP)
DA   October 12, 1990
BT1 Rotational Windstorms
 BT2 Act of Nature
RT   Dust Devils
RT   Subvortices
RT   Tornadoes
SO   Natural Phenomenon
DEF (SMRP) Fast-rotating columns of air over water. The formative stage of a waterspout is a dark spot above the water surface seen from the air. A funnel cloud pendant from a convective cloud is often observed from the ground. A mature waterspout induces a spray vortex, leaving a wake of foam along its path. Waterspouts commonly form over warm tropical and subtropical waters. When a waterspout moves inland it must be identified, by definition, as a tornado. Likewise, a tornado is identified as a waterspout during its passage over a large body of water. About 75% of waterspouts form beneath towering cumuli while 25% are spawned by cumulonimbus clouds. 90% of waterspouts are F1 (112 mph) or weaker; F4 (207 mph) or stronger waterspouts are rare and exceptional.

WEAK STORIES
(SEA)
DA   October 12, 1990
BT1 Stories
SO   Construction
DEF (SEA) Stories in which individual story strength is less than 80 percent of that of the story above.

WEAPON SIMULATORS
(DOE Order 5480.16)
DA   October 12, 1990
NT1 Law Simulators
RT   Blank Ammunition

RT   Engagement Simulation Systems
RT   Firearms
SO   Firearms
DEF (DOE Order 5480.16) Devices that simulate the function of actual weapons without emitting projectiles or detonating large explosive charges. Generally they will emit some fragments and heat, flame, or smoke and are capable of injuring personnel.

WEED EATER
(2981 WEED EATER)
DA   January 3, 1991
BT1 Tools - Powered
 BT2 Tools (DOE FRASE Vocabulary)
  BT3 Equipment
SO   DOE FRASE VOCABULARY

WEEDS
(USC)
DA   November 15, 1990
BT1 Plants
DEF (USC) Plants which grow where they are not wanted.

WEIGHT OF EVIDENCE
(EPA)
DA   October 12, 1990
SO   Environmental Protection Agency
       Glossary
DEF (EPA) Tissue-specific and represents the fraction of the total health risk resulting from uniform, whole-body irradiation that could be contributed to that particular tissue. The weighting factors recommended by the ICRP (Publication 26) and used here are Gonads (0.25), Breasts (0.15), Red Bone Marrow (0.12), Lungs (0.12), Thyroid (0.03), Bone Surfaces (0.03), Remainder (0.30) [Remainder means the five other organs with the highest dose (e.g., liver, kidney, spleen, thymus, adrenal, pancreas, stomach, small intestine, or upper and lower large intestine, but excluding skin, lens of the eye, and extremities). The weighting factor for each of these organs is 0.06]. (EPA) An EPA classification system for characterizing the extent to which the available data indicate that an agent is a human carcinogen.

WEIGHTING FACTORS
(DOE Orders 5400.5, 5480.11; ESH; IAEA)
DA   October 12, 1990
RT   Critical Organs
RT   Effective Dose Equivalent
RT   Irradiation
RT   Risks
SO   Industrial Hygiene
SO   Radiation
DEF (ESH) Tissue-specific and

represent the fraction of the total health risk resulting from uniform, whole-body irradiation that could be contributed to that particular tissue. They are used in the calculation of annual and committed effective dose equivalent to equate the risk arising from the irradiation of tissue T to the total risk when the whole body is uniformly irradiated. The weighting factors recommended by the ICRP (Publication 26) and used here are: gonands, 0.25; breasts, 0.15; red bone marrow,0.12;lungs,0.12; thyroid,0.03; bone surfaces,0.03; remainder,0.03 (remainder means the five other organs with the next highest risk, including liver, kidney, spleen, thymus, adrenals, pancreas, stomach, small intestine, or upper and lower intestine, but excluding skin, lens of the eye, and extremities. The weighting factor for each such organ is 0.06.

WELDER'S HOOD
(2693 WELDER HOOD)
DA   January 3, 1991
BT1 Head Protection
 BT2 Personal Protective Equipment
  BT3 Equipment/Parts - Personal
       Protective (DOE FRASE
       Vocabulary)
   BT4 Equipment
SO   DOE FRASE VOCABULARY

WELDER/SOLDERER
(0771 WELDER)
DA   November 28, 1990
BT1 Precision/Production Personnel
 BT2 Occupations
 BT2 Personnel
SO   DOE FRASE VOCABULARY

WELL INJECTION
(SDWA; CFR; EPA)
DA   October 12, 1990
SY   Underground Injection
RT   Injection Wells
SO   Environmental Management
SO   Environmental Protection Agency
       Glossary
SO   Water Pollution
DEF The subsurface emplacement of fluids through a bored, drilled or driven well; or through a dug well, where the depth of the dug well is greater than the largest surface dimension. (EPA) The subsurface emplacement of fluids in a well. (ES&H Audit Manual) Disposal of liquid wastes through a hole or shaft into a subsurface stratum.

WELL MONITORING
(SDWA; CFR; ESH)

DA   October 12, 1990
BT1 Monitoring
 BT2 Activities
RT   Wells
SO   Water Pollution
DEF (ESH) The measurement, by
      on-site instruments or laboratory
      methods, of the quality of water in
      a well.

WELL PAD
(1627 WELL PAD)
DA   December 10, 1990
BT1 Site (DOE FRASE Vocabulary)
 BT2 Sites/Areas
SO   DOE FRASE VOCABULARY

WELL PLUGS
(SDWA; CFR; EPA)
DA   October 12, 1990
RT   Packers
RT   Plugging
SO   Environmental Protection Agency
      Glossary
SO   Water Pollution
DEF Watertight and gastight seals
      installed in a borehole or well to
      prevent movement of fluids.

WELL STIMULATION
(SDWA; CFR)
DA   October 12, 1990
BT1 Processes
RT   Injection Wells
SO   Water Pollution
DEF Several processes used to clean
      the well bore, enlarge channels,
      and increase pore space in the
      interval to be injected thus making
      it possible for wastewater to move
      more readily into the formation,
      and includes (1) surging, (2)
      jetting, (3) blasting, (4) acidizing,
      (5) hydraulic fracturing.

WELL WORKOVERS
(SDWA; CFR)
DA   October 12, 1990
RT   Injection Wells
RT   Wells
SO   Water Pollution
DEF Any reentries of an injection well;
      including, but not limited to, the
      pulling of tubular goods,
      cementing or casing repairs; and
      excluding any routine maintenance
      (e.g. reseating the packer at the
      same depth, or repairs to surface
      equipment).

WELLS
(SWDA; RCRA; SDWA; CFR; EPA)
DA   October 12, 1990
NT1 Abandoned Wells
NT1 Disposal Wells
NT1 Injection Wells
 NT2 Class II Wells
  NT3 Existing Class II Wells
  NT3 New Class II Wells

 NT2 Existing Injection Wells
NT1 Monitoring Wells
RT   Casings
RT   Current Plugging and
      Abandonment Cost Estimates
RT   Injection
RT   Plugging
RT   Surface Casings
RT   Underground Sources of Drinking
      Water
RT   Water Pollution
RT   Well Monitoring
RT   Well Workovers
SO   Environmental Management
SO   Environmental Protection Agency
      Glossary
SO   Water Pollution
DEF Bored, drilled or driven shafts, or
      dug holes, whose depth is greater
      than the largest surface
      dimension.

WERF
DA   October 12, 1990
SEE Waste Experimental Reduction
      Facility
SO   Acronyms

**WEST VALLEY NUCLEAR SERVICES
CO., INC.**
DA   January 11, 1991
BT1 Companies
 BT2 Commercial Organizations
  BT3 Organizations
BT1 DOE Contractors
 BT2 Potentially Responsible Parties
RT   Idaho Operations Office

**WESTERN AREA POWER
ADMINISTRATION**
DA   May 15, 1991
BT1 Power Marketing Administrations
SO   Management
DEF (U.S. Government Manual) The
      Administration is responsible for
      the federal electric
      power-marketing and transmission
      functions in 15 central and
      western States, encompassing a
      geographic area of 1.3 million
      square miles. The Administration
      sells power to 532 customers,
      consisting of cooperatives,
      municipalities, public utility
      districts, private utilities, Federal
      and State Agencies, and irrigation
      districts. These wholesale power
      customers provide service to
      consumers in Arizona, California,
      Colorado, Iowa, Kansas,
      Minnesota, Montana, Nebraska,
      Nevada, New Mexico, North
      Dakota, South Dakota, Texas,
      Utah, and Wyoming.

WESTINGHOUSE ELECTRIC CORP.
DA   January 11, 1991
BT1 Companies
 BT2 Commercial Organizations

 BT3 Organizations
BT1 DOE Contractors
 BT2 Potentially Responsible Parties
RT   Albuquerque Operations Office
RT   Naval Reactors
RT   Waste Isolation Pilot Plant

**WESTINGHOUSE HANFORD
COMPANY**
DA   January 8, 1991
SF   *WHC*
BT1 Companies
 BT2 Commercial Organizations
  BT3 Organizations
BT1 DOE Contractors
 BT2 Potentially Responsible Parties
RT   Richland Operations Office

**WESTINGHOUSE IDAHO NUCLEAR
CO., INC.**
DA   January 11, 1991
BT1 Companies
 BT2 Commercial Organizations
  BT3 Organizations
BT1 DOE Contractors
 BT2 Potentially Responsible Parties
RT   Idaho Chemical Processing Plant
RT   Idaho Operations Office

**WESTINGHOUSE MATERIALS
COMPANY OF OHIO**
DA   January 8, 1991
SF   *WMCO*
BT1 Companies
 BT2 Commercial Organizations
  BT3 Organizations
BT1 DOE Contractors
 BT2 Potentially Responsible Parties
RT   Feed Materials Production Center
RT   Oak Ridge Operations Office

**WESTINGHOUSE SAVANNAH RIVER
COMPANY**
DA   January 11, 1991
BT1 Companies
 BT2 Commercial Organizations
  BT3 Organizations
BT1 DOE Contractors
 BT2 Potentially Responsible Parties
RT   Savannah River Operations Office
RT   Savannah River Site

WETLANDS
(CWA; RHA; CFR; EPA)
DA   October 12, 1990
BT1 Sites/Areas
BT1 Surface Water Resources
 BT2 Natural Resources
NT1 Bogs
NT1 Bottomland Hardwoods
NT1 Fens
NT1 Marshes
 NT2 Tidal Marshes
NT1 Swamps
RT   Muck Soils
RT   Notice of Involvement
RT   Surface Waters
RT   Unacceptable Adverse Effects

SO Environmental Protection Agency
Glossary
SO Water Pollution
DEF Those areas that are inundated or
saturated by surface or ground
water at a frequency and duration
sufficient to support, and that
under normal circumstances do
support, a prevalence of
vegetation typically adapted for life
in saturated soil conditions.
Wetlands generally include
swamps, marshes, bogs,
estuaries, and similar areas. See
also 10 CFR 1022.

WHC
DA October 12, 1990
SEE Westinghouse Hanford Company
SO Acronyms

WHOLE BODY
(CAA; CFR)
DA October 12, 1990
NT1 Critical Organs
RT Effective Dose Equivalent
RT Extremities
RT Human Body Parts
RT Radiation Safety
SO Air Pollution
DEF All organs or tissues exclusive of
the integumentary system (skin)
and the cornea.

WHOLE BODY COUNTER
(EMER)
DA February 1, 1991
BT1 Radiation Detectors
BT2 Equipment
SO Emergency Preparedness
DEF (EMER) A device used to identify
and measure the radioactivity in
the body of human beings or
animals; it uses heavy shielding to
keep out background radiation and
measures the activity with
ultrasensitive scintillation detectors
and electronic equipment. This
instrument is also referred to as
an in-vivo counter.

WHOLE BODY DOSE
(EMER)
DA February 1, 1991
BT1 Doses
SO Emergency Preparedness
DEF (EMER) The dose of radiation
received by the body in its
entirety, as distinct from a dose to
a limited area of the body.

WHPP
DA October 12, 1990
SEE Waste Handling and Packaging
Plant
SO Acronyms

WILDLIFE
(ESA; CFR)

DA October 19, 1990
RT Animals
SO Endangered Species
DEF Any member of the animal
kingdom, including without
limitation any mammal, fish, bird
(including any migratory,
nonmigratory, or endangered bird
for which protection is also
afforded by treaty or other
international agreement),
amphibian, reptile, mollusk,
crustacean, arthropod or other
invertebrate, and includes any
part, product, egg, or offspring
thereof, or the dead body or parts
thereof.

WILDLIFE REFUGES
(EPA)
DA October 12, 1990
BT1 Habitats
NT1 Manatee Refuges
RT Critical Habitats
SO Environmental Protection Agency
Glossary
DEF (EPA) Areas designated for the
protection of wild animals, within
which hunting and fishing are
either prohibited or strictly
controlled.

WIN
DA October 12, 1990
SEE Waste Information Network
SO Acronyms
SO Environmental Management
SO Wastes

WIND DAMAGE
(1541 WIND DAMAGE)
DA November 29, 1990
BT1 Damage
BT2 Nature of Property Damage
SO DOE FRASE VOCABULARY

WIPE OUT
(SSDC)
DA October 12, 1990
SO System Safety Development Center
Glossary
DEF (SSDC) Destroy the total entry
(used to purge an incorrect entry
on a computer or data stores
system).

WIPP
DA October 12, 1990
SEE Waste Isolation Pilot Plant
SO Acronyms

WIRING
(2434 WIRING)
DA January 3, 1991
BT1 Equipment/Parts - Electrical (DOE
FRASE Vocabulary)
BT2 Equipment
SO DOE FRASE VOCABULARY

WITNESSES
(SSDC)
DA October 12, 1990
RT Accidents
SO System Safety Development Center
Glossary
DEF (SSDC) Persons who have
firsthand knowledge of some fact
related directly or indirectly to an
accident or incident.

WL
DA October 12, 1990
SEE Working Level
SO Acronyms

WLM
DA October 12, 1990
SEE Working Level Month
SO Acronyms

WMCO
DA October 12, 1990
SEE Westinghouse Materials Company
of Ohio
SO Acronyms

WMin
DA October 12, 1990
SEE Waste Minimization
SO Acronyms
SO Environmental Management
SO Wastes

WMO
DA October 12, 1990
SEE Waste Management Operations
SO Acronyms
SO Environmental Management
SO Wastes

WOOD-BURNING STOVE POLLUTION
(EPA)
DA October 12, 1990
BT1 Air Pollution
BT2 Pollution
SO Environmental Protection Agency
Glossary
DEF (EPA) Air pollution caused by
emissions of particulate matter,
carbon monoxide, total suspended
particulates, and polycyclic organic
matter from wood-burning stoves.

WOOD RESIDUES
(CAA; CFR)
DA October 12, 1990
BT1 Residues
BT2 Hazardous Materials
BT3 Materials
RT Fossil Fuel and Wood
Residue-Fired Steam Generating
Unit
SO Air Pollution
DEF Bark, sawdust, slabs, chips,
shavings, mill trim, and other wood
products derived from wood

SY-Synonymous Terms

SO-Source/Subject Category

SF-See From

processing and forest
management operations.

WORK ENVIRONMENT
(DOE Order 6430.1A)
DA   October 12, 1990
BT1 Environment
RT   Arbitration
RT   Human Factors Engineering
RT   Interfaces
RT   Occupational Injuries
RT   Occupational Exposure
RT   Working Conditions
SO   Construction
DEF (DOE Order 6430.1A) The
     surroundings in which systems
     operate. Includes all of the
     conditions that may affect one or
     more system components, e.g.,
     temperature/humidity, noise, light,
     vibration, toxic materials,
     radioactive materials.

WORK PROCESSES
(SSDC)
DA   October 12, 1990
BT1 Administrative Processes
 BT2 Processes
RT   Arbitration
RT   Standard Job Procedures
SO   System Safety Development Center
     Glossary
DEF (SSDC) All of the components
     involved to get work done. Work
     processes consist of three major
     elements; (1) procedures and
     management system control; (2)
     hardware and equipment in the
     process; and (3) personnel
     involved in the process.

WORK STATION
(1726 WORK STATION)
DA   December 10, 1990
BT1 Site (DOE FRASE Vocabulary)
 BT2 Sites/Areas
BT1 Station
 BT2 Site (DOE FRASE Vocabulary)
  BT3 Sites/Areas
SO   DOE FRASE VOCABULARY

WORKING CONDITION A
(IAEA)
DA   November 20, 1990
BT1 Working Conditions
 BT2 Conditions
RT   Dose Limit
RT   Special Health Supervision
SO   Radiation
DEF (IAEA) Conditions where the
     annual exposures might exceed
     three-tenths of the dose equivalent
     limits.

WORKING CONDITION B
(IAEA)
DA   November 20, 1990
BT1 Working Conditions
 BT2 Conditions

RT   Dose Limit
SO   Radiation
DEF (IAEA) Conditions where it is most
     unlikely that the annual exposures
     will exceed three-tenths of the
     dose equivalent limits.

WORKING CONDITIONS
(IAEA)
DA   October 12, 1990
BT1 Conditions
NT1 Working Condition A
NT1 Working Condition B
RT   Occupational Exposure
RT   Radiation Safety
RT   Work Environment
SO   Radiation
DEF (IAEA) Conditions under which
     workers are occupationally
     exposed to ionizing radiation.

WORKING LEVEL
(EPA)
DA   October 12, 1990
SF   WL
BT1 Radiation Units
 BT2 Units of Measure
RT   Detailed Operating Procedures
RT   Radon Progeny
RT   Working Level Month
SO   Environmental Protection Agency
     Glossary
SO   Radiation
DEF (EPA) A unit of measure for
     documenting exposure to radon
     decay products. One working level
     is equal to approximately 200
     picocuries per liter. (IAEA) A unit
     for potential alpha energy
     concentration (i.e. the sum of the
     total energy per unit volume of air
     carried by alpha particles emitted
     during the complete decay of each
     atom and its progeny in a unit
     volume of air).

WORKING LEVEL MONTH
(EPA; IAEA; NCRP)
DA   October 12, 1990
SF   WLM
BT1 Radiation Units
 BT2 Units of Measure
NT1 Cumulative Working Level Months
RT   Exposure
RT   Radon Progeny
RT   Working Level
SO   Environmental Protection Agency
     Glossary
SO   Radiation
DEF A unit of measure used to
     determine cumulative exposure to
     radon. A unit of exposure to air
     concentrations of potential alpha
     energy released from short-lived
     radon progeny. One working level
     month is defined as the cumulative
     exposure equivalent to exposure
     at one working level for a working
     month of 170 hours.

WRAP
DA   October 12, 1990
SEE Waste Receiving and Processing
     Plant
SO   Acronyms

WRENCH(S)
(3071 WRENCH)
DA   January 3, 1991
BT1 Tools (DOE FRASE Vocabulary)
 BT2 Equipment
NT1 Allen Wrench
NT1 Impact Wrench
NT1 Power Impact Wrench(s)
NT1 Ratchet Wrench(s)
NT1 Torque Wrench(s)
SO   DOE FRASE VOCABULARY

WRIST(S)
(1099 WRIST)
DA   November 28, 1990
BT1 Joint(s)
 BT2 Human Body Parts
RT   Hand(s)
SO   DOE FRASE VOCABULARY

WRIST BAND
(2694 WRIST BAND)
DA   January 3, 1991
BT1 Hand Protection
 BT2 Personal Protective Equipment
  BT3 Equipment/Parts - Personal
       Protective (DOE FRASE
       Vocabulary)
   BT4 Equipment
SO   DOE FRASE VOCABULARY

WT
DA   October 12, 1990
SEE Water Tight
SO   Acronyms

X-RAY EQUIPMENT
(2435 X-RAY EQUIPM)
DA   January 3, 1991
BT1 Equipment/Parts - Electrical (DOE
     FRASE Vocabulary)
 BT2 Equipment
SO   DOE FRASE VOCABULARY

X-RAYS
(NIH; EMER)
DA   October 12, 1990
BT1 Ionizing Radiation
 BT2 Radiation
RT   Gamma Radiation
RT   Half Value Layer (Half Thickness)
SO   Emergency Preparedness
SO   Radiation
DEF (NIH) Penetrating electromagnetic
     radiations having wave lengths
     shorter than those of visible light.
     They are usually produced by
     bombarding a metallic target with
     fast electrons in a high vacuum. In
     nuclear reactions it is customary to
     refer to photons originating in the
     nucleus as gamma rays, and
     those originating in the

BT-Broader Term   NT-Narrower Term   RT Related Term

extranuclear part of the atom as x rays. These rays are sometimes called roentgen rays after their discoverer, W. C. Roentgen.

XENOBIOTIC
(EPA)
DA   October 12, 1990
RT Detergents
RT Pesticides
RT Synthetic Organic Chemicals
SO Environmental Protection Agency Glossary
DEF (EPA) Term for non-naturally occurring man-made substances found in the environment (i.e., synthetic material solvents, plastics).

Y-12 PLANT
DA   January 11, 1991
BT1 Government-Owned Contractor-Operated Facilities
 BT2 Federal Facilities
 BT3 Facilities
RT Martin Marietta Energy Systems
RT Oak Ridge Operations Office
DEF (Capsule Review of DOE Research and Development and Field Facilities, 1986) Built in 1943 as part of the Manhattan Project, the Oak Ridge Y-12 plant was established to separate uranium isotopes using the electromagnetic process. When the process was discontinued after World War II, the role of the Y-12 Plant changed to manufacturing and developmental engineering. The missions of the Y-12 Plant include production of nuclear weapon components and subassemblies, development and fabrication of test hardware for the weapon design laboratories, fabrication support for other Martin Marietta Energy Systems plants and support for federal agencies.

YARD AREA
(1628 YARD AREA)
DA   December 10, 1990
BT1 Area
 BT2 Sites/Areas
SO DOE FRASE VOCABULARY

Yards³
DA   October 12, 1990
SEE Cubic Yards
SO Acronyms

YEARS LIVING RADIOISOTOPES
(EDB)

DA   March 29, 1991

BT1 Radionuclides
 BT2 CERCLA Hazardous Substances
  BT3 Hazardous Substances
 BT2 Nuclides
NT1 Antimony 125
NT1 Cesium 134
NT1 Cobalt 60
NT1 Radium 226
NT1 Radium 228
NT1 Ruthenium 103
NT1 Strontium 90
NT1 Thorium 230
NT1 Tritium
NT1 Uranium 238
SO Radiation

YELLOWCAKE
(NCRP; EMER)
DA   October 12, 1990
RT Uranium
SO Emergency Preparedness
SO Radiation
DEF (NCRP) A product of uranium mills, concentrated in uranium content and suitable for shipment for further processing into fuel for reactors.

ZENER DIODES
(NFI)
DA   October 12, 1990
BT1 Equipment
SO Nuclear Facilities Incident Database
SO Radiation
DEF (NFI) Junction diodes, semiconductors.

ZINC
DA   January 8, 1991
SF *Zn (Periodic Element)*

ZINC 65
(EDB)
DA   March 29, 1991
BT1 Beta-Minus Decay Radioisotopes
 BT2 Beta Decay Radioisotopes
  BT3 Radionuclides
   BT4 CERCLA Hazardous Substances
    BT5 Hazardous Substances
   BT4 Nuclides
BT1 Days Living Radioisotopes
 BT2 Radionuclides
  BT3 CERCLA Hazardous Substances
   BT4 Hazardous Substances
  BT3 Nuclides
SO Radiation

ZION PROBABILISTIC SAFETY STUDY
DA   January 8, 1991
SF *ZPSS*
BT1 Studies

ZIRCONIUM 95
(EDB)

DA   March 29, 1991
BT1 Beta-Minus Decay Radioisotopes
 BT2 Beta Decay Radioisotopes
  BT3 Radionuclides
   BT4 CERCLA Hazardous Substances
    BT5 Hazardous Substances
   BT4 Nuclides
BT1 Days Living Radioisotopes
 BT2 Radionuclides
  BT3 CERCLA Hazardous Substances
   BT4 Hazardous Substances
  BT3 Nuclides
SO Radiation

Zn (Periodic Element)
DA   October 12, 1990
SEE Zinc
SO Acronyms

ZONES
DA   February 25, 1991
BT1 Sites/Areas
NT1 Aeration Zones
NT1 Buckled Zones
NT1 Buffer Zone
NT1 Coastal Zones
NT1 Confining Zones
NT1 Contiguous Zones
NT1 Contingency Planning Zones
NT1 Disturbed Zones
NT1 Emergency Planning Zones
NT1 Engineering Control Zones
NT1 Excavation Zone
NT1 Flow Zone
NT1 Hydrogeologic Units
NT1 Injection Zones
 NT2 Injection Interval
NT1 Inland Zones
NT1 Isolation Zones
NT1 Law Hazard Zone
NT1 Mixing Zones
NT1 Planning Zones
NT1 Radiation Zone
NT1 Saturated Zones
NT1 Treatment Zones
NT1 Unsaturated Zones
NT1 Vulnerable Zones

ZOOPLANKTON
(EPA)
DA   October 19, 1990
BT1 Plankton
RT Fish
SO Environmental Protection Agency Glossary
DEF (EPA) Tiny aquatic animals eaten by fish.

ZPSS
DA   October 12, 1990
SEE Zion Probabilistic Safety Study
SO Acronyms

Acronyms

A/E
SEE Architect Engineer

AACS
SEE Airborne Activity Confinement
Systems

ABLE
SEE Scram

ABOSFn
SEE Nominal Automatic Burnout Safety
Factor

ABS-SC
SEE Automatic Backup Shutdown of the
Safety Computer

AC
SEE Alternating Current

AC Motors
SEE Allis-Chalmers Motors

ACC
SEE Abnormal Condition Control

ACH
SEE Air Changes Per Hour

ACLs
SEE Alternate Concentration Limits

ACO
SEE Administrative Consent Order

ACR
SEE Area Control Room

ACRR
SEE Annular Core Research Reactor

ACRS
SEE Advisory Committee on Reactor
Safety

ADC
SEE Analog to Digital Converter

ADI
SEE Acceptable Daily Intake

ADM
SEE Action Description Memoranda

ADS (Activity Data Sheets)
SEE Activity Data Sheets

ADS (automatic depressurization system)
SEE Automatic Depressurization System

AEA
SEE Atomic Energy Act

AEC
SEE Atomic Energy Commission

AEOD
SEE NRS Office for Analysis and
Evaluation of Operational Data

AFB
SEE Air Force Base

AFESC
SEE Air Force Engineering and Services
Center

AFW
SEE Auxiliary Feedwater

AIA
SEE Automatic Incident Actions

AIC
SEE Acceptable Intake for Chronic
Exposure

AIF
SEE Atomic Industrial Forum

AIS
SEE Acceptable Intake for Subchronic
Exposure

ALAP
SEE As Low As Practicable

ALARA
SEE As Low As Reasonably Achievable

ALI
SEE Annual Limit on Intake

ALS
SEE Advanced Light Source

AMAD
SEE Activity-Median Aerodynamic
Diameter

ANL-E
SEE Argonne National Laboratory-East
(Chicago)

ANL-W
SEE Argonne National Laboratory-West
(At INEL)

ANO-1
SEE Arkansas Nuclear One-1

ANS
SEE American Nuclear Society

ANSI
SEE American National Standards
Institute

AO
SEE Auxiliary Orifices

AOT
SEE Allowable Outage Time

API
SEE Axial Power Indicator

API (American Petroleum Institute)
SEE American Petroleum Institute

APM
SEE Axial Power Monitors

ARARs
SEE Applicable or Relevant and
Appropriate Requirements

ARI
SEE Alternate Rod Insertion

ARPA
SEE Archaeological Resources
Protection Act of 1979

ARS
  SEE Alternate Removal Systems

ASME
  SEE American Society of Mechanical
    Engineers

ATS
  SEE Automatic Transfer Switch

ATWS
  SEE Anticipated Transient Without
    Scram

B&RC
  SEE Budget and Reporting Code

B/O
  SEE Blackout

BAAQMD
  SEE Bay Area Air Quality Management
    District

BACT
  SEE Best Available Control Technology

BART
  SEE Best Available Retrofit Technology

BAT
  SEE Best Available Technology

BBC
  SEE Balanced Biological Communities

BCL
  SEE Battelle Columbus Laboratories

BDAT
  SEE Best Demonstrated Available
    Technology

BEF
  SEE Bottom End Fitting

BFA
  SEE Blank Fire Adapter (BFA)

BFI
  SEE Bottom Fitting Insert

BG (Blanket Gas)
  SEE Blanket Gas

BG (Burial Ground)
  SEE Burial Grounds

BIF
  SEE Boiler/Industrial Furnaces

BIT
  SEE Boron Injection Tank

BLVR
  SEE Building Power Low Voltage Relay

BMP
  SEE Best Management Practices

BNL
  SEE Brookhaven National Laboratory

BOD (Biological Oxygen Demand)
  SEE Biological Oxygen Demand

BOP
  SEE Balance of Plant

BOR
  SEE Burnout Risk

BOSF
  SEE Burnout Safety Factor

BOSFn
  SEE Nominal BOSF

BPS
  SEE Boeing Petroleum Services

BPT
  SEE Best Practical Technology

BPV
  SEE Back Pressure Valve

Bq
  SEE Becquerel

BRC
  SEE Below Regulatory Concern

BTDR
  SEE Building Time Delay Relay

BTR
  SEE BPS Technical Representative

BTU
  SEE British Thermal Unit

BUSTR
  SEE Ohio Bureau of Underground
    Storage Tank Regulation

BWR
  SEE Boiling Water Reactor

BWST
  SEE Borated Water Storage Tank

BZ
  SEE Buckled Zones

C&D
  SEE Charge and Discharge

Ca
  SEE Calcium

CAA
  SEE Clean Air Act

CAD
  SEE Computer-Assisted Design

CAIRS
  SEE Computerized Accident/Incident
    Reporting System

CAMS
  SEE Continuous Air Monitors

CAR
  SEE Corrective Action Reporting

CARC
  SEE Containment Air Recirculation and
    Cooling

CAS
  SEE Condition Assessment Surveys

CATS
  SEE Computer Assisted Tracking
    System

CCDF
  SEE Complementary Cumulative
    Distribution Function

CCF
  SEE Common Cause Failure

CCR
  SEE Central Control Room

CCTV
  SEE Crane Control, Close Circuit TV

CCW
  SEE Component Cooling Water

Cd
  SEE Cadmium

CDC
  SEE Center for Disease Control

CDF
  SEE Core Damage Frequency

CDH
  SEE Colorado Department of Health

CDPM
  SEE Crane Drip Pan Monitor

CDR
  SEE Conceptual Design Report

CE
  SEE Combustion Engineering

CEARP
  SEE Comprehensive Environmental
    Assessment and Response
    Program

CEL
  SEE Channel Effluent Limit

CEQ
  SEE Council on Environmental Quality

CERCLA
  SEE Comprehensive Environmental
    Response, Compensation, etc.

CF
  SEE Containment Failure

CFEE
  SEE Conference of Federal
    Environmental Engineers

CFM
  SEE Cubic Feet Per Minute

CFR
  SEE Code of Federal Regulations

CFS
  SEE Core Flood System

CFT
  SEE Core Flood Tanks

CFTs
  SEE Californium Targets

CF_4
  SEE Carbon Tetrafluoride

CH (Chicago Operations Office)
  SEE Chicago Operations Office

CH (Component Handling)
  SEE Component Handling

CH (contact handled)
  SEE Contact Handled

CHEMS
  SEE Chemical Hazards Emergency
    Management System

CHR
  SEE Confinement Heat Removal

CIC
  SEE Compensated Ion Chamber

CIF
  SEE Consolidated Incinerator Facility

CIL
  SEE Critical Items List

CIL (Computer)
  SEE Computer Inoperative Limits

CILRT
  SEE Containment Integrated Leak Rate
    Test

CIP
  SEE Capital Improvement Project

CIT
  SEE Critical Incident Technique

Cl
  SEE Chlorine

CIF_3
  SEE Chlorine Trifluoride

CLP
  SEE Contract Laboratory Program

CM
  SEE Core Melt

CM/sec
  SEE Centimeters per second

CMA
  SEE Crane Maintenance Area

CMS
  SEE RCRA Corrective Measures Study

CMT
  SEE Crisis Management Teams

CNHR
  SEE Containment Atmosphere Heat
    Removal

CNRR
  SEE Containment Radioactivity Removal

CO (Carbon Monoxide)
  SEE Carbon Monoxide

CO (Contracting Officer)
  SEE Contracting Officers

COCA
  SEE Consent Order and Compliance
    Agreement

COD (Chemical Oxygen Demand)
  SEE Chemical Oxygen Demand

COD (Computer Operations Division)
  SEE Computer Operations Division

COGEMT
  SEE Continuity of Government
    Emergency Management Teams

COH
  SEE Coefficient of Haze

CO_2
  SEE Carbon Dioxide

CPAF
  SEE Cost Plus Awards Fee

CPL
  SEE Confinement Protection Limits

Cr
  SEE Chromium

CRD
  SEE Control Rod Drive

CRDCS
  SEE Control Rod Drive Control System

CRDM
  SEE Control Rod Drive Mechanism

CRM (Corrective Repair Maintenance)
  SEE Corrective Repair Maintenance

CRM (Count Rate Meter)
  SEE Count Rate Meter

CRT
  SEE Coolant Return Tank

Cr^{+6}
  SEE Chromium-hexavalent

CS
  SEE Containment Spray

CS&R
  SEE Codes, Standards, and Regulations

CSR
  SEE Scram Relay

CSS
  SEE Containment Spray System

CSSI
  SEE Containment Spray System
    (Post-accident Injection Phase)

CSSR
  SEE Containment Spray System
    (Post-accident Recirculation
    Phase)

CST
  SEE Condensate Storage Tank

CSWE
  SEE Central Service Works Engineering

CT
  SEE Current Transformer

CTA
  SEE Central Training Academy

CTG
  SEE Control Technique Guidelines

CTR
  SEE Critical Temperature Ratio

Cu
  SEE Copper

CVCS
  SEE Chemical Volume and Control
     System

CWA
  SEE Clean Water Act

CWGM
  SEE Cooling Water Gamma Monitor

CWLM
  SEE Cumulative Working Level Months

CWS
  SEE Contaminated Water Storage

CX
  SEE Categorical Exclusions

$C_2C_{12}F_4$
  SEE Freon-114

D
  SEE Absorbed Dose

D&D
  SEE Decontamination and
     Decommissioning

D&E
  SEE Discharge and Exit

d/m
  SEE Drips per Minute

D/M/G
  SEE Disintegrations per minute per gram

DA
  SEE DOE Design Agency

DAC
  SEE Derived Air Concentration

db (Decibel)
  SEE Decibel

DB (Dumbbell)
  SEE Dumbbell

DBA
  SEE Design Basis Accidents

DBE
  SEE Design Basis Earthquakes

DBF
  SEE Design Basis Fires

DBFL
  SEE Design Basis Floods

DBT
  SEE Design Basis Tornadoes

DC
  SEE Direct Current

DCG
  SEE Derived Concentration Guide

DCH
  SEE Direct Containment Heating

DCR
  SEE Design Change Request

DFDP
  SEE Defense Facility Decommissioning
     Program

DFM
  SEE Flow Monitor

DG
  SEE Diesel Generator

DHR
  SEE Decay Heat Removal

DI
  SEE Deionized

DIH
  SEE Differential in Hours

DIL
  SEE Daily Instruction Logs

DIV
  SEE Divisions

DMA
  SEE Diagnosis of Multiple Alarms
     System

DNO
  SEE Do Not Operate

DOE
  SEE U.S. Department of Energy

DOJ
  SEE U.S. Department of Justice

DOL
  SEE U.S. Department of Labor

DOP
  SEE Detailed Operating Procedures

DOT
  SEE U.S. Department of Transportation

DP (Defense Programs)
  SEE Defense Programs

dP (Differential Pressure)
  SEE Differential Pressure

DRE
  SEE Destruction Removal Efficiency

DSFM
  SEE Division of the State Fire Marshall

DST
  SEE Double-shell Tank

DUVAS
  SEE Derivative Ultraviolet Absorbtion
     Spectroscopy

DW
  SEE Drywell

DWMP
  SEE Defense Waste Management Plan

DWPF
  SEE Defense Waste Processing Facility

DWTM
  SEE Office of Defense Waste and
     Transportation Management

E&CF
  SEE Events and Causal Factors

E&S
  SEE Error and Sensitivity

E-MAD
  SEE Engine Maintenance Assembly and
     Disassembly

EA (Emergency Actions)
  SEE Emergency Actions

EA (Environmental Assessment)
  SEE Environmental Assessments

EAS
  SEE Engineering Assistance Section

EBF
 SEE Eccentric Braced Frames

EBWR
 SEE Experimental Boiling Water Reactor

ECC
 SEE Emergency Control Centers

ECCI
 SEE Emergency Core Cooling Injection

ECCS
 SEE Emergency Core Cooling System

ECP
 SEE Engineering Change Proposal

ECS (Emergency Control Station)
 SEE Emergency Control Stations

ECS (Emergency Cooling System)
 SEE Emergency Cooling System

ECT
 SEE Evaporator Condensate Tank

ECW
 SEE Effluent Cooling Water

ECWD
 SEE Effluent Cooling Water Drainage

EDB
 SEE Ethylene Dibromide

EDG
 SEE Emergency Diesel Generator

EED (construction)
 SEE Electroexplosive Devices

EED (Equipment Engineering
Department)
 SEE Equipment Engineering Department

EEM
 SEE Essential Equipment Monitor

EF
 SEE End Fitting

EFC
 SEE External Fission Counter

EFS
 SEE Emergency Feedwater System

EFΔP
 SEE End Fitting Delta Pressure

EG&G

EG (Emergency generators)
 SEE Emergency (Diesel) Generators

EG (Engine-Generators)
 SEE Engine-Generators

EH
 SEE Office of the Assistant Secretary for
      Environment, et. al.

EIS
 SEE Environmental Impact Statements

EMT
 SEE Emergency Medical Technician

EOC
 SEE Emergency Operations Centers

EOD
 SEE Explosives Ordnance Disposal

EOP
 SEE Emergency Operating Procedures

EP
 SEE Essential Power

EP Toxic
 SEE Extraction Procedure Toxic

EPA
 SEE U.S. Environmental Protection
      Agency

EPRI
 SEE Electric Power Research Institute

EQ
 SEE Environmental Qualification

1E (Electrical Equipment)
 SEE Class 1E Electrical Equipment

ER (Environmental Restoration)
 SEE Office of Environmental Restoration
      and Waste Management

ER (Office of Energy Research)
 SEE Office of Energy Research

ERAB
 SEE Energy Research Advisory Board

ERCWS
 SEE Emergency Raw Cooling Water
      System

ERDA
 SEE Energy Research and Development
      Administration

ERL
 SEE Emergency Reference Levels

ERT
 SEE Emergency Response Teams

ERV
 SEE Emergency Relief Valve

ES&H
 SEE Environment, Safety, and Health
      (ES&H) Program
 OR   Environment, Safety, and Health

ESA
 SEE Endangered Species Act

ESD
 SEE Emergency Shutdown Device

ESF
 SEE Engineered Safety Features
 OR   Engineered Safety Features
      (Shutdown)

ESFAS
 SEE Engineered Safety Features
      Actuation System

ESP
 SEE Electrostatic Precipitators

ESS
 SEE Engagement Simulation Systems

ESW
 SEE Emergency Service Water

ET
 SEE Event Tree

ETEC
 SEE Energy Technology Engineering
      Center (Canoga Park)

F-V
 SEE Fussell-Vesely

F/S
 SEE Frequency/Severity

FAA
 SEE Federal Aviation Administration

FAR
 SEE Federal Aviation Regulation

FCs
 SEE Fluorocarbons

FDA
 SEE Fuel Distribution Analyzer

Fe (Periodic Element)
 SEE Iron

FEFGC
SEE Fuel Element Failure Gas
Chromatograph

FFA
SEE Federal Facilities Agreement

FFCA
SEE Federal Facilities Compliance
Agreement

FFE
SEE Failed Fuel Element

FFTF
SEE Fast Flux Test Facility

FHAR
SEE Fire Hazard Analysis Report

FICI
SEE Failed Instrument Component
Inspection

FM
SEE Factory Mutual

FMEA
SEE Failure Mode and Effect Analysis

FMPC
SEE Feed Materials Production Center

FNSI
SEE Finding of No Significant Impact

FONSI
SEE Finding of No Significant Impact

FP
SEE Fission Products

FPTS
SEE Fire Protection Tracking System

FRACAS
SEE Failure Reporting Analysis and
Corrective Action System

FRGT
SEE Fast Response Gamma
Thermometer

FSAR
SEE Final Safety Analysis Report

FSH
SEE Fuel Sleeve Housing

FTA
SEE Fault Tree Analysis

FUSRAP
SEE Formerly Utilized Sites Remedial
Actions Program

FVC
SEE Filtered Vented Containment

FW
SEE Feedwater

FWA
SEE Flow Weighted Average (Delta T)

FY
SEE Fiscal Year

FZ
SEE Flow Zone

F_2
SEE Fluorine

G-M
SEE Geiger-Mueller Counters

G/T
SEE Activity per Ton

gal
SEE gallon

GAT
SEE Goodyear Atomic Corporation

GCEP
SEE Gas Centrifuge Enrichment Plant

GDC
SEE General Design Criteria

GDP
SEE Gaseous Diffusion Plant

GE
SEE General Electric

GGNS
SEE Grand Gulf Nuclear Station

GIDEP
SEE Government Industry Data
Exchange Program

GJPO
SEE Grand Junction Projects Office

GOCO
SEE Government-Owned
Contractor-Operated Facilities

GP
SEE Graphic Panel

GPP
SEE General Plant Projects

GPU
SEE Grapper Pick Up

GTCC
SEE Greater-Than-Class-C

GTF
SEE Grout Treatment Facility

GTM
SEE Gang Temperature Monitor

GVR
SEE Gas Volume Ratio

H
SEE Dose Equivalents

H&SM
SEE Health and Safety Manual,
PUB-3000

HATS
SEE Hazard Abatement Tracking
System

HAZRAP
SEE Hazardous Waste Remedial
Actions Program

HBCU
SEE Historically Black Colleges and
Universities

HC
SEE Hydrocarbons

HCF
SEE Hot Channel Factor

HDP
SEE High Delta Pressure

HE (Effective Dose)
SEE Effective Dose

HE (High Explosives)
SEE High Explosives

HE (Human Error)
SEE Human Error

HEPA
SEE High Efficiency Particulate Air
Filters

HF
SEE Hydrogen Fluoride

Hg
SEE Mercury

HIPO
  SEE High Potential Incidents

HLC
  SEE High Level Caves

HLFM
  SEE High Level Flux Monitor

HLLW
  SEE High-Level Liquid Waste

HLW
  SEE High-Level Wastes

HM
  SEE Health Monitoring

HOCs
  SEE Halogenated Organic Compounds

HP
  SEE Health Physics

HPCI
  SEE High Pressure Coolant Injection

HPCS
  SEE High Pressure Core Spray

HPIS
  SEE High Pressure Injection System

HPLC
  SEE High Performance Liquid
       Chromatography

HPPP
  SEE High Pressure Pump Pad

HPRS
  SEE High Pressure Recirculation
       System

HPSI
  SEE High Pressure Spray (Post-accident
       Injection Phase)

HPSR
  SEE High Pressure Spray (Post-accident
       Recirculation Phase)

HPSW
  SEE High Pressure Service Water

HQ
  SEE DOE Headquarters

HRM
  SEE Hoisting and Rigging Manual

HS
  SEE Hydrogen Sulfide

HSE
  SEE Health, Safety, and Environment
       Division

HSWA
  SEE Hazardous and Solid Waste
       Amendments

HTA
  SEE High Temperature Alarm

HTO
  SEE Tritium Oxides

HWVP
  SEE Hanford Waste Vitrification

HX
  SEE Heat Exchangers

I&C
  SEE Instrumentation and Control

I/M
  SEE Inspection and Maintenance

I/P
  SEE Circuit to Pressure Converter

IA (Incident Actions)
  SEE Incident Actions

IA (Instrument Air)
  SEE Instrument Air

IAG
  SEE Interagency Agreement

IBM
  SEE International Business Machines

IC (Incident Control)
  SEE Incident Control

IC (Isolation Condenser)
  SEE Isolation Condenser

ICP/MS
  SEE Inductively Coupled Plasma/Mass
       Spectrometer

ICPP
  SEE Idaho Chemical Processing Plant

ICS
  SEE Intermittent Control System

ICU
  SEE Interface Control Unit

ID
  SEE Idaho Operations Office

IDB
  SEE Integrated Data Base

IDCOR
  SEE Industry Degraded Core
       Rulemaking Program

IDLH
  SEE Immediately Dangerous to Life and
       Health

IE (Industrial Engineering)
  SEE Industrial Engineering

IE (Office of Inspection and Enforcement)
  SEE NRS Office of Inspection and
       Enforcement

IEEE
  SEE Institute of Electrical and Electronic
       Engineers

IFC
  SEE Internal Fission Counter

IFI
  SEE Inspector Follow-up Items

IGFM
  SEE Internal Gamma Flux Monitor

IH (Industrial Hygiene)
  SEE Industrial Hygiene

IH (Inner Housing)
  SEE Inner Housing

IHE
  SEE Insensitive High Explosives

ILRT
  SEE Integrated Leak Rate Test

ILS
  SEE Integrated Logistics Support

INEL
  SEE Idaho National Engineering
       Laboratory

INPO
  SEE Institute for Nuclear Power
       Operations

IORV
  SEE Inadvertently Opened Relief Valve

IP
  SEE EIS Implementation Plans

IPE
  SEE Individual Plant Examination

IPM
  SEE Integrated Pest Management

IREP
  SEE Interim Reliability Evalution
      Program

IRI
  SEE Industrial Risk Insurers

IRIS
  SEE Integrated Risk Information System

IRO
  SEE Insulated Uranium Oxide

IST
  SEE In-Situ Treatment

ISV
  SEE In-Situ Vitrification

IT
  SEE Inner Targets

ITRI
  SEE Inhalation Toxicology Research
      Institute

IX
  SEE Ion Exchange

JSA
  SEE Job Safety Analysis

KCP
  SEE Kansas City Plant

KCR
  SEE Crane Control Room

Kg/yr
  SEE Kilograms/year

Km
  SEE Kilometer

Km²
  SEE square kilometers

KOR
  SEE Knowledge of Results

KV
  SEE Kilovolt

LANL
  SEE Los Alamos National Laboratory

LAW
  SEE Light Antitank Weapon

LBL
  SEE Lawrence Berkeley Laboratory

lbs/year
  SEE Pounds/year

LC
  SEE Locked Closed

LCO
  SEE Limiting Condition for Operation

LDR
  SEE Land Disposal Restrictions

LEGM
  SEE Low Energy Gamma Monitor

LEHR
  SEE Laboratory for Energy-Related
      Health Research

LEL
  SEE Lower Explosive Limit

LEPC
  SEE Local Emergency Planning
      Committees

LER
  SEE Licensee Event Report

LET
  SEE Linear Energy Transfer

LLD
  SEE Lower Limit of Detection

LLRW
  SEE Low Level Radioactive Wastes

LLW
  SEE Low Level Wastes

LLWDDD
  SEE Low Level Waste Disposal
      Development Demonstration

LNP
  SEE Loss of Normal Power

LO
  SEE Locked Open

LOC
  SEE Level of Concern

LOCA
  SEE Loss of Coolant Accident

LOCV
  SEE Loss of Condenser Vacuum

LOOP
  SEE Loss of Offsite Power

LOP
  SEE Loss of Offsite Power

LOSP
  SEE Loss of Offsite Power

LOTA
  SEE Loss of Target Accident

LPCI
  SEE Low Pressure Coolant Injection

LPCR
  SEE Low Pressure Cooling Recirculation
      Phase

LPCS
  SEE Low Pressure Core Spray

LPI
  SEE Low Pressure Injection

LPP
  SEE Long Plenum Plugs

LPPP
  SEE Low Pressure Pump Pad

LPRS
  SEE Low Pressure Recirculation System

LPRSX
  SEE Low Pressure Recirculation System
      Heat Exchanger

LPSW
  SEE Low Pressure Service Water

LTA
  SEE Less than Adequate

LWC
  SEE Lost Workday Cases

LWR
  SEE Light Water Reactor

M
  SEE Moment Magnitude

M&O
  SEE Management and Operating
      Contractor for DOE Facility

M&T
  SEE Main and Trim

MAP
  SEE Mitigation Action Plans

MB
  SEE Machine Basin

MBA
  SEE Materials Balance Area

MBO
  SEE Management by Objective

MCA
  SEE Maximum Credible Accident

MCC
  SEE Motor Control Center

MCL
  SEE Maximum Contaminant Levels

MCLG
  SEE Maximum Contaminant Level Goal

MCLs
  SEE Maximum Concentration Limits

10 MDLM
  SEE 10-minute Delay Line Gamma
    Monitor

MDNR
  SEE Missouri Department of Natural
    Resources

MEPS
  SEE Multimedia Environmental Pollutant
    Assessment System

MFL
  SEE Maximum Foreseeable Loss

MFW
  SEE Main Feedwater

MFWCS
  SEE Main Feedwater and Condensate
    System

MG
  SEE Motor Generator

MG-MA
  SEE Motor Generator-Motor Alternator

mg/kg
  SEE milligram/killigram

mg/l
  SEE milligram/liter

mgallon
  SEE million gallons

MGD
  SEE Million Gallons Per Day

MgF_2
  SEE Magnesium Fluoride

MIAC
  SEE Maintenance Information and
    Control

MILES
  SEE Multiple Integrated Laser
    Engagement System

MIs
  SEE Minority Institutions

MIT
  SEE Massachusetts Institute of
    Technology

MMES
  SEE Martin Marietta Energy Systems

MNCR
  SEE Material Nonconformance Report

MOC
  SEE Margin of Control (Criticality)

MORT
  SEE Management Oversight and Risk
    Tree

MOU
  SEE Memorandum of Understanding

MOV
  SEE Motor Operated Valves

MP&L
  SEE Mississippi Power and Light

MPC
  SEE Maximum Permissible
    Concentration

MPD
  SEE Maximum Permissible Dose

MPL
  SEE Maximum Possible Loss

mR
  SEE Milliroentgen

mR/h
  SEE Microroentgen/hour

MRC
  SEE Maintenance Requirement Card

MRS
  SEE Moderator Recovery System

MSA
  SEE Major System Acquisition

MSDS
  SEE Material Safety Data Sheets

MSIV
  SEE Main Steam Isolation Valve

MTBF
  SEE Mean Time Between Failures

MTE
  SEE Measurement and Test Equipment

MTI
  SEE Maintenance Team Inspection

MTR
  SEE Materials Test Reactor

MTSAA
  SEE Multidiscipline Technical Safety
    Assurance Appraisal

MTTR
  SEE Mean Time to Repair

MUX
  SEE Multiplexor

μCi
  SEE Microcuries

MW
  SEE Mixed Wastes

MWMF
  SEE Mixed Waste Management Facility

MWO
  SEE Maintenance Work Order

MWt
  SEE Megawatt thermal

M_b
  SEE Body Wave Magnitude

M_L
  SEE Local Magnitude (M_L)

M_s
  SEE Surface Wave Magnitude

NAAQS
  SEE National Ambient Air Quality
    Standards

NAS
  SEE National Academy of Sciences

NASA
  SEE National Aeronautic and Space
    Administration

NC
  SEE Normally Closed

NCF
  SEE No Containment Failure

NCO
  SEE NEPA Compliance Officers

NCP
  SEE National Contingency Plan

NCR
  SEE Non-Conformance Report

NCSTATE
  SEE North Carolina State University

NDE
  SEE Non-Destructive Evaluation

NDT
  SEE Non-Destructive Test

NE
  SEE Nuclear Energy

NEC
  SEE National Electrical Code

NEPA
  SEE National Environmental Policy Act

NESHAPS
  SEE National Emissions Standards for
        Hazardous Air Pollutants

NFA
  SEE No Flow Assemblies

NFPA
  SEE National Fire Protection Association

NIM
  SEE Nuclear Incident Monitor

NIST
  SEE National Institute of Standards and
        Testing

NLU
  SEE Normal Latch Up

NMEID
  SEE New Mexico Environmental
        Improvement Division

NMWQCCR
  SEE New Mexico Water Quality Control
        Commission Regulations

NO (Nitric Oxide)
  SEE Nitric Oxides

NO (normally open)
  SEE Normally Open

NOA
  SEE Notice of Availability

NOAEL
  SEE Non-Observed Adverse Effect Level

NOD
  SEE Notice of Deficiency

NOHSCP
  SEE National Oil and Hazardous
        Substances Contingency Plan

NOI
  SEE Notice of Intent

NON
  SEE Notification of Non-Compliance

NOS
  SEE Not Otherwise Specified

NOV
  SEE Notice of Violation

NO$_2$
  SEE Nitrogen Dioxide (NO_2)

NPAR
  SEE Nuclear Plant Aging Research
        Program

NPDES
  SEE National Pollutant Discharge
        Elimination System

NPL
  SEE National Priorities List

NPP
  SEE Nuclear Power Plants

NPR
  SEE New Production Reactor

NPRDS
  SEE Nuclear Plant Reliability Data
        System

NPSH
  SEE Net Positive Suction Head

NQA-1
  SEE Nuclear Quality Assurance-1

NRC
  SEE Nuclear Regulatory Commission

NRC (NON-REUSEABLE CONTAINER)
  SY Non-Reuseable Containers

NRDC
  SEE Natural Resource Defense Council

NRR
  SEE NRC Office of Nuclear Reactor
        Regulation

NRT
  SEE National Response Teams

NSAC
  SEE Nuclear Safety Analysis Center

NSPS
  SEE New Source Performance
        Standards

NTA
  SEE Nitrilotriacetic Acid

NTG
  SEE Nuclear Test Gage

NTNCWS
  SEE Non-Transient Non-Community
        Water System

NTS
  SEE Nevada Test Site

NV
  SEE Nevada Operations Office

NWC
  SEE DOE Nuclear Weapons Complex

NWPA
  SEE Nuclear Waste Policy Act

O&M
  SEE Operations and Maintenance

OAC
  SEE Ohio Administrative Code

OAR
  SEE Operation Assessment and
        Readiness

OBA
  SEE Operating Basis Accident

OBE
  SEE Operating Basis Earthquake

OBES
  SEE Office of Basic Energy Sciences

OC
  SEE Outer Chuck

OCAW
  SEE Oil Chemical and Atomic Workers

OCB
  SEE Oil Circuit Breaker

OCM
  SEE Operational Change Memos

OCRWM
  SEE Office of Civilian Radioactive Waste
    Management

ODC
  SEE Ohio Department of Commerce

ODH
  SEE Ohio Department of Health

OEPA
  SEE Ohio Environmental Protection
    Agency

OER
  SEE Office of Energy Research

OH
  SEE Outer Housing

OHD
  SEE Occupational Health Division

OHER
  SEE Office of Health and Environmental
    Research

OL
  SEE Operating Limit

OLC
  SEE On-Line Computer

OM
  SEE Operating Methods

OMB
  SEE Office of Management and Budget

OP
  SEE Operating Procedures

OR
  SEE Oak Ridge Operations Office

ORAU
  SEE Oak Ridge Associated Universities

ORC
  SEE Ohio Revised Code

ORGDP
  SEE Oak Ridge Gaseous Diffusion Plant

ORM
  SEE Other Regulated Material

ORNL
  SEE Oak Ridge National Laboratory

ORO
  SEE Oak Ridge Operations Office

ORPS
  SEE Occurrence Reporting and
    Processing System

OSC
  SEE On-Scene Coordinators

OSHA
  SEE Occupational Safety and Health
    Administration

OSP *(Office of Special Projects)*
  SEE Office of Special Projects

OSP *(Operational Safety Procedures)*
  SEE Operational Safety Procedures

OSP *Coordinators*
  SEE Office of Special Projects
    Coordinators

OSR
  SEE Operational Safety Requirements

OTD
  SEE Office of Technology Development

OVEC
  SEE Ohio Valley Electric Company

OWP
  SEE Office of Weapons Production

OWR
  SEE Omega West Reactor

O_3
  SEE Ozone

P
  SEE Primary Body Waves

P&ID
  SEE Piping and Instrumentation Drawing

PA
  SEE DOE Production Agency

PA *(Preliminary Assessment)*
  SEE Preliminary Assessment

PAG
  SEE Protective Action Guides

Pan Am
  SEE Pan American World Services, INC.

PARD
  SEE Protect as Restricted Data

Pb *(Periodic Element)*
  SEE Lead

PBF
  SEE Power Burst Facility

PCBs
  SEE Polychlorinated Biphenyls

PCC
  SEE Primary Component Cooling

pCi/g
  SEE Picocuries/gram

pCi/L
  SEE Picocuries Per Liter

PCS
  SEE Power Conversion System

PDB
  SEE Plant Damage Bin

PDC
  SEE Program Development Computer

PDM
  SEE Power Density Monitor

PEARL
  SEE Personnel Expertise and Resource
    Listing

PECSS
  SEE Poisoned Emergency Cooling
    System Sampling

PEL
  SEE Permissible Exposure Limits

PEO
  SEE Program Enrichment Office

PERT
  SEE Program Evaluation and Review
    Technique

PETS
  SEE Procedures for Evaluating
    Technical Specifications Program

PFR
  SEE Phase Failure Relays

PG
  SEE Process Gas

PGDP
  SEE Paducah Gaseous Diffusion Plant
    (Paducah)

PJA
  SEE Proper Job Analysis

PJI
SEE Proper Job Instruction

PLWA
SEE Primary Light Water Addition

PM
SEE Preventive Maintenance

PMA
SEE Power Marketing Administrations

PMF
SEE Probable Maximum Flood

PNL
SEE Pacific Northwest Laboratory

PORC
SEE Plant Oversight Review Committee

PORTS
SEE Portsmouth Gaseous Diffusion
Plant

PORV
SEE Power Operated Relief Valve

pots
SEE Potentiometer

POTW
SEE Publicly Owned Treatment Works

PP
SEE Plenum Pressure

PPC
SEE Project Planning and Control

PPL
SEE Priority Problem List

PRA
SEE Probabilistic Risk Assessment

PREPP
SEE Process Experimental Pilot Plant

PRISIM
SEE Plant Risk Status Information
Management System

PRP
SEE Potentially Responsible Parties

PRV
SEE Pressure Regulator Valve

PSA
SEE Preliminary Safety Analysis

PSB
SEE Pump Shaft Break

PSD
SEE Prevention of Significant
Deterioration

PSI
SEE Pollutant Standard Index

PSI (Pounds per Square Inch)
SEE Pounds per Square Inch

PSIA
SEE Pounds per Square Inch Absolute

PSIG
SEE Pounds per Square Inch Gauge

PSO (Program Secretarial Officer)
SEE Program Secretarial Officers

PSO (Program Senior Official)
SEE Program Senior Officials

PSR
SEE Primary System Relief

PSS
SEE Probabilistic Safety Study

PTP
SEE Peak-To-Peak Ratio

PUEC
SEE Portsmouth Uranium Enrichment
Complex

PVC
SEE Polyvinyl Chloride

PW
SEE Process Water

PWA
SEE Process Waste Assessment

PWGM
SEE Process Water Gamma Monitor

PWR
SEE Pressurized Water Reactor

Q
SEE Quality Factors

QA
SEE Quality Assurance

QAOK
SEE Quality Assurance Acceptance

QC
SEE Quality Control

QIC
SEE Quality Inspection Control

QV
SEE Quality Verification

R
SEE Roentgen

R&D
SEE Research and Development

R&PM
SEE Regulations and Procedures
Manual, PUB-201

R&QA
SEE Reliability and Quality Assurance

RA
SEE Remedial Actions

RAAS
SEE Remedial Action Assessment
System

RACT
SEE Reasonably Available Control
Technology

RAD
SEE Radiation Absorbed Dose

RAM
SEE Reliability, Availability, and
Maintainability

RAP
SEE Remedial Action Program

RBC
SEE Reactor Building Cooling

RBCCW
SEE Reactor Building Closed Cooling
Water

RBCLCW
SEE Reactor Building Closed Loop
Cooling Water

RBCS
SEE Reactor Building Fan Coolers

RBE
SEE Relative Biological Effectiveness

RBS
SEE Reactor Building Spray

RBSI
SEE Reactor Building Spray Injection

RBSR
SEE Reactor Building Spray
Recirculation

RBSVS
  SEE Reactor Building Standby
    Ventilation System

RCB
  SEE Retrieval Containment Building

RCIC
  SEE Reactor Core Isolation Cooling

RCM
  SEE Reliability-Centered Maintenance

RCP
  SEE Reactor Coolant Pump

RCRA
  SEE Resource Conservation and
    Recovery Act

RCS
  SEE Reactor Coolant System

RCSI
  SEE Reactor Coolant System Integrity

RCW
  SEE Recirculated Cooling Water

RD
  SEE Remedial Design

RDDT&E
  SEE Research, Development,
    Demonstration, Testing and
    Evaluation

rDNA
  SEE Recombinant DNA

REDAC
  SEE Remote Detection and Control

REDS
  SEE Reactor Electric Distribution System

REECo
  SEE Reynolds Electrical and
    Engineering Company

REHR
  SEE Reactor Heat Removal

REM (Radiation Exposure Module)
  SEE Radiation Exposure Module

REM (Roentgen Equivalent Man)
  SEE Roentgen Equivalent Man

RER
  SEE Rod Equipment Room

RES
  SEE NRC Office of Nuclear Regulatory
    Research

RESC
  SEE Reactor Subcriticality

RFA
  SEE RCRA Facility Assessment

RFP
  SEE Rocky Flats Plant

RG
  SEE Regulatory Guide

RH
  SEE Remote Handled

RHR
  SEE Residual Heat Removal

RI
  SEE Remedial Investigations

RIG
  SEE Risk-Based Inspection Guide

RIM
  SEE Relative Importance Measure

RL

RMCG
  SEE Reactor Materials Control Group

RMCL
  SEE Recommended Maximum
    Contaminant Level

RMI
  SEE Reactive Metals, Inc.

RNA
  SEE Ribonucleic Acid

RO (Reactor Operator)
  SEE Reactor Operator

RO (Remains Open)
  SEE Remains Open

ROD (CERCLA Record of Decision)
  SEE Record of Decision (CERCLA)

ROD (NEPA Record of Decision)
  SEE Record of Decision (NEPA)

ROD (Report of Discrepancy)
  SEE Report of Discrepancy

RPIS
  SEE Real Property Inventory System

RPM (Radial Power Monitor)
  SEE Radial Power Monitor

RPM (Remedial Project Manager)
  SEE Remedial Project Managers

RPS
  SEE Reactor Protection System

RPT
  SEE Recirculation Pump Trip

RPV
  SEE Reactor Pressure Vessel

RQ
  SEE Reportable Quantities

RRI
  SEE RCRA Remedial Investigation

RRMT
  SEE Reactor and Reactor Material
    Technology

RRT
  SEE Regional Response Teams

RSE
  SEE Removal Site Evaluations

RSO (Reported Significant Observations)
  SEE Reported Significant Observations

RSPA
  SEE Research and Special Programs
    Administration

RSS
  SEE Reactor Safety Study

RSSMAP
  SEE Reactor Safety Study Methodology
    Application Program

RTD
  SEE Reactor Technology Department

RTR
  SEE Roof-Top-Ratio

RV
  SEE Reactor Vessel

RVC
  SEE Reactor Volume Control

RW (DOE Program Office)
  SEE Office of Civilian Radioactive Waste
    Management

RW (River Water)
  SEE River Water

RWE
  SEE Reactor Works Engineering

RWIS
  SEE Raw Water Intake Structure

RWL
  SEE Reactor Water Level

RWQCB
  SEE Regional Water Quality Control
      Board

RWST
  SEE Refueling Water Storage Tank

RX
  SEE Reactors

RZ
  SEE Radiation Zone

S&A
  SEE Sampling and Analysis

S&M
  SEE Surveillance and Maintenance

S Waves
  SEE Secondary Body Waves

S/R
  SEE Safety/Relief

SAIC
  SEE Science Applications International
      Corporation

SALP
  SEE Systematic Assessment of
      Licensee Performance

SAM
  SEE Site Availability Mode

SAN
  SEE San Francisco Operations Office

SAR
  SEE Safety Analysis Reports

SARA
  SEE Superfund Amendments and
      Reauthorization Act of 1986

SAS
  SEE Semi-annual soils

SASS
  SEE Safety Assurance System
      Summary

SAT
  SEE Semi-Automatic Trim

SAV
  SEE Semi-annual vegetation

SBLOCA
  SEE Small Break LOCA

SBO
  SEE Station Blackout

SC
  SEE Safety Computer

SCF
  SEE Standard Cubic Foot

SCG
  SEE Standby Diesel Generator

SCTI
  SEE Sodium Components Test
      Installation

SCW
  SEE Service Clarified Water

SDC
  SEE Shutdown Cooling

SDHX
  SEE Shutdown Heat Exchanger

SDS
  SEE Shutdown Sequencer

SDWA
  SEE Safe Drinking Water Act

SEP
  SEE Systematic Evaluation Program

SER
  SEE Safety Evaluation Report

SERC
  SEE State Emergency Response
      Commission

SFMP
  SEE Surplus Facilities Managment
      Program

SG
  SEE Steam Generator

SGTR
  SEE Steam Generator Tube Rupture

SGTS
  SEE Standby Gas Treatment System

SH
  SEE Sleeve Housings

SHT
  SEE Seal Head Tank

SI
  SEE Site Inspections

Si (Periodic Element)
  SEE Silicon

SICS
  SEE Safety Injection Control System

SIMS
  SEE Standards Information Management
      System

SIP
  SEE State Implementation Plans

SIS
  SEE Special Isotope Separator

SIT
  SEE Safety Injection Tank

SJP
  SEE Standard Job Procedures

SLAC
  SEE Stanford Linear Accelerator Center

SLB
  SEE Shallow Land Burial

SLCS
  SEE Standby Liquid Control System

SLI
  SEE Sandia Laboratory Instruction

SMC
  SEE Selection and Monitoring Chassis

SMCL
  SEE Secondary Maximum Contaminant
      Levels

SMOA
  SEE Superfund Memorandum of
      Agreement

SMRSF
  SEE Special Moment Resisting Space
      Frame

SNL
  SEE Sandia National Laboratories

SNLA
  SEE Sandia National
      Laboratories-Alburquerque

SNLL
  SEE Sandia National
      Laboratories-Livermore

SNM
  SEE Special Nuclear Materials

SOCs
  SEE Synthetic Organic Chemicals

SOP (Management)
  SEE Standard Operating Procedures

SOP (Standard Operating Power)
  SEE Standard Operating Power

SORV
  SEE Stuck-Open Relief Valve

SO_2
  SEE Sulfur Dioxide

SO_2F_2
  SEE Sulfuryl Fluoride

SP (Suppression Pool)
  SEE Suppression Pool

SPC
  SEE Suppression Pool Cooling

SPCC
  SEE Spill Prevention Control and
      Countermeasures Plan

SPMS
  SEE Safety Performance Measurement
      System

SPP
  SEE Standard Practice Procedures

SPR
  SEE Strategic Petroleum Reserve

SPR III
  SEE Sandia Pulse Reactor III

SPS
  SEE Safety Precedence Sequence

SR
  SEE Savannah River Operations Office

SR/VO
  SEE Safety/Relief Valve Open

SRC
  SEE Safety Review Committee

SRO
  SEE Senior Reactor Operator

SRS
  SEE Savannah River Site

SRV
  SEE Safety/Relief Valve

SRW
  SEE Savannah River Water

SS
  SEE Shift Supervisor

SSDC
  SEE System Safety Development Center

SSFI
  SEE Safety System Functional
      Inspection

SSOMI
  SEE Safety System Outage Modification
      Inspection

SSPS
  SEE Solid State Protection System

SSPSA
  SEE Seabrook Station Probabilistic
      Safety Assessment

SSR
  SEE Secondary System Relief

SSS (Safe Shutdown System)
  SEE Safe Shutdown System

SSS (Supplementary Safety System)
  SEE Supplementary Safety System

SST
  SEE Single Shell Tanks

ST
  SEE Sleeve Target

STA
  SEE Shift Technical Advisor

STC
  SEE Single Strip Containers

STGWG
  SEE State and Tribal Government
      Working Group

STI
  SEE Surveillance Test Interval

STM
  SEE Stack Tritium Monitor

SWAT
  SEE Solid Waste Assessment Test

SWDA
  SEE Solid Waste Disposal Act

SWGR
  SEE Switchgear Room

SWMU
  SEE Solid Waste Management Units

SWP
  SEE Safety Work Permit

SWS (Salt Water System)
  SEE Salt Water System

SWS (Service Water System)
  SEE Service Water System

SWSA
  SEE Solid Waste Storage Area

T&B
  SEE Top and Bottom

TA
  SEE Test Authorizations

TALM
  SEE Temperature Alarm Monitor

TAN
  SEE Test Area North

TAP
  SEE Toxic Air Pollutants

TB
  SEE Turbine Building

TBSCCW
  SEE Turbine Building Secondary Closed
      Cooling Water

TBSW
  SEE Turbine Building Service Water

TBV
  SEE Turbine Block Valve

Tc (Periodic element)
  SEE Technetium

TC (Thermocouple)
  SEE Thermocouple(s)

TCE
  SEE Trichloroethylene (TCE)

TCLP
  SEE Toxicity Characteristic Leaching
      Procedure

TD
  SEE Time Delay

TDC
SEE Tennessee Department of Commerce

THERP
SEE Technique for Human Error Rate Prediction

THM
SEE Trihalomethane

TIR
SEE Total Indicated Runout

TLD
SEE Thermoluminescent Dosimeters

TLV
SEE Threshold Limit Values

TOC
SEE Total Organic Carbon

TPC
SEE Temporary Procedure Change

TPL
SEE Transient Protection Limit

TPQ
SEE Threshold Planning Quantities

TR
SEE Transformer Room

TRA
SEE Test Reactor Area

TRAM
SEE TRAnsient Multiplexor

TRC
SEE Total Recordable Cases

TRE
SEE Turbidity Removal Evaporator

TRU
SEE Transuranic Nuclides

TS
SEE Technical Specifications

TSA
SEE Technical Safety Appraisals

TSCA
SEE Toxic Substances Control Act

TSCM
SEE Temperature Scram Circuit Monitor

TSD
SEE Treatment, Storage, and Disposal

TSH
SEE Target Sleeve Housing

TSS
SEE Total Suspended Solids

TTO
SEE Total Toxic Organics

TTR (Time to Repair)
SEE Time to Repair

TTR (Tonopah Test Range)
SEE Tonopah Test Range

TWA
SEE Time Weighted Average

TWF
SEE Transuranic Waste Facility

TWFM
SEE Traveling Wire Flux Monitor

TWTSF
SEE Transuranic Waste Treatment and Storage Facility

UC
SEE University of California

UCC
SEE Ultra Clean Coal

UCI
SEE University of California-Irvine

UDH
SEE Utah Department of Health

UE
SEE United Electric

UFSAR
SEE Updated Final Safety Analysis Report

UF$_4$
SEE Uranium Tetrafluoride

UF$_6$
SEE Uranium Hexafluoride

UIC
SEE Underground Injection Control

UMT
SEE Uranium Mill Tailings

UMTRA
SEE Uranium Mill Tailings Remedial Action

UMTRAP
SEE Uranium Mill Tailings Remedial Action Program

UORS (Reports)
SEE Unusual Occurrence Reports

UORS (System)
SEE Unusual Occurrence Reporting System

UPS
SEE Uninterruptible Power Supplies

USATHAMA
SEE U.S. Army Toxic and Hazardous Materials Center

USDW
SEE Underground Sources of Drinking Water

USH
SEE Universal Sleeve Housing

USRADS
SEE Ultrasonic Ranging and Data System

UST
SEE Underground Storage Tanks

UTENN
SEE University of Tennessee Center for Biotechnology

UV
SEE Undervoltage

VAC
SEE Volts Alternating Current

VAM
SEE Vibration and Acoustic Monitoring

VCAPCD
SEE Ventura County Air Pollution Control Districty

VCP
SEE Vitrified Clay Pipe

VDC
SEE Volts Direct Current

VHAP
SEE Volatile Hazardous Air Pollutant (VHAP)

VHT
SEE Very High Temperature

VLF
SEE Very Low Flow

VLFC
  SEE Very Low Flow Constant

VM
  SEE Volt Meter

VOCs
  SEE Volatile Organic Compounds

VRE
  SEE Vibrating Reed Electrometers

VTS
  SEE Vertical Tube Storage

VU
  SEE Veeder Units

WACC
  SEE Waste Acceptance Criteria
    Committee

WAG
  SEE Waste Area Grouping

WERF
  SEE Waste Experimental Reduction
    Facility

WHC
  SEE Westinghouse Hanford Company

WHPP
  SEE Waste Handling and Packaging
    Plant

WIN
  SEE Waste Information Network

WIPP
  SEE Waste Isolation Pilot Plant

WL
  SEE Working Level

WLM
  SEE Working Level Month

WMCO
  SEE Westinghouse Materials Company
    of Ohio

WMin
  SEE Waste Minimization

WMO
  SEE Waste Management Operations

WRAP
  SEE Waste Receiving and Processing
    Plant

WT
  SEE Water Tight

Yards³
  SEE Cubic Yards

Zn (Periodic Element)
  SEE Zinc

ZPSS
  SEE Zion Probabilistic Safety Study

Source Terms/Subject Categories

AIR POLLUTION
ST   Abandoned Areas
ST   Active Mines
ST   Acts
ST   Adequately Wetted
ST   Administrators
ST   Adverse Impact on Visibility
ST   Affected Facilities
ST   Agencies
ST   Agreement States
ST   Allowable Emissions
ST   Area Sources
ST   Arsenic Kitchens
ST   Arsenic Containing Glass Types
ST   As Expeditiously as Practicable
ST   Asbestos
ST   Asbestos Materials
ST   Asbestos Mills
ST   Asbestos Tailings
ST   Asbestos Waste from Control
        Devices
ST   Asbestos-Containing Waste
        Materials
ST   Banking (Air Pollution)
ST   Baseline Concentration
ST   Beryllium
ST   Beryllium Alloys
ST   Beryllium Ores
ST   Beryllium Propellants
ST   Beryllium-Containing Wastes
ST   Best Available Control Technology
ST   Best Available Retrofit Technology
ST   Blowing
ST   Bulk Resins
ST   Bulkheads
ST   Calciners
ST   Capacity Factor
ST   Capital Expenditure
ST   Cell Rooms
ST   Ceramic Plants
ST   Charging
ST   Closed Vent Systems
ST   Coal
ST   Coal Refuse
ST   Commercial Asbestos
ST   Commercial Arsenic
ST   Compliance Schedules
ST   Condenser Stack Gases
ST   Connectors
ST   Construction
ST   Continuous Disposal
ST   Control Strategies
ST   Control Devices
ST   Converter Arsenic Charging Rate
ST   Copper Converters
ST   Copper Converter Department
ST   Copper Mattes

ST   Critical Organs
ST   Cullet
ST   Curie
ST   Curtail
ST   Delayed Compliance Orders
ST   Demolition
ST   Denuders
ST   Dewatered
ST   Dispersion Techniques
ST   Dispersion Resins
ST   Dose Equivalents
ST   Double Block and Bleed System
ST   Drift
ST   Effects on Welfare
ST   Effective Date
ST   Effective Dose Equivalent
ST   Elemental Phosphorus Plants
ST   Emergency Renovation Operations
ST   Emission Limitation
ST   End Boxes
ST   End Box Ventilation System
ST   EPA
ST   Equipment
ST   Equivalent Methods
ST   Ethylene Dichloride Plants
ST   Ethylene Dichloride Purification
ST   Excess Emissions
ST   Exhaust Gases
ST   Existing Stationary Facilities
ST   Existing Sources
ST   Existing Tailings Piles
ST   Extraction Plants
ST   Fabricating
ST   Facilities
ST   Facility Components
ST   Federal Land Managers
ST   Field Gases
ST   First Attempt at Repair
ST   Fixed Capital Costs
ST   Fossil Fuels
ST   Fossil Fuel and Wood
        Residue-Fired Steam Generating
        Unit
ST   Foundries
ST   Friable Asbestos
ST   Fugitive Emissions
ST   Glass Melting Furnaces
ST   Hazardous Air Pollution
ST   Heat Input
ST   Hydraulic Lift Tanks
ST   Hydrogen Gas Streams
ST   In Heavy Liquid Service
ST   In Light Liquid Service
ST   In Liquid Service
ST   In Vacuum Service
ST   In VHAP Service
ST   In Vinyl Chloride Service

ST   In VOC Service
ST   In-Process Wastewater
ST   In-Situ Sampling Systems
ST   Inactive Mines
ST   Inactive Waste Disposal Sites
ST   Incinerators
ST   Increments of Progress
ST   Inorganic Arsenic
ST   Installations
ST   Integral Vista
ST   Interstate Air Pollution Control
        Agencies
ST   Intermittent Control System
ST   Latex Resins
ST   Lead Mattes
ST   Leaks
ST   Local Agencies
ST   Lowest Achievable Emission Rate
ST   Major Emitting Facilities
ST   Major Stationary Source
ST   Malfunctions
ST   Man-Made Air Pollution
ST   Mandatory Class I Federal Areas
ST   Mercury
ST   Mercury Chlor-alkali Cells
ST   Mercury Chlor-alkali Electrolyzers
ST   Mercury Ore
ST   Mercury Ore Processing Facilities
ST   Modification
ST   Monitoring Systems
ST   Motor Vehicles
ST   National Standards (Air Pollution)
ST   Natural Gas Liquids
ST   Natural Gas Processing Plants
ST   Natural Conditions
ST   New Sources
ST   New Source Performance
        Standards
ST   New Tailings
ST   New Tailings Impoundment
ST   Nitric Acid Plants
ST   Nodulizing Kilns
ST   Non-Attainment Areas
ST   Non-Fractionating Plants
ST   NRC-Licensed Facilities
ST   Opacity
ST   Open Ended Valve
ST   Open-Ended Line
ST   Operators
ST   Outside Air
ST   Owners
ST   Particulate Matter
ST   Particulate Matter Emissions
ST   Particulate Asbestos Materials
ST   Phased Disposal
ST   Planned Renovation Operations
ST PM_{10}

ST PM$_{10}$ Emissions
ST Point Sources
ST Polyvinyl Chloride Plants
ST Pot Furnaces
ST Potential to Emit
ST Pouring
ST Pressure Releases
ST Prevention of Significant
      Deterioration Regulations
ST Primary Standard Attainment Date
ST Primary Standards
ST Primary Copper Smelter
ST Primary Emission Control Systems
ST Process Emissions
ST Process Units
ST Process Unit Shutdown
ST Product Accumulator Vessels
ST Propellants
ST Propellant Plants
ST Radionuclides
ST Reactors (Chemical)
ST Reactor Opening Loss
ST Reasonable Further Progress
ST Reasonably Attributable
ST Reasonably Available Control
      Technology
ST Rebricking
ST Reconstruction
ST Reference Methods
ST Regions
ST Regional Office
ST Relief Valves
ST Relief Valve Discharge
ST Renovation
ST Resin Grade
ST Roadways
ST Roasting
ST Rocket Motor Test Sites
ST Run
ST Schedule of Compliance
ST Secondary Emissions
ST Secondary Standards
ST Secondary Hood Systems
ST Semiannual
ST Sensors
ST Shutdown (Emissions)
ST Significant Impairment
ST Skimming
ST Slip Gauges
ST Sludge
ST Sludge Dryer
ST Stacks
ST Standard of Performance
ST Standard Operating Procedures
ST Standard Pressure
ST Standard Temperature
ST Startup
ST State Agencies
ST State Implementation Plans
ST Stationary Sources
ST Strippers
ST Strip
ST Structural Members
ST Sulfuric Acid Plants
ST Temporarily Abandoned Areas
ST Theoretical Arsenic Emissions
      Factor
ST Total Suspended Particulates
ST Traceable
ST Transportation Control Measures
ST Types of Resin

ST Uncontrolled Total Arsenic
      Emissions
ST Underground Uranium Mines
ST Uranium By-Product Materials
ST Variances
ST Vinyl Chloride Plants
ST Visibility Impairments
ST Visible Emissions
ST Volatile Hazardous Air Pollutant
      (VHAP)
ST Wastewater Treatment Processes
ST Whole Body
ST Wood Residues

AVIATION SAFETY
ST Aircraft Accidents
ST Aircraft Incidents
ST Aircraft
ST Airports
ST Aviation Operations
ST Charter Operations
ST Civil Aircraft
ST Flight Crewmembers
ST Ground Crews
ST Heliports
ST Helipads
ST Modern Aircraft
ST Passengers (Aircraft)
ST Public Aircraft

COMPENSATION AND LIABILITY
ST Activation (Emergency)
ST Administrators
ST Alternative Water Supplies
ST Articles
ST Barrels
ST Biological Additives
ST Burning Agents
ST Chemical Agents
ST Claims
ST Claimant
ST Coastal Waters
ST Coastal Zones
ST Comprehensive Environmental
      Response, Compensation, etc.
ST Consumer Products
ST Contaminants
ST Contiguous Zones
ST Contractual Relationships
ST Discharges
ST Dispersants
ST Drinking Water Supplies
ST Environment
ST Extremely Hazardous Substances
ST Facilities
ST Feasibility Studies
ST Federally Permitted Releases
ST First Federal Officials
ST Full-time Employees
ST Groundwater
ST Guarantors
ST Hazardous Substances
ST Hazardous Chemicals
ST Hazardous Wastes
ST Hazard Categories
ST Incineration Vessels
ST Indian Tribes
ST Inland Waters
ST Inland Zones
ST Insurance

ST Lead Agencies
ST Listed Hazardous Substances
ST Management of Migrations
ST Manufacture
ST Material Safety Data Sheets
ST Mixtures
ST Natural Resources
ST Navigable Waters
ST Offshore Facilities
ST Oil Pollution Fund
ST Oils
ST On-Scene Coordinators
ST On-Shore Facilities
ST Operable Units
ST Operators
ST Otherwise Use
ST Owners
ST Pollutants
ST Pollution Liability
ST Primary Documents
ST Process (Toxic Chemical)
ST Purchasing Group
ST Reasonable Costs
ST Releases
ST Relevant and Appropriate
      Requirements
ST Remedial Investigations
ST Remedial Project Managers
ST Remedial Actions
ST Remedies
ST Removal
ST Reportable Quantities
ST Risk Retention Group
ST Secondary Documents
ST Senior Management Officials
ST Sinking Agents
ST Site Quality Assurance Plan
ST Size Classes of Discharges
ST Size Classes of Releases
ST Source Control Remedial Actions
ST Specified Ports and Harbors
ST States
ST Superfund
ST Surface Collecting Agents
ST Threshold Planning Quantities
ST Toxic Chemicals
ST Transportation
ST Transport (Hazardous Substances)
ST Trustees
ST Trust Fund (CERCLA)
ST United States
ST Unlisted Hazardous Substances
ST Vessels

CONSTRUCTION
ST Accidents (explosive)
ST Allowable Soil Bearing Capacity
ST Ambient
ST Anaerobic Digestion
ST Anticipated Operational
      Occurrences
ST Approved Storage Containers
ST Aquifers
ST As Low As Reasonably Achievable
ST Auxiliary Air Units
ST Ballast (Railroads)
ST Base Shear
ST Base Course
ST Bases
ST Bearing Wall Systems

ST Bearing Capacity
ST Bench Marks
ST Bentonite Clays
ST Best Available Technology
ST Bird Strike
ST Boiler/Industrial Furnaces
ST Borings
ST Boundary Elements
ST Braced Frame
ST Building Frame Systems
ST Building Acquisitions
ST Caisson Foundations
ST Cantilever Footings
ST Capillary Water
ST Cased Explosives
ST Classified Information
ST Classified Interests
ST Classified Matter
ST Classified Telecommunications
   Facilities
ST Cognizant DOE Authority
ST Collectors
ST Concentric Braced Frames
ST Concrete Encasement
ST Confinement Areas
ST Confinement Systems
ST Construction Joints
ST Construction Projects
ST Construction Project Planning
ST Cooper E
ST Covers
ST Credible Accidents
ST Criteria
ST Critical Areas
ST Critical Facilities
ST Criticality Incidents
ST Crossing Frogs
ST Crypto
ST Cultural Resource Sites
ST Curb Inlets
ST Curb Return
ST Datum
ST Dead Loads
ST Decommissioning
ST Decontamination
ST Dedicated Fire Water Systems
ST Deflagration
ST Deflection Angles
ST Department of Energy Resident
   Construction Contractors
ST Departmental-Approved Equipment
ST Departmental Elements
ST Department of Energy Project
   Construction Contractors
ST Design Basis Accidents
ST Design Basis Earthquakes
ST Design Basis Fires
ST Design Basis Floods
ST Design Basis Tornadoes
ST Design Floods
ST Detection
ST Detection Equipment
ST Detonators
ST Detonations
ST Diaphragms
ST Diaphragm Chord
ST Diaphragm Strut
ST Disturbed Zones
ST DOE Contractors
ST DOE Energy Management
   Coordinators

ST DOE Fire Protection Authorities
ST DOE Safeguards and Security
   Coordinators
ST Dual Systems
ST Duress Systems
ST Earth-Lined Channels
ST Eccentric Braced Frames
ST Effective Dose Equivalent
ST Effluents
ST Egress
ST Electroexplosive Devices
ST Emergency Control Centers
ST Emergency Control Stations
ST Emergency Operations Centers
ST Emergency Planning Zones
ST Emergency Power
ST Enclosures
ST Energy Monitoring and Control
   Systems
ST Energy Management Systems
ST Engineered Safety Features
ST Entry Control Points
ST Environmental Surveys
ST Environmental Audits
ST Essential Facilities
ST Exclusion Areas
ST Expansion Joints
ST Explosives
ST Explosives Activities
ST Explosives Bays
ST Explosives Buildings
ST Explosives Hazard Classes
ST Explosives Hazard, Class I
ST Explosives Hazard, Class II
ST Explosives Hazard, Class III
ST Explosives Hazard, Class IV
ST External Corrosion
ST Facilities
ST Facility Authority
ST Facility Boundaries
ST Fail-Safe
ST Federal Employee Occupational
   Safety and Health Program
ST Field Elements
ST Findings
ST Fissile Materials
ST Flexible Elements
ST Flexural Strength
ST Force Mains
ST Freeboard
ST Functional Appraisals
ST GDC Planning Board
ST General Design Criteria
ST Grade Beams
ST Halogenated
ST Hazardous Materials
ST High Efficiency Particulate Air
   Filters
ST High Explosives
ST High-Level Wastes
ST Holdup (Nuclear Material)
ST Horizontal Bracing Systems
ST Hot Lines
ST Human Factors
ST Human Factors Engineering
ST Hydraulic Structures
ST IHE Subassemblies
ST IHE Weapons
ST Improved Risk
ST In-Process Materials
ST In-Situ

ST In-Use Materials
ST Ingress
ST Inhabited Building Distance
ST Initiation Stimuli
ST Insensitive High Explosives
ST Installed Equipment
ST Internal Appraisals
ST Intermediate Moment Resisting
   Space Frame
ST Interfaces
ST Intraline Separation (Barricaded)
ST Intraline Separation (Unbarricaded)
ST Intrusion Alarm System (Perimeter
   or Interior)
ST Inverted Siphons
ST Ion Exchange
ST Isolation Zones
ST Joint Frequency Distribution
ST Karst Terrain
ST Land Application
ST Landfills
ST Lateral Force Resisting System
ST Leachates
ST Life Cycle Costs
ST Limited Areas
ST Line Organizations
ST Live Loads
ST LLW
ST Load Factor
ST Low Level Wastes
ST Magazine Separation
ST Magazines (Buildings)
ST Management Appraisals
ST Mass Concrete
ST Material Access Areas
ST Material Balance Area
ST Maximal Effective Pressure
ST Maximum Probable Flood
ST Moment Resisting Space Frame
ST Monumentation
ST Nationally Recognized Testing
   Laboratories
ST New Storage Facilities
ST Nuclear Facilities
ST Occupiable Area
ST Occupied Area (Explosives)
ST Operating Level
ST Operational Safety Requirements
ST Operating Area Compartment
ST Operating Basis Accident
ST Operational DBA
ST Ordinary Moment Resisting Space
   Frame
ST Orthogonal Effects
ST Overpressure
ST P-Delta Effect
ST Peak Positive Incident Pressure
ST Permafrost
ST Pervious
ST pH
ST Physical Protection (physical
   security)
ST Physically Separated
ST Plastic Yielding
ST Platforms
ST Plutonium Processing and Handling
   Facilities
ST Plutonium Storage Facilities
ST Point of Nearest Public Access
ST Portland Cement
ST Preliminary Safety Analysis Report

ST Primary Confinement Systems
ST Program Senior Officials
ST Project Design Criteria
ST Property Protection Area
ST Protected Areas
ST Public Travel Routes
ST Pyrophoric-Igniting Spontaneously
ST Quality Assurance
ST Quality Assurance Records
ST Quantity Distances
ST Radio Repeater Stations
ST Rational Method
ST Real Property Inventory System
ST Receiving Streams
ST Refractories
ST Regional Frequency Analysis
ST Reinforcement Ratio
ST REM (Roentgen Equivalent Man)
ST Remote Interrogation Points
ST Required Strength
ST Reservoir Routing
ST Response Time
ST Retaining Walls
ST Return Period
ST Roadway Crown
ST Routine Wastes
ST Safeguards
ST Safety Analysis Reports
ST Safety Analysis
ST Safety Class Items
ST Safety Limits
ST Sanitary Landfills
ST Sanitary Engineering Structures
ST Saturated Zones
ST Second Line Organization Level
ST Secure Communications Centers
ST Security
ST Security Areas
ST Security Interests
ST Seismic Category I
ST Service Magazine
ST Setback
ST Shear Walls
ST Sheet Piling
ST Shoring
ST Significant Modification
ST Single Failure
ST Site Boundaries
ST Site-Specific Safeguards and
      Security Plan
ST Slanting
ST SNM Vault
ST Soft Stories
ST Soil Resistivity
ST Soil Mechanics
ST Soil-Structure Resonance
ST Space Frames
ST Special Moment Resisting Space
      Frame
ST Special Nuclear Materials
ST Staging Bays (in-process)
ST Standby Power
ST Storage Area Compartments
ST Stories
ST Story Drifts
ST Story Drift Ratio
ST Story Shear
ST Strength
ST Structural Collapse
ST Subbase
ST Subcritical Flows

ST Subgrade Modulus
ST Subslabs
ST Substantial Construction
ST Supercritical Flows
ST Superelevation
ST Support Buildings
ST Surfactants
ST Tactical Response Forces
ST Technical Safety Appraisals
ST Tension Wires
ST TNT Equivalent
ST Towers
ST Transients (re: Explosives
      Facilities)
ST Transuranic Elements
ST Transverse
ST TRU Wastes
ST TSA
ST Unattended Openings
ST Underpinnings
ST Uninterruptible Power Supplies
ST Unirradiated Enriched Uranium
ST Unit Masonry
ST Unpackaging Rooms
ST Useful Life
ST Vault-type Rooms
ST Vectors
ST Vertical Load Carrying Space
      Frames
ST Vital Activities
ST Vital Areas
ST Vital Equipment
ST Vital Facilities
ST Vital Programs
ST Water Hammers
ST Weak Stories
ST Work Environment

EMERGENCY PREPAREDNESS
ST Absorbed Dose
ST Accidents
ST Acute Effects
ST Acute Exposure
ST Acute Toxicity
ST Affected Persons
ST Agencies
ST Agency Lead Officials
ST Agreement States
ST Airborne Response Teams
ST Airborne Releases
ST Alarms
ST Alert
ST Alert Signals
ST Alpha
ST Alpha Particles
ST Annual Limit on Intake
ST Asphyxiants
ST Assessment Actions
ST Assessments
ST Attack Warning Signals
ST Attention Signals
ST Beta Particles
ST Biological Half-Time
ST Bomb Incidents
ST Bomb Threats
ST Bravo
ST By-Product Materials
ST CAMS
ST Central Alarm Stations
ST Chain of Command

ST Charlie
ST Chronic Effects
ST Chronic Exposure
ST Chronic Toxicity
ST Classification
ST Classified Matter
ST COGEMT
ST Cognizant Federal Agencies
ST Cognizant Federal Agency Officials
ST Collective Dose Equivalent
ST Committed Dose Equivalent
ST Committed Effective Dose
      Equivalent
ST Communicators
ST Community Awareness and
      Emergency Response Programs
ST Comprehensive Environmental
      Response, Compensation, etc.
ST COMVAN
ST Concept of Operations
ST Confidential
ST Consequence Assessments
ST Consequences
ST Containment
ST Contamination
ST Containment Commanders
ST Continuity of Government
      Emergency Management Teams
ST Contingency Planning Zones
ST Continuous Air Monitors
ST Controlled Copies
ST Controlled Areas
ST Controllers
ST Core Melt Accidents
ST Corrective Actions
ST Corrosive Materials
ST Couriers
ST Covert Threats
ST Credibility Assessments
ST Crisis Managers
ST Critical Mass
ST Criticality
ST Criticality Accidents
ST Criticality Alarms
ST Curie
ST Damage Assessments
ST Decontamination
ST Defense Readiness Conditions
ST Denials
ST Derived Air Concentration
ST Derived Concentration Guide
ST Detection
ST Director of Emergency Operations
ST DOE Accident Response Group
      Team Leaders
ST DOE Emergency Operations Center
ST DOE Property
ST Dose Equivalents
ST Doses
ST Dosimeters
ST Dosimetry
ST Drills
ST Duress Systems
ST Duty Officers
ST Effective Dose Equivalent
ST Effective Half-Life
ST Emergency Management Team
      Directors
ST Emergency Equipment
ST Emergency Management
      Coordination Committee

ST Emergency Management Teams
ST Emergency Response
ST Emergency and Hazardous
   Chemical Inventory Forms
ST Emergency Briefing Centers
ST Emergency Broadcast System
ST Emergency Management
   Coordination Committee
   Secretariat
ST Emergency Management
   Information System
ST Emergency Management Systems
ST Emergency Plans
ST Emergency Planning,
   Preparedness, and Response
   Program
ST Emergencies
ST Emergency Procedures
ST Emergency Resources
ST Emergency Response Levels
ST Emergency Support Teams
ST Emergency Telecommunications
   Services
ST Emergency Control Centers
ST Emergency Control Stations
ST Emergency Operations Centers
ST Emergency Planning Zones
ST Emergency Systems
ST Emergency Response Teams
ST Energy Emergencies
ST Energy Emergency Management
   Teams
ST Environment, Safety, and Health
   (ES&H) Program
ST Exempt or Limited Quantities
ST Exercises
ST Explosive Ordnance Disposal
   Teams
ST Extremely Hazardous Substances
ST Extremely Hazardous Wastes
ST Facilities
ST Federal Coordinating Officers
ST Federal Field Exercises
ST Federal Radiological Emergency
   Response Plans
ST Federal Radiological Monitoring
   and Assessment Centers
ST Federal Radiological Monitoring
   and Assessment Plans
ST Federal Response Centers
ST Field Command Posts
ST Field Facility/Building Emergency
   Plans
ST Field Monitoring
ST Field Site Emergency Plans
ST First Federal Officials
ST Fissile
ST Fission Products
ST Fixed Nuclear Facilities
ST FRMAC Director
ST Full-Scale Exercises
ST Gamma Rays
ST General Emergencies
ST Half-Life
ST Hazardous Substances
ST Hazardous Chemicals
ST Hazardous Wastes
ST Hazard Analysis
ST Hazard and Operability Study
ST Hazardous Materials Emergencies
ST Hazardous Materials

ST Health Physics
ST Hostage Negotiation Teams
ST Hostage Throw Phone
ST Immediately Dangerous to Life or
   Health Values
ST Incidents
ST Ingestion Exposure Pathways
ST Inspection and Evaluation
ST Interagency Radiological
   Assistance Plans
ST Internal Security Reports
ST Ionization Chambers
ST Ionizing Radiation
ST Isotopes
ST Joint Information Center
ST Joint Nuclear Accident Coordination
   Center
ST Large Quantities
ST Lethal Concentration Low
ST Lethal Dose Low
ST Lethal Dose of Radiation
ST Level of Concern
ST Liaison Officers
ST Licenses
ST Limited Responses
ST Local Emergency Planning
   Committees
ST Local Governments
ST Low Level Wastes
ST Low Specific Activity Materials
ST Material Safety Data Sheets
ST Median Lethal Concentration
ST Median Lethal Dose
ST Mobile Command Vehicles
ST Monitoring
ST Mutual Assistance Agreements
ST National Response Centers
ST National Response Teams
ST National Defense Areas
ST National Emergencies
ST National Security Areas
ST National Security Information
ST National Warning System
ST National Contingency Plan
ST Natural Disasters
ST Non-Nuclear Facilities
ST Non-Penetrating Radiation
ST Normal Form
ST Notification
ST Nuclear Criticality Safety
ST Nuclear Facilities
ST Nuclear Accident Dosimeter
ST Nuclear By-product Material
   License
ST Nuclear Emergency Search Teams
ST Nuclear Fuel Cycle
ST Nuclear Materials Management and
   Standards System
ST Nuclear Regulatory Commission
   Licensed Activities
ST Nuclear Threat Incidents
ST Nuclear Weapons Accidents
ST Nuclear Regulatory Commission
ST Nuclides
ST Off-Site Federal Support
ST Off-site Notification/Warning
ST Off-site
ST Official Use Only
ST Officially Reportable Events
ST On-Scene Coordinators
ST On-Scene

ST On-Scene Commanders
ST On-Site Federal Support
ST On-Site Technical Directors
ST On-Site
ST Operational Emergencies
ST Operational Accidents
ST Operational Emergency
   Management Teams
ST Operational Emergency
   Preparedness Management Plans
ST Operational Emergency Response
   Levels
ST Owners
ST Packaging
ST PEL
ST Penetrating Radiation
ST Permissible Exposure Limits
ST Planning Zones
ST Plumes
ST Plume Exposure Pathways
ST Population Dose Projections
ST Post-Incident Activities
ST Potassium Iodide
ST Preparedness
ST Primary Emergency Plans
ST Program Senior Officials
ST Projected Dose
ST Protective Action Guides
ST Protect as Restricted Data
ST Protective Actions
ST Protective Action
   Recommendations
ST Protective Measures
ST Protons
ST Public Affairs (Personnel)
ST Public Information Officers
ST Quality Factors
ST Radiation Emergency Assistance
   Center/Training Sites
ST Radionuclides
ST Radioactivity
ST Radioactive Material Transportation
   Accidents
ST Radioactive Material Transportation
   Incidents
ST Radiological Accidents
ST Radiological Assistance Programs
ST Radiological Assistance Teams
ST Radiological Releases
ST Radiological Transportation
   Incidents
ST Reactor Facilities
ST Readiness Assurance
ST Recovery Actions
ST Recovery Plans
ST Reference Levels
ST Regional Response Teams
ST Regional Assistance Committees
ST Regional Coordinating Offices
ST Relative Biological Effectiveness
ST Releases
ST REM (Roentgen Equivalent Man)
ST Remedial Project Managers
ST Reportable Quantities
ST Resource Conservation and
   Recovery Act
ST Responses
ST Safeguards
ST Safeguards and Security Alerts
ST Safeguards and Security Alert I
   (Code Designator: Alpha)

ST Safeguards and Security Alert II
      (Code Designator: Bravo)
ST Safeguards and Security Alert III
      (Code Designator: Charlie)
ST Secure Communications Centers
ST Secure Automatic Communications
      Network
ST Security Communications Control
      Center
ST Security Events
ST Senior Controllers
ST Senior Federal Emergency
      Management Agency Officials
ST Senior Scientific Advisers
ST Senior Security Supervisors
ST Sensitive Compartmented
      Information
ST Sensitive Nuclear Material
      Production Information
ST Shelters
ST Short-Term Exposure Limit
ST Significant Incidents
ST Site (DOE FRASE Vocabulary)
ST Site Emergencies
ST Small Quantity Generators
ST Solid Wastes
ST Source Materials
ST Special Nuclear Materials
ST Special Form
ST Special Nuclear Materials
      Emergencies
ST Special Populations
ST Special Response Teams
ST Spent Fuel
ST Standard Operating Procedures
ST State Emergency Response
      Commission
ST State Coordinating Officers
ST Subcommittee on Federal
      Response
ST Superfund Amendments and
      Reauthorization Act of 1986
ST Surveillance and Nuclear Detection
      Systems
ST Suspicious Devices
ST Tabletop
ST Tactical Response Teams
ST Telecommunication
ST Terrorist
ST Terrorist Response Plans
ST Terrorist Threats
ST Terrorist Acts
ST Threats
ST Threshold Planning Quantities
ST Threshold Limit Values
ST Toxic Chemicals
ST Toxic Chemical Release Forms
ST Transuranic Wastes (TRU Waste)
ST Transuranic Elements
ST Transportation Incidents
ST Type A Packaging
ST Type B Packaging
ST Unified Dose Assessments
ST Unusual Events
ST Uranium Hexafluoride
ST Vital Records
ST Vulnerable Zones
ST Whole Body Counter
ST Whole Body Dose
ST X-rays
ST Yellowcake

ENDANGERED SPECIES
ST Actions
ST Action Areas
ST Acts
ST Adverse Modifications
ST Agency Actions
ST Agreements
ST Alaskan Natives
ST Alternative Courses of Action
ST Application for Federal Assistance
ST Assistant Administrator for
      Fisheries
ST Authorized Officers
ST Biological Assessments
ST Biological Opinion
ST Candidates
ST Captivity-bred
ST Captivity
ST Commercial Activities
ST Conferences
ST Conservation Recommendations
ST Critical Habitats
ST Designated Non-Federal
      Representatives
ST Destruction
ST Directors
ST Effects of the Action
ST Essential Experimental Populations
ST Experimental Populations
ST Federal Agencies
ST Fish
ST Foreign Commerce
ST Formal Consultation
ST Incidental Take (Taking)
ST Informal Consultation
ST Irreversible Commitment of
      Resources
ST License Applicants
ST List
ST Listed Species
ST Major Construction Activities
ST Manatee Protection Areas
ST Manatee Refuges
ST Manatee Sanctuaries
ST On-Going Projects
ST Permits
ST Permit Applicants
ST Plants
ST Plastrons
ST Populations
ST Preliminary Biological Opinion
ST Program
ST Project Segment
ST Proposed Critical Habitats
ST Proposed Species
ST Proposed Actions
ST Public Hearings
ST Reasonable and Prudent
      Alternatives
ST Reasonable and Prudent Measures
ST Recovery
ST Resident Species
ST Sea Turtles
ST Service
ST Species
ST Specimens
ST State Agencies
ST States
ST Subsistence
ST Threatened Species
ST Wasteful Manner

ST Waterborne Activities
ST Water Vehicles
ST Wildlife

ENVIRONMENTAL MANAGEMENT
ST Aboveground Tanks
ST Accident Response Capabilities
      Coordinating Committee
ST Accident Response Groups
ST Accidents
ST Action Levels
ST Activation (Emergency)
ST Active Portions
ST Active Ingredients
ST Acute Dermal LD 50
ST Acute LD 50
ST Acute Oral LD 50
ST Acute Toxicity
ST Additives
ST Aerial Measuring Systems
ST Agricultural Solid Wastes
ST Airborne Radioactive Materials
ST Aircraft Accidents
ST Aircraft Incidents
ST Alternate Emergency Operations
      Center
ST Ambient Air
ST Ancillary Equipment
ST Annual Limit on Intake
ST Anti-Microbial Agents
ST Applied Research
ST Approved Programs
ST Aquifers
ST Archaeological Resources
      Protection Act of 1979
ST Archaeological Resources
ST Area Sources
ST Articles
ST As Low As Reasonably Achievable
ST Asbestos
ST Asbestos Materials
ST Asbestos Mills
ST Asbestos Waste from Control
      Devices
ST Asbestos-Containing Waste
      Materials
ST Atmospheric Release Advisory
      Capability
ST Atomic Weapons
ST Atomic Energy Act Facilities
ST Attractants
ST Authorized Representatives
ST Background Soil Ph
ST Balers
ST Barrels
ST Barriers
ST Baseline Concentration
ST Batches
ST Below Regulatory Concern
ST Beryllium
ST Beryllium Alloys
ST Beryllium Ores
ST Beryllium-Containing Wastes
ST Best Available Control Technology
ST Bottom Ash
ST BRC
ST Burial Operations
ST By-products
ST CAIRS
ST Carriers

ST   Category A Reactors
ST   Categorical Exclusions
ST   CATS
ST   Cell
ST   Certification
ST   Certified Applicators
ST   Changed Use Pattern
ST   Chemical Substances
ST   Chemical Hazards Emergency
        Management System
ST   CHEMS
ST   Chronic Toxicity
ST   Claims
ST   Classified Matter
ST   Closed Portions
ST   Closure
ST   Coastal Waters
ST   Coastal Zones
ST   Cognizant Federal Agencies
ST   Cognizant Federal Agency Officials
ST   Commercial Asbestos
ST   Commercial Applicators
ST   Compliance Considerations
ST   Components
ST   Comprehensive Environmental
        Response, Compensation, etc.
ST   Computerized Accident/Incident
        Reporting System
ST   Computer Assisted Tracking
        System
ST   Conference of Federal
        Environmental Engineers
ST   Confined Aquifers
ST   Confinement Systems
ST   Consolidated Pre-Manufacture
        Notices
ST   Construction
ST   Contaminants
ST   Containers
ST   Containment
ST   Contamination
ST   Contiguous Zones
ST   Contingency Plans
ST   Continuity of Government
ST   Continuity of Government
        Emergency Management Teams
ST   Controlled Copies
ST   Controlled Areas
ST   Control Systems
ST   Control Substances
ST   Corrective Actions
ST   Corrective Action Management
        Units
ST   Corrosion Experts
ST   Cover Materials
ST   Criteria
ST   Critical Mass
ST   Daily Cover
ST   Damage Assessments
ST   Decommissioning
ST   Decontamination Areas
ST   Decontamination
ST   Defoliants
ST   Demolition
ST   Department of Energy Sites
ST   Depositories
ST   Derived Air Concentration
ST   Dikes
ST   Director of Emergency Operations
ST   Discharge of Pollutants
ST   Discharges

ST   Disposal
ST   Disposal Sites
ST   DOE Contractors
ST   DOE Operations
ST   Dose Equivalents
ST   Dose Commitments
ST   Draft Permits
ST   Drums
ST   Economic Poisons
ST   Effective Dose Equivalent
ST   Effluent Limitation Guidelines
ST   Effluents
ST   Effluent Monitoring
ST   EIS Implementation Plans
ST   Elementary Neutralization Units
ST   Emergency Management Team
        Directors
ST   Emergency Equipment
ST   Emergency Management
        Coordination Committee
ST   Emergency Management Teams
ST   Emergency Response
ST   Emergencies
ST   Emergency Control Stations
ST   Emissions
ST   Endangered Species Act
ST   Engineering Control Zones
ST   Environment
ST   Environmental Audits
ST   Environmental Protection Standards
ST   Environmental Monitoring
ST   Environmental Surveillance
ST   Environmental Assessments
ST   Environmental Impact Statements
ST   EOC
ST   EPA Hazardous Waste Numbers
ST   Equipment Room
ST   Equivalent Methods
ST   Exclusion Areas
ST   Exemptions
ST   Existing Sources
ST   Existing Hazardous Waste
        Management (HWM) Facilities
ST   Existing Tank Systems
ST   Facilities
ST   Factor Relationship and Sequence
        of Events Vocabulary
ST   Faults
ST   Feasibility Studies
ST   Federally Permitted Releases
ST   Federal Agencies
ST   Federal Facilities
ST   Federal Land Managers
ST   Field Organizations
ST   Field Elements
ST   Field Level Exemptions
ST   Final Closure
ST   Final Authorizations
ST   Final Cover
ST   Finding of No Significant Impact
ST   First Federal Officials
ST   Fissile Materials
ST   Fissile Classification
ST   Fissionable Materials
ST   Fissionable Materials Handlers
ST   Fixed Sources
ST   Floodplains
ST   Fly Ash
ST   Formula Quantities
ST   Free Liquids
ST   Freeboard

ST   Friable Asbestos Materials
ST   Fugitive Emissions
ST   Functional Appraisals
ST   Fungicides
ST   Generators (Pollution)
ST   Generic Significance
ST   General Purpose Facilities Projects
ST   Generic Exemptions
ST   General Plant Projects
ST   General Environment
ST   Geologic Repository Operations
        Areas
ST   Geologic Repositories
ST   Grate Siftings
ST   Groundwater
ST   Guide Specifications
ST   Halocarbons
ST   Halogenated Organic Compounds
ST   Hazardous Substances
ST   Hazardous Wastes
ST   Hazardous Waste Management
        Units
ST   Hazardous Waste Generation
ST   Hazardous Waste Management
ST   Hazardous Materials Response
        Teams
ST   Hazardous Constituents
ST   Hazards
ST   Hazardous Air Pollutants
ST   Hazardous Waste Management
        Facilities
ST   Hazardous Materials
ST   Health and Safety Studies
ST   Health Hazards
ST   Heavy Metals
ST   Herbicides
ST   High Radiation Areas
ST   High Efficiency Particulate Air
        Filters
ST   High-Grade Paper
ST   High-Level Wastes
ST   Hosts
ST   Human Environment
ST   HWM Facilities
ST   Hydrolysis
ST   Immediately Dangerous to Life and
        Health
ST   Imminent and Substantial
        Endangerment
ST   Imminent Danger
ST   Imminent Hazard
ST   Implementation Plans
ST   In Situ Volatization
ST   Inactive Portions
ST   Inactive Facilities
ST   Inactive Hazardous Waste Disposal
        Sites
ST   Incompatible Wastes
ST   Indirect Discharges
ST   Individual Generation Sites
ST   Industrial Furnaces
ST   Inert Ingredients
ST   Infectious Wastes
ST   Inground Tanks
ST   Inhalation LC 50
ST   Injection Wells
ST   Inland Waters
ST   Inland Zones
ST   Inner Liners
ST   Insecticides
ST   Installation Inspectors

ST Institutional Solid Wastes
ST Institution Control
ST Interim Authorizations
ST International Shipments
ST Internal Appraisals
ST Interagency Group on Energy
      Vulnerability
ST Interferences
ST Intermediate Cover
ST Intermediate Pre-Manufacture
      Notices
ST Intermediates
ST Interim (Permit) Status
ST Interim Actions
ST Inventory of Open Dumps
ST Laboratories
ST Land Disposal
ST Land Treatment Facilities
ST Land Disposal Units
ST Land Reclamation
ST Land Disposal Facilities
ST Landfill Cells
ST Landfills
ST Leachates
ST Lead Agencies
ST Leak Detection Systems
ST Licenses
ST Liners
ST Lithosphere
ST Manufacture
ST Material Safety Data Sheets
ST Maximum Normal Operating
      Pressure
ST Maximum Contaminant Levels
ST Metabolites
ST Mining Wastes
ST Miscellaneous Units
ST Mitigation
ST Mitigation Action Plans
ST Mixed Wastes
ST Mixtures
ST Modified Hazard Ranking System
ST Monitoring
ST Monthly NEPA Reports
ST Movement
ST Municipalities
ST Municipal Solid Wastes
ST National Pollutant Discharge
      Elimination System
ST National Environmental Policy Act
      Documents
ST National Capacity Variances
ST National Security
ST National Environmental Policy Act
ST Natural Resources
ST Natural Barriers
ST Natural Phenomena Emergencies
ST Natural Resource Damage
      Preassessment Screens
ST Navigable Waters
ST Near Surface Disposal
ST NEPA Compliance Guides
ST NEPA Compliance Officers
ST NEPA Documents
ST NEPA Status Reports
ST New Chemical Substances
ST New Hazardous Waste
      Management (HWM) Facilities
ST New Sources
ST New Tank Systems

ST Non-Contact Cooling Water
      Pollutants
ST Non-Contact Cooling Water
ST Non-Employee Radiation Workers
ST Non-Isolated Intermediates
ST Non-Stochastic Effects
ST Non-Target Organisms
ST Normal Form Radioactive Materials
ST Notice of Availability
ST Notice of Intent
ST Notice of Violation
ST Nuclear Regulatory Commission
ST Occupational Workers
ST Occupational Doses
ST Occurrences
ST Occurrence Reporting and
      Processing System
ST Office of Special Projects
ST Offshore Facilities
ST Oil Pollution Fund
ST On Ground Tanks
ST On-Scene Coordinators
ST On-Shore Facilities
ST On-Site Discharges
ST On-Site
ST Oncogenic
ST Open Burning
ST Open Dumps
ST Operable Units
ST Operators
ST Operations
ST Operational Emergencies
ST Operational Readiness Reviews
ST Operations Office Managers
ST Operating Basis Earthquake
ST OSC
ST Overfire Air
ST Owners
ST Oxygen Deficiencies
ST Packaging
ST Packages
ST Part B Permits
ST Part A Permits
ST Partial Closure
ST Performance Assessments
ST Permanent Exemptions
ST Permanent Variances
ST Permits
ST Permit (RCRA)
ST Permissible Exposure Limits
ST Personnel Expertise and Resource
      Listing
ST Pesticide Related Wastes
ST Pesticides
ST Physical Construction
ST Piles (Wastes)
ST Point Sources
ST Point of Compliance
ST Pollution
ST Pollutants
ST Polychlorinated Biphenyls
ST Positive Exposure
ST Post Consumer Wastes
ST Post Emergency Responses
ST POTW
ST Pre-Manufacture Notices
ST Precious Metals
ST Preliminary Designs
ST Pressure
ST Pretreatment
ST Primary Drinking Water Regulations

ST Private Applicators
ST Process (Toxic Chemical)
ST Process Waste Water
ST Process Waste Water Pollutants
ST Procurement Items
ST Procuring Agencies
ST Program Secretarial Officers
ST Program Senior Officials
ST Programmatic NEPA Documents
ST Propellants
ST PSO (Program Senior Official)
ST Public Water Systems
ST Published Exposure Levels
ST Publicly Owned Treatment Works
ST Public Lands
ST Pyrophoric Materials
ST Quality Assurance
ST Quality Factors
ST Quality Assurance Units
ST Quarterly Environmental
      Compliance Reports
ST Radiation Areas
ST Radionuclides
ST Radioactive Materials
ST Radioactive Wastes
ST Radioactive Mixed Wastes
ST Radiological Areas
ST Radioactivity
ST Raw Data
ST Reasonably Available Control
      Technology
ST Reclamation
ST Record of Decision (NEPA)
ST Record of Decision (CERCLA)
ST Recovered Resources
ST Reference Methods
ST Reference Standards
ST Reimbursement Period
ST Releases
ST Relevant and Appropriate
      Requirements
ST Remedial Investigations
ST Remedial Project Managers
ST Remedial Actions
ST Remote-Handled Transuranic
      Wastes
ST Removal
ST Renovation
ST Reportable Quantities
ST Representative Samples
ST Research and Development
ST Residential Solid Wastes
ST Resource Conservation
ST Resource Recovery
ST Resource Recovery Systems
ST Resource Recovery Facilities
ST Response Spectrum
ST Restricted Areas
ST Risks
ST Roentgen
ST Run-off
ST Run-on
ST Safe Mass
ST Safe Shutdown Earthquake 135
ST Safeguards and Security
      Emergencies
ST Safety Analysis Reports
ST Sanitary Landfills
ST Saturated Zones
ST Schedule of Compliance
ST Sealed Sources

ST Secondary Environmental Monitor
ST Secondary Drinking Water Regulations
ST Secretarial Officers
ST Section 5 Notices
ST Significant Environmental Compliance Issues
ST Significant Quantities
ST Significant Modification
ST Significance
ST SIMS
ST Site (DOE FRASE Vocabulary)
ST Site Characterization
ST Site Development and Facility Utilization Plans
ST Site-Limited Intermediates
ST Site-Wide NEPA Documents
ST Sludge
ST Small Quantity Generators
ST Solid Wastes
ST Solid Waste Management Facilities
ST Solid-Waste-Derived Fuel
ST Solvents
ST Source Materials
ST Specific Ionization
ST Special Form Radioactive Materials
ST Special Nuclear Material Scrap
ST Special Wastes
ST Specifications
ST Specimens
ST Special Nuclear Material of Low Strategic Significance
ST Specific Activity
ST Special Reviews
ST Special Nuclear Materials
ST Spent Materials
ST Spills
ST Standard of Performance
ST Standards
ST Standards Information Management System
ST State Coordination
ST State Notification
ST Strategic Special Nuclear Materials
ST Stratosphere
ST Subacute Dietary LC 50
ST Substantial
ST Sumps
ST Supplement Analyses
ST Supplemental EIS
ST Surface Impoundment
ST Surplus Facilities
ST Surveys
ST Tank Systems
ST Tanks
ST Technical Safety Appraisals
ST Technically Qualified Individuals
ST Temporary Exemptions
ST Temporary Variances
ST Test Substances
ST Test Systems
ST Thermal Treatment
ST Tiering
ST Totally Enclosed Treatment Facilities
ST Toxic Pollutants
ST Toxic Substances
ST Toxicity
ST Traceable
ST Transfer Facilities
ST Transport Vehicles

ST Transport Indices
ST Transuranic (TRU) Contaminated Materials
ST Transuranic Wastes (TRU Waste)
ST Transportation
ST Treatability Studies
ST Treatment
ST Treatment Zones
ST Treatment Facilities
ST Treatment Technique Requirements
ST Treatment Technologies
ST Trenching
ST Trust Fund (CERCLA)
ST Type A Quantities
ST Type B Quantities
ST Uncontrolled Hazardous Waste Sites
ST Underground Injection
ST Underground Tanks
ST Underground Injection Control
ST Underfire Air
ST Underground Sources of Drinking Water
ST Unsaturated Zones
ST Unusual Occurrences
ST Uppermost Aquifers
ST Validation
ST Variances
ST Vectors
ST Virgin Materials
ST Volatile Organic Compounds
ST Waste Management
ST Wastewater Treatment Units
ST Waste Containers
ST Waste Packages
ST Wastes
ST Waste Information Network
ST Waste Minimization
ST Waste Management Operations
ST Water Systems
ST Water Management Division Director
ST Water Table
ST Water Dumping
ST Well Injection
ST Wells
ST WIN
ST WMin
ST WMO

ENVIRONMENTAL PROTECTION AGENCY GLOSSARY
ST A-Scale Sound Levels
ST Abandoned Wells
ST Abatement
ST ABEL
ST Absorption (Waste)
ST Absorption (Radiation)
ST Absorption (Chemical)
ST Accelerators
ST Acceptable Daily Intake
ST Acceptable Intake for Chronic Exposure
ST Acceptable Intake for Subchronic Exposure
ST Accident Sites
ST Acclimatization
ST Acetylcholine
ST Acid Deposition
ST Acid Rain

ST Action Levels
ST Activated Carbon
ST Activated Sludges
ST Active Ingredients
ST Acute Exposure
ST Acute Toxicity
ST Adaptations
ST Add On Control Devices
ST Adhesion
ST Administrative Orders
ST Administrative Procedures Act
ST Adulterants
ST Advanced Waste Water Treatment
ST Advisory
ST Aeration
ST Aeration Tanks
ST Aerobic
ST Aerobic Treatment
ST Aerosols
ST Afterburners
ST Agent Orange
ST Agglomeration
ST Agglutination
ST Agricultural Pollution
ST Air Contaminants
ST Air Curtains
ST Air Mass
ST Air Monitoring
ST Air Pollutants
ST Air Pollution
ST Air Pollution Episodes
ST Air Quality Control Regions
ST Air Quality Criteria
ST Air Quality Standards
ST Airborne Particulates
ST Airborne Releases
ST Alachlor
ST Alar
ST Aldicarb
ST Algae
ST Algal Blooms
ST Alpha Particles
ST Alternate Methods
ST Ambient Air
ST Anadromous Fish
ST Anaerobic
ST Analytes
ST Antagonism
ST Antarctic "Ozone Hole"
ST Anti-Degradation Clause
ST Antibodies
ST Antigens
ST Aquifers
ST Arbitration
ST Area Sources
ST Area of Review
ST Asbestos
ST Asbestosis
ST Ashes
ST Assimilation
ST Atmosphere
ST Atomize
ST Attainment Areas
ST Attenuation
ST Attractants
ST Attrition
ST Autotrophic
ST Background Levels
ST Bacteria
ST Baffle Chambers
ST Baghouse Filters

ST Baling
ST Ballistic Separators
ST Band Application
ST Banking (Air Pollution)
ST Bar Screens
ST Barrier Coating(s)
ST Basal Application
ST BEN
ST Benthic Organisms (Benthos)
ST Benthic Region
ST Beryllium
ST Beta Particles
ST Bioaccumulative
ST Bioassays
ST Biochemical Oxygen Demand
ST Biodegradable
ST Biological Controls
ST Biological Magnification
ST Biological Oxidation
ST Biological Treatment
ST Biomass
ST Biomonitoring
ST Biosphere
ST Biostabilizers
ST Biotechnology
ST Biotic Communities
ST Black Lung
ST Blackwater
ST BOD5
ST Bogs
ST Booms
ST Botanical Pesticides
ST Bottle Bill
ST Bottomland Hardwoods
ST Brackish Waters
ST Broadcast Application
ST Bubble
ST Bubble Policy
ST Buffer Strips
ST Burial Grounds
ST By-products
ST Cadmium
ST Cancellations
ST Caps
ST Capture Efficiency
ST Carbon Adsorbers
ST Carbon Dioxide
ST Carbon Monoxide
ST Carboxyhemoglobin
ST Carcinogens
ST Carcinogen Risk Assessment
    Verification Endeavor Workgroup
ST Carrying Capacity
ST Casks
ST Catalytic Converters
ST Catalytic Incinerators
ST Catanadramous Fish
ST Categorical Exclusions
ST Categorical Pretreatment Standards
ST Cathodic Protection
ST Caustic Soda
ST CBOD5
ST Cells
ST Centrifugal Collectors
ST Cesium (Cs)
ST Channelization
ST Chemical Oxygen Demand
ST Chemical Treatments
ST Chemicals of Potential Concern
ST Chemosterilants
ST Chilling Effect

ST Chlorinated Hydrocarbons
ST Chlorinated Solvents
ST Chlorination
ST Chlorinators
ST Chlorine-Contact Chambers
ST Chlorofluorocarbons (CFCs)
ST Chlorosis
ST Chromium
ST Chronic RfD
ST Chronic Toxicity
ST Clarification
ST Clarifiers
ST Cleanup
ST Clear Cut
ST Cloning
ST Closed-Loop Recycling
ST Coagulation
ST Coastal Zones
ST Coefficient of Haze
ST Coffins
ST Coliform Index
ST Combined Sewers
ST Combustion
ST Combustion Products
ST Command Posts
ST Comment Periods
ST Comminuters
ST Comminution
ST Common Laboratory Contaminants
ST Community Water Systems
ST Community Relations
ST Compaction
ST Compliance Schedules
ST Compliance Coatings
ST Compost
ST Composting
ST Conditional Registration
ST Confined Aquifers
ST Consent Decrees
ST Conservation
ST Contaminants
ST Contact Pesticides
ST Contact Rate
ST Contingency Plans
ST Contour Plowing
ST Contract Laboratory Program
ST Contract Labs
ST Contract-Required Quantitation
    Limits
ST Contrails
ST Control Technique Guidelines
ST Conventional Pollutants
ST Conventional Systems
ST Coolants
ST Cooling Towers
ST Cooperative Agreements
ST Corrosion
ST Cost Recovery
ST Covers
ST Cover Materials
ST Crawl Spaces
ST Criteria
ST Criteria Pollutants
ST Cubic Feet Per Minute
ST Cultural Eutrophication
ST Cumulative Working Level Months
ST Curie
ST Cutie Pies
ST Cyclone Collectors
ST Data Call-In
ST DDT

ST Dechlorination
ST Decibel
ST Decomposition
ST Defoliants
ST Degradation
ST Delegated States
ST Delist
ST Denitrification
ST Depletion Curves
ST Depressurization
ST Dermal Toxicity
ST DES
ST Desalinization
ST Desiccants
ST Designated Pollutants
ST Designated Uses
ST Designer Bugs
ST Desulfurization
ST Detection Limits
ST Detergents
ST Developers
ST Developmental RfD
ST Diatomaceous Earth (Diatomite)
ST Diazinon
ST Dicofol
ST Differentiation
ST Diffused Air Process
ST Digesters
ST Digestion
ST Dikes
ST Dilution Ratio
ST Dinocap
ST Dinoseb
ST Dioxins
ST Direct Dischargers
ST Disinfectants
ST Dispersants
ST Disposal
ST Dissolved Oxygen
ST Dissolved Solids
ST Distillation
ST DNA
ST DNA Hybridization
ST Dose-Response Evaluation
ST Doses
ST Dosimeters
ST Dredging
ST Dumps
ST Dust
ST Dustfall Jars
ST Dystrophic Lakes
ST Ecological Impacts
ST Ecology
ST Economic Poisons
ST Ecosphere
ST Ecosystems
ST Effluent Limitations
ST Effluents
ST Electrodialysis
ST Electrostatic Precipitators
ST Eligible Costs
ST Eminent Domain
ST Emissions
ST Emission Factor
ST Emission Inventory
ST Emissions Trading
ST Endangered Species
ST Endangered Assessment
ST Enforcement Decision Documents
ST Enrichment
ST Environment

ST Environmental Audits
ST Environmental Assessments
ST Environmental Impact Statements
ST Environmental Response Teams
ST EPA
ST Epidemics
ST Epidemiology
ST Episodes (Pollution)
ST Equilibrium
ST Equivalent Methods
ST Erosion
ST Estuaries
ST Ethylene Dibromide
ST Eutrophic Lakes
ST Eutrophication
ST Evaporation Ponds
ST Evapotranspiration
ST Exclusionary
ST Exempt Solvents
ST Exposure
ST Exposure Assessment
ST Exposure Event
ST Exposure Pathways
ST Exposure Point
ST Exposure Point Concentration
ST Exposure Route
ST Fabric Filters
ST Feasibility Studies
ST Fecal Coliform Bacteria
ST Feedlots
ST Fens
ST Fermentation
ST Fertilizers
ST Field Sampling Plans
ST Filling
ST Filtration
ST Finding of No Significant Impact
ST First Draw
ST Flocculation
ST Flocs
ST Floor Sweeps
ST Flow Sand Filtration
ST Flue Gases
ST Flue Gas Desulfurization
ST Flumes
ST Fluorides
ST Fluorocarbons
ST Fluorosis
ST Flush
ST Fly Ash
ST Fogging
ST Food Chains
ST Formaldehyde
ST Formulation
ST Fresh Water
ST Fuel Economy Standard
ST Fugitive Emissions
ST Fumes
ST Fumigants
ST Functional Equivalent
ST Fungi
ST Fungicides
ST Game Fish
ST Gamma Radiation
ST Gasification
ST Geiger Counters
ST Gene Library
ST General Permits
ST Generators (Pollution)
ST Genes
ST Genetic Engineering

ST Germicides
ST Grain Loading
ST Granular Activated Carbon
     Treatment
ST Gray Water
ST Greenhouse Effect
ST Grinder Pumps
ST Gross Alpha Particle Activity
ST Gross Beta Particle Activity
ST Groundwater
ST Groundcovers
ST Habitats
ST Half-Life
ST Halogens
ST Halons
ST Hammermills
ST Hard Water
ST Hazardous Substances
ST Hazardous Wastes
ST Hazard Analysis
ST Hazard Wastes Characteristics
ST Hazard Quotient
ST Hazardous Air Pollutants
ST Hazardous Ranking System
ST Hazards Identification
ST Heat Island Effects
ST Heavy Metals
ST Heptachlor
ST Herbicides
ST Herbivores
ST Heterotrophic Organisms
ST High-Density Polyethylene
ST High-Level Radioactive Wastes
ST Holding Ponds
ST Hood Capture Efficiency
ST Hosts
ST Humus
ST Hybrid
ST Hybridoma
ST Hydrocarbons
ST Hydrogen Sulfide
ST Hydrogeology
ST Hydrology
ST Ignitability
ST Immediately Dangerous to Life and
     Health
ST Impoundment
ST In Vitro
ST In Vivo
ST Incinerators
ST Incineration
ST Incineration at Sea
ST Indirect Discharges
ST Indoor Air
ST Indoor Air Pollution
ST Indoor Climate
ST Inert Ingredients
ST Inertial Separators
ST Infiltration
ST Inflow
ST Influents
ST Information File
ST Injection Wells
ST Injection Zones
ST Inocula
ST Inorganic Chemicals
ST Insecticides
ST Inspection and Maintenance
ST Instream Use
ST Intake
ST Integrated Pest Management

ST Integrated Risk Information System
ST Interceptor Sewers
ST Interim (Permit) Status
ST Interstate Carrier Water Supplies
ST Interstate Waters
ST Interstitial Monitoring
ST Inversion
ST Ion Exchange Treatment
ST Ionization Chambers
ST Ionizing Radiation
ST Ions
ST Irradiated Food
ST Irradiation
ST Irrigation
ST Isotopes
ST Kinetic Rate Coefficient
ST Lagoons
ST Land Application
ST Land Farming (of waste)
ST Landfills
ST Lateral Sewers
ST LC 50
ST LD 0
ST LD 50
ST LD L0
ST Leachates
ST Leachate Collection Systems
ST Leaching
ST Lead
ST Lethal Concentration
ST Lethal Dose
ST Lifting Station
ST Lifts
ST Limestone Scrubbing
ST Limiting Factors
ST Limnology
ST Liners
ST Lipid Solubility
ST Liquefaction
ST List
ST Listed Wastes
ST LLW
ST Local Emergency Planning
     Committees
ST Low Level Radioactive Wastes
ST Lower Explosive Limit
ST Lowest Achievable Emission Rate
ST Lowest-Observed-Adverse-Effect
     Level
ST Major Modification
ST Major Stationary Source
ST Manufacturers Formulation
ST Marine Sanitation Devices
ST Marshes
ST Material Safety Data Sheets
ST Maximum Contaminant Levels
ST Mechanical Aeration
ST Mechanical Turbulence
ST Media
ST Mercury
ST Metabolites
ST Methane
ST Method 18
ST Method 24
ST Method 25
ST Microbes
ST Microbial Pesticides
ST Microorganisms
ST Million Gallons Per Day
ST Mists
ST Mitigation

ST Mixed Liquor
ST Mobile Sources
ST Model Plants
ST Modeling
ST Monitoring
ST Monitoring Wells
ST Monoclonal Antibodies
ST Muck Soils
ST Mulches
ST Multiple Uses
ST Mutagens
ST Mutate
ST National Pollutant Discharge Elimination System
ST National Ambient Air Quality Standards
ST National Emissions Standards for Hazardous Air Pollutants
ST National Oil and Hazardous Substances Contingency Plan
ST National Priorities List
ST National Response Centers
ST National Response Teams
ST Natural Gas
ST Natural Selection
ST Naturally Occurring Background Levels
ST Navigable Waters
ST Necrosis
ST Nematocides
ST Neutralization
ST New Sources
ST New Source Performance Standards
ST Nitrates
ST Nitric Oxides
ST Nitrification
ST Nitrilotriacetic Acid
ST Nitrites
ST Nitrogen Dioxide (NO$_2$)
ST Nitrogen Oxides (NO$_x$)
ST Nitrogenous Wastes
ST No-Observed-Effect-Level
ST Non-Attainment Areas
ST Non-Community Water Systems
ST Non-Conventional Pollutants
ST Non-detects
ST Non-Ionizing Electromagnetic Radiation
ST Non-Observed Adverse Effect Level
ST Non-Point Sources
ST Notice of Availability
ST Nuclear Power Plants
ST Nuclear Winter
ST Nutrients
ST Off-Site Facilities
ST Oil Fingerprinting
ST Oil Spills
ST Oligotrophic Lakes
ST On-Scene Coordinators
ST On-Site Facilities
ST Oncogenic
ST Opacity
ST Open Burning
ST Open Dumps
ST Operable Units
ST Operations and Maintenance
ST Organic
ST Organic Chemicals
ST Organic Matter
ST Organisms

ST Organophosphates
ST Organotins
ST Osmosis
ST Outfall
ST Overburden
ST Overland Flow
ST Overturn
ST Oxidants
ST Oxidation
ST Oxidation Ponds
ST Oxygenated Solvents
ST Ozonators
ST Ozone
ST Ozone Depletion
ST Packed Towers
ST Pandemic
ST Paraquat
ST Particulate Loading
ST Particulates
ST Pathogens
ST PCBs
ST Percolation
ST Permeability
ST Permits
ST Persistent Pesticides
ST Pesticides
ST Pesticide Tolerance
ST Pests
ST pH
ST Phenols
ST Pheromones
ST Phosphates
ST Phosphorus
ST Photochemical Smog
ST Photosynthesis
ST Physical and Chemical Treatments
ST Phytoplankton
ST Picocurie
ST Picocuries Per Liter
ST Pigs
ST Piles (Wastes)
ST Plankton
ST Plasmids
ST Plastics
ST Plugging
ST Plumes
ST Plutonium
ST Point Sources
ST Pollen
ST Pollution
ST Pollutants
ST Pollutant Standard Index
ST Polyelectrolytes
ST Polymers
ST Polyvinyl Chloride
ST Populations
ST Positive Data
ST Post-Closure
ST Potable Water
ST Potentially Responsible Parties
ST PPM/PPB
ST Precipitates
ST Precipitators
ST Precursors
ST Preliminary Assessment
ST Pressure Sewers
ST Pretreatment
ST Prevention of Significant Deterioration
ST Primary Drinking Water Regulations
ST Primary Waste Treatment

ST Process Weight
ST Proteins
ST Protoplasts
ST Public Water Systems
ST Publicly Owned Treatment Works
ST Pumping Stations
ST Putrescible
ST PVC
ST Pyrolysis
ST Quality Assurance Project Plans
ST Quality Control
ST Quantitation Limit
ST Quench Tanks
ST Radiation
ST Radiation Absorbed Dose
ST Radiation Standards
ST Radionuclides
ST Radioactive Substances
ST Radiobiology
ST Radon
ST Radon Decay Products
ST Rasps
ST Raw Sewage
ST Reasonably Available Control Technology
ST Receiving Waters
ST Recharge Areas
ST Recombinant DNA
ST Recombinant Bacteria
ST Recommended Maximum Contaminant Level
ST Reconstructed Sources
ST Record of Decision (NEPA)
ST Record of Decision (CERCLA)
ST Recycling
ST Red Borders
ST Red Tide
ST Reentry Intervals
ST Reference Dose (RfD)
ST Refuse Reclamation
ST Regeneration
ST Regional Response Teams
ST Registrants
ST Registration Standards
ST Registration
ST REM (Roentgen Equivalent Man)
ST Remedial Investigations
ST Remedial Project Managers
ST Remedial Actions
ST Remedial Design
ST Remedial Responses
ST Removal Actions
ST Reportable Quantities
ST Reregistrations
ST Reservoirs
ST Residual
ST Resource Recovery
ST Resource
ST Response Actions
ST Restoration
ST Restricted Use
ST Restriction Enzymes
ST Reuse
ST Reverse Osmosis
ST RFD
ST Ringlemann Chart
ST Riparian Habitats
ST Riparian Rights
ST Risk Assessment
ST Risk Management
ST Risk Communication

ST   ROD (NEPA Record of Decision)
ST   Rodenticides
ST   Roentgen Equivalent Man
ST   Rough Fish
ST   Routine Analytical Services
ST   Rubbish
ST   Run-off
ST   Salt Water Intrusion
ST   Salts
ST   Salvage
ST   Sample Management Offices
ST   Sampling and Analysis Plans
ST   Sand Filters
ST   Sanitary Surveys
ST   Sanitary Sewers
ST   Sanitation
ST   Saturated Zones
ST   Scrap
ST   Screening
ST   Scrubbers
ST   Secondary Treatment
ST   Secondary Drinking Water
       Regulations
ST   Secure Maximum Contaminant
       Levels
ST   Sedimentation
ST   Sedimentation Tanks
ST   Sediments
ST   Selective Pesticides
ST   Semi-Confined Aquifers
ST   Senescence
ST   Septic Tanks
ST   Service Connectors
ST   Settling Chamber
ST   Settling Tanks
ST   Settleable Solids
ST   Sewage
ST   Sewage Sludge
ST   Sewers
ST   Sewerage
ST   Shotgun
ST   Signal Words
ST   Significant Deterioration
ST   Significant Municipal Facilities
ST   Significant Violations
ST   Silt
ST   Silviculture
ST   Sinking
ST   Site Inspections
ST   Siting
ST   Skimming
ST   Slope Factor
ST   Sludge
ST   Slurries
ST   Smelters
ST   Smog
ST   Smoke
ST   Soft Detergents
ST   Soft Water
ST   Soil Adsorption Field
ST   Soil Conditioner
ST   Soil Gas
ST   Solder
ST   Sole Source Aquifer
ST   Solid Wastes
ST   Solid Waste Management
ST   Solid Waste Disposal
ST   Solidification and Stabilization
ST   Solvents
ST   Soot
ST   Sorption

ST   Species
ST   Special Analytical Services
ST   Special Reviews
ST   Spill Prevention Control and
       Countermeasures Plan
ST   Spoil
ST   Sprawl
ST   Stabilization
ST   Stable Air
ST   Stack Effect
ST   Stacks
ST   Stagnation
ST   Standards
ST   State Emergency Response
       Commission
ST   State Implementation Plans
ST   Stationary Sources
ST   Sterilization
ST   Storage
ST   Storm Sewers
ST   Stratosphere
ST   Strip Cropping
ST   Strip Mining
ST   Subchronic RfD (RfD)
ST   Sulfur Dioxide
ST   Sumps
ST   Superfund
ST   Surface Impoundment
ST   Surface Waters
ST   Surfactants
ST   Surveillance Systems
ST   Suspended Solids
ST   Suspension
ST   Suspension Cultures
ST   Swamps
ST   Synergism
ST   Synthetic Organic Chemicals
ST   Systemic Pesticides
ST   Tailings
ST   Target Compound List
ST   Technology-Based Standards
ST   Teratogens
ST   Terracing
ST   Tertiary Treatment
ST   Thermal Pollution
ST   Threshold Planning Quantities
ST   Threshold Limit Values
ST   Tidal Marshes
ST   Tolerances
ST   Total Exposure Points
ST   Total Suspended Solids
ST   Toxic Chemical Release Forms
ST   Toxic Clouds
ST   Toxic Pollutants
ST   Toxic Substances
ST   Toxicants
ST   Toxicity
ST   Transformation
ST   Transpiration
ST   Trash-To-Energy Plans
ST   Treatment, Storage, and Disposal
       Facilities
ST   Trichloroethylene (TCE)
ST   Trickling Filters
ST   Trihalomethane
ST   Troposphere
ST   Trust Fund (CERCLA)
ST   Tundra
ST   Turbidimeters
ST   Turbidity
ST   Ubiquitous Background Levels

ST   Ultra Clean Coal
ST   Ultraviolet Rays
ST   Underground Storage Tanks
ST   Underground Sources of Drinking
       Water
ST   Unsaturated Zones
ST   Uranium
ST   Urban Runoff
ST   Vaccines
ST   Vapors
ST   Vapor Capture Systems
ST   Vapor Dispersion
ST   Vapor Plumes
ST   Vaporization
ST   Variances
ST   Vectors
ST   Ventilation
ST   Vinyl Chloride
ST   Viruses
ST   Volatile
ST   Volatile Organic Compounds
ST   Volatile Synthetic Organic
       Chemicals
ST   Vulnerable Zones Radius
ST   Vulnerability Analyses
ST   Vulnerable Zones
ST   Wastes
ST   Waste Load Allocations
ST   Waste Treatment Plants
ST   Waste Treatment Stream
ST   Wastewater
ST   Wastewater Operations and
       Maintenance
ST   Water Pollution
ST   Water Quality Criteria
ST   Water Quality Standards
ST   Water Solubility
ST   Water Suppliers
ST   Water Supply System
ST   Water Table
ST   Watershed
ST   Weight of Evidence
ST   Well Injection
ST   Well Plugs
ST   Wells
ST   Wetlands
ST   Wildlife Refuges
ST   Wood-Burning Stove Pollution
ST   Working Level
ST   Working Level Month
ST   Xenobiotic
ST   Zooplankton

FIREARMS
ST   Ammunition
ST   Armorers
ST   Authorized Weapons
ST   Automatic Rifles
ST   Blank Ammunition
ST   Blank Fire Adapter (BFA)
ST   Bullet Containment Devices
ST   Central Training Academy
ST   Clearing Barrels
ST   Contractors
ST   Controllers
ST   Defective Firearm
ST   Diversionary Devices
ST   Double Action Semiautomatic
       Pistols
ST   Dry Firing

ST Duds
ST Engagement Simulation Systems
ST ESS Blanks
ST Field Elements
ST Firearms Ranges
ST Flares
ST Flash Grenades
ST Force-on-Force Exercises
ST Grenade Launchers
ST Guards
ST Handguns
ST Hangfires
ST Inhabited Building Distance
ST Laser Eye Safety Distance
ST Law Hazard Zone
ST Law Simulators
ST Light Antitank Weapon
ST Live Round Excluders
ST Machine Guns
ST Machine Pistols
ST Magazines
ST Misfires
ST Multiple Integrated Laser
   Engagement System
ST Munitions
ST Net Explosive Weight
ST Night Simulations
ST Night Vision Goggles
ST Pistols
ST Protective Force Personnel
ST Public Traffic Route Distance
ST Quantity Distances
ST Range Masters
ST Range Safety Officers
ST Ranges (Firearms)
ST Rifles
ST Safety Analysis Reports
ST Security Facilities
ST Security Inspectors
ST Semiautomatic Firearms
ST Senior Controllers
ST Shadow Forces
ST Shadow Force Weapons
ST Shotgun, Pump
ST Shotgun, Semiautomatic
ST Simulators
ST Single Action Semiautomatic Pistols
ST Small Arms
ST Smoke Grenades
ST Special Weapons
ST Submachine Guns, Closed Bolt
ST Submachine Guns, Open Bolt
ST Tagging
ST Unauthorized Discharges
ST Weapon Simulators

FIRES
ST Consultant Fire Protection Survey
   Program
ST Fire Protection
ST Fire Protection Systems
ST Improved Risk
ST Maximum Credible Loss
ST Maximum Possible Fire Loss
ST National Security
ST Property
ST Property Loss

HAZARDOUS MATERIALS
ST Administrators

ST Agencies
ST Allegations
ST Annual Document Logs
ST Annual Reports
ST Asbestos
ST Asbestos-containing Materials
ST Atmosphere Gases
ST Authorized Inspectors
ST Barges
ST Batches
ST Best Management Practices
ST Blasting Agents
ST Bottles
ST Break-bulk
ST Bulk Packaging
ST Bureau of Explosives
ST By-products
ST Capacitors
ST Captain of the Port
ST Carfloats
ST Cargo Aircraft Only
ST Cargo Tanks
ST Cargo Vessels
ST Carriers
ST Certification
ST Chemical Substances
ST Chemical Waste Landfills
ST Chemical Ammunition
ST Class A Explosives
ST Class B Explosives
ST Class C Explosives
ST Close Reflection By Water
ST Combustible Liquids
ST Commercial Storers of PCB Waste
ST Competent Authorities
ST Compressed Gases
ST Compressed Gases in Solution
ST Consumer Commodities
ST Contaminants
ST Containerships
ST Containment Vessels
ST Control Substances
ST Conventional Pollutants
ST Corrosive Materials
ST Cryogenic Liquids
ST Cylinders
ST Depleted Uranium
ST Designated Facilities
ST Discharge of Pollutants
ST Disposal
ST Disposal Packages
ST Disposer of PCB Waste
ST District Commander
ST Double Wash/Rinse
ST Effluent Limitations
ST Effluent Limitation Guidelines
ST Emergency Situations
ST Engines
ST Environment
ST EPA
ST EPA Guidance Document
ST EPA Identification Number
ST Etiologic Agents
ST Experimental Start Date
ST Experimental Termination Date
ST Explosive Projectiles
ST Explosive Bombs
ST Explosive Mines
ST Explosive Torpedoes
ST Facilities
ST Ferry Vessels

ST Fissile Materials
ST Fissile Classification
ST Fissile Class I
ST Fissile Class II
ST Fissile Class III
ST Flammable Compressed Gases
ST Flammable Liquids
ST Flammable Solids
ST Flash Point
ST Fluorescent Light Ballasts
ST Freight Containers
ST Friable Asbestos-Containing
   Materials
ST Friable Asbestos Materials
ST Fuel Tanks
ST Grenades
ST Gross Weight
ST Hazardous Substances
ST Hazardous Wastes
ST Health and Safety Studies
ST Hermetically Sealed
ST High Concentration PCBs
ST High-Contact Industrial Surfaces
ST Highway Route Controlled Quantity
ST Impurities
ST Incinerators
ST Industrial Buildings
ST Intermodal Portable Tanks
ST Irritating Materials
ST Jet Thrust Units
ST Known Human Effects
ST Large High Voltage Capacitors
ST Large Low Voltage Capacitors
ST Leaks
ST Limited Quantity
ST Liquefied Compressed Gases
ST Liquids
ST Local Educational Agencies
ST Low Concentration PCBs
ST Low Specific Activity Materials
ST Low Specific Activity
ST Magazine Vessels
ST Manned Control Centers
ST Manufacture
ST Market/Marketers
ST Marks
ST Maximum Normal Operating
   Pressure
ST Migration
ST Mineral Oil PCB Transformers
ST Mixtures
ST Mode
ST Moderators (Nuclear)
ST Motor Vehicles
ST Motor Fuels
ST Municipal Solid Wastes
ST Name of Contents
ST Navigable Waters
ST Net Weight
ST New Chemical Substances
ST New Sources
ST Non-bulk Packaging
ST Non-Flammable Compressed
   Gases
ST Non-fixed Radioactive
   Contamination
ST Non-Impervious Solid Surfaces
ST Non-Liquefied Compressed Gases
ST Non-PCB Transformers
ST Non-Restricted Access Areas
ST Normal Form Radioactive Materials

ST NOS
ST Not Otherwise Specified
ST NRC (Non-Reuseable Container)
ST Offshore Facilities
ST Oils
ST On-Shore Facilities
ST On-Site
ST Operators
ST Optimum Interspersed
   Hydrogenous Moderation
ST Organic Peroxides
ST Other Regulated Material
ST Other Restricted Access
   (nonsubstation) Locations
ST Outage
ST Outdoor Electrical Substations
ST Outside Container
ST Overpack
ST Owners
ST Oxidizing Materials
ST Packaging
ST Passenger-carrying Aircraft
ST Passenger Vessels
ST PCB Articles
ST PCB Containers
ST PCB Equipment
ST PCB Items
ST PCB Transformers
ST PCB Transformer Rupture
ST PCB Waste Generator
ST PCB Wastes
ST PCBs
ST Permits
ST Placarded Cars
ST Point Sources
ST Poisons
ST Pollution
ST Pollutants
ST Portable Tanks
ST Posing an Exposure Risk to Food
   or Feed
ST Pounds per Square Inch
ST Pounds per Square Inch Gauge
ST Preferred Routes
ST Preferred Highway
ST Primary Coolants
ST Private Track
ST Private Siding
ST Process (Toxic Chemical)
ST Process for Commercial Purposes
ST Proper Shipping Name
ST Public Vessels
ST Pyrophoric Liquids
ST Qualified Incinerators
ST Quality Assurance Units
ST Radioactive Materials
ST Radioactive Articles
ST Radioactive Contents
ST Radon
ST Rail Freight Cars
ST Raw Data
ST Recycled PCBs
ST Reference Substances
ST Removal
ST Reportable Quantities
ST Requirements
ST Research and Special Programs
   Administration
ST Residues
ST Response Actions
ST Responsible Parties

ST Retailers
ST Retrofill
ST Rocket Ammunition
ST Schools
ST Sheathing
ST Shipping Papers
ST Significant Adverse Reactions
ST Single Strip Containers
ST Small Capacitors
ST Small Manufacturers
ST Small Quantities for Research and
   Development
ST Small Arms Ammunition
ST Soils
ST Special Form Radioactive Materials
ST Specimens
ST Spills
ST Spill Area
ST Spill Boundaries
ST Spontaneously Combustible
   Materials
ST Standard of Performance
ST Standard Cubic Foot
ST Standard Wipe Test
ST Standards
ST Starter Cartridges
ST States
ST State-Designated Routes
ST State Routing Agencies
ST Stowage
ST Strong Outside Containers
ST Studies
ST Study Completion Date
ST Study Directors
ST Study Initiation Date
ST Technical Names
ST Test Substances
ST Test Systems
ST Testing Requirements Rule
ST Testing Requirements
ST Totally Enclosed Manner
ST Toxic Pollutants
ST Trailership
ST Trainships
ST Transfer Facilities
ST Transport Vehicles
ST Transport Indices
ST Transporter of PCB Waste
ST Ullage
ST Unit Load Devices
ST United States
ST Vehicles
ST Vessels
ST Virgin Materials
ST Viscous Liquids
ST Volatility
ST Waste Oils
ST Waters of the United States
ST Water Reactive Materials
ST Water Resistant

INDUSTRIAL HYGIENE
ST Absorbed Dose
ST ALI
ST Annual Effective Dose Equivalent
ST Annual Limit on Intake
ST As Low As Reasonably Achievable
ST Atomic Energy Act
ST Best Available Technology
ST Collective Dose Equivalent

ST Collective Effective Dose Equivalent
ST Committed Dose Equivalent
ST Committed Effective Dose
   Equivalent
ST Controlled Areas
ST Cumulative Annual Effective Dose
   Equivalent
ST Deep Dose Equivalent
ST Derived Air Concentration
ST Derived Concentration Guide
ST DOE Contractors
ST DOE Operations
ST Dose Equivalents
ST Effective Dose Equivalent
ST Environment, Safety, and Health
   (ES&H) Program
ST Environmental Surveys
ST Environmental Audits
ST Environment, Safety, and Health
   Overview
ST ES&H
ST Exceptions
ST Extremities
ST Federal Employee Occupational
   Safety and Health Program
ST Field Organizations
ST Generic Exemptions
ST Hazardous Waste Facilities
ST Headquarters Coordinating Teams
ST Laboratories
ST Lens of Eye Dose Equivalents
ST Line Organizations
ST Minimum Requirements and
   Standards
ST Monitoring
ST Non-Stochastic Effects
ST Occupational Workers
ST Occupational Medicine
ST Occupational Medical Program
ST Operating Level
ST PEL
ST Permissible Exposure Limits
ST Program Senior Officials
ST Public Doses
ST Quality Factors
ST Radiation Workers
ST Radioactive Wastes
ST Radioactive Mixed Wastes
ST Radiological Areas
ST Shallow Dose Equivalents
ST Stochastic Effects
ST Technical Safety Appraisals
ST Total Recordable Cases
ST Transuranic (TRU) Contaminated
   Materials
ST Transuranic Wastes (TRU Waste)
ST Triage
ST TSA
ST Unreviewed Safety Questions
ST Variances
ST Weighting Factors

INDUSTRIAL SAFETY
ST Category A Reactors
ST Complaints
ST Compliance Inspections
ST Contractors
ST Controls (Reactor)
ST Contracting Officers

ST Contracting Officer's
   Representative
ST Contractor Employees
ST Department of Energy Operations
ST Discrimination
ST Disposal Systems
ST Emergency Systems
ST Employees' Representative
ST Engineered Safety Features
ST Exceptions
ST Field Organizations
ST Field Elements
ST Government-Owned
   Contractor-Operated Facilities
ST Health Examinations
ST Imminent Danger
ST Initial Startup
ST Inspections (Nuclear)
ST Intruder Barriers
ST Modification
ST Monitored Workers
ST Monitored Personnel Locator File
ST National Institute for Occupational
   Safety and Health
ST Non-Employee Radiation Workers
ST Occupational Safety and Health
   Administration
ST Operators
ST Operable
ST PEL
ST Permanent Variances
ST Permissible Exposure Limits
ST Pressure
ST Program Offices
ST Program Secretarial Officers
ST Radiation Protection
ST Reactor Facilities
ST Reactor Operations
ST Reactor Operator
ST Reactor Projects
ST Reactor Supervisor
ST Risks
ST Safe Mass
ST Safety Analysis Reports
ST Safety Analysis
ST Safety Guides
ST Safety Documents
ST Safety Reviews
ST Safety and Health Directors
ST Senior Reactor Operator
ST Significant Quantities
ST Significant Modification
ST Standby
ST Supervisors
ST Technical Specifications
ST Temporary Variances
ST Total Recordable Cases
ST TS
ST Under Construction
ST Undue Risks
ST Unreviewed Safety Questions
ST Verification of Training and
   Retraining

MANAGEMENT
ST Action Plans
ST Action Description Memoranda
ST Activities
ST Administrative Consent Order
ST Alaska Power Administration

ST Aliquots
ST Argonne National Laboratory
ST Assessments
ST Assessment Team Leaders
ST Assessment Plans
ST Atomic Energy Defense Activities
ST Best Management Practice
   Findings
ST Best Practical Technology
ST Boards
ST Bonneville Power Administration
ST Budget and Reporting Code
ST Chemical Hazards Emergency
   Management System
ST Civilian Nuclear Activities
ST Code of Federal Regulations
ST Committees
ST Companies
ST Compliance Findings
ST Concerns
ST Condition Assessment Surveys
ST Contractors
ST Contracting Officers
ST Coordination Processes
ST Corrective Action Reporting
ST Corrective Repair Maintenance
ST Criteria
ST Critical Organs
ST Critical Population Group
ST Department of Energy Contractors
ST Department of Energy Sites
ST Detailed Operating Procedures
ST Divisions
ST DOE Contractors
ST DOE Facility Representatives
ST DOE Operations
ST DOE Representatives
ST Dose Equivalents
ST Dose Commitments
ST Effluents
ST Effluent Monitoring
ST Emergency Systems
ST Engineering Change Proposal
ST Environment, Safety, and Health
   (ES&H) Program
ST Environmental Detection Limits
ST Environmental Occurrences
ST Environmental Surveys
ST Environmental Audits
ST Environmental Protection Standards
ST Environmental Monitoring
ST Environmental Surveillance
ST Environmental Assessments
ST Environmental Impact Statements
ST Environment Assessments
ST EOP
ST ES&H
ST Exceptions
ST Facilities
ST Facility Managers
ST Failures
ST Field Organizations
ST Findings
ST Finding of No Significant Impact
ST Fiscal Year
ST Functional Appraisals
ST Functional Units
ST Generic Significance
ST General Design Criteria
ST Goals
ST Government Standards

ST Government-Owned
   Contractor-Operated Facilities
ST Graded Approach
ST Groups
ST Heads of Field Operations
ST Heads of Headquarters Elements
ST Implementation Plans
ST Information Systems
ST Installed Equipment
ST Institutes
ST Internal Appraisals
ST Interagency Committee on
   Standards Policy
ST Investigations
ST Investigation Report
ST Issue Identification
ST Landlords
ST Line Organizations
ST Lower Limit of Detection
ST Maintenance Backlog
ST Maintenance Management
ST Management Appraisals
ST Management System Factors
ST Management and Organization
   Assessment
ST Mandatory Standards
ST Manifests
ST Migration
ST Milestones
ST Monitored Visitors
ST National Environmental Policy Act
   Documents
ST Noteworthy Practices
ST Notice of Involvement
ST Notice of Violation
ST Notification
ST Occurrences
ST Occurrence Reports
ST Offices
ST Office of the Assistant Secretary for
   Nuclear Energy
ST Office of the Assistant Secretary for
   Fossil Energy
ST Office of the Assistant Secretary for
   Defense Programs
ST Office of the Assistant Secretary for
   Conservation et.al.
ST Office of the Assistant Secretary for
   Environment, et. al.
ST Office of Environmental Restoration
   and Waste Management
ST Office of New Production Reactors
ST Office of Defense Waste and
   Transportation Management
ST Office of Special Projects
ST Office of Basic Energy Sciences
ST Office of Civilian Radioactive Waste
   Management
ST Office of Energy Research
ST Office of Health and Environmental
   Research
ST Office of Management and Budget
ST Office of Technology Development
ST Office of Weapons Production
ST On-Site Discharges
ST Operators
ST Operating Level
ST Operations
ST Operating Procedures
ST Organizations
ST Other Equipment

ST Outlier
ST Performance Objectives
ST Permanent Exemptions
ST Personal Property
ST Positive Exposure
ST Post Removal Site Control
ST Potential Hazards
ST Power Marketing Administrations
ST Predictive Maintenance
ST Production Facilities
ST Program Offices
ST Program Secretarial Officers
ST Program Senior Officials
ST Program
ST Program Organizations
ST Programmatic Equipment
ST Public Notices
ST Radiation Records Repositories
ST Radiation Exposure Module
ST Real Property
ST Record of Decision (NEPA)
ST Recorded Doses
ST Reference Standards
ST Regulatory Organizations
ST Related Personal Property
ST Reliability-Centered Maintenance
ST Risks
ST ROD (NEPA Record of Decision)
ST Root Cause Analysis
ST Safety and Health Assessments
ST Sanitized
ST Second Line Organization Level
ST Self-Assessment
ST Senior Management Officials
ST Senior Controllers
ST Significant Environmental
      Compliance Issues
ST SIMS
ST Site Closure and Stabilization
ST SOP (Management)
ST Source Control Maintenance
      Measures
ST Southwestern Power Administration
ST Southeastern Power Administration
ST Specific Control Factors
ST Spiked Samples
ST Standard Operating Procedures
ST Standard Curve
ST Standard Deviation
ST Standards Information Management
      System
ST Standard Practice Procedures
ST Substantial
ST Supervisors
ST Superfund Memorandum of
      Agreement
ST Superfund State Contracts
ST Support Agencies
ST Support Agency Coordinators
ST Technical Experts
ST Technical Support
ST Temporary Exemptions
ST Terminated Employees
ST Test Authorizations
ST Test Area North
ST Tiger Team Assessments
ST Tiger Teams
ST Tiger Team Assessment Program
ST Tiger Team Assessment Reports
ST Tiger Team Leaders
ST Tiger Team Administrators

ST Total Recordable Cases
ST Trained Investigators
ST Training
ST Trending
ST Undue Risks
ST Universities
ST Unusual Occurrences
ST Unusual Occurrence Reports
ST User Agencies
ST Utilization Facilities
ST Voluntary Standards Bodies
ST Voluntary Standards
ST Western Area Power Administration

NATURAL PHENOMENON
ST Act of Nature
ST Act of God
ST Body Wave Magnitude
ST Compressional Waves
ST Differential Settlement
ST Dip-Slip
ST Divergent Windstorms
ST Downbursts
ST Downslope Wind
ST Dust Devils
ST Earthquake Magnitude
ST Flow Failures
ST Ground Cracks
ST Ground Failures
ST Gust Fronts
ST Horizontal Displacements
ST Landslides
ST Lateral Spreads
ST Liquefaction
ST Local Magnitude (M_L)
ST Love Seismic Waves
ST Microbursts
ST Moment Magnitude
ST Oblique Slip
ST Primary Body Waves
ST Quick Clays
ST Rayleigh Seismic Waves
ST Richter Magnitude (M_L)
ST Rotational Windstorms
ST Secondary Body Waves
ST Strike Slip
ST Subsidence
ST Subvortices
ST Surface Faulting
ST Surface Wave Magnitude
ST Tectonic Deformations
ST Tornadoes
ST Tsunamis
ST Uplifts
ST Waterspouts

NUCLEAR FACILITIES INCIDENT DATABASE
ST 10 MDLM
ST 3 AC Flow
ST Abnormal Condition Control
ST Action Charlie
ST Automatic Backup Shutdown of the
      Safety Computer
ST Backup Systems
ST Beetles
ST Blanket Assembly
ST Bottom End Fitting
ST Bottom Fitting Insert
ST Bypasses

ST CANDIS Computer Tapes
ST Carbon Dioxide Spaces
ST Check Valves
ST CLDTMP
ST Containment Storage Tanks
ST Control Rod Fault
ST Cutie Pies
ST D Machines
ST D&E Canal
ST Delta K Rods
ST Delta P
ST Delta T
ST DIL
ST DNO
ST Dog Houses
ST Dry Caves
ST ED Charge
ST Emergency Spray Water System
ST Failed Element Monitors
ST FLOCHK
ST Forests (Reactor)
ST G-O-S
ST Gangs
ST Golf Bags
ST Green Slugs
ST HARP
ST IMM
ST Inert Atmosphere
ST Ink
ST IR Circuit
ST Kanne Alarm
ST Leader Seater
ST Magneforming
ST Megger Testing
ST MSCANS
ST N-Unit
ST Noise Pollution Abatement
ST O-Unit
ST Occupational Exposure
ST One-Inch Components
ST OX2 Assembly
ST PAD
ST PAR Ponds
ST PILOT
ST Pistol Grip Switches
ST Plugging
ST Polybor
ST Power Losses
ST Rabbits
ST Raincoats
ST Reactor Shutdown
ST Septifoils
ST Shaft Break Limit
ST Shaft Guides
ST Sparger Assemblies
ST Spiders
ST Sweeping the Gas Plenum
ST Test Stations
ST Torpedoes
ST TRIAC
ST TSCANS
ST ZENER Diodes

QUALITY ASSURANCE
ST Accuracy
ST American National Metric Council
ST Analytical Batches
ST Appraisals
ST Blanks
ST Calibration Checks

ST Check Samples
ST Data Quality Assessments
ST Department's Procurements
ST Department's Metric Coordinator
ST DOE Programs
ST Field Operations Plans
ST Field Audits
ST Hard Conversion
ST Implementation Plans
ST Inch-Pound System of Units
ST International System of Units
ST Laboratory Audits
ST Matrix/Spike-Duplicate Analysis
ST Method Quantification Limits
ST Metric System
ST Metrication Operating Committee
ST Metric Transition Committee
ST Metric Transition Plan
ST Metrication
ST MQL
ST Precision
ST Quality Assurance Plans
ST Quality Assurance Overviews
ST Reagent Grade
ST Replicate Samples
ST Sampling Plans
ST Soft Conversion
ST Standard Curve
ST Surrogates
ST System Audits
ST Verification
ST Water

RADIATION
ST Abnormal Exposure Conditions
ST Absorption (Radiation)
ST Absorbed Dose
ST Absorbed Dose Rate
ST Accelerator Produced Radioactive
    Materials
ST Accidents
ST Action Charlie
ST Activation (Emergency)
ST Activation (Nuclear)
ST Activity (Nuclear)
ST Activity-Median Aerodynamic
    Diameter
ST Advanced Test Reactor
ST ALI
ST Alpha Decay Radioisotopes
ST Alpha Particles
ST Alpha Rays
ST Alternate Removal Systems
ST Annihilation (Electron)
ST Annual Dose Equivalent Limit
ST Annual Limit on Intake
ST Antimony 125
ST Approved Medical Practitioners
ST Atoms
ST Authorized Limits
ST Automatic Incident Actions
ST Autoradiographs
ST Axial Power Monitors
ST Background Radiation
ST Backup Systems
ST Becquerel
ST Beetles
ST Benefaction
ST Beryllium 7
ST Beta Decay Radioisotopes

ST Beta Decay
ST Beta Particles
ST Beta Rays
ST Beta-Minus Decay Radioisotopes
ST Bioassay Procedures
ST Biological Clearance Rate
ST Biological Half-Life
ST Blanket Assembly
ST Body Content
ST Bone Seekers
ST Bremsstrahlung
ST Bronchial Epithelium
ST Buckled Zones
ST Buildup Factor
ST Bypasses
ST Calibration
ST CANDIS Computer Tapes
ST Carbon Dioxide Spaces
ST Carrier Free
ST Cerium 144
ST Cerium 141
ST Cesium 137
ST Cesium 134
ST Charged Particle Equilibrium
ST Check Valves
ST Cobalt 60
ST Cobalt 58
ST Collective Effective Dose Equivalent
ST Collective Effective Dose
    Equivalent Commitment
ST Collective Effective Dose
    Equivalent Rate
ST Committed Dose Equivalent
ST Competent Authorities
ST Computer Inoperative Limits
ST Containment
ST Containment Storage Tanks
ST Contaminated Water Storage
ST Contamination
ST Controlled Areas
ST Control Systems
ST Control Rod Fault
ST Cost-Effective Analysis
ST Count (Radiation Measurement)
ST Critical Organs
ST Critical Group
ST Critical Pathways
ST Criticality
ST Criticality Accidents
ST Criticality Excursions
ST Curie
ST Cutie Pies
ST D Machines
ST D&E Canal
ST Daily Instruction Logs
ST Days Living Radioisotopes
ST de minimis
ST Decontamination
ST Decontamination Factor
ST Deionized Water
ST Delayed Neutron Precursors
ST Delayed Proton Precursors
ST Delta K Rods
ST Delta P
ST Delta T
ST Deposition
ST Derived Air Concentration
ST Derived Limits
ST Detriment
ST Differential Cost Benefit Analysis
ST Differential in Hours

ST Direct Bioassays
ST Directly Ionizing Particles
ST Do Not Operate Tags
ST Dog Houses
ST DOP Tested
ST Dose Equivalents
ST Dose Equivalent Index
ST Dose Limit
ST Dose Rate Meters
ST Dose Upper Bounds
ST Doses
ST Dosimeters
ST Dry Caves
ST ED Charge
ST Effective Dose Equivalent
ST Effective Dose Equivalent
    Commitment
ST Effective Half-Life
ST Efficiency (Counters)
ST Electrons
ST Electron Capture
ST Electron Volt
ST Element 104 Isotopes
ST Element 105 Isotopes
ST Element 106 Isotopes
ST Element 107 Isotopes
ST Element 108 Isotopes
ST Element 109 Isotopes
ST Emergency Exposure
ST Emergency Spray Water System
ST Emergency Reference Levels
ST Emergency Systems
ST Energy Fluence
ST Energy Fluence Rate
ST Energy Flux
ST Environmental Exposure
ST Equilibrium
ST Exposure
ST Exposure Rate
ST Exposure Pathways
ST Extraordinary Nuclear Occurrences
ST Failed Element Monitors
ST Failures
ST Fertile
ST Film Badges
ST Filters (Radiology)
ST Fissile
ST Fission Products
ST Forests (Reactor)
ST Gamma Rays
ST Gang Temperature Monitor
ST Gangs
ST Gas Volume Ratio
ST Geiger-Mueller Counters
ST Geometry
ST Golf Bags
ST Gray
ST Green Slugs
ST Half Value Layer (Half Thickness)
ST Half-Life
ST HARP
ST Health Physics
ST Heap-leach Extraction
ST Heavy Ion Decay Radioisotopes
ST Hours Living Radioisotopes
ST IMM
ST In-Situ Extraction
ST Indirectly Ionizing Particles
ST Induced Radioactivity
ST Inert Atmosphere
ST Ingestion

ST Inhalation
ST Ink
ST Intake
ST Intervention Levels
ST Inverse Square Law
ST Investigation Levels
ST Iodine 131
ST Ionization
ST Ionization Chambers
ST Ionizing Radiation
ST Ions
ST IR Circuit
ST Isomeric Transition Isotopes
ST Isotopic Tracers
ST Isotopes
ST Kanne Alarm
ST Kerma
ST Labeled Compounds
ST Limits
ST Linear Energy Transfer
ST Lung Classes
ST Magneforming
ST Main and Trim
ST Manganese 54
ST Mass Attenuation Coefficient
ST Mass Energy Absorption Coefficient
ST Mass Energy Transfer Coefficient
ST Maximum Permissible
   Concentration
ST Maximum Permissible Dose
ST Medical Exposure
ST Megger Testing
ST Microsec Living Radioisotopes
ST Milliroentgen
ST Millisec Living Radioisotopes
ST Minutes Living Radioisotopes
ST Monitoring
ST N-Unit
ST Nanosec Living Radioisotopes
ST Natural Background Radiation
ST Naturally Occurring Radioactive
   Material
ST Neutron-Deficient Isotopes
ST Neutron-Rich Isotopes
ST Neutron Poisons
ST Noise Pollution Abatement
ST Non-Stochastic Radiation Effects
ST Nuclear Regulatory Commission
ST Nuclear Poisons
ST Nuclides
ST O-Unit
ST Occupational Exposure
ST One-Inch Components
ST Operating Organizations
ST Operational (Radiation) Limits
ST OX2 Assembly
ST Packaging
ST PAD
ST PAR Ponds
ST Particle Fluence
ST Particle Flux
ST Peak-To-Peak Ratio
ST Phantoms
ST Phosphate Rocks
ST Piles (Nuclear)
ST Pistol Grip Switches
ST Planned Special Exposure
ST Plugging
ST Polybor
ST Power Losses
ST Power Burst Facility

ST Prescribed Limits
ST Primary Limits
ST Producing (Special Nuclear
   Materials)
ST Production Facilities
ST Protective Barriers
ST Proton Decay Radioisotopes
ST Quality Factors
ST Rabbits
ST Radiation Protection
ST Radiation Induced Hereditary
   Effects
ST Radiation Induced Genetic Effects
ST Radiant Energy
ST Radiation Protection Officers
ST Radiation Safety
ST Radiation Shields
ST Radiation Exposure Module
ST Radioactive Waste Management
ST Radionuclides
ST Radioactivity
ST Radionuclide Barriers (Natural or
   Engineered)
ST Radioactive Contamination
ST Radioactivity Decay Constant
ST Radioactive Effluents
ST Radioactive Half-Life
ST Radioactive Nuclide Intake
ST Radiological Monitoring
ST Radioactive Decay
ST Radioactive Series
ST Radioactive Source Terms
ST Radiological Surveys
ST Radiotoxicity
ST Radium 226
ST Radium 228
ST Radon
ST Radon Flux
ST Radon Progeny
ST Radon 222
ST Raffinate
ST Raincoats
ST Reactor Shutdown
ST Reactor Materials
ST Recording Levels
ST Reduction
ST Reference Levels
ST Reference Man
ST Relative Biological Effectiveness
ST REM (Roentgen Equivalent Man)
ST REM (Radiation Exposure Module)
ST Remedial Measures
ST Remote Detection and Control
   Systems
ST Reprocessing
ST Retention
ST Risks
ST Roentgen
ST Roof-Top-Ratio
ST Ruthenium 103
ST Sampling
ST Scintillation Counters
ST Sealed Sources
ST Seconds Living Radioisotopes
ST Secondary Limits
ST Selection and Monitoring Chassis
ST Selenium 75
ST Septifoils
ST Shaft Break Limit
ST Shaft Guides
ST Shielding Materials

ST Shutdown Cooling
ST Shutdown Heat Exchanger
ST Shutdown Sequencer
ST Side-Stream Extraction
ST Sievert
ST Smear
ST Somatic Radiation Effects
ST Source Materials
ST Sparger Assemblies
ST Specific Ionization
ST Special Health Supervision
ST Specific Activity
ST Spiders
ST Stochastic Radiation Effects
ST Supervised Areas
ST Surveillance
ST Sweeping the Gas Plenum
ST SWP Conditions
ST Tailings
ST Tellurium 132
ST Test Stations
ST Test and Evalution Facilities
ST Thermoluminescent Dosimeters
ST Thorium
ST Thorium 230
ST Tissue Equivalents
ST Tolerances
ST Torpedoes
ST Total Mass Stopping Power
ST Total Indicated Runout
ST TRAM Values
ST Transuranic Radioactive Wastes
ST Transuranic Radionuclides
ST Transuranic Nuclides
ST TRIAC
ST Tritium
ST Unattached Fractions
ST Uptake
ST Uranium
ST Uranium 238
ST Uranium Mill Tailings
ST Utilization Facilities
ST Very Low Flow Constant
ST Vibrating Reed Electrometers
ST Weighting Factors
ST Working Conditions
ST Working Level
ST Working Level Month
ST Working Condition A
ST Working Condition B
ST X-rays
ST Years Living Radioisotopes
ST Yellowcake
ST ZENER Diodes
ST Zinc 65
ST Zirconium 95

STANDARDS
ST Accredited
ST Accrediting Boards
ST Accreditation Review Teams
ST Compliance Considerations
ST Contractors
ST Contractor Self-Evaluation Reports
ST Critical Mass
ST DOE Representatives
ST Exceptions
ST Exemptions
ST Field Organizations
ST Field Level Exemptions

ST Fissionable Materials
ST Fissionable Materials Handlers
ST Government Standards
ST Interagency Committee on
   Standards Policy
ST Level 1 Compliance
ST Level 2 Compliance
ST Level 3 Compliance
ST Line Organizations
ST Mandatory Standards
ST Nuclear Criticality Safety
ST Nuclear Facilities
ST Nuclear Criticality
ST Occupational Safety and Health
   Administration
ST Operations
ST Operational Safety Requirements
ST Operational Readiness Reviews
ST Performance Testing Laboratories
ST Performance-Based Training
ST Permanent Exemptions
ST Personnel Dosimetry Programs
ST Personnel Dosimeters
ST Potential Hazards
ST Program Offices
ST Reference Standards
ST Self-Evaluation
ST Seriousness
ST Standards-developing Groups
ST Standby
ST Subject Matter Experts
ST Technical Experts
ST Temporary Exemptions
ST Training Accreditation Program
   Staff
ST Training Programs
ST Training Program Accreditation
   Plan
ST Voluntary Standards Bodies
ST Voluntary Standards

SYSTEM SAFETY DEVELOPMENT CENTER GLOSSARY

ST Acceptable Risks
ST Accidents
ST Accident Prone Situation
ST Accident Types
ST Act of Nature
ST Action Propensity
ST Actual Cash Value
ST Actuarial
ST Advanced Test Reactor
ST Amelioration
ST Analytical (Logic) Trees
ST Analyses
ST AND gate
ST As Low As Reasonably Achievable
ST As Low As Practicable
ST Asphyxiation
ST Assigned Risks
ST Assumed Risks
ST Attractive Nuisance
ST Audits (SSDC)
ST Backfit Reviews
ST Barriers
ST Basic Causes
ST Behavioral Stereotypes
ST Breakthrough
ST Catalogs
ST Causes

ST Change
ST Change Analyses
ST Codes, Standards, and Regulations
ST Computer Stores
ST Conditional Probability
ST Conditional AND gate
ST Conditional OR Gate
ST Configuration Control
ST Confidence
ST Consequential Losses
ST Constraints
ST Containment
ST Contingent Liability
ST Controls (Reactor)
ST Control Barriers
ST Contractual Liability
ST Contributing Causes
ST Control Systems
ST Coupling
ST Critical Incident Technique
ST Critical Jobs (task)
ST Cumulative Frequency (probability)
ST Danger
ST De Minimus Violations
ST Defects
ST Delphi Process
ST Density Function
ST Derived Projection
ST Detailed Operating Procedures
ST Diagrams
ST Difference Deviation
ST Disabilities
ST Disabling Injuries
ST Distribution
ST Drafting Break
ST Empirical Correlates
ST Energy
ST Energy Flow
ST Energy Trace
ST Ergonomics
ST Error Sampling
ST Events
ST Events and Causal Factors
ST Evidence
ST Extreme Value
ST Extreme Value Projection
ST Fact
ST Factory Mutual
ST Failure Mode and Effect Analysis
ST Faults
ST Fault Tree
ST Feedback
ST Feedback Loop
ST Fiduciaries
ST First Aid
ST Flow Charts
ST Frequency Distribution
ST Frequency/Severity
ST Functional Appraisals
ST Guidelines
ST Hazards
ST Hazard Analysis
ST Hazard Classifications
ST High Potential Incidents
ST Histograms
ST Human Engineering
ST Human Performance
ST Human Reliability
ST Human Factors
ST Human Factors Engineering
ST Immediate Causes

ST Imminent Danger
ST Incidence Rate, Total Recordable
   Cases (TRC)
ST Incidence Rate, Lost Workday
   Cases (LWC)
ST Incidence Rate, WDL
ST Incidence Rate, WDLR
ST Incidence Rate, LWD (Lost Work
   Days)
ST Incidents
ST Incurred Losses
ST Indemnify
ST Independent (Safety) Reviews
ST Innovation Diffusion
ST Inspections
ST Inspections (Nuclear)
ST Intangible Risks
ST Internal Audits
ST Investigations
ST Job Safety Analysis
ST Key Word
ST Knowledge of Results
ST Known Precedents
ST KOR
ST Less than Adequate
ST Line Organizations
ST Line of Balance
ST Line Management
ST Logic Gates (Boolean)
ST Logical Correlates
ST Logic Gates
ST Logic Trees
ST Loss Control Management
ST Loss Ratio
ST Management Appraisals
ST Management Oversight and Risk
   Tree
ST Management by Objective
ST Management Systems
ST Management
ST Maps
ST Matrices
ST Matrix Organizations
ST Maximum Foreseeable Loss
ST Maximum Probable Loss
ST Maximum Possible Loss
ST Mean Annual Loss
ST Mean
ST Medical Treatment
ST Minor Accidents
ST Mishaps
ST Mode
ST Monitoring
ST MORT
ST Negligence
ST Occupational Injuries
ST Occupational Illnesses
ST Occurrences
ST Operational Readiness
ST OR gate
ST Oversights and Omissions
ST Perceived Risks
ST Permanent Partial Disabilities
ST Permanent Total Disabilities
ST PERT
ST Physical Harm
ST Policies
ST Positive (objective) Trees
ST Preliminary Safety Analysis
ST Priority Problem List
ST Probable Causes

ST Probability
ST Procedures
ST Product Liability
ST Program Evaluation and Review
   Technique
ST Project Managers
ST Proper Job Analysis
ST Proper Job Instruction
ST Proximate Causes
ST Quality Assurance
ST Range (Statistical)
ST Recommendations
ST Redundance
ST Reflex Arc Responses
ST Reliability and Quality Assurance
ST Replacement Values
ST Reported Significant Observations
ST Residual Risks
ST Retention
ST Return Period
ST Reviews
ST Risk Analysis
ST Risk Assessment
ST Risk Management
ST Risks
ST Root Causes
ST Safe
ST Safe Work Permit
ST Safety Analysis Reports
ST Safety Appraisals
ST Safety Barriers
ST Safety
ST Safety (Hazard) Analysis
ST Safety Analysis Process
ST Safety Assurance System
   Summary
ST Safety Precedence Sequence
ST Schematics
ST Self-Insurance
ST Sensors
ST Serious Violations
ST Single Failure Analysis
ST Software
ST Staff Organization
ST Standard Deviation
ST Standard Job Procedures
ST Stimulus-Mediation-Response
ST Stimulus-Response
ST Subordinate Management
ST Summation Gates
ST Surveys
ST System (Safety) Analyses
ST System Safety
ST System Safety Development Center
ST Targets
ST Task Analysis
ST Taxonomy
ST Threshold Limit Values
ST Time Loss (T/I) Analysis
ST Time Weighted Average
ST Tolerance Limits
ST Top Management
ST Torts
ST Total Recordable Cases
ST Trained Investigators
ST Trigger
ST Trigger Event
ST Uncertainty
ST Upstream Processes
ST Values
ST Variances

ST Vocabulary
ST Wipe Out
ST Witnesses
ST Work Processes

WASTES
ST Aboveground Tanks
ST Aboveground Releases
ST Absorption (Waste)
ST Accessible Environment
ST Accidental Occurrences
ST Active Life
ST Active Portions
ST Active Maintenance
ST Acts
ST Administrators
ST Aeration Zones
ST Ancillary Equipment
ST Anticipated Processes and Events
ST Applicable Standards and
   Limitations
ST Approved Programs
ST Appropriate Act and Regulations
ST Approved States
ST Aquifers
ST Assets
ST Authorized Representatives
ST Belowground Releases
ST Below Regulatory Concern
ST Bottom Ash
ST BRC
ST Cathodic Protection
ST Cathodic Protection Testers
ST Certification
ST Clean Water Act
ST Closed Portions
ST Closure Plans
ST Closure
ST Closure Period
ST Compatible
ST Components
ST Confined Aquifers
ST Connected Piping
ST Construction
ST Consumptive Use
ST Containers
ST Contingency Plans
ST Cooperative Agreements
ST Corrosion Experts
ST Current Assets
ST Current Liabilities
ST Current Closure Cost Estimate
ST Current Post-Closure Cost Estimate
ST Current Plugging and
   Abandonment Cost Estimates
ST Daily Cover
ST Demonstration
ST Depositories
ST Designated Facilities
ST Dielectric Materials
ST Dikes
ST Directors
ST Discarded Materials
ST Discharges
ST Disposal Facilities
ST Disposal
ST Disposal Areas
ST Disposal Packages
ST Disposal Systems
ST Disturbed Zones

ST Draft Permits
ST Effluent Monitoring
ST Electrical Equipment
ST Elementary Neutralization Units
ST Engineering Control Zones
ST Engineered Barrier Systems
ST Environmental Surveillance
ST EPA
ST EPA Hazardous Waste Numbers
ST EPA Identification Number
ST EPA Region
ST Equivalent Methods
ST Excavation Zone
ST Existing Hazardous Waste
   Management (HWM) Facilities
ST Existing Tank Systems
ST Existing Facilities
ST Extraordinary Nuclear Occurrences
ST Facilities
ST Facility Mailing Lists
ST Farm Tanks
ST Federal Agencies
ST Federal Land Managers
ST Final Closure
ST Final Authorizations
ST Flow-Through Process Tanks
ST Fly Ash
ST Food-chain Crops
ST Free Liquids
ST Free Products
ST Freeboard
ST Functionally Equivalent
   Components
ST Gathering Lines
ST General Permits
ST Generators (Pollution)
ST General Environment
ST Groundwater
ST Halogenated Organic Compounds
ST Hazardous Waste Management
   Units
ST Hazardous Substance UST System
ST Hazardous Waste Generation
ST Hazardous Waste Management
ST Hazardous Constituents
ST Hazardous Waste Discharge
ST Hazardous Waste Management
   Facilities
ST Heating Oils
ST HWM Facilities
ST Hydraulic Lift Tanks
ST Hydrogeologic Units
ST Implementing Agencies
ST Impoundment
ST Inactive Portions
ST Incinerators
ST Incompatible Wastes
ST Inconsistencies
ST Indian Tribes
ST Indian Governing Bodies
ST Individual Generation Sites
ST Industrial Furnaces
ST Infectious Wastes
ST Inground Tanks
ST Injection Interval
ST Injection Wells
ST Inner Liners
ST Installation Inspectors
ST Interim Authorizations
ST International Shipments
ST Interstate Agencies

ST	Intermunicipal Agencies	ST	PSD Permits	ST	Transmissive Faults
ST	Intruder Barriers	ST	Publicly Owned Treatment Works	ST	Transmissive Fractures
ST	Land Disposal	ST	Radioactive Waste Management	ST	Transportation
ST	Land Treatment Facilities	ST	Radioactive Wastes	ST	Transporters
ST	Land Disposal Facilities	ST	Radioactive Mixed Wastes	ST	Treatability Studies
ST	Landfills	ST	Radioactivity	ST	Treatment
ST	Leachates	ST	Re-refined Oils	ST	Treatment Zones
ST	Leak Detection Systems	ST	Re-refined Oil	ST	Treatment Technologies
ST	Legal Defense Costs	ST	Recoverable	ST	U.S. Environmental Protection
ST	Liabilities	ST	Recovered Materials		Agency
ST	Liners	ST	Recovered Resources	ST	Underground Storage Tanks
ST	Liquid Traps	ST	Recycled Oils	ST	Underground Areas
ST	Lithosphere	ST	Reference Man	ST	Underground Injection
ST	Long Term Contracts	ST	Regional Authority	ST	Underground Releases
ST	Lubricating Oils	ST	Regional Administrator	ST	Underground Tanks
ST	Maintenance	ST	Regulated Substances	ST	Underground Sources of Drinking
ST	Major Facilities	ST	Regulated Activities		Water
ST	Major PSD Stationary Source	ST	Releases	ST	United States
ST	Management	ST	Release Detection	ST	Unsaturated Zones
ST	Manifest Document Number	ST	Release of Property	ST	Upgrades
ST	Manifests	ST	Remedial Actions	ST	Uppermost Aquifers
ST	Medical Wastes	ST	Representative Samples	ST	Used Oils
ST	Mining Wastes	ST	Residential Tanks	ST	UST Systems
ST	Mixed Wastes	ST	Residual Radioactive Materials	ST	Vessels
ST	Motor Fuels	ST	Resource Conservation	ST	Virgin Materials
ST	Movement	ST	Resource Recovery	ST	Waste Management
ST	Municipalities	ST	Resource Recovery Systems	ST	Wastewater Treatment Units
ST	National Pollutant Discharge	ST	Resource Recovery Facilities	ST	Waste Isolation Pilot Plant
	Elimination System	ST	Run-off	ST	Waste Water Collection Systems
ST	Near Surface Disposal Facilities	ST	Run-on	ST	Wastewater Treatment Tanks
ST	Net Working Capital	ST	Sanitary Landfills	ST	Waste Minimization
ST	Net Worth	ST	Saturated Zones	ST	Waste Management Operations
ST	New Facilities	ST	Schedule of Compliance	ST	Water(bulk shipment)
ST	New Hazardous Waste	ST	Septic Tanks	ST	WIN
	Management (HWM) Facilities	ST	Settleable Solids	ST	WMin
ST	New Tank Systems	ST	Sewers	ST	WMO
ST	New Tank Components	ST	Sewerage		
ST	Non-Operational Storage Tanks	ST	Sludge		
ST	Non-Sudden Accidental	ST	Small Quantity Generators		
	Occurrences	ST	Soil Column	**WATER POLLUTION**	
ST	Off-site	ST	Solid Wastes	ST	Abandoned Wells
ST	On Ground Tanks	ST	Solid Waste Management	ST	Acts
ST	On-Site	ST	Solid Waste Management Facilities	ST	Administrators
ST	Open Burning	ST	Solid Waste Planning	ST	Agencies
ST	Open Dumps	ST	Source Control Maintenance	ST	Algae Bloom
ST	Operational Life		Measures	ST	Appropriate Act and Regulations
ST	Operators	ST	Special Source of Ground Water	ST	Approved State Primacy Program
ST	Overfill Releases	ST	States	ST	Aquatic Environment
ST	Owners	ST	State Director	ST	Aquifers
ST	Parent Corporations	ST	State Hazardous Wastes	ST	Artificial Reefs
ST	Partial Closure	ST	State Authority	ST	Assets
ST	PCBs	ST	State/EPA Agreements	ST	Best Available Technology
ST	Permits	ST	Stochastic Effects	ST	BIA (Bureau of Indian Affairs)
ST	Permit-by-Rule	ST	Storage	ST	Casings
ST	Permits Necessary to Begin	ST	Storm Water	ST	Catastrophic Collapses
	Physical Construction	ST	Substantial Business Relationships	ST	Cementing
ST	Personnel	ST	Sudden Accidental Occurrences	ST	Class II Wells
ST	Petroleum UST Systems	ST	Sumps	ST	Clean Water Act
ST	Petroleum	ST	Superfund Memorandum of	ST	Coagulation
ST	Physical Construction		Agreement	ST	Community Water Systems
ST	Piles (Wastes)	ST	Surface Impoundment	ST	Confining Beds
ST	Pipeline Facilities	ST	Tangible Net Worth	ST	Confining Zones
ST	Piping	ST	Tank Systems	ST	Confluent Growth
ST	Point Sources	ST	Tanks	ST	Contaminants
ST	Pollutant Discharge Elimination	ST	Test and Evalution Facilities	ST	Contaminant Carriers
	System	ST	Thermal Treatment	ST	Conventional Filtration Treatment
ST	Polychlorinated Biphenyls	ST	Totally Enclosed Treatment	ST	Conventional Mines
ST	Post Removal Site Control		Facilities	ST	CTcalc
ST	Procurement Items	ST	Transfer Stations	ST	Cultivating
ST	Procuring Agencies	ST	Transfer Facilities	ST	Current Assets
ST	Protective Action Guides	ST	Transport Vehicles	ST	Current Liabilities

ST Current Plugging and
       Abandonment Cost Estimates
ST Dams
ST Deny
ST Diatomaceous Earth Filtration
ST Dikes
ST Directors
ST Direct Filtration
ST Discharge of Dredged Material
ST Discharge of Fill Material
ST Discharge Point
ST Discretionary Authority
ST Disinfectants
ST Disinfectant Contact Time
ST Disinfection
ST Disposal Sites
ST Disposal Wells
ST Dose Equivalents
ST Dredged Materials
ST Drinking Water Coolers
ST EPA
ST Exempted Aquifers
ST Existing Class II Wells
ST Existing Injection Wells
ST Extraction Sites
ST Facilities
ST Faults
ST Federal Agencies
ST Fill Materials
ST Filtration
ST Flocculation
ST Flow Rates
ST Fluids
ST Formations
ST Formation Fluid
ST General Permits
ST Generators (Pollution)
ST Gross Alpha Particle Activity
ST Gross Beta Particle Activity
ST Groundwater
ST Halogens
ST Hazardous Wastes
ST Hazardous Waste Management
       Facilities
ST Headwaters
ST High Tide Line
ST HWM Facilities
ST Hydrolysis
ST Hydrogeologic Units
ST Indian Tribes
ST Individual Permits
ST Injection Interval
ST Injection Wells

ST Injection Zones

ST Injection
ST Interstate Agencies
ST Lakes
ST Lead Free
ST Legionella
ST Letter of Permission
ST Liabilities
ST Lithology
ST Local Educational Agencies
ST Man-made Beta Particle and
       Photon Emitters
ST Maximum Contaminant Levels
ST Maximum Contaminant Level Goal
ST Maximum Total Trihalomethane
       Potential (MTP)
ST Minor Drainage
ST Mixing Zones
ST Municipalities
ST Net Working Capital
ST Net Worth
ST New Class II Wells
ST Non-Community Water Systems
ST Non-Transient Non-Community
       Water System
ST Operators
ST Ordinary High Water Mark
ST Owners
ST Packers
ST Parent Corporations
ST Performance Evaluation Samples
ST Permits
ST Permitting Authorities
ST Picocurie
ST Plugging Records
ST Plugging and Abandonment Plans
ST Plugging
ST Point of Disinfectant Application
ST Point-of-entry Treatment Devices
ST Point-of-use Treatment Devices
ST Pollution
ST Pollutants
ST Practicable
ST Pressure (Surface)
ST Pretreatment Facilities
ST Primary Drinking Water Regulations
ST Primary Enforcement Responsibility
ST Principal Source Aquifers
ST Prohibit Specification
ST Public Water Systems
ST Publicly Owned Treatment Works
ST Radioactive Wastes
ST Regional Administrator

ST REM (Roentgen Equivalent Man)
ST Residual Disinfectant Concentration

ST Safe Drinking Water Act
ST Sanitary Surveys
ST Schedule of Compliance
ST Schools
ST Secondary Drinking Water
       Regulations
ST Secondary Maximum Contaminant
       Levels
ST Sedimentation
ST Slow Sand Filtration
ST Special Aquatic Sites
ST Special Source of Ground Water
ST Standard Sample
ST States
ST State Director
ST State Primary Drinking Water
       Regulation
ST State Program Revisions
ST Strata
ST Subsidence
ST Surface Casings
ST Surface Waters
ST Tangible Net Worth
ST TBT Paints (Trybutilin)
ST Territorial Seas
ST Tidal Waters
ST Total Dissolved Solids
ST Total Trihalomethanes (TTHM)
ST Transmissive Faults
ST Transmissive Fractures
ST Treatment Technique Requirements
ST Treatment Technologies
ST Trihalomethane
ST U.S. Environmental Protection
       Agency
ST UIC
ST Unacceptable Adverse Effects
ST Underground Injection
ST Underground Injection Control
ST Underground Sources of Drinking
       Water
ST Viruses
ST Waters of the United States
ST Water Systems
ST Waterborne Disease Outbreaks
ST Water Bloom
ST Well Injection
ST Well Monitoring
ST Well Plugs
ST Well Stimulation
ST Well Workovers
ST Wells
ST Wetlands

DOE FRASE Vocabulary

ABDOMEN
(1115 ABDOMEN)
DA November 28, 1990
BT1 Trunk
 BT2 Human Body Parts
RT Groin
SO DOE FRASE VOCABULARY

ABRASION
(1300 AB)
DA November 28, 1990
BT1 Injuries
RT Razor(s)
SO DOE FRASE VOCABULARY

ACID SUIT
(2650 ACID SUIT)
DA January 3, 1991
BT1 Anticontamination Clothing
 BT2 Clothing
 BT2 Personal Protective Equipment
 BT3 Equipment/Parts - Personal
  Protective (DOE FRASE
  Vocabulary)
 BT4 Equipment
SO DOE FRASE VOCABULARY

**ACTIVITY TYPES (DOE FRASE
VOCABULARY)**
(DOE FRASE Vocabulary Numeric Keys
 1225-1299)
DA November 28, 1990
BT1 Activities
NT1 Building/Equip Maint/Repair
  Activity
NT1 Classified Activity
NT1 Construction Activity
NT1 Decommissioning Activity
NT1 Decontamination Activity
NT1 Emergency Response Activity
NT1 Equipment Installation Activity
NT1 Food Service Activity
NT1 Fuel Handling Activity
NT1 Grounds Maintenance Activity
NT1 Inspection/Monitoring Activity
NT1 Janitorial/Housekeeping Activity
NT1 Material Handling Activity
NT1 Mining/Drilling Activity
 NT2 Strip Mining
NT1 No Activity
NT1 Office Activity
NT1 Other Non-Task Activity
NT1 Physical Fitness Training Activity
NT1 Pre Start-Up/ Calibration Activity
NT1 Production/Operation Activity
NT1 Reactor Refueling Activity

NT1 Recreation/Break Activity
NT1 Research/Testing Activity
NT1 Security Activity
NT1 Training/Education Activity
NT1 Transportation Activity
NT1 Travel Activity
NT1 Unknown/Undetermined Activity
NT1 Vehicle Maint/Repair Activity
RT DOE FRASE Categories
SO DOE FRASE VOCABULARY
DEF A subject category used with the
  DOE FRASE Vocabulary.

ADAPTER(S)
(2375 ADAPTER)
DA January 3, 1991
BT1 Equipment/Parts - Electrical (DOE
  FRASE Vocabulary)
 BT2 Equipment
SO DOE FRASE VOCABULARY

**ADMIN. SUPPORT/CLERICAL
EMPLOYEE**
(0450 OFFICE;CLERK)
DA November 28, 1990
BT1 Occupations
BT1 Personnel
NT1 Firefighter
NT1 Food Service Employee
NT1 Janitor
NT1 Misc Service Employee
NT1 Security Guard
RT Office Activity
SO DOE FRASE VOCABULARY

AGITATOR
(2100 AGITATOR)
DA December 10, 1990
BT1 Machines (DOE FRASE
  Vocabulary)
 BT2 Equipment
SO DOE FRASE VOCABULARY

AGRICULTURAL MACHINE
(2101 AGRICULTURAL)
DA December 10, 1990
BT1 Machines (DOE FRASE
  Vocabulary)
 BT2 Equipment
SO DOE FRASE VOCABULARY

AGRICULTURE PERSONNEL
DA November 29, 1990
BT1 Occupations
BT1 Personnel

NT1 Forest Worker
NT1 Groundskeeper
NT1 Misc Agriculture Employee
SO DOE FRASE VOCABULARY

AIR COMPRESSOR
(2102 AIR COMPRESS)
DA December 10, 1990
BT1 Compressor
 BT2 Machines (DOE FRASE
  Vocabulary)
 BT3 Equipment
SO DOE FRASE VOCABULARY

AIR CONDITIONER
(2475 AIR CONDITIO)
DA January 3, 1991
SO DOE FRASE VOCABULARY

AIR DRYER
(2103 AIR DRYER)
DA December 10, 1990
BT1 Machines (DOE FRASE
  Vocabulary)
 BT2 Equipment
SO DOE FRASE VOCABULARY

AIR HOIST
(2300 AIR HOIST)
DA December 10, 1990
BT1 Hoist(s)
 BT2 Hoisting Apparatus
 BT3 Material Handling Device
 BT4 Devices
 BT4 Equipment/Parts - Material
  Handling (DOE FRASE
  Vocabulary)
 BT5 Equipment
SO DOE FRASE VOCABULARY

AIR LOCK ROOM
(1675 AIR LOCK ROO)
DA December 10, 1990
BT1 Room
 BT2 Sites/Areas
SO DOE FRASE VOCABULARY

AIR MASK
(2651 AIR MASK)
DA January 3, 1991
BT1 Personal Protective Equipment
 BT2 Equipment/Parts - Personal
  Protective (DOE FRASE
  Vocabulary)
 BT3 Equipment

RT Dust Mask
RT Nose
SO DOE FRASE VOCABULARY

AIR TAMPER
(2950 AIR TAMPER)
DA   January 3, 1991
BT1 Tools - Powered
 BT2 Tools (DOE FRASE Vocabulary)
  BT3 Equipment
SO DOE FRASE VOCABULARY

AIRCRAFT PILOT
(0825 AIR PILOT)
DA   November 28, 1990
BT1 Transport Personnel
 BT2 Occupations
 BT2 Personnel
RT Aircraft
SO DOE FRASE VOCABULARY

ALLEN WRENCH
(3025 ALLEN WRENCH)
DA   January 3, 1991
BT1 Wrench(s)
 BT2 Tools (DOE FRASE Vocabulary)
  BT3 Equipment
SO DOE FRASE VOCABULARY

ALLEY
(1600 ALLEY)
DA   December 10, 1990
BT1 Site (DOE FRASE Vocabulary)
 BT2 Sites/Areas
SO DOE FRASE VOCABULARY

AMMETER(S)
(2750 AMMETER)
DA   January 3, 1991
BT1 Equipment/Parts -
     Instrumentation/Measuring (DOE
     FRASE Voc.)
 BT2 Equipment
BT1 Indicator(s)
 BT2 Instrument(s)
  BT3 Equipment/Parts -
       Instrumentation/Measuring
       (DOE FRASE Voc.)
   BT4 Equipment
SO DOE FRASE VOCABULARY

AMPLIFIER(S)
(2376 AMPLIFIER)
DA   January 3, 1991
BT1 Equipment/Parts - Electrical (DOE
     FRASE Vocabulary)
 BT2 Equipment
SO DOE FRASE VOCABULARY

AMPUTATION
(1301 AMPUTATION)
DA   November 28, 1990
BT1 Injuries
RT Avulsion
SO DOE FRASE VOCABULARY

ANIMAL BITE
(1302 ANIMAL BITE)

DA   November 28, 1990
BT1 Injuries
NT1 Snake Bite
SO DOE FRASE VOCABULARY

ANKLE(S)
(1126 ANKLE)
DA   November 28, 1990
BT1 Joint(s)
 BT2 Human Body Parts
RT Foot/Feet
RT Leg(s)
SO DOE FRASE VOCABULARY

ANKLE PROTECTION
(2652 ANKLE PROTEC)
DA   January 3, 1991
BT1 Foot Protection
 BT2 Personal Protective Equipment
  BT3 Equipment/Parts - Personal
       Protective (DOE FRASE
       Vocabulary)
   BT4 Equipment
SO DOE FRASE VOCABULARY

ANTENNA
(2377 ANTENNA)
DA   January 3, 1991
BT1 Equipment/Parts - Electrical (DOE
     FRASE Vocabulary)
 BT2 Equipment
SO DOE FRASE VOCABULARY

ANTICONTAMINATION CLOTHING
(2695 ANTI C CLOTH)
DA   January 3, 1991
BT1 Clothing
BT1 Personal Protective Equipment
 BT2 Equipment/Parts - Personal
      Protective (DOE FRASE
      Vocabulary)
  BT3 Equipment
NT1 Acid Suit
NT1 Lab Coat
NT1 Radiation Suit
SO DOE FRASE VOCABULARY

APPLIANCE(S)
(2378 APPLIANCE)
DA   January 3, 1991
BT1 Equipment/Parts - Electrical (DOE
     FRASE Vocabulary)
 BT2 Equipment
NT1 Heating Appliance(s)
SO DOE FRASE VOCABULARY

AREA
(1601 AREA)
DA   December 10, 1990
BT1 Sites/Areas
NT1 Canal Area
NT1 Construction Area
NT1 Customer Service Area
NT1 Data Processing Area
NT1 Decontamination Areas
NT1 Eating Area
 NT2 Cafeteria
NT1 Field Area
NT1 Maintenance Area

NT1 Office Area
NT1 Pit Area
NT1 Production/Operations Area
NT1 Pump Area
NT1 Recreation Area
NT1 Sand Blasting Area
NT1 Storage Area
 NT2 Stock Room
 NT2 Warehouse
NT1 Sump Area
NT1 Test Area
 NT2 Fast Flux Test Facility
 NT2 Nevada Test Site
 NT2 Rocket Motor Test Sites
 NT2 Test Reactor Facility
 NT2 Test Stations
 NT2 Test Reactor Area
 NT2 Tonopah Test Range
NT1 Vehicle Service Area
NT1 Yard Area
SO DOE FRASE VOCABULARY

AREA AROUND EYE
(1083 AREA EYE)
DA   November 28, 1990
BT1 Face
 BT2 Head
  BT3 Human Body Parts
NT1 Eyelid(s)
RT Eye Protection
RT Eye(s)
SO DOE FRASE VOCABULARY

ARM(S)
(1095 ARM)
DA   November 28, 1990
BT1 Human Body Parts
NT1 Forearm(s)
NT1 Upper Arm
RT Elbow(s)
RT Forearm Protection
RT Hand(s)
SO DOE FRASE VOCABULARY

ASPHYXIA
(1303 ASPHYXIA)
DA   November 28, 1990
BT1 Injuries
SO DOE FRASE VOCABULARY

AUTOCLAVE
(2476 AUTOCLAVE)
DA   January 3, 1991
BT1 Heater(s)
 BT2 Equipment/Parts - Heating (DOE
      FRASE Vocabulary)
  BT3 Equipment
SO DOE FRASE VOCABULARY

AUTOMATIC REACTOR SCRAM
(1543 ARS)
DA   November 29, 1990
BT1 Nature of Property Damage
BT1 Scram
 BT2 Reactor Shutdown
RT Manual Reactor Scram
RT Nuclear Facility
RT Unscheduled Shutdown

RT Violations
SO DOE FRASE VOCABULARY

AVULSION
(1304 AV)
DA   November 28, 1990
BT1 Injuries
RT Amputation
SO DOE FRASE VOCABULARY

AX(S)
(3026 AX)
DA   January 3, 1991
BT1 Tools - Manual
 BT2 Tools (DOE FRASE Vocabulary)
  BT3 Equipment
RT Hatchet(s)
SO DOE FRASE VOCABULARY

BACK
(1107 BACKP)
DA   November 28, 1990
BT1 Trunk
 BT2 Human Body Parts
NT1 Lower Back
NT1 Upper Back
RT Ruptured Disk
RT Spinal Cord
SO DOE FRASE VOCABULARY

BALCONY
(1677 BALCONY)
DA   December 10, 1990
BT1 Site (DOE FRASE Vocabulary)
 BT2 Sites/Areas
SO DOE FRASE VOCABULARY

BALL MILL
(2104 BALL MILL)
DA   December 10, 1990
BT1 Milling Machine
 BT2 Machines (DOE FRASE
      Vocabulary)
  BT3 Equipment
SO DOE FRASE VOCABULARY

BAND SAW
(2105 BAND SAW)
DA   December 10, 1990
BT1 Machines (DOE FRASE
     Vocabulary)
 BT2 Equipment
BT1 Saws
SO DOE FRASE VOCABULARY

BASEMENT
(1678 BASEMENT)
DA   December 10, 1990
BT1 Site (DOE FRASE Vocabulary)
 BT2 Sites/Areas
SO DOE FRASE VOCABULARY

BATTERY(S)
(2379 BATTERY)
DA   January 3, 1991
BT1 Equipment/Parts - Electrical (DOE
      FRASE Vocabulary)

 BT2 Equipment
SO DOE FRASE VOCABULARY

**BENIGN AND UNSPECIFIED
NEOPLASM**
(1376 BENIGN AND U)
DA   November 28, 1990
BT1 Illnesses
SO DOE FRASE VOCABULARY

BIRDCAGE
(2550 BIRDCAGE)
DA   January 3, 1991
BT1 Equipment/Parts - Nuclear (DOE
      FRASE Vocabulary)
 BT2 Equipment
 BT2 Reactor Components
SO DOE FRASE VOCABULARY

BLISTER(S)
(1305 BLISTER)
DA   November 28, 1990
BT1 Injuries
SO DOE FRASE VOCABULARY

BLOOD
(1140 BLOOD)
DA   November 28, 1990
BT1 Circulatory System
 BT2 Body System(s)
  BT3 Human Body Parts
RT Blood Clot
SO DOE FRASE VOCABULARY

BLOOD CLOT
(1306 BLOOD CLOT)
DA   November 28, 1990
BT1 Injuries
RT Blood
SO DOE FRASE VOCABULARY

BODY OF WATER
(1603 BODY OF WATE)
DA   December 10, 1990
BT1 Site (DOE FRASE Vocabulary)
 BT2 Sites/Areas
RT Dam
RT Drowning
RT Pond
RT River/Creek
SO DOE FRASE VOCABULARY

BODY SYSTEM(S)
(1135 BODY SYSTEM)
DA   November 28, 1990
BT1 Human Body Parts
NT1 Circulatory System
 NT2 Blood
NT1 Digestive System
 NT2 Mouth
  NT3 Tooth/Teeth
 NT2 Rectum
NT1 Excretory System
NT1 Nervous System
 NT2 Brain
 NT2 Spinal Cord
NT1 Respiratory System
 NT2 Bronchial Epithelium

NT2 Nose
NT2 Throat
SO DOE FRASE VOCABULARY

BOILER(S)
(2479 BOILER)
DA   January 3, 1991
BT1 Heater(s)
 BT2 Equipment/Parts - Heating (DOE
       FRASE Vocabulary)
  BT3 Equipment
NT1 Boiler Part(s)
RT Boiler/Industrial Furnaces
SO DOE FRASE VOCABULARY

BOILER PART(S)
(2478 BOILER PART)
DA   January 3, 1991
BT1 Boiler(s)
 BT2 Heater(s)
  BT3 Equipment/Parts - Heating (DOE
        FRASE Vocabulary)
   BT4 Equipment
RT Boiler/Industrial Furnaces
SO DOE FRASE VOCABULARY

BONE(S)
(1136 BONE)
DA   November 28, 1990
BT1 Human Body Parts
NT1 Rib(s)
NT1 Spine
 NT2 Coccyx
RT Fracture
SO DOE FRASE VOCABULARY

BOOM
(2301 BOOM)
DA   December 10, 1990
BT1 Equipment/Parts - Material
      Handling (DOE FRASE
      Vocabulary)
 BT2 Equipment
RT Derrick
SO DOE FRASE VOCABULARY

BOOM CRANE
(2302 BOOM CRANE)
DA   December 10, 1990
BT1 Crane(s)
 BT2 Material Handling Device
  BT3 Devices
  BT3 Equipment/Parts - Material
        Handling (DOE FRASE
        Vocabulary)
   BT4 Equipment
SO DOE FRASE VOCABULARY

BORING MACHINE
(2106 BORING MACHI)
DA   December 10, 1990
BT1 Machines (DOE FRASE
      Vocabulary)
 BT2 Equipment
SO DOE FRASE VOCABULARY

BRAIN
(1079 BRAIN)

DA   November 28, 1990
BT1 Nervous System
 BT2 Body System(s)
  BT3 Human Body Parts
RT Head
RT Skull (Frase)
SO DOE FRASE VOCABULARY

BRAIN DAMAGE
(1307 BRAIN DAMA)
DA   November 28, 1990
BT1 Injuries
NT1 Concussion
NT1 Contusion(S)
RT Skull (Frase)
SO DOE FRASE VOCABULARY

BREAST(S)
(1110 BREAST)
DA   November 28, 1990
BT1 Human Body Parts
RT Chest
SO DOE FRASE VOCABULARY

BRIDGE
(1850 BRIDGE)
DA   December 10, 1990
BT1 Structures (DOE FRASE
    Vocabulary)
SO DOE FRASE VOCABULARY

BRIDGE CRANE
(2303 BRIDGE CRANE)
DA   December 10, 1990
BT1 Crane(s)
 BT2 Material Handling Device
  BT3 Devices
  BT3 Equipment/Parts - Material
      Handling (DOE FRASE
      Vocabulary)
   BT4 Equipment
BT1 Equipment/Parts -
    Instrumentation/Measuring (DOE
    FRASE Voc.)
 BT2 Equipment
SO DOE FRASE VOCABULARY

BROOM(S)
(3027 BROOM)
DA   January 3, 1991
BT1 Tools - Manual
 BT2 Tools (DOE FRASE Vocabulary)
  BT3 Equipment
SO DOE FRASE VOCABULARY

BRUSH(S)
(3028 BRUSH)
DA   January 3, 1991
BT1 Tools - Manual
 BT2 Tools (DOE FRASE Vocabulary)
  BT3 Equipment
SO DOE FRASE VOCABULARY

**BUILDING (DOE FRASE
VOCABULARY)**
(1775 BUILDING; DOE Order 4330.4A)
DA   December 10, 1990

BT1 Facilities and Buildings (DOE
    FRASE Vocabulary)
 BT2 Facilities
NT1 Containment Building
NT1 Demineralizer Plant
NT1 Fire Station
NT1 Garage
NT1 Guard Station
NT1 Hospital
NT1 Hot Shop
NT1 Machine Shop
NT1 Maintenance Shop
NT1 Motel/Hotel
NT1 Research & Development
    Laboratory
NT1 Restaurant
NT1 Sewage Plant
NT1 Steam Plant
SO DOE FRASE VOCABULARY
DEF (DOE Order 4330.4A) A roofed
    structure that is suitable for
    housing people, material, or
    equipment. Also included are
    sheds and other roofed structures
    that provide partial protection from
    the weather.

**BUILDING/EQUIP MAINT/REPAIR
ACTIVITY**
(1228 BEM ACTIVITY)
DA   November 28, 1990
BT1 Activity Types (DOE FRASE
    Vocabulary)
 BT2 Activities
BT1 Maintenance
 BT2 Activities
SO DOE FRASE VOCABULARY

BUMP CAP
(2653 BUMP CAP)
DA   January 3, 1991
BT1 Head Protection
 BT2 Personal Protective Equipment
  BT3 Equipment/Parts - Personal
      Protective (DOE FRASE
      Vocabulary)
   BT4 Equipment
SO DOE FRASE VOCABULARY

BURIAL BOX(S)
(2551 BURIAL BOX)
DA   January 3, 1991
BT1 Equipment/Parts - Nuclear (DOE
    FRASE Vocabulary)
 BT2 Equipment
 BT2 Reactor Components
SO DOE FRASE VOCABULARY

BURN(S)
(1308 BURN;N)
DA   November 28, 1990
BT1 Injuries
NT1 Chemical Burn(s)
NT1 Electrical Burn(s)
NT1 Flash Burn(s)
SO DOE FRASE VOCABULARY

BURNER(S)
(2480 BURNER)

DA   January 3, 1991
BT1 Equipment/Parts - Heating (DOE
    FRASE Vocabulary)
 BT2 Equipment
BT1 Heating Equipment
SO DOE FRASE VOCABULARY

BURSITIS
DA   November 28, 1990
BT1 Injuries
SO DOE FRASE VOCABULARY

BUS DRIVER
(0821 BUS DRIVER)
DA   November 28, 1990
BT1 Transport Personnel
 BT2 Occupations
 BT2 Personnel
SO DOE FRASE VOCABULARY

BUTTOCK(S)
(1119 BUTTOCK)
DA   November 28, 1990
BT1 Trunk
 BT2 Human Body Parts
RT Hip(s)
SO DOE FRASE VOCABULARY

CAFETERIA
(1680 CAFETERIA)
DA   December 10, 1990
BT1 Eating Area
 BT2 Area
  BT3 Sites/Areas
BT1 Site (DOE FRASE Vocabulary)
 BT2 Sites/Areas
RT Kitchen
RT Restaurant
SO DOE FRASE VOCABULARY

CANAL AREA
(1604 CANAL AREA)
DA   December 10, 1990
BT1 Area
 BT2 Sites/Areas
SO DOE FRASE VOCABULARY

CAPACITOR(S)
(2380 CAPACITOR)
DA   January 3, 1991
BT1 Equipment/Parts - Electrical (DOE
    FRASE Vocabulary)
 BT2 Equipment
SO DOE FRASE VOCABULARY

CARPENTER
(0642 CARPENTER)
DA   November 28, 1990
BT1 Repair/Construction Personnel
 BT2 Occupations
 BT2 Personnel
SO DOE FRASE VOCABULARY

CART(S)
(2304 CART)
DA   December 10, 1990
BT1 Hand Truck(s)
 BT2 Material Handling Device

BT3 Devices
BT3 Equipment/Parts - Material
Handling (DOE FRASE
Vocabulary)
BT4 Equipment
SO DOE FRASE VOCABULARY

CATWALK
(1681 CATWALK)
DA December 10, 1990
BT1 Site (DOE FRASE Vocabulary)
BT2 Sites/Areas
SO DOE FRASE VOCABULARY

CELL
(1682 CELL)
DA December 10, 1990
BT1 Site (DOE FRASE Vocabulary)
BT2 Sites/Areas
SO DOE FRASE VOCABULARY
SO Environmental Management

CENTRIFUGE
(2110 CENTRIFUGE)
DA December 10, 1990
BT1 Machines (DOE FRASE
Vocabulary)
BT2 Equipment
SO DOE FRASE VOCABULARY

CHAIN SAW
(2951 CHAIN SAW)
DA January 3, 1991
BT1 Saws
BT1 Tools - Powered
BT2 Tools (DOE FRASE Vocabulary)
BT3 Equipment
SO DOE FRASE VOCABULARY

CHAMBER
(1683 CHAMBER)
DA December 10, 1990
BT1 Site (DOE FRASE Vocabulary)
BT2 Sites/Areas
RT Room
SO DOE FRASE VOCABULARY

CHEMICAL BURN(S)
(1310 CHEMICAL BUR)
DA November 28, 1990
BT1 Burn(s)
BT2 Injuries
SO DOE FRASE VOCABULARY

CHEMICAL CONTAMINATION
(1525 CHEMICAL CON)
DA November 28, 1990
BT1 Contamination
RT Environmental Release
RT Hazardous Spill
SO DOE FRASE VOCABULARY

CHEMICAL REACTION
(1377 CHEMICAL REA)
DA November 28, 1990
BT1 Illnesses
RT Insulin Reaction
SO DOE FRASE VOCABULARY

CHEST
(1109 CHEST)
DA November 28, 1990
BT1 Trunk
BT2 Human Body Parts
RT Breast(s)
RT Respiratory System
RT Rib(s)
SO DOE FRASE VOCABULARY

CHIN
(1091 CHIN)
DA November 28, 1990
BT1 Head
BT2 Human Body Parts
RT Jaw
SO DOE FRASE VOCABULARY

CHIN STRAP
(2654 CHIN STRAP)
DA January 3, 1991
BT1 Personal Protective Equipment
BT2 Equipment/Parts - Personal
Protective (DOE FRASE
Vocabulary)
BT3 Equipment
SO DOE FRASE VOCABULARY

CHISEL(S)
(3029 CHISEL)
DA January 3, 1991
BT1 Tools - Manual
BT2 Tools (DOE FRASE Vocabulary)
BT3 Equipment
SO DOE FRASE VOCABULARY

CIRCUIT(S)
(2382 CIRCUIT)
DA January 3, 1991
BT1 Equipment/Parts - Electrical (DOE
FRASE Vocabulary)
BT2 Equipment
SO DOE FRASE VOCABULARY

CIRCUIT BREAKER(S)
(2381 CIRCIUT BREA)
DA January 3, 1991
BT1 Equipment/Parts - Electrical (DOE
FRASE Vocabulary)
BT2 Equipment
SO DOE FRASE VOCABULARY

CIRCULATORY SYSTEM
(1141 CIRCULATORY)
DA November 28, 1990
BT1 Body System(s)
BT2 Human Body Parts
NT1 Blood
SO DOE FRASE VOCABULARY

CLADDING
(2552 CLADDING)
DA January 3, 1991
BT1 Fuel Plate
BT2 Fuel Element(s)
BT3 Equipment/Parts - Nuclear (DOE
FRASE Vocabulary)
BT4 Equipment

BT4 Reactor Components
SO DOE FRASE VOCABULARY
DEF (NRC Glossary of Terms: Nuclear
Power and Radiation) The
thin-walled metal tube that forms
the outer jacket of a nuclear fuel
rod. It prevents corrosion of the
fuel by the coolant and the release
of fission products into the coolant.
Aluminum, stainless steel and
zirconium alloys are common
cladding materials.

CLASSIFIED ACTIVITY
(1229 CLASSIFIED A)
DA November 28, 1990
BT1 Activity Types (DOE FRASE
Vocabulary)
BT2 Activities
RT Classification
SO DOE FRASE VOCABULARY

CLOTHING
DA January 30, 1991
NT1 Anticontamination Clothing
NT2 Acid Suit
NT2 Lab Coat
NT2 Radiation Suit
NT1 Flame Retardant Clothing
RT Equipment/Parts - Personal
Protective (DOE FRASE
Vocabulary)
SO DOE FRASE VOCABULARY

COCCYX
(1116 COCCYX)
DA November 28, 1990
BT1 Spine
BT2 Bone(s)
BT3 Human Body Parts
SO DOE FRASE VOCABULARY

COLD TRAP
(2553 COLD TRAP)
DA January 3, 1991
BT1 Equipment/Parts - Nuclear (DOE
FRASE Vocabulary)
BT2 Equipment
BT2 Reactor Components
SO DOE FRASE VOCABULARY

COMPACTOR
(2112 COMPACTOR)
DA December 10, 1990
BT1 Machines (DOE FRASE
Vocabulary)
BT2 Equipment
SO DOE FRASE VOCABULARY

**COMPLICATIONS PECULIAR TO MED.
CARE**
(1378 COMPLICATION)
DA November 28, 1990
BT1 Illnesses
SO DOE FRASE VOCABULARY

COMPRESSOR
(2113 COMPRESSOR)

DA   December 10, 1990
BT1 Machines (DOE FRASE
         Vocabulary)
BT2 Equipment
NT1 Air Compressor
SO   DOE FRASE VOCABULARY

COMPUTER(S)
(2384 COMPUTER)
DA   January 3, 1991
BT1 Equipment/Parts - Electrical (DOE
         FRASE Vocabulary)
BT2 Equipment
RT   Data Processing Area
SO   DOE FRASE VOCABULARY

COMPUTER TERMINAL(S)
(2383 COMPUTER TER)
DA   January 3, 1991
BT1 Equipment/Parts - Electrical (DOE
         FRASE Vocabulary)
BT2 Equipment
RT   Data Processing Area
SO   DOE FRASE VOCABULARY

CONCRETE SAW
(2114 CONCRETE SAW)
DA   December 10, 1990
BT1 Machines (DOE FRASE
         Vocabulary)
BT2 Equipment
BT1 Saws
SO   DOE FRASE VOCABULARY

CONCUSSION
(1311 CONCUSSION)
DA   November 28, 1990
BT1 Brain Damage
BT2 Injuries
RT   Contusion(S)
RT   Head
RT   Head Protection
RT   Loss of Consciousness
RT   Skull (Frase)
SO   DOE FRASE VOCABULARY

**COND OF RESPIRATORY SYS.
NON-TOXIC**
(1379 COND OF RESP)
DA   November 28, 1990
BT1 Diseases
BT1 Illnesses
NT1 Upper Respiratory Condition
NT1 Upper Respiratory Disease
NT2 Tuberculosis
RT   Respiratory System
SO   DOE FRASE VOCABULARY

CONDENSER(S)
(2385 CONDENSER)
DA   January 3, 1991
BT1 Equipment/Parts - Electrical (DOE
         FRASE Vocabulary)
BT2 Equipment
SO   DOE FRASE VOCABULARY

CONDUCTOR(S)
(2386 CONDUCTOR)

DA   January 3, 1991
BT1 Equipment/Parts - Electrical (DOE
         FRASE Vocabulary)
BT2 Equipment
SO   DOE FRASE VOCABULARY

CONDUIT(S)
(2387 CONDUIT)
DA   January 3, 1991
BT1 Equipment/Parts - Electrical (DOE
         FRASE Vocabulary)
BT2 Equipment
SO   DOE FRASE VOCABULARY

CONJUNCTIVITIS
(1380 CONJUNCTIVIT)
DA   November 28, 1990
BT1 Contagious or Infectious Disease
BT2 Diseases
BT2 Illnesses
BT1 Illnesses
RT   Eyelid(s)
SO   DOE FRASE VOCABULARY

CONSTANT AIR MONITOR
(2751 CONSTANT AIR)
DA   January 3, 1991
BT1 Equipment/Parts -
         Instrumentation/Measuring (DOE
         FRASE Voc.)
BT2 Equipment
BT1 Monitors
BT2 Equipment
SO   DOE FRASE VOCABULARY

CONSTRUCTION ACTIVITY
(1227 CON ACTIVITY)
DA   November 28, 1990
BT1 Activity Types (DOE FRASE
         Vocabulary)
BT2 Activities
RT   Jackhammer
RT   Misc Repair/Construction
         Employee
SO   DOE FRASE VOCABULARY

CONSTRUCTION AREA
(1605 CONST AREA)
DA   December 10, 1990
BT1 Area
BT2 Sites/Areas
SO   DOE FRASE VOCABULARY

**CONTAGIOUS OR INFECTIOUS
DISEASE**
(1382 CONTAGIOUS O)
DA   November 28, 1990
BT1 Diseases
BT1 Illnesses
NT1 Conjunctivitis
SO   DOE FRASE VOCABULARY

CONTAINMENT BUILDING
(1776 CONTAIN BLDG)
DA   December 10, 1990
BT1 Building (DOE FRASE Vocabulary)
BT2 Facilities and Buildings (DOE
         FRASE Vocabulary)

BT3 Facilities
SO   DOE FRASE VOCABULARY

CONTROL PANEL(S)
(2554 CONTROL PANE)
DA   January 3, 1991
BT1 Equipment/Parts - Nuclear (DOE
         FRASE Vocabulary)
BT2 Equipment
BT2 Reactor Components
SO   DOE FRASE VOCABULARY

CONTROL ROD(S)
(2556 CONTROL ROD)
DA   January 3, 1991
BT1 Equipment/Parts - Nuclear (DOE
         FRASE Vocabulary)
BT2 Equipment
BT2 Reactor Components
NT1 Gangs
NT1 Raincoats
RT   Control Rod Drive Assembly
RT   Core(s)
SO   DOE FRASE VOCABULARY

CONTROL ROD DRIVE ASSEMBLY
(2555 CRDA)
DA   January 3, 1991
BT1 Control Rod Drive Control System
BT2 Control Systems
BT3 Controls
BT3 Systems
BT2 Reactor Components
BT1 Equipment/Parts - Nuclear (DOE
         FRASE Vocabulary)
BT2 Equipment
BT2 Reactor Components
RT   Control Rod(s)
RT   Rod Drive
SO   DOE FRASE VOCABULARY

CONTROL ROOM
(1684 CONTROL ROOM)
DA   December 10, 1990
BT1 Room
BT2 Sites/Areas
SO   DOE FRASE VOCABULARY
DEF The area in a nuclear power plant
         from which most of the plant
         power production and emergency
         safety equipment can be operated
         by remote control.

CONTUSION(S)
(1312 CO)
DA   November 28, 1990
BT1 Brain Damage
BT2 Injuries
RT   Concussion
RT   Head
RT   Head Protection
RT   Loss of Consciousness
RT   Skull (Frase)
SO   DOE FRASE VOCABULARY

CONVEYOR(S)
(2310 CONVEYOR)
DA   December 10, 1990
BT1 Material Handling Device

BT2 Devices
BT2 Equipment/Parts - Material
    Handling (DOE FRASE
    Vocabulary)
BT3 Equipment
SO DOE FRASE VOCABULARY

COOLANT RETURN PUMP
(2557 CR PUMP)
DA January 3, 1991
BT1 Equipment/Parts - Nuclear (DOE
    FRASE Vocabulary)
BT2 Equipment
BT2 Reactor Components
BT1 Pump(s)
BT2 Machines (DOE FRASE
    Vocabulary)
BT3 Equipment
SO DOE FRASE VOCABULARY

COOLING SYSTEM
(2026 COOLING SYST)
DA December 10, 1990
BT1 Systems (DOE FRASE Vocabulary)
BT2 Systems
SO DOE FRASE VOCABULARY

CORE(S)
(2558 CORE)
DA January 3, 1991
BT1 Equipment/Parts - Nuclear (DOE
    FRASE Vocabulary)
BT2 Equipment
BT2 Reactor Components
RT Control Rod(s)
RT Fuel Element(s)
RT Reactor(s)
SO DOE FRASE VOCABULARY

CORRIDOR/HALL
(1685 CORRIDOR)
DA December 10, 1990
BT1 Site (DOE FRASE Vocabulary)
BT2 Sites/Areas
SO DOE FRASE VOCABULARY

COUNT
(2780 COUNT)
DA January 3, 1991
BT1 Equipment/Parts -
    Instrumentation/Measuring (DOE
    FRASE Voc.)
BT2 Equipment
SO DOE FRASE VOCABULARY

CRANE(S)
(2313 CRANE)
DA December 10, 1990
BT1 Material Handling Device
BT2 Devices
BT2 Equipment/Parts - Material
    Handling (DOE FRASE
    Vocabulary)
BT3 Equipment
NT1 Boom Crane
NT1 Bridge Crane
NT1 Mobile Crane
NT1 Overhead Crane

RT Derrick
SO DOE FRASE VOCABULARY

CRANE BRIDGE
(2312 CRANE BRIDGE)
DA December 10, 1990
SO DOE FRASE VOCABULARY

CRITICALITY ALARM SYSTEM
(2027 CAS)
DA December 10, 1990
BT1 Alarms
BT2 Devices
BT1 Emergency Systems
BT2 Systems
BT1 Systems (DOE FRASE Vocabulary)
BT2 Systems
RT Violation of Criticality Specifictn
SO DOE FRASE VOCABULARY

CROWBAR(S)
(3030 CROWBAR)
DA January 3, 1991
BT1 Tools - Manual
BT2 Tools (DOE FRASE Vocabulary)
BT3 Equipment
SO DOE FRASE VOCABULARY

CRUSHING MACHINE
(2115 CRUSHING MAC)
DA December 10, 1990
BT1 Machines (DOE FRASE
    Vocabulary)
BT2 Equipment
SO DOE FRASE VOCABULARY

CUSTOMER SERVICE AREA
(1687 CS AREA)
DA December 10, 1990
BT1 Area
BT2 Sites/Areas
SO DOE FRASE VOCABULARY

CUTTER(S)
(3031 CUTTER)
DA January 3, 1991
BT1 Tools - Manual
BT2 Tools (DOE FRASE Vocabulary)
BT3 Equipment
RT Scissors
RT Snips
SO DOE FRASE VOCABULARY

DAM
(1852 DAM)
DA December 10, 1990
BT1 Structures (DOE FRASE
    Vocabulary)
RT Body of Water
RT Pond
SO DOE FRASE VOCABULARY

DAMAGE
(1526 DAMAGE;N)
DA November 28, 1990
BT1 Nature of Property Damage
NT1 Electronic Damage
NT1 Fire Damage

NT1 Heat Damage
NT1 Mechanical Damage
NT1 Smoke Damage
NT1 Structural Damage
NT1 Water Damage
NT1 Wind Damage
SO DOE FRASE VOCABULARY

DAMAGE TO PROSTHETIC DEVICE
(1313 DAMAGE TO PR)
DA November 28, 1990
BT1 Injuries
SO DOE FRASE VOCABULARY

DATA PROCESSING AREA
(1688 DP AREA)
DA December 10, 1990
BT1 Area
BT2 Sites/Areas
RT Computer Terminal(s)
RT Computer(s)
SO DOE FRASE VOCABULARY

DEATH
(1314 DEATH; TSCA; CFR)
DA November 28, 1990
BT1 Injuries
SO DOE FRASE VOCABULARY

DECOMMISSIONING ACTIVITY
(1250 DEC ACTIVITY)
DA November 28, 1990
BT1 Activity Types (DOE FRASE
    Vocabulary)
BT2 Activities
RT Nuclear Facility
SO DOE FRASE VOCABULARY

DECONTAMINATION ACTIVITY
(1255)
DA November 28, 1990
BT1 Activity Types (DOE FRASE
    Vocabulary)
BT2 Activities
RT Nuclear Facility
SO DOE FRASE VOCABULARY

DECONTAMINATION AREAS
(1689 DECON AREA; TSCA; CFR)
DA December 10, 1990
BT1 Area
BT2 Sites/Areas
SO DOE FRASE VOCABULARY
SO Environmental Management

DELAY OF EXPERIMENT
(1575 DELAY OF EXP)
DA November 28, 1990
BT1 Delays
BT1 Nature of Programmatic Impact
SO DOE FRASE VOCABULARY

DELAY OF RESPONSE
(1576 DELAY OF RES)
DA November 28, 1990
BT1 Delays
BT1 Nature of Programmatic Impact
SO DOE FRASE VOCABULARY

DELAY OF SAMPLING
(1577 DELAY OF SAM)
DA   November 28, 1990
BT1 Delays
BT1 Nature of Programmatic Impact
SO   DOE FRASE VOCABULARY

DELAYS
DA   January 30, 1991
NT1 Delay of Experiment
NT1 Delay of Response
NT1 Delay of Sampling
NT1 Program Significant Delay
RT   Nature of Programmatic Impact
SO   DOE FRASE VOCABULARY

DELUGE SYSTEM
(2028 DELUGE SYSTE)
DA   December 10, 1990
BT1 Systems (DOE FRASE Vocabulary)
 BT2 Systems
SO   DOE FRASE VOCABULARY

DEMINERALIZER PLANT
(1796 DEMINERALIZE)
DA   December 10, 1990
BT1 Building (DOE FRASE Vocabulary)
 BT2 Facilities and Buildings (DOE
       FRASE Vocabulary)
  BT3 Facilities
SO   DOE FRASE VOCABULARY

DENT
(1527 DENT)
DA   November 28, 1990
BT1 Nature of Property Damage
SO   DOE FRASE VOCABULARY

DENTAL INJURY
(1315 DENTAL INJUR)
DA   November 28, 1990
BT1 Injuries
RT   Mouth
RT   Tooth/Teeth
SO   DOE FRASE VOCABULARY

DERMATITIS
(1382 DERMATITIS)
DA   November 28, 1990
BT1 Illnesses
RT   Other Skin Conditions
RT   Skin
SO   DOE FRASE VOCABULARY

DERRICK
(2314 DERRICK)
DA   December 10, 1990
BT1 Hoisting Apparatus
 BT2 Material Handling Device
  BT3 Devices
  BT3 Equipment/Parts - Material
       Handling (DOE FRASE
       Vocabulary)
   BT4 Equipment
RT   Boom
RT   Crane(s)
SO   DOE FRASE VOCABULARY

DEVICES
DA   January 30, 1991
NT1 Accelerators
NT1 Alarms
 NT2 Attack Warning Signals
 NT2 Attention Signals
 NT2 Criticality Alarm System
 NT2 Criticality Alarms
 NT2 Fire Alarm System
 NT2 High Temperature Alarm
 NT2 Intrusion Alarm System (Perimeter
      or Interior)
 NT2 Kanne Alarm
 NT2 Level Alarm
 NT2 Radiation Alarm System
NT1 Bar Screens
NT1 Blank Fire Adapter (BFA)
NT1 Bullet Containment Devices
 NT2 Clearing Barrels
NT1 Catalytic Converters
NT1 Control Devices
 NT2 Add On Control Devices
 NT2 Afterburners
 NT2 Arsenic Kitchens
 NT2 Carbon Adsorbers
 NT2 Catalytic Incinerators
 NT2 Cyclone Collectors
 NT2 Packed Towers
 NT2 Precipitators
  NT3 Electrostatic Precipitators
 NT2 Primary Emission Control Systems
 NT2 Secondary Hood Systems
 NT2 Vapor Capture Systems
  NT3 Floor Sweeps
NT1 Crossing Frogs
NT1 Drinking Water Coolers
NT1 Electroexplosive Devices
 NT2 Detonators
NT1 Elementary Neutralization Units
NT1 Emergency Shutdown Device
NT1 Fall Protection Device
 NT2 Safety Belt
 NT2 Safety Line
 NT2 Safety Net
NT1 Filters
 NT2 Baghouse Filters
 NT2 Fabric Filters
 NT2 Filters (Radiology)
 NT2 Hepa Filter
 NT2 High Efficiency Particulate Air
      Filters
 NT2 Sand Filters
 NT2 Trickling Filters
NT1 Flares
NT1 Golf Bags
NT1 Graphic Panel
NT1 Grapper Pick Up
NT1 Grinder Pumps
NT1 Inertial Separators
 NT2 Cyclone Collectors
NT1 Lifting Station
NT1 Live Round Excluders
NT1 Load Securing Device
NT1 Marine Sanitation Devices
NT1 Material Handling Device
 NT2 Conveyor(s)
 NT2 Crane(s)
  NT3 Boom Crane
  NT3 Bridge Crane
  NT3 Mobile Crane
  NT3 Overhead Crane

 NT2 Earth Moving Equipment
  NT3 Dredge(s)
 NT2 Hand Truck(s)
  NT3 Cart(s)
  NT3 Dolly
 NT2 Hoisting Apparatus
  NT3 Derrick
  NT3 Elevator
  NT3 Forklift(s)
  NT3 Hoist(s)
   NT4 Air Hoist
  NT3 Lift Bucket
  NT3 Manlift(s)
  NT3 Scissor Lift
NT1 Ozonators
NT1 Packers
NT1 Point-of-entry Treatment Devices
NT1 Point-of-use Treatment Devices
NT1 Pumping Stations
NT1 Reset Device
NT1 Safety Device/System
NT1 Scrubbers
NT1 Sensors
NT1 Sludge Dryer
NT1 Suspicious Devices
NT1 Torpedoes
NT1 Turbidimeters
NT1 Valves
 NT2 Back Pressure Valve
 NT2 Check Valves
 NT2 Main Steam Isolation Valve
 NT2 Motor Operated Valves
 NT2 Open Ended Valve
 NT2 Pressure Regulator Valve
 NT2 Relief Valves
  NT3 Emergency Relief Valve
  NT3 Inadvertently Opened Relief
       Valve
  NT3 Power Operated Relief Valve
  NT3 Safety/Relief Valve
  NT3 Stuck-Open Relief Valve
 NT2 Turbine Block Valve
NT1 Wastewater Treatment Units
 NT2 Digesters
 NT2 Wastewater Treatment Tanks
RT   Equipment
RT   Production Facilities
SO   DOE FRASE VOCABULARY

DIGESTIVE SYSTEM
(1142 DIGESTIVE SY)
DA   November 28, 1990
BT1 Body System(s)
 BT2 Human Body Parts
NT1 Mouth
 NT2 Tooth/Teeth
NT1 Rectum
SO   DOE FRASE VOCABULARY

DIODE(S)
(2388 DIODE)
DA   January 3, 1991
BT1 Equipment/Parts - Electrical (DOE
      FRASE Vocabulary)
 BT2 Equipment
SO   DOE FRASE VOCABULARY

DISCHARGE MACHINE
(2116 DISCHARGE MA)
DA   December 10, 1990

BT1 Machines (DOE FRASE
   Vocabulary)
 BT2 Equipment
SO DOE FRASE VOCABULARY

**DISEASE OF CENTRAL NERVOUS
SYSTEM**
(1384 DISEASE OF C)
DA November 28, 1990
BT1 Diseases
BT1 Illnesses
SO DOE FRASE VOCABULARY

DISEASES OF BLOOD
(1385 DISEASES OF)
DA November 28, 1990
BT1 Diseases
BT1 Illnesses
NT1 Hepatitis
SO DOE FRASE VOCABULARY

DISK SANDER(S)
(2952 DISK SANDER)
DA January 3, 1991
BT1 Tools - Powered
 BT2 Tools (DOE FRASE Vocabulary)
 BT3 Equipment
SO DOE FRASE VOCABULARY

DISLOCATION
(1316 DISLOCATION)
DA November 28, 1990
BT1 Injuries
SO DOE FRASE VOCABULARY

DOCTOR/NURSE
(0260 DOCTOR;NURSE)
DA November 28, 1990
BT1 Professional Personnel
 BT2 Occupations
 BT2 Personnel
RT Hospital
SO DOE FRASE VOCABULARY

DOE FRASE CATEGORIES
DA January 31, 1991
RT Activity Types (DOE FRASE
   Vocabulary)
RT Equipment/Parts - Electrical (DOE
   FRASE Vocabulary)
RT Equipment/Parts - Heating (DOE
   FRASE Vocabulary)
RT Equipment/Parts -
   Instrumentation/Measuring (DOE
   FRASE Voc.)
RT Equipment/Parts - Material
   Handling (DOE FRASE
   Vocabulary)
RT Equipment/Parts - Nuclear (DOE
   FRASE Vocabulary)
RT Equipment/Parts - Personal
   Protective (DOE FRASE
   Vocabulary)
RT Facilities and Buildings (DOE
   FRASE Vocabulary)
RT FRASE Vocabulary
RT Human Body Parts
RT Illnesses
RT Injuries

RT Machines (DOE FRASE
   Vocabulary)
RT Nature of Programmatic Impact
RT Nature of Property Damage
RT Occupations
RT Sites/Areas
RT Structures (DOE FRASE
   Vocabulary)
RT Systems (DOE FRASE Vocabulary)
RT Tools - Manual
RT Tools - Powered
SO DOE FRASE VOCABULARY
DEF Categories under which FRASE
   (Factor Relationship And
   Sequence of Events) terms and
   phrases are grouped according to
   broader concepts or things. See
   individual category entries for a list
   (and structure) of terms and
   phrases under that heading.

DOLLY
(2315 DOLLY)
DA December 10, 1990
BT1 Hand Truck(s)
 BT2 Material Handling Device
 BT3 Devices
 BT3 Equipment/Parts - Material
   Handling (DOE FRASE
   Vocabulary)
 BT4 Equipment
SO DOE FRASE VOCABULARY

DREDGE(S)
(2916 DREDGE)
DA December 10, 1990
BT1 Earth Moving Equipment
 BT2 Material Handling Device
 BT3 Devices
 BT3 Equipment/Parts - Material
   Handling (DOE FRASE
   Vocabulary)
 BT4 Equipment
SO DOE FRASE VOCABULARY

DRILL RIG(S)
(2317 DRILL RIG)
DA December 10, 1990
BT1 Equipment/Parts - Material
   Handling (DOE FRASE
   Vocabulary)
 BT2 Equipment
SO DOE FRASE VOCABULARY

DRILLING MACHINE
(2117 DRILLING MAC)
DA December 10, 1990
BT1 Machines (DOE FRASE
   Vocabulary)
 BT2 Equipment
SO DOE FRASE VOCABULARY

DRINKING WATER SYSTEM
(2029 DRINKING WAT)
DA December 10, 1990
BT1 Public Water Systems
 BT2 Water Systems
 BT3 Publicly Owned Treatment Works
 BT4 Treatment Facilities

 BT5 Facilities
 BT3 Systems
BT1 Systems (DOE FRASE Vocabulary)
 BT2 Systems
RT Raw Water System
RT Waste Water System
SO DOE FRASE VOCABULARY

DROWNING
(1317 DROWNING)
DA November 28, 1990
BT1 Injuries
RT Body of Water
RT Pool
SO DOE FRASE VOCABULARY

DRYING MACHINE
(2119 DRYING MACHI)
DA December 10, 1990
BT1 Machines (DOE FRASE
   Vocabulary)
 BT2 Equipment
SO DOE FRASE VOCABULARY

DUST MASK
(2655 DUST MASK)
DA January 3, 1991
BT1 Personal Protective Equipment
 BT2 Equipment/Parts - Personal
   Protective (DOE FRASE
   Vocabulary)
 BT3 Equipment
RT Air Mask
RT Nose
SO DOE FRASE VOCABULARY

DWELLING
(1729 DWELLING)
DA December 10, 1990
BT1 Site (DOE FRASE Vocabulary)
 BT2 Sites/Areas
SO DOE FRASE VOCABULARY

EAR(S)
(1085 EAR)
DA November 28, 1990
BT1 Head
 BT2 Human Body Parts
RT Ear Muffs
RT Ear Plug(s)
RT Hearing Impairment
RT Hearing Protection
SO DOE FRASE VOCABULARY

EAR MUFFS
(2656 EAR MUFFS)
DA January 3, 1991
RT Ear Plug(s)
RT Ear(s)
RT Hearing Impairment
RT Hearing Protection
SO DOE FRASE VOCABULARY

EAR PLUG(S)
(2657 EAR PLUG)
DA January 3, 1991
BT1 Hearing Protection
 BT2 Personal Protective Equipment

BT3 Equipment/Parts - Personal
Protective (DOE FRASE
Vocabulary)
 BT4 Equipment
RT Ear Muffs
RT Ear(s)
RT Hearing Impairment
SO DOE FRASE VOCABULARY

EARTH MOVING EQUIPMENT
(2318 EARTH MOVING)
DA December 10, 1990
BT1 Material Handling Device
 BT2 Devices
 BT2 Equipment/Parts - Material
Handling (DOE FRASE
Vocabulary)
 BT3 Equipment
NT1 Dredge(s)
SO DOE FRASE VOCABULARY

EATING AREA
(1691 EATING AREA)
DA December 10, 1990
BT1 Area
 BT2 Sites/Areas
NT1 Cafeteria
RT Food Service Activity
RT Kitchen
RT Restaurant
SO DOE FRASE VOCABULARY

EFFECTS
DA January 30, 1991
SY Consequences
NT1 Acute Effects
NT1 Chilling Effect
NT1 Chronic Effects
NT1 Ecological Impacts
NT1 Effects on Welfare
NT1 Effects of the Action
NT1 Frostbite/Other Low Temp Effects
NT1 Genetic Radiation Effects
NT1 Greenhouse Effect
NT1 Heat Stroke/Other High Temp
Effect
NT1 Heat Island Effects
NT1 Ionizing Radiation Effects
NT1 Known Human Effects
NT1 Non-Stochastic Effects
 NT2 Non-Stochastic Radiation Effects
NT1 Orthogonal Effects
NT1 P-Delta Effect
NT1 Somatic Radiation Effects
NT1 Stack Effect
NT1 Stochastic Effects
 NT2 Radiation Induced Hereditary
Effects
 NT2 Radiation Induced Genetic Effects
 NT2 Stochastic Radiation Effects
NT1 Systemic Effects
NT1 Toxic Effects to Single System
NT1 Unacceptable Adverse Effects
RT Illnesses
RT Injuries
SO DOE FRASE VOCABULARY

ELBOW(S)
(1097 ELBOW)

DA November 28, 1990
BT1 Joint(s)
 BT2 Human Body Parts
RT Arm(s)
RT Forearm Protection
SO DOE FRASE VOCABULARY

ELECTRIC SHOCK
(1318 ELECTRIC SHO)
DA November 28, 1990
BT1 Injuries
RT Electrical Burn(s)
RT Electrocution
RT Electrical Short
SO DOE FRASE VOCABULARY

ELECTRICAL APPARATUS
(2389 ELEC APPARAT)
DA January 3, 1991
BT1 Equipment/Parts - Electrical (DOE
FRASE Vocabulary)
 BT2 Equipment
SO DOE FRASE VOCABULARY

ELECTRICAL BURN(S)
(1319 ELEC BURN)
DA November 28, 1990
BT1 Burn(s)
 BT2 Injuries
RT Electric Shock
RT Electrocution
SO DOE FRASE VOCABULARY

ELECTRICAL BUS
(2390 ELEC BUS)
DA January 3, 1991
BT1 Equipment/Parts - Electrical (DOE
FRASE Vocabulary)
 BT2 Equipment
SO DOE FRASE VOCABULARY

ELECTRICAL INSULATOR(S)
(2391 ELEC INSULAT)
DA January 3, 1991
BT1 Equipment/Parts - Electrical (DOE
FRASE Vocabulary)
 BT2 Equipment
SO DOE FRASE VOCABULARY

ELECTRICAL OFFICE MACHINE
(2120 ELEC OFFICE)
DA December 10, 1990
BT1 Machines (DOE FRASE
Vocabulary)
 BT2 Equipment
SO DOE FRASE VOCABULARY

ELECTRICAL SHORT
(1528 ELEC SHORT)
DA November 28, 1990
BT1 Nature of Property Damage
RT Electric Shock
RT Electronic Damage
SO DOE FRASE VOCABULARY

ELECTRICIAN
(0643 ELECTRICIAN)
DA November 28, 1990

BT1 Repair/Construction Personnel
 BT2 Occupations
 BT2 Personnel
SO DOE FRASE VOCABULARY

ELECTROCUTION
(1320 ELECTROCUTIO)
DA November 28, 1990
BT1 Injuries
RT Electric Shock
RT Electrical Burn(s)
SO DOE FRASE VOCABULARY

ELECTRODE(S)
(2392 ELECTRODE)
DA January 3, 1991
BT1 Equipment/Parts - Electrical (DOE
FRASE Vocabulary)
 BT2 Equipment
SO DOE FRASE VOCABULARY

**ELECTROMECHANICAL
MANIPULATOR(S)**
(2559 EM)
DA January 3, 1991
BT1 Manipulator
 BT2 Equipment/Parts - Nuclear (DOE
FRASE Vocabulary)
 BT3 Equipment
 BT3 Reactor Components
SO DOE FRASE VOCABULARY

ELECTRONIC DAMAGE
(1529 ELECTRONIC D)
DA November 28, 1990
BT1 Damage
 BT2 Nature of Property Damage
RT Electrical Short
SO DOE FRASE VOCABULARY

ELEVATOR
(2320 ELEVATOR)
DA December 10, 1990
BT1 Hoisting Apparatus
 BT2 Material Handling Device
 BT3 Devices
 BT3 Equipment/Parts - Material
Handling (DOE FRASE
Vocabulary)
 BT4 Equipment
RT Escalator
SO DOE FRASE VOCABULARY

**EMERGENCY CORE COOLING
SYSTEM**
(2030 ECCS)
DA December 10, 1990
SF ECCS
BT1 Emergency Systems
 BT2 Systems
BT1 Reactor Protection System
 BT2 Emergency Systems
 BT3 Systems
BT1 Systems (DOE FRASE Vocabulary)
 BT2 Systems
SO DOE FRASE VOCABULARY
DEF Reactor system components
(pumps, valves, heat exchangers,
tanks and piping) that are

specifically designed to remove residual heat from the reactor fuel rods should the normal core cooling system fail.

EMERGENCY DIESEL GENERATOR
(2121 EDG)
DA   December 10, 1990
SF   *EDG*
BT1 Machines (DOE FRASE Vocabulary)
 BT2 Equipment
RT   Emergency Power System
SO   DOE FRASE VOCABULARY

EMERGENCY POWER SYSTEM
(2031 EPS)
DA   December 10, 1990
BT1 Emergency Systems
 BT2 Systems
BT1 Systems (DOE FRASE Vocabulary)
 BT2 Systems
RT   Emergency Diesel Generator
SO   DOE FRASE VOCABULARY

EMERGENCY RESPONSE ACTIVITY
(1230 ER ACTIVITY)
DA   November 28, 1990
BT1 Activity Types (DOE FRASE Vocabulary)
 BT2 Activities
SO   DOE FRASE VOCABULARY

EN ROUTE
(1606 EN ROUTE)
DA   December 10, 1990
BT1 Site (DOE FRASE Vocabulary)
 BT2 Sites/Areas
SO   DOE FRASE VOCABULARY

ENGINEER
(0160 ENGINEER)
DA   November 28, 1990
BT1 Professional Personnel
 BT2 Occupations
 BT2 Personnel
SO   DOE FRASE VOCABULARY

ENGINEERING TECHNICIAN
(0370 ENGR TECH)
DA   November 28, 1990
BT1 Technicians
 BT2 Professional Personnel
  BT3 Occupations
  BT3 Personnel
SO   DOE FRASE VOCABULARY

ENVIRONMENTAL RELEASE
(1563 ENV RELEASE)
DA   November 29, 1990
BT1 Nature of Property Damage
BT1 Releases
RT   Chemical Contamination
RT   Hazardous Spill
RT   Leak Detector
SO   DOE FRASE VOCABULARY

EQUIPMENT INSTALLATION ACTIVITY
(1231 EI ACTIVITY)
DA   November 28, 1990
BT1 Activity Types (DOE FRASE Vocabulary)
 BT2 Activities
SO   DOE FRASE VOCABULARY

EQUIPMENT OPERATOR
(0830 EQUIPMENT OP)
DA   November 28, 1990
BT1 Transport Personnel
 BT2 Occupations
 BT2 Personnel
SO   DOE FRASE VOCABULARY

EQUIPMENT ROOM
(1693 EQUIPMENT RO)
DA   December 10, 1990
BT1 Room
 BT2 Sites/Areas
SO   DOE FRASE VOCABULARY
SO   Environmental Management

EQUIPMENT/PARTS - ELECTRICAL (DOE FRASE VOCABULARY)
(DOE FRASE Vocabulary Numeric Keys 2375-2474)
DA   January 17, 1991
BT1 Equipment
NT1 Adapter(s)
NT1 Amplifier(s)
NT1 Antenna
NT1 Appliance(s)
 NT2 Heating Appliance(s)
NT1 Battery(s)
NT1 Capacitor(s)
NT1 Circuit Breaker(s)
NT1 Circuit(s)
NT1 Computer Terminal(s)
NT1 Computer(s)
NT1 Condenser(s)
NT1 Conductor(s)
NT1 Conduit(s)
NT1 Diode(s)
NT1 Electrical Apparatus
NT1 Electrical Bus
NT1 Electrical Insulator(s)
NT1 Electrode(s)
NT1 Fuse(s)
NT1 Ground Fault Interrupter(s)
NT1 Heat Coil
NT1 Insulator
NT1 Jumper
NT1 Laser
NT1 Light(s)
NT1 Magnetic/Electrolytic Apparatus
NT1 Motor(s)
NT1 Outlet/Receptacle(s)
NT1 Power Cord(s)
NT1 Power Key(s)
NT1 Power Lead
NT1 Power Line(s)
NT1 Power Pole(s)
NT1 Power Supply
 NT2 Uninterruptible Power Supplies
NT1 Preamplifier(s)
NT1 Pull Box
NT1 Radio
NT1 Rectifier(s)

NT1 Regulator
NT1 Relay(s)
 NT2 Building Power Low Voltage Relay
 NT2 Building Time Delay Relay
 NT2 Phase Failure Relays
 NT2 Scram Relay
NT1 Reset Device
NT1 Resistor(s)
NT1 Rheostat(s)
NT1 Shut-Down Circuit(s)
NT1 Solar Collector(s)
NT1 Solenoid(s)
NT1 Starter
NT1 Switch Box
NT1 Switchboard
NT1 Switchgear
NT1 Thermocouple(s)
NT1 Transducer(s)
NT1 Transformer(s)
NT1 Transmitter(s)
NT1 Trouble Light
NT1 Ultraviolet Equipment
NT1 Voltage
NT1 Voltage Regulator
NT1 Wiring
NT1 X-Ray Equipment
RT   DOE FRASE Categories
SO   DOE FRASE VOCABULARY
DEF A subject category used with the DOE FRASE Vocabulary.

EQUIPMENT/PARTS - HEATING (DOE FRASE VOCABULARY)
(DOE FRASE Vocabulary Numeric Keys 2475-2549)
DA   January 17, 1991
SY   Heating Equipment
BT1 Equipment
NT1 Burner(s)
NT1 Evaporator(s)
NT1 Heater(s)
 NT2 Autoclave
 NT2 Boiler(s)
  NT3 Boiler Part(s)
 NT2 Feed Water Heater(s)
 NT2 Furnace(s)
  NT3 Boiler/Industrial Furnaces
 NT2 Oven(s)
 NT2 Stove(s)
NT1 Incinerators
 NT2 Catalytic Incinerators
 NT2 Qualified Incinerators
NT1 Pilot Light(s)
NT1 Retort
RT   DOE FRASE Categories
RT   Heating Appliance(s)
SO   DOE FRASE VOCABULARY
DEF A subject category used with the DOE FRASE Vocabulary.

EQUIPMENT/PARTS - INSTRUMENTATION/MEASURING (DOE FRASE VOC.)
(DOE FRASE Vocabulary Numeric Keys 2750-2824)
DA   January 17, 1991
BT1 Equipment
NT1 Ammeter(s)
NT1 Bridge Crane
NT1 Constant Air Monitor

NT1 Count
NT1 Fast Response Gamma
    Thermometer
NT1 Instrument(s)
NT2 Gauge(s)
NT3 Scale(s)
NT3 Straight Edge Ruler
NT3 Thermostat(s)
NT3 Time Clock
NT2 Graphitar(s)
NT2 Indicator(s)
NT3 Ammeter(s)
NT3 Flow Meter(s)
NT3 Ohmmeter(s)
NT3 Potentiometer
NT3 Spectrometer(s)
NT3 Tachometer(s)
NT3 Thermoluminescent Dosimeters
NT3 Thermometer(s)
NT3 Volt Meter
NT2 Level Alarm
NT2 Log N Recorder
NT2 Ram(s)
NT2 Recorder
NT2 Testing Equipment
NT3 Heat Detector
NT3 Leak Detector
NT3 Oscilloscope(s)
NT3 Scintillation Probe(s)
NT3 Sensor(s)
RT DOE FRASE Categories
SO DOE FRASE VOCABULARY
DEF A subject category used with the
    DOE FRASE Category.

**EQUIPMENT/PARTS - MATERIAL
HANDLING (DOE FRASE
VOCABULARY)**
(DOE FRASE Vocabulary Numeric Keys
2300-2374)
DA January 17, 1991
BT1 Equipment
NT1 Boom
NT1 Drill Rig(s)
NT1 Heavy Mobile Equipment
NT1 Highway Construction Equipment
NT1 Hook(s)
NT1 Load Securing Device
NT1 Manipulator Tape
NT1 Material Handling Device
NT2 Conveyor(s)
NT2 Crane(s)
NT3 Boom Crane
NT3 Bridge Crane
NT3 Mobile Crane
NT3 Overhead Crane
NT2 Earth Moving Equipment
NT3 Dredge(s)
NT2 Hand Truck(s)
NT3 Cart(s)
NT3 Dolly
NT2 Hoisting Apparatus
NT3 Derrick
NT3 Elevator
NT3 Forklift(s)
NT3 Hoist(s)
NT4 Air Hoist
NT3 Lift Bucket
NT3 Manlift(s)
NT3 Scissor Lift

NT1 Reel(s)
NT1 Scraper
NT1 Sling(s)
NT1 Tine(s)
RT DOE FRASE Categories
SO DOE FRASE VOCABULARY
DEF A subject category used with the
    DOE FRASE Vocabulary.

**EQUIPMENT/PARTS - NUCLEAR (DOE
FRASE VOCABULARY)**
(DOE FRASE Vocabulary Numeric Keys
2550-2649)
DA January 17, 1991
BT1 Equipment
BT1 Reactor Components
NT1 Birdcage
NT1 Burial Box(s)
NT1 Cold Trap
NT1 Control Panel(s)
NT1 Control Rod Drive Assembly
NT1 Control Rod(s)
NT2 Gangs
NT2 Raincoats
NT1 Coolant Return Pump
NT1 Core(s)
NT1 Exhaust Stack
NT1 Flux Channel(s)
NT1 Fuel Element(s)
NT2 Fuel Plate
NT3 Cladding
NT3 Plutonium Fuel Plate(s)
NT2 Fuel Rod(s)
NT2 Piles (Nuclear)
NT1 Fuel Handling Equipment
NT2 Fuel Canister(s)
NT2 Fuel Drawer(s)
NT1 Fuel Tool(s)
NT1 Glove Box
NT1 Heat Exchanger
NT2 Low Pressure Recirculation
    System Heat Exchanger
NT2 Shutdown Heat Exchanger
NT1 Hepa Filter
NT1 Ion Exchange Column
NT1 Ion Pump
NT1 Log N Channel
NT1 Manipulator
NT2 Electromechanical Manipulator(s)
NT1 Neutron Source
NT1 Neutron(s)
NT1 Plutonium Process Hood
NT1 Plutonium Recovery Process
NT1 Pressure Vessel Part
NT1 Pressure Vessel(s)
NT1 Reactor Fuel
NT1 Reactor Waste
NT1 Reactor(s)
NT1 Rod Drive
NT1 Shield Plug(s)
NT1 Sodium Scrubber(s)
NT1 Test Train
NT1 Transient Operation
NT1 Transient Test
RT DOE FRASE Categories
SO DOE FRASE VOCABULARY
DEF A subject category used with the
    DOE FRASE Vocabulary.

EQUIPMENT/PARTS - PERSONAL

**PROTECTIVE (DOE FRASE
VOCABULARY)**
(DOE FRASE Vocabulary Numeric Keys
2650-2749)
DA January 17, 1991
BT1 Equipment
NT1 Personal Protective Equipment
NT2 Air Mask
NT2 Anticontamination Clothing
NT3 Acid Suit
NT3 Lab Coat
NT3 Radiation Suit
NT2 Chin Strap
NT2 Dust Mask
NT2 Eye Protection
NT3 Goggles
NT3 Safety Glasses
NT4 Safety Glasses W Side Shields
NT4 Tinted Safety Glasses
NT3 Side Shields
NT2 Fall Protection Device
NT3 Safety Belt
NT3 Safety Line
NT3 Safety Net
NT2 Flame Retardant Clothing
NT2 Foot Protection
NT3 Ankle Protection
NT3 Metatarsal Protection
NT3 Safety Boots
NT3 Safety Shoe(s)
NT3 Shoe Cover(s)
NT4 Metal Shoe Cover
NT2 Hand Protection
NT3 Glove(s)
NT3 Wrist Band
NT2 Head Protection
NT3 Bump Cap
NT3 Faceshield
NT3 Hard Hat
NT3 Helmet
NT3 Sand Blaster's Hood
NT3 Welder's Hood
NT2 Hearing Protection
NT3 Ear Plug(s)
NT2 Leggings
NT2 Padding
NT2 Respirator(s)
NT2 Seat Belt(s)
RT Barriers
RT Clothing
RT DOE FRASE Categories
SO DOE FRASE VOCABULARY
DEF A subject category used with the
    DOE FRASE Vocabulary.

ESCALATOR
(1694 ESCALATOR)
DA December 10, 1990
BT1 Site (DOE FRASE Vocabulary)
BT2 Sites/Areas
RT Elevator
SO DOE FRASE VOCABULARY

ESCAPE HATCH
(1695 ESCAPE HATCH)
DA December 10, 1990
BT1 Site (DOE FRASE Vocabulary)
BT2 Sites/Areas
SO DOE FRASE VOCABULARY

EVAPORATOR(S)
(2481 EVAPORATOR)
DA   January 3, 1991
BT1 Equipment/Parts - Heating (DOE
     FRASE Vocabulary)
 BT2 Equipment
 BT1 Heating Equipment
 SO   DOE FRASE VOCABULARY

EXCRETORY SYSTEM
(1143 EXCRETORY SY)
DA   November 28, 1990
BT1 Body System(s)
 BT2 Human Body Parts
 SO   DOE FRASE VOCABULARY

EXHAUST STACK
(2560 EXHAUST STAC)
DA   January 3, 1991
SY   Stack Effect
BT1 Equipment/Parts - Nuclear (DOE
     FRASE Vocabulary)
 BT2 Equipment
 BT2 Reactor Components
 SO   DOE FRASE VOCABULARY

EXIT
(1696 EXIT;N)
DA   December 10, 1990
BT1 Site (DOE FRASE Vocabulary)
 BT2 Sites/Areas
 SO   DOE FRASE VOCABULARY

EXPLOSION SUPPRESSION SYSTEM
(2032 ESS)
DA   December 10, 1990
BT1 Emergency Systems
 BT2 Systems
BT1 Systems (DOE FRASE Vocabulary)
 BT2 Systems
 SO   DOE FRASE VOCABULARY

EYE(S)
(1084 EYE)
DA   November 28, 1990
BT1 Face
 BT2 Head
  BT3 Human Body Parts
 RT   Area Around Eye
 RT   Eye Protection
 RT   Eyelid(s)
 RT   Goggles
 RT   Safety Glasses
 RT   Safety Glasses W Side Shields
 RT   Side Shields
 SO   DOE FRASE VOCABULARY

EYE PROTECTION
(2658 EYE PROTECTI)
DA   January 3, 1991
BT1 Personal Protective Equipment
 BT2 Equipment/Parts - Personal
      Protective (DOE FRASE
      Vocabulary)
  BT3 Equipment
 NT1 Goggles
 NT1 Safety Glasses
  NT2 Safety Glasses W Side Shields
  NT2 Tinted Safety Glasses

NT1 Side Shields
 RT   Area Around Eye
 RT   Eye(s)
 RT   Eyelid(s)
 SO   DOE FRASE VOCABULARY

EYELID(S)
(1082 EYELID)
DA   November 28, 1990
BT1 Area Around Eye
 BT2 Face
  BT3 Head
   BT4 Human Body Parts
 RT   Conjunctivitis
 RT   Eye Protection
 RT   Eye(s)
 SO   DOE FRASE VOCABULARY

FACE
(1081 FACE)
DA   November 28, 1990
BT1 Head
 BT2 Human Body Parts
 NT1 Area Around Eye
  NT2 Eyelid(s)
 NT1 Eye(s)
 NT1 Lip(s)
 NT1 Nose
 RT   Faceshield
 RT   Mouth
 SO   DOE FRASE VOCABULARY

FACESHIELD
(2659 FACESHIELD)
DA   January 3, 1991
BT1 Head Protection
 BT2 Personal Protective Equipment
  BT3 Equipment/Parts - Personal
       Protective (DOE FRASE
       Vocabulary)
   BT4 Equipment
 RT   Face
 RT   Forehead
 SO   DOE FRASE VOCABULARY

**FACILITIES AND BUILDINGS (DOE
FRASE VOCABULARY)**
(DOE FRASE Vocabulary Numeric Keys
1775-1849)
DA   January 17, 1991
BT1 Facilities
NT1 Building (DOE FRASE Vocabulary)
 NT2 Containment Building
 NT2 Demineralizer Plant
 NT2 Fire Station
 NT2 Garage
 NT2 Guard Station
 NT2 Hospital
 NT2 Hot Shop
 NT2 Machine Shop
 NT2 Maintenance Shop
 NT2 Motel/Hotel
 NT2 Research & Development
      Laboratory
 NT2 Restaurant
 NT2 Sewage Plant
 NT2 Steam Plant
NT1 Facility (DOE FRASE Vocabulary)
 NT2 Fuel Fabrication Facility

 NT2 Fusion Facility
 NT2 Hot Cell Facility
 NT2 Irradiated Fuel Process Facility
 NT2 Irradiated Fuel Storage Facility
 NT2 Low Power Reactor Facility
 NT2 Nuclear Facility
 NT2 Particle Accelerator Facility
 NT2 Production Reactor Facility
 NT2 Rad Waste Treatment/Storage
      Facility
 NT2 Test Reactor Facility
 NT2 Waste Monitoring Facility
 RT   DOE FRASE Categories
 RT   Targets
 SO   DOE FRASE VOCABULARY
DEF A subject category used with the
    DOE FRASE Vocabulary.

FACILITY (DOE FRASE VOCABULARY)
(1778 Facility)
DA   January 31, 1991
BT1 Facilities and Buildings (DOE
     FRASE Vocabulary)
 BT2 Facilities
 NT1 Fuel Fabrication Facility
 NT1 Fusion Facility
 NT1 Hot Cell Facility
 NT1 Irradiated Fuel Process Facility
 NT1 Irradiated Fuel Storage Facility
 NT1 Low Power Reactor Facility
 NT1 Nuclear Facility
 NT1 Particle Accelerator Facility
 NT1 Production Reactor Facility
 NT1 Rad Waste Treatment/Storage
      Facility
 NT1 Test Reactor Facility
 NT1 Waste Monitoring Facility
 SO   DOE FRASE VOCABULARY

**FACTOR RELATIONSHIP AND
SEQUENCE OF EVENTS
VOCABULARY**
DA   February 28, 1991
SY   DOE FRASE VOCABULARY
SY   FRASE Vocabulary
BT1 Vocabulary
 RT   Safety Performance Measurement
      System
 SO   Environmental Management
DEF A controlled vocabulary designed
    for searching the narrative
    description fields of records on the
    Safety Performance Measurement
    System (SPMS).

FALL PROTECTION DEVICE
(2660 FALL PROTECT)
DA   January 3, 1991
BT1 Devices
BT1 Personal Protective Equipment
 BT2 Equipment/Parts - Personal
      Protective (DOE FRASE
      Vocabulary)
  BT3 Equipment
 NT1 Safety Belt
 NT1 Safety Line
 NT1 Safety Net
 SO   DOE FRASE VOCABULARY

FEED WATER HEATER(S)
(2482 FEED WATER H)
DA   January 3, 1991
BT1 Heater(s)
 BT2 Equipment/Parts - Heating (DOE
       FRASE Vocabulary)
  BT3 Equipment
 SO   DOE FRASE VOCABULARY

FIELD AREA
(1607 FIELD AREA)
DA   December 10, 1990
BT1 Area
 BT2 Sites/Areas
 SO   DOE FRASE VOCABULARY

FILE(S)
(3032 FILE)
DA   January 3, 1991
BT1 Tools - Manual
 BT2 Tools (DOE FRASE Vocabulary)
  BT3 Equipment
 SO   DOE FRASE VOCABULARY

FINGER(S)
(1104 FINGER)
DA   November 28, 1990
BT1 Hand(s)
 BT2 Human Body Parts
 RT   Fingernail(s)
 RT   Thumb(s)
 SO   DOE FRASE VOCABULARY

FINGERNAIL(S)
(1105 FINGERNAIL)
DA   November 28, 1990
BT1 Skin
 BT2 Human Body Parts
 RT   Finger(s)
 SO   DOE FRASE VOCABULARY

FIRE ALARM SYSTEM
(2033 FIRE ALARM S)
DA   December 10, 1990
BT1 Alarms
 BT2 Devices
BT1 Fire Protection Systems
 BT2 Emergency Systems
  BT3 Systems
BT1 Systems (DOE FRASE Vocabulary)
 BT2 Systems
 RT   Fire Escape
 RT   Fire Station
 RT   Fire Suppression System
 RT   Firefighter
 SO   DOE FRASE VOCABULARY

FIRE DAMAGE
(1530 FIRE DAMAGE)
DA   November 29, 1990
BT1 Damage
 BT2 Nature of Property Damage
 RT   Heat Damage
 RT   Smoke Damage
 SO   DOE FRASE VOCABULARY

FIRE ESCAPE
(1697 FIRE ESCAPE)

DA   December 10, 1990
BT1 Site (DOE FRASE Vocabulary)
 BT2 Sites/Areas
 RT   Fire Alarm System
 RT   Fire Suppression System
 SO   DOE FRASE VOCABULARY

FIRE STATION
(1779 FIRE STATION)
DA   December 10, 1990
BT1 Building (DOE FRASE Vocabulary)
 BT2 Facilities and Buildings (DOE
       FRASE Vocabulary)
  BT3 Facilities
BT1 Station
 BT2 Site (DOE FRASE Vocabulary)
  BT3 Sites/Areas
 RT   Fire Alarm System
 RT   Firefighter
 SO   DOE FRASE VOCABULARY

FIRE SUPPRESSION SYSTEM
(2058 FS SYSTEM)
DA   December 10, 1990
BT1 Fire Protection Systems
 BT2 Emergency Systems
  BT3 Systems
BT1 Systems (DOE FRASE Vocabulary)
 BT2 Systems
 RT   Fire Escape
 RT   Fire Alarm System
 RT   Foam System
 RT   Sprinkler System
 SO   DOE FRASE VOCABULARY

FIREFIGHTER
(0512 FIREFIGHTER)
DA   November 28, 1990
BT1 Admin. Support/Clerical Employee
 BT2 Occupations
 BT2 Personnel
 RT   Fire Station
 RT   Fire Alarm System
 SO   DOE FRASE VOCABULARY

FIRING RANGE
(1608 FIRING RANGE)
DA   December 10, 1990
BT1 Site (DOE FRASE Vocabulary)
 BT2 Sites/Areas
 SO   DOE FRASE VOCABULARY

FLAME RETARDANT CLOTHING
(2661 FLAME RETARD)
DA   January 3, 1991
BT1 Clothing
BT1 Personal Protective Equipment
 BT2 Equipment/Parts - Personal
       Protective (DOE FRASE
       Vocabulary)
  BT3 Equipment
 SO   DOE FRASE VOCABULARY

FLASH BURN(S)
(1321 FLASH BURN)
DA   November 28, 1990
BT1 Burn(s)
 BT2 Injuries
 SO   DOE FRASE VOCABULARY

FLOOR OPENING
(1698 FLOOR OPENIN)
DA   December 10, 1990
BT1 Site (DOE FRASE Vocabulary)
 BT2 Sites/Areas
 SO   DOE FRASE VOCABULARY

FLOW METER(S)
(2752 FLOW METER)
DA   January 3, 1991
BT1 Indicator(s)
 BT2 Instrument(s)
  BT3 Equipment/Parts -
       Instrumentation/Measuring
       (DOE FRASE Voc.)
   BT4 Equipment
 SO   DOE FRASE VOCABULARY

FLUX CHANNEL(S)
(2561 FLUX CHANNEL)
DA   January 3, 1991
BT1 Equipment/Parts - Nuclear (DOE
       FRASE Vocabulary)
 BT2 Equipment
 BT2 Reactor Components
 SO   DOE FRASE VOCABULARY

FOAM SYSTEM
(2034 FOAM SYSTEM)
DA   December 10, 1990
BT1 Systems (DOE FRASE Vocabulary)
 BT2 Systems
 RT   Fire Suppression System
 SO   DOE FRASE VOCABULARY

FOOD SERVICE ACTIVITY
(1232 FS ACTIVITY)
DA   November 28, 1990
BT1 Activity Types (DOE FRASE
       Vocabulary)
 BT2 Activities
 RT   Eating Area
 SO   DOE FRASE VOCABULARY

FOOD SERVICE EMPLOYEE
(0521 FOOD SERVICE)
DA   November 28, 1990
BT1 Admin. Support/Clerical Employee
 BT2 Occupations
 BT2 Personnel
 RT   Kitchen
 SO   DOE FRASE VOCABULARY

FOOT PROTECTION
(2662 FOOT PROTECT)
DA   January 3, 1991
BT1 Personal Protective Equipment
 BT2 Equipment/Parts - Personal
       Protective (DOE FRASE
       Vocabulary)
  BT3 Equipment
 NT1 Ankle Protection
 NT1 Metatarsal Protection
 NT1 Safety Boots
 NT1 Safety Shoe(s)
 NT1 Shoe Cover(s)
  NT2 Metal Shoe Cover
 RT   Foot/Feet
 RT   Heel(s)

RT Sole(s)
RT Toe(s)
SO DOE FRASE VOCABULARY

FOOT/FEET
(1127 FOOTP)
DA November 28, 1990
BT1 Human Body Parts
NT1 Heel(s)
NT1 Sole(s)
NT1 Toe(s)
RT Ankle(s)
RT Foot Protection
RT Leg(s)
RT Safety Boots
SO DOE FRASE VOCABULARY

FOREARM(S)
(1098 FOREARM)
DA November 28, 1990
BT1 Arm(s)
 BT2 Human Body Parts
RT Forearm Protection
SO DOE FRASE VOCABULARY

FOREARM PROTECTION
(2663 FOREARM PROT)
DA January 3, 1991
RT Arm(s)
RT Elbow(s)
RT Forearm(s)
SO DOE FRASE VOCABULARY

FOREHEAD
(1080 FOREHEAD)
DA November 28, 1990
BT1 Head
 BT2 Human Body Parts
RT Faceshield
RT Head Protection
SO DOE FRASE VOCABULARY

FOREST WORKER
(0570 FOREST WORKE)
DA November 28, 1990
BT1 Agriculture Personnel
 BT2 Occupations
 BT2 Personnel
SO DOE FRASE VOCABULARY

FORKLIFT(S)
(2321 FORKLIFT)
DA December 10, 1990
BT1 Hoisting Apparatus
 BT2 Material Handling Device
 BT3 Devices
 BT3 Equipment/Parts - Material
    Handling (DOE FRASE
    Vocabulary)
 BT4 Equipment
SO DOE FRASE VOCABULARY

FRACTURE
(1322 FR)
DA November 28, 1990
BT1 Injuries
RT Bone(s)
SO DOE FRASE VOCABULARY

FRASE VOCABULARY
DA February 28, 1991
SY DOE FRASE VOCABULARY
SY Factor Relationship and Sequence
    of Events Vocabulary
BT1 Vocabulary
RT DOE FRASE Categories
RT Key Word
RT Safety Performance Measurement
    System
DEF A controlled vocabulary designed
    for use in searching the narrative
    description fields of records on the
    Safety Performance Measurement
    System (SPMS).

**FROSTBITE/OTHER LOW TEMP
EFFECTS**
(1386 FROSTBITE)
DA November 28, 1990
BT1 Effects
BT1 Illnesses
SO DOE FRASE VOCABULARY

FUEL CANISTER(S)
(2562 FUEL CANISTE)
DA January 3, 1991
BT1 Fuel Handling Equipment
 BT2 Equipment/Parts - Nuclear (DOE
    FRASE Vocabulary)
 BT3 Equipment
 BT3 Reactor Components
RT Fuel Element(s)
SO DOE FRASE VOCABULARY

FUEL DRAWER(S)
(2563 FUEL DRAWER)
DA January 3, 1991
BT1 Fuel Handling Equipment
 BT2 Equipment/Parts - Nuclear (DOE
    FRASE Vocabulary)
 BT3 Equipment
 BT3 Reactor Components
SO DOE FRASE VOCABULARY

FUEL ELEMENT(S)
(2564 FUEL ELEMENT)
DA January 3, 1991
BT1 Equipment/Parts - Nuclear (DOE
    FRASE Vocabulary)
 BT2 Equipment
 BT2 Reactor Components
NT1 Fuel Plate
 NT2 Cladding
 NT2 Plutonium Fuel Plate(s)
NT1 Fuel Rod(s)
NT1 Piles (Nuclear)
RT Core(s)
RT Fuel Canister(s)
RT Fuel Handling Equipment
RT Reactor Fuel
SO DOE FRASE VOCABULARY
DEF (NRC Glossary of Terms: Nuclear
    Power and Radiation) A cluster of
    fuel rods (or plates). Also called a
    fuel assembly. Many fuel
    assemblies make up a reactor
    core.

FUEL FABRICATION FACILITY
(1797 FUEL FABRICA)
DA December 10, 1990
BT1 Facility (DOE FRASE Vocabulary)
 BT2 Facilities and Buildings (DOE
    FRASE Vocabulary)
 BT3 Facilities
SO DOE FRASE VOCABULARY

FUEL HANDLING ACTIVITY
(1233 FH ACTIVITY)
DA November 28, 1990
BT1 Activity Types (DOE FRASE
    Vocabulary)
 BT2 Activities
SO DOE FRASE VOCABULARY

FUEL HANDLING EQUIPMENT
(2565 FH EQUIPMENT)
DA January 3, 1991
BT1 Equipment/Parts - Nuclear (DOE
    FRASE Vocabulary)
 BT2 Equipment
 BT2 Reactor Components
NT1 Fuel Canister(s)
NT1 Fuel Drawer(s)
RT Fuel Element(s)
SO DOE FRASE VOCABULARY

FUEL PLATE
(2566 FUEL PLATE)
DA January 3, 1991
BT1 Fuel Element(s)
 BT2 Equipment/Parts - Nuclear (DOE
    FRASE Vocabulary)
 BT3 Equipment
 BT3 Reactor Components
NT1 Cladding
NT1 Plutonium Fuel Plate(s)
SO DOE FRASE VOCABULARY

FUEL ROD(S)
(2567 FUEL ROD)
DA January 3, 1991
BT1 Fuel Element(s)
 BT2 Equipment/Parts - Nuclear (DOE
    FRASE Vocabulary)
 BT3 Equipment
 BT3 Reactor Components
SO DOE FRASE VOCABULARY

FUEL TOOL(S)
(2568 FUEL TOOL)
DA January 3, 1991
BT1 Equipment/Parts - Nuclear (DOE
    FRASE Vocabulary)
 BT2 Equipment
 BT2 Reactor Components
SO DOE FRASE VOCABULARY

FURNACE(S)
(2483 FURNACE)
DA January 3, 1991
BT1 Heater(s)
 BT2 Equipment/Parts - Heating (DOE
    FRASE Vocabulary)
 BT3 Equipment
NT1 Boiler/Industrial Furnaces
SO DOE FRASE VOCABULARY

FUSE(S)
(2394 FUSE)
DA January 3, 1991
BT1 Equipment/Parts - Electrical (DOE
     FRASE Vocabulary)
BT2 Equipment
SO DOE FRASE VOCABULARY

FUSION FACILITY
(1798 FUSION FACIL)
DA December 10, 1990
BT1 Facility (DOE FRASE Vocabulary)
BT2 Facilities and Buildings (DOE
     FRASE Vocabulary)
BT3 Facilities
SO DOE FRASE VOCABULARY

GALLERY
(1700 GALLERY)
DA December 10, 1990
BT1 Site (DOE FRASE Vocabulary)
BT2 Sites/Areas
SO DOE FRASE VOCABULARY

GARAGE
(1780 GARAGE)
DA December 10, 1990
BT1 Building (DOE FRASE Vocabulary)
BT2 Facilities and Buildings (DOE
     FRASE Vocabulary)
BT3 Facilities
RT Maintenance Area
RT Maintenance Shop
RT Mechanic/Repairer
RT Parking Space
RT Parking Lot
RT Vehicle Service Area
RT Vehicle Maint/Repair Activity
SO DOE FRASE VOCABULARY

GAUGE(S)
(2753 GAUGE)
DA January 3, 1991
BT1 Instrument(s)
BT2 Equipment/Parts -
     Instrumentation/Measuring (DOE
     FRASE Voc.)
BT3 Equipment
NT1 Scale(s)
NT1 Straight Edge Ruler
NT1 Thermostat(s)
NT1 Time Clock
SO DOE FRASE VOCABULARY

GENITALS
(1118 GENITALS)
DA November 28, 1990
BT1 Human Body Parts
RT Groin
SO DOE FRASE VOCABULARY

GLOVE(S)
(2664 GLOVE)
DA January 3, 1991
BT1 Hand Protection
BT2 Personal Protective Equipment
BT3 Equipment/Parts - Personal
     Protective (DOE FRASE
     Vocabulary)

BT4 Equipment
RT Hand(s)
RT Knuckle(s)
RT Thumb(s)
SO DOE FRASE VOCABULARY

GLOVE BOX
(2569 GLOVE BOX)
DA January 3, 1991
BT1 Equipment/Parts - Nuclear (DOE
     FRASE Vocabulary)
BT2 Equipment
BT2 Reactor Components
SO DOE FRASE VOCABULARY

GOGGLES
(2665 GOGGLES)
DA January 3, 1991
BT1 Eye Protection
BT2 Personal Protective Equipment
BT3 Equipment/Parts - Personal
     Protective (DOE FRASE
     Vocabulary)
BT4 Equipment
RT Eye(s)
SO DOE FRASE VOCABULARY

GRAPHITAR(S)
(2754 GRAPHITAR)
DA January 3, 1991
BT1 Instrument(s)
BT2 Equipment/Parts -
     Instrumentation/Measuring (DOE
     FRASE Voc.)
BT3 Equipment
SO DOE FRASE VOCABULARY

GRINDING MACHINE
(2123 GRINDING MAC)
DA December 10, 1990
BT1 Machines (DOE FRASE
     Vocabulary)
BT2 Equipment
SO DOE FRASE VOCABULARY

GROIN
(1117 GROIN)
DA November 28, 1990
BT1 Trunk
BT2 Human Body Parts
RT Abdomen
RT Genitals
RT Thigh(s)
SO DOE FRASE VOCABULARY

GROUND
(1609 GROUND;N)
DA December 10, 1990
BT1 Site (DOE FRASE Vocabulary)
BT2 Sites/Areas
SO DOE FRASE VOCABULARY

GROUND FAULT INTERRUPTER(S)
(2396 GFI)
DA January 3, 1991
BT1 Equipment/Parts - Electrical (DOE
     FRASE Vocabulary)

BT2 Equipment
SO DOE FRASE VOCABULARY

GROUNDS MAINTENANCE ACTIVITY
(1234 GM ACTIVITY)
DA November 28, 1990
BT1 Activity Types (DOE FRASE
     Vocabulary)
BT2 Activities
BT1 Maintenance
BT2 Activities
RT Groundskeeper
SO DOE FRASE VOCABULARY

GROUNDSKEEPER
(0562 GROUNDSKEEPE)
DA November 28, 1990
BT1 Agriculture Personnel
BT2 Occupations
BT2 Personnel
RT Grounds Maintenance Activity
SO DOE FRASE VOCABULARY

GUARD STATION
(1781 GUARD STATIO)
DA December 10, 1990
BT1 Building (DOE FRASE Vocabulary)
BT2 Facilities and Buildings (DOE
     FRASE Vocabulary)
BT3 Facilities
BT1 Station
BT2 Site (DOE FRASE Vocabulary)
BT3 Sites/Areas
RT Plant Protection System
RT Security Guard
SO DOE FRASE VOCABULARY

HACK SAW
(3033 HACK SAW)
DA January 3, 1991
BT1 Saws
BT1 Tools - Manual
BT2 Tools (DOE FRASE Vocabulary)
BT3 Equipment
SO DOE FRASE VOCABULARY

HAIR
(1076 HAIR)
DA November 28, 1990
BT1 Skin
BT2 Human Body Parts
RT Scalp
SO DOE FRASE VOCABULARY

HALON SYSTEM
(2035 HALON SYSTEM)
DA December 10, 1990
BT1 Systems (DOE FRASE Vocabulary)
BT2 Systems
SO DOE FRASE VOCABULARY

HAMMER
(3034 HAMMER)
DA January 3, 1991
BT1 Tools - Manual
BT2 Tools (DOE FRASE Vocabulary)
BT3 Equipment

NT1 Sledgehammer(s)
SO DOE FRASE VOCABULARY

HAND(S)
(1100 HAND)
DA   November 28, 1990
BT1 Human Body Parts
NT1 Finger(s)
NT1 Palm(s)
NT1 Thumb(s)
RT Arm(s)
RT Glove(s)
RT Hand Protection
RT Knuckle(s)
RT Wrist(s)
SO DOE FRASE VOCABULARY

HAND IRON(S)
(3035 HAND IRON)
DA   January 3, 1991
BT1 Tools - Manual
 BT2 Tools (DOE FRASE Vocabulary)
 BT3 Equipment
SO DOE FRASE VOCABULARY

HAND PROTECTION
(2666 HAND PROTECT)
DA   January 3, 1991
BT1 Personal Protective Equipment
 BT2 Equipment/Parts - Personal
      Protective (DOE FRASE
      Vocabulary)
 BT3 Equipment
NT1 Glove(s)
NT1 Wrist Band
RT Hand(s)
RT Knuckle(s)
RT Palm(s)
RT Thumb(s)
SO DOE FRASE VOCABULARY

HAND SAW
(3036 HAND SAW)
DA   January 3, 1991
BT1 Saws
BT1 Tools - Manual
 BT2 Tools (DOE FRASE Vocabulary)
 BT3 Equipment
SO DOE FRASE VOCABULARY

HAND STAPLER
(3037 HAND STAPLER)
DA   January 3, 1991
BT1 Tools - Manual
 BT2 Tools (DOE FRASE Vocabulary)
 BT3 Equipment
SO DOE FRASE VOCABULARY

HAND TRUCK(S)
(2323 HAND TRUCK)
DA   December 10, 1990
BT1 Material Handling Device
 BT2 Devices
 BT2 Equipment/Parts - Material
      Handling (DOE FRASE
      Vocabulary)
 BT3 Equipment
NT1 Cart(s)

NT1 Dolly
SO DOE FRASE VOCABULARY

HANDLER/LABORER/HELPER
(0850 LABORER)
DA   November 28, 1990
BT1 Occupations
BT1 Personnel
SO DOE FRASE VOCABULARY

HARD HAT
(2667 HARD HAT)
DA   January 3, 1991
BT1 Head Protection
 BT2 Personal Protective Equipment
  BT3 Equipment/Parts - Personal
      Protective (DOE FRASE
      Vocabulary)
  BT4 Equipment
RT Helmet
SO DOE FRASE VOCABULARY

HATCH
(1701 HATCH)
DA   December 10, 1990
BT1 Site (DOE FRASE Vocabulary)
 BT2 Sites/Areas
SO DOE FRASE VOCABULARY

HATCHET(S)
(3038 HATCHET)
DA   January 3, 1991
BT1 Tools - Manual
 BT2 Tools (DOE FRASE Vocabulary)
 BT3 Equipment
RT Ax(s)
SO DOE FRASE VOCABULARY

HAZARDOUS SPILL
(1531 HAZARDOUS SP)
DA   November 29, 1990
BT1 Nature of Property Damage
RT Chemical Contamination
RT Environmental Release
SO DOE FRASE VOCABULARY

HEAD
(1075 HEAD)
DA   November 28, 1990
BT1 Human Body Parts
NT1 Chin
NT1 Ear(s)
NT1 Face
 NT2 Area Around Eye
  NT3 Eyelid(s)
 NT2 Eye(s)
 NT2 Lip(s)
 NT2 Nose
NT1 Forehead
NT1 Skull (Frase)
 NT2 Jaw
RT Brain
RT Concussion
RT Contusion(S)
RT Mouth
RT Scalp
SO DOE FRASE VOCABULARY

HEAD PROTECTION
(2668 HEAD PROTECT)
DA   January 3, 1991
BT1 Personal Protective Equipment
 BT2 Equipment/Parts - Personal
      Protective (DOE FRASE
      Vocabulary)
 BT3 Equipment
NT1 Bump Cap
NT1 Faceshield
NT1 Hard Hat
NT1 Helmet
NT1 Sand Blaster's Hood
NT1 Welder's Hood
RT Concussion
RT Contusion(S)
RT Forehead
RT Skull (Frase)
SO DOE FRASE VOCABULARY

HEALTH PHYSICIST
(0184 HEALTH PHYS)
DA   November 28, 1990
BT1 Scientist
 BT2 Professional Personnel
  BT3 Occupations
  BT3 Personnel
SO DOE FRASE VOCABULARY
DEF An individual engaged in the study
    of science concerned with
    recognition, evaluation and control
    of health hazards from ionizing
    radiation.

HEALTH TECHNICIAN
(0360 HEALTH TECH)
DA   November 28, 1990
BT1 Technicians
 BT2 Professional Personnel
  BT3 Occupations
  BT3 Personnel
SO DOE FRASE VOCABULARY

HEARING IMPAIRMENT
(1387 HEARING IMPA)
DA   November 28, 1990
BT1 Illnesses
RT Ear Muffs
RT Ear Plug(s)
RT Ear(s)
SO DOE FRASE VOCABULARY

HEARING PROTECTION
(2669 HEARING PROT)
DA   January 3, 1991
BT1 Personal Protective Equipment
 BT2 Equipment/Parts - Personal
      Protective (DOE FRASE
      Vocabulary)
 BT3 Equipment
NT1 Ear Plug(s)
RT Ear Muffs
RT Ear(s)
SO DOE FRASE VOCABULARY

HEART ATTACK
(1389 HEART ATTACK)
DA   November 28, 1990

BT1 Illnesses
SO DOE FRASE VOCABULARY

HEAT COIL
(2397 HEAT COIL)
DA January 3, 1991
BT1 Equipment/Parts - Electrical (DOE
   FRASE Vocabulary)
BT2 Equipment
SO DOE FRASE VOCABULARY

HEAT DAMAGE
(1532 HEAT DAMAGE)
DA November 29, 1990
BT1 Damage
BT2 Nature of Property Damage
RT Fire Damage
RT Water Damage
SO DOE FRASE VOCABULARY

HEAT DETECTOR
(2755 HEAT DETECTO)
DA January 3, 1991
BT1 Testing Equipment
BT2 Instrument(s)
BT3 Equipment/Parts -
   Instrumentation/Measuring
   (DOE FRASE Voc.)
BT4 Equipment
SO DOE FRASE VOCABULARY

HEAT EXCHANGER
(2570 HEAT EXCHANG)
DA January 3, 1991
BT1 Equipment/Parts - Nuclear (DOE
   FRASE Vocabulary)
BT2 Equipment
BT2 Reactor Components
NT1 Low Pressure Recirculation System
   Heat Exchanger
NT1 Shutdown Heat Exchanger
SO DOE FRASE VOCABULARY
DEF (NRC Glossary of Terms: Nuclear
   Power and Radiation) Any device
   that transfers heat from one fluid
   (liquid or gas) to another fluid or to
   the environment.

**HEAT STROKE/OTHER HIGH TEMP
EFFECT**
(1323 HEAT STROKE)
DA November 28, 1990
BT1 Effects
BT1 Injuries
RT Sunburn
SO DOE FRASE VOCABULARY

HEATER(S)
(2485 HEATER)
DA January 3, 1991
BT1 Equipment/Parts - Heating (DOE
   FRASE Vocabulary)
BT2 Equipment
NT1 Autoclave
NT1 Boiler(s)
NT2 Boiler Part(s)
NT1 Feed Water Heater(s)
NT1 Furnace(s)
NT2 Boiler/Industrial Furnaces

NT1 Oven(s)
NT1 Stove(s)
RT Boiler/Industrial Furnaces
SO DOE FRASE VOCABULARY

HEATING APPLIANCE(S)
DA January 3, 1991
BT1 Appliance(s)
BT2 Equipment/Parts - Electrical (DOE
   FRASE Vocabulary)
BT3 Equipment
RT Equipment/Parts - Heating (DOE
   FRASE Vocabulary)
RT Heating Equipment
SO DOE FRASE VOCABULARY

HEATING EQUIPMENT
(2486 HEATING EQUI)
DA January 3, 1991
SY Equipment/Parts - Heating (DOE
   FRASE Vocabulary)
NT1 Burner(s)
NT1 Evaporator(s)
NT1 Incinerators
NT2 Catalytic Incinerators
NT2 Qualified Incinerators
NT1 Pilot Light(s)
NT1 Retort
RT Heating Appliance(s)
SO DOE FRASE VOCABULARY

HEAVY MOBILE EQUIPMENT
(2324 HM EQUIPMENT)
DA December 10, 1990
BT1 Equipment/Parts - Material
   Handling (DOE FRASE
   Vocabulary)
BT2 Equipment
SO DOE FRASE VOCABULARY

HEEL(S)
(1128 HEEL)
DA November 28, 1990
BT1 Foot/Feet
BT2 Human Body Parts
RT Foot Protection
SO DOE FRASE VOCABULARY

HELMET
(2670 HELMET)
DA January 3, 1991
BT1 Head Protection
BT2 Personal Protective Equipment
BT3 Equipment/Parts - Personal
   Protective (DOE FRASE
   Vocabulary)
BT4 Equipment
RT Hard Hat
SO DOE FRASE VOCABULARY

HEMORRHOIDS
(1390 HEMORRHOIDS)
DA November 28, 1990
BT1 Illnesses
SO DOE FRASE VOCABULARY

HEPA FILTER
(2571 HEPA FILTER)

DA January 3, 1991
BT1 Equipment/Parts - Nuclear (DOE
   FRASE Vocabulary)
BT2 Equipment
BT2 Reactor Components
BT1 Filters
BT2 Devices
SO DOE FRASE VOCABULARY

HEPATITIS
(1392 INFECTIOUS H)
DA November 28, 1990
BT1 Diseases of Blood
BT2 Diseases
BT2 Illnesses
SO DOE FRASE VOCABULARY

HERNIA
(1324 HERNIA)
DA November 28, 1990
BT1 Injuries
SO DOE FRASE VOCABULARY

HIGHBAY
(1703 HIGHBAY)
DA December 10, 1990
BT1 Site (DOE FRASE Vocabulary)
BT2 Sites/Areas
SO DOE FRASE VOCABULARY

HIGHWAY
(1854 HIGHWAY)
DA December 10, 1990
BT1 Routes (Transportation)
BT1 Structures (DOE FRASE
   Vocabulary)
RT Intersection
SO DOE FRASE VOCABULARY

**HIGHWAY CONSTRUCTION
EQUIPMENT**
(2325 HIGHWAY CONS)
DA December 10, 1990
BT1 Equipment/Parts - Material
   Handling (DOE FRASE
   Vocabulary)
BT2 Equipment
SO DOE FRASE VOCABULARY

HILL
(1610 HILL)
DA December 10, 1990
BT1 Site (DOE FRASE Vocabulary)
BT2 Sites/Areas
SO DOE FRASE VOCABULARY

HIP(S)
(1122 HIP)
DA November 28, 1990
BT1 Joint(s)
BT2 Human Body Parts
RT Buttock(s)
SO DOE FRASE VOCABULARY

HOE(S)
(3039 HOE)
DA January 3, 1991
BT1 Tools - Manual

BT2 Tools (DOE FRASE Vocabulary)
  BT3 Equipment
  SO DOE FRASE VOCABULARY

HOIST(S)
(2327 HOIST)
DA December 10, 1990
BT1 Hoisting Apparatus
  BT2 Material Handling Device
    BT3 Devices
    BT3 Equipment/Parts - Material
      Handling (DOE FRASE
      Vocabulary)
      BT4 Equipment
NT1 Air Hoist
SO DOE FRASE VOCABULARY

HOISTING APPARATUS
(2328 HOIST APPAR)
DA December 10, 1990
BT1 Material Handling Device
  BT2 Devices
  BT2 Equipment/Parts - Material
    Handling (DOE FRASE
    Vocabulary)
    BT3 Equipment
NT1 Derrick
NT1 Elevator
NT1 Forklift(s)
NT1 Hoist(s)
  NT2 Air Hoist
NT1 Lift Bucket
NT1 Manlift(s)
NT1 Scissor Lift
SO DOE FRASE VOCABULARY

HOOK(S)
(2330 HOOK;N)
DA December 10, 1990
BT1 Equipment/Parts - Material
  Handling (DOE FRASE
  Vocabulary)
  BT2 Equipment
SO DOE FRASE VOCABULARY

HOSPITAL
(1782 HOSPITAL)
DA December 10, 1990
BT1 Building (DOE FRASE Vocabulary)
  BT2 Facilities and Buildings (DOE
    FRASE Vocabulary)
    BT3 Facilities
RT Doctor/Nurse
RT Illnesses
RT Injuries
RT Mental Disorders
SO DOE FRASE VOCABULARY

HOT CELL FACILITY
(1799 HOT CELL)
DA December 10, 1990
BT1 Facility (DOE FRASE Vocabulary)
  BT2 Facilities and Buildings (DOE
    FRASE Vocabulary)
    BT3 Facilities
  SO DOE FRASE VOCABULARY

HOT SHOP
(1783 HOT SHOP)

DA December 10, 1990
BT1 Building (DOE FRASE Vocabulary)
  BT2 Facilities and Buildings (DOE
    FRASE Vocabulary)
    BT3 Facilities
SO DOE FRASE VOCABULARY

HUMAN BODY PARTS
(DOE FRASE Vocabulary Numeric Keys
1075-1174)
DA November 29, 1990
NT1 Arm(s)
  NT2 Forearm(s)
  NT2 Upper Arm
NT1 Body System(s)
  NT2 Circulatory System
    NT3 Blood
  NT2 Digestive System
    NT3 Mouth
      NT4 Tooth/Teeth
    NT3 Rectum
  NT2 Excretory System
  NT2 Nervous System
    NT3 Brain
    NT3 Spinal Cord
  NT2 Respiratory System
    NT3 Bronchial Epithelium
    NT3 Nose
    NT3 Throat
NT1 Bone(s)
  NT2 Rib(s)
  NT2 Spine
    NT3 Coccyx
NT1 Breast(s)
NT1 Extremities
NT1 Foot/Feet
  NT2 Heel(s)
  NT2 Sole(s)
  NT2 Toe(s)
NT1 Genitals
NT1 Hand(s)
  NT2 Finger(s)
  NT2 Palm(s)
  NT2 Thumb(s)
NT1 Head
  NT2 Chin
  NT2 Ear(s)
  NT2 Face
    NT3 Area Around Eye
      NT4 Eyelid(s)
    NT3 Eye(s)
    NT3 Lip(s)
    NT3 Nose
  NT2 Forehead
  NT2 Skull (Frase)
    NT3 Jaw
NT1 Joint(s)
  NT2 Ankle(s)
  NT2 Elbow(s)
  NT2 Hip(s)
  NT2 Knee(s)
  NT2 Knuckle(s)
  NT2 Shoulder(s)
  NT2 Wrist(s)
NT1 Leg(s)
  NT2 Lower Leg
  NT2 Thigh(s)
NT1 Multiple Body Parts
NT1 Muscle/Tendon(s)
NT1 Neck

NT1 Skin
  NT2 Fingernail(s)
  NT2 Hair
  NT2 Scalp
NT1 Trunk
  NT2 Abdomen
  NT2 Back
    NT3 Lower Back
    NT3 Upper Back
  NT2 Buttock(s)
  NT2 Chest
  NT2 Groin
  NT2 Rib(s)
RT DOE FRASE Categories
RT Targets
RT Whole Body
SO DOE FRASE VOCABULARY

HVAC SYSTEM
(2036 HVAC SYSTEM)
DA December 10, 1990
SY Heating, Ventilation, Air
  Conditioning
BT1 Systems (DOE FRASE Vocabulary)
  BT2 Systems
SO DOE FRASE VOCABULARY

ILLNESS
(1391 ILLNESS)
DA November 28, 1990
RT Injury
SO DOE FRASE VOCABULARY

ILLNESSES
(DOE FRASE Vocabulary Numeric Keys
1375-1449)
DA January 17, 1991
NT1 Asbestosis
NT1 Benign and Unspecified Neoplasm
NT1 Chemical Reaction
NT1 Complications Peculiar to Med.
  Care
NT1 Cond of Respiratory Sys. Non-Toxic
  NT2 Upper Respiratory Condition
  NT2 Upper Respiratory Disease
    NT3 Tuberculosis
NT1 Conjunctivitis
NT1 Contagious or Infectious Disease
  NT2 Conjunctivitis
NT1 Dermatitis
NT1 Disease of Central Nervous System
NT1 Diseases of Blood
  NT2 Hepatitis
NT1 Fluorosis
NT1 Frostbite/Other Low Temp Effects
NT1 Hearing Impairment
NT1 Heart Attack
NT1 Hemorrhoids
NT1 Insulin Reaction
NT1 Ionizing Radiation Effects
NT1 Malignant Neoplasm
NT1 Mental Disorders
NT1 Occupational Illnesses
NT1 Other Complications
NT1 Other Skin Conditions
NT1 Physical Harm
NT1 Poisoning
  NT2 Systemic Poisoning
NT1 Sunburn
NT1 Synovitis

NT1 Systemic Effects
NT1 Tendonitis
NT1 Tetanus
NT1 Toxic Effects to Single System
NT1 Ulcer(s)
RT Diseases
RT DOE FRASE Categories
RT Effects
RT Hospital
RT Injuries
SO DOE FRASE VOCABULARY
DEF Equivalent to the DOE FRASE
    Vocabulary Category "Nature of
    Illness".

IMPACT WRENCH
(3040 IMPACT WRENC)
DA January 3, 1991
BT1 Wrench(s)
 BT2 Tools (DOE FRASE Vocabulary)
  BT3 Equipment
SO DOE FRASE VOCABULARY

INDICATOR(S)
(2756 INDICATOR)
DA January 3, 1991
BT1 Instrument(s)
 BT2 Equipment/Parts -
     Instrumentation/Measuring (DOE
     FRASE Voc.)
  BT3 Equipment
NT1 Ammeter(s)
NT1 Flow Meter(s)
NT1 Ohmmeter(s)
NT1 Potentiometer
NT1 Spectrometer(s)
NT1 Tachometer(s)
NT1 Thermoluminescent Dosimeters
NT1 Thermometer(s)
NT1 Volt Meter
RT Trouble Light
SO DOE FRASE VOCABULARY

INFECTION
(1325 INFECTION)
DA November 28, 1990
BT1 Injuries
RT Inflammation
RT Irritation
SO DOE FRASE VOCABULARY

INFLAMMATION
(1326 INFLAMMATION)
DA November 28, 1990
BT1 Injuries
RT Infection
RT Irritation
SO DOE FRASE VOCABULARY

INJURIES
(DOE FRASE Vocabulary Numeric Keys
 1300-1374)
DA January 17, 1991
NT1 Abrasion
NT1 Amputation
NT1 Animal Bite
 NT2 Snake Bite
NT1 Asphyxia
NT1 Avulsion

NT1 Blister(s)
NT1 Blood Clot
NT1 Bodily Injuries
NT1 Brain Damage
 NT2 Concussion
 NT2 Contusion(S)
NT1 Burn(s)
 NT2 Chemical Burn(s)
 NT2 Electrical Burn(s)
 NT2 Flash Burn(s)
NT1 Bursitis
NT1 Damage to Prosthetic Device
NT1 Death
NT1 Dental Injury
NT1 Disabling Injuries
NT1 Dislocation
NT1 Drowning
NT1 Electric Shock
NT1 Electrocution
NT1 Fracture
NT1 Heat Stroke/Other High Temp
     Effect
NT1 Hernia
NT1 Infection
NT1 Inflammation
NT1 Insect Sting
NT1 Internal Deposition
NT1 Irritation
NT1 Laceration
NT1 Loss of Consciousness
NT1 Loss of Condenser Vacuum
NT1 Multiple Injuries
NT1 No Personnel Injury
NT1 Nosebleed
NT1 Occupational Injuries
NT1 Physical Harm
NT1 Pinched Nerve
NT1 Puncture
NT1 Radiation Exposure
NT1 Ruptured Disk
NT1 Sprain
NT1 Strain
NT1 Strangulation
NT1 Torn Cartilage
RT Diseases
RT DOE FRASE Categories
RT Effects
RT Hospital
RT Illnesses
SO DOE FRASE VOCABULARY
DEF Equivalent to the DOE FRASE
    Vocabulary Category "Nature of
    Injury".

INJURY
(1327 INJURY)
DA November 28, 1990
RT Illness
SO DOE FRASE VOCABULARY

INSECT STING
(1328 INSECT STING)
DA November 28, 1990
BT1 Injuries
SO DOE FRASE VOCABULARY

INSPECTION/MONITORING ACTIVITY
(1235 IM ACTIVITY)
DA November 28, 1990

BT1 Activity Types (DOE FRASE
     Vocabulary)
 BT2 Activities
SO DOE FRASE VOCABULARY

INSTRUMENT(S)
(2757 INSTRUMENT)
DA January 3, 1991
BT1 Equipment/Parts -
     Instrumentation/Measuring (DOE
     FRASE Voc.)
 BT2 Equipment
NT1 Gauge(s)
 NT2 Scale(s)
 NT2 Straight Edge Ruler
 NT2 Thermostat(s)
 NT2 Time Clock
NT1 Graphitar(s)
NT1 Indicator(s)
 NT2 Ammeter(s)
 NT2 Flow Meter(s)
 NT2 Ohmmeter(s)
 NT2 Potentiometer
 NT2 Spectrometer(s)
 NT2 Tachometer(s)
 NT2 Thermoluminescent Dosimeters
 NT2 Thermometer(s)
 NT2 Volt Meter
NT1 Level Alarm
NT1 Log N Recorder
NT1 Ram(s)
NT1 Recorder
NT1 Testing Equipment
 NT2 Heat Detector
 NT2 Leak Detector
 NT2 Oscilloscope(s)
 NT2 Scintillation Probe(s)
 NT2 Sensor(s)
SO DOE FRASE VOCABULARY

INSTRUMENT AIR SYSTEM
(2037 IAS)
DA December 10, 1990
BT1 Systems (DOE FRASE Vocabulary)
 BT2 Systems
SO DOE FRASE VOCABULARY

INSULATOR
(2436 INSULATOR)
DA January 3, 1991
BT1 Equipment/Parts - Electrical (DOE
     FRASE Vocabulary)
 BT2 Equipment
SO DOE FRASE VOCABULARY

INSULIN REACTION
(1393 INSULIN REAC)
DA November 28, 1990
BT1 Illnesses
RT Chemical Reaction
SO DOE FRASE VOCABULARY

INTERLOCK SYSTEM
(2038 INTERLOCK SY)
DA December 10, 1990
BT1 Systems (DOE FRASE Vocabulary)
 BT2 Systems
SO DOE FRASE VOCABULARY

INTERNAL DEPOSITION
(1344 INT DEPOSITI)
DA November 28, 1990
BT1 Injuries
SO DOE FRASE VOCABULARY

INTERSECTION
(1611 INTERSECTION)
DA December 10, 1990
BT1 Site (DOE FRASE Vocabulary)
 BT2 Sites/Areas
RT Highway
SO DOE FRASE VOCABULARY

ION EXCHANGE COLUMN
(2574 ION EXCHANGE)
DA January 3, 1991
BT1 Equipment/Parts - Nuclear (DOE
    FRASE Vocabulary)
 BT2 Equipment
 BT2 Reactor Components
SO DOE FRASE VOCABULARY

ION PUMP
(2575 ION PUMP)
DA January 3, 1991
BT1 Equipment/Parts - Nuclear (DOE
    FRASE Vocabulary)
 BT2 Equipment
 BT2 Reactor Components
BT1 Pump(s)
 BT2 Machines (DOE FRASE
    Vocabulary)
 BT3 Equipment
SO DOE FRASE VOCABULARY

IONIZING RADIATION EFFECTS
(1394 IONIZ RAD EF)
DA November 28, 1990
BT1 Effects
BT1 Illnesses
SO DOE FRASE VOCABULARY

**IRRADIATED FUEL PROCESS
FACILITY**
(1800 IFP FACILITY)
DA December 10, 1990
BT1 Facility (DOE FRASE Vocabulary)
 BT2 Facilities and Buildings (DOE
    FRASE Vocabulary)
 BT3 Facilities
SO DOE FRASE VOCABULARY

**IRRADIATED FUEL STORAGE
FACILITY**
(1801 IRS FACILITY)
DA December 10, 1990
BT1 Facility (DOE FRASE Vocabulary)
 BT2 Facilities and Buildings (DOE
    FRASE Vocabulary)
 BT3 Facilities
SO DOE FRASE VOCABULARY

IRRITATION
(1329 IRRITATION)
DA November 28, 1990
BT1 Injuries
RT Infection

RT Inflammation
SO DOE FRASE VOCABULARY

ISOSTATIC PRESS
(2124 ISOSTATIC PR)
DA December 10, 1990
BT1 Press
 BT2 Machines (DOE FRASE
    Vocabulary)
 BT3 Equipment
SO DOE FRASE VOCABULARY

JACKHAMMER
(2953 JACKHAMMER)
DA January 3, 1991
BT1 Tools - Powered
 BT2 Tools (DOE FRASE Vocabulary)
 BT3 Equipment
RT Construction Activity
SO DOE FRASE VOCABULARY

JANITOR
(0524 JANITOR)
DA November 28, 1990
BT1 Admin. Support/Clerical Employee
 BT2 Occupations
 BT2 Personnel
RT Janitorial/Housekeeping Activity
SO DOE FRASE VOCABULARY

**JANITORIAL/HOUSEKEEPING
ACTIVITY**
(1236 JH ACTIVITY)
DA November 28, 1990
BT1 Activity Types (DOE FRASE
    Vocabulary)
 BT2 Activities
RT Janitor
SO DOE FRASE VOCABULARY

JAW
(1090 JAW)
DA November 28, 1990
BT1 Skull (Frase)
 BT2 Head
 BT3 Human Body Parts
RT Chin
SO DOE FRASE VOCABULARY

JOINT(S)
(1137 JOINT)
DA November 28, 1990
BT1 Human Body Parts
NT1 Ankle(s)
NT1 Elbow(s)
NT1 Hip(s)
NT1 Knee(s)
NT1 Knuckle(s)
NT1 Shoulder(s)
NT1 Wrist(s)
SO DOE FRASE VOCABULARY

JUMPER
(2399 JUMPER)
DA January 3, 1991
BT1 Equipment/Parts - Electrical (DOE
    FRASE Vocabulary)

 BT2 Equipment
SO DOE FRASE VOCABULARY

KITCHEN
(1705 KITCHEN)
DA December 10, 1990
BT1 Site (DOE FRASE Vocabulary)
 BT2 Sites/Areas
RT Cafeteria
RT Eating Area
RT Food Service Employee
SO DOE FRASE VOCABULARY

KNEE(S)
(1124 KNEE)
DA November 28, 1990
BT1 Joint(s)
 BT2 Human Body Parts
RT Leg(s)
RT Leggings
SO DOE FRASE VOCABULARY

KNIFE(S)
(3041 KNIFE)
DA January 3, 1991
BT1 Tools - Manual
 BT2 Tools (DOE FRASE Vocabulary)
 BT3 Equipment
NT1 Pocket Knife(s)
RT Laceration
SO DOE FRASE VOCABULARY

KNUCKLE(S)
(1102 KNUCKLE)
DA November 28, 1990
BT1 Joint(s)
 BT2 Human Body Parts
RT Glove(s)
RT Hand Protection
RT Hand(s)
SO DOE FRASE VOCABULARY

LAB COAT
(2671 LAB COAT)
DA January 3, 1991
BT1 Anticontamination Clothing
 BT2 Clothing
 BT2 Personal Protective Equipment
 BT3 Equipment/Parts - Personal
    Protective (DOE FRASE
    Vocabulary)
 BT4 Equipment
SO DOE FRASE VOCABULARY

LABORATORY
(1706 LABORATORY)
DA December 10, 1990
BT1 Site (DOE FRASE Vocabulary)
 BT2 Sites/Areas
SO DOE FRASE VOCABULARY

LACERATION
(1330 LA)
DA November 28, 1990
BT1 Injuries
RT Knife(s)
RT Puncture
SO DOE FRASE VOCABULARY

LASER
(2400 LASER)
DA   January 3, 1991
BT1 Equipment/Parts - Electrical (DOE
     FRASE Vocabulary)
 BT2 Equipment
SO   DOE FRASE VOCABULARY

LATHE
(2125 LATHE)
DA   December 10, 1990
BT1 Machines (DOE FRASE
     Vocabulary)
 BT2 Equipment
SO   DOE FRASE VOCABULARY

LEAK DETECTOR
(2759 LEAK DETECTO)
DA   January 3, 1991
BT1 Leak Detection Systems
 BT2 Systems
BT1 Testing Equipment
 BT2 Instrument(s)
  BT3 Equipment/Parts -
       Instrumentation/Measuring
       (DOE FRASE Voc.)
   BT4 Equipment
RT   Environmental Release
SO   DOE FRASE VOCABULARY

LEG(S)
(1121 LEGP)
DA   November 28, 1990
BT1 Human Body Parts
NT1 Lower Leg
NT1 Thigh(s)
RT   Ankle(s)
RT   Foot/Feet
RT   Knee(s)
RT   Leggings
SO   DOE FRASE VOCABULARY

LEGGINGS
(2672 LEGGINGS)
DA   January 3, 1991
BT1 Personal Protective Equipment
 BT2 Equipment/Parts - Personal
      Protective (DOE FRASE
      Vocabulary)
  BT3 Equipment
RT   Knee(s)
RT   Leg(s)
RT   Lower Leg
RT   Thigh(s)
SO   DOE FRASE VOCABULARY

LEVEL ALARM
(2760 LEVEL ALARM)
DA   January 3, 1991
BT1 Alarms
 BT2 Devices
BT1 Instrument(s)
 BT2 Equipment/Parts -
      Instrumentation/Measuring (DOE
      FRASE Voc.)
  BT3 Equipment
SO   DOE FRASE VOCABULARY

LIFT BUCKET
(2331 LIFT BUCKET)
DA   December 10, 1990
BT1 Hoisting Apparatus
 BT2 Material Handling Device
  BT3 Devices
  BT3 Equipment/Parts - Material
       Handling (DOE FRASE
       Vocabulary)
   BT4 Equipment
SO   DOE FRASE VOCABULARY

LIGHT(S)
(2401 LIGHT;N)
DA   January 3, 1991
BT1 Equipment/Parts - Electrical (DOE
     FRASE Vocabulary)
 BT2 Equipment
SO   DOE FRASE VOCABULARY

LIP(S)
DA   November 28, 1990
BT1 Face
 BT2 Head
  BT3 Human Body Parts
RT   Mouth
SO   DOE FRASE VOCABULARY

LOAD SECURING DEVICE
(2332 LOAD SECURIN)
DA   December 11, 1990
BT1 Devices
BT1 Equipment/Parts - Material
     Handling (DOE FRASE
     Vocabulary)
 BT2 Equipment
RT   Barriers
RT   Controls
SO   DOE FRASE VOCABULARY

LOADING DOCK
(1855 LOADING DOCK)
DA   December 10, 1990
BT1 Structures (DOE FRASE
     Vocabulary)
SO   DOE FRASE VOCABULARY

LOBBY
(1707 LOBBY)
DA   December 10, 1990
BT1 Site (DOE FRASE Vocabulary)
 BT2 Sites/Areas
SO   DOE FRASE VOCABULARY

LOCKER/SHOWER ROOM
(1708 LOCKER ROOM)
DA   December 10, 1990
BT1 Room
 BT2 Sites/Areas
SO   DOE FRASE VOCABULARY

LOG N CHANNEL
(2577 LOG N CHANNE)
DA   January 3, 1991
BT1 Equipment/Parts - Nuclear (DOE
     FRASE Vocabulary)
 BT2 Equipment

 BT2 Reactor Components
SO   DOE FRASE VOCABULARY

LOG N RECORDER
(2761 LOG N RECORD)
DA   January 3, 1991
BT1 Instrument(s)
 BT2 Equipment/Parts -
      Instrumentation/Measuring (DOE
      FRASE Voc.)
  BT3 Equipment
SO   DOE FRASE VOCABULARY

LOSS OF CONSCIOUSNESS
(1331 LOSS OF CONS)
DA   November 28, 1990
BT1 Injuries
BT1 Losses
RT   Concussion
RT   Contusion(S)
SO   DOE FRASE VOCABULARY

LOSS OF EXPERIMENT
(1547 LOSS OF EXPE)
DA   November 29, 1990
BT1 Losses
BT1 Nature of Property Damage
SO   DOE FRASE VOCABULARY

LOSS OF INSTRUMENT AIR
(1548 LOSS OF INST)
DA   November 29, 1990
BT1 Losses
BT1 Nature of Property Damage
SO   DOE FRASE VOCABULARY

LOSS OF MATERIAL
(1533 LOSS OF MATE)
DA   November 29, 1990
BT1 Losses
BT1 Nature of Property Damage
SO   DOE FRASE VOCABULARY

LOSS OF OPERATING TIME
(1578 LOSS OF OPER)
DA   November 28, 1990
BT1 Losses
BT1 Nature of Programmatic Impact
BT1 Time Designations
SO   DOE FRASE VOCABULARY

LOSS OF PRODUCTION
(1579 LOSS OF PROD)
DA   November 28, 1990
BT1 Losses
BT1 Nature of Programmatic Impact
SO   DOE FRASE VOCABULARY

LOSS OF SPECIMEN
(1551 LOSS OF SPEC)
DA   November 29, 1990
BT1 Losses
BT1 Nature of Property Damage
SO   DOE FRASE VOCABULARY

LOSS OF TRANSMISSION
(1552 LOSS OF TRAN)
DA   November 29, 1990

BT1 Losses
BT1 Nature of Property Damage
SO DOE FRASE VOCABULARY

LOW POWER REACTOR FACILITY
(1802 LOW POWER RE)
DA December 10, 1990
BT1 Facility (DOE FRASE Vocabulary)
  BT2 Facilities and Buildings (DOE
      FRASE Vocabulary)
    BT3 Facilities
BT1 Reactor Facilities
  BT2 Nuclear Facilities
    BT3 Facilities
SO DOE FRASE VOCABULARY

LOWER BACK
(1114 LOWER BACK)
DA November 28, 1990
BT1 Back
  BT2 Trunk
    BT3 Human Body Parts
SO DOE FRASE VOCABULARY

LOWER LEG
(1125 LOWER LEG)
DA November 28, 1990
BT1 Leg(s)
  BT2 Human Body Parts
RT Leggings
SO DOE FRASE VOCABULARY

MACHINE SETUP/OPERATOR
(0710 MACHINE OPER)
DA November 28, 1990
BT1 Precision/Production Personnel
  BT2 Occupations
  BT2 Personnel
SO DOE FRASE VOCABULARY

MACHINE SHOP
(1784 MACHINE SHOP)
DA December 10, 1990
BT1 Building (DOE FRASE Vocabulary)
  BT2 Facilities and Buildings (DOE
      FRASE Vocabulary)
    BT3 Facilities
SO DOE FRASE VOCABULARY

MACHINES (DOE FRASE VOCABULARY)
(DOE FRASE Vocabulary Numeric Keys
2100-2199)
DA December 10, 1990
BT1 Equipment
NT1 Agitator
NT1 Agricultural Machine
NT1 Air Dryer
NT1 Band Saw
NT1 Biostabilizers
NT1 Boring Machine
NT1 Centrifuge
NT1 Comminuters
NT1 Compactor
NT1 Compressor
  NT2 Air Compressor
NT1 Concrete Saw
NT1 Crushing Machine
NT1 Discharge Machine

NT1 Drilling Machine
NT1 Drying Machine
NT1 Electrical Office Machine
NT1 Emergency Diesel Generator
NT1 Grinding Machine
NT1 Hammermills
NT1 Lathe
NT1 Milling Machine
  NT2 Ball Mill
  NT2 Roll Mill
NT1 Mixing Machine
NT1 Polishing Machine
NT1 Press
  NT2 Isostatic Press
NT1 Printing Machine
NT1 Pump(s)
  NT2 Coolant Return Pump
  NT2 Ion Pump
  NT2 Reactor Coolant Pump
  NT2 Sump Pump
NT1 Recirculator
NT1 Sanding Machine
NT1 Slicing Machine
NT1 Stitching/Sewing Machine
NT1 Table Saw
NT1 Turbine
NT1 Vacuum Cleaner
RT DOE FRASE Categories
SO DOE FRASE VOCABULARY
DEF A subject category used with the
    DOE FRASE Vocabulary.

MACHINIST
(0681 MACHINIST)
DA November 28, 1990
BT1 Precision/Production Personnel
  BT2 Occupations
  BT2 Personnel
SO DOE FRASE VOCABULARY

MAGNETIC/ELECTROLYTIC APPARATUS
(2402 MEA)
DA January 3, 1991
BT1 Equipment/Parts - Electrical (DOE
    FRASE Vocabulary)
  BT2 Equipment
SO DOE FRASE VOCABULARY

MAINTENANCE AREA
(1612 MAINT AREA)
DA December 10, 1990
BT1 Area
  BT2 Sites/Areas
RT Garage
RT Mechanic/Repairer
SO DOE FRASE VOCABULARY

MAINTENANCE SHOP
(1785 MAINT SHOP)
DA December 10, 1990
BT1 Building (DOE FRASE Vocabulary)
  BT2 Facilities and Buildings (DOE
      FRASE Vocabulary)
    BT3 Facilities
RT Garage
RT Mechanic/Repairer
SO DOE FRASE VOCABULARY

MALIGNANT NEOPLASM
(1395 MALIGNANT NE)
DA November 28, 1990
BT1 Illnesses
SO DOE FRASE VOCABULARY

MANAGER/ADMINISTRATOR
(0110 MANAGER)
DA November 28, 1990
SO DOE FRASE VOCABULARY

MANIPULATOR
(2578 MANIPULATOR)
DA January 3, 1991
BT1 Equipment/Parts - Nuclear (DOE
    FRASE Vocabulary)
  BT2 Equipment
  BT2 Reactor Components
NT1 Electromechanical Manipulator(s)
RT Manipulator Tape
SO DOE FRASE VOCABULARY

MANIPULATOR TAPE
(2333 MANIP TAPE)
DA December 11, 1990
BT1 Equipment/Parts - Material
    Handling (DOE FRASE
    Vocabulary)
  BT2 Equipment
RT Manipulator
SO DOE FRASE VOCABULARY

MANLIFT(S)
(2334 MANLIFT)
DA December 10, 1990
BT1 Hoisting Apparatus
  BT2 Material Handling Device
    BT3 Devices
    BT3 Equipment/Parts - Material
        Handling (DOE FRASE
        Vocabulary)
      BT4 Equipment
SO DOE FRASE VOCABULARY

MANUAL REACTOR SCRAM
(1553 MANUAL REACT)
DA November 29, 1990
SY Action Charlie
BT1 Nature of Property Damage
BT1 Scram
  BT2 Reactor Shutdown
RT Automatic Reactor Scram
RT Nuclear Facility
RT Unscheduled Shutdown
RT Violations
SO DOE FRASE VOCABULARY

MASON
(0641 MASON)
DA November 28, 1990
BT1 Repair/Construction Personnel
  BT2 Occupations
  BT2 Personnel
SO DOE FRASE VOCABULARY

MATERIAL HANDLING ACTIVITY
(1237 MH ACTIVITY)
DA November 28, 1990

BT1 Activity Types (DOE FRASE
    Vocabulary)
 BT2 Activities
SO DOE FRASE VOCABULARY

MATERIAL HANDLING DEVICE
(2335 MH DEVICE)
DA December 10, 1990
BT1 Devices
BT1 Equipment/Parts - Material
    Handling (DOE FRASE
    Vocabulary)
 BT2 Equipment
NT1 Conveyor(s)
NT1 Crane(s)
 NT2 Boom Crane
 NT2 Bridge Crane
 NT2 Mobile Crane
 NT2 Overhead Crane
NT1 Earth Moving Equipment
 NT2 Dredge(s)
NT1 Hand Truck(s)
 NT2 Cart(s)
 NT2 Dolly
NT1 Hoisting Apparatus
 NT2 Derrick
 NT2 Elevator
 NT2 Forklift(s)
 NT2 Hoist(s)
  NT3 Air Hoist
 NT2 Lift Bucket
 NT2 Manlift(s)
 NT2 Scissor Lift
SO DOE FRASE VOCABULARY

MECHANIC/REPAIRER
(0610 MECHANIC)
DA November 28, 1990
BT1 Repair/Construction Personnel
 BT2 Occupations
 BT2 Personnel
RT Garage
RT Maintenance Area
RT Maintenance Shop
SO DOE FRASE VOCABULARY

MECHANICAL DAMAGE
(1534 MECH DAMAGE)
DA November 29, 1990
BT1 Damage
 BT2 Nature of Property Damage
SO DOE FRASE VOCABULARY

MENTAL DISORDERS
(1396 MENTAL DISOR)
DA November 28, 1990
BT1 Illnesses
RT Hospital
SO DOE FRASE VOCABULARY

METAL SHOE COVER
(2673 METAL SHOE C)
DA January 3, 1991
BT1 Shoe Cover(s)
 BT2 Foot Protection
  BT3 Personal Protective Equipment
   BT4 Equipment/Parts - Personal
       Protective (DOE FRASE
       Vocabulary)

 BT5 Equipment
SO DOE FRASE VOCABULARY

METATARSAL PROTECTION
(2674 METATARSAL P)
DA January 3, 1991
BT1 Foot Protection
 BT2 Personal Protective Equipment
  BT3 Equipment/Parts - Personal
      Protective (DOE FRASE
      Vocabulary)
   BT4 Equipment
RT Toe(s)
SO DOE FRASE VOCABULARY

MILITARY PERSONNEL
(0910 MILITARY)
DA November 28, 1990
BT1 Occupations
BT1 Personnel
SO DOE FRASE VOCABULARY

MILLING MACHINE
(2128 MILLING MACH)
DA December 10, 1990
BT1 Machines (DOE FRASE
    Vocabulary)
 BT2 Equipment
NT1 Ball Mill
NT1 Roll Mill
SO DOE FRASE VOCABULARY

MINE
(1856 MINE)
DA December 10, 1990
BT1 Structures (DOE FRASE
    Vocabulary)
SO DOE FRASE VOCABULARY

MINER/DRILLER
(0650 MINER;DRILLE)
DA November 28, 1990
BT1 Repair/Construction Personnel
 BT2 Occupations
 BT2 Personnel
RT Mining/Drilling Activity
SO DOE FRASE VOCABULARY

MINING/DRILLING ACTIVITY
(1251 MD ACTIVITY)
DA November 28, 1990
BT1 Activity Types (DOE FRASE
    Vocabulary)
 BT2 Activities
NT1 Strip Mining
RT Miner/Driller
SO DOE FRASE VOCABULARY

MISC AGRICULTURE EMPLOYEE
(0580 MISC AGRICUL)
DA November 28, 1990
BT1 Agriculture Personnel
 BT2 Occupations
 BT2 Personnel
BT1 Misc Employee
 BT2 Occupations
 BT2 Personnel
SO DOE FRASE VOCABULARY

MISC EMPLOYEE
(0990 MISC EMPLOYE)
DA November 28, 1990
BT1 Occupations
BT1 Personnel
NT1 Misc Service Employee
NT1 Misc Agriculture Employee
NT1 Misc Repair/Construction
    Employee
NT1 Misc Precision/Production
    Employee
NT1 Misc Transport Employee
SO DOE FRASE VOCABULARY

**MISC PRECISION/PRODUCTION
EMPLOYEE**
(0780 MISC PRECIS)
DA November 28, 1990
BT1 Misc Employee
 BT2 Occupations
 BT2 Personnel
BT1 Precision/Production Personnel
 BT2 Occupations
 BT2 Personnel
SO DOE FRASE VOCABULARY

MISC PROFESSIONAL
(0200 MISC PROF)
DA November 28, 1990
BT1 Professional Personnel
 BT2 Occupations
 BT2 Personnel
SO DOE FRASE VOCABULARY

**MISC REPAIR/CONSTRUCTION
EMPLOYEE**
(0660 MISC REPAIR)
DA November 28, 1990
BT1 Misc Employee
 BT2 Occupations
 BT2 Personnel
BT1 Repair/Construction Personnel
 BT2 Occupations
 BT2 Personnel
RT Construction Activity
SO DOE FRASE VOCABULARY

MISC SERVICE EMPLOYEE
(0525 MISC SERVICE)
DA November 28, 1990
BT1 Admin. Support/Clerical Employee
 BT2 Occupations
 BT2 Personnel
BT1 Misc Employee
 BT2 Occupations
 BT2 Personnel
SO DOE FRASE VOCABULARY

MISC TRANSPORT EMPLOYEE
(0840 MISC TRANSPO)
DA November 28, 1990
BT1 Misc Employee
 BT2 Occupations
 BT2 Personnel
BT1 Transport Personnel
 BT2 Occupations
 BT2 Personnel
RT Transportation Activity
SO DOE FRASE VOCABULARY

MISCELLANEOUS TECHNICIAN
(0390 MISC TECH)
DA   November 28, 1990
BT1 Technicians
 BT2 Professional Personnel
  BT3 Occupations
  BT3 Personnel
SO DOE FRASE VOCABULARY

MIXER(S)
(2954 MIXER)
DA   January 3, 1991
BT1 Tools - Powered
 BT2 Tools (DOE FRASE Vocabulary)
  BT3 Equipment
SO DOE FRASE VOCABULARY

MIXING MACHINE
(2130 MIXING MACHI)
DA   December 10, 1990
BT1 Machines (DOE FRASE
     Vocabulary)
 BT2 Equipment
SO DOE FRASE VOCABULARY

MOBILE CRANE
(2337 MOBILE CRANE)
DA   December 10, 1990
BT1 Crane(s)
 BT2 Material Handling Device
  BT3 Devices
  BT3 Equipment/Parts - Material
       Handling (DOE FRASE
       Vocabulary)
   BT4 Equipment
SO DOE FRASE VOCABULARY

MOP(S)
(3043 MOP)
DA   January 3, 1991
BT1 Tools - Manual
 BT2 Tools (DOE FRASE Vocabulary)
  BT3 Equipment
SO DOE FRASE VOCABULARY

MOTEL/HOTEL
(1786 MOTEL)
DA   December 10, 1990
BT1 Building (DOE FRASE Vocabulary)
 BT2 Facilities and Buildings (DOE
       FRASE Vocabulary)
  BT3 Facilities
SO DOE FRASE VOCABULARY

MOTOR(S)
(2403 MOTOR)
DA   January 3, 1991
BT1 Equipment/Parts - Electrical (DOE
     FRASE Vocabulary)
 BT2 Equipment
SO DOE FRASE VOCABULARY

MOUTH
(1088 MOUTH)
DA   November 28, 1990
BT1 Digestive System
 BT2 Body System(s)
  BT3 Human Body Parts

NT1 Tooth/Teeth
RT Dental Injury
RT Face
RT Head
RT Lip(s)
SO DOE FRASE VOCABULARY

MULTIPLE BODY PARTS
(1134 MULTIPLE BP)
DA   November 28, 1990
BT1 Human Body Parts
SO DOE FRASE VOCABULARY

MULTIPLE INJURIES
(1332 MULTIPLE INJ)
DA   November 28, 1990
BT1 Injuries
SO DOE FRASE VOCABULARY

MUSCLE/TENDON(S)
(1138 MUSCLE)
DA   November 28, 1990
BT1 Human Body Parts
RT Tendonitis
SO DOE FRASE VOCABULARY

NAIL GUN
(2955 NAIL GUN)
DA   January 3, 1991
BT1 Tools - Powered
 BT2 Tools (DOE FRASE Vocabulary)
  BT3 Equipment
SO DOE FRASE VOCABULARY

NATURE OF PROGRAMMATIC IMPACT
(DOE FRASE Vocabulary Numeric Keys
   1575-1599)
DA   January 17, 1991
NT1 Delay of Experiment
NT1 Delay of Response
NT1 Delay of Sampling
NT1 Loss of Operating Time
NT1 Loss of Production
NT1 No Programmatic Impact
NT1 Unscheduled Shutdown
NT1 Unscheduled Evacuation
RT Damage Assessments
RT Delays
RT DOE FRASE Categories
SO DOE FRASE VOCABULARY
DEF A subject category used with the
     DOE FRASE Vocabulary.

NATURE OF PROPERTY DAMAGE
(DOE FRASE Vocabulary Numeric Keys
   1525-1574)
DA   January 17, 1991
NT1 Automatic Reactor Scram
NT1 Damage
 NT2 Electronic Damage
 NT2 Fire Damage
 NT2 Heat Damage
 NT2 Mechanical Damage
 NT2 Smoke Damage
 NT2 Structural Damage
 NT2 Water Damage
 NT2 Wind Damage
NT1 Dent
NT1 Electrical Short

NT1 Environmental Release
NT1 Hazardous Spill
NT1 Loss of Transmission
NT1 Loss of Material
NT1 Loss of Specimen
NT1 Loss of Instrument Air
NT1 Loss of Experiment
NT1 Manual Reactor Scram
NT1 Power Outage
NT1 Power Shutdown
NT1 Radioactive Contamination
 NT2 Non-fixed Radioactive
      Contamination
NT1 Vehicle Accident
NT1 Violation of Working Limit
NT1 Violation of Operational Safety Req
NT1 Violation of Tech Specification
NT1 Violation of Occup Safety
     Regulation
NT1 Violation of Criticality Specifictn
RT Damage Assessments
RT DOE FRASE Categories
RT Violations
SO DOE FRASE VOCABULARY
DEF A subject category used with the
     DOE FRASE Vocabulary.

NECK
(1093 NECK)
DA   November 28, 1990
BT1 Human Body Parts
RT Spine
RT Throat
SO DOE FRASE VOCABULARY

NERVOUS SYSTEM
(1144 NERVOUS SYST)
DA   November 28, 1990
BT1 Body System(s)
 BT2 Human Body Parts
NT1 Brain
NT1 Spinal Cord
SO DOE FRASE VOCABULARY

NEUTRON(S)
(2581 NEUTRON)
DA   January 3, 1991
BT1 Equipment/Parts - Nuclear (DOE
     FRASE Vocabulary)
 BT2 Equipment
 BT2 Reactor Components
RT Neutron Source
SO DOE FRASE VOCABULARY
DEF (NRC Glossary of Terms: Nuclear
     Power and Radiation) Uncharged
     elementary particles with a mass
     slightly greater than that of the
     proton, and found in the nucleus of
     every atom heavier than hydrogen.

NEUTRON SOURCE
(2579 NEUTRON SOUR)
DA   January 3, 1991
BT1 Equipment/Parts - Nuclear (DOE
     FRASE Vocabulary)
 BT2 Equipment
 BT2 Reactor Components
RT Neutron(s)

SO DOE FRASE VOCABULARY
DEF (NRC Glossary of Terms: Nuclear
Power and Radiation) A
radioactive material (decays by
neutron emission) that can be
inserted into a reactor to ensure
that a sufficient quantity of
neutrons is available to start a
chain reaction and register on
neutron detection equipment.

NO ACTIVITY
(1254 NO ACTIVITY)
DA November 28, 1990
BT1 Activity Types (DOE FRASE
Vocabulary)
BT2 Activities
SO DOE FRASE VOCABULARY

NO PERSONNEL INJURY
(1333 NO PERSONNEL)
DA November 28, 1990
BT1 Injuries
SO DOE FRASE VOCABULARY

NO PROGRAMMATIC IMPACT
(1582 NO PROG IMPA)
DA November 28, 1990
BT1 Nature of Programmatic Impact
SO DOE FRASE VOCABULARY

NO PROPERTY DAMAGE OR INJURY
(1535 NO PROP DAM)
DA November 29, 1990
SO DOE FRASE VOCABULARY

NON-POWERED HANDTOOL(S)
(3044 NP HANDTOOL)
DA January 3, 1991
BT1 Tools - Manual
BT2 Tools (DOE FRASE Vocabulary)
BT3 Equipment
SO DOE FRASE VOCABULARY

NOSE
(1086 NOSE)
DA November 28, 1990
BT1 Face
BT2 Head
BT3 Human Body Parts
BT1 Respiratory System
BT2 Body System(s)
BT3 Human Body Parts
RT Air Mask
RT Dust Mask
SO DOE FRASE VOCABULARY

NOSEBLEED
(1334 NOSEBLEED)
DA November 28, 1990
BT1 Injuries
SO DOE FRASE VOCABULARY

NUCLEAR FACILITY
(1787 NUCLEAR FAC)
DA December 10, 1990
BT1 Facility (DOE FRASE Vocabulary)

BT2 Facilities and Buildings (DOE
FRASE Vocabulary)
BT3 Facilities
RT Automatic Reactor Scram
RT Decommissioning Activity
RT Decontamination Activity
RT Manual Reactor Scram
RT Particle Accelerator Facility
RT Production Reactor Facility
SO DOE FRASE VOCABULARY

OCCUPATION (UNK)
(0001 OCCUP;UNK)
DA November 28, 1990
SO DOE FRASE VOCABULARY

OCCUPATIONS
(DOE FRASE Vocabulary Numeric Keys
0001-0999)
DA November 29, 1990
NT1 Admin. Support/Clerical Employee
NT2 Firefighter
NT2 Food Service Employee
NT2 Janitor
NT2 Misc Service Employee
NT2 Security Guard
NT1 Agriculture Personnel
NT2 Forest Worker
NT2 Groundskeeper
NT2 Misc Agriculture Employee
NT1 Handler/Laborer/Helper
NT1 Military Personnel
NT1 Misc Employee
NT2 Misc Service Employee
NT2 Misc Agriculture Employee
NT2 Misc Repair/Construction
Employee
NT2 Misc Precision/Production
Employee
NT2 Misc Transport Employee
NT1 Precision/Production Personnel
NT2 Machinist
NT2 Machine Setup/Operator
NT2 Misc Precision/Production
Employee
NT2 Operator, Plant/System/Utility
NT2 Sheet Metal Worker
NT2 Welder/Solderer
NT1 Professional Personnel
NT2 Doctor/Nurse
NT2 Engineer
NT2 Misc Professional
NT2 Scientist
NT3 Health Physicist
NT2 Technicians
NT3 Engineering Technician
NT3 Health Technician
NT3 Miscellaneous Technician
NT3 Radiation Monitor/Technician
NT3 Science Technician
NT1 Repair/Construction Personnel
NT2 Carpenter
NT2 Electrician
NT2 Mason
NT2 Mechanic/Repairer
NT2 Miner/Driller
NT2 Misc Repair/Construction
Employee
NT2 Painter
NT2 Pipe Fitter

NT1 Sales Person
NT1 Transport Personnel
NT2 Aircraft Pilot
NT2 Bus Driver
NT2 Equipment Operator
NT2 Misc Transport Employee
NT2 Truck Driver
RT DOE FRASE Categories
SO DOE FRASE VOCABULARY
DEF A subject category used with the
DOE FRASE Vocabulary.

OFF GAS SYSTEM
(2039 OFF GAS SYST)
DA December 10, 1990
BT1 Systems (DOE FRASE Vocabulary)
BT2 Systems
SO DOE FRASE VOCABULARY

OFFICE ACTIVITY
(1226 O ACTIVITY)
DA November 28, 1990
BT1 Activity Types (DOE FRASE
Vocabulary)
BT2 Activities
RT Admin. Support/Clerical Employee
RT Office Area
SO DOE FRASE VOCABULARY

OFFICE AREA
(1709 OFFICE AREA)
DA December 10, 1990
BT1 Area
BT2 Sites/Areas
RT Office Activity
SO DOE FRASE VOCABULARY

OHMMETER(S)
(2762 OHMMETER)
DA January 3, 1991
BT1 Indicator(s)
BT2 Instrument(s)
BT3 Equipment/Parts -
Instrumentation/Measuring
(DOE FRASE Voc.)
BT4 Equipment
SO DOE FRASE VOCABULARY

OPERATOR, PLANT/SYSTEM/UTILITY
(0690 OP;UTIL)
DA November 28, 1990
BT1 Precision/Production Personnel
BT2 Occupations
BT2 Personnel
SO DOE FRASE VOCABULARY

OSCILLOSCOPE(S)
(2763 OSCILLOSCOPE)
DA January 3, 1991
BT1 Testing Equipment
BT2 Instrument(s)
BT3 Equipment/Parts -
Instrumentation/Measuring
(DOE FRASE Voc.)
BT4 Equipment
SO DOE FRASE VOCABULARY

OTHER COMPLICATIONS
(1397 OTHER COMPLI)
DA November 28, 1990
BT1 Illnesses
SO DOE FRASE VOCABULARY

OTHER NON-TASK ACTIVITY
(1253 ONT ACTIVITY)
DA November 28, 1990
BT1 Activity Types (DOE FRASE
   Vocabulary)
 BT2 Activities
SO DOE FRASE VOCABULARY

OTHER SKIN CONDITIONS
(1398 OTHER SKIN C)
DA November 28, 1990
BT1 Illnesses
RT Dermatitis
RT Skin
SO DOE FRASE VOCABULARY

OUTLET/RECEPTACLE(S)
(2404 OUTLET)
DA January 3, 1991
BT1 Equipment/Parts - Electrical (DOE
   FRASE Vocabulary)
 BT2 Equipment
SO DOE FRASE VOCABULARY

OVEN(S)
(2488 OVEN)
DA January 3, 1991
BT1 Heater(s)
 BT2 Equipment/Parts - Heating (DOE
   FRASE Vocabulary)
 BT3 Equipment
SO DOE FRASE VOCABULARY

OVERHEAD CRANE
(2338 OVERHEAD CRA)
DA December 10, 1990
BT1 Crane(s)
 BT2 Material Handling Device
 BT3 Devices
 BT3 Equipment/Parts - Material
   Handling (DOE FRASE
   Vocabulary)
 BT4 Equipment
SO DOE FRASE VOCABULARY

PADDING
(2675 PADDING)
DA January 3, 1991
BT1 Personal Protective Equipment
 BT2 Equipment/Parts - Personal
   Protective (DOE FRASE
   Vocabulary)
 BT3 Equipment
SO DOE FRASE VOCABULARY

PAINTER
(0644 PAINTER)
DA November 28, 1990
BT1 Repair/Construction Personnel
 BT2 Occupations
 BT2 Personnel
SO DOE FRASE VOCABULARY

PALM(S)
(1101 PALM)
DA November 28, 1990
BT1 Hand(s)
 BT2 Human Body Parts
RT Hand Protection
RT Skin
SO DOE FRASE VOCABULARY

PARKING LOT
(1857 PARKING LOT)
DA December 10, 1990
BT1 Structures (DOE FRASE
   Vocabulary)
RT Garage
RT Parking Space
SO DOE FRASE VOCABULARY

PARKING SPACE
(1613 PARKING SPAC)
DA December 10, 1990
BT1 Site (DOE FRASE Vocabulary)
 BT2 Sites/Areas
RT Garage
RT Parking Lot
SO DOE FRASE VOCABULARY

PARTICLE ACCELERATOR FACILITY
(1803 PARTICLE ACC)
DA December 10, 1990
BT1 Facility (DOE FRASE Vocabulary)
 BT2 Facilities and Buildings (DOE
   FRASE Vocabulary)
 BT3 Facilities
RT Nuclear Facility
SO DOE FRASE VOCABULARY

PERSONAL PROTECTIVE EQUIPMENT
(2676 PPE)
DA January 3, 1991
BT1 Equipment/Parts - Personal
   Protective (DOE FRASE
   Vocabulary)
 BT2 Equipment
NT1 Air Mask
NT1 Anticontamination Clothing
 NT2 Acid Suit
 NT2 Lab Coat
 NT2 Radiation Suit
NT1 Chin Strap
NT1 Dust Mask
NT1 Eye Protection
 NT2 Goggles
 NT2 Safety Glasses
  NT3 Safety Glasses W Side Shields
  NT3 Tinted Safety Glasses
 NT2 Side Shields
NT1 Fall Protection Device
 NT2 Safety Belt
 NT2 Safety Line
 NT2 Safety Net
NT1 Flame Retardant Clothing
NT1 Foot Protection
 NT2 Ankle Protection
 NT2 Metatarsal Protection
 NT2 Safety Boots
 NT2 Safety Shoe(s)
 NT2 Shoe Cover(s)
  NT3 Metal Shoe Cover

NT1 Hand Protection
 NT2 Glove(s)
 NT2 Wrist Band
NT1 Head Protection
 NT2 Bump Cap
 NT2 Faceshield
 NT2 Hard Hat
 NT2 Helmet
 NT2 Sand Blaster's Hood
 NT2 Welder's Hood
NT1 Hearing Protection
 NT2 Ear Plug(s)
NT1 Leggings
NT1 Padding
NT1 Respirator(s)
NT1 Seat Belt(s)
SO DOE FRASE VOCABULARY

**PHYSICAL FITNESS TRAINING
ACTIVITY**
(1238 PFT ACTIVITY)
DA November 28, 1990
BT1 Activity Types (DOE FRASE
   Vocabulary)
 BT2 Activities
SO DOE FRASE VOCABULARY

PICK(S)
(3045 PICK;N)
DA January 3, 1991
BT1 Tools - Manual
 BT2 Tools (DOE FRASE Vocabulary)
 BT3 Equipment
SO DOE FRASE VOCABULARY

PIER
(1858 PIER)
DA December 10, 1990
BT1 Structures (DOE FRASE
   Vocabulary)
SO DOE FRASE VOCABULARY

PILOT LIGHT(S)
(2489 PILOT LIGHT)
DA January 3, 1991
BT1 Equipment/Parts - Heating (DOE
   FRASE Vocabulary)
 BT2 Equipment
BT1 Heating Equipment
SO DOE FRASE VOCABULARY

PINCHED NERVE
(1335 PINCH NERV)
DA November 28, 1990
BT1 Injuries
RT Spinal Cord
RT Upper Back
SO DOE FRASE VOCABULARY

PIPE FITTER
(0645 PIPE FITTER)
DA November 28, 1990
BT1 Repair/Construction Personnel
 BT2 Occupations
 BT2 Personnel
SO DOE FRASE VOCABULARY

PIT AREA
(1614 PIT AREA)
DA   December 10, 1990
BT1 Area
 BT2 Sites/Areas
SO DOE FRASE VOCABULARY

PLANE(S)
(3046 PLANE)
DA   January 3, 1991
BT1 Tools - Manual
 BT2 Tools (DOE FRASE Vocabulary)
 BT3 Equipment
SO DOE FRASE VOCABULARY

PLANT PROTECTION SYSTEM
(2040 PPS)
DA   December 10, 1990
BT1 Emergency Systems
 BT2 Systems
BT1 Systems (DOE FRASE Vocabulary)
 BT2 Systems
RT Guard Station
RT Security Guard
RT Security Activity
SO DOE FRASE VOCABULARY

PLIERS
(3047 PLIERS)
DA   January 3, 1991
BT1 Tools - Manual
 BT2 Tools (DOE FRASE Vocabulary)
 BT3 Equipment
SO DOE FRASE VOCABULARY

PLUTONIUM FUEL PLATE(S)
(2582 PU FUEL PLAT)
DA   January 3, 1991
BT1 Fuel Plate
 BT2 Fuel Element(s)
 BT3 Equipment/Parts - Nuclear (DOE
      FRASE Vocabulary)
 BT4 Equipment
 BT4 Reactor Components
SO DOE FRASE VOCABULARY

PLUTONIUM PROCESS HOOD
(2583 PU PROC HOOD)
DA   January 3, 1991
BT1 Equipment/Parts - Nuclear (DOE
      FRASE Vocabulary)
 BT2 Equipment
 BT2 Reactor Components
RT Barriers
RT Plutonium Recovery Process
SO DOE FRASE VOCABULARY

PLUTONIUM RECOVERY PROCESS
(2584 PLUT PROC HOOD)
DA   January 3, 1991
BT1 Equipment/Parts - Nuclear (DOE
      FRASE Vocabulary)
 BT2 Equipment
 BT2 Reactor Components
RT Plutonium Process Hood
SO DOE FRASE VOCABULARY

POCKET KNIFE(S)
(3048 POCKET KNIFE)
DA   January 3, 1991
BT1 Knife(s)
 BT2 Tools - Manual
 BT3 Tools (DOE FRASE Vocabulary)
 BT4 Equipment
SO DOE FRASE VOCABULARY

POISONING
(1399 POISONING)
DA   November 28, 1990
BT1 Illnesses
NT1 Systemic Poisoning
RT Toxic Effects to Single System
SO DOE FRASE VOCABULARY

POLISHING MACHINE
(2133 POLISHING MA)
DA   December 10, 1990
BT1 Machines (DOE FRASE
      Vocabulary)
 BT2 Equipment
SO DOE FRASE VOCABULARY

POND
(1859 POND)
DA   December 10, 1990
BT1 Structures (DOE FRASE
      Vocabulary)
RT Body of Water
RT Dam
SO DOE FRASE VOCABULARY

POOL
(1727 POOL)
DA   December 10, 1990
BT1 Site (DOE FRASE Vocabulary)
 BT2 Sites/Areas
RT Drowning
SO DOE FRASE VOCABULARY

POST DRIVER(S)
(2956 POST DRIVER)
DA   January 3, 1991
BT1 Tools - Powered
 BT2 Tools (DOE FRASE Vocabulary)
 BT3 Equipment
SO DOE FRASE VOCABULARY

POTENTIOMETER
(2764 POTENTIOMETE)
DA   January 3, 1991
SF *pots*
BT1 Indicator(s)
 BT2 Instrument(s)
 BT3 Equipment/Parts -
      Instrumentation/Measuring
      (DOE FRASE Voc.)
 BT4 Equipment
SO DOE FRASE VOCABULARY

POWER ACTUATED TOOL(S)
(2957 POWER ACTUA)
DA   January 3, 1991
BT1 Tools - Powered
 BT2 Tools (DOE FRASE Vocabulary)

 BT3 Equipment
SO DOE FRASE VOCABULARY

POWER BUFFER(S)
(2958 POWER BUFFER)
DA   January 3, 1991
BT1 Tools - Powered
 BT2 Tools (DOE FRASE Vocabulary)
 BT3 Equipment
SO DOE FRASE VOCABULARY

POWER CHISEL(S)
(2959 POWER CHISEL)
DA   January 3, 1991
BT1 Tools - Powered
 BT2 Tools (DOE FRASE Vocabulary)
 BT3 Equipment
SO DOE FRASE VOCABULARY

POWER CORD(S)
(2405 POWER CORD)
DA   January 3, 1991
BT1 Equipment/Parts - Electrical (DOE
      FRASE Vocabulary)
 BT2 Equipment
SO DOE FRASE VOCABULARY

POWER CUTTER(S)
(2960 POWER CUTTER)
DA   January 3, 1991
BT1 Tools - Powered
 BT2 Tools (DOE FRASE Vocabulary)
 BT3 Equipment
SO DOE FRASE VOCABULARY

POWER DRILL(S)
(2961 POWER DRILL)
DA   January 3, 1991
BT1 Tools - Powered
 BT2 Tools (DOE FRASE Vocabulary)
 BT3 Equipment
SO DOE FRASE VOCABULARY

POWER FILE(S)
(2962 POWER FILE)
DA   January 3, 1991
BT1 Tools - Powered
 BT2 Tools (DOE FRASE Vocabulary)
 BT3 Equipment
SO DOE FRASE VOCABULARY

POWER GRINDER(S)
(2963 POWER GRINDE)
DA   January 3, 1991
BT1 Tools - Powered
 BT2 Tools (DOE FRASE Vocabulary)
 BT3 Equipment
SO DOE FRASE VOCABULARY

POWER HAMMER(S)
(2964 POWER HAMMER)
DA   January 3, 1991
BT1 Tools - Powered
 BT2 Tools (DOE FRASE Vocabulary)
 BT3 Equipment
SO DOE FRASE VOCABULARY

POWER IMPACT WRENCH(S)
(2965 POWER IMPACT)
DA January 3, 1991
BT1 Tools - Powered
  BT2 Tools (DOE FRASE Vocabulary)
  BT3 Equipment
BT1 Wrench(s)
  BT2 Tools (DOE FRASE Vocabulary)
  BT3 Equipment
SO DOE FRASE VOCABULARY

POWER KEY(S)
(2406 POWER KEY)
DA January 3, 1991
BT1 Equipment/Parts - Electrical (DOE
    FRASE Vocabulary)
  BT2 Equipment
SO DOE FRASE VOCABULARY

POWER LEAD
(2407 POWER LEAD)
DA January 3, 1991
BT1 Equipment/Parts - Electrical (DOE
    FRASE Vocabulary)
  BT2 Equipment
SO DOE FRASE VOCABULARY

POWER LINE(S)
(2408 POWER LINE)
DA January 3, 1991
BT1 Equipment/Parts - Electrical (DOE
    FRASE Vocabulary)
  BT2 Equipment
SO DOE FRASE VOCABULARY

POWER OUTAGE
(1562 POWER OUTAGE)
DA November 29, 1990
BT1 Nature of Property Damage
RT Emergency Situations
RT Power Shutdown
SO DOE FRASE VOCABULARY

POWER POLE(S)
(2409 POWER POLE)
DA January 3, 1991
BT1 Equipment/Parts - Electrical (DOE
    FRASE Vocabulary)
  BT2 Equipment
SO DOE FRASE VOCABULARY

POWER POLISHER(S)
(2967 POWER POLISH)
DA January 3, 1991
BT1 Tools - Powered
  BT2 Tools (DOE FRASE Vocabulary)
  BT3 Equipment
SO DOE FRASE VOCABULARY

POWER RIVETER(S)
(2968 POWER RIVETE)
DA January 3, 1991
BT1 Tools - Powered
  BT2 Tools (DOE FRASE Vocabulary)
  BT3 Equipment
SO DOE FRASE VOCABULARY

POWER SANDBLASTER(S)
(2969 POWER SANDBL)
DA January 3, 1991
BT1 Tools - Powered
  BT2 Tools (DOE FRASE Vocabulary)
  BT3 Equipment
SO DOE FRASE VOCABULARY

POWER SANDER(S)
(2970 POWER SANDER)
DA January 3, 1991
SO DOE FRASE VOCABULARY

POWER SCREWDRIVER(S)
(2972 POWER SCREWD)
DA January 3, 1991
BT1 Tools - Powered
  BT2 Tools (DOE FRASE Vocabulary)
  BT3 Equipment
SO DOE FRASE VOCABULARY

POWER SHUTDOWN
(1554 POWER SHUTDO)
DA November 29, 1990
BT1 Nature of Property Damage
RT Power Outage
SO DOE FRASE VOCABULARY

POWER SPRAY GUN(S)
(2973 POWER SPRAY)
DA January 3, 1991
BT1 Tools - Powered
  BT2 Tools (DOE FRASE Vocabulary)
  BT3 Equipment
SO DOE FRASE VOCABULARY

POWER STAPLER(S)
(2975 POWER STAPLE)
DA January 3, 1991
BT1 Tools - Powered
  BT2 Tools (DOE FRASE Vocabulary)
  BT3 Equipment
SO DOE FRASE VOCABULARY

POWER SUPPLY
(2410 POWER SUPPLY)
DA January 3, 1991
BT1 Equipment/Parts - Electrical (DOE
    FRASE Vocabulary)
  BT2 Equipment
NT1 Uninterruptible Power Supplies
SO DOE FRASE VOCABULARY

POWER WAXER(S)
(2976 POWER WAXER)
DA January 3, 1991
BT1 Tools - Powered
  BT2 Tools (DOE FRASE Vocabulary)
  BT3 Equipment
SO DOE FRASE VOCABULARY

POWERED HANDTOOL(S)
(2977 POWERED HT)
DA January 3, 1991
BT1 Tools - Powered
  BT2 Tools (DOE FRASE Vocabulary)
  BT3 Equipment
SO DOE FRASE VOCABULARY

**PRE START-UP/ CALIBRATION
ACTIVITY**
(1239 PSC ACTIVITY)
DA November 28, 1990
BT1 Activity Types (DOE FRASE
    Vocabulary)
  BT2 Activities
RT Start-Up Procedure
SO DOE FRASE VOCABULARY

PREAMPLIFIER(S)
(2411 PREAMPLIFIER)
DA January 3, 1991
BT1 Equipment/Parts - Electrical (DOE
    FRASE Vocabulary)
  BT2 Equipment
SO DOE FRASE VOCABULARY

**PRECISION/PRODUCTION
PERSONNEL**
DA November 29, 1990
BT1 Occupations
BT1 Personnel
NT1 Machinist
NT1 Machine Setup/Operator
NT1 Misc Precision/Production
    Employee
NT1 Operator, Plant/System/Utility
NT1 Sheet Metal Worker
NT1 Welder/Solderer
SO DOE FRASE VOCABULARY

PRESS
(2134 PRESS;N)
DA December 10, 1990
BT1 Machines (DOE FRASE
    Vocabulary)
  BT2 Equipment
NT1 Isostatic Press
RT Printing Machine
SO DOE FRASE VOCABULARY

PRESSURE RELIEF SYSTEM
(2041 PRESSURE REL)
DA December 10, 1990
BT1 Systems (DOE FRASE Vocabulary)
  BT2 Systems
SO DOE FRASE VOCABULARY

PRESSURE VESSEL(S)
(2586 PRESSURE VES)
DA January 3, 1991
BT1 Equipment/Parts - Nuclear (DOE
    FRASE Vocabulary)
  BT2 Equipment
  BT2 Reactor Components
RT Pressure Vessel Part
RT Reactor(s)
SO DOE FRASE VOCABULARY
DEF (NRC Glossary of Terms: Nuclear
    Power and Radiation) A
    strong-walled container housing
    the core of most types of power
    reactors; it usually also contains
    the moderator, neutron reflector,
    thermal shield and control rods.

PRESSURE VESSEL PART
(2585 PRESSURE VP)

DA   January 3, 1991
BT1 Equipment/Parts - Nuclear (DOE
        FRASE Vocabulary)
 BT2 Equipment
 BT2 Reactor Components
RT   Pressure Vessel(s)
SO   DOE FRASE VOCABULARY

PRINTING MACHINE
(2135 PRINTING MAC)
DA   December 10, 1990
BT1 Machines (DOE FRASE
        Vocabulary)
 BT2 Equipment
RT   Press
SO   DOE FRASE VOCABULARY

PRODUCTION REACTOR FACILITY
(1804 PROD REACTOR)
DA   December 10, 1990
BT1 Facility (DOE FRASE Vocabulary)
 BT2 Facilities and Buildings (DOE
        FRASE Vocabulary)
  BT3 Facilities
BT1 Reactor Facilities
 BT2 Nuclear Facilities
  BT3 Facilities
RT   Nuclear Facility
SO   DOE FRASE VOCABULARY

PRODUCTION/OPERATION ACTIVITY
(1240 PO ACTIVITY)
DA   November 28, 1990
BT1 Activity Types (DOE FRASE
        Vocabulary)
 BT2 Activities
RT   Production/Operations Area
SO   DOE FRASE VOCABULARY

PRODUCTION/OPERATIONS AREA
(1710 P;O AREA)
DA   December 10, 1990
BT1 Area
 BT2 Sites/Areas
RT   Production/Operation Activity
SO   DOE FRASE VOCABULARY

PROFESSIONAL PERSONNEL
DA   November 29, 1990
BT1 Occupations
BT1 Personnel
NT1 Doctor/Nurse
NT1 Engineer
NT1 Misc Professional
NT1 Scientist
 NT2 Health Physicist
NT1 Technicians
 NT2 Engineering Technician
 NT2 Health Technician
 NT2 Miscellaneous Technician
 NT2 Radiation Monitor/Technician
 NT2 Science Technician
SO   DOE FRASE VOCABULARY

PROGRAMMATIC IMPACT
(1583 PROGRAMMATIC)
DA   November 28, 1990
SO   DOE FRASE VOCABULARY

PROTECTIVE RELAY SYSTEM
(2042 PROTECTIVE R)
DA   December 10, 1990
BT1 Systems (DOE FRASE Vocabulary)
 BT2 Systems
SO   DOE FRASE VOCABULARY

PUBLIC ADDRESS SYSTEM
(2043 PA SYSTEM)
DA   December 10, 1990
BT1 Systems (DOE FRASE Vocabulary)
 BT2 Systems
SO   DOE FRASE VOCABULARY

PULL BOX
(2412 PULL BOX)
DA   January 3, 1991
BT1 Equipment/Parts - Electrical (DOE
        FRASE Vocabulary)
 BT2 Equipment
SO   DOE FRASE VOCABULARY

PUMP(S)
(2137 PUMP;N)
DA   December 10, 1990
BT1 Machines (DOE FRASE
        Vocabulary)
 BT2 Equipment
NT1 Coolant Return Pump
NT1 Ion Pump
NT1 Reactor Coolant Pump
NT1 Sump Pump
SO   DOE FRASE VOCABULARY

PUMP AREA
(1615 PUMP AREA)
DA   December 10, 1990
BT1 Area
 BT2 Sites/Areas
SO   DOE FRASE VOCABULARY

PUNCH
(3049 PUNCH)
DA   January 3, 1991
BT1 Tools - Manual
 BT2 Tools (DOE FRASE Vocabulary)
  BT3 Equipment
SO   DOE FRASE VOCABULARY

PUNCTURE
(1336 PUNCTURE)
DA   November 28, 1990
BT1 Injuries
RT   Laceration
SO   DOE FRASE VOCABULARY

PUNCTURE PROTECTION
(2677 PUNCTURE PRO)
DA   January 3, 1991
SO   DOE FRASE VOCABULARY

PURGE SYSTEM
(2044 PURGE SYSTEM)
DA   December 10, 1990
BT1 Systems (DOE FRASE Vocabulary)
 BT2 Systems
SO   DOE FRASE VOCABULARY

**RAD WASTE TREATMENT/STORAGE
FACILITY**
(1806 RWT FACILITY)
DA   December 10, 1990
BT1 Facility (DOE FRASE Vocabulary)
 BT2 Facilities and Buildings (DOE
        FRASE Vocabulary)
  BT3 Facilities
SO   DOE FRASE VOCABULARY

RADIATION ALARM SYSTEM
(2045 RADIATION AS)
DA   December 10, 1990
BT1 Alarms
 BT2 Devices
BT1 Emergency Systems
 BT2 Systems
BT1 Systems (DOE FRASE Vocabulary)
 BT2 Systems
SO   DOE FRASE VOCABULARY

RADIATION EXPOSURE
(1343 RAD EXPOSURE)
DA   November 28, 1990
BT1 Injuries
RT   Absorption (Radiation)
RT   Radiation Exposure Module
SO   DOE FRASE VOCABULARY

RADIATION MONITOR/TECHNICIAN
(0383 RAD TECH)
DA   November 28, 1990
BT1 Technicians
 BT2 Professional Personnel
  BT3 Occupations
  BT3 Personnel
SO   DOE FRASE VOCABULARY

RADIATION SUIT
(2678 RADIATION SU)
DA   January 3, 1991
BT1 Anticontamination Clothing
 BT2 Clothing
 BT2 Personal Protective Equipment
  BT3 Equipment/Parts - Personal
        Protective (DOE FRASE
        Vocabulary)
   BT4 Equipment
SO   DOE FRASE VOCABULARY

RADIO
(2437 RADIO)
DA   January 3, 1991
BT1 Equipment/Parts - Electrical (DOE
        FRASE Vocabulary)
 BT2 Equipment
SO   DOE FRASE VOCABULARY

RAKE(S)
(3050 RAKE)
DA   January 3, 1991
BT1 Tools - Manual
 BT2 Tools (DOE FRASE Vocabulary)
  BT3 Equipment
SO   DOE FRASE VOCABULARY

RAM(S)
(2765 RAM)

DA   January 3, 1991
BT1 Instrument(s)
  BT2 Equipment/Parts -
      Instrumentation/Measuring (DOE
      FRASE Voc.)
  BT3 Equipment
SO   DOE FRASE VOCABULARY

RATCHET WRENCH(S)
(3051 RATCHET WREN)
DA   January 3, 1991
BT1 Wrench(s)
  BT2 Tools (DOE FRASE Vocabulary)
  BT3 Equipment
SO   DOE FRASE VOCABULARY

RAW WATER SYSTEM
(2046 RAW WATER SY)
DA   December 10, 1990
BT1 Systems (DOE FRASE Vocabulary)
  BT2 Systems
RT   Drinking Water System
SO   DOE FRASE VOCABULARY

RAZOR(S)
(3052 RAZOR)
DA   January 3, 1991
BT1 Tools - Manual
  BT2 Tools (DOE FRASE Vocabulary)
  BT3 Equipment
RT   Abrasion
SO   DOE FRASE VOCABULARY

REACTOR(S)
(2589 REACTOR)
DA   January 3, 1991
BT1 Equipment/Parts - Nuclear (DOE
    FRASE Vocabulary)
  BT2 Equipment
  BT2 Reactor Components
RT   Core(s)
RT   Pressure Vessel(s)
SO   DOE FRASE VOCABULARY

REACTOR FUEL
(2587 REACTOR FUEL)
DA   January 3, 1991
BT1 Equipment/Parts - Nuclear (DOE
    FRASE Vocabulary)
  BT2 Equipment
  BT2 Reactor Components
BT1 Fuels
BT1 Reactor Components
RT   Buckled Zones
RT   Core Melt Accidents
RT   Fuel Element(s)
SO   DOE FRASE VOCABULARY

REACTOR REFUELING ACTIVITY
(1248 RR ACTIVITY)
DA   November 28, 1990
BT1 Activity Types (DOE FRASE
    Vocabulary)
  BT2 Activities
RT   Reactor Components
SO   DOE FRASE VOCABULARY

REACTOR WASTE
(2588 REACTOR WAST)
DA   January 3, 1991
BT1 Equipment/Parts - Nuclear (DOE
    FRASE Vocabulary)
  BT2 Equipment
  BT2 Reactor Components
SO   DOE FRASE VOCABULARY

RECIRCULATOR
(2138 RECIRCULATOR)
DA   December 10, 1990
BT1 Machines (DOE FRASE
    Vocabulary)
  BT2 Equipment
SO   DOE FRASE VOCABULARY

RECORDER
(2766 RECORDER)
DA   January 3, 1991
BT1 Instrument(s)
  BT2 Equipment/Parts -
      Instrumentation/Measuring (DOE
      FRASE Voc.)
  BT3 Equipment
SO   DOE FRASE VOCABULARY

RECREATION AREA
(1616 RECREATION A)
DA   December 10, 1990
BT1 Area
  BT2 Sites/Areas
RT   Recreation/Break Activity
SO   DOE FRASE VOCABULARY

RECREATION/BREAK ACTIVITY
(1241 RB ACTIVITY)
DA   November 28, 1990
BT1 Activity Types (DOE FRASE
    Vocabulary)
  BT2 Activities
RT   Recreation Area
SO   DOE FRASE VOCABULARY

RECTIFIER(S)
(2413 RECTIFIER)
DA   January 3, 1991
BT1 Equipment/Parts - Electrical (DOE
    FRASE Vocabulary)
  BT2 Equipment
SO   DOE FRASE VOCABULARY

RECTUM
(1120 RECTUM)
DA   November 28, 1990
BT1 Digestive System
  BT2 Body System(s)
  BT3 Human Body Parts
SO   DOE FRASE VOCABULARY

REEL(S)
(2339 REEL)
DA   December 10, 1990
BT1 Equipment/Parts - Material
    Handling (DOE FRASE
    Vocabulary)
  BT2 Equipment
SO   DOE FRASE VOCABULARY

REFRIGERATION ROOM
(1711 REFRIG ROOM)
DA   December 10, 1990
BT1 Room
  BT2 Sites/Areas
SO   DOE FRASE VOCABULARY

REGULATOR
(2414 REGULATOR)
DA   January 3, 1991
BT1 Equipment/Parts - Electrical (DOE
    FRASE Vocabulary)
  BT2 Equipment
SO   DOE FRASE VOCABULARY

RELAY(S)
(2415 RELAY)
DA   January 3, 1991
BT1 Equipment/Parts - Electrical (DOE
    FRASE Vocabulary)
  BT2 Equipment
NT1 Building Power Low Voltage Relay
NT1 Building Time Delay Relay
NT1 Phase Failure Relays
NT1 Scram Relay
SO   DOE FRASE VOCABULARY

**REPAIR/CONSTRUCTION
PERSONNEL**
DA   November 29, 1990
BT1 Occupations
BT1 Personnel
NT1 Carpenter
NT1 Electrician
NT1 Mason
NT1 Mechanic/Repairer
NT1 Miner/Driller
NT1 Misc Repair/Construction
    Employee
NT1 Painter
NT1 Pipe Fitter
SO   DOE FRASE VOCABULARY

**RESEARCH & DEVELOPMENT
LABORATORY**
(1805 R&D LABORATO)
DA   December 10, 1990
BT1 Building (DOE FRASE Vocabulary)
  BT2 Facilities and Buildings (DOE
      FRASE Vocabulary)
  BT3 Facilities
BT1 Laboratories
  BT2 Research and Development
      Organizations
  BT3 Organizations
SO   DOE FRASE VOCABULARY

RESEARCH/TESTING ACTIVITY
(1242 RT ACTIVITY)
DA   November 28, 1990
BT1 Activity Types (DOE FRASE
    Vocabulary)
  BT2 Activities
RT   Scientist
RT   Test Area
RT   Test Reactor Facility
RT   Testing Equipment
RT   Transient Test
SO   DOE FRASE VOCABULARY

RESET DEVICE
(2416 RESET DEVICE)
DA January 3, 1991
BT1 Devices
BT1 Equipment/Parts - Electrical (DOE
    FRASE Vocabulary)
 BT2 Equipment
SO DOE FRASE VOCABULARY

RESISTOR(S)
(2417 RESISTOR)
DA January 3, 1991
BT1 Equipment/Parts - Electrical (DOE
    FRASE Vocabulary)
 BT2 Equipment
SO DOE FRASE VOCABULARY

RESPIRATOR(S)
(2679 RESPIRATOR)
DA January 3, 1991
BT1 Personal Protective Equipment
 BT2 Equipment/Parts - Personal
     Protective (DOE FRASE
     Vocabulary)
  BT3 Equipment
SO DOE FRASE VOCABULARY

RESPIRATORY SYSTEM
(1145 RESP SYSTEM)
DA November 28, 1990
BT1 Body System(s)
 BT2 Human Body Parts
NT1 Bronchial Epithelium
NT1 Nose
NT1 Throat
RT Chest
RT Cond of Respiratory Sys. Non-Toxic
RT Upper Respiratory Condition
RT Upper Respiratory Disease
SO DOE FRASE VOCABULARY

REST ROOM
(1712 REST ROOM)
DA December 10, 1990
BT1 Room
 BT2 Sites/Areas
SO DOE FRASE VOCABULARY

RESTAURANT
(1789 RESTAURANT)
DA December 10, 1990
BT1 Building (DOE FRASE Vocabulary)
 BT2 Facilities and Buildings (DOE
     FRASE Vocabulary)
  BT3 Facilities
RT Cafeteria
RT Eating Area
SO DOE FRASE VOCABULARY

RETORT
(2491 RETORT)
DA January 3, 1991
BT1 Equipment/Parts - Heating (DOE
    FRASE Vocabulary)
 BT2 Equipment
BT1 Heating Equipment
SO DOE FRASE VOCABULARY

RHEOSTAT(S)
(2418 RHEOSTAT)
DA January 3, 1991
BT1 Equipment/Parts - Electrical (DOE
    FRASE Vocabulary)
 BT2 Equipment
SO DOE FRASE VOCABULARY

RIB(S)
(1111 RIB)
DA November 28, 1990
BT1 Bone(s)
 BT2 Human Body Parts
BT1 Trunk
 BT2 Human Body Parts
RT Chest
SO DOE FRASE VOCABULARY

RIVER/CREEK
(1617 RIVER)
DA December 10, 1990
BT1 Surface Water Resources
 BT2 Natural Resources
RT Body of Water
SO DOE FRASE VOCABULARY

ROAD
(1861 ROAD)
DA December 10, 1990
BT1 Structures (DOE FRASE
    Vocabulary)
SO DOE FRASE VOCABULARY

ROD DRIVE
(2590 ROD DRIVE)
DA January 3, 1991
BT1 Equipment/Parts - Nuclear (DOE
    FRASE Vocabulary)
 BT2 Equipment
 BT2 Reactor Components
RT Control Rod Drive Assembly
SO DOE FRASE VOCABULARY

ROLL MILL
(2139 ROLL MILL)
DA December 10, 1990
BT1 Milling Machine
 BT2 Machines (DOE FRASE
     Vocabulary)
  BT3 Equipment
SO DOE FRASE VOCABULARY

ROOM
(1713 Room)
DA January 18, 1991
BT1 Sites/Areas
NT1 Air Lock Room
NT1 Central Control Room
NT1 Clean Rooms
NT1 Control Room
NT1 Equipment Room
NT1 Locker/Shower Room
NT1 Refrigeration Room
NT1 Rest Room
NT1 Shipping Room
NT1 Stock Room
RT Chamber
SO DOE FRASE VOCABULARY

ROUTER(S)
(2978 ROUTER)
DA January 3, 1991
BT1 Tools - Powered
 BT2 Tools (DOE FRASE Vocabulary)
  BT3 Equipment
SO DOE FRASE VOCABULARY

RUPTURED DISK
(1337 RUPTURED DIS)
DA November 28, 1990
BT1 Injuries
RT Back
RT Spinal Cord
RT Upper Back
SO DOE FRASE VOCABULARY

SAFETY BELT
(2680 SAFETY BELT)
DA January 3, 1991
BT1 Fall Protection Device
 BT2 Devices
 BT2 Personal Protective Equipment
  BT3 Equipment/Parts - Personal
      Protective (DOE FRASE
      Vocabulary)
   BT4 Equipment
SO DOE FRASE VOCABULARY

SAFETY BOOTS
(2681 SAFETY BOOTS)
DA January 3, 1991
BT1 Foot Protection
 BT2 Personal Protective Equipment
  BT3 Equipment/Parts - Personal
      Protective (DOE FRASE
      Vocabulary)
   BT4 Equipment
RT Foot/Feet
SO DOE FRASE VOCABULARY

SAFETY GLASSES
(2682 SAF GLASS)
DA January 3, 1991
BT1 Eye Protection
 BT2 Personal Protective Equipment
  BT3 Equipment/Parts - Personal
      Protective (DOE FRASE
      Vocabulary)
   BT4 Equipment
NT1 Safety Glasses W Side Shields
NT1 Tinted Safety Glasses
RT Eye(s)
SO DOE FRASE VOCABULARY

SAFETY GLASSES W SIDE SHIELDS
(2683 SAF GLASS SS)
DA January 3, 1991
BT1 Safety Glasses
 BT2 Eye Protection
  BT3 Personal Protective Equipment
   BT4 Equipment/Parts - Personal
       Protective (DOE FRASE
       Vocabulary)
    BT5 Equipment
RT Eye(s)
SO DOE FRASE VOCABULARY

SAFETY LINE
(2684 SAFETY LINE)
DA   January 3, 1991
BT1 Fall Protection Device
BT2 Devices
BT2 Personal Protective Equipment
BT3 Equipment/Parts - Personal
      Protective (DOE FRASE
      Vocabulary)
BT4 Equipment
SO   DOE FRASE VOCABULARY

SAFETY NET
(2685 SAFETY NET)
DA   January 3, 1991
BT1 Fall Protection Device
BT2 Devices
BT2 Personal Protective Equipment
BT3 Equipment/Parts - Personal
      Protective (DOE FRASE
      Vocabulary)
BT4 Equipment
SO   DOE FRASE VOCABULARY

SAFETY SHOE(S)
(2686 SAFETY SHOE)
DA   January 3, 1991
BT1 Foot Protection
BT2 Personal Protective Equipment
BT3 Equipment/Parts - Personal
      Protective (DOE FRASE
      Vocabulary)
BT4 Equipment
SO   DOE FRASE VOCABULARY

SALES PERSON
(0400 SALES PERSON)
DA   November 28, 1990
BT1 Occupations
BT1 Personnel
SO   DOE FRASE VOCABULARY

SAND BLASTER'S HOOD
(2687 SAND BL HOOD)
DA   January 3, 1991
BT1 Head Protection
BT2 Personal Protective Equipment
BT3 Equipment/Parts - Personal
      Protective (DOE FRASE
      Vocabulary)
BT4 Equipment
RT   Sand Blasting Area
SO   DOE FRASE VOCABULARY

SAND BLASTING AREA
(1714 SB AREA)
DA   December 10, 1990
BT1 Area
BT2 Sites/Areas
RT   Sand Blaster's Hood
SO   DOE FRASE VOCABULARY

SANDING MACHINE
(2140 SANDING MACH)
DA   December 10, 1990
BT1 Machines (DOE FRASE
      Vocabulary)
BT2 Equipment
SO   DOE FRASE VOCABULARY

SANITARY TREATMENT SYSTEM
(2048 SANITARY TS)
DA   December 10, 1990
BT1 Systems (DOE FRASE Vocabulary)
BT2 Systems
RT   Sewage Plant
SO   DOE FRASE VOCABULARY

SAWS
DA   January 30, 1991
NT1 Band Saw
NT1 Chain Saw
NT1 Concrete Saw
NT1 Hack Saw
NT1 Hand Saw
NT1 Table Saw
SO   DOE FRASE VOCABULARY

SCALE(S)
(2767 SCALE)
DA   January 3, 1991
BT1 Gauge(s)
BT2 Instrument(s)
BT3 Equipment/Parts -
      Instrumentation/Measuring
      (DOE FRASE Voc.)
BT4 Equipment
SO   DOE FRASE VOCABULARY

SCALP
(1077 SCALP)
DA   November 28, 1990
BT1 Skin
BT2 Human Body Parts
RT   Hair
RT   Head
SO   DOE FRASE VOCABULARY

SCALPEL(S)
(3053 SCALPEL)
DA   January 3, 1991
BT1 Tools - Manual
BT2 Tools (DOE FRASE Vocabulary)
BT3 Equipment
SO   DOE FRASE VOCABULARY

SCIENCE TECHNICIAN
(0380 SCIENCE TECH)
DA   November 28, 1990
BT1 Technicians
BT2 Professional Personnel
BT3 Occupations
BT3 Personnel
SO   DOE FRASE VOCABULARY

SCIENTIST
(0170 SCIENTIST)
DA   November 28, 1990
BT1 Professional Personnel
BT2 Occupations
BT2 Personnel
NT1 Health Physicist
RT   Research/Testing Activity
SO   DOE FRASE VOCABULARY

SCINTILLLATION PROBE(S)
(2768 SCINTIL PROB)
DA   January 3, 1991

BT1 Testing Equipment
BT2 Instrument(s)
BT3 Equipment/Parts -
      Instrumentation/Measuring
      (DOE FRASE Voc.)
BT4 Equipment
SO   DOE FRASE VOCABULARY

SCISSOR LIFT
(2340 SCISSOR LIFT)
DA   December 10, 1990
BT1 Hoisting Apparatus
BT2 Material Handling Device
BT3 Devices
BT3 Equipment/Parts - Material
      Handling (DOE FRASE
      Vocabulary)
BT4 Equipment
SO   DOE FRASE VOCABULARY

SCISSORS
(3054 SCISSORS)
DA   January 3, 1991
BT1 Tools - Manual
BT2 Tools (DOE FRASE Vocabulary)
BT3 Equipment
RT   Cutter(s)
RT   Snips
SO   DOE FRASE VOCABULARY

SCRAPER
(2341 SCRAPER)
DA   December 10, 1990
BT1 Equipment/Parts - Material
      Handling (DOE FRASE
      Vocabulary)
BT2 Equipment
SO   DOE FRASE VOCABULARY

SCREWDRIVER(S)
(3055 SCREWDRIVER)
DA   January 3, 1991
BT1 Tools - Manual
BT2 Tools (DOE FRASE Vocabulary)
BT3 Equipment
SO   DOE FRASE VOCABULARY

SEAT BELT(S)
(2688 SEAT BELT)
DA   January 3, 1991
BT1 Personal Protective Equipment
BT2 Equipment/Parts - Personal
      Protective (DOE FRASE
      Vocabulary)
BT3 Equipment
RT   Vehicle Accident
SO   DOE FRASE VOCABULARY

SECURITY ACTIVITY
(1243 SEC ACTIVITY)
DA   November 28, 1990
BT1 Activity Types (DOE FRASE
      Vocabulary)
BT2 Activities
RT   Plant Protection System
RT   Security Guard
SO   DOE FRASE VOCABULARY

SECURITY GUARD
(0513 SECURITY GUA)
DA   November 28, 1990
BT1 Admin. Support/Clerical Employee
 BT2 Occupations
 BT2 Personnel
RT   Guard Station
RT   Plant Protection System
RT   Security Activity
SO   DOE FRASE VOCABULARY

SENSOR(S)
(2769 SENSOR)
DA   January 3, 1991
BT1 Testing Equipment
 BT2 Instrument(s)
  BT3 Equipment/Parts -
      Instrumentation/Measuring
      (DOE FRASE Voc.)
   BT4 Equipment
SO   DOE FRASE VOCABULARY

SEPARATOR
(2142 SEPARATOR)
DA   December 10, 1990
SO   DOE FRASE VOCABULARY

SEWAGE PLANT
(1790 SEWAGE PLANT)
DA   December 10, 1990
BT1 Building (DOE FRASE Vocabulary)
 BT2 Facilities and Buildings (DOE
     FRASE Vocabulary)
  BT3 Facilities
RT   Sanitary Treatment System
RT   Sewer System
SO   DOE FRASE VOCABULARY

SEWER SYSTEM
(2050 SEWER SYSTEM)
DA   December 10, 1990
BT1 Systems (DOE FRASE Vocabulary)
 BT2 Systems
RT   Sewage Plant
SO   DOE FRASE VOCABULARY

SHEAR(S)
(3056 SHEAR)
DA   January 3, 1991
BT1 Tools - Manual
 BT2 Tools (DOE FRASE Vocabulary)
  BT3 Equipment
SO   DOE FRASE VOCABULARY

SHEET METAL WORKER
(0682 METAL WORKER)
DA   November 28, 1990
BT1 Precision/Production Personnel
 BT2 Occupations
 BT2 Personnel
SO   DOE FRASE VOCABULARY

SHIELD PLUG(S)
(2591 SHIELD PLUG)
DA   January 3, 1991
BT1 Equipment/Parts - Nuclear (DOE
     FRASE Vocabulary)
 BT2 Equipment

 BT2 Reactor Components
SO   DOE FRASE VOCABULARY

SHIPPING ROOM
(1715 SHIPPING ROO)
DA   December 10, 1990
BT1 Room
 BT2 Sites/Areas
SO   DOE FRASE VOCABULARY

SHOE COVER(S)
(2690 SHOE COVER)
DA   January 3, 1991
BT1 Foot Protection
 BT2 Personal Protective Equipment
  BT3 Equipment/Parts - Personal
      Protective (DOE FRASE
      Vocabulary)
   BT4 Equipment
NT1 Metal Shoe Cover
SO   DOE FRASE VOCABULARY

SHOULDER(S)
(1094 SHOULDER)
DA   November 28, 1990
BT1 Joint(s)
 BT2 Human Body Parts
RT   Upper Arm
SO   DOE FRASE VOCABULARY

SHOVEL(S)
(3057 SHOVEL)
DA   January 3, 1991
BT1 Tools - Manual
 BT2 Tools (DOE FRASE Vocabulary)
  BT3 Equipment
SO   DOE FRASE VOCABULARY

SHUT-DOWN CIRCUIT(S)
(2419 SHUT-DOWN CI)
DA   January 3, 1991
BT1 Equipment/Parts - Electrical (DOE
     FRASE Vocabulary)
 BT2 Equipment
SO   DOE FRASE VOCABULARY

SIDE SHIELDS
(2691 SIDE SHIELDS)
DA   January 3, 1991
BT1 Eye Protection
 BT2 Personal Protective Equipment
  BT3 Equipment/Parts - Personal
      Protective (DOE FRASE
      Vocabulary)
   BT4 Equipment
RT   Eye(s)
SO   DOE FRASE VOCABULARY

SILO
(1863 SILO)
DA   December 10, 1990
BT1 Structures (DOE FRASE
     Vocabulary)
SO   DOE FRASE VOCABULARY

SITE (DOE FRASE VOCABULARY)
(1618 SITE; EMER)
DA   December 10, 1990

BT1 Sites/Areas
NT1 Alley
NT1 Balcony
NT1 Basement
NT1 Body of Water
NT1 Cafeteria
NT1 Catwalk
NT1 Cell
NT1 Chamber
NT1 Corridor/Hall
NT1 Dwelling
NT1 En Route
NT1 Escalator
NT1 Escape Hatch
NT1 Exit
NT1 Fire Escape
NT1 Firing Range
NT1 Floor Opening
NT1 Gallery
NT1 Ground
NT1 Hatch
NT1 Highbay
NT1 Hill
NT1 Intersection
NT1 Kitchen
NT1 Laboratory
NT1 Lobby
NT1 Parking Space
NT1 Pool
NT1 Stall/Booth(s)
NT1 Station
 NT2 Fire Station
 NT2 Guard Station
 NT2 Work Station
NT1 Substation
NT1 Switchyard
NT1 Tank Farm
NT1 Vault
 NT2 SNM Vault
NT1 Waste Disposal Site
NT1 Well Pad
NT1 Work Station
SO   DOE FRASE VOCABULARY
SO   Emergency Preparedness
SO   Environmental Management

SITES/AREAS
(DOE FRASE Vocabulary Numeric Keys
16000-1774)
DA   January 17, 1991
NT1 Abandoned Areas
 NT2 Temporarily Abandoned Areas
NT1 Accident Sites
NT1 Action Areas
NT1 Active Portions
NT1 Air Quality Control Regions
 NT2 Federal Class I areas
  NT3 Mandatory Class I Federal Areas
 NT2 Regions
NT1 Area
 NT2 Canal Area
 NT2 Construction Area
 NT2 Customer Service Area
 NT2 Data Processing Area
 NT2 Decontamination Areas
 NT2 Eating Area
  NT3 Cafeteria
 NT2 Field Area
 NT2 Maintenance Area
 NT2 Office Area

NT2 Pit Area
NT2 Production/Operations Area
NT2 Pump Area
NT2 Recreation Area
NT2 Sand Blasting Area
NT2 Storage Area
NT3 Stock Room
NT3 Warehouse
NT2 Sump Area
NT2 Test Area
NT3 Fast Flux Test Facility
NT3 Nevada Test Site
NT3 Rocket Motor Test Sites
NT3 Test Reactor Facility
NT3 Test Stations
NT3 Test Reactor Area
NT3 Tonopah Test Range
NT2 Vehicle Service Area
NT2 Yard Area
NT1 Area of Review
NT1 Assessment Area
NT1 Attainment Areas
NT1 Burial Grounds
NT1 Central Alarm Stations
NT1 Commercial Areas
NT1 Confinement Areas
NT1 Controlled Areas
NT2 Disturbed Zones
NT1 Conventional Mines
NT2 Active Mines
NT2 Inactive Mines
NT2 Underground Uranium Mines
NT1 Corrective Action Management
    Units
NT2 Disposal Units
NT3 Land Disposal Units
NT4 Landfill Cells
NT4 Landfills
NT5 Chemical Waste Landfills
NT5 Sanitary Landfills
NT5 Specially Designated Landfills
NT2 Hazardous Waste Management
    Units
NT3 Miscellaneous Units
NT2 Solid Waste Management Units
NT1 Crane Maintenance Area
NT1 Crawl Spaces
NT1 Critical Areas
NT1 Cultural Resource Sites
NT1 Department of Energy Sites
NT1 Depositories
NT1 Discharge Point
NT1 Disposal Sites
NT2 Inactive Waste Disposal Sites
NT3 Inactive Hazardous Waste
    Disposal Sites
NT1 Disposal Areas
NT1 DOE Property
NT1 DOE Reservation
NT1 Dog Houses
NT1 Dry Caves
NT1 Dumps
NT2 Open Dumps
NT1 Emergency Control Stations
NT1 Entry Control Points
NT1 EPA Region
NT1 Evaporation Ponds
NT1 Extraction Sites
NT1 Feedlots
NT1 Field Command Posts
NT1 Firearms Ranges

NT1 Floodplains
NT1 Geologic Repositories
NT2 Geologic Repository Operations
    Areas
NT1 High-Contact Residential Surfaces
NT1 High-Contact Commercial Surfaces
NT1 High-Contact Industrial Surfaces
NT1 Impoundment
NT2 New Tailings Impoundment
NT2 PAR Ponds
NT2 Reservoirs
NT2 Surface Impoundment
NT1 Inactive Portions
NT1 Indian Lands
NT1 Individual Generation Sites
NT1 Licensed Sites
NT1 Manatee Protection Areas
NT2 Manatee Refuges
NT2 Manatee Sanctuaries
NT1 Material Access Areas
NT1 National Defense Areas
NT1 National Security Areas
NT1 Non-Attainment Areas
NT1 Non-Restricted Access Areas
NT1 Occupiable Area
NT1 Occupied Area (Explosives)
NT1 Off-site
NT1 On-Scene
NT1 On-Site
NT1 Operating Area Compartment
NT1 Other Restricted Access
    (nonsubstation) Locations
NT1 Outdoor Electrical Substations
NT1 Outfall
NT1 Point of Disinfectant Application
NT1 Point of Compliance
NT1 Point of Nearest Public Access
NT1 Property Protection Area
NT1 Protected Areas
NT1 Public Lands
NT1 Radiation Areas
NT2 High Radiation Areas
NT1 Radiation Emergency Assistance
    Center/Training Sites
NT1 Radiological Areas
NT1 Ranges (Firearms)
NT1 Recharge Areas
NT1 Remote Interrogation Points
NT1 Residential Areas
NT1 Restricted Areas
NT1 River Basins
NT1 Room
NT2 Air Lock Room
NT2 Central Control Room
NT2 Clean Rooms
NT2 Control Room
NT2 Equipment Room
NT2 Locker/Shower Room
NT2 Refrigeration Room
NT2 Rest Room
NT2 Shipping Room
NT2 Stock Room
NT1 Security Areas
NT2 Exclusion Areas
NT2 Limited Areas
NT2 SNM Vault
NT2 Vault-type Rooms
NT2 Vital Areas
NT1 Site (DOE FRASE Vocabulary)
NT2 Alley
NT2 Balcony

NT2 Basement
NT2 Body of Water
NT2 Cafeteria
NT2 Catwalk
NT2 Cell
NT2 Chamber
NT2 Corridor/Hall
NT2 Dwelling
NT2 En Route
NT2 Escalator
NT2 Escape Hatch
NT2 Exit
NT2 Fire Escape
NT2 Firing Range
NT2 Floor Opening
NT2 Gallery
NT2 Ground
NT2 Hatch
NT2 Highbay
NT2 Hill
NT2 Intersection
NT2 Kitchen
NT2 Laboratory
NT2 Lobby
NT2 Parking Space
NT2 Pool
NT2 Stall/Booth(s)
NT2 Station
NT3 Fire Station
NT3 Guard Station
NT3 Work Station
NT2 Substation
NT2 Switchyard
NT2 Tank Farm
NT2 Vault
NT3 SNM Vault
NT2 Waste Disposal Site
NT2 Well Pad
NT2 Work Station
NT1 Site Boundaries
NT1 Soil Adsorption Field
NT1 Solid Waste Storage Area
NT1 Specified Ports and Harbors
NT1 Special Aquatic Sites
NT1 Spill Area
NT1 Spill Boundaries
NT1 Stabilization Ponds
NT1 Storage Area Compartments
NT1 Supervised Areas
NT1 Test Area North
NT1 Transfer Stations
NT1 Unattended Openings
NT1 Uncontrolled Hazardous Waste
    Sites
NT1 Underground Areas
NT1 Unpackaging Rooms
NT1 Watershed
NT1 Wetlands
NT2 Bogs
NT2 Bottomland Hardwoods
NT2 Fens
NT2 Marshes
NT3 Tidal Marshes
NT2 Swamps
NT1 Zones
NT2 Aeration Zones
NT2 Buckled Zones
NT2 Buffer Zone
NT2 Coastal Zones
NT2 Confining Zones
NT2 Contiguous Zones

NT2 Contingency Planning Zones
NT2 Disturbed Zones
NT2 Emergency Planning Zones
NT2 Engineering Control Zones
NT2 Excavation Zone
NT2 Flow Zone
NT2 Hydrogeologic Units
NT2 Injection Zones
 NT3 Injection Interval
NT2 Inland Zones
NT2 Isolation Zones
NT2 Law Hazard Zone
NT2 Mixing Zones
NT2 Planning Zones
NT2 Radiation Zone
NT2 Saturated Zones
NT2 Treatment Zones
NT2 Unsaturated Zones
NT2 Vulnerable Zones
RT DOE FRASE Categories
RT Facilities
RT Real Property
RT Siting
RT Structures (DOE FRASE
     Vocabulary)
RT Targets
SO DOE FRASE VOCABULARY
DEF Designations of particular locations.
     Also, A subject category used with
     the DOE FRASE Vocabulary.

SKIN
(1132 SKIN)
DA November 28, 1990
BT1 Human Body Parts
NT1 Fingernail(s)
NT1 Hair
NT1 Scalp
RT Dermatitis
RT Other Skin Conditions
RT Palm(s)
SO DOE FRASE VOCABULARY

SKULL (FRASE)
(1078 SKULL)
DA November 28, 1990
BT1 Head
 BT2 Human Body Parts
NT1 Jaw
RT Brain
RT Brain Damage
RT Concussion
RT Contusion(S)
RT Head Protection
SO DOE FRASE VOCABULARY

SLEDGEHAMMER(S)
(3059 SLEDGEHAMMER)
DA January 3, 1991
BT1 Hammer
 BT2 Tools - Manual
  BT3 Tools (DOE FRASE Vocabulary)
   BT4 Equipment
SO DOE FRASE VOCABULARY

SLICING MACHINE
(2144 SLICING MACH)
DA December 10, 1990

BT1 Machines (DOE FRASE
     Vocabulary)
 BT2 Equipment
SO DOE FRASE VOCABULARY

SLING(S)
(2342 SLING)
DA December 10, 1990
BT1 Equipment/Parts - Material
     Handling (DOE FRASE
     Vocabulary)
 BT2 Equipment
SO DOE FRASE VOCABULARY

SMOKE DAMAGE
(1537 SMOKE DAMAGE)
DA November 29, 1990
BT1 Damage
 BT2 Nature of Property Damage
RT Fire Damage
RT Water Damage
SO DOE FRASE VOCABULARY

SNAKE BITE
(1338 SNAKE BITE)
DA November 28, 1990
BT1 Animal Bite
 BT2 Injuries
SO DOE FRASE VOCABULARY

SNIPS
(3060 SNIPS)
DA January 3, 1991
BT1 Tools - Manual
 BT2 Tools (DOE FRASE Vocabulary)
  BT3 Equipment
RT Cutter(s)
RT Scissors
SO DOE FRASE VOCABULARY

SNUBBER(S)
(3061 SNUBBER)
DA January 3, 1991
BT1 Tools - Manual
 BT2 Tools (DOE FRASE Vocabulary)
  BT3 Equipment
SO DOE FRASE VOCABULARY

SODIUM SCRUBBER(S)
(2593 SODIUM SCRUB)
DA January 3, 1991
BT1 Equipment/Parts - Nuclear (DOE
     FRASE Vocabulary)
 BT2 Equipment
 BT2 Reactor Components
SO DOE FRASE VOCABULARY

SOLAR COLLECTOR(S)
(2420 SOLAR COLLEC)
DA January 3, 1991
BT1 Equipment/Parts - Electrical (DOE
     FRASE Vocabulary)
 BT2 Equipment
SO DOE FRASE VOCABULARY

SOLDERING IRON(S)
(2979 SOLDERING IR)
DA January 3, 1991

BT1 Tools - Powered
 BT2 Tools (DOE FRASE Vocabulary)
  BT3 Equipment
SO DOE FRASE VOCABULARY

SOLE(S)
(1130 SOLE)
DA November 28, 1990
BT1 Foot/Feet
 BT2 Human Body Parts
RT Foot Protection
SO DOE FRASE VOCABULARY

SOLENOID(S)
(2421 SOLENOID)
DA January 3, 1991
BT1 Equipment/Parts - Electrical (DOE
     FRASE Vocabulary)
 BT2 Equipment
SO DOE FRASE VOCABULARY

SPATULA(S)
(3062 SPATULA)
DA January 3, 1991
BT1 Tools - Manual
 BT2 Tools (DOE FRASE Vocabulary)
  BT3 Equipment
SO DOE FRASE VOCABULARY

SPECTROMETER(S)
(2770 SPECTROMETER)
DA January 3, 1991
BT1 Indicator(s)
 BT2 Instrument(s)
  BT3 Equipment/Parts -
     Instrumentation/Measuring
     (DOE FRASE Voc.)
   BT4 Equipment
SO DOE FRASE VOCABULARY

SPINAL CORD
(1113 SPINAL CORD)
DA November 28, 1990
BT1 Nervous System
 BT2 Body System(s)
  BT3 Human Body Parts
RT Back
RT Pinched Nerve
RT Ruptured Disk
RT Spine
RT Upper Back
SO DOE FRASE VOCABULARY

SPINE
(1112 SPINE)
DA November 28, 1990
BT1 Bone(s)
 BT2 Human Body Parts
NT1 Coccyx
RT Neck
RT Spinal Cord
SO DOE FRASE VOCABULARY

SPRAIN
(1339 SPRAIN)
DA November 28, 1990
BT1 Injuries

RT Strain
SO DOE FRASE VOCABULARY

SPRAY GUN(S)
(2980 SPRAY GUN)
DA January 3, 1991
BT1 Tools - Powered
 BT2 Tools (DOE FRASE Vocabulary)
  BT3 Equipment
SO DOE FRASE VOCABULARY

SPRINKLER SYSTEM
(2051 SPRINKLER SY)
DA December 10, 1990
BT1 Dedicated Fire Water Systems
 BT2 Fire Protection Systems
  BT3 Emergency Systems
   BT4 Systems
BT1 Systems (DOE FRASE Vocabulary)
 BT2 Systems
RT Fire Suppression System
SO DOE FRASE VOCABULARY

STALL/BOOTH(S)
(1716 STALL;N)
DA December 10, 1990
BT1 Site (DOE FRASE Vocabulary)
 BT2 Sites/Areas
SO DOE FRASE VOCABULARY

START-UP PROCEDURE
(1244 SU PROCEDURE)
DA November 28, 1990
RT Pre Start-Up/ Calibration Activity
SO DOE FRASE VOCABULARY

STARTER
(2422 STARTER)
DA January 3, 1991
BT1 Equipment/Parts - Electrical (DOE
    FRASE Vocabulary)
 BT2 Equipment
SO DOE FRASE VOCABULARY

STATION
(1717 STATION)
DA December 10, 1990
BT1 Site (DOE FRASE Vocabulary)
 BT2 Sites/Areas
NT1 Fire Station
NT1 Guard Station
NT1 Work Station
RT Substation
SO DOE FRASE VOCABULARY

STEAM PLANT
(1807 STEAM PLANT)
DA December 10, 1990
BT1 Building (DOE FRASE Vocabulary)
 BT2 Facilities and Buildings (DOE
     FRASE Vocabulary)
  BT3 Facilities
SO DOE FRASE VOCABULARY

STITCHING/SEWING MACHINE
(2145 STITCHING MA)
DA December 10, 1990

BT1 Machines (DOE FRASE
    Vocabulary)
 BT2 Equipment
SO DOE FRASE VOCABULARY

STOCK ROOM
(1719 STOCK ROOM)
DA December 10, 1990
BT1 Room
 BT2 Sites/Areas
BT1 Storage Area
 BT2 Area
  BT3 Sites/Areas
SO DOE FRASE VOCABULARY

STORAGE AREA
(1619 STORAGE AREA)
DA December 10, 1990
BT1 Area
 BT2 Sites/Areas
NT1 Stock Room
NT1 Warehouse
SO DOE FRASE VOCABULARY

STOVE(S)
(2492 STOVE)
DA January 3, 1991
BT1 Heater(s)
 BT2 Equipment/Parts - Heating (DOE
     FRASE Vocabulary)
  BT3 Equipment
SO DOE FRASE VOCABULARY

STRAIGHT EDGE RULER
(2771 STRAIGHT EDG)
DA January 3, 1991
BT1 Gauge(s)
 BT2 Instrument(s)
  BT3 Equipment/Parts -
      Instrumentation/Measuring
      (DOE FRASE Voc.)
   BT4 Equipment
SO DOE FRASE VOCABULARY

STRAIN
(1340 STRAIN)
DA November 28, 1990
BT1 Injuries
RT Sprain
SO DOE FRASE VOCABULARY

STRANGULATION
(1341 STRANGULATIO)
DA November 28, 1990
BT1 Injuries
SO DOE FRASE VOCABULARY

STRUCTURAL DAMAGE
(1538 STRUCTURAL D)
DA November 29, 1990
BT1 Damage
 BT2 Nature of Property Damage
RT Structures (DOE FRASE
    Vocabulary)
SO DOE FRASE VOCABULARY

**STRUCTURES (DOE FRASE
VOCABULARY)**
(DOE Order 4330.4A; DOE FRASE
 Vocabulary Numeric Keys 1850-1924)
DA January 17, 1991
NT1 Artificial Reefs
NT1 Bridge
NT1 Cooling Towers
NT1 Dam
NT1 Earth-Lined Channels
NT1 Explosives Buildings
 NT2 Explosives Bays
  NT3 Staging Bays (in-process)
NT1 Facility Boundaries
NT1 Highway
NT1 Hydraulic Structures
 NT2 Dams
 NT2 Dikes
  NT3 Terracing
NT1 Loading Dock
NT1 Mine
NT1 Parking Lot
NT1 Pier
NT1 Platforms
NT1 Pond
NT1 Raw Water Intake Structure
NT1 Road
NT1 Silo
NT1 Space Frames
 NT2 Building Frame Systems
  NT3 Bearing Wall Systems
  NT3 Braced Frame
   NT4 Concentric Braced Frames
   NT4 Eccentric Braced Frames
  NT3 Horizontal Bracing Systems
  NT3 Lateral Force Resisting System
  NT3 Shoring
 NT2 Moment Resisting Space Frame
  NT3 Intermediate Moment Resisting
      Space Frame
  NT3 Ordinary Moment Resisting
      Space Frame
  NT3 Special Moment Resisting Space
      Frame
 NT2 Vertical Load Carrying Space
     Frames
NT1 Trailer Building
NT1 Trench/Ditch
RT DOE FRASE Categories
RT Facilities
RT Sites/Areas
RT Structural Damage
RT Targets
RT Towers
SO DOE FRASE VOCABULARY
DEF (DOE Order 4330.4A) Any fixed
    real property improvement
    constructed on or in the land that
    is not a building or utility (e.g.,
    bridges, towers, and tanks).
    Denotes something constructed or
    built. Also, a subject category used
    with the DOE FRASE Vocabulary.

SUBSTATION
(1620 SUBSTATION)
DA December 10, 1990
BT1 Site (DOE FRASE Vocabulary)
 BT2 Sites/Areas

RT   Station
SO   DOE FRASE VOCABULARY

SUMP AREA
(1621 SUMP AREA)
DA   December 10, 1990
BT1   Area
 BT2   Sites/Areas
SO   DOE FRASE VOCABULARY

SUMP PUMP
(2146 SUMP PUMP)
DA   December 10, 1990
BT1   Pump(s)
 BT2   Machines (DOE FRASE
        Vocabulary)
  BT3   Equipment
SO   DOE FRASE VOCABULARY

SUNBURN
(1400 SUNBURN)
DA   November 28, 1990
BT1   Illnesses
RT   Heat Stroke/Other High Temp
      Effect
SO   DOE FRASE VOCABULARY

SWITCH BOX
(2423 SWITCH BOX)
DA   January 3, 1991
BT1   Equipment/Parts - Electrical (DOE
        FRASE Vocabulary)
 BT2   Equipment
SO   DOE FRASE VOCABULARY

SWITCHBOARD
(2424 SWITCHBOARD)
DA   January 3, 1991
BT1   Equipment/Parts - Electrical (DOE
        FRASE Vocabulary)
 BT2   Equipment
SO   DOE FRASE VOCABULARY

SWITCHGEAR
(2425 SWITCHGEAR)
DA   January 3, 1991
BT1   Equipment/Parts - Electrical (DOE
        FRASE Vocabulary)
 BT2   Equipment
SO   DOE FRASE VOCABULARY

SWITCHYARD
(1622 SWITCHYARD)
DA   December 10, 1990
BT1   Site (DOE FRASE Vocabulary)
 BT2   Sites/Areas
SO   DOE FRASE VOCABULARY

SYNOVITIS
(1401 SYNOVITIS)
DA   November 28, 1990
BT1   Illnesses
RT   Tendonitis
SO   DOE FRASE VOCABULARY

SYSTEMIC EFFECTS
(1402 SYSTEMIC EFF)
DA   November 28, 1990

BT1   Effects
BT1   Illnesses
SO   DOE FRASE VOCABULARY

SYSTEMIC POISONING
(1403 SYSTEMIC POI)
DA   November 28, 1990
BT1   Poisoning
 BT2   Illnesses
SO   DOE FRASE VOCABULARY

**SYSTEMS (DOE FRASE
VOCABULARY)**
(DOE FRASE Vocabulary Numeric Keys
  2025-2099)
DA   December 10, 1990
BT1   Systems
NT1   Cooling System
NT1   Criticality Alarm System
NT1   Deluge System
NT1   Drinking Water System
NT1   Emergency Core Cooling System
NT1   Emergency Power System
NT1   Explosion Suppression System
NT1   Fire Alarm System
NT1   Fire Suppression System
NT1   Foam System
NT1   Halon System
NT1   HVAC System
NT1   Instrument Air System
NT1   Interlock System
NT1   Off Gas System
NT1   Plant Protection System
NT1   Pressure Relief System
NT1   Protective Relay System
NT1   Public Address System
NT1   Purge System
NT1   Radiation Alarm System
NT1   Raw Water System
NT1   Sanitary Treatment System
NT1   Sewer System
NT1   Sprinkler System
NT1   Vacuum System
NT1   Ventilation System
 NT2   End Box Ventilation System
 NT2   Reactor Building Standby
        Ventilation System
NT1   Waste Water System
 NT2   Waste Water Collection Systems
RT   Controls
RT   DOE FRASE Categories
SO   DOE FRASE VOCABULARY
DEF A subject category used with the
      DOE FRASE Vocabulary.

TABLE SAW
(2147 TABLE SAW)
DA   December 10, 1990
BT1   Machines (DOE FRASE
        Vocabulary)
 BT2   Equipment
BT1   Saws
SO   DOE FRASE VOCABULARY

TACHOMETER(S)
(2773 TACHOMETER)
DA   January 3, 1991
BT1   Indicator(s)
 BT2   Instrument(s)

 BT3   Equipment/Parts -
        Instrumentation/Measuring
        (DOE FRASE Voc.)
  BT4   Equipment
SO   DOE FRASE VOCABULARY

TANK FARM
(1623 TANK FARM)
DA   December 10, 1990
BT1   Site (DOE FRASE Vocabulary)
 BT2   Sites/Areas
SO   DOE FRASE VOCABULARY

TECHNICIANS
DA   November 29, 1990
BT1   Professional Personnel
 BT2   Occupations
 BT2   Personnel
NT1   Engineering Technician
NT1   Health Technician
NT1   Miscellaneous Technician
NT1   Radiation Monitor/Technician
NT1   Science Technician
SO   DOE FRASE VOCABULARY

TENDONITIS
(1404 TENDONITIS)
DA   November 28, 1990
BT1   Illnesses
RT   Muscle/Tendon(s)
RT   Synovitis
SO   DOE FRASE VOCABULARY

TEST AREA
(1624 TEST AREA)
DA   December 10, 1990
BT1   Area
 BT2   Sites/Areas
NT1   Fast Flux Test Facility
NT1   Nevada Test Site
NT1   Rocket Motor Test Sites
NT1   Test Reactor Facility
NT1   Test Stations
NT1   Test Reactor Area
NT1   Tonopah Test Range
RT   Research/Testing Activity
SO   DOE FRASE VOCABULARY

TEST REACTOR FACILITY
(1808 TEST REACTOR)
DA   December 10, 1990
BT1   Facility (DOE FRASE Vocabulary)
 BT2   Facilities and Buildings (DOE
        FRASE Vocabulary)
  BT3   Facilities
BT1   Reactor Facilities
 BT2   Nuclear Facilities
  BT3   Facilities
BT1   Test Area
 BT2   Area
  BT3   Sites/Areas
RT   Research/Testing Activity
SO   DOE FRASE VOCABULARY
DEF (NWPA) An at-depth, prototypic,
      underground cavity with
      subsurface lateral excavations
      extending from a central shaft that
      is used for research and
      development purposes, including

the development of data and
experience for the safe handling
and disposal of solidified high-level
radioactive waste, transuranic
waste, or spent nuclear fuel.

TEST TRAIN
(2594 TEST TRAIN)
DA January 3, 1991
BT1 Equipment/Parts - Nuclear (DOE
     FRASE Vocabulary)
  BT2 Equipment
  BT2 Reactor Components
SO DOE FRASE VOCABULARY

TESTING EQUIPMENT
(2774 TEST EQUIPMT)
DA January 3, 1991
BT1 Instrument(s)
  BT2 Equipment/Parts -
       Instrumentation/Measuring (DOE
       FRASE Voc.)
    BT3 Equipment
NT1 Heat Detector
NT1 Leak Detector
NT1 Oscilloscope(s)
NT1 Scintillation Probe(s)
NT1 Sensor(s)
RT Research/Testing Activity
SO DOE FRASE VOCABULARY

TETANUS
(1406 TETANUS)
DA November 28, 1990
BT1 Illnesses
SO DOE FRASE VOCABULARY

THERMOCOUPLE(S)
(2426 THERMOCOUPLE)
DA January 3, 1991
SF TC (Thermocouple)
BT1 Equipment/Parts - Electrical (DOE
     FRASE Vocabulary)
  BT2 Equipment
SO DOE FRASE VOCABULARY

THERMOMETER(S)
(2776 THERMOMETER)
DA January 3, 1991
BT1 Indicator(s)
  BT2 Instrument(s)
    BT3 Equipment/Parts -
         Instrumentation/Measuring
         (DOE FRASE Voc.)
      BT4 Equipment
SO DOE FRASE VOCABULARY

THERMOSTAT(S)
(2777 THERMOSTAT)
DA January 3, 1991
BT1 Gauge(s)
  BT2 Instrument(s)
    BT3 Equipment/Parts -
         Instrumentation/Measuring
         (DOE FRASE Voc.)
      BT4 Equipment
SO DOE FRASE VOCABULARY

THIGH(S)
(1123 THIGH)
DA November 28, 1990
BT1 Leg(s)
  BT2 Human Body Parts
RT Groin
RT Leggings
SO DOE FRASE VOCABULARY

THREADER(S)
(3063 THREADER)
DA January 3, 1991
BT1 Tools - Manual
  BT2 Tools (DOE FRASE Vocabulary)
    BT3 Equipment
SO DOE FRASE VOCABULARY

THROAT
(1092 THROAT)
DA November 28, 1990
BT1 Respiratory System
  BT2 Body System(s)
    BT3 Human Body Parts
RT Neck
SO DOE FRASE VOCABULARY

THUMB(S)
(1103 THUMB)
DA November 28, 1990
BT1 Hand(s)
  BT2 Human Body Parts
RT Finger(s)
RT Glove(s)
RT Hand Protection
SO DOE FRASE VOCABULARY

TIME CLOCK
(2778 TIME CLOCK)
DA January 3, 1991
BT1 Gauge(s)
  BT2 Instrument(s)
    BT3 Equipment/Parts -
         Instrumentation/Measuring
         (DOE FRASE Voc.)
      BT4 Equipment
SO DOE FRASE VOCABULARY

TINE(S)
(2343 TINE)
DA December 10, 1990
BT1 Equipment/Parts - Material
     Handling (DOE FRASE
     Vocabulary)
  BT2 Equipment
SO DOE FRASE VOCABULARY

TINTED SAFETY GLASSES
(2692 TINTED SAFET)
DA January 3, 1991
BT1 Safety Glasses
  BT2 Eye Protection
    BT3 Personal Protective Equipment
      BT4 Equipment/Parts - Personal
           Protective (DOE FRASE
           Vocabulary)
        BT5 Equipment
SO DOE FRASE VOCABULARY

TOE(S)
(1131 TOE)
DA November 28, 1990
BT1 Foot/Feet
  BT2 Human Body Parts
RT Foot Protection
RT Metatarsal Protection
SO DOE FRASE VOCABULARY

TONGS
(3064 TONGS)
DA January 3, 1991
BT1 Tools - Manual
  BT2 Tools (DOE FRASE Vocabulary)
    BT3 Equipment
SO DOE FRASE VOCABULARY

TOOLS (DOE FRASE VOCABULARY)
(3065 TOOL)
DA January 3, 1991
BT1 Equipment
NT1 Leader Seater
NT1 Tools - Manual
  NT2 Ax(s)
  NT2 Broom(s)
  NT2 Brush(s)
  NT2 Chisel(s)
  NT2 Crowbar(s)
  NT2 Cutter(s)
  NT2 File(s)
  NT2 Hack Saw
  NT2 Hammer
    NT3 Sledgehammer(s)
  NT2 Hand Iron(s)
  NT2 Hand Saw
  NT2 Hand Stapler
  NT2 Hatchet(s)
  NT2 Hoe(s)
  NT2 Knife(s)
    NT3 Pocket Knife(s)
  NT2 Mop(s)
  NT2 Non-Powered Handtool(s)
  NT2 Pick(s)
  NT2 Plane(s)
  NT2 Pliers
  NT2 Punch
  NT2 Rake(s)
  NT2 Razor(s)
  NT2 Scalpel(s)
  NT2 Scissors
  NT2 Screwdriver(s)
  NT2 Shear(s)
  NT2 Shovel(s)
  NT2 Snips
  NT2 Snubber(s)
  NT2 Spatula(s)
  NT2 Threader(s)
  NT2 Tongs
  NT2 Torch(s)
  NT2 Tweezers
  NT2 Vise
  NT2 Visegrips
NT1 Tools - Powered
  NT2 Air Tamper
  NT2 Chain Saw
  NT2 Disk Sander(s)
  NT2 Jackhammer
  NT2 Mixer(s)
  NT2 Nail Gun
  NT2 Post Driver(s)

NT2 Power Actuated Tool(s)
NT2 Power Buffer(s)
NT2 Power Chisel(s)
NT2 Power Cutter(s)
NT2 Power Drill(s)
NT2 Power File(s)
NT2 Power Grinder(s)
NT2 Power Hammer(s)
NT2 Power Impact Wrench(s)
NT2 Power Polisher(s)
NT2 Power Riveter(s)
NT2 Power Sandblaster(s)
NT2 Power Screwdriver(s)
NT2 Power Spray Gun(s)
NT2 Power Stapler(s)
NT2 Power Waxer(s)
NT2 Powered Handtool(s)
NT2 Router(s)
NT2 Soldering Iron(s)
NT2 Spray Gun(s)
NT2 Weed Eater
NT1 Wrench(s)
 NT2 Allen Wrench
 NT2 Impact Wrench
 NT2 Power Impact Wrench(s)
 NT2 Ratchet Wrench(s)
 NT2 Torque Wrench(s)
SO DOE FRASE VOCABULARY

TOOLS - MANUAL
(DOE FRASE Vocabulary Numeric Keys
  3025-3099)
DA January 17, 1991
BT1 Tools (DOE FRASE Vocabulary)
 BT2 Equipment
NT1 Ax(s)
NT1 Broom(s)
NT1 Brush(s)
NT1 Chisel(s)
NT1 Crowbar(s)
NT1 Cutter(s)
NT1 File(s)
NT1 Hack Saw
NT1 Hammer
 NT2 Sledgehammer(s)
NT1 Hand Iron(s)
NT1 Hand Saw
NT1 Hand Stapler
NT1 Hatchet(s)
NT1 Hoe(s)
NT1 Knife(s)
 NT2 Pocket Knife(s)
NT1 Mop(s)
NT1 Non-Powered Handtool(s)
NT1 Pick(s)
NT1 Plane(s)
NT1 Pliers
NT1 Punch
NT1 Rake(s)
NT1 Razor(s)
NT1 Scalpel(s)
NT1 Scissors
NT1 Screwdriver(s)
NT1 Shear(s)
NT1 Shovel(s)
NT1 Snips
NT1 Snubber(s)
NT1 Spatula(s)
NT1 Threader(s)
NT1 Tongs

NT1 Torch(s)
NT1 Tweezers
NT1 Vise
NT1 Visegrips
RT DOE FRASE Categories
RT Tools - Powered
SO DOE FRASE VOCABULARY
DEF A subject category used with the
    DOE FRASE Vocabulary.

TOOLS - POWERED
(DOE FRASE Vocabulary Numeric Keys
  2950-3024)
DA January 17, 1991
BT1 Tools (DOE FRASE Vocabulary)
 BT2 Equipment
NT1 Air Tamper
NT1 Chain Saw
NT1 Disk Sander(s)
NT1 Jackhammer
NT1 Mixer(s)
NT1 Nail Gun
NT1 Post Driver(s)
NT1 Power Actuated Tool(s)
NT1 Power Buffer(s)
NT1 Power Chisel(s)
NT1 Power Cutter(s)
NT1 Power Drill(s)
NT1 Power File(s)
NT1 Power Grinder(s)
NT1 Power Hammer(s)
NT1 Power Impact Wrench(s)
NT1 Power Polisher(s)
NT1 Power Riveter(s)
NT1 Power Sandblaster(s)
NT1 Power Screwdriver(s)
NT1 Power Spray Gun(s)
NT1 Power Stapler(s)
NT1 Power Waxer(s)
NT1 Powered Handtool(s)
NT1 Router(s)
NT1 Soldering Iron(s)
NT1 Spray Gun(s)
NT1 Weed Eater
RT DOE FRASE Categories
RT Tools - Manual
SO DOE FRASE VOCABULARY
DEF A subject category used with the
    DOE FRASE Vocabulary.

TOOTH/TEETH
(1089 TOOTH)
DA November 28, 1990
BT1 Mouth
 BT2 Digestive System
  BT3 Body System(s)
   BT4 Human Body Parts
RT Dental Injury
SO DOE FRASE VOCABULARY

TORCH(S)
(3066 TORCH)
DA January 3, 1991
BT1 Tools - Manual
 BT2 Tools (DOE FRASE Vocabulary)
  BT3 Equipment
SO DOE FRASE VOCABULARY

TORN CARTILAGE
(1342 TORN CARTILA)
DA November 28, 1990
BT1 Injuries
SO DOE FRASE VOCABULARY

TORQUE WRENCH(S)
(3067 TORQUE WRENC)
DA January 3, 1991
BT1 Wrench(s)
 BT2 Tools (DOE FRASE Vocabulary)
  BT3 Equipment
SO DOE FRASE VOCABULARY

TOXIC EFFECTS TO SINGLE SYSTEM
(1407 TOXIC EFFECT)
DA November 28, 1990
BT1 Effects
BT1 Illnesses
RT Poisoning
SO DOE FRASE VOCABULARY

TRAILER BUILDING
(1866 TRAILER; BLD)
DA December 10, 1990
BT1 Structures (DOE FRASE
    Vocabulary)
SO DOE FRASE VOCABULARY

TRAINING/EDUCATION ACTIVITY
(1245 TE ACTIVITY)
DA November 28, 1990
BT1 Activity Types (DOE FRASE
    Vocabulary)
 BT2 Activities
SO DOE FRASE VOCABULARY

TRANSDUCER(S)
(2427 TRANSDUCER)
DA January 3, 1991
BT1 Equipment/Parts - Electrical (DOE
    FRASE Vocabulary)
 BT2 Equipment
SO DOE FRASE VOCABULARY

TRANSFORMER(S)
(2428 TRANSFORMER)
DA January 3, 1991
BT1 Equipment/Parts - Electrical (DOE
    FRASE Vocabulary)
 BT2 Equipment
SO DOE FRASE VOCABULARY

TRANSIENT OPERATION
(2596 TRANSIENT OP)
DA January 3, 1991
BT1 Equipment/Parts - Nuclear (DOE
    FRASE Vocabulary)
 BT2 Equipment
 BT2 Reactor Components
SO DOE FRASE VOCABULARY

TRANSIENT TEST
(2597 TRANSIENT TE)
DA January 3, 1991
BT1 Equipment/Parts - Nuclear (DOE
    FRASE Vocabulary)
 BT2 Equipment

BT2 Reactor Components
RT Research/Testing Activity
SO DOE FRASE VOCABULARY

TRANSMITTER(S)
(2429 TRANSMITTER)
DA January 3, 1991
BT1 Equipment/Parts - Electrical (DOE
    FRASE Vocabulary)
BT2 Equipment
SO DOE FRASE VOCABULARY

TRANSPORT PERSONNEL
DA November 29, 1990
BT1 Occupations
BT1 Personnel
NT1 Aircraft Pilot
NT1 Bus Driver
NT1 Equipment Operator
NT1 Misc Transport Employee
NT1 Truck Driver
RT Transportation Activity
RT Travel Activity
SO DOE FRASE VOCABULARY

TRANSPORTATION ACTIVITY
(1246 TRS ACTIVITY)
DA November 28, 1990
BT1 Activity Types (DOE FRASE
    Vocabulary)
BT2 Activities
RT Misc Transport Employee
RT Transport Personnel
SO DOE FRASE VOCABULARY

TRAVEL ACTIVITY
(1249 TRA ACTIVITY)
DA November 28, 1990
BT1 Activity Types (DOE FRASE
    Vocabulary)
BT2 Activities
RT Transport Personnel
SO DOE FRASE VOCABULARY

TRENCH/DITCH
(1867 TRENCH)
DA December 10, 1990
BT1 Structures (DOE FRASE
    Vocabulary)
SO DOE FRASE VOCABULARY

TROUBLE LIGHT
(2430 TROUBLE LIGH)
DA January 3, 1991
BT1 Equipment/Parts - Electrical (DOE
    FRASE Vocabulary)
BT2 Equipment
RT Indicator(s)
SO DOE FRASE VOCABULARY

TRUCK DRIVER
(0820 TRUCK DRIVER)
DA November 28, 1990
BT1 Transport Personnel
BT2 Occupations
BT2 Personnel
SO DOE FRASE VOCABULARY

TRUNK
(1106 TRUNK)
DA November 28, 1990
BT1 Human Body Parts
NT1 Abdomen
NT1 Back
NT2 Lower Back
NT2 Upper Back
NT1 Buttock(s)
NT1 Chest
NT1 Groin
NT1 Rib(s)
SO DOE FRASE VOCABULARY

TUBERCULOSIS
(1409 TUBERCULOSIS)
DA November 28, 1990
BT1 Upper Respiratory Disease
BT2 Cond of Respiratory Sys.
    Non-Toxic
BT3 Diseases
BT3 Illnesses
RT Upper Respiratory Condition
SO DOE FRASE VOCABULARY

TURBINE
(2148 TURBINE)
DA December 10, 1990
BT1 Facility Components
BT1 Machines (DOE FRASE
    Vocabulary)
BT2 Equipment
SO DOE FRASE VOCABULARY
DEF (NRC Glossary of Terms: Nuclear
    Power and Radiation) A rotary
    engine made with a series of
    curved vanes on a rotating shaft.
    Usually turned by water or steam.
    Turbines are considered to be the
    most economical means to turn
    large electrical generators.

TWEEZERS
(3068 TWEEZERS)
DA January 3, 1991
BT1 Tools - Manual
BT2 Tools (DOE FRASE Vocabulary)
BT3 Equipment
SO DOE FRASE VOCABULARY

ULCER(S)
(1410 ULCER)
DA November 28, 1990
BT1 Illnesses
SO DOE FRASE VOCABULARY

ULTRAVIOLET EQUIPMENT
(2431 ULTRAVIOLET)
DA January 3, 1991
BT1 Equipment/Parts - Electrical (DOE
    FRASE Vocabulary)
BT2 Equipment
SO DOE FRASE VOCABULARY

UNKNOWN/UNDETERMINED
ACTIVITY
(1252 UU ACTIVITY)
DA November 28, 1990

BT1 Activity Types (DOE FRASE
    Vocabulary)
BT2 Activities
SO DOE FRASE VOCABULARY

UNSCHEDULED EVACUATION
(1581 UNSCHED EVAC)
DA November 28, 1990
BT1 Nature of Programmatic Impact
SO DOE FRASE VOCABULARY

UNSCHEDULED SHUTDOWN
(1580 UNSCH SHUTDN)
DA November 28, 1990
BT1 Nature of Programmatic Impact
RT Automatic Reactor Scram
RT Manual Reactor Scram
RT Violations
SO DOE FRASE VOCABULARY

UPPER ARM
(1096 UPPER ARM)
DA November 28, 1990
BT1 Arm(s)
BT2 Human Body Parts
RT Shoulder(s)
SO DOE FRASE VOCABULARY

UPPER BACK
(1108 UPPER BACK)
DA November 28, 1990
BT1 Back
BT2 Trunk
BT3 Human Body Parts
RT Pinched Nerve
RT Ruptured Disk
RT Spinal Cord
SO DOE FRASE VOCABULARY

UPPER RESPIRATORY CONDITION
(1411 UPPER RC)
DA November 28, 1990
BT1 Cond of Respiratory Sys. Non-Toxic
BT2 Diseases
BT2 Illnesses
RT Respiratory System
RT Tuberculosis
SO DOE FRASE VOCABULARY

UPPER RESPIRATORY DISEASE
(1412 UPPER RD)
DA November 28, 1990
BT1 Cond of Respiratory Sys. Non-Toxic
BT2 Diseases
BT2 Illnesses
NT1 Tuberculosis
RT Respiratory System
SO DOE FRASE VOCABULARY

VACUUM CLEANER
(2150 VACUUM CLEAN)
DA December 10, 1990
BT1 Machines (DOE FRASE
    Vocabulary)
BT2 Equipment
RT Vacuum System
SO DOE FRASE VOCABULARY

VACUUM SYSTEM
(2054 VAC SYSTEM)
DA   December 10, 1990
BT1 Systems (DOE FRASE Vocabulary)
 BT2 Systems
RT Vacuum Cleaner
SO DOE FRASE VOCABULARY

VAULT
(1723 VAULT)
DA   December 10, 1990
BT1 Site (DOE FRASE Vocabulary)
 BT2 Sites/Areas
NT1 SNM Vault
SO DOE FRASE VOCABULARY

VEHICLE ACCIDENT
(1539 VEHICLE ACC)
DA   November 29, 1990
BT1 Nature of Property Damage
RT Seat Belt(s)
SO DOE FRASE VOCABULARY

VEHICLE MAINT/REPAIR ACTIVITY
(1247 VMR ACTIVITY)
DA   November 28, 1990
BT1 Activity Types (DOE FRASE
        Vocabulary)
 BT2 Activities
BT1 Maintenance
 BT2 Activities
RT Garage
SO DOE FRASE VOCABULARY

VEHICLE SERVICE AREA
(1625 VS AREA)
DA   December 10, 1990
BT1 Area
 BT2 Sites/Areas
RT Garage
SO DOE FRASE VOCABULARY

VENTILATION SYSTEM
(2055 VENTIL SYST)
DA   December 10, 1990
BT1 Systems (DOE FRASE Vocabulary)
 BT2 Systems
NT1 End Box Ventilation System
NT1 Reactor Building Standby
        Ventilation System
RT Ventilation
SO DOE FRASE VOCABULARY

**VIOLATION OF CRITICALITY
SPECIFICTN**
(1555 VIOLATION CS)
DA   November 29, 1990
BT1 Nature of Property Damage
BT1 Violations
RT Criticality Alarm System
SO DOE FRASE VOCABULARY

**VIOLATION OF OCCUP SAFETY
REGULATION**
(1556 VIOLATION SR)
DA   November 29, 1990
BT1 Nature of Property Damage

BT1 Violations
SO DOE FRASE VOCABULARY

**VIOLATION OF OPERATIONAL
SAFETY REQ**
(1558 VIOLATIO OSR)
DA   November 29, 1990
BT1 Nature of Property Damage
BT1 Violations
SO DOE FRASE VOCABULARY

VIOLATION OF TECH SPECIFICATION
(1557 VIOLATION TS)
DA   November 29, 1990
BT1 Nature of Property Damage
BT1 Violations
SO DOE FRASE VOCABULARY

VIOLATION OF WORKING LIMIT
(1559 VI WRK LIMIT)
DA   November 29, 1990
BT1 Nature of Property Damage
BT1 Violations
SO DOE FRASE VOCABULARY

VIOLATIONS
DA   January 30, 1991
NT1 De Minimus Violations
NT1 Serious Violations
NT1 Significant Violations
NT1 Violation of Working Limit
NT1 Violation of Operational Safety Req
NT1 Violation of Tech Specification
NT1 Violation of Occup Safety
        Regulation
NT1 Violation of Criticality Specifictn
RT Automatic Reactor Scram
RT Manual Reactor Scram
RT Nature of Property Damage
RT Unscheduled Shutdown
SO DOE FRASE VOCABULARY

VISE
(3069 VISE)
DA   January 3, 1991
BT1 Tools - Manual
 BT2 Tools (DOE FRASE Vocabulary)
  BT3 Equipment
SO DOE FRASE VOCABULARY

VISEGRIPS
(3070 VISEGRIPS)
DA   January 3, 1991
BT1 Tools - Manual
 BT2 Tools (DOE FRASE Vocabulary)
  BT3 Equipment
SO DOE FRASE VOCABULARY

VOCABULARY
(SSDC)
DA   October 12, 1990
NT1 DOE FRASE VOCABULARY
NT1 Factor Relationship and Sequence
        of Events Vocabulary
NT1 FRASE Vocabulary
NT1 Radioactive Source Terms
RT Catalogs

SO System Safety Development Center
        Glossary
DEF (SSDC) The set of words employed
        by a language, group, individual,
        or work, or in relation to a subject.

VOLT METER(S)
(2779 VOLT METER)
DA   January 3, 1991
SO DOE FRASE VOCABULARY

VOLTAGE
(2432 VOLTAGE)
DA   January 3, 1991
BT1 Equipment/Parts - Electrical (DOE
        FRASE Vocabulary)
 BT2 Equipment
SO DOE FRASE VOCABULARY

VOLTAGE REGULATOR
(2433 VOLTAGE REG)
DA   January 3, 1991
BT1 Equipment/Parts - Electrical (DOE
        FRASE Vocabulary)
 BT2 Equipment
SO DOE FRASE VOCABULARY

WAREHOUSE
(1791 WAREHOUSE)
DA   December 10, 1990
BT1 Storage Area
 BT2 Area
  BT3 Sites/Areas
SO DOE FRASE VOCABULARY

WASTE DISPOSAL SITE
(1626 WD SITE)
DA   December 10, 1990
BT1 Site (DOE FRASE Vocabulary)
 BT2 Sites/Areas
RT Depositories
SO DOE FRASE VOCABULARY

WASTE MONITORING FACILITY
(1792 WM FACILITY)
DA   December 10, 1990
BT1 Facility (DOE FRASE Vocabulary)
 BT2 Facilities and Buildings (DOE
        FRASE Vocabulary)
  BT3 Facilities
SO DOE FRASE VOCABULARY

WASTE WATER SYSTEM
(2057 WW SYSTEM)
DA   December 10, 1990
BT1 Systems (DOE FRASE Vocabulary)
 BT2 Systems
BT1 Water Systems
 BT2 Publicly Owned Treatment Works
  BT3 Treatment Facilities
   BT4 Facilities
 BT2 Systems
NT1 Waste Water Collection Systems
RT Drinking Water System
SO DOE FRASE VOCABULARY

WATER DAMAGE
(1540 WATER DAMAGE)

DA   November 29, 1990
BT1 Damage
 BT2 Nature of Property Damage
RT   Heat Damage
RT   Smoke Damage
SO   DOE FRASE VOCABULARY

WEED EATER
(2981 WEED EATER)
DA   January 3, 1991
BT1 Tools - Powered
 BT2 Tools (DOE FRASE Vocabulary)
  BT3 Equipment
SO   DOE FRASE VOCABULARY

WELDER'S HOOD
(2693 WELDER HOOD)
DA   January 3, 1991
BT1 Head Protection
 BT2 Personal Protective Equipment
  BT3 Equipment/Parts - Personal
      Protective (DOE FRASE
      Vocabulary)
   BT4 Equipment
SO   DOE FRASE VOCABULARY

WELDER/SOLDERER
(0771 WELDER)
DA   November 28, 1990
BT1 Precision/Production Personnel
 BT2 Occupations
 BT2 Personnel
SO   DOE FRASE VOCABULARY

WELL PAD
(1627 WELL PAD)

DA   December 10, 1990

BT1 Site (DOE FRASE Vocabulary)
 BT2 Sites/Areas
SO   DOE FRASE VOCABULARY

WIND DAMAGE
(1541 WIND DAMAGE)
DA   November 29, 1990
BT1 Damage
 BT2 Nature of Property Damage
SO   DOE FRASE VOCABULARY

WIRING
(2434 WIRING)
DA   January 3, 1991
BT1 Equipment/Parts - Electrical (DOE
     FRASE Vocabulary)
 BT2 Equipment
SO   DOE FRASE VOCABULARY

WORK STATION
(1726 WORK STATION)
DA   December 10, 1990
BT1 Site (DOE FRASE Vocabulary)
 BT2 Sites/Areas
BT1 Station
 BT2 Site (DOE FRASE Vocabulary)
  BT3 Sites/Areas
SO   DOE FRASE VOCABULARY

WRENCH(S)
(3071 WRENCH)
DA   January 3, 1991
BT1 Tools (DOE FRASE Vocabulary)
 BT2 Equipment
NT1 Allen Wrench

NT1 Impact Wrench
NT1 Power Impact Wrench(s)

NT1 Ratchet Wrench(s)
NT1 Torque Wrench(s)
SO   DOE FRASE VOCABULARY

WRIST(S)
(1099 WRIST)
DA   November 28, 1990
BT1 Joint(s)
 BT2 Human Body Parts
RT   Hand(s)
SO   DOE FRASE VOCABULARY

WRIST BAND
(2694 WRIST BAND)
DA   January 3, 1991
BT1 Hand Protection
 BT2 Personal Protective Equipment
  BT3 Equipment/Parts - Personal
      Protective (DOE FRASE
      Vocabulary)
   BT4 Equipment
SO   DOE FRASE VOCABULARY

X-RAY EQUIPMENT
(2435 X-RAY EQUIPM)
DA   January 3, 1991
BT1 Equipment/Parts - Electrical (DOE
     FRASE Vocabulary)
 BT2 Equipment
SO   DOE FRASE VOCABULARY

YARD AREA
(1628 YARD AREA)
DA   December 10, 1990
BT1 Area
 BT2 Sites/Areas
SO   DOE FRASE VOCABULARY

Printed and bound by CPI Group (UK) Ltd, Croydon, CR0 4YY

23/10/2024

01778245-0020